HUMAN ANATOMY

HUMAN ANATOMY

AN ILLUSTRATED GUIDE TO THE STRUCTURE AND FUNCTION OF THE BODY

GENERAL EDITOR:
JANE DE BURGH

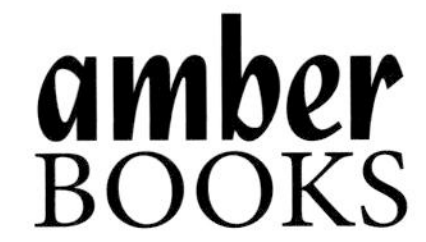

First published in 2025

Published by
Amber Books Ltd
United House
North Road
London N7 9DP
United Kingdom

www.amberbooks.co.uk
Facebook: amberbooks
YouTube: amberbooksltd
Instagram: amberbooksltd
X(Twitter): @amberbooks

ISBN: 978-1-83886-488-0

Project Editor: Michael Spilling
Designers: Mark Batley, Lewis Hughes-Batley and Keren Harragan
Picture Research: Terry Forshaw

Printed in China

CONTENTS

Introduction

The human body, with its complex network of interdependent systems and structures, has presented us with an enduring challenge – to discover and unravel the intricate workings of each of its parts and to use this knowledge to enhance and develop new ways to fight illness and disease.

The body is made up of trillions of microscopic units, each with its own unique function, yet all collaborating together to create one highly functioning entity. The study of the structure of these cells, tissues and organs is known as anatomy (from the Greek word 'anatome' meaning dissection). The science behind how organs and cells function is known as physiology.

The study of anatomy and physiology has been at the forefront of medical practice for centuries, which is why we can now understand the normal functions of the body and diagnose disease. In the past, the only way to gain anatomical knowledge was through the dissection of dead bodies, or cadavers, but historically there have been many obstacles to the practice of human dissection, primarily for religious and ethical reasons. Today, individuals can still choose to donate their cadavers for anatomical study and research, although dissection is seldom performed at universities as other more advanced, cost-effective methods for visualizing the body have become available.

Anatomy through the centuries

Although dissection for the purpose of education is recorded as having been carried out as early as 300 BC, it was not until the 2nd-century AD that Greek physician Galen made significant progress in the study of anatomy. Dissection was forbidden in Rome where he lived and worked, so he encouraged his students to dissect local monkeys, which he believed had similar anatomy to humans. Galen supported the then-popular but inaccurate theory that the body contains four humours: blood, phlegm, yellow bile and red bile. It was widely believed that an imbalance of these humours caused illness. Despite these obsolete beliefs, Galen made important advances in understanding anatomy and physiology through his animal dissections and he wrote influential treatises that drew parallels with the inner functions of the human body.

Anatomical drawings

In the Middle Ages, from the 10th to 15th centuries, a period of relative stagnation followed these advances. Medicine continued to be studied at universities under a master physician, but students relied heavily on ancient texts and the observations of ancient physicians like Galen. In the 15th century, however, the Italian artist Leonardo da Vinci recognized the importance of anatomical investigation to increasing medical knowledge. He believed that human dissection was the only way to fully understand the body and during his lifetime he carried out about 30 such dissections. His highly detailed sketches of the internal organs of the human body were not well publicized, however, and in 1513, the Pope banned da Vinci from conducting further dissections.

During the Renaissance in Europe, in the early 16th century, interest in anatomy began to grow again and some of the early Greek and Roman texts, such as Galen's *On Anatomical Procedures,* were rediscovered. By the mid-1500s, progress was being made by Andreas Vesalius, a Flemish professor of anatomy and surgery. Vesalius had trained at Padua University in Italy, which boasted a newly established anatomical theatre for human

Opposite: The artist Leonardo da Vinci performed dissections to inform his anatomical drawings of the human body.

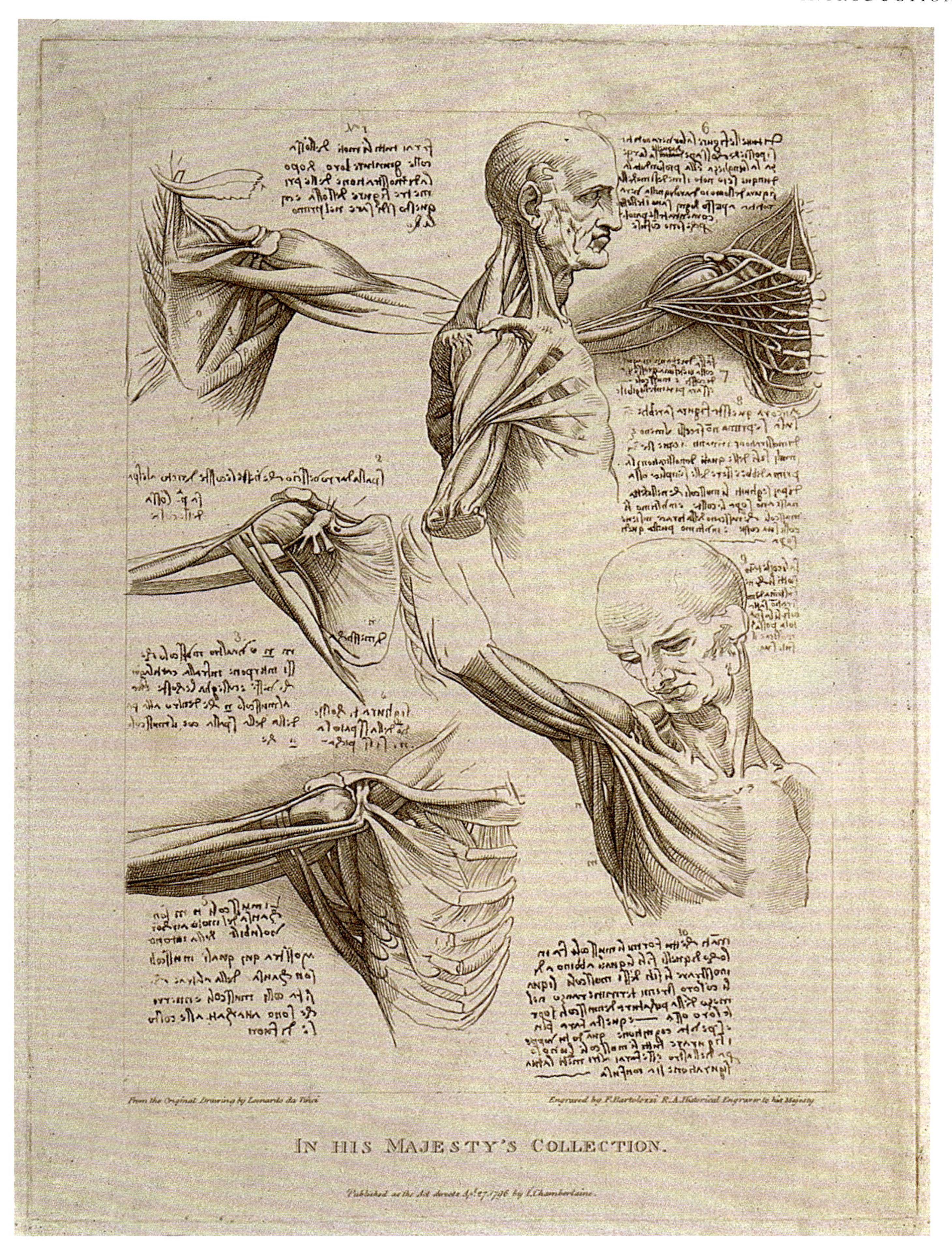
From the Original Drawing by Leonardo da Vinci
Engraved by F.Bartolozzi R.A. Historical Engraver to his Majesty
IN HIS MAJESTY'S COLLECTION.
Published as the Act directs Apl. 27. 1796 by J.Chamberlaine.

Above: *The Anatomy Lesson of Dr Nicolaes Tulp* (1632), by Rembrandt, depicts a group of doctors observing an arm dissection.

dissection. In 1543 Vesalius challenged previously held beliefs in his text *De humani corporis fabrica* (On the Structure of the Human Body). Based on dissections of human bodies, his treatise demonstrated how human anatomy differed from that of animals, which hugely influenced the science of anatomy.

In 1628, the English physician William Harvey published his revolutionary text *An Anatomical Study of the Motion of the Heart and Blood in Animals*. In this, he described how the heart works as a muscle, causing the arteries to pulsate and constantly move blood around the body. His text also demonstrated the importance of valves in the circulation system.

Over the last three centuries, countless other scientists and doctors have contributed to our knowledge of anatomy. With the invention of the microscope in the late 17th century, new frontiers for anatomical research opened up, as cells could now be visualized. A new era in medical technology began, laying the foundations for major breakthroughs in the treatment of disease that continued well into the 20th century. In 1895, the German physicist Wilhelm Röntgen discovered X-rays, which offered unprecedented opportunities for examining the inner workings of the human body. From 1800, the arrival of anaesthetics transformed surgical operations, and in 1901 Austrian physician Karl Landsteiner's discovery of the four blood groups paved the way for some of the most important advances in the 20th century.

Imaging techniques

Improvements in Rontgen's X-ray machine made it an increasingly powerful diagnostic tool, and in the 1950s, Ian Donald, a professor of midwifery at Scotland's University of Glasgow, pioneered the use of ultrasound in obstetrics and gynaecology, thus reducing the potentially harmful radiation of X-rays. The technique was initially used only in pregnancy, but it soon transformed the visualization of internal organs across other medical specialities and improved the early detection of abnormalities. One of the 20th century's most significant contributions to medicine has been – and remains – the constant refinement of imaging technologies. These have allowed scientists to observe the inner workings of the body and have revolutionized the diagnosis and treatment of many serious conditions. More recent imaging technologies include MRI, which uses magnetic fields and radio waves to create images, and the computed tomography (CT) scan, which is a highly accurate three-dimensional form of X-ray. A further innovation in the field of imaging has been the use of endoscopes, flexible viewing instruments that allow internal examinations without surgery.

Discovering DNA

One of the most monumental landmarks in medical history came in 1953, when American James Watson and Englishman Francis Crick deciphered the structure of deoxyribonucleic acid (DNA). The pair claimed that they had found 'the secret of life', not an unfounded claim since the double helix structure of the DNA molecule contains the genes that form the basis of all living tissue. The effects of this discovery continue to resonate into the 21st century, and decades later resulted in the international scientific research programme known as the Human Genome Project.

Of all the medical advances, gene therapy and stem cell research promise to offer the greatest potential. In 1990 the Human Genome Project was spearheaded by scientists in the USA. The aim was to identify all 25,000 genes in the human genome and to unlock the sequence of the 3 billion chemical bases that make up DNA. In 2003 the project was completed. With every gene in the body mapped out, scientists are able to compare the genes in healthy and non-healthy individuals and identify problem genes. Gene therapy involves replacing the faulty gene with a normal gene, thus eliminating a disease. Athough gene therapy remains in its infancy, trials are progressing internationally.

Virtual anatomy

Historically, the teaching of anatomy to medical students has involved cadaver dissections, textbooks, photographs, as well as preserved specimens. These are effective methods, but they are often not practical or cost-effective. For example, during the Covid-19 pandemic, online learning was paramount. With medical advances in imaging techniques, 3D-imaging allows almost lifelike images of the body's organs, while sophisticated computerized imaging can produce dissection-style interactive images. Increasingly, artificial intelligence (AI)

Below: Modern CT scanners provide detailed cross-sectional images of inside the body using X-rays and computers.

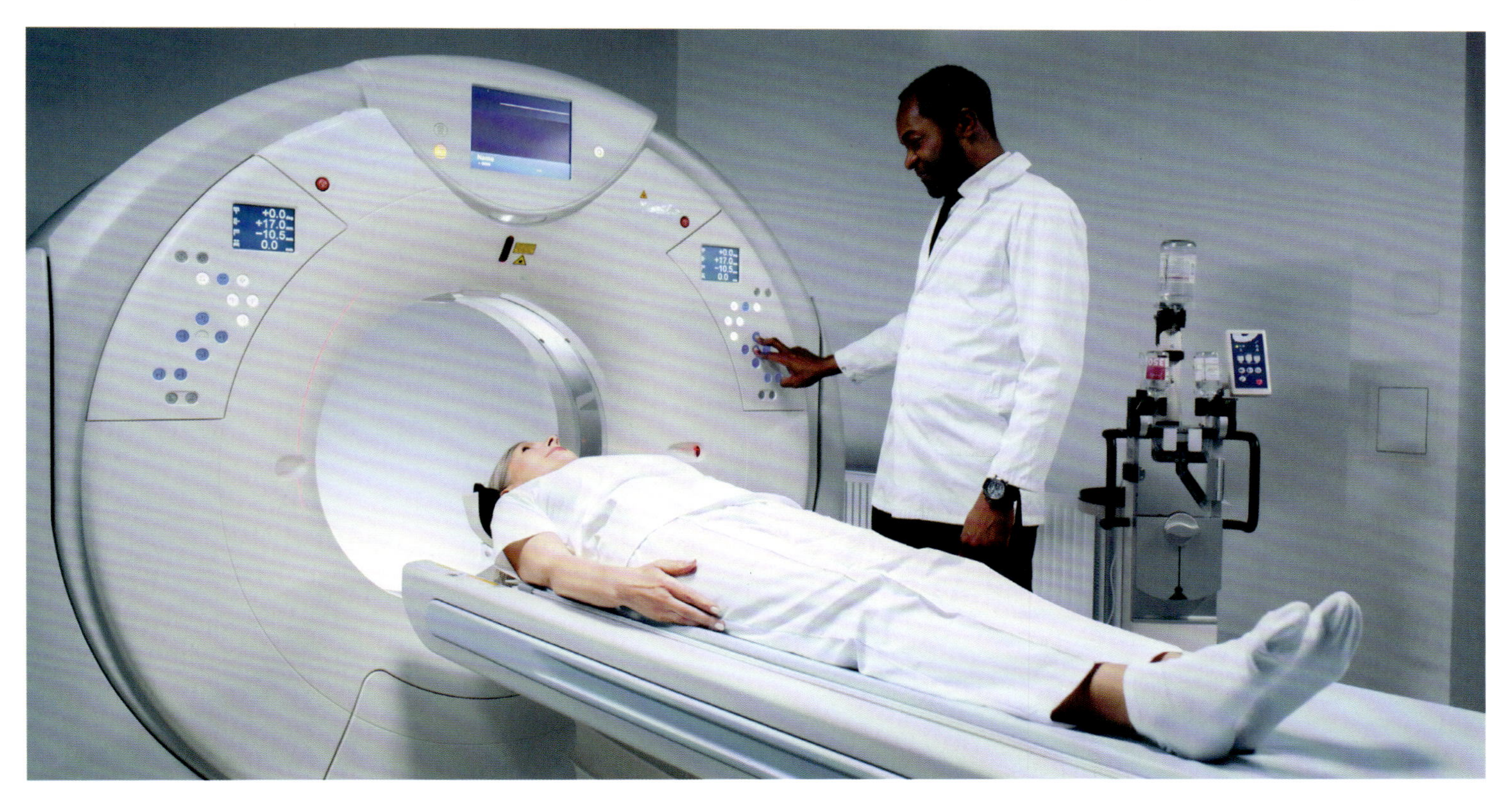

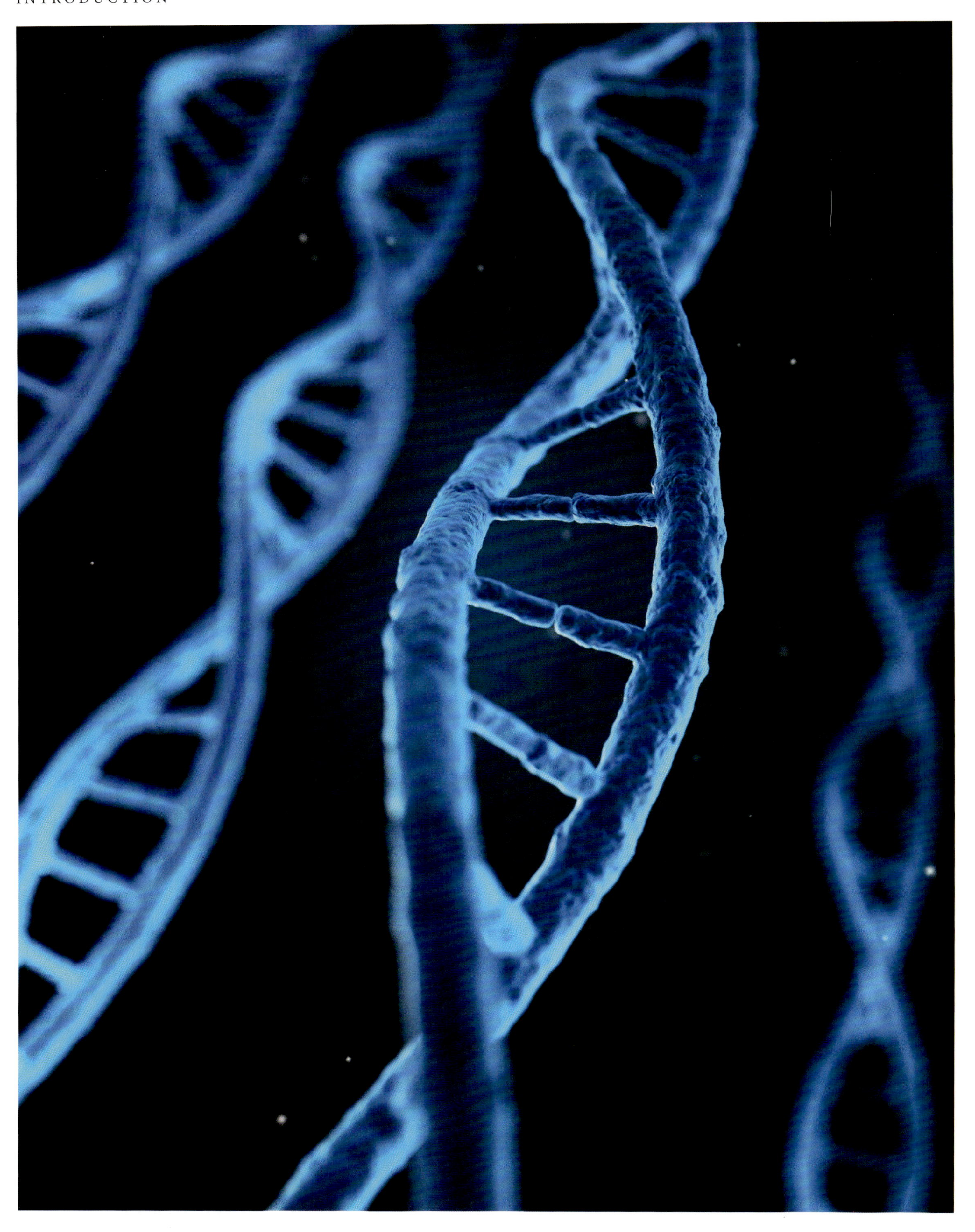

is being used to teach anatomy, with virtual dissections and interactive tools adding to the learning experience.

Approaches to anatomy

There are two approaches to the study of human anatomy. The first uses a regional, or segmental, approach, which divides the body up into sections and examines the organs and tissues within each segment, usually the head, the thorax, the abdomen, the pelvis and upper and lower limbs. The second approach looks at the 'systems' of the body, for example, the digestive system, which processes food and drink from the moment of 'ingestion' to the moment of 'excretion'.

Human Anatomy adopts a systems-based approach, detailing the anatomy of each system as well as describing how the organs function:

CELLS AND TISSUES
This section describes the basic units within the body and explains how they communicate with each other.

SKIN, HAIR AND NAILS
The skin is the largest organ in the body and this chapter describes its many functions.

THE MUSCULOSKELETAL SYSTEM
In this system muscles, tendons and ligaments throughout the body attach to the skeleton to enable movement.

THE NERVOUS SYSTEM
This system is the control centre of the body, comprising the central nervous system (the brain and spinal cord) and the peripheral nervous system (the nerves that carry messages to and from the brain).

THE SENSES
The special senses help us to make sense of the world around us and enable us to communicate our thoughts.

THE CARDIOVASCULAR SYSTEM
This system ensures that oxygen and nutrients are delivered to all cells in the body via the blood. The heart acts as a powerful pump to circulate blood through the vessels.

THE RESPIRATORY SYSTEM
Without oxygen, we would cease to exist. Our lungs take in oxygenated air, absorb vital oxygen and excrete carbon dioxide.

THE DIGESTIVE SYSTEM
All cells and tissues require nutrients to survive. The digestive system processes food and absorbs the necessary nutrients, excreting waste products as faeces.

THE URINARY SYSTEM
Comprising the kidneys, ureters, bladder and urethra, this system is responsible for fluid and electrolyte balance in the body and also excretes waste in the form of urine.

THE LYMPHATIC AND IMMUNE SYSTEMS
These systems work together to protect the body against foreign invaders and infection. Blood cells, lymph glands and bone marrow all form part of these systems.

THE ENDOCRINE SYSTEM
This network produces important chemicals known as hormones. These hormones are vital to life and are manufactured by specialized glands and tissues in the body collectively known as the endocrine system.

THE REPRODUCTIVE SYSTEM
This system enables reproduction and the roles of both male and female systems are described here.

AGEING AND DEATH
The book finishes with a section on the ultimate degeneration of cells and tissues at the end of life and describes what happens to our bodies after death.

Opposite: The discovery of the double helix (or twisted ladder) structure of DNA provided the basis for future research into human genes.

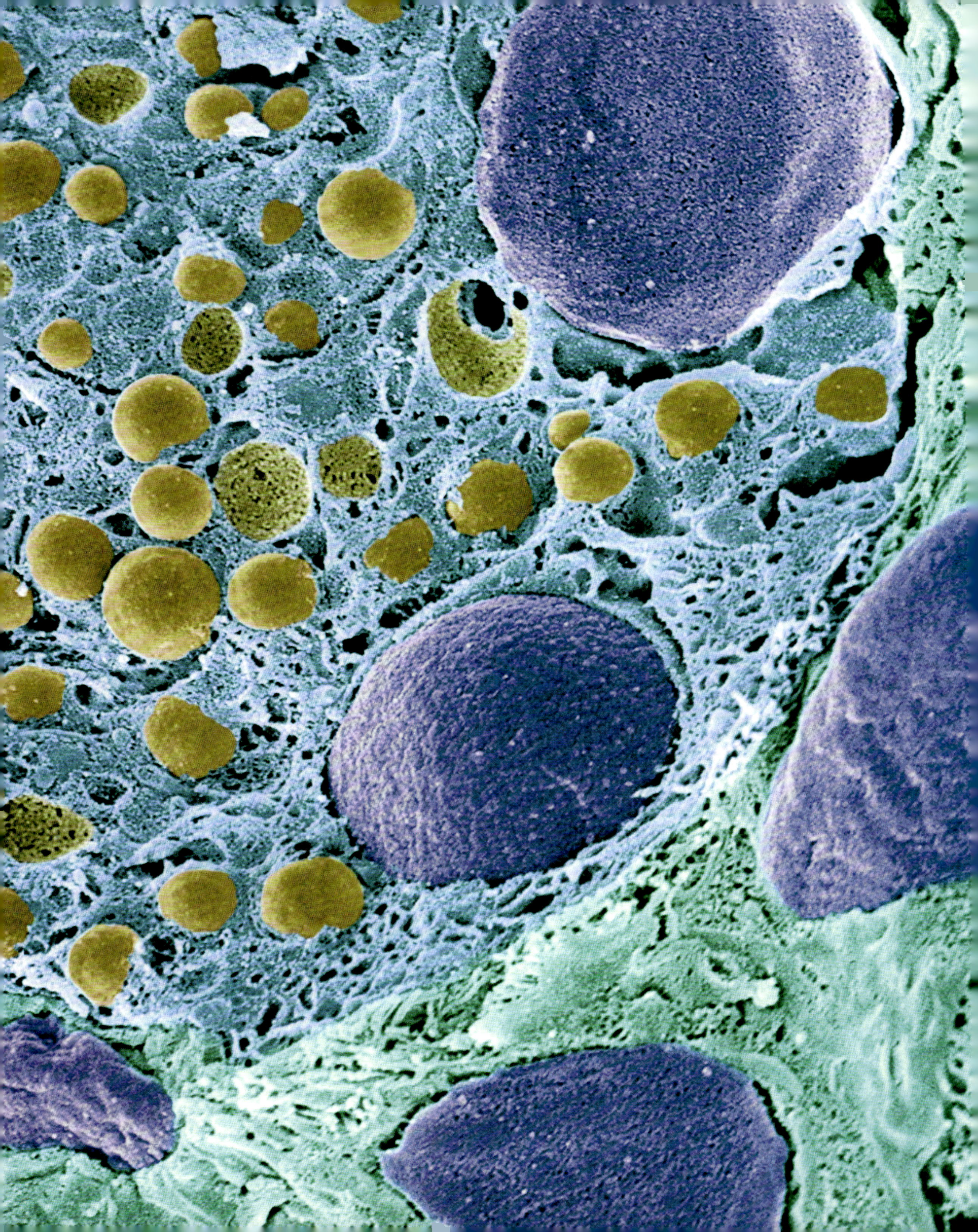

CHAPTER 1

Cells and Tissues

Our bodies are made up of trillions of cells, each with its own specialized function. Cells are the smallest units within the body and the fundamental building blocks for the muscles, bones, tissues, blood, nerves and skin that make up our body systems. At the centre of each cell is the nucleus. This contains the DNA that holds the cell's genetic code and which is the blueprint for making proteins, essential for the development and growth of all body structures.

This chapter details the structure and function of human cells, explaining how cells communicate and work together and how the genes carried within the cells dictate the unique characteristics of each individual.

Opposite: This magnified image of pancreatic tissue shows zymogen granules that secrete digestive enzymes. Cell nuclei are shown in the blue.

Cell Function

All the living tissue in the body is made up of cells – bounded compartments filled with a concentrated solution of chemicals. Cells are the smallest living unit in the body.

Every tissue in the body is made up of groups of cells performing specialized functions and linked by intricate systems of communication. These cells are microscopic, so not visible to the naked eye. Most animal cells are approximately 0.01–0.10mm (0.0003–0.003in) in size.

There are over 200 different types of cells in the body. Although enormously complex, the final structure of the human body is generated by a limited repertoire of cell activities. Most cells grow, divide and die while performing functions particular to their tissue type, for example, the contraction of muscle cells.

Organelles

Typically, cells contain structural elements called organelles, which are involved in the cell's metabolism and life cycle. This includes the uptake of nutrients, cell division and synthesis of proteins – the molecules responsible for most of the cell's enzymatic, metabolic and structural functions. Organelles are highly complex units, which are themselves held in place by their own membranes.

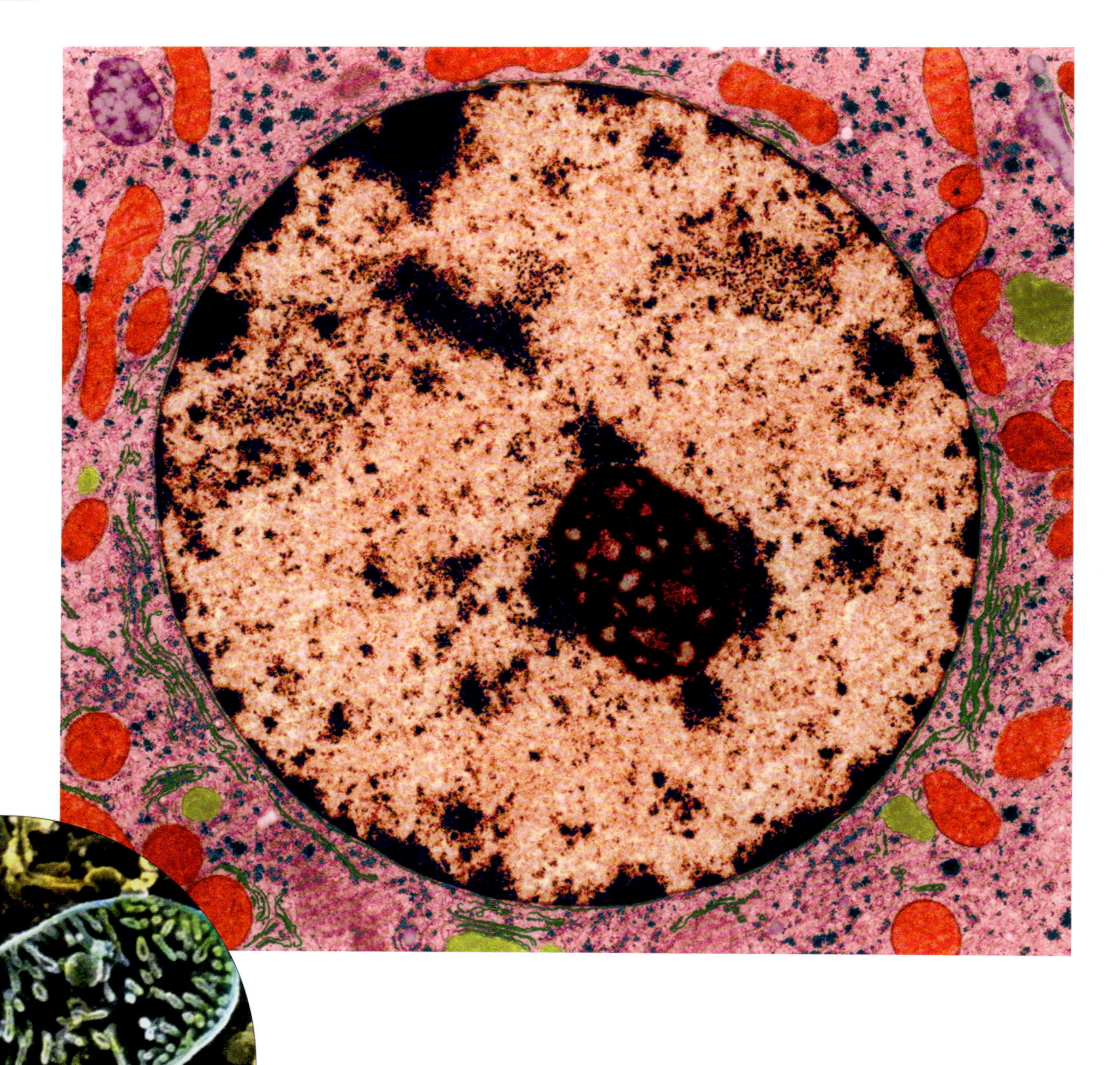

Right: A hepatocyte is seen on a micrograph. This is a specialized liver cell that performs several functions.

Below: All living cells are powered by 'batteries' known as mitochondria.

Structure of a Cell

Cell structure can be divided into the outer membrane, the DNA-containing nucleus and organelles. Each component has a specific function, such as energy production, storage or the synthesis of proteins.

Plasma Membrane
The plasma membrane surrounds each cell and separates it from its external environment. Contained within the membrane is a solution of proteins, electrolytes and carbohydrates called the cytosol, as well as membrane-bound subcellular structures called organelles. Spanning across the membrane are proteins responsible for communication with the external environment and for transport of nutrients and waste.

Nucleus
The nucleus is in the centre of the cell, and contains the cell's DNA arranged into chromosomes, as well as structural proteins for coiling and protecting the DNA. The nucleus is surrounded by a membrane containing large pores, allowing for movement of molecules between the nucleus and the cytosol, while retaining the chromosomes inside the nucleus.

Organelles
The cytoplasm is the inner contents of the cell, not including the nucleus, which is made up of fluid (the cytosol) and large numbers of organelles. The organelles include:

■ **Mitochondria**
Responsible for energy production. Nutrients in the form of sugars and fats are broken down in the presence of oxygen to make ATP (adenosine triphosphate), a source of energy used by a cell.

■ **Ribosomes**
Ribosomes carry out the production of proteins, using the blueprint recorded in the genetic material of the cell.

■ **Endoplasmic reticulum**
This is a vast network of tubes, sacs and sheets of membrane that runs throughout the cell. It allows for the transport and storage of molecules.

■ **Golgi apparatus**
The Golgi apparatus is a stack of flattened sacs, critical in the modification, packaging and sorting of large molecules in the cell.

■ **Vesicles and vacuoles**
Vesicles are membrane-bound areas within a cell for specialized processes or storage. Vacuoles resemble 'holes' under the microscope and are typically regions of storage or digestion surrounded by a membrane.

■ **Cytoskeleton**
The cytoskeleton is the fine meshwork of protein filaments that is used to maintain the cell's shape, to anchor components in place and to provide a basis for the cell's movements.

Inside the cell

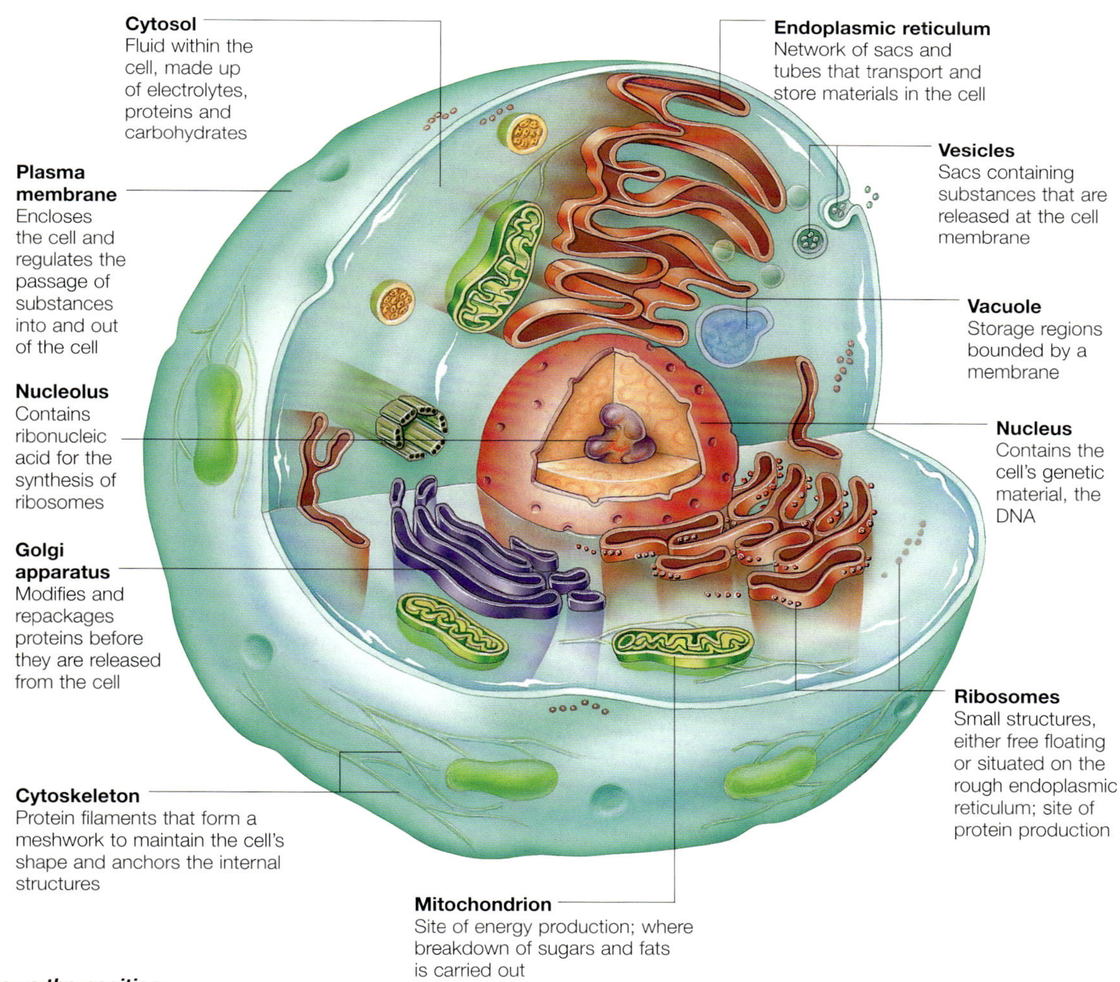

This cut-away shows the position of the nucleus as well as many organelles, the functioning structures of the cell.

Cell Division

The vast majority of cells that make up the human body divide on a regular basis. This occurs not only during periods of growth, but also when worn-out cells need to be replaced.

All tissues are made up of cells, microscopic membrane-bounded compartments. New cells are made by cell division, during which a cell replicates its genetic material and then separates its contents into two daughter cells. The process of cell division occurs continuously throughout the body, both during the development of the fetus and throughout adulthood.

Replacement

Cells divide when body tissue is growing, or when the cells in a particular area wear out and need to be replaced. Division is carefully regulated and must occur in accordance with the needs of the surrounding tissue, as well as in time with the internal cell growth cycle.

Embryonic Cells

The most prolific cell division occurs during early embryonic development; over nine months, a fertilized egg (one cell) develops into an embryo and, subsequently, a fetus of over 10 thousand million cells.

As development proceeds, many cells switch from dividing to performing a specialized function (such as becoming pacemaker cells in the heart), a process known as differentiation.

In almost all tissues there are stem cells, which are cells that are not fully differentiated, but which can divide and differentiate in response to stimuli or tissue damage.

Cell Cycle

The cell division cycle is the process by which one cell doubles its genetic material and then divides into two identical daughter cells. The cycle has two main stages: interphase, when the cell's components are replicated, and mitosis, when the cell divides into two.

Interphase is divided into two gap phases (G_1 and G_2) and a synthesis (S) phase. During the first gap phase (G_1), the cell produces carbohydrates, lipids and proteins. Slow-growing cells, such as liver cells, may remain in this phase for years, whereas fast-growing cells, such as those in bone marrow, spend only 16–24 hours in the G_1 phase.

If a cell is not actively dividing, it exits the cell cycle during G_1 and enters a state called G_0. For example, in adults many highly specialized cells, such as neurones (nerve cells) and heart muscle cells, do not divide and remain in phase G_0. This makes healing and regeneration in these tissues slow and at times impossible.

Replicating Chromosomes

The next period of interphase – which is known as the S phase – sees the replication of the chromosomes so that the cell temporarily has 92 chromosomes instead of the normal 46. Proteins are also synthesized during the S phase, including those that form the spindle structures that pull the chromosomes apart. In most human cells, the S phase lasts between 8 and 10 hours. Additional proteins are synthesized in the second gap phase, G_2.

The cell divides into two daughter cells during a process called mitosis. Mitosis is subdivided into four phases: prophase, metaphase, anaphase and telophase. The duration of a complete cell cycle varies from a day to a year, depending on the type of cell.

Examples of the rates of replacement for different cell types are:

- Liver cells: 12 months
- Red blood cells: 80–120 days
- Skin cells: 14–28 days
- Intestinal mucosa: 3–5 days.

Below: Once an egg cell is fertilized, it divides progressively; this human embryo is at the four cell stage.

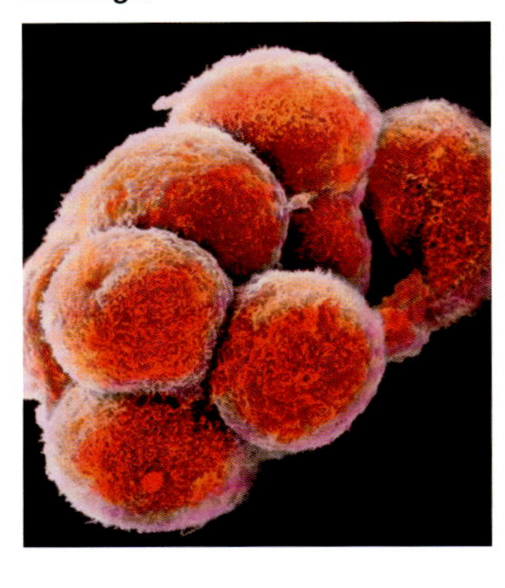

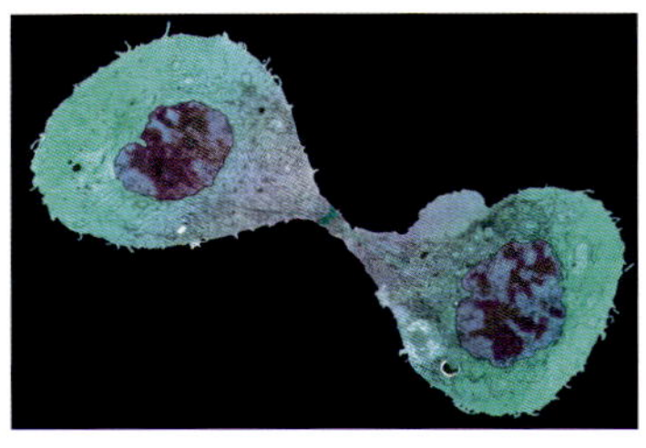

Above: During the process of cell division, the chromosomes, which contain genetic material, separate into each of the two new (daughter) cells.

Right: Cell division is divided into two stages: interphase (purple), when the cell contents are replicated, and mitosis (orange), when the cell divides. Interphase is subdivided into G_1, S and G_2 phases. Mitosis is subdivided into prophase, metaphase, anaphase and telophase.

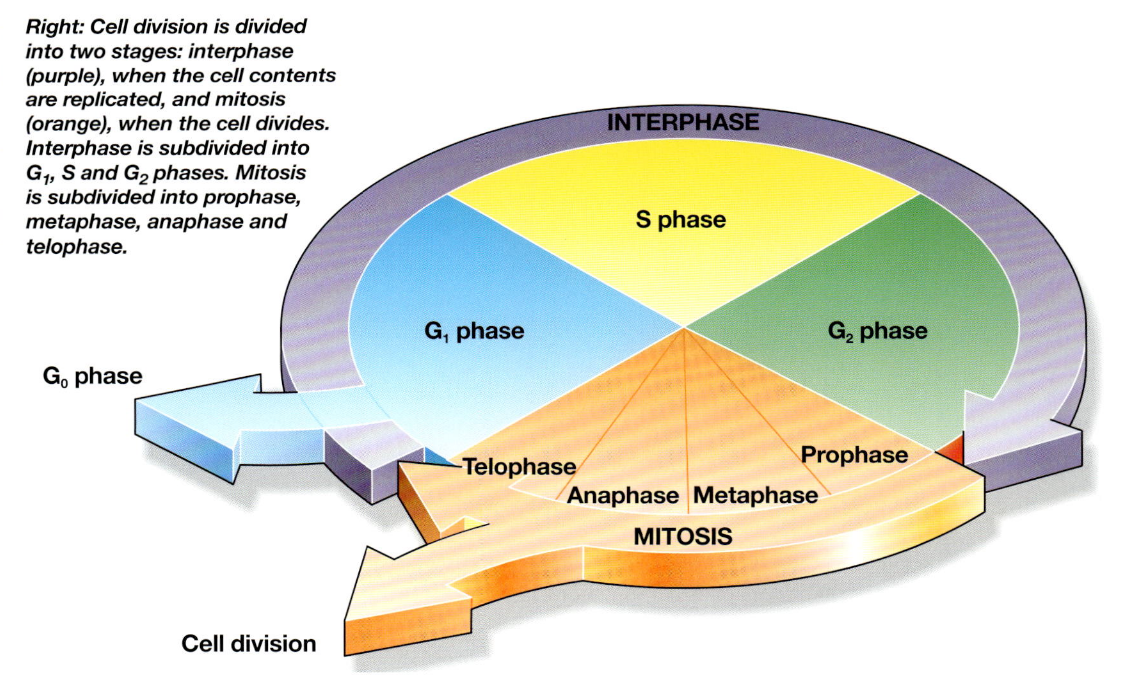

Mitosis

The four stages of mitosis

1 Prophase
During prophase, the DNA condenses into recognizable chromosomes, the nucleus disbands and the nuclear contents enter the cytoplasm.

Right: This scanning electron micrograph (SEM) shows the condensed chromosomes (red), nuclear membrane (orange) and cytoplasm (green).

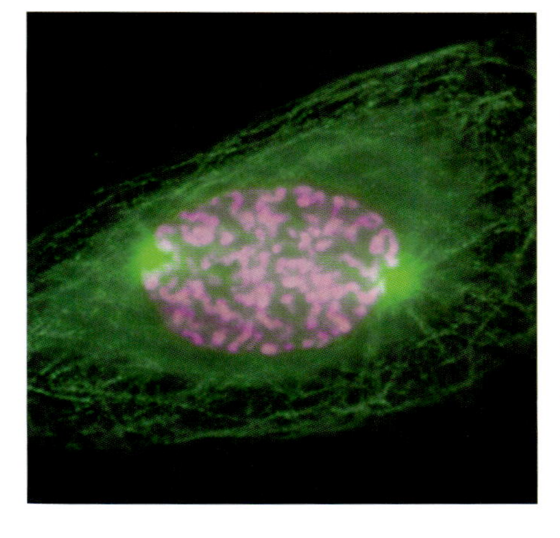

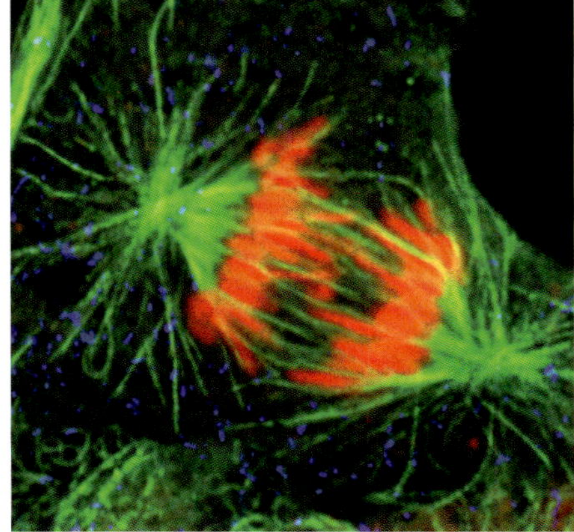

2 Metaphase
During metaphase the chromosomes attach to the mitotic apparatus, a series of specially synthesized protein filaments anchored to opposite sides of the cell.

Left: Here, the cell is in late metaphase: the nuclear membrane has disappeared and the chromosomes (red) are aligned along the centre of the cell.

3 Anaphase
During anaphase, the chromosomes are pulled away from each other by the mitotic apparatus in the cell. Half of the chromosomes go to each side of the cell.

Right: This SEM shows the first stages of anaphase, when the chromosomes separate and the cell membrane becomes indented.

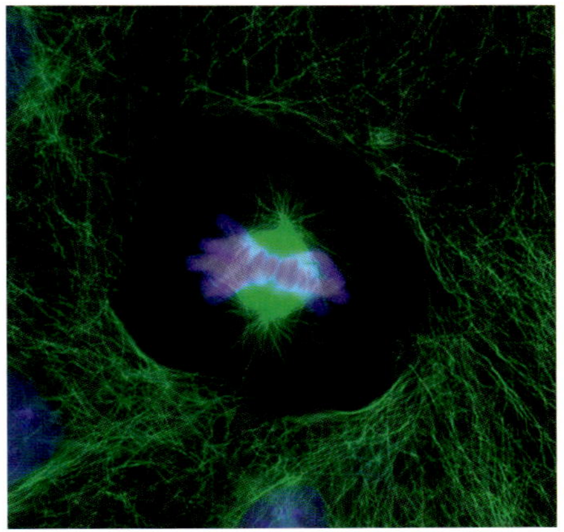

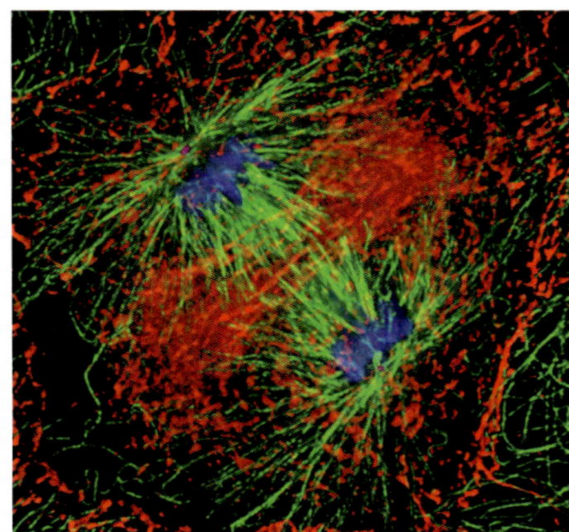

4 Telophase
During telophase, the nuclear membranes reform, the cell's contents are redistributed and the membrane 'pinches off' to form two cells.

Left: The cell is in late telophase in this SEM: the two newly formed cells are still joined by a narrow bridge containing elements of the mitotic apparatus.

Cell death

Necrosis
When the body is exposed to mechanical or chemical damage, cells may die simply because they are no longer able to function properly. This process, called necrosis, happens when the cell's integrity is violated, or when molecules or structures essential to the cell's survival are no longer available or functional.

For example, after a person dies, nutrients and oxygen are unavailable to all of the cells of the body, which then undergo necrosis and die. Gangrene is another example of necrosis – the dead tissue turns black due to the action of certain bacteria on haemoglobin, which is broken down to produce dark iron sulphide deposits.

Apoptosis
The majority of cells have an inbuilt programme that leads to eventual apoptosis. Scientists believe that this programme is as intrinsic to the cell as mitosis. There are two main reasons why a cell would undergo apoptosis. First, programmed cell death is often needed for the proper development of the human body. For example, the formation of the fingers and toes of the fetus requires the removal of the tissue between them by apoptosis.

Second, apoptosis may be needed to destroy cells that represent a threat to the organism. For example, defensive T-lymphocyte cells kill virus-infected cells by inducing apoptosis in them.

In the early stages of fetal development, fingers are connected, giving them a webbed appearance. This webbing disappears (as shown) as the fetus develops, due to apoptosis.

Cell Communication

For the body to act in a coordinated manner, it is essential that cells communicate with each other. They do this either by releasing chemical messengers or by electrically exciting neighbouring cells.

The human body contains a total of around 10,000,000,000,000 (10 trillion) cells, made from just over 200 different cell types. However, the benefits of having specialized cells can only be realized if this multicellular organization behaves in a coordinated manner.

- Internal stimuli – The body must be able to respond to changes in its internal environment. For example, cells in the pancreas detect the rise in blood glucose concentration after a meal; they release a hormone – insulin – which makes the cells of other tissues absorb glucose from the blood to provide energy.
- External stimuli – Similarly, the body must also be able to detect and respond to external stimuli. For example, visual information provides stimuli to other cells to react, for example, in the case of the 'fight-or-flight' mechanism.

Both internal and external stimuli are detected by specialized chemicals (normally proteins) called receptors which transduce (convert) information into a form that can be relayed to other cells within the body. Usually communication between the body's cells is accomplished using either chemical messengers or electrical currents.

Electrical Communication

Most electrical messages are carried by nerve cells (although heart cells also communicate electrically) that are specially adapted for carrying nerve impulses from one region of the body to another, sometimes along nerves that are up to a metre in length.

The main advantage of electrical communication is the speed at which information can be communicated; some nerves are able to transmit nerve impulses at rates of 120m (390ft) per second. As the 'wiring' of neurones is especially precise, information can be delivered to very specific locations.

Chemical Communication

Many chemical messengers, such as hormones, are released into the bloodstream and can affect a wide number of cells. However, their effect is slower.

For example, when a person is exposed to a stressful situation, the 'fight-or-flight' hormone adrenaline does not activate for 15–30 seconds. This is because the process is more complex; adrenaline molecules diffuse from the adrenal gland (located just above the kidneys) into the bloodstream and are carried around the body to the target organs (such as the heart, increasing both its rate and strength of beating).

Below: Nerve cells communicate by releasing chemical messengers that affect the electrical excitability of neighbouring cells.

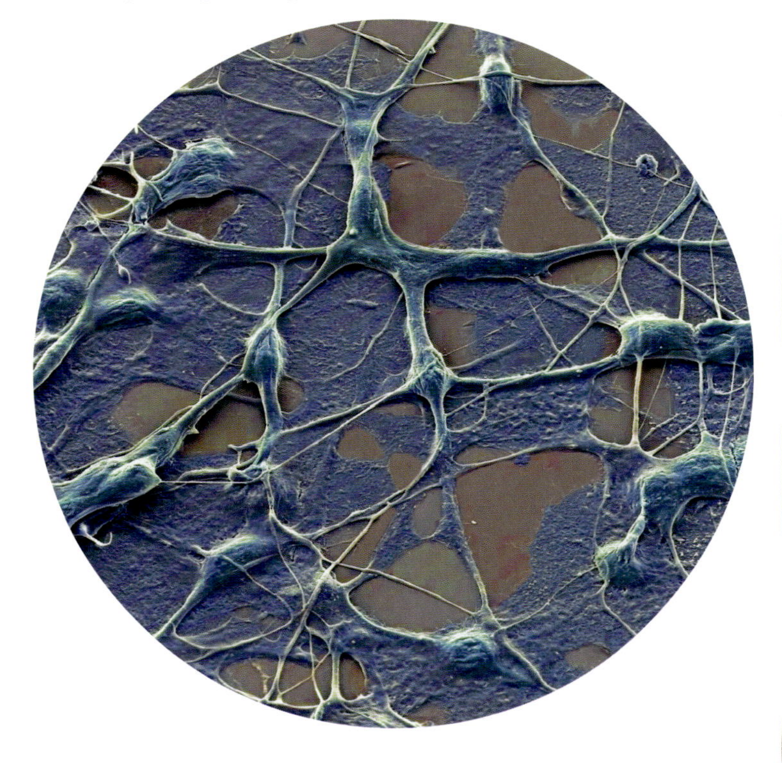

Heart cells communicate electrically

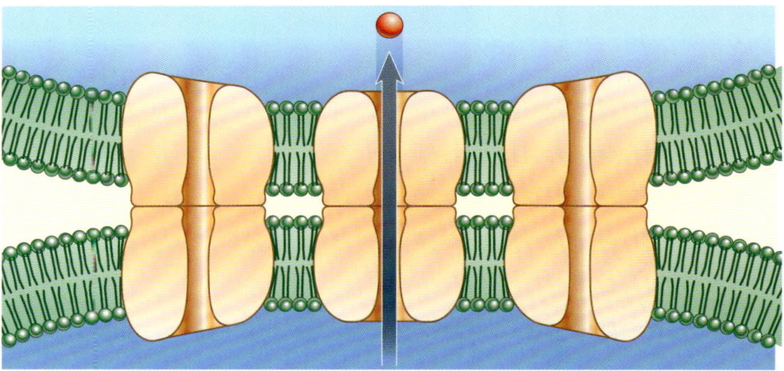

Above: Heart cells are joined by protein pores that allow electrically charged ions to cross the cell membrane. These enable a wave of electrical excitation to travel through the heart.

Left: Heart cells (green) communicate with each other electrically. Chemicals (for example, adrenaline) released by distant tissues can affect their behaviour.

Types of chemical communication

Hormones

Hormones are chemicals that are released by an endocrine gland into the bloodstream, which then carries them to distant sites throughout the body. They have either a specific site of action or affect a wide variety of different cells, simultaneously regulating a large number of different bodily processes. However, the 'target cells' are specifically programmed to respond to that particular hormone.

The hormone adrenaline, for example, is released into the blood from the adrenal medulla, the central region of each of the adrenal gland that lie above the kidneys. Adrenaline has a broad range of actions, which include constriction of the blood vessels, increased cardiac activity, dilation of the pupils in the eye and inhibition of the gastrointestinal tract.

Hormone specificity

Hormones are released by endocrine glands only in relation to the body's need at any given time. In a healthy person there is never an over- or under-production of hormones.

For a hormone to affect the internal biochemistry of a cell (a cell's 'behaviour'), the cell must have an appropriate protein receptor embedded within its cell membrane. These receptors bind with the hormone to act on the cell.

Hormones are chemical messengers that are released by a gland into the bloodstream, which then carries the hormone to distant tissues.

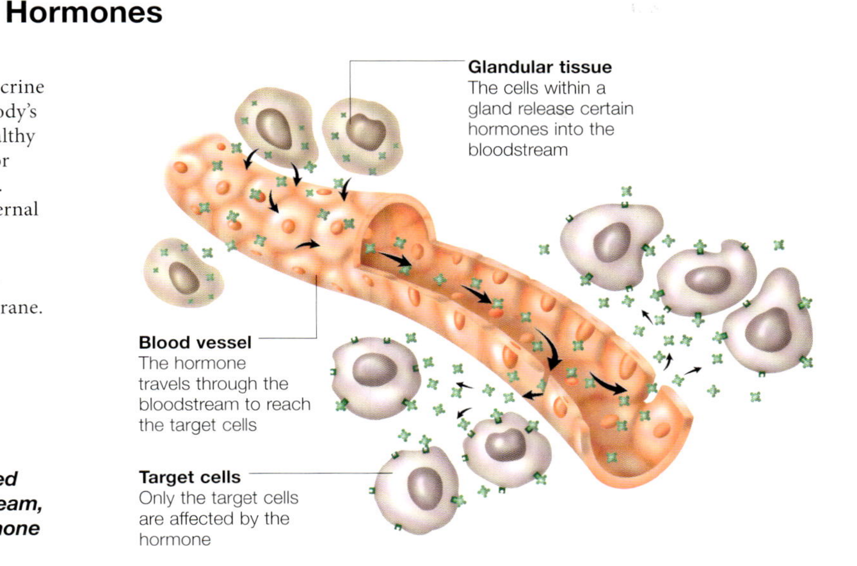

Paracrine and autocrine factors

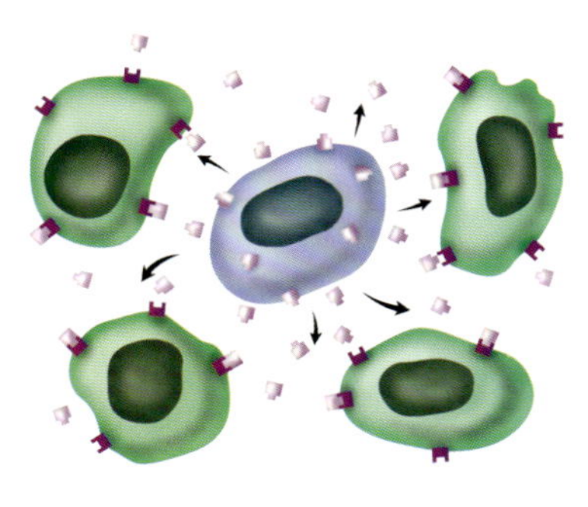

Paracrine factors are released into the water-filled space between the cells. They affect cells of a different type to the one that released them.

The second group of chemical messengers differ from hormones in that they are not transported by the bloodstream to their target cells.

Rather, these chemicals are released into the watery space that lies between the cells to affect either the same type of cell that released them (autocrine factors – 'auto' meaning 'self') or different, though nearby, cells (paracrine factors).

It should be noted, however, that a chemical can be both an autocrine and paracrine factor.

Paracrine factors

One of the most common paracrine factors is the chemical histamine. Histamine is released from specialized cells called mast cells that are present in most tissues. It is involved in allergic reactions and in some of the inflammatory chemical pathways that are initiated when a tissue is damaged. Antihistamines work by preventing mast cells from releasing this paracrine factor.

Autocrine factors

Autocrine factors affect the same type of tissue that released them. For example, most cells release autocrine factors that inhibit their own cell division and that of similar nearby cells. Cancerous cells are thought to either not release, or not respond to, these inhibitors resulting in cell division proceeding unabated.

Autocrine factors are chemical messengers that only affect the same type of cell that originally released them.

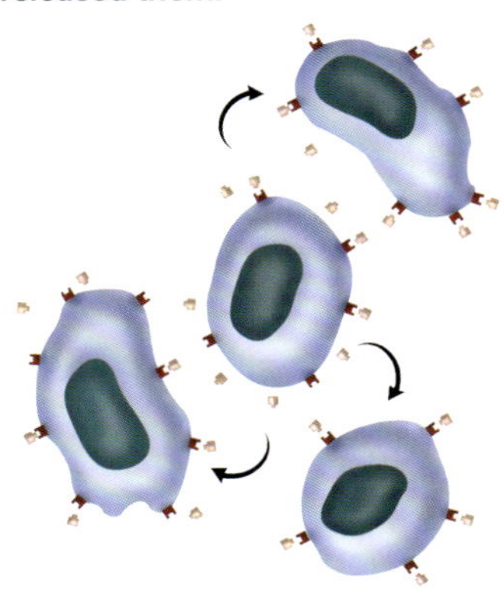

Neurohormones

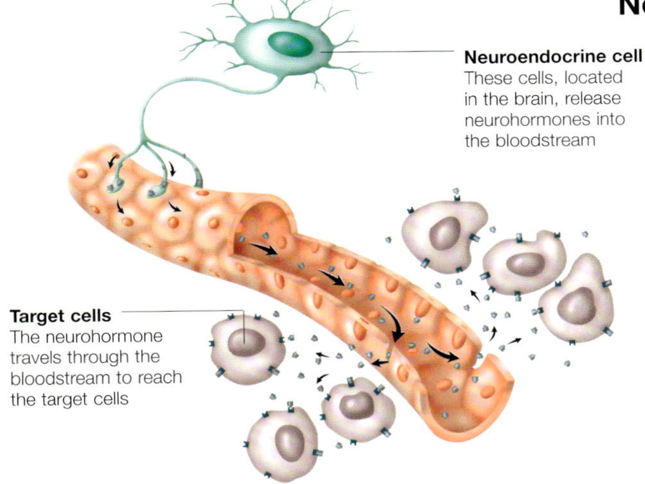

Most neurones communicate with each other by releasing a chemical messenger that diffuses between the gap (called a synapse) that separates them.

However, some neurones do not synapse with another nerve cell. Rather, their synaptic terminals are located near blood vessels; when these neurones are stimulated they release a neurohormone into the bloodstream, which is then carried to distant target organs in much the same way that a hormone is released from a gland.

Neurohormones are released by specialized nerve cells called neuroendocrine cells. These chemicals are carried in the bloodstream to the target cells.

Oxytocin

Oxytocin is a neurohormone that is released into the bloodstream by neuroendocrine cells located in the hypothalamus. This occurs in response to the stimulation of sensory nerves in the mother's nipple by a suckling infant. The blood carries the neurohormone to the mammary gland where it causes milk to be ejected from the nipple.

Structure of the Cell Membrane

The cell membrane separates one cell from another and from the external environment. Since it is only permeable to certain molecules, the internal environment of the cell can be tightly controlled.

Every cell is covered in a membrane, made mainly of phospholipids (phosphate-containing fat molecules) and proteins, which acts as a barrier. Some molecules can pass freely through it, whereas others have either restricted or no access.

The cell membrane, also known as the plasma membrane, surrounds the whole of the cell including the organelles (the subcellular components). It is much more than a simple protective covering; by determining which chemicals are allowed to pass into and out of the cell, the cell is able to tightly control its internal environment, as well as communicate with other cells.

Chemical Construction

The cell membrane is composed of four groups of chemicals: proteins (55 per cent); phospholipids (25 per cent); cholesterol (15 per cent); and carbohydrates and other lipids (5 per cent)

- Phospholipid molecules are arranged in two layers (known as a 'bilayer')
- Proteins provide a means by which water-soluble molecules can enter and leave a cell. They also allow cells to communicate with, recognize and adhere to each other
- Cholesterol molecules are, in a sense, 'dissolved' within the phospholipid bilayer. Cholesterol reduces the fluidity of the membrane by interfering with the lateral movement of the phospholipid tails.
- Carbohydrates are attached to proteins (glycoproteins) and to lipids (glycolipids). They invariably protrude on the outside surface of the membrane and are important for cell adhesion and communication.

The cell membrane allows molecules through to the interior of the cell if they: are small, such as water and amino acids; can dissolve easily in lipids, for example, oxygen, carbon dioxide and steroids; have an opposite ionic charge to the membrane; or contain specialized carrier proteins.

Chemical construction

The cell membrane is a complex structure that separates the cell from other cells and the external environment.

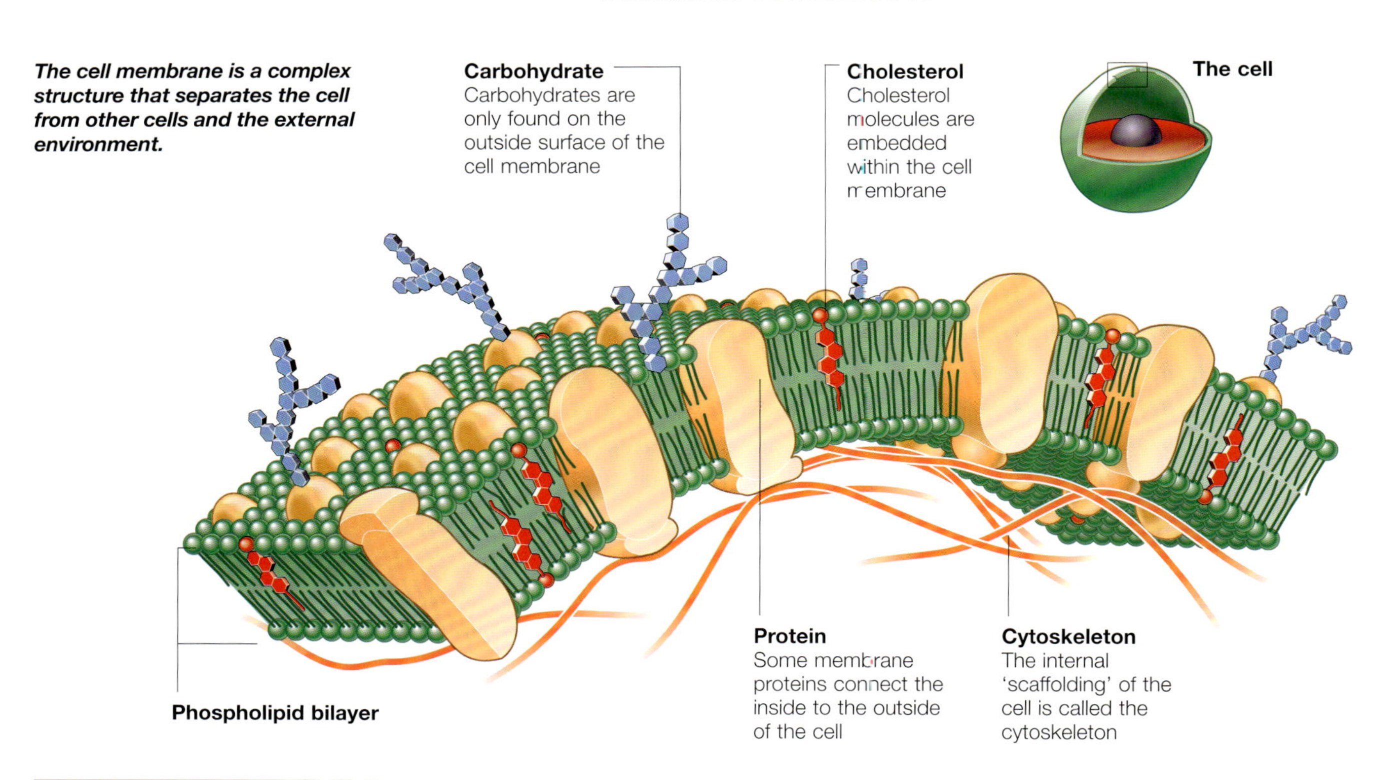

Specializations of the membrane

Cell membranes are not the same in all cells. This is because of the huge variety of functions that cells perform in different parts of the body.

Microvilli

Microvilli are specialized infoldings of the cell membrane that greatly increase its total surface area. This is especially useful in cells whose main role is to absorb chemicals from the outside to the inside of the cell. For example, around 1,000 microvilli are found in each intestinal epithelial cell, which act to absorb nutrients from the gastrointestinal tract. Each of these microvilli is about one-thousandth of a millimetre in length, and so increases the surface area available for uptake of nutrients by up to 20 times.

Adhesion between cells

While some cells, such as blood cells and sperm, are independent entities with some degree of movement, the majority of the body's cells are knitted together to form tissues; the body's cells are joined together by specialized membrane junctions.

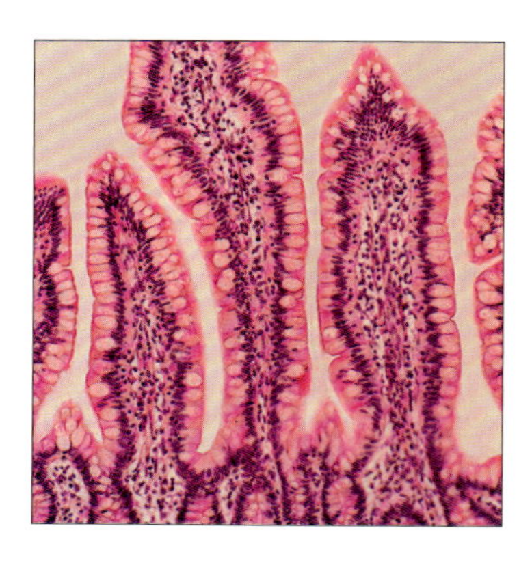

Microvilli in the lining of the small intestine increase the surface area to absorb nutrients. They also contain specialized cells that produce anti-bacterial chemicals.

Membrane proteins

Proteins embedded in the cell membrane play an important role in many cellular functions. Some span the cell membrane, connecting the internal cell to the external environment, allowing cells to communicate with each other chemically.

Membrane proteins are responsible for most of the specialized functions of the cell membrane. They can be broken down into two main groups:

- Integral proteins – while some integral proteins protrude through the cell membrane on one side only, the majority cross the membrane and so are exposed to both the inside and outside of the cell.

  These 'trans-membrane' proteins often allow substances to be exchanged between the internal and external environments, either by providing a pore through the membrane, or by physically ferrying the molecules across.
- Integral proteins also provide binding sites for chemicals released by other cells; this allows cells, including neurones (nerve cells), to communicate with each other.
- Peripheral proteins – these are not embedded in the phospholipid bilayer but are usually attached to the internal side of integral proteins. They can act as enzymes, which speed up chemical reactions inside the cell and are involved in changing the cell's shape, for example, during division.

Functions of membrane proteins

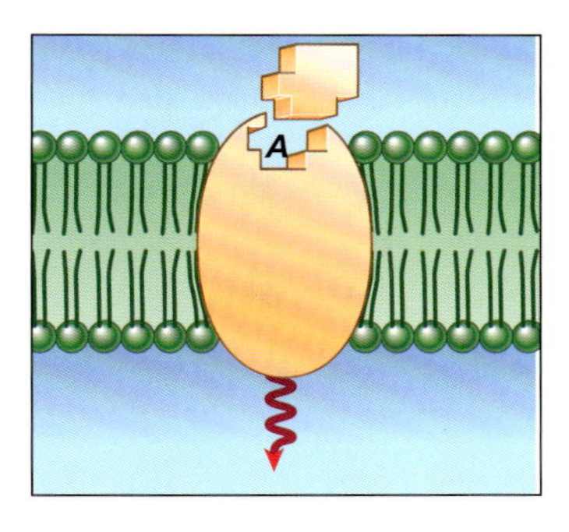

The external surface of some proteins provides a 'binding site' (A) for chemical messengers released from other cells.

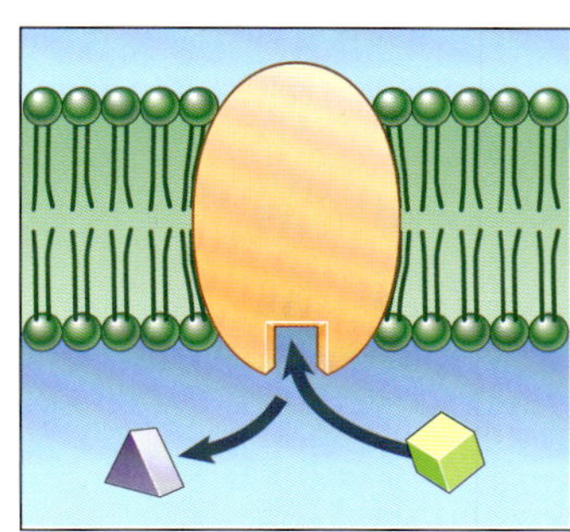

The internal surface of some proteins acts as an enzyme, speeding up chemical reactions that occur inside the cell.

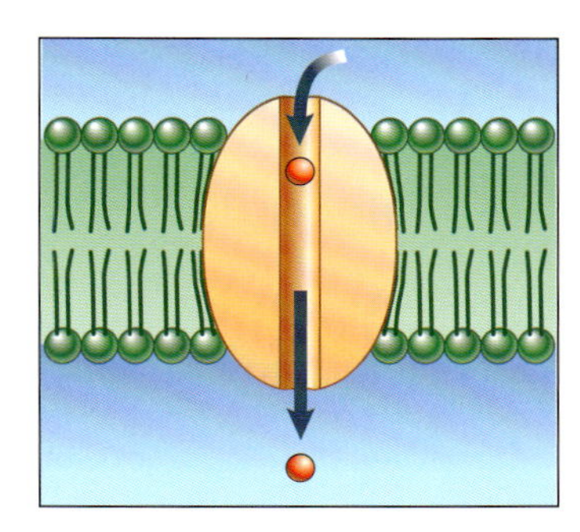

Transport proteins span the membrane, providing a pore for chemicals to travel either into, or out of, the cell.

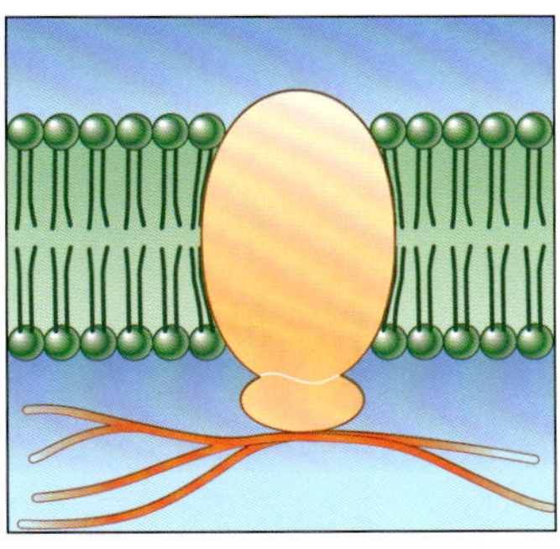

The internal scaffolding of the cell (cytoskeleton – red strands) attaches to the internal surface of membrane proteins.

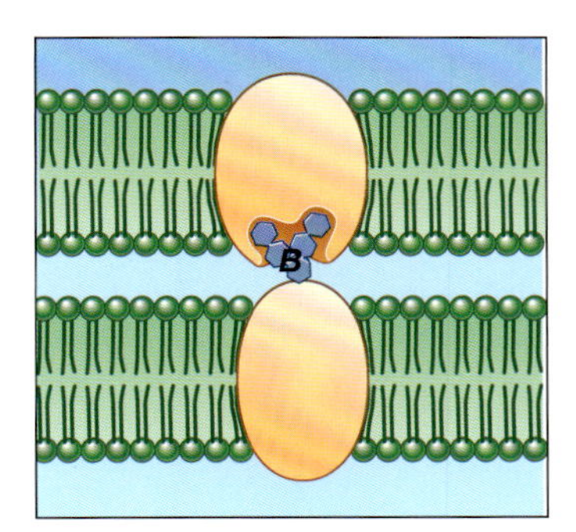

Some glycoproteins (molecules made of proteins joined to carbohydrates) act as 'identification tags' (B).

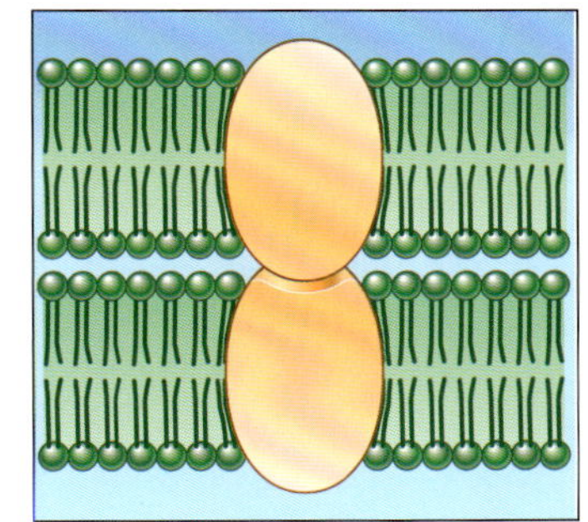

Membrane proteins of adjacent cells may join together, providing various kinds of junctions between the two cells.

Phospholipids

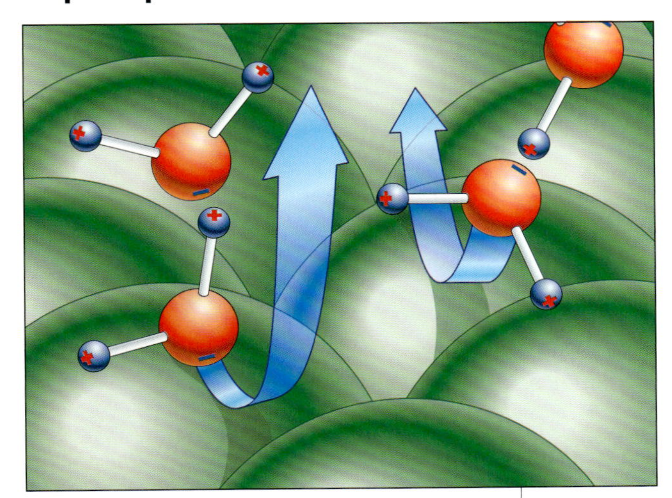

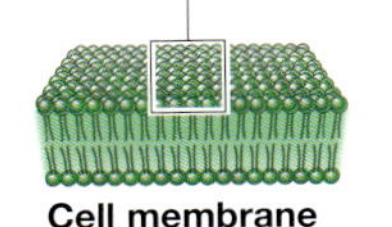

Water molecules are composed of two hydrogen atoms attached to one oxygen atom (H_2O). Although they have no net electrical charge, the oxygen at one end of the molecule tends to be slightly negative and the two hydrogens at the other tend to be slightly positive. Water is thus said to be a 'polar' molecule since, like a magnet, it has two electrical 'poles'. This results in water molecules interacting with each other electrically: the negative oxygen in one water molecule is attracted to the positive hydrogens of neighbouring water molecules. The degree of this attraction, which depends on the temperature, determines whether the molecules form ice, water or steam. Water also interacts with other polar molecules, such as glucose. However, non-polar molecules, which include fats, are insoluble in water.

Water molecules are made of a slightly negative oxygen atom (red), attached to two slightly positive hydrogen atoms (blue). Since the phospholipid 'tails' are non-polar, the cell membrane is impermeable to water.

Phospholipids and water

Phospholipids are ideal building blocks for the membrane, which is designed to separate the contents of the cell from the outside environment. A phospholipid molecule is made of a phosphorus-containing, 'water-loving' head attached to a lipid-containing, 'water-hating' tail. When phospholipids are mixed with water, the 'water-loving' heads mix with, while the 'water-hating' tails avoid, the water molecules. Thus water molecules can only pass through the cell membrane through pores in proteins that are embedded within it.

Passive Transport of Molecules

Cells in the body must control their internal environment in order to function properly. The cell (or plasma) membrane provides a barrier that regulates the passage of chemicals passing into and out of the cell.

Each cell in the body is surrounded by a cell, or plasma, membrane. This is an important barrier that separates the cells' internal environment from their external environment. The cell membrane plays a vital role in regulating the contents of cells so they are able to function.

Semi-permeable Membranes

However, the cell membrane is not an impenetrable barrier. It allows some substances to pass freely across it, while restricting, or totally preventing, the passage of other chemicals; hence the term 'semi-permeable membrane'.

For example, glucose is an essential molecule that provides the body with energy and is able to move easily through the cell membrane. Unused glucose is converted into glucose-6-phosphate, a chemical that is unable to travel back across the cell membrane.

Other molecules that can pass through the cell membrane easily include oxygen, which is used in the metabolism of glucose; and carbon dioxide, a waste product that diffuses out of the cell.

Protein Pores

Other particles, such as sodium ions or amino acids, can only pass through the membrane via 'pores' provided by specific membrane proteins; many of these act like gates – opening or closing only in response to a predetermined chemical signal.

For example, some protein pores only open when a chemical released from another cell (for example a hormone) binds to its external surface. Other protein pores open in response to a change in electrical voltage.

Passive transport

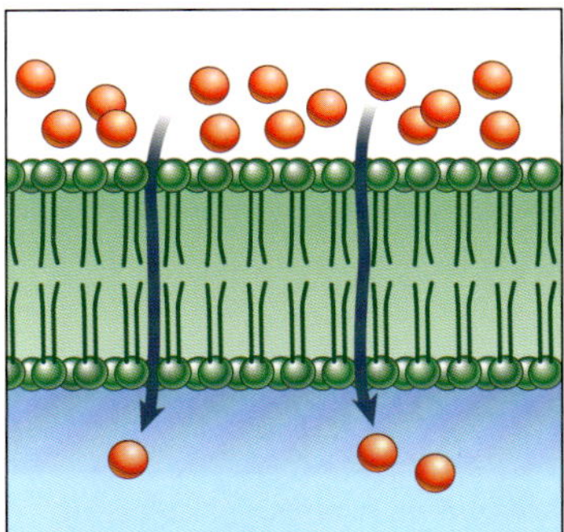

SIMPLE DIFFUSION
Some molecules, such as steroids, can travel freely across the cell membrane (green) into the cell.

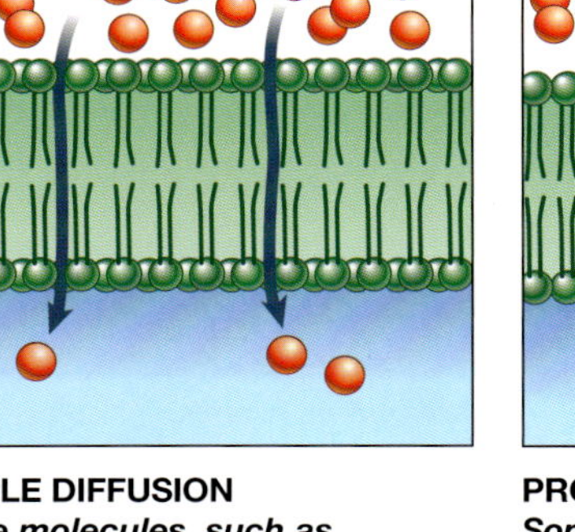

PROTEIN PORES
Some proteins provide 'pores' that allow small chemicals, such as sodium atoms, to cross the cell membrane.

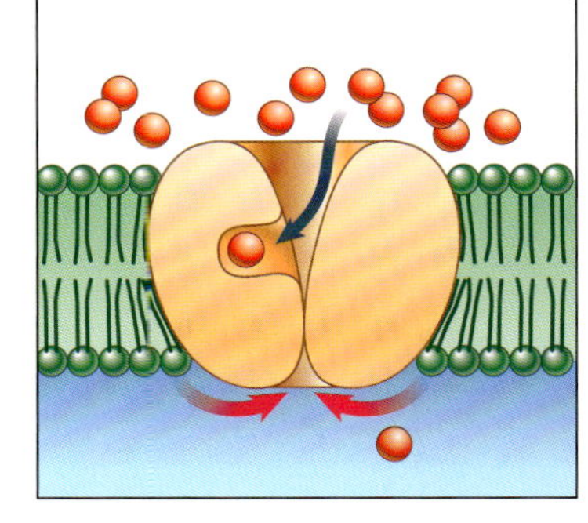

FACILITATED TRANSPORT 1
Larger molecules, such as glucose, are 'ferried' across the membrane. These differ from pores as they are not permanently open.

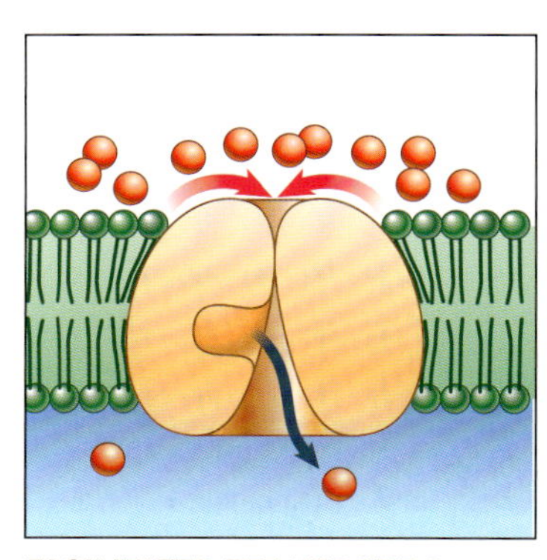

FACILITATED TRANSPORT 2
The protein channel alters its shape slightly after the chemical has 'docked' with it, allowing the chemical to be released on the inside of the cell membrane.

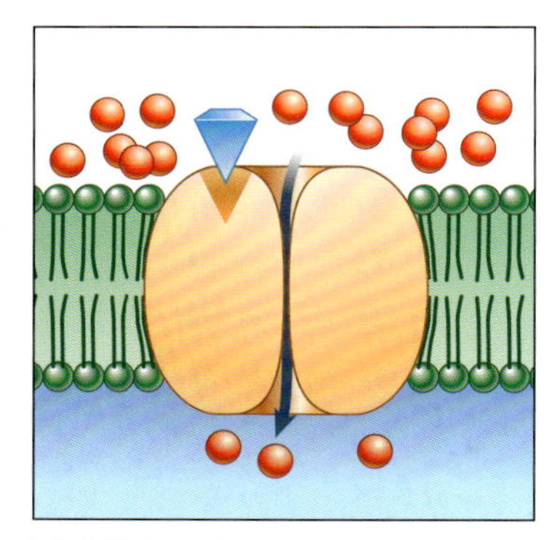

RECEPTOR-MEDIATED OPENING OF PROTEIN PORE 1
Some protein channels only open when a messenger molecule (often released by another cell) binds to the outer surface of the protein, like a key.

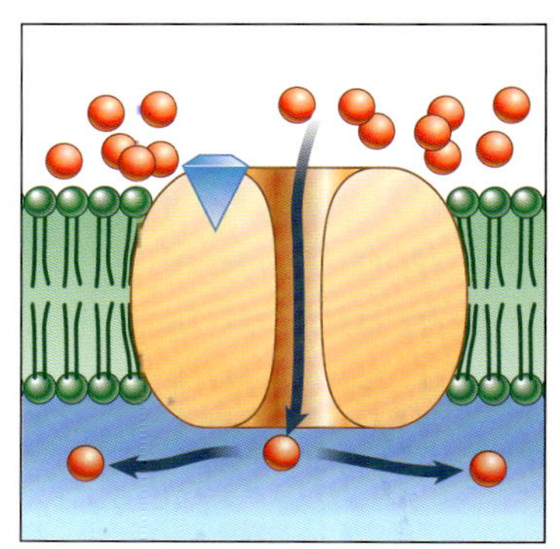

RECEPTOR-MEDIATED OPENING OF PROTEIN PORE 2
These 'receptor proteins' are especially important in neurones (nerve cells), as they allow one neurone to influence the internal environment of another.

Osmosis

Osmosis is a passive process by which water molecules move across a membrane from an area of high concentration to one of low concentration.

This process is illustrated below, with a two-compartment water vessel separated centrally by a semi-permeable membrane that has small pores allowing the passage of water but not larger molecules. The left side contains water and the right side a sugar solution indicated by the red dots. Water molecules are 'pulled' across the membrane into the sugar solution by osmotic pressure.

Osmosis plays an important role in the human body. For example, blood volume can be controlled by altering the concentration of sodium in the urine; water flows into the urinary tract as a result of osmosis and is then excreted, thus lowering the blood volume.

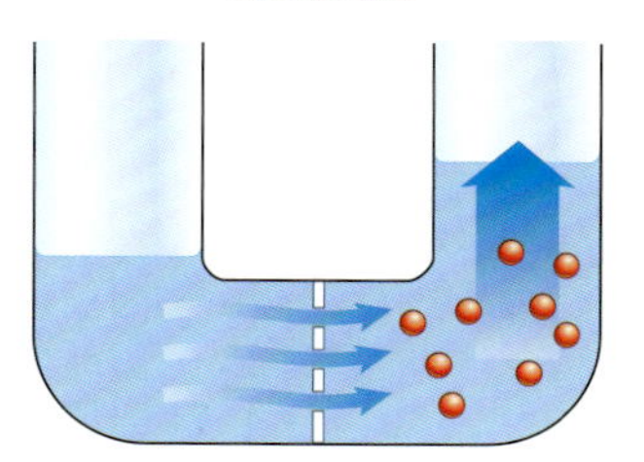

The membrane is permeable to water, but not sugar molecules; therefore water travels across the membrane to the sugar solution.

Active Transport

Some molecules are unable to cross the cell membrane unaided. This may be because they are insoluble in the cell membrane or too large to pass through the protein pores.

When cells require substances to move across the membrane, they can use active transport to achieve this against a concentration gradient, that is from an area of low to high concentration. Unlike diffusion or osmosis, active transport requires the cell to use energy.

There are two main types of active transport: membrane pumps and vesicular transport. Membrane pumps are proteins that run through the membrane and are able to propel a small number of molecules across the membrane. Vesicular transport, in contrast, involves the transport of many molecules.

Membrane pumps

Simple active transport

One way of carrying a chemical against its concentration gradient (in this case from inside to outside a cell) is for a protein to ferry it; this process requires energy.

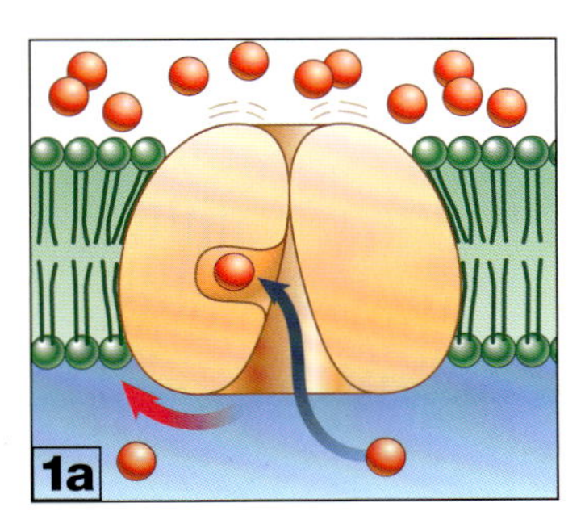

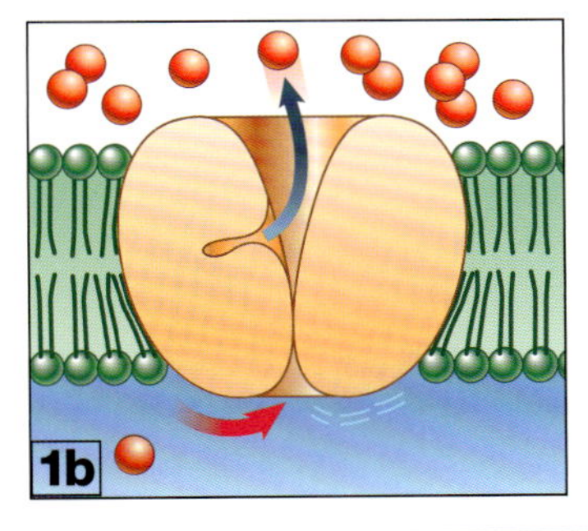

Co-transport

If one molecule is diffusing down its concentration gradient (red spheres), another molecule (purple triangles) can 'hitch a ride' and so travel against its concentration gradient.

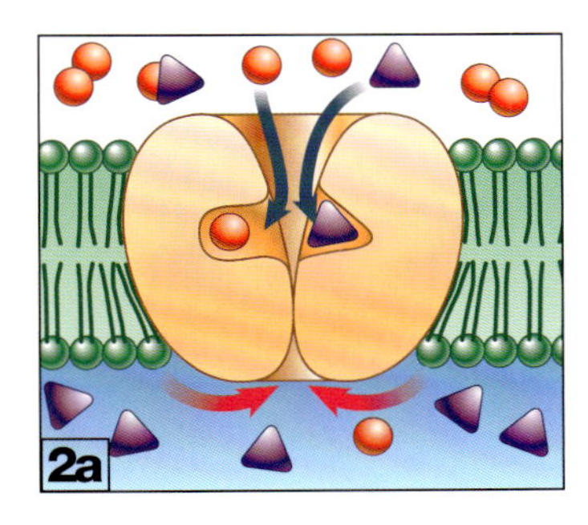

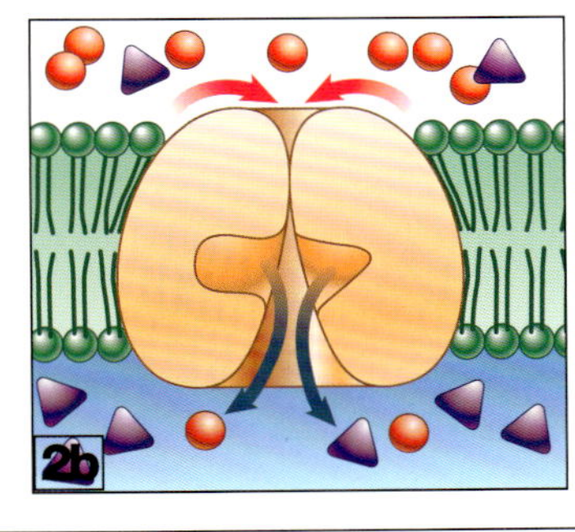

Counter-transport

Counter-transport is similar to co-transport in that the 'downhill' movement of one chemical provides the energy for another chemical to be transported 'uphill'.

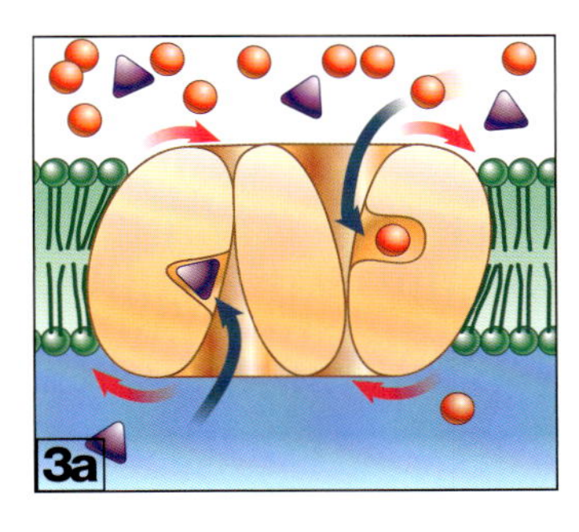

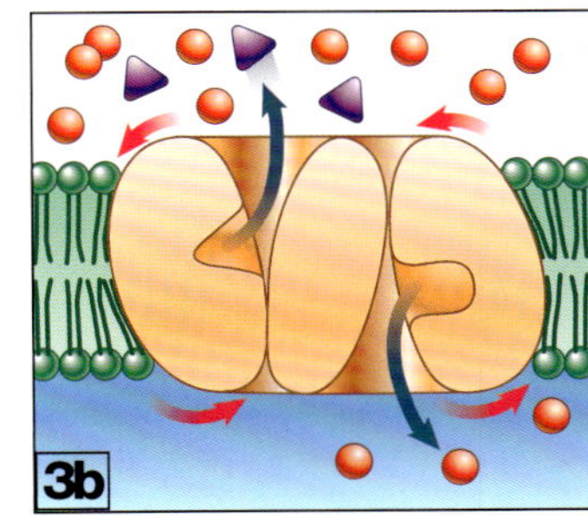

Endocytosis

Endocytosis is in many ways the complete opposite of exocytosis; it is a process that allows substances to be taken into the cell from the external environment. There are three main types of endocytosis:

- Phagocytosis (literally 'cell eating'): large, solid material (for example, a bacterium) is engulfed by the cell's membrane and taken into the cell where it is digested
- Pinocytosis (literally 'cell drinking'): droplets of fluid from outside the cell, containing dissolved molecules, are engulfed
- Receptor-mediated endocytosis: when only very specific molecules are engulfed.

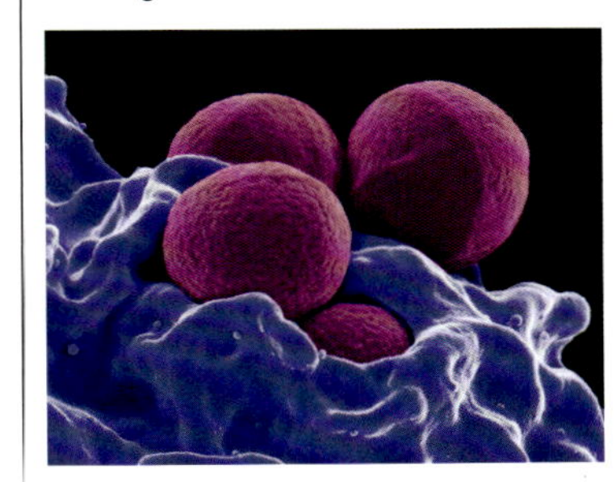

A white blood cell in this magnified image engulfs staphylococcus aureus bacteria (in pink).

Vesicular transport – exocytosis

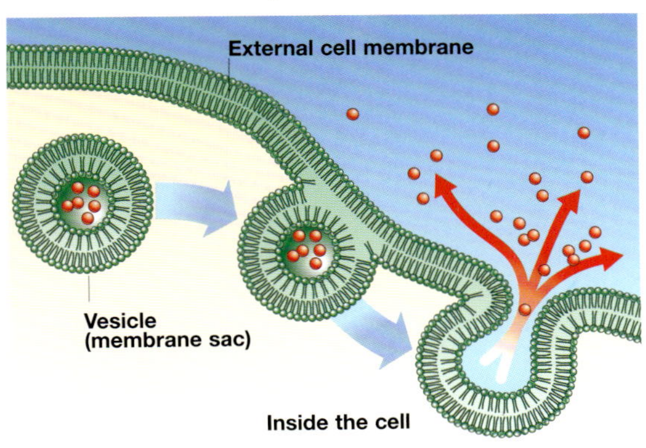

The chemical to be released from the cell is packaged in a membrane sac (vesicle). This sac fuses with the cell membrane, releasing its contents.

Exocytosis is a process by which large quantities of a substance are transported from the inside to the outside of the cell. Like membrane pumps, exocytosis requires energy. This mode of transport is responsible for the secretion of hormones from endocrine glands and for the release of neurotransmitters from nerve cells. As a result, exocytosis plays a major role in allowing communication between cells.

Process of exocytosis

The substance to be released from the cell is first packaged within a sac, called a 'vesicle', composed of phospholipids and proteins, just like those found in the plasma membrane. The vesicle then moves to the cell membrane. Proteins on the vesicle recognize, and bind with, proteins on the membrane causing the two membranes to fuse and finally rupture, spilling the contents of the vesicle outside of the cell.

The cell membrane does not increase in size as more of these vesicles dock with it. Instead these vesicles are constantly recycled.

Deoxyribonucleic Acid (DNA)

DNA is the genetic material of all organisms, located in the nucleus of every cell. The discovery of its chemical structure revolutionized the biological sciences and our understanding of human genetics.

The chemical properties of DNA (deoxyribonucleic acid) allow it to carry out two highly important functions:

- It provides the body's cells with the 'recipes' needed to build proteins from the 20 essential amino acids found in proteins
- It is able to make copies of itself, and so provides the means by which these protein 'recipes' can be transmitted from one generation to the next; this means that characteristics such as eye colour or facial features can be passed from parent to child.

Chromosomes

The vast majority of human DNA is packaged into 23 different pairs of chromosomes, which are stored within the cell's nucleus; one set of 23 chromosomes is inherited from the father and one set is inherited from the mother.

The exceptions to this rule are sperm cells and egg cells, which contain only one set of 23 chromosomes; and red blood cells, which contain no chromosomes at all.

Genes

Useful DNA – as opposed to so-called 'junk' DNA (see below) – is packaged within the chromosomes into what are known as genes, of which there are thought to be around 100,000 in the human body. Each of these genes provides the 'recipe' that tells the cell how to make a specific protein.

However, while each of the body's cells contains a copy of every protein recipe, not all of these recipes are 'switched on'. This is what differentiates a heart cell from, for example, a liver cell – each produces its own set of proteins.

'Junk' DNA

Most DNA is so-called 'junk' DNA, which serves no known purpose in the human body. Much of this DNA is inherited from our distant ancestors and their parasites, dating back to when life on earth began around four billion years ago.

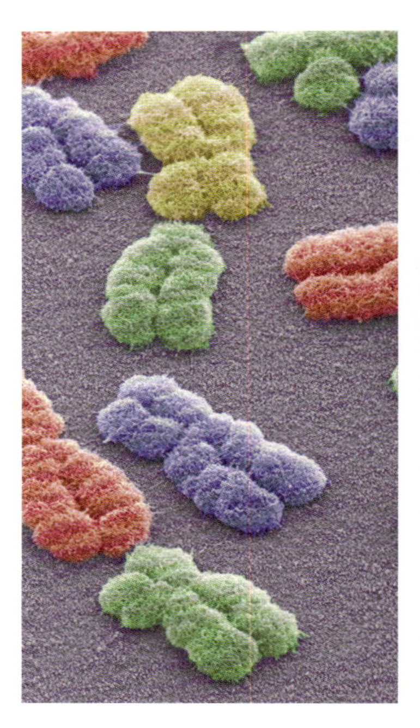

In humans, DNA is packaged within 23 pairs of chromosomes. These 'X' shaped structures replicate during cell division.

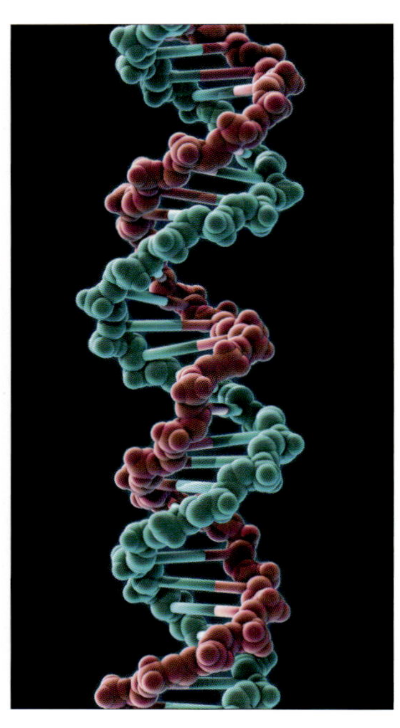

DNA is composed of two strands of nucleotides

DNA function

The ability of DNA to make copies of itself is the direct result of its chemical structure. A molecule of DNA consists of two interlinked strands of nucleotides, which are exact mirror images of each other, and which run in opposite directions.

James Watson (left) and Francis Crick (right) were awarded a Nobel Prize for their contribution to the discovery of DNA structure.

These two strands are each made up of a sugar-phosphate 'backbone', to which are attached specialized molecules, called bases. These bases are: adenine, guanine, cytosine and thymine (abbreviated to the letters A, G, C and T). What gives DNA its special properties is that the four bases will only pair off in the following combinations: A with T and G with C. Thus a strand of DNA with the sequence of 'TGATCG' will only bind with a complementary strand with the sequence 'ACTAGC'.

DNA is made of two strands, which run in opposite directions. The two strands are joined by special molecules called bases.

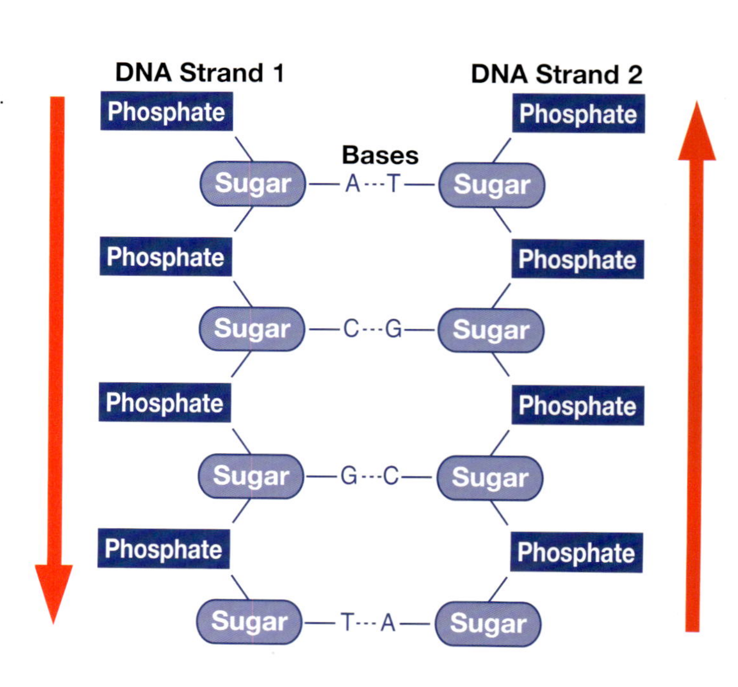

DNA structure

DNA Replication

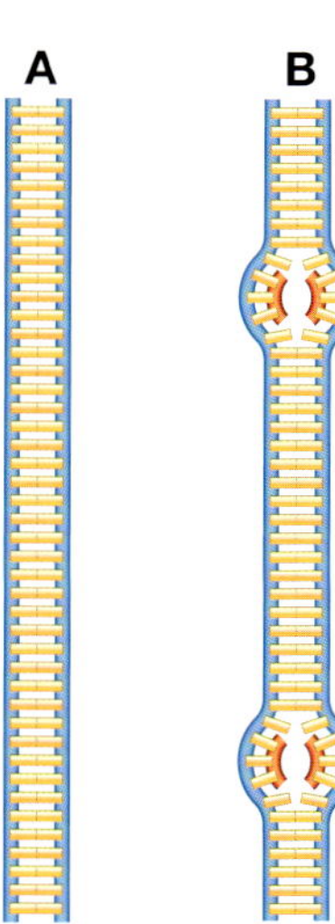

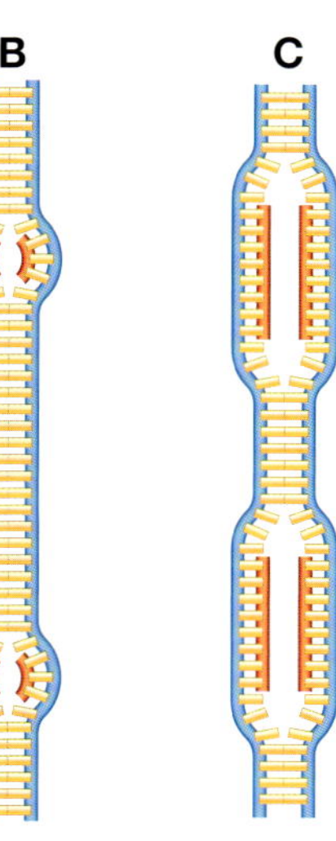

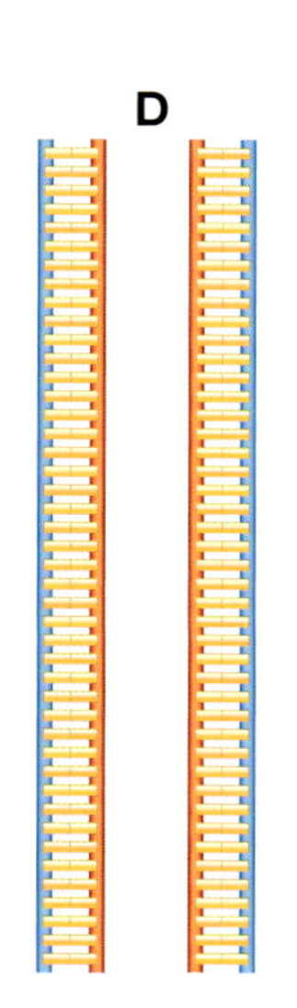

DNA makes a copy of itself during a process called replication. First, the original double helix 'parent' DNA is unwound, so separating and exposing the base pairs. Since each base will only bind with one of the three others (such as A with T, but not with G or C), a complementary strand can then be built up on each of the parent strands. Thus one DNA double helix becomes two identical double helices.

DNA (A) is simultaneously unwound at a number of points (B). A new strand (red) is built on each parent strand (B and C) forming two DNA molecules (D). The two strands of a DNA molecule are separated from each other and a new strand is built onto each one. Thus, two identical strands can be made from the original.

DNA molecule

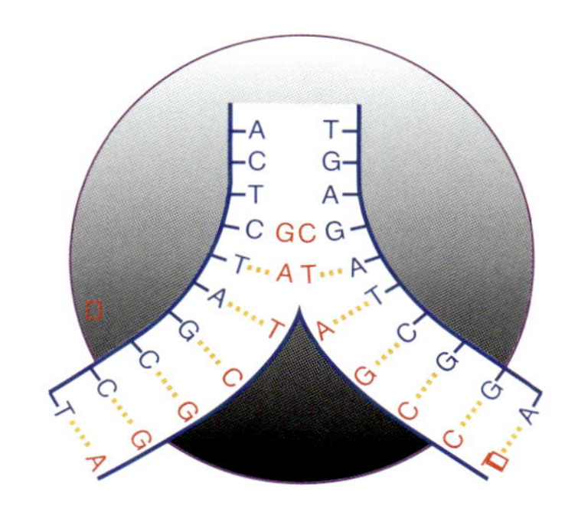

Ribosomes

In many ways DNA is similar to a language. However, unlike the English language, there are only 64 three-letter 'words' that can be formed using the four letters A, G, C and T. Geneticists call these 'words' codons as each codes for a specific amino acid, the building blocks of proteins.

Proteins are made on structures called ribosomes in the cytoplasm. However, since DNA is unable to leave the nucleus, first of all one strand of DNA is 'transcribed' onto a single-stranded messenger molecule, with a very similar structure to a DNA strand, called messenger RNA (mRNA), which can cross the nuclear membrane. The mRNA is then 'translated' on the ribosomes in the cytoplasm so that the correct amino acids are joined together in the correct order.

Proteins are built from amino acids according to a template copied from DNA. This process occurs in subcellular structures called ribosomes, which are found inside the cell.

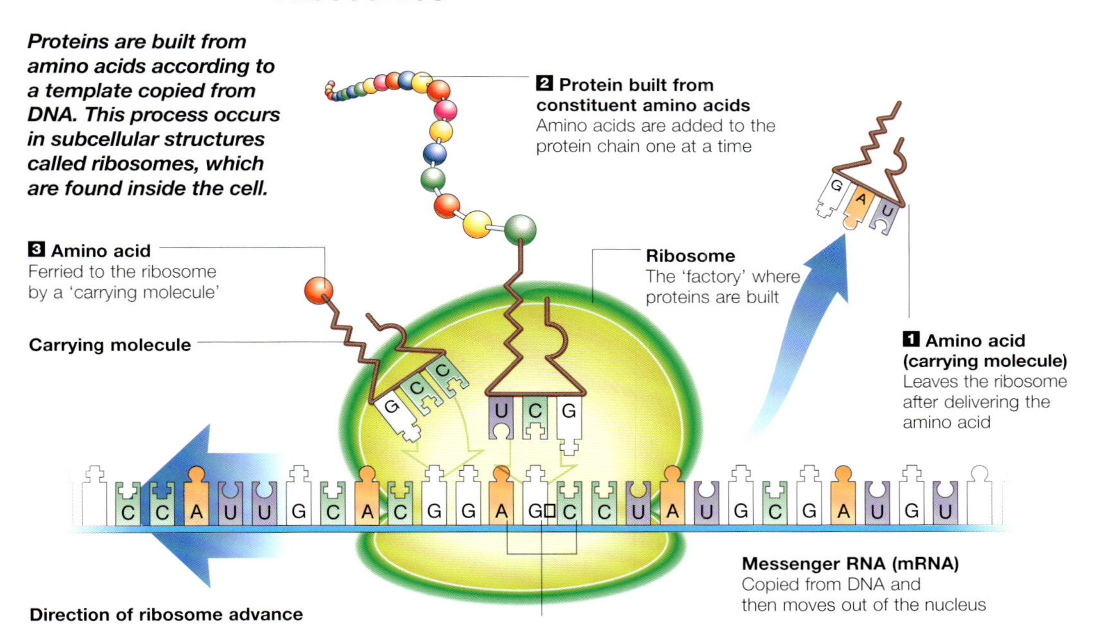

Genes

Faulty genes do not always lead to disease. It is possible for people who are normal to carry abnormal genes and only learn of the problem when they have an affected child by another, equally normal carrier.

An observable characteristic in a person is called a phenotype. This could be a disease, a blood group, eye colour, nose shape or any other such attribute. The genetic information that gives rise to a phenotype is known as the genotype.

Alleles

A gene locus is the site on a chromosome at which a gene for a particular trait lies. The different forms of a gene that may be present at a gene locus are called alleles. If there are two alleles – 'A' and 'a' – for a given gene locus, then three genotypes may be formed. These are: 'AA', 'Aa' or 'aa'. 'Aa' is known as a heterozygote and 'aa' and 'AA' are homozygotes.

If the 'A' allele is dominant, it will mask the effects of the recessive 'a' allele, and produce a recognizable phenotype in a heterozygous (Aa) individual. Recessive alleles only result in a recognizable phenotype in the homozygous state (aa).

Co-dominance

If both alleles are recognized in the heterozygous state (Aa), they are seen as co-dominant. The expression of the ABO blood groups is an example of the effects of co-dominance.

Dominant and recessive genes

In these family trees, the parents have a dominant allele 'A' and recessive allele 'a'. If 'A' represents the phenotype of brown eyes, and 'a' represents blue eyes, only those with the 'aa' genotype will have blue eyes. Otherwise, the dominant 'A' allele dictates the phenotype.

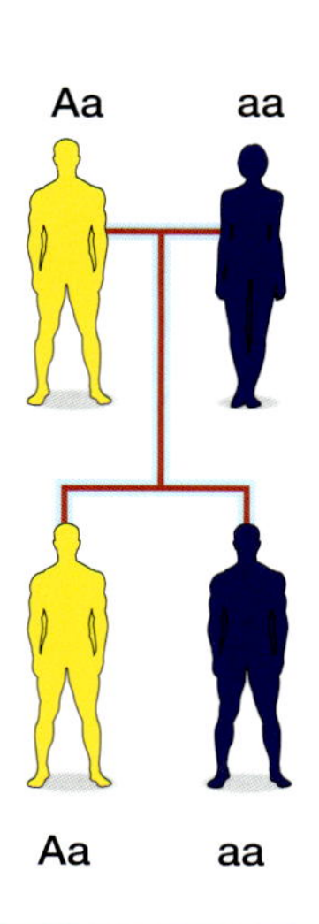

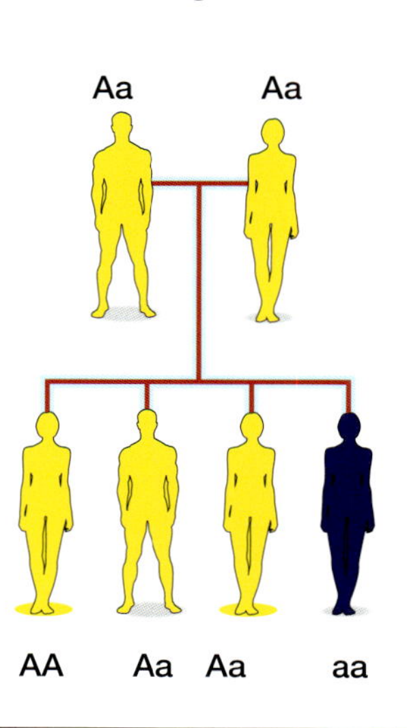

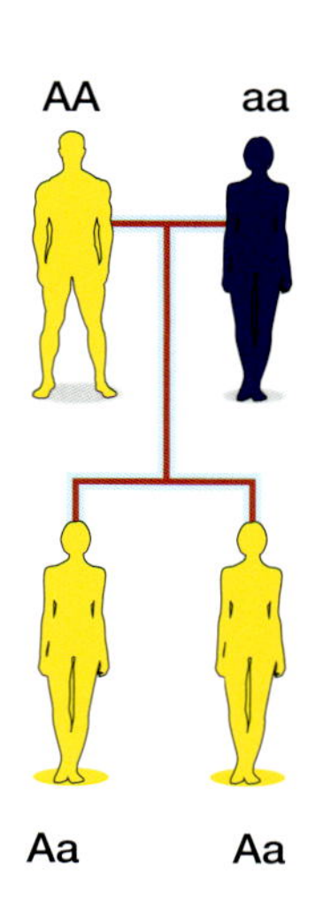

Autosomal dominant conditions

Affected individuals (male or female) who carry a dominant abnormal gene have a 50 per cent chance of producing affected offspring with a normal partner. Only people who inherit a copy of the gene will be affected. Achondroplasia (dwarfism) is an autosomal dominant condition.

An individual is either normal or affected by an autosomal dominant condition. When their partner is unaffected there is a 50 per cent chance that a child will be affected.

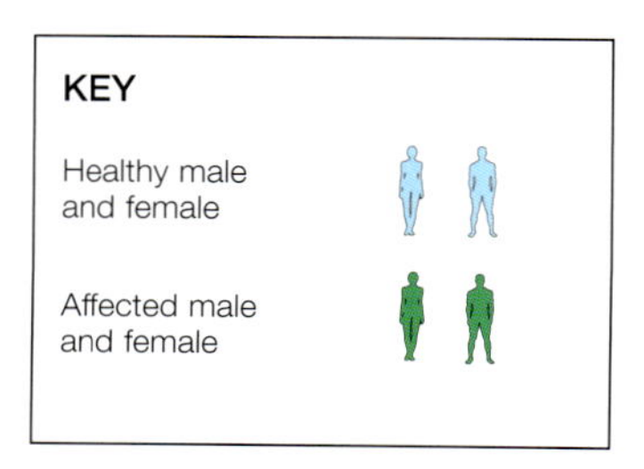

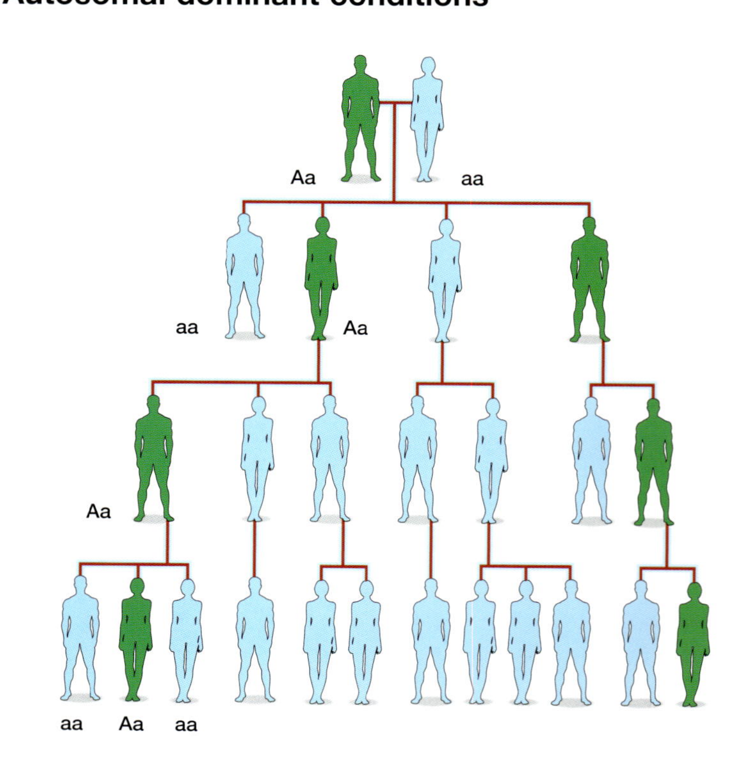

Autosomal recessive conditions

Unaffected individuals of either sex may be carriers. When two carriers (Aa; heterozygotes) have a child, there is a one in four (25 per cent) risk that the child will be affected. An example of an autosomal recessive condition is sickle cell disease, a disease of the blood that mainly affects people of African ancestry.

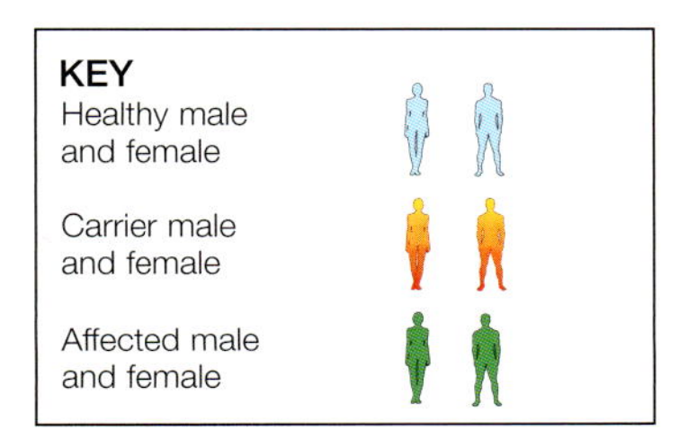

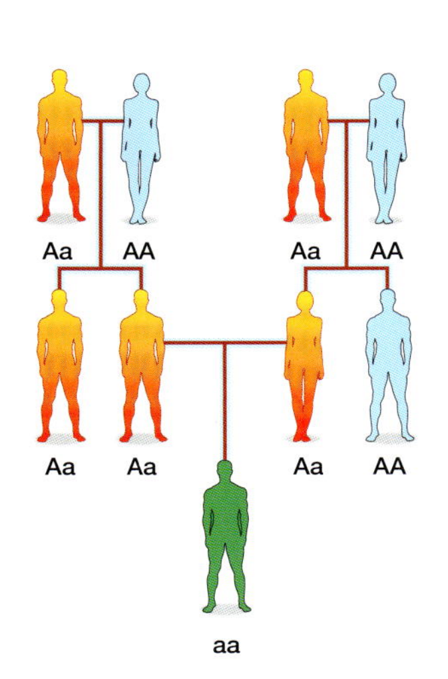

Sickle cell disease affects haemoglobin, causing red blood cells to distort into a distinctive sickle shape. The condition occurs when an individual inherits the sickle cell gene from both parents.

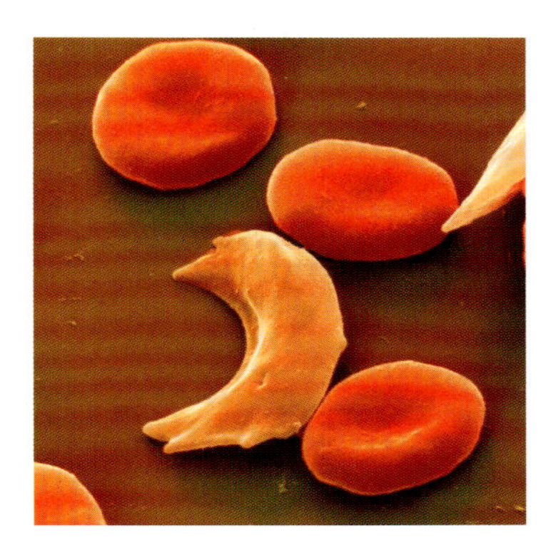

Sex-linked conditions

In these conditions, the abnormal trait is carried on the sex chromosomes (X and Y). Males only have one X chromosome, therefore all daughters inherit their father's X chromosome. They will also inherit one of the two X chromosomes from their mother. A son will inherit his father's Y chromosome and one of his mother's two X chromosomes.

If one of the two X chromosomes from the mother contains a gene that can give rise to a disorder, she is referred to as a 'carrier'. Half the sons of a carrier female are likely to be affected. Half the daughters will be carriers for the gene. Males are clinically affected because they carry only one X chromosome, while females are unaffected because they have two X chromosomes. Affected males can only inherit the gene through a female line.

A well-known example of an X-linked disorder is the haemophilia that affected Queen Victoria's family line (see right). Baldness can also be X-linked.

Y-linked traits include genes for sex determination and male development. Father-to-son transmission is only possible in Y-liked traits as the Y chromosome is only inherited by sons.

This family tree shows the sex-linked recessive inheritance of haemophilia in the royal families of Europe. All affected individuals can trace their inheritance back to Queen Victoria in the 19th century, who was a carrier of the disease. The current British royal family is unaffected as they are descended from an unaffected individual (Victoria's son, King Edward VII).

Figures represent multiple siblings

Queen Victoria

2

Frederick III of Germany

Edward VII

Leopold of Albany

Maurice of Battenburg

Frederick of Hesse

Tsar Nicholas II of Russia

Alfonso XIII of Spain

Leopold of Battenburg

4

2

Waldemar of Prussia

Henry of Prussia

Tsarevich Alexis of Russia

Rupert

Prince Alfonso of Spain

Gonzalo of Spain

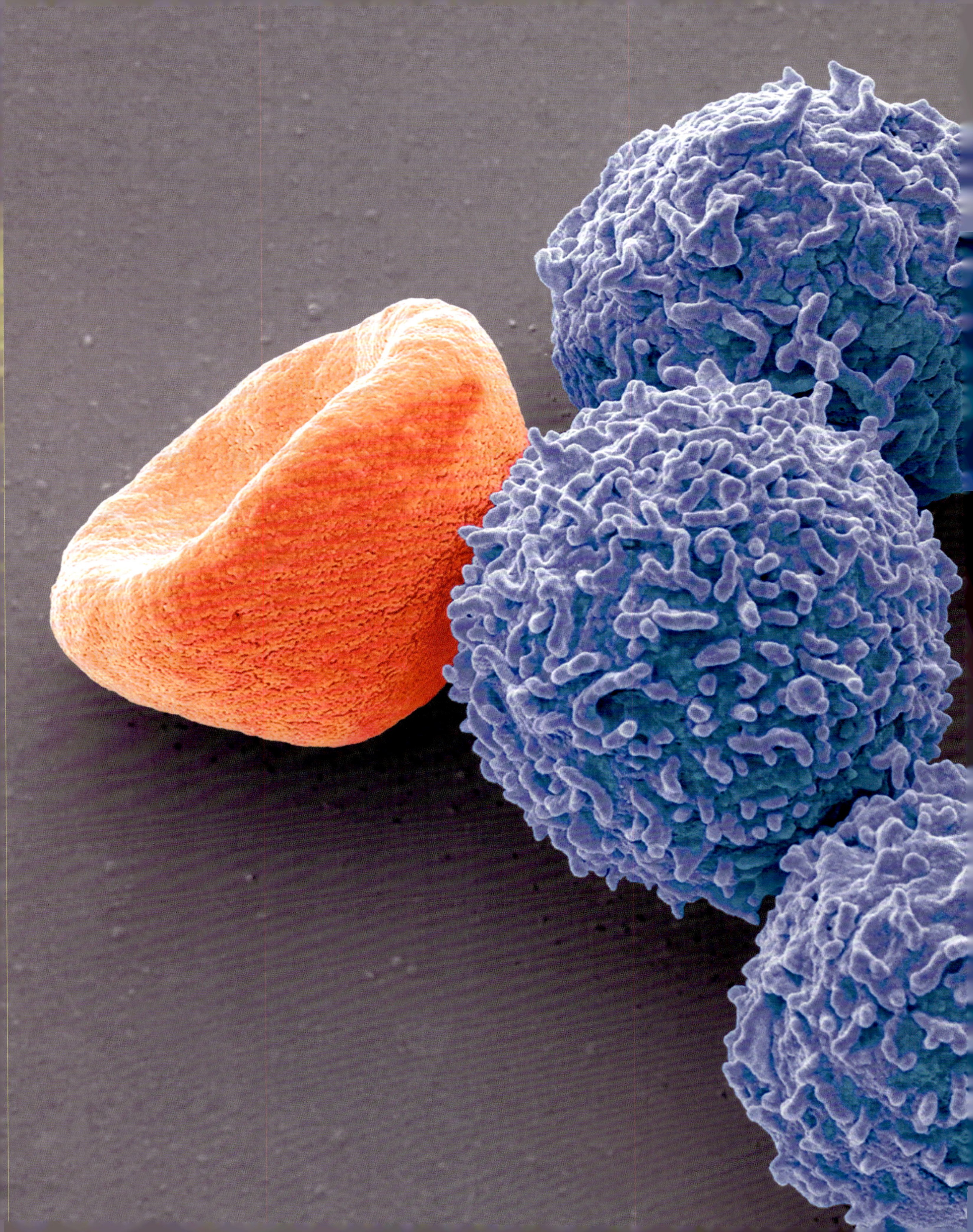

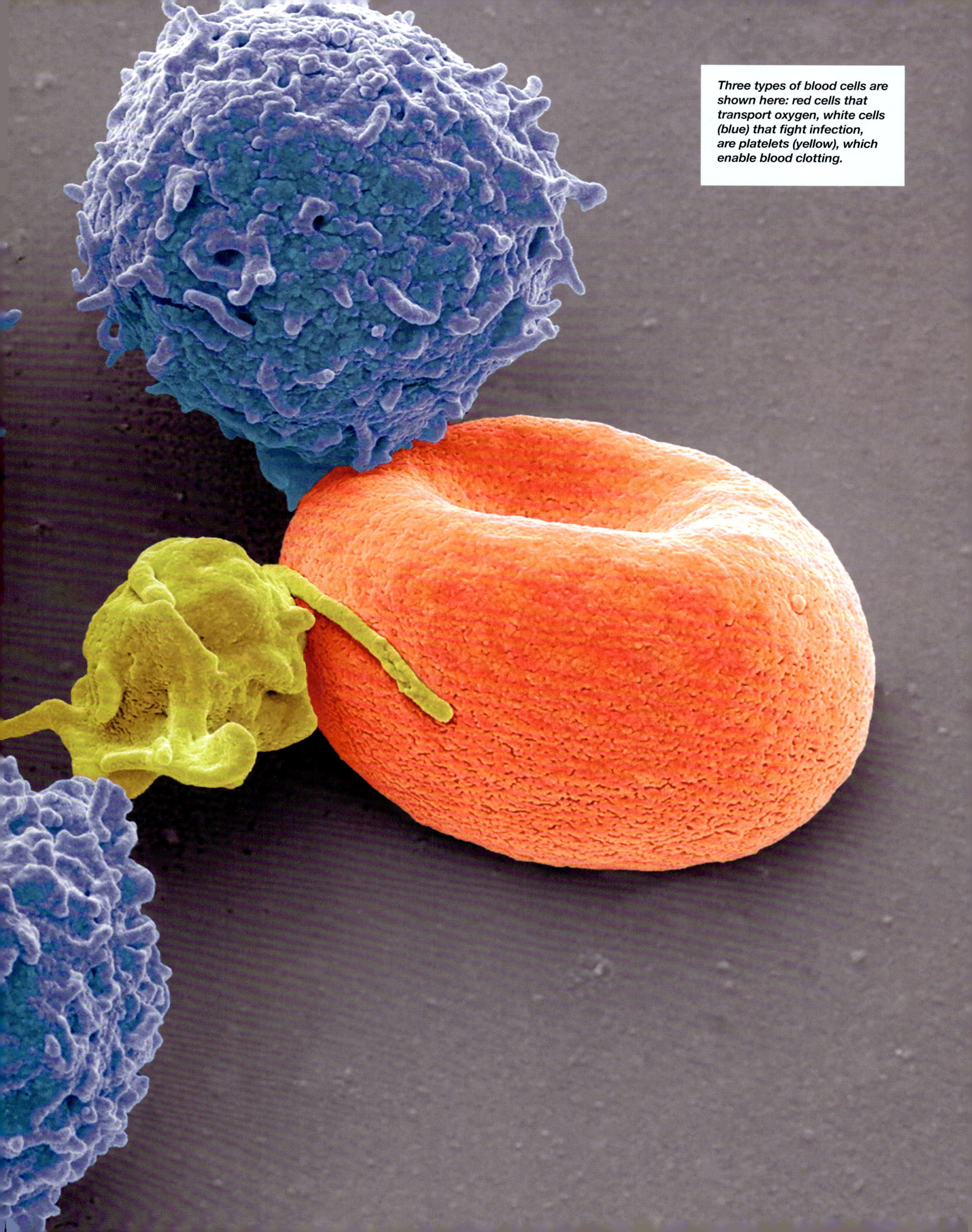

Three types of blood cells are shown here: red cells that transport oxygen, white cells (blue) that fight infection, are platelets (yellow), which enable blood clotting.

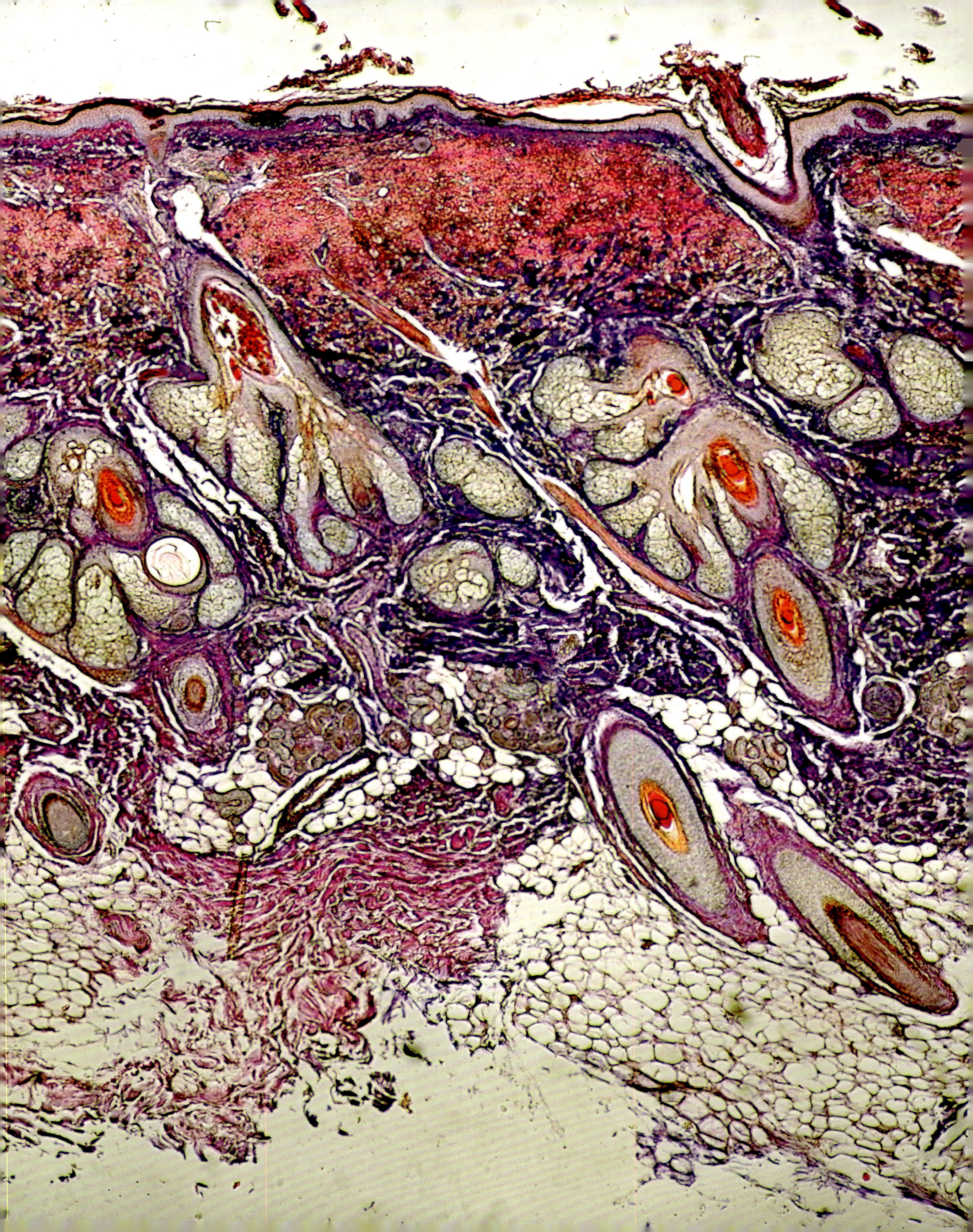

CHAPTER 2

Skin, Hair and Nails

The skin and its glands, hair and nails are collectively known as the integumentary system. Together they form a barrier that separates the external environment from the complex internal structures of the human body. The skin is one of the largest organs in the body, and in an average sized adult, has a surface area of up to 2sq m (21sq ft). The skin has numerous functions, including protection from infection and sunlight, maintaining body temperature, and processing vitamin D, which is vital for health. Hair develops from follicles in the skin and also has a protective role against the sun, insects, dust and other foreign bodies. Most of the skin is covered with hair, with some exceptions, such as the palms and soles. Nails are formed from the outer layer of skin, the epidermis, that hardens into a protein called keratin. They develop on the fingers and toes, buffering them against injury and enabling us to grasp objects.

Opposite: A cross section of the skin from the scalp shows hair follicles, sweat glands and the main layers of skin.

Skin

The skin, together with our hair and nails, makes up the integumentary system. Functions of the skin include heat regulation and defence against microbial attack.

The skin covers the entire human body and has a surface area of about 1.5–2 sq m (16–21sq ft). It accounts for about 7 per cent of the weight of the body and weighs around 4kg (10lb).

Two Layers

Skin is composed of two layers – the epidermis and dermis.

- Epidermis – this is the thinner of the two layers of skin and serves as a tough protective covering for the underlying dermis. It is made up of numerous layers of cells, the innermost of which consist of living cube-shaped cells that divide rapidly, providing cells for the outer layers.

By the time these cells reach the outer layers, they have died and become flattened, before being 'sloughed off' by abrasion. The epidermis has no blood supply of its own, and depends upon diffusion of nutrients from the plentiful supply of blood to the dermis below.

- Dermis – this is the thicker layer of skin, which lies protected under the epidermis. It is composed of connective tissue and has elastic fibres for flexibility, and collagen fibres for strength. The dermis contains a rich supply of blood vessels as well as numerous sensory nerve endings. Lying within this layer are the other important structures of the integumentary system, including hair follicles and oil (sebaceous) and sweat glands.

Cross-section of skin

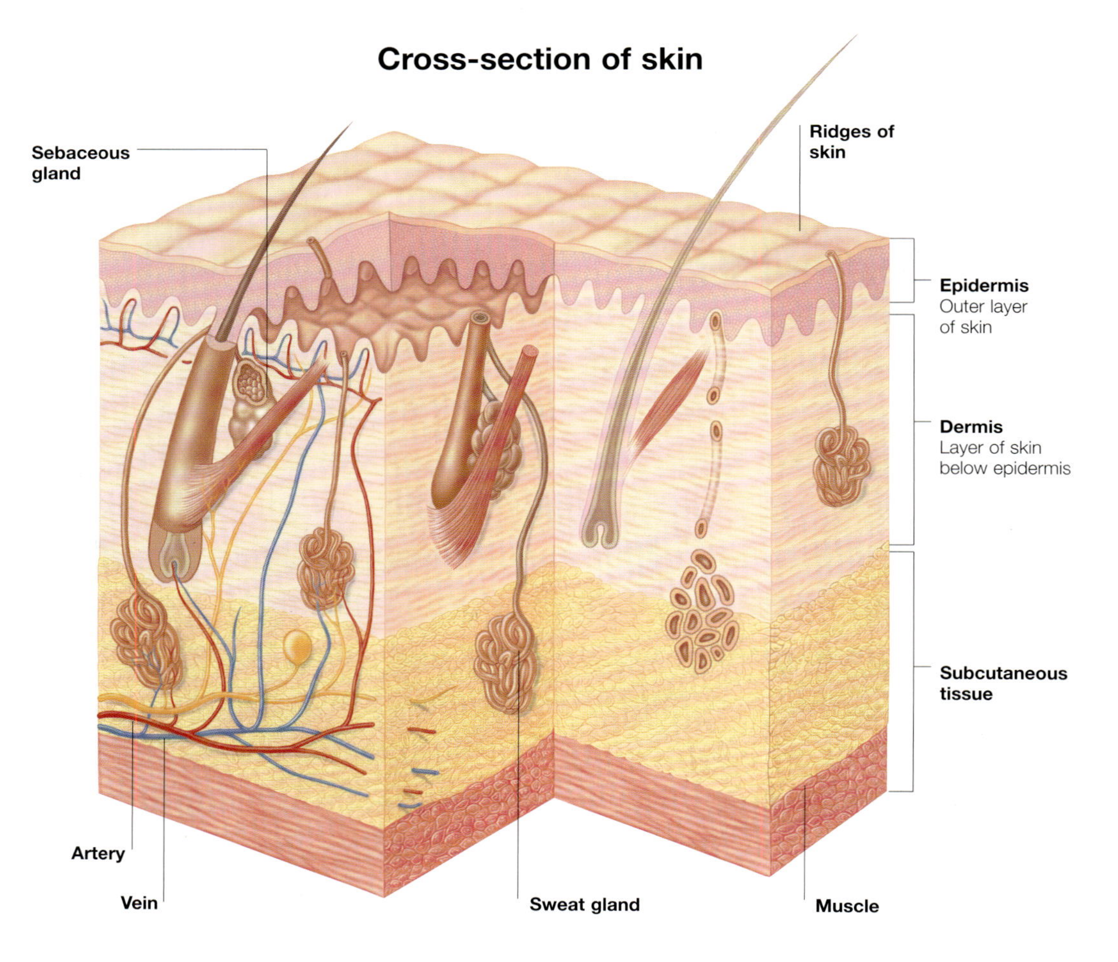

The skin is composed of two main layers: the epidermis and the dermis. The epidermis is nourished indirectly by the blood vessels in the dermis.

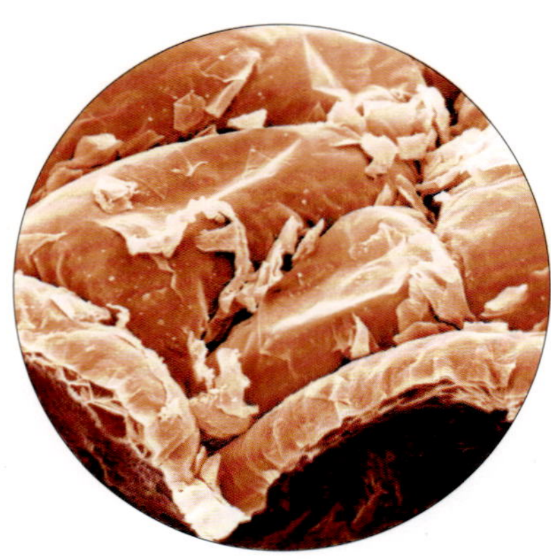

The surface of the skin, or epidermis, is formed of squamous epithelial cells. These cells die and slough off to be replaced by new cells.

The role of skin

The skin is a remarkable organ, covering the entire surface area of the body. Skin plays a number of important roles in protecting the body and also helps to control body temperature.

Skin performs a number of important roles. These include:

- Protection – the collagen fibres of the dermis give the skin strength and resistance.

- Regulation of temperature – through vasoconstriction (narrowing) and vasodilation (widening) of blood vessels in the dermis. The production of sweat also helps to cool the body.

- Barrier against bacterial infection – large numbers of micro-organisms are naturally present on the surface of the skin. These compete with harmful bacteria, preventing them from invading the body.

- Sensitivity to touch and pain – the dermis contains a dense network of nerve endings that are sensitive to pain and pressure. These nerves provide the brain with vital information about the body in relation to its environment, allowing it to act accordingly, for example, retracting the hand when something hot is touched.

- Prevention of unregulated water loss – the sebaceous glands of the dermis secrete an oily substance known as sebum. This coats the skin, making it effectively waterproof. Collagen fibres within the dermis also hold water.

- Protection against ultraviolet (UV) radiation – the pigment melanin (produced by melanocytes in the epidermis) acts as a filter to the harmful ultraviolet radiation produced by the sun.

- Manufacture of vitamin D – this is produced in response to sunlight and helps to regulate the metabolism of calcium.

Anatomy of the skin

The skin acts as a barrier, protecting the internal environment of the body from heat, infection and dehydration. It also contains receptors to touch and pain.

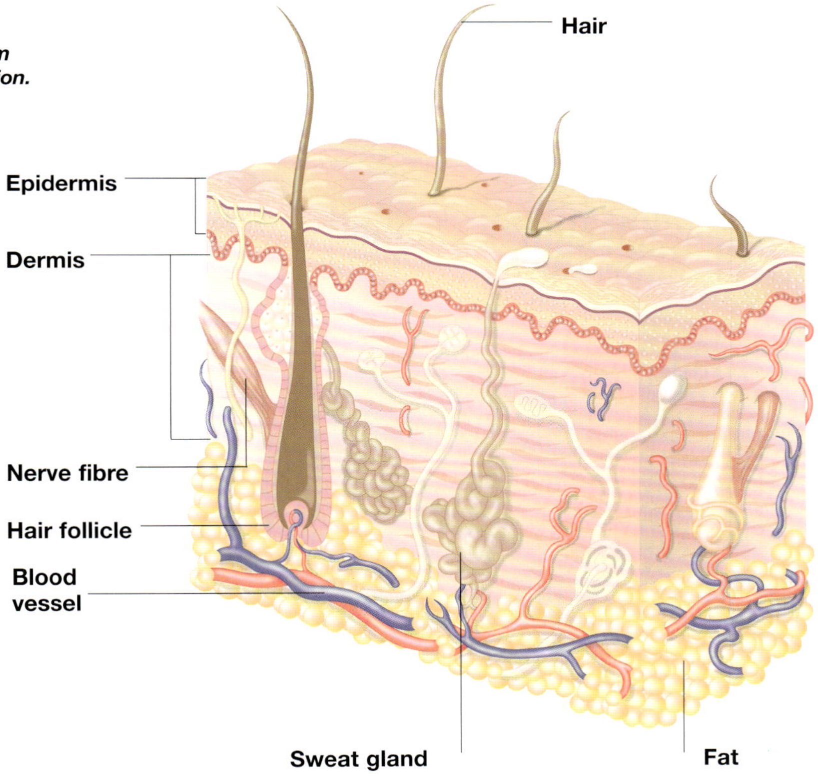

Hair
Epidermis
Dermis
Nerve fibre
Hair follicle
Blood vessel
Sweat gland
Fat

Skin pigmentation and repair

Skin colour largely depends on the presence of melanin. Production of this pigment protects the skin from harmful radiation produced by the sun.

The colour of the skin depends on a combination of factors, such as skin thickness, blood flow and pigment concentration.

Pigments

In areas where the skin is very thin and blood flow is good, it will appear much darker (such as over the lips) due to the red colour of the pigment haemoglobin in the blood.

The pigments that affect skin colour are carotene (found mainly in the dermis) and melanin (produced by melanocyte cells present in the epidermis). In general, the production of melanin determines the colour of the skin. Melanocyte numbers are approximately equal in everyone so skin colour is determined by the amount of melanin that is produced.

Sun exposure

Skin responds to ultraviolet rays in sunlight by producing greater amounts of melanin. As levels of melanin increase, the skin darkens forming a filter against the harmful radiation produced by the sun.

Freckles are another example of the skin's reaction to the sun, representing concentrated areas of melanin-producing cells.

Sunburn

If sun exposure does not take place gradually, however, the skin is unable to produce melanin fast enough to filter out the suns harmful rays.

As a result the skin burns, becoming inflamed and very tender. Prolonged exposure to UV radiation can permanently damage the skin cells leading to premature ageing of the skin and, sometimes, skin cancer.

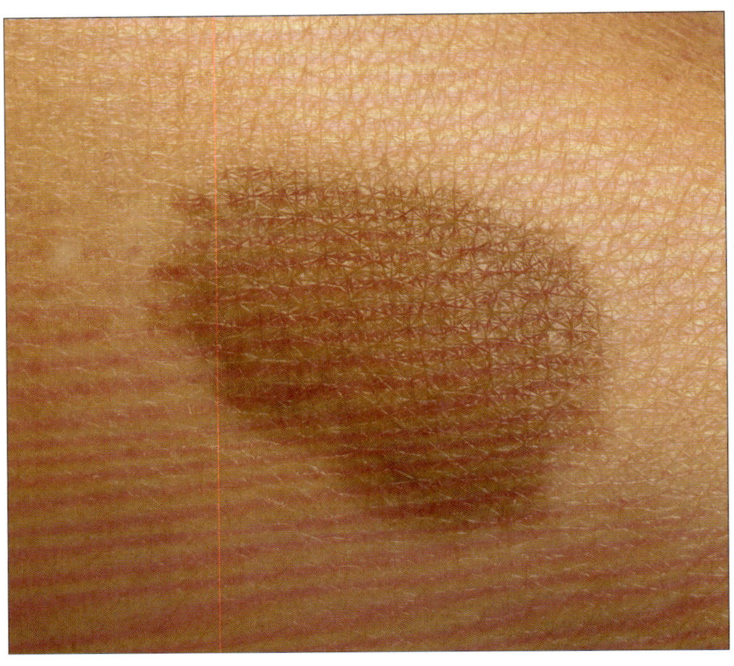

Birthmarks are formed of skin pigmentation (darkening) and are present at birth. They are usually harmless.

Skin repair

When skin is cut, for example, by surgical incision, the sides of the wound will grow back together if they are held in place with stitches. Where there is tissue loss, however, a remarkable process occurs during which new skin is regenerated.

Skin cells adjacent to the wound break away from the cells below, migrate to the wounded area and enlarge to fill the space. Other cells surrounding the wound multiply rapidly to replace the cells lost.

Eventually migrating cells from all sides of the wound converge. Once the wound is entirely covered, cell migration stops. The wound will continue to heal as epithelial cells multiply, and normal thickness is restored.

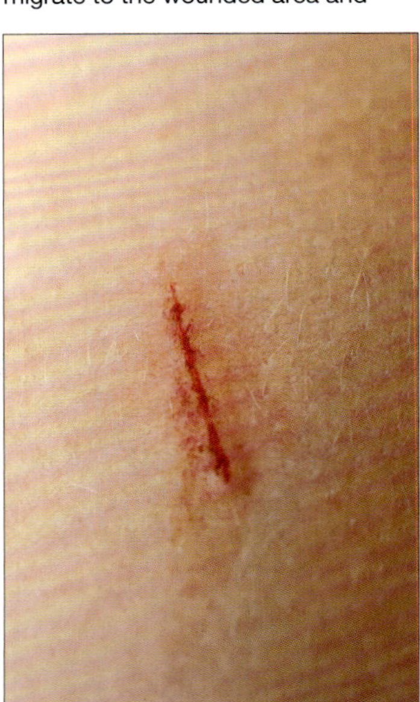

Left: Following skin damage, the surrounding tissue becomes red and inflamed as white blood cells accumulate.

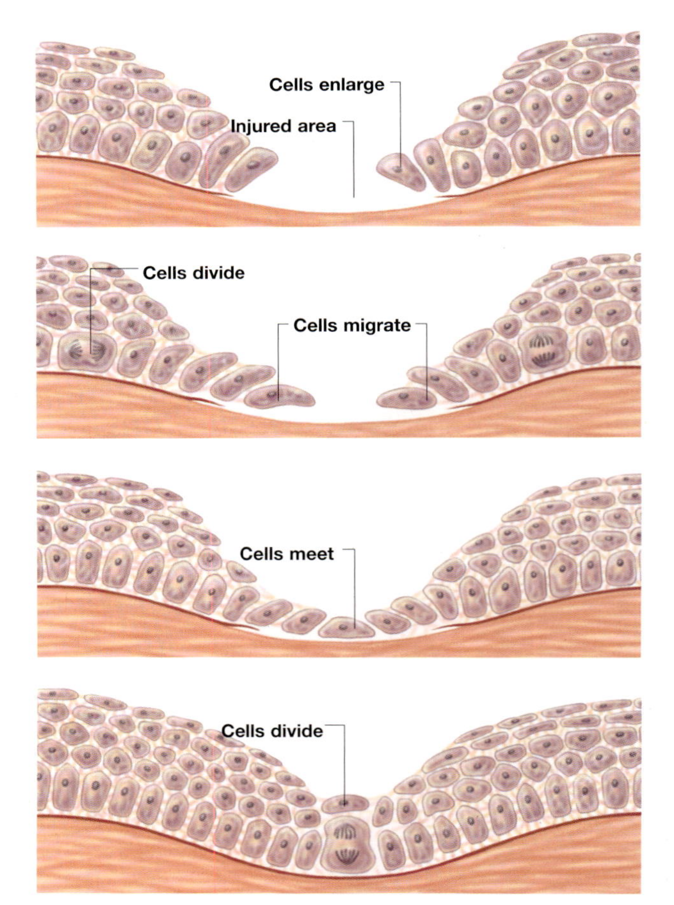

Right: When skin is wounded, the surrounding cells move to the site of the wound, and multiply until the area is covered.

Sweat glands

In the dermis of the skin are glands that produce sweat. This vital function enables the body to cool as sweat on the surface of the skin evaporates into the atmosphere taking with it excess body heat.

Cross-section of the dermis

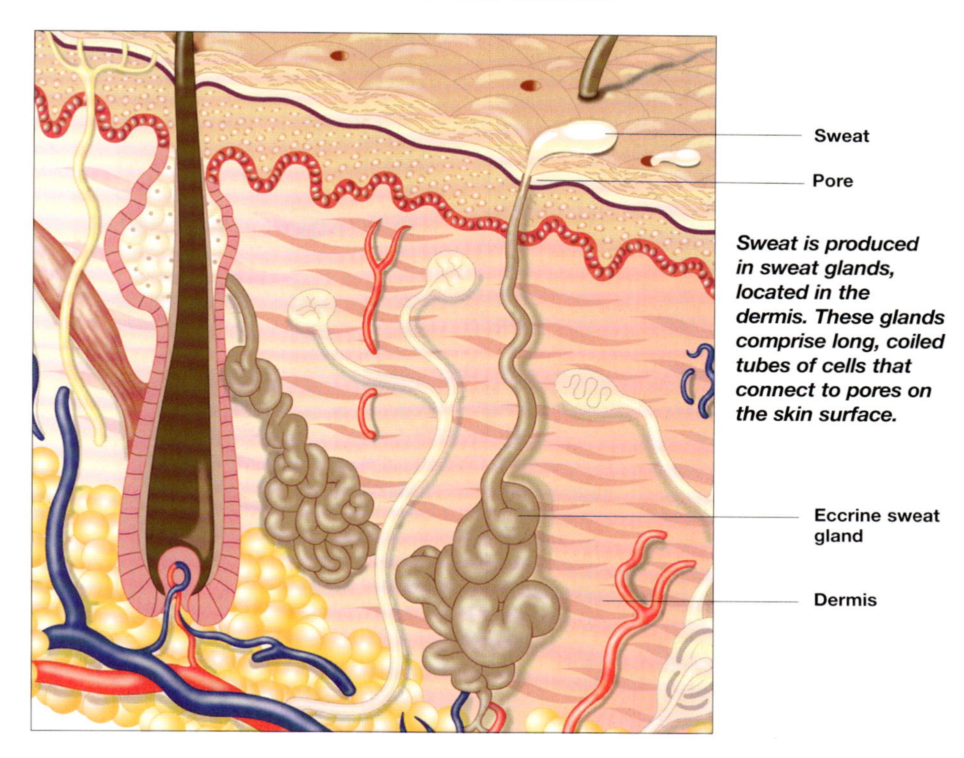

Sweat is produced in sweat glands, located in the dermis. These glands comprise long, coiled tubes of cells that connect to pores on the skin surface.

The body constantly produces sweat. This process is the body's main way of ridding itself of excess heat. The amount of sweat the body produces depends upon the state of emotion and physical activity. Sweat can be produced in response to stress, high temperature and exercise.

Sweat glands
Sweat is manufactured in the sweat glands, also known as sudoriferous glands. These are located in the dermis of the skin, along with nerve endings and hair follicles. On average, each person has around 2.6 million sweat glands, which are distributed over the entire body, with the exception of the lips, nipples and areas of the genitals.

How sweat is produced
Sweat glands consist of long, coiled, hollow tubes of cells. The coiled portion in the dermis is where sweat is produced. The long portion is a duct that connects the gland to tiny openings (pores) located on the outer surface of the skin. Nerve cells from the sympathetic nervous system (a division of the autonomic nervous system) connect to the sweat glands. There are two types of gland:
- Eccrine – these are the most numerous type of sweat gland. They are found all over the body, particularly on the palms of the hands, soles of the feet and forehead. Eccrine glands are active from birth.
- Apocrine – these sweat glands are mostly confined to the armpits and around the pubic region. Usually, they end in hair follicles rather than pore openings. Apocrine sweat glands are larger than eccrine glands, and typically only develop and become active once puberty has begun.

Sweat production

Stimulation of an eccrine gland causes the cells lining the gland to secrete a fluid that is similar to plasma, but without the fatty acids and proteins. Sweat is mostly water with high concentrations of sodium and chloride (salts) and a low concentration of potassium.

Fluid originates in the spaces between cells (interstitial spaces), which are provided with fluid by the blood vessels (capillaries) in the dermis.

The fluid passes from the coiled portion and up through the straight duct. What happens to this fluid when it reaches the straight portion of the sweat duct depends upon the rate of sweat production.

- Low sweat flow – at rest and in a cool environment, the sweat glands are not stimulated to produce much sweat. The cells of the straight duct have time to reabsorb most of the water and salts, so not much fluid actually reaches the surface of the skin as sweat.

The composition of this sweat is different from that of its primary source: it contains less sodium and chloride, and more potassium.

- High sweat flow – this occurs in higher temperatures or during exercise. Cells in the straight portion of the sweat duct do not have time to reabsorb all the water, sodium and chloride from the primary secretion. As a result, higher levels of sweat reach the surface of the skin, and its composition is similar to that of the primary secretion.

Apocrine sweat
Sweat is produced in the apocrine glands in a similar way, but apocrine differs from eccrine sweat in that it contains fatty acids and proteins. For this reason, apocrine sweat is thicker and milky-yellow in colour.

Odour
Sweat itself has no odour, but when bacteria present on the hair and skin metabolize the proteins and fatty acids present in apocrine sweat, an unpleasant odour is produced.

The constituents of sweat vary according to temperature and activity. If sweat production is minimal, then the sweat contains less salts.

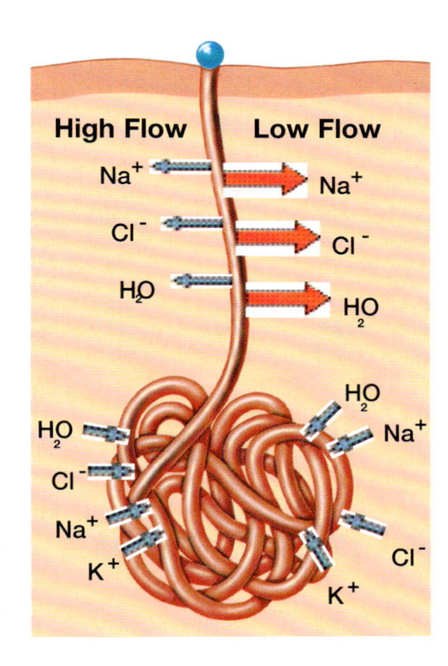

H_2O	Water
K^+	Potassium
Na^+	Sodium
Cl^-	Chloride

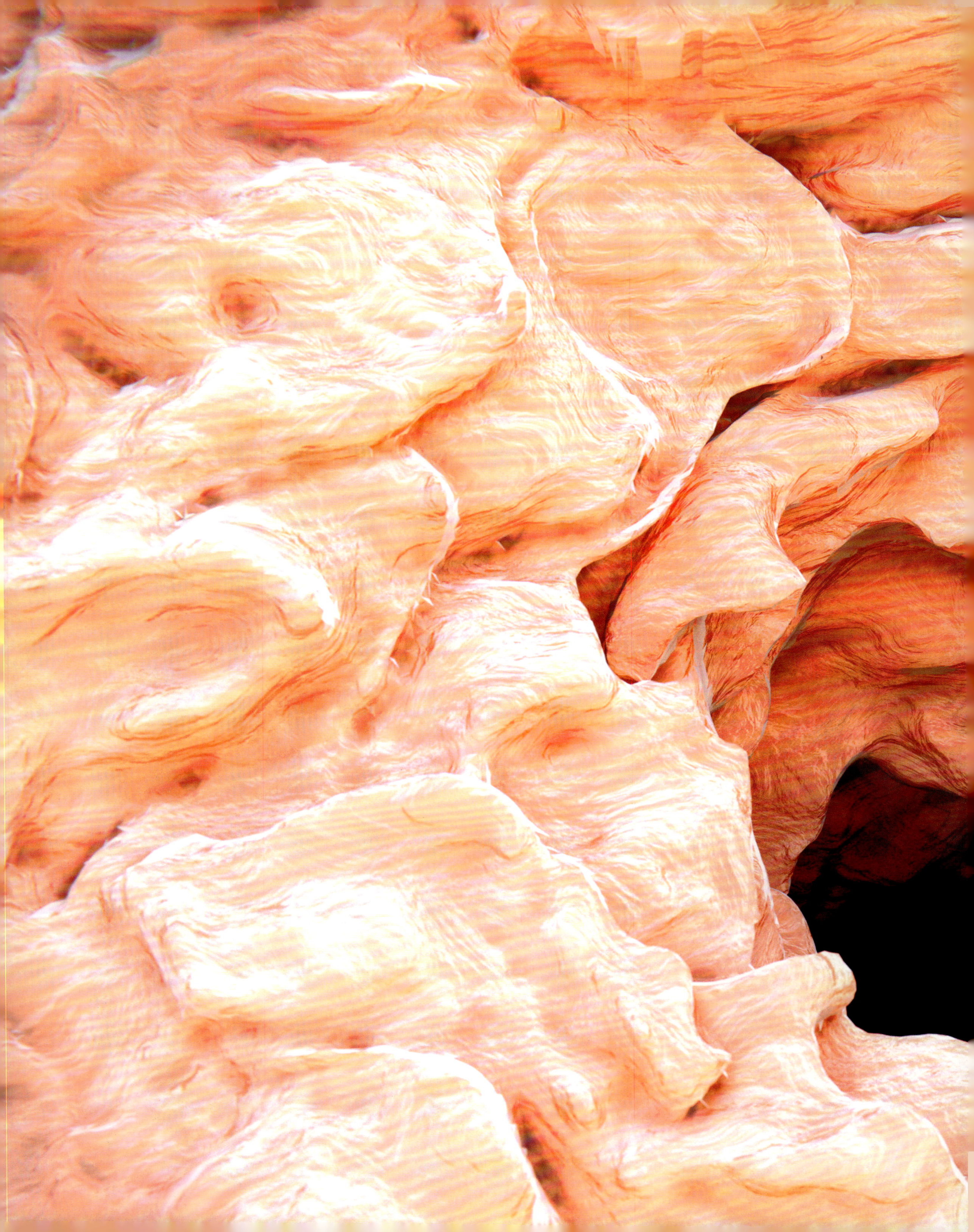

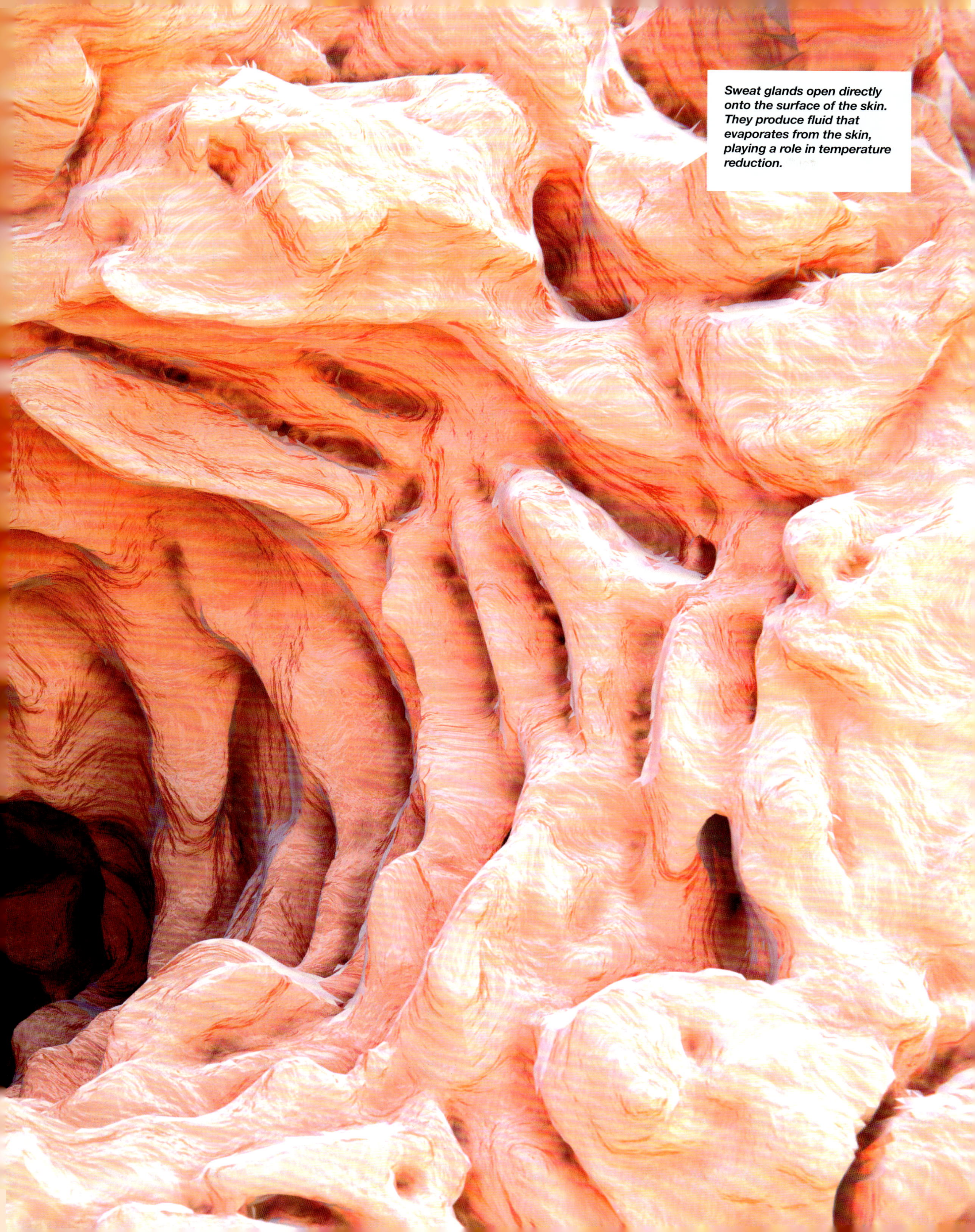

Sweat glands open directly onto the surface of the skin. They produce fluid that evaporates from the skin, playing a role in temperature reduction.

Hair

There are two main types of human hair: vellus and terminal. Only terminal hair has a central core or medulla and responds to the male sex hormone, testosterone.

Cuticle
This is formed from a single layer of overlapping cells. When this layer wears away at the end of long hairs, 'split ends' may occur

Cortex
This consists of several layers of flattened cells that contain various amounts of the pigment melanin – this produces different hair colours

Hair root

Medulla
The central core, which is present only in coarse, terminal hair

Hair bulb

Enlarged cross-section

A hair consists of three layers. The cuticle is the outer covering; the cortex makes up the bulk of the hair; and the medulla is the central core.

The surface of the human body is covered with millions of hairs. They are most noticeable on the head, around the external genitalia and under the arm. The only parts of the body without hair are the nipples, lips, parts of the external genitalia and the palms and soles. The function of hair is primarily protection from injury of the scalp, from heat and cold and to screen the eyes, ears and nostrils from insects and foreign particles.

Structure of a hair
Hair is composed of flexible strands of the hard protein, keratin. It is produced by hair follicles within the dermis. Each hair is made up of three concentric layers:

- The cuticle – the outermost layer, this is composed of a single layer of hard keratin cells that overlap one another like roof tiles.
- The medulla (the central core) – consists of large cells containing soft keratin that are partially separated by air spaces.
- The cortex – the bulky layer surrounding the medulla, consists of several layers of flattened, hard keratin-containing cells. This outer layer of the hair contains the most keratin, and strengthens and protects the hair, helping to keep the inner layers compacted.

Types of Hair
Vellus hair describes the soft hair that covers most of the body in women and children. It is short, fine and usually light in colour. Vellus hair shafts do not have a central medulla. Terminal hair is much coarser. It occurs on the head, as eyelashes and eyebrows and as public and axillary hair. Terminal hair has a central medulla within its shaft and grows in response to male sex hormone hormones such as testosterone.

Colour and Texture
The exact composition of the keratin produced by the body is determined by our genes and differs between individuals. Since keratin is responsible for the texture of the hair shaft, this can vary greatly. The shape of the hair shaft determines whether the hair is straight or curly; the rounder the shaft in cross-section, the straighter the hair.

The colour of hair depends upon the presence of the pigment melanin, produced by melanocytes in the bulb of the hair follicle, and then transferred to the cortex. Dark hair contains true melanin, like that found in the skin, while blond and red hair results from types of melanin that contain sulphur and iron. Grey or white hair results from decreased melanin production (genetically triggered) and from the replacement of melanin by air bubbles in the shaft.

Adults have around 120,000 hairs on their head. Those with red hair tend to have fewer hairs, while those with blonde hair have more.

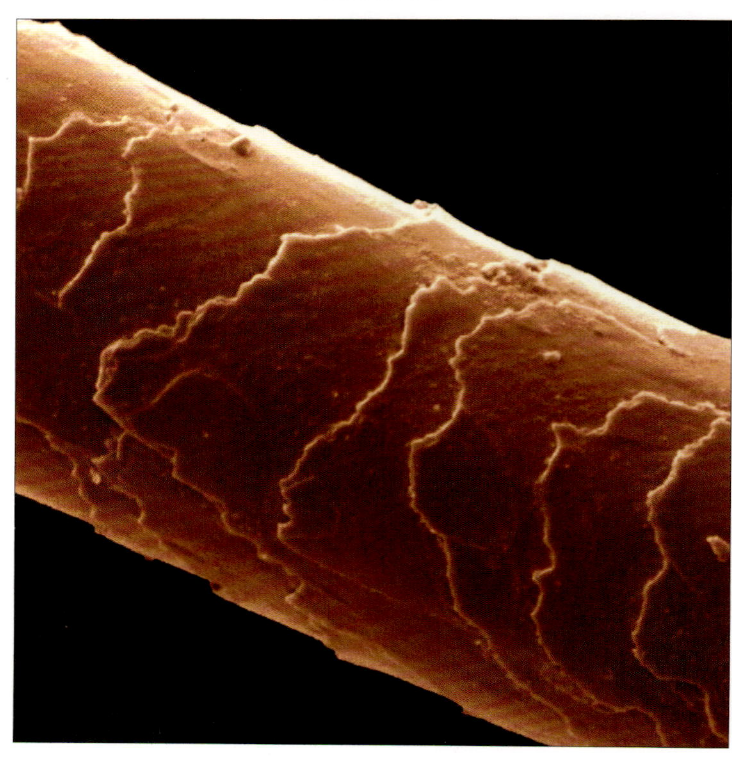

The hair cuticle is shown here, greatly magnified. The single layer of overlapping dead cells strengthen the hair.

Hair follicle and sebaceous glands

Hairs are produced within hair follicles on the skin's surface. A number of other structures are associated with follicles, including sebaceous glands, nerve endings and tiny muscles that pull the hair erect.

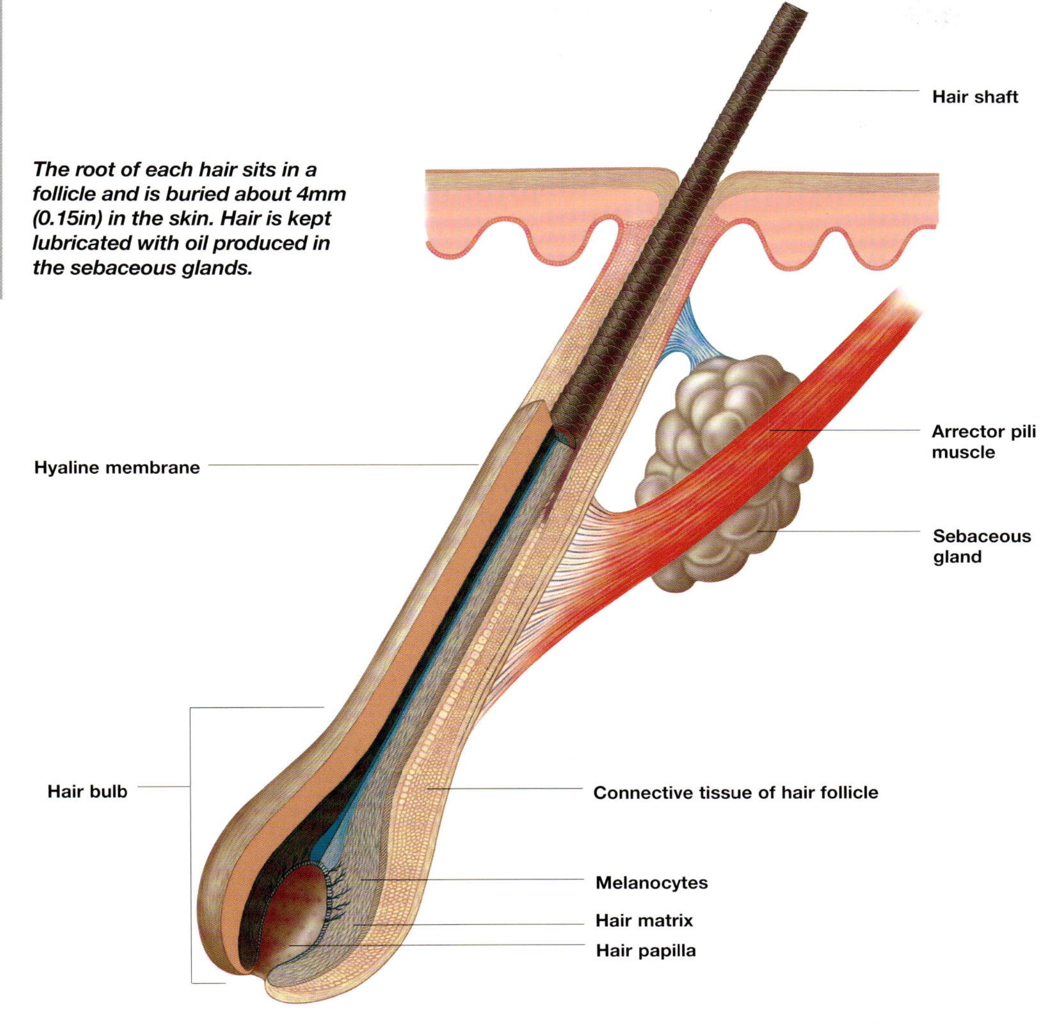

The root of each hair sits in a follicle and is buried about 4mm (0.15in) in the skin. Hair is kept lubricated with oil produced in the sebaceous glands.

Surrounding the root of each hair is a follicle, the base of which has an expanded end known as the hair bulb . The bulb contains the papilla, which is an indentation of loose connective tissue. The papilla contains a knot of tiny blood vessels that provide nutrients for the hair. Within the papilla are also the hair production cells, known as the matrix. Each time a hair is shed, the matrix divides to form a new hair.

Sebaceous glands

Sebaceous, or oil, glands lie alongside hair follicles wherever they are on the surface of the body. They produce an oily substance, known as sebum, which drains out of the gland through a sebaceous duct into the hair follicle. The sebum then passes out around the emerging hair shaft to reach the surface of the body.

The amount of sebum produced depends upon the size of the sebaceous gland, which in turn depends upon the levels of circulating hormones, especially androgens (male sex hormones). The largest sebaceous glands are on the head, neck, and back and front of the chest.

The function of sebum is to soften and lubricate the skin and hair, and to prevent the skin from drying out. It also contains substances that kill bacteria, which might otherwise cause infection of the skin and hair follicle.

Nerve endings

A network of tiny nerve endings lie around the bulb of the hair follicle. These nerves are stimulated by any movement of the base of the hair. If the hair is bent by pressure somewhere along its shaft, these nerve endings will fire, sending signals to the brain. This is what happens, for instance, when an insect alights on the skin; the slight bending of hairs it causes sets off a chain of events, resulting in a reflex action to remove it before it stings. In this way hair contributes to our sense of touch.

Arrector pili muscle

Each hair follicle is attached to a tiny muscle called an arrector pili, which literally means 'raiser of hair'. When this muscle contracts, it causes the hair to move from its normal, angled position to a vertically erect one.

When this occurs within many hair follicles, we see (and feel) 'goose pimples' which are commonly stimulated by either cold or fear.

Stages of growth

Hair is produced in cycles that involve a growth phase and a resting phase. During the growth phase the hair is formed and extends as cells are added at the base of the root. This phase can last from around two to six years. As hair grows approximately 10cm (4in) a year, any individual hair is unlikely to grow more than a metre long.

Resting phase

Eventually, cell division pauses (the resting phase) and growth of the hair stops. The hair follicle shrinks to one sixth of its normal length, and the dermal papilla, responsible for the nourishment of new hair cells, breaks away from the root bulb. During this phase the dead hair is held in place. It is these hairs that seem to come out in handfuls when hair is washed or brushed. Eventually a new cycle starts, and the hair is shed from the hair follicle as the production of a new hair begins.

Hair types

The length of each phase depends on the type of body hair: scalp hairs tend to grow for a period of three years and rest for one or two years, while eyelash hair, which is much shorter, will grow for around 30 days, and rest for 105 days before being shed. At any one time around 90 per cent of scalp hairs will be in the growing stage, and there is a normal loss of around 100 scalp hairs per day.

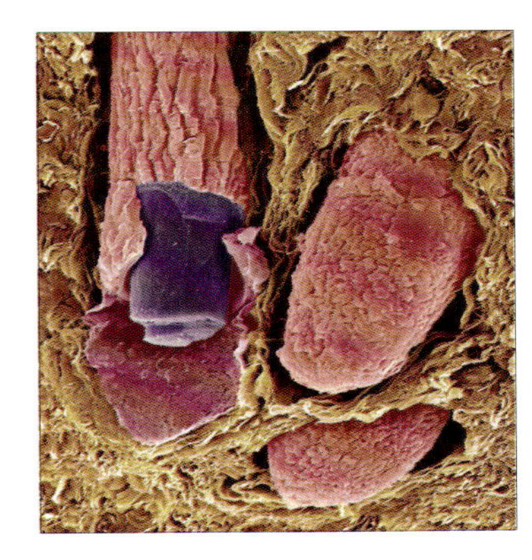

The hair root sits in a follicle, within which are tiny blood vessels.

Nails

Nails lie on the dorsal (back) surfaces of the ends of the fingers and toes, overlying the terminal phalanx (final bone) of each, providing a hard protective covering.

The nails are flattened, elastic structures that begin to grow on the upper surface of the tips of the fingers and toes in the third month of fetal development. The different parts of the nail include:

- Nail plate – each nail is composed of a plate of hard keratin (the same substance as is found in hair), which is continuously produced at its root
- Lateral nail fold – except for the free edge of the nail, at its furthermost end, the nail is surrounded and overlapped by folds of skin (nail folds)
- Free edge – the nail separates from the underlying surface at its furthermost point to form a free edge. The extent of this nail at the free edge depends upon personal preference and wear and tear
- Root, or matrix – this lies at the base of the nail beneath the nail itself and the nail fold. This part of the nail is closest to the skin, and it is here that the hard keratin of the nail is produced by cell division. If the root of the nail is destroyed the nail cannot grow back
- Lunula – the paler, crescent-shaped area at the base of the nail where the matrix is visible through the nail
- Cuticle (eponychium). This covers the proximal (near) end of the nail and extends over the nail plate to help protect the matrix from infection by invading micro-organisms.
- Nail bed – this is the area underlying the entire nail.

Functions

Like skin and hair, the nails are composed of keratin, a tough protein. This acts as a shock absorber, protecting the tips of the fingers and toes. In addition, the fingernails are useful tools for tasks such as undoing a shoelace, picking up small objects or scratching an itch.

Despite the fact that the nails lack nerves, they also serve as excellent 'antennae', since they are embedded within sensitive tissue that detects any impact when the nail touches an object.

Cross-section of nail

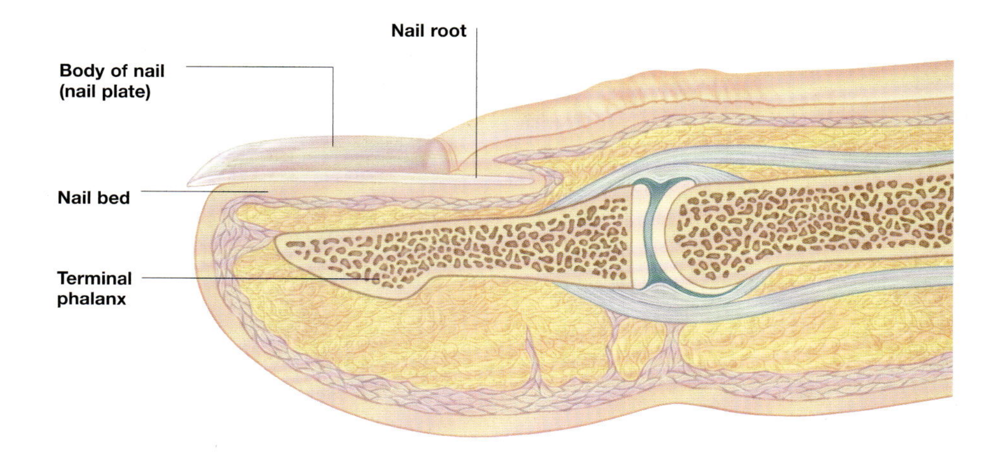

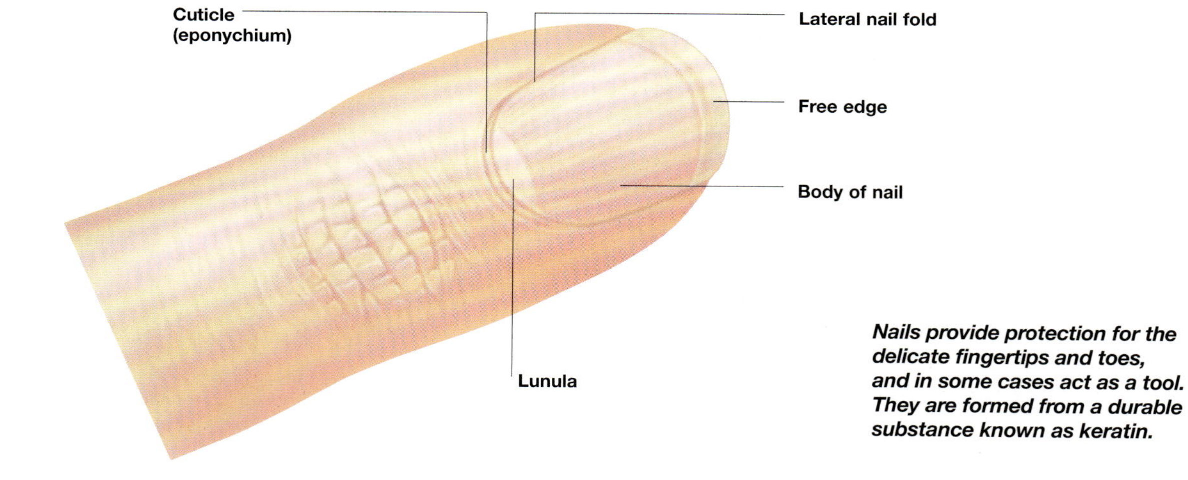

Nails provide protection for the delicate fingertips and toes, and in some cases act as a tool. They are formed from a durable substance known as keratin.

How nails grow

The nails are extensions of the outer skin layer, and grow continuously throughout life.

Nail growth takes place continually throughout life. The area where nail growth occurs in both the fingers and toes is the germinal, or nail, matrix, which is located at the lower part of the nail bed, beneath the root of the nail. Here, epidermal cells divide and become enriched with keratin, which thickens to become nail. As the cells of the matrix are transformed into nail cells, they are pushed forward over the nail bed.

Blood and nerve supply

For a normal rate of growth, and to produce normal, pink, healthy nails, there must be a good blood supply to the root of the nail; nails look pink because of the large number of blood vessels in the dermis. The nail itself is has no blood vessels or nerves, but the underlying tissue, the nail bed, is rich in capillaries and has a good nerve supply. The crescent-shaped lunula at the lower end of the nail is whitish because the blood enriched tissue below does not show through.

Rate of growth

On average it takes around three to six months for a nail to grow from its base to the tip of the finger. Fingernails grow much quicker than toenails and the corresponding time for a toenail may be up to two years. In most people, finger and toe nails are kept short due to cutting and abrasion. Without this, nails are capable of growing to a great length.

Nail hydration

The nails are very porous and they can hold up to 100 times as much water as the equivalent weight of skin. In this way, the nails limit the amount of water entering the tissues of the fingertips. Water taken up by the nails is eventually lost through evaporation as they dry out and resume their normal size.

Nail injury

Injury to a nail is common, as fingers may be trapped in doors or hit with a blunt instrument. Following an injury, nail growth tends to accelerate until it has recovered. However, if the germinal matrix of the nail is destroyed, the nail will cease to grow.

The nails consist of curved plates of hard keratin. Beneath the lunula area lies the nail matrix – this is responsible for the growth of the nail.

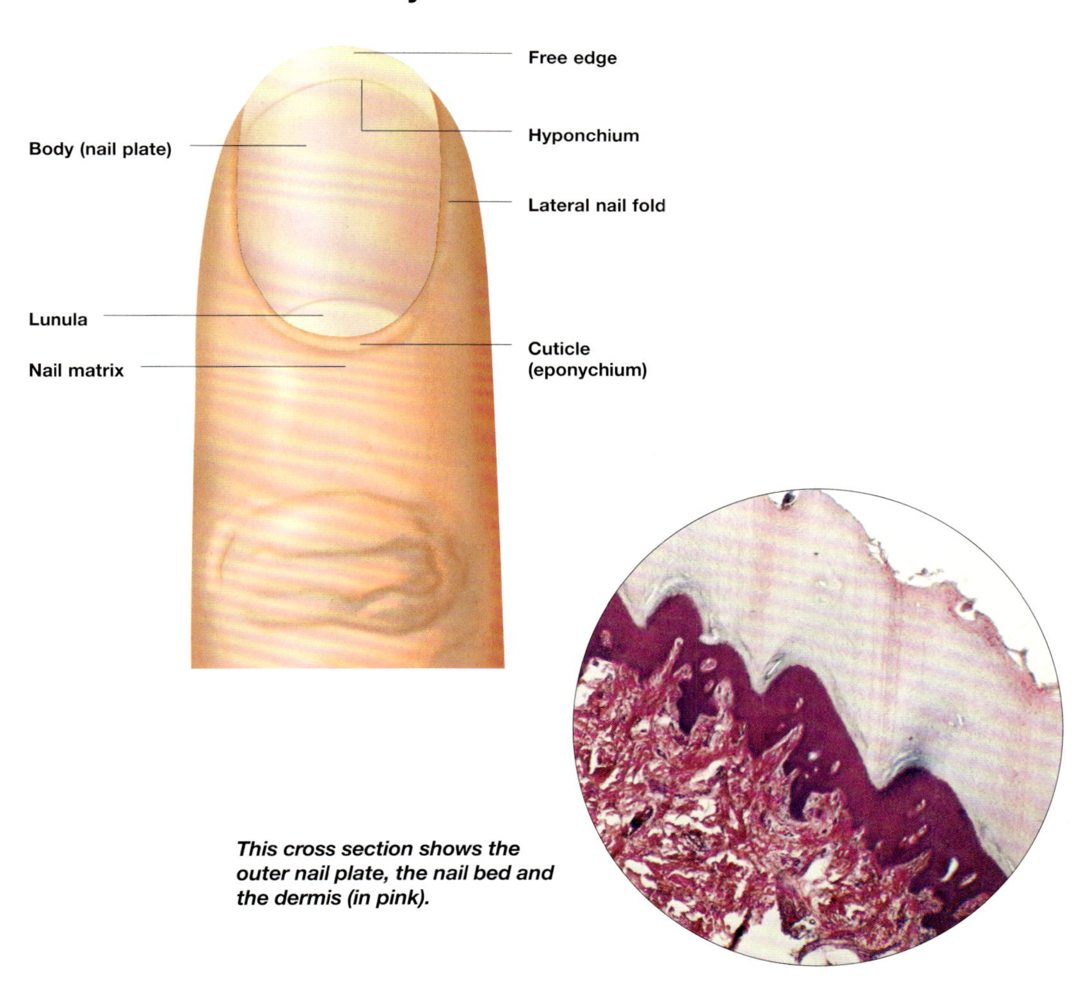

This cross section shows the outer nail plate, the nail bed and the dermis (in pink).

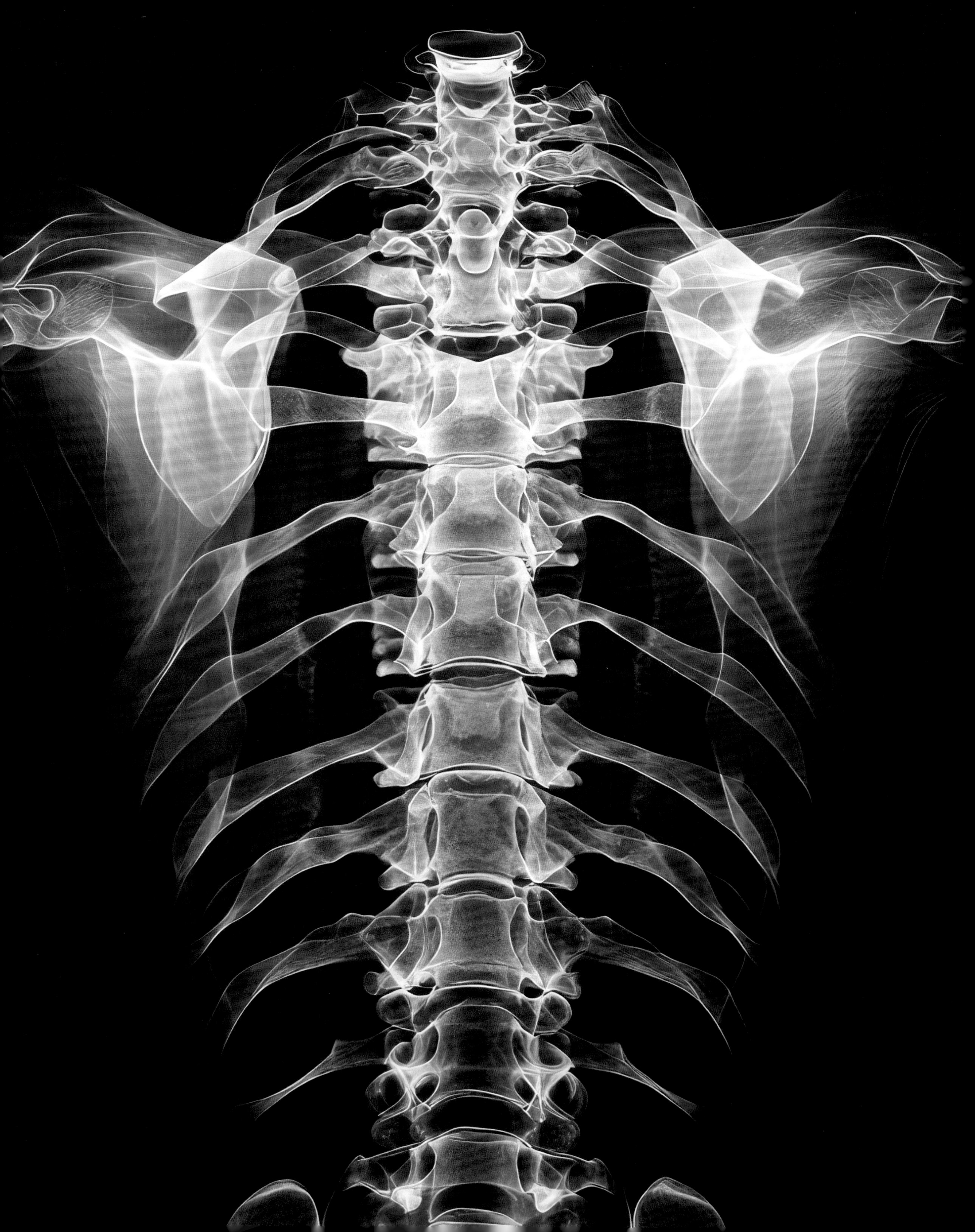

CHAPTER 3

The Musculoskeletal System

The most extensive system in the body, the musculoskeletal system, consists of a mechanized network of bones, muscles, ligaments and tendons that work together to enable movement, or locomotion. Around 206 bones slot together to form the skeleton, which provides support for the body and protects vital organs. The tendons and ligaments, sometimes known as the articular system, attach the muscles to the bones and it is at these attachment sites, known as joints, that movement occurs, for example at the elbow or knee. Muscles, of which there are several types, are also responsible for movement but they have several other functions. Specialized cardiac muscle is the primary tissue of the heart walls, and smooth muscle forms part of the walls of the blood vessels and hollow organs such as the intestines. This section describes the structure of the components and how they function together.

Opposite: The spinal column, shown here with the ribs and pectoral girdle, provides stability and support to the entire skeleton.

Skeleton

The skeleton is made up of bone and cartilage and accounts for one-fifth of the body's weight. Over 200 bones form a living structure, superbly designed to support and protect the body.

The human skeleton provides a stable yet flexible framework for the other tissues of the body. Cartilage is more flexible than bone and so is found in the places where movement occurs.

Functions of Bone
The bones of the skeleton have a number of vital functions:

- Support – bones support the body when standing, and hold soft internal organs in place
- Protection – the brain and spinal cord are protected by the skull and vertebral column, while the ribcage protects the heart and lungs
- Movement – throughout the body, muscles attach to bones to give them the leverage to bring about movement
- Storage of minerals – calcium and phosphate ions are stored in bone to be drawn upon when necessary
- Blood cell formation – the marrow cavity of some bones, such as the sternum, is a site of production of red blood cells.

Formation
The bony skeleton is formed in fetal life, but grows throughout childhood. A fetus of six weeks has a skeleton made of fibrous membranes and hyaline cartilage, which converts into bone during pregnancy. After birth, and until the end of adolescence, the skeleton grows in weight and length as well as being remodelled.

Anterior view of the human skeleton

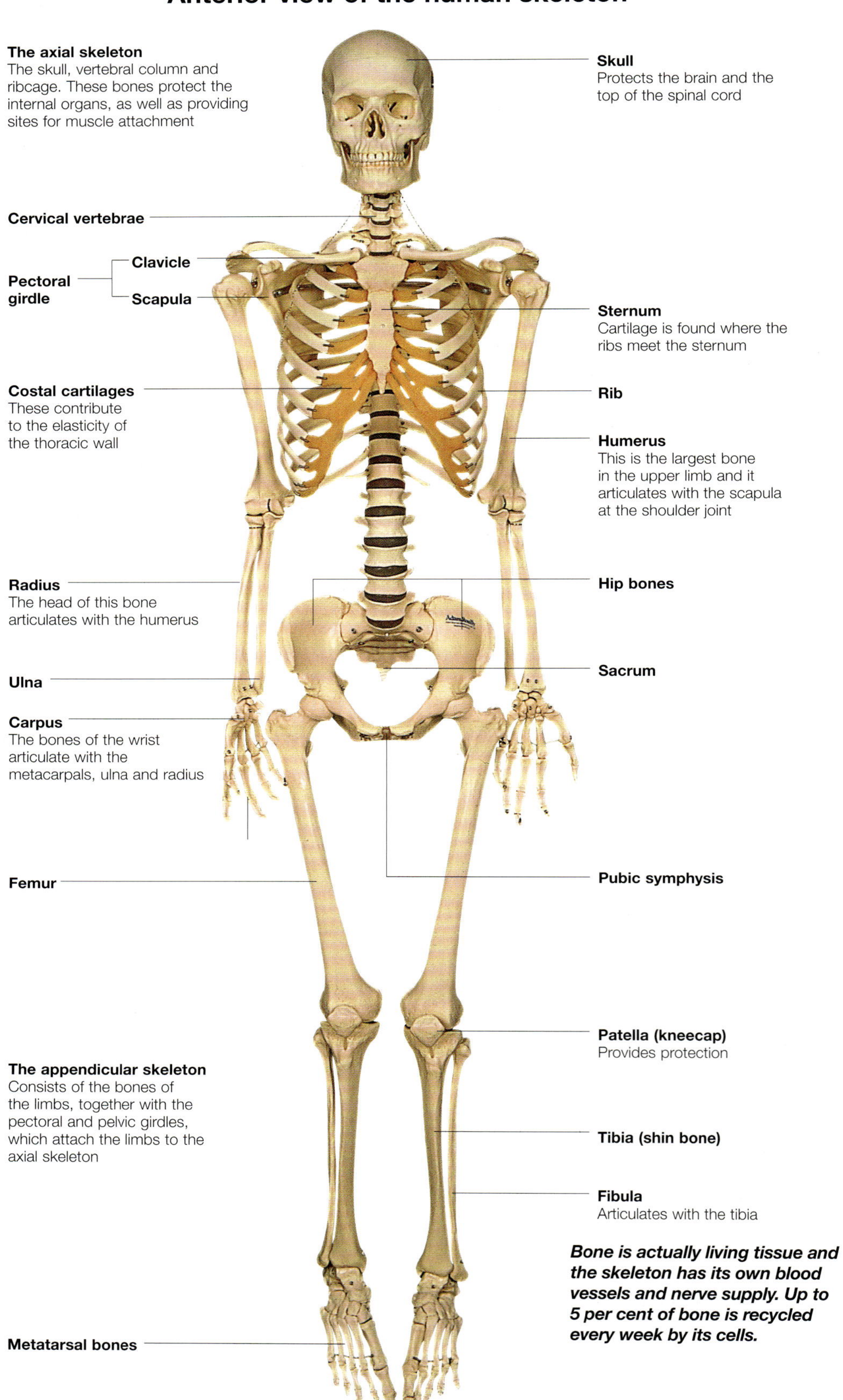

Bone is actually living tissue and the skeleton has its own blood vessels and nerve supply. Up to 5 per cent of bone is recycled every week by its cells.

Bone markings and features

Each bone of the skeleton is shaped to fulfil its functions. Bones bear marks, ridges and notches that relate to other structures with which they come into contact.

Projections
Projections on the surface of a bone often occur where muscles, tendons or ligaments are attached or where a joint is formed. Examples include:
- Condyle – rounded projection at a joint (such as the femoral condyle at the knee)
- Epicondyle – the raised area above a condyle (such as on the lower humerus, at the elbow)
- Crest – prominent ridge of bone (such as the iliac crest of the pelvic bone)
- Tubercle – small raised area (such as the greater tubercle at the top of the humerus)
- Line – long, narrow raised ridge (such as the soleal line at the back of the tibia).

Depressions and grooves
Depressions, holes and grooves are usually found where blood vessels and nerves must pass through or around bones. Examples include:
- Fossa – a shallow, bowl-like depression (such as the infraspinous fossa of the shoulder blade or the iliac fossa, which is a depression found on the ilium)
- Foramen – a hole in a bone to allow the passage of a particular vessel or nerve (such as the jugular foramen in the skull that allows passage of the internal jugular vein)
- Notch – an indentation that is found at the edge of a bone (such as the greater sciatic notch, which is partly formed by the ilium)
- Groove – a furrow or elongated depression that marks the route of a vessel or nerve along a bone (such as the oblique radial groove at the back of the humerus).

Posterior view of the human skeleton

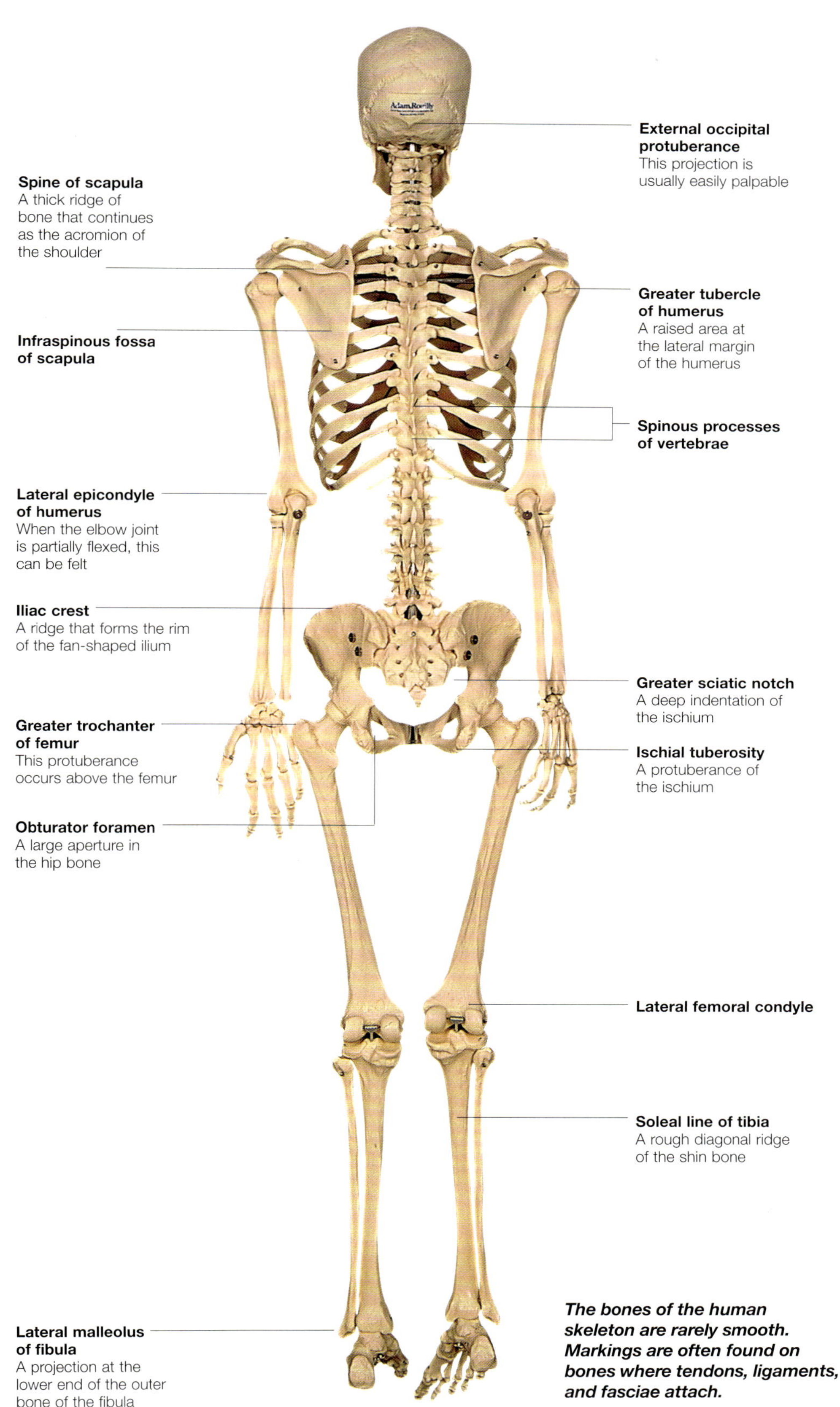

The bones of the human skeleton are rarely smooth. Markings are often found on bones where tendons, ligaments, and fasciae attach.

Structure of Bone Tissue

Bones are living tissue and are in a constant state of renewal. They form the basis of the skeleton and are responsible for movement as well as containing bone marrow and vital minerals.

Bone is composed of a calcified matrix in which bone cells are embedded. The matrix is made up of flexible collagen fibres in which crystals of hydroxyapatite (a calcium salt) are deposited. Three principal bone cell types are found within this matrix: osteoblasts (bone forming cells); osteoclasts (bone 'eating' cells; and osteocytes (bone cells that have fully matured).

Bone Tissue

Bone tissue exists in two forms: compact (or cortical) bone and spongy (or cancellous) bone.

Compact Bone

Compact bone makes up the outer covering of all bones and is thickest in the places that receive the greatest stress. It is made up of a series of canals and passageways; these provide a route for the nerves, blood vessels and lymphatic vessels that extend through each bone.

The structural units of compact bones (osteons) are elongated cylinders which lie parallel to the long axis of the bone. Osteons are composed of a group of lamellae (hollow tubes) of bone matrix arranged concentrically.

The lamellae are organized in such a way that the collagen fibres in adjacent lamellae run in opposite directions; this is intended to reinforce the bone against twisting forces. Each osteon is nourished by blood vessels and served by nerve fibres that run throughout its centre, known as Haversian canals.

Volkmann's canals connect the blood vessels and nerve of the periosteum (membrane around the bone) to those in the central canals and medullary cavity (which contains bone marrow).

Mature bone cells (osteocytes) are located in the small cavities (lacunae) between each lamella.

Spongy Bone

Spongy bone makes up the inner part of most bones and is much lighter and less dense than compact bone. This is due to the fact that it contains a number of cavities that are filled with marrow. Spongy bone is strengthened by a criss-cross network of bony supports, known as trabeculae.

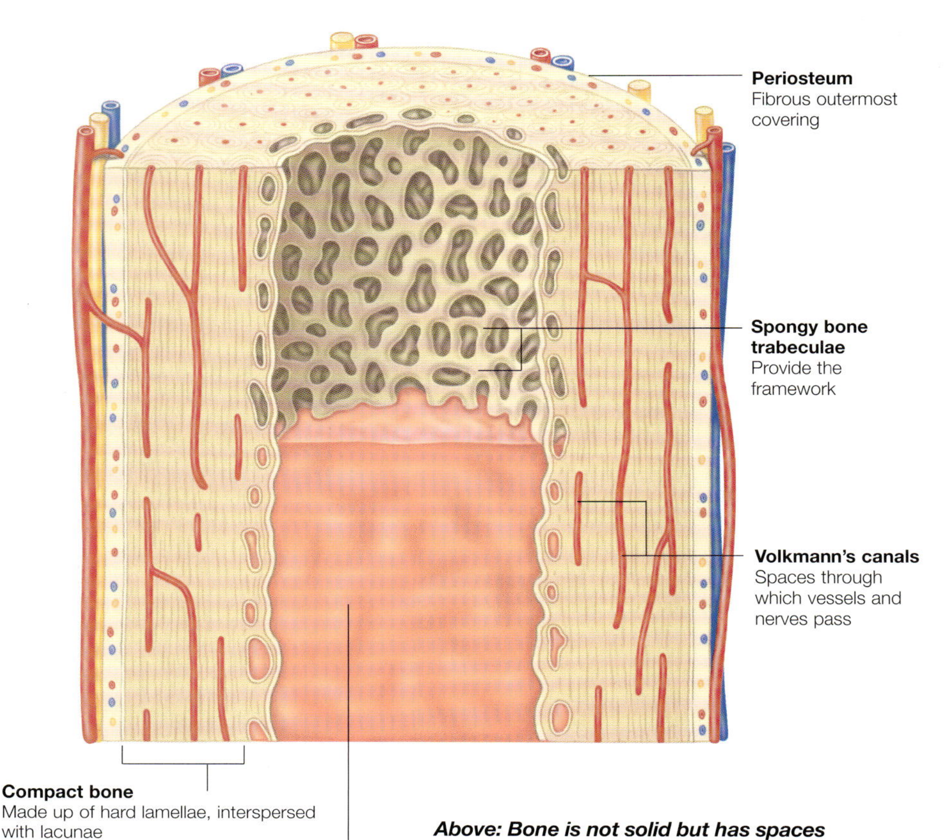

Above: Bone is not solid but has spaces between the hard components. The size and distribution of these spaces dictates whether bone is compact or spongy.

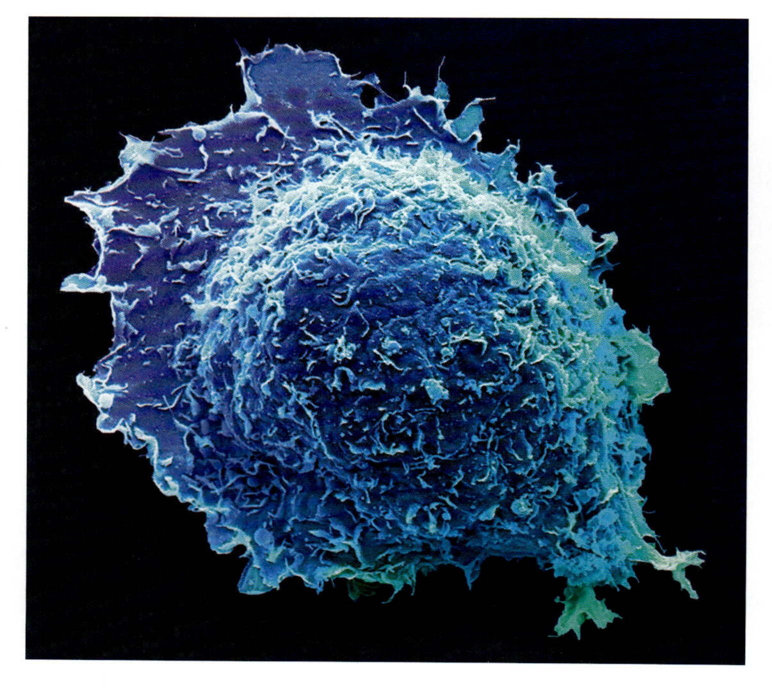

Left: Osteoblast bone cell. Coloured scanning electron micrograph (SEM) of an osteoblast bone cell. Osteocytes are osteoblasts (bone-producing cells) that have become trapped within bone cavities (lacunae, depressions in bone). They produce and secrete the organic matrix of the bone (osteoid). As soon as osteoid is formed, calcium salts crystallise inside it to form mineralised bone.

Formation of bones

Bone formation begins in the embryo and continues for the first 20 years of life. Development takes place from a number of ossification centres and once these are fully calcified, no further elongation can occur.

The skeleton is made up of a variety of different bones, ranging from the flat bones found in the skull to the long bones of the limbs. Each bone is designed for a different function.

Long bones

The longest bones within the body are those of the upper and lower limbs. Each long bone consists of three main components:

- Diaphysis – a hollow shaft, composed of compact bone
- Epiphysis – at each end of the bone; the site of articulation between bones
- Epiphyseal (growth) plate – composed of spongy bone and the site of bone elongation.

Protective membrane

The entire bone is covered by the two-layered periosteum. The outer layer of this membrane consists of fibrous connective tissue. The inner layer of the periosteum contains osteoblasts and osteoclasts, the cells that are responsible for the constant replenishment of the bone.

Bone development

Skeletal development begins in the embryo and continues for around two decades. It is a complex process under genetic control, and is modulated by endocrine, physical and biological processes. A template of the skeleton forms in the embryo from the primitive embryonic tissue. As the embryo develops, this tissue becomes recognizable as cartilage (soft, elastic connective tissue) and individual 'bones' begin to be seen.

Ossification

Normal bone then forms within these templates by a process known as ossification. This takes place either directly around the early bone-forming cells of the fetus (intramembranous ossification) or by replacing a cartilage model with bone (endochondral ossification).

The formation of compact bone commences at sites in the bone shafts known as primary ossification centres. Osteoblasts within the cartilage secrete a gelatinous substance called osteoid, which is hardened by mineral salts to form bone. The cartilage cells die and are replaced by further osteoblasts.

Ossification of long bones continues until only a thin strip of cartilage remains at either end. This cartilage (the epiphyseal plate) is the site of secondary bone growth up to late adolescence.

The sequence of the formation of ossification centres follows a prescribed pattern, allowing experts to age skeletons by the extent of ossification.

Mature bone

Once the bone has reached full length, the shaft, growth plate and epiphyses are all ossified and fuse to form continuous bone. No further elongation can take place after this time.

Structure of a typical long bone

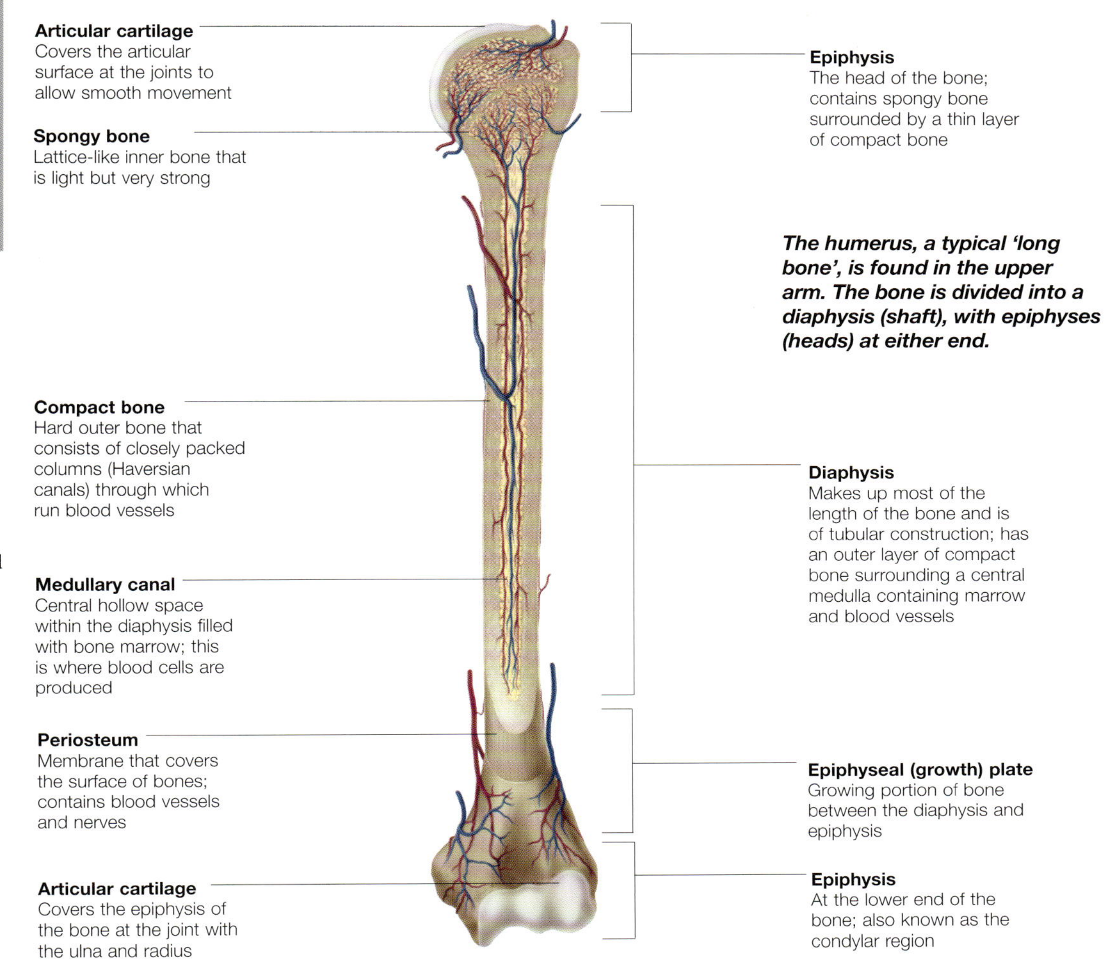

The humerus, a typical 'long bone', is found in the upper arm. The bone is divided into a diaphysis (shaft), with epiphyses (heads) at either end.

Long bone of a newborn

Epiphysis (bone end)
Growth plate
Blood vessel
Marrow cavity
Diaphysis (shaft)

Long bone of a child

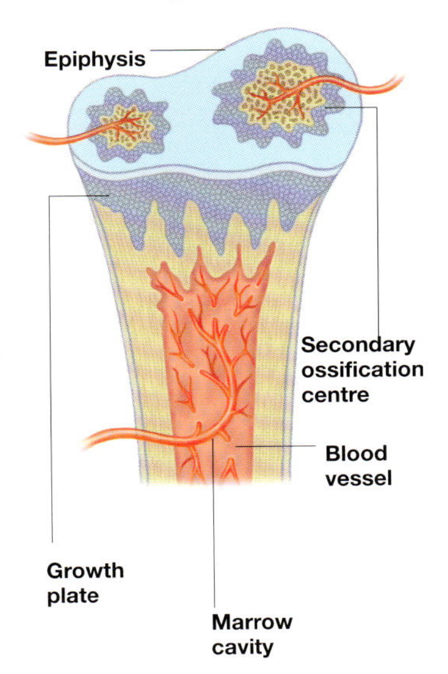

In a newborn baby, the shaft is mostly bone, while the bone ends consist of cartilage. In a child, new bone forms from secondary ossification centres at the bone ends.

Bone Remodelling

Although bones cease to grow after late adolescence, bone is a very dynamic tissue. Bone is continually being reabsorbed and regenerated as its structure is constantly changing.

Bone has an excellent ability to reshape itself. This process, known as remodelling, occurs during growth and continues throughout life.

Bone Remodelling
Remodelling occurs continuously, organizing the bone into orderly units that enable the bone mass to withstand mechanical forces. Old bone is removed by osteoclasts, while osteoblasts form new bone.

Bone Reabsorption
Osteoclasts secrete enzymes that break down the bone matrix, as well as acids that convert the resulting calcium salts into a soluble form (which can enter the bloodstream). Osteoclast activity takes place behind the epiphyseal growth zone to reduce expanded ends to the width of the lengthening shaft. Osteoclasts also act within the bone in order to clear the long tubular spaces that will accommodate bone marrow.

Hormonal Regulation
While the osteoclasts reabsorb bone, osteoblasts make new bone to maintain the skeletal structure. This process is regulated by hormones, growth factors and vitamin D. During childhood, bone formation outweighs bone destruction, resulting in gradual growth. After skeletal maturity has been reached, however, the two processes occur in equilibrium so that growth proceeds more gradually.

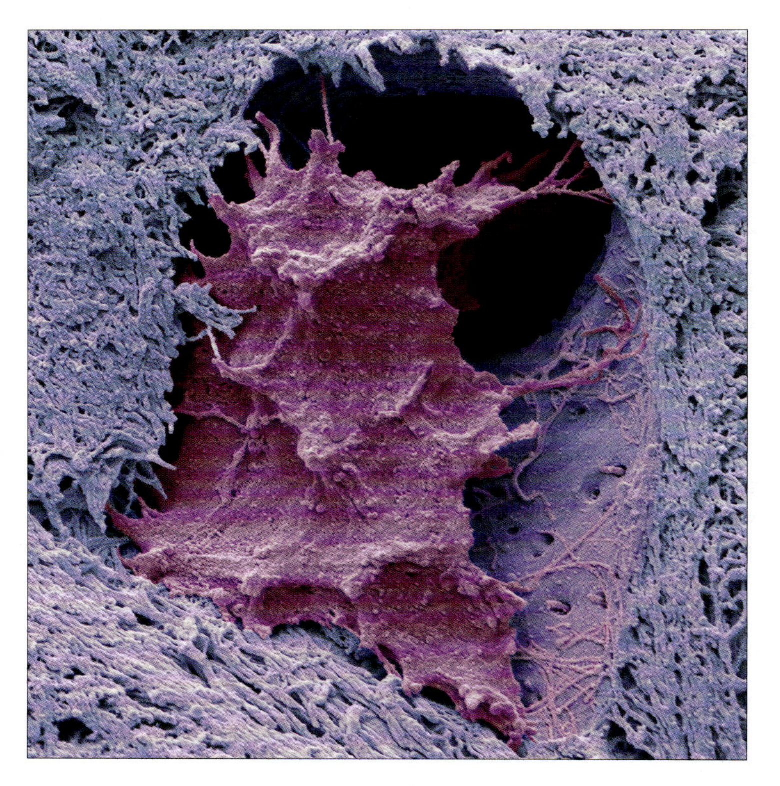

This micrograph shows an osteoblast (pink) surrounded by bone tissue (blue). Osteoblasts form a new bone by producing bone matrix.

Long Bones
The process of remodelling is especially important for the long bones that support the limbs. These are wider at each end than in the middle, providing extra strength at the joint. As osteoclasts destroy the old epiphyseal swellings of the bone, osteoblasts within the growth zone create a new epiphysis.

Within each of the tubular spaces cleared by osteoclasts inside the bone, the osteoblasts follow along, laying down a layer of new bone.

Rates of Remodelling
Bone remodelling is not a uniform process; it takes place at different rates throughout the skeleton. Bone remodelling takes place more in areas where the bone undergoes the greatest stress. This means that the bones that receive the most stress are subject to much remodelling. The femur, for example, one of the load-bearing bones of the leg, is effectively replaced every five to six months.

Bone that is subject to increased stress is constantly remodelled. The femur, the top of which is shown here, is effectively replaced every six months.

Calcium regulation

Bone remodelling not only alters the structure of the bone, but also helps to regulate the levels of calcium ions in the blood. Calcium is necessary for healthy nerve transmission, the formation of cellular membranes and the ability of the blood to clot.

Bone contains about 99 per cent of the body's calcium. When body fluid calcium levels fall too low, parathyroid hormone stimulates osteoclast activity and calcium is released into the bloodstream. When body fluid calcium levels become too high, calcitonin hormone inhibits reabsorption, restricting the release of calcium from the bones.

Bone Repair

If bone is subjected to a force beyond its strength it will fracture. New bone must be formed and remodelled for the fracture to heal.

One of the processes which is dependent on the remodelling of bone is the repair mechanism that takes place after a fracture.

Bone fractures

Fractures occur when a bone experiences a force greater than its resistance or strength.

These can occur as the result of a spontaneous force, or after years of continued stress upon a bone. Bones are particularly susceptible to fractures later in life when they are less elastic and bone mineral density declines. Bone repair takes place in four main stages.

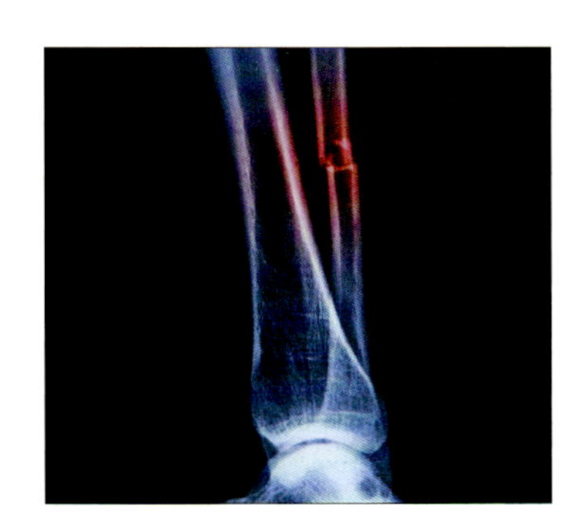

Bone fractures are common throughout life but the body's repair system is very effective. This fracture of the fibula should heal within six weeks.

Blood clot formation

1 A fracture of the bone causes the blood vessels in the area (mainly those of the periosteum, the protective covering of the bone) to rupture.

As these vessels bleed, a clot is formed at the site of the fracture giving rise to the characteristic swelling that often accompanies a broken bone. Very soon, bone cells deprived of nutrition begin to die and the site becomes extremely painful.

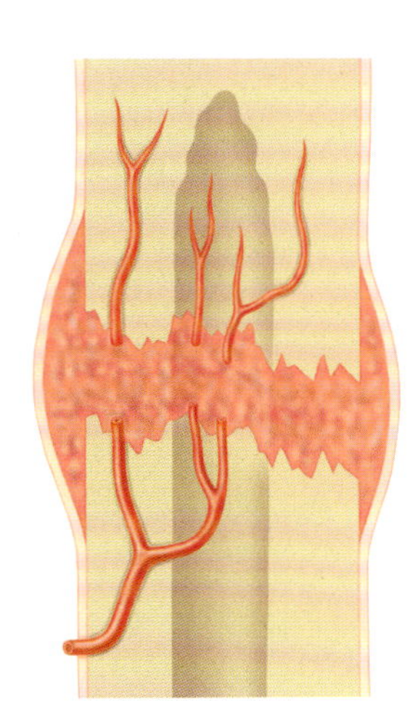

Blood vessels at the site of the fracture rupture, causing a blood clot to form. The nerves lining the periosteum are also severed, causing much pain.

Fibrocartilage callus formation

2 Several days after the injury, blood vessels and undifferentiated cells from surrounding tissues invade the area. Some of these cells develop into fibroblasts, which produce a network of collagen fibres between the bone fragments. Other cells form chondroblasts, which secrete cartilage matrix.

This zone of tissue repair between the two ends is known as a fibrocartilage callus.

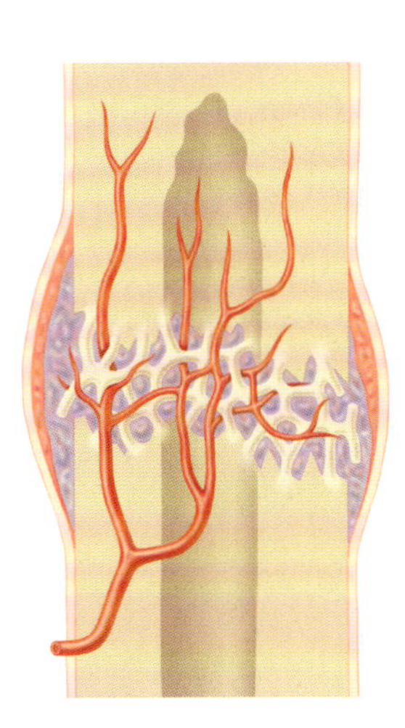

Blood vessels and cells invade the site of the fracture. The cells produce a matrix of collagen fibres and cartilage, forming a fibrocartilage callus.

Bony callus formation

3 Osteoblasts and osteoclasts migrate towards the affected area multiplying rapidly within the fibrocartilage callus.

Osteoblasts within the callus secrete osteoid, converting it into a bony callus.

This bony callus is composed of two portions: an external callus located around the outside of the fracture and an internal callus located between the broken bone fragments.

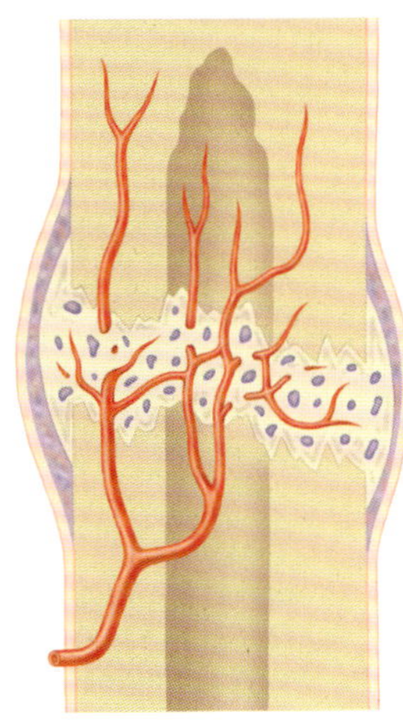

Within the fibrous callus, osteoblasts secrete a substance called osteoid, which hardens, forming a bony callus.

Bone remodelling

4 Bone formation is usually complete within four to six weeks of injury.

Once the new bone has been formed it will slowly be remodelled to form compact and spongy bone.

Total healing may require up to several months depending on the nature of the fracture and the specific function of the limb – weight-bearing limbs take longer to repair.

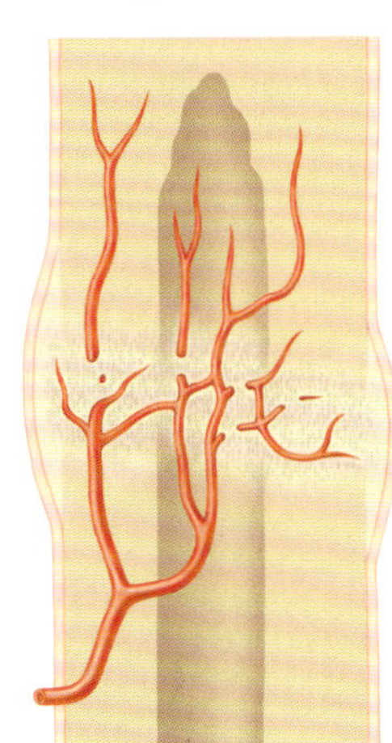

As the new bone is formed it is remodelled by osteoclasts. In this way the bony callus is smoothed out, and the bone regains its original structure.

Types of Joints

A joint is formed where two or more bones meet. Some allow movement and so give mobility to the body, whereas others protect and support the body by holding the bones rigid against each another.

The joints of the body can be divided into three main structural groups, according to the tissues that lie between the bones. These groups are known as fibrous, cartilaginous and synovial.

Fibrous Joints

Where two bones are connected by a fibrous joint, they are held together with collagen (a protein). Collagen fibres allow little, if any, movement. Fibrous joints are located in the body where the movement of one bone upon the other should be prevented, such as in the skull.

Cartilaginous Joints

The ends of the bones in a cartilaginous joint are covered with a thin layer of hyaline (glass-like) cartilage, with the bones being connected by tough fibrocartilage. The whole joint is covered by a fibrous capsule.

Cartilaginous joints do not allow much movement but they can 'relax' under pressure, therefore giving flexibility to structures such as the spinal column.

Synovial Joints

Most joints of the body are synovial, and allow easy movement between the bones. In a synovial joint, the bones are covered by hyaline cartilage and separated by fluid. The joint cavity is lined by a synovial membrane and the whole joint is enclosed by a fibrous capsule.

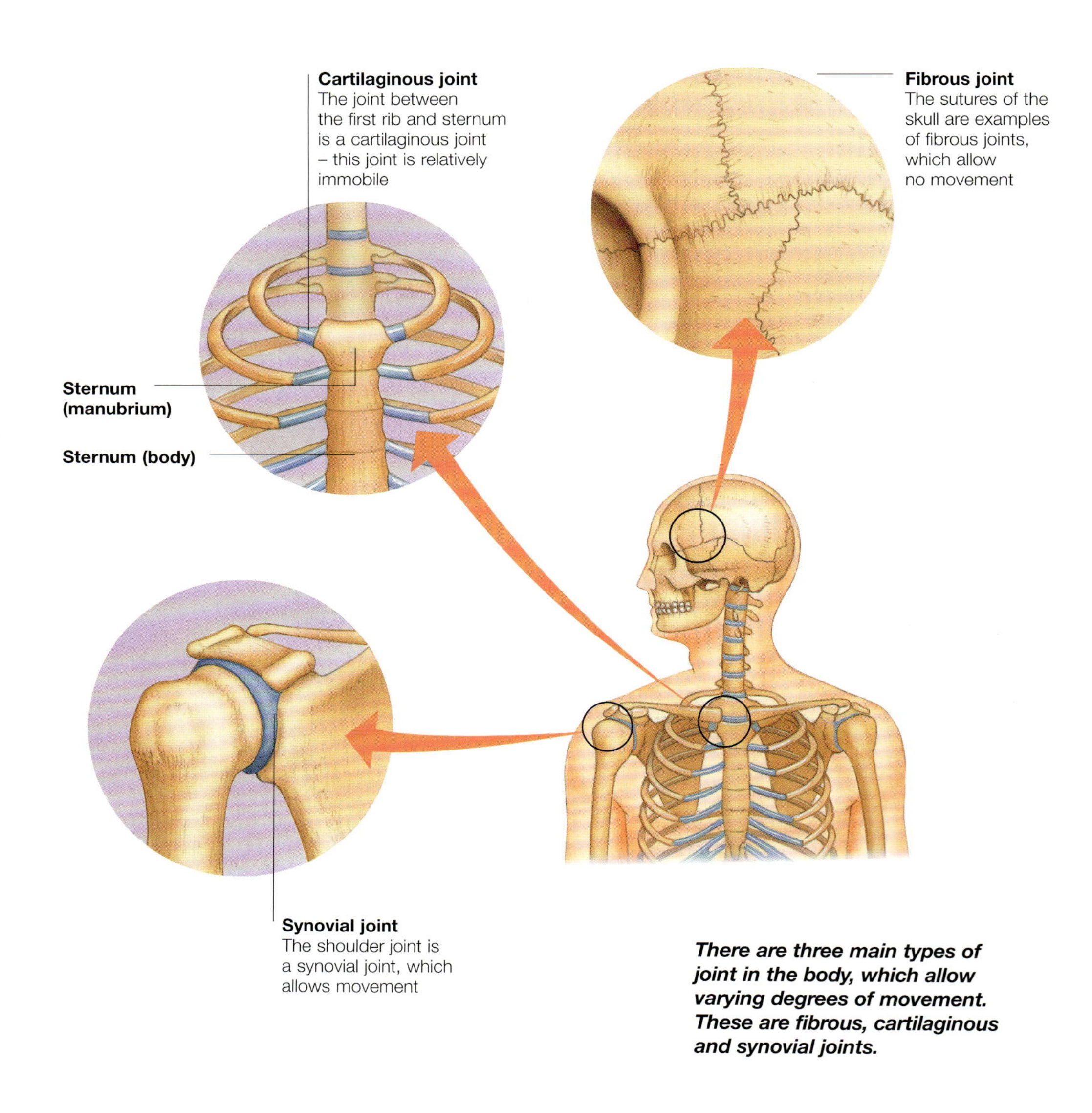

There are three main types of joint in the body, which allow varying degrees of movement. These are fibrous, cartilaginous and synovial joints.

Functional groups of joints

The functional classification of joints refers to the level of movement they allow; some permit little or no movement, and others free movement.

The classification of joints shown opposite is based on the structure of the tissues that make up the joint. Joints can also be grouped according to their function. Perhaps the most important function of a joint is to allow or prevent movement. On this basis, there are three groups:

- Synarthroses – these are joints that allow no movement. They are found predominantly within the axial skeleton (the central skeleton, excluding the limbs), where bones are more likely to fulfil the functions of support and protection than mobility. An example of synarthroses are the fibrous joints (sutures) of the skull
- Amphiarthroses – these are joints that allow slight movement. They are found in areas where some flexibility is needed, but a greater degree of movement would be unsuitable. Examples of amphiarthroses include the vertebral joints in the spine and the fibrous interosseous membrane in the forearm
- Diarthroses – these are joints that allow free movement. They predominate in the limbs, where mobility and movement are the prime functions. Some examples of diarthroses are the hip, shoulder and elbow joints.

Synarthrosis

Synarthrosis is a type of joint that allows no movement. These immovable joints are typically found in areas like the skull, where bones are tightly connected by fibrous tissue or cartilage, providing strength and protection.

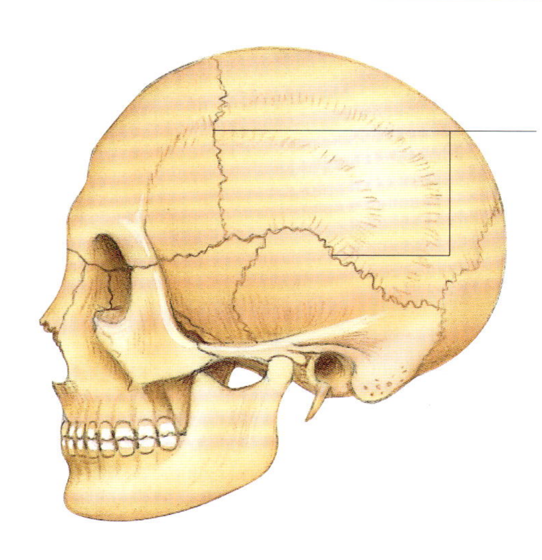

Diarthrosis

The elbow is a diarthrotic joint, which allows flexibility. The articular capsule allows plenty of freedom for extending the elbow joint.

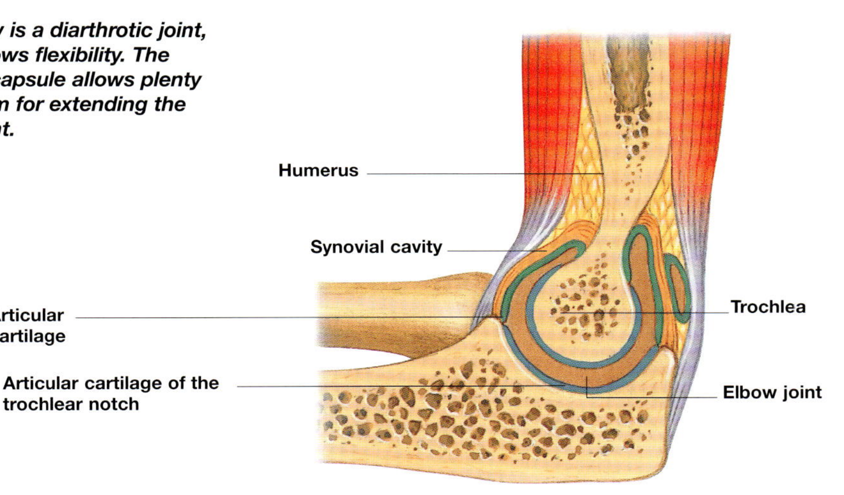

Fibrous Joints

Fibrous joints are so named after the tissue from which they are formed. There is little or no movement at a fibrous joint.

The bones of a fibrous joint are connected solely by long collagen fibres; there is no cartilage and no fluid-filled joint cavity. Because of its structure, a fibrous joint does not allow much real movement of the bones against each other. What little movement there is, is determined by the length of the collagen fibres.

Groups of Fibrous Joints

Fibrous joints can be further subdivided into three groups:

- Sutures – literally meaning 'seams', sutures are the tough fibrous joints between the interlocking bones of the skull. Short collagen fibres allow no side-to-side movement of these bones upon each other although there may be some slight 'springing' of the bones if pressure is applied. The presence of fixed fibrous joints in the skull gives great protection to the vulnerable brain tissue that lies beneath
- Syndesmoses – here, the bones are connected by a sheet of fibrous connective tissue, and the length of the fibres varies from joint to joint. These may also be known as interosseous membranes and are a feature of the forearm and the lower leg, where two bones lie side by side, acting as a unit. Syndesmoses tend to have longer fibres than sutures and so allow a little more movement
- Gomphoses – this is a very specialized type of fibrous joint with only one example in the human body, the tooth socket. In a gomphosis, a peg-like process sits in a depression, or socket, and is held in place by fibrous tissue, in this case the periodontal ligament. Movement is generally abnormal but micromovement is essential to eating to allow adjustment of the pressure of the bite.

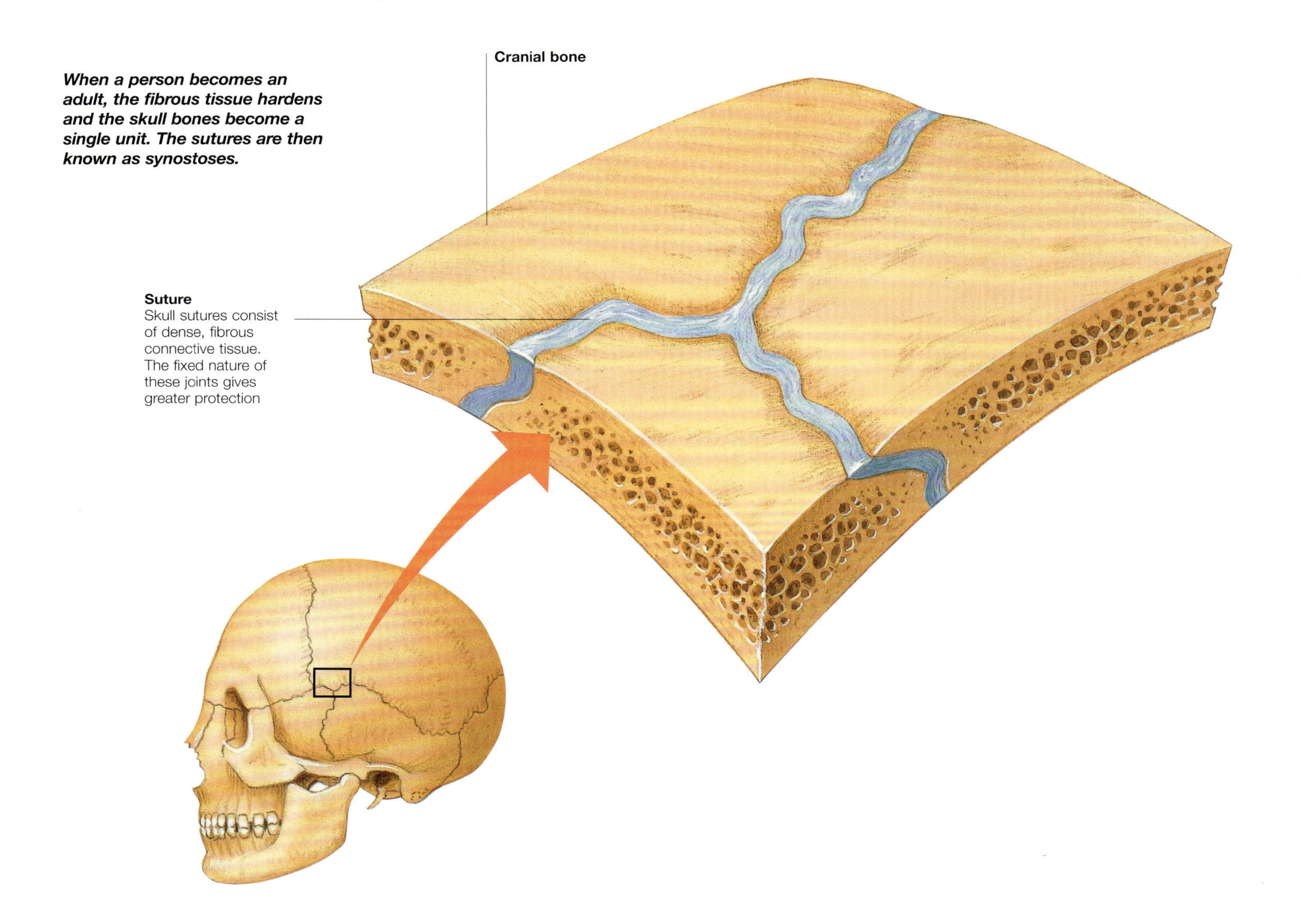

When a person becomes an adult, the fibrous tissue hardens and the skull bones become a single unit. The sutures are then known as synostoses.

Cartilaginous Joints

Cartilaginous joints are formed either of cartilage or of cartilage and fibrous tissue. They allow slight movement and are primarily found between the vertebrae.

In a cartilaginous joint, the bone ends are covered by hyaline cartilage. In some cases, there is a plate of tough fibrocartilage between the bones. The joint is usually enclosed within a fibrous capsule. They allow more movement than fibrous joints but much less than a synovial joint. There are two types of cartilaginous joints:

- Primary cartilaginous joints – those joints where two ends of bone are connected by a plate of hyaline cartilage. Primary cartilaginous joints are mostly present in the growing long bones of children. However, there are a few that exist in the mature skeleton, for example, between the first rib and the top of the sternum (breastbone).

- Secondary cartilaginous joints or 'symphyses' – these are joints in which a plate of tough fibrocartilage lies between the bones. They are strong, slightly movable joints that often perform the function of shock absorbers. An example of a secondary cartilaginous joint is found in the vertebral column, in which the individual vertebrae are covered with hyaline cartilage and are connected to each other by resilient fibrocartilaginous intervertebral discs. The pubic symphysis in the pelvis is also a cartilaginous joint.

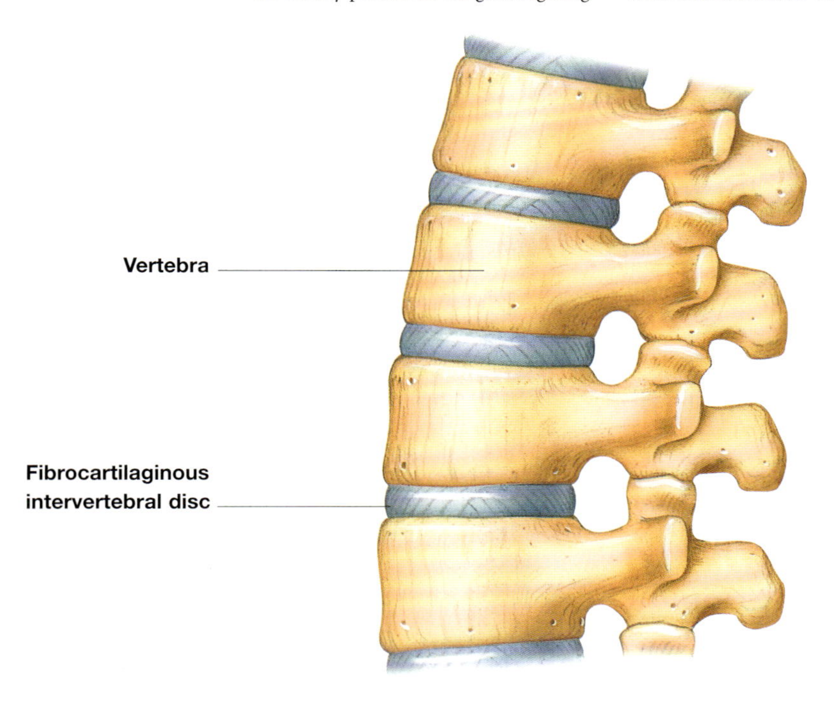

Fibrocartilaginous discs are in the joints in the vertebrae and act as shock absorbers. These tough joints allow a small amount of movement.

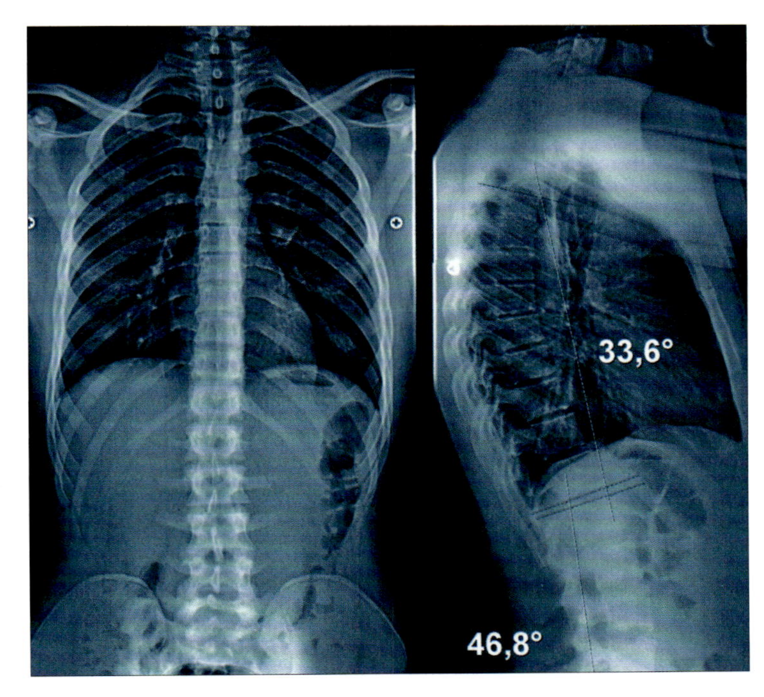

X-ray of the whole spinal column, standing, from the front and side, with vertebrae angle measurements.

Synovial Joints

Synovial joints allow much greater movement than fibrous and cartilaginous joints. They are classified into six groups according to their structure and movement.

Synovial joints are the most common type of joint in the body. They have a fluid-filled joint cavity between the articulating bones that allows a great deal of free movement, which is why many of the joints of the limbs are synovial.

Common Features:

- Joint cavity – the distinctive feature of a synovial joint is the presence of a joint cavity between the articulating bones. This is filled with synovial fluid.
- Synovial membrane – this is a layer of connective tissue that is rich in blood vessels. It lines the joint cavity except where cartilage covers the bone ends.
- Synovial fluid – this fills the joint cavity and is produced by the synovial membrane. In a healthy joint there is only a fine layer of this thick and viscous fluid, which is ideal for lubricating the joint.
- Articular cartilage – this is hyaline (transparent), flexible and slightly spongy so that it can act as a shock absorber for the bones, helping to reduce wear and tear and friction.
- Articular capsule – this tough, fibrous capsule surrounds and protects the joint. It is continuous with the periosteum (protective covering) of the bones above and below the joint.

Schematic representation of a synovial joint

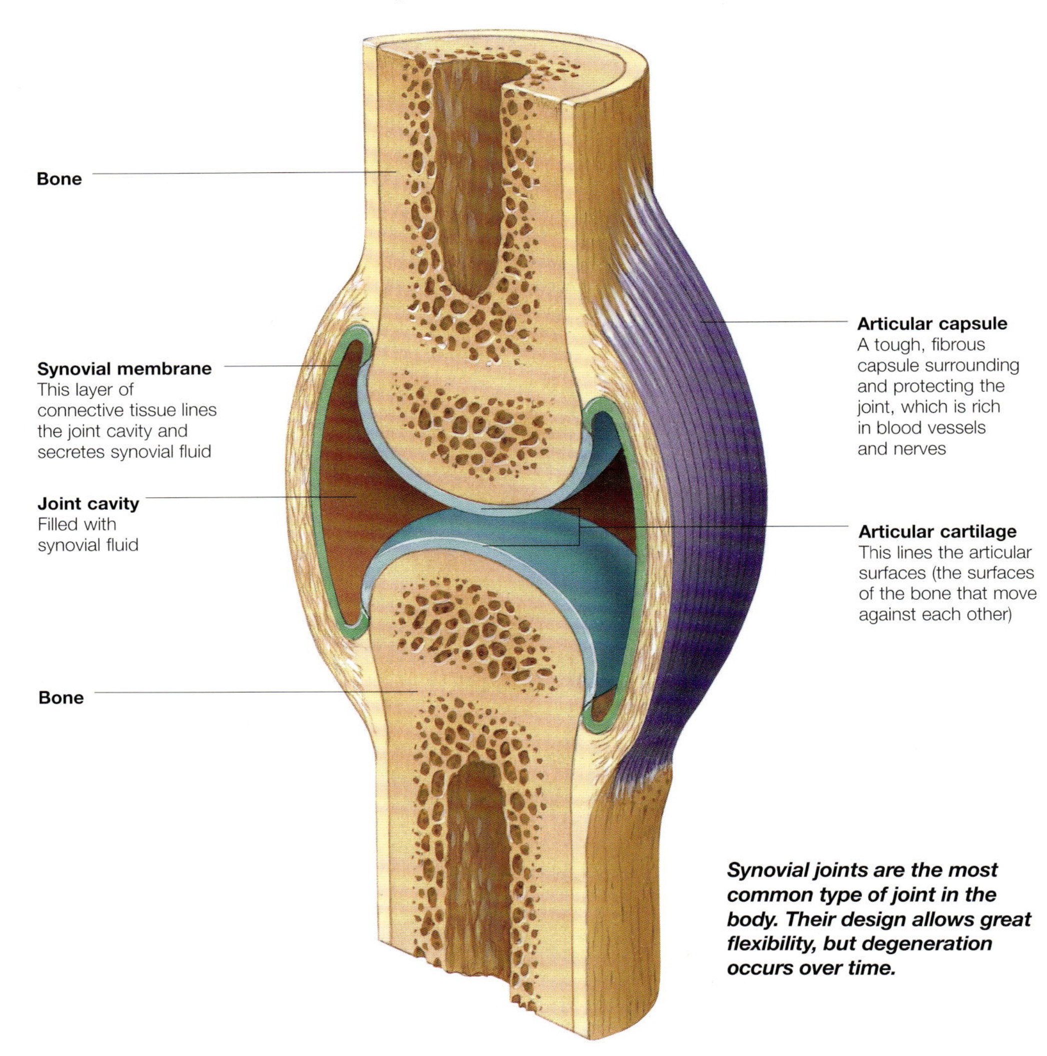

Synovial joints are the most common type of joint in the body. Their design allows great flexibility, but degeneration occurs over time.

Stability of synovial joints

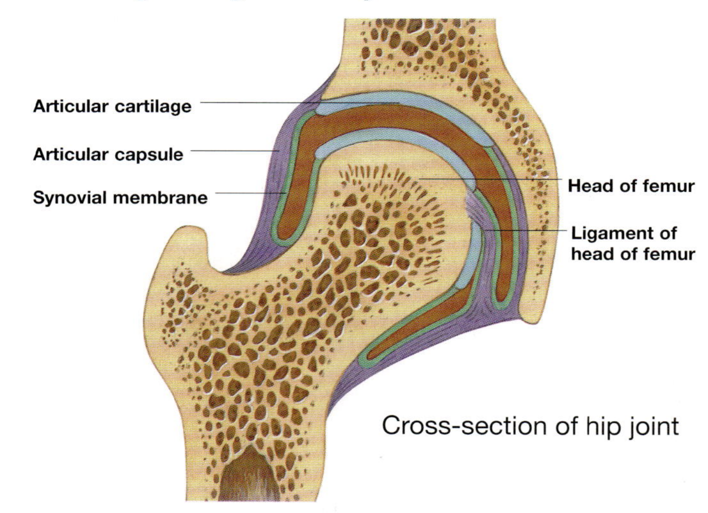

Cross-section of hip joint

To prevent instability and possible dislocation, a synovial joint depends upon three factors – the shape of the bone surfaces, the presence of ligaments and the tone of the surrounding muscle:

- Articular surfaces – in some cases the shape of the articular surfaces is important in the stability of a joint. The ball and deep socket arrangement of the hip, for example, greatly adds to this joint's stability.
- Ligaments – these support and strengthen the articular capsule of synovial joints and help to stabilize them by preventing excessive movement. But ligaments can be damaged by overstretching if the joint is not stabilized by other means
- Muscle tone – the tone in the joint's surrounding muscles is the crucial factor in its stabilization. The gentle background contraction of muscle fibres, even when the muscle is relaxed, acts to provide stability. The muscle both holds the joint ends together and brings them back into alignment after movement.

The free movement of synovial joints means that they rely for stability on the surrounding ligaments and muscles. The joint shape also plays a part.

Types of synovial joints

Although synovial joints have many structural features in common, they can be divided into six distinct groups according to the shape of their articular surfaces and the type of movement the joints allow.

Plane joints
In plane joints the articular surfaces are flat and usually allow movement in one plane only. Examples include the acromioclavicular joint between the shoulder blade and the collarbone, and the joints between the articular processes of the vertebrae in the spine.

Hinge joints
Hinge joints act like the hinge on a door: the articular surfaces can move in only one plane around one axis. The best example of a hinge joint is the elbow where only flexion (bending) and extension (straightening) are allowed.

Pivot joints
In a pivot joint a rounded or conical process of one bone inserts into a sleeve or ring of another. Rotation is the only permitted movement, such as is illustrated by the atlas and axis vertebrae, which allow the head to be turned from side to side.

Ball and socket joints
In a ball and socket joint one articular surface is rounded and sits within a cup-shaped socket. This is the most mobile of joint types and allows movement in all directions. Examples include the hip and the shoulder joints.

Saddle joints
The saddle-shaped articular surfaces of this joint allow movement in two different planes, such as occurs at the base of the thumb where the first metacarpal articulates with the trapezium of the wrist.

Condyloid joints
Condyloid joints have oval articular surfaces that allow a range of movements including flexion, extension, side-to-side movement and the circular movement of circumduction. One example of a condyloid joint is the 'knuckles', or metacarpophalangeal joints.

Types of synovial joint

Each type of synovial joint allows a particular kind of movement. This ranges from movement in one plane to that over several.

Plane joint

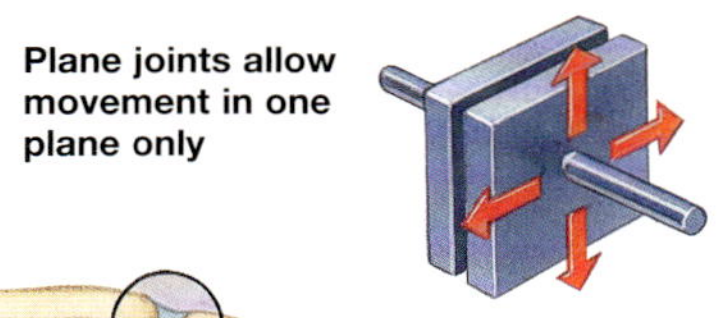

Hinge joint

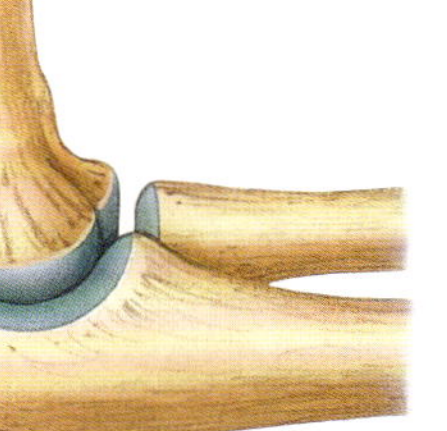

Pivot joint

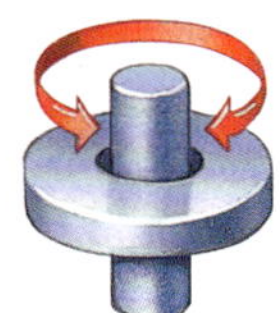

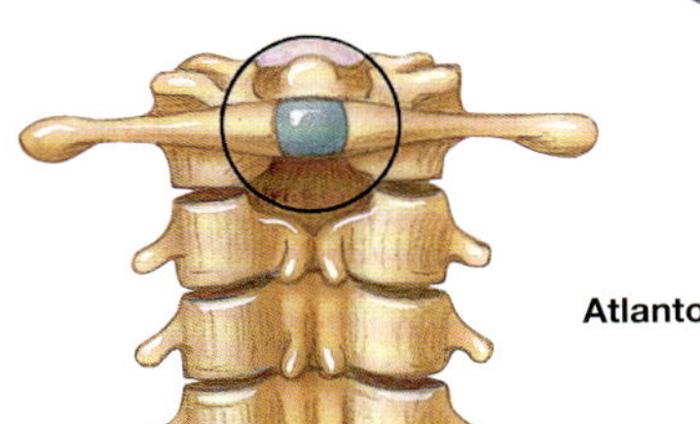

Ball and socket joint

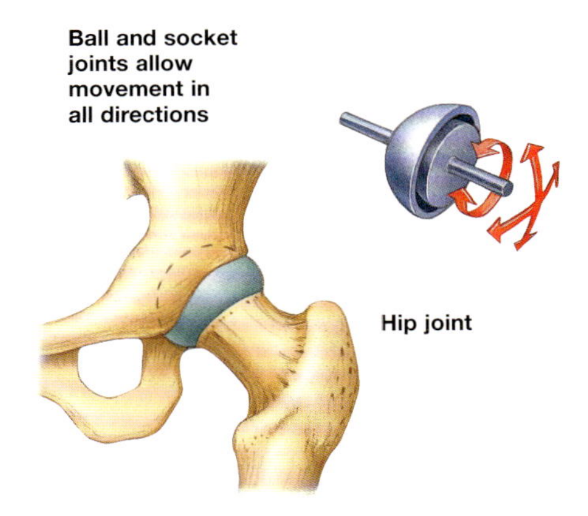

Types of Muscle

There are three main types of muscle in the body – skeletal muscle is used for voluntary movement, smooth muscle controls internal organs and cardiac muscle keeps the heart beating.

Skeletal muscle
The most familiar muscles in the body are the skeletal muscles (also known as striated or voluntary muscles) that are attached to bone and enable movement. Voluntary muscles are under conscious control, and can contract powerfully exerting a great deal of force, or can perform delicate movements such as picking up a small object.

Smooth muscle
So called for its lack of striations when viewed under a microscope, smooth muscle does not come under voluntary control. Instead it is controlled by the autonomic nervous system, which regulates the body's internal environment. Smooth muscle is found in the walls of hollow structures such as the intestines and bladder, and acts to regulate the size of those organs.

Cardiac muscle
Cardiac muscle is a specialized form of striated muscle that is found only in the heart and walls of the great vessels adjoining it, such as the aorta and superior vena cava. This type of muscle makes up almost all of the mass of the heart walls (the myocardium). It is not under conscious control but is regulated by the autonomic nervous system.

Skeletal muscle

Muscle fibres are bound together by connective tissue (epimysium) and divided into groups (fascicles) by a sheath, or perimysium.

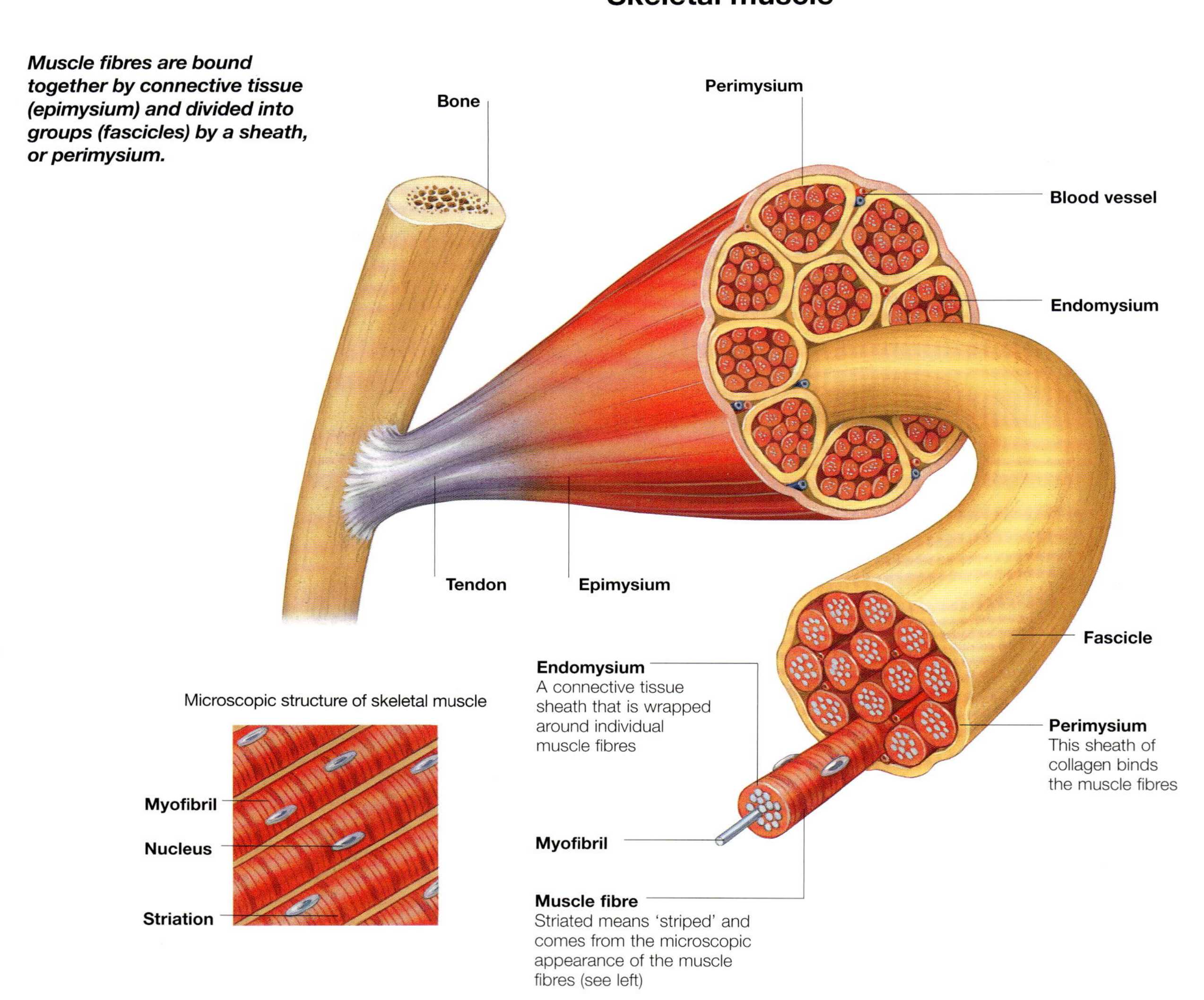

Shapes of skeletal muscle

Although all skeletal muscles are made up of fascicles, or groups of muscle fibres, the arrangement of these fascicles may vary. This variation leads to different muscle shapes throughout the body.

There are several ways of describing the various shapes of muscle, including:

- Flat – muscles, such as the external oblique in the abdominal wall, may be flat, yet fairly broad. They may cover a wide area and sometimes insert into an aponeurosis (a broad sheet of connective tissue)
- Fusiform – many muscles are of this 'spindle-shaped' form, where the rounded belly tapers at each end. Examples include the biceps and triceps muscles of the upper arm, which have more than one head
- Pennate – these muscles are named for their similarity to a feather (the word 'penna' means feather). They may be described as being unipennate (for example, extensor digitorum longus), bipennate (such as rectus femoris) or multipennate (for example, the deltoid). Multipennate muscles resemble a number of feathers placed next to one another
- Circular – these muscles, also known as sphincteral muscles, surround body openings. Contraction of these muscles, where the fibres are arranged in concentric rings, closes the opening. Circular muscles within the face include the orbicularis oculi, which closes the eye
- Convergent – these muscles are fan-shaped and the muscle fibres arise from a wide origin and converge on a narrow tendon. In some cases, these muscles take on a triangular shape. Examples include the large pectoral muscles.

Function

The arrangement of the fascicles in a muscle influences that muscle's action and power. When muscle fibres contract, they shorten to about 70 per cent of their relaxed length. If the muscle is long with parallel fibres, such as the sartorius muscle in the leg, it can shorten a great deal but has little strength.

If the degree of shortening is not as important as the power it can produce, the muscle may have numerous fibres packed tightly together and converging on a single point. This is the arrangement in multipennate muscles such as the deltoid in the shoulder.

Fascicle arrangement in relation to muscle structure

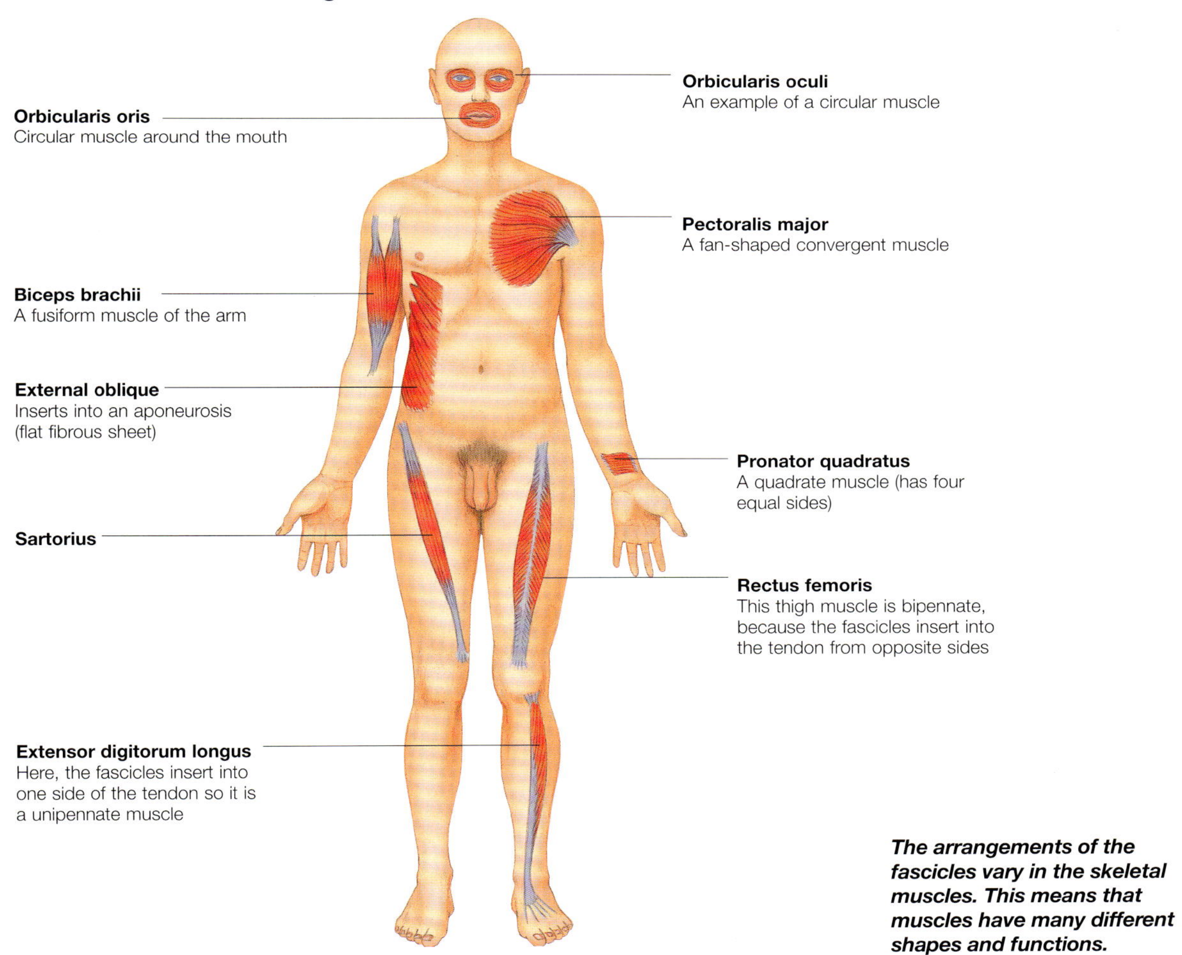

The arrangements of the fascicles vary in the skeletal muscles. This means that muscles have many different shapes and functions.

Muscle Contraction

Muscle tissue accounts for about half of the body's total mass, and is constantly at work, whether articulating the skeleton, enabling the heart to beat or passing food through the gut.

Muscle is tissue that is capable of contracting. The main types of muscle are voluntary and involuntary muscle. The contraction of voluntary (or skeletal) muscle can be consciously controlled, and this type of muscle is linked to parts of the skeleton to produce movement.

Voluntary Muscle

Muscle that moves bones is known as striated muscle due to its striped appearance. It consists of bundles of fibres bound together, with each fibre made up of a single long, multi-nucleated cell that stretches from one end of the muscle to the other. Each fibre consists of many long thin strands, known as myofibrils. These consist of two kinds of tiny, overlapping protein filaments made of actin and myosin, giving the myofibril a banded appearance. The bands of neighbouring myofibrils line up so that the whole fibre appears to be striped.

Involuntary Muscle

Involuntary muscle is not under the brain's conscious control. It is controlled automatically by the nervous system and is found in non-skeletal parts of the body. The heart, for example, is made up of involuntary muscle, beating without conscious effort.

Muscle structure

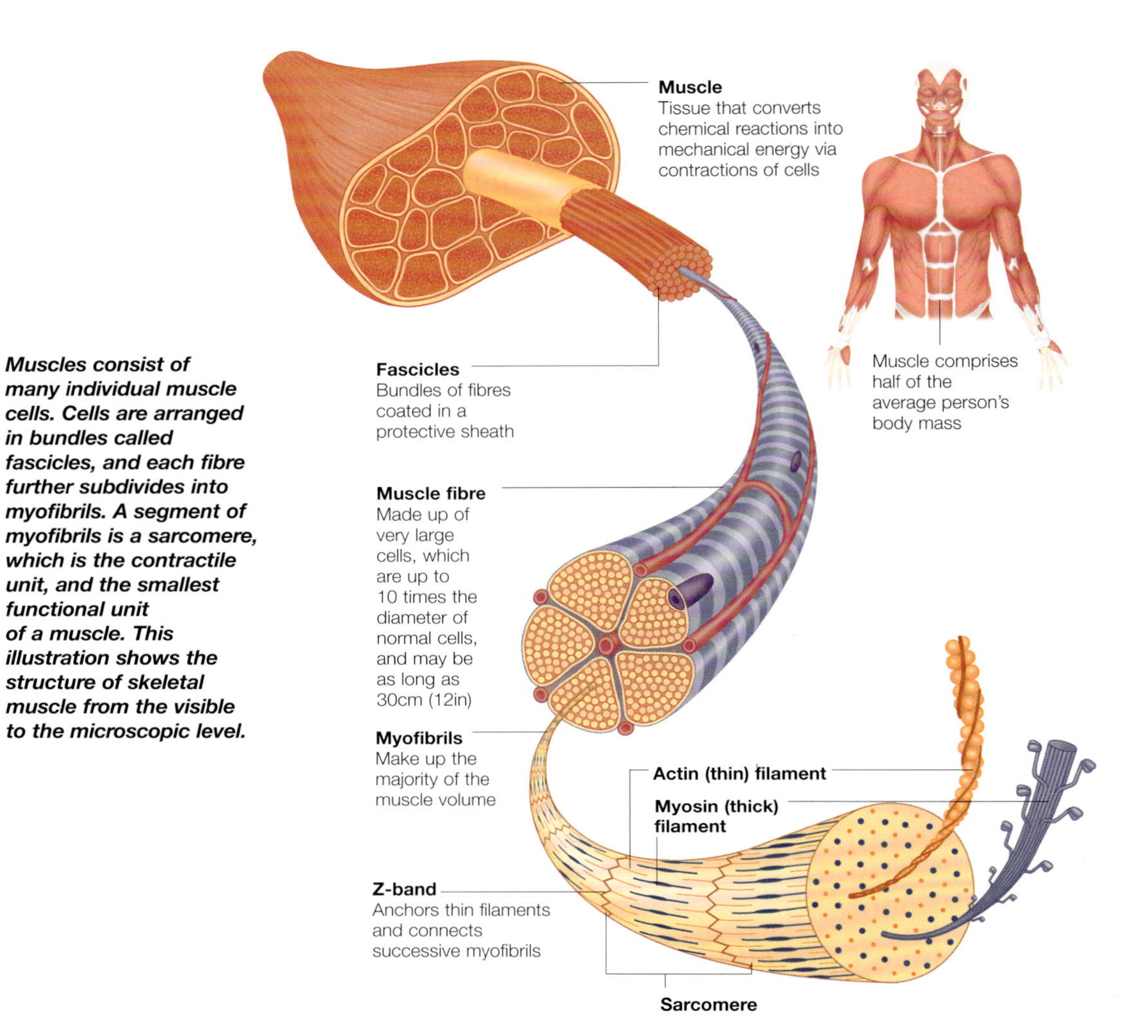

Muscles consist of many individual muscle cells. Cells are arranged in bundles called fascicles, and each fibre further subdivides into myofibrils. A segment of myofibrils is a sarcomere, which is the contractile unit, and the smallest functional unit of a muscle. This illustration shows the structure of skeletal muscle from the visible to the microscopic level.

How muscles contract

A muscle contracts when it is stimulated by a nerve impulse, which causes complex chemical changes to take place in the muscle fibres. Each group of filaments lies in a small chamber (sarcomere) in which the thin actin filaments are attached to each end. The thick myosin filaments lie between the actin filaments in the middle of the sarcomere.

When provided with energy – usually obtained from glycogen ('animal starch') stored in the muscle – the myosin filaments form chemical bonds with the actin filaments, and these bonds are repeatedly broken and remade further along. In this way, the myosin filaments work their way along the actin filaments like ratchets, with the result that the whole sarcomere becomes shorter and fatter.

When the muscle is no longer being stimulated, the chemical action ceases. The bonds between the filaments are no longer formed and the muscle relaxes.

Contraction of an opposing muscle stretches the filaments apart, and this is triggered by a chemical called acetylcholine, which is released by the nerve endings and alights on special receptive areas in the muscle. As long as acetylcholine is present in these areas, the muscle remains contracted.

Relaxed

Z-band

Myosin filament

Actin filament

Fully contracted

Contraction is achieved by myosin fibres rapidly breaking and reforming bonds with the actin fibres, in a ratchet-like manner.

How involuntary muscles move

The body contains two types of involuntary muscle (muscle not under the conscious control of the brain) – smooth muscle and cardiac muscle.

Smooth muscle

Smooth muscle and cardiac muscle are both able to contract involuntarily, without conscious control. They are controlled by nerve impulses from the autonomic (unconscious) nervous system. The nervous system uses hormones, neurotransmitters and other receptors to control smooth muscle spontaneously and when necessary.

Smooth muscle is found in many parts of the body, notably the gut, but also in such places as the lungs, bladder and sex organs. It consists of spindle-shaped cells, whose average length is only a fraction of a millimetre.

The cells are tapered at both ends, have single nuclei and are arranged in bundles held together by a substance that acts as a cement. These bundles are grouped into larger bundles or flattened bands, held together by connective tissue. The arrangement of the cells is much looser than the regular pattern found in striped muscle, but the contraction of smooth muscle still results from the movement of filaments, which are found in the walls of the cells.

Contraction

Contraction of smooth muscle is generally slower than that of striated muscle, and does not necessarily take place throughout the whole muscle.

An action typical of smooth muscle is found in the intestines, where a band of muscle usually contracts over a certain part of its length, then relaxes while another part contracts, thus producing waves of contraction down the muscle, called peristalsis. This enables food to be passed down the digestive tract, into the stomach and through the intestines.

Cardiac muscle cells are less elongated than skeletal muscle. Adjacent cells are closely attached to each other by proteins called intercalated discs. Structures called desmosomes form junctions, allowing electrical signals to be transmitted between cells.

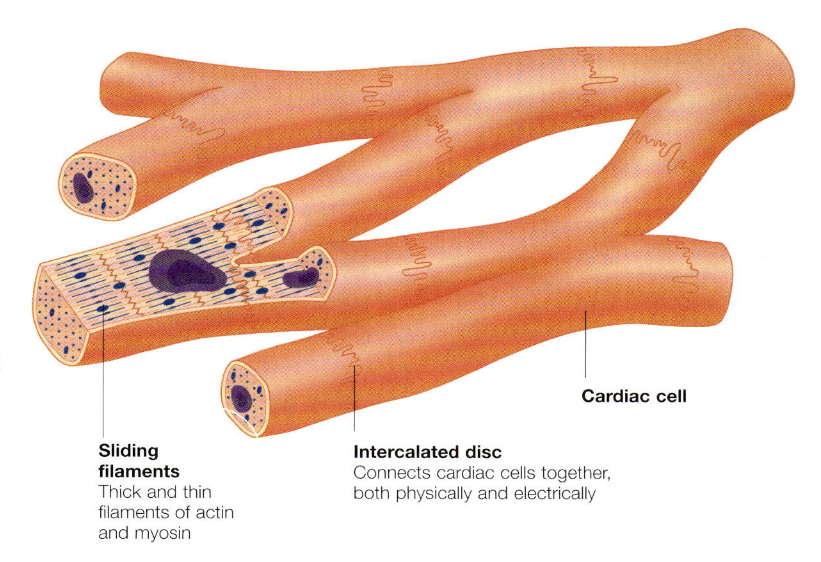

Cardiac muscle

Cardiac muscle is only found in the heart, and its structure is somewhere between that of striated muscle and smooth muscle. It has a striped appearance when viewed through a microscope, but the cells are shorter and more box-shaped than striated muscle fibres. Most of the cells are divided at the ends, and the subdivisions form connections with cells that lie alongside. In this way, a resilient network of fibres is formed with the ability to act in unison, and it is this structure that gives heart muscle its toughness.

In an average lifetime, the heart beats over two billion times and pumps some 550,000 tonnes of blood. In order to keep the heart contracting steadily and regularly throughout this time, the heartbeat is controlled by electrical impulses that are generated from within the heart itself by specialized nodes.

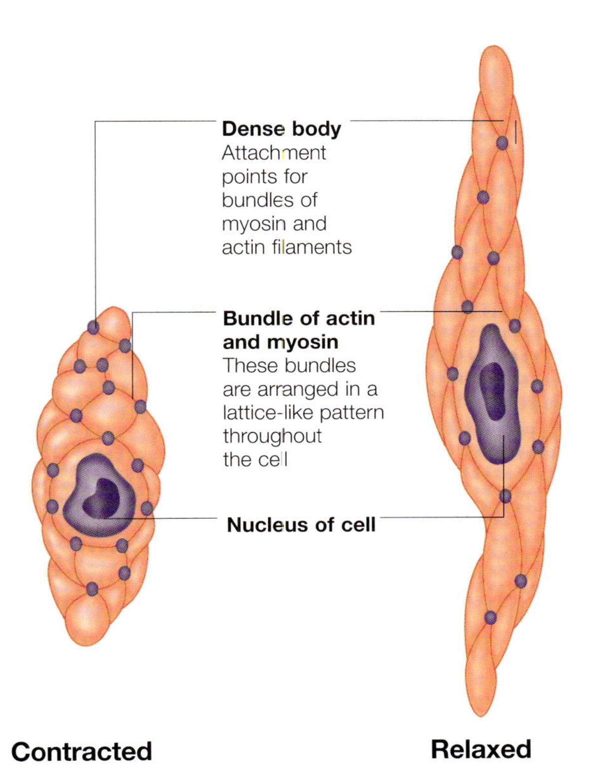

Smooth muscle surrounds hollow body structures, such as the oesophagus, bladder, uterus and blood vessels. The rate of contraction of its cells is relatively slow, but they are more energy-efficient and able to maintain contraction for a longer time.

Ligaments

Ligaments are bundles of tough fibres that hold and stablize adjacent bones at joints. They are formed of connective tissue.

Ligaments usually bind bones together at joints and are found throughout the body. They consist of dense, regular connective tissue that contains closely packed bundles of collagen and elastic fibres, together with fibroblasts (specialized cells that produce this tissue). The fibres all run in the same direction and allow this strong flexible tissue to withstand considerable pulling forces in that direction.

Shape and Stability

Being 'wavy', collagen fibres have a certain amount of give in them which, together with the elastic fibres, allows the ligament to stretch a little under tension and then return to its previous length. This helps a joint to retain its original shape and stability before, during and after movement.

Some ligaments, such as the ligamenta nuchae in the back of the neck, have an unusually high number of elastic fibres, which enables them to stretch further than other ligaments without tearing. This allows a greater degree of movement between the bones to which they are attached.

Abdominal and Pelvic Ligaments

Not all ligaments are connected to bones. For example, some internal organs are secured in place, and to each other, by ligaments. The oesophagus and upper part of the stomach are held to the diaphragm by the phrenicoesophageal ligaments. These ligaments are extensions of the fascia, a connective tissue that covers the diaphragm's surface. In the pelvis, the broad ligament helps to keep the uterus in position.

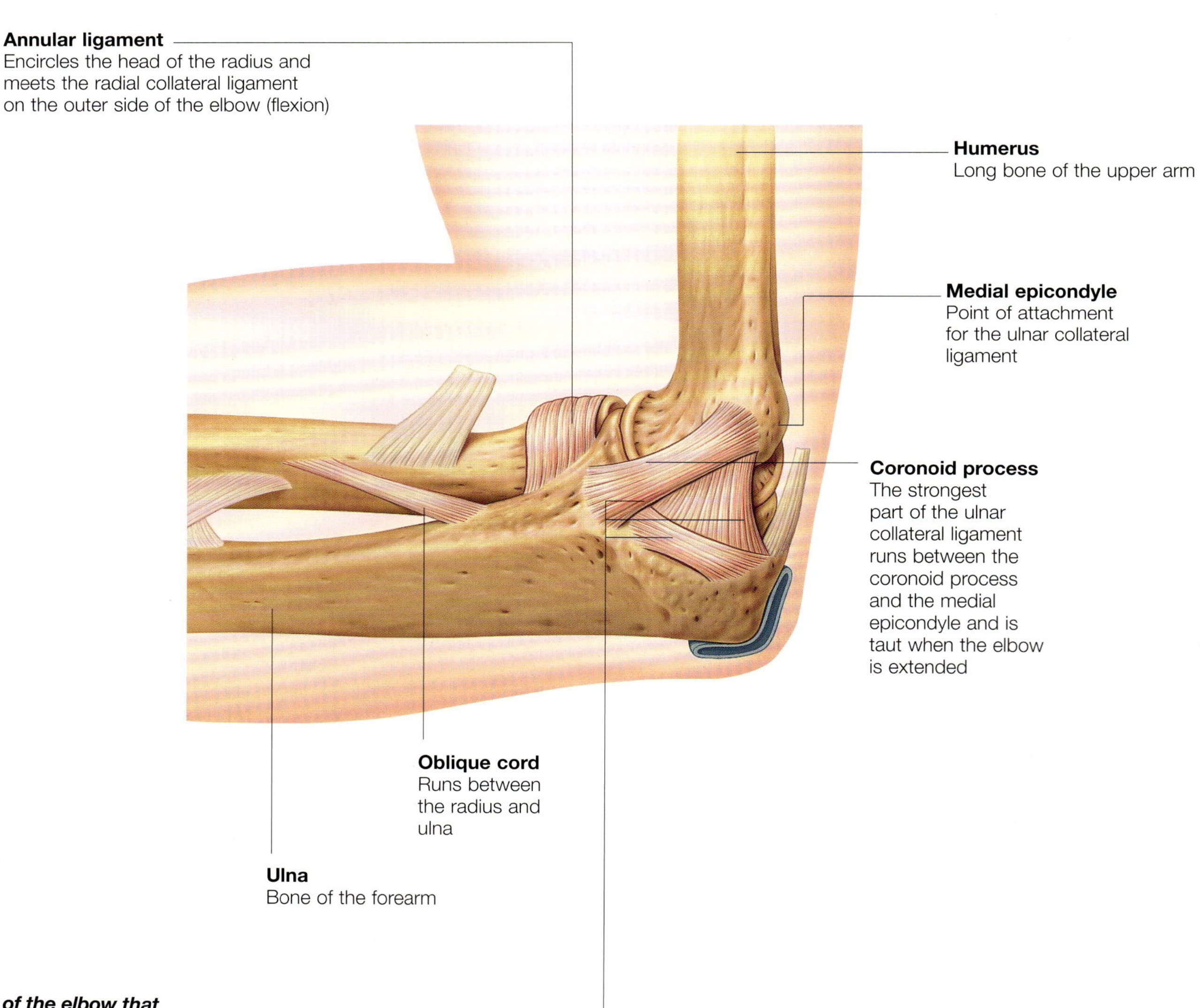

The ligaments of the elbow that strengthen the joint and hold the bones in place are seen in this medial view (inner side) of the joint bent to 90 degrees.

Tendons

Tendons attach muscles to bones. They are composed of dense connective tissue formed of collagen, which is arranged in parallel bundles.

There are approximately 4000 tendons located throughout the human body. They are strong, silvery-white cords formed of fibrous connective tissue. Tendons attach skeletal bones to muscle, enabling movement and providing stability. When a muscle contracts, the tendon acts as a lever and pulls back the attached bone, causing it to move.

Layers of the Tendon

The three layers of the muscle – epimysium (fibrous connective tissue), perimysium and endomysium – are continuous with the connective tissue beyond the muscle cells to form the tendon. Tendons run from the skeletal muscle at the myotendinous junction to the periosteum of a bone at the osteotendinous junction.

Tendon Sheath

Some tendons, for example, those in the wrist and ankle, are surrounded by fibrous tubes of connective tissue called sheaths. This sheath is composed of three layers: an outer or parietal layer, an inner or visceral layer and a central space that contains synovial fluid. The sheath stabilizes the tendon and also allows smooth movement.

Size and Shape of Tendons

Tendons vary in shape and size. For example, the calcaneus (Achilles) tendon at the back of the ankle is the largest tendon in the body and can be up to 25cm (10in) in length. A thin, flat tendon is known as an aponeurosis. The aponeurosis in the scalp attaches two muscles: the occipitalis and the frontalis. The long thin tendons of the hand (the extensor tendons) are vital in enabling the fingers to straighten and bend.

Tendons in the foot

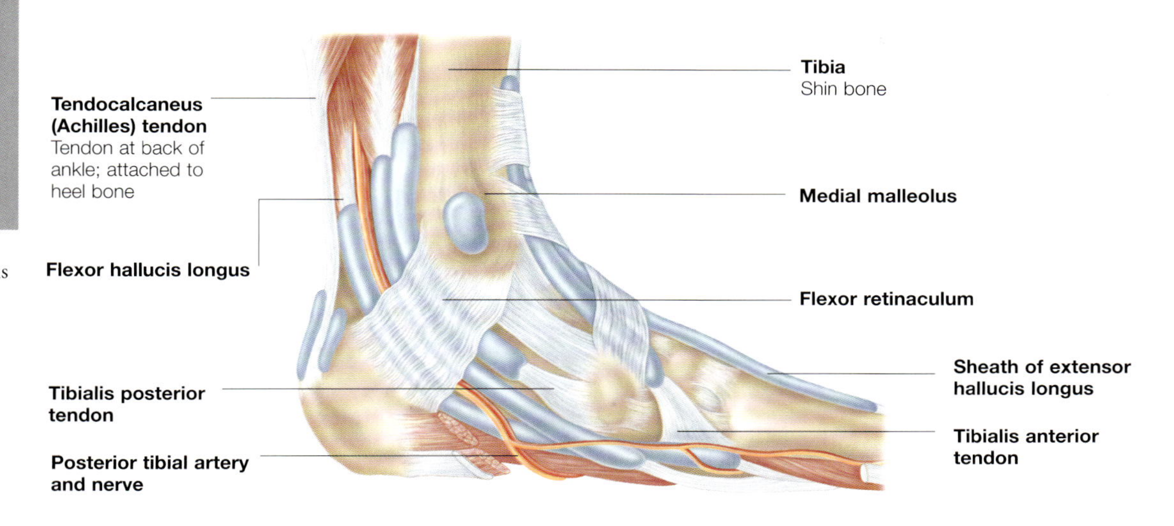

Tendons in the scalp

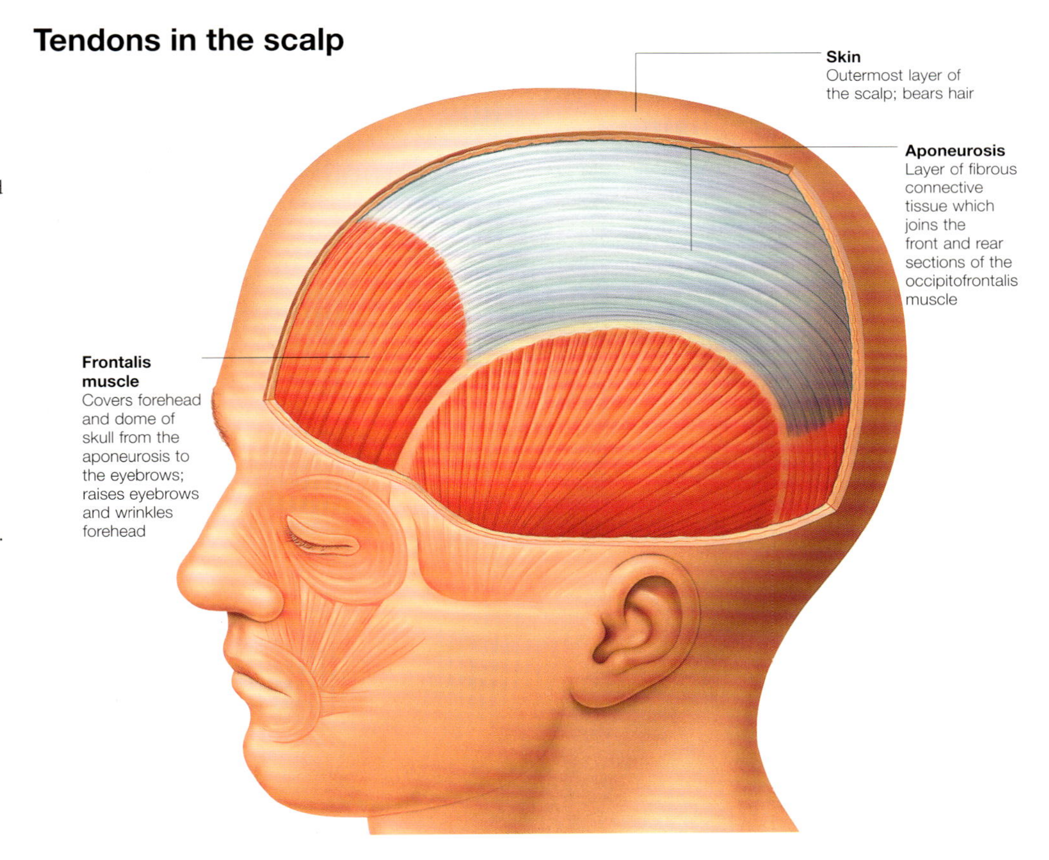

Skull

The skull is the head's natural crash helmet, protecting the brain and sense organs from damage. It is made up of 28 separate bones and is the most complex element of the human skeleton.

The skull, or cranium, is the bony covering of the head, which protects the brain, sections of the cranial nerves and the organs of special sense such as the eyes. It also provides attachment for many of the muscles of the neck and head.

Although often thought of as a single bone, the skull is made up of 28 separate bones. It can be divided into two main sections: the cranium and the mandible. Whereas most of the bones of the skull articulate by relatively fixed joints, the mandible (jawbone) is easily detached. The cranium is then subdivided into a number of smaller regions, including:

- Cranial vault (upper dome part of the skull)
- Cranial base
- Facial skeleton
- Upper jaw
- Acoustic cavities
- Cranial cavities (interior of skull housing the brain).

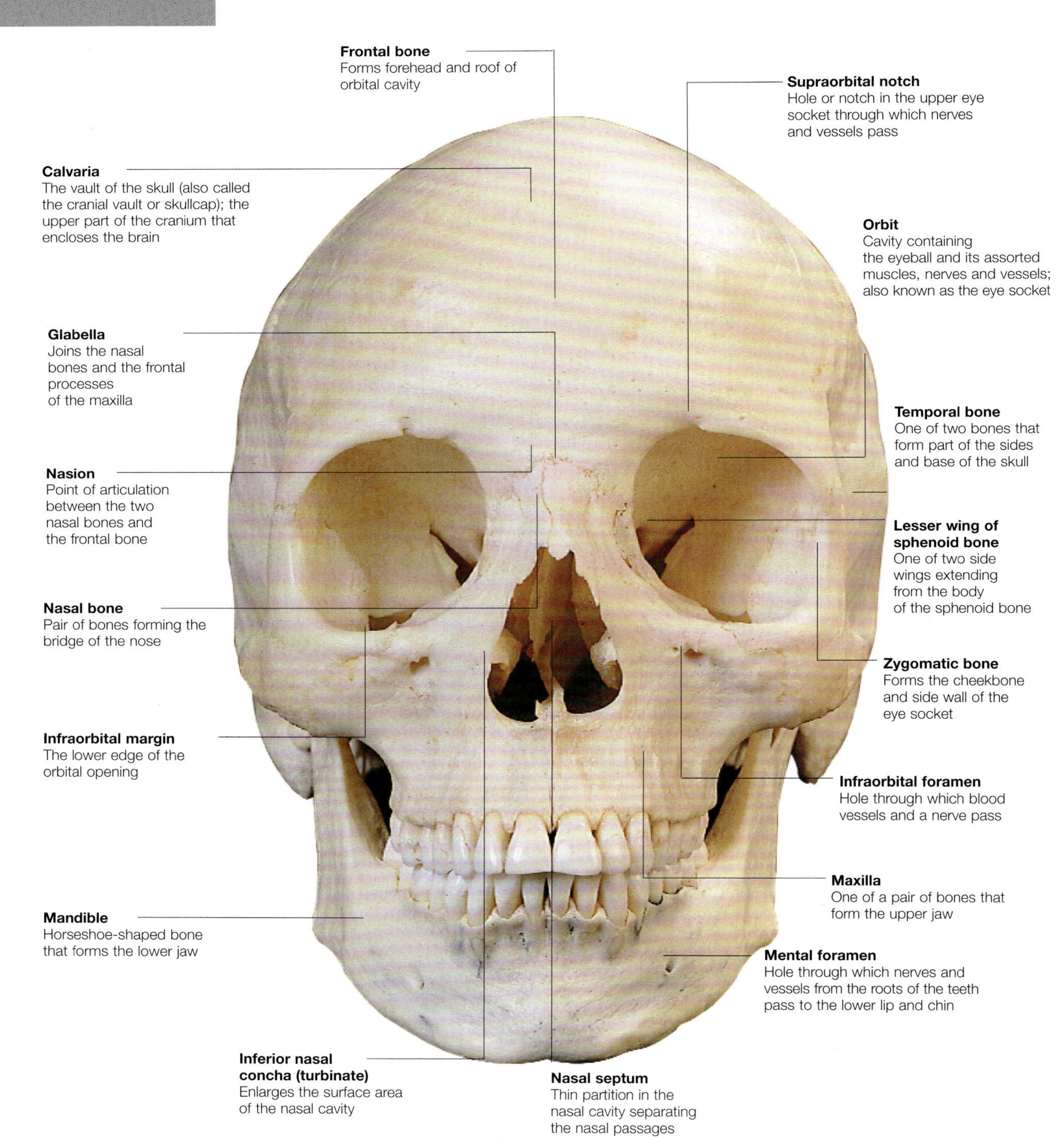

Side of the skull

A lateral or side view of the skull clearly reveals the complexity of the structure, which has many separate bones as well as the joints between them.

Several of the bones of the skull are paired, with one on either side of the midline of the head. The nasal, zygomatic, parietal and temporal bones all conform to this symmetry. Others, such as the ethmoid and sphenoid bones, occur singly along the midline. Some bones develop in two separate halves and then fuse at the midline, namely the frontal bone and the mandible (lower jaw).

The bones of the skull constantly undergo a process of remodelling: new bone develops on the outer surface of the skull, while the excess on the inside is reabsorbed into the bloodstream. This dynamic process is facilitated by the presence of numerous cells and a good blood supply.

Joints of the skull: sutures

The only moveable skull joint is the temporomandibular joint, where the jaw hinges against the cranium, and which allows all the actions of chewing and speech. All the other bones are fixed to each other by joints known as sutures, which are only found in the skull. In the adult, these comprise thin zones of unmineralized fibrous tissue bonding the irregular, interlocking margins of adjacent bones.

The purpose of sutures in the skull of the developing infant is to allow for growth at right angles to their alignment. For example, the coronal suture allows growth in length and the squamosal allows for increase in height of the skull.

During the rapid period of cranial growth, from baby to child, the enlarging brain forces the bones apart at their sutures, and new bone is then deposited at the edge of the sutures, stabilizing the skull at its new size. By the age of seven, the sutural growth has slowed, and the skull enlarges at a slower rate by bone remodelling.

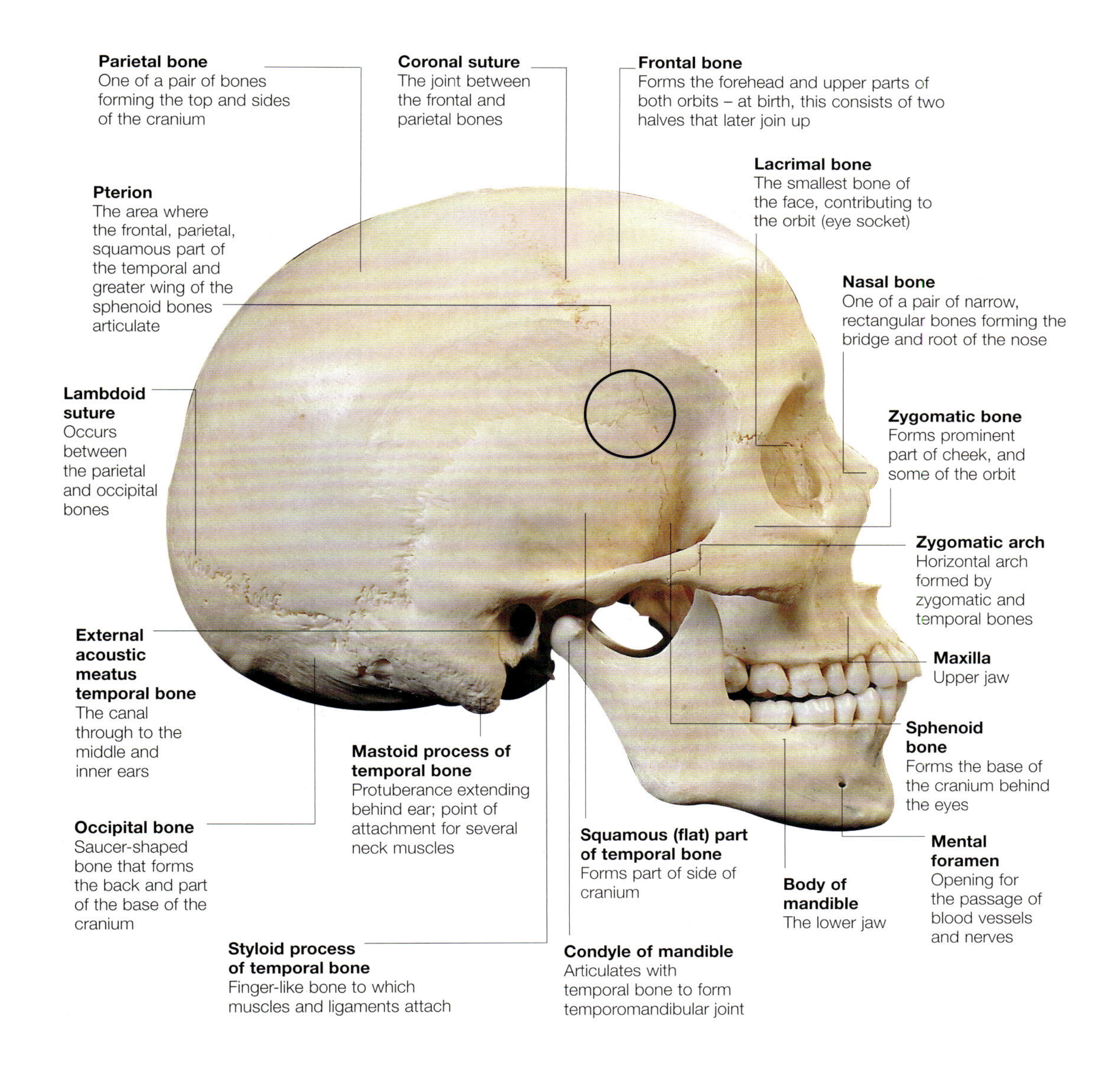

Top of the skull

The calvaria (also known as the vault of the skull or the skullcap) is the upper section of the cranium, surrounding and protecting the brain.

The four bones that make up the calvaria are the frontal bone, the two parietals and a portion of the occipital bone.

These bones are formed by a process in which the original soft connective tissue membrane ossifies (hardens) into bone substance, without going through the intermediate cartilage stage, as happens with some other bones of the skull.

Points of interest in the calvaria include:

- The sagittal suture running longitudinally from the lambdoid suture at the back of the head to the coronal suture
- The vertex (highest point) of the skull; the central uppermost part, along the sagittal suture
- The distance between the two parietal tuberosities is the widest part of the cranium.
- The complex, interlocking nature of the sutures, which enable substantial skull growth in the formative years, and provide strength and stability in the adult skull.

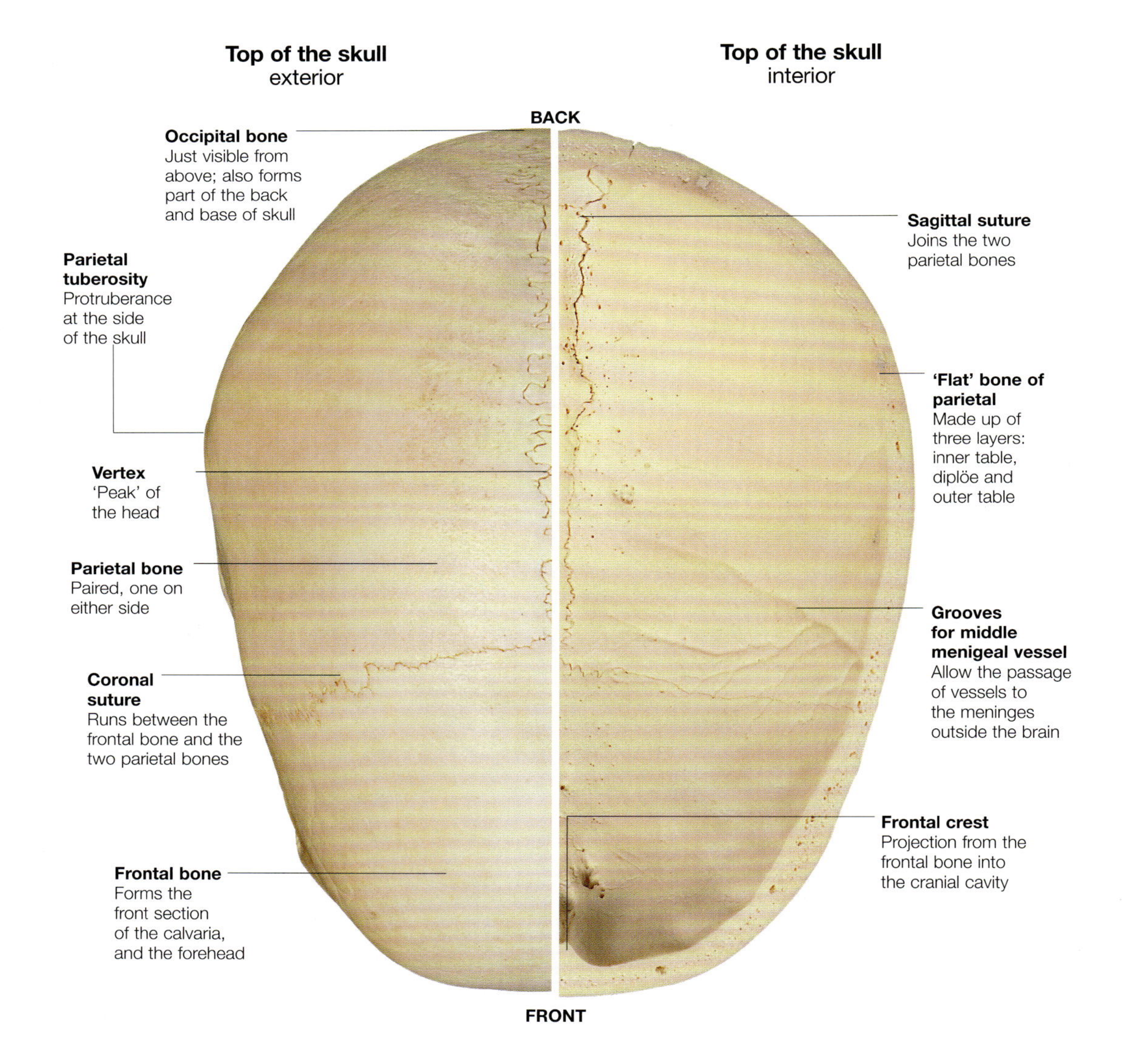

Base of the skull

This unusual view of the skull is from below. The upper jaw and the hole through which the spinal cord passes can be seen.

The bones found in the midline region of the base of the skull (the ethmoid, sphenoid and part of the occipital bone) develop in a different way from those of the vault of the skull. They are derived from an earlier cartilaginous structure in a process called endochondral ossification.

The maxillae are the two tooth-bearing bones of the upper jaw, one on each side. The palatine processes of the maxillae and the horizontal plates of the palatine bones form the hard palate.

Foramina

Foramina is the plural form of the Latin word 'foramen' meaning hole or opening. These openings are the numerous canals through the bones of the skull that allow blood vessels and the 12 pairs of cranial nerves to pass in and out of the cranial cavity.

Other small, less regular, channels link the external veins of the skull with those on the inside. They are termed emissary veins and the openings are emissary foramina. Such pathways can allow the spread of infection from outside the skull to a more serious infection inside. The most important foramina are:

- Foramen magnum, where the spinal cord joins the brain stem
- Foramen lacerum, between the petrous part of the temporal bone and the sphenoid bone
- Foramen ovale, (one on either side) for the mandibular branch of the trigeminal nerve
- Foramen spinosum, allows the middle meningeal artery to pass into the interior of the cranial cavity
- Stylomastoid foramen, allows the seventh cranial nerve to pass
- Jugular foramen, aperture for the sigmoid and inferior petrosal sinus and three of the cranial nerves
- Carotid canal, for the passage of the carotid artery (main artery in the neck) and associated nerve fibres.

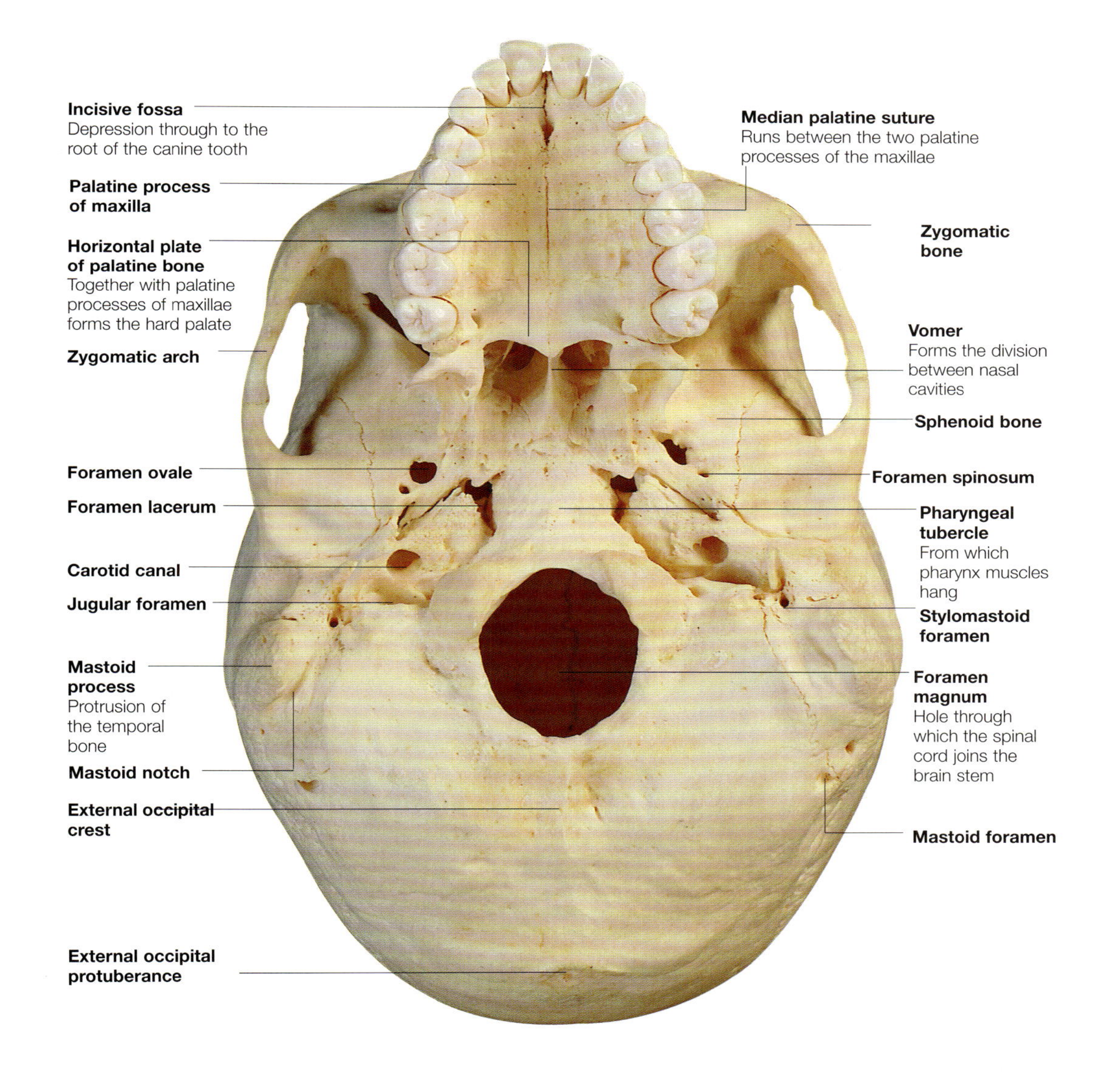

Paranasal Sinuses

The paranasal sinuses are air-filled cavities in the bones around the nasal cavity that warm inhaled air, lighten the weight of the skull and produce mucus that flows into the nasal cavity.

There are four pairs of paranasal sinuses that are named according to the bones in which they are situated. These four pairs are:

- Maxillary sinuses
- Ethmoidal sinuses
- Frontal sinuses
- Sphenoidal sinuses.

Each sinus opens into its half of the nasal cavity through a tiny opening called an ostium on the side of the nasal cavity. The paranasal sinuses are very small, or even absent, at the time of birth, and remain small until puberty when the sinuses enlarge fairly rapidly.

Frontal Sinuses

The frontal sinuses are situated within the frontal bone (the bone of the forehead). Each is variable in size, corresponding to an area just above the inner part of the eyebrow. The frontal sinuses are situated above the opening into the nasal cavity in the middle meatus.

Mucus Production

The paranasal sinuses have an inner lining of mucous membrane containing cells that secrete fluid. Some of the cells possess hair-like projections that propel secretions into the nasal cavity through the ostia (openings). Drainage of mucous secretions is efficient and aided by gravity.

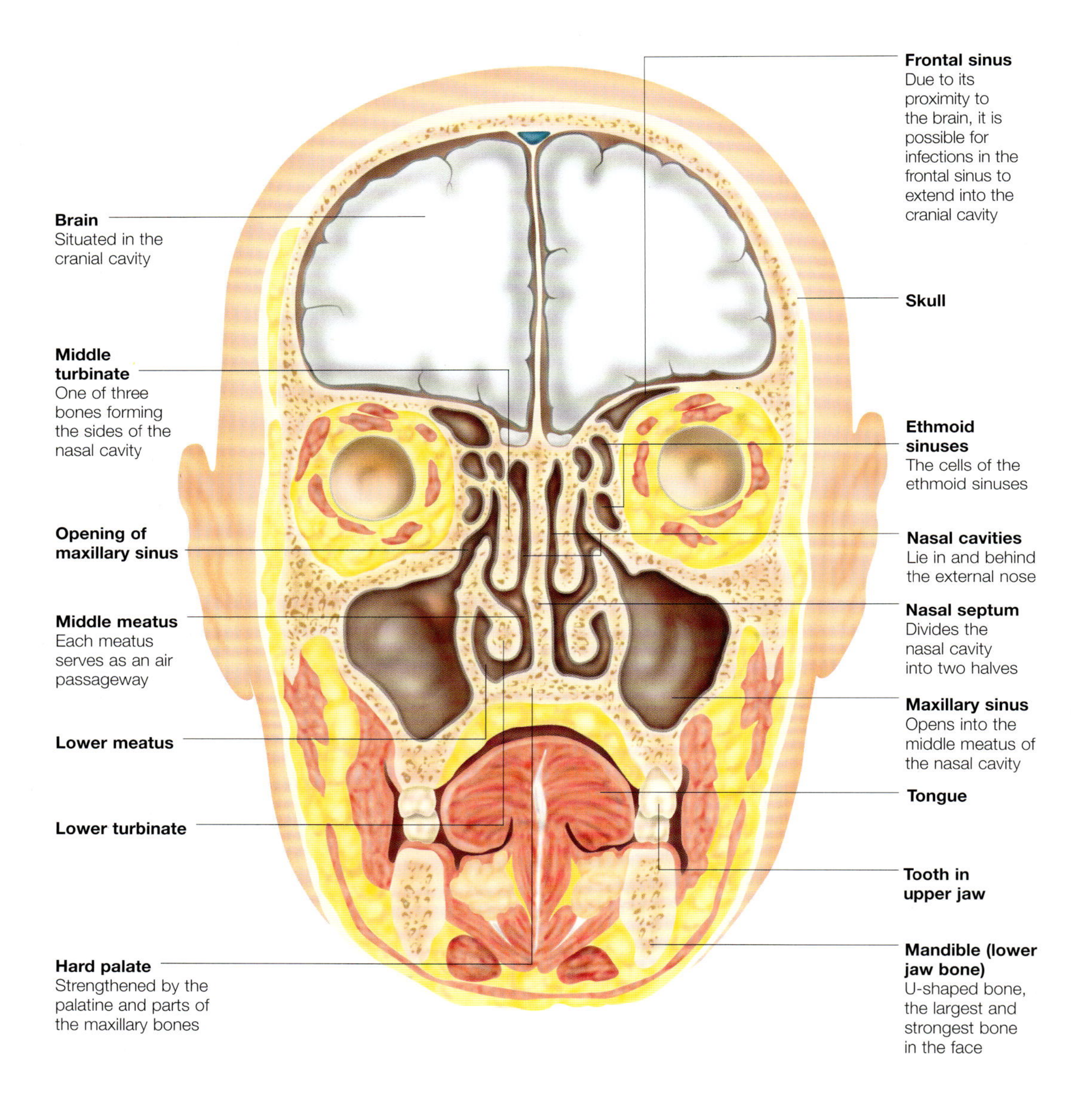

Inside the sinuses

The efficiency of mucous drainage from each of the four pairs of sinuses depends on their location. Effective drainage lessens the risk of sinus infection.

Sphenoidal sinuses
Each of the paired sphenoidal sinuses are behind the roof of the nasal cavity, within the sphenoid bone. The two sphenoidal sinuses lie side by side, separated by a thin, vertical, bony partition (the septum).

Each sphenoidal sinus opens into the uppermost part of the side wall of the nasal cavity (immediately above the upper turbinate) and also drains fairly efficiently into the nasal cavity.

Ethmoidal sinuses
Each ethmoidal sinus is situated between the thin, inner wall of the orbit (eye socket) and the side wall of the nasal cavity. Unlike the other paranasal sinuses, these sinuses are made up of multiple communicating cavities called ethmoid air cells. These cells are subdivided into front, middle and back groups. The front and middle groups of air cells open into the middle meatus, while the back group opens into the upper meatus. Drainage into the nasal cavity is moderately efficient.

Maxillary sinuses
The largest of the pairs of sinuses are the maxillary sinuses, situated within the maxillae (cheekbones). Infections and inflammation are more common here than in any of the other paranasal sinuses. This is because the drainage of mucous secretions from this sinus to the nasal cavity is not very efficient.

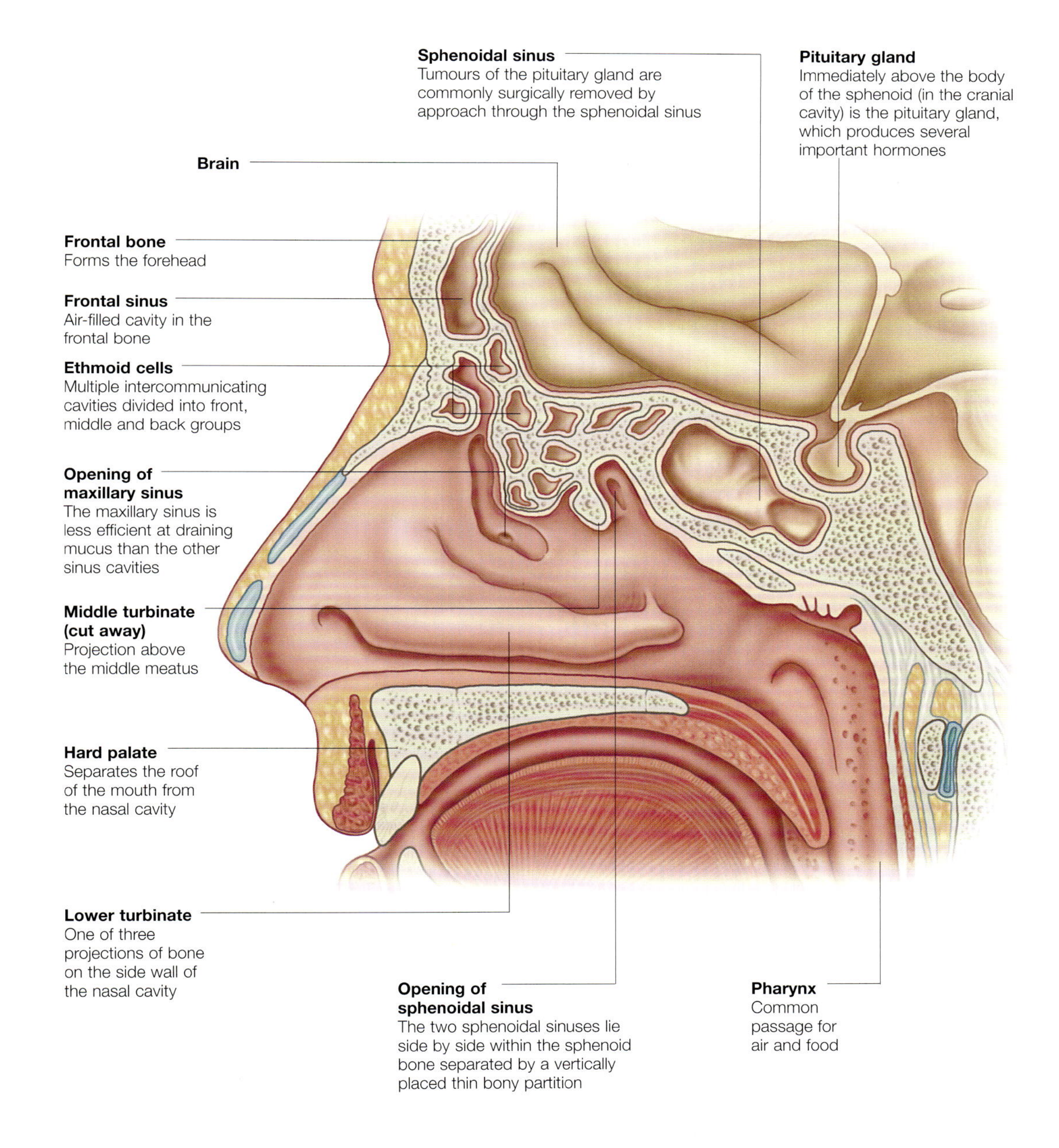

Infratemporal Fossa

The infratemporal fossa (a fossa is a depression or hollow) is a region at the side of the head that contains a number of important nerves, blood vessels and muscles involved in mastication (chewing).

The infratemporal fossa is located below the base of the skull, between the pharynx and the ramus, which is the upper side of the mandible (lower jawbone).

Dental Importance
The region is of particular importance to dental surgeons, not only because many of its components are essential to the process of mastication, but also as many of the nerves and blood vessels supplying the mouth pass through it.

Anatomy of the Fossa
The region is largely defined by the skeletal boundaries of the infratemporal fossa. The anterior boundary is the posterior surface of the maxillary bone, and the posterior boundary is the styloid process of the temporal bone and the carotid sheath. The midline boundary is formed by the lateral pterygoid plate of the sphenoid bone; the lateral boundary is the ramus of the mandible and the roof is the base of the greater wing of the sphenoid bone. The infratemporal fossa has no floor, and is continuous with the neck.

Contents of the Fossa
The fossa contains the pterygoid muscles, branches of the mandibular nerve, the chorda tympani branch of the facial nerve, the otic ganglion (part of the autonomic nervous system), the maxillary artery and the pterygoid venous plexus (the vessels surrounding pterygoid muscles).

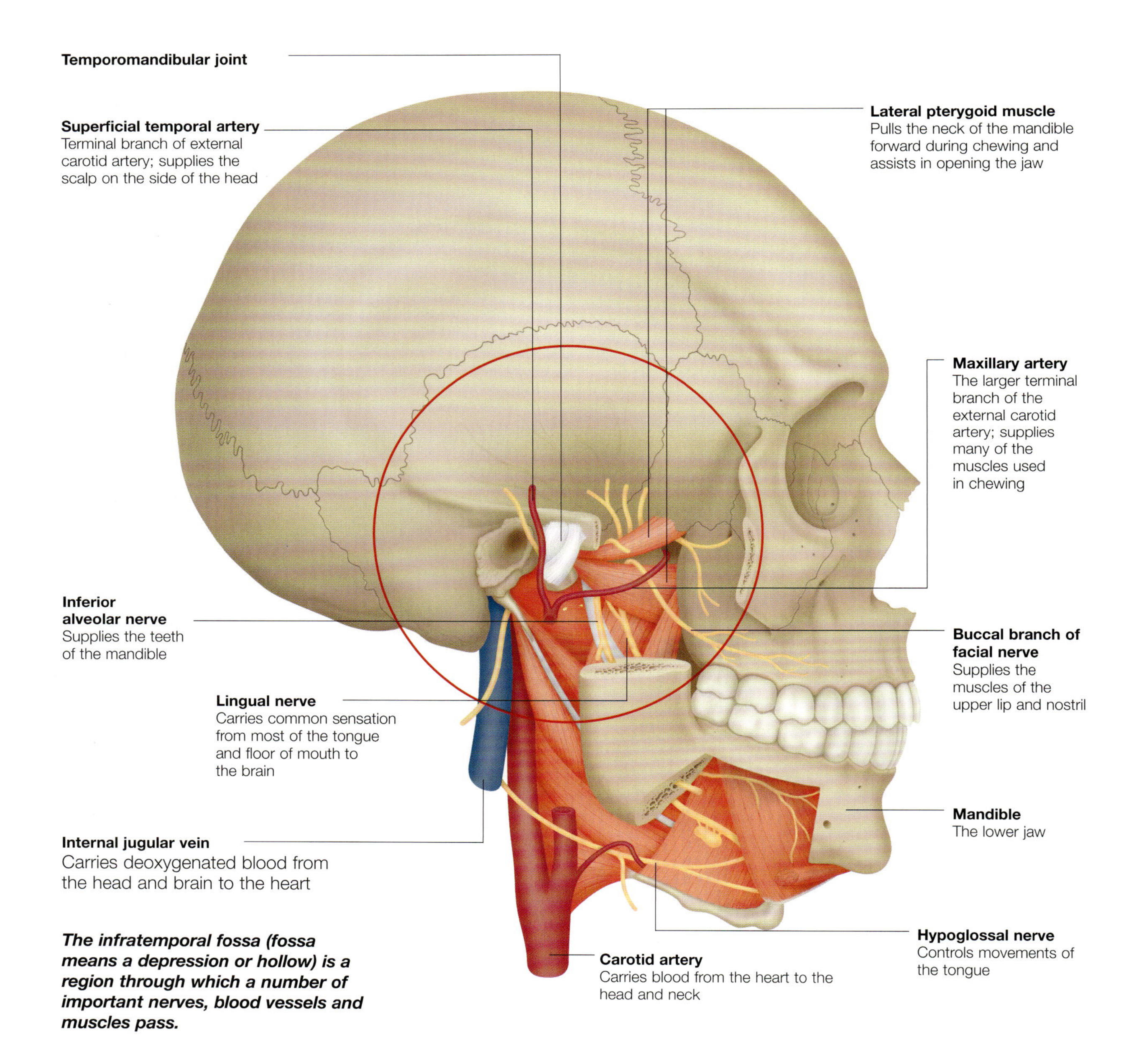

The infratemporal fossa (fossa means a depression or hollow) is a region through which a number of important nerves, blood vessels and muscles pass.

Pterygopalatine Fossa

The pterygopalatine fossa is a funnel-shaped space between the bones of the head. It contains important nerves and blood vessels that supply the eyes, mouth, nose and face.

The fossa can be located via the pterygomaxillary fissure, a narrow gap between the pterygoid plates of the sphenoid bone, and the back of the upper jaw (maxilla). This leads to the lateral part of the fossa.

Location of the Fossa

The fossa is a small funnel-shaped space that tapers downwards and lies below the back of the orbit, the socket of the eyeball. It is located behind the maxilla and its back wall is formed by the pterygoid plates and the greater wing of the sphenoid bone. The palatine bone forms its midline and its floor. It is a very important distribution centre as it communicates with all of the important regions of the head including the mouth, nose, eyes and face, infratemporal fossa and also with the brain.

The main components of the pterygopalatine fossa are the maxillary artery and nerve (branch of the trigeminal nerve) and the pterygopalatine ganglion. These enter and exit the region through the sphenopalatine foramen. The artery supplies oxygenated blood to the maxillary teeth, the hard and soft palate, the nasal cavity, the paranasal sinuses and the skin of the lower eyelid, the nose and the upper lip. The maxillary nerve supplies sensation to large areas of the face.

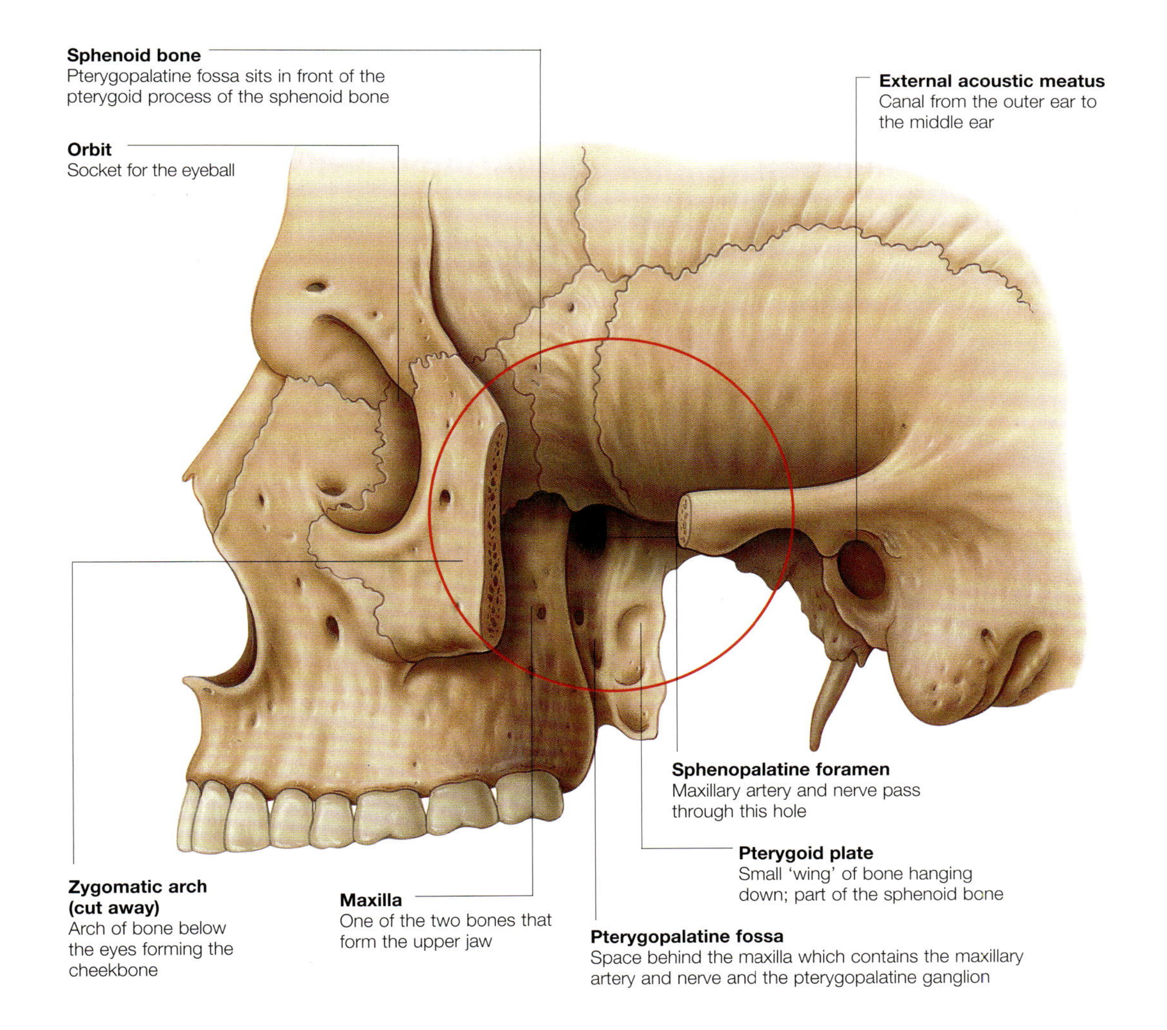

Muscles of the Scalp

The muscles of the scalp lie below the skin and a layer of connective tissue. They act to move the skin of the forehead and the jaw while chewing.

The occipitofrontalis is a large muscle formed of two sections, one at the front and one at the back of the scalp. These are connected by a thin, tough, fibrous sheet (the aponeurosis). The occipitalis is the section of muscle that arises from the top of the back of the neck and passes forwards to the aponeurosis. It acts to pull the scalp backwards. The frontalis is the section of muscle over the forehead, arising from the skin overlying the eyebrow and passing back to become continuous with the aponeurosis. This muscle acts to raise the eyebrows, thus wrinkling the forehead or pulling the scalp forwards, as when frowning.

The temporalis muscle lies at the side of the scalp, above the ears, and runs from the skull down to the lower jaw. It is involved in chewing.

Loose Connective Tissue

The fourth layer, underlying the muscle and aponeurosis, is a layer of loose connective tissue that allows the layers above to move relatively freely over the layer below. The pericranium is the fifth layer of the scalp and is the tough membrane that covers the bone of the skull itself.

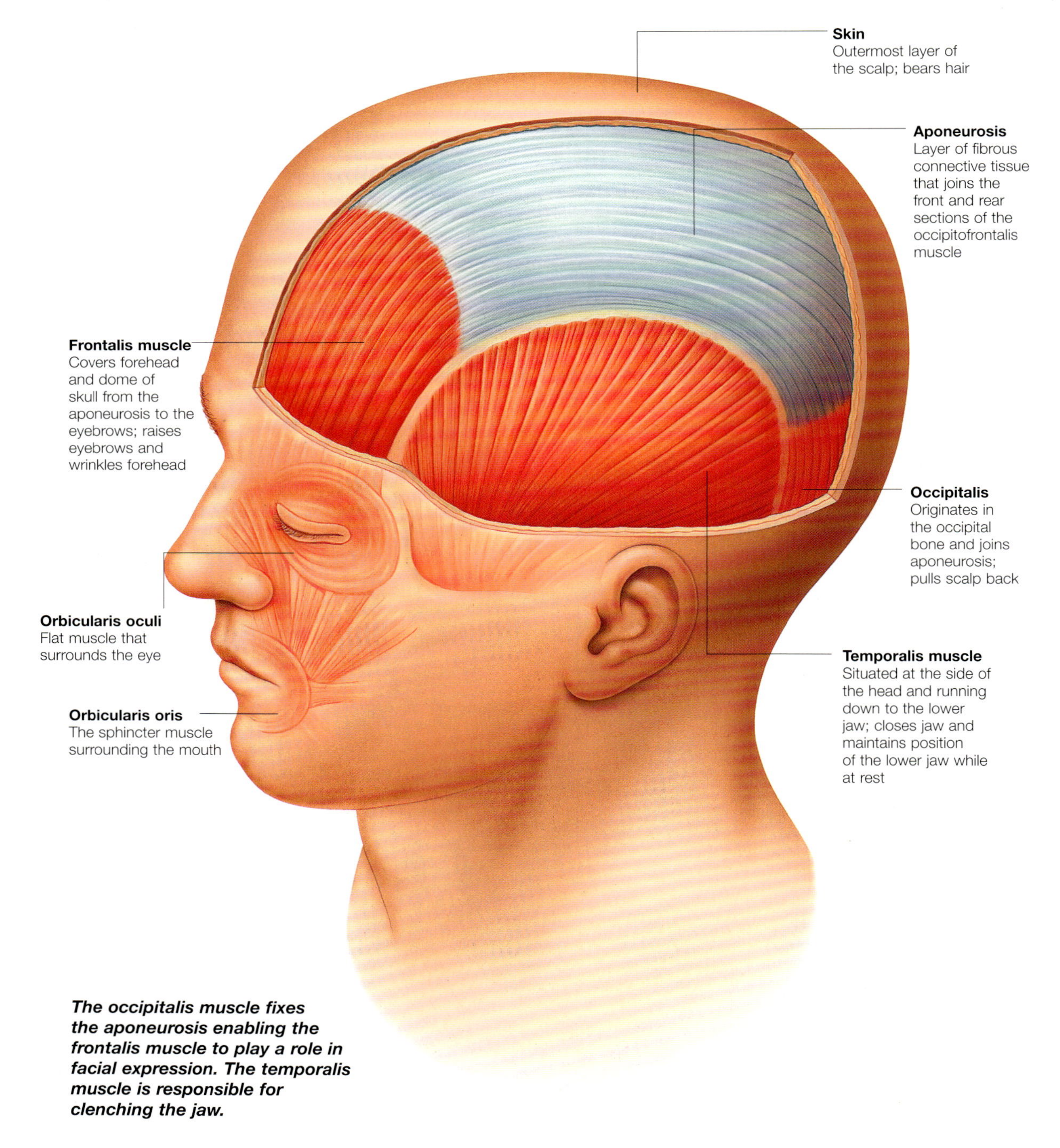

The occipitalis muscle fixes the aponeurosis enabling the frontalis muscle to play a role in facial expression. The temporalis muscle is responsible for clenching the jaw.

Muscles of mastication

The muscles of mastication help us chew our food but also play a part in speech, breathing and yawning.

The muscles of mastication are the muscles that move the mandible (jawbone) up and down, and forwards and backwards, resulting in the opening and closing of the mouth. This action is used in activities such as speaking, breathing through the mouth and in yawning. The closing action is also part of the movements necessary for biting off and chewing up food (mastication), when side-to-side slewing of the jaw is also employed.

Moving the jaw
All jaw movements take place at the pair of temporomandibular joints, which lie in front of the ears.

The bones forming the joint are the head of the mandible (the rounded section at the top of the jawbone) and the mandibular fossa of the temporal bone (the hollow in the skull in which the head of the mandible sits).

The hinge-like action allows up and down movements of the jaw.

Additionally, the head of the mandible is covered with a closely fitting disc of cartilage, which allows forward and backward rocking movements. This latter movement enables the lower jaw to be slewed across the upper jaw on opening, and so provides the sideways force necessary to grind up hard food on closing the mouth and chewing.

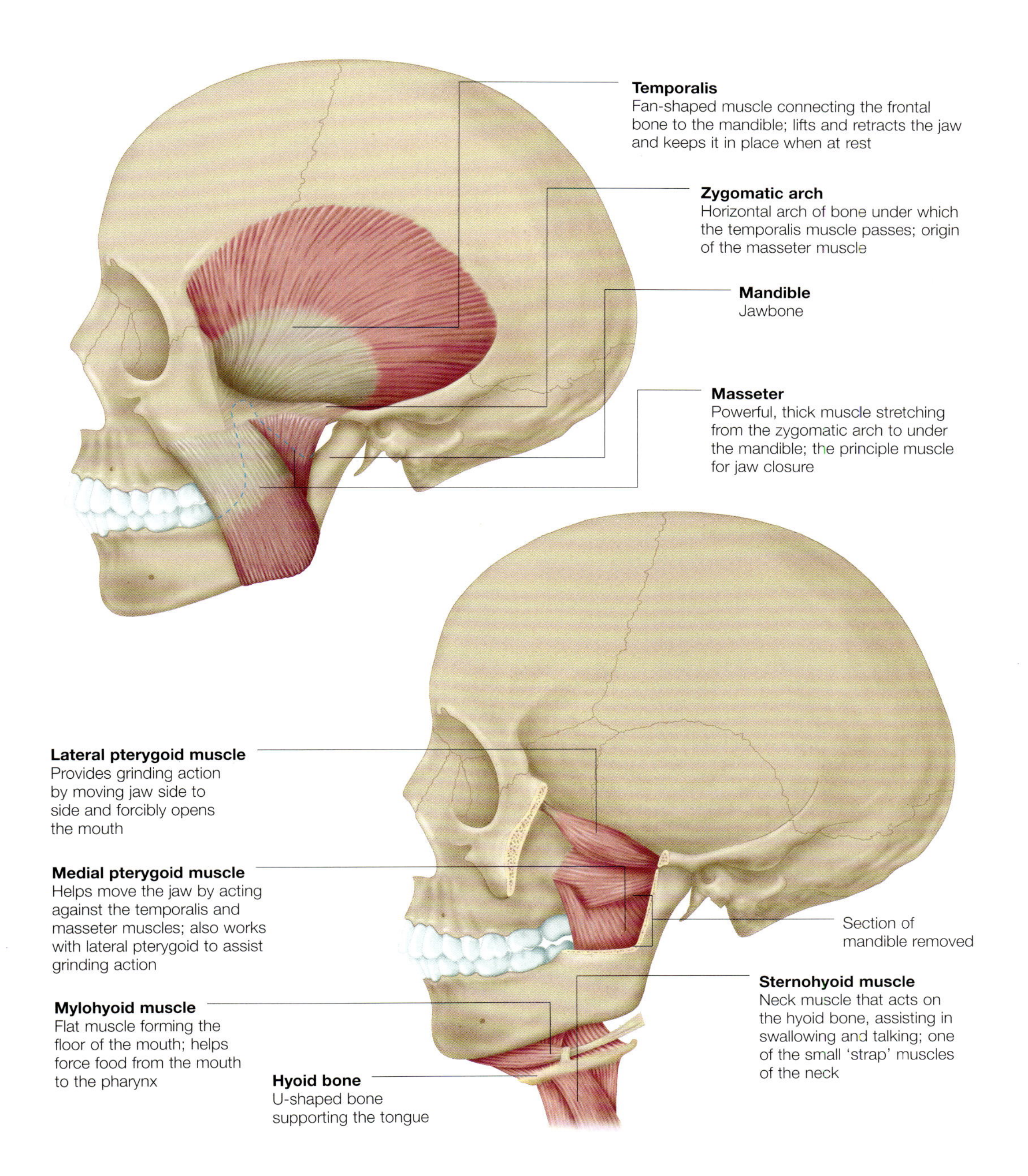

Facial Muscles

One of the features that distinguishes humans from animals is our ability to communicate using a wide range of facial expressions that are facilitated by a complex system of facial muscles.

Just under the skin of the scalp and face lies a group of thin muscles, collectively known as the muscles of facial expression. These muscles play a crucial role in a number of ways:

- they alter facial expression
- they provide a means of non-verbal communication by transmitting a range of emotional information
- they are one of the means of articulating speech
- they form sphincters that open and close the orifices of the face – the eyes and mouth.

Skin and bone

The majority of facial muscles are attached to a skull bone at one end and to a deep layer of skin (dermis) at the other. From these attachments, it can be seen see how the numerous muscles alter facial expression, and also how they eventually cause creases and wrinkles in the overlying skin.

A number of small muscles called 'dilators' open the mouth. They radiate out from the corners of the mouth and lips, where they have an attachment to bone. The mouth and lips can also be pulled up, pushed down and moved from side to side.

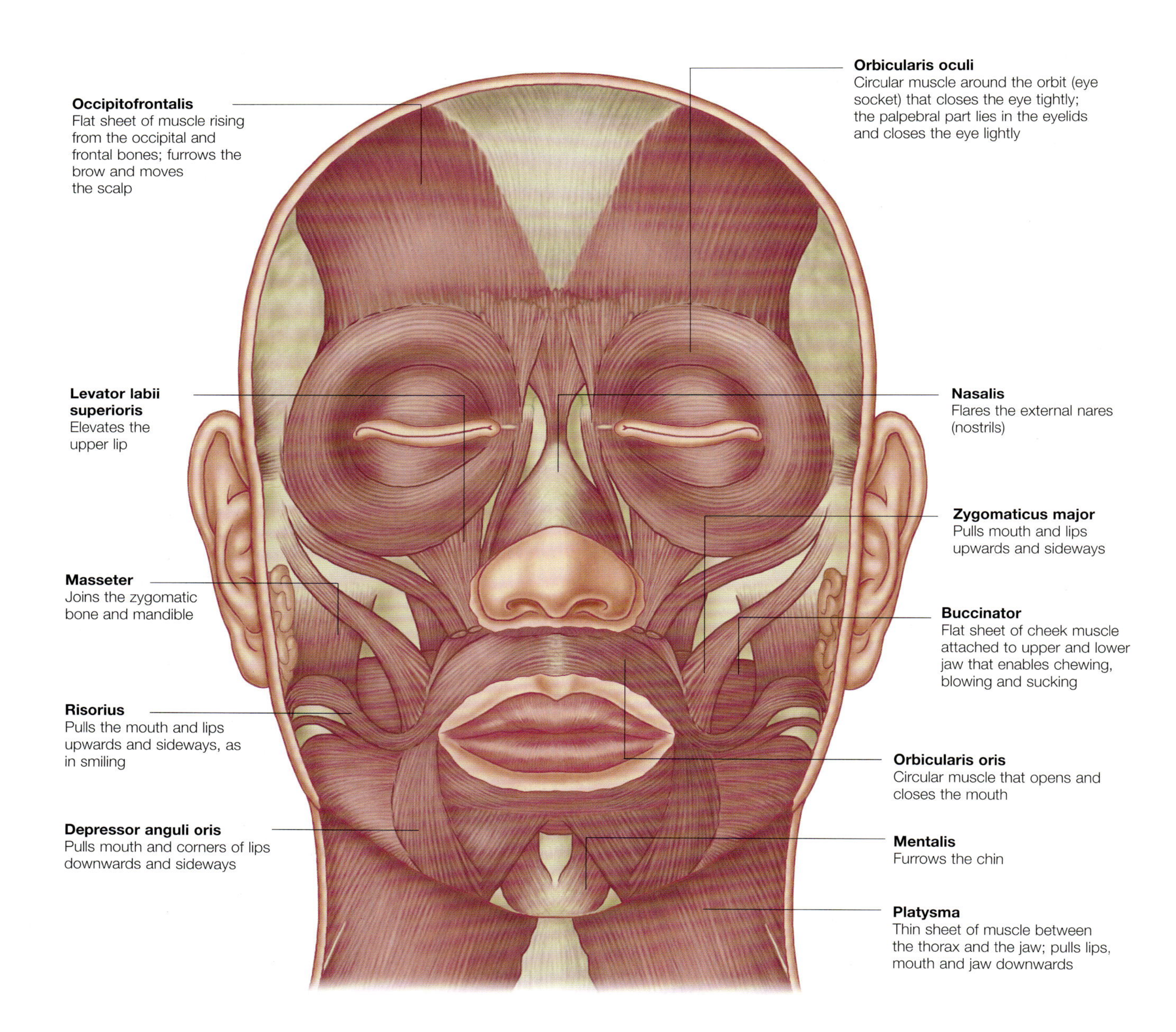

Muscles of the eye

The eyelids communicate a range of non-verbal signals. The eyelids are also vital for cleaning and lubricating the eyes.

The orbicularis oculi is the muscle responsible for the closing of the eye. This flat sphincter muscle lines the rim of the orbit (eye socket), and various sections of it can be manipulated individually.

Part of the orbicularis oculi lies in the eyelid (the palpebral part). This section of the muscle closes the eye lightly, as in sleeping or blinking. This action also aids the flow of lacrimal secretion (tears) across the conjunctiva (the membrane covering the eyes) to keep it clean, free of foreign bodies and lubricated.

Opening the lids

A larger part of the orbicularis oculi consists of concentrically arranged fibres that cover the front of the eye socket. The role of this part of the muscle is to close the eye tightly to protect against a blow or bright light. The second orbital muscle is the levator palpebrae superioris. This small muscle pulls on the upper lid to open the eye. Unlike the larger orbicularis, this muscle lies within the eye socket.

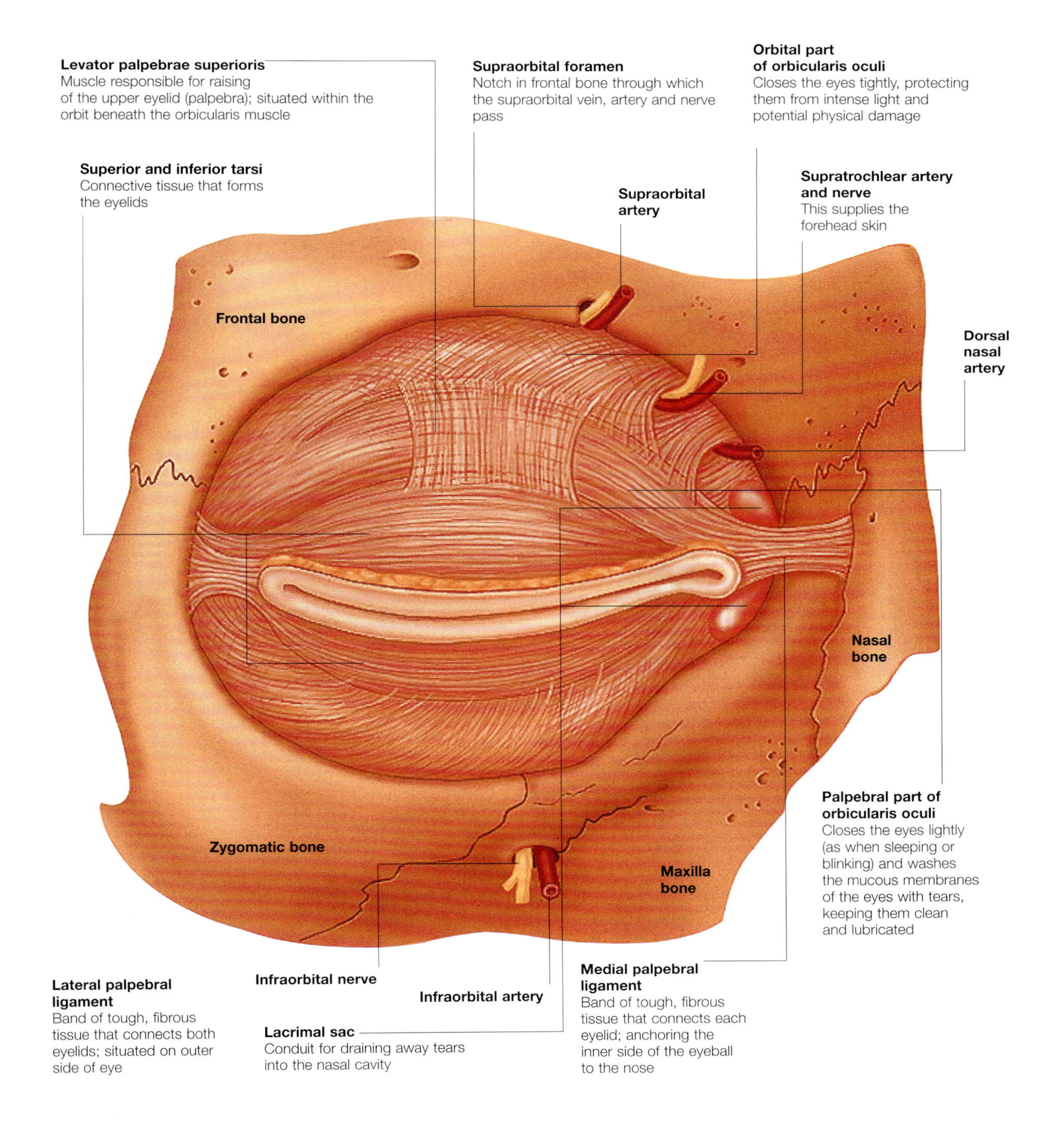

Muscles of the Neck

Two groups of muscles run vertically down the front of the neck from the mandible to the sternum: the suprahyoid and infrahyoid groups. They attach to the hyoid bone and act to raise and lower it and the larynx during swallowing.

These muscles are concerned with movements of the jaw, hyoid bone and larynx and aid swallowing. The hyoid bone separates these two groups into the suprahyoid (above) and the infrahyoid (below).

The Suprahyoid Muscles
This is a group of paired muscles, lying between the jaw and the hyoid bone. The digastric muscle has two spindle-shaped bellies connected by a tendon. The anterior belly is attached to the mandible near the midline, while the posterior belly arises from the base of the skull. The connecting tendon slides freely through a fibrous 'sling' that is attached to the hyoid bone. The stylohyoid is a small muscle that passes from the styloid process in the base of the skull forwards and downwards to the hyoid bone. Arising from the back of the mandible, the paired mylohyoid muscles unite in the midline to form the floor of the mouth. Posteriorly, they attach to the hyoid bone. The geniohyoid is a narrow muscle that runs along the floor of the mouth from the back of the mandible in the midline to the hyoid bone below.

The Infrahyoid Muscles
These are the sternohyoid, omohyoid, thyrohyoid and sternothyroid muscles that lie between the hyoid bone and the sternum. After swallowing the infrahyoid muscles act to return the hyoid bone and larynx to their previous positions.

Infrahyoid and suprahyoid muscles

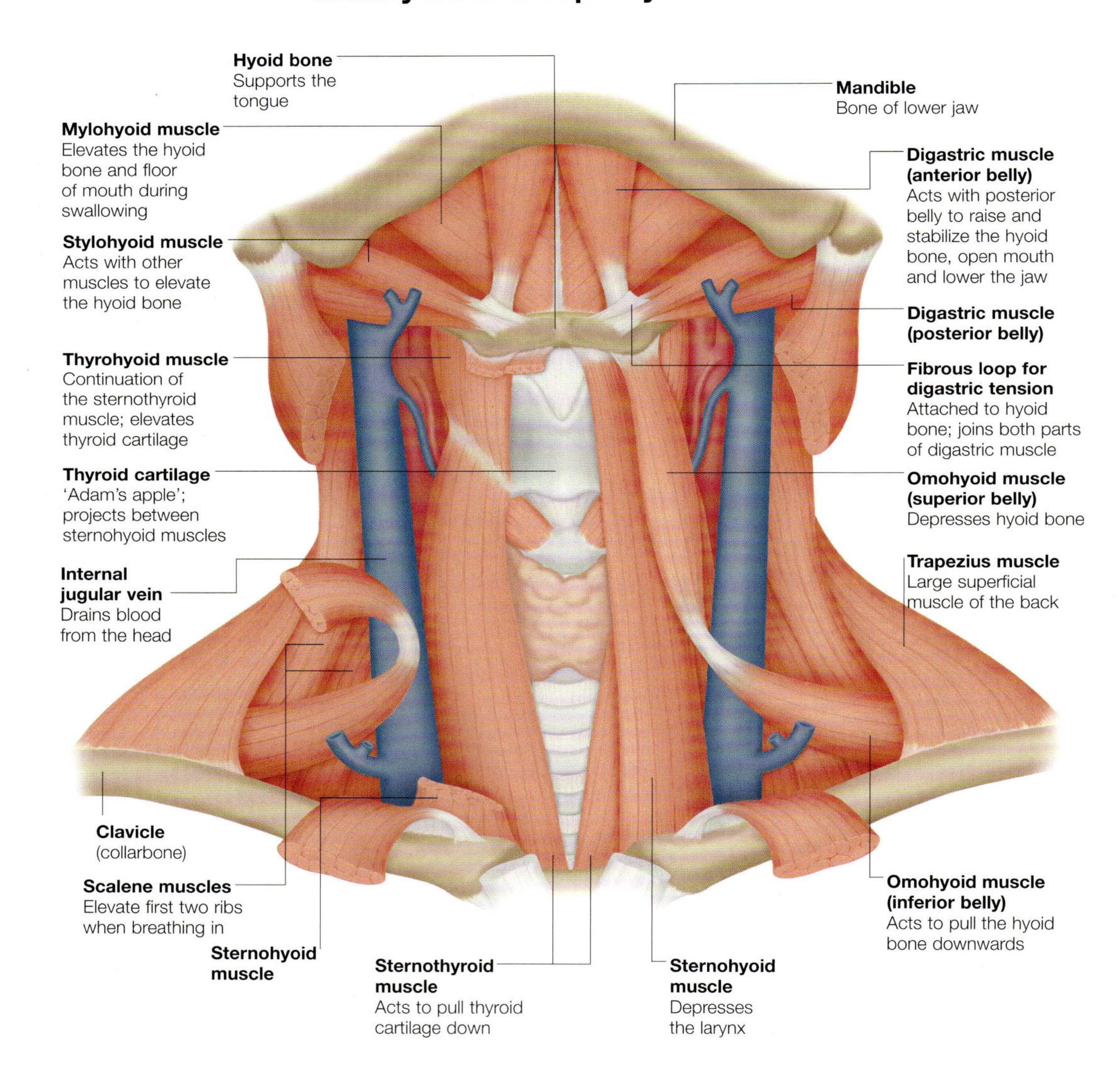

Action of the neck muscles

The suprahyoid and infrahyoid groups of muscles have opposing actions on the larynx and hyoid bone. This enables us to swallow.

The small U-shaped hyoid bone is attached to the larynx and mandible by tendons and muscles. Swallowing is enabled by the upward and forward movement of the hyoid and the larynx, which widens the pharynx (gullet) and helps to close off the trachea so that food cannot enter the respiratory tract.

Suprahyoid muscles

The mylohyoid, geniohyoid and the anterior belly of the digastric muscle act together to pull the hyoid and the larynx forwards and up during swallowing. They also enable the mouth to be opened against resistance. The stylohyoid and the posterior belly of the digastric muscle together lift and pull back the hyoid bone and the larynx.

Infrahyoid muscles

The infrahyoid group of muscles act together to pull the hyoid and the larynx back down to their normal positions, as at the end of the act of swallowing. When contracted, the infrahyoid muscles lower and fix the hyoid bone so that the suprahyoid muscles can pull against it to open the mouth.

Action of the suprahyoid and infrahyoid muscles

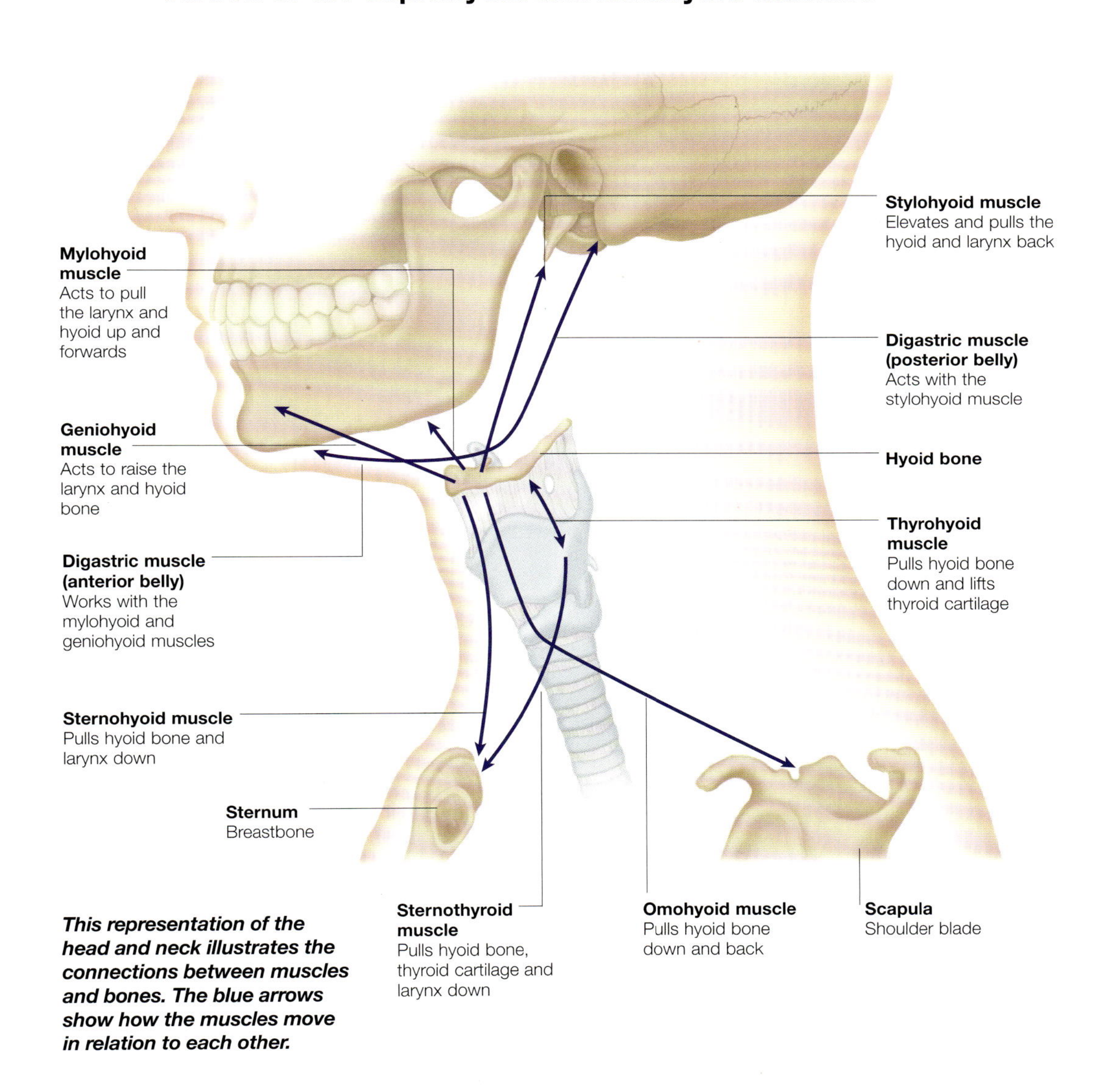

This representation of the head and neck illustrates the connections between muscles and bones. The blue arrows show how the muscles move in relation to each other.

Neck Flexor Muscles

The flexor muscles of the neck work to keep the head stable and upright on the spine. These muscles also enable flexion of the neck and head and raise the first two ribs during inspiration.

The centre of gravity of the head lies in front of the spinal column. As a result, constant activity in the muscles and ligaments of the back of the neck is necessary to keep the head from falling forwards.

Much of the forward and lateral flexion of the head and neck is achieved by the coordinated action of three types of neck flexor muscles: the scalenes, the prevertebral muscles and the powerful sternocleidomastoid.

Scalene Muscles

Three muscles run from the transverse processes either side of the cervical vertebrae to attach to the first and second ribs. The scalenus anterior and the scalenus medius both originate from the third to the sixth cervical vertebrae (C3 to C6) and attach to the first rib. The scalenus posterior muscle may be absent or may be part of the scalenus medius muscle. It passes down to the second rib.

Prevertebral Muscles

These lie in front of the cervical vertebrae and extend from the skull down to the junction of the neck and chest. Rectus capitis anterior and lateralis muscles are short muscles that run from the skull to the first cervical vertebra. The longus capitis muscles are longer muscles that lie in line with the tendons of origin of the scalenus anterior muscle. The longus colli connects the vertebrae to each other so they move as a unit.

Front view of scalene and prevertebral muscles

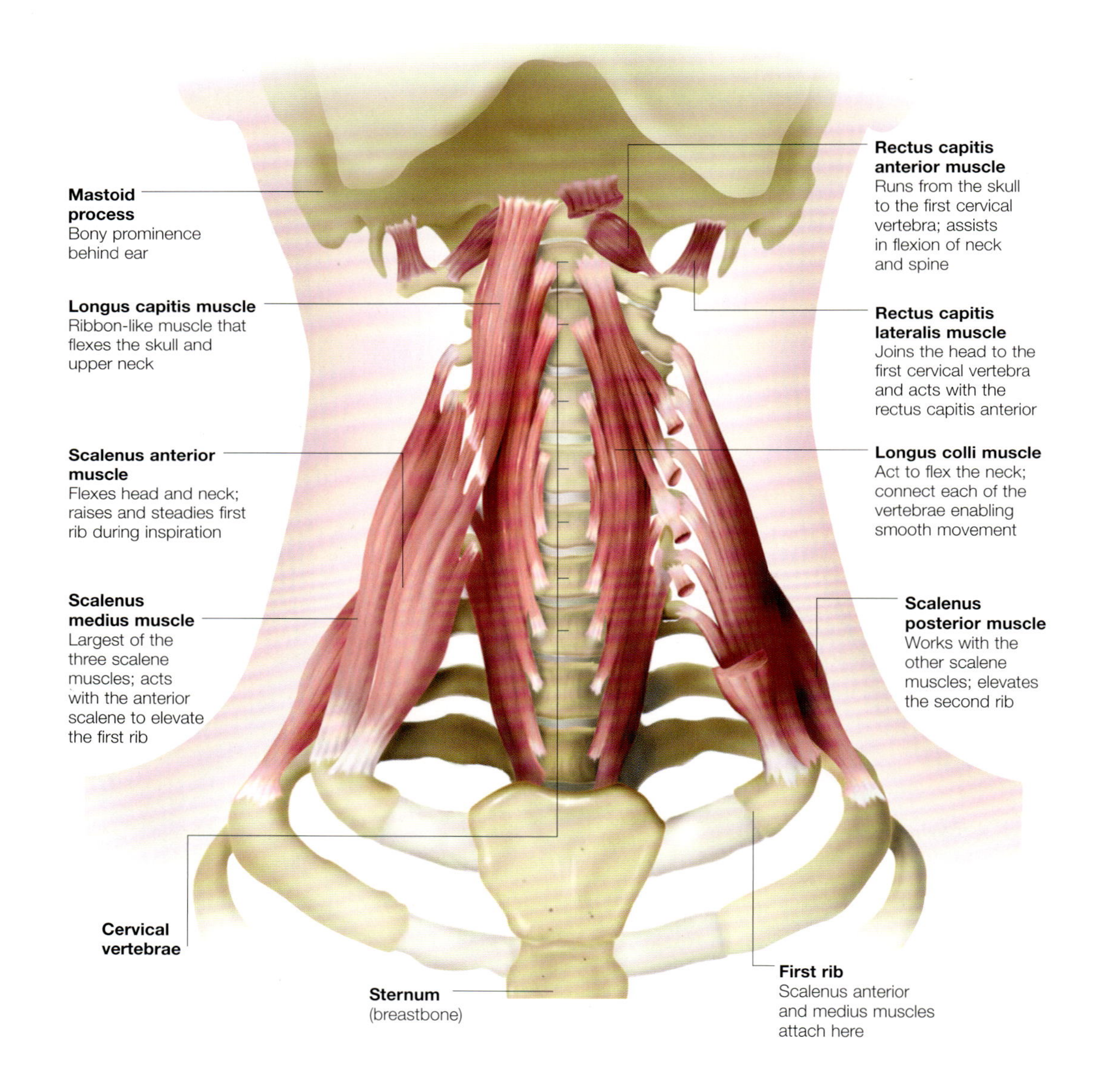

Sternocleidomastoid muscles

The large sternocleidomastoid muscles are powerful flexors of the head and cervical spine. They work in conjunction with the scalene and prevertebral muscles.

The sternocleidomastoid muscles are the major head flexor muscles. These paired muscles can be seen very prominently under the skin on either side of the front of the neck. They run from the mastoid process (a prominence on the base of the skull) down and forwards to the sternum (breastbone) and clavicle (collarbone). At this lower end, each splits into two segments; one part attaches to the front of the upper sternum, while the second, deeper part, attaches to the clavicle.

Rotation and protraction

When the sternocleidomastoid contracts on one side of the neck only, it causes the face to be turned to the opposite side and inclined slightly upwards. If other muscles hold the neck vertical at the same time, simple rotation of the head results. The muscle can be felt under the skin when the head is rotated.

While the sternocleidomastoid and other muscles contract, the muscles at the back of the neck relax, allowing the head to be pulled forwards. This movement is known as protraction – when the head is moved forward in relation to the body while keeping it vertical and maintaining a horizontal gaze. An example is the action of the head when someone tries to look over their shoulder.

Side view of sternocleidomastoid muscle

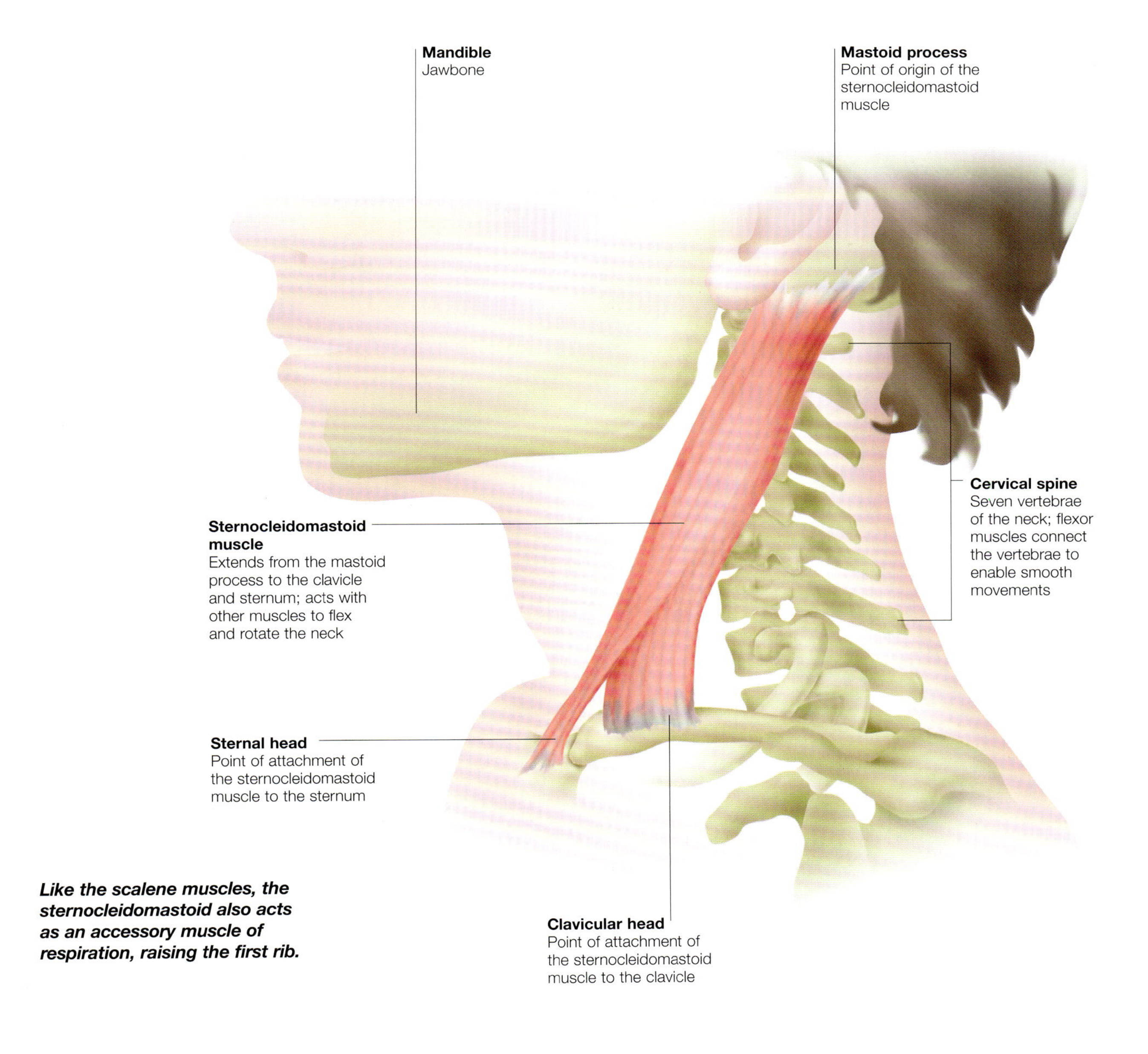

Like the scalene muscles, the sternocleidomastoid also acts as an accessory muscle of respiration, raising the first rib.

Vertebral Column

The vertebral column gives our bodies flexibility and keeps us upright. It also protects the spinal cord. This part of the skeleton is known as the axial skeleton, as it is on the longitudinal axis.

The vertebral column forms the part of the skeleton known as the backbone or spine. The spine supports the skull and gives attachment to the pelvic girdle, supporting the lower limbs. As well as its obvious role in posture and locomotion, the vertebral column surrounds and protects the spinal cord. Like all bones, its marrow is a source of blood cells, and it acts as a reservoir for calcium ions.

Curvatures
The spine exhibits four curvatures when viewed from the side. The cervical and lumbar curvatures are convex anteriorly (forwards). The thoracic and sacral curvatures are convex posteriorly (backwards). The cervical curvature develops in infancy as the baby learns to hold its head upright; similarly, the lumbar curvature forms as the baby learns to walk.

Intervertebral discs
Between each vertebra is a disc of connective tissue with a soft jelly-like centre and a tougher surrounding tissue known as the annulus fibrosus. These discs account for about 25 per cent of the length of the vertebral column, and act as shock absorbers, protecting the vertebrae and allowing a wide range of different movements.

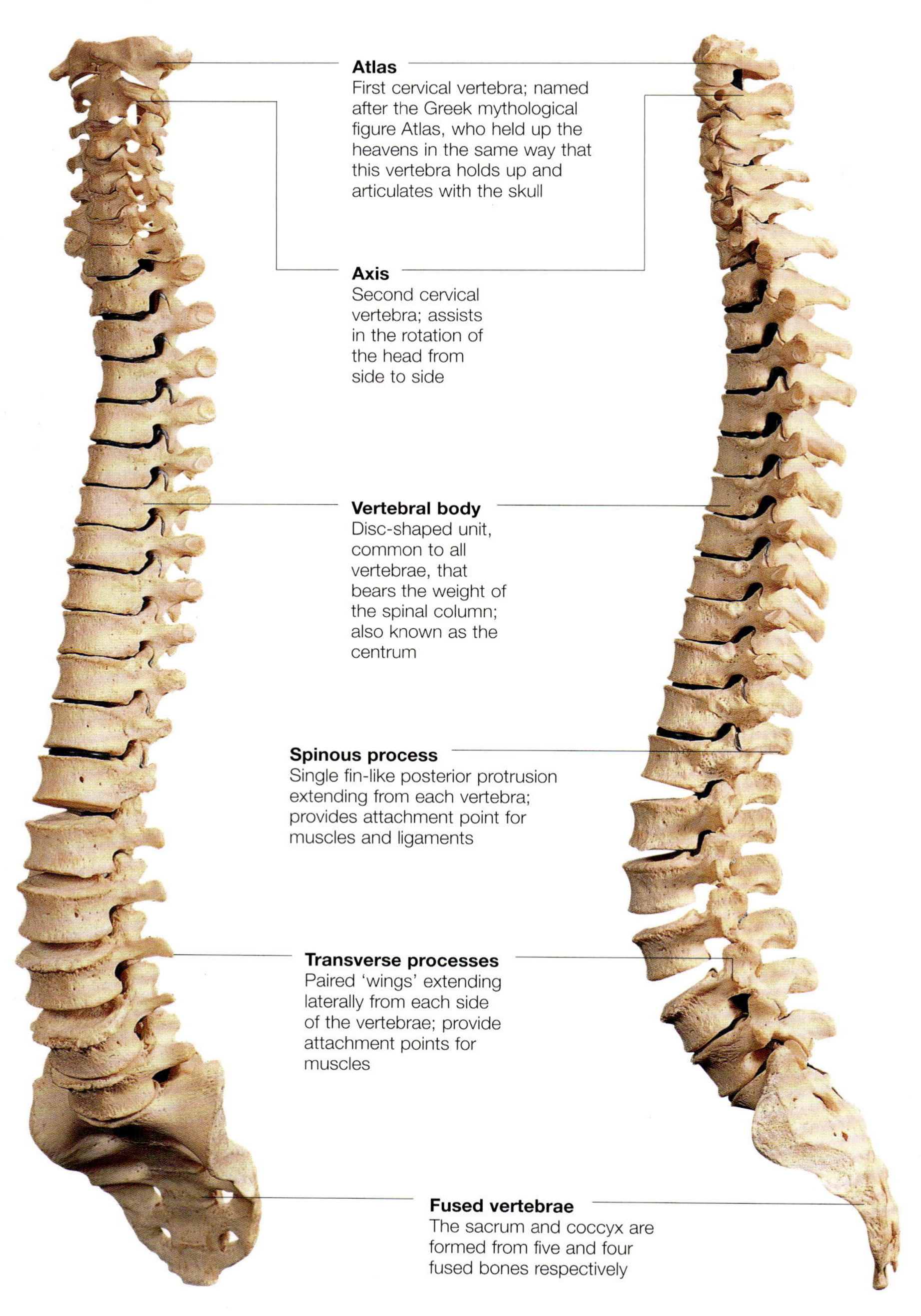

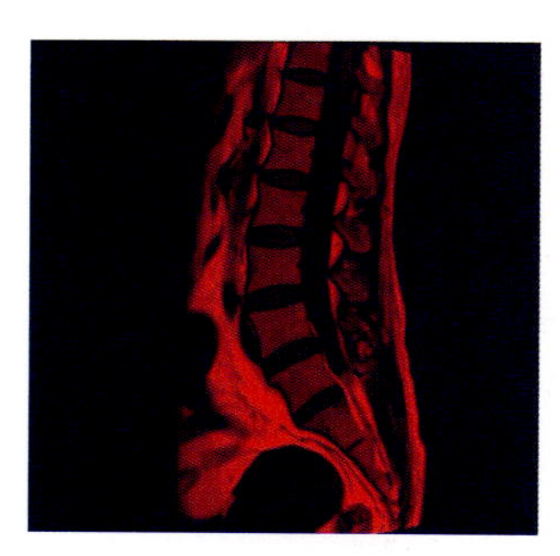

The sacral and lumbar curvatures of the lower spine can be clearly seen in this MRI scan.

Structure of the Spine

The spine is divided into five main sections, each of which has a specific function. Together they maintain the stability of the skeleton as a whole.

The adult vertebral column consists of a series of 33 bones, 26 of which are known as the vertebrae. There are seven cervical vertebrae in the neck and upper back region, 12 thoracic at the rear of the thoracic cavity, and five lumbar in the region of the lower back. The five sacral vertebrae are fused into a single bone called the sacrum. Similarly, the four coccygeal vertabrae are fused to form the coccyx.

Structure

Each vertebra consists of a body in front and a neural arch at the back that surrounds and protects the spinal cord. From the neural arch arise the transverse processes and spines that provide attachments for muscles and ligaments. Adjacent vertebrae articulate at joints, allowing movement. Movements between adjacent vertebrae are relatively small but, when accumulated over the whole length of the vertebral column, give the trunk considerable mobility.

The nerves leaving and entering the spinal cord do so through gaps, called intervertebral foramina, between adjacent vertebrae.

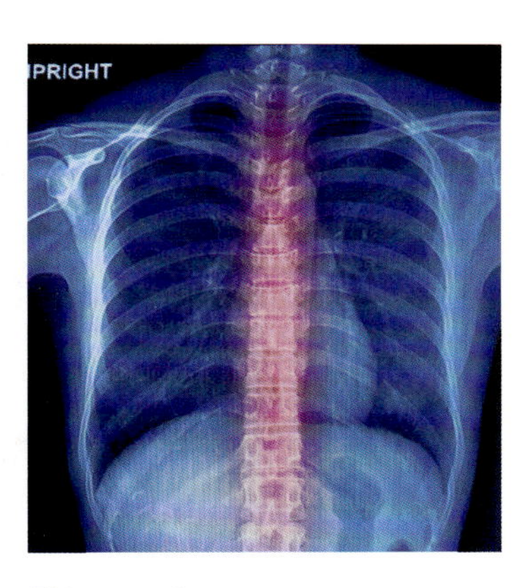

This x-ray image shows the spinal column, formed of 33 separate bones.

Rear view

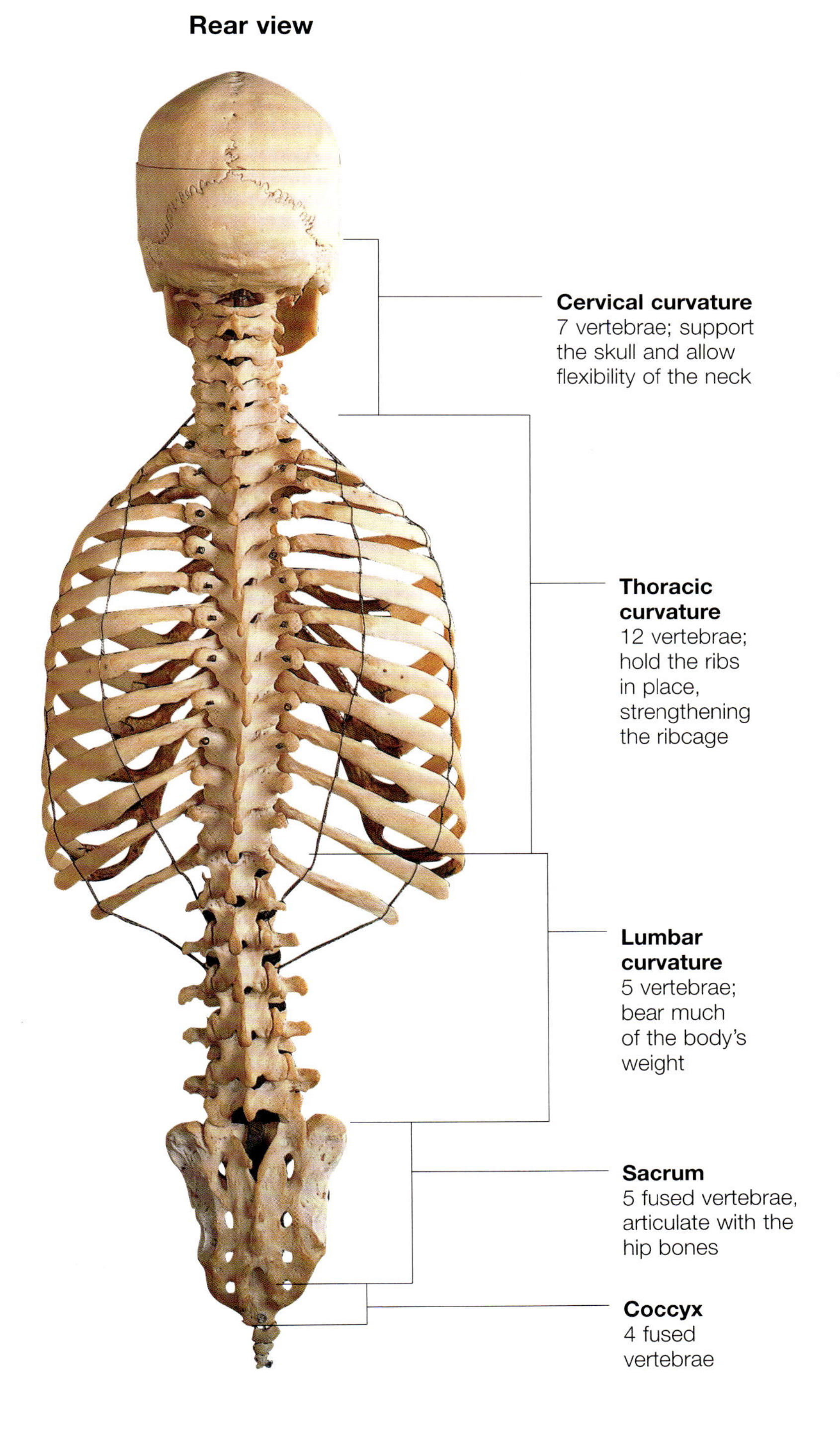

Cervical Vertebrae

There are seven cervical vertebrae, which together make up the skeletal structure of the neck. These vertebrae protect the spinal cord, support the skull and allow a range of movement.

Of the seven cervical vertebrae, the lower five appear similar, although the seventh has distinctive features. The first cervical vertebra (atlas) and the second cervical vertebra (axis) show specializations related to how the vertebral column articulates with the skull.

Typical cervical vertebra
The third to the sixth cervical vertebrae are formed of two main components, a body towards the front and a vertebral arch at the rear. These surround the vertebral foramen (hole) which, as part of the vertebral column, forms the vertebral canal. The body of a cervical vertebra is small compared with those in other areas, and is nearly cylindrical.

Vertebral Arch
The vertebral arch can be subdivided into two main elements. The pedicles, by which it is attached to the body, contain notches that allow the passage of spinal nerves. The laminae are thin plates of bone that are directed backwards and which fuse in the midline, forming a bifid (divided) spine.

Associated with each vertebral arch are a pair of transverse processes. These are sites for muscle attachment, allowing movement and containing foramina through which blood vessels run.

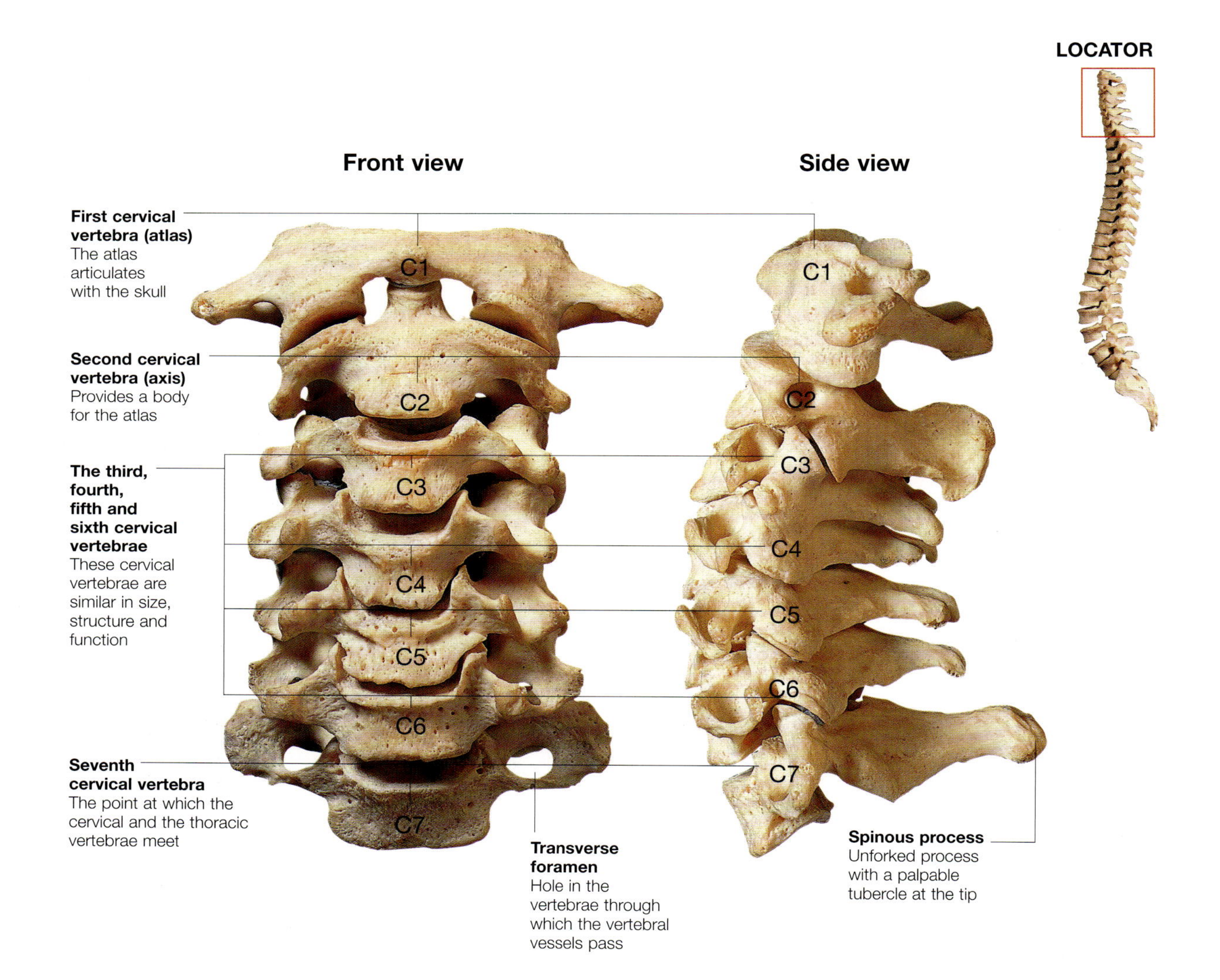

First, second and seventh vertebrae

The first, second and seventh cervical vertebrae differ structurally from the others, in relation to their unique functions.

First cervical vertebra

The first cervical vertebra, the atlas, is the vertebra that articulates with the skull. Unlike the other vertebrae, it does not have a body, this being incorporated into the second cervical vertebra as the dens. It also has no spine. Instead, the atlas takes the form of a thin ring of bone with anterior and posterior arches, the surface of which show grooves related to the vertebral arteries before they enter the skull through the foramen magnum.

Second cervical vertebra

The second cervical vertebra, known as the axis, can be distinguished from other cervical vertebrae by the presence of a tooth-like process called the dens (odontoid process). The dens articulates with the facet on the bottom surface of the anterior (front) arch of the atlas. Rotation of the head occurs at this joint.

The body of the axis vertebra resembles the bodies of the other cervical vertebrae.

Seventh cervical vertebra

This vertebra has the largest spine of any cervical vertebra and, as the first to be easily felt along the spine, it has been termed the vertebra prominens.

The transverse processes are also larger than those of the other cervical vertebrae and the oval foramen transversarium transmits an accessory vertebral vein.

First cervical vertebra (atlas)

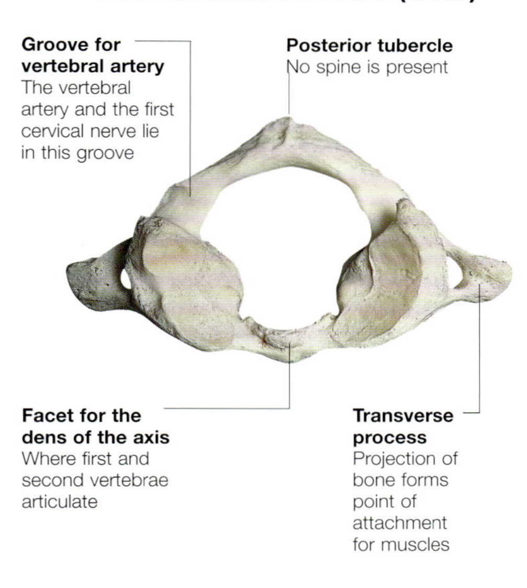

Second cervical vertebra (axis)

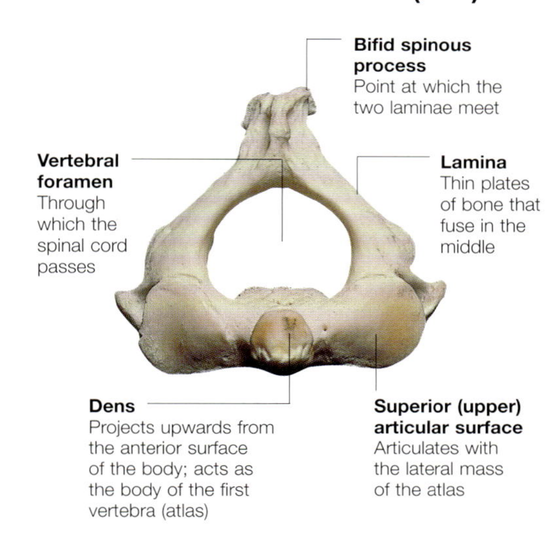

Fifth (typical) cervical vertebra

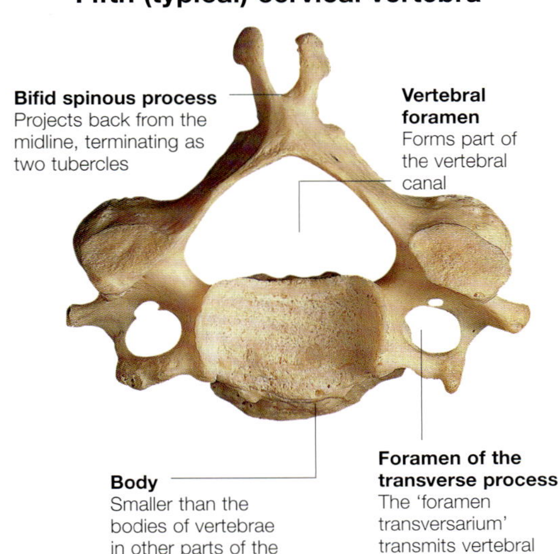

Seventh cervical vertebra

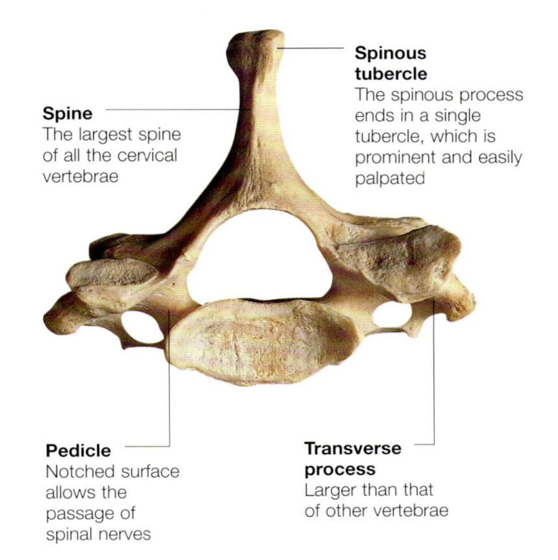

Cervical Ligaments

The cervical ligaments play an essential role in securing and stabilizing the vertebrae of the neck. They consist of elastic fibres that enable smooth flexion of the head and neck.

The ligaments of the neck have a vital role in binding the cervical vertebrae together while allowing a wide range of movements. Some of the cervical ligaments are the continuation of those found further down the spine but other, more specialized ligaments are needed at the point where the spine articulates with the skull.

Ligament function

The first two cervical vertebrae, the atlas and the axis, are adapted to the role of supporting and allowing movement of the skull on the spine. The axis, below the atlas, has a spike of bone – the dens – projecting up into the front of the vertebral canal. If the dens moves within the vertebral canal, it could damage the nervous tissue.

A group of ligaments bind the dens to the atlas and the skull:

- The cruciform ligament – this cross-shaped ligament has a very thick central band passing from one side of the atlas to the other, holding the dens forwards and away from the spinal cord; an upper and a lower band attach the axis to the skull
- The apical ligament – a small ligament that passes from the tip of the dens to the skull
- The alar ligaments – two strong bands that pass up and outwards from the dens to the skull.

Posterior view of cervical ligaments

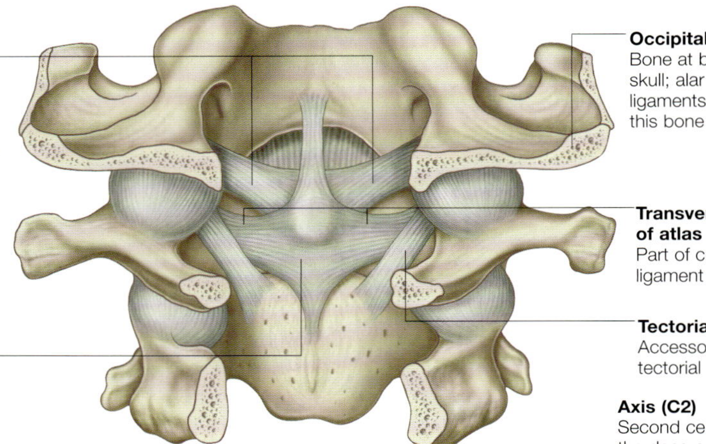

Posterior view of dens and deep cervical ligaments

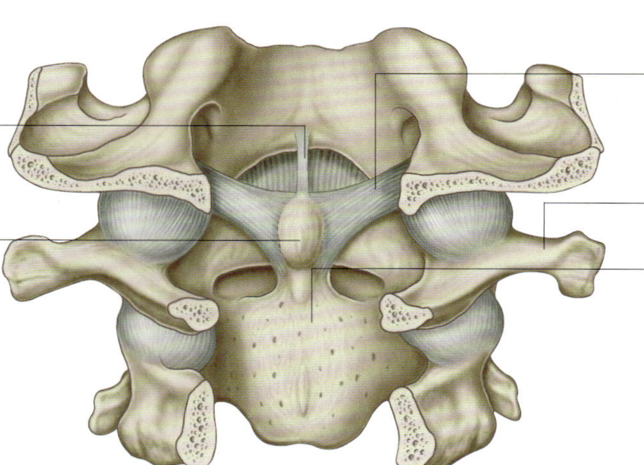

Together, the cervical ligaments restrict the movement of the dens, guarding particularly against backward movement into the vertebral canal.

Longitudinal ligaments

The vertebrae of the spine are attached by ligaments running longitudinally. These ligaments support the spine, and its elastic fibres enable a range of movement.

The vertebrae of the spine are linked and supported by two ligaments that pass from skull to sacrum, the anterior and the posterior longitudinal ligaments.

Longitudinal ligaments

The anterior longitudinal ligament is a broad, strong, fibrous band attached to the front of the intervertebral discs and to the vertebrae. It connects the vertebrae and attaches to the skull as the anterior atlanto-occipital membrane.

The posterior longitudinal ligament is narrower and weaker and connects the backs of the vertebral bodies up to the atlas where it becomes the tectorial membrane attached to the skull.

There are many other short ligaments joining each vertebra to the one above and below. The ligamenta flava, for example, contains elastic tissue and helps preserve the curvatures of the spine and assists with straightening the spine after bending forwards. The ligamenta flava attaches to the skull as the posterior atlanto-occipital membrane.

The nuchal ligament

The nuchal ligament is a strong, thick band lying in the midline of the back of the neck. It attaches to the spines of the cervical vertebrae and is used as an attachment for muscles such as the trapezius. It is elastic and helps support the weight of the head; it also stretches while the head and neck are flexed.

Right lateral view of cervical ligaments

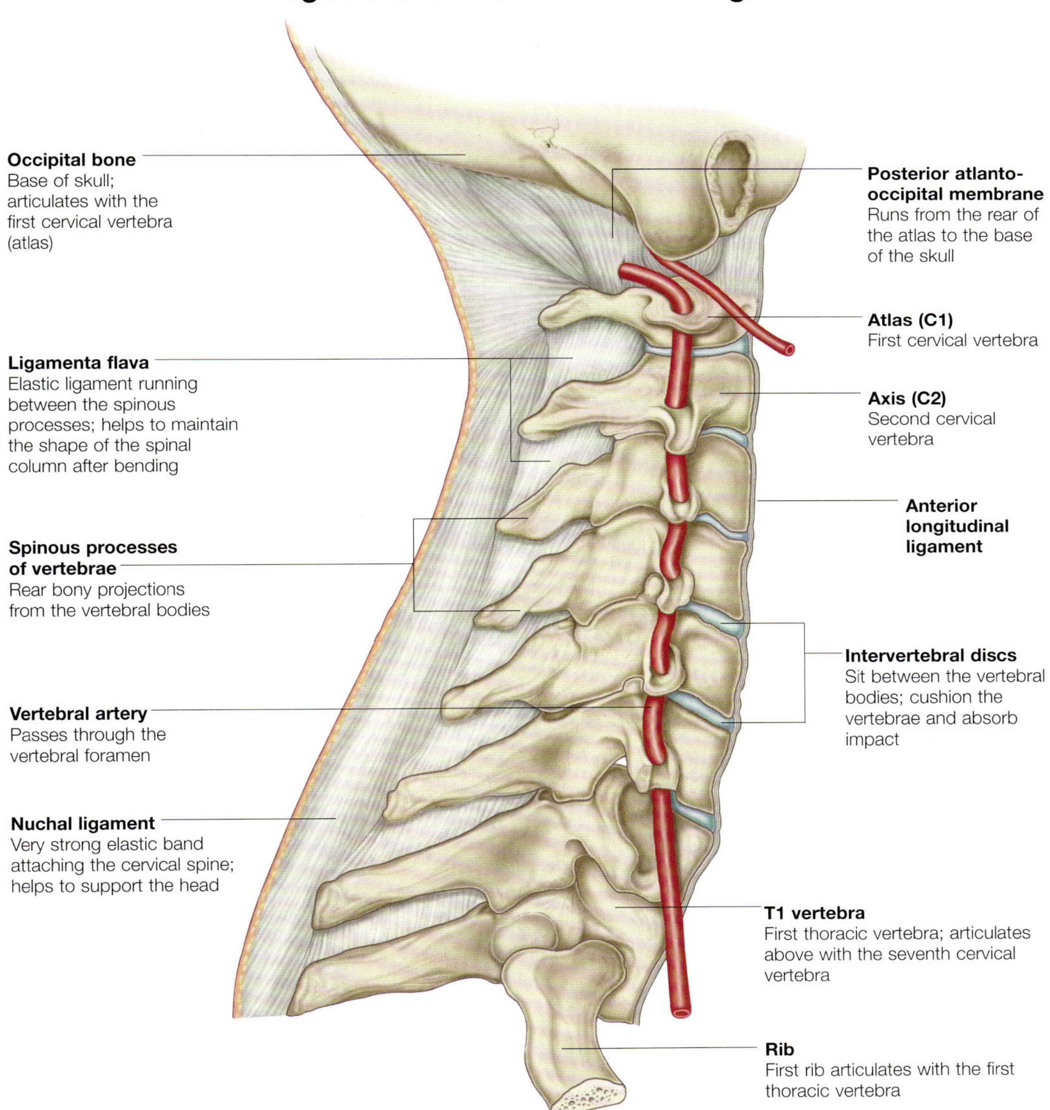

Thoracic Vertebrae

The 12 thoracic vertebrae are the bones of the spinal column to which the ribs are attached. The thoracic vertebrae sit between the cervical vertebrae of the neck and the lumbar vertebrae of the lower back.

Each thoracic vertebrae has two components: a cylindrical body in front and a vertebral arch behind. The body and vertebral arch surround a gap, called the vertebral foramen, which is rounded.

When all the vertebrae are articulated together, the space formed by the linked vertebral foramina forms the vertebral canal. This houses the spinal cord, which is surrounded by three protective layers known as the meninges.

Bony processes
The part of the vertebral arch that attaches to the body on each side is called the pedicle and the arch is completed behind by two laminae that meet in the midline to form the spinous process. These processes project downwards (like the tiles of a roof), with that of the eighth being the longest and most vertical. At the junction of the pedicles and laminae are the projecting transverse processes. These decrease in size from the top down.

Muscle attachments
Muscles and ligaments are attached to the spines and transverse processes. The thoracic vertebrae articulate with each other at the intervertebral joints. Between the vertebral bodies are discs that act as shock absorbers. Each vertebra has four surfaces (facets), which form moveable synovial joints with the adjacent vertebrae – one pair of facets articulates with the vertebra above, the other pair with the vertebra below. All of these joints are strengthened by ligaments.

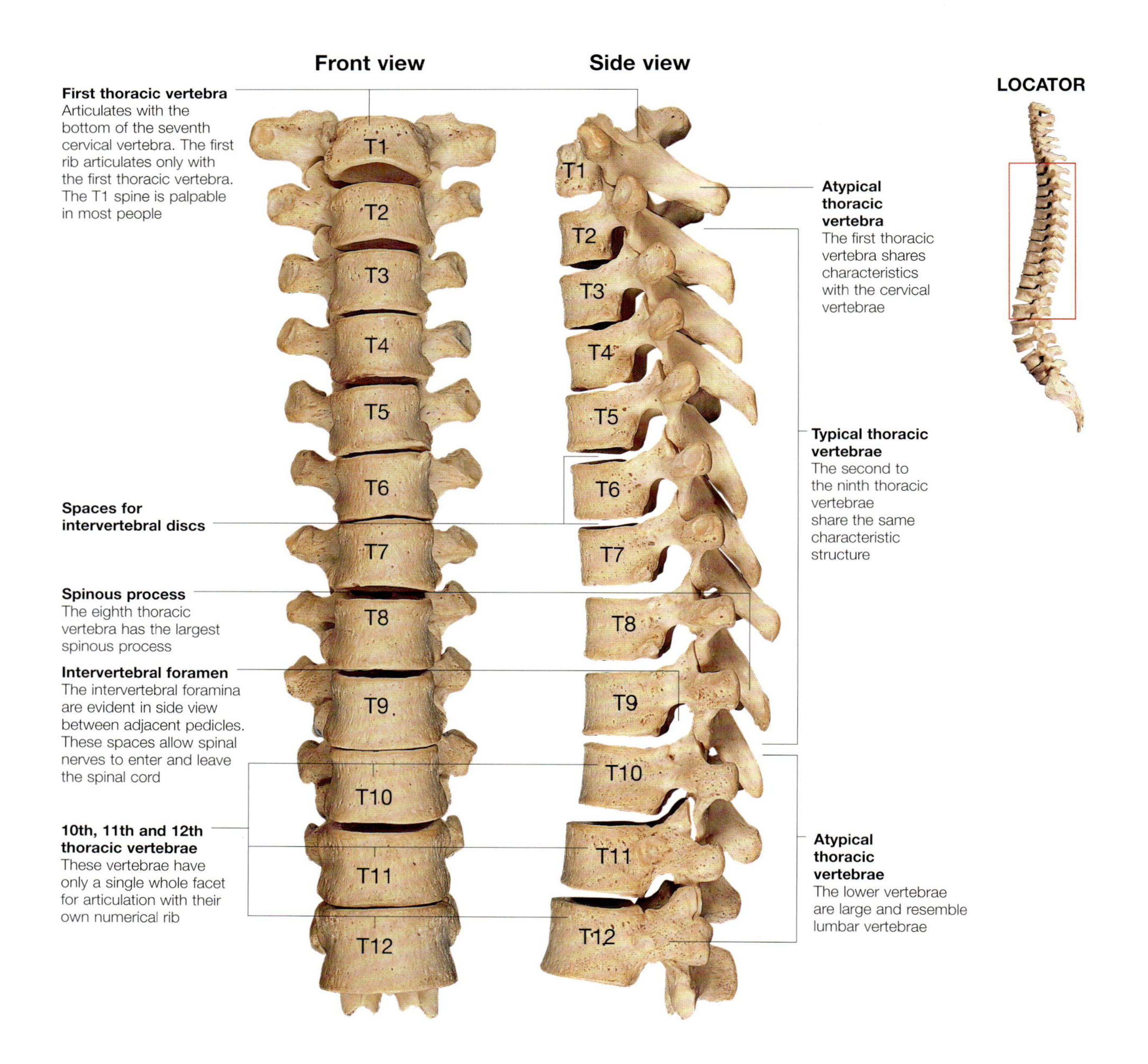

Features of the thoracic vertebrae

The thoracic vertebrae can be easily distinguished from the typical cervical vertebrae.

The thoracic vertebrae differ from the cervical vertebrae in several ways:

- An absence of the transverse process foramen (the foramen transversarium, through which nerves and blood vessels pass in the cervical vertebrae)
- A single, rather than bifid (two-part) spine
- The vertebral canal, through which the spinal cord runs, is smaller and more circular
- The presence of facets enabling the ribs to articulate with the spine. Each typical thoracic vertebra has six facets for rib articulation, three on each side.

The head of the rib lies in the region of the intervertebral disc, at the back, and has two hemi-facets that articulate with its own numbered vertebra (upper border) and the vertebra immediately above (lower border).

The exceptions to the above are the first, 10th, 11th and 12th thoracic vertebrae. In the first thoracic vertebra, the facet on the upper border is a whole facet (rather than a half), as the first rib articulates only with its own vertebra.

Each of the 10th, 11th and 12th vertebrae has only one whole facet to articulate with its own numerical rib. The 11th and 12th vertebrae have no articulation with the tubercle of the corresponding rib (and therefore no articular facet). The last two ribs are called 'floating ribs' as they have no connections to the ribs above.

Fifth (typical) thoracic vertebra (front view)

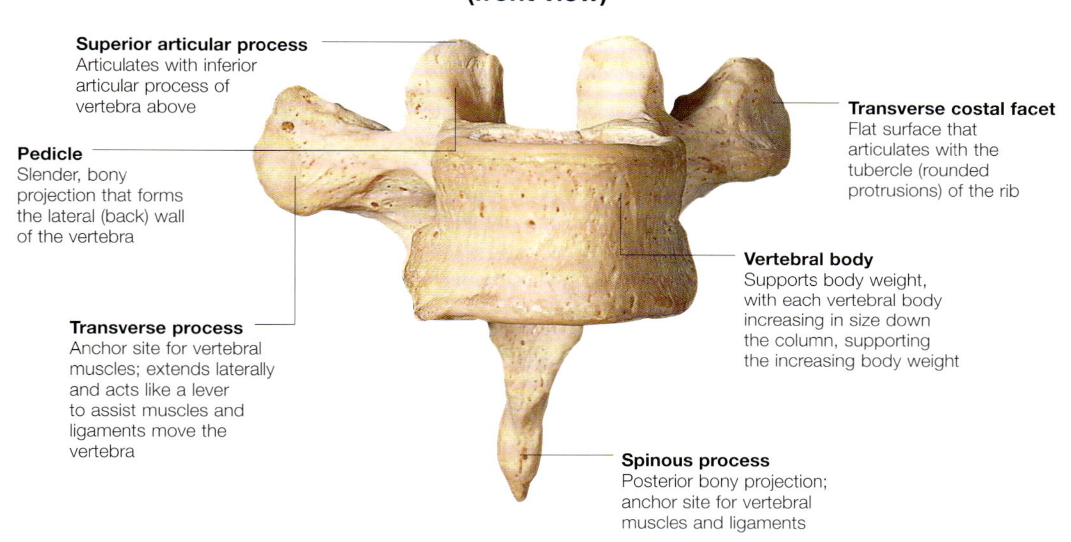

First (atypical) thoracic vertebra (side view)

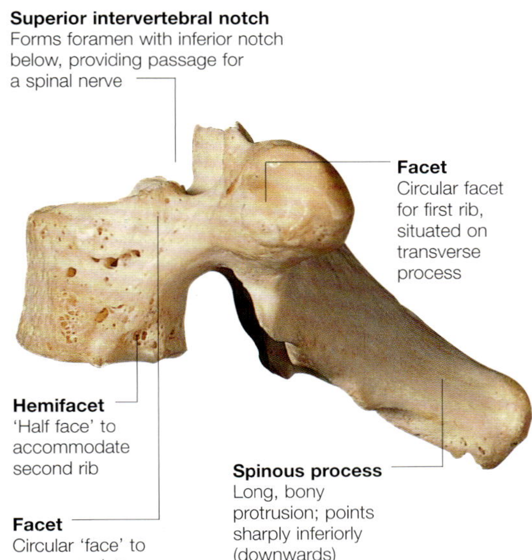

12th (atypical) thoracic vertebra (side view)

Body of vertebra
Structure of the lower thoracic vertebrae begins to resemble that of the lumbar vertebrae; only one round facet is present each side

Transverse process
11th and 12th thoracic vertebrae lack facet on transverse process

Inferior intervertebral notch
Forms the intervertebral foramen, through which a spinal nerve passes

Spinous process
At the base of the thoracic vertebrae, the spinous processes are small and rounded, resembling those of the lumbar vertebrae

Muscles of the Back

The muscles of the back give us our upright posture and allow flexibility and mobility of the spine. The superficial back muscles also act with other muscles to move the shoulders and upper arms.

The deep muscles of the back aid support and movement of the spine, whereas the superficial back muscles act to move the arm and shoulder.

Superficial muscles

The trapezius is a large, fan-shaped muscle whose top edge forms the visible slope from neck to shoulder. It attaches to the skull and helps to hold up and rotate the head and enables us to brace the shoulders back. The latissimus dorsi, the largest and most powerful back muscle, is attached to the spine from above the lower edge of the trapezius and runs down to the back of the pelvis.

Smaller muscles contribute to this superficial muscle layer. Levator scapulae, rhomboid major and rhomboid minor run between the spine and the scapula and act to move the scapula up and inwards. The 'rotator cuff' is a group of muscles that run between the scapula and head of the humerus (upper arm bone) at the shoulder joint. They hold the head of the humerus into the shoulder joint.

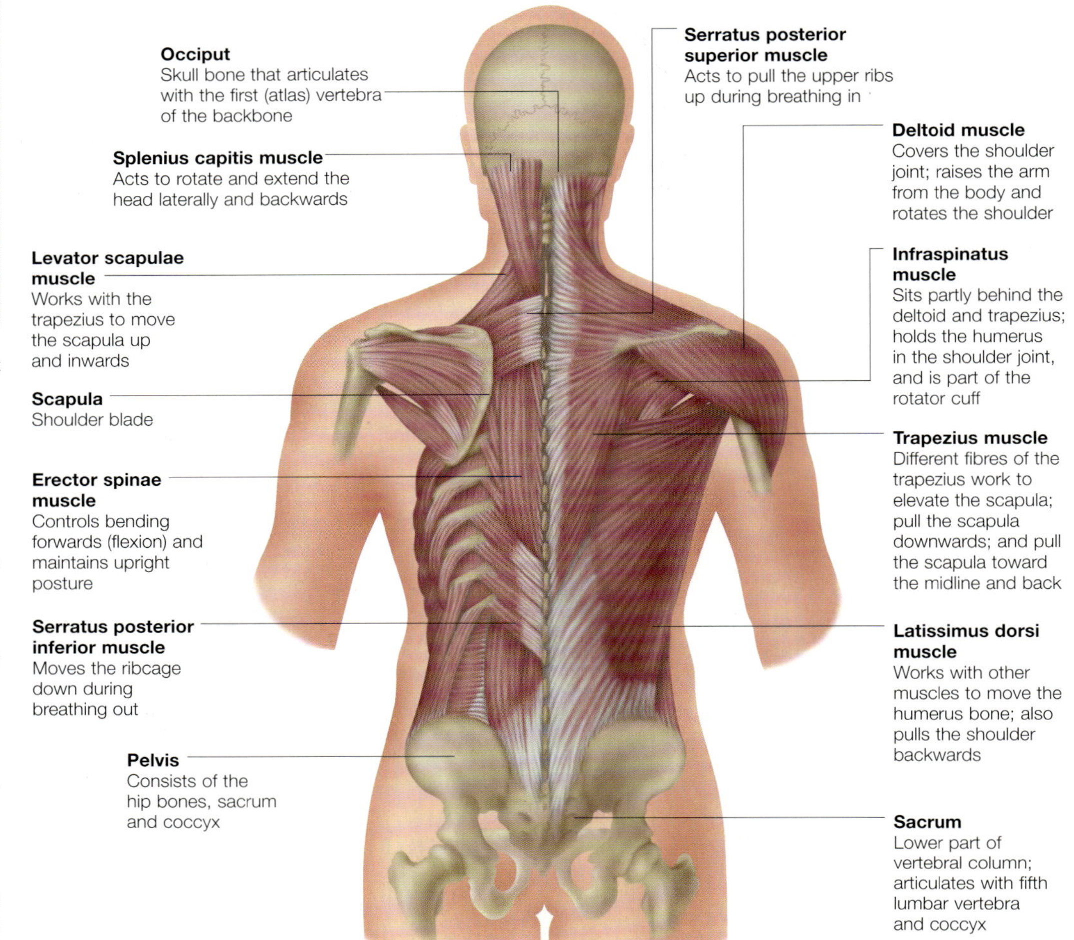

Supporting the head and neck

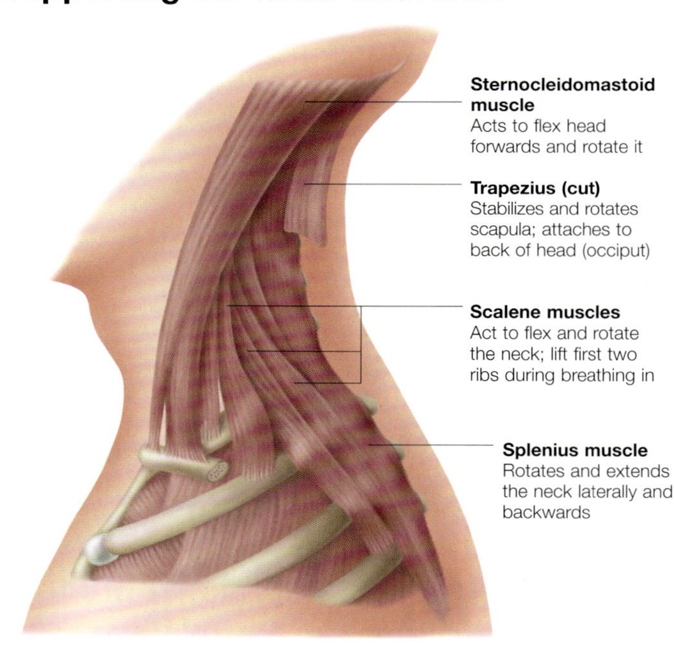

The deep muscles at the top of the spine are attached to the skull and act to keep the neck extended and the head upright. The centre of gravity of the head is in front of the spine and so constant contraction of the muscles at the back of the neck is needed to prevent the head from falling forwards

Sternocleidomastoid muscles

The sternocleidomastoid muscles are large muscles on either side of the neck. They act as the major muscles of flexion of the head and can be braced to support the head when it is elevated – as happens when rising from a lying position. The sternocleidomastoid muscles are aided in these tasks by several other deep muscles.

A lateral view of the muscles of the neck reveals some of the major muscles responsible for supporting and flexing the head and neck.

Splenius and trapezius muscles

The splenius muscles are broad sheets of muscle fibres that wrap around and over the deeper muscles of the neck. The splenius muscles originate in the cervical vertebrae (in the neck) and insert in the occipital bone at the back of the skull. When the muscles of one side of the neck are used alone, the neck and head is extended laterally or rotated; when used together and in collaboration with other muscles, the splenius muscles help to extend the neck backwards.

Extension of the neck is also aided by the action of the trapezius muscle, which attaches to the occipital bone, thoracic vertebrae and scapula.

Deep muscles of the back

The deep muscles of the back attach to underlying bones of the spine, pelvis and ribs. They act together to allow smooth movements of the spine.

Muscles must be attached to bone to provide the leverage they need to perform their functions. The bony attachments of the deep muscles of the back include the vertebrae, the ribs, the base of the skull and the pelvis.

Deep muscle layers

The deep muscles of the back are built up in layers; the most deeply located muscles are very short, running from each vertebra obliquely to the one above. Over these lie muscles that are longer and run vertically between several vertebrae and the ribs. More superficially, the muscles become longer and some are attached to the pelvic bones and the occiput (back of the base of the skull) as well as to the vertebrae.

There are numerous muscles in these layers. Each muscle is individually named according to its position, and they all act in varying combinations rather than individually. Together, they form the large group of deep muscles that lie on either side of the spine and act in conjunction to maintain the spine in an S-shaped curve, enabling fluid movement of the back.

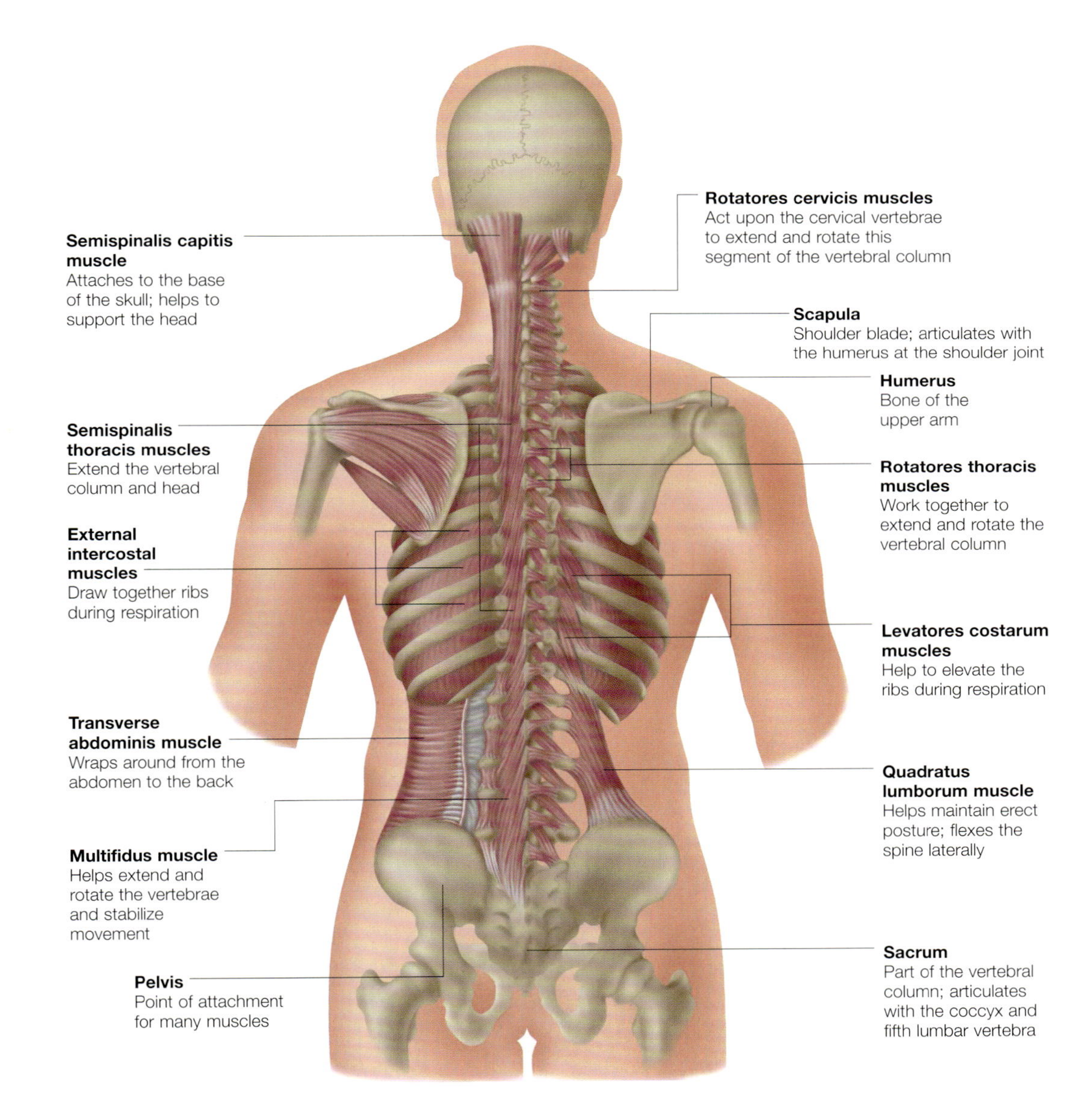

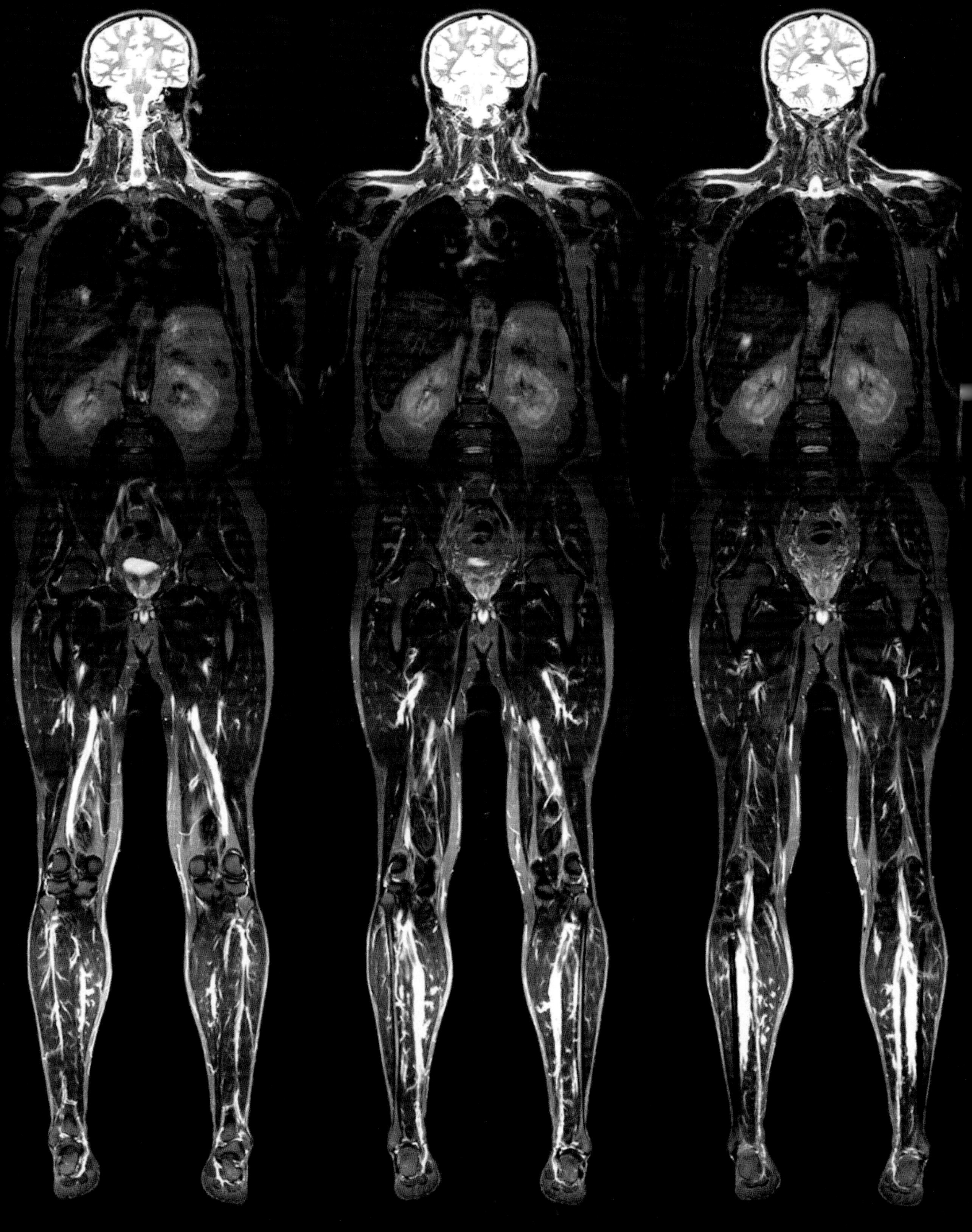

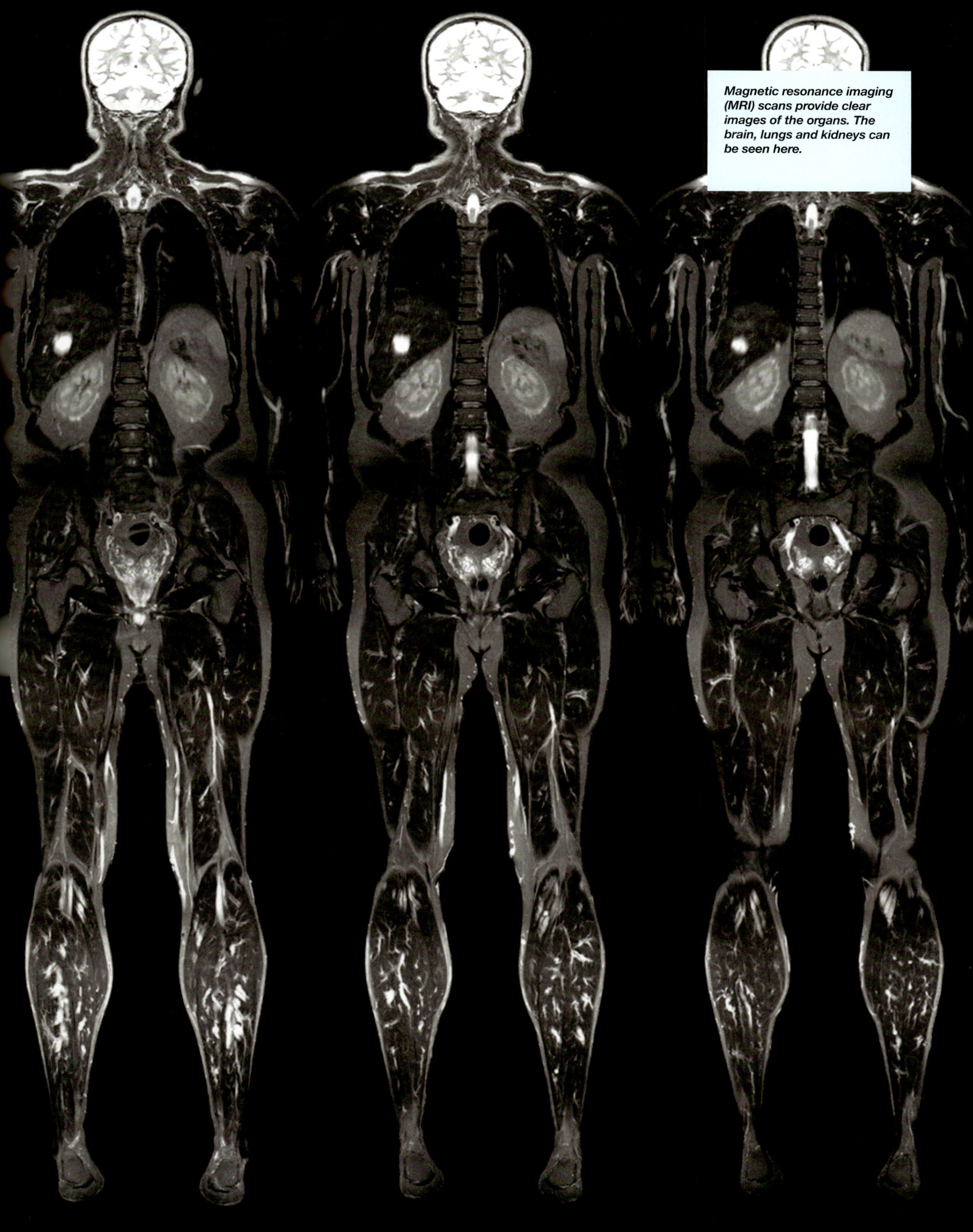

Magnetic resonance imaging (MRI) scans provide clear images of the organs. The brain, lungs and kidneys can be seen here.

Abdominal Wall

The abdominal cavity lies between the diaphragm and the pelvis. The abdominal wall at the front and sides of the body consists of different muscular layers, surrounding and supporting the cavity.

The posterior (rear) abdominal wall is formed by the lower ribs, the spine and accompanying muscles, while the anterolateral (front and side) wall consists entirely of muscle and fibrous sheets (aponeuroses).

Under the skin and subcutaneous fat layer there lie the muscle layers of the abdominal wall. The muscles here lie in three broad sheets: the external oblique, the internal oblique and the transversus abdominis, which give the abdomen support in all directions. In addition, there is a wide band of muscle, the rectus abdominis, which runs vertically from the front of the ribcage down to the front of the pelvis.

External Oblique

The external oblique muscle forms the most superficial layer of abdominal muscle. It is in the form of a broad, thin sheet whose fibres run down and inwards.

The muscle arises from the under-surfaces of the lower ribs. The fibres fan out into a wide sheet of tough connective tissue known as the external oblique aponeurosis. At the lower end the fibres insert into the top of the pubic bones.

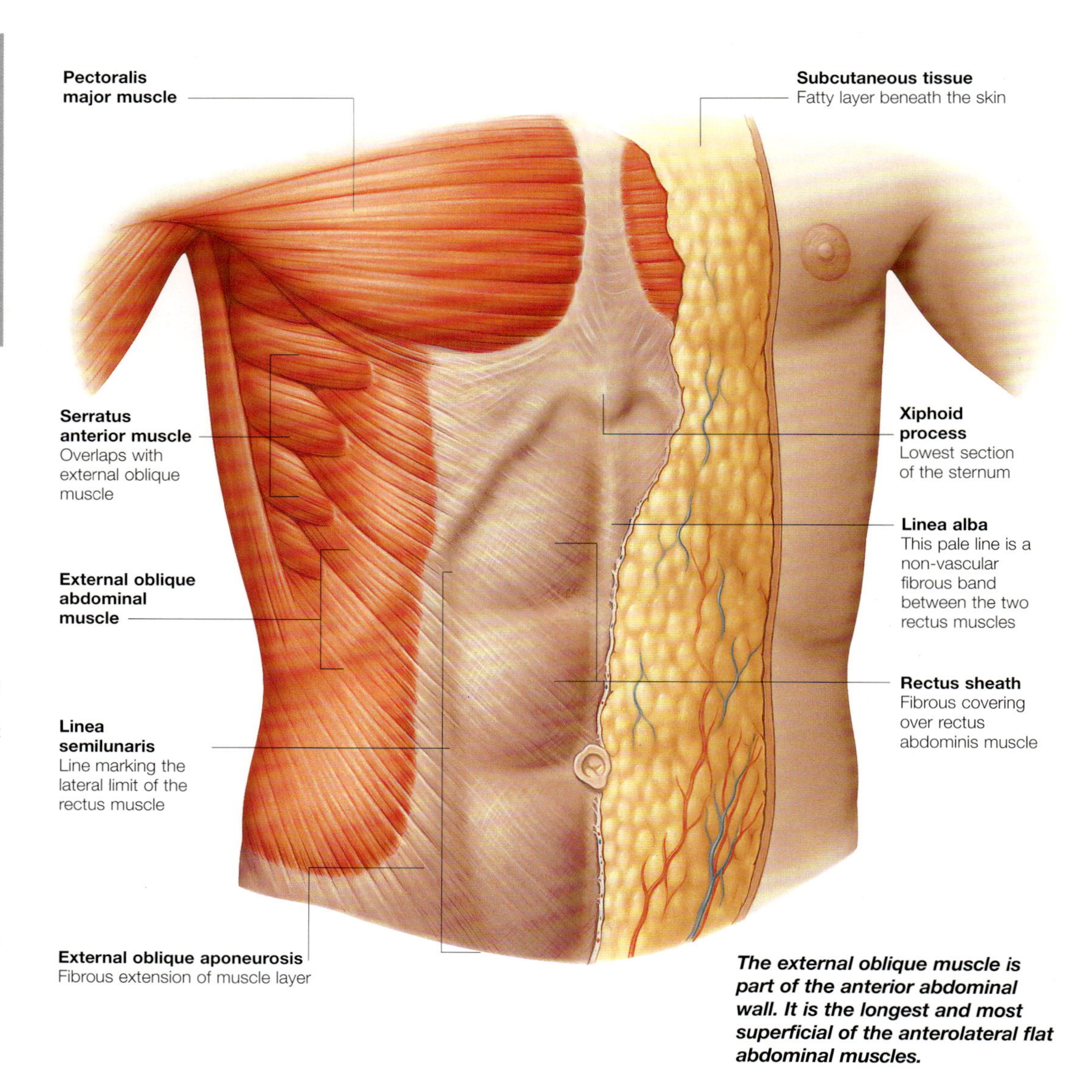

The external oblique muscle is part of the anterior abdominal wall. It is the longest and most superficial of the anterolateral flat abdominal muscles.

Layers of the abdominal wall

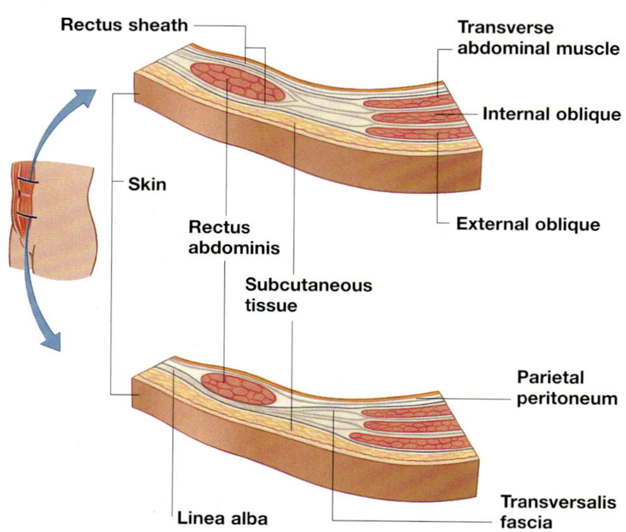

The layers of the abdominal wall include the following:

- Skin – the natural lines of cleavage of the skin lie horizontally over the greater part of the abdominal wall
- Superficial fatty layer, or Camper's fascia – this can be very thick and, in obese people, may cause the abdominal wall to lie in folds
- Superficial membranous layer, or Scarpa's fascia – this thin layer is continuous with the superficial fascia in adjoining parts of the body
- Three muscle layers: external and internal oblique, and transversus abdominis
- Deep fascia layers lying between, and separating, these muscle layers
- Transversalis fascia — this firm, membranous sheet lines most of the abdominal wall, merging with the tissues lining the underside of the diaphragm above and the pelvis below
- Fat, lying between the transversalis fascia and the peritoneum
- Peritoneum — this is the delicate, lubricating membrane that lines the abdominal cavity and covers the surfaces of many abdominal organs.

These transverse sections of the abdominal wall show its layered structure. The fibrous layers from the muscle interweave around the rectus abdominis.

Deeper muscles of the abdominal wall

Beneath the external oblique muscle lie two more layers of sheet-like muscle – the internal oblique and the transversus abdominis. The rectus abdominis (or the 'abs') run vertically down the centre.

The internal oblique muscle is a broad, thin sheet that lies deep to the external oblique. Its fibres originate from the lumbar fascia (a layer of connective tissue on either side of the spine), the iliac crest of the pelvis and the inguinal ligament in the groin. Like the external oblique, the internal oblique muscle inserts into a tough, broad aponeurosis, which splits to enclose the rectus abdominis muscle (rectus sheath).

Transversus abdominis
This is the innermost of the three sheets of muscle that support the abdominal contents. Its fibres run horizontally to insert into an aponeurosis that lies behind the rectus abdominis muscle for much of its length.

Rectus abdominis
These two strap-like muscles run vertically down the front of the abdominal wall. The upper part of each muscle is wider and thinner than the lower part. Lying between the muscles is a thin, tendinous band of tough connective tissue, the linea alba.

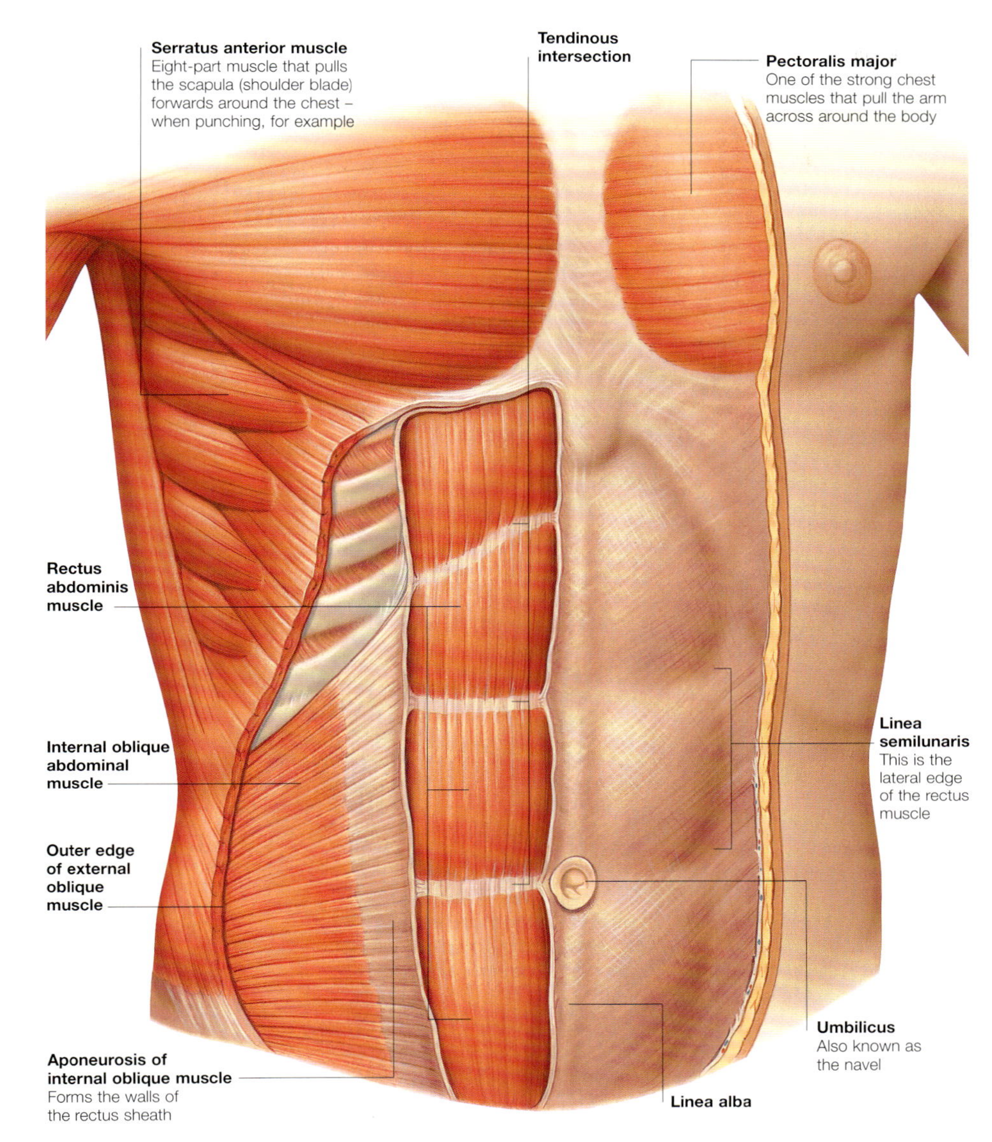

The rectus sheath

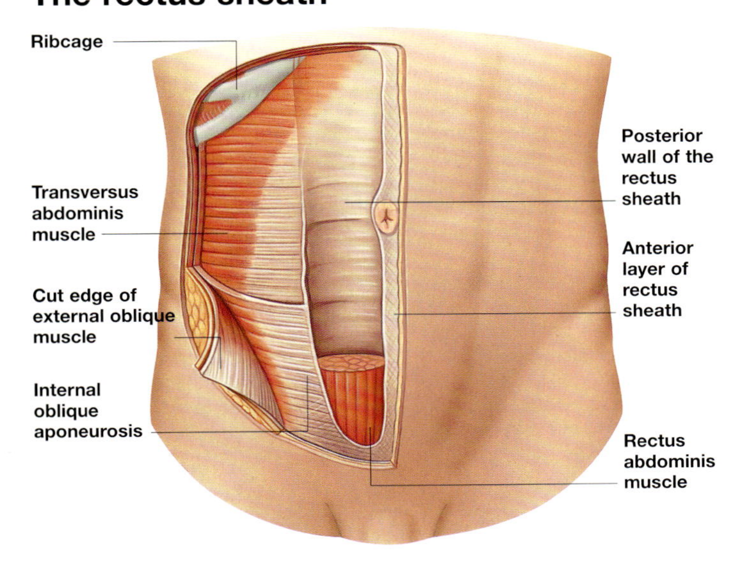

The rectus abdominis muscle is enclosed within a sheath of connective tissue formed by the coming together of the aponeuroses of the three muscle sheets of the abdominal wall. (An aponeurosis is a thin but strong sheet of fibrous tissue.)

■ The upper three-quarters of the rectus sheath differ from the lower quarter due to the way in which the three aponeuroses interweave.

The rectus abdominis muscle extends along the front of the abdomen. It is enclosed by connective tissue known as the rectus sheath.

■ The upper rectus sheath
The anterior wall is formed by the external oblique aponeurosis and half of the internal oblique aponeurosis, while the posterior wall is formed by the remaining half of the internal oblique and the transversus aponeurosis.

■ The lower rectus sheath
Three aponeuroses lie in front of the rectus abdominis muscle, which lies, therefore, directly on the transversalis fascia beneath.

The three aponeuroses meet in the midline to form the tough linea alba. As well as the rectus abdominis muscle, the rectus sheath contains blood vessels that lie deep to the muscle.

Ribcage

The ribcage protects the vital organs of the thorax, as well as providing sites for the attachment of muscles of the back, chest and shoulders. It is also light enough to move during breathing.

The ribcage is supported at the back by the 12 thoracic vertebrae of the spinal column and is formed by the 12 paired ribs, the costal cartilages and the sternum, or breastbone, at the front. Each of the 12 pairs of ribs attach posteriorly (at the back) to the corresponding numbered thoracic vertebra. The ribs then curve down and around the chest towards the anterior (front) surface of the body.

The ribs can be divided into two groups according to their anterior (front) site of attachment:

■ **True (vertebrosternal) ribs**
The first seven pairs of ribs attach anteriorly directly to the sternum via individual costal cartilages.

■ **False (vertebrochondral) ribs**
These do not attach directly to the sternum. Rib pairs eight to 10 attach indirectly to the sternum via fused costal cartilages. Rib pairs 11 and 12 do not have attachments to bone or cartilage and so are known as 'floating' ribs. Their anterior ends lie buried within the musculature of the lateral abdominal wall.

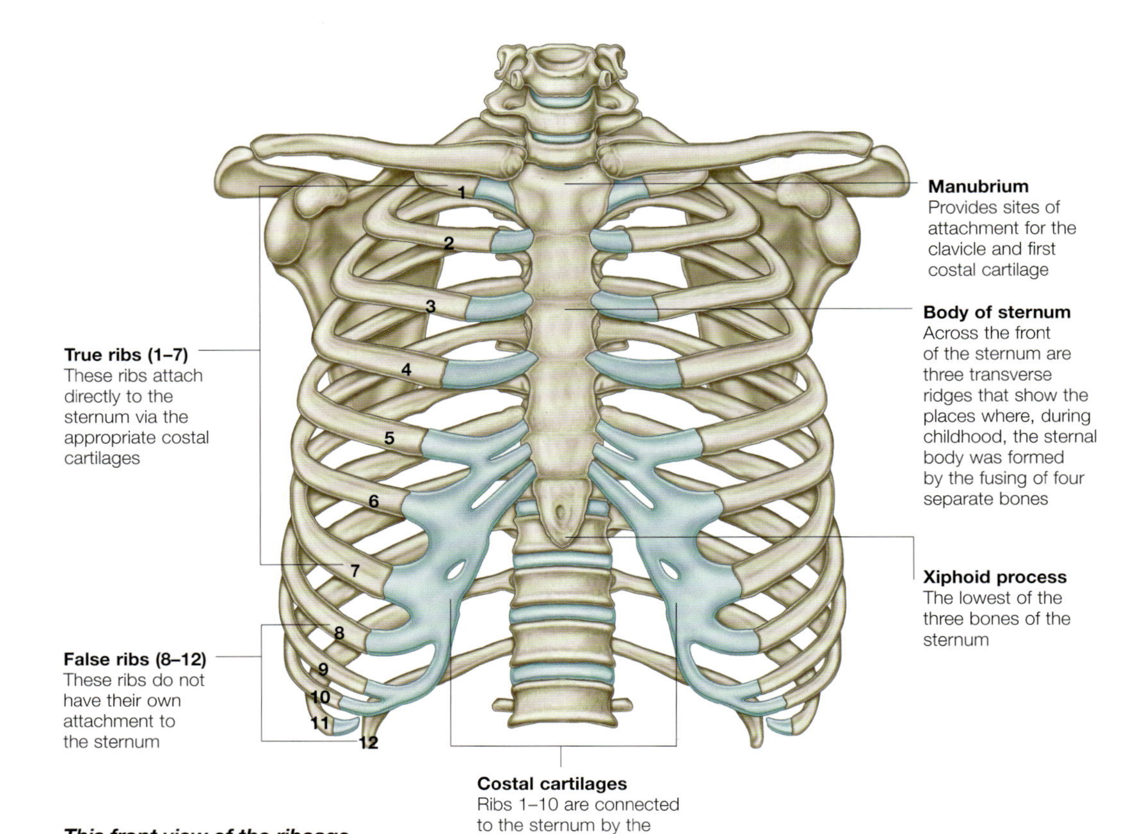

This front view of the ribcage shows the sternum and the 12 pairs of ribs and their associated costal cartilages.

Rib structure

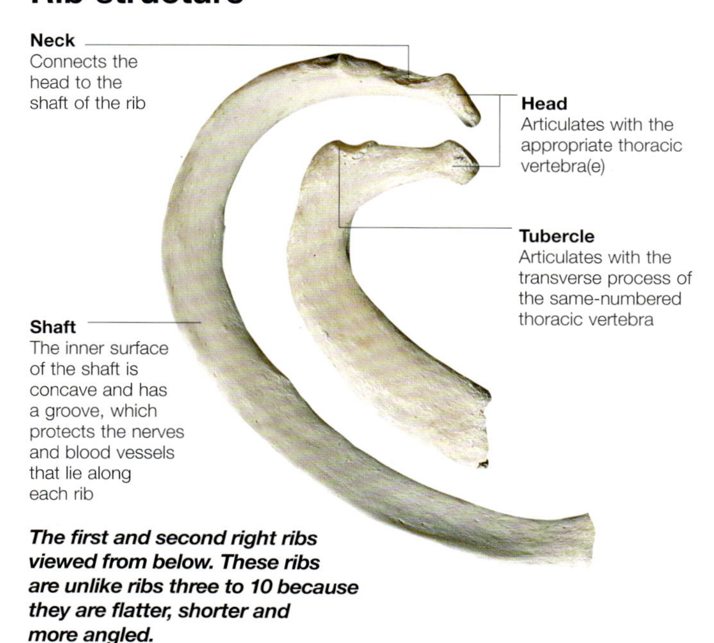

The first and second right ribs viewed from below. These ribs are unlike ribs three to 10 because they are flatter, shorter and more angled.

While they all vary slightly in their structure, ribs three to 10 are similar enough to be described as 'typical ribs'. They consist of the following parts:

■ **Head.** This connects to the thoracic vertebra with the same numeric value and the one immediately above that (for example, the fourth rib attaches to both the third and fourth thoracic vertebrae)

■ **Neck.** This narrowed length of rib connects the head to the shaft or body

■ **Tubercle.** This raised, roughened area lies at the junction of neck and shaft and bears a facet for articulation with the transverse process of the thoracic vertebra

■ **Shaft.** The rib continues as a flattened, curved bone that bends around at the 'costal angle' to encircle the thorax.

Dissimilar ribs

■ **First rib.** This is the widest, shortest and most flattened rib; it has only one facet on its head for articulation with the first thoracic vertebra. On its upper surface it has a prominent 'scalene tubercle'

■ **Second rib.** This rib is thinner than the first, its shaft being more like that of a typical rib. Halfway down the shaft it has a second prominent tubercle for the attachment of muscles

■ **11th and 12th ribs (floating ribs).** These have only a single facet on their heads and do not have a point of articulation between their tubercle and the transverse process of the corresponding thoracic vertebrae. The ends of their shafts carry only a cap of cartilage and do not connect with any of the other ribs.

Sternum

The sternum (breastbone) is a long, flat bone that lies vertically at the centre of the anterior (front) surface of the ribcage.

The sternum has three parts:

- **The manubrium.** This bone forms the upper part of the sternum and is in the shape of a rough triangle with a prominent, and easily palpable, notch in the centre of its superior surface, the 'suprasternal notch'
- **The body.** The manubrium and the body of the sternum lie in slightly different planes, angled so that their junction, the manubriosternal joint, projects forwards forming the 'sternal angle of Louis'. The body of the sternum is longer than the manubrium, forming the greater length of the breastbone
- **The xiphoid process.** This is a small pointed bone that projects downwards and slightly backwards from the lower end of the body of the sternum. In young people it may be cartilaginous, but it usually becomes completely ossified (changed to bone) by 40–50 years of age.

Anterior view

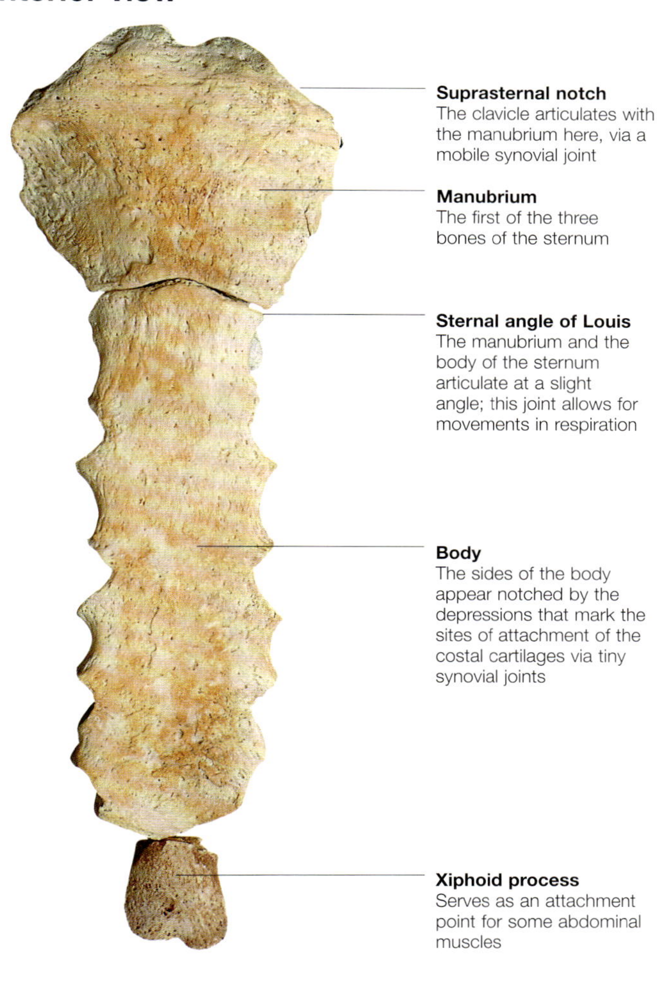

The sternum (breastbone) consists of three parts: the manubrium, the body and the xiphoid process.

Costal cartilages

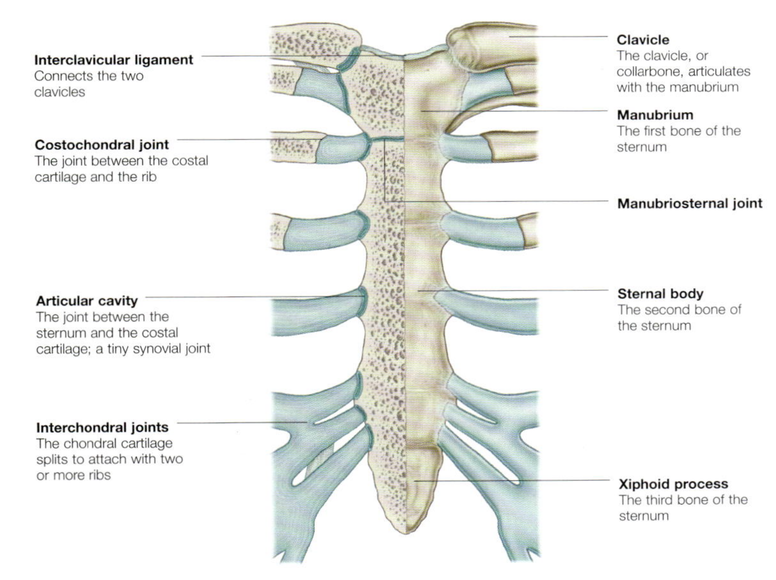

The ribs are connected to the sternum, or breastbone, by the costal cartilages. These flexible and resilient structures are made of 'hyaline' cartilage, which is tough but elastic, and so their presence contributes to the mobility of the thoracic wall.

During the process of respiration, the costal cartilages can stretch and twist to allow the ribcage to lift and expand as air is breathed into the lungs, and afterwards 'spring back' to regain their shape and position.

The first seven costal cartilages attach directly to the sternum, the next three attach to the costal cartilage directly above, while the last two are really just caps of cartilage on the ends of the rib shafts, which attach only to the soft tissues of the lateral abdominal wall.

The costal cartilages connect the sternum to the upper 10 pairs of ribs. These flexible structures provide much of the mobility of the thoracic cage.

Muscles of the Ribcage

The bony skeleton of the ribcage is sheathed in several layers of muscle. These include many of the powerful muscles of the upper limbs and back, as well as those that act upon the ribcage alone.

The muscles of the ribcage are concerned with respiration (breathing). They attach only to the ribcage and the thoracic spine. They form the structure of the thoracic wall, enclosing and protecting the vital internal organs of the thorax.

Intercostal Muscles

The intercostal muscles occupy the 11 intercostal spaces between the ribs. They lie in three layers:

- **External intercostal muscles**
These are the most superficial and the fibres of each external intercostal muscle run downwards and forwards to the rib below. Their contraction acts to lift the ribs during inspiration.
- **Internal intercostal muscles**
The internal intercostal muscles lie just under the external intercostals and at right angles to them; their fibres run downwards and backwards from the upper to the lower rib. Like the external intercostal muscles, they act to assist in inspiration.
- **Innermost intercostal muscles**
These lie deep to the internal intercostal muscles, their fibres running in the same direction. They are separated from the internal intercostals by connective tissue containing nerves and blood vessels.

Internal front view of the chest wall

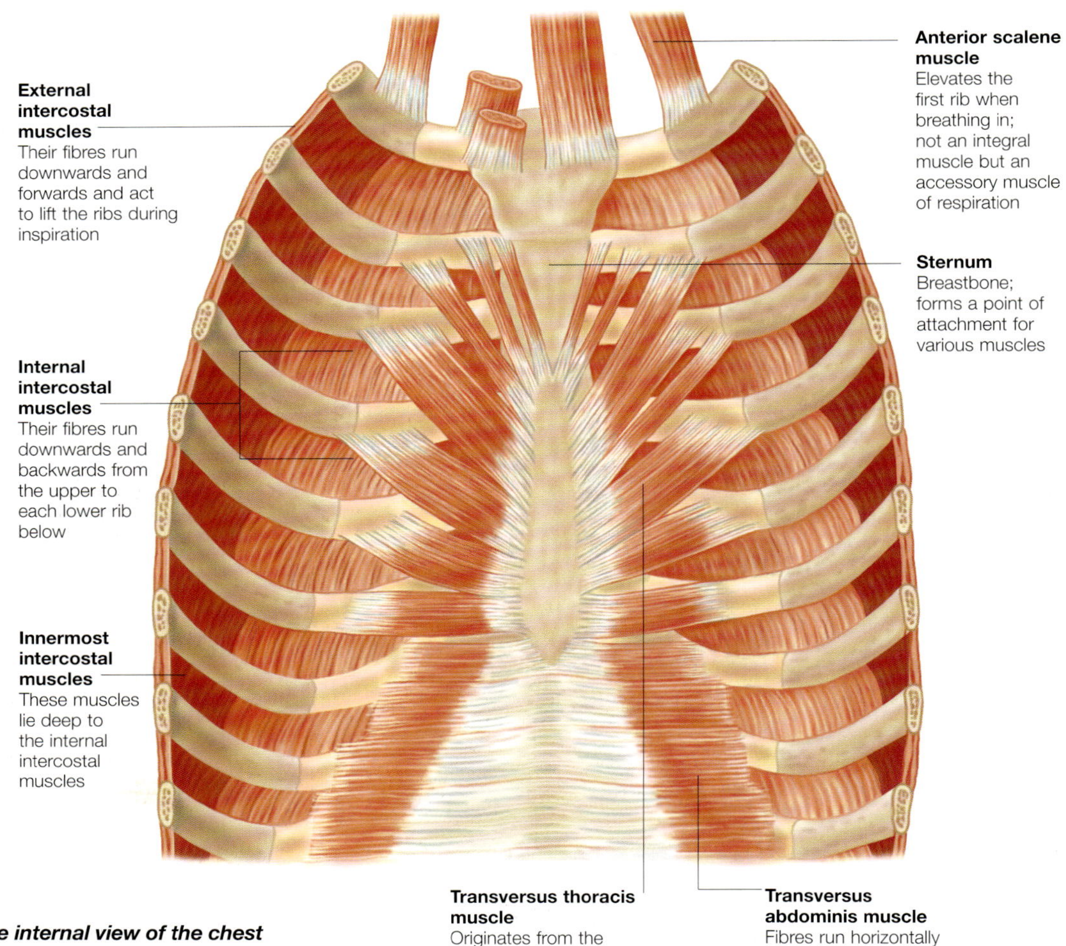

The internal view of the chest wall reveals the various muscles that attach to the sternum and the layers of intercostal muscles.

Integral muscles of the ribcage

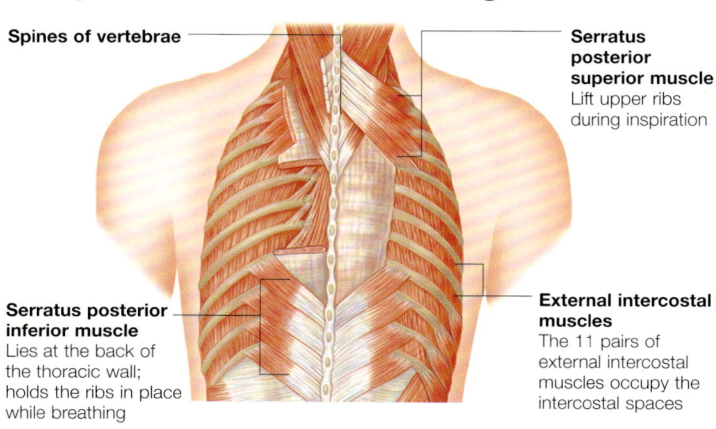

The view of the back of the thoracic wall reveals the external intercostal muscles and serratus inferior and superior muscles.

The integral muscles that form the structure of the ribcage include:

- **Intercostal muscles**
These lie in three layers filling each intercostal space between the ribs
- **Subcostal muscles**
These are small muscles that run down on the inner surface of the posterior thoracic wall between the lower ribs. Their fibres run in the same direction as those of the internal intercostal muscles and act to help in elevation of the ribs
- **Transversus thoracis muscles**
Small muscles that may vary in size and shape lying on the inside of the front of the chest
- **Serratus posterior**
These lie at the back of the thoracic wall in two parts: superior, which lifts the upper ribs during inspiration; and inferior, which holds the ribs in place during breathing.

Lumbar Vertebrae

The five lumbar vertebrae of the lower back are the largest and strongest vertebrae of the spinal column.

The lumbar vertebrae are the largest and strongest in the vertebral column as the lower the position of the bones of the spinal column, the more body weight they must bear. The lumbar vertebral joints are designed to allow maximum flexion (bend forwards), and some lateral flexion (reach sideways), but little rotation (this occurs at the thoracic level).

Basic Structure
Each lumbar vertebra has the same basic structure as the cervical and thoracic vertebrae, consisting of a cylindrical body in the front and a vertebral arch behind which enclose a space, the vertebral foramen. Each vertebral arch comprises a number of processes. There are two laterally projecting transverse processes, a centrally positioned spinous process and two pairs of articular facets. The transverse processes and spines are shorter and thicker than those of other vertebrae and are well adapted for the attachment of the large back muscles and strong ligaments.

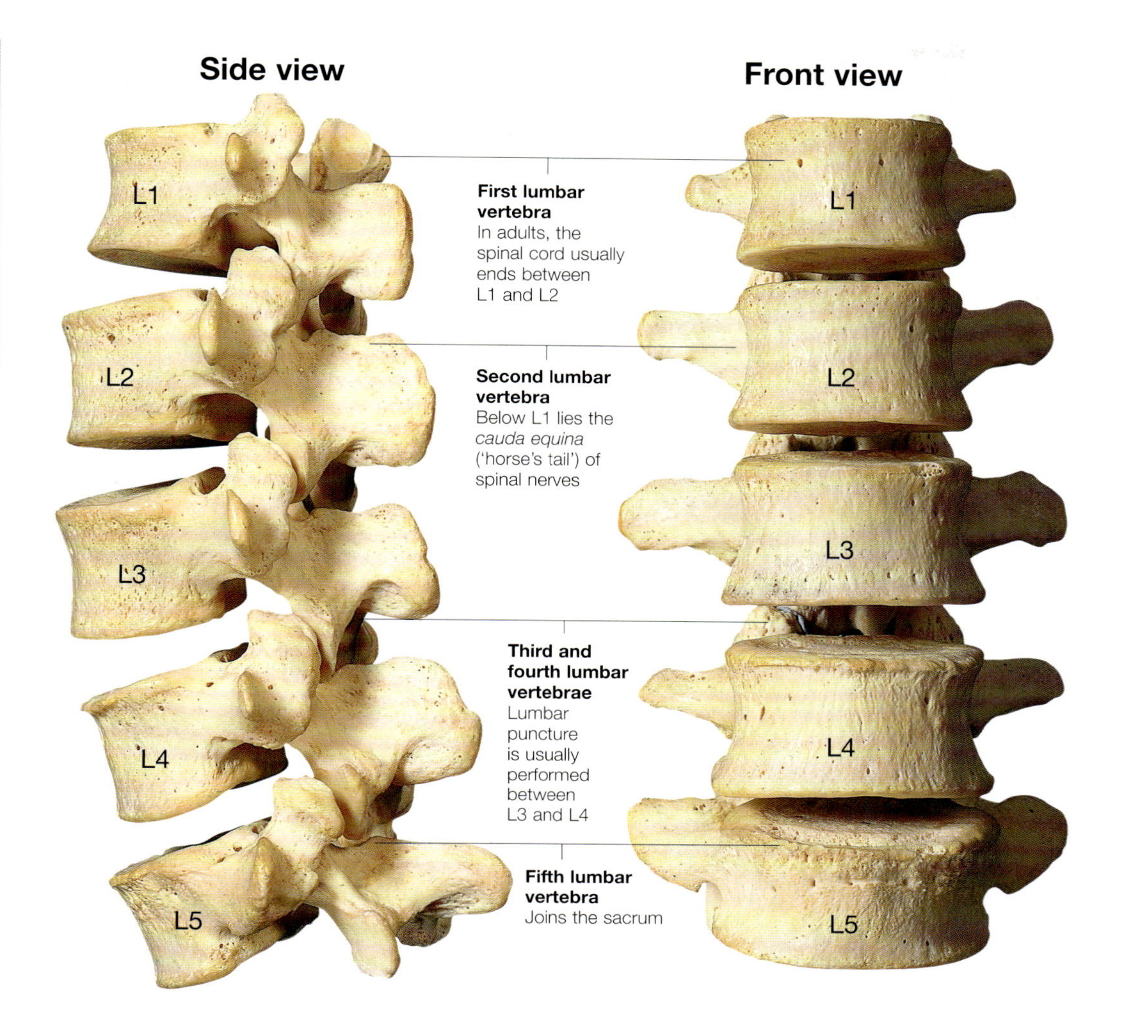

The front of the lumbar vertebrae form a convex curve when viewed from the side, known as lumbar lordosis. This increases strength and helps to absorb shock.

The five lumbar vertebrae are subject to greater vertical compression forces than the rest of the spine. For this reason, these vertebrae are large and strong.

Typical lumbar vertebrae

The lumbar vertebrae have no articulations for ribs and the orientation of the articular processes prevent rotation of this part of the spine.

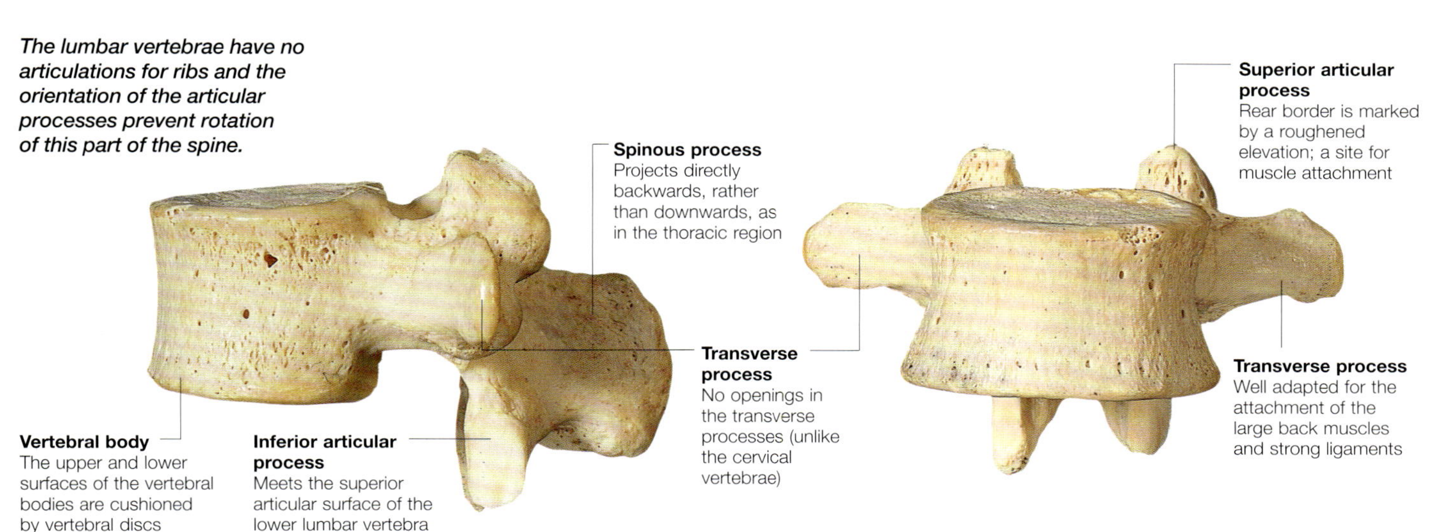

Lumbar Ligaments

The intervertebral discs and connecting ligaments support the bones of the spine. They act as shock absorbers, reducing wear on the vertebrae.

The intervertebral discs link the bones of adjacent vertebrae, prevent dislocation of the vertebral column and also act as shock absorbers between the vertebrae. Intervertebral discs contribute about one-fifth of the length of the vertebral column, and are thickest in the lumbar region where the vertical compression forces are greatest.

Strength and Stability

To reinforce stability, the vertebral bodies are strengthened by tough, longitudinally running ligaments, consisting of fibrous tissue, in the front and rear. These ligaments are firmly attached to the intervertebral discs and adjacent edges of the vertebral body, but loosely attached to the rest of the body. They are known as the supraspinous ligament, the interspinous ligament and the anterior longitudinal ligament.

Synovial Joints

Movement between vertebrae is the result of the action of muscles attached to the processes of the vertebral arches. The joints associated with the articular processes are synovial joints, allowing the adjacent surfaces to glide smoothly over each other. Each synovial joint is surrounded by a loose joint capsule. The joints of the vertebral arches are strengthened by various ligaments. The ligamenta flava join the laminae of the adjacent vertebra and contain elastic tissue.

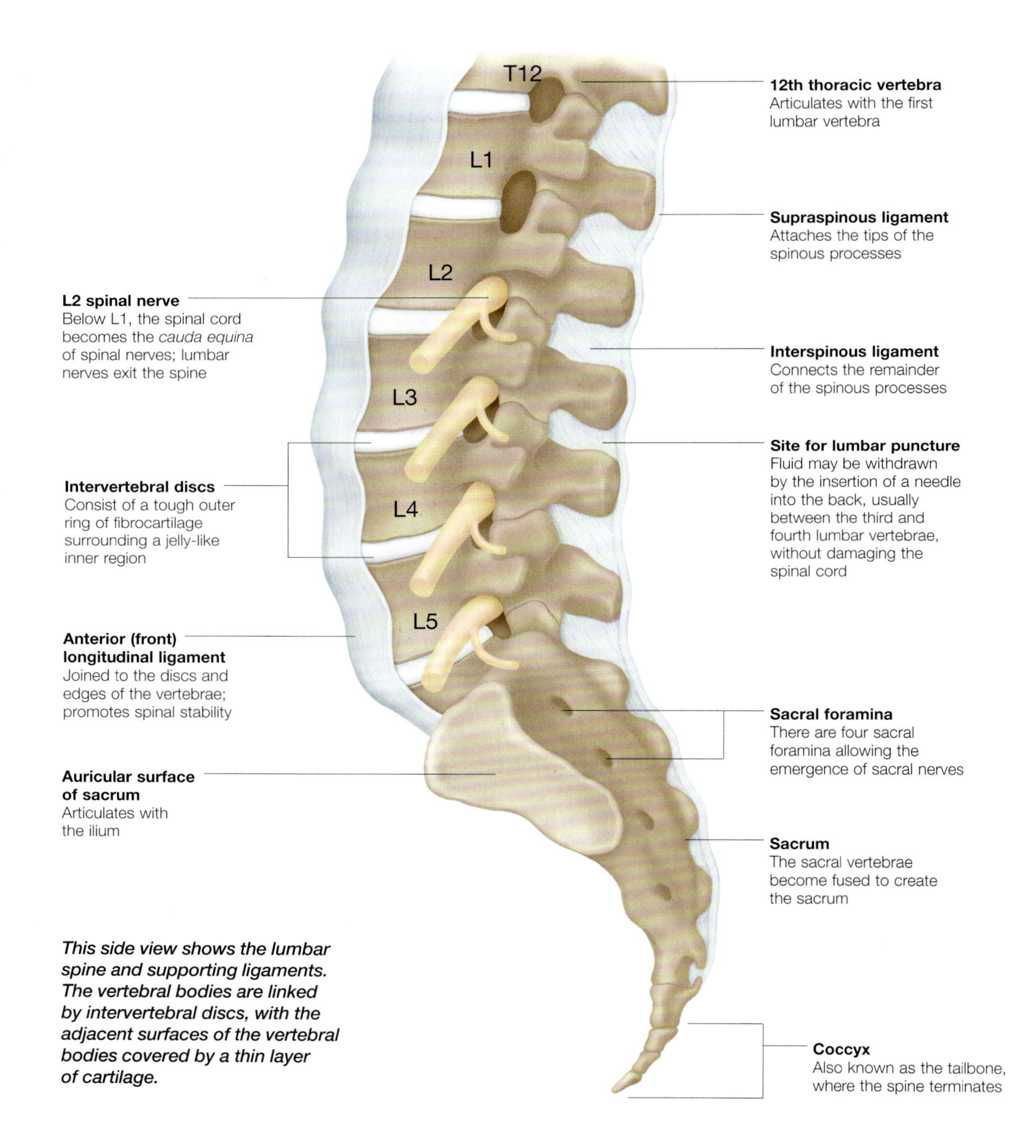

This side view shows the lumbar spine and supporting ligaments. The vertebral bodies are linked by intervertebral discs, with the adjacent surfaces of the vertebral bodies covered by a thin layer of cartilage.

Sacrum and Coccyx

The sacrum and coccyx form the tail end of the spinal column. Both are formed from fused vertebrae, allowing attachment for weight-bearing ligaments and muscles, and helping to protect pelvic organs.

The sacrum is composed of five fused sacral vertebrae. It performs several functions: it attaches the vertebral column to the pelvic girdle, supporting the body's weight and transmitting it to the legs; it protects pelvic organs, such as the uterus and bladder; and it allows attachment of muscles that move the thigh.

The sacrum is shaped like an upside-down triangle, the five fused vertebral bodies diminishing in size from the wide base above (formed by the first sacral vertebra and the sacral alae, or 'wings') towards the apex below, where the coccyx is attached.

Centrally, horizontal bony ridges indicate the junctions between vertebrae; these are the remnants of intervertebral discs. On either side, sacral foramina allow the passage of the ventral sacral motor nerve roots.

The Coccyx

Attached to the base of the sacrum, the coccyx is the remains of the tail seen in our primate ancestors. It consists of a small, pyramid-shaped bone formed from four fused vertebrae, and allows the attachment of ligaments and muscles, forming the anal sphincter.

Sacro-iliac Joint

On either side the sacrum articulates with the pelvic bones at the sacro-iliac joints. The joint surface is covered with the type of cartilage found in free-moving joints. This allows flexibility, for example, during childbirth.

Pelvic (inner) surface of sacrum

Side view of sacrum

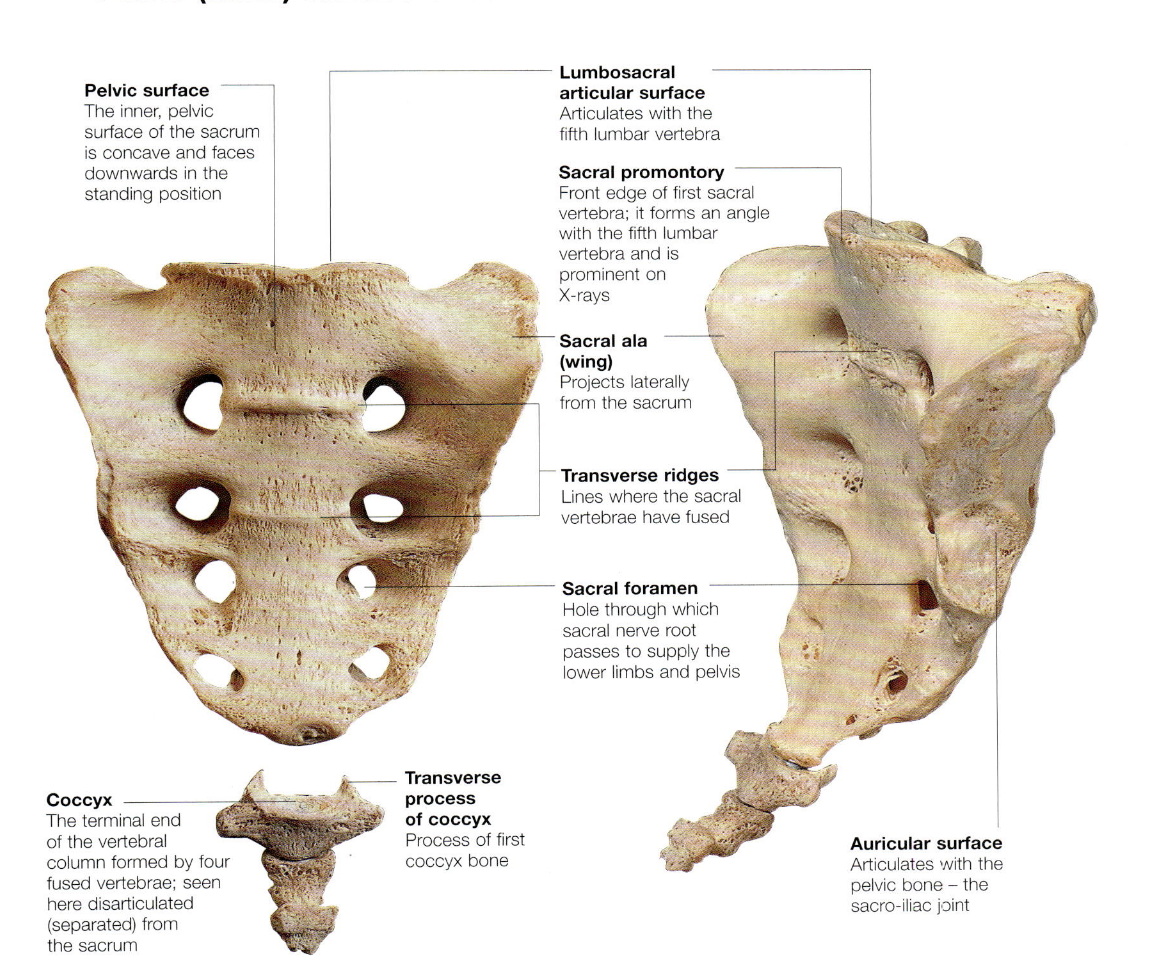

Pectoral Girdle

The pectoral (shoulder) girdle is the bony structure that articulates with and supports the upper limb. It consists of the clavicles at the front of the chest and the scapulae that lie flat against the back.

The upper limb is connected to the skeleton by the pectoral or shoulder girdle, made up of the clavicle (collarbone) and the scapula (shoulder blade). The pectoral girdle has only one joint with the central skeleton, at the inner end of the clavicle where it articulates with the sternum.

The Clavicle

The clavicle is an S-shaped bone that lies horizontally at the upper border of the chest. The front and upper surfaces of the clavicle are mostly smooth, while the under surfaces are roughened and grooved by the attachments of muscles and ligaments.

The medial (inner) end of the clavicle has a large oval facet that connects with the sternum at the sternoclavicular joint. A smaller facet lies at the other end where the clavicle articulates with the acromion (a bony prominence of the scapula) at the acromioclavicular joint.

The clavicle acts as a strut to brace the upper limb away from the body, thereby allowing a range of free movement. Along with the scapula and its muscular connections, it also transmits the force of impacts on the upper limb to the skeleton.

Pectoral girdle from above

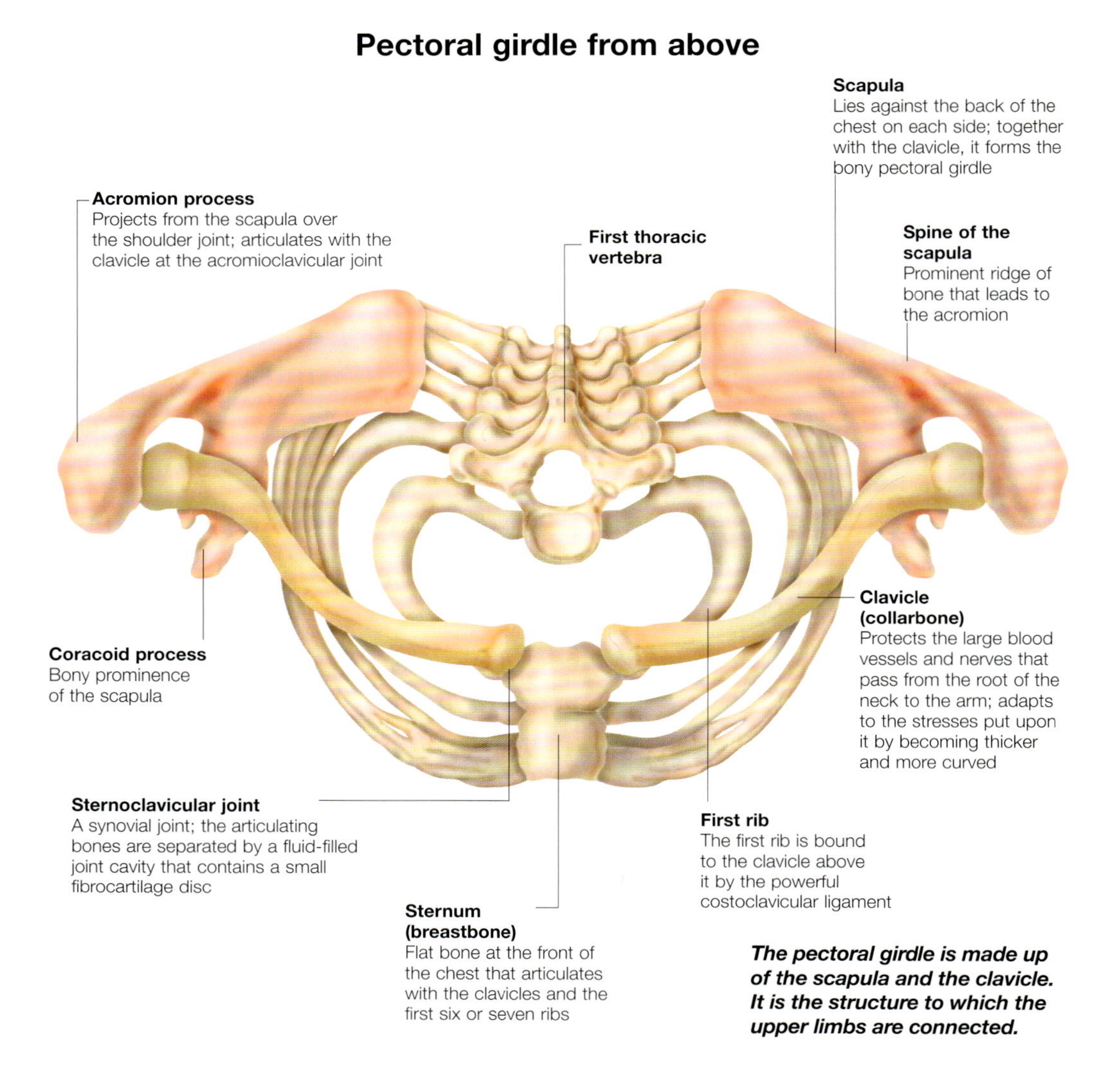

The pectoral girdle is made up of the scapula and the clavicle. It is the structure to which the upper limbs are connected.

Joints of the clavicle

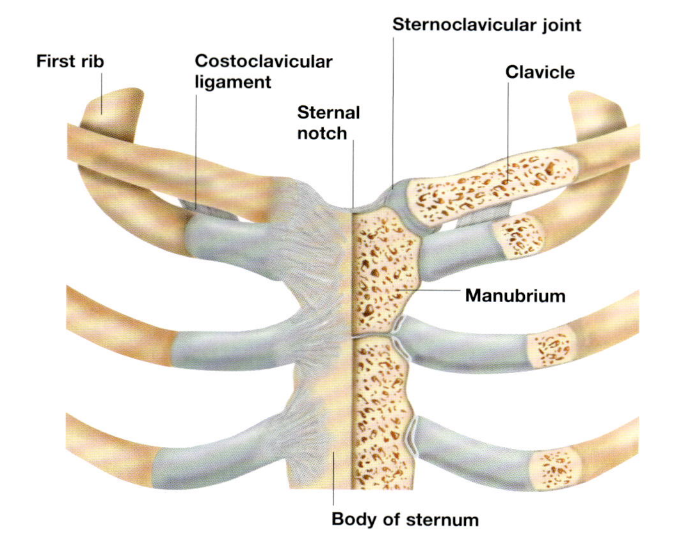

A tough fibrous capsule (sheath), together with strong surrounding ligaments, holds the sternoclavicular joint firmly in place.

The sternoclavicular joint is the only bony connection between the pectoral girdle and the rest of the skeleton. It can be felt under the skin, as the sternal end of the clavicle is fairly large and extends above the top of the manubrium (the top of the sternum), both sides together forming the familiar 'sternal notch' at the base of the neck.

The cavity is divided into two by an articular disc made of fibrocartilage, which improves the fit of the bones and keeps the joint stable. The joint is further stabilized by the costoclavicular ligament, which anchors its underside to the first rib. Only a small degree of movement is possible at the sternoclavicular joint; the outer end of the clavicle can move upwards, as when shrugging the shoulders, or forwards when the arm reaches out to pick up something in front of the body.

The acromioclavicular joint is formed between the outer end of the clavicle and the acromion of the scapula. The acromio-clavicular joint rotates the scapula on the clavicle under the influence of muscles that attach the scapula to the rest of the skeleton.

Scapula

The scapula is a flat, triangular-shaped bone that lies against the back of the chest. Together with the clavicle (collarbone), it forms the bony pectoral girdle.

The scapula, or shoulder blade, lies against the back of the chest on each side overlying the second to seventh ribs. As a rough triangle, the scapula has three borders: medial (inner), lateral (outer) and superior, with three angles between them.

Anterior surface

The scapula has two surfaces: anterior (front) and posterior (back). The anterior or costal (rib) surface lies against the ribs at the back of the chest and is concave, having a large hollow called the subscapular fossa that provides a large surface area for the attachment of muscles.

Posterior surface

The posterior surface is divided by a prominent spine. The supraspinous fossa is the small area above the spine, while the infraspinous fossa lies below. These hollows provide s ites of attachment for muscles of the same name.

Bony processes

The spine of the scapula is a thick projecting ridge, continuous with the bony outcrop called the acromion. This is a flattened prominence that forms the tip of the shoulder.

The lateral angle, the thickest part of the scapula, contains the glenoid cavity, the depression into which t he head of the humerus fits at the shoulder joint. The coracoid process is an important site of attachment of muscles and ligaments.

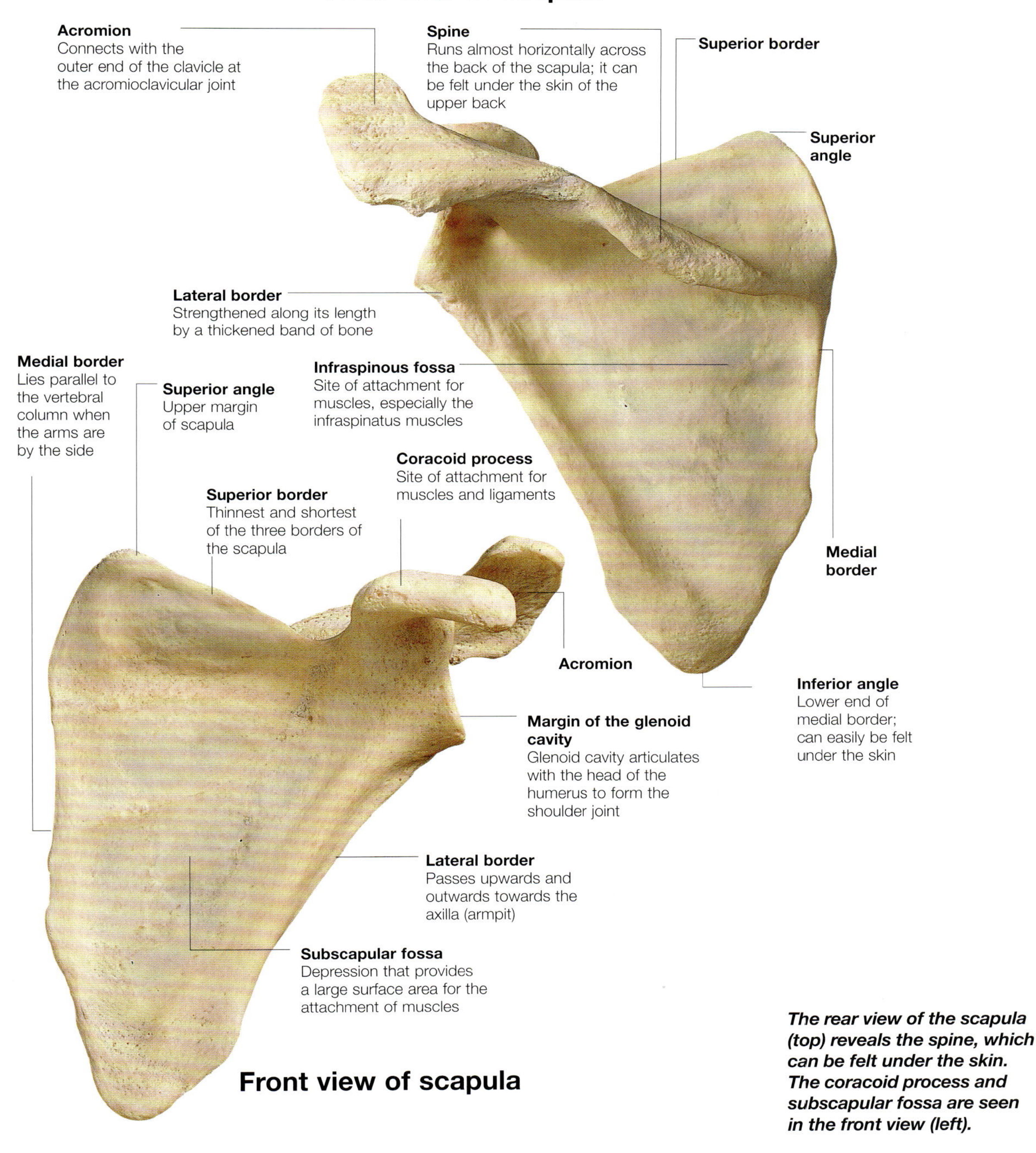

The rear view of the scapula (top) reveals the spine, which can be felt under the skin. The coracoid process and subscapular fossa are seen in the front view (left).

Muscles of the Pectoral Girdle

The pectoral girdle consists of the scapulae and clavicles, and is responsible for attaching the upper limbs to the central skeleton. The pectoral girdle muscles hold the scapulae and clavicles in place.

The pectoral girdle is defined as the structure that attaches the upper limbs to the axial skeleton. The muscles of the pectoral girdle attach primarily to the scapulae and clavicles; however, there are a few muscles that connect the upper limb directly to the central skeleton, and cause indirect movements of the pectoral girdle. This group of muscles lie superficially on the trunk; pectoralis major at the front and latissimus dorsi at the back.

Pectoralis Major

Pectoralis major arises via two heads, one from the sternum (breastbone) and adjacent rib (costal) cartilages, and another from the middle third of the clavicle. Its tendon twists anti-clockwise as it courses towards the outer lip of the bicipital groove, on the upper end of the humerus. This twisting gives the clavicular head a greater mechanical advantage during flexion of the arm.

The sternocostal head of pectoralis major is a powerful adductor of the arm (pulling the limb towards the body). If the arm is kept fixed, this muscle can elevate the ribs as an accessory muscle of breathing.

Latissimus Dorsi

The latissimus dorsi acts in conjunction with the pectoralis major to perform some arm movements, such as paddling a canoe.

Beneath the Pectoralis Major

Lying beneath the pectoralis major are the subclavius and pectoralis minor musclea. The subclavius is an insignificant muscle that may help to stabilize the clavicle during movements of the pectoral girdle.

Pectoralis minor arises from the second, third, fourth and fifth ribs, and attaches to the coracoid process of the scapula. It assists in pulling the scapula against and around the trunk wall. This action (protraction) is necessary to 'throw' a punch. The main muscle that performs protraction is serratus anterior. This muscle wraps itself around the wall of the ribcage, to attach to the scapula's inner edge. The lower four digitations converge on the inferior angle of the scapula, and are involved in assisting trapezius during scapular rotations.

The sternocleidomastoid muscle, which arises from the medial third of the clavicle and manubrium sternum, is mainly involved with movements of the head and neck.

Anterior muscles of the pectoral girdle

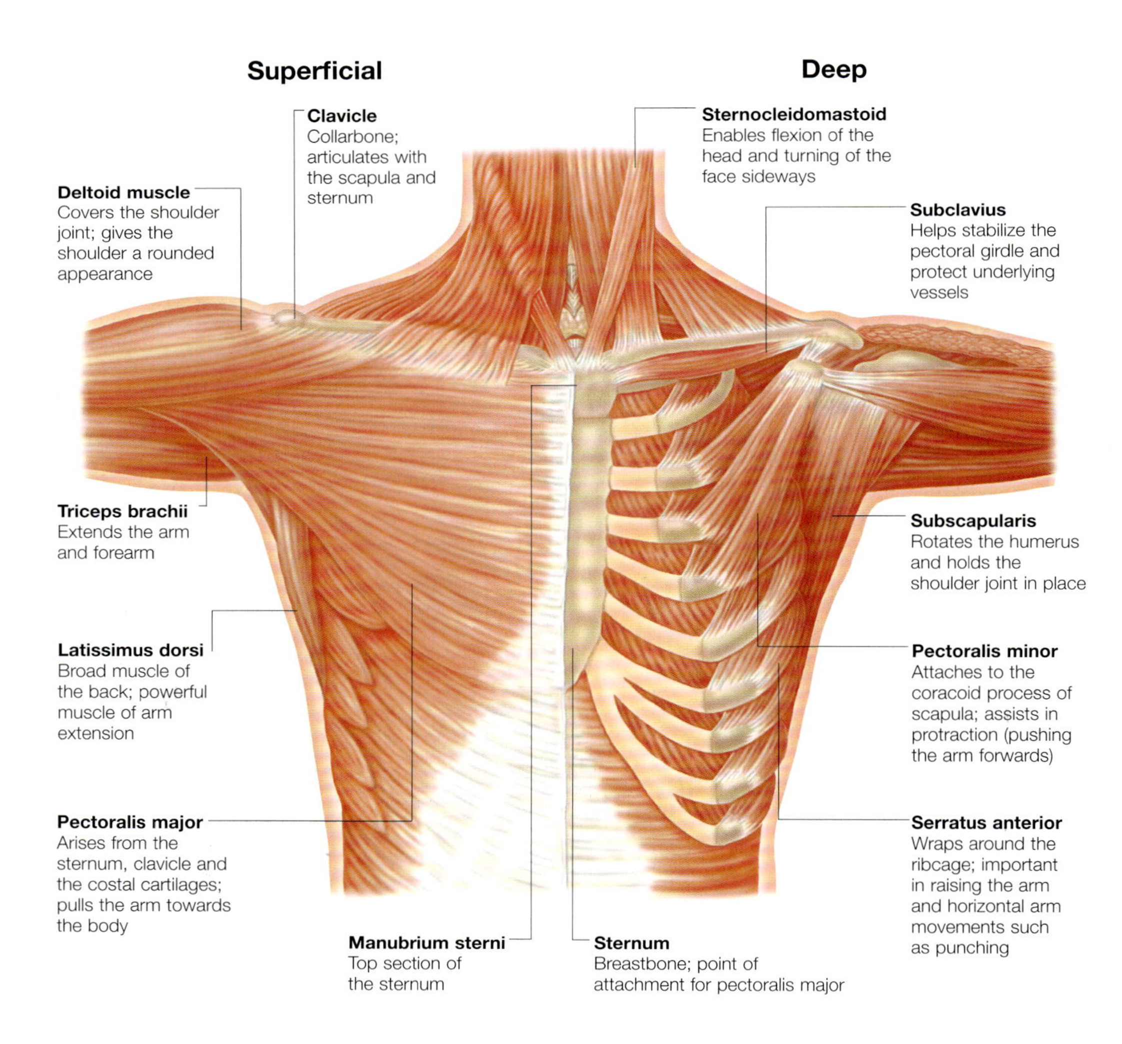

Pectoral girdle from the back

The latissimus dorsi and the large trapezius are superficial muscles of the back that attach to and influence the movement of the pectoral girdle.

Latissimus dorsi arises from the lower thoracic and the lumbar and sacral vertebrae. It also arises from the thoracolumbar fascia and posterior part of the iliac crest, with a few fibres attaching to the lower four ribs. From this broad base (latissimus means 'broadest' in Latin), it converges onto the floor of the bicipital groove at the upper end of the humerus.

This muscle assists pectoralis major in pulling the arm towards the body (adduction). Since the muscle wraps itself around the lower ribs, it assists during forceful expiration (breathing out), for example, during coughing.

Trapezius

Partly overlapping the latissimus dorsi is the lower part of the trapezius muscle. The trapezius also has a broad origin from the base of the skull (occipital protuberance) to the spines of the 12 thoracic vertebrae. The lower fibres attach to the spine of the scapula; intermediate fibres to the acromion process; and upper fibres to the outer third of the clavicle. The upper part serves to shrug the shoulders; the middle and lower parts serve to laterally rotate the scapula.

Deep muscles

The scapula is pulled backwards (retracted) by the action of the rhomboids (rhomboid major and minor). These attach the inner edge of the scapula to the vertebral column. These muscles allow the shoulder to be 'braced'; as seen before a punch is thrown, or before a forceful push, since this maximizes the force in protraction (forwards).

Since these muscles lie deep to the trapezius, they are difficult to see and feel. However, should they become paralyzed on one side, the scapula on that side would be displaced further away from the midline. Rhomboid major and minor, together with the levator scapulae, also act to rotate the scapula medially, and hence counteract the actions of the trapezius and serratus anterior.

Scapula

Movements of the scapula are essential in providing the widest range of motion at the shoulder joint. Although there is no anatomical joint between the scapula and the trunk, clinicians often refer to a scapulothoracic 'joint', since there is a great deal of movement between the two.

These movements are also transferred to the sternoclavicular joint through the clavicle. The sternoclavicular joint is the sole connecting link between the pectoral girdle and the trunk.

Posterior muscles of the pectoral girdle

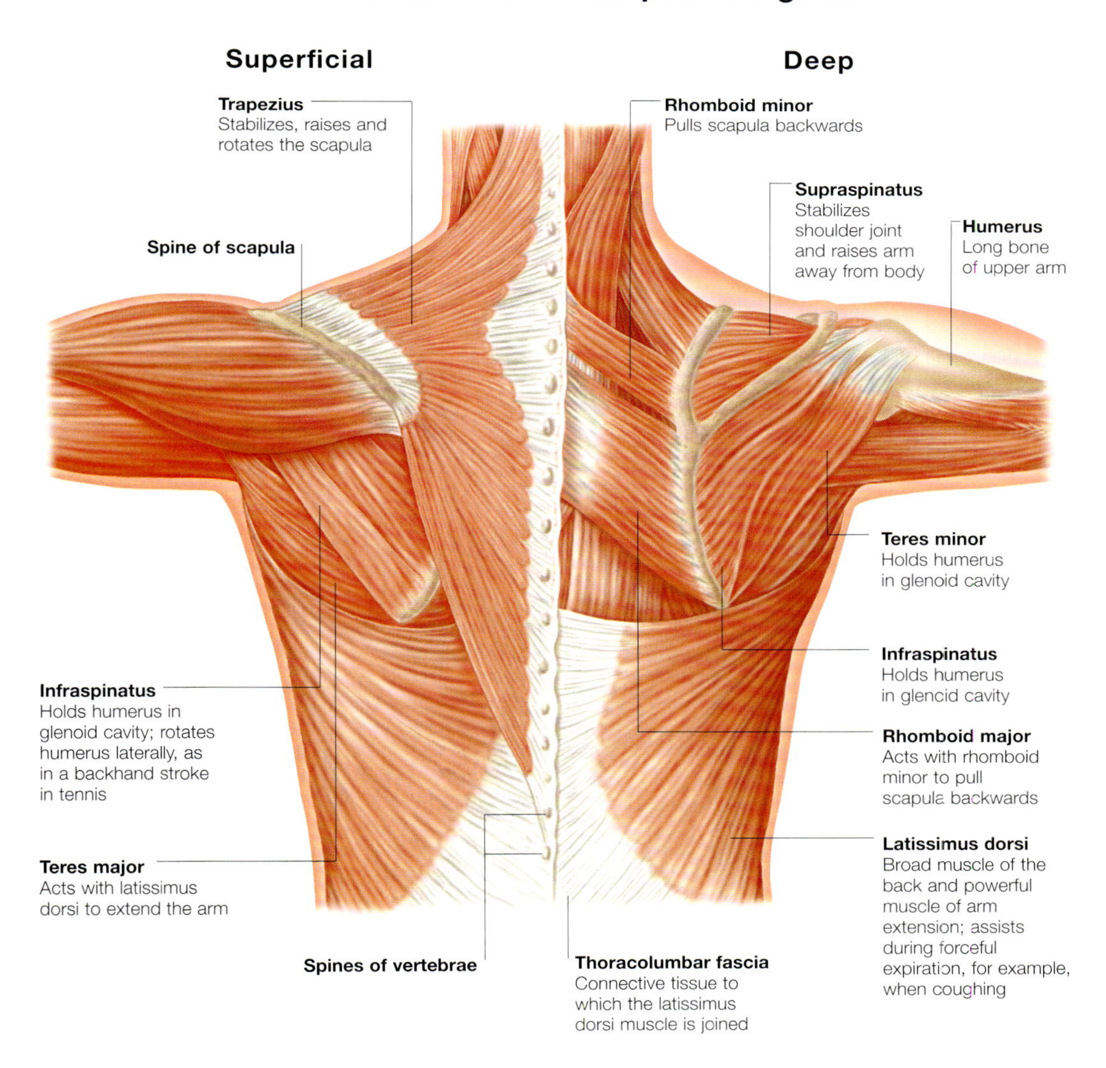

Shoulder Joint

The glenohumeral (shoulder joint) is a ball-and-socket joint at the point of articulation of the humerus and the scapula. The construction of this joint allows the arm a wide range of movement.

The glenohumeral, or shoulder joint, is the point of articulation between the glenoid cavity of the scapula and the head of the humerus. It is a ball-and-socket synovial joint that allows the upper limb a wide range of movement.

Articular Surface

To permit a wide range of movement, the head of the humerus provides a large articular surface. The glenoid cavity of the scapula, deepened by a ring of tough fibrocartilage (the glenoid labrum), offers only a shallow socket. Consequently, the joint needs to be held firmly together by the surrounding muscles and ligaments. A thin layer of smooth articular cartilage allows the bones to slip over each other with minimum friction.

Joint Capsule

The joint is surrounded by a loose capsule of fibrous tissue. This capsule is lined by a synovial membrane that covers all the inner surfaces of the joint except those covered with cartilage. The cells of the synovial membrane secrete synovial fluid, that lubricates and nourishes the joint.

Anterior shoulder joint

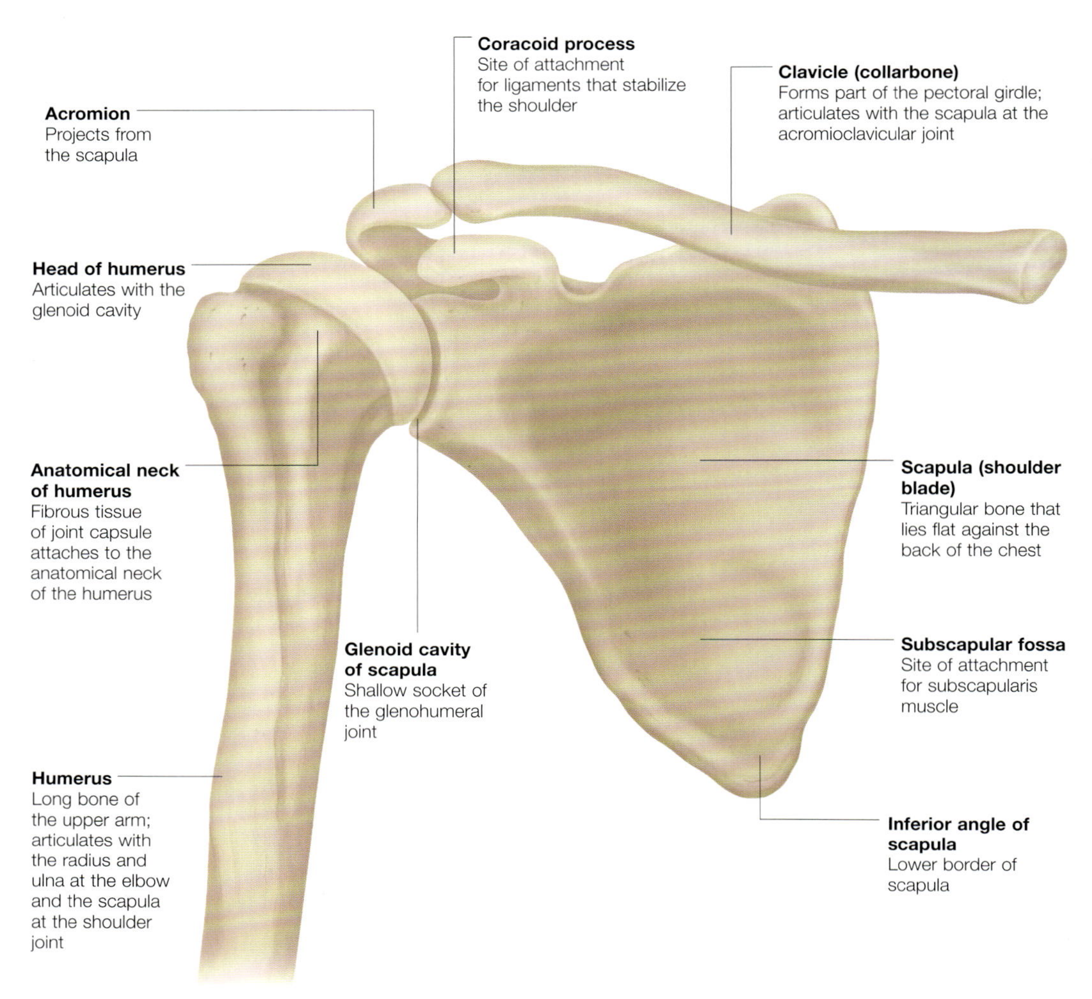

Bursae of the shoulder joint

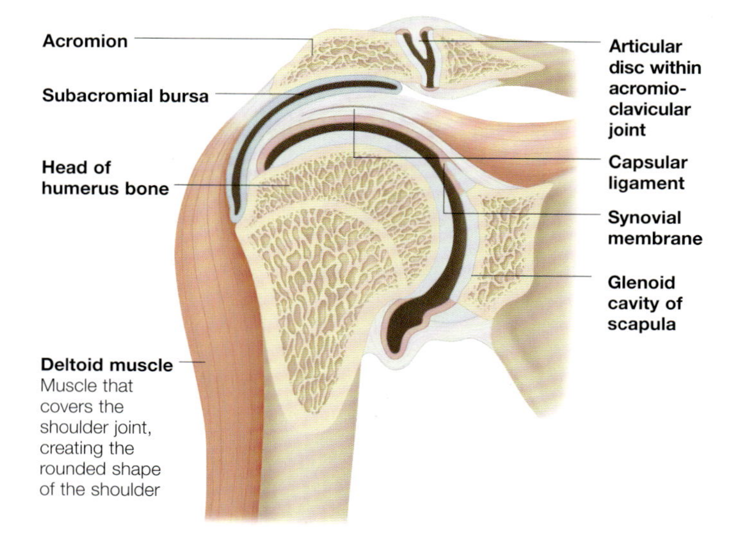

A bursa (*plural*: bursae) is a flattened fibrous sac lined with synovial membrane that contains a small amount of viscous synovial fluid. Bursae act to reduce the friction between structures that necessarily rub against each other during normal movement. Bursae are located at various points around the body where ligaments, muscle and tendons rub against bone.

Bursae may become abnormally enlarged at a point of unusual pressure, such as occurs when a bunion develops at the base of the big toe where a shoe rubs.

A coronal section through the glenohumeral joint shows the position of the bursa of the shoulder joint.

The shoulder joint has several important bursae:

- **Subscapular bursa (not shown)**
This protects the tendon of the subscapularis muscle as it passes over the neck of the scapula. It usually has an opening, which leads into the joint cavity and so may actually be thought of as an outpouching of that cavity.
- **Subacromial bursa**
This lies above the glenohumeral joint beneath the acromion and the coracoacromial ligament. It allows free movement of the muscles that pass beneath it. It is usually a true bursa, with no connection to the joint cavity of the shoulder.

Ligaments of the shoulder joint

The ligaments of the shoulder joint, along with the surrounding muscles, are crucial for the stability of this shallow ball-and-socket joint.

The ligaments around any joint contribute to its stability by holding the bones firmly together. In the shoulder joint, the main stabilizers are the surrounding muscles, but ligaments also play a role.

Stabilizing ligaments

The fibrous joint capsule has ligaments within it that help to strengthen the joint:

- The glenohumeral ligaments are three weak, fibrous bands that reinforce the front of the capsule
- The coracohumeral ligament is a strong, broad band that strengthens the upper aspect of the capsule. Although not actually part of the glenohumeral joint itself, the coracoacromial ligament is important as it spans the gap between the acromion and the coracoid process of the scapula. The arch of bone and ligament is so strong that even if the humerus is forcibly pushed up, it will not break; the clavicle or the humerus will give way first.
- The transverse humeral ligament runs from the greater to the lesser tuberosity of the humerus, creating a tunnel for the passage of the biceps brachii tendon within its synovial sheath.

Posterior shoulder joint

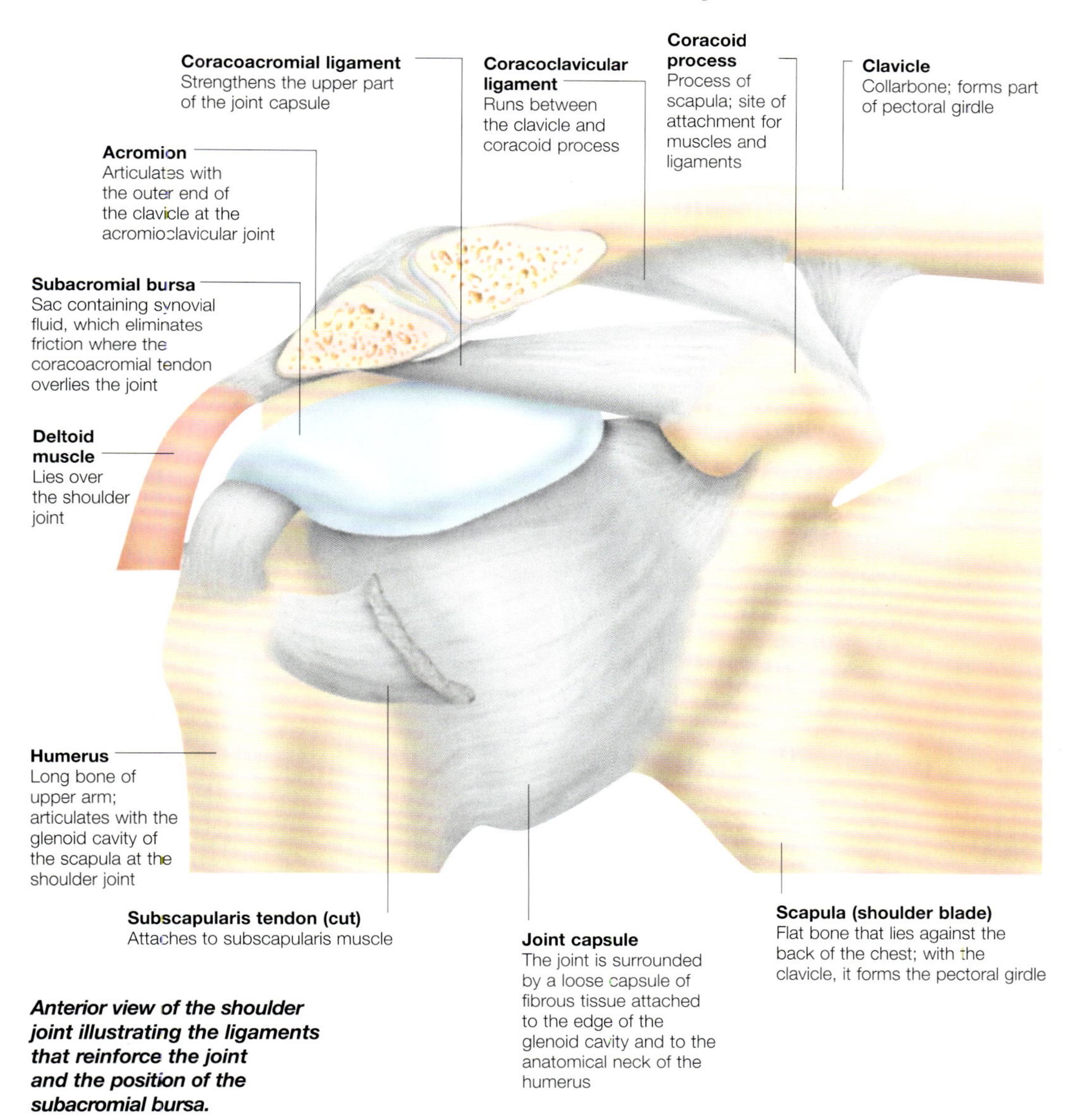

Anterior view of the shoulder joint illustrating the ligaments that reinforce the joint and the position of the subacromial bursa.

Movements of the Shoulder Joint

The shoulder joint is a ball-and-socket joint that allows 360 degrees of movement to give maximum flexibility. In addition to enabling these movements, the muscles of the pectoral girdle add stability.

The movements of the shoulder joint take place around three axes: a horizontal axis through the centre of the glenoid fossa; a second axis perpendicular to this (front-back) through the humeral head; and a third running vertically through the shaft of the humerus. These give the axes of flexion and extension, adduction and abduction, and medial (internal) and lateral (external) rotation respectively. A combination of these movements can allow the circular motion of the limb called circumduction.

Muscles of shoulder movement
Many of the muscles involved in these movements are attached to the pectoral girdle. The scapula has muscles attached to its rear and front surfaces and the coracoid process, a bony projection. Some muscles arise directly from the trunk (pectoralis major and latissimus dorsi). Other muscles influence the movement of the humerus even though they are not attached to it directly (such as trapezius). They do this by moving the scapula, and hence the shoulder joint.

Anterior view of muscles of the shoulder

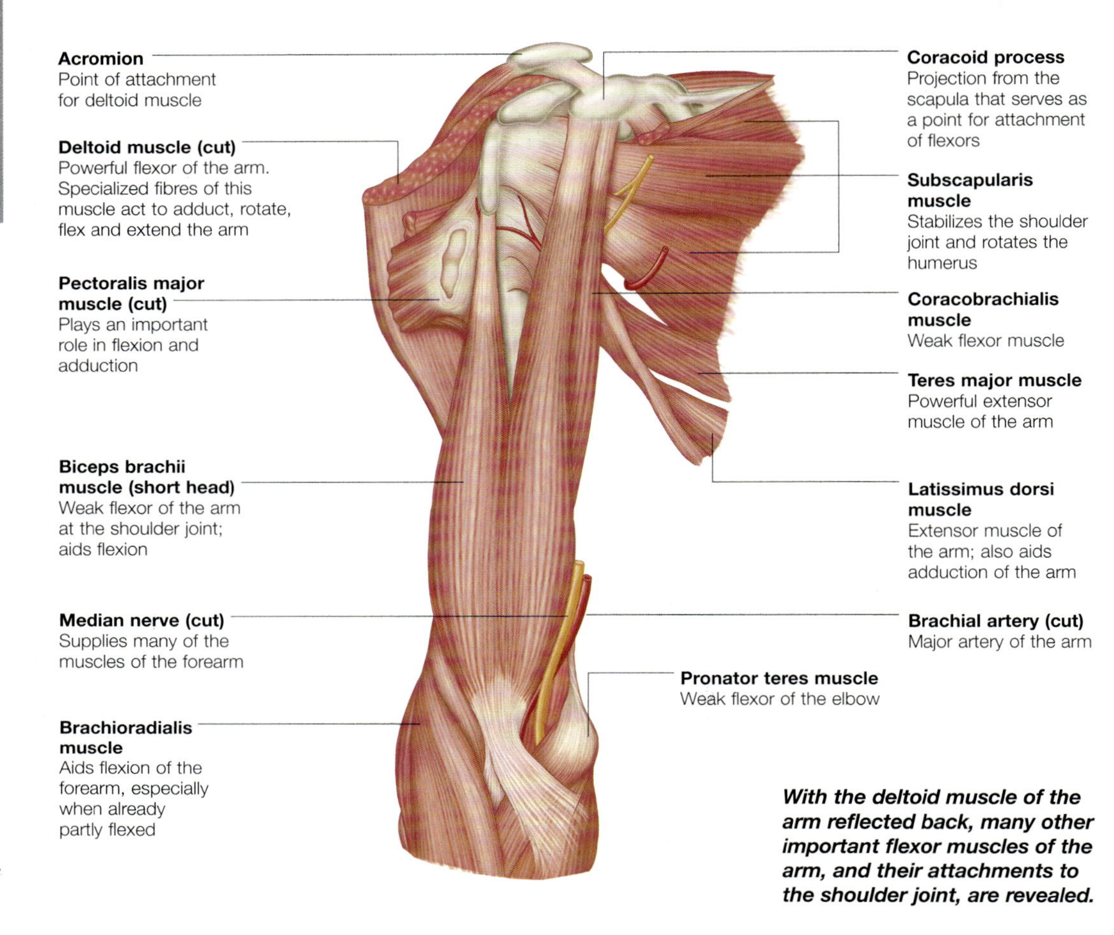

With the deltoid muscle of the arm reflected back, many other important flexor muscles of the arm, and their attachments to the shoulder joint, are revealed.

Movements of the shoulder joint

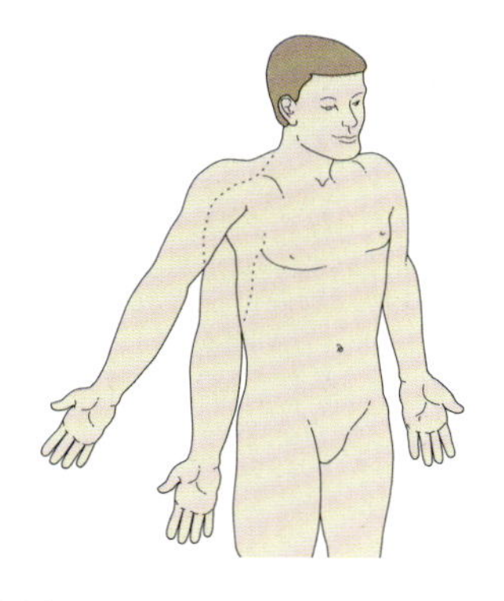

Adduction (towards the body) of the arm is brought about by pectoralis major and latissimus dorsi muscles; abduction (away from the body) by supraspinatus and the deltoid.

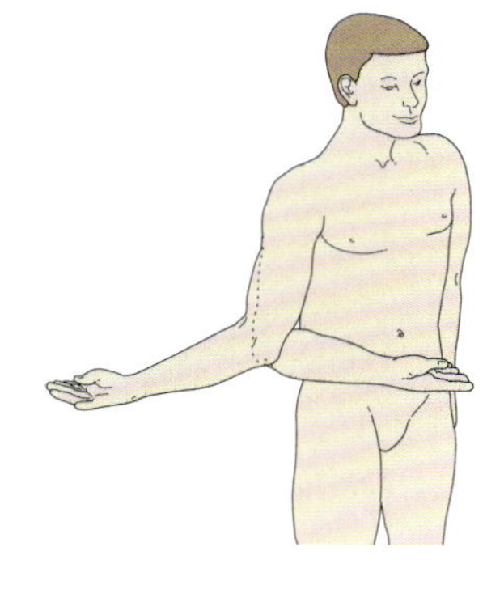

Muscles that bring about lateral rotation are infraspinatus, teres minor and the posterior fibres of the deltoid muscle. Medial rotators form the 'rotator cuff' group of muscles.

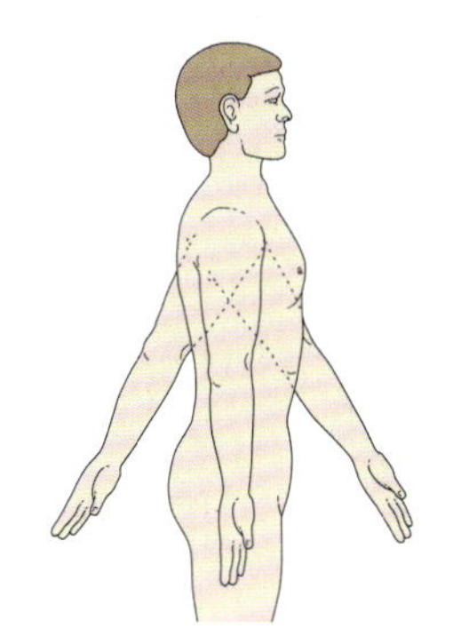

Flexion (forward movement) is due to biceps, coracobrachialis, deltoid and pectoralis major. Extension (backward movement) is due to rear fibres of deltoid, latissimus dorsi and teres major.

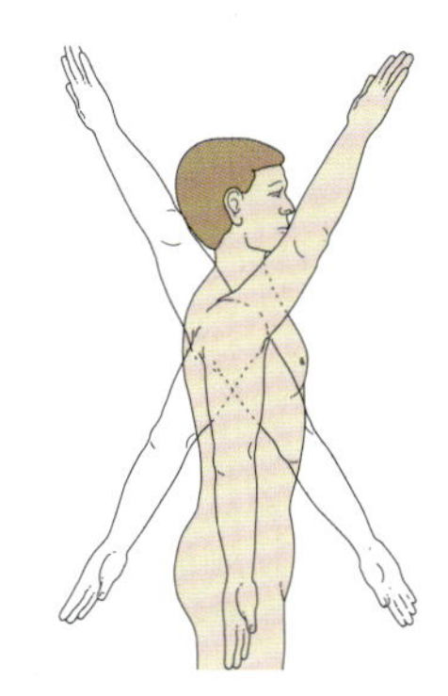

Circumduction is a combination of these movements. It is dependent on the clavicle holding the shoulder joint in the glenoid cavity and contractions of different muscle groups.

Rotation of the arm and the 'rotator cuff'

The shoulder muscles work together to rotate the arm in a circular motion. They also act individually to move the humerus and upper arm.

The arm can move in a circular movement, or 'rotate', using a number of shoulder muscles. The pectoralis major, anterior fibres of deltoid, teres major and latissimus dorsi muscles all cause medial rotation of the humerus. The most powerful medial rotator, however, is subscapularis. This muscle occupies the entire front surface of the scapula, and attaches to the joint capsule around the lesser tuberosity of the humerus.

Rotator cuff

Subscapularis is one of four short muscles known as the 'rotator cuff', which attach to and strengthen the joint capsule. They also pull the humerus into the socket of the joint (glenoid fossa), increasing contact of the bony elements. This is the key factor contributing to the stability of the joint. Injury to the rotator cuff muscles is disabling as the stability of the humerus in the joint is lost.

Muscles of shoulder movement (front)

Muscles of shoulder movement (back)

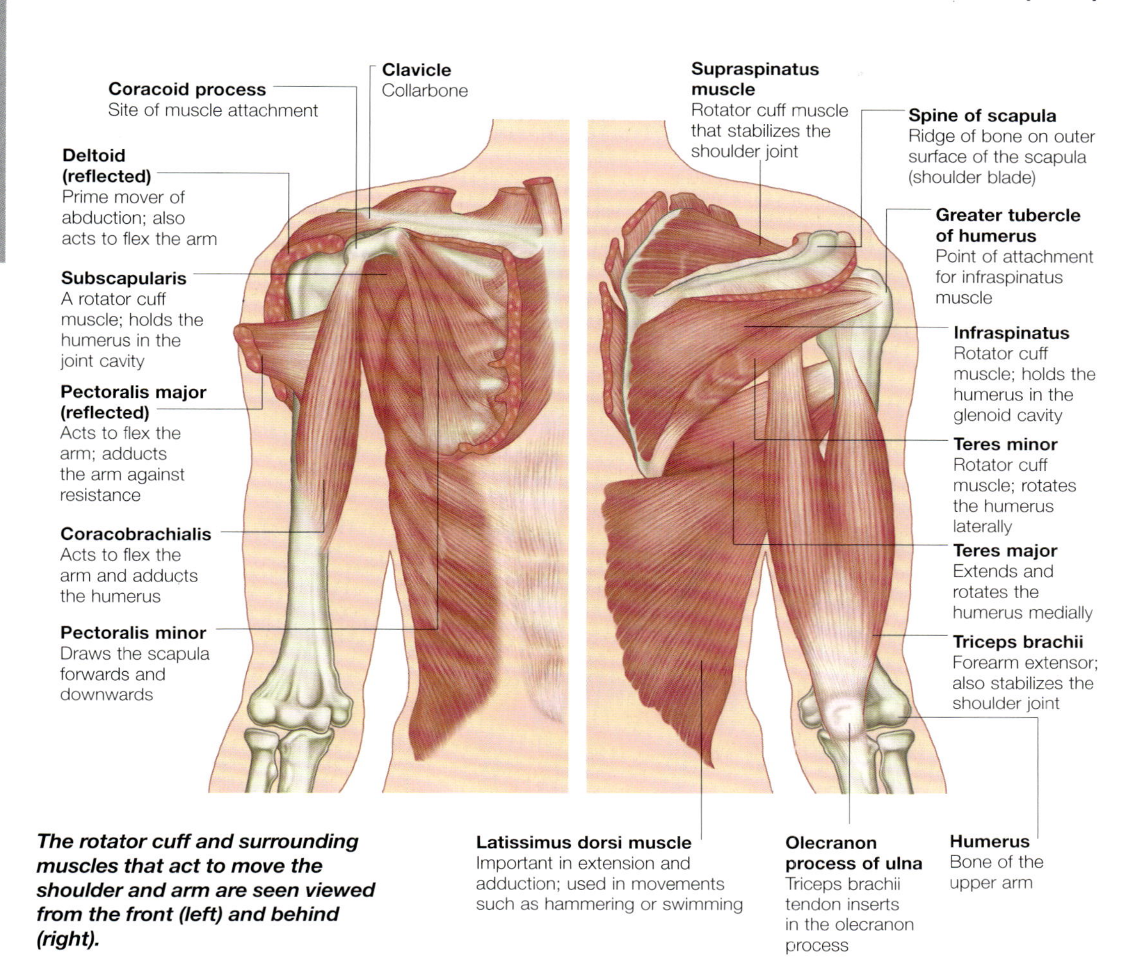

The rotator cuff and surrounding muscles that act to move the shoulder and arm are seen viewed from the front (left) and behind (right).

Abduction of the arm

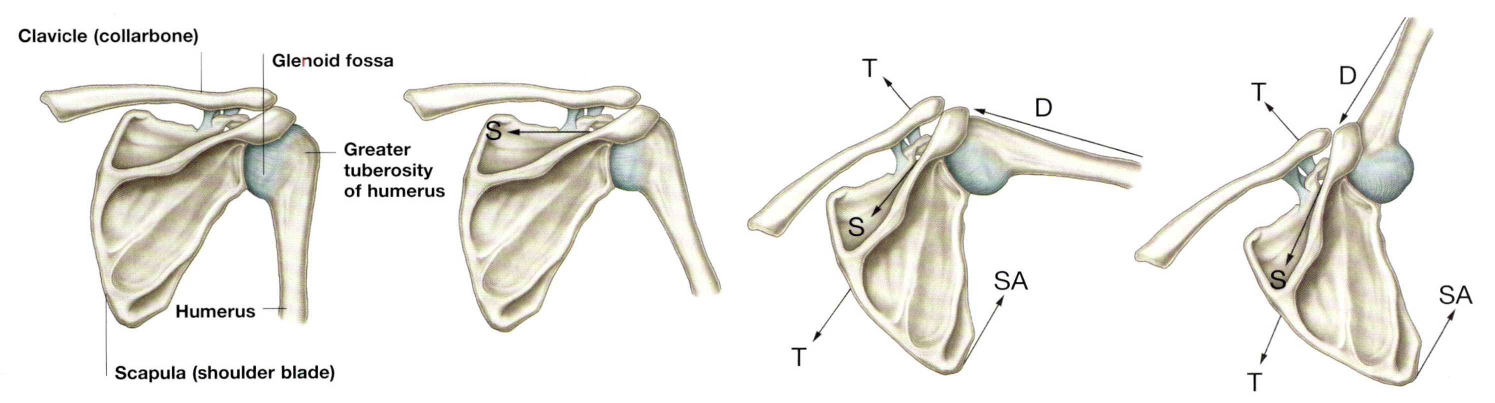

Abduction (movement away from the body) is a weak action performed by supraspinatus (S) and the acromial (middle) part of deltoid (D) muscle.

KEY		
S	=	Supraspinatus
D	=	Deltoid
T	=	Trapezius
SA	=	Serratus anterior

From the resting position, the deltoid muscle can pull the humerus upwards but not outwards. The supraspinatus muscle is in a good mechanical position to initiate abduction. Once the motion has started, the deltoid muscle takes over and continues the movement.

An obstacle to full abduction is the bony contact of the greater tuberosity of the humerus and the acromion process of the scapula. This bony contact would prevent raising of the arm higher than horizontal (for example, arms held straight outwards at shoulder height).

Rotating the scapula using the trapezius muscle (T) allows the hands to be raised above the head. The scapula rotates so that the glenoid fossa points upwards, taking the acromion process with it. The humerus rotates so that articular contact of the joint is maintained.

Structure of the Humerus

The humerus, a typical 'long bone', is found in the upper arm. It has a long shaft with expanded ends that connect with the scapula at the shoulder joint and the radius and ulna at the elbow.

At the top of the humerus (the proximal end) lies the smooth, hemispherical head that fits into the glenoid cavity of the scapula at the shoulder joint. Behind the head is a shallow constriction known as the 'anatomical neck' of the humerus, which separates the head from two bony prominences, the greater and the lesser tuberosities. These are sites for muscle attachment and are separated by the intertubercular (or bicipital) groove.

The shaft

At the upper end of the shaft is the slightly narrowed 'surgical neck' of the humerus – a common site for fractures. The relatively smooth shaft has two distinctive features. About half way down the shaft, on the lateral (outer) side, lies the deltoid tuberosity, a raised site of attachment of the deltoid muscle. The second feature is the radial (or spiral) groove that runs across the back of the middle part of the shaft. This depression marks the path of the radial nerve and the profunda brachii artery.

Ridges at each side of the lower shaft pass down to end in the prominent medial (inner) and lateral epicondyles. There are two main parts to the articular surface: the trochlea, which articulates with the ulna; and the capitulum, which articulates with the radius.

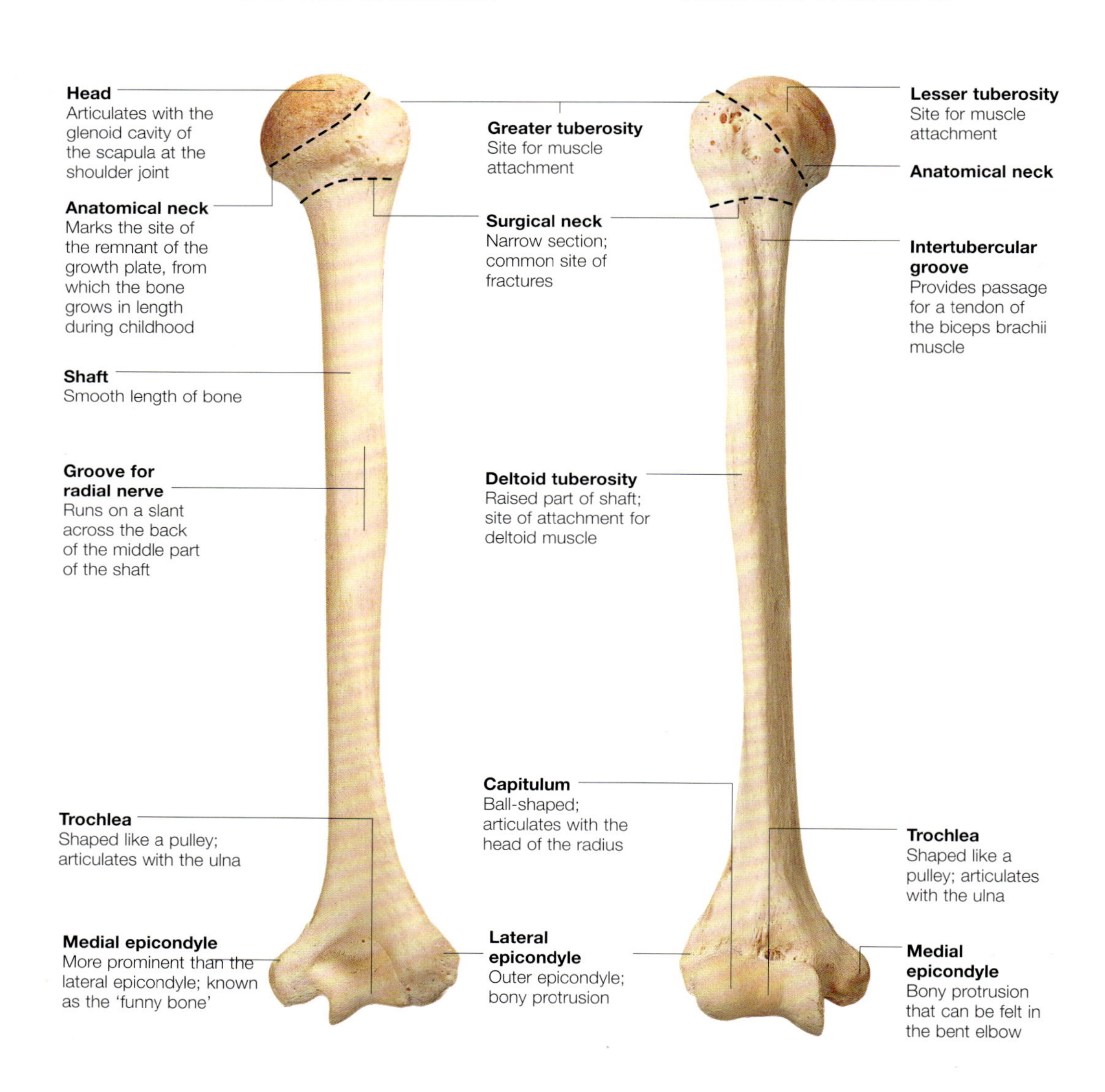

Inside the humerus

The structure of the humerus is typical of the long bones. The bone is divided into the diaphysis (shaft) and the epiphysis (head) at either end.

Long bones are elongated in shape and longer than they are wide. Most of the bones of the limbs are long bones, even the small bones of the fingers, and as such they have many features in common with the humerus.

The humerus consists of a diaphysis, or shaft, with an epiphysis (expanded head) at each end. The diaphysis is of tubular construction with an outer layer of dense, thick bone surrounding a central medulla (inner region) containing fat cells. The epiphyses of the humerus are, at the upper end, the head and at the lower end the condylar region. These are composed of a thin layer of compact bone covering cancellous (spongy) bone, which makes up the greater volume.

Bone surface

The surface of the humerus (and all long bones) is covered by a thick membrane, the periosteum. The articular surfaces at the joints are the only parts of the bone not covered by the periosteum. These surfaces are covered by tough articular (or hyaline) cartilage that is smooth, allowing the bones to glide over each other.

The outer compact bone receives its blood supply from the arteries of the periosteum, and will die if that periosteum is stripped off, while the inner parts of the bone are supplied by occasional nutrient arteries that pierce the compact bone.

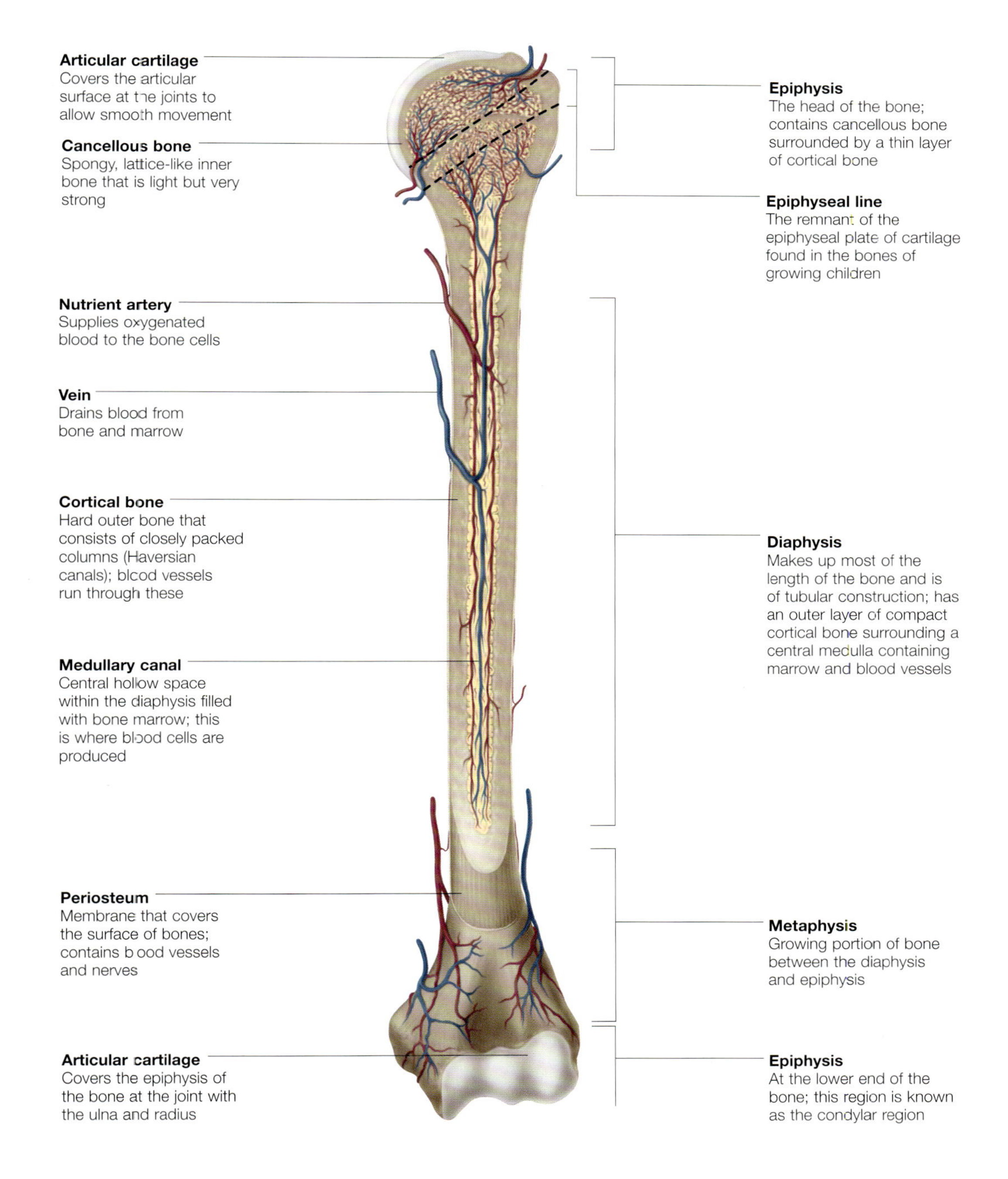

The Ulna

The ulna and the radius are the long bones of the forearm. They articulate with the humerus and the wrist bones and are uniquely adapted to enable rotation of the hand and forearm.

The ulna and radius are the two parallel long bones of the forearm and lie between the elbow and wrist joints. The ulna lies on the same side as the little finger (medially), while the radius lies on the same side as the thumb (laterally). The radio-ulnar joints allow the ulna and radius to rotate around each other in the movements peculiar to the forearm known as 'pronation' (rotating the forearm so that the palm faces down), and 'supination' (rotating the forearm so that the palm faces up).

The Ulna

The ulna is longer than the radius and is the main stabilizing bone of the forearm with a long shaft and two expanded ends. The upper end has two prominent projections, the olecranon and the coronoid process. These are separated by the deep trochlear notch, which articulates with the trochlea of the humerus.

On the outer side of the coronoid process, there is a small, rounded recess (the radial notch), which is the site of articulation of the upper end of the ulna with the neighbouring head of the radius. The head of the ulna is separated from the wrist joint by an articular disc and does not play much part in the wrist joint itself.

Interosseus Membrane

Formed of tough connective tissue, the interosseus membrane binds the radius and ulna together. The membrane is broad so allows movement between the bones and provides sites of attachment for some of the deep muscles of the forearm.

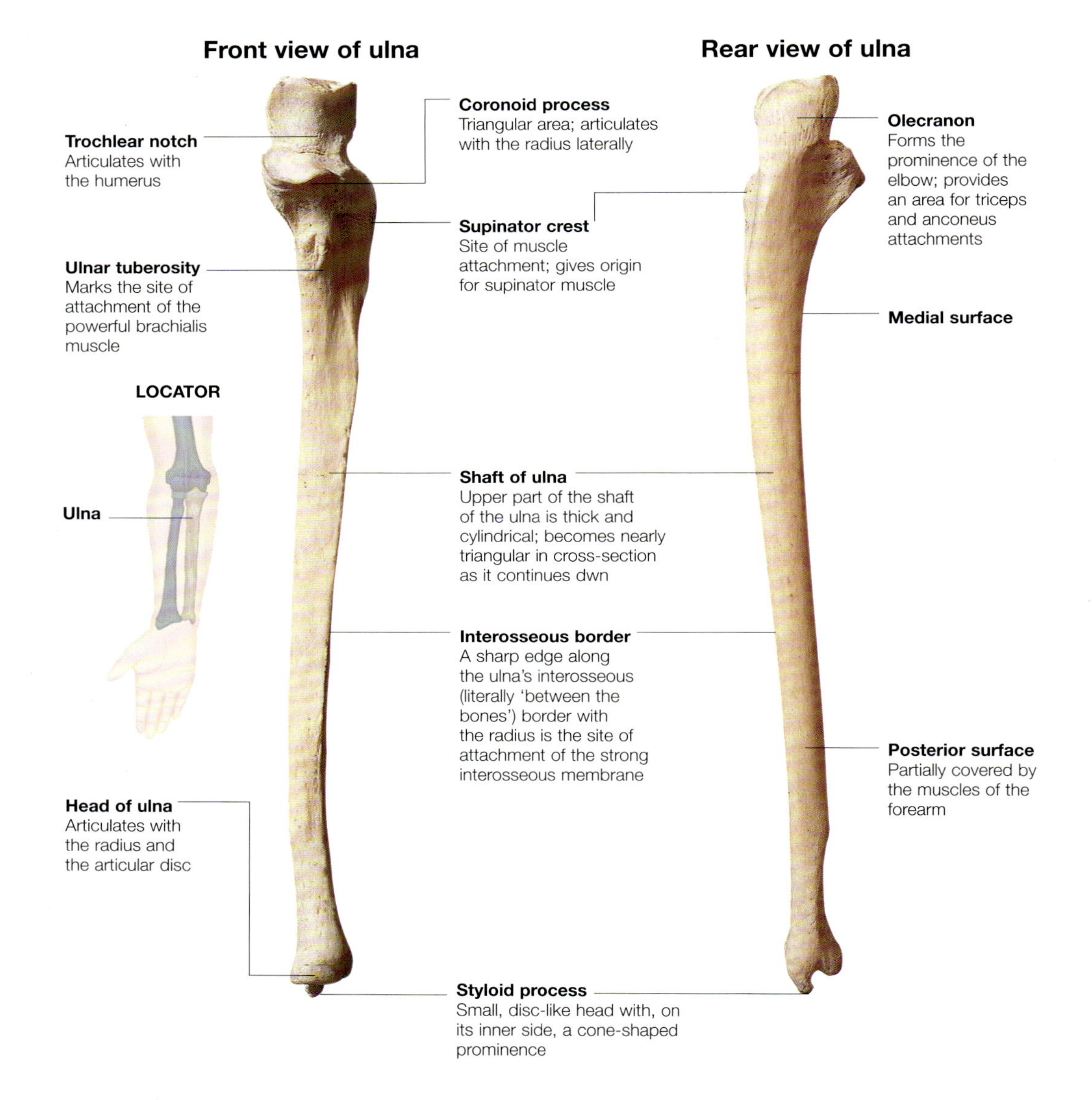

The Radius

The radius is the shorter of the two bones of the forearm and articulates with the wrist. It is joined firmly to the ulna by a tough layer of connective tissue.

Like the ulna, the radius has a long shaft with upper and lower expanded ends. While the ulna is the forearm bone that contributes most to the elbow, the radius plays a major part in the wrist joint.

Head of the Radius

The disc-like head of the radius is concave above where it articulates with the capitulum of the humerus in the elbow joint. The cartilage that covers this concavity continues down over the head, especially on the side of the ulna, to allow the smooth articulation of the head of the radius with the radial notch at the upper end of the ulna.

The Shaft

The shaft of the radius becomes progressively thicker as it continues down to the wrist. It also has a sharp edge for attachment of the interosseous membrane. On the inner side, next to the ulna, there is a concavity (the ulnar notch), which is the site for articulation with the head of the ulna.

Extending from the opposite side is the radial styloid process, a blunt cone that projects a little further down than the ulnar styloid process. At the back of the end of the radius, and easily felt at the back of the wrist, is the dorsal tubercle.

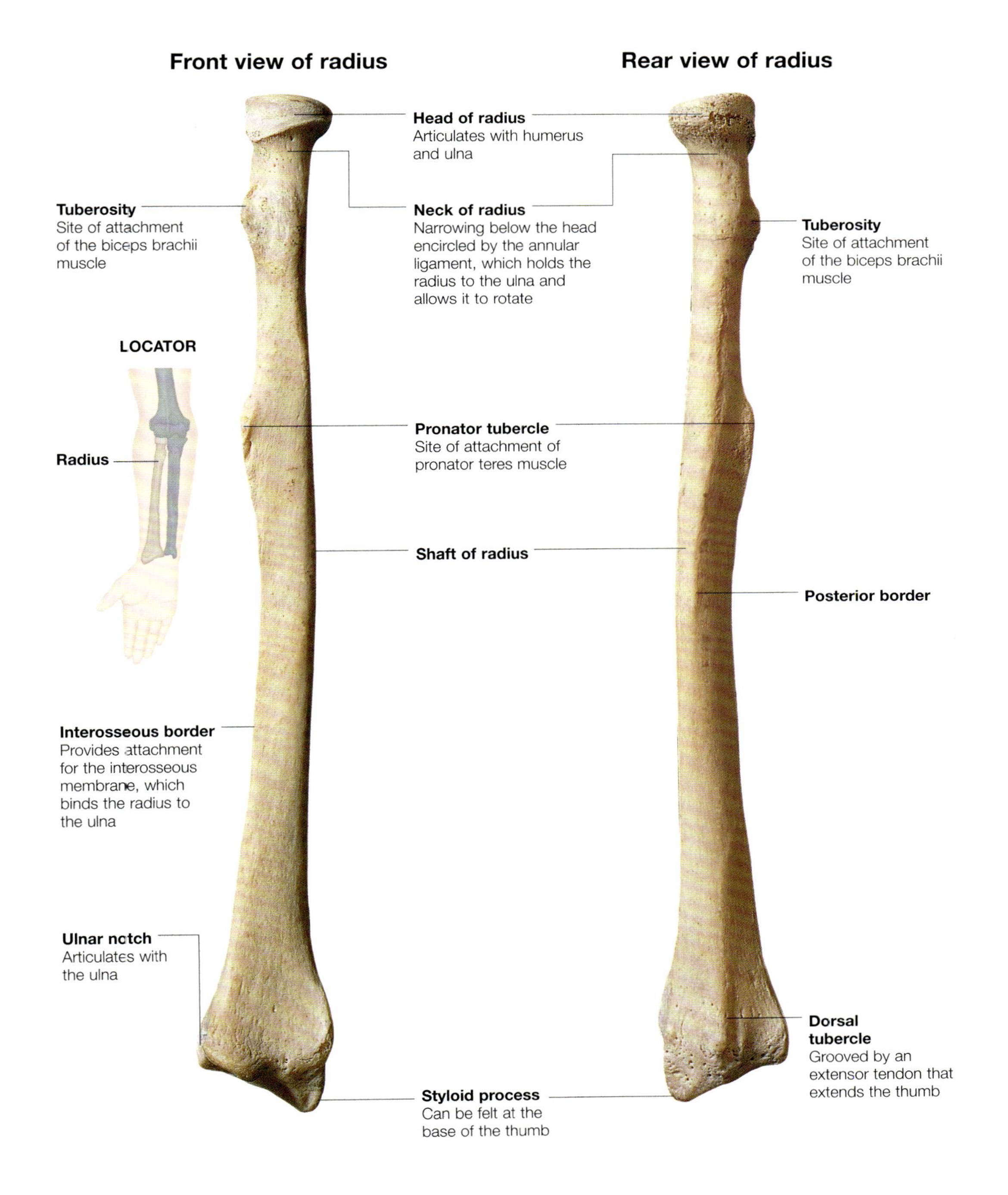

Elbow

The elbow is the fluid-filled joint at which the humerus of the upper arm and the radius and ulna of the forearm articulate. This joint structure only allows hinge-like movement but is extremely stable.

The elbow is a synovial joint between the lower end of the humerus and the upper ends of the ulna and radius. It is a 'hinge' joint, where the only movements possible are flexion and extension (bending and straightening).

Structure

The pulley-shaped trochlea of the lower end of the humerus articulates with the deep trochlear notch of the ulna, while its hemispherical capitulum articulates with the head of the radius. All the opposing joint surfaces are covered by smooth articular cartilage (hyaline) to reduce friction between bony surfaces during movement. The whole joint is surrounded by a fibrous capsule that extends from the articular surfaces of the humerus to the upper end of the ulna. The capsule is loose at the back of the elbow to allow flexion and extension. The capsule is lined with synovial membrane that secretes thick synovial fluid filling the joint cavity. This fluid nourishes the joint and acts as a lubricant. The joint cavity is continuous with that of the superior radioulnar joint below.

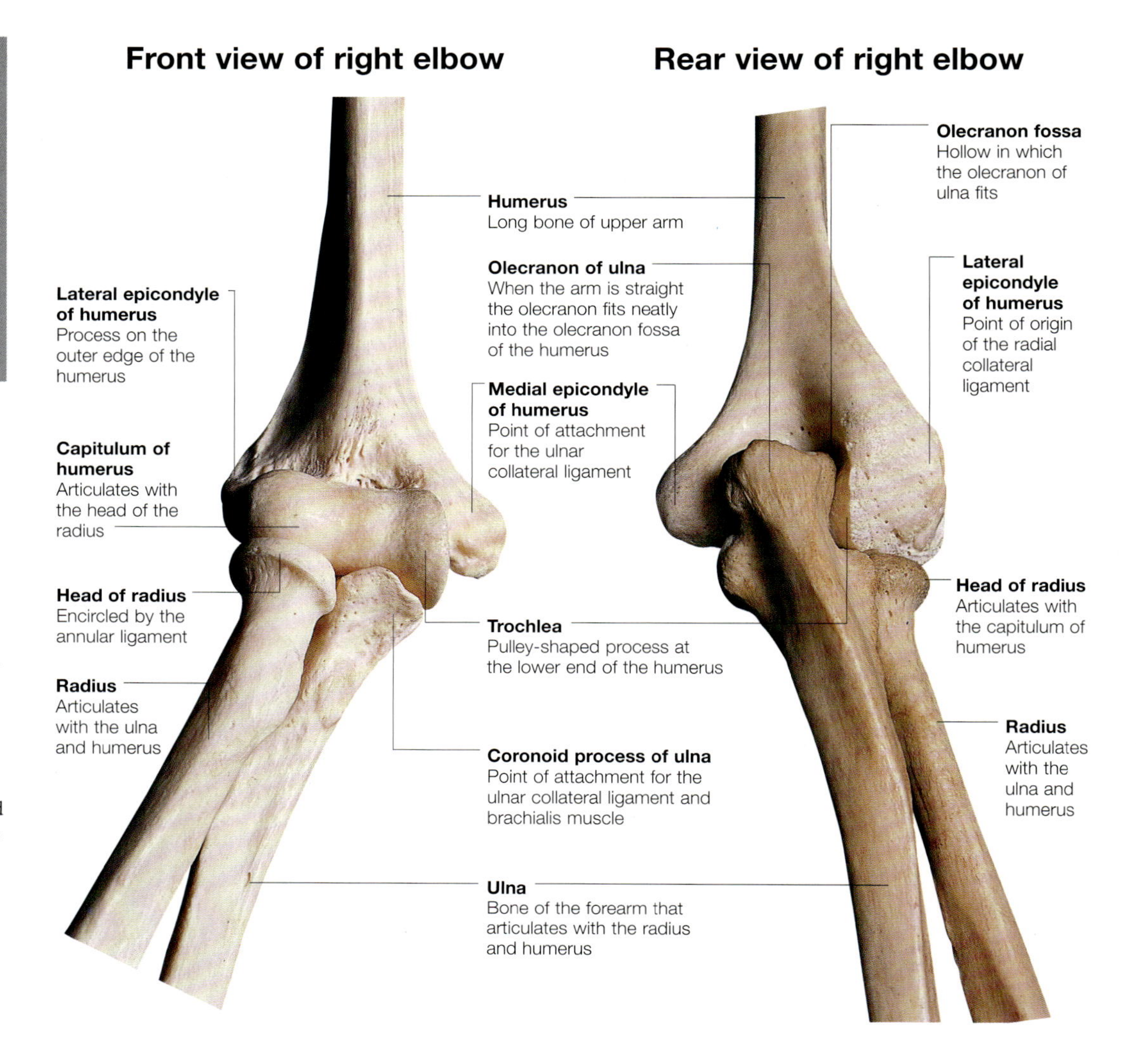

Stability and movement of the elbow

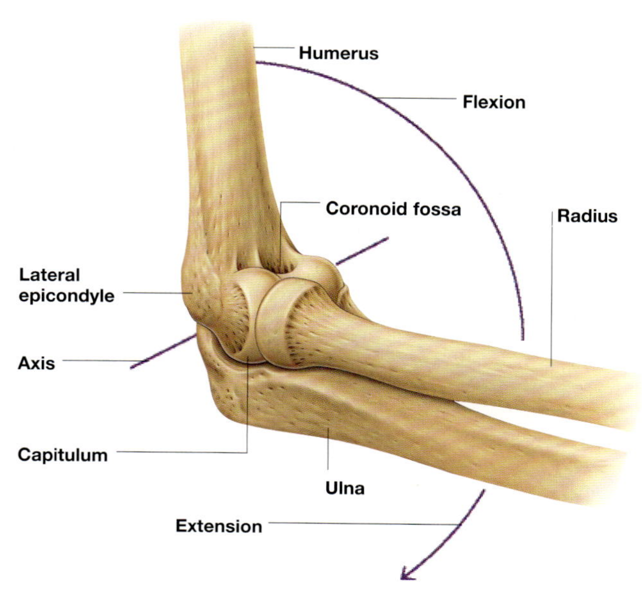

The main stability of the elbow comes from the size and depth of the trochlear notch of the ulna, which effectively grips the lower end of the humerus like a wrench. The depth of this bony notch is increased by the presence of a band of the medial collateral ligament. Because of the shape of this joint, and the presence of the strong collateral ligaments on each side, the elbow can move only as a hinge.

The elbow joint is capable of two movements: flexion (bending) and extension (straightening), as indicated by the purple arrow.

Flexion

Flexion (bending) of the elbow is achieved by contraction of the powerful muscles at the front of the upper arm such as brachialis and the well-known biceps brachii. The movement is limited at its fullest extent by the coming together of the forearm and the upper arm.

Extension

Extension (straightening) of the elbow is mainly achieved by contraction of the triceps muscle at the back of the upper arm, assisted by gravity. At full extension, with the arm straight, the olecranon of the ulna fits neatly into the olecranon fossa (hollow) of the lower end of the back of the humerus. This fitting together of the two bones prevents over-extension of the elbow, and so adds to its stability.

Ligaments of the elbow

The elbow is supported and strengthened at each side by the strong collateral ligaments. These are thickenings of the joint capsule.

The radial collateral ligament is a fan-shaped ligament that originates from the lateral epicondyle – a bony prominence on the outer side of the lower end of the humerus – and runs down to blend with the annular ligament, which encircles the head of the radius. It is not attached to the radius itself so does not restrict movement of the radius during pronation (when the forearm is rotated so that the palm faces down) and supination (when the forearm is rotated so that the palm faces up).

The ulnar collateral ligament runs between the medial (inner) epicondyle of the humerus and the upper end of the ulna. It is in three parts that form a rough triangle.

Carrying angle

When the arm is fully extended downwards with the palm facing forwards, the long axis of the forearm is not in line with the long axis of the upper arm, but deviates slightly outwards.

The angle so formed at the elbow is known as the 'carrying angle' and is greater in women than in men (by about 10 degrees), possibly to accommodate the wider hips of the female body. The carrying angle disappears when the forearm is pronated (turned so the palm faces in to the body).

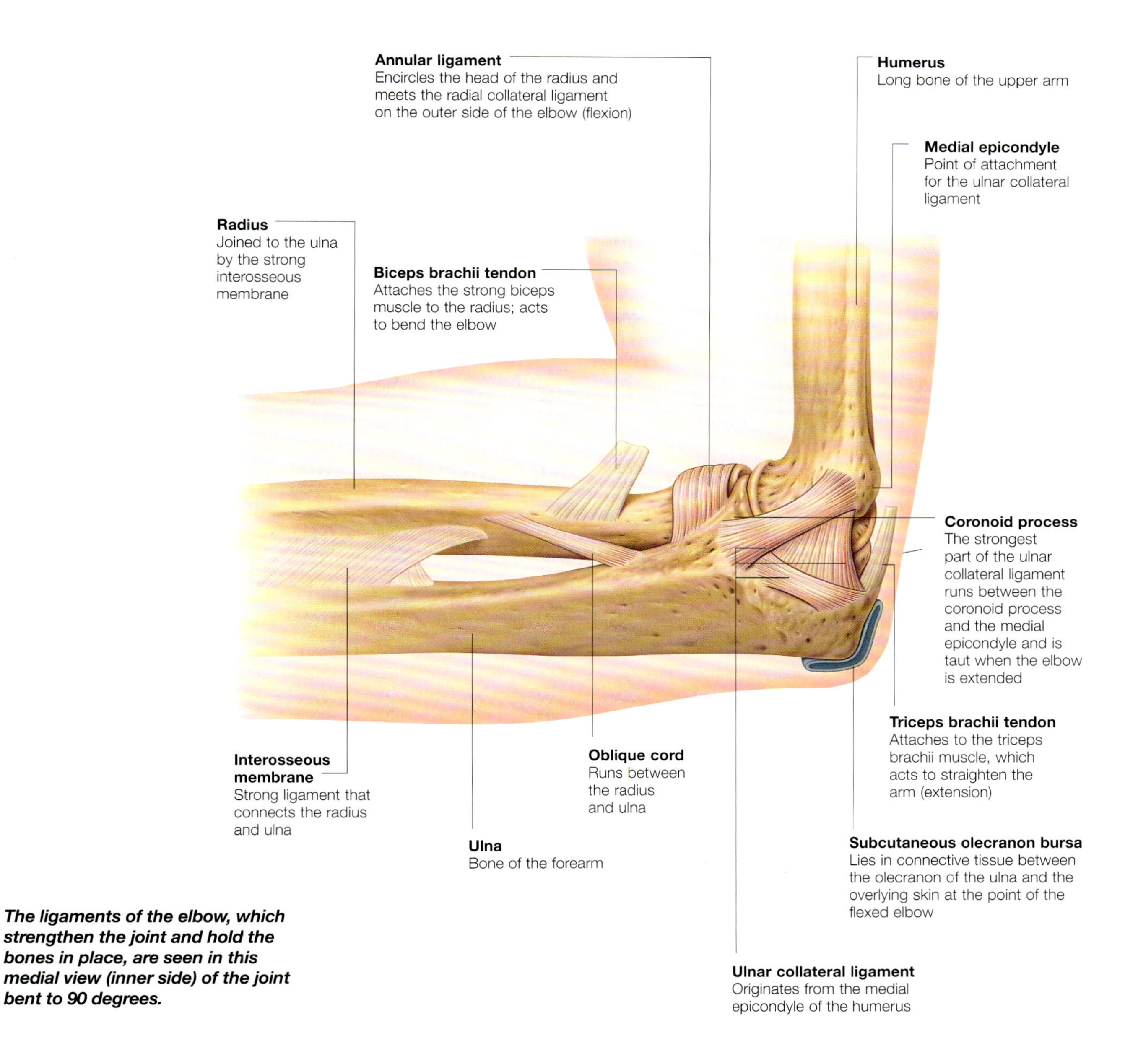

The ligaments of the elbow, which strengthen the joint and hold the bones in place, are seen in this medial view (inner side) of the joint bent to 90 degrees.

Forearm Movements

The movements of pronation (palm facing down or back) and supination (palm facing up or forwards) are peculiar to the movements of the forearm.

Pronation and supination are achieved by contraction of muscles that rotate the radius around the stationary ulna. The upper and lower radio-ulnar joints act as pivots.

The muscles responsible for the actions of pronation and supination generally lie deep to the other muscles of the forearm which are mainly concerned with movements of the wrist and fingers.

Muscles of pronation

The muscles that enable pronation include:

- Pronator teres. This muscle lies in front of the bones of the forearm and acts to pronate the forearm and flex (bend) the elbow. It arises from two heads that are attached to the coronoid process of the ulna and the medial epicondyle of the humerus (a bony prominence on the inner side of the elbow). It passes down and outwards to the middle of the outer edge of the radius where it can exert maximum leverage on the bone.
- Pronator quadratus. This is a small muscle that connects the front of the lower quarter of the ulna to the radius and acts to help the interosseous membrane tie the two bones together, as well as to pronate the forearm.

Muscles of supination

The muscles that enable supination include:

- Supinator. This is a deep muscle that lies behind the forearm bones just below the elbow. It arises from the outer edge of the elbow and from a raised line on the ulna, the supinator crest. It then passes down and around the outer edge of the upper radius to insert into the upper third of the shaft.
- Biceps brachii. This powerful muscle acts to supinate the forearm as well as its better known function of flexing the elbow. It is the muscle involved when powerful supination is needed, such as driving in a screw
- Brachioradialis. This is a muscle that runs from the outer part of the humerus just above the elbow down to the outer side of the lower end of the radius. It acts to pull the forearm back to the position of rest, midway between pronation and supination; if the forearm is fully pronated it will act as a supinator; if fully supinated, it will act as a pronator.

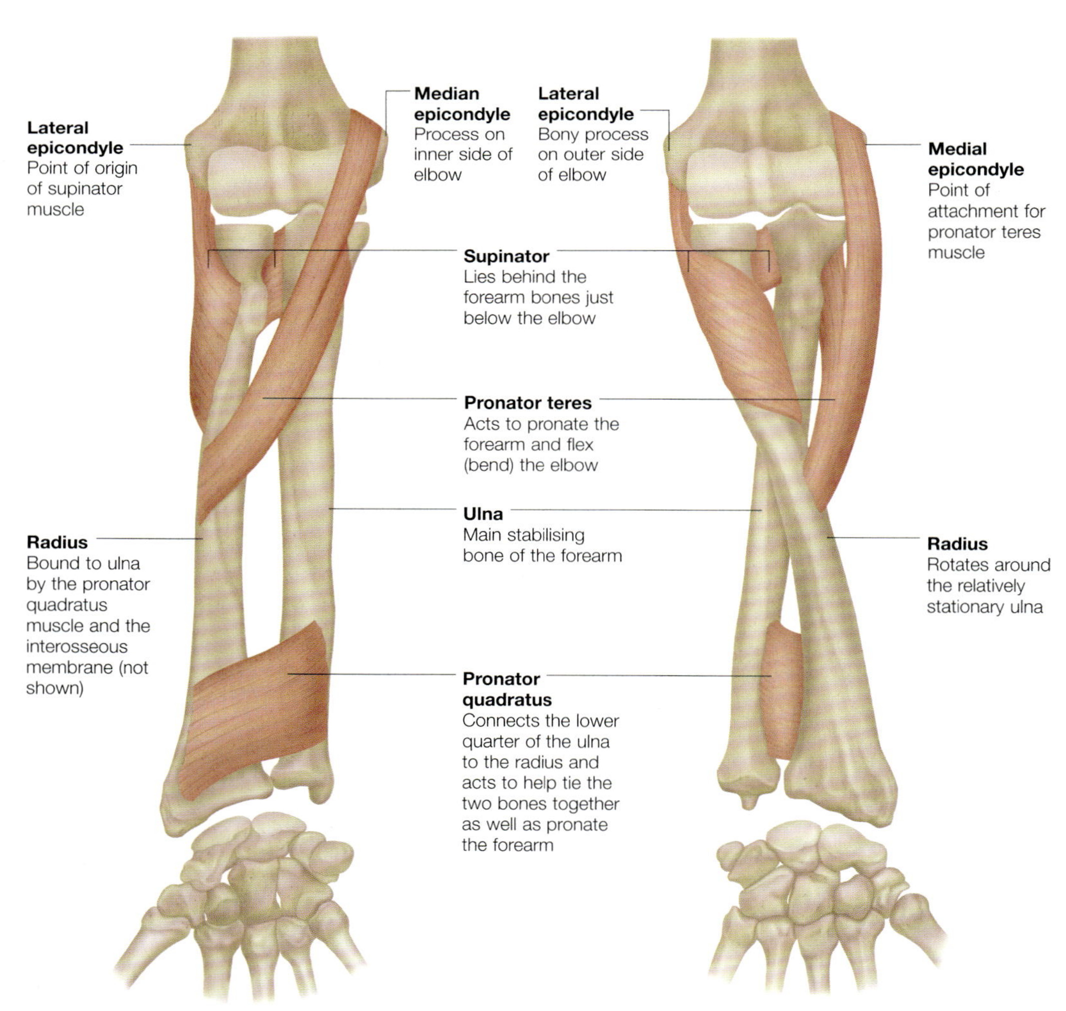

Radio-ulnar joints

As well as the articulations at the elbow and wrist, the radius and ulna articulate with each other at both their upper and lower ends.

The radio-ulnar joints at both ends of the forearm allow for pronation (the hand turning palm down) and supination (the hand turning palm up) as the radius rotates around the ulna.

Upper (Superior) Radio-ulnar Joint
This is a synovial joint that acts like a pivot. The head of the radius articulates with the radial notch of the upper end of the ulna and is held in place by a strong annular ligament.

The whole joint is surrounded and supported by the fibrous joint capsule (lined by synovial membrane, which produces the lubricating synovial fluid). During pronation and supination, the head of the radius rotates within the circle formed by the annular ligament and the radial notch.

Lower (inferior) Radio-ulnar Joint
This is also a pivot joint where the lower end of the radius rotates around the relatively stationary head of the ulna. The rounded head of the ulna articulates with the ulnar notch on the inner side of the lower end of the radius. The bones are bound together by the 'triangular ligament', which is a tough fibrocartilaginous disc separating this joint from the joint cavity of the wrist.

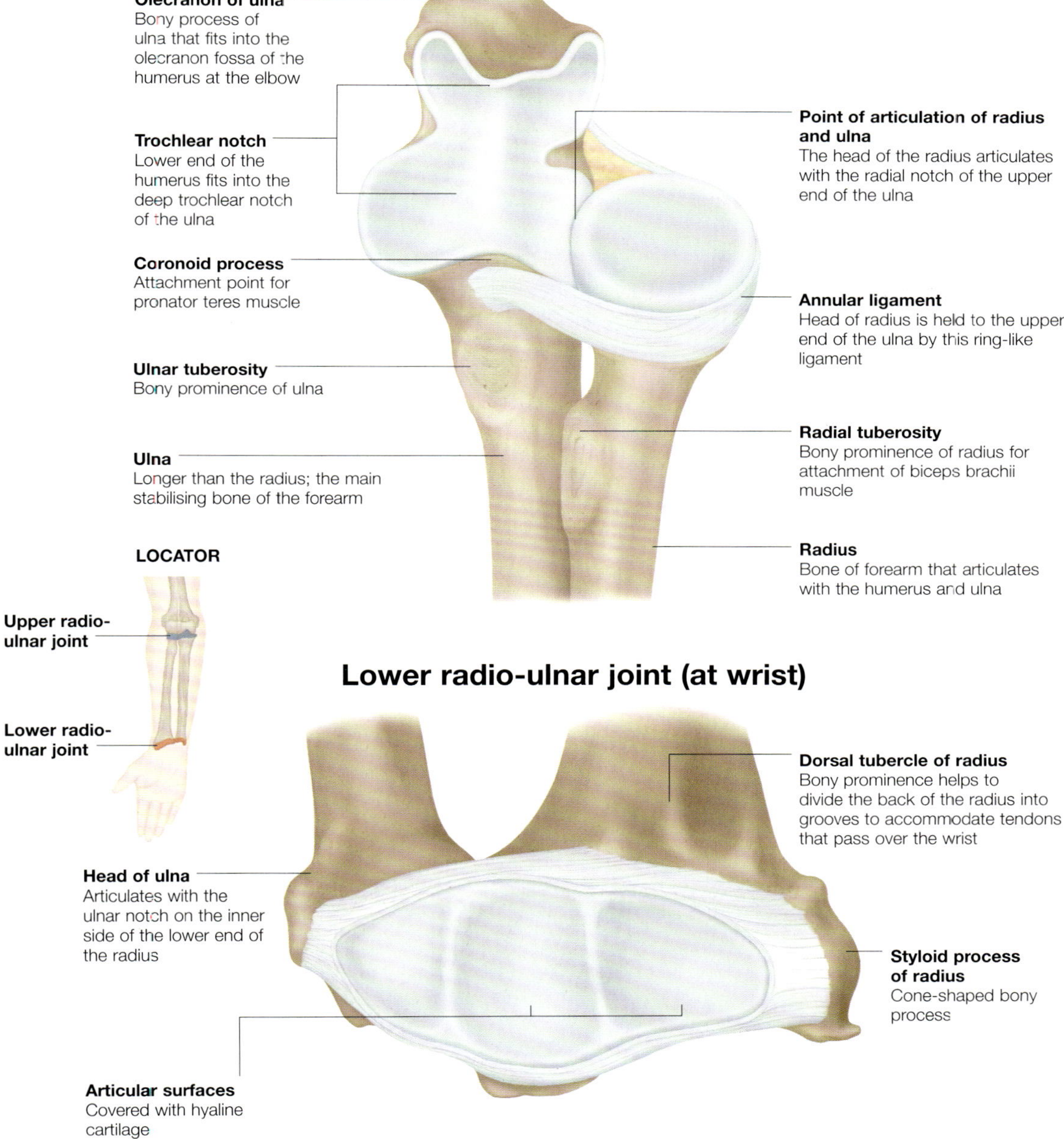

Muscles of the Upper Arm

The musculature of the upper arm is divided into two distinct compartments. The muscles of the anterior compartment act to flex the arm and the muscles of the posterior compartment extend it.

The muscles of the anterior (front) compartment of the upper arm are all flexors, that is they bend the arm:

- **Biceps brachii.** Arises from two heads that join together to form the body of the muscle. The bulging body then tapers as it runs down to form the strong tendon of insertion. When the elbow is straight, the biceps acts to flex the forearm. However, when the elbow is already bent the biceps muscle is a powerful supinator of the forearm, rotating the forearm so that the hand is palm up.
- **Brachialis**. Arises from the lower half of the anterior surface of the humerus and passes down to cover the front of the elbow joint, its tendon inserting into the coronoid process and tuberosity of the ulna. Brachialis is the main flexor muscle of the elbow, whatever the position of the forearm.
- **Coracobrachialis**. Arises from the tip of the coracoid process of the scapula and runs down and outwards to insert into the inner surface of the humerus. This muscle helps to flex the upper arm at the shoulder and to pull it back into line with the body (adduction).

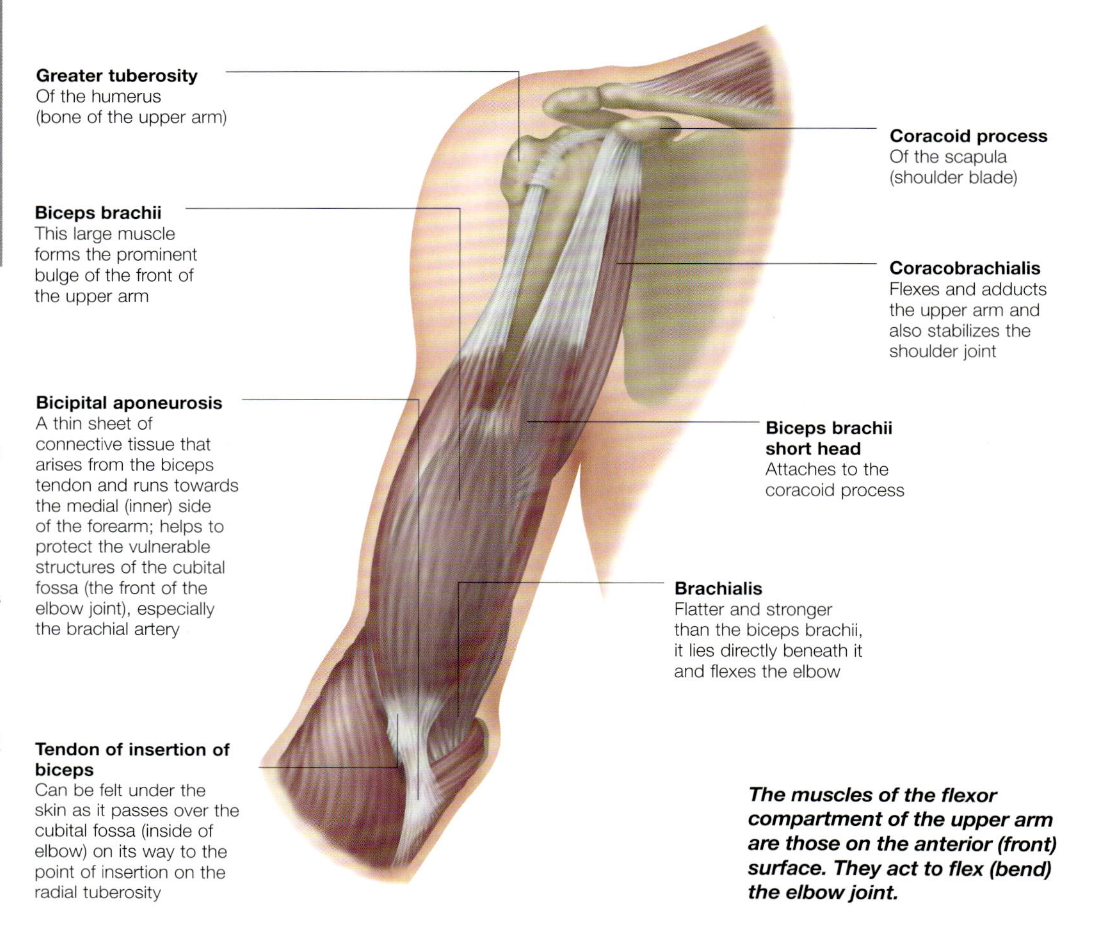

The muscles of the flexor compartment of the upper arm are those on the anterior (front) surface. They act to flex (bend) the elbow joint.

The long head of biceps

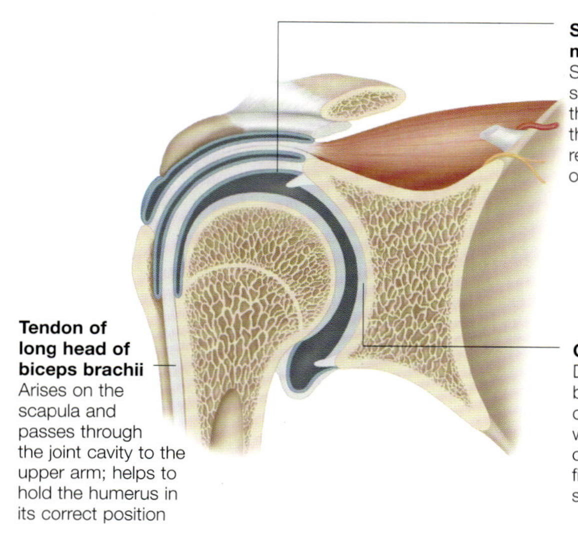

The biceps brachialis has two heads (the long head and the short head) that join together to form the large muscle. The short head originates at the apex of the coracoid process of the scapula. The long head of biceps brachii arises from a point on the scapula just above the glenoid fossa. The rounded tendon crosses the head of the humerus actually within the cavity of the shoulder joint before emerging in the upper arm.

This section through the shoulder joint is angled to show the tendon of the long head of biceps. The name biceps means 'having two heads'.

Tendon of the long head
As it passes out of the shoulder joint cavity, the tendon of the long head runs in the 'bicipital groove' between the lesser and greater tubercles of the humerus, surrounded by a sheath of synovial membrane (fluid-secreting connective tissue). The fluid acts to lubricate the tendon, reducing friction on movement. The position of attachment of this head of the biceps muscle allows it to help in stabilizing the shoulder joint, as well as in flexing the arm.

Muscles of the posterior compartment

The muscles of the back of the upper arm act to extend the elbow, so straightening the whole arm.

The posterior compartment has one major muscle, the triceps brachii. This is a large, bulky muscle that lies posterior to the humerus and, as its name implies, has three heads: the long head; the lateral head; and the medial head. These converge in the middle of the upper arm on a wide, flattened tendon that passes down over a small bursa to attach to the olecranon process of the ulna. The main action of the triceps is to straighten the elbow joint. The long head of the triceps muscle also helps to stabilize the shoulder joint.

Anconeus

The small triangular shaped anconeus muscle lies behind and below the elbow joint. As with the triceps, it extends the elbow and also has a function in the stabilization of the elbow joint.

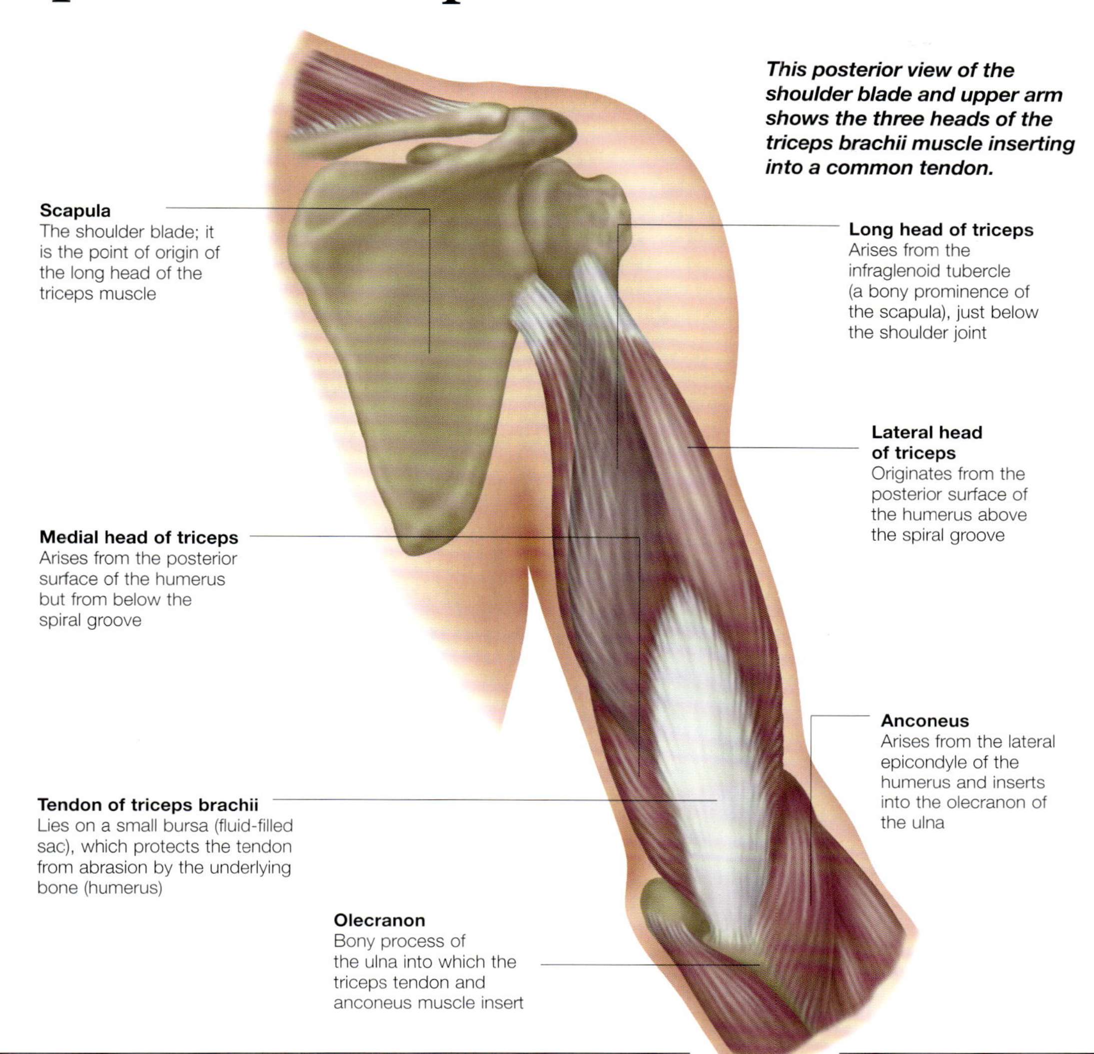

This posterior view of the shoulder blade and upper arm shows the three heads of the triceps brachii muscle inserting into a common tendon.

Cross-section of the upper arm

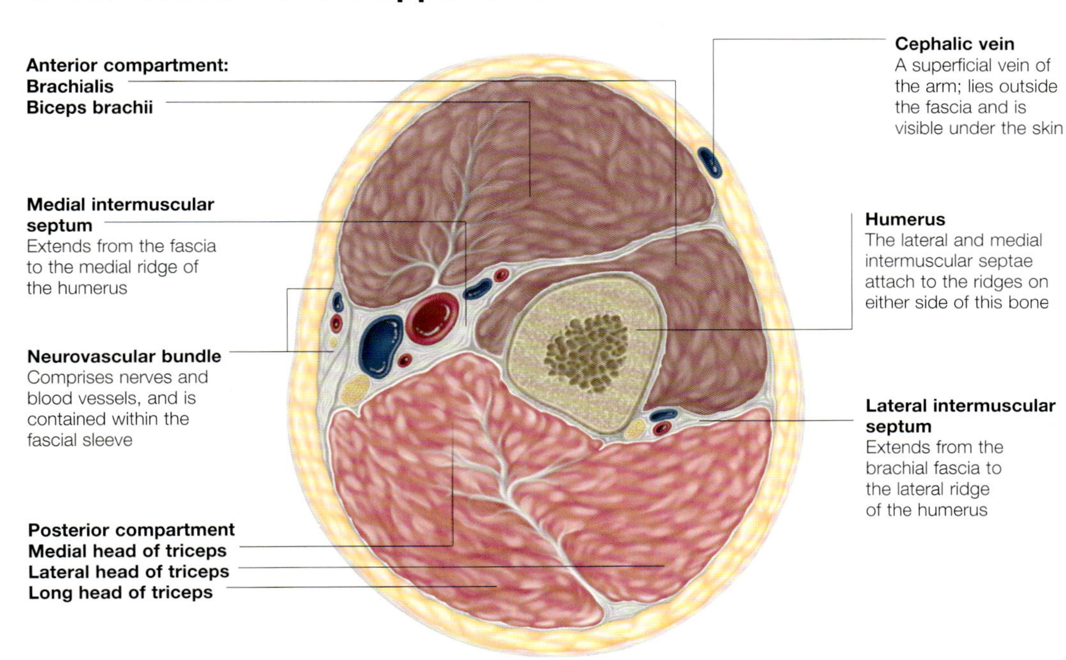

The arm is divided into distinct compartments, surrounded by sheets of connective tissue known as fascia. The brachial fascia is like a sleeve under the skin within which the major structures of the arm lie. Further fascial divisions, the lateral and medial intermuscular septae, arise from this brachial fascia and attach to the ridges on either side of the humerus, dividing the upper arm into the anterior and posterior muscular compartments. Also contained within the fascial 'sleeve' are groups of nerves and blood vessels (neurovascular bundles).

The muscles of the anterior (darker coloured) and posterior (lighter coloured) compartments are separated by the medial and lateral intermuscular septae.

Muscles of the Forearm

The flexor muscles of the front compartment of the forearm act to flex the hand, wrist and fingers. They are divided into superficial and deep muscles of the flexor and extensor compartments.

Superficial flexor muscles

This compartment, or section, of the forearm lies in the front of the forearm and contains muscles that flex the wrist and fingers as well as some that pronate the forearm (turn the hand palm down). They are subdivided into superficial and deep layers according to position.

The superficial group contains five muscles that all originate at the medial epicondyle of the humerus, where their fibres merge to form the 'common flexor tendon':

- Pronator teres – pronates the forearm and flexes the elbow
- Flexor carpi radialis – acts to produce flexion and abduction (bending away from the midline of the body) of the wrist
- Palmaris longus – this small muscle is absent in 14 per cent of people; it acts to flex the wrist
- Flexor carpi ulnaris – this muscle flexes and adducts the wrist (bends away from the midline of the body); unlike the other muscles of the flexor compartment, this muscle is innervated by the ulnar nerve
- Flexor digitorum superficialis – this is the largest superficial muscle of the forearm and it acts, as its name suggests, to flex the fingers, or digits.

Pronator teres
Pronates the forearm and flexes the elbow

Palmaris longus
A weak wrist-flexor muscle

Flexor carpi radialis
Flexes the wrist and bends it away from the midline of the body

Flexor carpi ulnaris
Flexes the wrist and bends it towards the midline of the body

Flexor digitorum superficialis
Flexes the fingers

The five main superficial flexor muscles of the forearm are shown in this illustration. These muscles originate from the humerus bone of the upper arm.

Deep flexor muscles

The deep layer of the flexor compartment consists of three muscles:

Flexor digitorum profundus
This bulky muscle originates from a wide area of the ulna and neighbouring interosseous membrane (a strong sheet of tissue connecting the radius and ulna). It is the only muscle that flexes the last joint of the fingers and so acts with its more superficial counterpart to curl the fingers.

Like the flexor digitorum superficialis muscle, this deeper muscle divides into four tendons that pass through the carpal tunnel within the same synovial sheath. The tendons insert into the bases of the distal (far end) phalanges of the four fingers.

Flexor pollicis longus
This muscle flexes the thumb. Its long, flat tendon passes through the carpal tunnel within its own synovial sheath and inserts into the base of the distal phalanx of the thumb (which, unlike the fingers, has only two phalanges).

Pronator quadratus
The deepest muscle of the anterior compartment, the pronator quadratus acts to pronate the forearm and is the only muscle that attaches solely to the radius and ulna. It also assists the interosseous membrane in binding the radius and ulna tightly together.

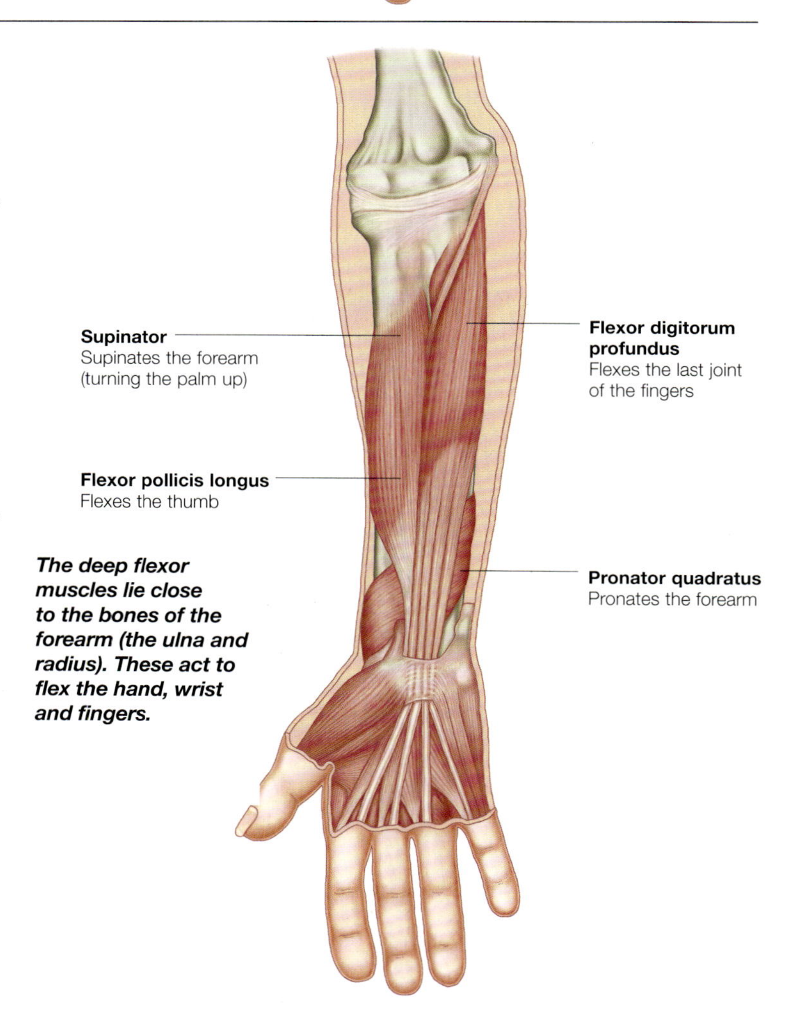

The deep flexor muscles lie close to the bones of the forearm (the ulna and radius). These act to flex the hand, wrist and fingers.

Flexing the hand

The muscles of the forearm are divided into front and rear compartments. The front flexor muscles bend the wrist and fingers and the rear extensor muscles act to straighten them again.

The muscles of the forearm are roughly divided into two groups, according to their function. These two groups are isolated from each other by the radius and ulna bones and by fascial layers (sheets of connective tissue) to form the 'anterior flexor compartment' and the 'posterior extensor compartment' of the forearm.

Opposing actions

The flexor muscles act to flex (bend) the wrist joint and the fingers, while the extensors act to extend (straighten) the same joints. Within these two groups are both deep and superficial muscles that act together to give the wide range of movements that are characteristic of the wrist and hand.

Forearm tendons

So that the wrist and hand can move flexibly, the bulk of muscle around the lower end of the upper limb is kept to a minimum. This is achieved by using long tendons from muscles higher up in the forearm to work the wrist, and the fingers.

Many of the muscles concerned with hand movements originate at the lower end of the humerus, so they are longer and more efficient. The humerus has developed two projections called the medial (inner) and lateral (outer) epicondyles. The flexor muscles are attached to the medial epicondyle while the extensor muscles are attached to the lateral epicondyle.

Cross-section through the forearm

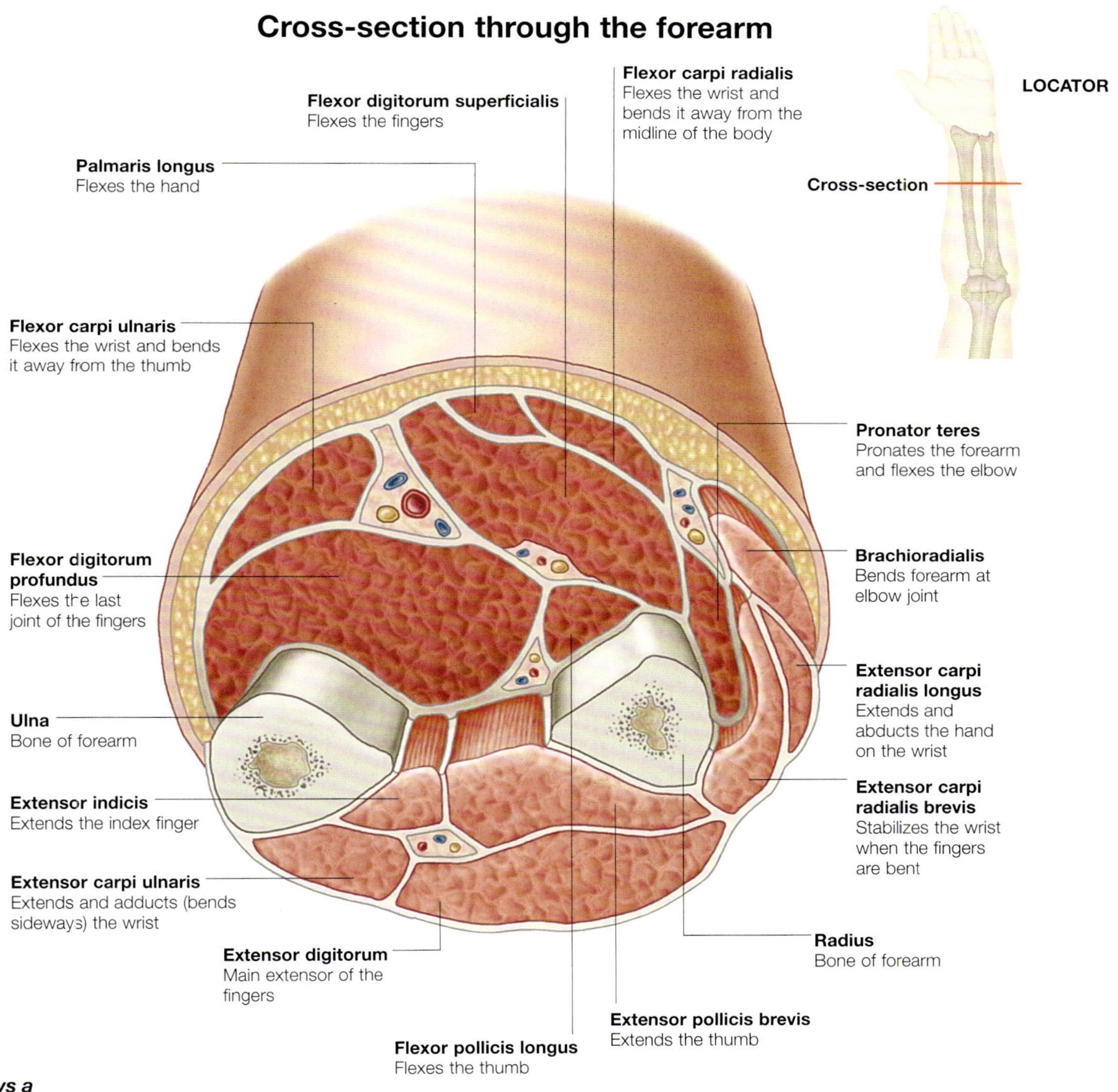

This illustration shows a cross-section through the forearm looking towards the hand with the palm upturned.

Extensor Muscles

Working together with the flexor muscles, the extensor muscles of the forearm enable a range of actions that allow the considerable mobility of the wrist, hand, fingers and thumb.

The posterior (rear) extensor compartment of the forearm contains muscles that act to straighten and pull back the wrist and fingers. These are separated from the flexor muscles at the front of the forearm by the radius and ulna bones, the strong membrane between them and the enveloping layer of thin connective tissue, the fascia of the forearm.

Extensor Actions

The extensors act to give the wide range of mobility required by the wrist and hand. The extensor muscles can be divided into three groups according to their functions:

- Muscles that move the hand or the wrist – straighten the wrist, pull the hand back or allow the hand to bend sideways
- Muscles that straighten the fingers, excluding the thumb
- Muscles that act only upon the thumb to extend it or pull it out sideways.

Superficial Extensors

Extensor carpi radialis longus
It acts to extend and abduct the hand on the wrist (bend sideways away from the little finger).

Extensor carpi radialis brevis
This muscle acts together with the extensor carpi radialis longus to stabilize the wrist joint when the four fingers are being flexed (bent into the palm).

Extensor carpi ulnaris
This long, slender muscle lies along the inner edge of the forearm and extends and adducts the wrist and is also needed for the action of clenching the fist.

Extensor digitorum
This muscle is the main extensor of the four fingers and its bulk makes up a good proportion of the back of the forearm.

Extensor digiti minimi
This runs down alongside the extensor digitorum and acts to help extend the little finger.

Brachioradialis
Although in the 'extensor compartment' of the forearm, this moderately powerful muscle acts to flex (bend) the forearm at the elbow joint. It returns a pronated or supinated forearm to the working position (thumb up).

Superficial extensor muscles

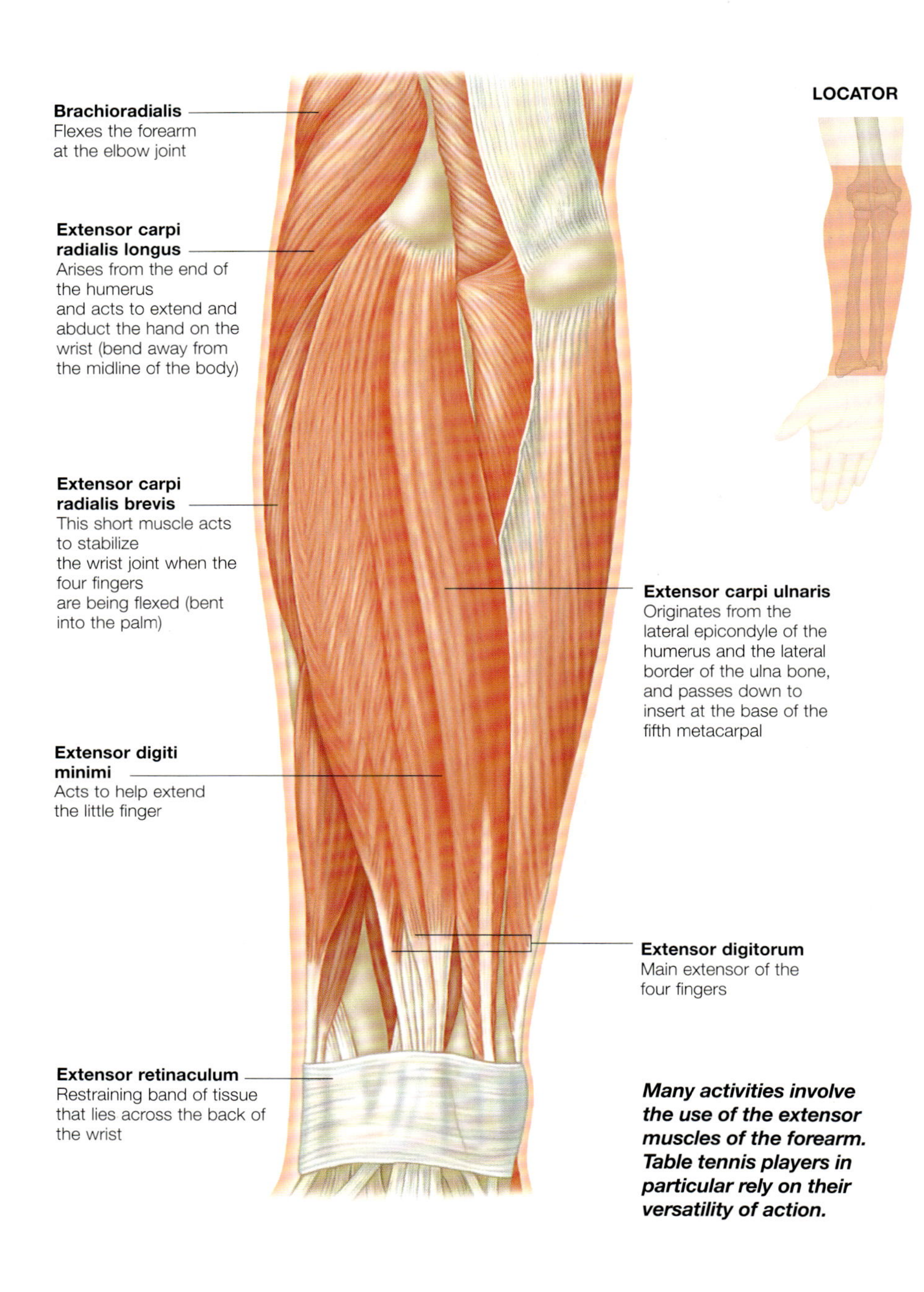

Many activities involve the use of the extensor muscles of the forearm. Table tennis players in particular rely on their versatility of action.

Deep extensor muscles

Lying closer to the underlying bones, the deep extensor layer includes muscles that act upon the thumb and the little finger individually.

The muscles of the extensor compartment, including the deep extensor muscles, receive their nerve supply from the radial nerve, which runs down the forearm alongside them. The deep extensor muscles include:

Extensor indicis
This muscle can act together with the extensor digitorum, or alone to extend the index finger. The index finger can be extended alone (as in the action of pointing) without the other fingers following, and this allows greater versatility to hand movement.

Abductor pollicis longus
This long muscle abducts the thumb, giving the action of lifting the thumb up and away from the plane of the palm. It can also extend the thumb, which is the action of moving the thumb out away from the other fingers sideways in the same plane. It originates from the back of the ulna, radius and adjoining interosseous membrane and runs down to insert at the base of the first metacarpal.

Extensor pollicis brevis
This short muscle extends the whole thumb, originating from the back of the radius and interosseous membrane and inserting at the base of the proximal phalanx of the thumb (the first bone in the thumb).

Extensor pollicis longus
This extensor of the thumb is larger than pollicis brevis.

Supinator
Together with the brachioradialis this muscle forms the floor of the cubital fossa. It is the major muscle involved in the action of supinating the forearm, which it does by rotating the radius. It is supplied by the deep branch of the radial nerve.

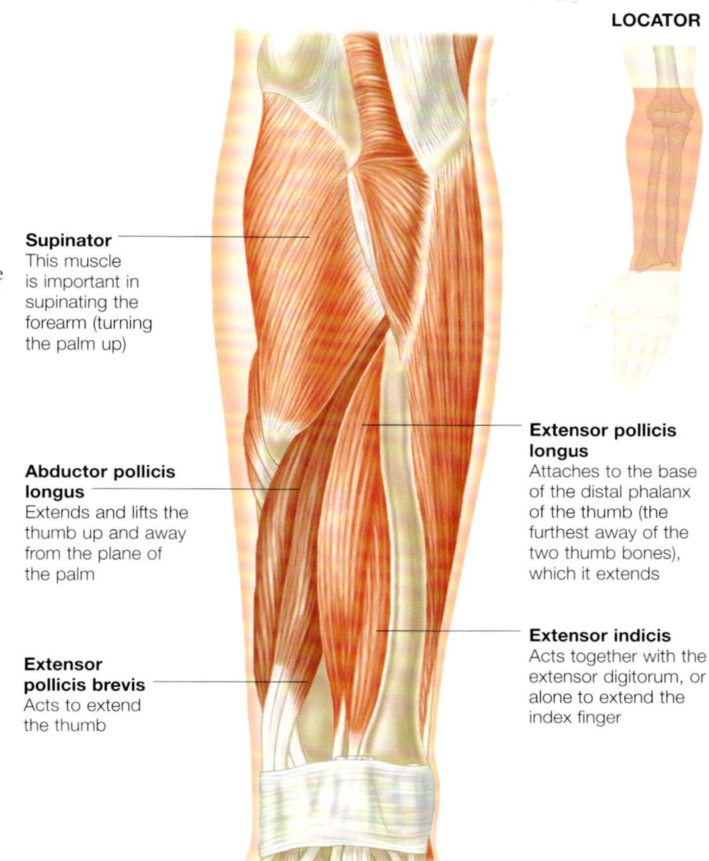

Together with the flexor muscles of the front of the forearm, the extensors act to give the wide range of mobility required by the wrist and hand.

Attachments of extensor tendons

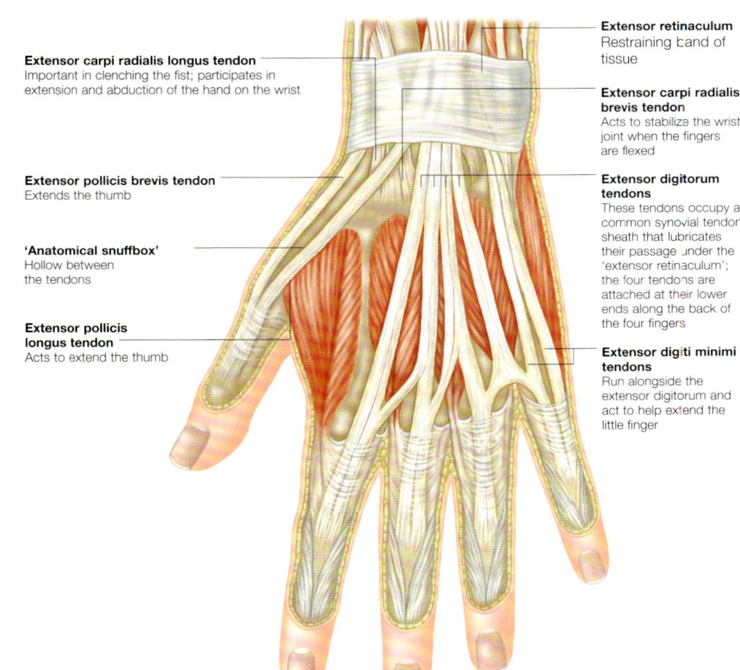

Most of the muscles of the posterior extensor compartment of the forearm terminate in long tendons that pass down over the back of the wrist to attach to the bones of the hands and fingers. In this way, the muscles that lie within the upper forearm can bring about extension (straightening and bending back) of the hand and fingers by 'remote control', so allowing the hand itself to be less bulky than it would otherwise have to be.

The site of attachment of each tendon, and whether it is to a hand- or finger-bone, will determine which joint of the hand will be straightened when that muscle contracts.

The extensor muscles are attached to the fingers and thumb of the hand by strong tendons protected in synovial sheaths.

Protecting the tendon

As the extensor tendons pass over the back of the wrist they pass under the 'extensor retinaculum', a horizontal restraining band of strong connective tissue that holds them in place against the joint as the hand moves.

To protect the tendons where they may rub against the underlying bones, and to lubricate their passage, the tendons lie within a fluid-filled synovial sheath.

Deep Fascia of the Arm

Between the subcutaneous tissue and muscles of the arm lies the deep fascia. This thin layer of connective tissue runs around the upper arm (brachial fascia) and the forearm (antebrachial fascia).

The deep fascia of the arm is a layer of thin, tough connective tissue that lies between the subcutaneous tissue and the muscles, enveloping the limb like a sleeve.

Brachial Fascia

The deep fascia of the upper arm is also called the brachial fascia. It is continuous above with the pectoral fascia, which encloses the pectoral muscles of the front of the chest, and the axillary fascia, which forms the floor of the armpit. Around the elbow the brachial fascia is attached to the epicondyles of the humerus (bony prominences at either side of the lower end of this bone), and to the olecranon of the ulna (one of the forearm bones).

Antebrachial Fascia

The deep fascia of the forearm is also known as the antebrachial fascia. It is continuous with the brachial fascia at the elbow and runs down to enclose the tissues of the forearm, forming the horizontal bands of the extensor and flexor retinacula at the wrist.

The bicipital aponeurosis is a membranous band that runs across the front of the elbow to merge with the antebrachial fascia. It arises from the tendon of the biceps muscle and inserts, with the deep fascia, into the subcutaneous border of the ulna. The deep fascia forms partitions (septa) that pass down and attach to the bones beneath, dividing the arm into spaces or compartments.

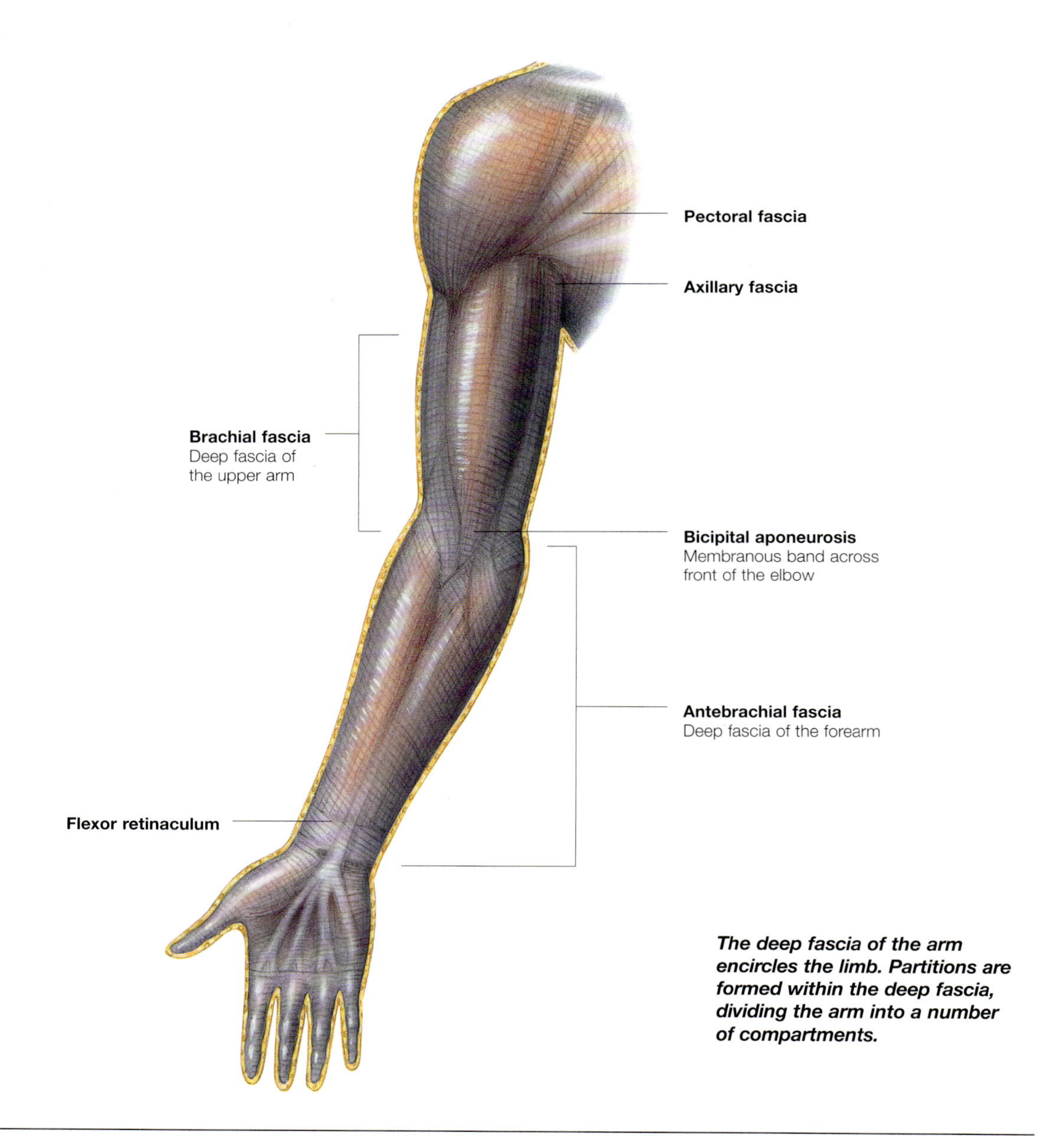

The deep fascia of the arm encircles the limb. Partitions are formed within the deep fascia, dividing the arm into a number of compartments.

Extensor retinaculum

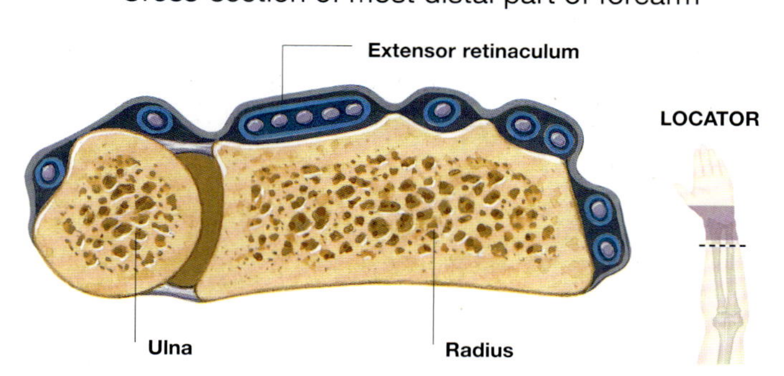

A retinaculum is a band of fascia that stretches across a joint to hold down or 'retain' tendons against the joint as it moves. The deep fascia of the arm gives rise to the extensor retinaculum of the wrist, which performs this function for the long tendons passing over the back of the wrist to reach the bones of the hand and fingers.

The extensor retinaculum is the continuation of the fascia over the wrist. It keeps the tendons against the joint as it moves.

Forearm bones

The extensor retinaculum stretches over the ends of the radius and ulna (the two bones of the forearm), and gives off fibrous septa that attach to these bones in several places.

The extensor tendons lie between the retinaculum and the bones and are divided by these attachments into six groups. The presence of the extensor retinaculum ensures that the long tendons of the back of the hand remain in position as the wrist moves rather than slipping sideways or 'bowstringing'.

Deep Fascia of the Hand

The continuation of the antebrachial fascia over the wrist forms the deep fascia of the hand. This fascia forms the palmar aponeurosis, a central thickened area. Partitions divide the palm into compartments.

The deep fascia of the hand is the continuation of the antebrachial fascia running down over the wrist to enclose the hand like a glove.

Palmar Aponeurosis

In the palm of the hand, the deep fascia is thin at the sides, but thickened centrally to form a tough, fibrous sheet known as the palmar aponeurosis.

The triangular palmar aponeurosis lies subcutaneously over the long flexor tendons and other soft tissues of the hand. One of its roles is to provide a firm site of attachment for the overlying skin, which helps to improve grip in the hand.

Flexor retinaculum

The proximal (wrist) end of the palmar aponeurosis is attached to the flexor retinaculum, the horizontal band of fascia that runs across the wrist to hold down the long flexor tendons.

Distally (towards the fingers), four longitudinal bands within the palmar aponeurosis pass up to the fingers, helping to form the fibrous sheaths that enclose the tendons there.

Compartments

Medial and lateral septa from the deep fascia divide the muscles of the hand into a number of compartments. These compartments comprise:

- The hypothenar compartment – this contains the muscles associated with the little finger, nerves and blood vessels. It is enclosed within the hypothenar fascia.
- The thenar compartment – this is surrounded by the thenar fascia, and houses some of the muscles that act upon the thumb.
- The adductor compartment – this is named for the action of its muscle, namely the adduction (pulling in) of the thumb.
- The central compartment – this lies under the palmar aponeurosis and contains flexor tendons, nerves and blood vessels.

Cross-section of the hand

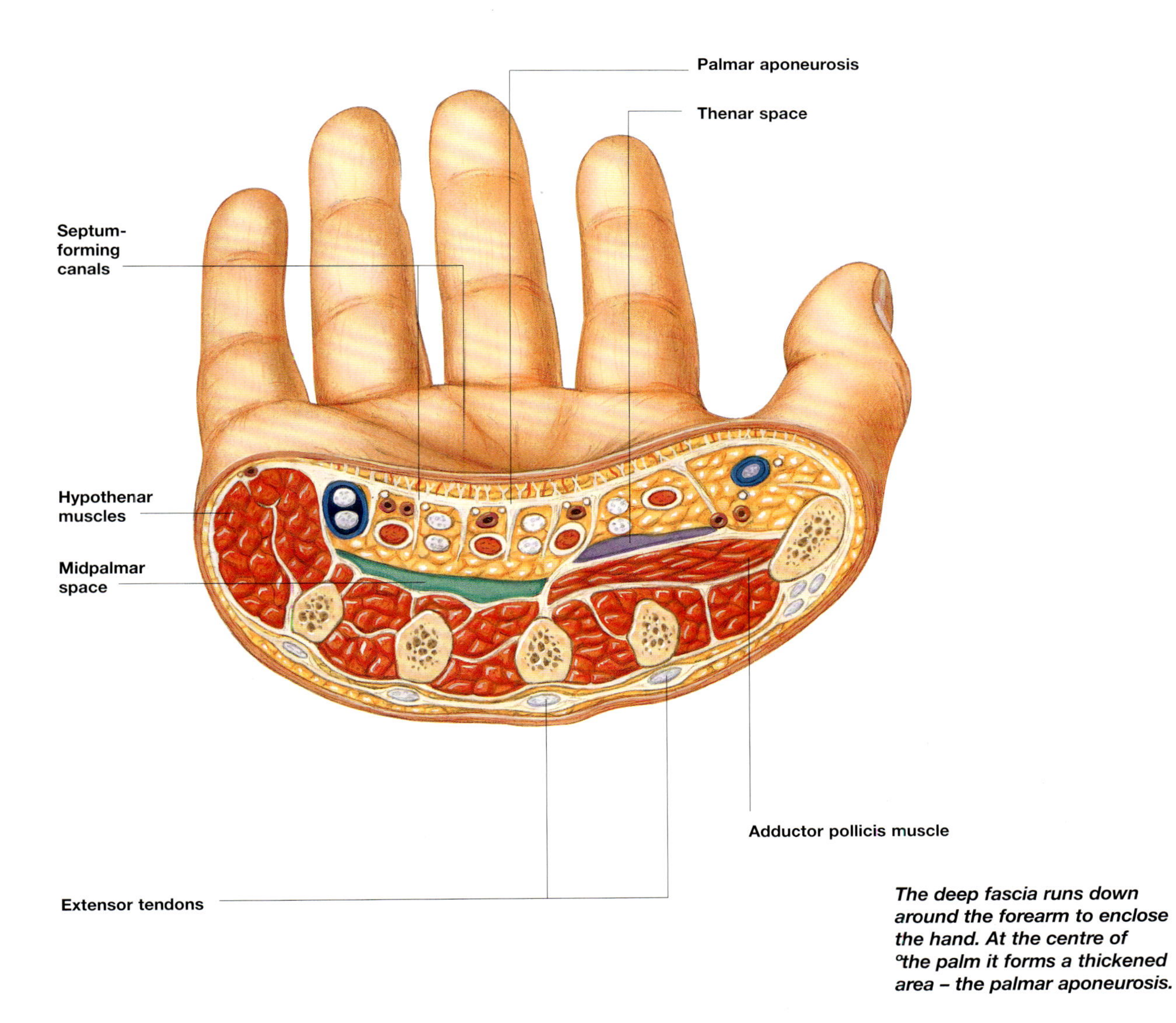

The deep fascia runs down around the forearm to enclose the hand. At the centre of the palm it forms a thickened area – the palmar aponeurosis.

Bones of the Wrist

The wrist lies between the radius and ulna of the forearm and the bones of the fingers. It is made up of eight marble-sized bones that move together to allow flexibility of the wrist joint and the hand.

The flexible wrist is formed of eight bones held together by ligaments. The carpal bones form two rows of four bones each – the proximal row (nearer the forearm) and the distal row (nearer the fingers). The main joint of the wrist is between the first of these two rows and the lower end of the radius.

The Proximal Row

- **Scaphoid** – a 'boat-shaped' bone with a large facet for articulation with the lower end of the radius; it articulates with three bones of the distal row
- **Lunate** – a moon-shaped bone that articulates with the end of the radius
- **Triquetral** – this pyramid-shaped bone articula tes with the disc of the inferior radioulnar joint and the pisiform bone
- **Pisiform** – this small bone plays no role in the wrist joint. It is about the size and shape of a pea and is a 'sesamoid' bone, a bone that lies within a muscle tendon.

The Distal Row

- **Trapezium** – a four-sided bone that lies between the scaphoid and the first metacarpal, with which it articulates
- **Trapezoid** – a small wedge-shaped bone that lies between the far end of the scaphoid and the second metacarpal
- **Capitate** – the largest of the carpal bones named after its large rounded head
- **Hamate** – a triangular bone with a hook-like process on its palmar surface.

Bones of the left wrist viewed from above

The way that the bones of the wrist sit in relation to each other is seen in this image. The top (distal) bones closest to the fingers are tinted orange; the bottom (proximal) bones closest to the forearm are purple.

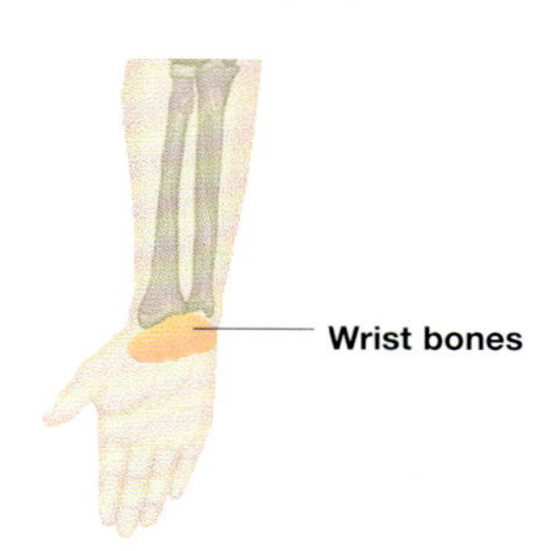

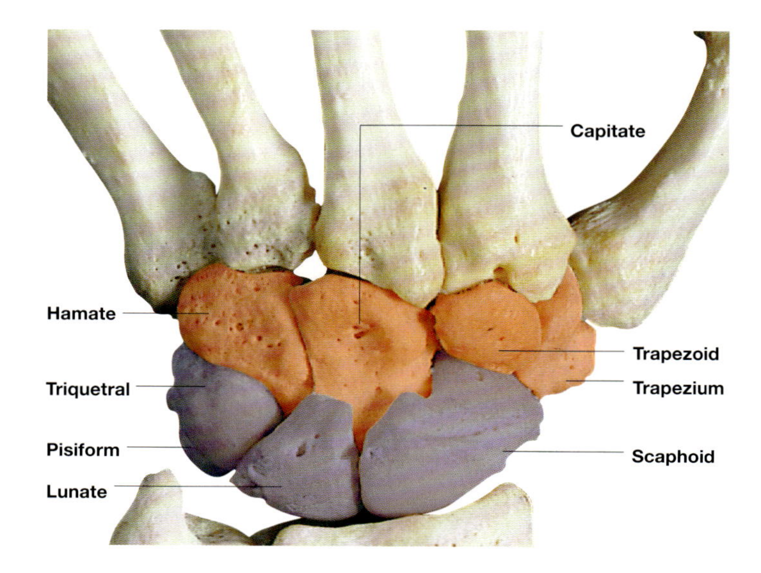

Bottom (proximal) row of wrist bones

The proximal row of carpal bones includes two bones that can be easily felt: the pisiform and scaphoid bones.

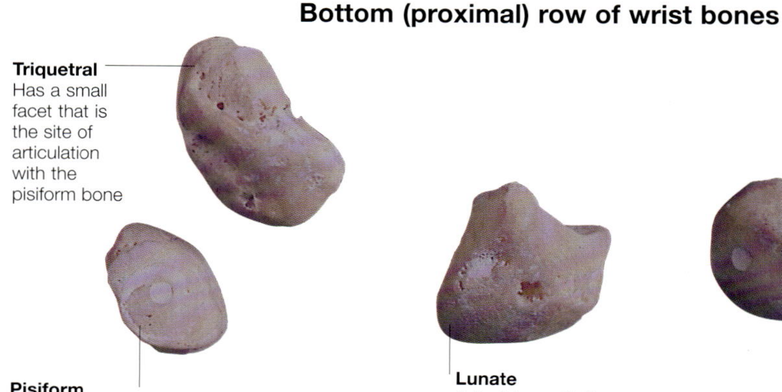

Triquetral
Has a small facet that is the site of articulation with the pisiform bone

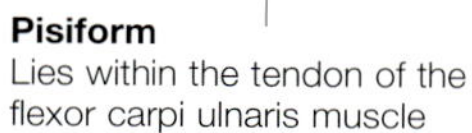

Pisiform
Lies within the tendon of the flexor carpi ulnaris muscle

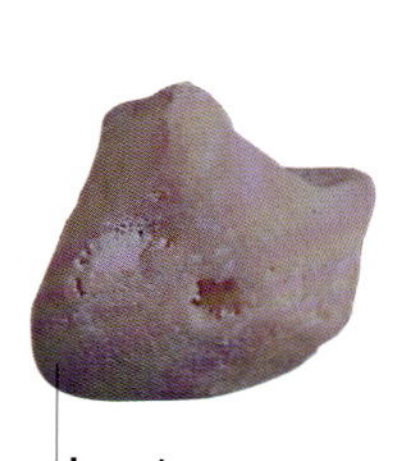

Lunate
Articulates with the lower end of the radius

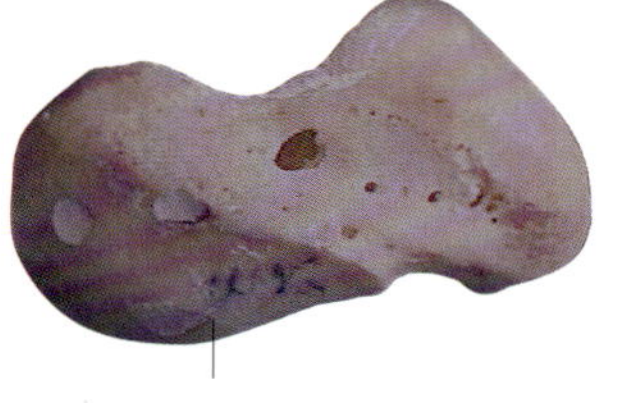

Scaphoid
Has a narrowed 'waist' and may be the site of a fracture

Top (distal) row of wrist bones

Hamate
Articulates with the lunate and the triquetral bones

Capitate
Largest of the carpal bones

Trapezoid
Lies between the far end of the scaphoid and the second metacarpal bone of the hand

The top (distal) row of carpal bones lies between the bones of the hand and the bottom (proximal) row. Both of these rows are held together by ligaments.

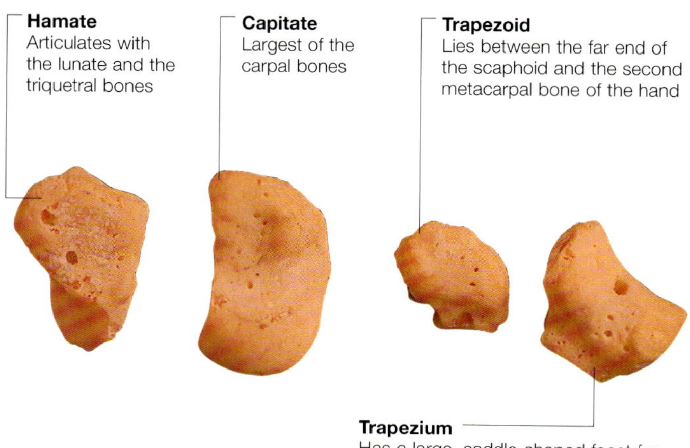

Trapezium
Has a large, saddle-shaped facet for articulation with the first metacarpal

Wrist joint

The bones of the wrist are covered in cartilage and enclosed by a synovial membrane. This secretes a viscous fluid that allows the bones to move in relation to one another with minimum friction.

The wrist, or radiocarpal, joint is a synovial (fluid-filled) joint. On one side lie the lower end of the radius and the articular disc of the inferior radioulnar joint; while on the other are three bones of the first row of carpal (wrist) bones: the scaphoid, lunate and triquetral bones. The fourth bone of this first row, the pisiform, plays no part in the wrist joint.

The radiocarpal joint

The radiocarpal joint is made up of three areas, which are separated by two low ridges:

- A lateral area (outer, on the same side as the thumb), which is formed by the lateral half of the end of the radius as it articulates with the scaphoid bone.
- A middle area where the medial (inner) half of the end of the radius articulates with the lunate bone.
- A medial area (inner, on the side of the little finger) where the articular disc that separates the wrist joint from the inferior radioulnar joint articulates with the triquetral bone.

Articular surfaces

All the articular surfaces are covered with smooth, articular (hyaline) cartilage for reduction of friction during movement. The joint is lined with synovial membrane that secretes thick, lubricating synovial fluid, and is surrounded by a fibrous capsule that is strengthened by ligaments.

The articular surfaces as a whole form an ellipsoid shape, with the long axis of the ellipse lying across the width of the wrist. The shape of the articular surfaces of a joint helps to determine the range of movements; an ellipsoid shape does not allow rotation of that joint. The joint is convex towards the hand.

The intercarpal joints

As well as the joint between the lower end of the forearm and the first row of carpal bones, there is articulation between the carpal bones themselves. There is a large, irregular, 'midcarpal' joint that lies between the two rows of carpal bones. It is a synovial joint with a joint cavity that extends into the gaps between the eight bones and allows them to glide over one another, giving the flexibility that is needed in the wrist.

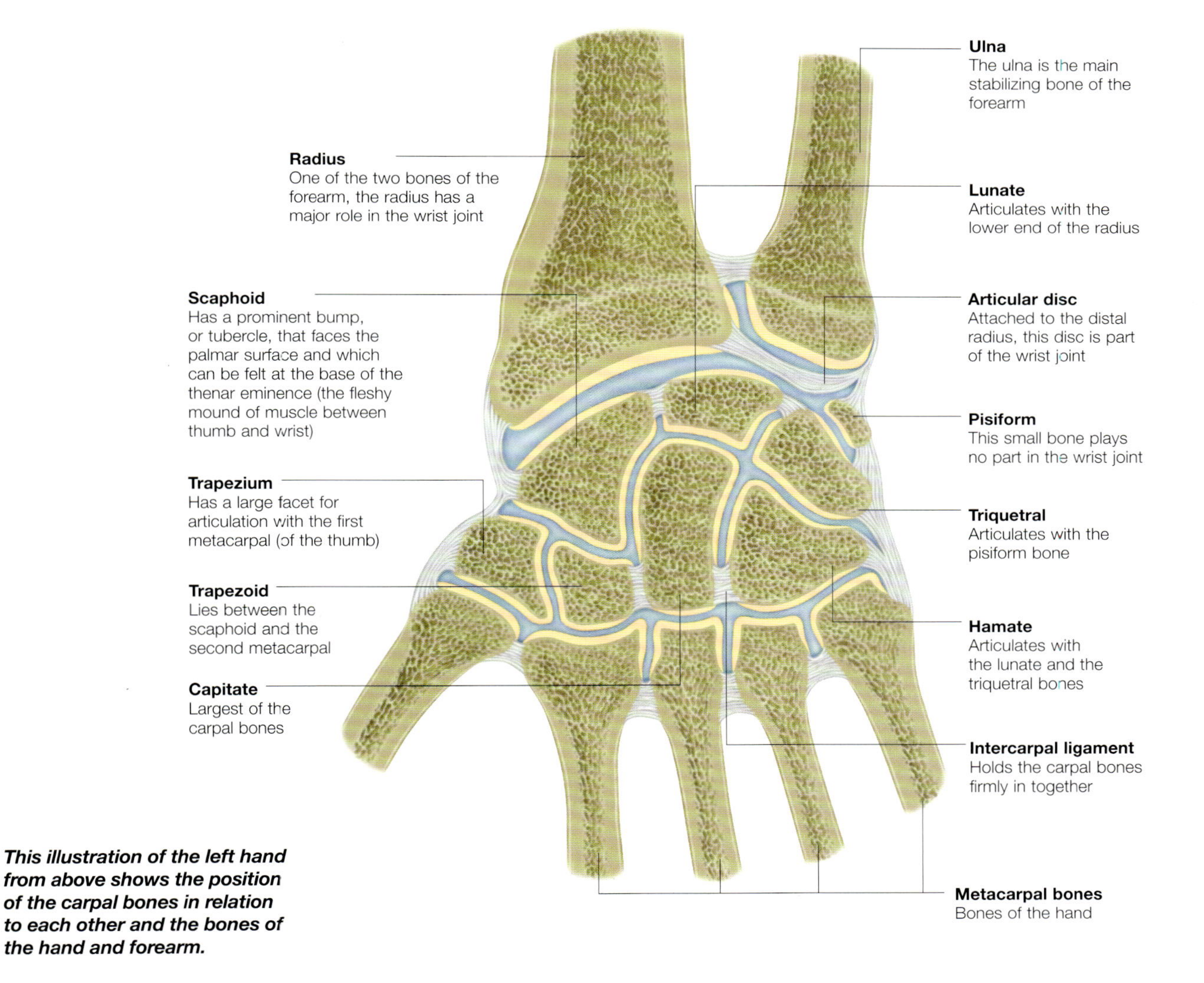

This illustration of the left hand from above shows the position of the carpal bones in relation to each other and the bones of the hand and forearm.

Carpal Tunnel

The strong ligaments of the wrist bind together the carpal bones, allowing stability and flexibility. Within the wrist is a fibrous band through which important tendons and nerves run – the carpal tunnel.

Tendons of the wrist

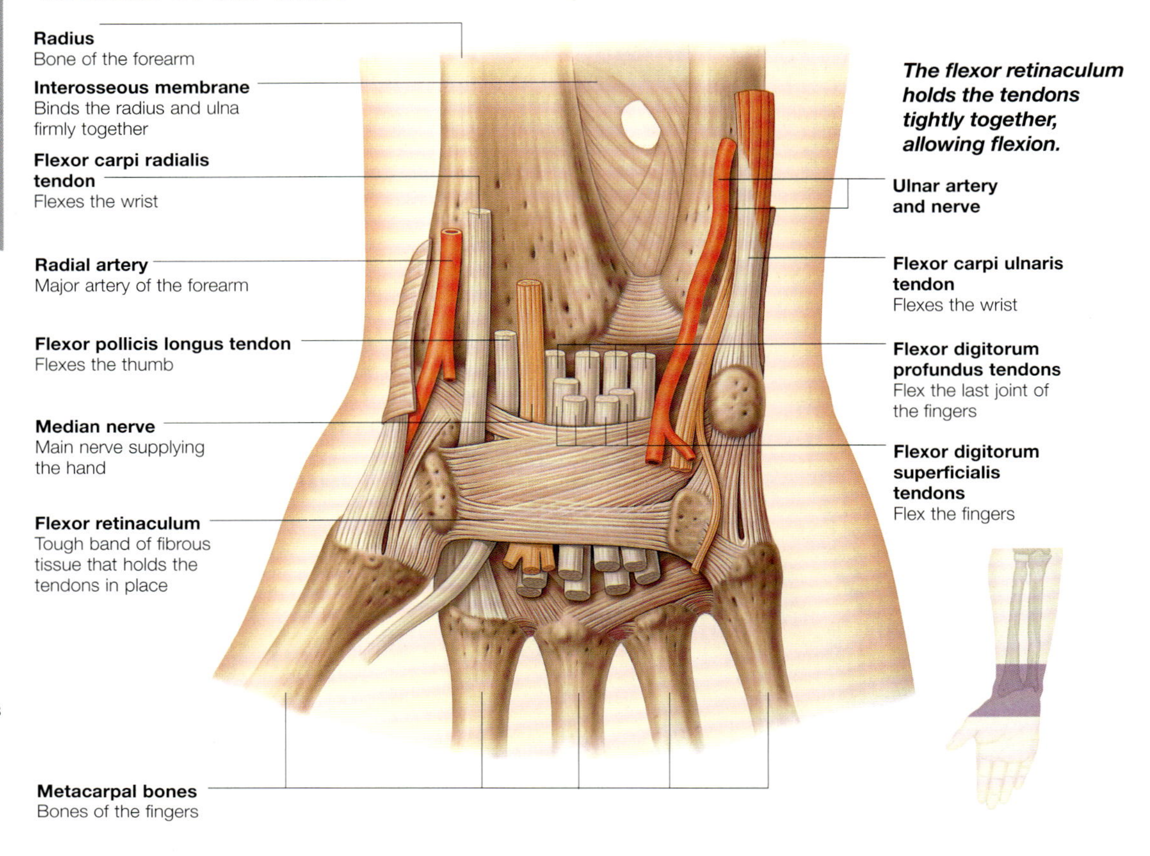

The flexor retinaculum holds the tendons tightly together, allowing flexion.

The eight carpal bones fit together in the wrist to form the shape of an arch. The arch is deepened on the palmar aspect of the wrist by the presence of the prominent tubercles of the scaphoid and trapezium bones on one side and the hook of the hamate and the pisiform bone on the other.

Carpal Tunnel

This bony arch is converted into a tunnel, known as the carpal tunnel, by a tough band of fibrous tissue, the flexor retinaculum. This band lies across the palmar surface and is attached on each side to the bony projections. It ensures that the tendons enclosed in the carpal tunnel are held close to the wrist even when the wrist is bent, so allowing flexion of the fingers at every position of the wrist.

Tendons

Running through the carpal tunnel are nine long tendons attached to muscles in the forearm. The four flexor digitorum profundus tendons and four flexor digitorum superficialis tendons are surrounded by a sheath and act to flex the wrist and fingers. The flexor pollicus tendon flexes the thumb. The flexor carpi radialis tendon lies outside the tunnel and bends the wrist away from the body.

Cross-section of the right wrist

The carpal tunnel (shown in blue) encloses the important tendons that enable the flexion of the wrist and fingers.

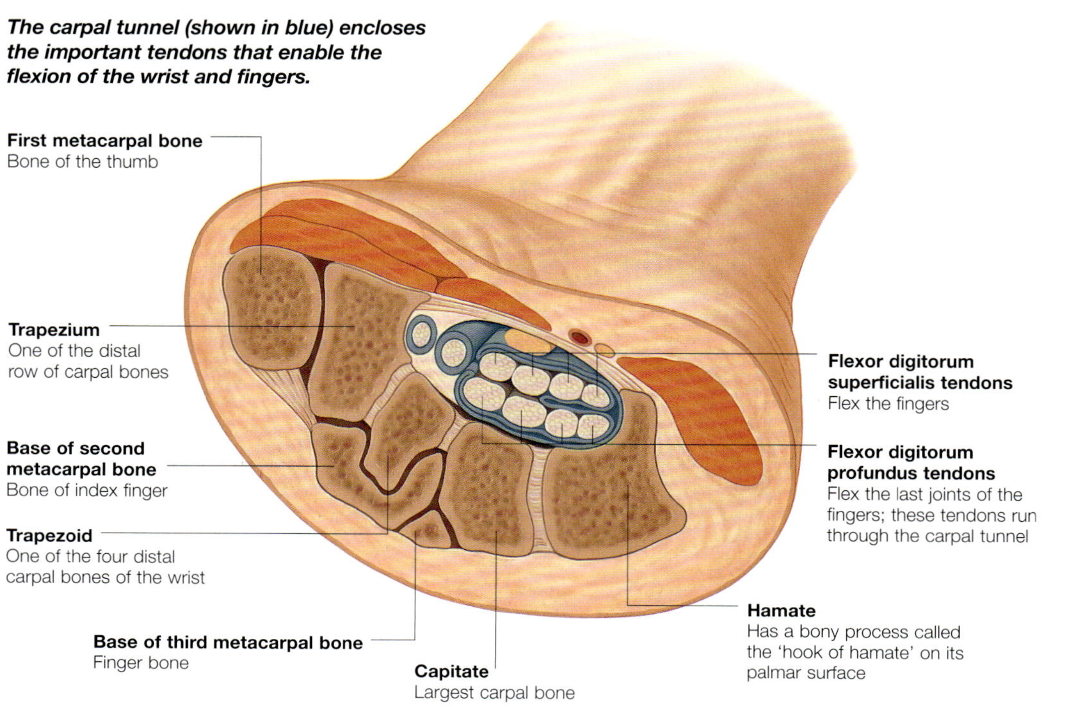

Ligaments of the Wrist

The ligaments of the wrist joint are thickenings of the joint capsule that help tie the wrist strongly to the lower ends of the radius and ulna.

The wrist joint itself cannot rotate and so rotation of the hand is achieved by pronation and supination of the forearm. The strong ligaments between the carpal bones and the radius are important as they 'carry' the hand round with the forearm during these actions. These ligaments include:

■ **Palmar radiocarpal ligaments**
Run from the radius to the carpal bones on the palm side of the hand. The fibres are directed so that the hand will go with the forearm during supination.

■ **Dorsal radiocarpal ligaments**
Run at the back of the wrist from the radius to the carpal bones and carry the hand back during pronation.

Collateral Ligaments

Strong collateral ligaments run down each side of the wrist to strengthen the joint capsule and add to the stability of the wrist. These limit the movements of the wrist joint when it is bent.

■ **Radial collateral ligament**
Runs between the styloid process of the radius and the scaphoid bone in the wrist and the trapezium bone in the carpus.

■ **Ulnar collateral ligament**
Runs between the styloid process of the ulna and the triquetral bone.

Dorsal view of the ligaments of the wrist

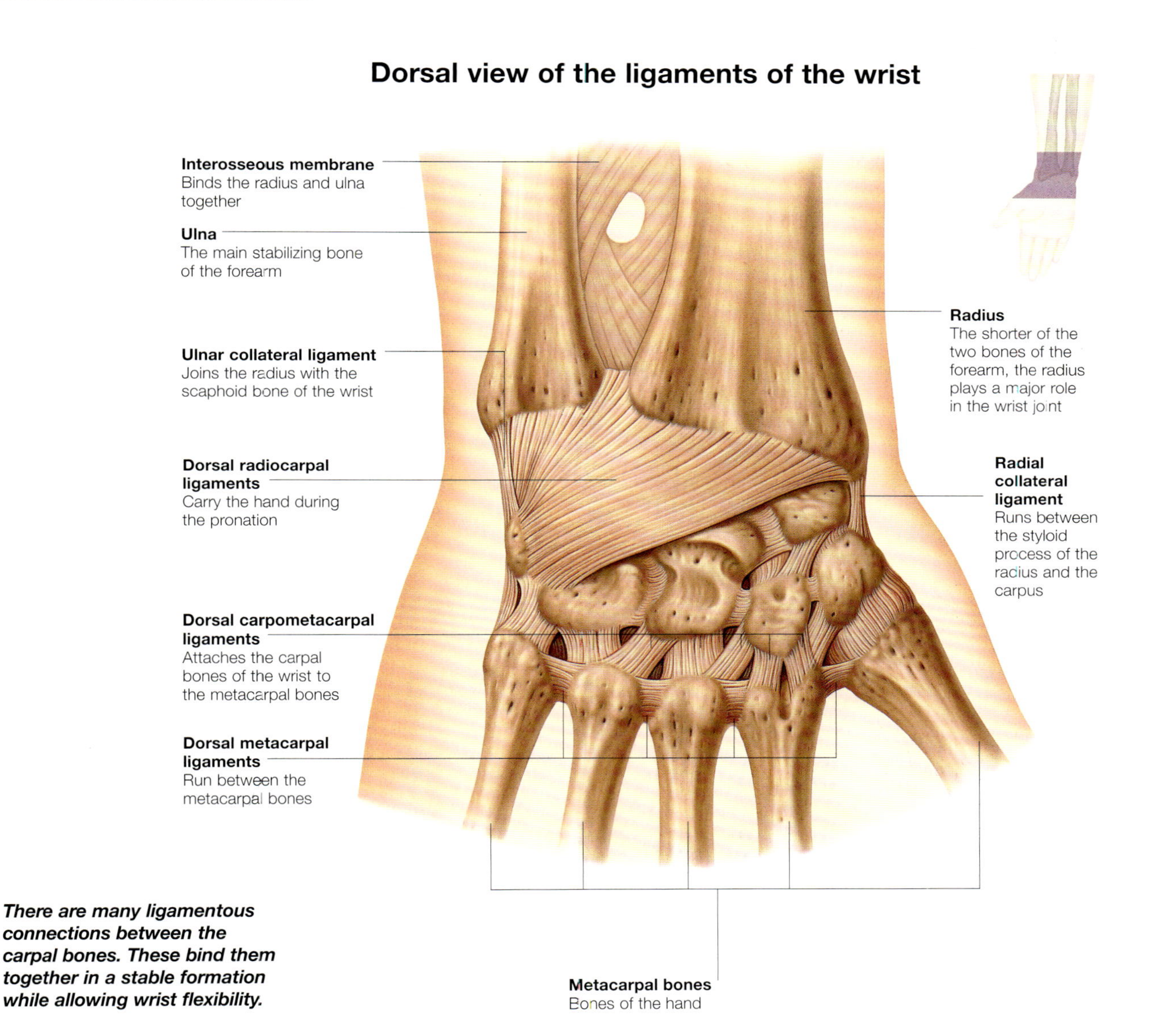

There are many ligamentous connections between the carpal bones. These bind them together in a stable formation while allowing wrist flexibility.

Bones of the Hand

The bones of the hand are separated into the metacarpal bones that support the palm and the phalanges or finger bones. The joints of these bones allow the fingers and the thumb great mobility.

The hand is formed of eight carpal bones of the wrist, five metacarpal bones that support the palm, and 14 phalanges, or finger bones.

The Metacarpals

Five slender bones radiate out from the wrist bones towards the fingers to form the support of the palm of the hand. They are numbered from one to five, starting at the thumb.

Each of the metacarpals is made up of a body, or shaft, and two slightly bulbous ends. The proximal end (near to the wrist), or base, articulates with one of the carpal bones. The distal end (away from the wrist), or head, articulates with the first phalanx of the corresponding finger. The heads of the metacarpals are known as the knuckles.

The Thumb

The first metacarpal, at the base of the thumb, is the shortest and thickest of the five bones and is rotated slightly out of line. It is extremely mobile, allowing a wider range of movement to the thumb than to the fingers, including the action of opposition whereby the thumb can touch the tips of each of the fingers.

The metacarpal bones

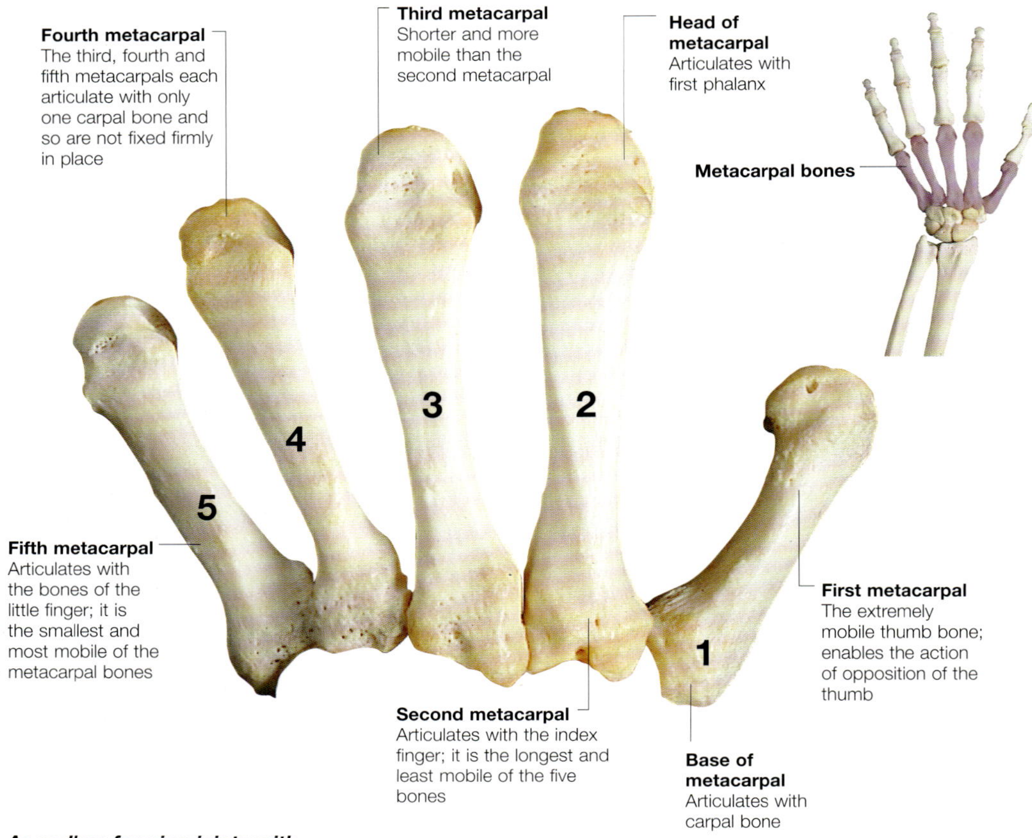

As well as forming joints with the carpal (wrist) bones, the five metacarpals articulate with each other laterally (at the sides) at their bases.

The phalanges

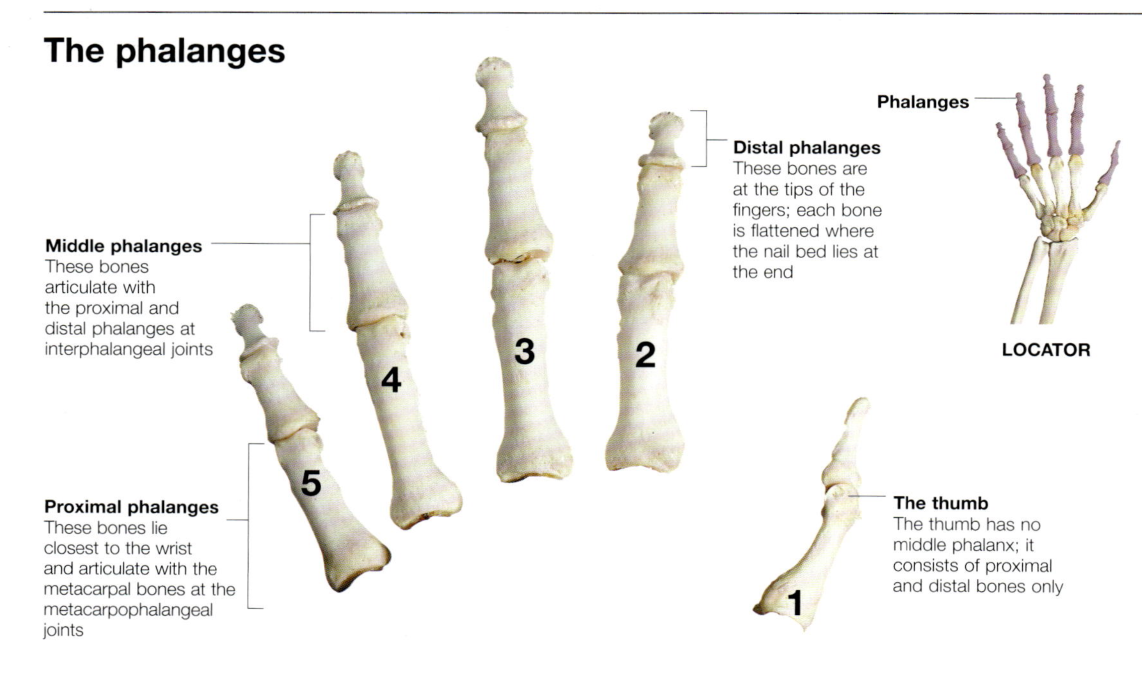

The phalanges (singular: 'phalanx') are the bones of the fingers, or digits. The digits are numbered one to five, with the thumb being number one. The first digit, the thumb, has only two phalanges, the other four digits each have three.

Each phalanx is a miniature long bone with a slender shaft, or body, and two expanded ends. In each digit the first, or proximal, phalanx is the largest while the end, or distal, phalanx is the smallest. The phalanges of the thumb are shorter and thicker than those in the other digits. Each of the small distal phalanges is characteristically flattened at the tip to form the skeletal support of the nail bed.

Each of the fingers, with the exception of the thumb, consists of three bones. These articulate with each other and with the metacarpal bones.

Finger joints

The joints between the phalanges are surrounded by fibrous capsules, lined with synovial membrane and supported by strong collateral ligaments.

The joints of the metacarpal bones with the carpal bones of the wrist – the carpometacarpal joints – are synovial (fluid-filled). The thumb has a saddle-shaped joint with the trapezium allowing a wide range of movement while the other metacarpals form 'plane' joints where the articulating surfaces are flat; they therefore have a limited range of movements.

Carpometacarpal joints

These are are surrounded by fibrous joint capsules lined with synovial membrane, which secretes the lubricating synovial fluid that fills the joint cavity. In most people there is a single, continuous joint cavity for the second to fifth carpometacarpal joints. The joint of the first metacarpal with the trapezium has its own separate joint cavity.

Metacarpophalangeal joints

The joints between the metacarpals and the proximal phalanges are 'condyloid' synovial joints – this shape allows movement in two planes. The fingers may flex and extend (bend and straighten), or abduct and adduct (move apart and together sideways, spreading the fingers). This adds to the mobility and versatility of the hand, as the fingers can be placed in a wide variety of positions.

Interphalangeal joints

The joints between each phalanx and the next are simple hinge-shaped joints that allow flexion and extension only.

Each finger has two interphalangeal joints, where the phalanges articulate with each other. These joints allow flexion and extension.

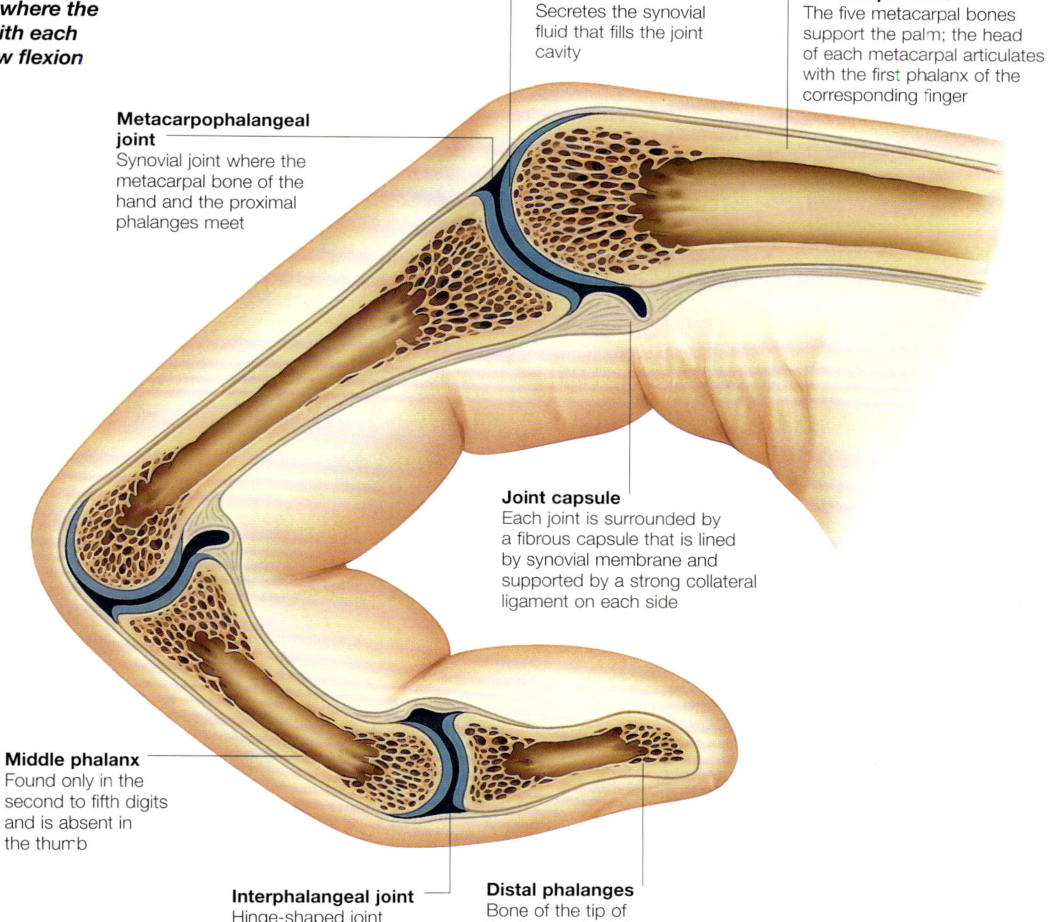

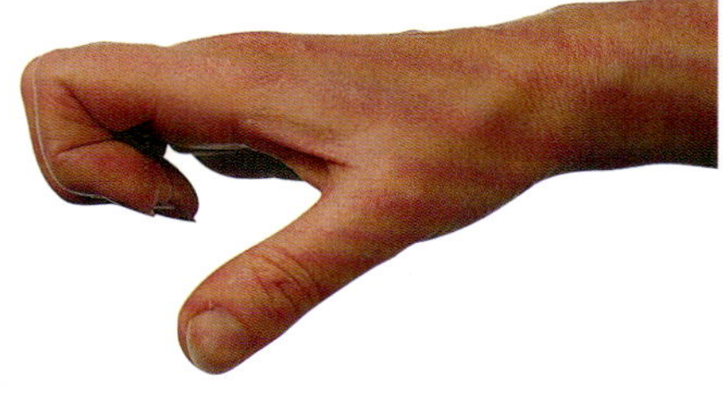

The hinge-shaped joints of the fingers – the interphalangeal joints – enable flexion and extension. The fingers may be flexed without flexing the hand.

Muscles of the Hand

The human hand is a very versatile structure, capable of both powerful and delicate movements. These are enabled by the actions and interactions of numerous muscles.

Many of the powerful movements of the hand, which need the contractile strength of muscle tissue, are controlled by the action of muscles in the forearm via tendons. Precise and delicate actions are produced by small, or 'intrinsic', muscles. These can be divided into three groups:

- The muscles of the thenar eminence between the base of the thumb and the wrist, which move the thumb
- The muscles of the hypothenar eminence (between the little finger and the wrist), which move the little finger
- The short muscles that run deep in the palm of the hand.

There are two groups of muscles that run longitudinally deep within the hand – the lumbricals and the interossei.

The Lumbricals

There are four lumbrical muscles arising in the palm from the tendons of the flexor digitorum profundus, a powerful muscle of the forearm. These muscles pass around the thumb side of the corresponding digit and insert into the area on the back of the finger, which contains the extensor tendons.

The lumbrical muscles

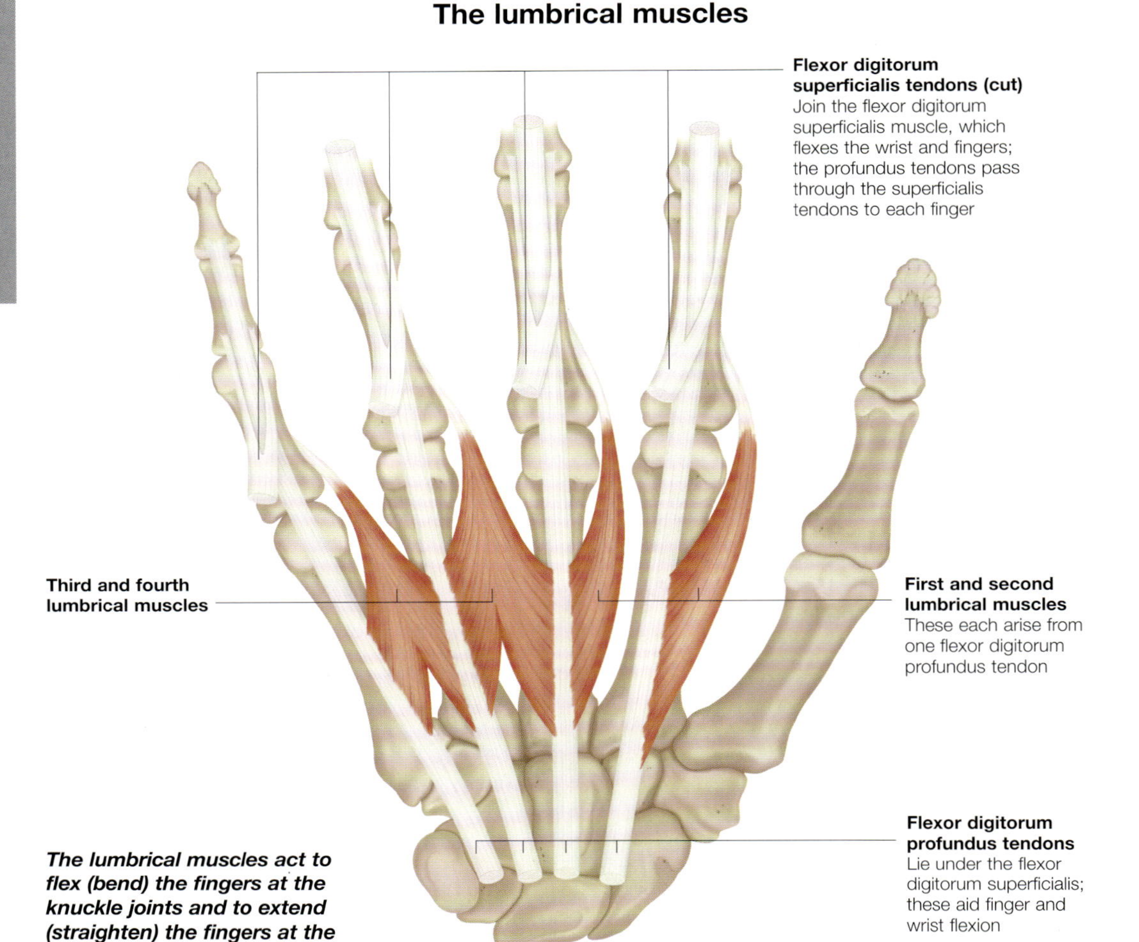

The lumbrical muscles act to flex (bend) the fingers at the knuckle joints and to extend (straighten) the fingers at the interphalangeal joints.

Interosseous muscles

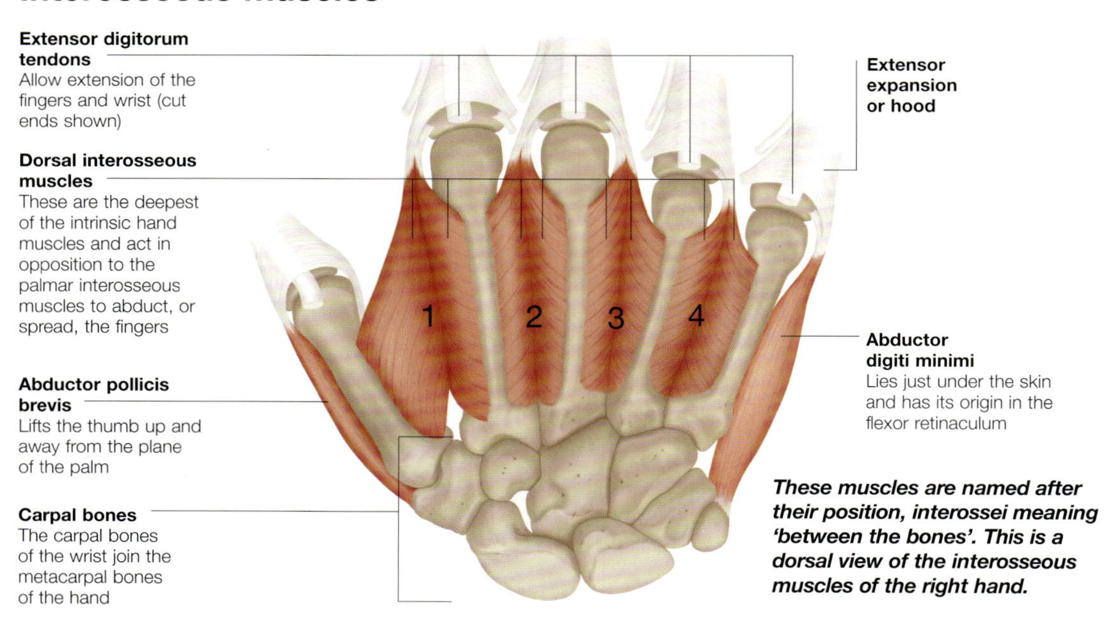

These muscles are named after their position, interossei meaning 'between the bones'. This is a dorsal view of the interosseous muscles of the right hand.

The interosseous muscles lie in two layers, those near the palm, the 'palmar interossei', and the deeper layer, the 'dorsal interossei'.

- Palmar interossei

These small muscles arise from the palmar surface of the metacarpals (excluding the third). The first two pass round the medial side of each digit before inserting into the dorsal (back) surface. Those which pass to digits four and five pass around the lateral (thumb) side. Contraction of these muscles pulls the fingers in together to give the action of adduction.

- Dorsal interossei

These larger muscles lie between the metacarpal bones of the hand, deep to the palmar interossei. Each arises from the sides of the metacarpal bones and act to spread the fingers.

Moving the thumb and the little finger

The muscles that move the thumb are contained in the thenar eminence at the base of the thumb; those that move the little finger are found in the hypothenar eminence, between the little finger and the wrist.

The four small muscles of the thenar eminence act together to allow the thumb to move in the manner that is so important to humans. This action is known as 'opposition' and is the action whereby the tip of the thumb is brought around to touch the tip of any of the fingers.

The muscles of the thenar eminence that move the thumb include:

Abductor pollicis brevis
Abductor pollicis brevis literally means the short muscle. It abducts the thumb (lifts the thumb up away from the palm).

Flexor pollicis brevis
Flexor pollicis brevis lies near to the centre of the palm and flexes the thumb.

Opponens pollicis
Opponens pollicis (the muscle that opposes the thumb) originates in the flexor retinaculum and the trapezium bone of the wrist and it inserts into the outer border of the first metacarpal bone.

Adductor pollicis
Adductor pollicis brings the abducted thumb back in line with the palm. This is a deeply placed muscle that has two heads of origin separated by the radial artery, which join to form a tendon. This tendon often contains a 'sesamoid' bone, a small bone that lies completely within the tendon and makes no connections with other bones.

The hypothenar eminence

The muscles of the smaller hypothenar eminence form the swelling that lies between the little finger and the wrist. These muscles act together to move the little finger towards the thumb in the action of cupping the hand, or when gripping the lid of a jar to twist it off.

Abductor digiti minimi
This muscle lies just under the skin and has its origin in the flexor retinaculum and the pisiform bone of the wrist. It inserts into the side of the base of the little finger.

Flexor digiti minimi
This short flexor muscle lies alongside the previous muscle but nearer to the centre of the palm. It originates in the flexor retinaculum and the hamate bone of the wrist, and inserts into the base of the little finger.

Opponens digiti minimi
This muscle, which opposes the little finger, lies underneath the more superficial muscles of the hypothenar eminence.

Palmaris brevis muscle
This short muscle has no attachments to bone, but originates in the palmar aponeurosis (connective tissue sheet that lies in the palm) and inserts into the skin overlying the hypothenar eminence. It acts to wrinkle the skin, which is believed to aid grip.

Palmar view of the right hand

The muscles that activate the thumb and little finger are shown in this illustration.

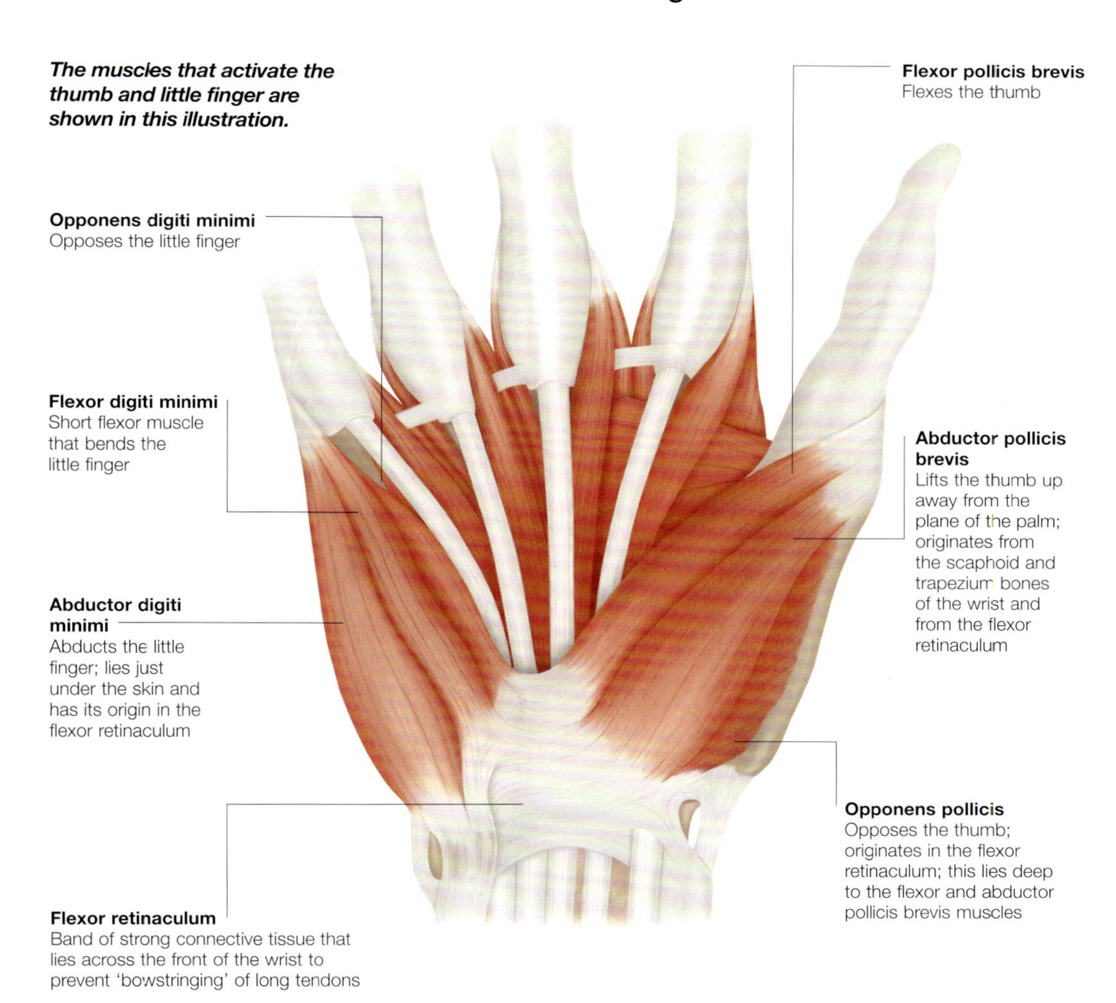

Bones of the Pelvis

The basin-like pelvis is formed by the hip bones, sacrum and coccyx. The pelvic bones provide sites of attachment for many important muscles, and also help to protect the vital pelvic organs.

Anterior view of the adult female pelvis

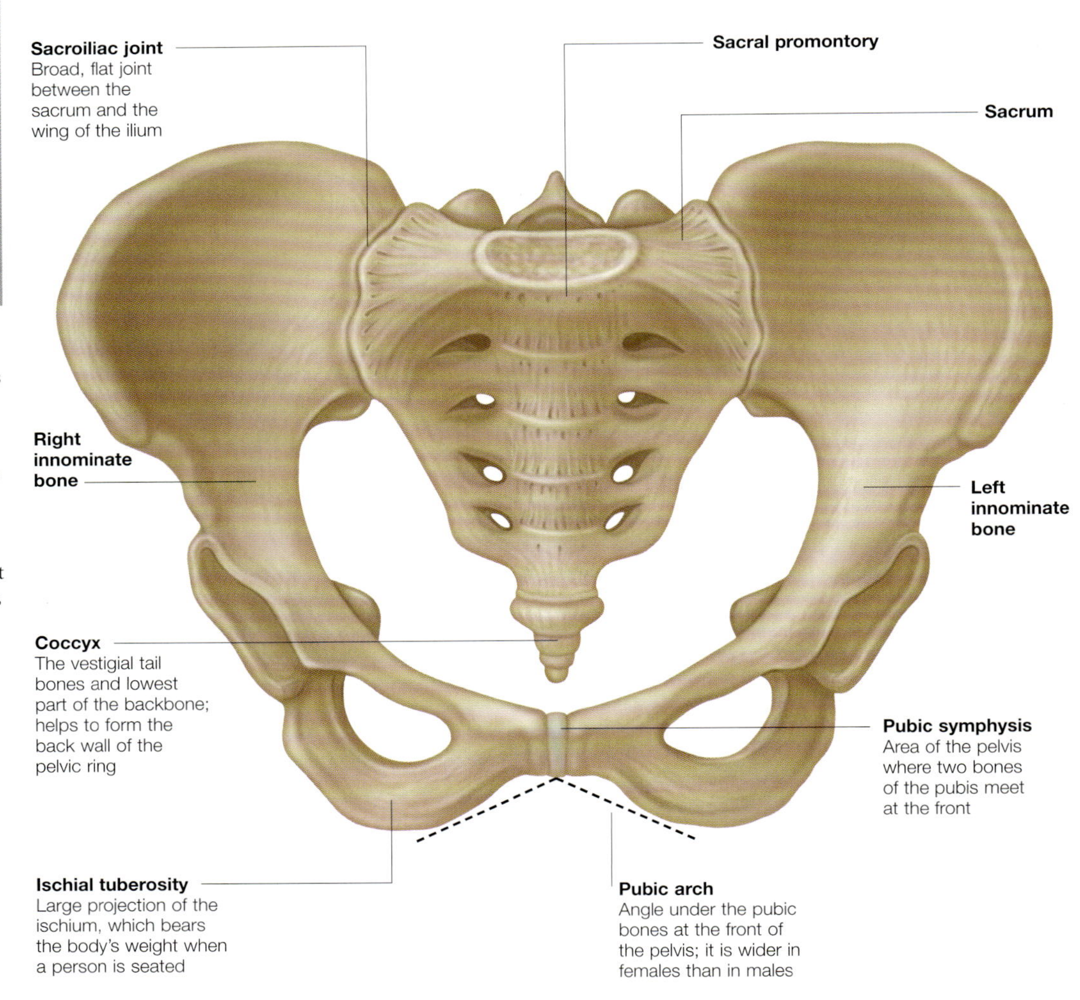

The bones of the pelvis form a ring that connects the spine to the legs and protects pelvic contents such as the reproductive organs and bladder. The pelvic bones, to which many strong muscles are attached, allow the body weight to be transferred to the legs for great stability.

The basin-like pelvis consists of the innominate (hip) bones, the sacrum and the coccyx. The innominate bones meet at the pubic symphysis anteriorly. Posteriorly, these two bones are joined to the sacrum. Extending down from the sacrum at the back of the pelvis is the coccyx.

False and True Pelvis

The pelvis can be said to be divided into two parts by an imaginary line passing through the sacral promontory and the pubic symphysis:

- Above the sacral promontory, the false pelvis flares out and supports the lower abdominal contents.
- Below this plane lies the true pelvis; in females, it forms the birth canal through which the baby passes. The female pelvis also has a wider pelvic inlet. The male pelvis is heavier with thicker bones.

Differences between the male and female pelvis

Adult male pelvis from the front

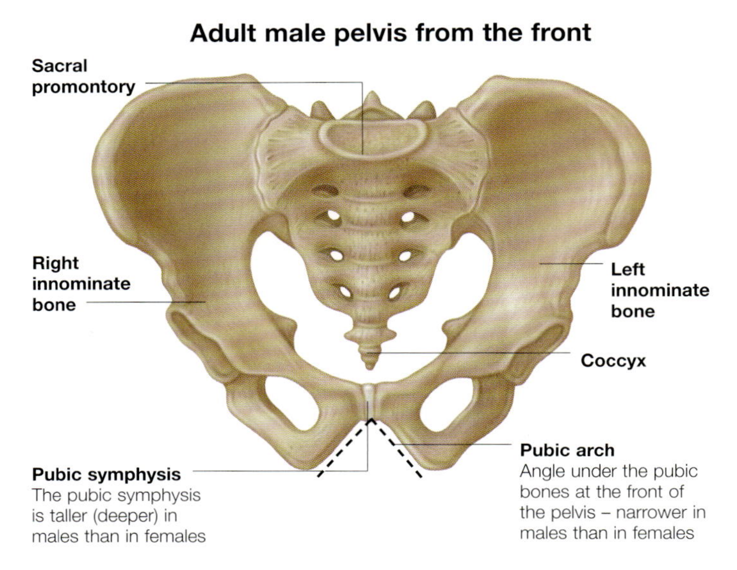

The skeletons of men and women differ in a number of places, but nowhere is this more marked than in the pelvis.

Physical variations

The differences between the male and female pelvis can be attributed to two factors: the requirements of childbirth and the fact that, in general, men are heavier and more muscular than women. Some of the more obvious differences are:

- General structure – the male pelvis is heavier, with thicker bones
- Pelvic inlet – the 'way into' the true pelvis is a wide oval in females but narrower and heart-shaped in males
- Pelvic canal – the 'way through' the true pelvis is roughly cylindrical in females, whereas in males it tapers
- Pubic arch – the angle under the pubic bones at the front of the pelvis is wider in females (100 degrees or more) than in males (90 degrees or less).

These differences, together with other, more subtle, measurements, may be used by forensic pathologists and anthropologists to determine the sex of a skeleton.

The male pelvis differs from the female pelvis in being heavier, with thicker bones. The pubic arch is narrower and the pubic symphysis deeper in males.

Hip bone

The two hip bones are fused together at the front and join with the sacrum at the back. They each consist of three bones – the ilium, ischium and pubis.

The two innominate (hip) bones constitute the greater part of the pelvis, joining with each other at the front and with the sacrum at the back.

Structure

The hip bone is large and strong, as it transmits forces between the legs and the spine. As with most bones, it has areas that are raised or roughened for the attachment of muscle or ligaments. The hip bone is formed by the fusion of three bones: the ilium, the ischium and the pubis. In children, these three bones are joined only by cartilage. At puberty, they fuse to form the single innominate, or hip, bone on each side.

Features

The upper margin of the hip bone is formed by the widened iliac crest. Further down the hip bone becomes the ischial tuberosity, a projection of the ischium. The obturator foramen lies below and slightly in front of the acetabulum, the latter receiving the head of the femur (thigh bone).

Right hip bone, lateral view

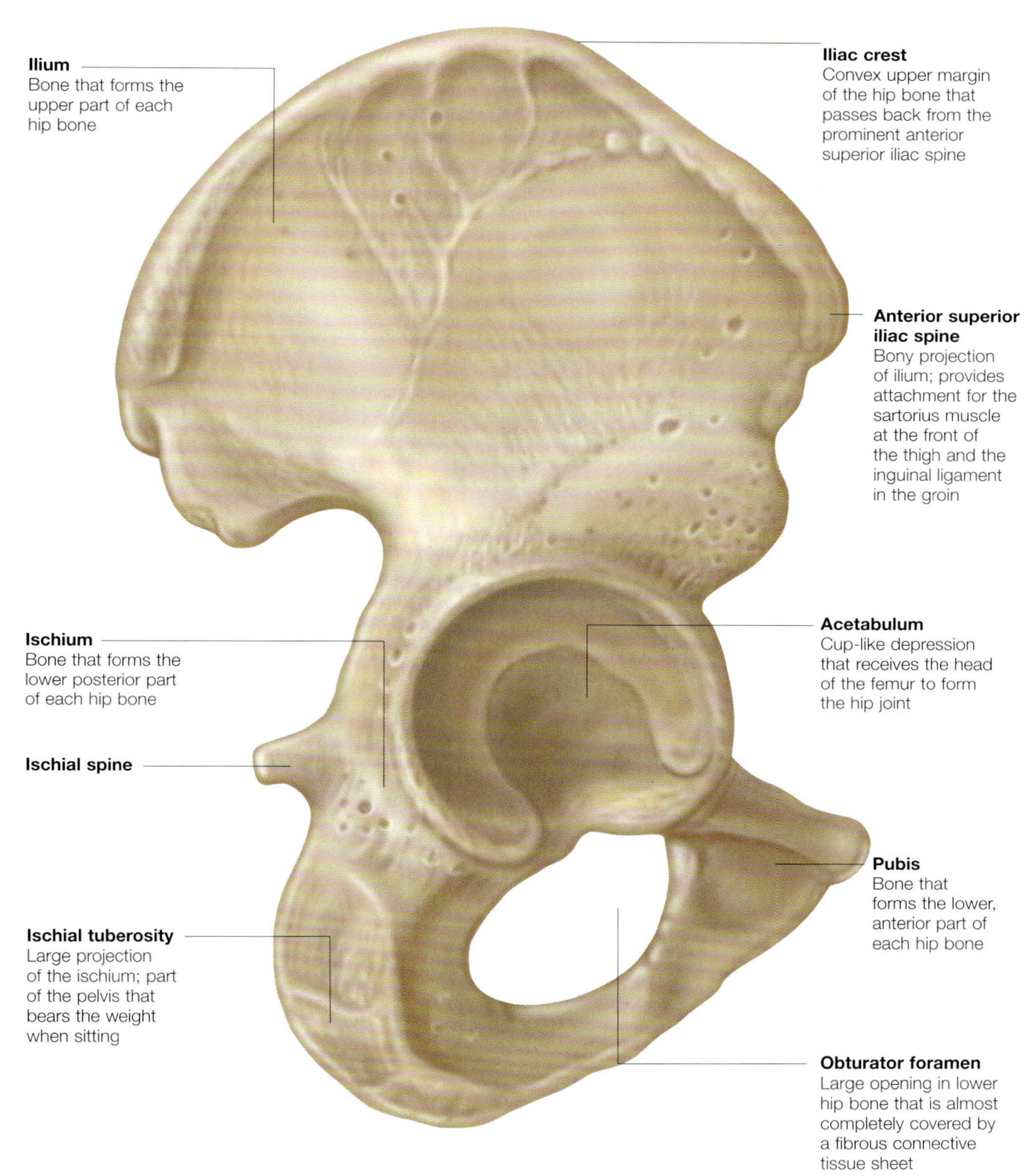

This lateral view of the hip bone clearly shows its constituent parts of ilium, ischium and pubis. These three bones fuse together at puberty.

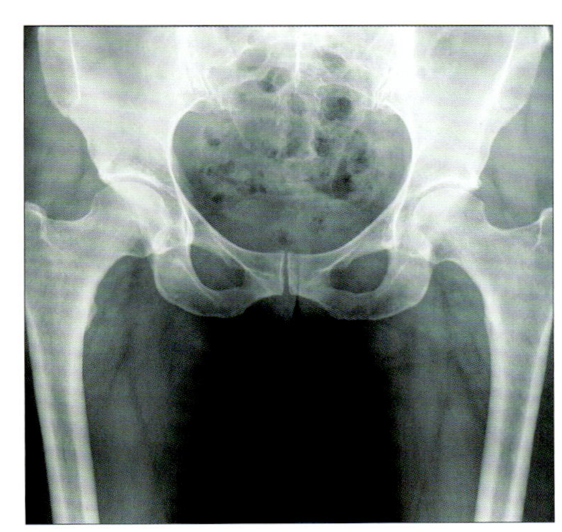

This plain x-ray shows the front view of the hip bones, with the head of the femor resting in the acetabulum.

Ligaments of the Pelvis

The pelvic bones are connected by joints that are bound together by ligaments to form a solid structure. The ligaments of the pelvis are some of the strongest in the body.

The pelvis must be structurally strong in order to perform its functions of transferring weight to the legs, and supporting the abdominal contents. The pelvic bones are themselves thick and strong but their overall stability is assured by the presence of a series of tough pelvic ligaments that bind them together.

Structure of Pelvis

The pelvis is formed from the paired innominate (hip) bones together with the sacrum and coccyx. These bones have joints between them and the pelvic ligaments are arranged so that they hold these joints together when resisting forces that would otherwise pull them apart.

Front View

The major ligaments of the pelvis are generally named after the two areas of bone that they link. Those most visible on an anterior (frontal) view of the pelvis are:

- Iliolumbar ligament
- Anterior sacroiliac ligament
- Sacrospinous ligament
- Anterior longitudinal ligament.

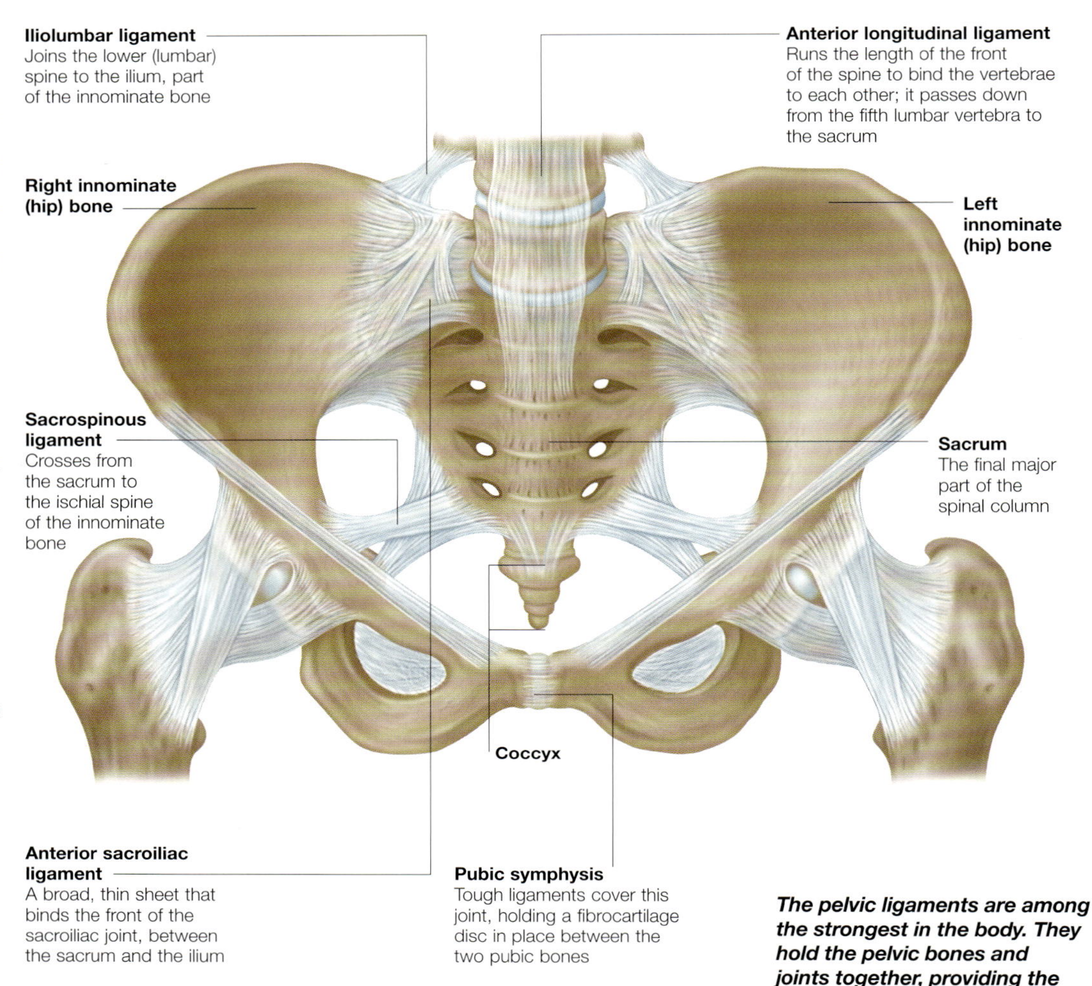

The pelvic ligaments are among the strongest in the body. They hold the pelvic bones and joints together, providing the necessary structural stability.

Posterior pelvic ligaments

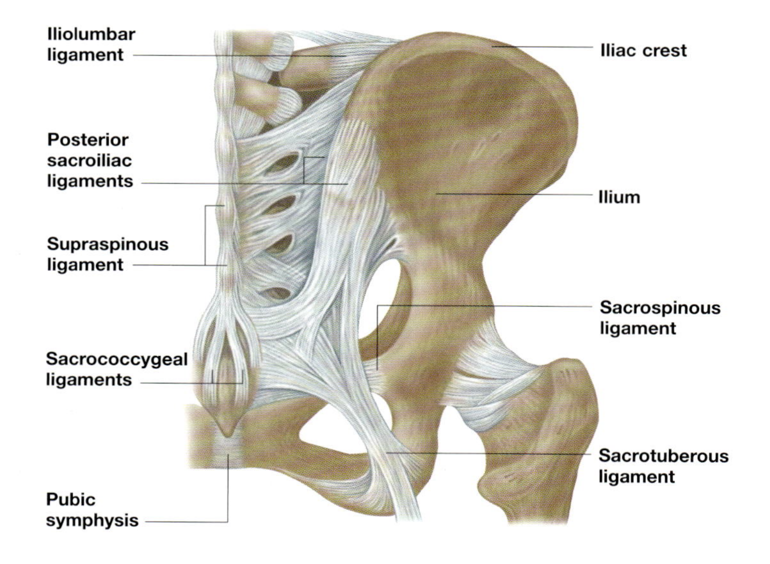

Viewing the pelvis from the rear shows the arrangement of ligaments that bind the bones at the back of the pelvis.

Function of ligaments

The posterior sacroiliac ligament crosses the sacroiliac joint as it passes downwards and inwards from the ilium to the sacrum. This ligament is stronger than the relatively thin anterior sacroiliac ligament, taking much of the strain of keeping the ilium connected to the sacrum on each side.

At the rear of the pelvis there are a number of ligaments. Each ligament strengthens the link between the different pelvic bones at the site of the joints.

The large and powerful sacrotuberous ligament can be clearly seen as it passes from the sacrum down to the roughened ischial tuberosity. Together with the sacrospinous ligament, which lies just in front of it, the sacrotuberous ligament acts to resist the rotational force on the sacrum generated by the weight of the body.

Just as the anterior longitudinal ligament joins the front of the spinal column, the tough supraspinous ligament stabilizes the vertebrae from the rear, joining their spinous processes. It terminates by forming the sacrococcygeal ligaments.

Joints of the Pelvis

The pelvis is a ring of bone on which the weight of the body is carried. Where bone meets bone, the pelvic joints are formed, bound together by the pelvic ligaments.

The pelvis is formed by the paired innominate bones and the sacrum with the coccyx attached. The pelvic joints that connect these bones, unlike the elbow or knee joints, are not designed to allow movement. The joints of the pelvis are bound strongly together by the pelvic ligaments to form a single, solid structure.

The Sacroiliac Joint

The sacroiliac joint is the largest and most important joint of the pelvis, between the sacrum (part of the lower spine) and the ilium (part of the large innominate bone). This joint needs to be strong as it bears the weight of the body. The shape of the bones at the joint surface contributes to the stability of this joint, having irregular indentations that can partially interlock.

However, the main stabilizing factor is the presence of the very strong posterior sacroiliac ligaments and the interosseous ligaments. These ligaments 'suspend' the sacrum between the two iliac bones, bearing the weight of the upper body.

Joint Movement

The sacroiliac joint belongs to the group of synovial, or fluid-filled, joints such as those of the elbow, shoulder and knee. However, unlike these joints there is very little movement in the sacroiliac pelvic joint to allow for stability.

Section through the sacroiliac joints and ligaments

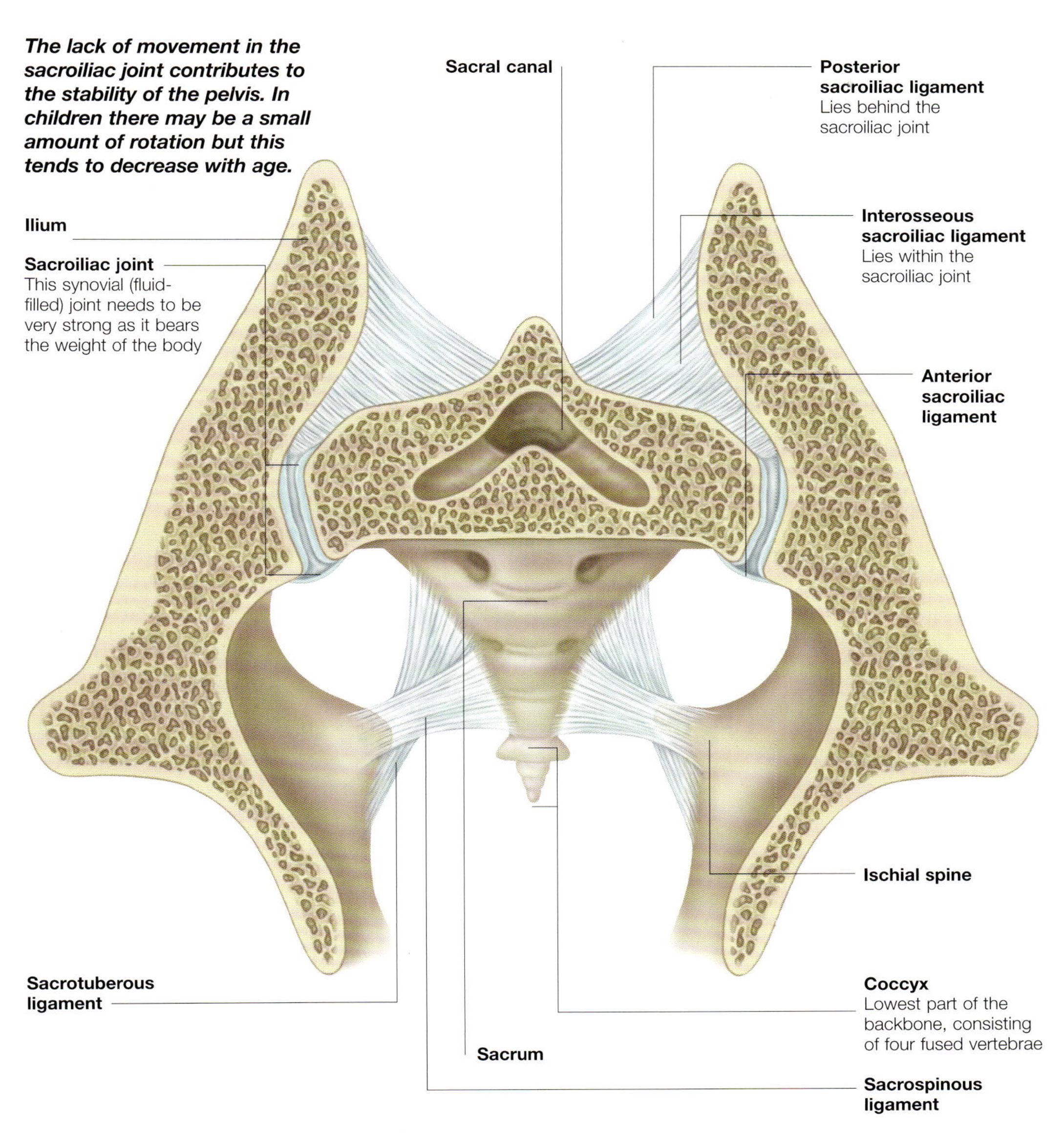

The lack of movement in the sacroiliac joint contributes to the stability of the pelvis. In children there may be a small amount of rotation but this tends to decrease with age.

Pubic symphysis

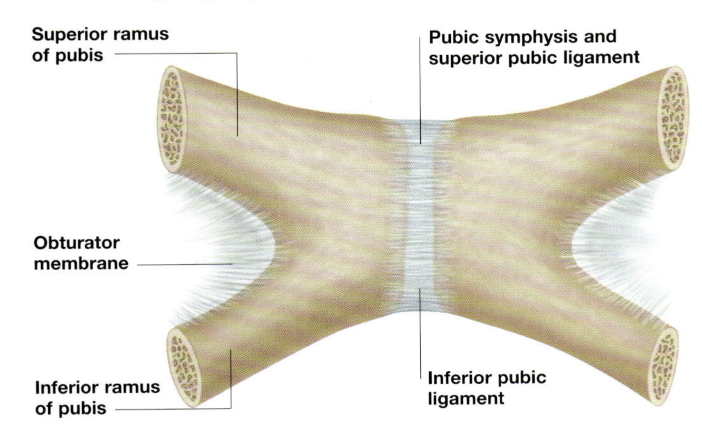

The pubic symphysis is the joint between the two pubic bones at the front of the pelvis. This is a very strong and stable joint, allowing almost no movement between the two pubic bones.

Cartilage

The pubic symphysis is a cartilaginous joint, the two bone surfaces being covered with a layer of hyaline cartilage and connected by fibrous ligaments.

Disc

Between the two bones within the joint there is fibrocartilage, which has a small cavity in the midline. This tissue tends to be more extensive in women than in men.

The joint, and especially the springy fibrocartilage, serves to act as a 'shock absorber', helping to reduce the chance of fracture of the bones when the pelvis receives sudden forces either directly or from the legs.

The pubic symphysis connects the two pubic bones. The bones are held in place by ligaments, and the bone surface is lined with cartilaginous tissue.

Pelvic Floor Muscles

The muscles of the pelvic floor play a vital role in supporting the abdominal and pelvic organs. They also help to regulate the processes of defecation and urination.

The pelvic floor muscles play a key role in supporting the abdominal and pelvic organs. In pregnancy, these muscles help to carry the growing weight of the uterus, and in childbirth they support the baby's head as the cervix dilates.

The muscles of the pelvic floor are attached to the inside of the ring-shaped pelvic skeleton, and slope downwards to form a rough funnel. The levator ani is the largest muscle of the pelvic floor. It is a wide, thin sheet made up of three parts:

- Pubococcygeus – the main part of the levator ani muscle
- Puborectalis – joins with its counterpart on the other side to form a U-shaped sling around the rectum
- Iliococcygeus – the posterior fibres of the levator ani. A second muscle, the coccygeus (or ischiococcygeus), lies behind the levator ani.

Perineal Body

The perineal body is a small mass of fibrous tissue that lies within the pelvic floor, in front of the anal canal. This provides a site for the attachment of many of the pelvic floor and perineal muscles, allowing paired muscles to pull against each other, normally one of the functions of bone. It also supports the internal organs of the pelvis.

Female pelvic diaphragm from above

The pelvic floor muscles are known as the pelvic diaphragm. The levator ani is the most important muscle and is named for its action in lifting the anus.

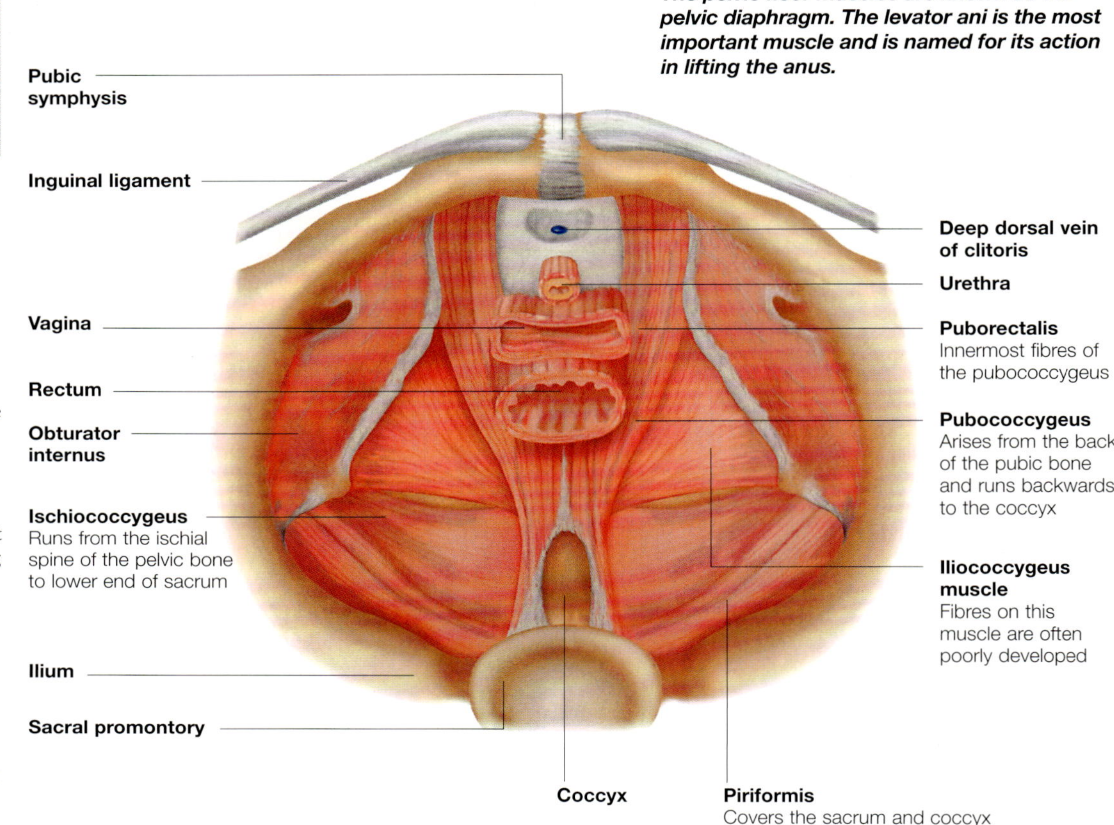

Perineal body

Female pelvis

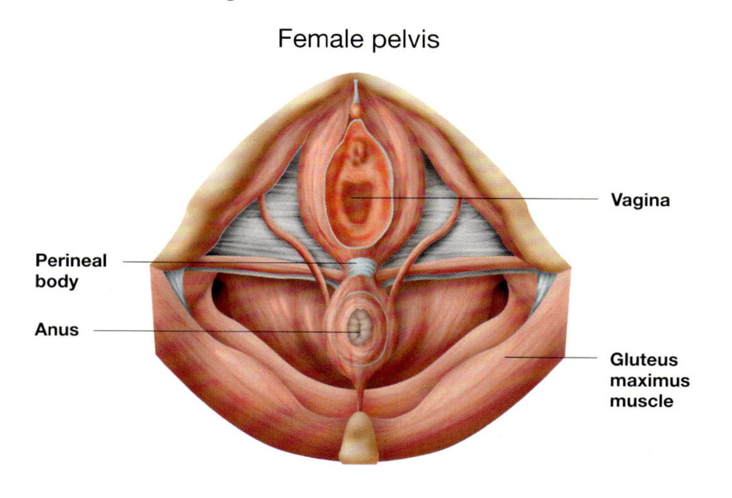

Episiotomy

The perineal body may become damaged during childbirth, either by stretching or tearing as the baby's head passes through the pelvic floor. Loss of the perineal body's support of the posterior vaginal wall may eventually lead to vaginal prolapse.

To prevent damage to the perineal body during childbirth, an obstetrician may perform an episiotomy. This deliberate incision into the muscle behind the vaginal opening enlarges this opening and avoids damage to the perineal body.

Although the perineal body is small and tucked away, it is a very important structure. It supports the organs of the pelvis that lie above it.

Openings of the pelvic floor

The pelvic floor resembles the diaphragm in the chest in that it forms a nearly continuous sheet with openings to allow important structures to pass through it.

From below, the pelvic floor can be seen to assume a funnel shape. The muscles of the pelvic floor are so arranged that there are two main openings:

- Anorectal hiatus – this opening, or hiatus, allows the rectum and anal canal to pass through the sheet of pelvic floor muscles to reach the anus beneath. The U-shaped fibres of the puborectalis muscle form the posterior edge of this hiatus.
- Urogenital hiatus – lying in front of the anorectal hiatus there is an opening in the pelvic floor for the urethra, which carries urine from the bladder out of the body. In females, the vagina also passes through the pelvic diaphragm within this opening, just behind the urethra.

Pelvic floor functions

The functions of the pelvic floor include:

- Supporting the internal organs of the abdomen and pelvis
- Helping to resist rises in pressure within the abdomen, such as during coughing and sneezing, which would otherwise cause the bladder/bowel to empty
- Assisting in the control of defecation and urination
- Helping to brace the trunk during forceful movements of the upper limbs, such as lifting heavy loads.

Male pelvic diaphragm from below

Pubic symphysis

Deep dorsal veins of penis

Urethra

Obturator internus

Rectum

Pubococcygeus

Iliococcygeus

Tip of coccyx

Gluteus maximus

Sacrum

Puborectalis
Has U-shaped fibres that form the posterior edge of the anorectal hiatus

Male pelvic diaphragm from above

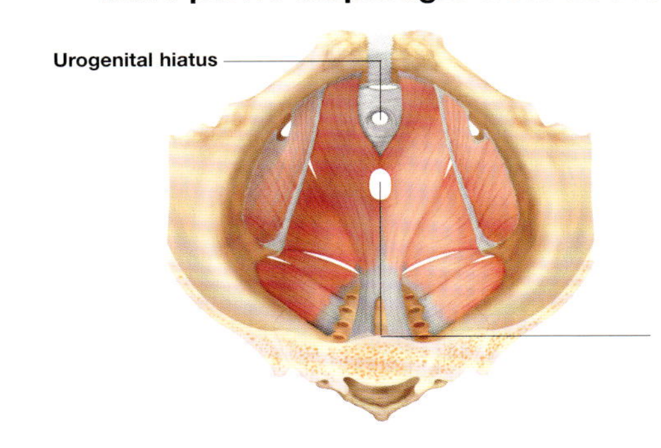

The pelvic floor muscles play a vital supporting role. Without them, the internal organs of the abdomen and pelvis would sink through the bony pelvic ring.

Anorectal hiatus
Normally a small opening, this must be able to expand during a bowel movement.

Ischioanal fossae

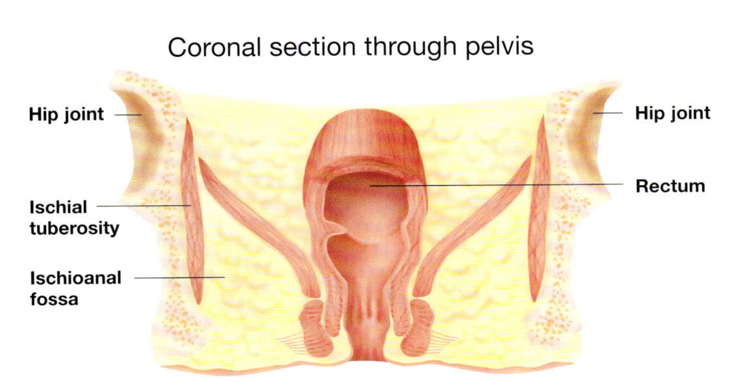

The ischioanal, or ischiorectal, fossae are spaces formed between the outside of the pelvic diaphragm and the skin around the anus.

The ischioanal fossae are filled with fat. This fat is divided into sections and supported by bands of connective tissue. The fat in the ischioanal fossae acts as a soft packing material that accommodates changes in the size and position of the anus during a bowel movement.

The ischioanal fossae are wedge-shaped, being narrowest at the top and widest at the bottom. The fossae are filled with sections of fat.

Muscles of the Gluteal Region

The gluteus maximus is the largest and heaviest of all the gluteal muscles and is situated in the buttock region. This strong, thick muscle plays an important part in enabling humans to stand.

The gluteal, or buttock, region lies behind the pelvis. The shape is formed by a number of large muscles that help to stabilize and move the hip joint. A layer of fat covers these muscles.

Gluteus Maximus

This is one of the largest muscles in the human body. It covers the other gluteal muscles with the exception of about one-third of the smaller gluteus medius.

The gluteus maximus arises from the ilium (a part of the bony pelvis), the back of the sacrum and the coccyx. Its fibres run down and outwards at a 45 degree angle towards the femur. Most of the fibres then insert into a band (the iliotibial tract).

Actions

The main function of the gluteus maximus is to extend (straighten) the leg as in standing from a sitting position.

When the leg is extended, as in standing, the gluteus maximus covers the bony ischial tuberosity. This bears the weight of the body when sitting. However, we never sit on the gluteus maximus muscle itself as it moves up and away from the ischial tuberosity when the leg is flexed (bent forward).

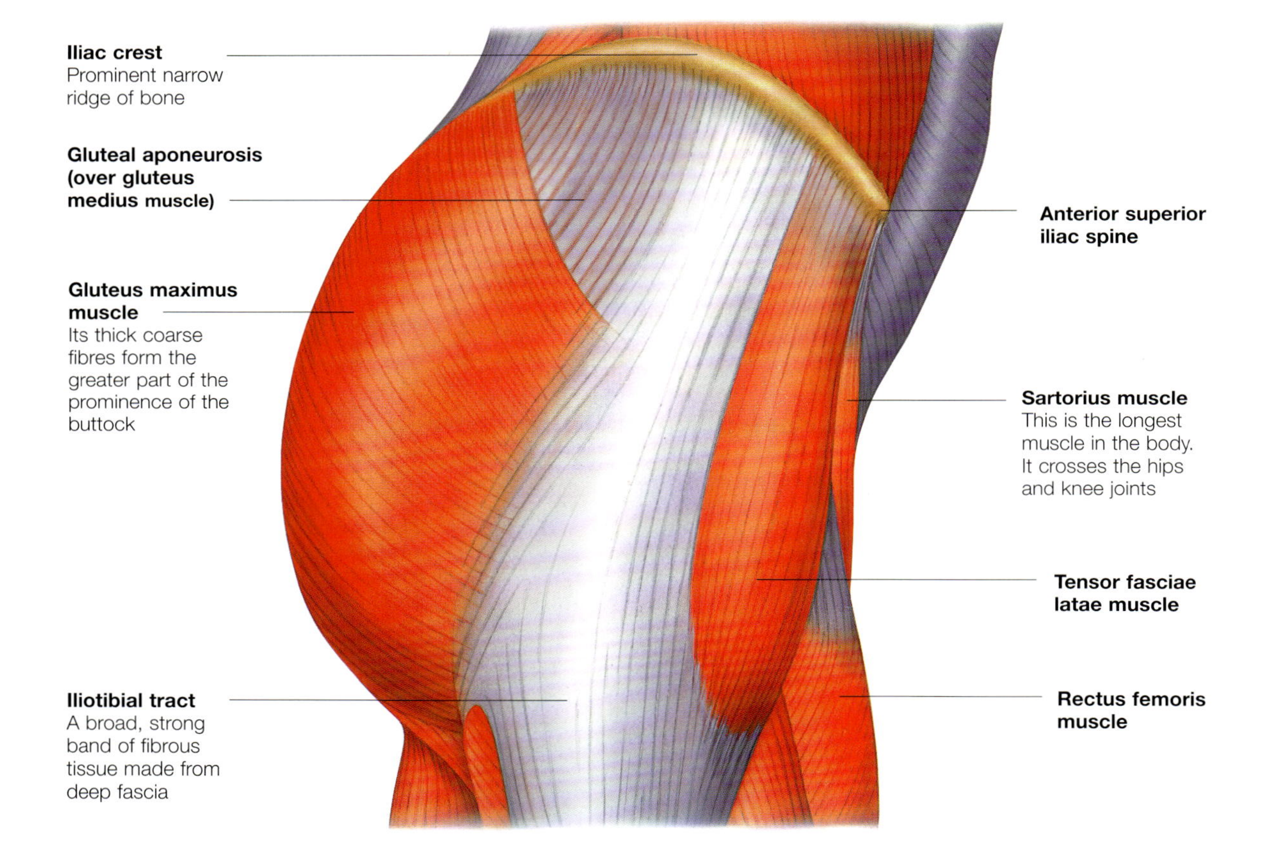

The gluteus maximus muscle is not very active during normal walking but comes into play during forceful actions such as running or walking upstairs.

Deeper muscles of the gluteal region

The muscles that lie deep to the gluteus maximus region play an important role in walking. They keep the pelvis level as each foot is lifted off the ground.

Quadratus femoris
This rectangular-shaped muscle extends laterally from the pelvis

Ischial tuberosity
The ischial tuberosities are the strongest parts of the hip bones

Gluteus maximus

Greater trochanter

Gluteus medius muscle
This is a thick muscle largely covered by the gluteus maximus

Gluteus minimus muscle
The smallest and deepest of the gluteal muscles

Piriformis muscle

Superior gemellus muscle

Obturator internus

Inferior gemellus muscle

The gluteus medius and minimus muscles rotate the thigh laterally and stabilize the hip.

Beneath the gluteus maximus lie a number of other muscles that act to stabilize the hip joint and move the lower limb.

Gluteus medius and minimus

The gluteus medius and gluteus minimus muscles lie deep to the gluteus maximus. They are both fan-shaped muscles with fibres that run in the same direction. Gluteus medius lies directly beneath gluteus maximus, with only about one-third of it not covered by this larger muscle. Its fibres originate from the external surface of the ilium (part of the pelvis) and insert into the greater trochanter, a protuberance of the femur.

Gluteus minimus lies directly beneath gluteus medius and is of a similar fan-like shape. Its fibres also originate from the ilium and insert into the greater trochanter.

Role in walking

The gluteus medius and gluteus minimus together have an essential role in the action of walking. These muscles act to hold the pelvis level when one foot is lifted from the ground, rather than letting it sag to that side. This allows the non-weight-bearing foot to clear the ground before being swung forward. A number of other muscles lie within this region, acting mainly to help certain movements of the lower limb at the hip. These include:

- Piriformis – this muscle, which is named for its pear shape, lies below gluteus minimus. It acts to rotate the thigh laterally, a movement that results in the foot turning outwards.
- Obturator internus, superior and inferior gemelli – these three muscles together form a composite three-headed muscle that lies below the piriformis muscle. These muscles rotate the thigh laterally and stabilize the hip joint
- Quadratus femoris – this short, thick muscle rotates the thigh laterally and helps to stabilize the hip joint.

Bursae of the gluteal region

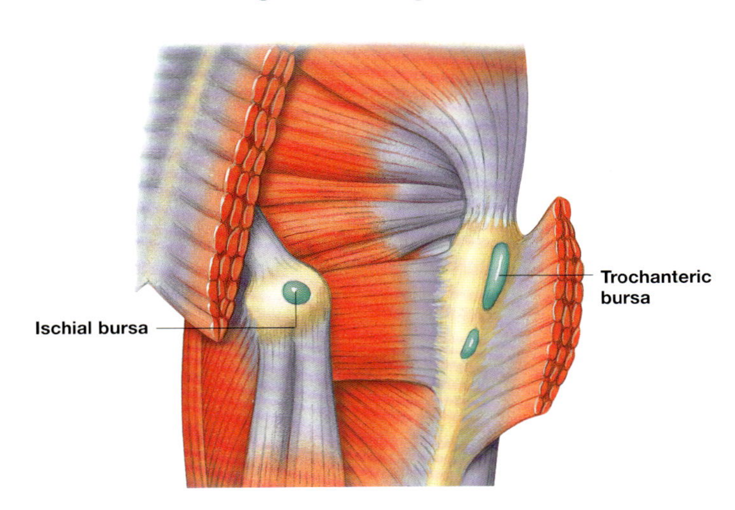

A bursa is a small fluid-filled sac rather like an underfilled water bottle. In many places in the body a bursa will be found where two structures, usually bone and tendon, move against each other.

Protection

The bursa lies between bone and muscle, protecting them from wear and tear.

There are three main groups of bursae in the gluteal region:

The gluteal region contains three main groups of bursae. The bursae help to ease the movement of the bones and tendons upon each other.

- The trochanteric bursae
These large bursae lie between the thick, upper fibres of the gluteus maximus muscle and the greater trochanter of the upper femur (thigh bone).

- The ischial bursa
This bursa, if present, lies between the lower fibres of the gluteus maximus muscle and the ischial tuberosity, the part of the pelvis that bears our weight during sitting.

- The gluteofemoral bursa
This bursa lies on the outer side of the leg, between the gluteus maximus and vastus lateralis muscles.

Hip Joint

The hip joint is the strong ball-and-socket joint that connects the lower limb to the pelvis. Of all the body's joints, the hip is second only to the shoulder in the range of movements it allows.

In the hip joint, the head of the femur (thigh bone) is the 'ball' that fits tightly into the 'socket' formed by the cup-like acetabulum of the hip bone.

Synovial Joint

The articular surfaces – the parts of the bone that come into contact with each other – are covered by a protective layer of hyaline cartilage, which is very smooth and slippery. The hip joint is a synovial joint, which means that movement is further lubricated by a thin layer of synovial fluid. The fluid lies between the articular surfaces within the synovial cavity and is secreted by the synovial membrane.

Acetabular Labrum

The depth of the socket formed by the acetabulum is increased by the presence of a ring of cartilage called the acetabular labrum. This structure brings greater stability to the joint, allowing the almost spherical femoral head to rest deep within the joint. The cartilage-covered articular surface of the acetabulum is not a continuous cup, or even a ring, but is horseshoe-shaped. There is a gap, the acetabular notch, at the lowest point, which is bridged by the complete ring of the acetabular labrum. The open centre of the 'horseshoe' is filled with a cushioning pad of fat.

Cross-section of the right hip joint

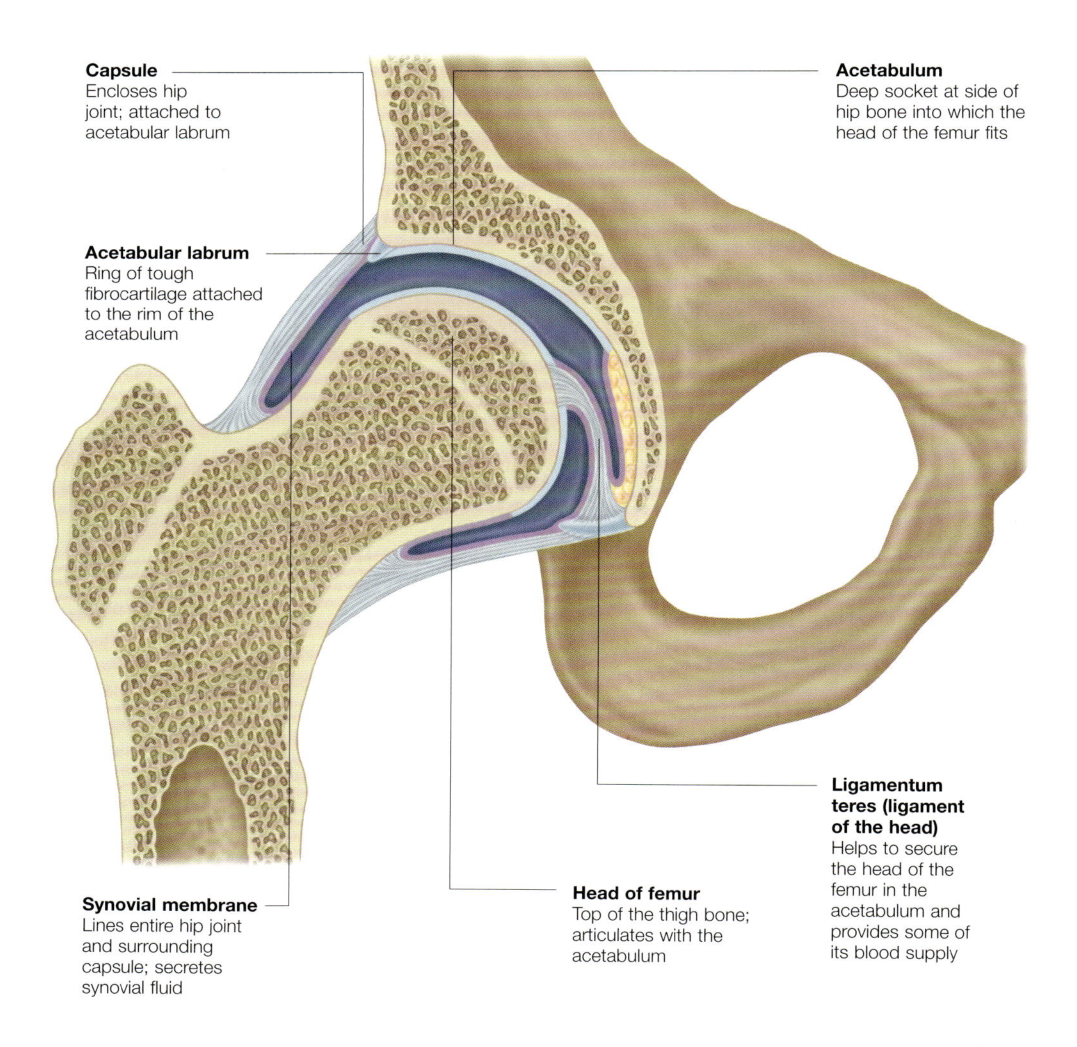

The hip joint is the ball-and-socket joint between the head of the femur and the hip bone. The joint is capable of a wide range of movement.

Ligaments of the hip joint

The hip joint is enclosed and protected by a thick, fibrous capsule. The capsule is flexible enough to allow the joint a wide range of movements but is strengthened by a number of tough ligaments.

The ligaments of the hip joint are thickened parts of the joint capsule, which extends from the rim of the acetabulum down to the neck of the femur.

These ligaments, which generally follow a spiral path from the hip bone to the femur, are named according to the parts of the bone to which they attach:

- Iliofemoral ligament
- Pubofemoral ligament
- Ischiofemoral ligament.

Movement and stability

The ball-and-socket nature of the hip joint allows it great mobility, second only to the shoulder joint in its range of movement. Unlike the shoulder, however, it needs to be very stable as it is a major weight-bearing joint. The hip joint is capable of the following movements:

- Flexion (bending forward, the knee coming up)
- Extension (bending the leg back behind the body)
- Abduction (moving the leg out to the side)
- Adduction (bringing the leg back to the midline)
- Rotation, which is greatest when the leg is flexed.

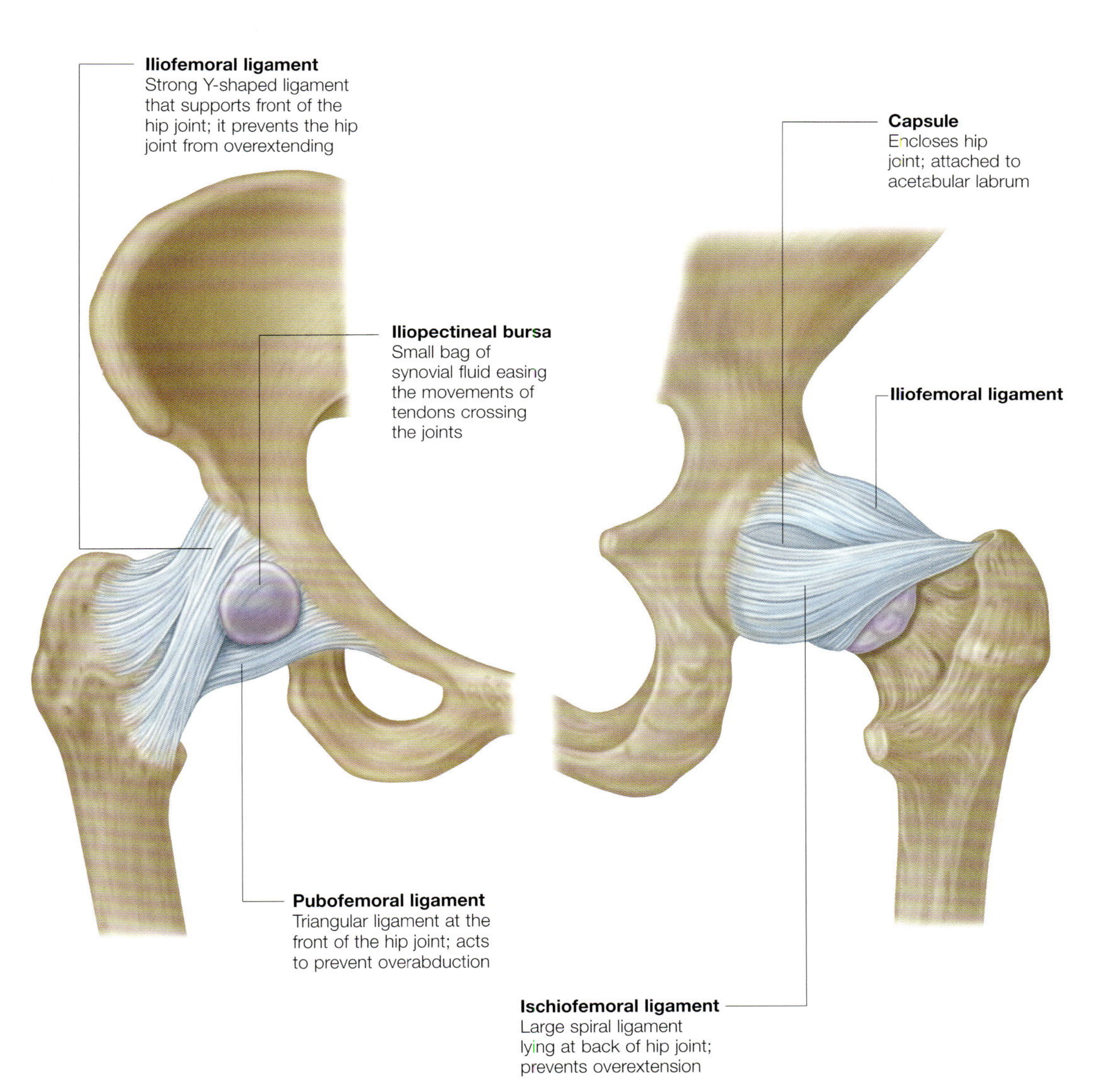

A fibrous capsule encloses the hip joint. This capsule is reinforced by a number of ligaments that spiral down from the hip bone to the femur.

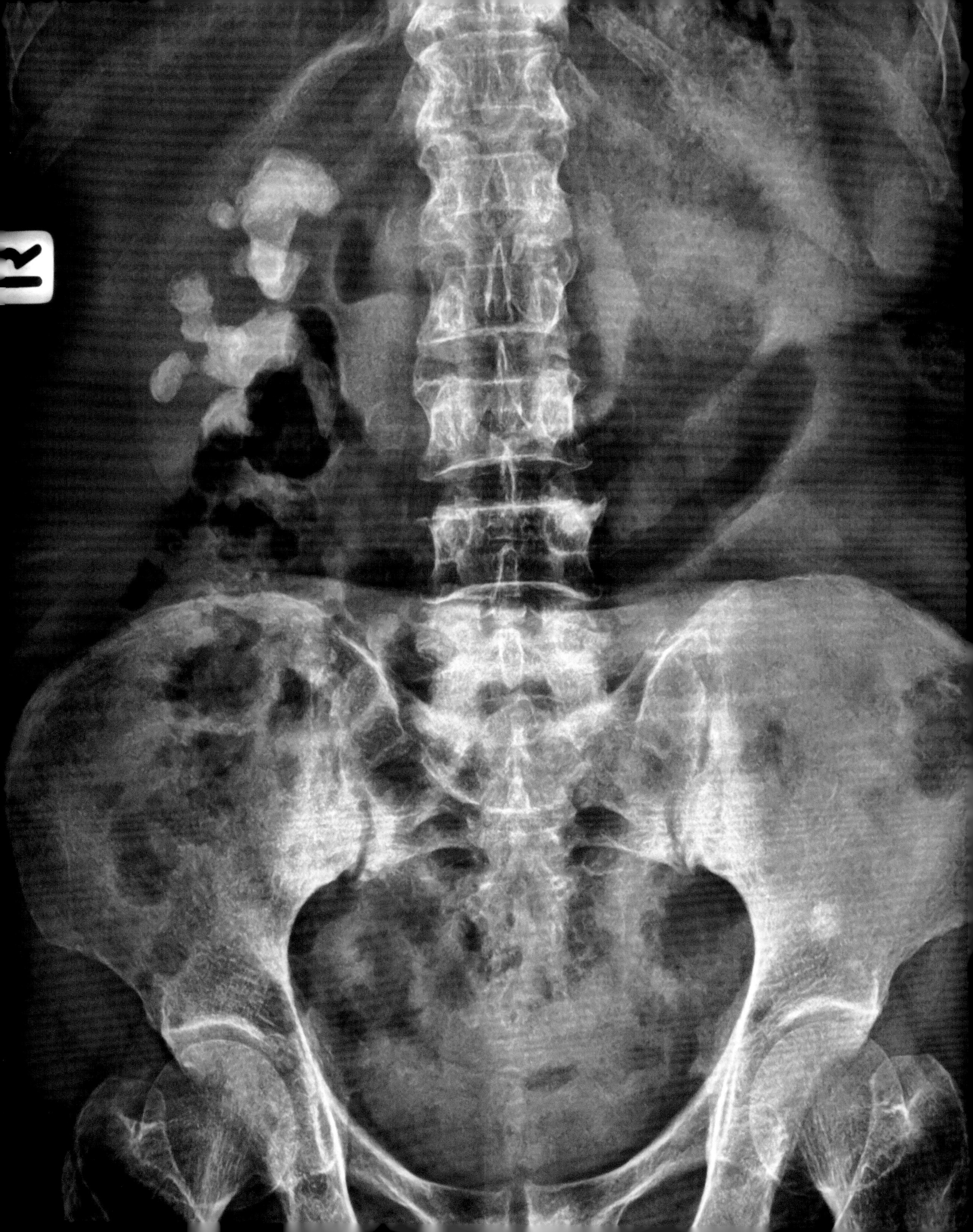
R

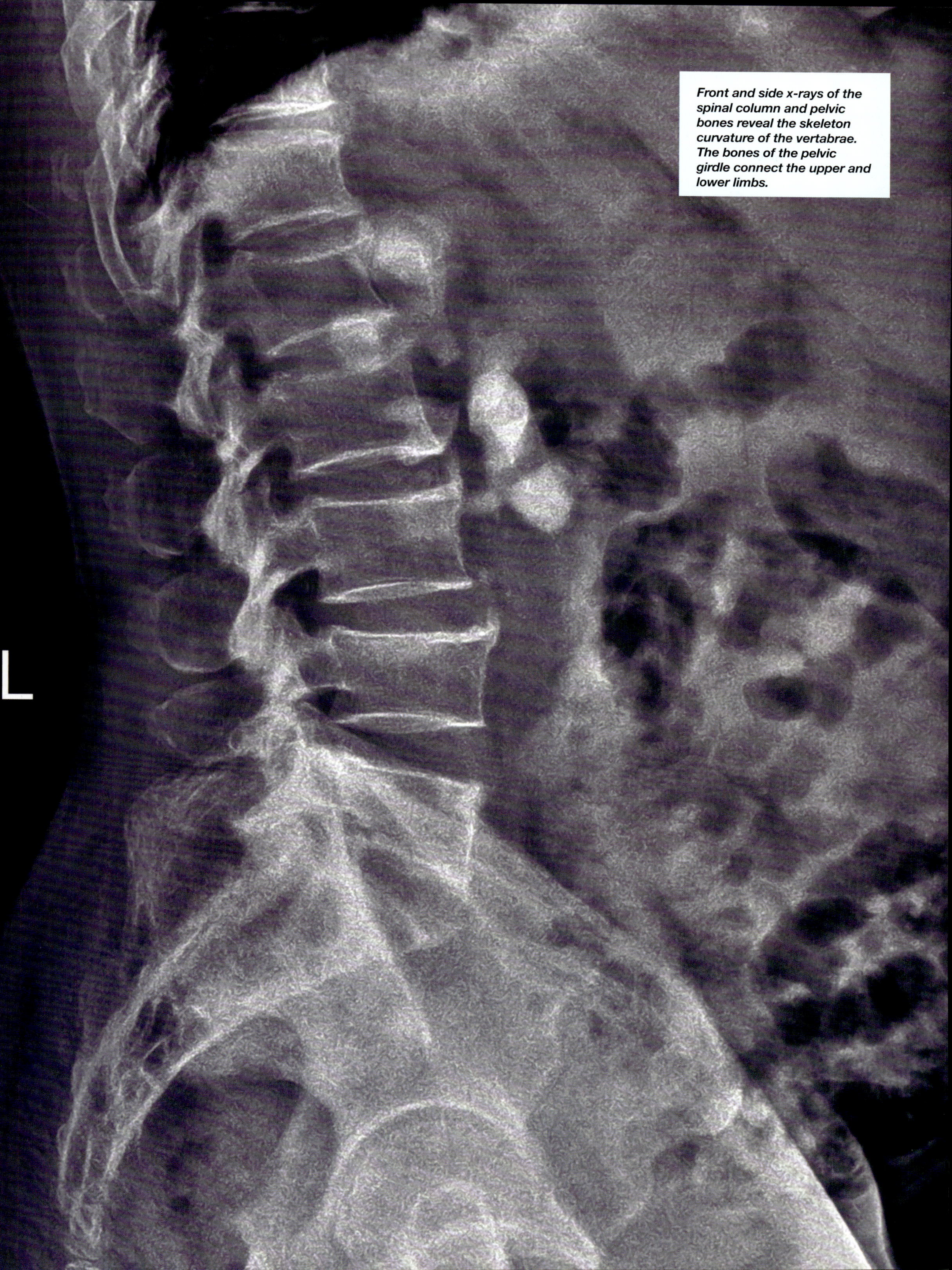

Front and side x-rays of the spinal column and pelvic bones reveal the skeleton curvature of the vartabrae. The bones of the pelvic girdle connect the upper and lower limbs.

Femur

The femur, or thigh bone, is the longest and heaviest bone in the body. Measuring about 45cm (18in) in length in adult males, the femur makes up about one quarter of a person's total height.

The femur has a long, thick shaft with two expanded ends. The upper end articulates with the pelvis to form the hip joint while the lower end articulates with the tibia and patella to form the knee joint.

Upper End

The femur upper end includes:

- Head – this is the near-spherical projection that forms the 'ball' of the ball and socket hip joint
- Neck – this is the narrowed area that connects the head to the body of the femur
- Greater and lesser trochanters – projections of bone allowing the attachment of muscles.

Shaft

The long central shaft of the femur is slightly bowed, being concave on its posterior surface. For much of its length the femur appears cylindrical, with a circular cross-section.

Lower End

The lower end of the femur is made up of two enlarged bony processes, the medial and lateral femoral condyles. These carry the smooth, curved surfaces that articulate with the tibia and patella to form the knee joint. The shape of the femoral condyles is outlined when the leg is viewed with the knee bent.

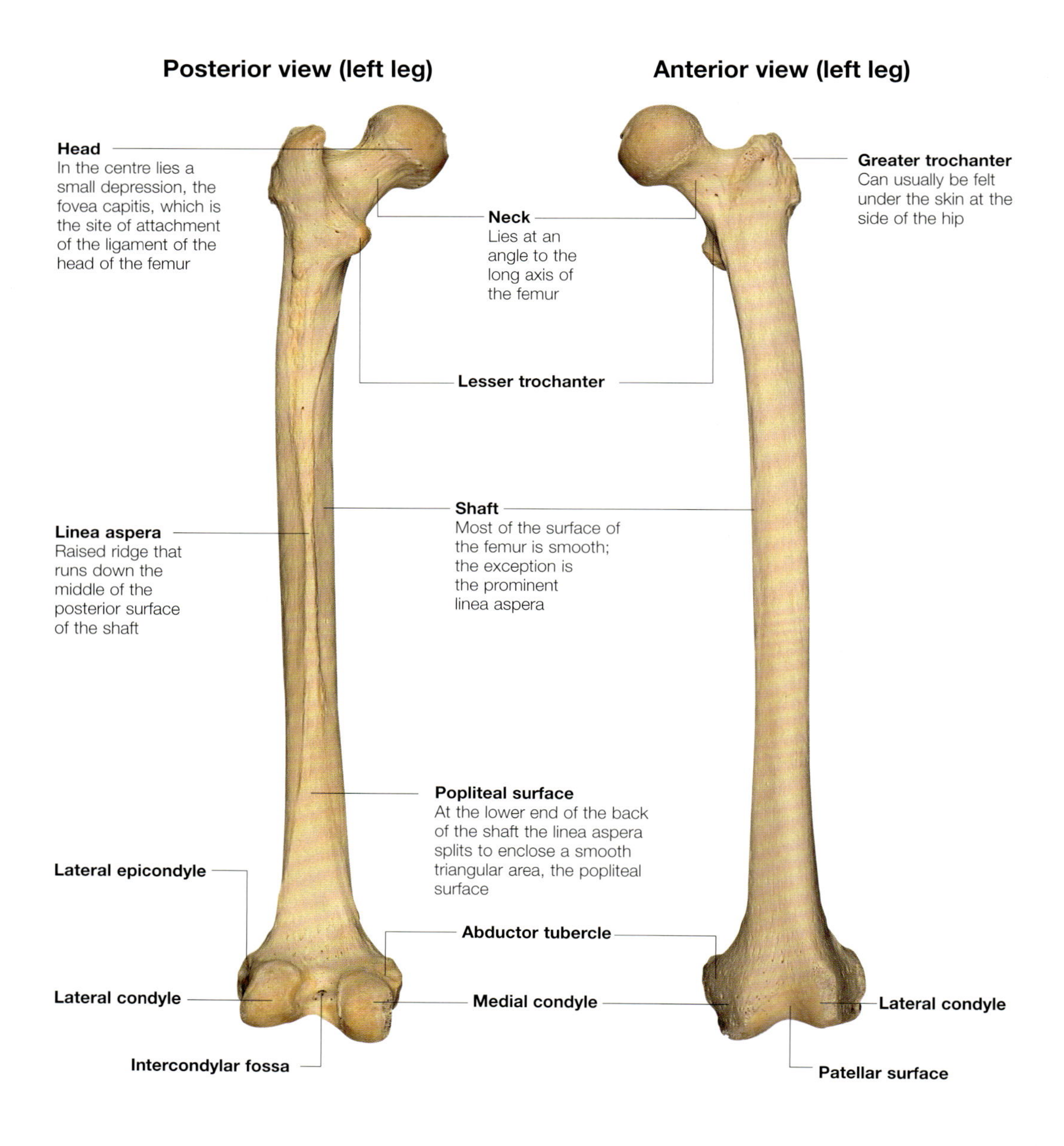

The femur runs from the hip joint to the knee. It is the longest bone in the body and is very strong.

Muscle attachments of the femur

The femur is a very strong bone that provides sites of attachment for many of the muscles of locomotion in the hip joint and legs.

Muscle origins
Some muscles, such as the powerful gluteus muscles, have their origins on the pelvic bones and so cross the hip joint to insert into the femur. When these muscles contract they cause the hip joint to move, allowing the leg to bend, straighten or move sideways.

Other muscles originate on the femur itself and pass down across the knee joint to insert on the tibia or the fibula, the two bones of the lower leg. These muscles allow the knee to bend or straighten. Together these muscles bring about movements of the legs such as in climbing, or rising from a sitting position.

Bony processes
Where muscle is attached to bone it causes a projection, or bony process, to arise. If the muscle is powerful, or if a number of muscles attach at the same site, the bony process can be pronounced. This is the case in the femur. The surface of the bone at the site of muscle attachment can also become quite roughened, unlike the smooth bone surface in between.

Points of attachment

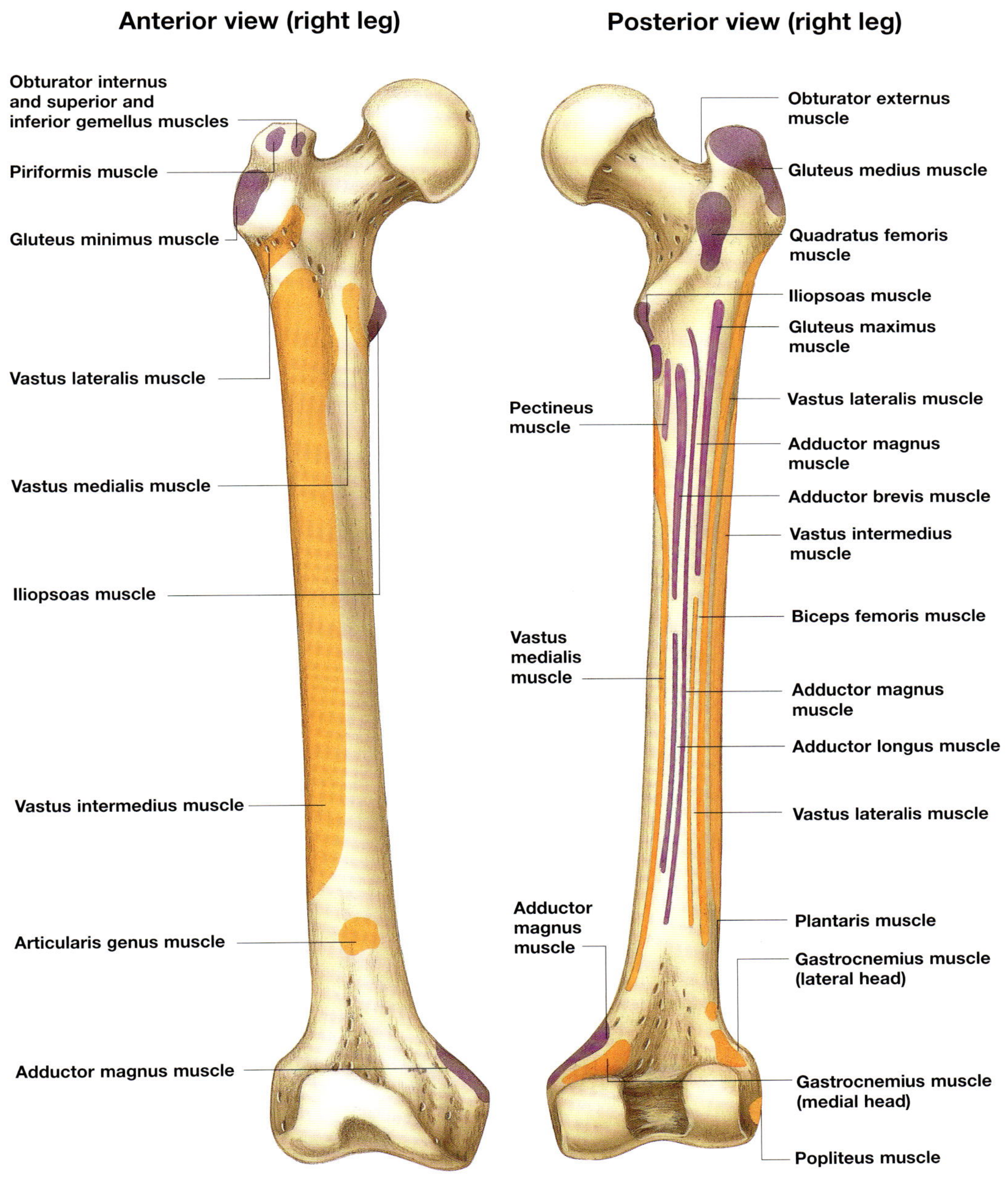

Anterior view (right leg)

Obturator internus and superior and inferior gemellus muscles
Piriformis muscle
Gluteus minimus muscle
Vastus lateralis muscle
Vastus medialis muscle
Iliopsoas muscle
Vastus intermedius muscle
Articularis genus muscle
Adductor magnus muscle

Posterior view (right leg)

Obturator externus muscle
Gluteus medius muscle
Quadratus femoris muscle
Iliopsoas muscle
Gluteus maximus muscle
Vastus lateralis muscle
Pectineus muscle
Adductor magnus muscle
Adductor brevis muscle
Vastus intermedius muscle
Biceps femoris muscle
Vastus medialis muscle
Adductor magnus muscle
Adductor longus muscle
Vastus lateralis muscle
Plantaris muscle
Adductor magnus muscle
Gastrocnemius muscle (lateral head)
Gastrocnemius muscle (medial head)
Popliteus muscle

This illustration shows the locations at which the particular muscles are attached. The areas of bone at the attachments become roughened.

Tibia and Fibula

The tibia and fibula together form the skeleton of the lower leg. The tibia is much larger and stronger than the fibula as it must bear the weight of the body.

Second only to the femur (thigh bone) in size, the tibia (shin bone) has the shape of a typical long bone, with an elongated shaft and two expanded ends. The tibia lies alongside the fibula, on the medial (inner) side, and articulates with the fibula at its upper and lower ends.

Tibial Condyles

The upper end of the tibia is expanded to form the medial and lateral tibial condyles that articulate with the femoral condyles at the knee joint. The lower end of the tibia is less pronounced than the upper end. It articulates with both the talus (ankle bone) and the lower end of the fibula.

Fibula

The fibula is a long, narrow bone that has none of the strength of the tibia. It lies next to the tibia, on its lateral (outer) side, and articulates with that bone. The fibula plays no part in the knee joint, but is an important support for the ankle.

The shaft of the fibula is narrow and bears the grooves and ridges associated with its major role as a site of attachment for leg muscles.

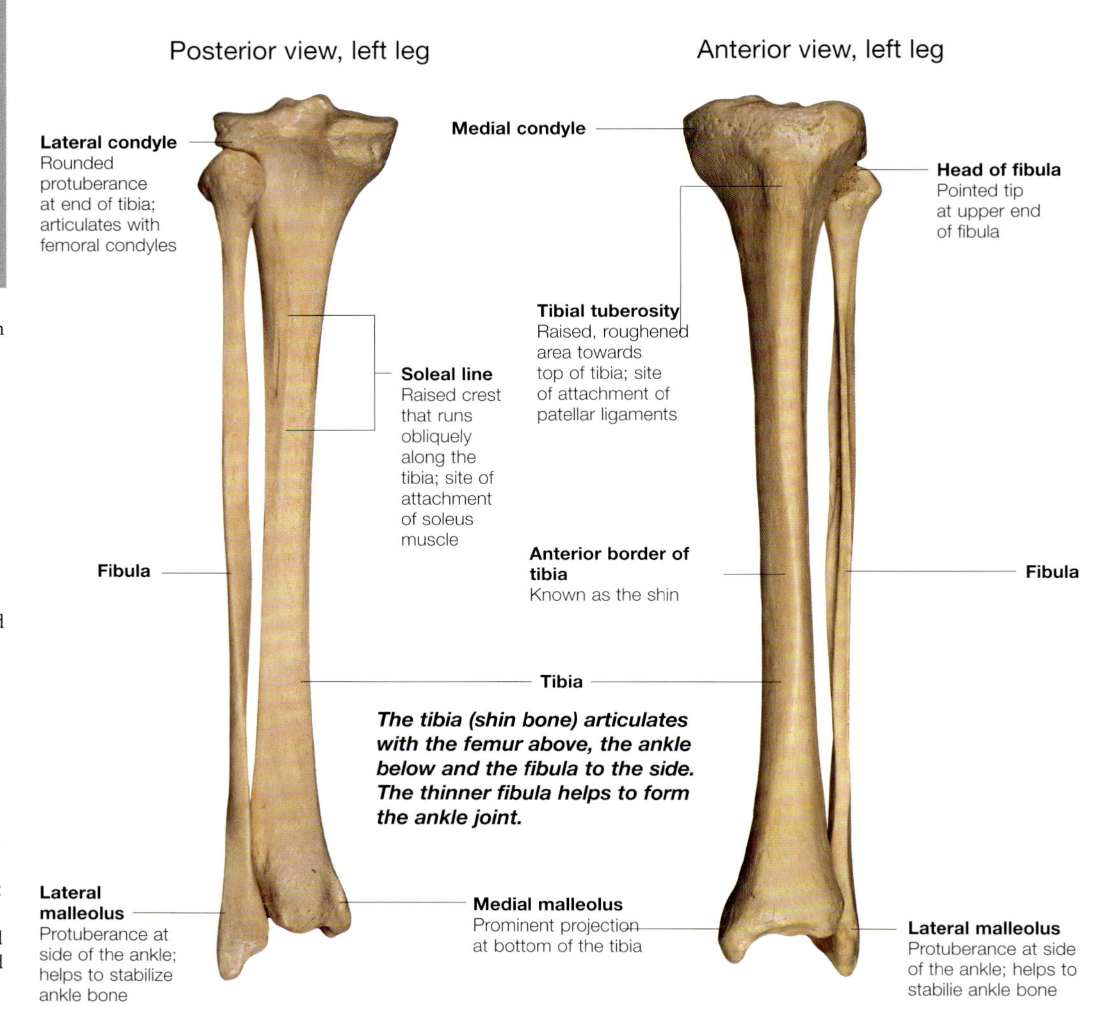

The tibia (shin bone) articulates with the femur above, the ankle below and the fibula to the side. The thinner fibula helps to form the ankle joint.

Cross-section of tibia and fibula

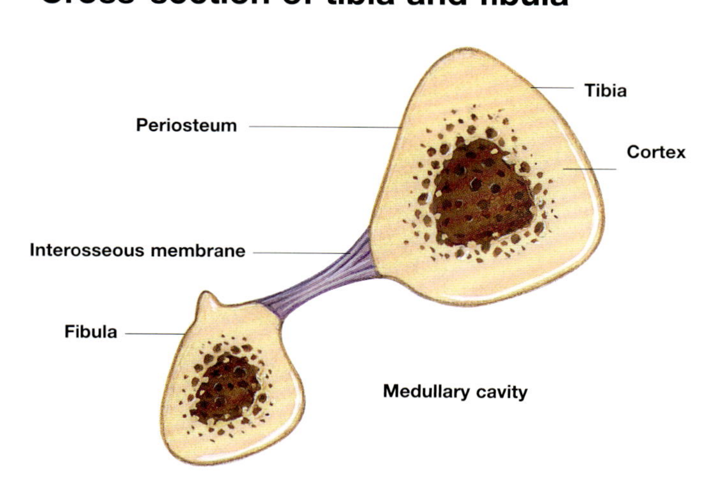

The shafts of the tibia and fibula are roughly triangular in cross-section. The tibial shaft is much greater in diameter than that of the fibula as it is the main weight-bearing element of the lower leg. The fibula acts as a strut, increasing the stability of the lower leg under load.

Long bones

The tibia and fibula have a typical long bone structure, with a thick, tubular outer cortex surrounding a spongy medullary cavity. Their hollow structure provides maximal mechanical strength with minimal support material, namely dense, cortical bone.

The shape of the bones is genetically determined, but is modified by the pull of developing muscles during childhood and into adulthood. In this way the bony ridges, such as the soleal line and tuberosities, form.

The tibia and fibula are enveloped in the periosteum (a tough connective tissue layer). The periosteum from the lateral border of the tibia and the medial border of the fibula blends into the interosseous membrane.

In cross-section, the tibial and fibular shafts are triangular in shape. The two bones are anchored together by the interosseous membrane.

Ligaments of the tibia and fibula

The ligaments are strong fibrous bands that surround the tibia and fibula and bind the two bones to each other and to the other leg bones with which they articulate.

There are a number of ligaments that surround the tibia and the fibula; they bind the two bones to each other and to other bones of the leg.

Proximal (upper) end

Just under the knee is the upper joint between the head of the fibula and the underside of the lateral tibial condyle. The joint is surrounded and protected by a fibrous joint capsule, which is strengthened by the anterior and posterior tibiofibular ligaments. The anterior ligament of the head of the fibula runs from the front of the fibular head across to the front of the lateral tibial condyle. The posterior ligament of the head of the fibula runs in a similar fashion behind the fibular head. The cruciate ligaments connect the tibia to the femur within the centre of the knee joint.

Other ligaments bind the bones of the lower leg to the femur. The strongest of these are the medial and lateral collateral ligaments of the knee joint, which run vertically down from the femur to the corresponding bone (tibia or fibula) beneath.

Distal (lower) end

The joint between the lower ends of the tibia and fibula allows no movement of one bone upon the other. Instead, the fibula is bound tightly to the tibia by fibrous ligaments to maintain the stability of the ankle joint. The main ligaments concerned are the anterior and posterior inferior (lower) tibiofibular ligaments. Other ligaments around the ankle bind the tibia and fibula to the foot bones.

Interosseous membrane

The fibres of the dense interosseous membrane run obliquely from the sharp interosseous border of the tibia across to the front of the fibula, binding the two bones together.

Anterior view of left leg with ligament attachments

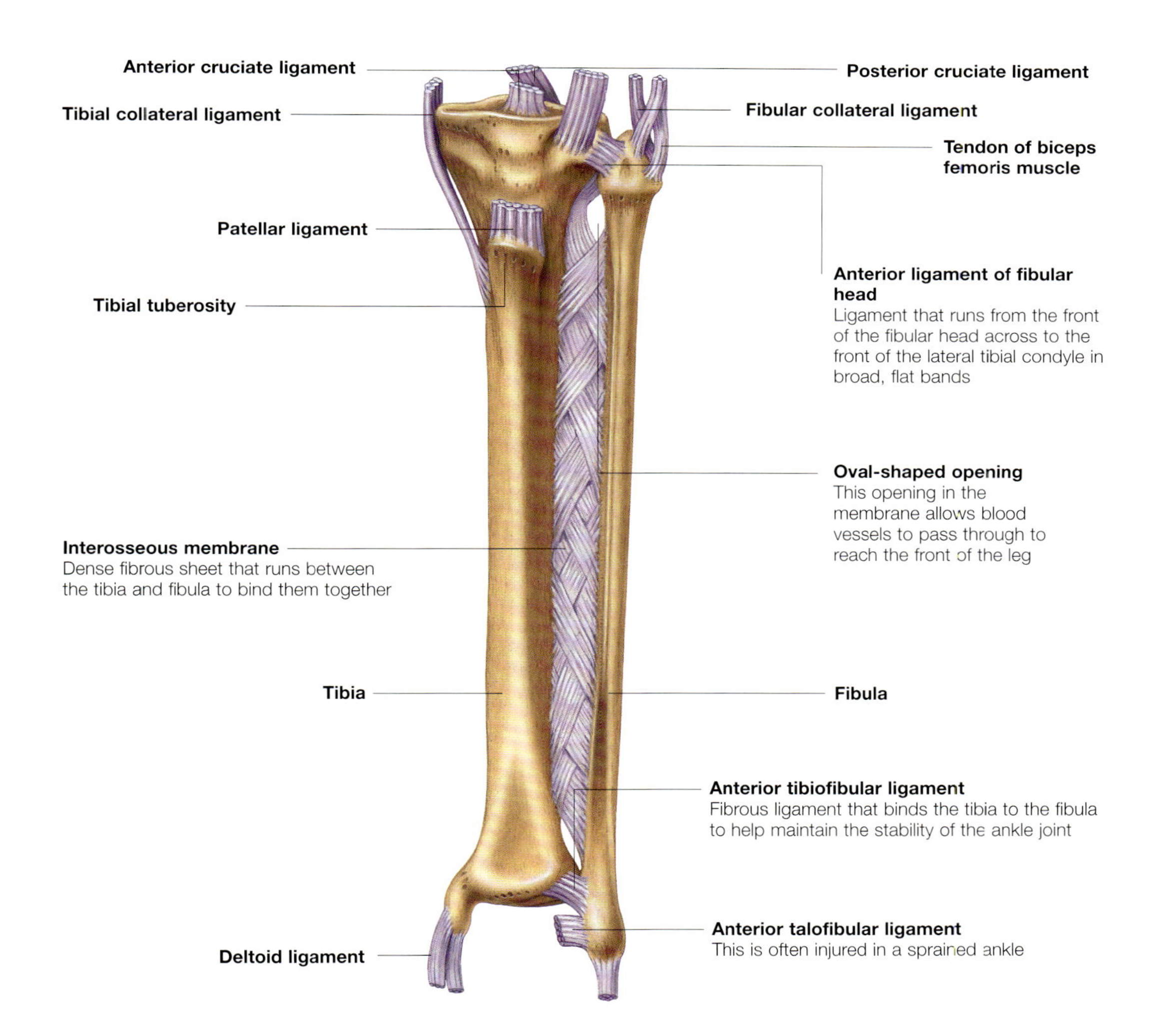

The tibia and fibula of the lower leg are surrounded by a number of ligaments. Ligaments are tough fibrous bands of connective tissue that bind bones together where they articulate at a joint.

Knee Joint

The knee is the joint between the end of the thigh bone and the top of the tibia. In front of the knee is the patella (kneecap), the convex surface of which can readily be felt under the skin.

The knee is the joint between the lower end of the femur (thigh bone) and the upper end of the tibia (the largest bone of the lower leg). The fibula (the smaller of the two lower leg bones) plays no part in the joint.

Structure

The knee is a synovial joint – one in which movement is lubricated by synovial fluid that is secreted by a membrane lining the joint cavity. Although the knee tends to be thought of as a single joint it is, in fact, the most complex joint in the body, being made up of three separate joints that share a common joint cavity:

- The joint between the patella (kneecap) and the lower end of the femur. Classified as a plane joint, this allows one bone to slide over the other.
- A joint on either side between the femoral condyles (the large bulbous ends of the femur) and the corresponding part of the upper tibia. These are said to be hinge joints as the movement they allow is akin to the movement of a door on its hinges.

Stability of the Knee

Although the 'fit' between the femoral condyles and the upper end of the tibia is not perfect, the knee is actually a reasonably stable joint. It relies heavily on the surrounding muscles and ligaments for its stability.

Sagittal section of knee

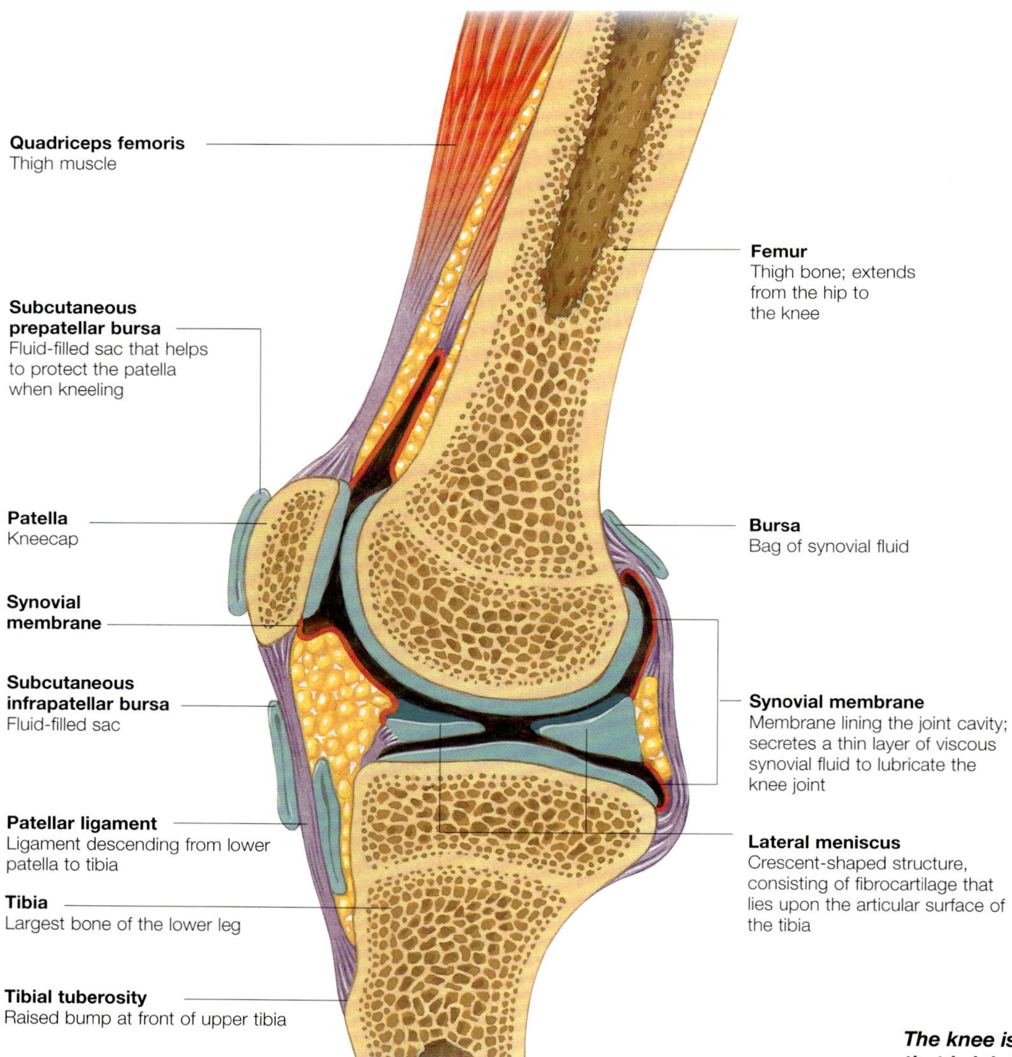

The knee is a synovial joint (one that is lubricated by a viscous synovial fluid). It is a stable but complex joint that is particularly susceptible to injury.

Menisci and patella

The menisci are crescent-shaped plates of tough fibrocartilage lying on the articular surface of the tibia. They act as 'shock absorbers' and prevent sideways movement of the femur.

The menisci are plates of tough fibrocartilage that lie upon the articular surface of the tibia, deepening the depression into which the femoral condyles fit. They act as shock absorbers and help to prevent the side-to-side rocking of the joint. Menisci are wedge-shaped in cross-section, their external margins being widest. Centrally, they taper to a thin, unattached edge. Anteriorly, the two menisci are attached to each other by the transverse ligament of the knee; the outer edges are firmly attached to the joint capsule. Attachment of the medial meniscus to the tibial collateral ligament is of significance as the meniscus can be damaged when this ligament is injured.

Patella

The patella is the largest sesamoid bone in the body, and protects the tendons from wear and tear. The patella is flattened with a convex surface at the front. Between the patella and the skin is a fluid-filled sac, or bursa, that acts as a buffer when kneeling. The back of the patella articulates with the lower end of the femur in a synovial joint. Fibres from the quadriceps thigh muscle insert into the upper border of the patella. The patellar ligament passes from the lower patella to the tibial tuberosity.

Superior view of knee (tibial plateau)

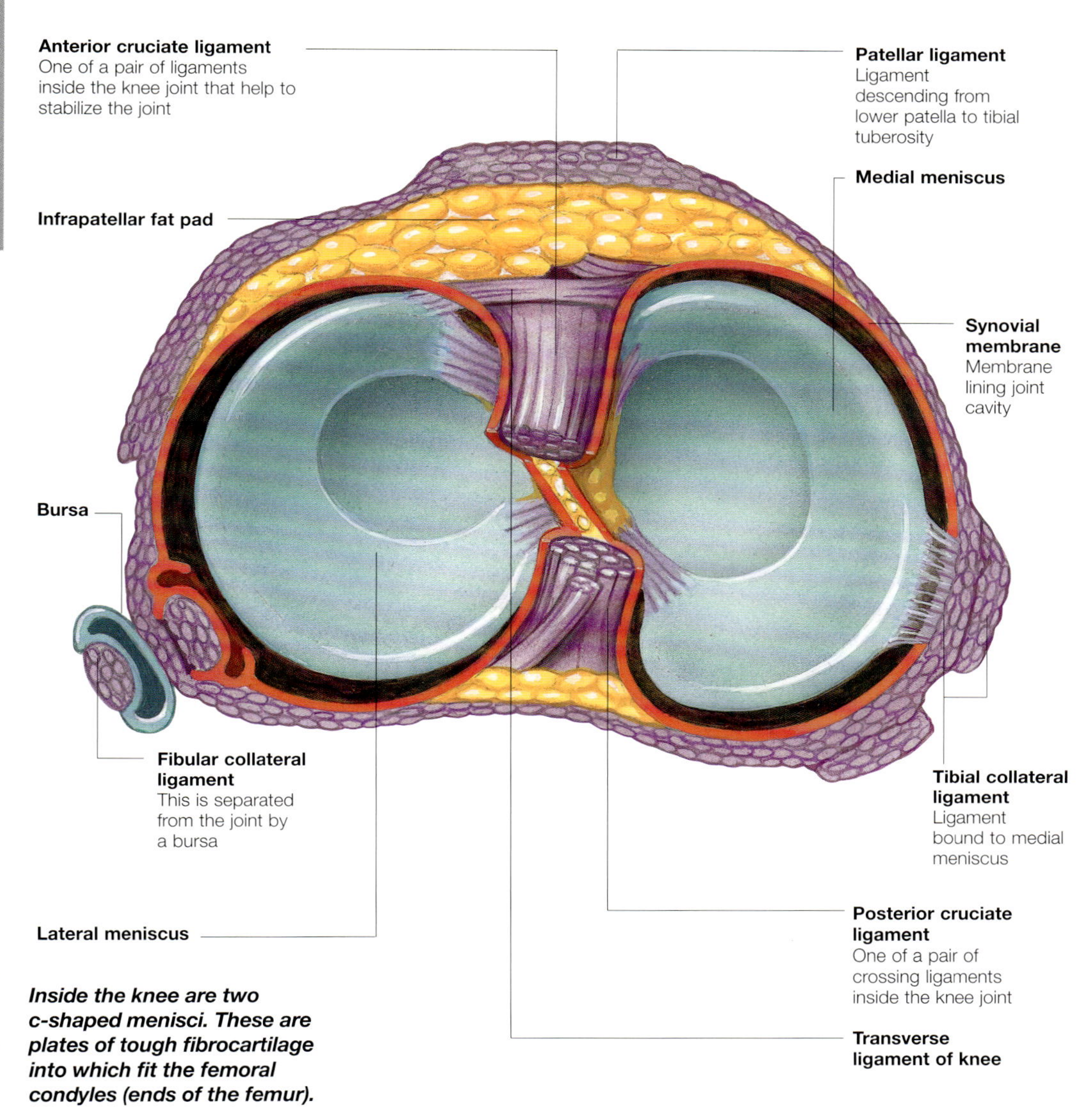

Inside the knee are two c-shaped menisci. These are plates of tough fibrocartilage into which fit the femoral condyles (ends of the femur).

The patella

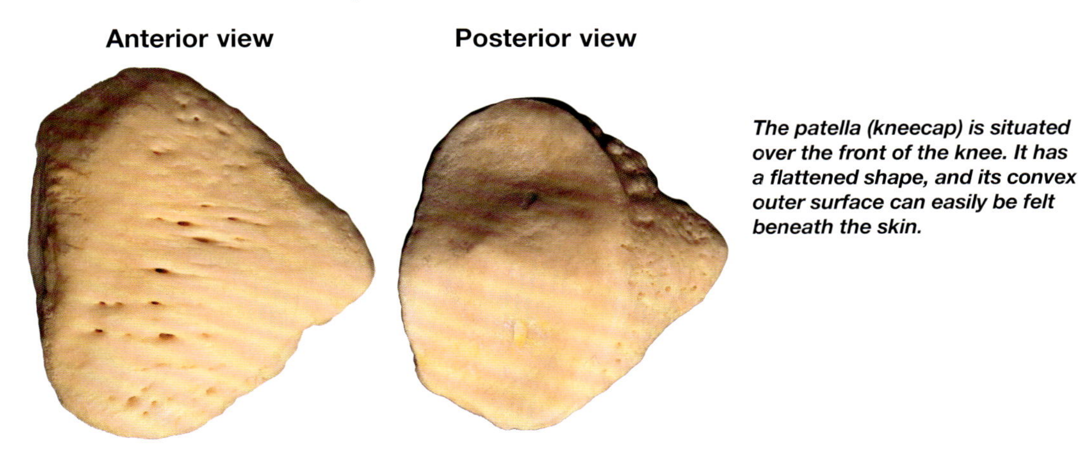

The patella (kneecap) is situated over the front of the knee. It has a flattened shape, and its convex outer surface can easily be felt beneath the skin.

Ligaments of the Knee

The knee joint is only partially enclosed in a capsule and relies on ligaments for its stability. Bursae are situated around the knee and allow smooth movement.

The stability of the knee depends on the ligaments and muscles that surround it. The ligaments can be divided into two groups, depending on their relationship to the fibrous capsule enclosing the knee joint, either outside it or within it.

Extracapsular Ligaments
These lie outside the capsule and prevent the lower leg bending forward at the knee. They include:

- Quadriceps tendon – extends from the tendon of the quadriceps femoris muscle. Supports the front of the knee.
- Fibular (or lateral) collateral ligament – a strong cord that binds the lower end of the outer femur to the head of the fibula.
- Tibial (or medial) collateral ligament – a strong flat band that runs from the lower end of the inner femur down to the tibia. It is weaker than the fibular collateral ligament and is more easily damaged.
- Oblique popliteal ligament – (not shown) this strengthens the capsule at the back.
- Arcuate popliteal ligament (not shown) – adds strength to the back of the knee.

Anterior view of flexed left knee

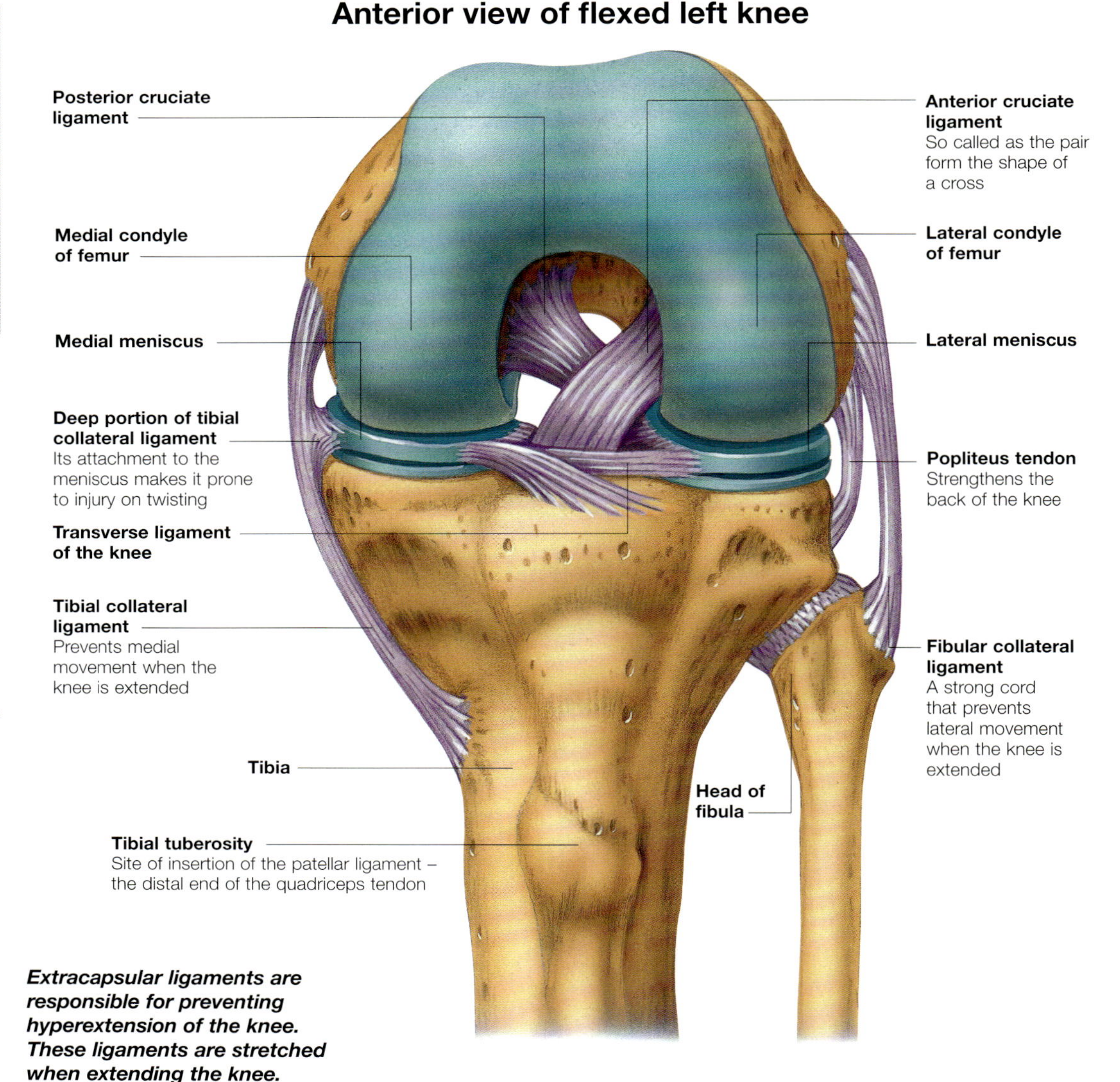

Extracapsular ligaments are responsible for preventing hyperextension of the knee. These ligaments are stretched when extending the knee.

Intracapsular ligaments

Lateral view of intracapsular ligaments

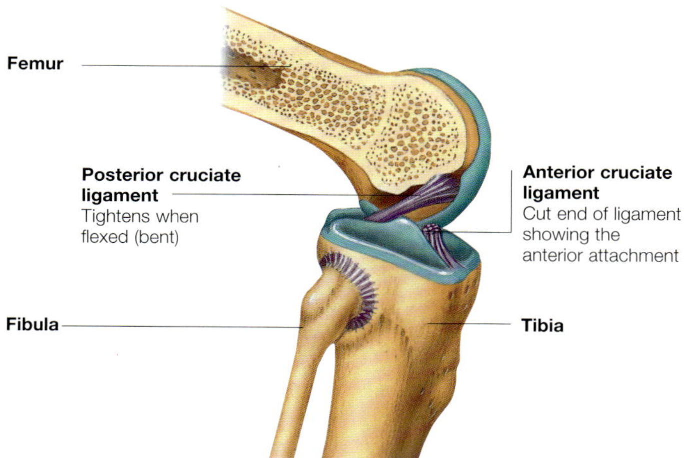

The intracapsular ligaments connect the tibia to the femur within the centre of the knee joint and prevent forward and backward displacement of the knee.

Cruciates
The two main intracapsular ligaments are known as the cruciate ligaments, as they form the shape of a cross.

- The anterior cruciate ligament – this is the weaker of the two cruciates, and is slack when the knee is flexed, taut when the knee is extended (straightened).
- The posterior cruciate ligament – this ligament tightens during flexion (bending) and is very important for the stability of the knee when bearing weight in a flexed position (for example, when walking downhill).

The intracapsular ligaments act to prevent anterior–posterior displacement and to stabilise articulating bones.

Bursae of the Knee

The bursae of the knee are small sacs filled with synovial fluid. They protect the structures inside the knee, reducing friction as they slide over each other when the joint is moving.

Bursae are small fluid-filled sacs found between two structures, usually bone and tendon, that regularly move against each other. The bursae protect the structures from wear and tear.

There are a number of bursae around the knee that protect the tendons during movement or allow easy movement of the skin across the patella.

Suprapatellar Bursa

Some of the bursae located around the knee joint are continuous with the joint cavity, the fluid-filled space between the articular surfaces. The suprapatellar bursa lies above the joint cavity between the lower end of the femur and the powerful quadriceps femoris muscle.

Prepatellar and Infrapatellar Bursae

These bursae surround the patella and the patellar ligament. The prepatellar bursa allows the skin to move freely over the patella during movement. The superficial and deep infrapatellar bursae lie around the lower end of the patellar ligament where it attaches to the tibial tuberosity.

There are about a dozen bursae located around the knee joint. They allow the structures of the knee to move freely over one another, reducing friction.

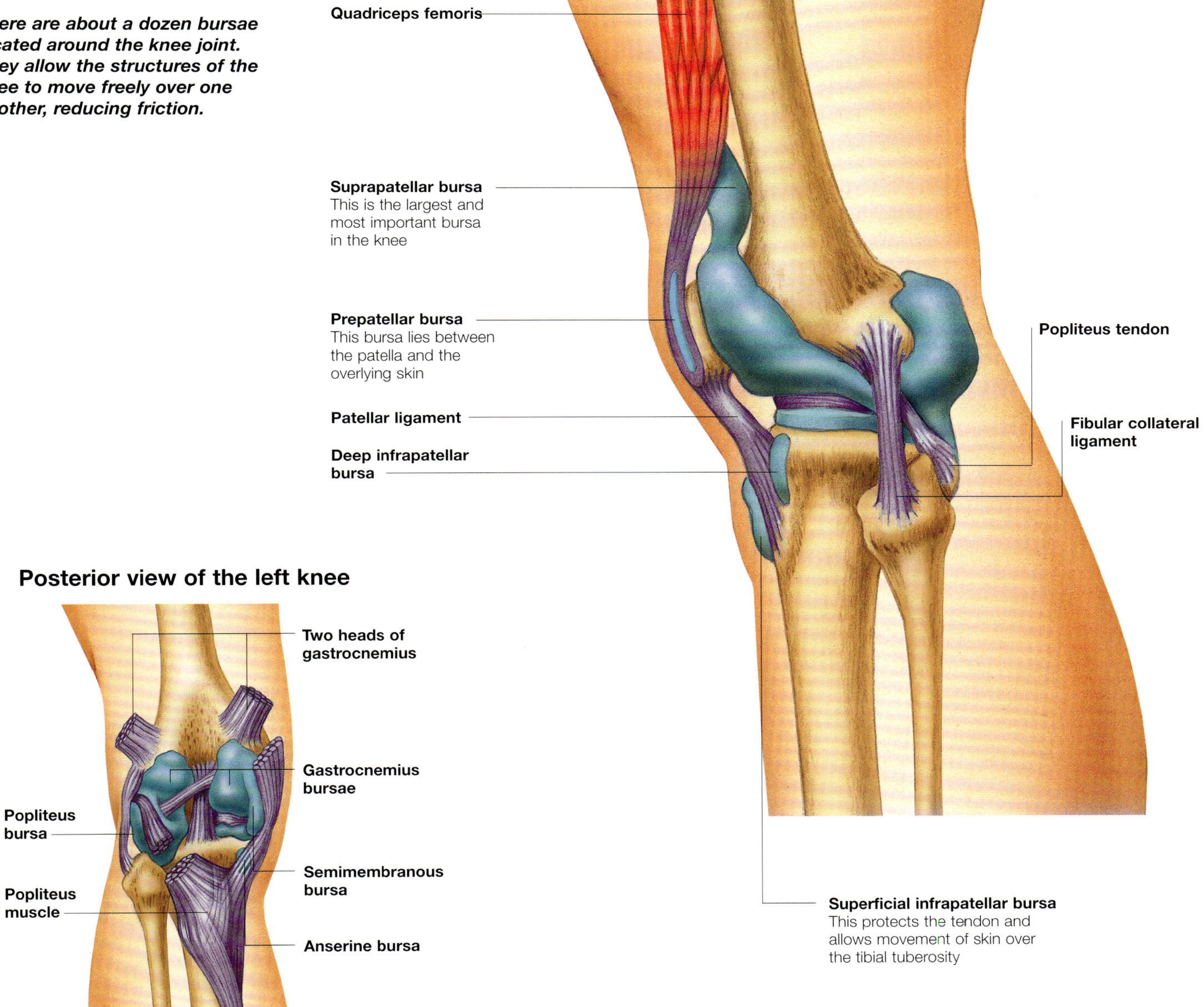

Muscles of the Thigh

The thigh is composed mainly of groups of large muscles that act to move the hip and the knee joint. Muscles that effect the movements of the thigh are among the strongest in the body.

The muscles of the thigh are divided into three groups; the anterior muscles lie in front of the femur, the posterior muscles lie behind, and the medial muscles (adductors) run between the inner femur (thigh) and the pelvis.

Anterior Muscles

These muscles flex or bend the hip, and extend or straighten the knee, the actions associated with walking. The muscles of this group include:

- **Iliopsoas**. This large muscle arises partly from the inside of the pelvis and partly from the lumbar vertebrae. Its fibres insert into the projection of the upper femur, the lesser trochanter. The iliopsoas is the most powerful of the muscles that flex the thigh, bringing the knee up and forwards.
- **Tensor fasciae latae**. Inserts into the strong band of connective tissue that runs down the outside of the leg to the tibia below the knee.
- **Sartorius**. The longest muscle in the body, which runs as a flat strap across the thigh from the anterior superior iliac spine of the pelvis. It crosses both the hip and knee joint before it inserts into the inner side of the top of the tibia.
- **Quadriceps femoris.** A large four-headed muscle.

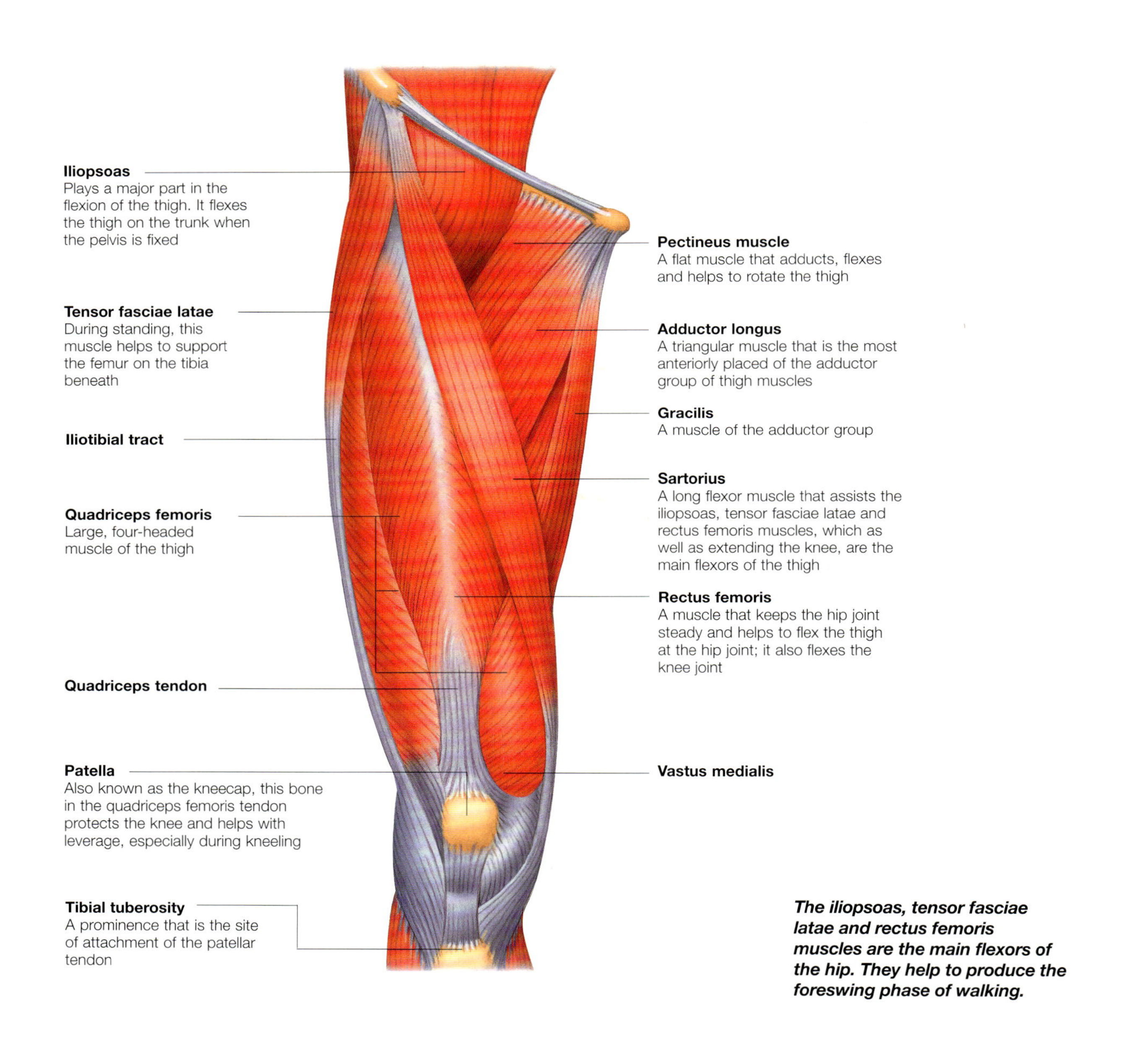

The iliopsoas, tensor fasciae latae and rectus femoris muscles are the main flexors of the hip. They help to produce the foreswing phase of walking.

Posterior thigh muscles

The three large muscles of the posterior thigh are together known as the hamstrings. These three muscles are the biceps femoris, semitendinosus and semimembranosus.

The hamstring muscles can both extend the hip and flex the knee. However, they cannot do both fully at the same time.

Biceps femoris
The biceps femoris muscle has two heads. The long head arises from the ischial tuberosity of the pelvis and the short head arises from the back of the femur. The rounded tendon of the biceps femoris can easily be felt and seen behind the outer side of the knee, especially if the knee is flexed against resistance.

Semitendinosus
Like the biceps femoris muscle, the semitendinosus arises from the ischial tuberosity of the pelvis. It is named for its long tendon, which begins about two-thirds of the way down its course. This tendon attaches to the inner side of the upper tibia.

Semimembranosus
This hamstring muscle arises from a flattened, membranous attachment to the ischial tuberosity of the pelvis. The muscle runs down the back of the thigh, deep to the semitendinosus. It inserts into the inner side of the upper tibia.

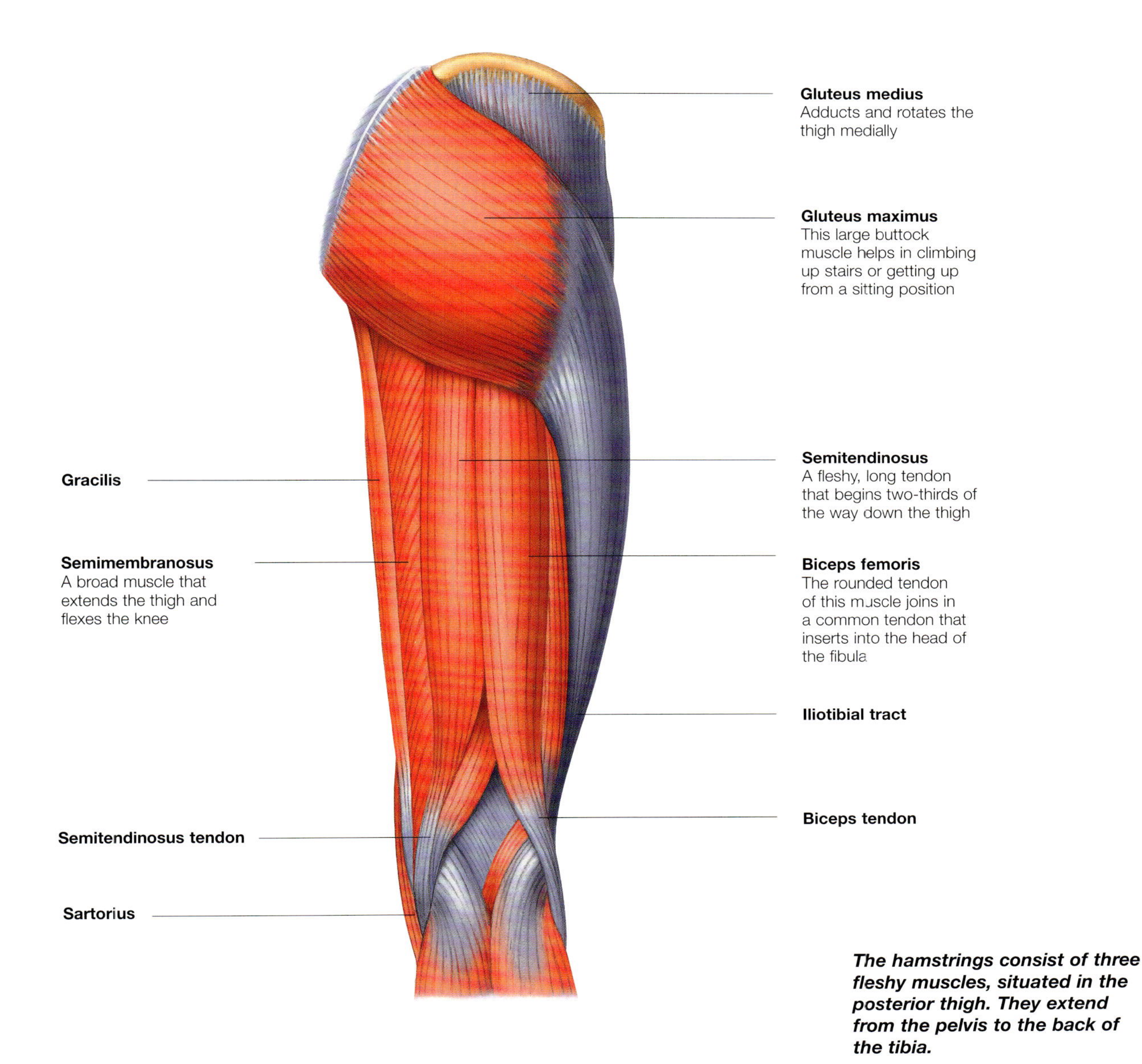

The hamstrings consist of three fleshy muscles, situated in the posterior thigh. They extend from the pelvis to the back of the tibia.

Quadriceps Femoris

The quadriceps thigh muscle is a powerful extensor that is used in running, jumping and climbing. It helps to straighten the knee when standing from a sitting position.

This large, four-headed muscle makes up the bulk of the thigh and is one of the strongest muscles in the body. It consists of four major parts whose tendons combine to form the strong quadriceps tendon. This inserts into the top of the patella and then continues down, as the patella tendon, to the front of the top of the tibia. The quadriceps femoris acts to straighten the knee.

The four parts of the quadriceps femoris are:

- Rectus femoris – a straight muscle, overlying the other parts. It helps to flex the hip joint and straighten the knee
- Vastus lateralis – the largest part of the quadriceps muscle
- Vastus medialis – lies on the inner side of the thigh
- Vastus intermedius – lies centrally, underneath the rectus femoris muscle.

A few small slips of the vastus intermedius pass down to the joint capsule of the knee. They ensure that folds of the capsule do not become trapped when the knee is straightened.

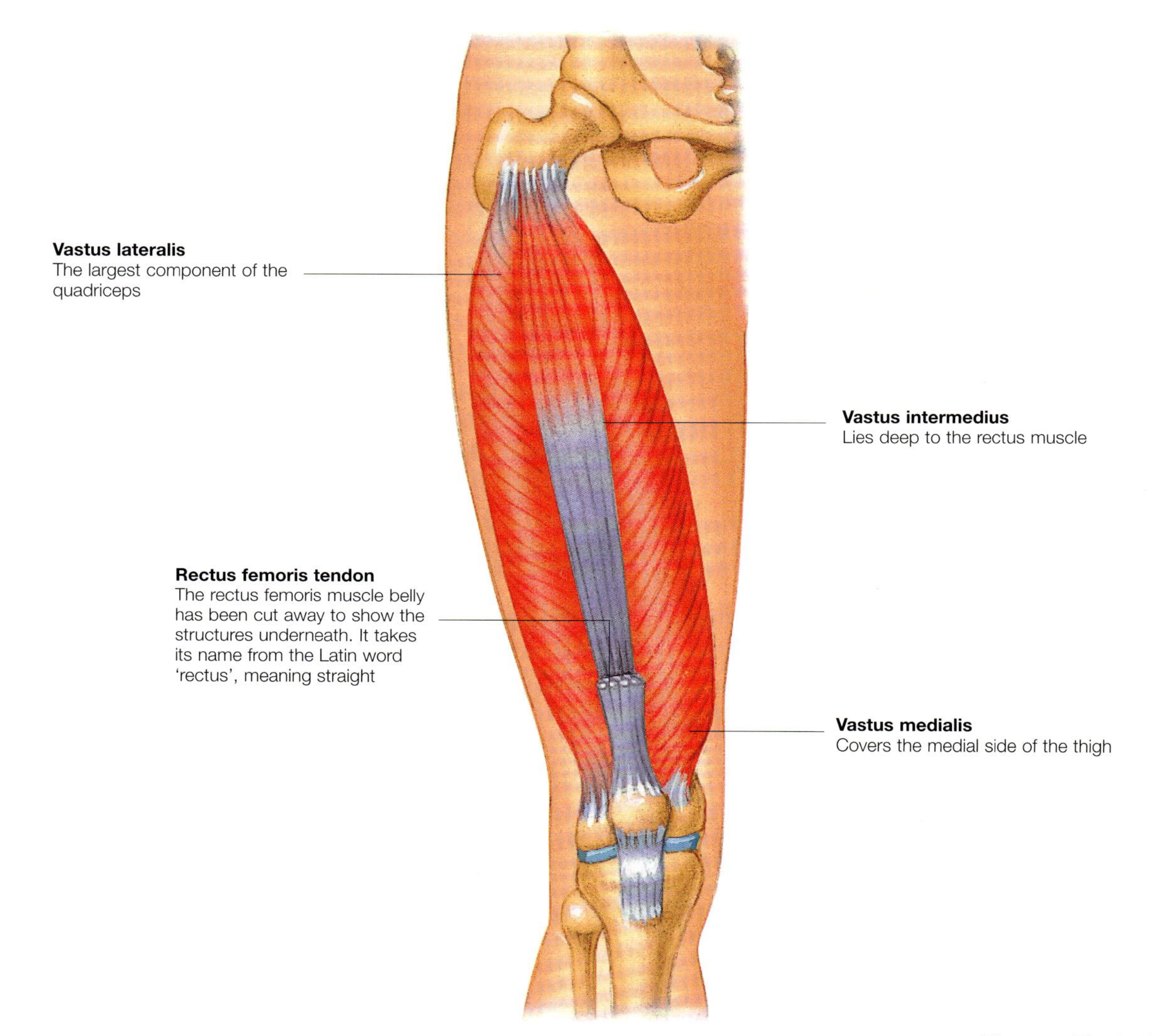

The quadriceps femoris is a large, four-headed muscle that runs the length of the thigh from the top of the femur to the patella.

Adductors

The inner thigh muscles are known as adductors and are responsible for moving the limb into the midline.

The muscles of the inner thigh are known as adductors and allow adduction of the thigh. This action moves the lower limb in towards the midline as when gripping the sides of a horse when riding. Adductor muscles arise from the lower part of the pelvis and insert into the femur at various levels. The muscles of this group include:

- Adductor longus – a large, fan-shaped muscle that lies in front of the other adductors and has a palpable tendon in the groin
- Adductor brevis – a shorter muscle that lies under the adductor longus
- Adductor magnus – a large triangular muscle that fulfils the function of both an adductor and a hamstring muscle
- Gracilis – a strap-like muscle that runs vertically down the inner thigh (not shown)
- Obturator externus – a small muscle that lies deeply within this group of adductors.

Adductor muscles may become strained during sporting activities, leading to a groin injury.

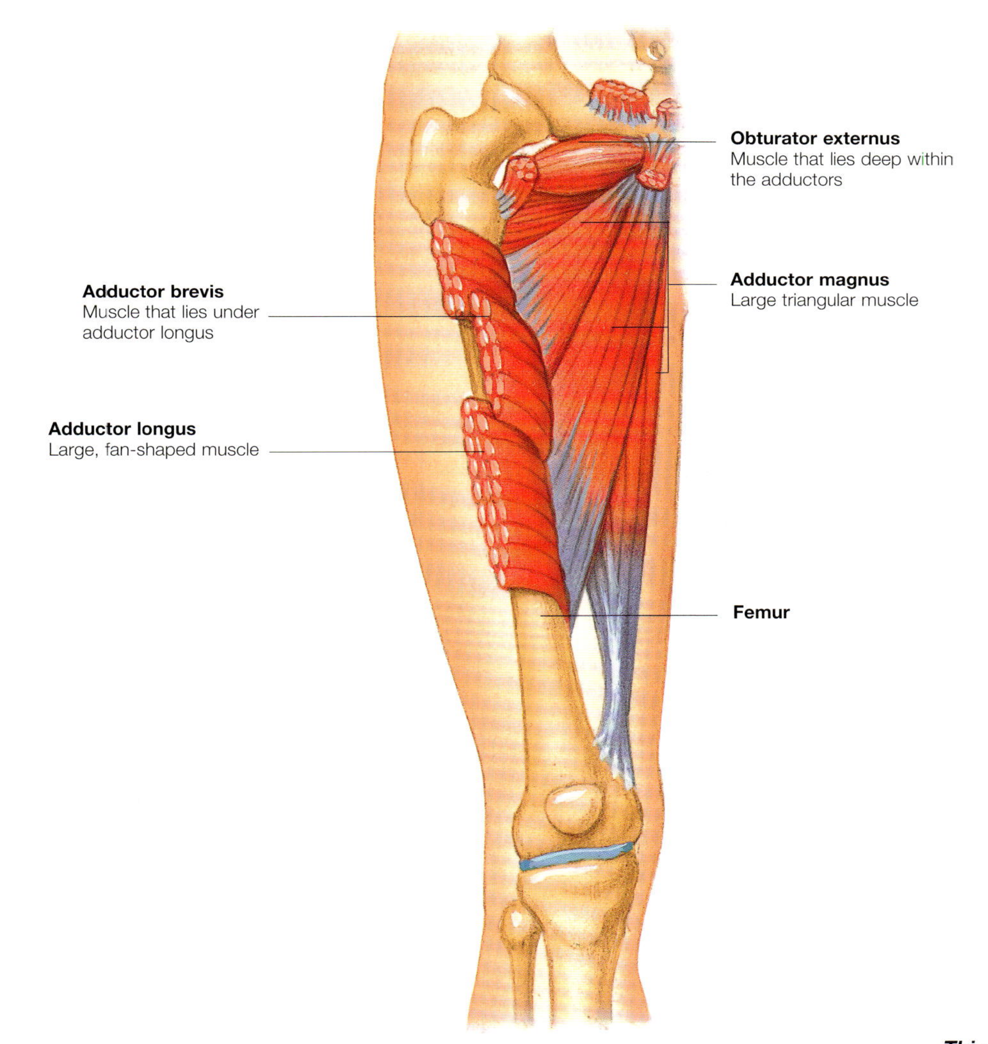

This medial group of thigh muscles run from the pelvic region to the lower femur. They enable the movement of adduction.

Muscles of the Lower Leg

There are three groups of muscles in the lower leg. Depending where they lie, they support and flex the ankle and foot, extend the toes and assist in lifting the body weight at the heel.

The muscles of the lower leg can be divided into three groups: the anterior group that lie in front of the tibia, the lateral group on the outer side of the lower leg and the posterior group.

Anterior Muscles

The anterior muscles of the lower leg include:

- Tibialis anterior – this muscle runs alongside the edge of the tibia
- Extensor digitorum longus – this muscle lies under the tibialis anterior and attaches to the outer four toes
- Peroneus (fibularis) tertius – this muscle is not always present but, when it is, it may join the extensor digitorum longus muscle. It inserts into the fifth metatarsal bone near the little toe
- Extensor hallucis longus – this thin muscle runs down to insert into the end of the hallux (big toe).

Dorsiflexors

These muscles all have a similar action in that they are dorsiflexors; when contracted they bend the ankle bringing the toes up and the heel down.

Anterior muscles of the lower leg

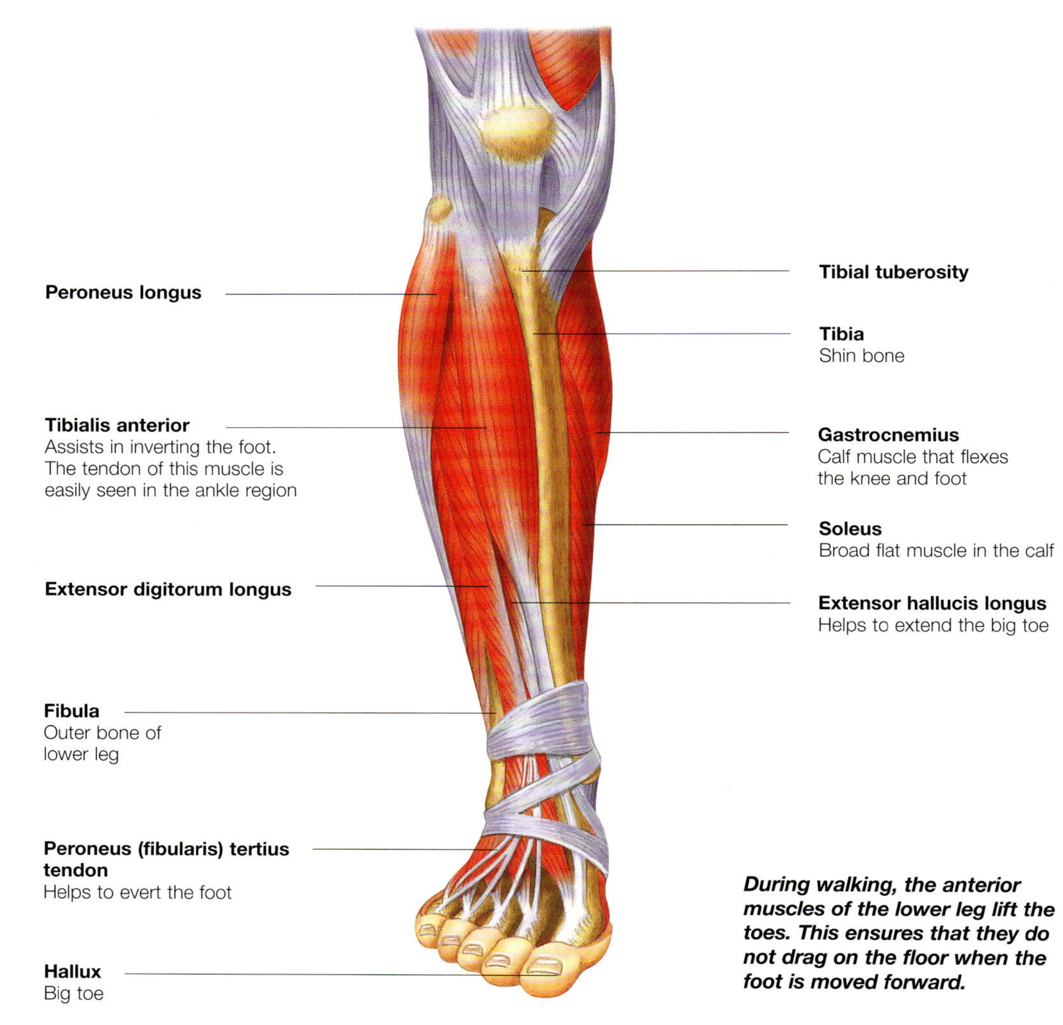

During walking, the anterior muscles of the lower leg lift the toes. This ensures that they do not drag on the floor when the foot is moved forward.

Lateral muscles of the lower leg

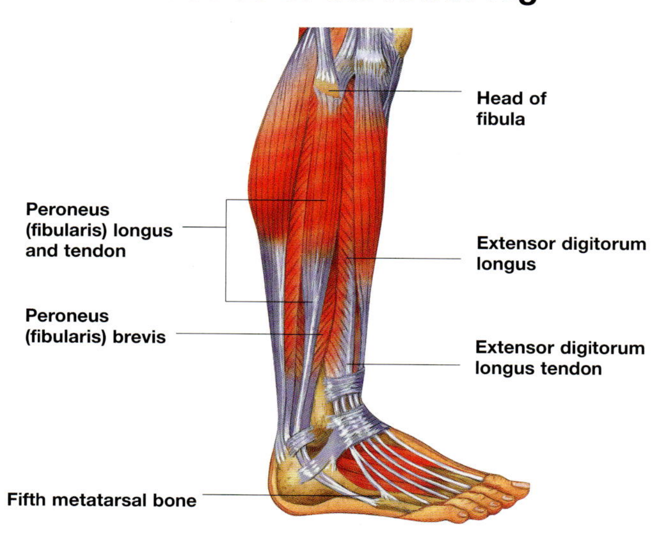

The muscles of the lateral (outer) compartment lie alongside the smaller of the two lower leg bones, the fibula. There are two muscles in this group:

- Peroneus (fibularis) longus

This muscle is the longer of the two and lies more superficially. It arises from the head and the upper portion of the narrow fibula and runs down to the sole of the foot.

- Peroneus (fibularis) brevis

As its name suggests, this is a short muscle, which lies underneath the peroneus longus muscle. It arises from the lower portion of the fibula and has a broad tendon that runs down to insert into the base of the fifth metatarsal bone of the foot.

Action of the lateral muscles

These two muscles together cause the foot to plantar flex, when the toes point down, and to evert, which means bending so that the sole faces outwards. In practice, these muscles help to support the ankle by resisting the movement of inversion (the sole facing inwards), which is when the joint is weakest.

The lateral muscles protect the ankle by resisting inversion of the foot. When the foot is inverted, the ankle is in a very vulnerable position.

Posterior muscles of the lower leg

The posterior group of muscles of the lower leg form the mound of the calf. Together, these muscles are strong and heavy, enabling them to work together to flex the foot and to support the weight of the body.

The posterior muscles, also known as calf muscles, are the largest group of the lower leg. They can be divided into superficial and deep layers.

Superficial calf muscles
These strong muscles form the bulk of the rounded calf and act to plantar flex the foot, lifting the heel and pointing the toes downwards. They include:

- Gastrocnemius – this large fleshy muscle is the most superficial. It has a distinctive shape with two heads that arise from the medial and lateral condyles of the femur. Its fibres run mainly vertically, which allows for the rapid and strong contractions needed in running and jumping.
- Soleus – this is a large and powerful muscle that lies under the gastrocnemius. It takes its name from its shape, being flat like a sole (a type of flatfish). Contraction of the soleus muscle is important for maintaining balance when standing.
- Plantaris – this muscle is sometimes absent, and when it is present it is rather small and thin. It is relatively unimportant in the lower leg.

Gastrocnemius and soleus have a single, common tendon, the large and powerful Achilles tendon that runs down from the lower edge of the calf to the heel.

Deep calf muscles
There are four muscles that make up the deep calf muscles:

- Popliteus – a thin triangular muscle that lies at the back of the knee in the popliteal fossa and acts to rotate the knee slightly to allow the straightened leg to be bent.
- Flexor digitorum longus – this muscle has long tendons that pass down to the outer four toes to allow them to curl under, or flex.
- Flexor hallucis longus – a powerful muscle attached to the base of the big toe. Acts to 'push off' during walking.
- Tibialis posterior – acts to invert the foot so that the sole faces inwards.

Posterior muscles of the lower leg

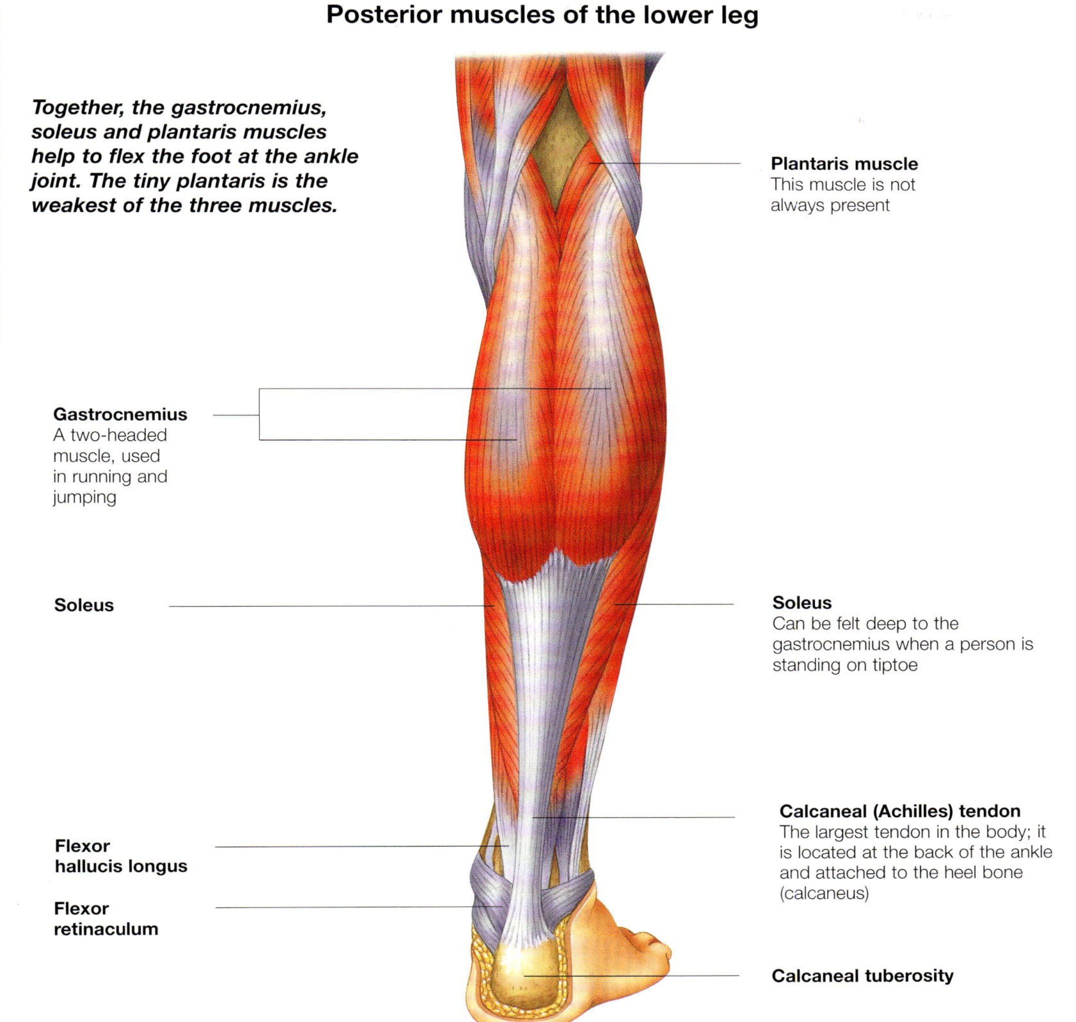

Together, the gastrocnemius, soleus and plantaris muscles help to flex the foot at the ankle joint. The tiny plantaris is the weakest of the three muscles.

Deep calf muscles

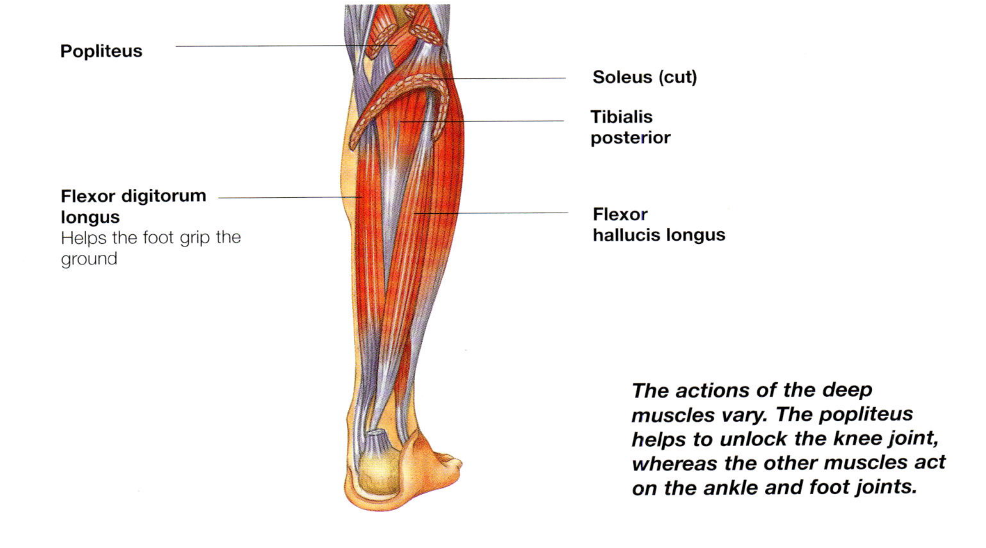

The actions of the deep muscles vary. The popliteus helps to unlock the knee joint, whereas the other muscles act on the ankle and foot joints.

Deep Fascia of the Leg

Lying just below the subcutaneous tissue, the deep fascia of the leg forms a strong, circular sheath around the muscles, bone and blood vessels. The fascia partitions the leg into three compartments.

Fascia is the name given to sheets of connective tissue in the body that enclose and bind structures such as muscles. Lying under the subcutaneous tissue, but above the muscles, the deep fascia of the leg is a membranous sheath that envelops the limb. It also forms partitions, which run down the leg until they meet bone, dividing the thigh and lower leg into a series of compartments.

Iliotibial Tract

On the outer side of the thigh, the deep fascia is thickened to form a tough vertical band known as the iliotibial tract. The tensor fasciae latae muscle inserts into this band, as does the greater part of the large and powerful gluteus maximus muscle.

Saphenous Opening

The saphenous opening is a gap in the deep fascia of the thigh. It allows the great saphenous vein to pass through to enter the femoral vein.

Venous Return

The deep fascia aids the return of venous blood by acting as a 'muscle pump'. The fascia is tough and inelastic and so resists the bulging of muscles as they contract. This causes them to put pressure upon the valved deep veins within the leg, so forcing venous blood back up the body.

Anterior and lateral view of the leg fascia

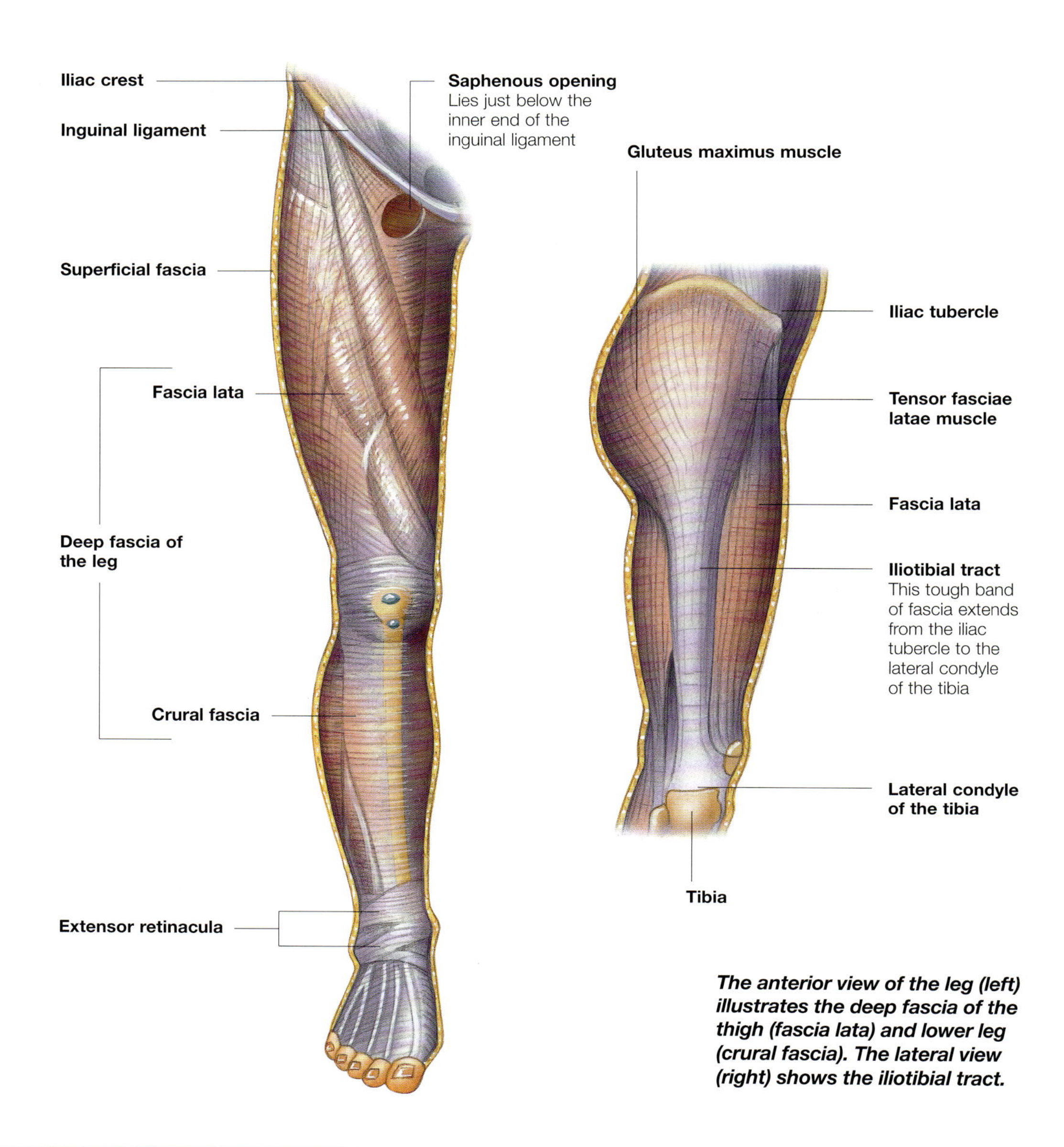

The anterior view of the leg (left) illustrates the deep fascia of the thigh (fascia lata) and lower leg (crural fascia). The lateral view (right) shows the iliotibial tract.

Compartments of the thigh

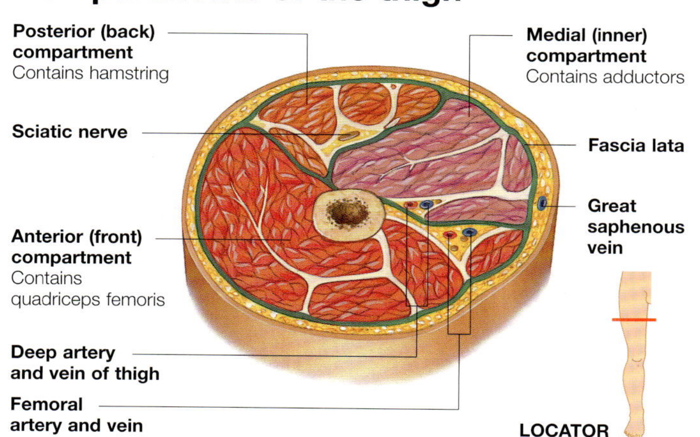

Partitions, or septa, arise from the fascia lata and run down to the femur. These divide the thigh into three compartments, each of which contain a group of muscles with similar actions and a similar nerve and blood supply:

- The anterior compartment – this contains muscles that flex the hip and extend the knee, supplied mainly by the femoral nerve and artery.
- The medial compartment – this is also known as the adductor compartment as it contains muscles that adduct the thigh (bring it in to the midline). These muscles are supplied by the obturator nerve and the profunda femoris and obturator arteries.
- The posterior compartment – this contains the powerful hamstring muscles that extend the hip and flex the knee. These muscles are supplied by the sciatic nerve and the profunda femoris artery.

The fascia lata envelops the limb, encircling the muscles, bone and blood vessels. The limb also has sheets of connective tissue running down to the femur.

Compartments of the lower leg

The deep fascia of the lower leg is also known as the crural fascia. It is a thick, membranous sheath, capable of binding muscles, arteries and veins within separate spaces: the anterior, lateral and posterior compartments.

The crural fascia of the lower leg attaches to the front and inner borders of the tibia below the knee, where it is continuous with its periosteum, the tough membrane that encloses bone. In the upper part of the lower leg, the crural fascia is thick and provides a site of attachment for muscle. Lower down it thins, except where it forms the horizontal retinacula at the ankle. Partitions arise from the deep surface of the crural fascia and attach to the fibula beneath to divide the lower leg into three compartments.

Anterior compartment
The anterior compartment contains muscles that dorsiflex the foot and extend, or straighten, the toes.

Lateral compartment
The lateral (outer) compartment lies alongside the fibula bone. It contains two muscles that plantarflex the foot and evert the foot.

Posterior compartment
The posterior compartment of the lower leg is subdivided by another partition, the transverse intermuscular septum, into deep and superficial layers. The powerful muscles here act to flex the foot and provide the main forward propulsive force in walking.

Retinacula
Around the ankle the deep fascia forms thickened horizontal bands known as the extensor retinacula. These are tough restraining bands which keep the underlying tendons in position against the ankle as the foot changes position.

Cross-section of the calf

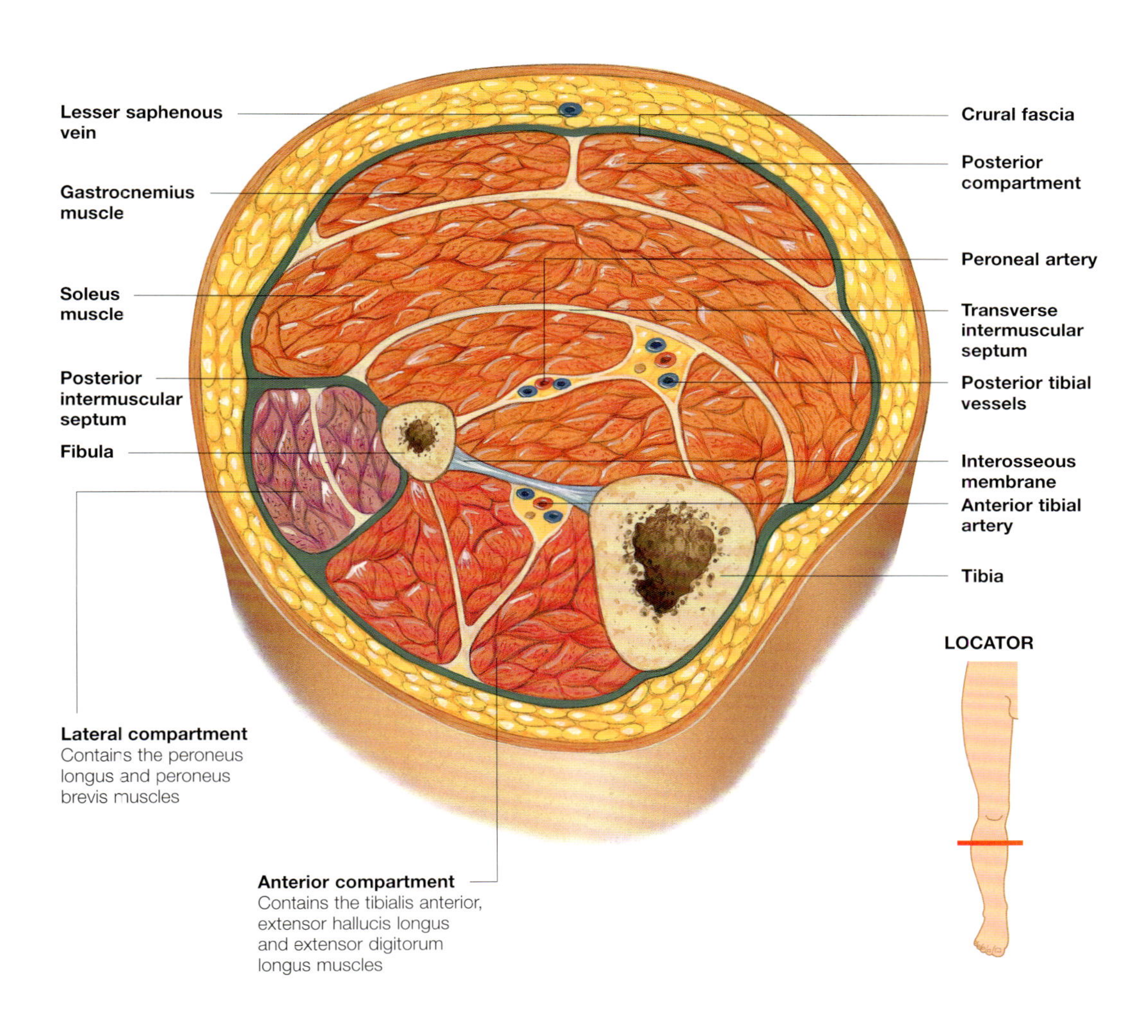

Each compartment houses particular muscles. The posterior compartment is subdivided by the transverse intermuscular septum.

Ankle and Ligaments

The ankle is the joint between the lower ends of the tibia and fibula, and the upper surface of the large foot bone, the talus. It is an example of a hinge joint.

At the ankle, a deep socket is formed by the lower ends of the tibia and fibula, the bones of the lower leg. Into this socket fits the pulley-shaped upper surface of the talus. The shape of the bones and the presence of strong supporting ligaments make the ankle very stable, an important feature for a major weight-bearing joint.

The Joint

The articular surfaces of the ankle joint – those parts of the bone that move against each other – are covered with a layer of smooth hyaline cartilage. This cartilage is surrounded by a thin synovial membrane that secretes a viscous fluid and lubricates the joint.

The articular surfaces

The articular surfaces of the ankle joint consist of the:

- Inside of the lateral malleolus, the expanded lower end of the fibula. This carries a facet (depression) that articulates with the outer side of the upper surface of the talus.
- Undersurface of the lower end of the tibia. This forms the roof of the socket, which articulates with the talus.
- Inside of the medial malleolus, the projection at the lower end of the tibia. This moves against the inner side of the upper surface of the talus.
- Trochlea of the talus. Named for its pulley shape, this upper part of the talus fits into the ankle joint, and articulates with the lower ends of the tibia and fibula.

Anterior view, left ankle

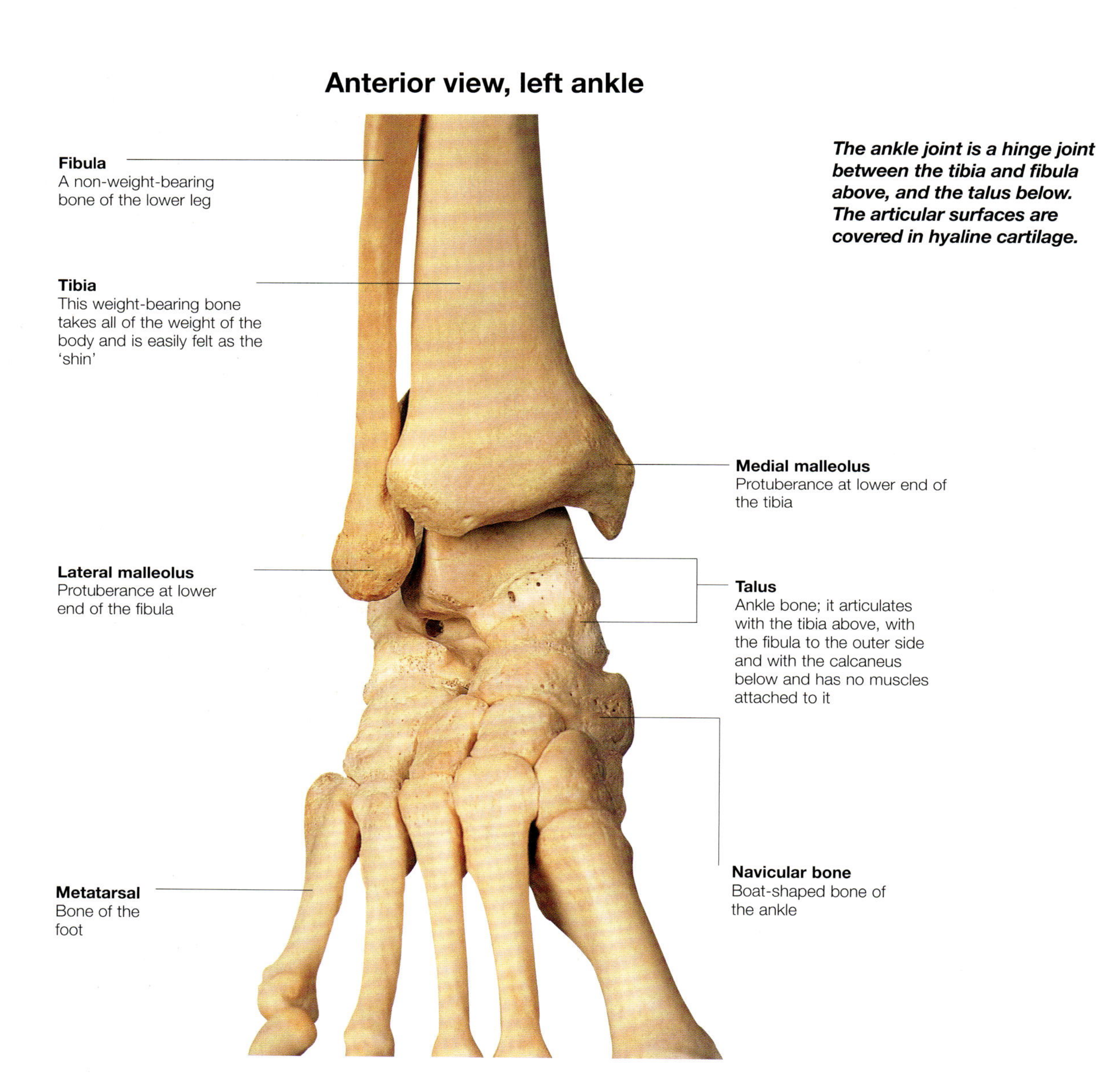

The ankle joint is a hinge joint between the tibia and fibula above, and the talus below. The articular surfaces are covered in hyaline cartilage.

Ligaments of the ankle

The ankle is supported by strong ligaments that help to stabilize this important weight-bearing joint.

The ankle joint must be stable as it bears the weight of the body. The presence of a variety of strong ligaments around the ankle helps to maintain this stability, while still allowing the necessary freedom of movement.

Like most joints, the ankle is enclosed within a tough fibrous capsule. Although the capsule is thin in front and behind, it is reinforced on each side by the strong medial (inner) and lateral (outer) ankle ligaments.

Medial ligament

Also known as the deltoid ligament, the medial ligament is a very strong structure that fans out from the tip of the medial malleolus of the tibia. It is usually described in three parts, each named for the bones that they connect:

- Anterior and posterior tibiotalar ligaments. Lying close against the bones, these parts of the medial ligament connect the tibia to the medial sides of the talus beneath.
- Tibionavicular ligament. More superficially, this part of the ligament runs between the tibia and the navicular, one of the bones of the foot.
- Tibiocalcaneal ligament. This strong ligament runs just under the skin from the tibia to the sustentaculum tali, a projection of the calcaneus (heel bone).

Together, these parts of the medial ligament support the ankle joint during the movement of eversion (the foot is turned out to the side).

Lateral ligament

The lateral ligament is weaker than the medial ligament, and is made up of three distinct bands:

- Anterior talofibular ligament. This runs forward from the lateral malleolus of the fibula to the talus.
- Calcanofibular ligament. This passes down from the tip of the lateral malleolus to the side of the talus
- Posterior talofibular ligament. This is a thick, stronger band that passes back from the lateral malleolus to the talus behind.

Lateral view (right foot)

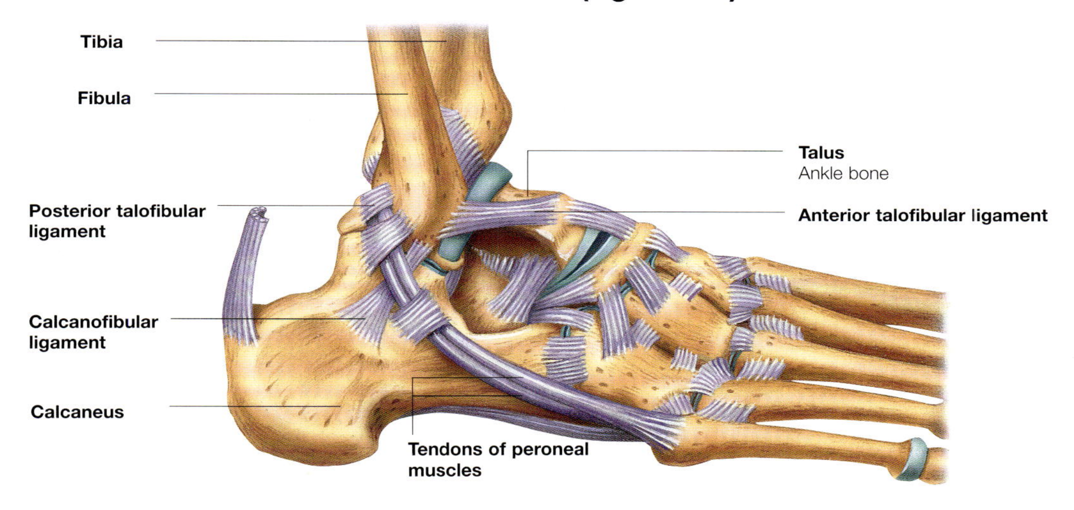

Medial view (right foot)

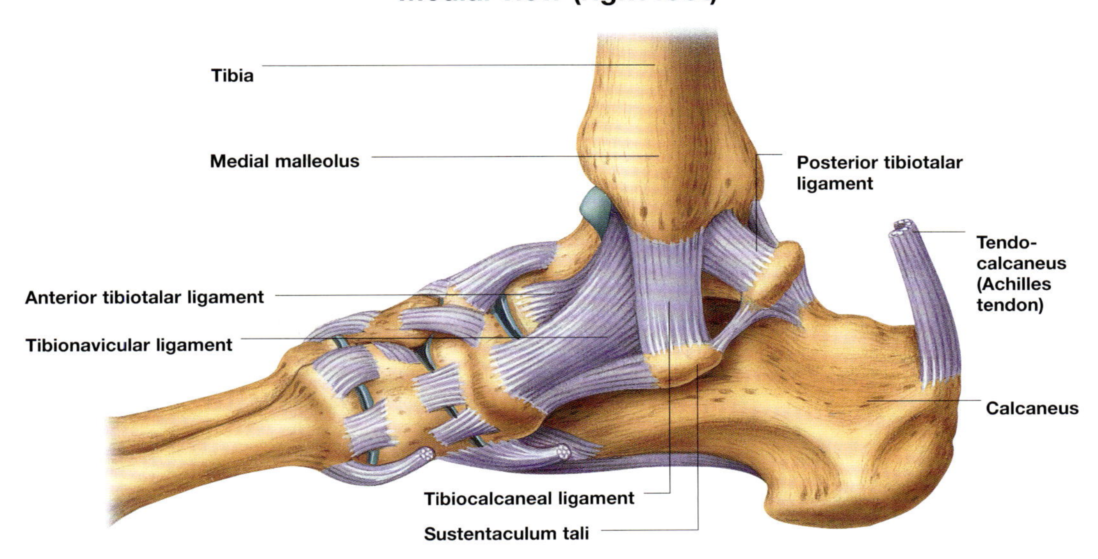

Bones of the Foot

The human foot contains 26 bones in total: seven larger, irregular tarsal bones; five metatarsals running the length of the foot; and 14 phalanges forming the skeleton of the toes.

The tarsal bones in the foot are equivalent to the carpal bones in the wrist, but there are seven tarsals as opposed to eight wrist bones. The tarsal bones also differ in their arrangement, reflecting the different functions of the hand and the foot.

Tarsal bones

The tarsal bones consist of:

- The talus – articulates with the tibia and fibula at the ankle joint. It bears the full weight of the body, transferred down from the tibia. Its shape is such that it can then spread this weight by passing the force backwards and downwards, and forwards to the front of the foot.
- The calcaneus – the heel bone and the largest bone in the foot. This weight bearing bone is the site of attachment of the Achilles tendon and has the role of transmitting the weight of the body from the talus to the ground.
- The navicular – a relatively small bone, named for its boat-like appearance. It has a projection, the navicular tuberosity.
- The cuboid – a bone roughly the shape of a cube. It lies on the outer side of the foot, and has a groove on its under surface to allow passage of a muscle tendon.
- The three cuneiforms – bones named according to their positions: medial, intermediate and lateral. The medial cuneiform is the largest of these three wedge-shaped bones.

Tarsal bones

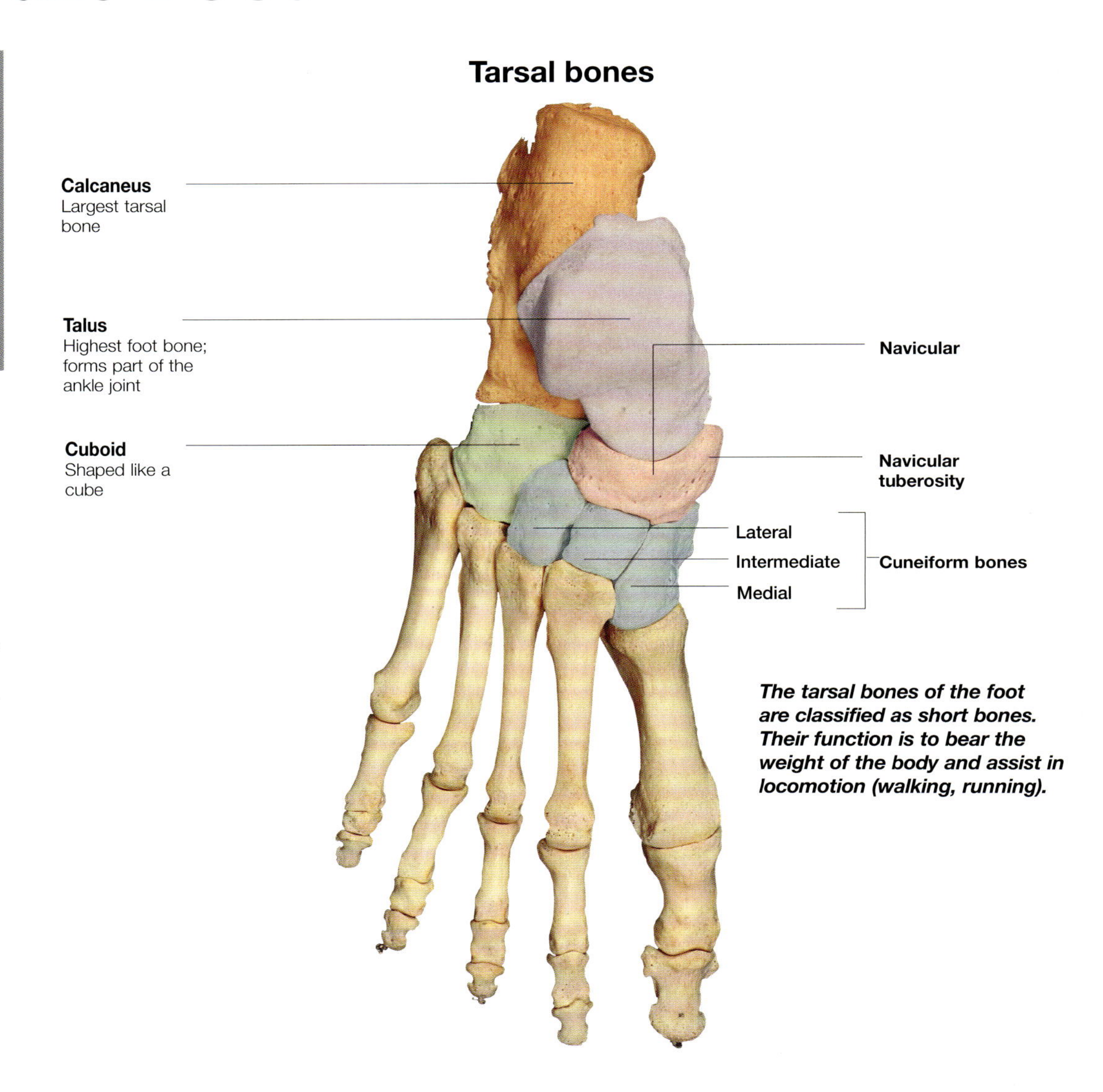

The tarsal bones of the foot are classified as short bones. Their function is to bear the weight of the body and assist in locomotion (walking, running).

View from above the calcaneus bone

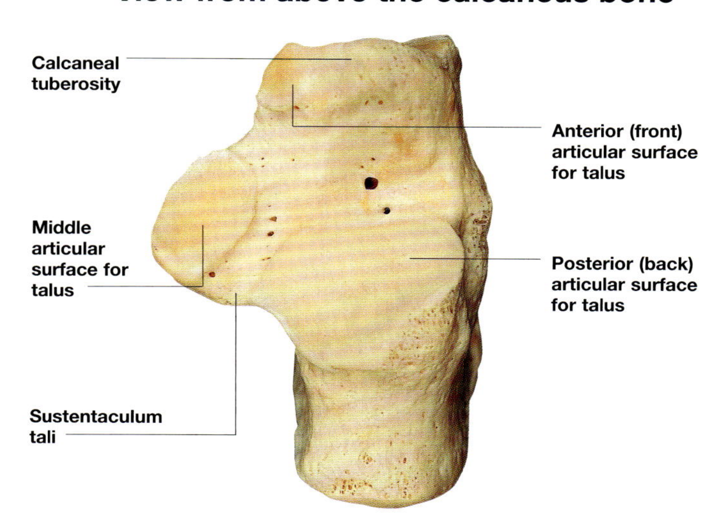

Metatarsals and phalanges

The metatarsals and phalanges in the foot are miniature long bones, consisting of a base, shaft and head.

Like the metacarpals in the hand, there are five metatarsals in the foot. They resemble the metacarpals in structure but their arrangement is slightly different. This is because the hallux lies in the same plane as the other toes and is not opposable like the thumb.

Metatarsals

Each metatarsal has a long shaft with two expanded ends, the base and the head. The bases of the metatarsals articulate with the tarsal bones in the middle of the foot. The heads articulate with the phalanges of the corresponding toes.

The metatarsals are numbered from one to five starting with the most medial, which lies behind the big toe. The first metatarsal is shorter and sturdier than the rest. It articulates with the first phalanx of the hallux.

Phalanges

The phalanges of the toe resemble the carpals of the fingers. There are 14 phalanges in the foot; the hallux has two and the other four toes have three each. The base of the first phalanx of each toe articulates with the head of the corresponding metatarsal. The phalanges of the big toe are thicker than those of the other toes.

Sesamoid bones of the foot

Sesamoid bones develop within the tendon of a muscle to protect it from wear and tear. Two sesamoid bones lie under the head of the first metatarsal within the two heads of the flexor hallucis brevis muscle and bear the weight of the body, especially as the toe pushes off in walking.

Lateral view of the foot

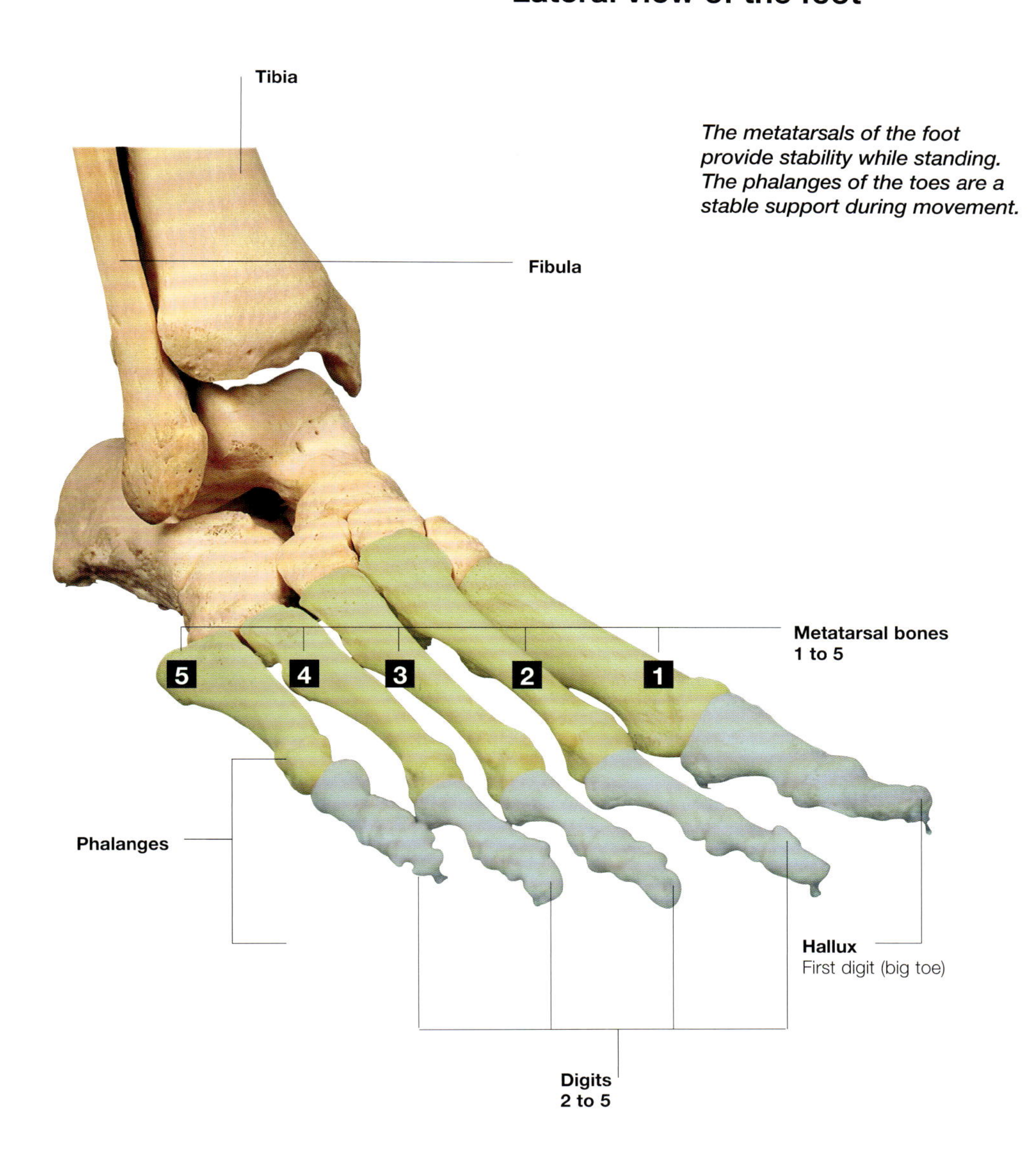

The metatarsals of the foot provide stability while standing. The phalanges of the toes are a stable support during movement.

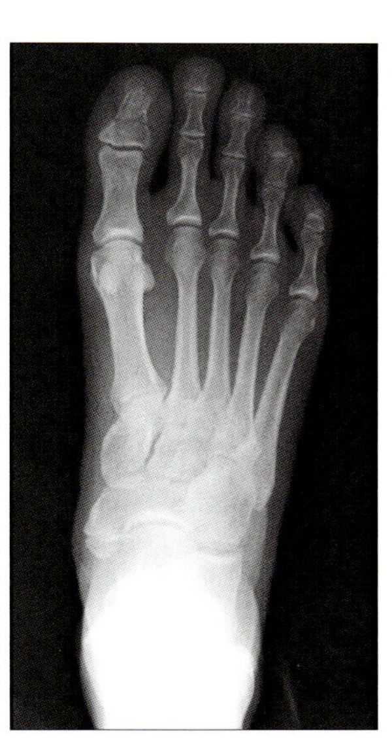

The 26 bones of the foot are clearly shown in this plain x-ray.

Ligaments and Joints of the Foot

The bones of the foot are arranged in such a way that they form bridge-like arches. These bones are supported by the presence of a number of strong ligaments.

The main supportive ligaments of the foot lie on the plantar (under) surface of the bones:

- The plantar calcaneonavicular, or spring ligament – stretches from the sustentaculum tali, a projection of the calcaneus, to the back of the navicular bone. This ligament helps maintain the longitudinal arch of the foot.
- The long plantar ligament – runs from under the calcaneus to the cuboid bone and the bases of the metatarsals. It maintains the arches of the foot.
- The plantar calcaneocuboid, or short plantar, ligament lies under the long plantar ligament and runs from the front of the undersurface of the calcaneus forward to the cuboid.

The metatarsals, phalanges and tarsals are bound to to each other by other ligaments running across the foot on the dorsal and plantar surface.

Joints of the Foot

The ankle joint only allows the foot to move up and down. Movements of the foot such as turning outwards (eversion) and inwards (inversion) are enabled by two joints:

- The transverse tarsal joint – a complex joint formed by parts of the calcaneus, talus, navicular and cuboid.
- The subtalar joint – where the talus moves against the calcaneus.

There are many other small synovial joints located in the foot where bone meets bone. These joints are held tightly together by tough ligaments and so little movement is possible. The joints between the phalanges allow movement of the toes, although the range is less than that of the fingers.

Ligaments of foot (plantar view)

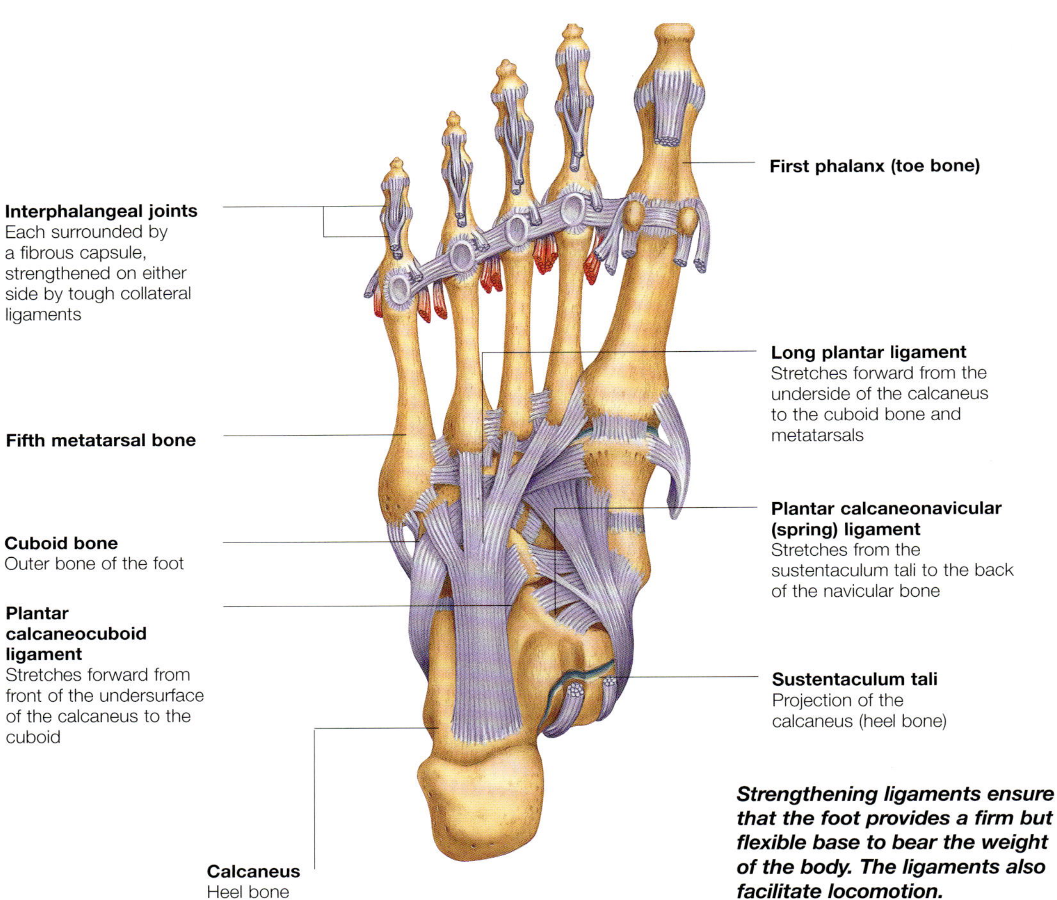

Strengthening ligaments ensure that the foot provides a firm but flexible base to bear the weight of the body. The ligaments also facilitate locomotion.

Joints of the foot

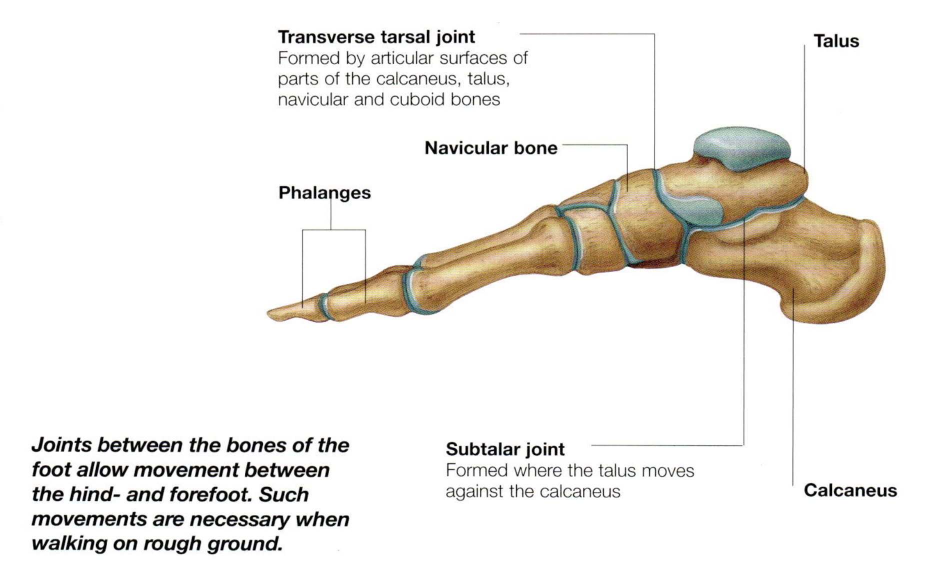

Joints between the bones of the foot allow movement between the hind- and forefoot. Such movements are necessary when walking on rough ground.

Arches of the Foot

A distinctive feature of the foot is that its bones are arranged in bridge-like arches. This allows the foot to be flexible enough to cope with uneven ground, while still being able to bear the weight of the body.

The arched shape of the foot can be illustrated by looking at a footprint. Only the heel, the outer edge of the foot, the pads under the metatarsal heads and the tips of the toes leave an impression. The rest of the foot is lifted away from the ground.

Three Arches

The foot has two longitudinal arches (medial and lateral) running along its length, and a transverse arch lying across it:

- Medial longitudinal arch. This is the higher and more important of the two longitudinal arches. The bones involved are the calcaneus, the talus, the navicular bone, the three cuneiform bones and the first three metatarsals. The head of the talus supports this arch.
- Lateral longitudinal arch. This arch is much lower and flatter, the bones resting on the ground when standing. The lateral arch is formed by the calcaneus, the cuboid and the fourth and fifth metatarsal bones.
- Transverse arch. This arch runs across the foot, supported on either side by the longitudinal arches, and is made up of the bases of the metatarsal bones, the cuboid and the three cuneiform bones.

Bones forming medial longitudinal arch of foot

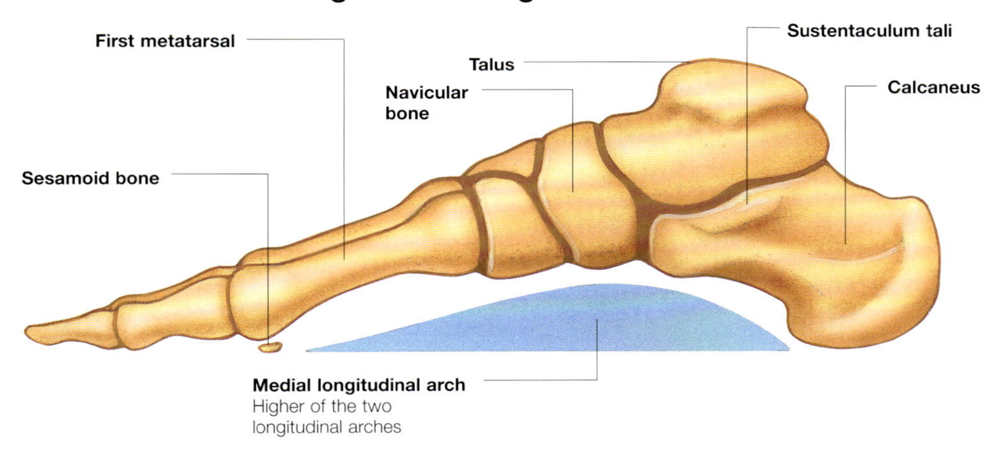

Bones forming lateral longitudinal arch of foot

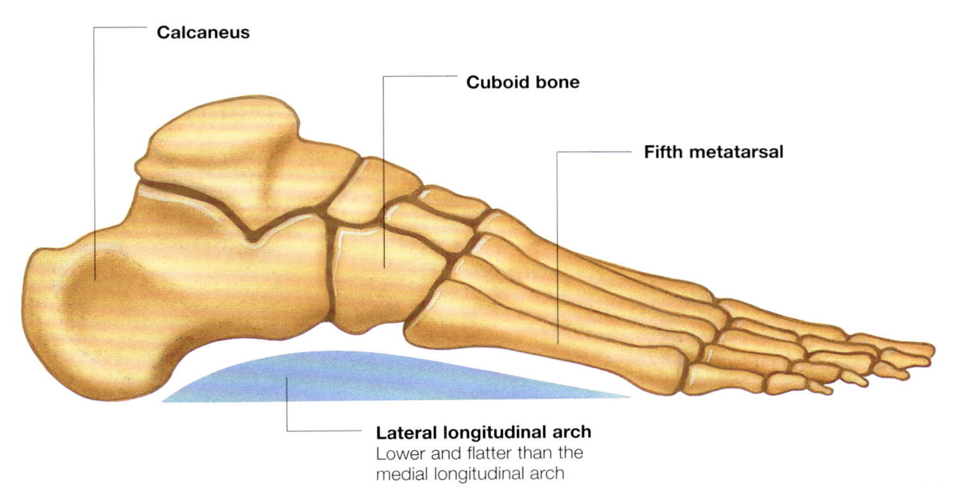

The bones of the foot form bridge-like arches. These are maintained by the shape of the bones and the strength of the ligaments and muscle tendons.

Tendons and Retinacula of the Foot

Many of the muscles that move the foot lie in the lower leg, rather than in the foot itself. Long tendons connect the leg muscles to the foot muscles, while retinacula bind the tendons firmly in place.

The leg muscles that move the bones in the foot have very long tendons that cross the ankle joint. At the joint they are held firmly in place by a series of retaining bands, or retinacula.

Retinacula of the Foot

There are four main retinacula:

- **Superior extensor retinaculum.** Lies above the ankle joint and retains the long tendons of the extensor muscles.
- **Inferior extensor retinaculum.** Lies beneath the ankle joint. It also retains extensor muscles.
- **Peroneal retinaculum.** Lies on the outer side of the ankle. It is in two parts, upper and lower, and retains the long peroneal muscle tendons.
- **Flexor retinaculum.** Lies on the inner side of the ankle and retains the long flexor tendons as they pass under the medial malleolus to reach the sole of the foot.

Long Tendons

The long tendons lie within lubricating synovial sheaths that protect them from damage as they run back and forth over the bones of the ankle. Some, for example the extensor digitorum longus and peroneus tertius, share a common synovial sheath.

Lateral view

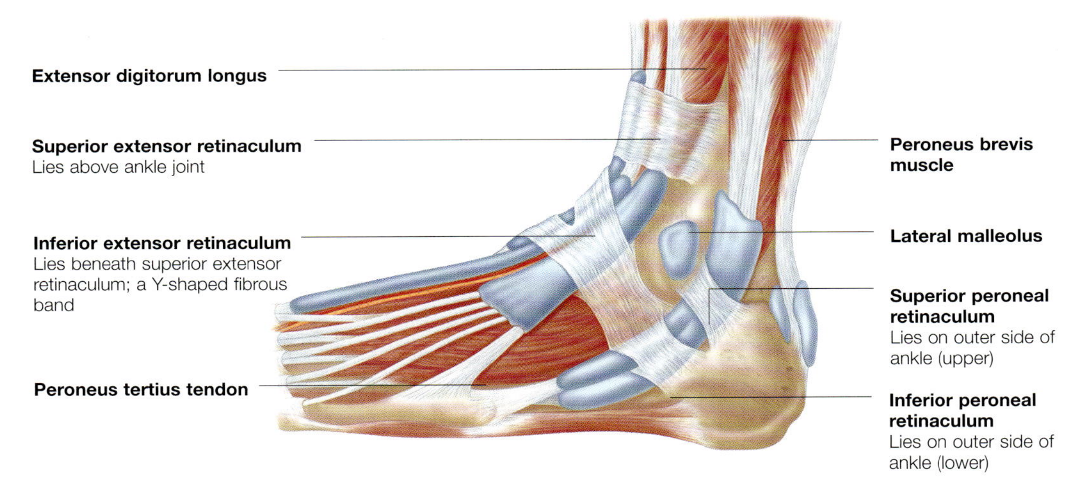

Medial view

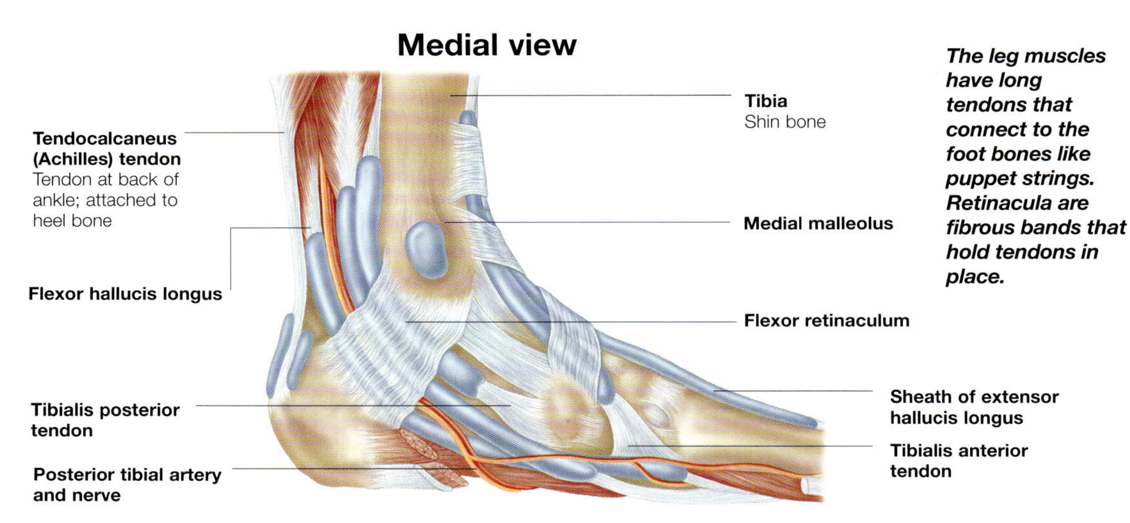

The leg muscles have long tendons that connect to the foot bones like puppet strings. Retinacula are fibrous bands that hold tendons in place.

Long tendons around ankle joint

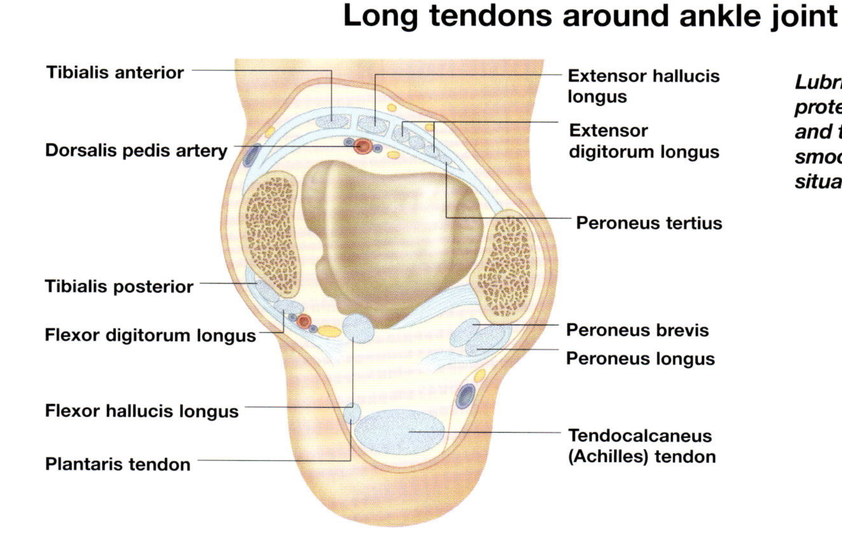

Lubricating tendon sheaths protect the tendons from wear and tear and help them run smoothly. The tendons are situated under the retinacula.

Muscles of the Top of the Foot

Although they are not especially powerful, the muscles that lie over the top of the foot play an important part in helping to extend the toes. The extensor digitorum brevis muscle tends to be used when the foot is already pointing upwards.

Most of the muscles that lie within the foot, the intrinsic muscles, are in the sole. The top, or dorsal surface, of the foot has just two muscles: the extensor digitorum brevis and the extensor hallucis brevis.

Muscles of the dorsal surface

- **Extensor digitorum brevis.** As its name suggests, this is a short muscle that extends (straightens or pulls upwards) the toes. It arises from the upper surface of the calcaneus, or heel bone, and the inferior extensor retinaculum. This muscle divides into three parts, each with a tendon that joins the corresponding long extensor tendon to insert into the second, third and fourth toes.
- **Extensor hallucis brevis.** This short muscle is really part of the extensor digitorum brevis. It runs down to insert into the big toe, or 'hallux', from which its name derives

Action of the Muscles

Together these two muscles assist the long extensor tendons in extending the first four toes. Although they do not have a particularly powerful action, they are useful in extending the toes when the foot itself is already pointing up, or dorsiflexed, as in this position the long extensors are unable to act further.

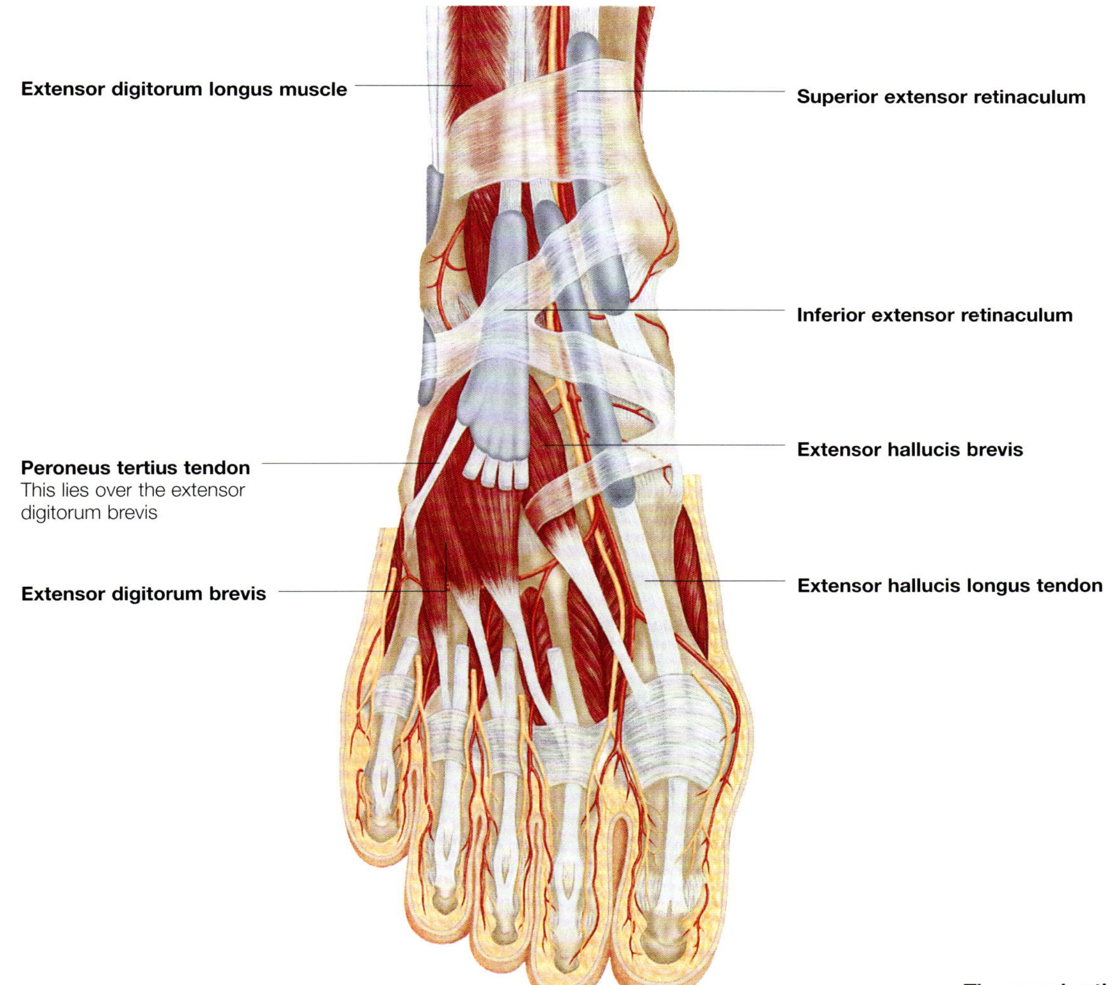

The muscles that lie at the top of the foot help to extend the toes. They assist the long extensors when the foot is dorsiflexed.

Muscles of the Soles of the Foot

Many of the movements of the bones and joints of the feet are brought about by muscles in the lower leg. However, there are also many small 'intrinsic' muscles that lie entirely within the foot.

The sole of the foot has four layers of intrinsic muscles. These work with the extrinsic (external to the foot) muscles to meet the varying demands placed upon the foot during standing, walking, running and jumping. They also help to support the bony arches of the foot and to allow us to stand on sloping or uneven ground.

First Muscle Layer

The first layer of sole muscles is the most superficial, lying just under the thick plantar aponeurosis. The muscles of this layer include:

- **Abductor hallucis** – this muscle lies along the medial (inner) border of the sole. It acts to abduct the big toe, or 'hallux', which means moving it away from the midline. It also flexes, or bends down, the hallux.
- **Flexor digitorum brevis** – this fleshy muscle lies down the centre of the sole and inserts into each of the lateral four toes. Contraction of this muscle causes those toes to flex
- **Abductor digiti minimi** – lying along the lateral (outer) border of the sole within this first layer, this muscle acts to abduct and flex the little toe.

These muscles are similar to the corresponding muscles in the hand, but their individual function is less important because the toes do not have such a wide range of movement as the fingers.

The muscles of the sole lie in four layers ranging from superficial to deep. The muscles of the first layer of the sole help to flex, abduct and adduct the toes.

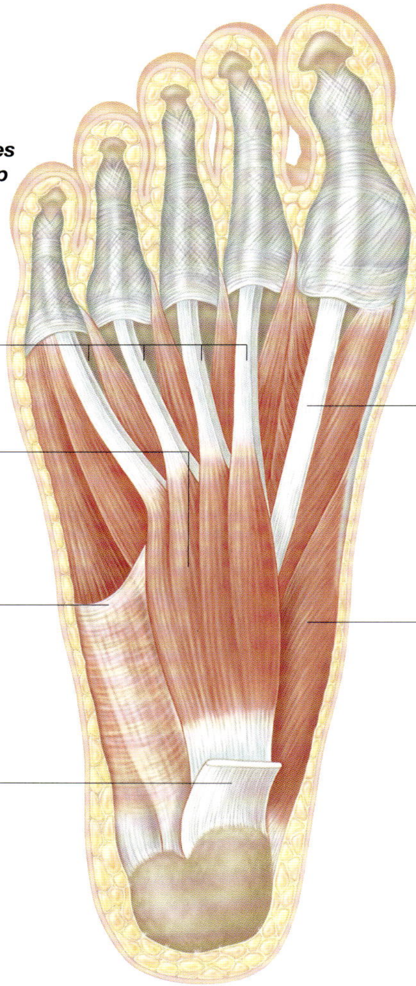

Flexor digitorum brevis tendons
Attached to the four toes

Flexor digitorum brevis
Assists in flexing the lateral four toes

Abductor digiti minimi
Helps to abduct the little toe and is the most lateral of the superficial sole muscles

Plantar aponeurosis (cut)

Flexor hallucis longus tendon

Abductor hallucis
Helps to flex and abduct the big toe and supports the medial bony arch

Plantar aponeurosis

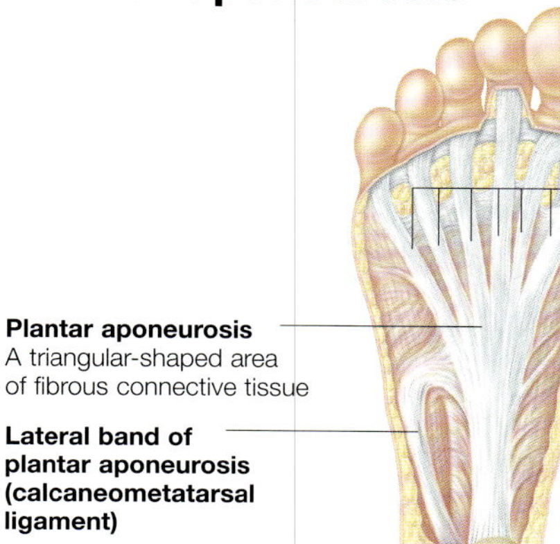

The skin of the sole is thick and overlies a layer of shock-absorbing fat pads. Under this layer lies a sheet of tough connective tissue called the plantar aponeurosis.

The plantar aponeurosis is the thickened central portion of the plantar fascia, the connective tissue that surrounds and encloses the muscles of the sole. The plantar aponeurosis consists of bands of strong fibrous tissue that run the length of the sole and insert into each of the toes. It also attaches to the skin above it and to the deeper tissues that lie below.

The plantar aponeurosis is a strong sheet of connective tissue. 'Plantar' refers to the sole of the foot just as 'palmar' refers to the palm of the hand.

Action

The plantar aponeurosis acts to hold together the parts of the foot and helps to protect the sole of the foot from injury. It also helps to support the bony arches of the foot.

Deep muscle layers of the sole

The muscles of the sole of the foot are made up of four different layers. All of these muscles act together, to help keep the bony arches of the feet stable.

Beneath the superficial layer of intrinsic muscles of the sole lie three further layers. These all have a contribution to make to the stability and flexibility of the foot, both at rest and in motion.

Although the deep muscles each have individual actions, their main role is to act together to maintain the stability of the bony arches of the feet.

Second muscle layer of the sole
The second muscle layer includes some tendons from the extrinsic muscles, as well as some smaller intrinsic muscles. Muscles and tendons that are included within this second layer of the sole are:

- **Quadratus plantae** (or flexor accessorius) muscle – this wide, rectangular muscle arises from two heads on either side of the heel. It inserts into the edge of the tendon of flexor digitorum longus where it acts by pulling backwards on this tendon and so stabilizes it as it flexes the toes.
- **Tendons of flexor hallucis longus and flexor digitorum longus** – these tendons enter the second muscular layer of the sole after winding around the medial malleolus (inner 'ankle bone').
- **The four lumbrical muscles** – named for their worm-like appearance, these four muscles arise from the tendons of flexor digitorum longus. These muscles act to extend the toes while the long tendons are flexing them, which helps to prevent the toes 'buckling under' when walking or running.

Flexor digitorum longus tendons

Sesamoid bones

Lumbrical muscles
Help with flexing and extending the toes

Flexor digiti minimi brevis muscle
Assists in flexing the little toe

Flexor hallucis longus tendon
Used to 'push off' in walking or jumping

Flexor digitorum longus tendon
Branches to each of the four lateral toes

Quadratus plantae muscle (flexor accessorius muscle)
Helps to flex the lateral four toes and is also able to flex the toes on its own

Tuberosity of calcaneus

Muscles from the second layer of the sole of the foot help to extend and flex the toes. They also help to stabilize the tendons during flexion of the toes.

Third and fourth muscle layers

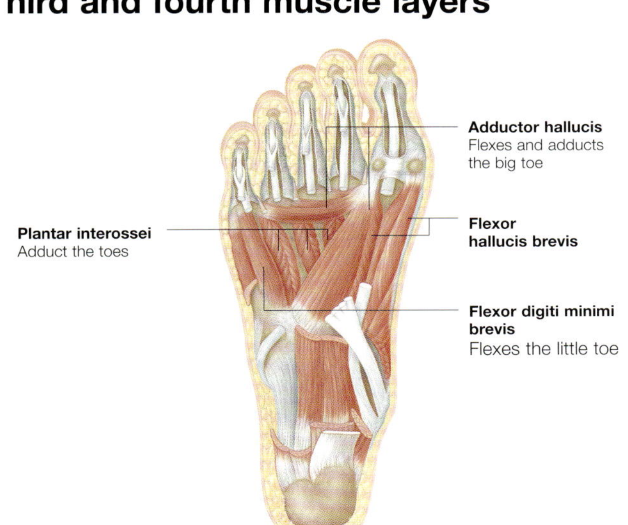

Lying deep to the long flexor tendons, the third muscle layer of the sole is made up of three small muscles:

- Flexor hallucis brevis. This is a short muscle that flexes the big toe. It arises from the cuboid and lateral cuneiform bones and then splits into two parts. Each of these two parts has a tendon that inserts into the base of the big toe. The two sesamoid bones of the foot lie within these tendons.
- Adductor hallucis. This muscle arises from two heads; an oblique head and a transverse head. They join to insert into the base of the big toe.
- Flexor digiti minimi brevis. This small muscle runs along the outer border of the foot to the little toe, which it helps to flex.

The three small muscles in the deep muscle layer of the sole help to flex the toes. Even deeper, between the bones, lie the seven interossei muscles.

Fourth muscle layer of the sole
The muscles of the fourth, and deepest layer of the sole are called the 'interossei' muscles, which literally means 'between the bones'. Unlike the hand, which has eight, there are only seven interossei in the foot. The four dorsal interossei muscles (not shown) abduct the toes, whereas the three plantar interossei adduct them.

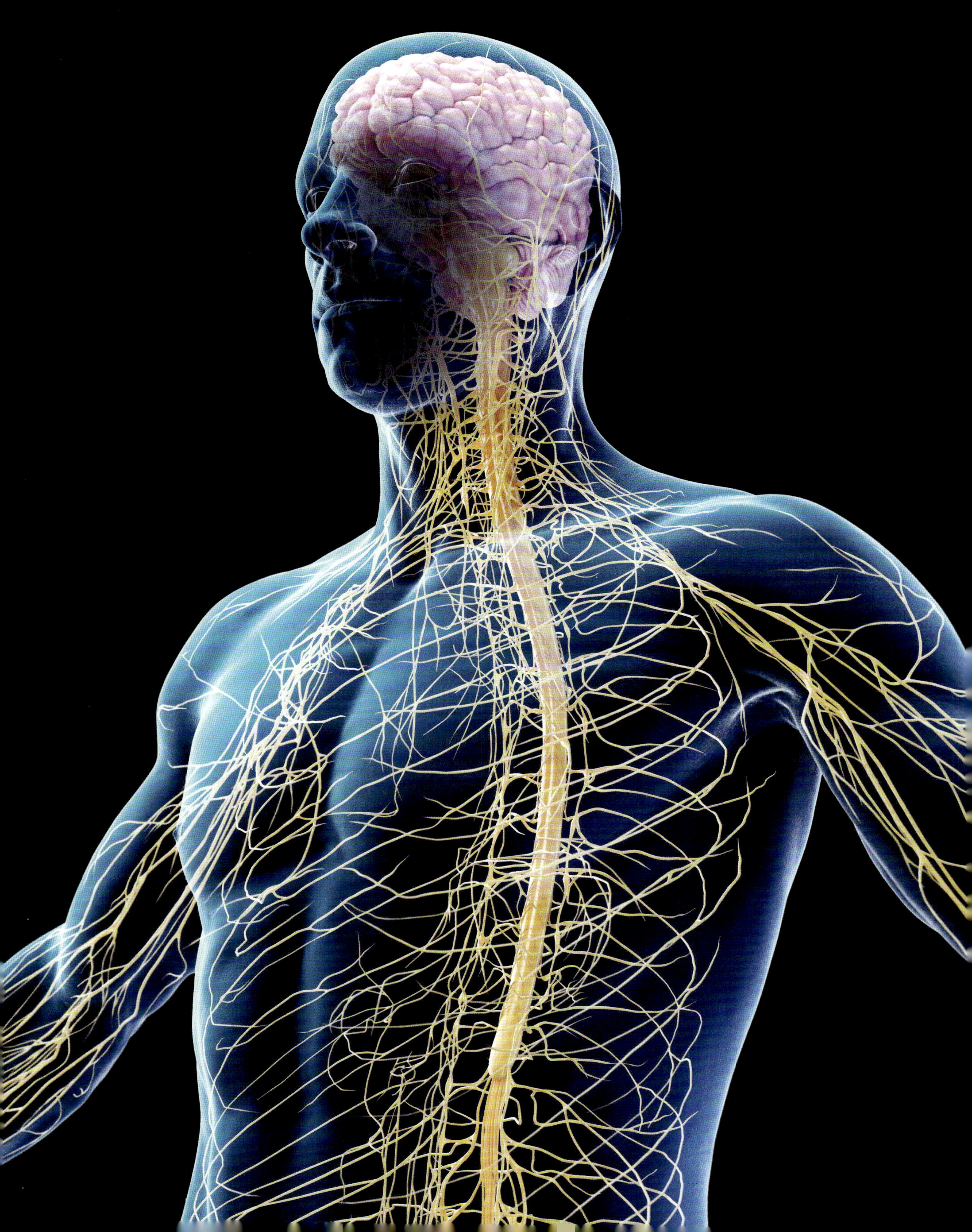

CHAPTER 4

The Nervous System

A complex network of nerve pathways connects the whole body to the brain, the central coordinating centre, enabling movement, thought, system functioning, breathing and much more. This section will describe the nervous system, outline the structure and function of the brain and explain how messages reach their destination via the spinal cord and peripheral nerves.

The brain and the spinal cord form the central nervous system. The brain is the most specialized organ in the body and contains around 100 billion neurones – the functioning units of the nervous system. Millions of messages travel constantly to and from the brain as electric signals, traversing the spinal cord from which branch hundreds of nerves that supply the whole body. These branching nerves form the peripheral nervous system and they supply tissues and organs, for example, the skin, muscles and digestive system, and enable us to move, feel, and function overall. In this section, the nerve supply to the head and limbs is described but the nerves of individual organs are covered in the relevant chapters.

Opposite: The brain is the control centre of the nervous system. Messages travel to and from the brain via the spinal cord and the network of peripheral nerves.

Neurone (nerve cell)

A neurone is a specialized cell of the nervous system. The main function of neurones is to carry information in the form of electrical impulses from one part of the body to the other.

The tissues of the nervous system are made up of two types of cells: neurones, or nerve cells, which transmit information in the form of electrical signals; and the smaller supporting cells (glial cells) that surround them. A nerve is a bundle of neurones supported by a sheath.

Common Features

Neurones are the large, highly specialized cells of the nervous system, whose function is to receive information and transmit it throughout the body.

Although variable in structure, neurones have some features in common:

- Cell body – the neurone possesses a single cell body from which a variable number of branching processes emerge
- Dendrites – these are the thin, branching processes of the neurone, which are extensions of the cell body
- Axon – each neurone has an axon carrying electrical impulses away from the cell body.

Types of Neurones

There are three main types of neurone:

- Interneural – found in the brain and spinal cord
- Sensory neurones – responsible for taste, smell, hearing, sight and touch
- Motor neurones – involved in movement, both voluntary and involuntary.

Structure of a motor neurone

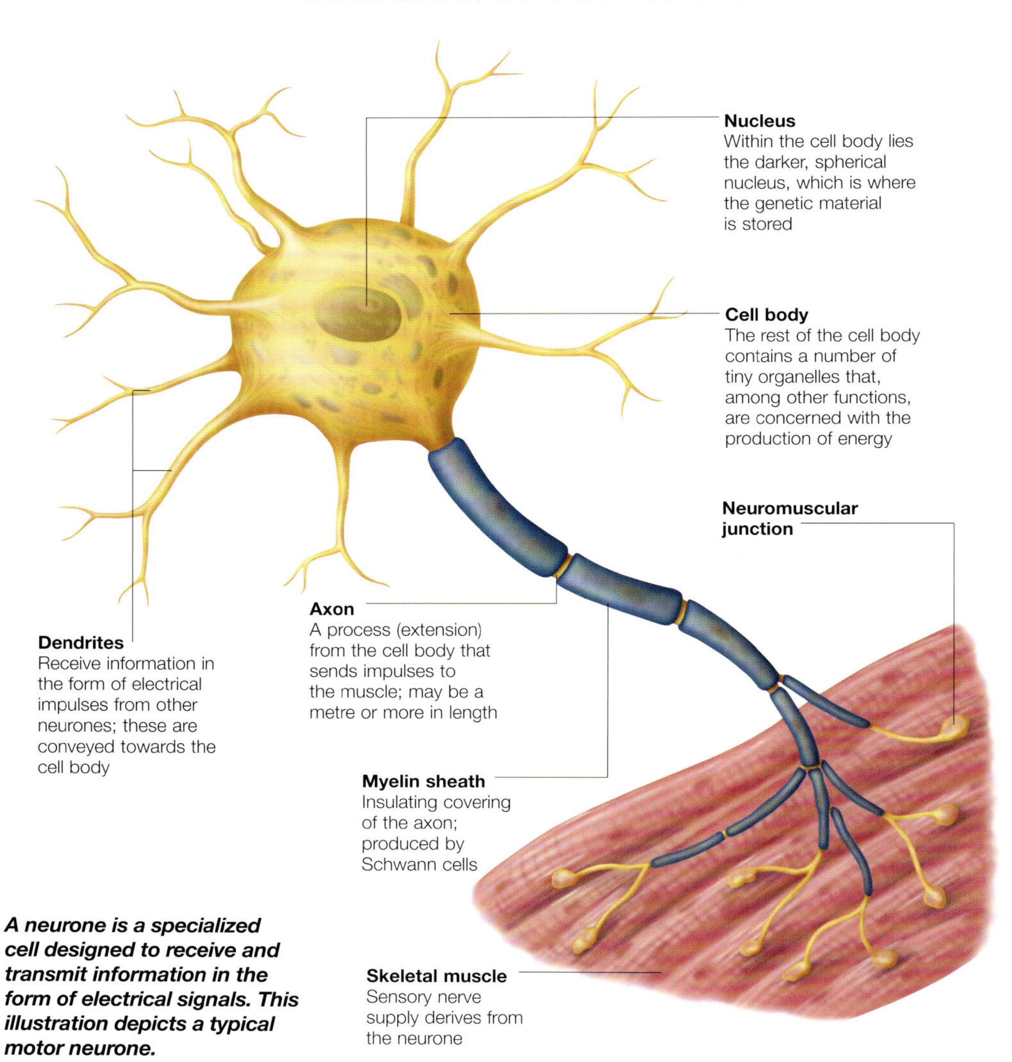

A neurone is a specialized cell designed to receive and transmit information in the form of electrical signals. This illustration depicts a typical motor neurone.

Myelin sheath

The speed of an electrical signal along a neurone's axon is increased by the presence of a myelin sheath – a layer of fatty insulation.

The myelin sheath is formed differently according to where it is located.

- In the peripheral nervous system (those nerves lying outside the brain and spinal cord), the myelin sheath is produced by specialized Schwann cells. These wrap themselves around the axon of a nerve cell to form a sheath of concentric circles of their cell membranes.
- In the central nervous system, neurones are given their myelin sheath by cells known as oligodendrocytes, which can myelinate more than one nerve axon at a time.

Appearance

Nerve fibres with myelin sheaths tend to look whiter than unmyelinated ones, which have a grey tinge. The 'white matter' of the brain is composed of dense collections of myelinated nerve fibres, whereas the 'grey matter' is made up of nerve cell bodies and unmyelinated fibres.

The gap between the cells, where there is no myelin, is known as the node of Ranvier. As an electrical signal passes down the nerve it must 'hop' from one node to the other, which makes it travel faster overall than if no myelin sheath were present.

Insulation of a peripheral nerve

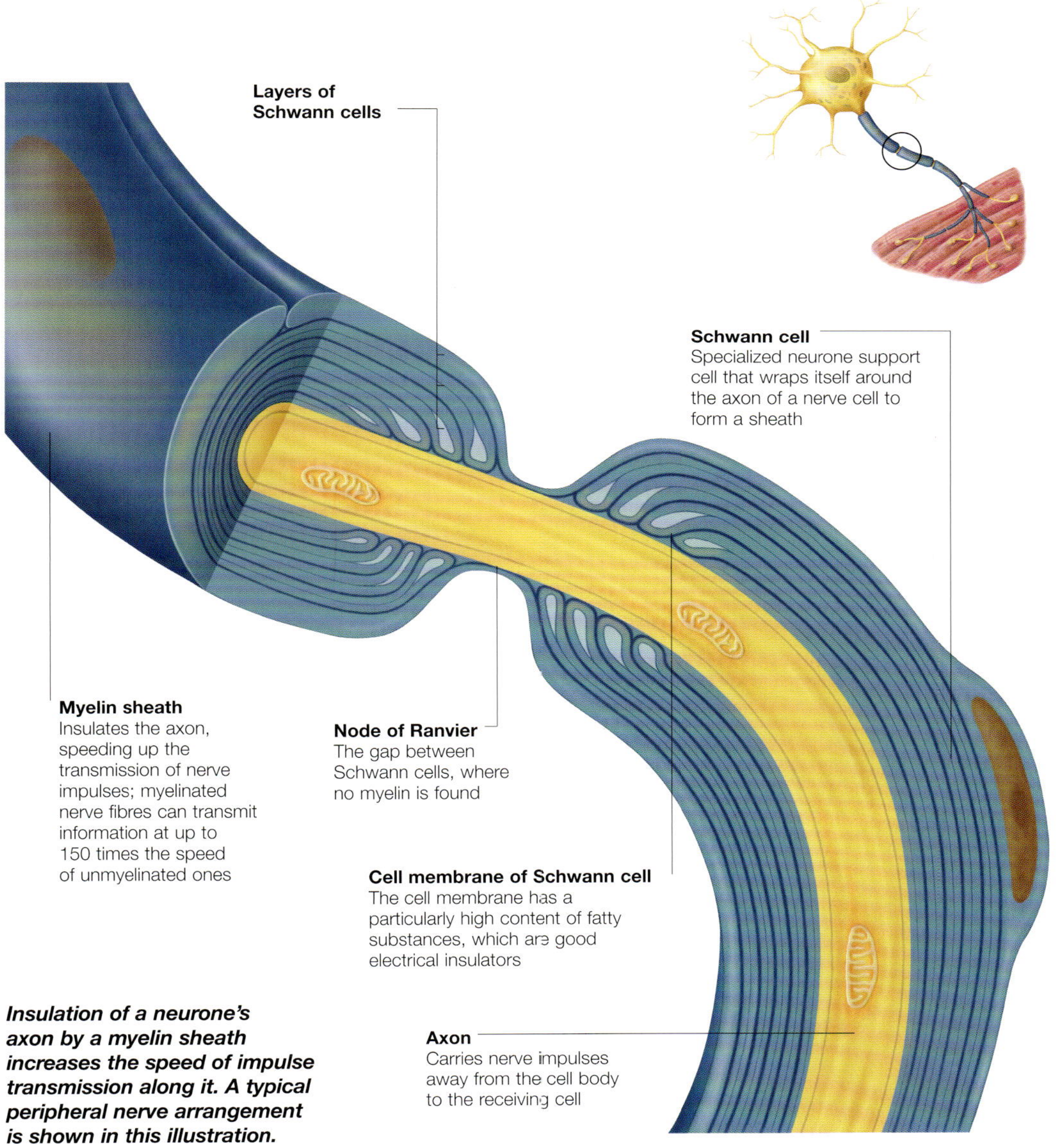

Insulation of a neurone's axon by a myelin sheath increases the speed of impulse transmission along it. A typical peripheral nerve arrangement is shown in this illustration.

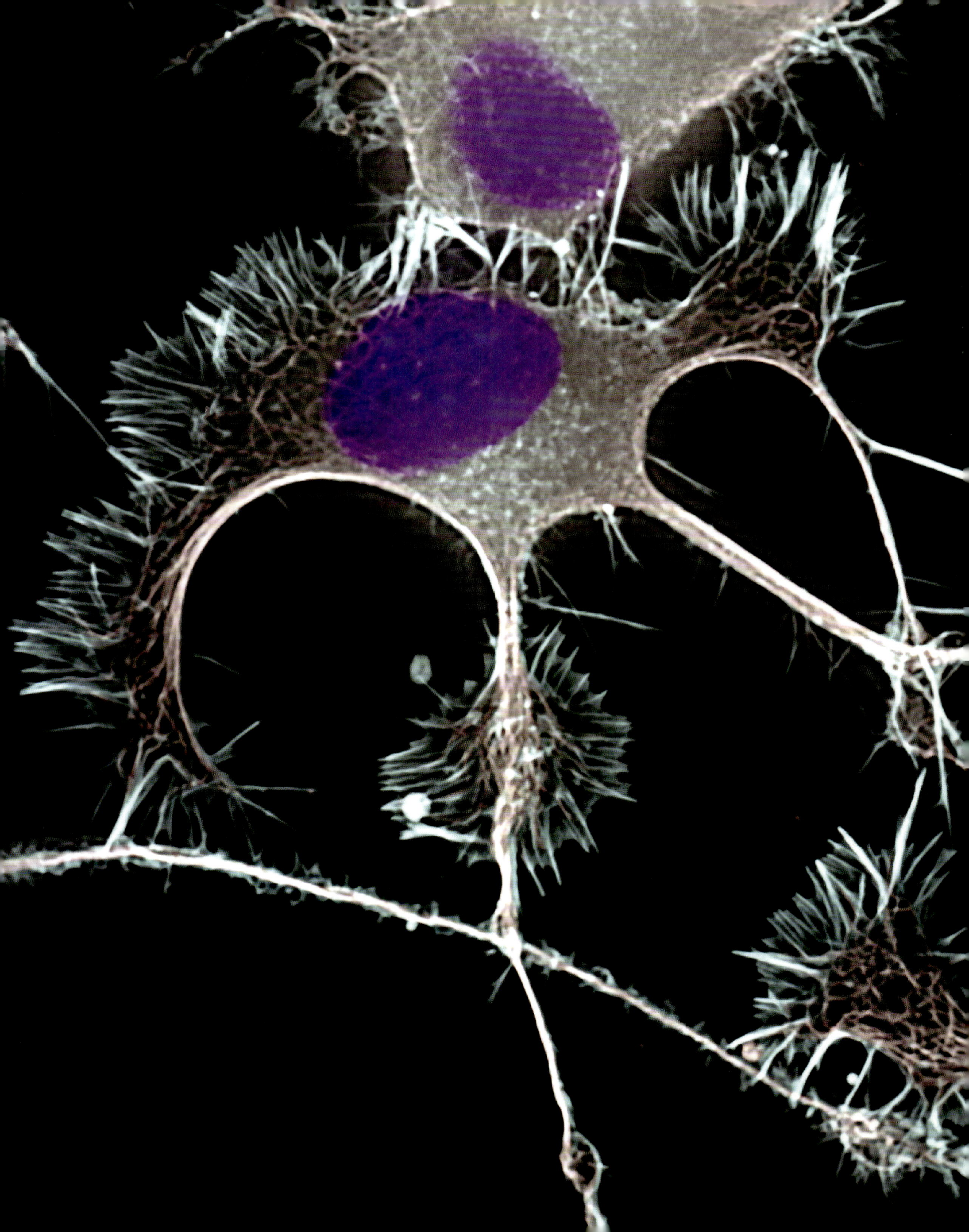

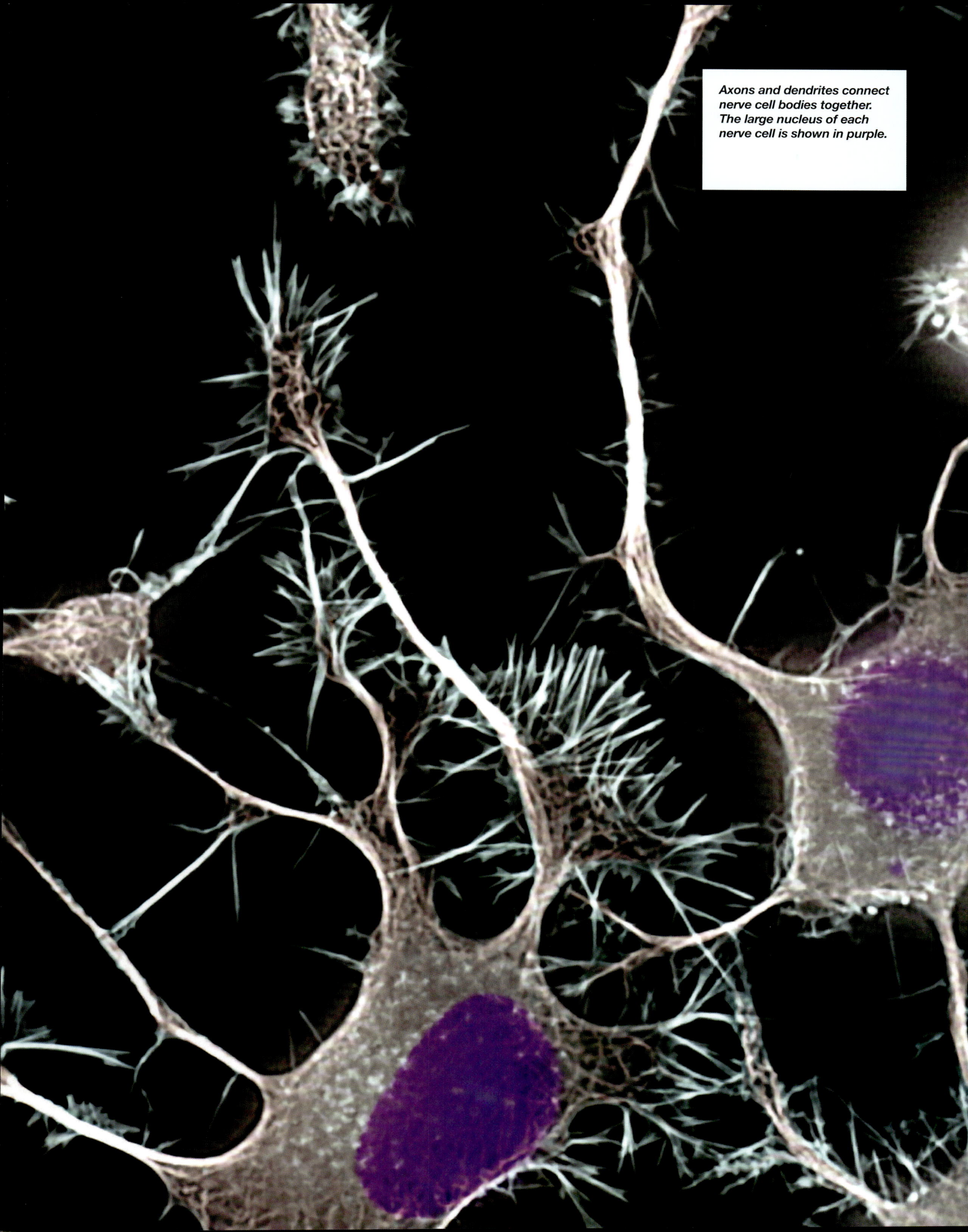

Axons and dendrites connect nerve cell bodies together. The large nucleus of each nerve cell is shown in purple.

How Neurones Work

Neurones generate nerve impulses, electrical messages that travel from one end of a nerve cell to the other. This ability is essential for us to interact successfully with the world around us.

The human central nervous system contains at least two hundred billion neurones; on average, each neurone communicates with thousands of other nerve cells. This complexity allows the brain to interpret the rich sensory input that it receives from the five senses and to react accordingly.

Neuroanatomy
Although neurones from different regions of the nervous system can look very dissimilar, they all contain the same three basic elements: dendrites, a cell body and an axon.

- Dendrites (from the Greek 'dendros', meaning tree) are branch-like protrusions of the cell membrane that provide a large surface area to receive neurotransmitters released from other neurones. The dendrites transduce (convert) this chemical information into small electrical impulses, which are then conveyed to the cell body.
- The greater part of a neurone is made up of the cell body which, like the majority of the body's cells, contains a nucleus. A region of the cell body called the axon hillock collates all the small nervous impulses generated by the many dendrites and initiates action potentials (nerve impulses) accordingly.
- The axon of a neurone carries nerve impulses from the cell body to the synaptic terminals – specialized endings that release neurotransmitters to communicate with other neurones.

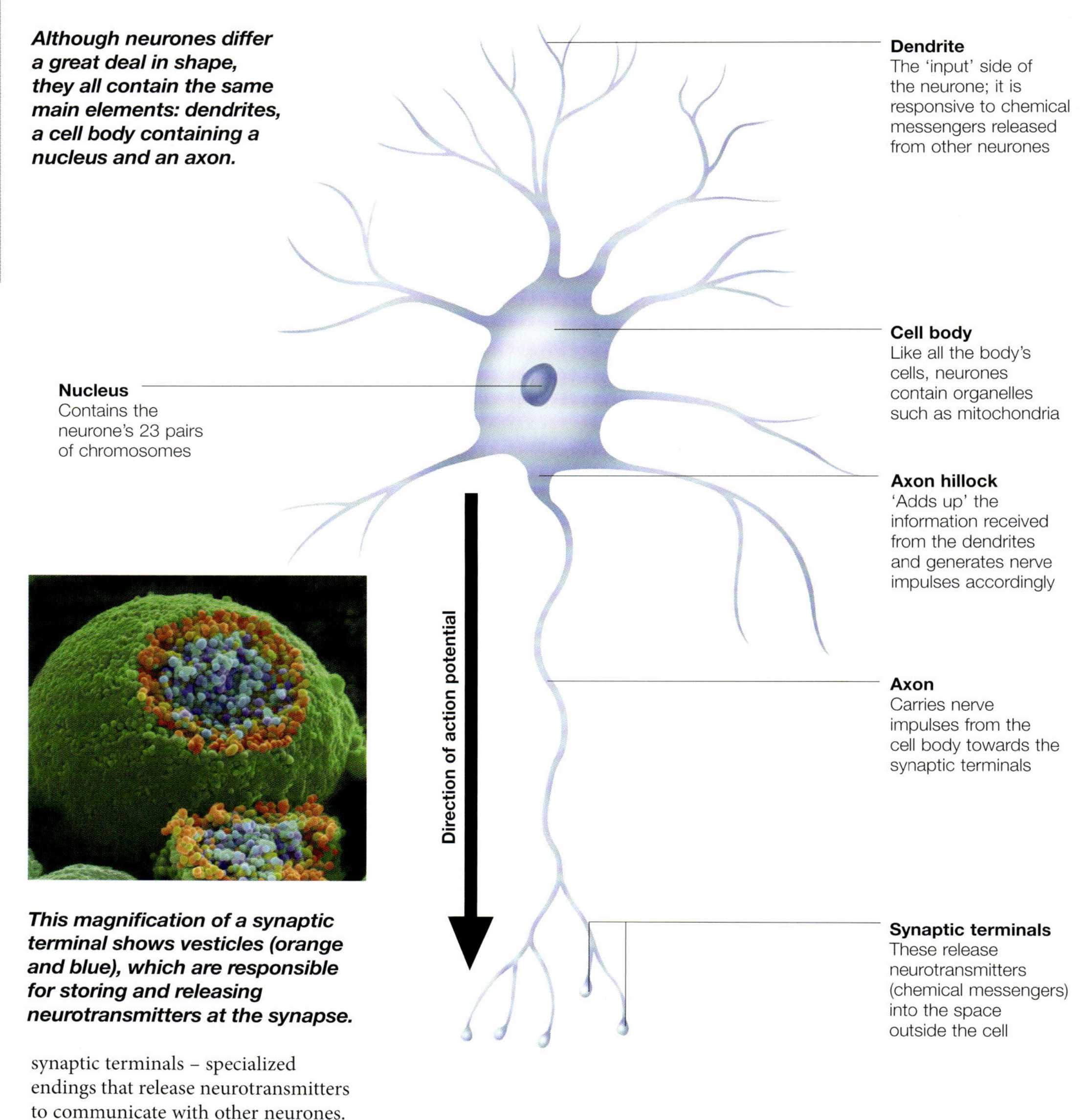

Although neurones differ a great deal in shape, they all contain the same main elements: dendrites, a cell body containing a nucleus and an axon.

This magnification of a synaptic terminal shows vesticles (orange and blue), which are responsible for storing and releasing neurotransmitters at the synapse.

What makes neurones different from other cells in the body?

The chemical composition of the fluid inside a cell (called the cytosol) is different from the composition of the fluid outside the cell (extracellular fluid).

Compared to the extracellular fluid, the cytosol has fewer positive charges and more negative charges; this means that the inside of the cell is slightly negative compared to the outside of the cell. This electrical charge across the membrane is called the membrane potential, and in most cells is about minus 70 millivolts (thousandths of a volt).

What makes neurones so special is that they can alter the electrical charge across their membrane to generate nerve impulses. They are able to do this because their cell membrane contains gated protein pores that allow electrically charged ions (sodium, potassium, calcium or chloride ions) to cross it transiently, so altering the neurone's membrane potential.

Other cells in the body do not have these protein pores and so their membrane potential stays relatively constant.

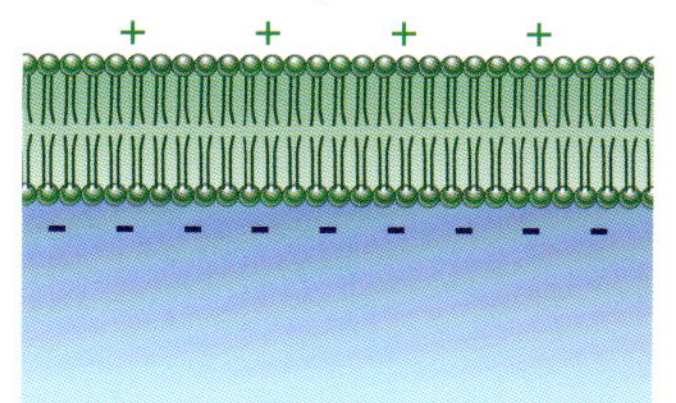

Unlike neurones, the majority of the body's cells do not have protein pores that can open and close in response to a pre-determined signal.

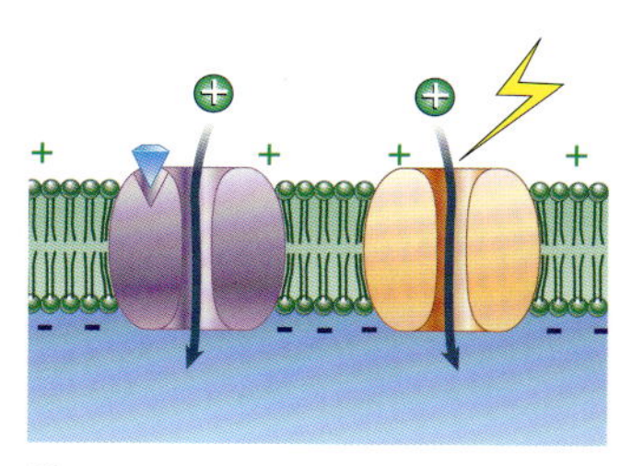

Neurones are able to generate nerve impulses because their membrane contains gated channels that respond either to chemical messengers (left) or to a change in voltage (right).

How nerve impulses are generated

Neurones generate nerve impulses by altering the charge across their membrane. If this is reduced, for example, by cooling, production of impulses is decreased.

A neurone can only generate an action potential (nerve impulse) when it has been adequately stimulated. Neurotransmitters (chemical messengers), which are released from nearby neurones, cause receptor proteins in the dendritic membrane to open. This allows positive sodium ions to flow into the dendrite, causing the membrane potential to become slightly less negative.

Threshold potential

If sufficient sodium ions enter the neurone to raise the membrane potential to the 'threshold potential', other voltage-dependent protein pores open, which allows even more positive sodium ions to enter the neurone. Action potentials are not graded in amplitude (that is, they do not vary in 'strength'). Rather, when the threshold potential is reached the membrane potential increases to its maximum level.

Action potential recovery

When the membrane potential reaches its maximum level, the sodium channels close and other channels, which are permeable to positive potassium ions, open in response to the high membrane potential (the potassium channels only open in response to a high voltage). The positive potassium ions flow out of the cell, bringing the membrane potential back towards its resting value.

Speed of nerve impulses

Each axon transmits nerve impulses at a constant speed. However there is a degree of variability in the speed at which different axons do this. For example, conduction speeds vary between 0.5 and 120m (1.6 and 394ft) per second. The speed depends on the diameter of the nerve (larger nerves conduct more quickly) and the degree to which the nerve is insulated – myelinated nerves are more efficient.

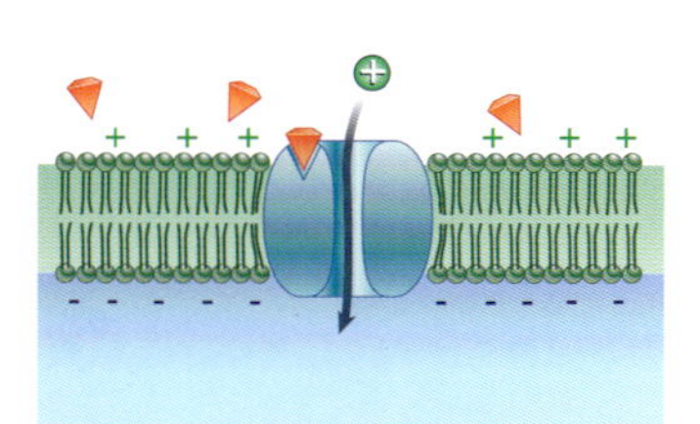

1. Some neurotransmitters open sodium channels in the dendritic membrane, thereby allowing positive sodium ions to flow into the cell.

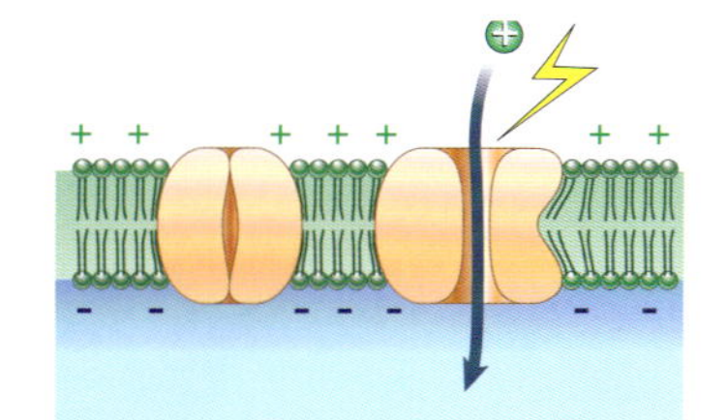

2. When the voltage reaches a threshold value, sodium channels open, and more positive ions enter the cell.

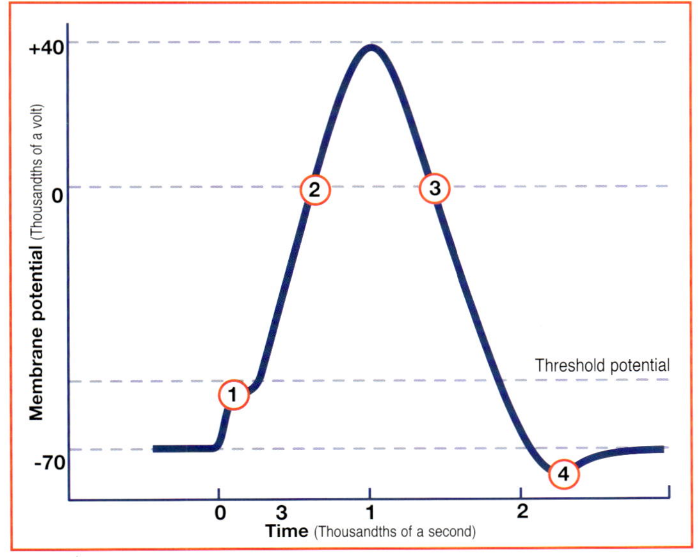

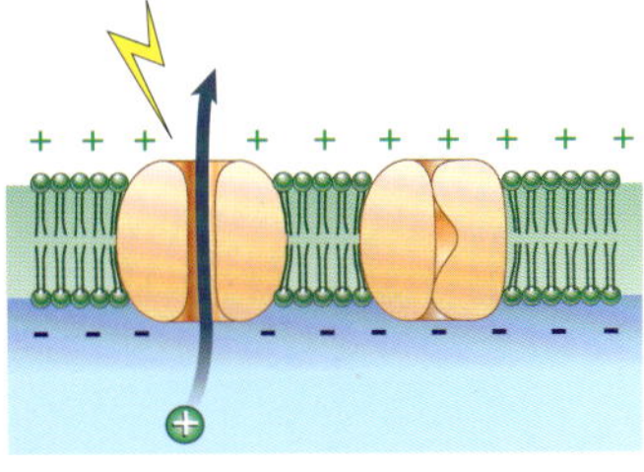

3. Sodium channels close and potassium channels open, allowing positive potassium ions to leave the cell; both these events act to lower the voltage.

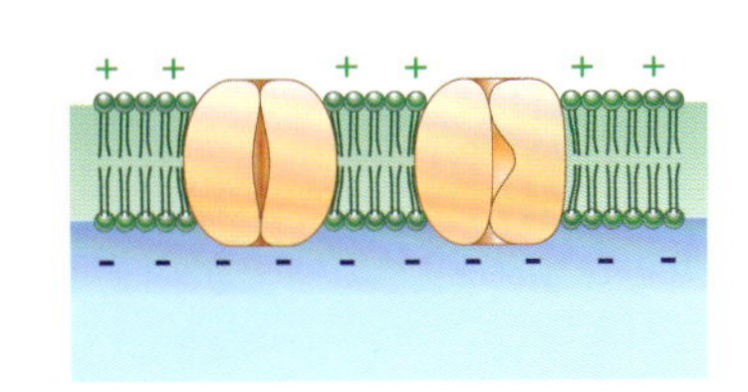

4. The sodium and potassium channels deactivate, meaning the neurone is at rest and no further action potentials can occur.

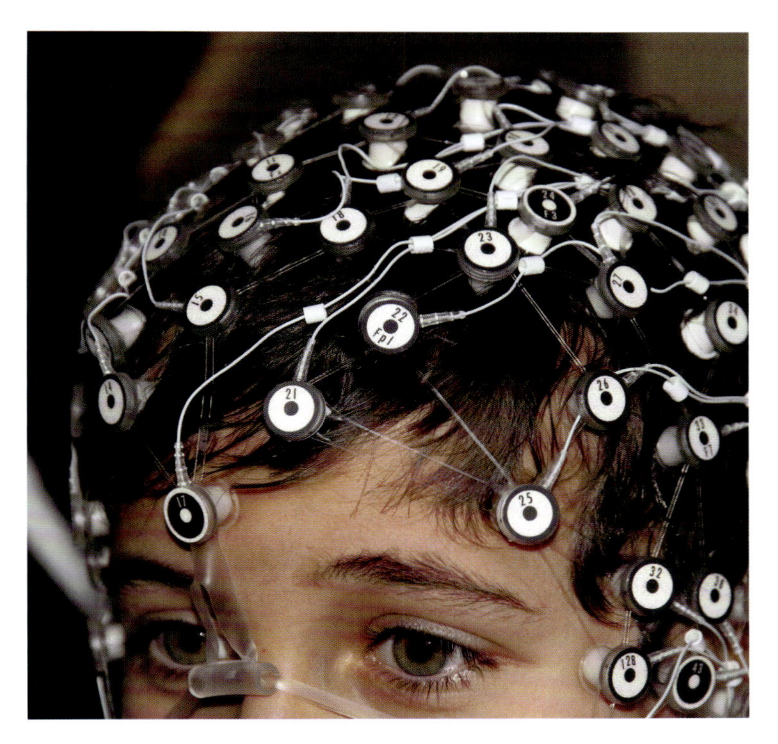

Electroencephalopathy (ECG) is a medical test used to monitor the electrical signals produced by the brain. Electrodes are attached to the brain and activity is recorded.

How Neurones Communicate

Neurones communicate with each other by releasing chemical messengers called neurotransmitters.

Nerve cells do not make direct contact with each other. There is a small gap (synaptic gap) that separates the nerve cell sending the information (pre-synaptic neurone) from the neurone receiving the information (post-synaptic neurone). This gap ensures that an electrical impulse cannot flow directly from one neurone to the next. Instead, when an impulse reaches the synaptic terminals, the sudden change in voltage causes calcium ions to flow into the pre-synaptic cell.

Release of Neurotransmitters

Calcium ions cause vesicles (small membrane-bound sacs containing chemical messengers called neurotransmitters) to move towards and dock with the pre-synaptic cell membrane, releasing their contents into the synaptic gap. The neurotransmitter molecules diffuse across to the post-synaptic cell and activate receptor proteins located within its membrane. This has the effect of exciting or inhibiting the post-synaptic cell (depending on the neurotransmitter and its associated receptor), increasing or decreasing the likelihood of an action potential being generated respectively.

After a neurotransmitter has bound with and activated its receptor on the post-synaptic membrane, it rapidly disengages and is either broken down by enzymes floating in the synaptic gap or is taken up into the pre-synaptic terminal where it is repackaged into another vesicle. This ensures that the effect of the transmitter on a receptor molecule is short-lived.

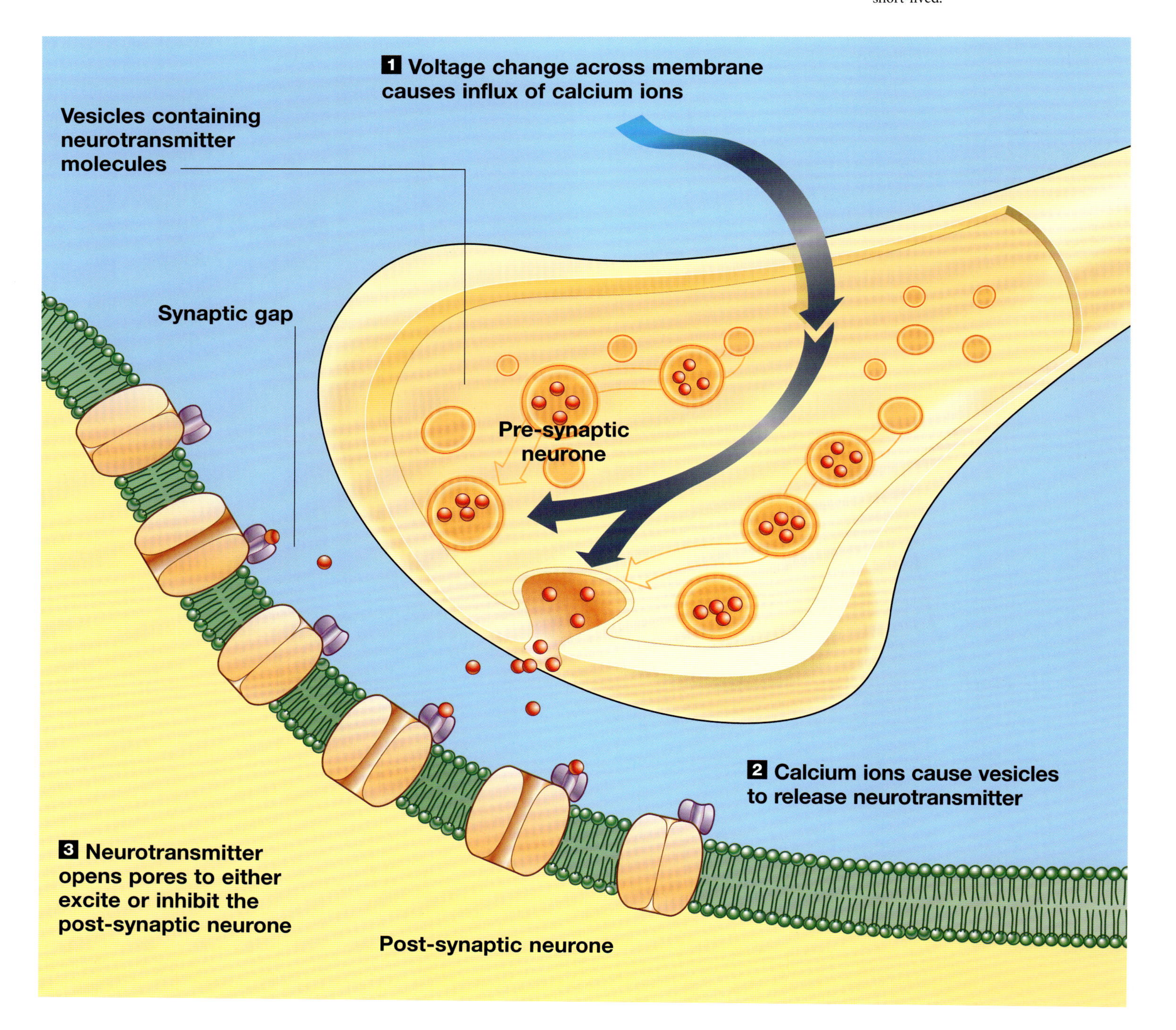

Neural processing

The brain is an incredibly complex structure; each of its neurones is connected to thousands of others located throughout the nervous system.

Since nerve impulses do not vary in strength, information is encoded in the frequency of nerve impulses (that is, the number of action potentials that a neurone generates per second), in a similar way to Morse code.

One of the big problems that neuroscientists face today is to try to understand how this relatively simple encoding system produces; for example, the emotional responses that we feel when a friend or relative dies, or the ability to throw a ball with such accuracy that it hits a target 20m (65ft) away.

It is clear that information is not transmitted from one neurone to another in a linear fashion. Rather, a single neurone is likely to receive synaptic inputs from many other neurones (called convergence) and be able to influence a large number (up to 100,000) of other neurones (called divergence).

It has been calculated that the number of possible routes for nerve impulses to take through this vast neural network is greater than the number of subatomic particles contained in the entire universe.

Information transfer does not occur in a linear fashion. A single neurone can thus influence and be influenced by thousands of other such cells.

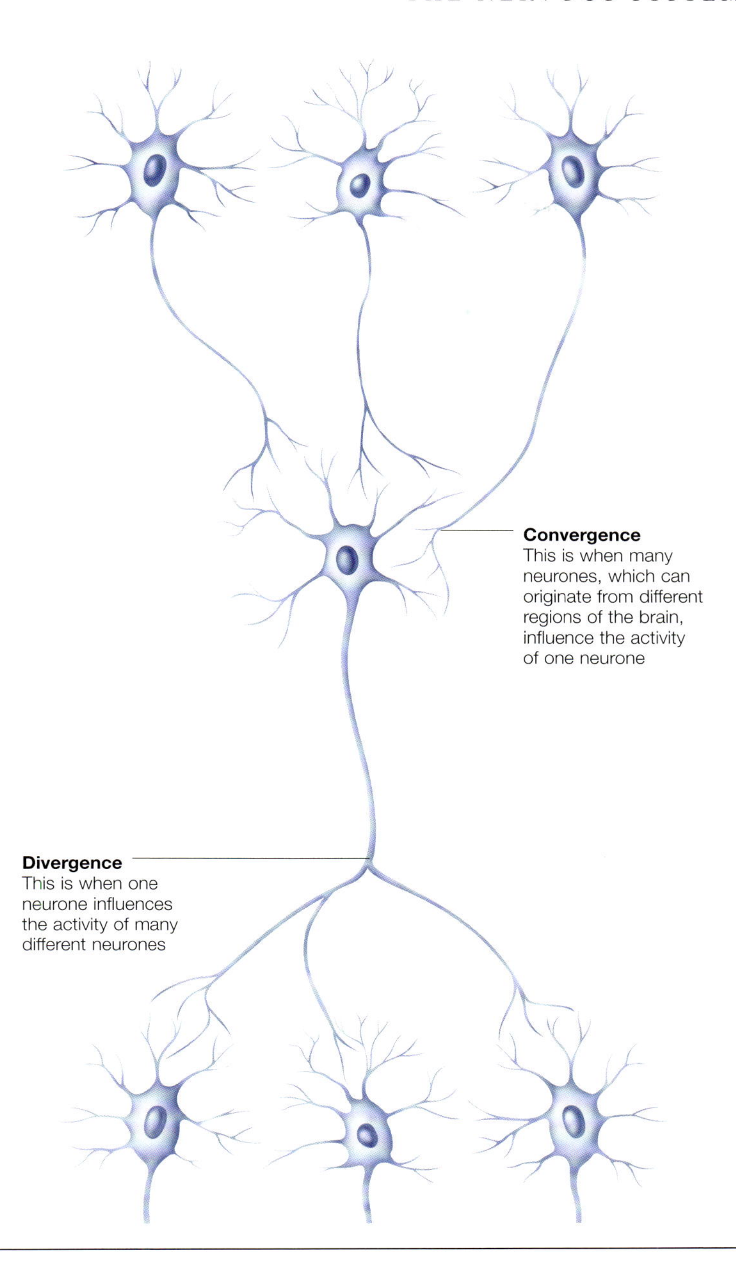

Types of synapse

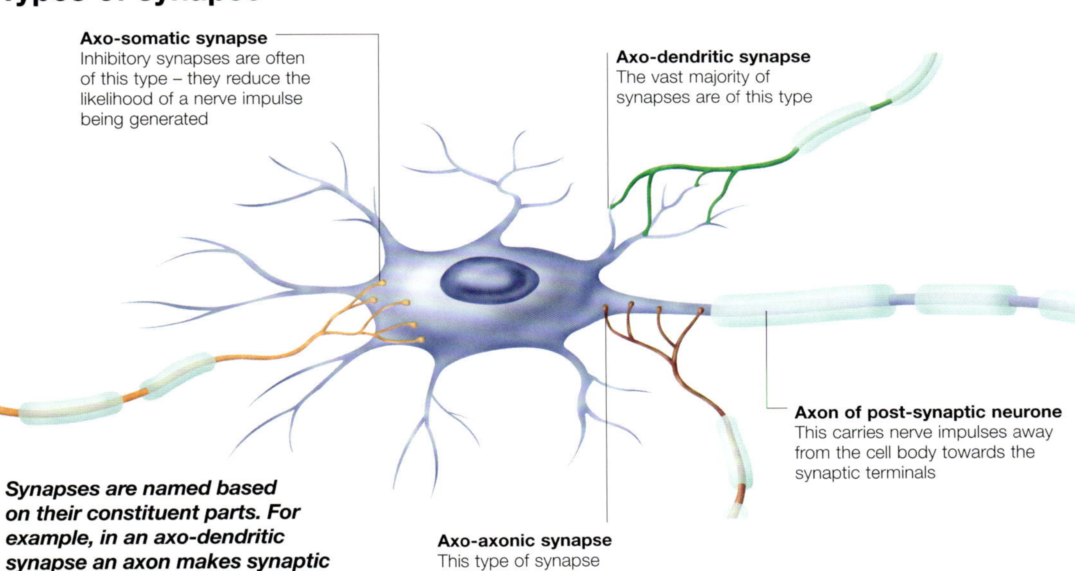

Synapses are named based on their constituent parts. For example, in an axo-dendritic synapse an axon makes synaptic contact with a dendrite.

There are two main types of synapse: those that cause the post-synaptic neurone to become excited, and those that cause it to become inhibited (this depends to a large degree on the type of neurotransmitter that is released). A neurone will only fire a nerve impulse when the excitatory inputs outweigh the inhibitory ones.

Strength of synapses

Each neurone receives a large number of both excitatory and inhibitory inputs. Each of the synapses present will have a greater or lesser effect in determining whether an action potential is initiated.

For example, synapses that have the most powerful effect are generally those close to the nerve impulse-initiating zone in the cell body (soma).

Brain

The brain is the key organ of the central nervous system and lies inside the skull. It controls many body functions including our heart rate, the ability to walk and run, and the creation of our thoughts.

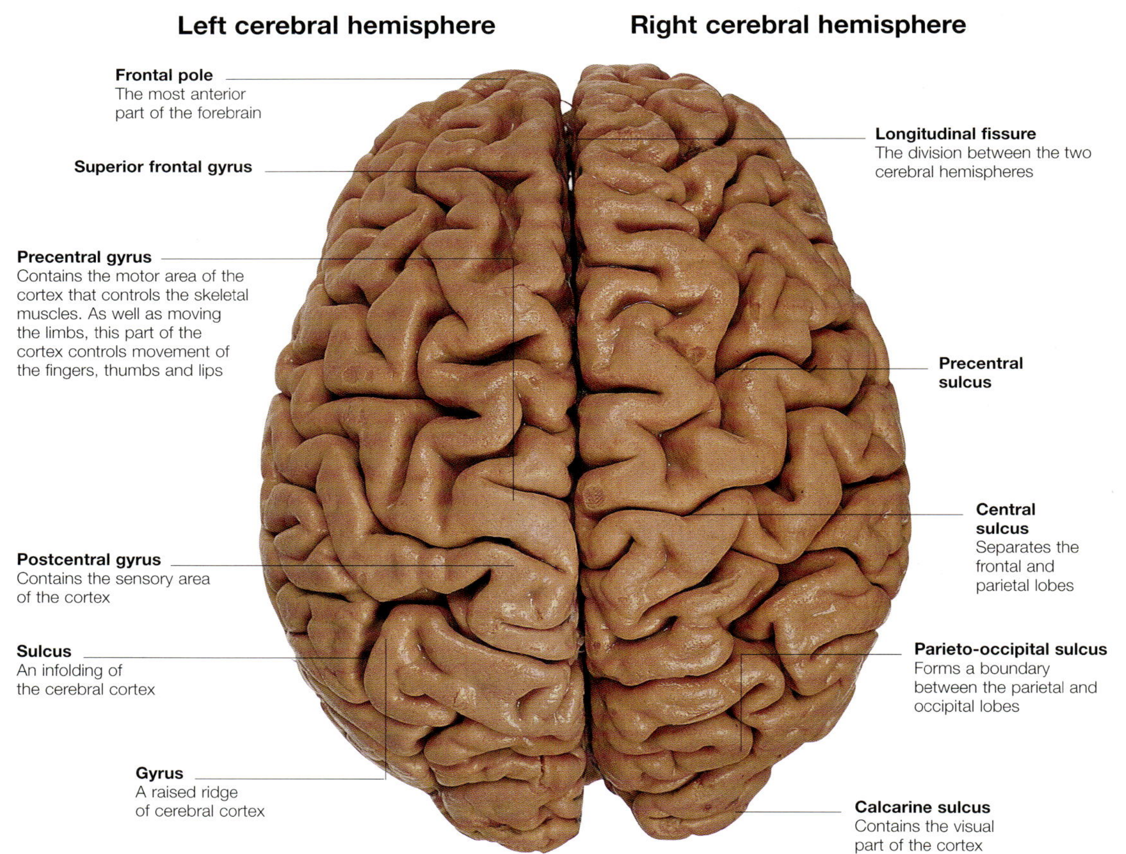

The brain has three major parts: forebrain, midbrain and hindbrain. The forebrain is divided into two halves, forming the left and right cerebral hemispheres.

Hemispheres

The cerebral hemispheres form the largest part of the forebrain. Their outer surface is folded into a series of gyri (ridges) and sulci (furrows) that greatly increases its surface area. Most of the surface of each hemisphere is hidden in the depths of the sulci.

Each hemisphere is divided into frontal, parietal, occipital and temporal lobes, named after the closely related bones of the skull. Connecting the two hemispheres is the corpus callosum, a large bundle of fibres deep in the longitudinal fissure.

Grey and White Matter

The hemispheres consist of an outer cortex of grey matter and an inner mass of white matter.

- Grey matter contains nerve cell bodies, and is found in the cortex of the cerebral and cerebellar hemispheres and in groups of subcortical nuclei.
- White matter comprises nerve fibres found below the cortex. They form the communication network of the brain, and can project to other areas of the cortex and spinal cord.

Gyri (ridges) and sulci (furrows)

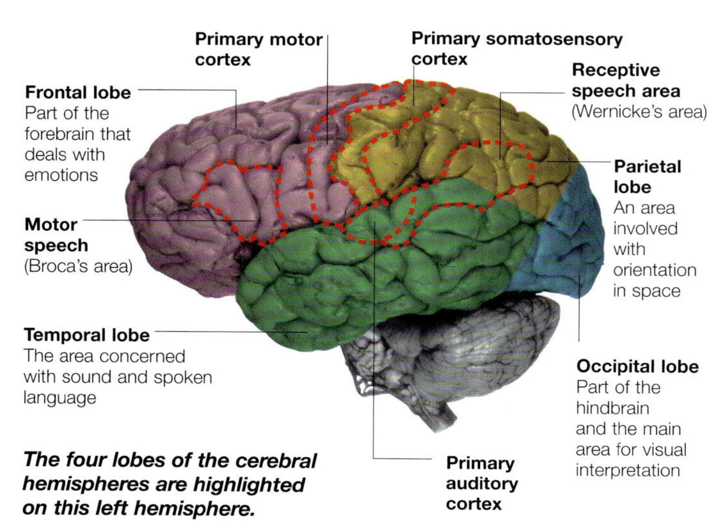

The four lobes of the cerebral hemispheres are highlighted on this left hemisphere.

The central sulcus runs from the longitudinal fissure to the lateral fissure, and marks the boundary between the frontal and parietal lobes. The precentral gyrus runs parallel to and in front of the central sulcus and contains the primary motor cortex, where voluntary movement is initiated.

The postcentral gyrus contains the primary somatosensory cortex that perceives bodily sensations. The parieto-occipital sulcus (on the medial surface of both hemispheres) marks the border between the parietal and occipital lobes.

The calcarine sulcus marks the position of the primary visual cortex, where visual images are perceived. The primary auditory cortex is located towards the posterior (back) end of the lateral fissure.

On the medial surface of the temporal lobe, at the rostral (front) end of the most superior gyrus, lies the primary olfactory cortex, which is involved with smell. Internal to the parahippocampal gyrus lies the hippocampus, which is part of the limbic system and is involved in memory formation.

The areas responsible for speech are located in the dominant hemisphere (usually the left) in each individual. The motor speech area (Broca's area) lies in the inferior frontal gyrus and is essential for the production of speech.

Inside the brain

This midline section through the two cerebral hemispheres reveals the main structures that control a vast number of activities in the body, including movement, sensation, speech and sleep.

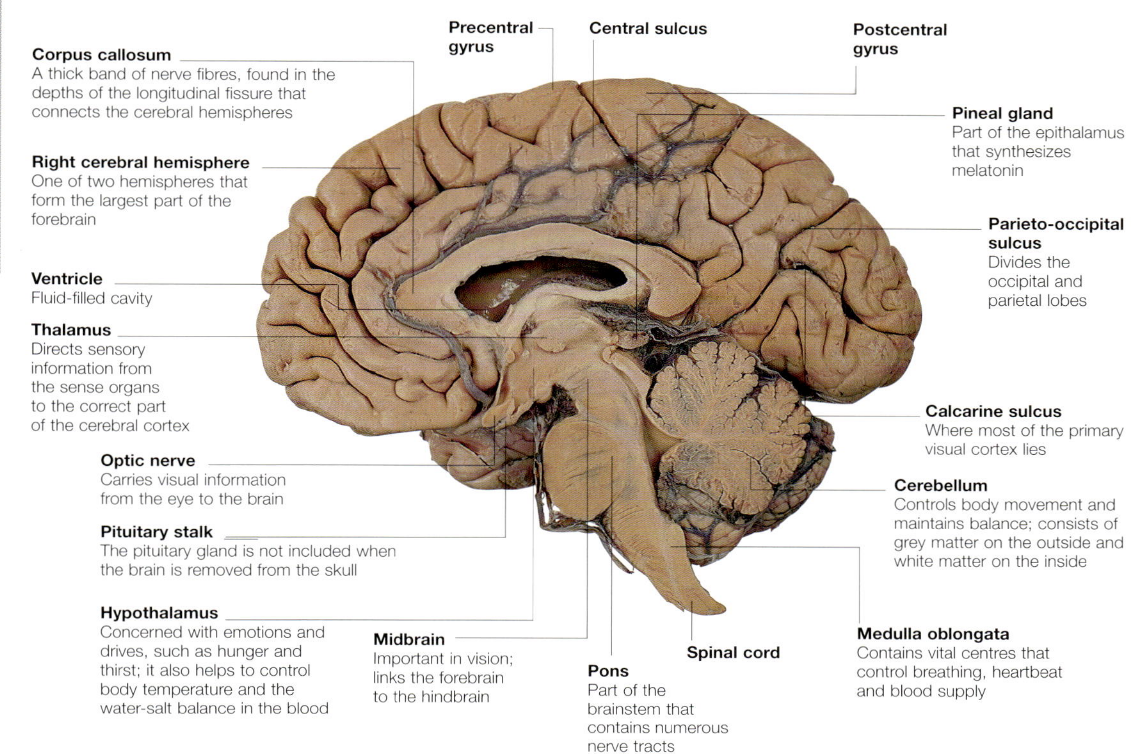

Each area of the brain plays a unique and complex function:

- Cerebellum – involved with movement and balance
- Brainstem (the midbrain, the pons and the medulla oblongata) – plays a crucial role in breathing and circulation
- Prefrontal cortex – has high-order cognitive functions, including abstract thinking, social behaviour and decision-making
- Basal ganglia – masses of grey matter situated within the white matter of the cerebral hemispheres. Involved in aspects of motor function, including movement programming, and motor memory retrieval
- Diencephalon – forms the central part of the brain and surrounds the third ventricle. Composed of the thalamus, hypothalamus, epithalamus and subthalamus
- Thalamus – the last relay station for information from the brainstem and spinal cord before it reaches the cortex
- Hypothalamus – involved in a variety of homeostatic mechanisms, and influences the sympathetic and parasympathetic nervous systems, controls body temperature, appetite and wakefulness
- Epithalamus – includes the pineal gland, which synthesizes melatonin and is involved in the control of the sleep/wake cycle
- Subthalamus – helps control movement
- Pituitary gland – secretes substances that influence the thyroid, adrenal glands and the gonads and produces growth factors. Also produces hormones that increase blood pressure, decrease urine production and cause uterine contraction.

Brainstem and cerebellum

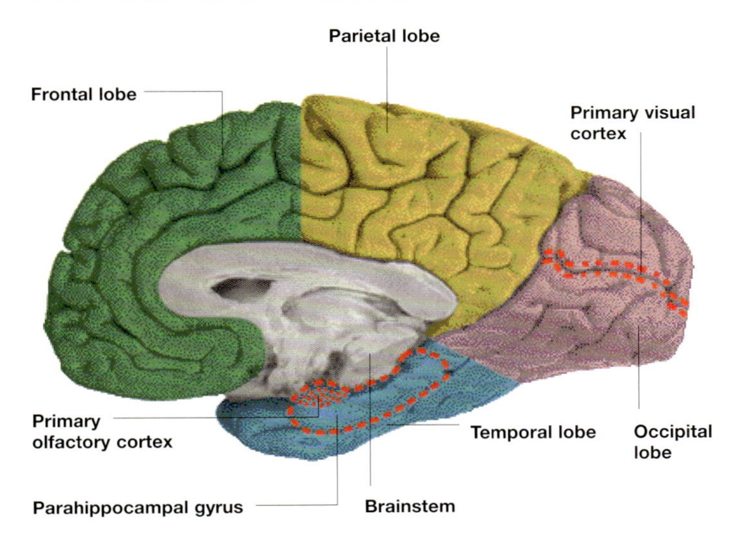

The posterior part of the diencephalon is connected to the midbrain, which is followed by the pons and medulla oblongata of the hindbrain. The midbrain and hindbrain contain the nerve fibres connecting the cerebral hemispheres to the cranial nerve nuclei, to lower centres within the brainstem and to the spinal cord. They also contain the cranial nerve nuclei.

Most of the reticular formation, a network of nerve pathways, lies in the midbrain and hindbrain. This system contains the important respiratory, cardiac and vasomotor centres.

The cerebellum lies posterior to the hindbrain and is attached to it by three pairs of narrow stalk-like structures called peduncles. Connections with the rest of the brain and spinal cord are established via these peduncles. The cerebellum functions at an unconscious level to coordinate movements initiated in other parts of the brain. It also controls the maintenance of balance and influences posture and muscle tone.

A view of the medial surface of the right hemisphere, with the brainstem removed, allowing the lower hemisphere to be seen.

Meninges

In anatomy the meninges are three membranes that cover and protect the brain and spinal cord.

The meninges cover the brain and spinal cord, and serve to protect these important structures. There are three layers: the dura, arachnoid and pia maters.

Dura Mater

The dura mater is a thick fibrous tissue that lines the inner layer of the skull. It deviates from the contours of the skull by forming a double fold (falx cerebri) that dips down between the cerebral hemispheres, and into the gaps between the cerebral hemispheres and the cerebellum (tentorium cerebelli).

Arachnoid Mater

The arachnoid mater is an impermeable membrane that follows the contours of the dura mater. It is separated from the dura mater by a small gap called the subdural space. The arachnoid mater is connected to the pia mater by bands of delicate web-like tissue known as trabeculae.

Pia Mater

The pia mater covers the surface of the brain and spinal cord. The space between the arachnoid and pia maters (subarachnoid space) is filled with cerebrospinal fluid (CSF). The brain and spinal cord are suspended in this fluid, which gives the most vital means of physical protection for the brain.

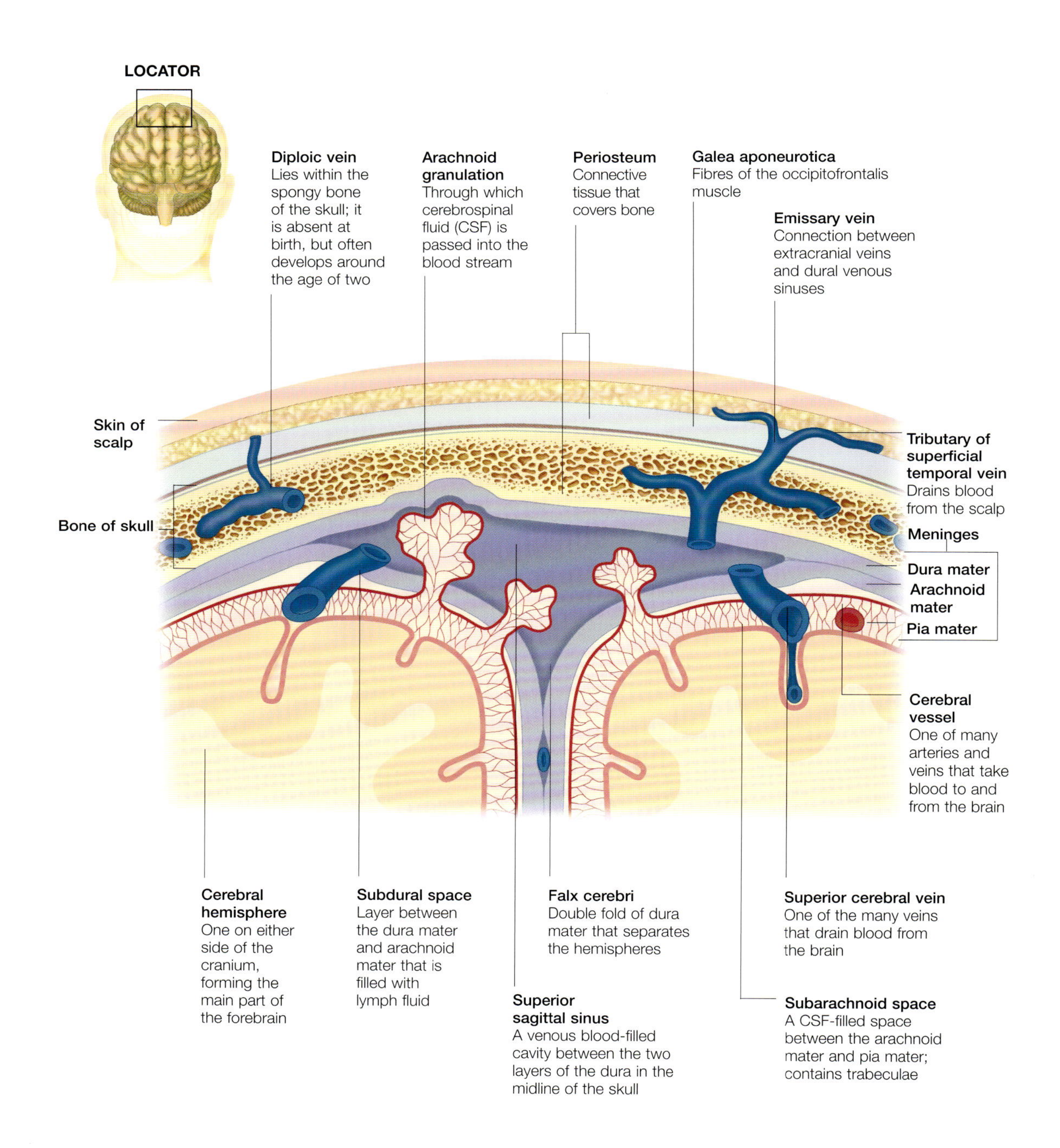

Dural Venous Sinuses

Dural sinuses are endothelial-lined spaces between two folds of the dura mater that play a role in the circulation and drainage of the blood and fluids that protect and bathe the brain.

There are 15 dural venous sinuses – blood-filled cavities between double folds of dura mater. Venous sinuses are lined by endothelium, but unlike other veins, they have no muscular layer. They are therefore very delicate, relying on surrounding tissues for support.

Venous Circulation

There are two sets of dural venous sinuses, those in the upper part of the skull and those on the floor of the skull. They receive blood draining from the brain via the cerebral and cerebellar veins, the red bone marrow of the skull via diploic veins, and the scalp via the emissary veins. They are crucial to the reabsorption of cerebrospinal fluid.

CSF Drainage

Cerebrospinal fluid is a liquid that bathes and protects the brain. To avoid a buildup of fluid, there is a system of balancing production of the fluid with drainage back into the venous system. CSF drains through the dura mater into the dural venous sinuses and from there into the venous system.

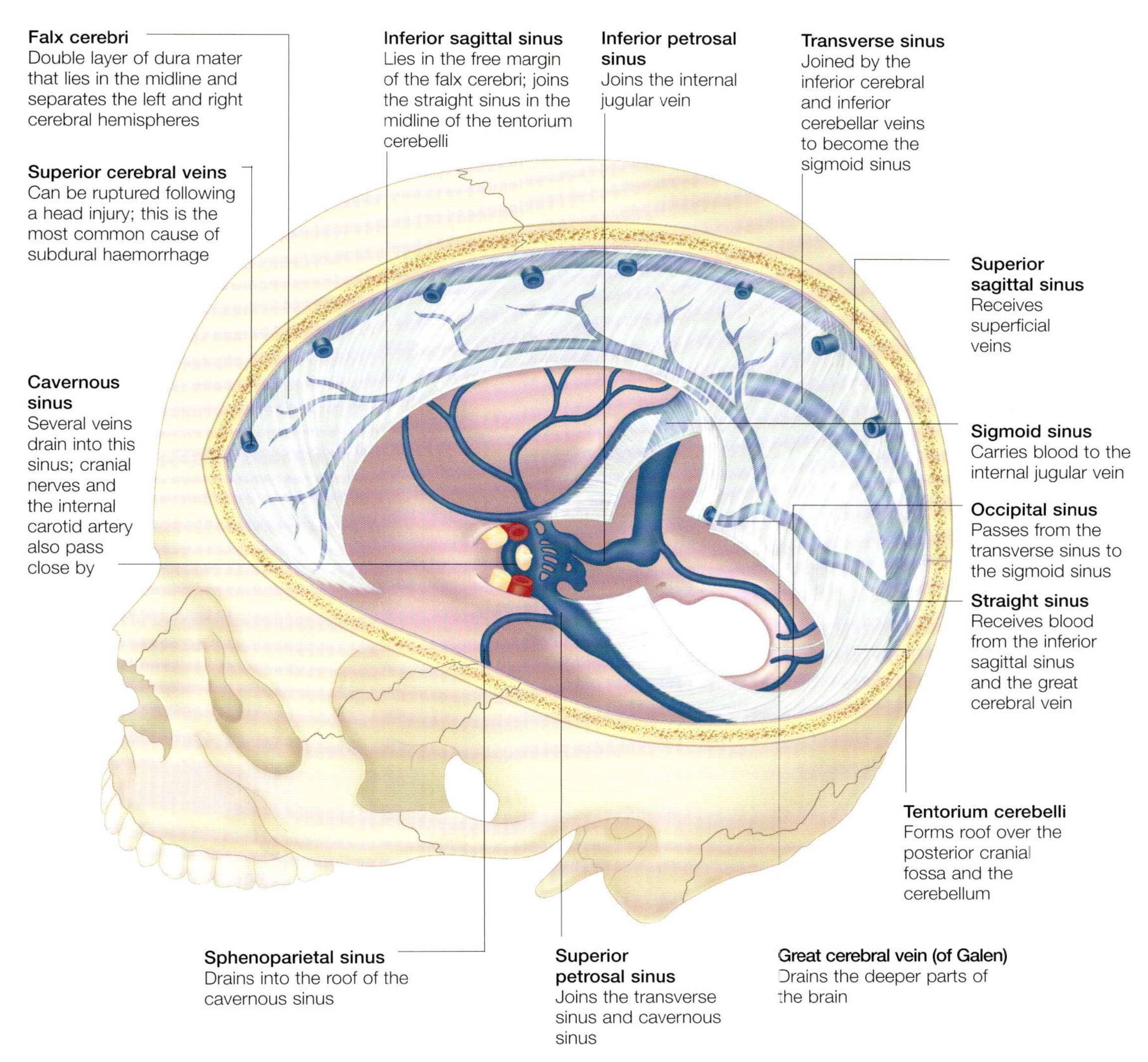

Sinuses in base of the skull

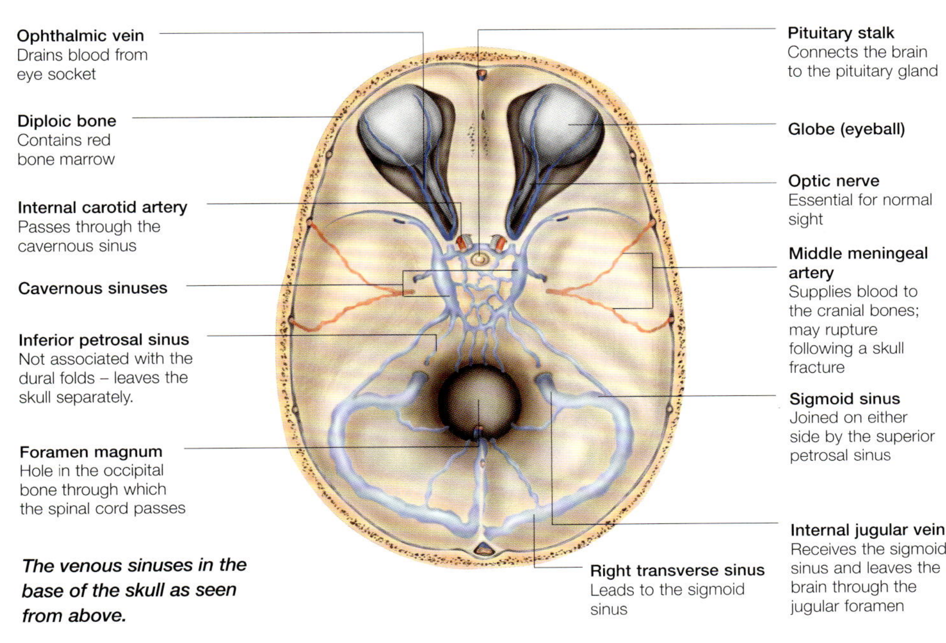

The venous sinuses in the base of the skull as seen from above.

There are seven pairs of sinuses on the floor of the skull. These include the transverse, inferior petrosal sinus, superior petrosal sinus, cavernous, sigmoid, sphenoparietal and occipital sinuses.

CAVERNOUS SINUSES

The cavernous sinuses lie on either side of the pituitary gland. The roof of each sinus is continuous with the dural sheet (diaphragma sellae) that covers the pituitary gland, surrounding the pituitary stalk.

Several important structures lie close to the cavernous sinuses. These include the internal carotid artery, three nerves supplying eye movement, and branches of the trigeminal nerve, which supplies sensation to the skin of the face and enables movement of the muscles of mastication.

Arteries of the Brain

The cerebral arteries provide the brain tissues with a rich supply of oxygenated blood.

The brain weighs about 1.4kg (3lb) and accounts for 2 per cent of our total body weight. However, it requires 15–20 per cent of the heart's output to be able to function properly. If the blood supply to the brain is cut off for as little as 10 seconds, loss of consciousness can result. Unless blood flow is quickly restored, it is only a matter of minutes before the damage becomes irreversible.

Arterial Network

Blood reaches the brain via two pairs of arteries. The internal carotid arteries originate from the common carotid arteries in the neck, enter the skull via the carotid canal and then branch to supply the cerebral cortex. The two main branches of the internal carotid are the middle and anterior cerebral arteries.

The vertebral arteries arise from the subclavian arteries, enter the skull via the foramen magnum and supply the brainstem and cerebellum. They join, forming the basilar artery, which then divides to produce the two posterior cerebral arteries that supply, among other areas, the occipital or visual cortex at the back of the brain.

These two sources of blood to the brain are linked by other arteries to form a circuit at the base of the brain. This is known as the 'circle of Willis'.

The arteries provide the brain with a rich supply of oxygenated blood

Inferior (from below) view of the brain

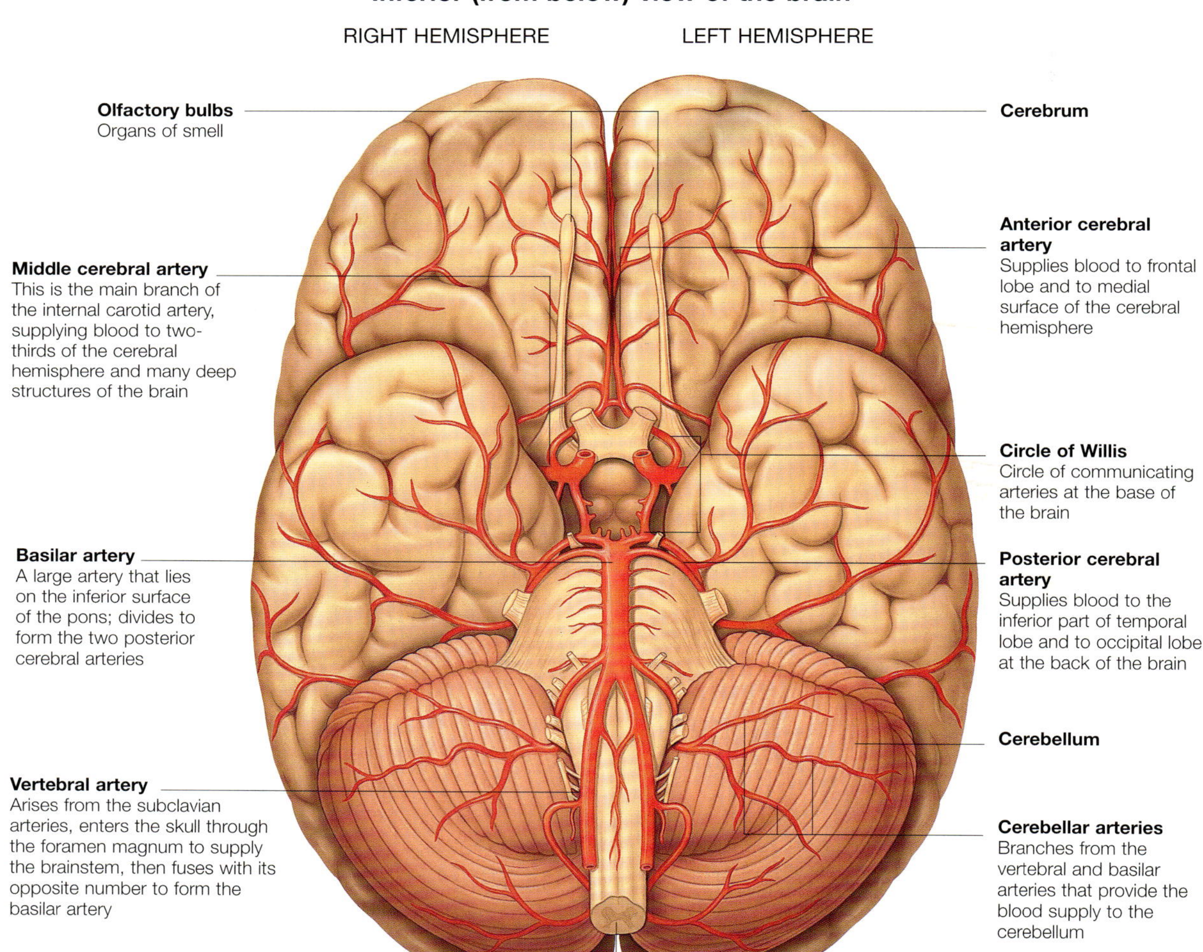

Veins of the Brain

Deep and superficial veins drain blood from the brain into a complex system of sinuses. These sinuses rely on gravity to return blood to the heart as, unlike other veins, they do not have valves.

The veins of the brain can be divided into deep and superficial groups. These veins, none of which have valves, drain into the venous sinuses of the skull.

Sinuses

The sinuses are formed between layers of dura mater, the tough outer membrane covering the brain, and they are unlike the veins in the rest of the body in that they have no muscular tissue in their walls.

The superficial veins have a variable arrangement on the surface of the brain and many of them are highly interconnected. Most superficial veins drain into the superior sagittal sinus.

Vein of Galen

By contrast, most of the deep veins, associated with structures within the body of the brain, drain into the straight sinus via the great cerebral vein (known as the vein of Galen).

The straight sinus and the superior sagittal sinus converge. Blood flows through the transverse and sigmoid sinuses and exits the skull through the internal jugular vein before flowing back towards the heart.

Beneath the brain, on either side of the sphenoid bone, are the cavernous sinuses. These drain blood from the orbit (eye socket) and the deep parts of the face.

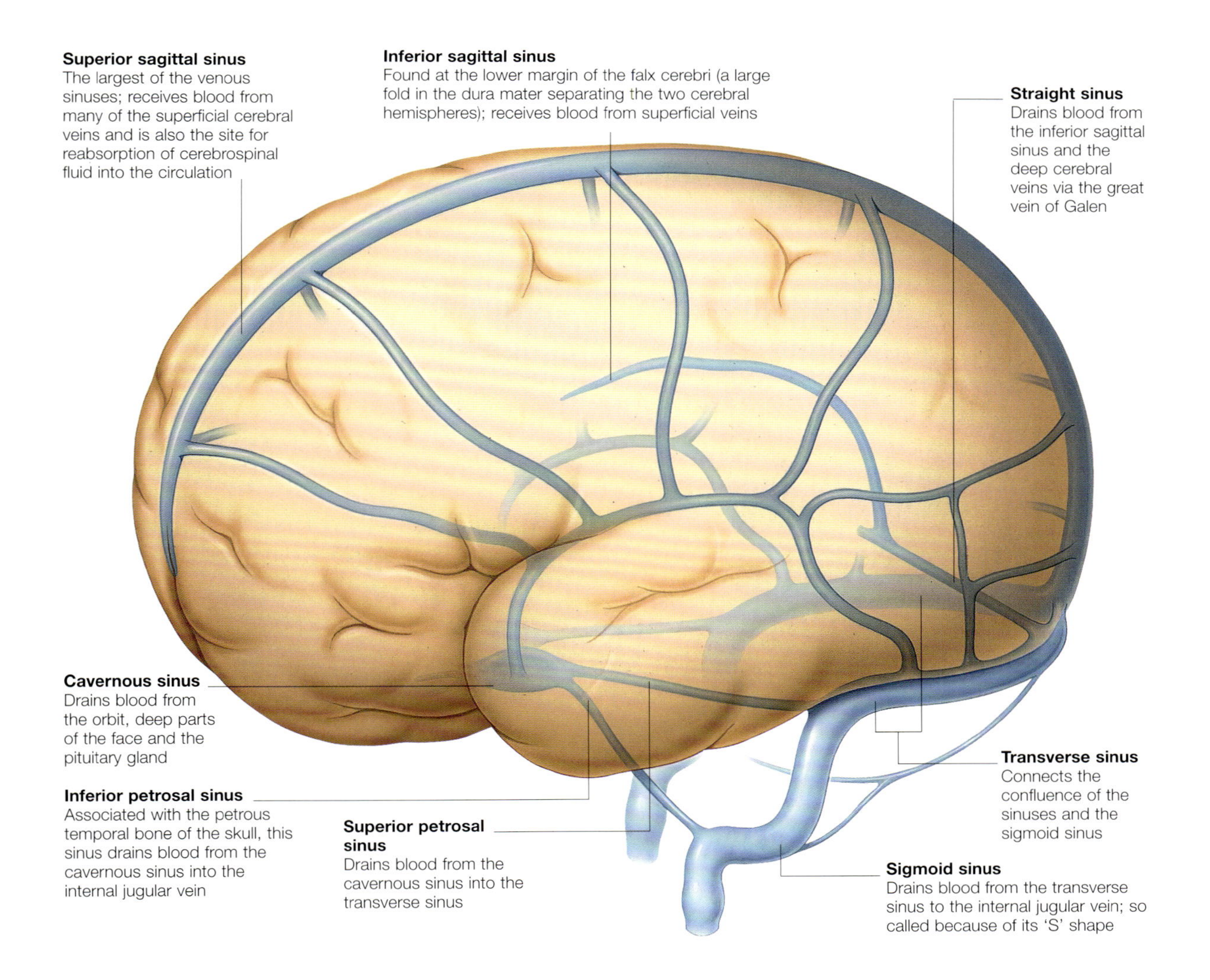

Ventricles of the Brain

The brain 'floats' in a protective layer of cerebrospinal fluid – the watery liquid produced in a system of cavities within the brain and brainstem.

The brain contains a system of communicating cavities known as the ventricles. There are four ventricles within the brain and brainstem, each of which contains a network of cells called the choroid plexus.

Cerebrospinal Fluid

These cells secrete cerebrospinal fluid (CSF), the fluid that surrounds the brain and spinal cord, protecting them from injury and infection.

Forebrain Ventricles

Three of the ventricles – the two (paired) lateral ventricles and the third ventricle – lie within the forebrain. The lateral ventricles are the largest, and lie within each cerebral hemisphere.

Each ventricle consists of a 'body' and three 'horns' – anterior (situated in the frontal lobe), posterior (occipital lobe) and inferior (temporal lobe). The third ventricle is a narrow cavity between the thalamus and hypothalamus.

Hindbrain Ventricle

The fourth ventricle is situated in the hindbrain, beneath the cerebellum. It is continuous with the third ventricle via a narrow channel called the cerebral aqueduct of the midbrain. The roof of the fourth ventricle is incomplete, allowing it to communicate with the subarachnoid space.

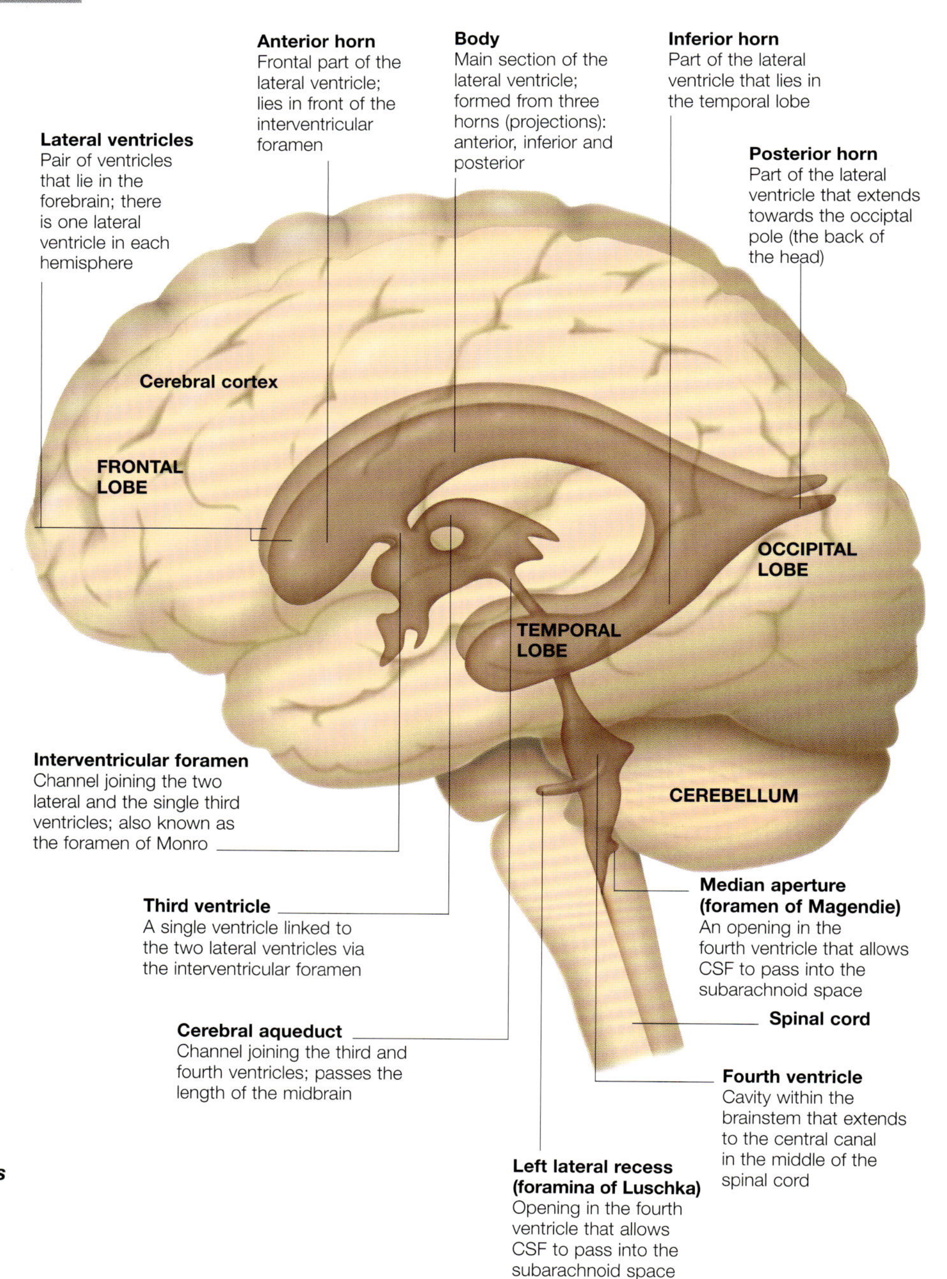

This sagittal section of the brain and brainstem reveals the four ventricles and the foramina and aqueducts that connect them.

Circulation of cerebrospinal fluid

Cerebrospinal fluid (CSF) is produced by the choroid plexus within the lateral, third and fourth ventricles.

The choroid plexuses are a rich system of blood vessels, epithelial cells and connective tissue originating from the pia mater, the innermost tissue surrounding the brain. The plexuses contain numerous folds (villous processes) projecting into the ventricles, from which cerebrospinal fluid is produced.

From the choroid plexuses in the two lateral ventricles, CSF passes to the third ventricle via the interventricular foramen. Together with additional fluid produced by the choroid plexus in the third ventricle, CSF then passes through the cerebral aqueduct of the midbrain and into the fourth ventricle. Additional fluid is produced by the choroid plexus in the fourth ventricle.

Subarachnoid space

From the fourth ventricle, CSF passes out into the subarachnoid space surrounding the brain. It does this through openings in the fourth ventricle – a median opening (foramen of Magendie) and two lateral ones (foramina of Luschka). Once in the subarachnoid space, the CSF circulates to surround the central nervous system. As CSF is produced constantly, it needs to be drained continuously to prevent buildup of pressure. This is achieved by passage of the CSF into the venous sinuses of the brain through protrusions (arachnoid granulations). These are particularly evident in the superior sagittal sinus.

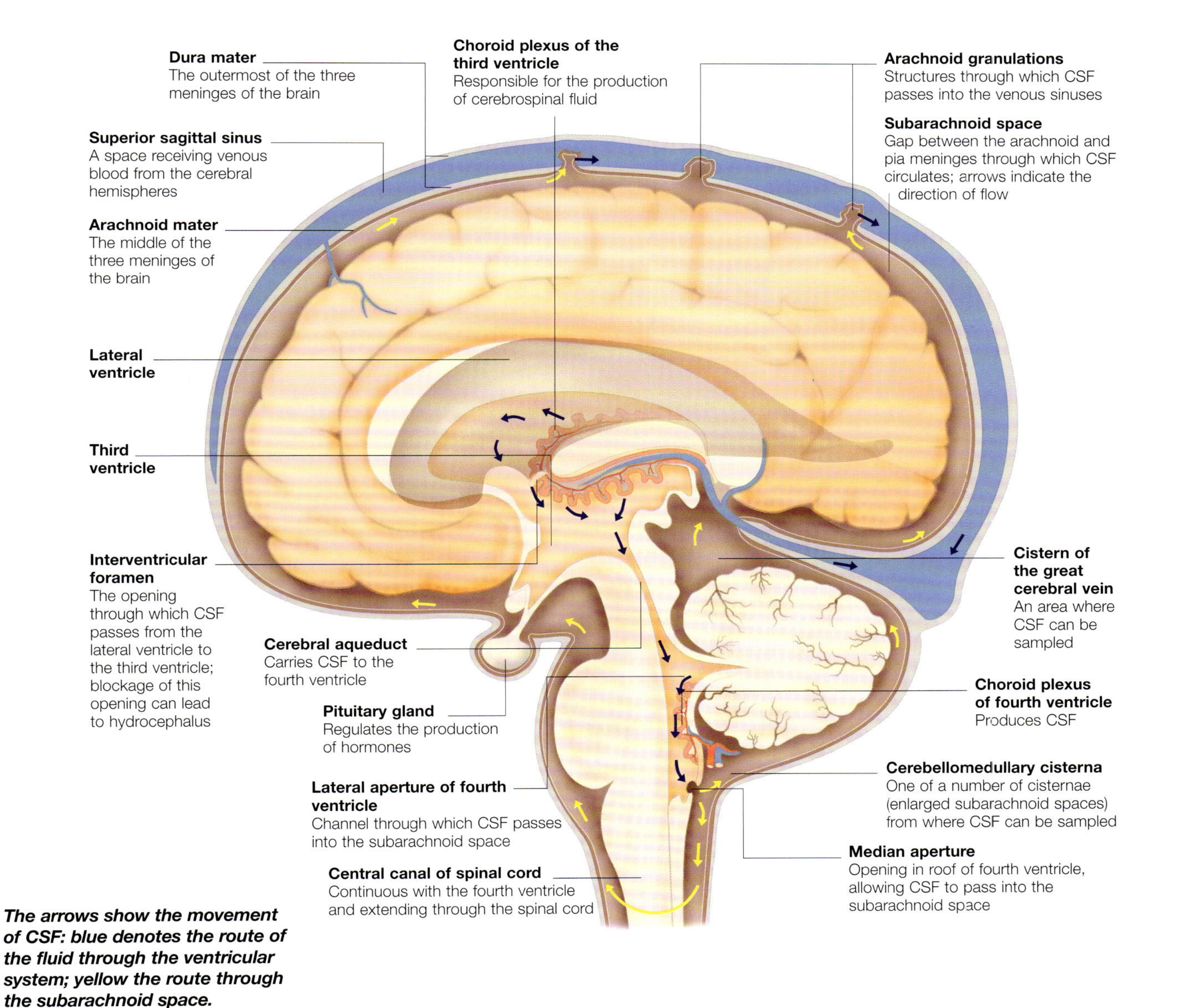

The arrows show the movement of CSF: blue denotes the route of the fluid through the ventricular system; yellow the route through the subarachnoid space.

Cerebral Hemispheres

The cerebral hemispheres are the largest part of the brain. In humans, they have developed out of proportion to the other parts of the brain, distinguishing our brains from those of other animals.

The left and right cerebral hemispheres are separated by the longitudinal fissure and have an outer layer of grey matter known as the cerebral cortex. Looking at the surface of the hemispheres from the top and side, there is a prominent groove running downwards, beginning about 1cm (0.3in) behind the midpoint between the front and back of the brain. This is the central sulcus or rolandic fissure. Further down on the lateral side there is a second large groove, the lateral sulcus or sylvian fissure.

Lobes of the Brain

The cerebral hemispheres are divided into lobes, named after the bones of the skull that lie over them:

- The frontal lobe lies in front of the rolandic fissure and above the sylvian fissure.
- The parietal lobe lies behind the rolandic fissure and above the back part of the sylvian fissure; it extends back as far as the parieto-occipital sulcus, a groove separating it from the occipital lobe, which is at the back of the brain.
- The temporal lobe is the area below the sylvian fissure and extends backwards to meet the occipital lobe.

At the bottom of the sylvian fissure there is another distinct area known as the insula or island of Reil. This triangular region has been buried by the growth of the adjacent parts of the brain and is not normally visible unless the sylvian fissure is spread open.

Lobes of the cerebral hemispheres

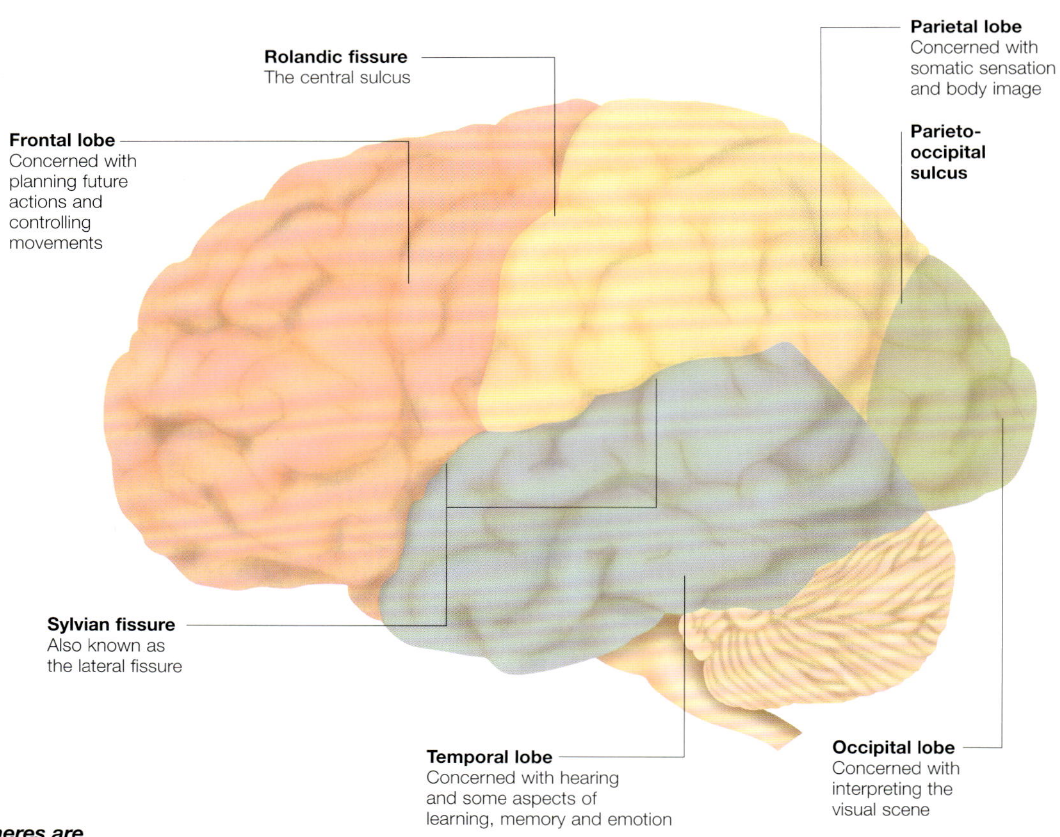

The cerebral hemispheres are each divided into four lobes. They are named after the bones of the skull that lie over them.

Functions of the cerebral cortex

The outer layer of the cerebral hemispheres is grey matter, which is known as the cerebral cortex. Different regions of the cerebral cortex have distinct and highly specialized functions.

The cerebral cortex is divided into:

- Motor areas that initiate and control movement. The primary motor cortex controls voluntary movement of the opposite side of the body. Just in front of the primary motor cortex is the premotor cortex and a third area, the supplementary motor area, lies on the inner surface of the frontal lobe. All of these areas work with the basal ganglia and cerebellum to allow us to perform complex sequences of finely controlled movements.
- Sensory areas that receive and integrate information from sensory receptors around the body. The primary somatosensory area receives information from sensory receptors on the opposite side of the body about touch, pain, temperature and the position of joints and muscles (proprioception).
- Association areas that are involved with the integration of more complex brain functions. These include the higher mental processes of learning, memory, language, judgement and reasoning, emotion and personality.

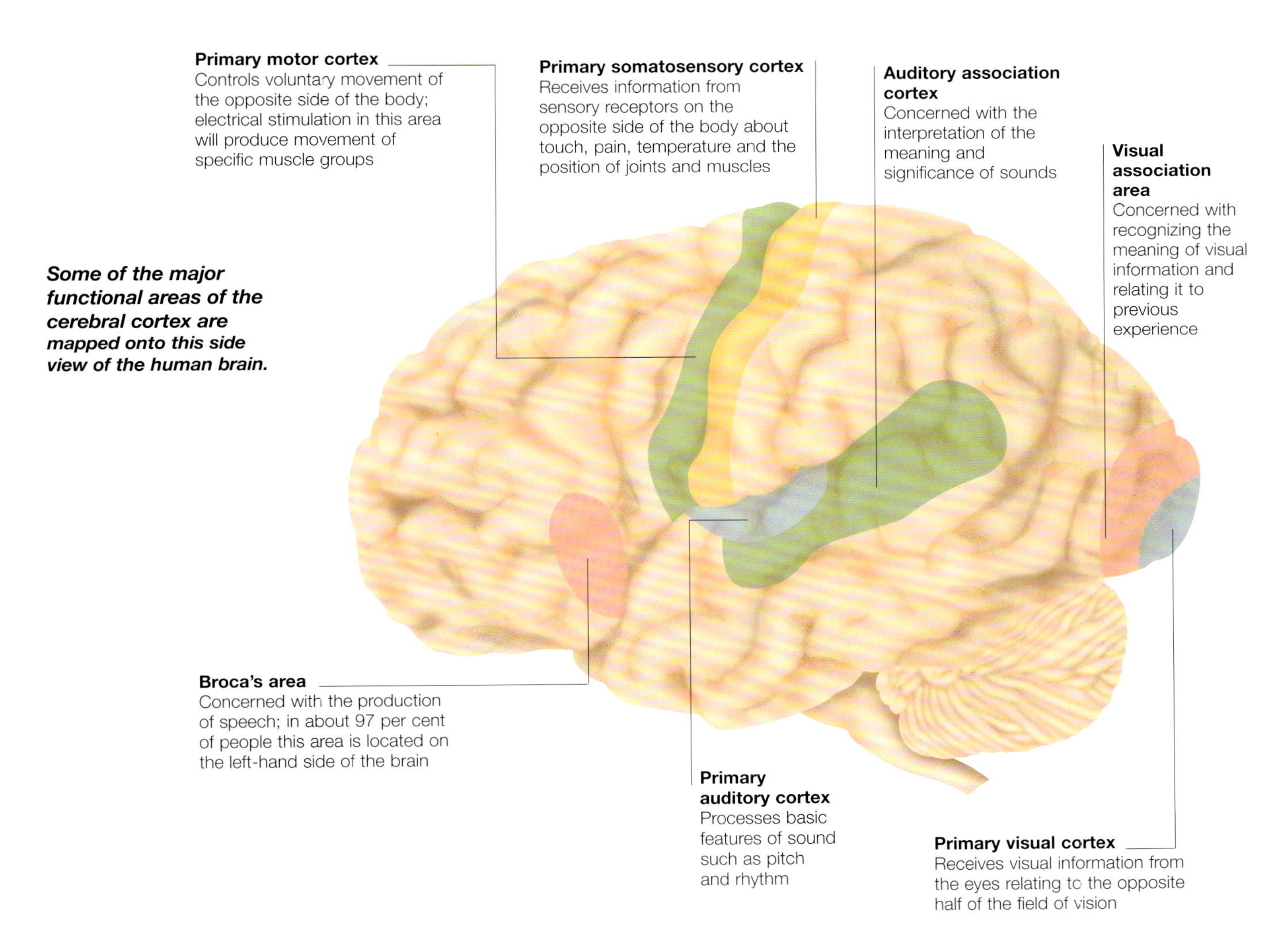

Some of the major functional areas of the cerebral cortex are mapped onto this side view of the human brain.

Structure of Cerebral Hemispheres

The cerebral cortex consists of two distinct layers: grey matter, a thin layer of nerve and glial cells about 2–4mm (0.07–0.15in) in thickness; and white matter, consisting of nerve fibres (axons) and glial cells.

On the surface of the hemispheres is the grey matter, which ranges in thickness from about 2 to 4mm (0.07 to 0.15in). The grey matter is made up of nerve cells (neurones) together with supporting glial cells. In most parts of the cortex, six separate layers of cells can be distinguished under a microscope.

Cortical Neurones

The cell bodies (which contain the cell's nucleus) of cortical neurones differ markedly in shape, although two primary types of cell can be distinguished:

- Pyramidal cells are so called because their cell body is shaped like a pyramid. Their axons (nerve fibres) project out of the cortex, carrying information to other regions of the brain.
- Non-pyramidal cells, in contrast, have a smaller and rounder cell body and are involved in receiving and analyzing input from other sources.

Layers of cells of the cortex (grey matter)

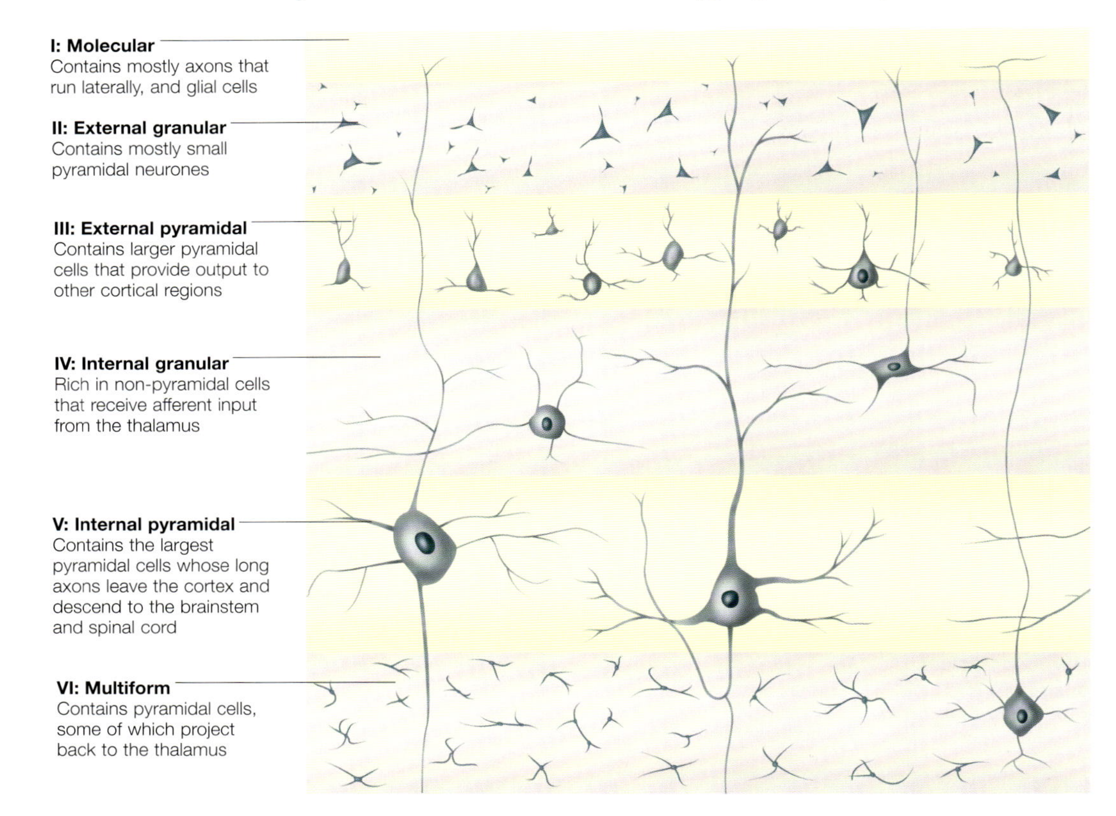

The grey matter of the cerebral hemispheres can be subdivided into six layers of cells, based on the type of brain cell present.

Brodmann areas

The thickness of the six individual layers varies between different brain regions. A German neurologist called Korbinian Brodmann (1868–1918) examined these differences by staining nerve cells and looking at them under a microscope.

In this way, Brodmann was able to classify the cerebral cortex into over 50 distinct areas according to predefined anatomical criteria. Subsequent work has shown that each of Brodmann's anatomical areas has its own physiological function and characteristic pattern of connections.

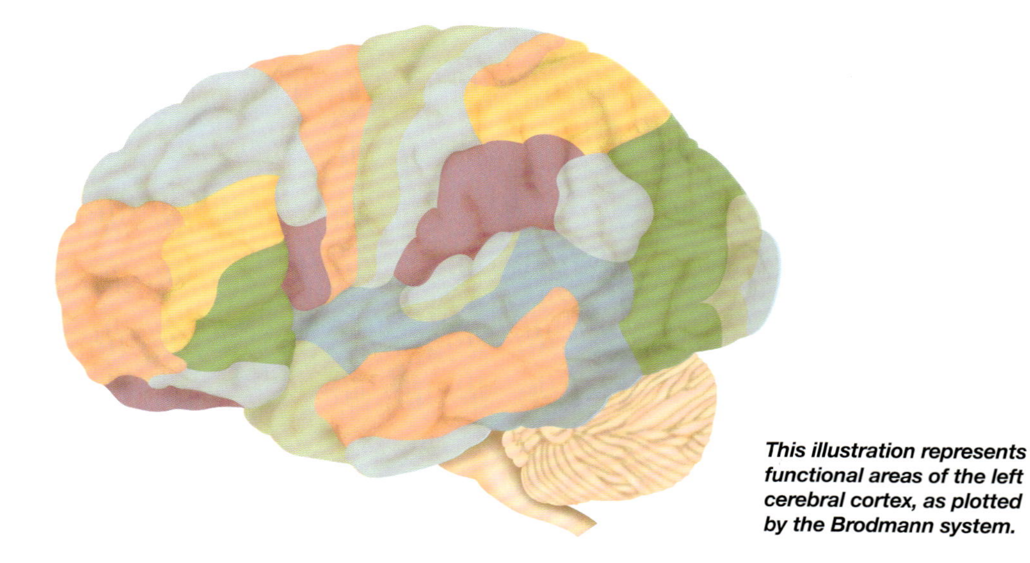

This illustration represents functional areas of the left cerebral cortex, as plotted by the Brodmann system.

White matter

The white matter is composed of nerve fibres, known as axons, which connect different regions of the brain.

Underneath the cerebral cortex (grey matter) is the white matter, which makes up the bulk of the inside of the cerebral hemispheres. It is arranged into bundles or tracts of three types:

■ Commissural fibres
These cross between the hemispheres and connect corresponding regions on the two sides. The corpus callosum is the largest of these tracts.

■ Association fibres
These connect different areas within the same hemisphere. Short association fibres connect adjacent gyri and long association fibres interconnect more widely separated regions of the cortex.

■ Projection fibres
These connect the cerebral cortex with the deeper underlying regions of the brain, the brainstem and the spinal cord. Projection fibres enable the cortex to receive incoming information from the rest of the body and to send out instructions controlling movement and other bodily functions.

Distribution of the major nerve fibre tracts

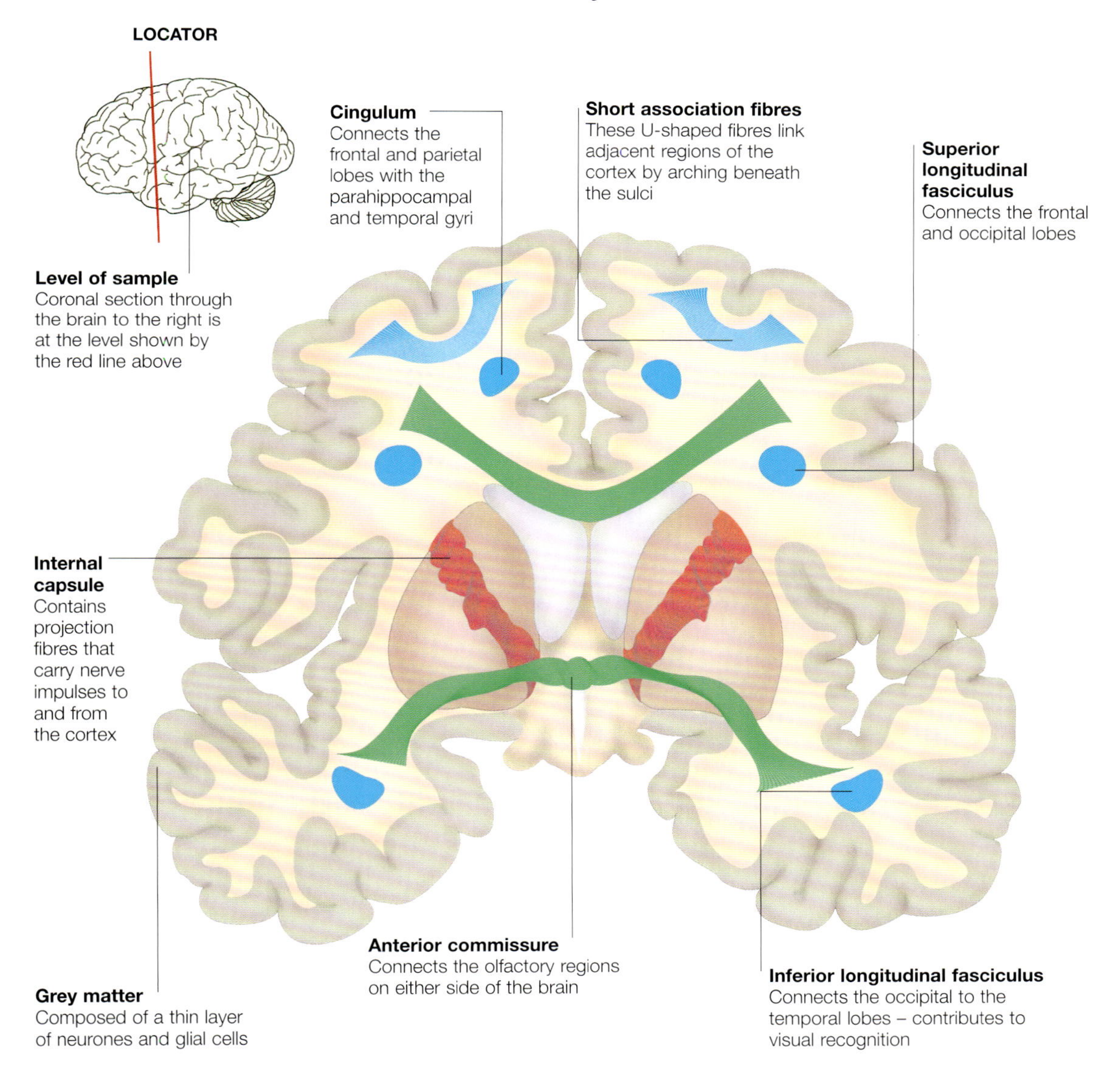

Nerve bundles can be classified into three groups – commissural fibres (green), association fibres (blue) and projection fibres (red).

Limbic System

The limbic system is a ring of interconnected structures that lies deep within the brain. It makes connections with other parts of the brain, and is associated with mood and memory.

The limbic system is a collection of structures deep within the brain that is associated with the perception of emotions and the body's response to them.

The limbic system is not one discrete part of the brain, but a ring of interconnected structures surrounding the top of the brainstem. The connections between these structures are complex, often forming loops or circuits and, as with much of the brain, their exact role is not fully understood.

Structure

The limbic system includes all or parts of the following brain structures:

- Amygdala – this almond-shaped nucleus appears to be linked to feelings of fear and aggression.
- Hippocampus – this structure seems to play a part in learning and memory.
- Thalamus – the anterior thalamic nuclei are collections of nerve cells that form part of the thalamus. One of their roles seems to lie in the control of instinctive drives.
- Cingulate gyrus – this connects the limbic system to the cerebral cortex, the part of the brain that carries conscious thoughts.
- Hypothalamus – this regulates the body's internal environment, including blood pressure, heart rate and hormone levels. The limbic system generates its effects on the body by sending messages to the hypothalamus.

Medial view of the limbic system within the brain

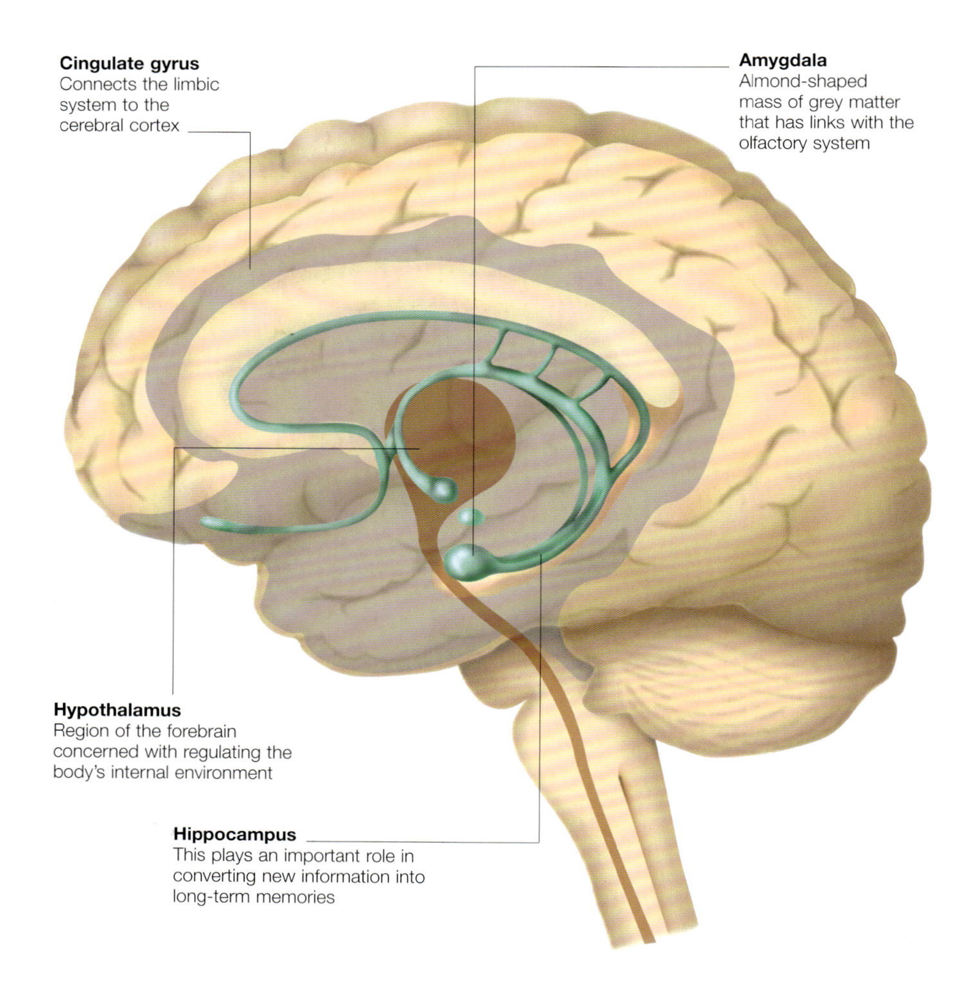

The limbic system connections encircle the upper part of the brainstem. They link with other parts of the brain and are associated with emotion.

Connections of the limbic system

The limbic system has connections with the higher centres in the cortex, and with the more primitive brainstem. It allows our emotions to influence the body, and enables the response to be regulated.

The human brain can be considered to be made up of three parts. These parts have evolved one after another over the millennia.

Brainstem
The 'oldest' part of the brain, in evolutionary terms, is the brainstem, which is concerned largely with unconscious control of the internal state of the body. The brainstem can be seen as a 'life support system'.

Limbic system
With the evolution of mammals came another 'layer' of brain, the limbic system. The limbic system allowed the development of feelings and emotions in response to sensory information. It is also associated with the development of newer – in evolutionary terms – behaviours, such as caring for and closeness to offspring (maternal bonding).

Cerebral cortex
This outer layer of the human brain is shared to some extent with higher mammals. It is the part of the brain that allows humans to think and reason. The cortex enables individuals to perceive the outside world and make conscious decisions about their behaviour and actions.

Role of the limbic system
The limbic system lies between the cortex and the brainstem and makes connections with both. Through its connections with the brainstem, the limbic system provides a way in which an individual's emotional state can influence the internal state of the body. For example, this may prepare the body for an act of self-preservation such as running away when experiencing fear, or for a sexual encounter.

Emotional response
The extensive connections between the limbic system and the cerebral cortex allow humans to use their knowledge of the outside world to regulate their response to emotions. The cerebral cortex can thus 'override' the more primitive limbic system when necessary.

The developing brain

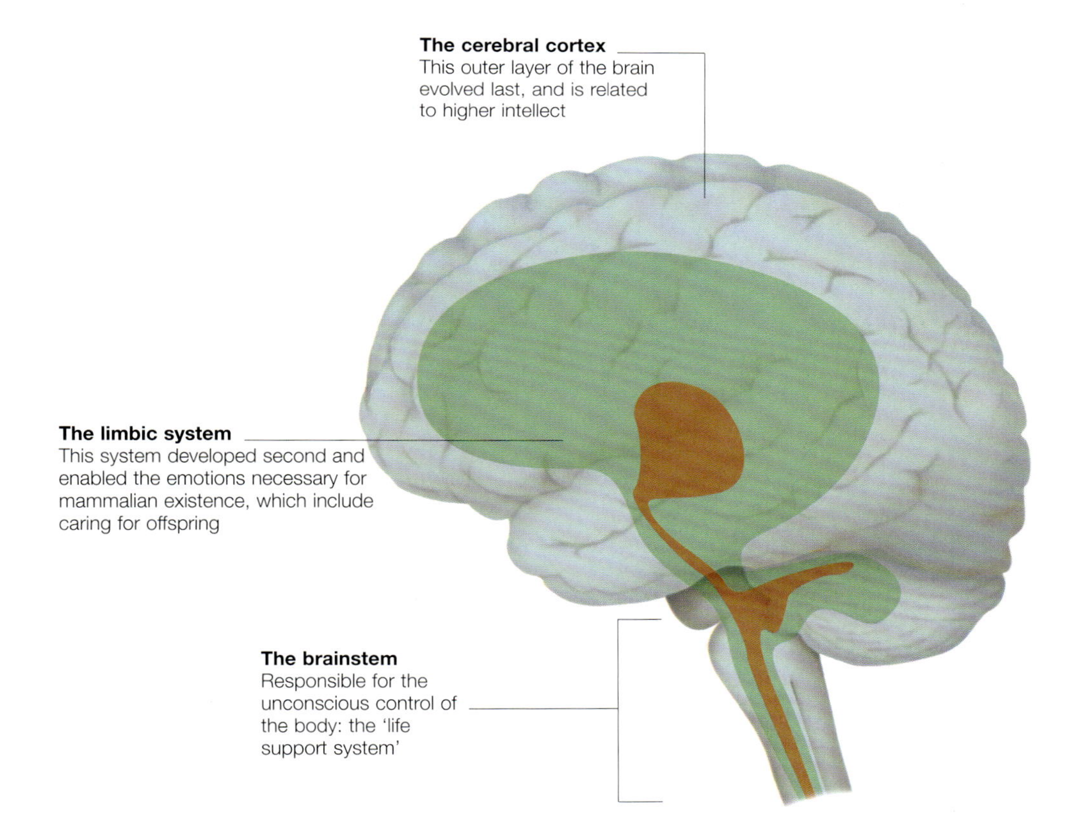

The three layers of the brain evolved one by one over thousands of years. Each is responsible for different bodily and intellectual functions.

Thalamus

The thalamus is a major sensory relay and integrating centre in the brain, lying deep within its central core. It consists of two halves, and receives sensory inputs of all types, except smell.

The thalamus is made up of paired egg-shaped masses of mostly grey matter (bodies of nerve cells) 3–4cm (1.1–1.5in) long and 1.5cm (0.6in) wide. It is located in the deep central core of the brain known as the diencephalon, or 'between brain'.

The thalamus makes up about 80 per cent of the diencephalon and lies on either side of the fluid-filled third ventricle. The right and left parts of the thalamus are connected by a bridge of grey matter – the massa intermedia, or interthalamic adhesion.

Neuroanatomy

The front end of the thalamus is rounded and is narrower than the back, which expands into the pulvinar. The upper surface of the thalamus is covered with a thin layer of white matter – the stratum zonale.

Medullary Laminae

A second layer of white matter – the external medullary lamina – covers the lateral surface. Its structure is very complex and it contains more than 25 distinct nuclei (collections of nerve cells with a common function).

These thalamic nuclear groups are separated by a vertical Y-shaped sheet of white matter – the internal medullary lamina. The anterior nucleus lies in the fork of the Y, and the tail divides the medial and lateral nuclei and splits to enclose the intralaminar nuclei.

Higher Brain Control

Each thalamic nucleus is linked to a distinct region of the cerebral cortex. These connections are made via a nerve fibre bundle called the internal capsule. Some thalamic nuclei relay information received from different sensory modalities – including physical sensation, vision and hearing – to the motor regions of the cortex.

Other thalamic nuclei are involved in transmitting information about body movement from the cerebellum to the motor regions of the frontal cortex. The thalamus is also involved in autonomic functions including the maintenance of consciousness.

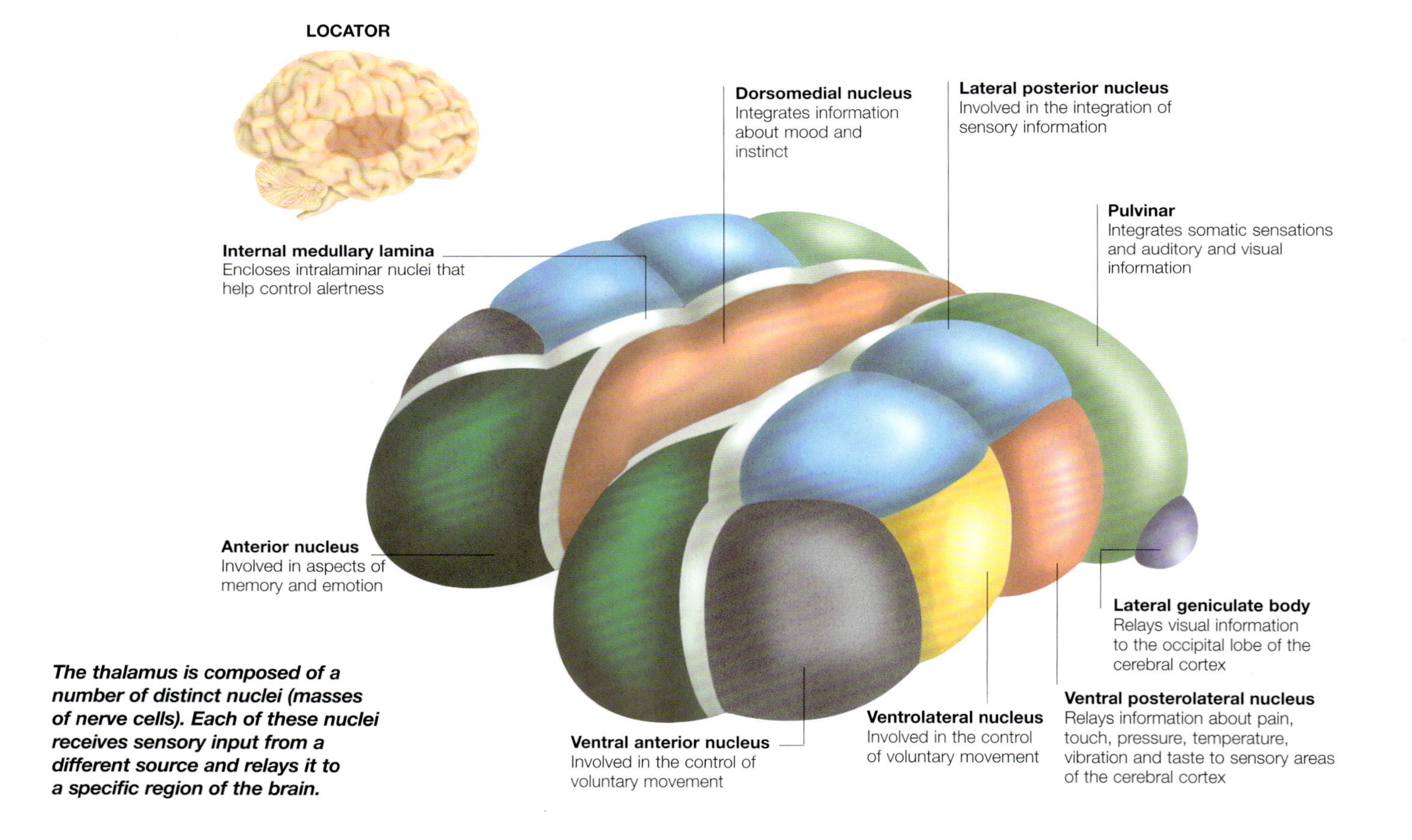

The thalamus is composed of a number of distinct nuclei (masses of nerve cells). Each of these nuclei receives sensory input from a different source and relays it to a specific region of the brain.

Hypothalamus

The hypothalamus is a complex structure located in the deep core of the brain. It regulates key aspects of body function, and is critical for homeostasis – the maintenance of equilibrium in the body's internal environment.

The hypothalamus is a small region of the diencephalon and is the size of a thumbnail. It lies below the thalamus and is separated from it by a shallow groove, the hypothalamic sulcus.

The hypothalamus is just behind the optic chiasm, the point at which the two optic nerves cross over as they travel from the eyes towards the back of the brain. Several distinct structures stand out on its undersurface:

- The mammillary bodies – two small, pea-like projections that are involved in the sense of smell
- The infundibulum or pituitary stalk – a hollow structure connecting the hypothalamus with the posterior part of the pituitary gland
- The tuber cinereum or median eminence – a greyish-blue, raised region surrounding the base of the infundibulum.

Hypothalamic control

The hypothalamus regulates a wide range of basic processes:

- Controls pituitary gland function
- Connects to the autonomic nervous system, influencing heart rate and blood pressure, contraction of the gut and bladder and sweating
- Influences hunger and thirst
- Acts as a thermostat for body temperature
- Contributes to daily patterns of sleeping and waking
- Associated with ability to retain new information.

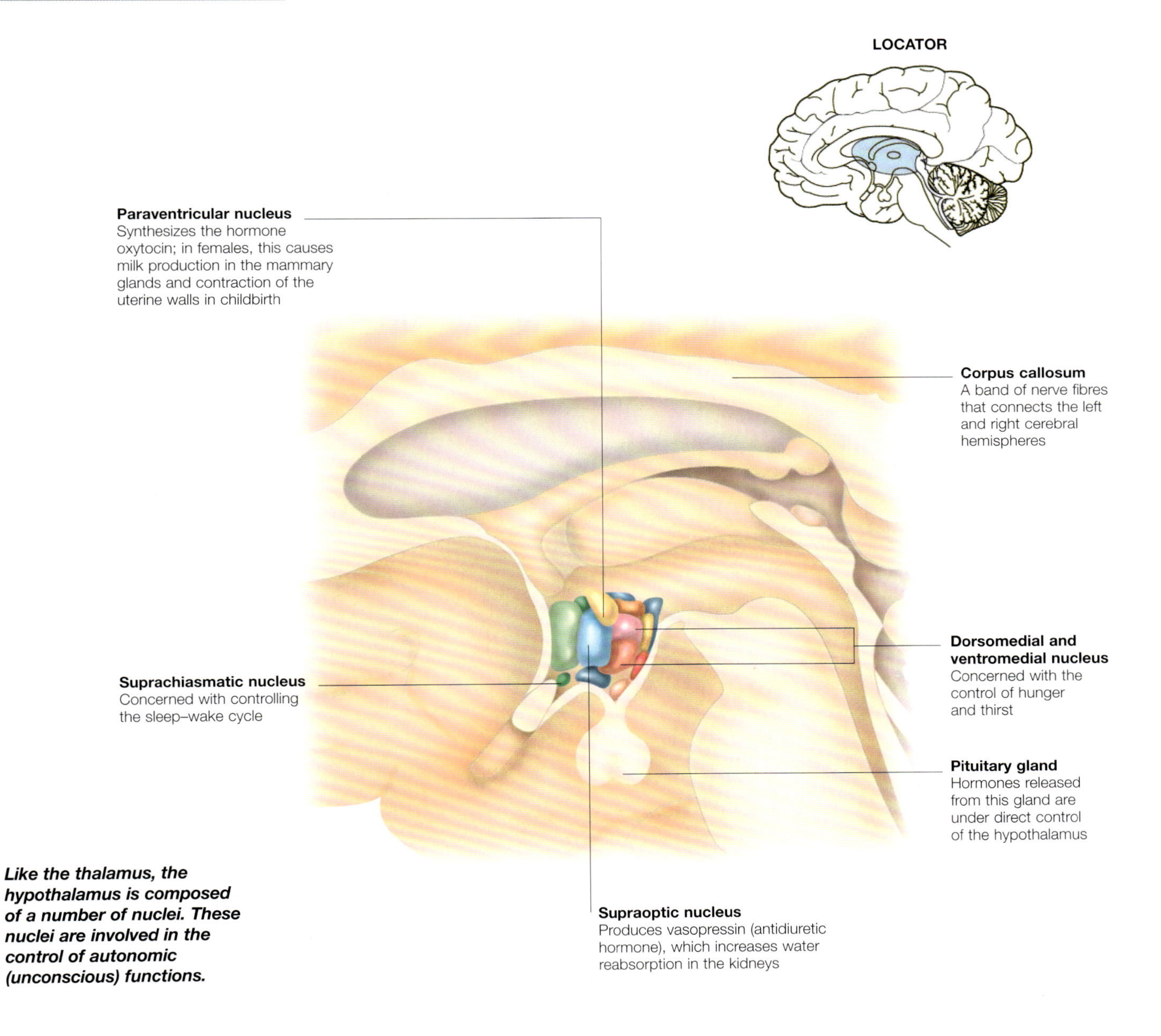

Like the thalamus, the hypothalamus is composed of a number of nuclei. These nuclei are involved in the control of autonomic (unconscious) functions.

Basal Ganglia

The basal ganglia lie deep within the white matter of the cerebral hemispheres. They are collections of nerve cell bodies that are involved in the control of movement.

The basal ganglia, also known as basal nuclei, are clusters of nerve cells that work closely with the cerebral cortex. These cells receive and return signals to and from the cortex and help to initiate and control complex movements, for example, writing.

Components

There are a number of component parts to the basal ganglia, which are closely related to each other. These include:

- Putamen. Together with the caudate nucleus, the putamen receives input from the cortex.
- Caudate nucleus. Named for its shape, as it has a long tail, this nucleus is continuous with the putamen at the anterior (front) end.
- Globus pallidus. This nucleus relays information from the putamen to the pigmented area of the midbrain known as the substantia nigra, with which it bears many similarities.

Grouping

Various names are associated with different groups of the basal ganglia. The term 'corpus striatum' (striped body) refers to the whole group of basal ganglia, whereas the 'striatum' includes only the putamen and caudate nuclei. Another term, the 'lentiform nucleus', refers to the putamen and the globus pallidus which, together, form a lens-shaped mass.

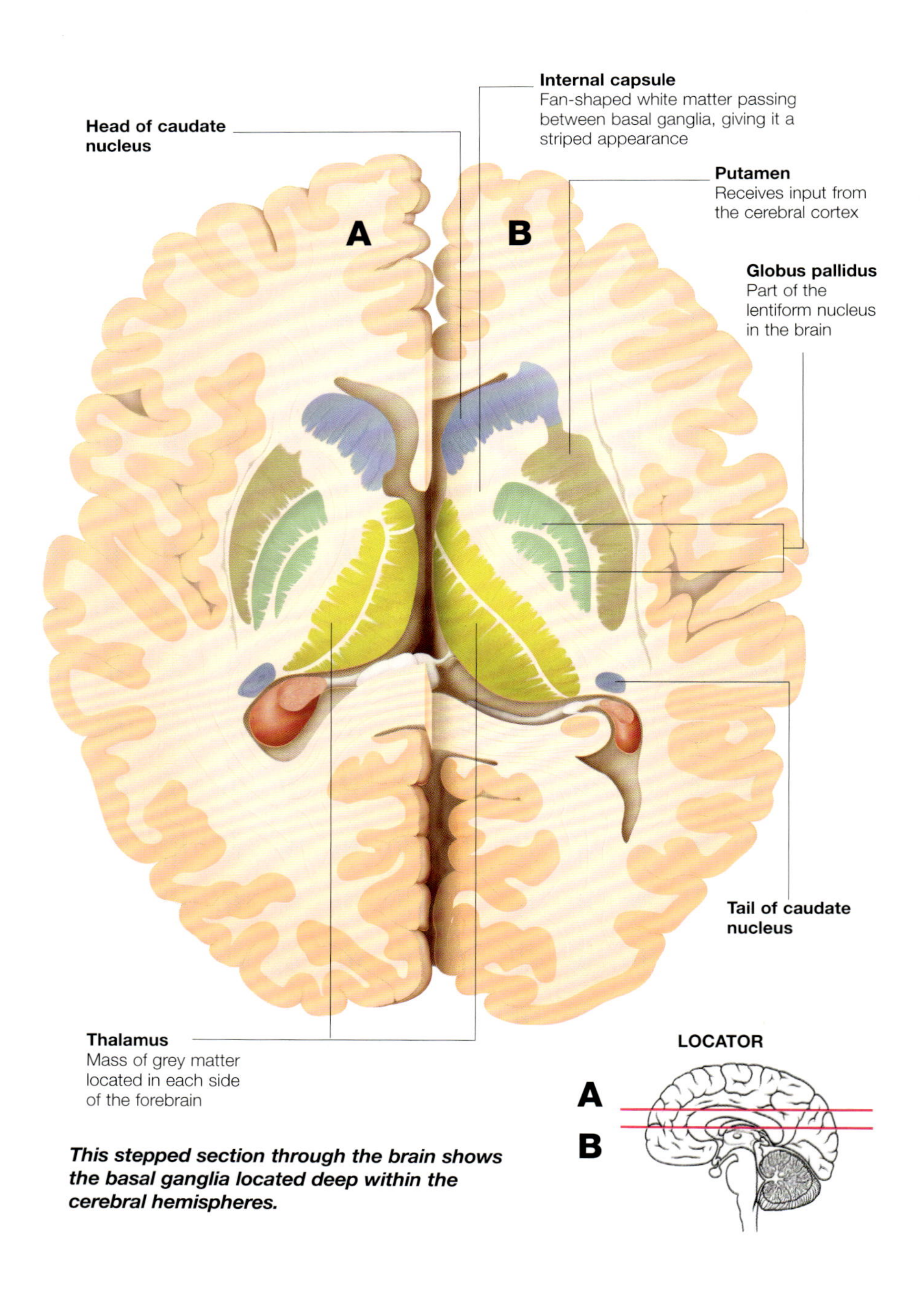

This stepped section through the brain shows the basal ganglia located deep within the cerebral hemispheres.

Structure and role of the basal ganglia

The overall shape of the basal ganglia is complex, and is hard to imagine by looking at two-dimensional cross-sections.

When seen in a three-dimensional view, the size and shape of the basal ganglia, together with their position within the brain as a whole, can be appreciated more easily.

Caudate nucleus shape

In particular, the shape of the caudate nucleus can now be understood – it connects at its head with the putamen, then bends back to arch over the thalamus before turning forwards again. The tip of the tail of the caudate nucleus ends as it merges with the amygdala, part of the limbic system that is concerned with unconscious, autonomic functions.

Role of the basal ganglia

The functions of the basal ganglia have been difficult to study because they lie deep within the brain and are therefore relatively inaccessible. Much of what is known about their function derives from the study of those patients who have disorders of the basal ganglia that lead to particular disruptions of movement and posture, such as Parkinson's disease.

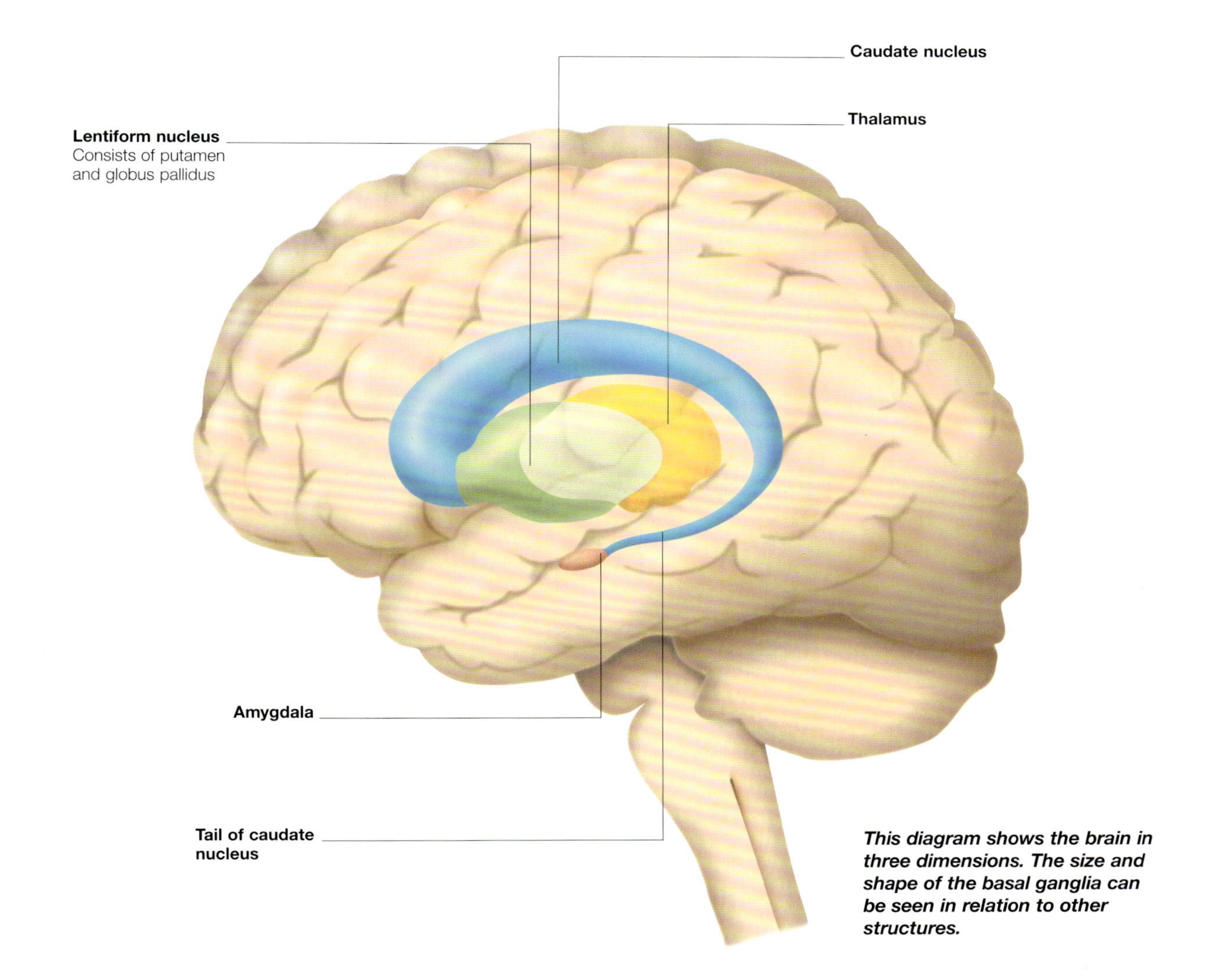

This diagram shows the brain in three dimensions. The size and shape of the basal ganglia can be seen in relation to other structures.

Cerebellum

The cerebellum (meaning 'little brain') lies under the occipital lobes of the cerebral cortex at the back of the brain. It is important in the control of fast muscular activity, such as running or playing the piano.

The cerebellum lies under the occipital lobes of the cerebral cortex at the back of the head. Its vital roles include movement coordination and the maintenance of balance and posture. The cerebellum works closely with the basal ganglia and cerebral cortex.

Structure

The cerebellum has two hemispheres that are bridged in the midline by the vermis. They extend laterally and posteriorly from the midline to form the bulk of the cerebellum.

Distinctive Appearance

The surface of the cerebellum has a very distinctive appearance. In contrast to the large folds of the cerebral hemispheres, it is made up of numerous fine folds (folia). Between the folia of the cerebellar surface lie deep fissures that divide it into three lobes: the anterior, posterior and flocculonodular lobe.

Cerebellar Peduncles

The cerebellum is connected to the brainstem by three pairs of nerve fibre tracts that make up the superior, middle and inferior cerebellar peduncles. All information to and from the cerebellum goes through the peduncles.

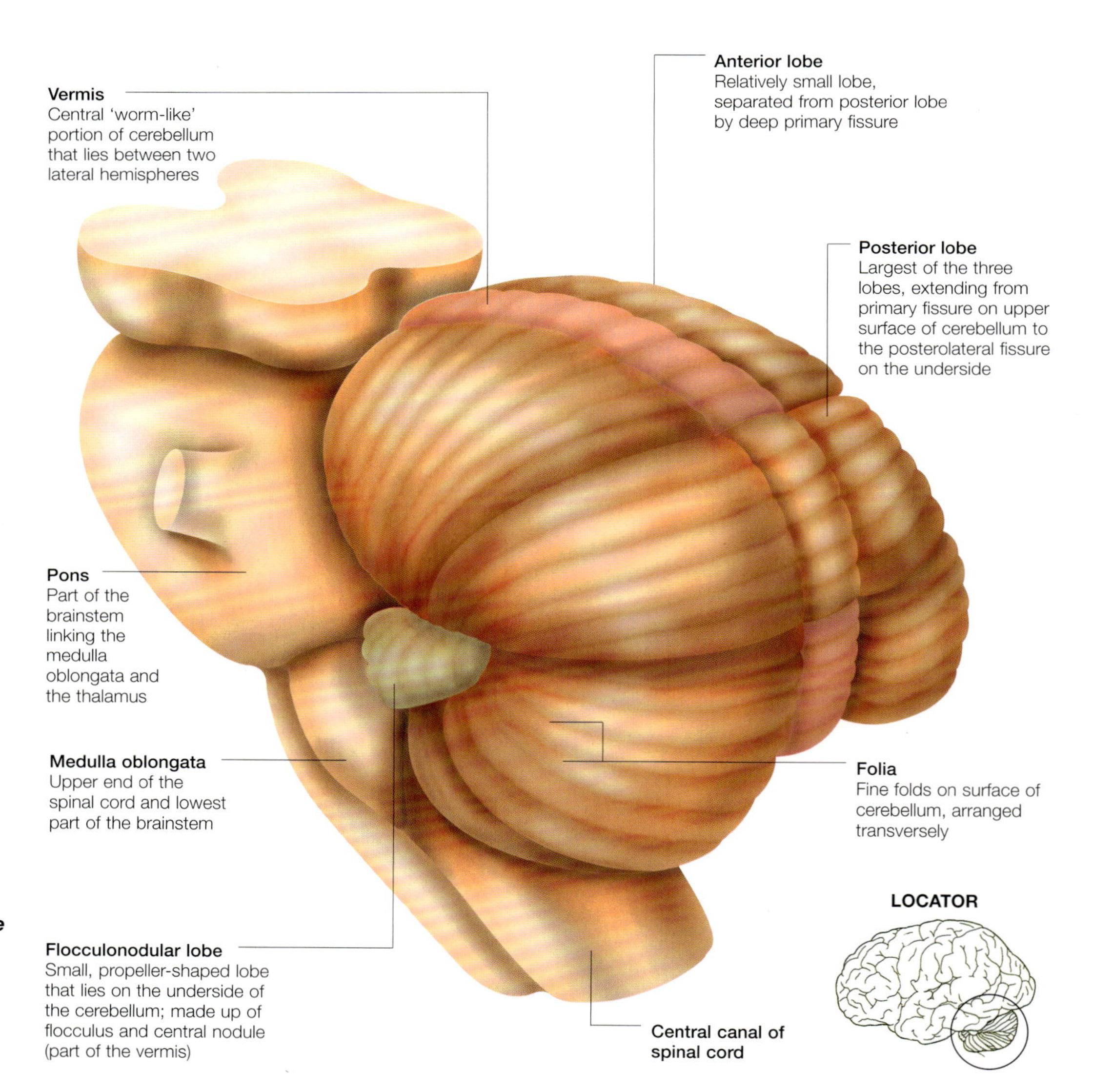

The cerebellum has two hemispheres, one on either side of the worm-like vermis. The surface of the cerebellum is made up of thin folds (folia).

Internal structure of the cerebellum

The cerebellum has an outer grey cortex and a core of nerve fibres, or white matter. Deep within the white matter lie four pairs of cerebellar nuclei: the fastigial, globose, emboliform and dentate nuclei.

The cerebellum has a surface layer of nerve cell bodies, or grey matter, which overlies a core of nerve fibres, or white matter. Deep within the white matter lie the cerebellar nuclei.

Cerebellar cortex

The cortex has an extensive surface area due to numerous fine folia (folds) on its surface. It is formed of the cell bodies and dendrites of most of the cerebellar neurones. The cells of the cortex receive information from outside the cerebellum via the cerebellar peduncles and make frequent connections between themselves within the cortex.

The cortex has several layers: a molecular layer that is rich in nerve cell bodies; a Purkinje cell layer that receives signals via their dendrites to pass to the cerebellar nuclei: and the granular cell layer that receive information from the peduncles and send signals through their axons up to the molecular layer.

Transfer of information

Signals from the cerebellar cortex are mainly conveyed in the fibres of the white matter down to the cerebellar nuclei. It is from here that information leaves the cerebellum to be carried to the rest of the central nervous system.

Cerebellar nuclei

The four pairs of cerebellar nuclei are known from the midline outwards as the fastigial, globose, emboliform and dentate nuclei.

Cross-section through cerebellum

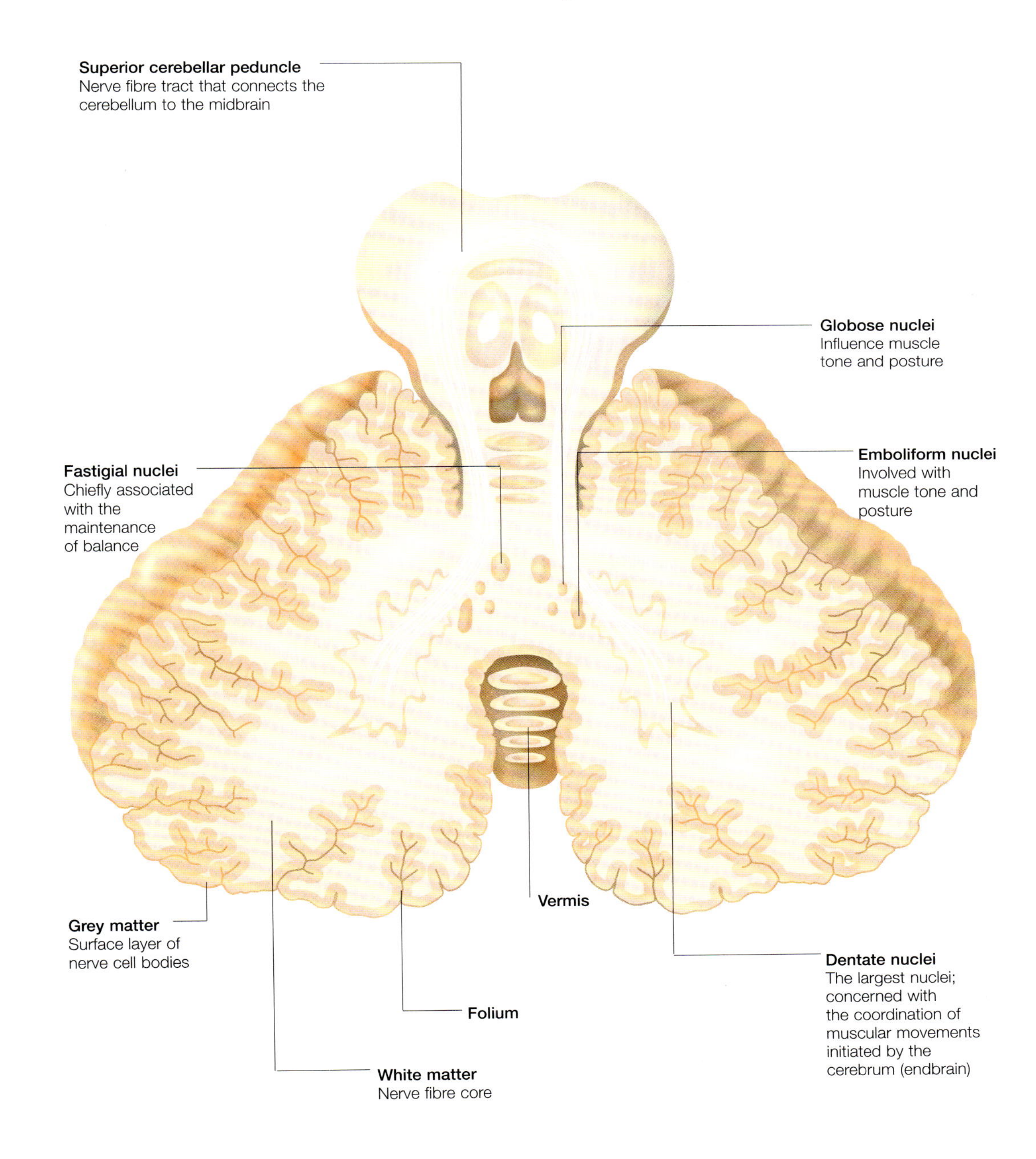

Brainstem

The brainstem lies at the junction of the brain and spinal cord. It helps to regulate breathing and blood circulation as well as having an effect upon a person's level of consciousness.

The brainstem is made up of three distinct parts: the midbrain, the pons, and the medulla oblongata. The midbrain connects with the higher brain above; the medulla is continuous with the spinal cord below.

Brainstem Appearance

The three parts of the brainstem can be viewed from underneath:

- The medulla oblongata – a bulge at the top of the spinal column. Pyramids, or columns, lie at either side of the midline. Nerve fibres within these columns carry messages from the cerebral cortex to the body. Raised areas known as the olives lie either side of the pyramids.
- The pons – contains a system of nerve fibres that originate in the nerve cell bodies deep within the pons.
- The midbrain – appears as two large columns, the cerebral crura, separated in the midline by a depression.

Cranial Nerves

Also present in the brainstem are some of the cranial nerves that supply much of the head. These nerves carry fibres associated with the cranial nerve nuclei, collections of grey matter that lie inside the brainstem.

Ventral surface of the brainstem

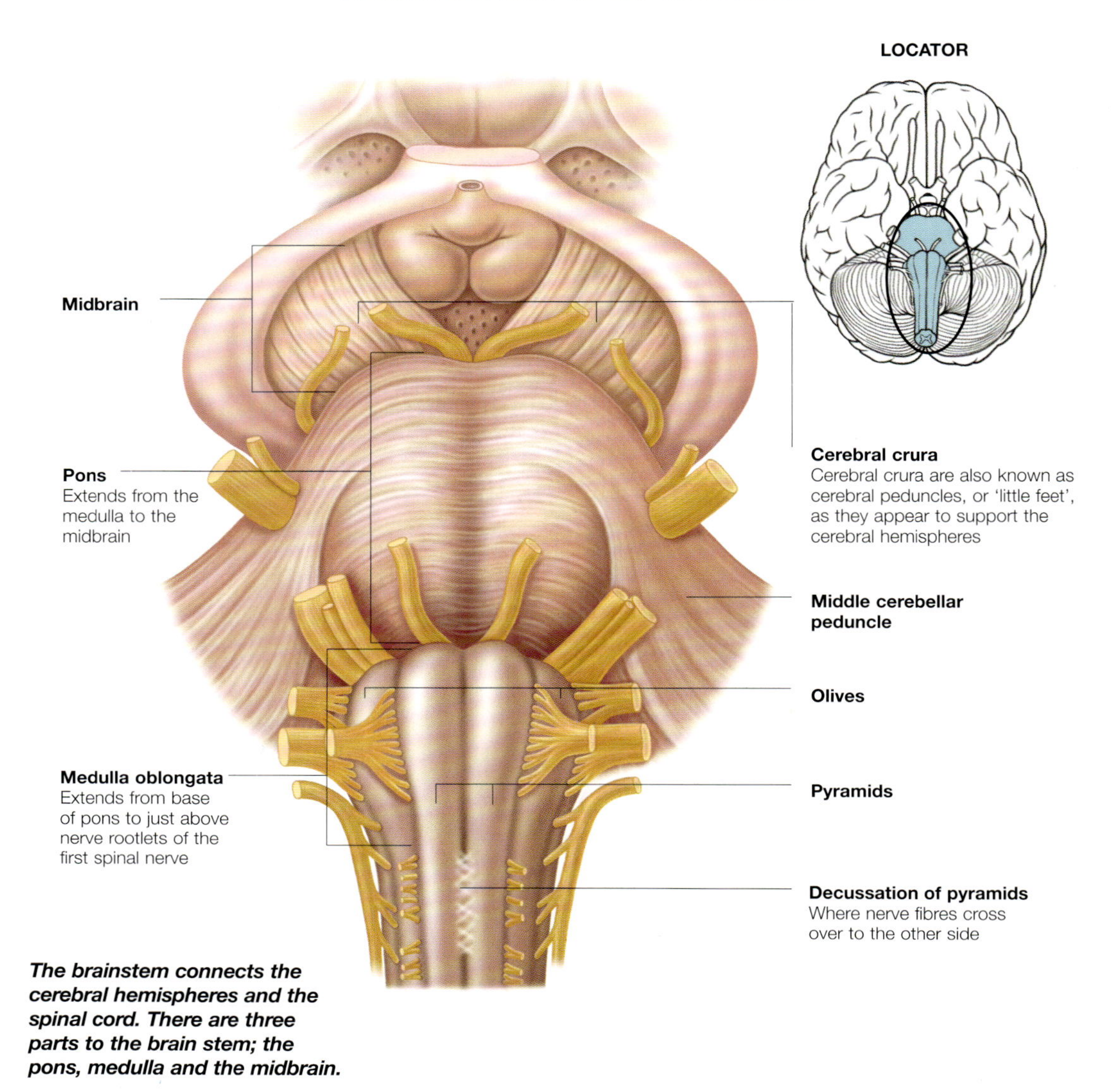

The brainstem connects the cerebral hemispheres and the spinal cord. There are three parts to the brain stem; the pons, medulla and the midbrain.

Relationships of the brainstem

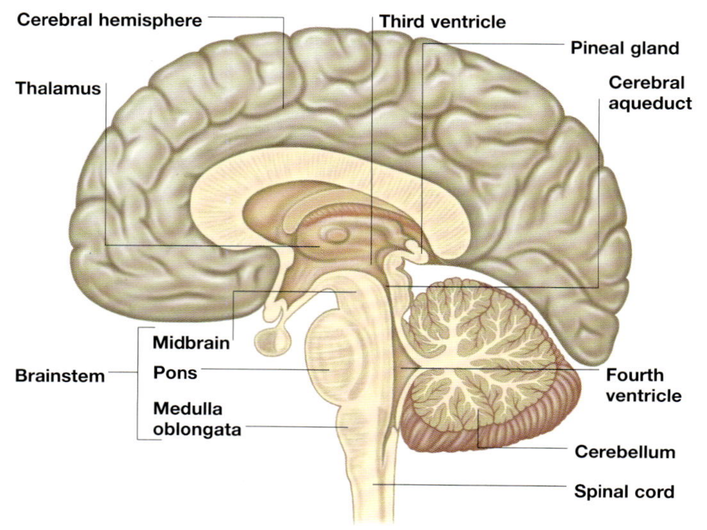

A sagittal section through the brain shows the position of the brainstem in relation to the other parts of the brain and spinal cord:

- The medulla oblongata – arises as a widening of the spinal cord at the level of the foramen magnum, the large hole in the bottom of the skull. The central canal of the spinal cord widens into the fourth ventricle allowing cerebrospinal fluid (CSF) to circulate between the brain and the spinal cord.

A sagittal view of the brain and spinal cord shows the location of the various structures. The brainstem is located in front of the cerebellum.

- The pons – lies above the medulla, at the level of the cerebellum. Above the pons lies the midbrain, encircling the cerebral aqueduct, which connects the fourth ventricle to the third ventricle.
- The midbrain – is the shortest part of the the brainstem and lies under the thalamus, the brain's central core, which is surrounded by the cerebral hemispheres. It thus lies below the thalamus and hypothalamus, and the tiny pineal gland.

Internal structure of the brainstem

The brainstem contains many areas of neural tissue that have a variety of functions vital to life and health. Responses to visual and auditory stimuli that influence head movement are also controlled here.

Cross-sections through the brainstem reveal its internal structure and the arrangement of white and grey matter, which differs according to the level at which the section is taken.

Medulla

The features of a section through the medulla are:

- The inferior olivary nucleus – a bag-like collection of grey matter that lies just under the olives. Other nuclei lying within the medulla include some belonging to the cranial nerves, such as the hypoglossal and the vagus nerves.
- The vestibular nuclear complex – an area that receives information from the ear and is concerned with balance and equilibrium.
- The reticular formation – a complex network of neurones, which is seen here and throughout the brainstem. It has a number of functions fundamental to life such as the control of respiration and circulation. The reticular formation is present in the midbrain as are several of the cranial nerve nuclei.

Midbrain

A cross-section through the midbrain shows the presence of:

- The cerebral aqueduct – the channel that connects the third and fourth ventricles
- Above the aqueduct lies an area called the tectum, while below it lie the large cerebral peduncles
- The cerebral peduncles – within these lie two structures on either side; the red nucleus and the substantia nigra
- The red nucleus is involved in control of movement, while damage to the substantia nigra is associated with Parkinson's disease.

Pons

The pons (not shown) is divided into upper and lower parts:

- The lower part – mostly made up of transverse nerve fibres, running across from the nuclei of the pons to the cerebellum
- The upper portion – contains a number of cranial nerve nuclei. The pons also contains part of the reticular formation.

Cross-sections of the brainstem

The numerous nuclei and tracts that are within the brainstem can be seen in these cross-sections. They are involved in most functions of the brain.

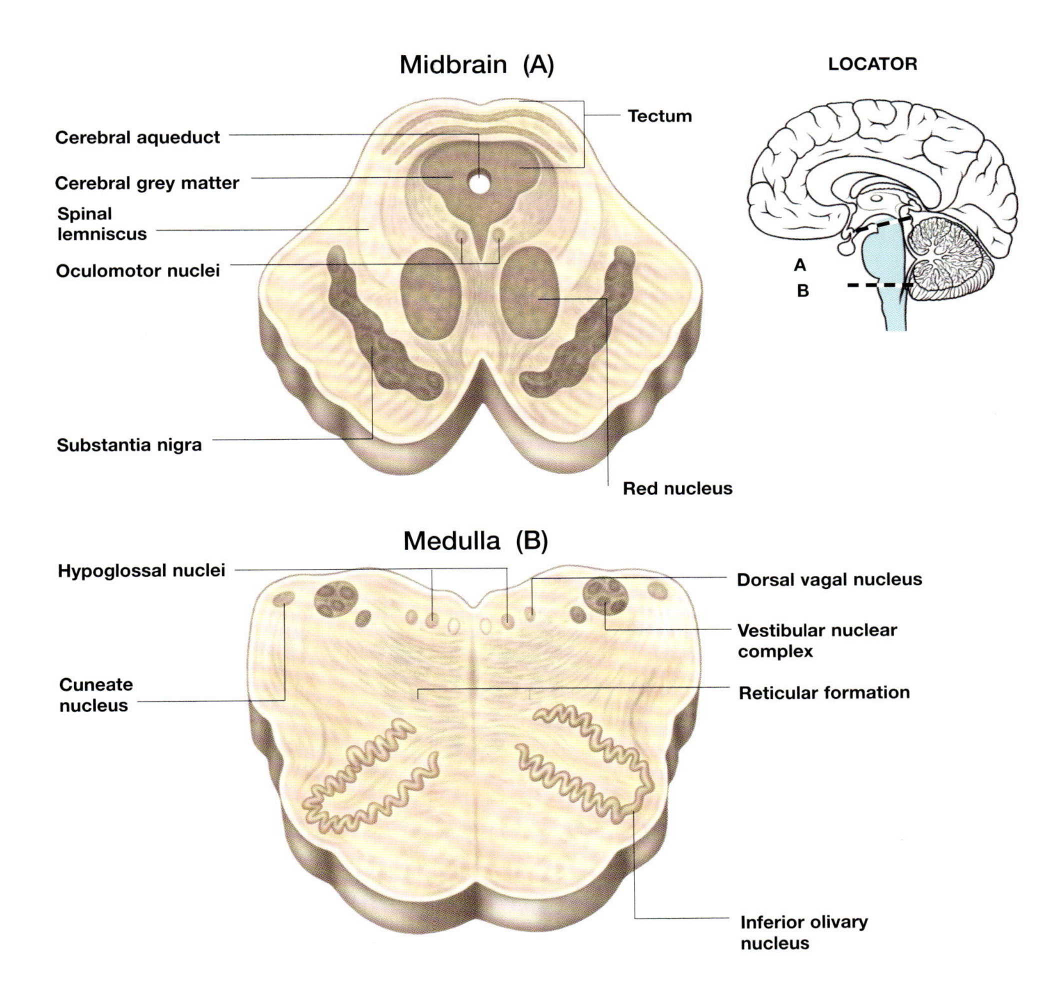

Cranial Nerves

There are 12 pairs of cranial nerves that emerge directly from either side of the brain passing through openings (foramina) in the skull. The cranial nerves carry information to and from the brain.

The cranial nerves all originate in the brain, and all but the first two are considered to be part of the peripheral nervous system as they are either sensory or mixed (sensory and motor fibres) nerves.

Numbered Pairs

There are 12 pairs of these important cranial nerves, which are named after the structures that they supply and each are numbered with Roman numerals. For example, the olfactory nerve (I) supplies the nose and enables smell. The first two pairs of cranial nerves originate in the forebrain whereas the remainder emerge from the brainstem.

Nerve Fibres

Cranial nerves are made up of sensory and motor nerve fibres and so carry information both to and from the central nervous system:

- Sensory nerve fibres carry information such as pain, touch and temperature sensations from the face as well as the senses of taste, vision and hearing
- Motor fibres send instructions to the head, neck and face muscles, allowing various facial expressions and eye movements
- Autonomic nerve fibres allow the subconscious control of internal structures such as the salivary glands, the iris and some of the major organs of the chest and abdomen.

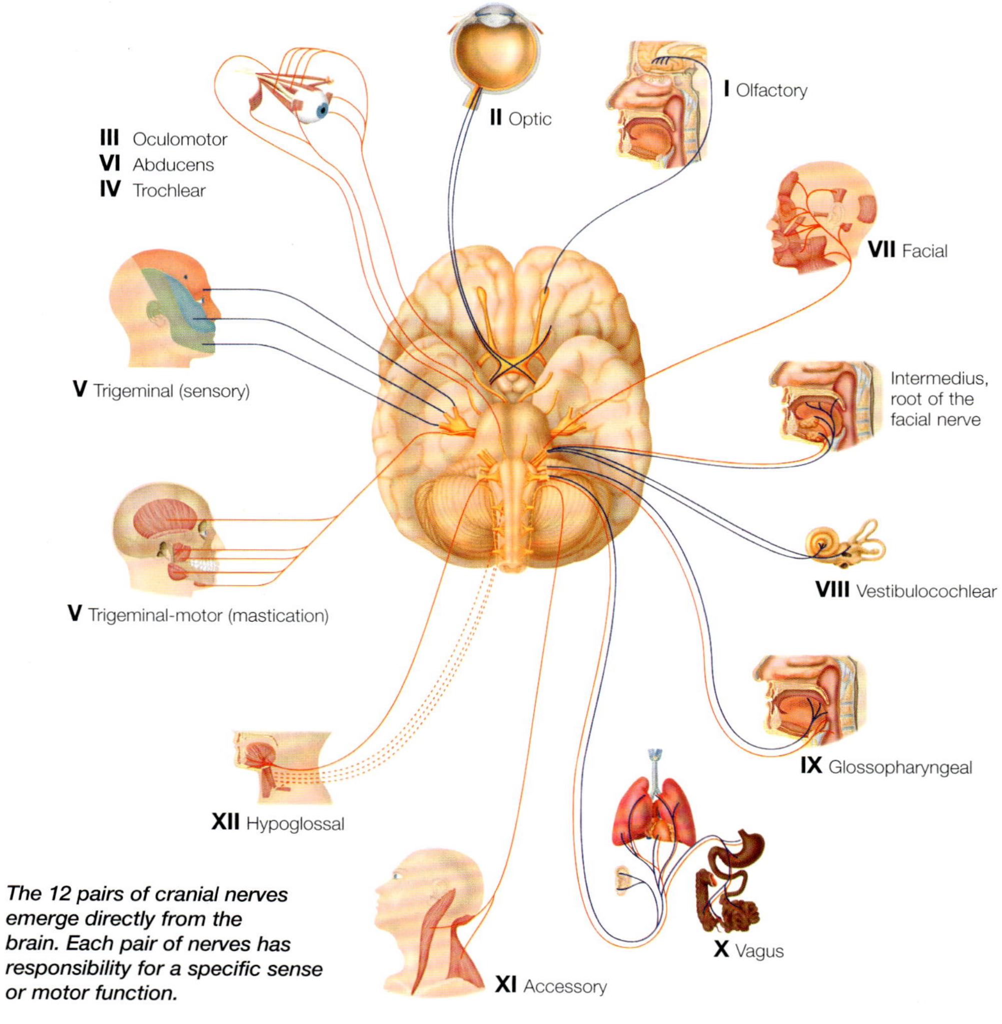

The 12 pairs of cranial nerves emerge directly from the brain. Each pair of nerves has responsibility for a specific sense or motor function.

Number	Name	Comments
I	Olfactory nerve	The sensory nerve of smell
II	Optic nerve	The sensory nerve of vision
III	Oculomotor nerve	Supplies four of the six muscles that move the eyeball
IV	Trochlear nerve	Supplies a muscle that moves the eyeball
V	Trigeminal nerve	Carries sensation from the face and moves the muscles in chewing
VI	Abducens nerve	Supplies a muscle that moves the eyeball
VII	Facial nerve	Moves the muscles of facial expression
VIII	Vestibulocochlear nerve	The sensory nerve for hearing and balance
IX	Glossopharyngeal nerve	Helps to innervate the tongue and pharynx (gullet)
X	Vagus nerve	Supplies many structures including organs in the thorax and abdomen
XI	Accessory nerve	Supplies structures in the throat and some neck muscles
XII	Hypoglossal nerve	Supplies the tongue muscles.

Olfactory Nerves I

The olfactory nerves are the tiny sensory nerves of smell. They run from the nasal mucosa to the olfactory bulbs in the brain.

The olfactory nerves carry the special sense of smell from the receptor cells in the nasal passages to the brain.

Olfactory Epithelium

The olfactory epithelium is the part of the lining of the nasal cavity that carries special receptor cells for the sense of smell. It is found in the upper part of the nasal cavity and the septum, the partition between the two sides of the cavity.

The olfactory receptors are specialized neurones, or nerve cells, that are able to detect odorous substances present in the form of minute droplets in the air.

Olfactory Nerves

Information from the olfactory receptor neurones is passed up to the brain through their long processes, or axons, which group together to form about 20 bundles. These bundles are the true olfactory nerves, which pass up through a thin perforated layer of bone, the cribriform plate of the ethmoid bone, to reach the olfactory bulbs in the cranial cavity. The fibres of the olfactory nerves make connections (synapse) with the neurones within the olfactory bulb.

Olfactory Bulbs

The paired olfactory bulbs are actually part of the brain, extended out on stalks called olfactory tracts, which contain fibres linking them to the cerebral hemispheres. Large specialized neurones, known as mitral cells, connect with the olfactory nerves within the olfactory bulb. This connection permits information about smell to be passed on from the olfactory nerves.

The axons of these mitral cells then carry this information to the olfactory centre of the brain via the olfactory tracts.

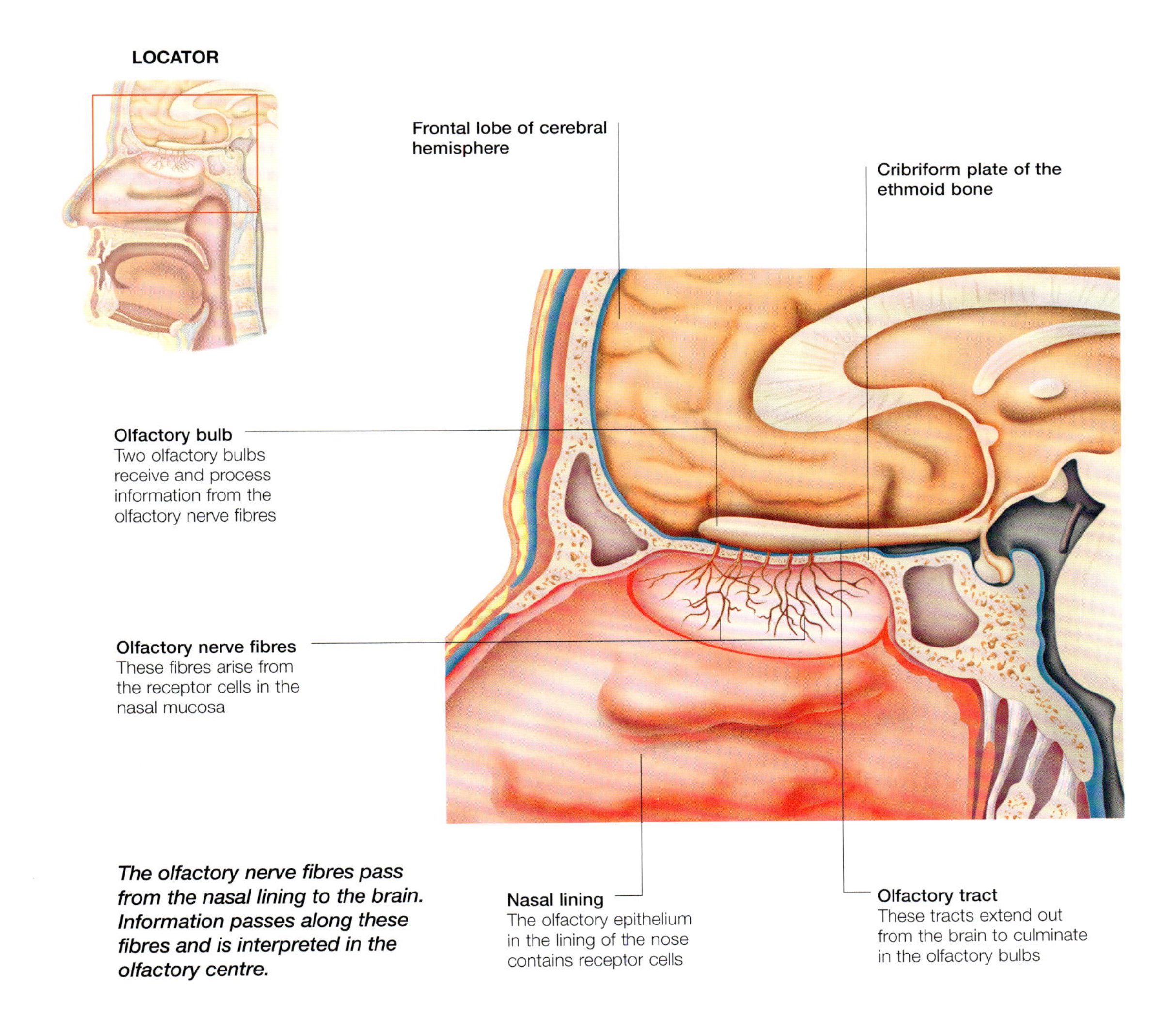

The olfactory nerve fibres pass from the nasal lining to the brain. Information passes along these fibres and is interpreted in the olfactory centre.

Optic Nerves II

Cranial nerves II, III, IV and VI are responsible for visual processes as well as for eye movements. The optic nerve (cranial nerve II) is the nerve of sight.

The optic nerve carries information from the retina at the back of the eye to the brain. Unlike some of the other cranial nerves, the optic nerve is solely sensory, which means that it only takes information to the brain, not from it.

The optic nerve is formed from the axons, or long processes, of the retinal cells at the back of the eye. These join together to form the nerve, which leaves the back of the eyeball at a point known as the optic disc. The optic nerve is covered by layers of meninges, the membranes that protect the brain. Running within the optic nerve are blood vessels that serve the retina.

Optic Nerve Fibres

The nerve fibres that originate in the retina enter the optic nerve, which then passes back through the eye socket to the optic canal, an opening in the skull. Entering the cranial cavity through this opening, the optic nerve fibres converge to form the optic chiasma where some of them cross to the other side. This exchange of nerve fibres allows for binocular vision.

The optic nerve fibres continue until they reach the lateral geniculate body of the brain. From here optic nerve fibres radiate back to the visual cortex where their information is processed.

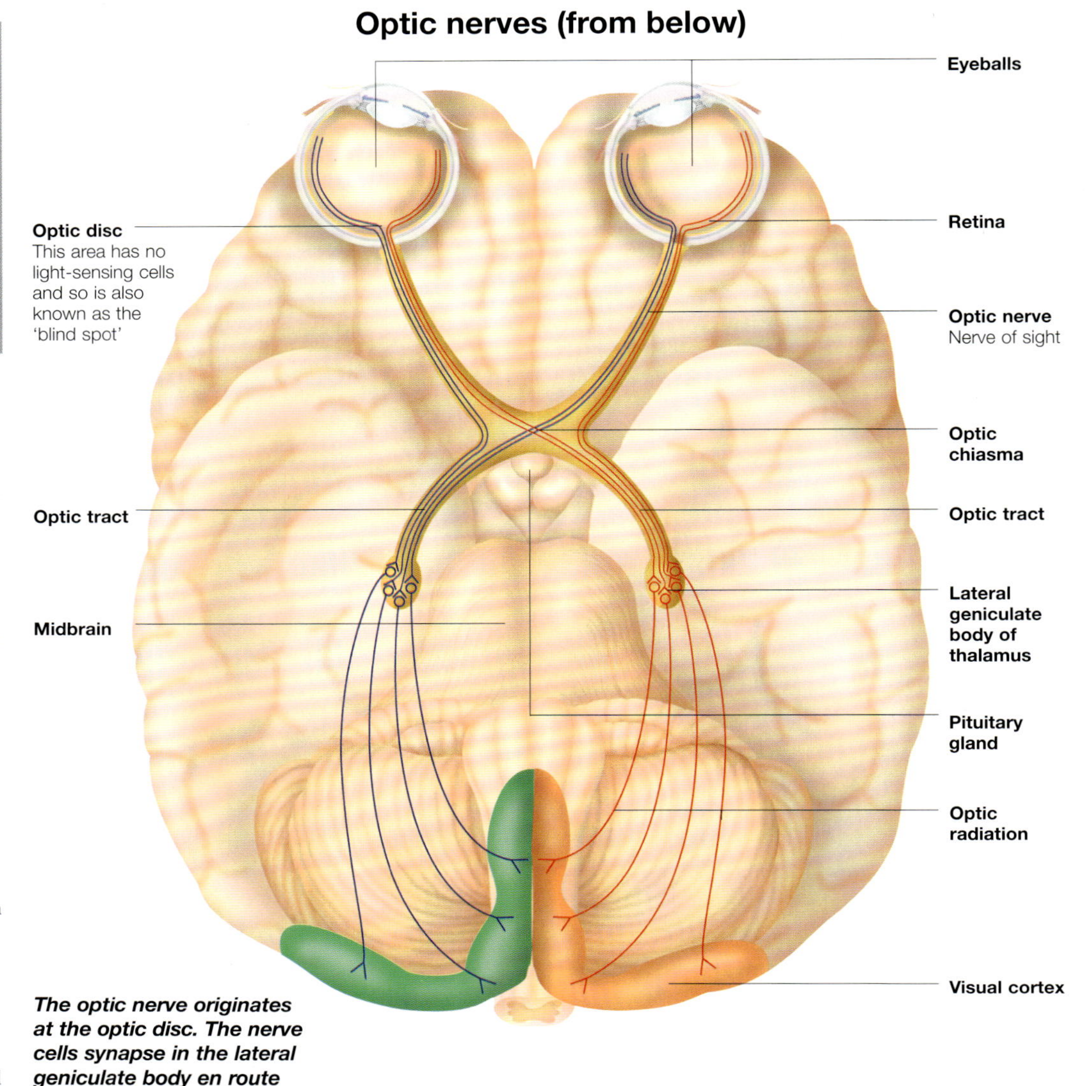

The optic nerve originates at the optic disc. The nerve cells synapse in the lateral geniculate body en route to the visual cortex.

Nerve cells of the retina

The optic nerve carries information from the retina, the delicate innermost layer at the back of the eyeball that responds to the presence of light.

Beneath a layer of pigmented cells lies the neural layer. This structure contains cells that can convert light energy into information in a form that the brain can process and understand.

Layers of nerve cells

The retina has three layers of nerve cells. Photoreceptor cells (rods and cones) lie at the deepest level and are stimulated by light. Above these cells is a layer of bipolar cells that make connections between the rods and cones and the uppermost layer of ganglion cells.

Ganglion cells receive information from the rods and cones via the bipolar cells. The axons of these ganglion cells run along the surface of the neural layer and converge to form the optic nerve.

Before reaching the rods and cones, light has to pass through two layers of cells. These are the ganglion and bipolar cell layers.

Structure of the retina

Locator
Axon of retinal ganglion cells
Ganglion cell layer
Bipolar cell layer
Optic disc
Layer of rods and cones
Optic nerve

Cranial Nerves III, IV and VI

The oculomotor, trochlear and abducens nerves are usually grouped together because between them they supply the six muscles that move the eye. These nerves do not carry any information relating to sight.

There are three cranial nerves that supply the muscles responsible for eye movement.

Oculomotor Nerve

The oculomotor nerve mainly carries motor fibres to the eye muscles, but it also has fibres that carry sensory information back to the brain concerning the position of those muscles. In addition, it contains some fibres from the autonomic nervous system that constrict the pupil and alter the lens shape.

Fibres of the oculomotor nerve originate in the midbrain, part of the brainstem, and leave the cranial cavity to enter the eye socket by passing through the superior orbital fissure. The nerve then splits into superior and inferior divisions.

Trochlear Nerve

The small trochlear nerve supplies only one of the muscles of eye movement, carrying both motor information to the muscle and sensory information from it. The fibres originate in the midbrain and then take a long course around the brainstem to enter the eye socket through the superior orbital fissure.

Abducens Nerve

The abducens (or abducent) nerve supplies the lateral rectus muscle of eye movement and has both motor fibres and sensory fibres. Its fibres originate in the pons, a part of the brainstem. The nerve then arrives at the eye socket by passing through the superior orbital fissure.

Oculomotor, trochlear and abducens nerves

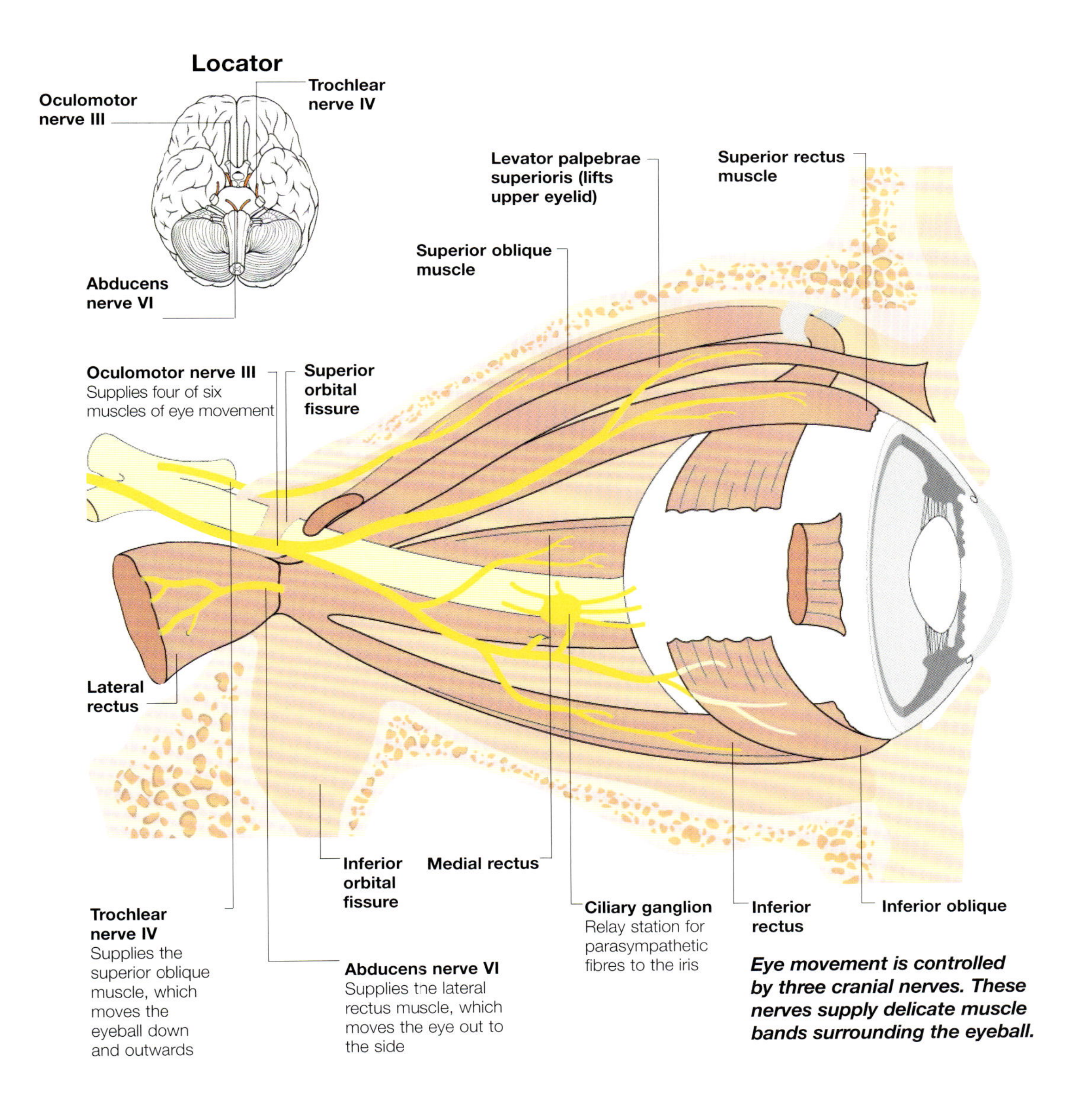

Eye movement is controlled by three cranial nerves. These nerves supply delicate muscle bands surrounding the eyeball.

Trigeminal Nerve V

The trigeminal nerve – the fifth and largest of the cranial nerves – is the main sensory nerve of the face. It has three branches: the ophthalmic, maxillary and mandibular nerves.

As its name implies, the trigeminal nerve has three branches, each of which supplies a part of the face:

- Ophthalmic branch – this carries purely sensory fibres and supplies structures in the upper part of the face and scalp on each side. The nerve emerges from the cranium through a gap (superior orbital fissure) in the back of the eye socket.
- Maxillary branch – this exits the cranium through a small hole (foramen rotundum) and runs forward to supply the central part of the face on each side.
- Mandibular branch – the mandibular nerve has both sensory and motor fibres. It passes out of the cranium through the foramen ovale to reach the lower third of the face and jaw.

Trigeminal Ganglion

Sensory information carried in the trigeminal nerve returns to the pons to be processed. On the way, information passes through the trigeminal ganglion, an expansion of the trigeminal nerve that houses the cell bodies of the nerve cells whose peripheral processes make up the ophthalmic, maxillary and mandibular nerves. The central processes of these nerve cells leave the ganglion to pass back to the brainstem as the solitary trigeminal nerve.

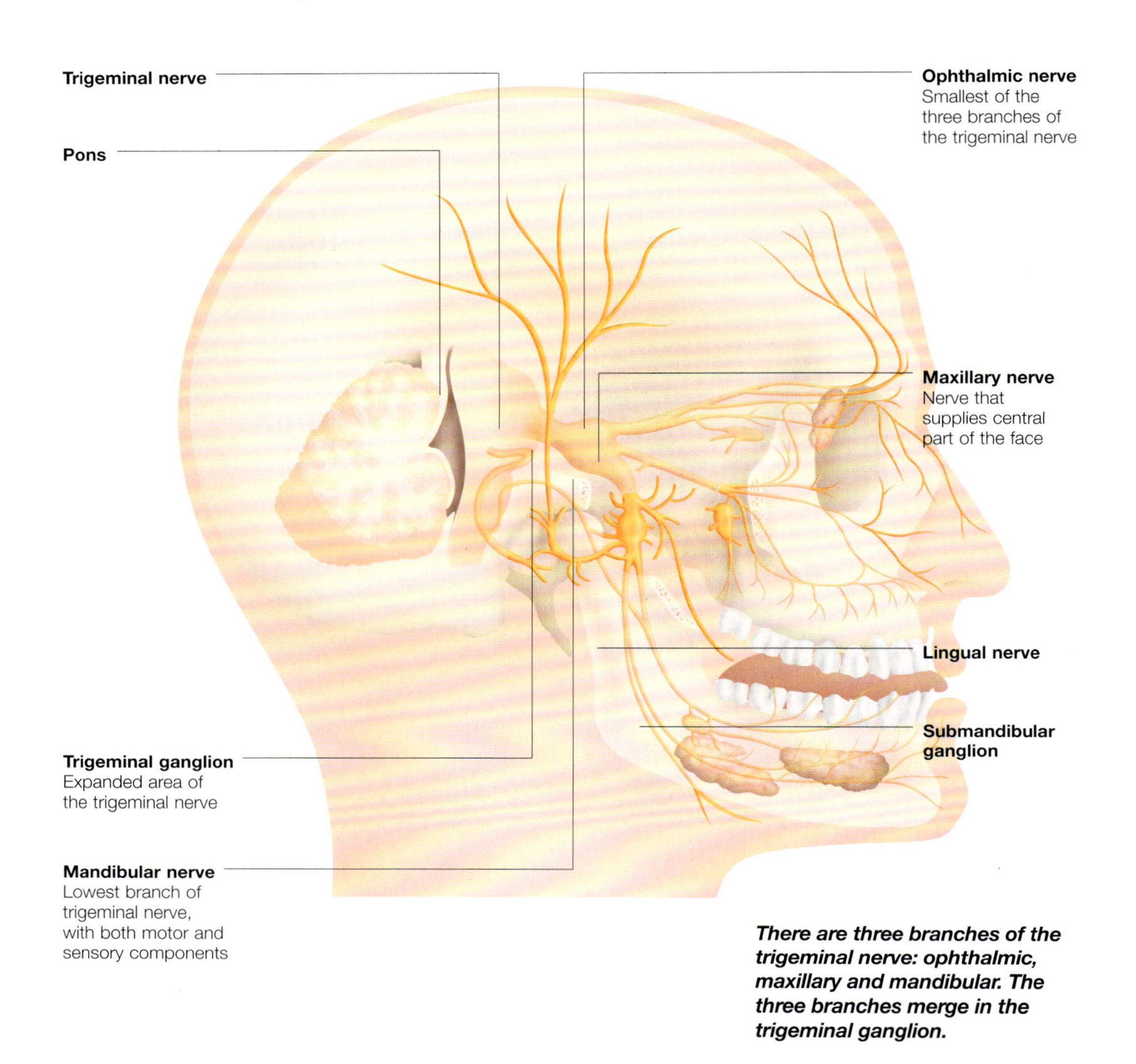

There are three branches of the trigeminal nerve: ophthalmic, maxillary and mandibular. The three branches merge in the trigeminal ganglion.

Sensory supply to the skin

Information that passes back through the trigeminal nerve includes the sensations of touch, pressure, pain and temperature from the face and scalp, cornea and the nasal and oral cavities.

Three divisions

The sensory supply to the face and scalp can be divided into three areas, corresponding to the three main branches of the trigeminal nerve:

- The ophthalmic nerve supplies the scalp, the skin of the upper eyelid and the centre of the nose
- The maxillary nerve supplies the lower eyelid, the side of the nose, the upper lip and cheek
- The mandibular nerve supplies the skin over the chin, lower lip and side of the face in front of and above the ear.

The skin of the rest of the face and head is supplied by branches of the first few spinal nerves of the cervical region.

The sensory supply to the face and scalp can be divided into three separate areas. These are the ophthalmic, maxillary and mandibular divisions.

Distribution of sensory fibres of face and scalp

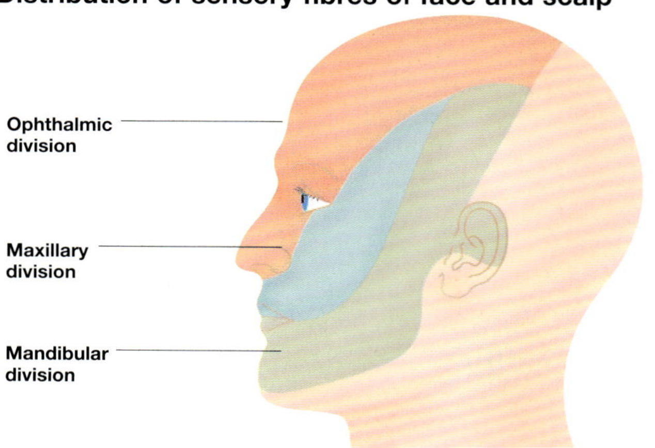

Motor branches of the trigeminal nerve

As well as receiving sensory information from the face, the trigeminal nerve provides the motor supply for some important muscles located around the jaw. Some of these muscles include those involved in chewing.

The large trigeminal nerve (cranial nerve V) is predominantly a sensory nerve, that carries information from the face and scalp back to the brain.

Muscle control

The trigeminal nerve also has a role to play in the control of some important muscles, including those involved in mastication (chewing). The majority of the muscles of the face, many of which are involved in facial expression, are supplied by another nerve, the facial nerve (cranial nerve VII).

Nerve supply

Only the mandibular nerve carries motor fibres. The muscles that receive their nerve supply from the motor fibres of the mandibular nerve include:

- Masseter – this strong muscle runs just under the skin, from the angle of the jaw up to the cheek. It lifts and extends the lower jaw during chewing and biting, and closes the jaw.
- Temporalis – this fan-shaped muscle can be felt in front of and above the ear as it contracts during chewing.
- Medial and lateral pterygoids – these strong muscles run between the lower jaw and the skull and enable chewing movements, including side-to-side grinding.
- Tensor veli palatini – this is one of the muscles of the soft palate.
- Mylohyoid – arising from the back of the mandible, the mylohyoid muscles of each side unite in the midline to form the floor of the mouth.
- Anterior belly of digastric – this muscle acts with others to pull the hyoid bone and the larynx forwards and upwards during swallowing.
- Tensor tympani – this tiny muscle tenses the eardrum to protect it from loud sounds.

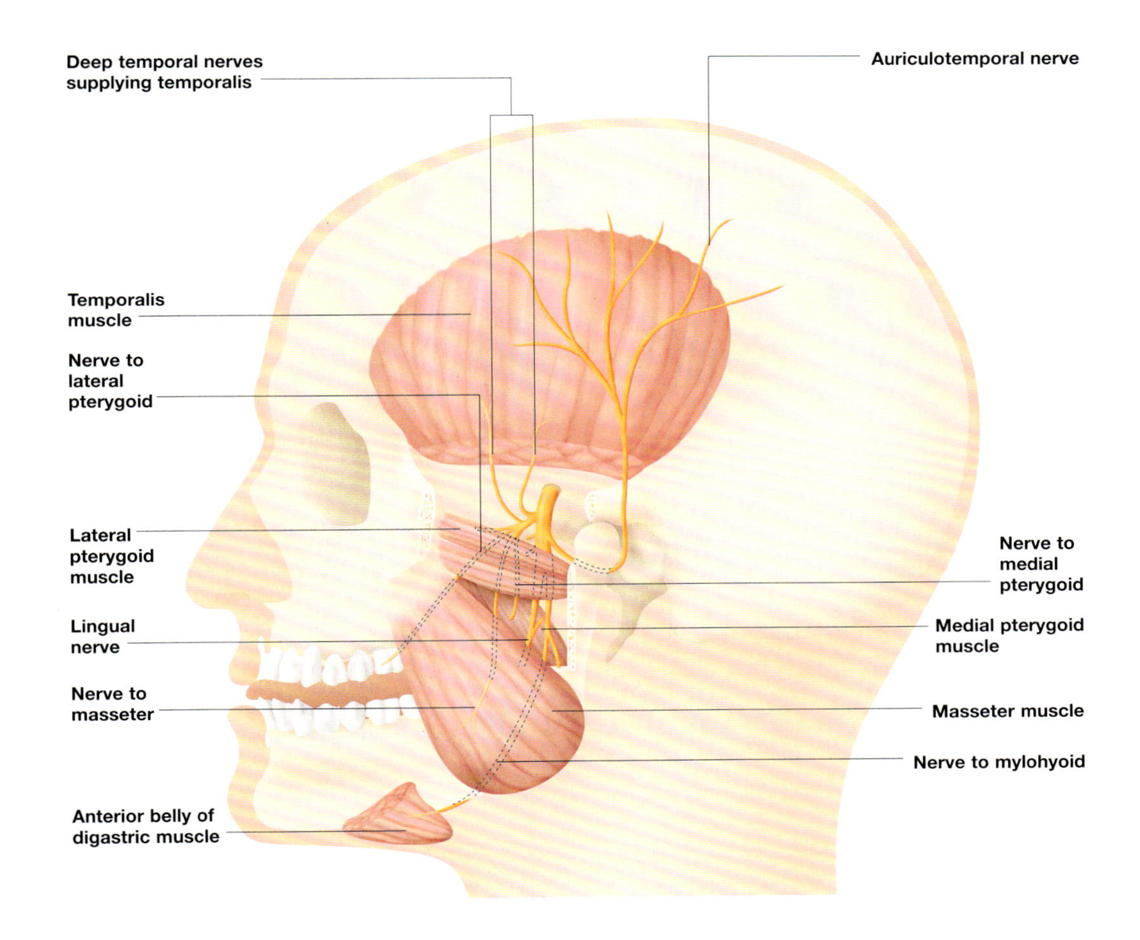

The trigeminal nerve plays an important role in controlling a number of muscles around the jaw. The muscles receive a nerve supply from the motor fibres of the mandibular nerve.

Facial Nerve VII

The seventh cranial nerve is the facial nerve, which supplies the muscles of facial expression. It also carries autonomic parasympathetic information to the lacrimal (tear) and salivary glands and conveys sensory signals.

The facial nerve emerges from the brainstem at the junction of the pons and medulla. There are two roots: the larger root carries motor fibres that transmit instructions to muscles; and the smaller root (nervus intermedius) that carries sensory information and parasympathetic fibres.

The facial nerve travels through the temporal bone, enclosed within the facial canal. The nerve comes into very close contact with the inner ear, before it emerges from the skull through a gap known as the stylomastoid foramen. The nerve then passes through the parotid gland (the salivary gland of the cheek), where it divides into its terminal branches.

There are six branches of the facial nerve:

- Posterior auricular – runs up behind the ear
- Temporal – passes up over the temples towards the forehead and upper eyelid
- Zygomatic – takes its name from the zygoma (cheekbone)
- Buccal – supplies the region around the mouth (buccal cavity)
- Mandibular – runs along the mandible (lower jaw)
- Cervical – runs downwards to the neck area.

Muscle Control

The facial nerve contains three types of fibres: motor, sensory and autonomic. The motor fibres control a number of important muscles, which include:

- Muscles of facial expression, including those that allow smiling and frowning
- Scalp muscles, including the occipitalis and auricular muscle, which give some degree of mobility to the scalp
- Posterior belly of digastric, which helps to raise the hyoid bone during swallowing and speaking
- Stylohyoid, a small muscle that also lifts the hyoid bone
- Stapedius muscle, which lies within the middle ear.

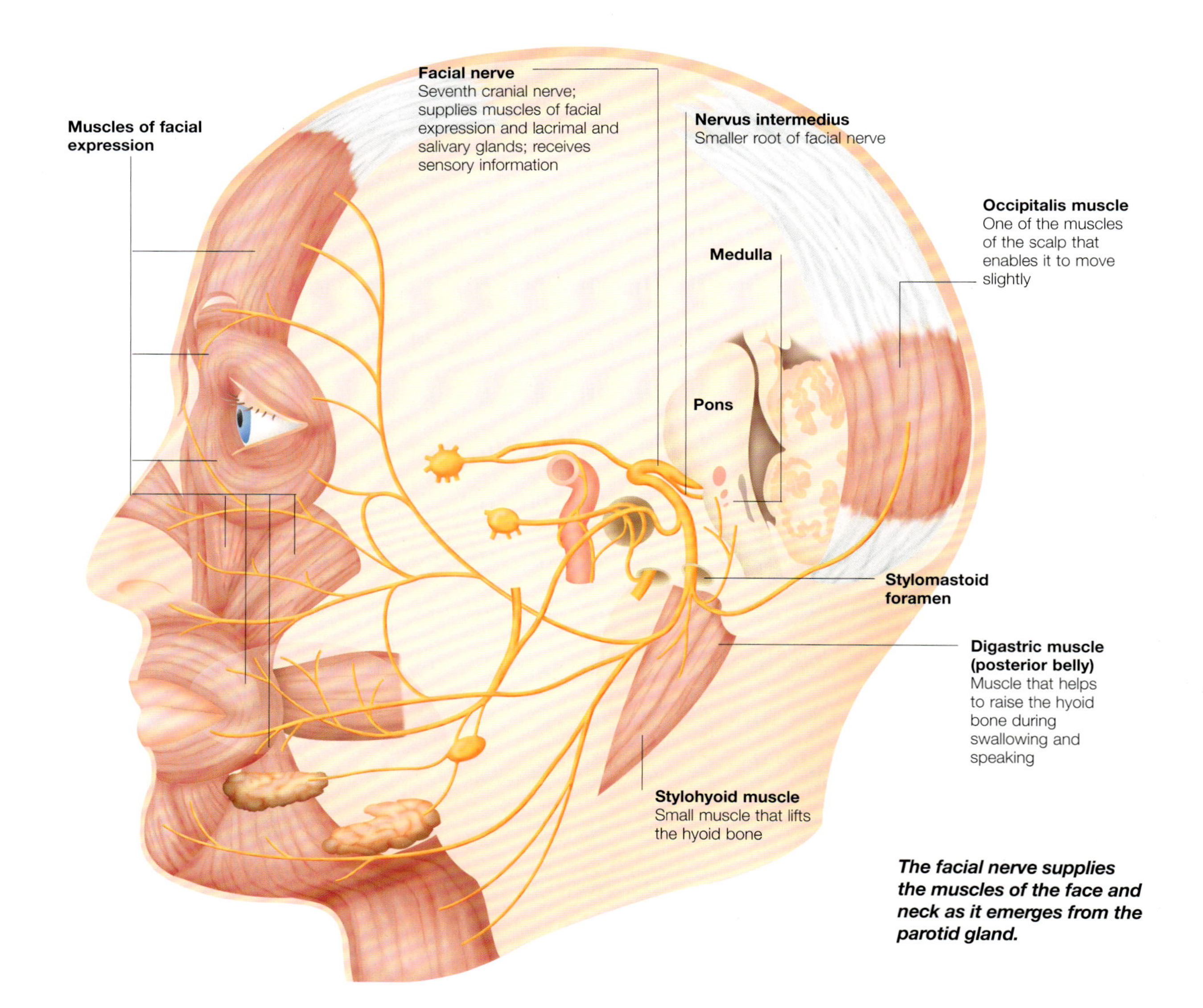

The facial nerve supplies the muscles of the face and neck as it emerges from the parotid gland.

Sensory and autonomic branches

As well as providing many muscles of the face with a nerve supply, the facial nerve carries sensory information, and transports fibres of the autonomic nervous system to the lacrimal and salivary glands.

The facial nerve carries sensory information back to the brain from:

- The tongue – the sense of taste from the front two-thirds of the tongue and soft palate is carried by fibres that become part of the facial nerve.
- A small area of skin around the external auditory meatus, the entrance to the ear canal.

The cell bodies of these sensory neurones lie in the geniculate ganglion, a swelling of the facial nerve.

The parasympathetic system is the part of the autonomic nervous system that is concerned with the unconscious control and regulation of the body's internal environment. The facial nerve contains some important fibres that provide a parasympathetic supply for the lacrimal (tear) glands, and the sublingual (under the tongue) and submandibular (under the jaw) salivary glands.

Parasympathetic fibres that leave the brain in the facial nerve run forward to form connections in two groups of neurone cell bodies called ganglia. These ganglia comprise:

- Pterygopalatine ganglion
Lying at the level of the cheekbone, this ganglion receives fibres that will go on to innervate the lacrimal, nasal and palatine glands.
- Submandibular ganglion
Lying nearer to the angle of the jaw, this ganglion receives fibres that will innervate the sublingual and submandibular salivary glands.

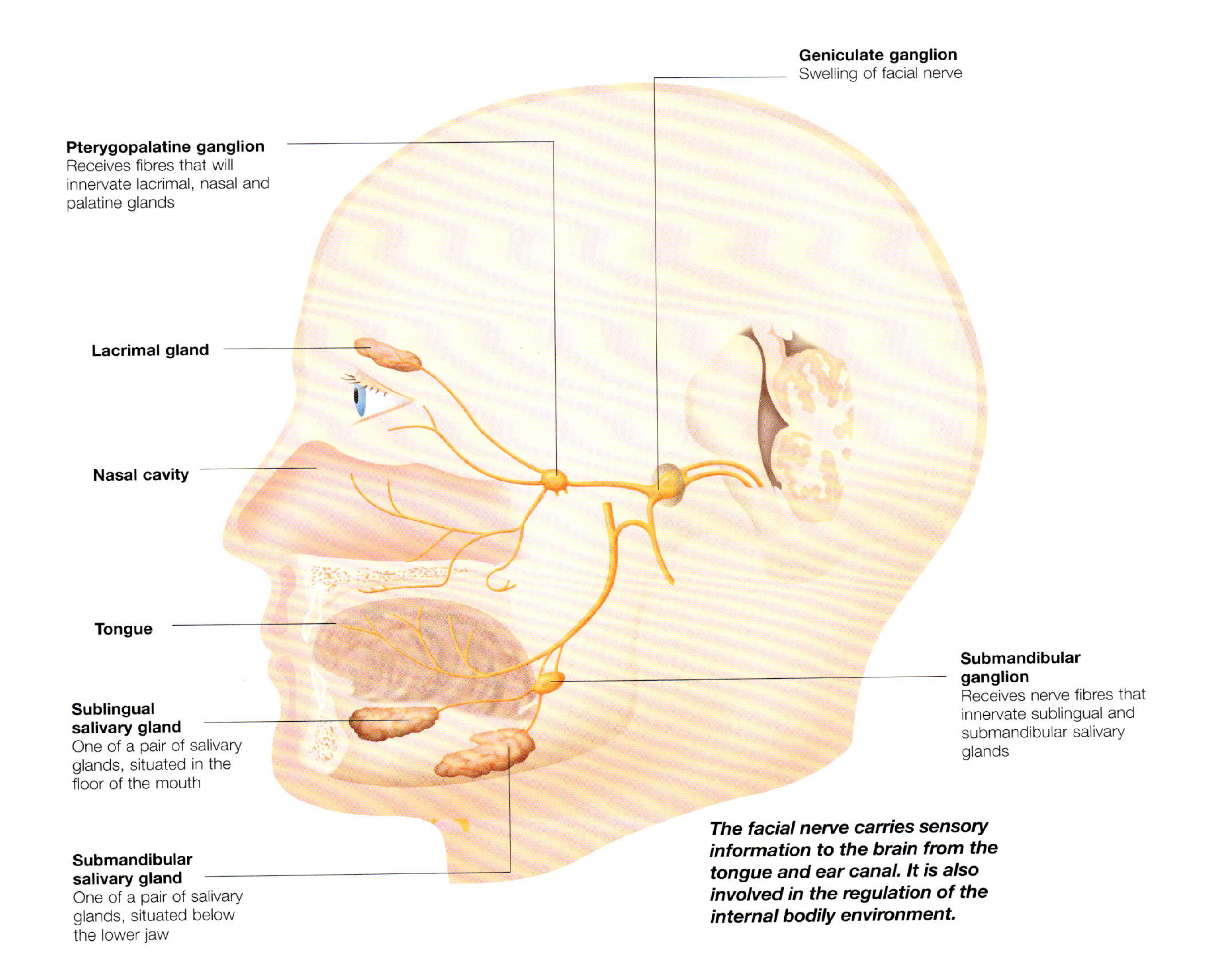

The facial nerve carries sensory information to the brain from the tongue and ear canal. It is also involved in the regulation of the internal bodily environment.

Vestibulocochlear Nerve VIII

The vestibulocochlear nerve is the eighth cranial nerve. It is responsible for relaying information about balance and hearing from the inner ear to the brain.

The vestibulocochlear nerve is the eighth of the 12 cranial nerves. It is a sensory nerve that carries information about balance and hearing from the inner ear to the brain.

The nerve is made up of two parts, the vestibular and the cochlear nerves, which correspond to these two functional areas of the inner ear.

Vestibular Nerve

The vestibular nerve carries information about the position and movement of the head from the semicircular canals and vestibule of the inner ear. These delicate structures contain sensory hair cells that are sensitive to head movement. Information from the hair cells is relayed via the vestibular nerve to the vestibular nuclei – four areas of grey matter within the brainstem.

Cochlear Nerve

The cochlear nerve carries information from hearing receptors in the cochlea, an organ of the inner ear. Fibres of the cochlear nerve receive information from the hair cells of the organ of Corti (spiral organ) within the cochlea. Information about sound is then carried away from the inner ear to the brainstem where it is processed further.

As they leave the inner ear, the vestibular and cochlear nerves join together to form the vestibulocochlear nerve. This nerve then passes through the internal auditory meatus to reach the brainstem.

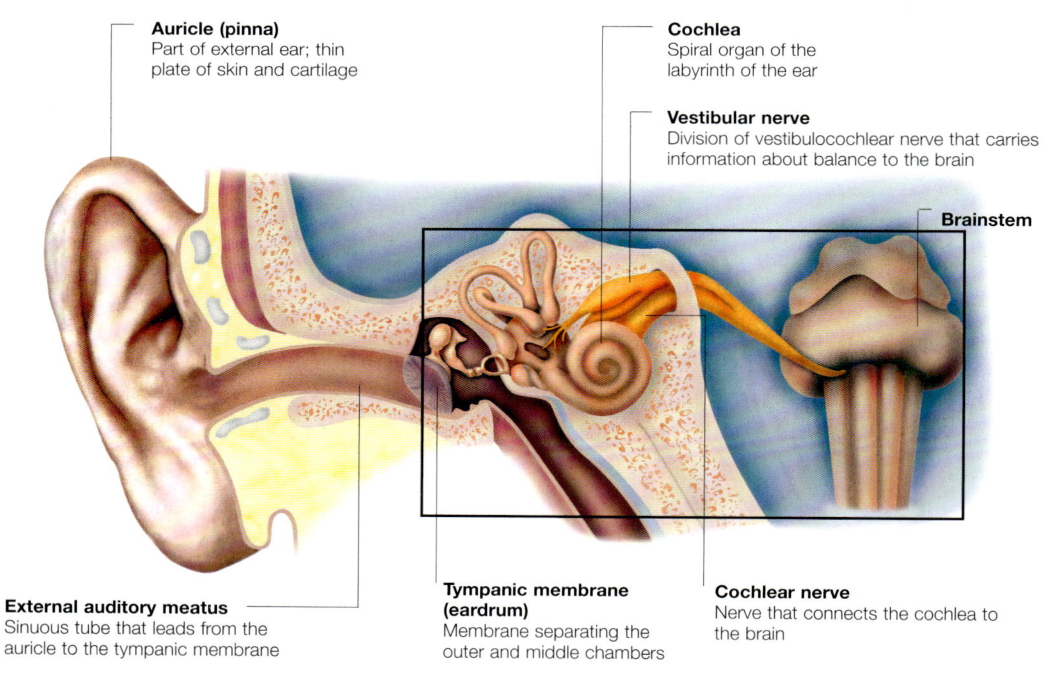

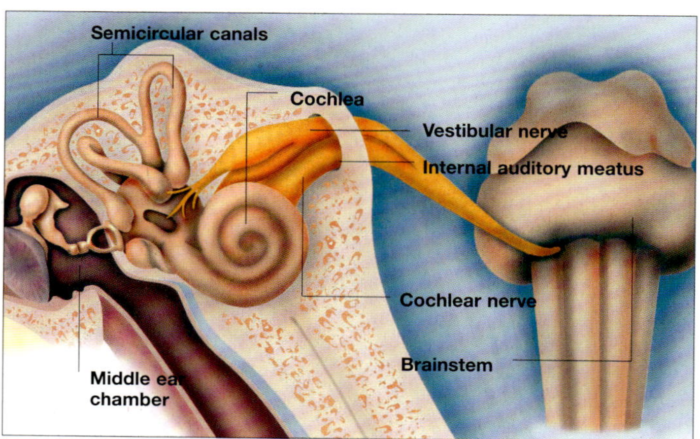

The vestibulocochlear nerve is vital to hearing, balance and posture. The nerve has two parts: the vestibular and cochlear nerves.

The information from the inner ear is transmitted through the internal auditory meatus to the brainstem. Sound is then analyzed in the auditory cortex.

Auditory pathway

The perception of sound involves the passage of information along a fairly complex path. This auditory pathway runs from the inner ear to the highest level of the brain, the cortex.

The pathway of hearing begins at the cochlea of the inner ear, which is stimulated by sound. The information is then carried to the brain along the vestibulocochlear nerve. Within the brain, the information passes through a number of connections before reaching the auditory cortex, where sound is analyzed.

Stages on the pathway

Points along the pathway include:

- Cochlea – the hair cells of the organ of Corti within the cochlea perceive sounds and convert them into electrical information.
- Vestibulocochlear nerve – nerve fibres receive data from the cochlea and pass it to the brainstem in the cochlear part of the vestibulocochlear nerve. The nerve cell bodies lie in the spiral ganglion.
- Cochlear nuclei – the fibres of the cochlear nerve make connections in the dorsal and ventral cochlear nuclei.
- Superior olivary nucleus – from the cochlear nuclei, auditory information passes up to the superior olivary nucleus, from where some fibres pass back to the cochlea to influence its perception of sound.
- Lateral lemniscus – nerve fibres from the superior olivary nucleus ascend the tract of the lateral lemniscus to reach the inferior colliculus. Some of the fibres in this area are thought to make connections that lead to the reflex contraction of small muscles in the ear in response to loud sounds.
- Medial geniculate nucleus of thalamus – fibres from the inferior colliculus ascend to the thalamus to the medial geniculate nucleus, the last staging point along the pathway of sound to the auditory cortex.
- Auditory cortex – the part of the cortex that receives information about sound is the primary auditory cortex in the temporal lobe. The area around the auditory cortex is known as Wernicke's area and it is here that information about sound is analyzed and interpreted.

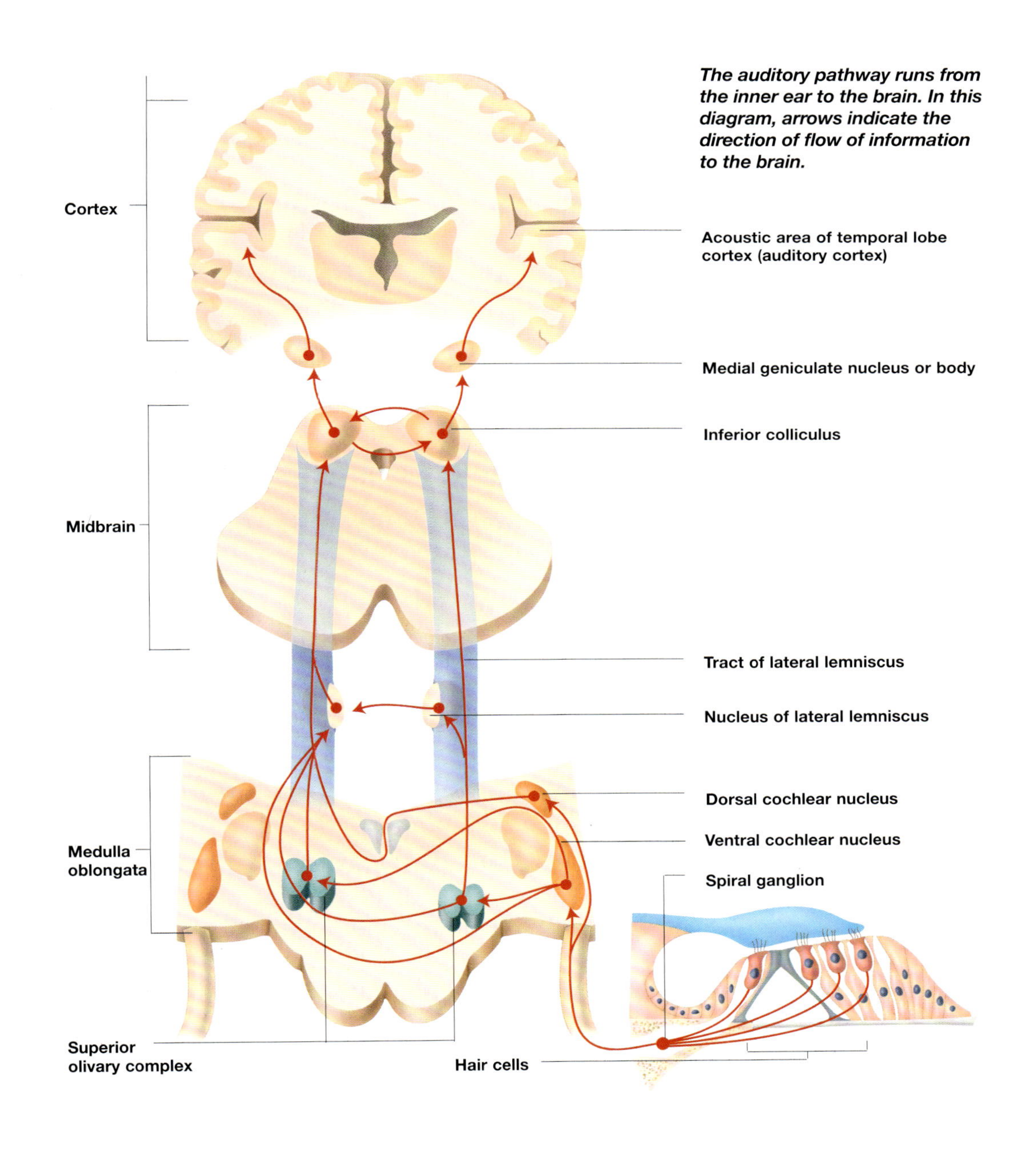

The auditory pathway runs from the inner ear to the brain. In this diagram, arrows indicate the direction of flow of information to the brain.

Vagus Nerve X

The vagus nerve is the tenth and largest of the cranial nerves, stretching down from the head to the abdomen. Its role in monitoring and controlling breathing and digestion makes it vital to life.

The tenth cranial nerve, the vagus, is the most extensive of the 12 cranial nerves.

Functions of the Vagus

This mixed nerve contains both sensory and motor fibres that provide:

- Sensation from the pharynx, larynx and organs of the chest and abdomen
- Taste from the root of the tongue and back of the throat
- Motor innervation of the muscles of the soft palate, pharynx and internal larynx
- The parasympathetic nerve supply to the internal organs of the chest and abdomen, which helps to monitor and control these organs subconsciously.

The vagus nerve has an important role in swallowing and speaking, as well as a vital part to play in the control of the heart, breathing and digestion.

Branches of the Vagus

Numerous branches include:

- A small branch from the base of the skull to the back of the brain, and a branch to a small area of skin around the ear
- Branches from the neck to the lower pharynx, larynx and the heart
- Branches from the thorax that form the plexuses serving the heart, lungs and oesophagus
- The anterior and posterior vagal trunks that supply the stomach and gut.

Schematic representation of branches of the vagus nerve

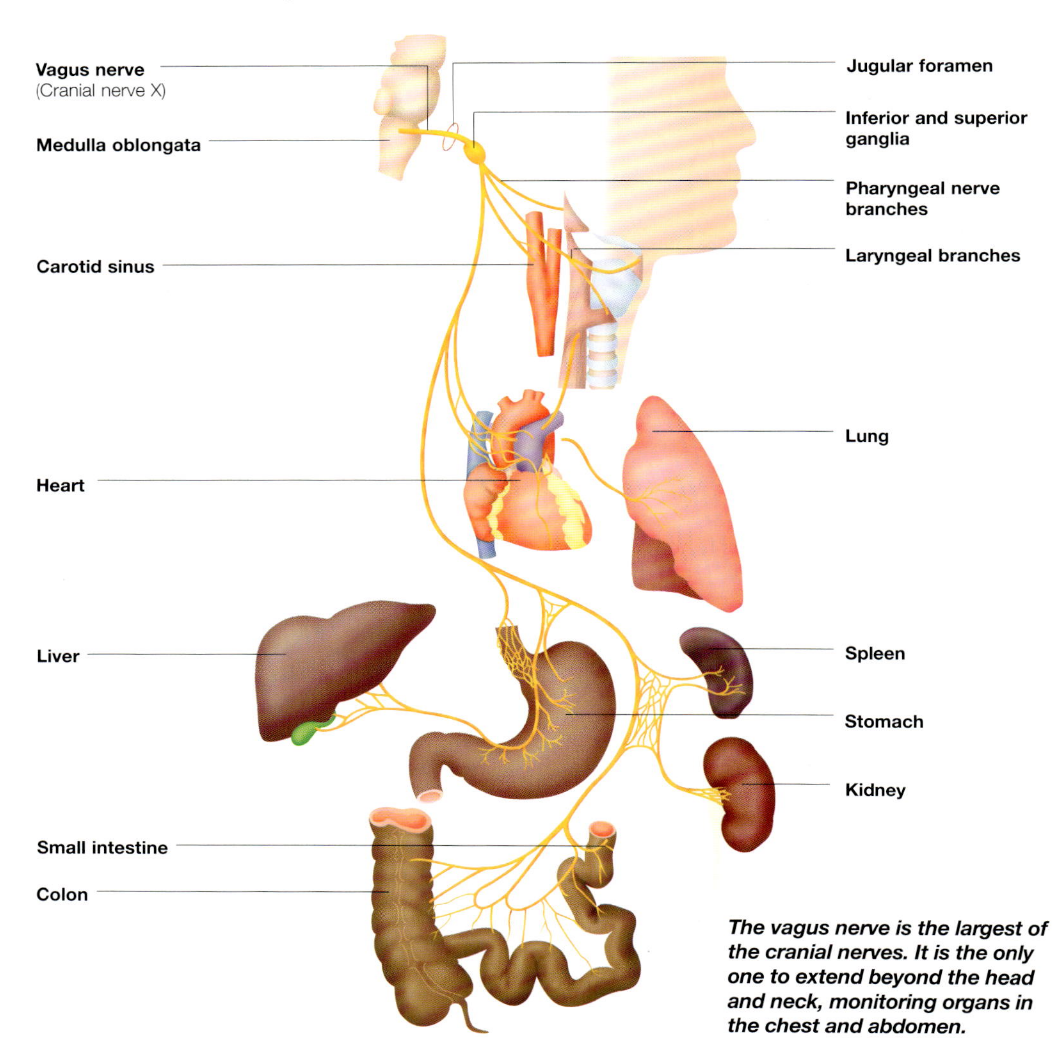

The vagus nerve is the largest of the cranial nerves. It is the only one to extend beyond the head and neck, monitoring organs in the chest and abdomen.

Course of the vagus in the neck

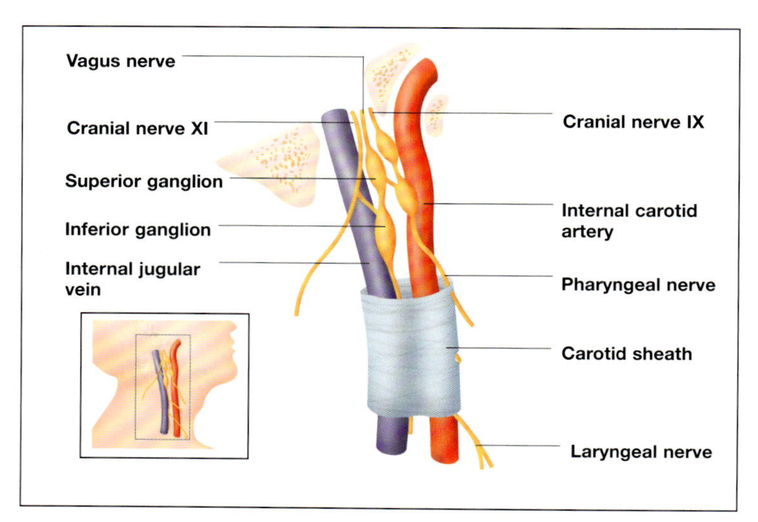

Fibres of the paired vagus nerves leave the medulla oblongata and exit the cranial cavity by passing through the jugular foramen together with cranial nerves IX and XI.

At this level the vagus has two ganglia (swellings containing nerve cell bodies): a small superior and a larger inferior ganglion. There are also connections here with cranial nerves IX and XI, and with fibres of the sympathetic nervous system.

The vagus nerve travels through the neck between the internal jugular vein and the carotid artery. They are all protected by the carotid sheath.

Through the neck

As the vagus emerges from the jugular foramen, it lies between the large carotid blood vessels within the carotid sheath (a protective sleeve of connective tissue). It continues down with them, through the neck to enter the thoracic cavity below.

Within the neck the vagus provides part of the nerve supply to the carotid sinus and the carotid bodies. These are specialized areas that detect changes in the pressure and chemical composition of blood.

Vagus in the thorax and abdomen

The vagus is the only cranial nerve to extend beyond the head and neck. Reaching into the thoracic and abdominal cavities, it has a vital role to play in the control of the internal organs of the chest and abdomen, including the heart.

The right and left vagal nerves follow a similar course in the head and neck. However, as they reach the root of the neck and enter the thoracic cavity their courses begin to differ:

- On the right, the vagus passes in front of the right subclavian artery, where it gives off the right recurrent laryngeal nerve. This runs back up to the larynx. The vagus then continues downwards, passing behind the large superior vena cava and the root of the right lung.
- On the left side, the vagus nerve enters the thoracic cavity, lying between the left common carotid and the left subclavian arteries. As it reaches and curves around the large aortic arch, it gives off the left recurrent laryngeal nerve, which loops under the aortic arch before returning to the larynx. The left vagus then descends behind the root of the left lung.

Vagal trunks

As they descend, the vagus nerves become part of a series of nerve networks, or plexuses, that supply the lungs, heart and oesophagus. Fibres from the oesophageal plexus converge to form two nerves, the anterior and posterior vagal trunks:

- The anterior vagal trunk is derived mainly from the left vagus nerve. It lies on the anterior surface of the oesophagus and extends towards the lesser curvature of the stomach, giving off branches to the stomach, liver and duodenum.
- The larger posterior vagal trunk is derived mainly from the right vagus. It gives off numerous branches to both sides of the stomach as well as contributing to the coeliac plexus.

The vagus helps to control and encourage the digestive processes through these nerves.

Descent of the vagus nerve through the thorax and abdomen

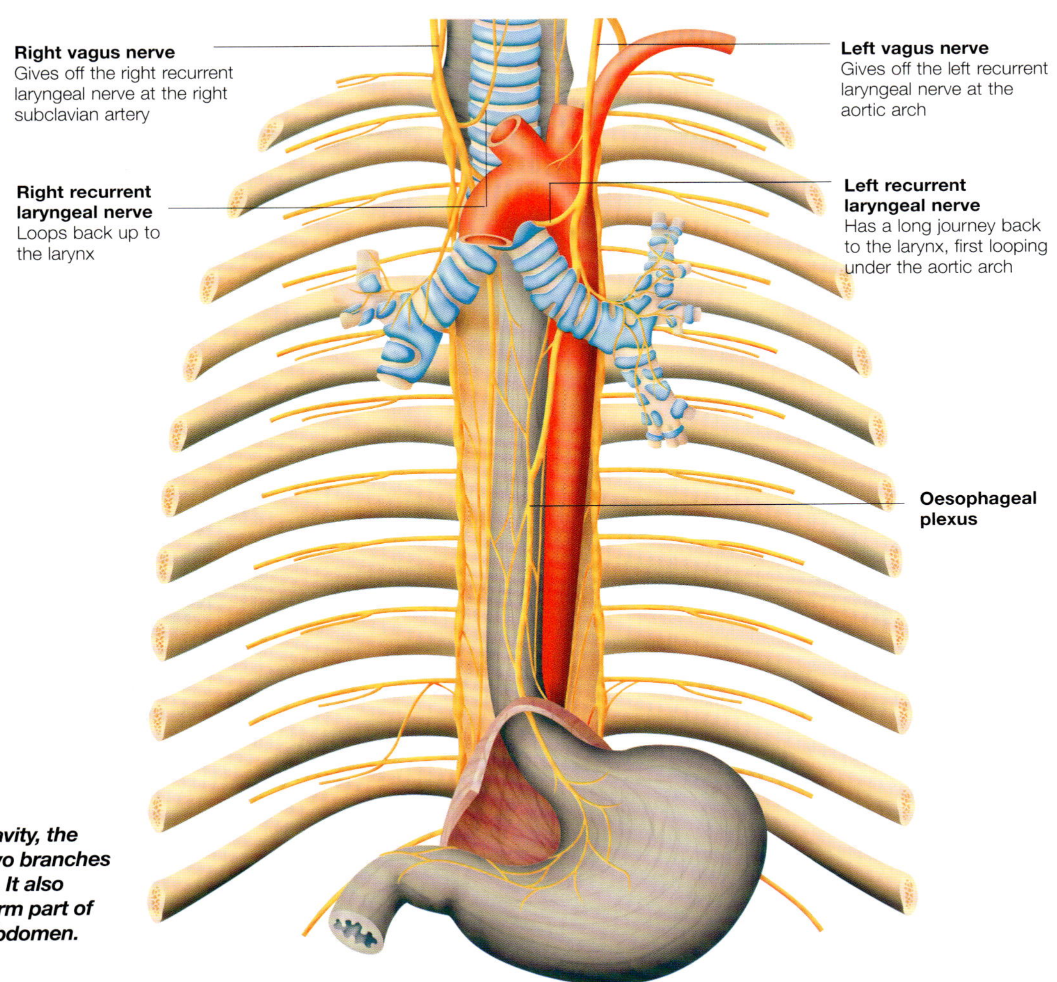

Within the thoracic cavity, the vagus nerve sends two branches back up to the larynx. It also continues down to form part of the plexuses of the abdomen.

Glossopharyngeal Nerve IX

The glossopharyngeal nerve (IX) carries sensory information from the throat and tongue to the brain. The accessory (XI) and hypoglossal (XII) nerves supply muscles within the throat and mouth.

The term 'glossal' relates to the tongue, and the glossopharyngeal nerve is named according to the main areas it serves: the tongue and the pharynx (throat).

Functions

The glossopharyngeal nerve is mixed, and carries both motor and sensory nerve fibres. It also carries fibres of the parasympathetic branch of the autonomic nervous system. Sensory information carried back to the brain by the glossopharyngeal nerve includes:

- Taste from the back third of the tongue
- Sensation from the lining of the pharynx, back third of the tongue and auditory (Eustachian) tube
- Blood oxygen and carbon dioxide levels from the carotid body (tissue within the carotid artery), and blood pressure is monitored by the carotid sinus.

Motor fibres of the nerve carry impulses to the stylopharyngeus muscle – one of the longitudinal muscles in the pharynx used in swallowing and speaking.

The glossopharyngeal nerve emerges from the medulla and runs forwards to leave the skull through the jugular foramen together with the tenth and eleventh cranial nerves. The nerve then travels down alongside the stylopharyngeus muscle towards the pharynx and the back of the tongue.

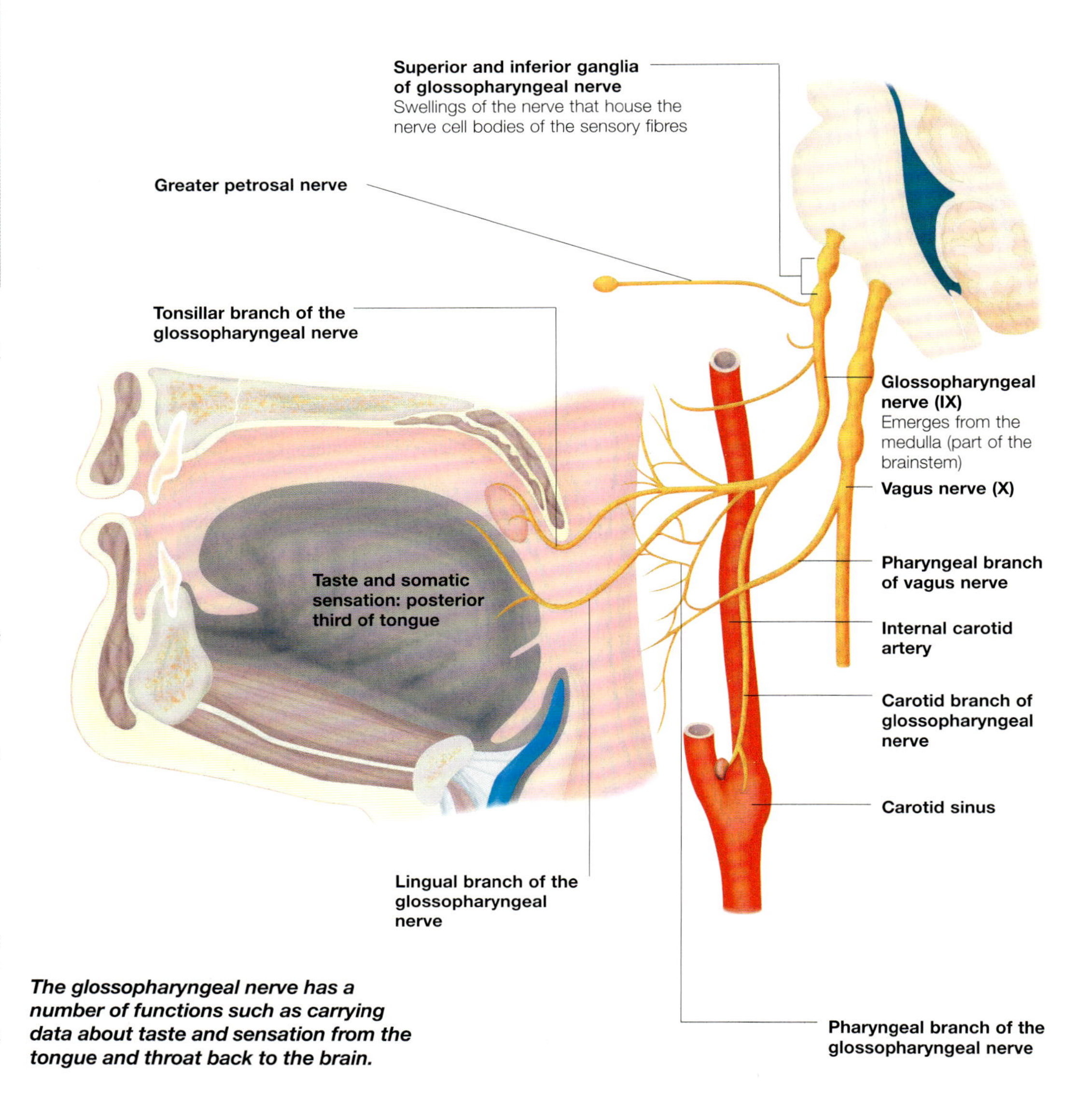

The glossopharyngeal nerve has a number of functions such as carrying data about taste and sensation from the tongue and throat back to the brain.

Accessory Nerve XI

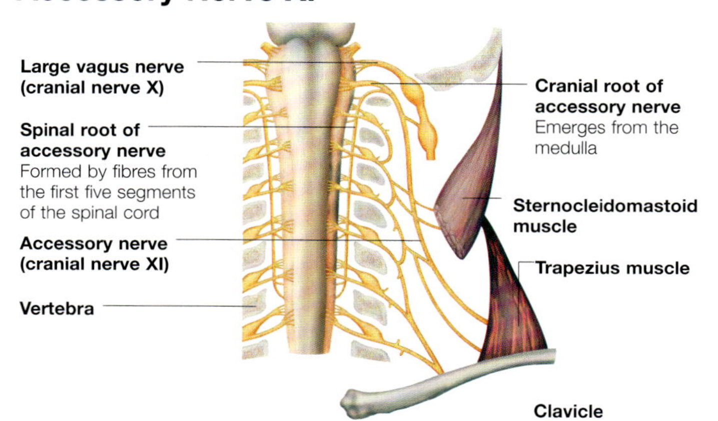

The accessory nerve is unique among the cranial nerves in that it has a spinal root as well as a cranial root. Together, these form the accessory nerve, which exits the skull through the jugular foramen.

Separate functions

Once through the jugular foramen, the two roots of the accessory nerve separate again to fulfil their functions.

The accessory nerve has both spinal and cranial roots. These join to exit the skull and then separate to perform separate functions in the body.

Fibres from the cranial root then join with the large vagus nerve and go on to supply the muscles of the soft palate, pharynx, larynx and oesophagus.

Fibres from the spinal root run down as the accessory nerve, lying alongside the internal carotid artery to reach the sternocleidomastoid muscle, which they supply. The spinal accessory nerve then continues on its journey to supply the large trapezius muscle at the back of the neck.

Hypoglossal Nerve XII

The twelfth cranial nerve is the hypoglossal (meaning 'under the tongue'), which supplies the muscles of the tongue. It has an important role in the actions of chewing, swallowing and speaking.

The hypoglossal nerve provides a motor nerve supply for many of the tongue's muscles including the three extrinsic muscles:

- The styloglossus muscle
- The hyoglossus muscle
- The genioglossus muscle.

The hypoglossal nerve is also joined by fibres from the first cervical nerve that go on to supply other structures. These include muscles attached to the hyoid bone in the neck, which provides a base for tongue movements. They also carry sensory information from the dura lining the rear part of the brain.

Course of the Hypoglossal

The hypoglossal nerve arises from each side of the medulla of the brainstem, usually as four separate roots that unite as they pass through an opening in the skull called the hypoglossal canal.

The paired nerves then pass down and outwards, between the internal carotid artery and the internal jugular vein, to the angle of the jaw. Here, the nerves curve forwards to run under the tongue, as the name of the nerve implies. The hypoglossal nerves end in a series of branches within the substance of the tongue.

Branches of the Hypoglossal

The branches of the hypoglossal nerve, together with the fibres from the cervical nerve that join it, include:

- A meningeal branch that returns through the hypoglossal canal to innervate (supply) the dura of the brain (not shown)
- A descending branch that joins the ansa cervicalis (a loop of nerves that supply the muscles below the hyoid)
- Terminal branches that provide a nerve supply to all the intrinsic and most of the extrinsic muscles of the tongue.

The hypoglossal nerve joins with the first cervical nerve to supply the muscles of the tongue. It arises in the medulla of the brainstem and ends within the tongue.

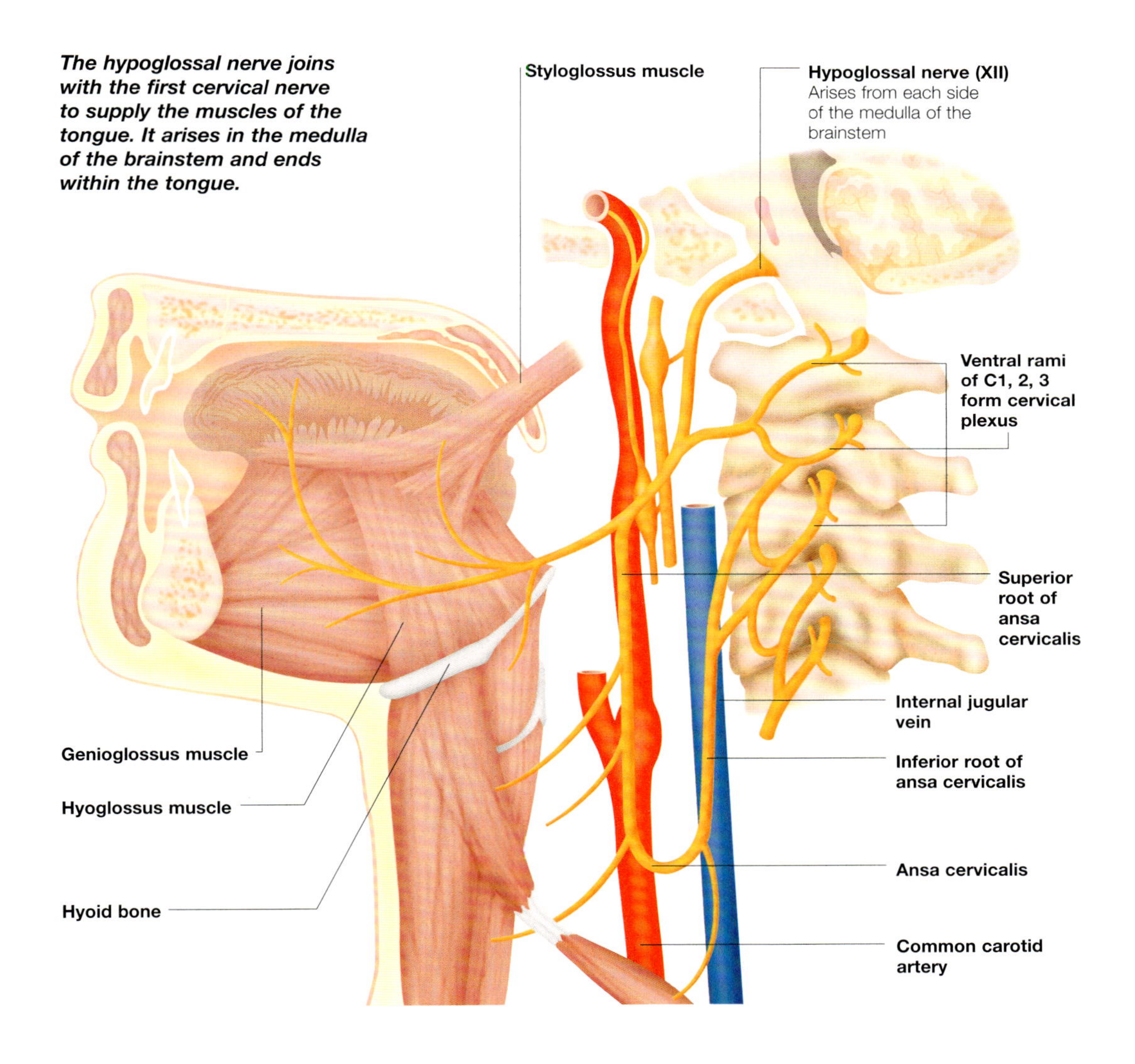

Spinal Cord

The spinal cord is the communication pathway between the brain and the body. It allows signals to pass down to control body function and up to carry data from the body to the brain.

The spinal cord is a slightly flattened cylindrical structure of 42–45cm (16–17in) length in adults, with an average diameter of about 2.5cm (1in). It begins as a continuation of the medulla oblongata, the lowest part of the brainstem, at the level of the foramen magnum, the largest opening in the base of the skull. It then runs down the length of the neck and back in the vertebral canal, protected by the bony vertebrae.

Development

Up to the third month of development in the womb, the spinal cord runs the entire length of the vertebral column. Later on, however, the vertebral column outgrows the cord, which by birth ends at the level of the third lumbar vertebra. This more rapid growth of the vertebral column continues so that in the adult, the spinal cord ends at about the level of the disc between the first and second lumbar vertebrae.

Anatomy of the Cord

The cord is enlarged in the regions of the neck and lower back (cervical and lumbar enlargements). The lower end of the cord tapers into a cone-shaped region – the conus medullaris. From this, the filum terminale – a thin strand of modified pia mater (a membrane that surrounds the brain and spinal cord) – continues downwards and is attached to the coccyx, anchoring the spinal cord.

Enlargements

The upper cervical enlargement extends from the level of the third cervical to the second thoracic vertebra and corresponds to the origin of the nerves to the arms. The lower lumbar enlargement extends from the ninth to the 12th thoracic vertebrae and gives rise to the nerve supply to the legs.

On the front surface of the spinal cord there is a deep, wide groove known as the anterior median fissure. On the back a narrower, shallow indentation – is called the posterior median sulcus. These two demarcations divide the spinal cord vertically into two halves.

Posterior view of spinal cord

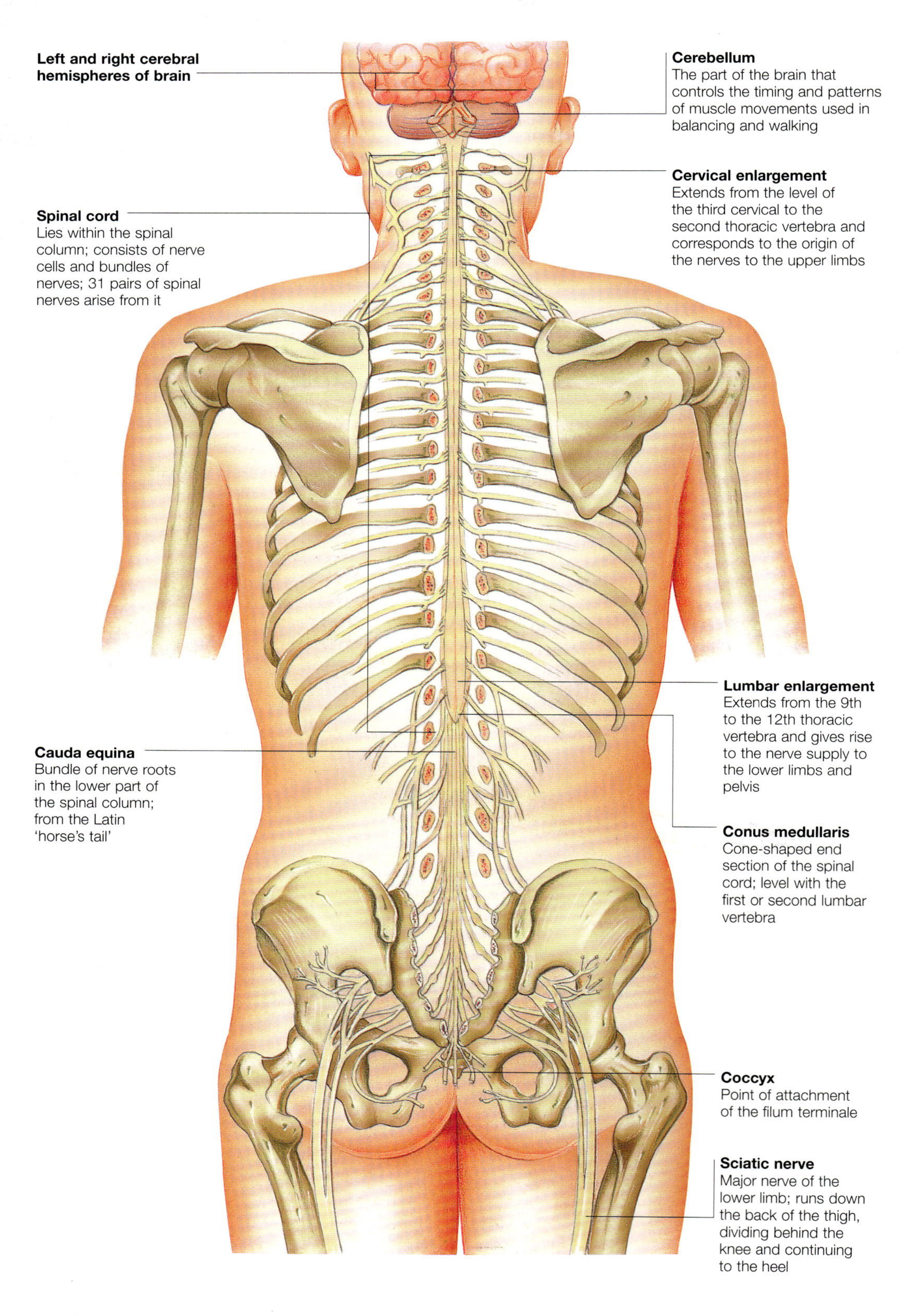

Cross-sections through the spinal cord

The appearance of the spinal cord varies at different levels, according to the amount of muscle supplied by the nerves that emanate from it.

The spinal cord is made up of an inner core of grey matter, which consists mainly of nerve cells and their supporting cells (neuroglia). These cells are surrounded by white matter, made up primarily of myelinated nerve fibres – nerves with an insulating sheath of the fatty substance myelin.

In cross-section, the grey matter typically has the shape of a letter H or a butterfly, with two anterior columns or horns, two posterior columns and a thin grey commissure connecting the grey matter in the two halves. There is a small central canal containing cerebrospinal fluid, which at its uppermost limit runs into the fourth ventricle in the region of the lower brainstem and cerebellum.

Variations in appearance

There is some variation in the appearance of a cross-section of the spinal cord at different levels. The amount of grey matter corresponds to the bulk of muscle whose nerve supply comes off at that level.

- Cervical: the cord is relatively large in the neck and has a more oval shape, known as an enlargement. Grey matter is prominent, corresponding to the cervical enlargement supplying the upper limbs. The posterior grey column is relatively narrow while the anterior grey horn is quite broad and expanded.

White matter is also plentiful, since all of the pathways (tracts) passing up from the various parts of the body to the brain have been gathered together. At the same time, all of the tracts descending from the brain are still grouped before they have given off their branches to the different regions of the body.

- Thoracic: the cord is almost circular and has a smaller diameter. There is an intermediate amount of white matter. Compared to the cervical and lumbar levels, the grey matter is not as prominent here and the anterior and posterior columns are quite narrow.
- Lumbar: the cord has a larger diameter again, corresponding to the increased amount of grey matter in the lumbar enlargement supplying the lower limbs. The white matter is less prominent, because many of the descending tracts have already branched off to their destinations, while only a few of the ascending tracts have been gathered together yet.
- Sacral: in the region of the conus medullaris, the grey matter takes the form of two oval-shaped masses that occupy most of the cord with very little white matter.

In the thoracic and upper lumbar and sacral regions of the cord a small lateral grey column projects between the anterior and dorsal columns, giving rise to fibres to the autonomic nervous system.

Tracts in the spinal cord

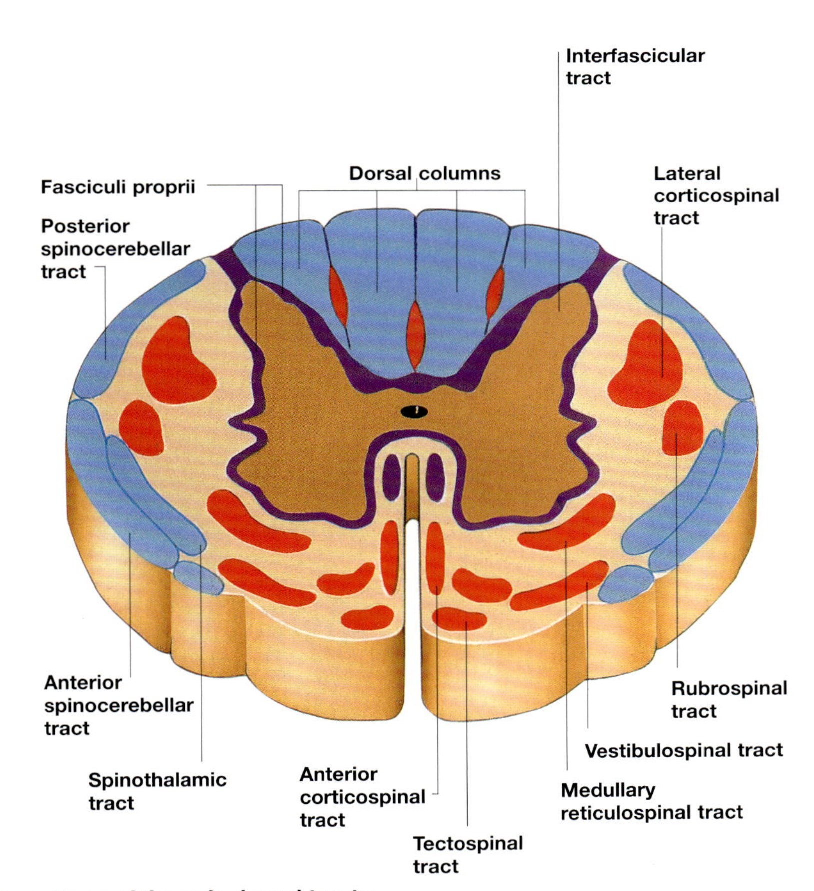

Locations of the spinal cord tracts

Blue: Ascending pathways

Red: Descending pathways

Purple: Fibres that pass in both directions

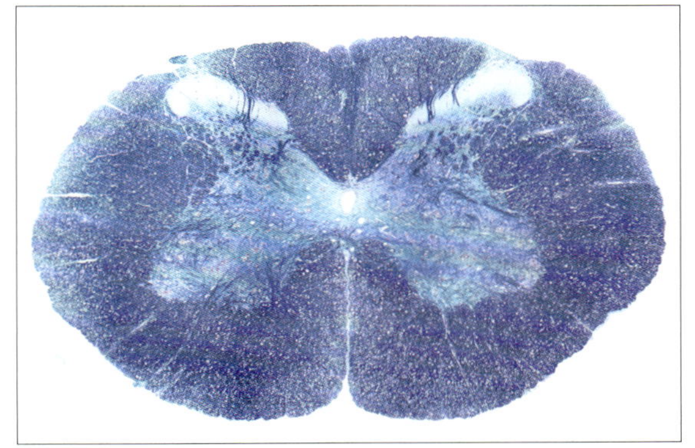

Cervical: the cord is relatively large and has an oval shape. Grey matter (dark red) is prominent, corresponding to the cervical enlargement supplying the upper limbs.

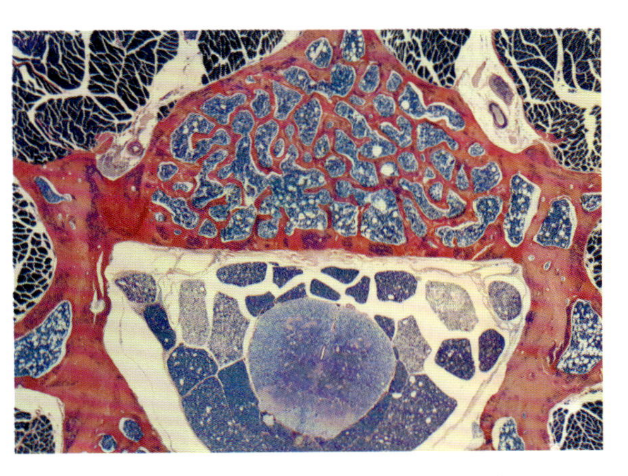

Sacral: in the region of the conus medullaris, the grey matter takes the form of two oval-shaped masses that occupy most of the cord with very little white matter.

Blood supply and membranes of the spinal cord

The spinal cord is supplied by a complex arrangement of arteries. This blood supply is vital for the normal functioning of the nervous system.

The anterior spinal arteries originate from the two vertebral arteries at the base of the brain and join together to form a single artery that runs down the front of the spinal cord in the anterior median fissure. Segmental branches of this artery supply the anterior two-thirds of the spinal cord.

The posterior spinal arteries also arise from the vertebral arteries and split into two descending branches that run either side of the cord, one behind and one in front of the attachment of the dorsal roots. These vessels supply the posterior third of the cord.

Radicular arteries

There is additional supply from radicular arteries that originate from the deep cervical arteries in the neck, the intercostal arteries in the chest and the lumbar arteries in the lower back. These vessels enter through the intervertebral foramina alongside the spinal nerves.

Usually, one of the anterior radicular arteries is larger than the others and is referred to as the artery of Adamkiewicz. It most commonly arises on the left-hand side from a branch of the descending aorta in the upper lumbar or lower thoracic region. This branch is the main blood supply to the lower two-thirds of the spinal cord.

Membranes that protect the spinal cord

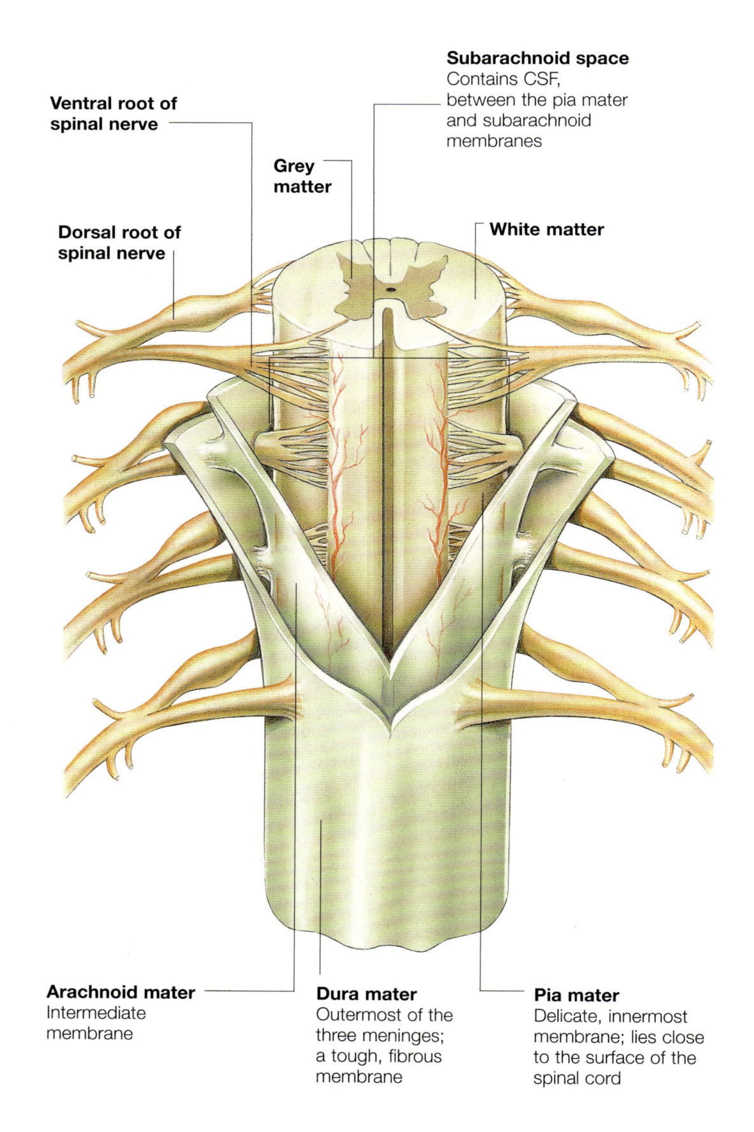

Like the brain, the spinal cord is surrounded and protected by three membranes. These are the meninges – dura mater, arachnoid mater and pia mater.

Spine protection

The bones of the vertebral column provide the main protection for the spinal cord, just as the skull does for the brain. However, like the brain, the cord has additional protection from three membranes, which continue down through the foramen magnum from inside the skull.

Dura mater

The dura mater is the tough, fibrous, outer membrane. The extradural or epidural space separates the dura from the bone of the vertebral bodies and contains fatty tissue and a plexus of veins.

Arachnoid mater

The middle membrane is the arachnoid mater, which is much thinner and more delicate, with an arrangement of connective tissue fibres resembling a spider's web. There is a potential subdural space between the dura and the arachnoid, normally containing only a very thin film of fluid.

Pia mater

The innermost membrane is the fine pia mater, which is closely applied to the surface of the spinal cord. It is transparent and richly supplied with fine blood vessels, which carry oxygen and nutrients to the cord.

Cerebrospinal fluid

Between the arachnoid and the pia is the subarachnoid space, which contains cerebrospinal fluid (CSF). This cushions the spinal cord, as well as helping to remove chemical waste products produced by nerve activity and metabolism. CSF is formed by the choroid plexuses inside the cerebral ventricles and circulates around the brain and spinal cord.

Denticulate ligaments

About 21 triangular extensions of the pia – the denticulate ligaments – pass outwards between the anterior and posterior nerve roots to join with the arachnoid and inner surface of the dura. The spinal cord is suspended by these in its dural sheath.

Spinal Nerves

There are 31 pairs of spinal nerves, arranged on each side of the spinal cord along its length. The pairs are grouped by region: eight cervical, 12 thoracic, five lumbar, five sacral and one coccygeal.

Each spinal nerve has two roots. The anterior, or ventral, root contains the axons of motor nerves that send impulses to control muscle movement. The posterior, or dorsal, root contains the axons of sensory nerves that send sensory information from the body into the spinal cord on its way to the brain.

Segments

Each root is formed by a series of small rootlets that attach it to the cord. The portion of the spinal cord that provides the rootlets for one dorsal root is referred to as a segment. In the lumbar and cervical regions, the rootlets are bunched closely, with the cord segments being about 1cm (0.3in) long. In the thoracic region they are more spread out, with segments more than 2cm (0.8in) long.

Nerve Formation

The ventral and dorsal roots join to form a single spinal nerve within the intervertebral foramina – small openings between the vertebrae through which the spinal nerves pass.Just before the point of fusion with the ventral root, there is an enlargement of each dorsal root. This enlargement is known as the dorsal root ganglion and is a collection of cell bodies of sensory nerves.

Rami

After passing through its intervertebral foramen, each spinal nerve divides into several branches, or rami, including:

- Ventral ramus: supplies the limbs and front and sides of the trunk
- Dorsal ramus: supplies the deep muscles and skin of the back
- Rami communicantes: part of the autonomic nervous system.

Cauda Equina

As the spinal cord is shorter than the vertebral column, the lower spinal nerve roots exit and travel down at an oblique angle. The lumbosacral nerve roots bunch together and pass down almost vertically. This led to the name cauda equina (Latin for horse's tail), which these lower nerve roots resemble.

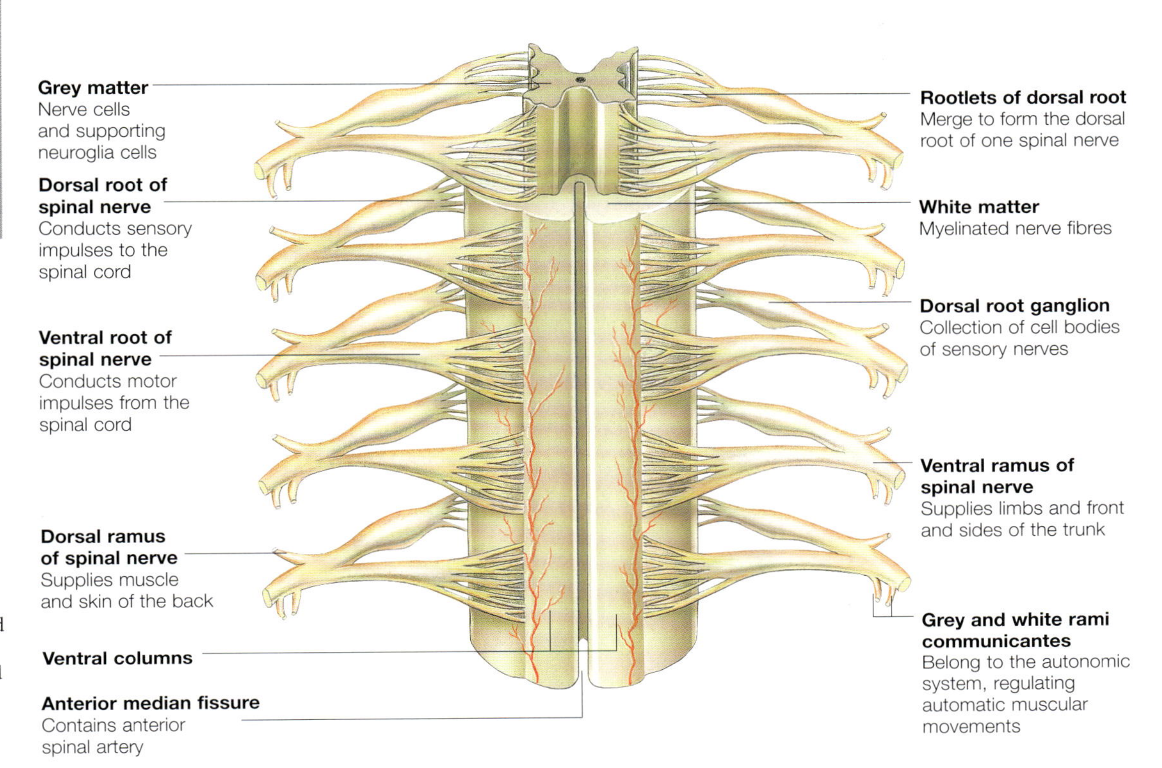

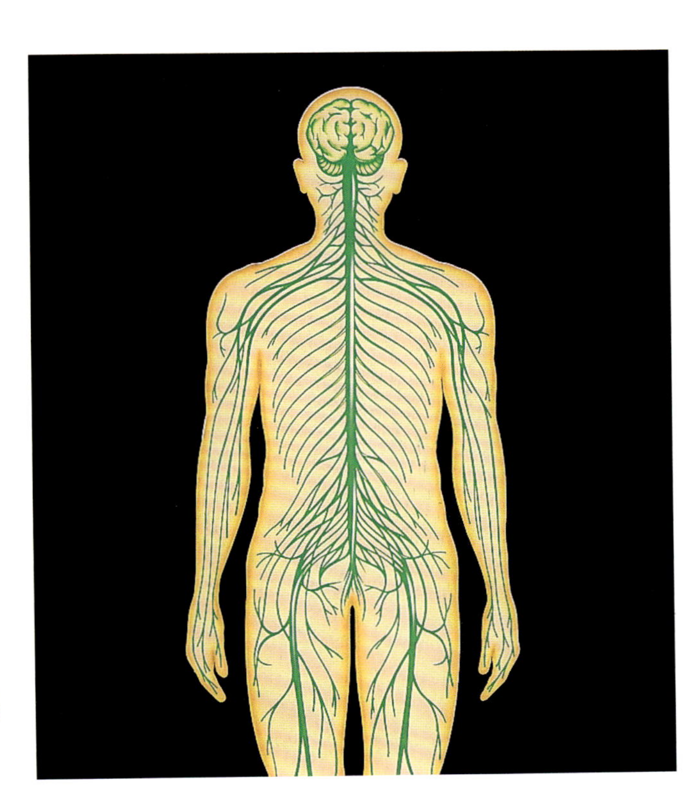

The 31 pairs of spinal nerves emerge from the spinal cord to supply data in the form of nerve impulses between the brain and the rest of the body.

Brachial Plexus

Lying within the root of the neck, and extending into the axilla, the brachial plexus is a complicated network of nerves from which arise the major nerves supplying the upper limbs.

At the level of each vertebra of the spine emerges a 'spinal nerve' that divides into dorsal and ventral parts, called 'rami'. The brachial plexus is formed by the joining and intermixing of the ventral rami at the level of the fifth to eighth cervical vertebrae and most of the ventral rami from the level of the first thoracic vertebra. Ventral rami are known as the 'roots' of the brachial plexus.

Structure

The roots of the brachial plexus join to form three 'trunks': superior, middle and inferior. As the complexity of the brachial plexus increases, each of the three trunks divides into an anterior (front) and a posterior (back) 'division'. In general, nerve fibres within the anterior divisions are those that go on to supply the anterior structures of the upper limb, while the fibres of the posterior divisions supply posterior upper limb structures.

From the six divisions, three 'cords' are formed, named for their positions in relation to the axillary artery to which they lie adjacent: lateral, medial and posterior.

Origins of the brachial plexus

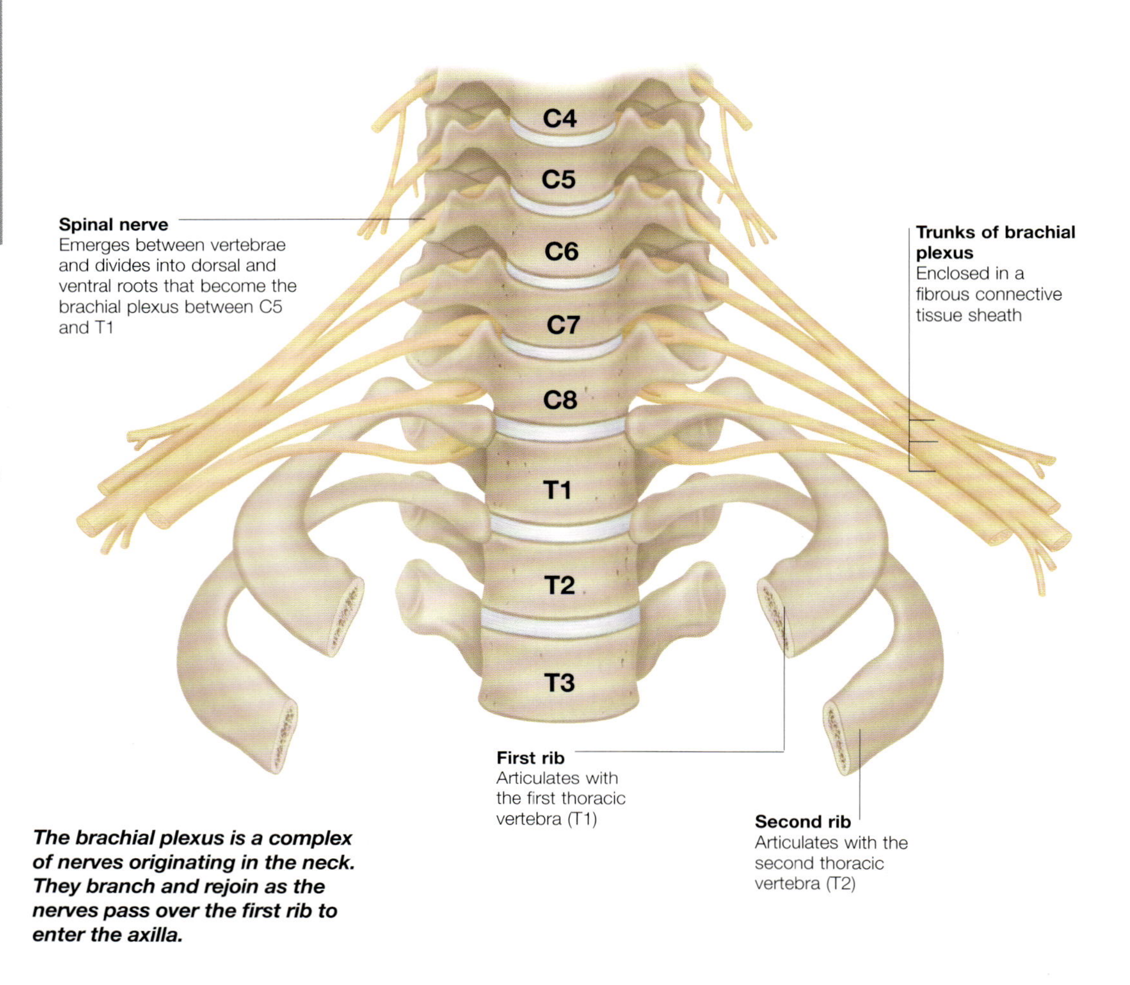

The brachial plexus is a complex of nerves originating in the neck. They branch and rejoin as the nerves pass over the first rib to enter the axilla.

Parts of the brachial plexus

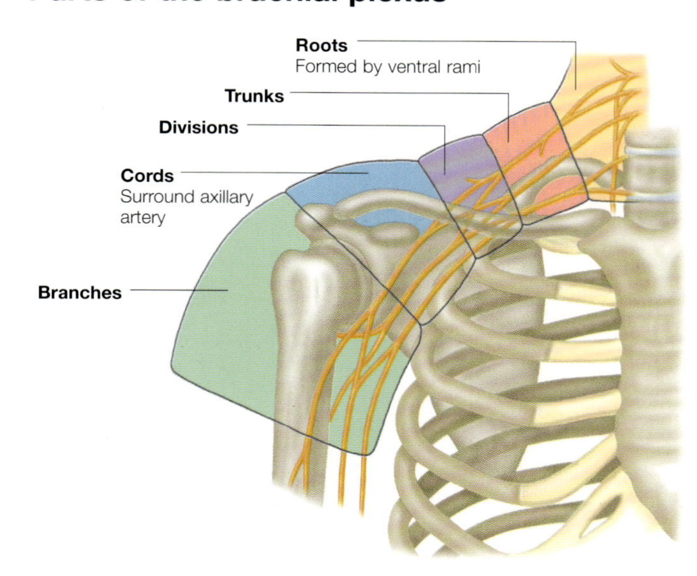

Anatomically, the brachial plexus is divided into sections which, starting from the spine, are known as roots, trunks, divisions, cords and branches.

Orientation

- Roots – the ventral rami of C5 to T1, lying within the neck to either side of the spinal column
- Three trunks – lie above the clavicle
- Divisions – arise from the trunks and pass behind the clavicle, entering the axilla
- Three cords – lie alongside the second part of the axillary artery within the axilla and inside the protective covering of the connective tissue of the axillary sheath
- The terminal branches of the brachial plexus leave the axilla as they pass into the upper limb.

The coloured blocks on this illustration show the anatomical levels of each different section of the brachial plexus.

Injuries to the brachial plexus

If the brachial plexus is injured, the effect upon function of the upper limb will vary, according to the level of the plexus at which the damage occurs. The nearer the injury is to the spine, the more generalized the resulting damage will be.

Sacral Plexus

The genitals, buttocks and lower limbs are supplied by nerve roots that emerge from the lumbar and sacral spine.

The sensory and motor nerve supply to and from the pelvis and legs is derived from a network of nerve roots called the sacral plexus. This lies on the rear wall of the pelvic cavity in front of the piriformis muscle. Contributions to the sacral plexus come from the lumbosacral trunk, representing the fourth and fifth lumbar nerve roots and the sacral nerve roots.

At the sacral plexus these nerve roots exchange nerve fibres and re-form into major nerves. These include the superior and inferior gluteal nerves, supplying the buttocks, and the sciatic nerve, which supplies the muscles of the leg. The parasympathetic splanchnic nerves (S1, S2, S3) regulate urination and defecation by controlling the internal sphincters, and also erection by dilating penile arterioles.

Sacral Foramina

The convex outer sacral surface has a ridge called the median crest in the midline, where the spinous processes fuse. The four posterior sacral foramina transmit the dorsal nerve roots. Nerves pass down the sacrum through the sacral canal.

A normal defect in the fusion of the fifth sacral vertebra posteriorly causes the canal to open out at the sacral hiatus.

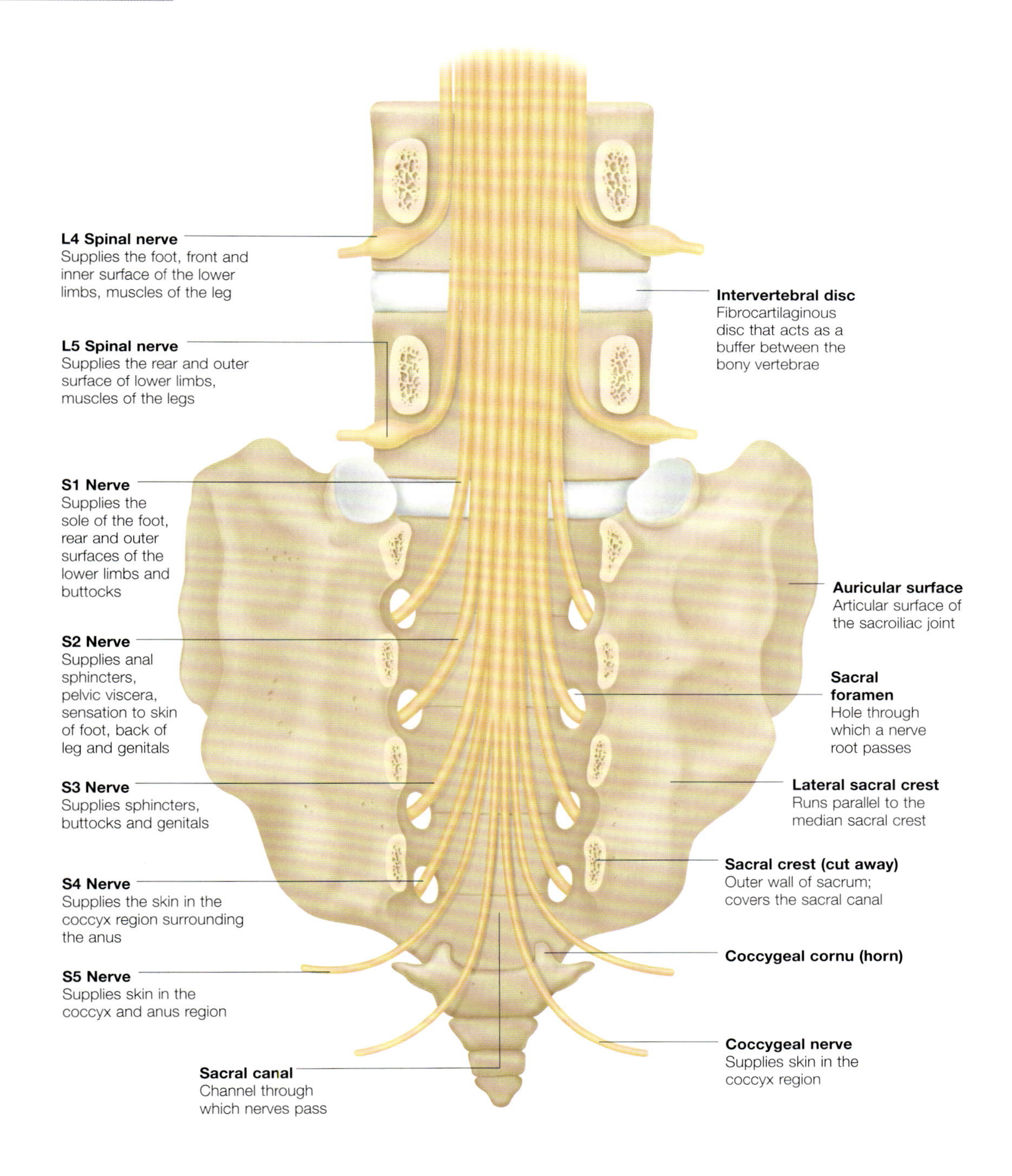

Peripheral Nervous System

The peripheral nervous system includes all the body's nerve tissue that is not in the brain and spinal cord. Its principal anatomical components are the cranial and spinal nerves.

The nervous system of the human body consists of the central nervous system (CNS), the brain and the spinal cord, and the peripheral nervous system (PNS).
The PNS consists of:
- **Sensory receptors** – specialized nerve endings that receive information about temperature, touch, pain, muscle stretching, and taste
- **Peripheral nerves** – bundles of nerve fibres that carry information to and from the CNS
- Motor nerve endings – specialized nerve endings that cause the muscle on which they lie to contract in response to a signal from the CNS.

There are two types of peripheral nerves:
- **Cranial nerves**

These emerge from the brain but are sometimes considered to be part of the PNS. They are concerned with receiving information from, and allowing control of, the head and neck. There are 12 pairs of cranial nerves.
- **Spinal nerves**

These arise from the spinal cord, each containing thousands of nerve fibres. Many of the 31 pairs of spinal nerves enter one of the complex networks, such as the brachial plexus that serves the upper limb, before becoming part of a large peripheral nerve.

The nerve supply to different organs and tissues of the body is discussed in each relevant section.

Major nerves of the peripheral nervous system

The peripheral nerves form the connection between the central nervous system – the brain and spinal cord – and the rest of the body.

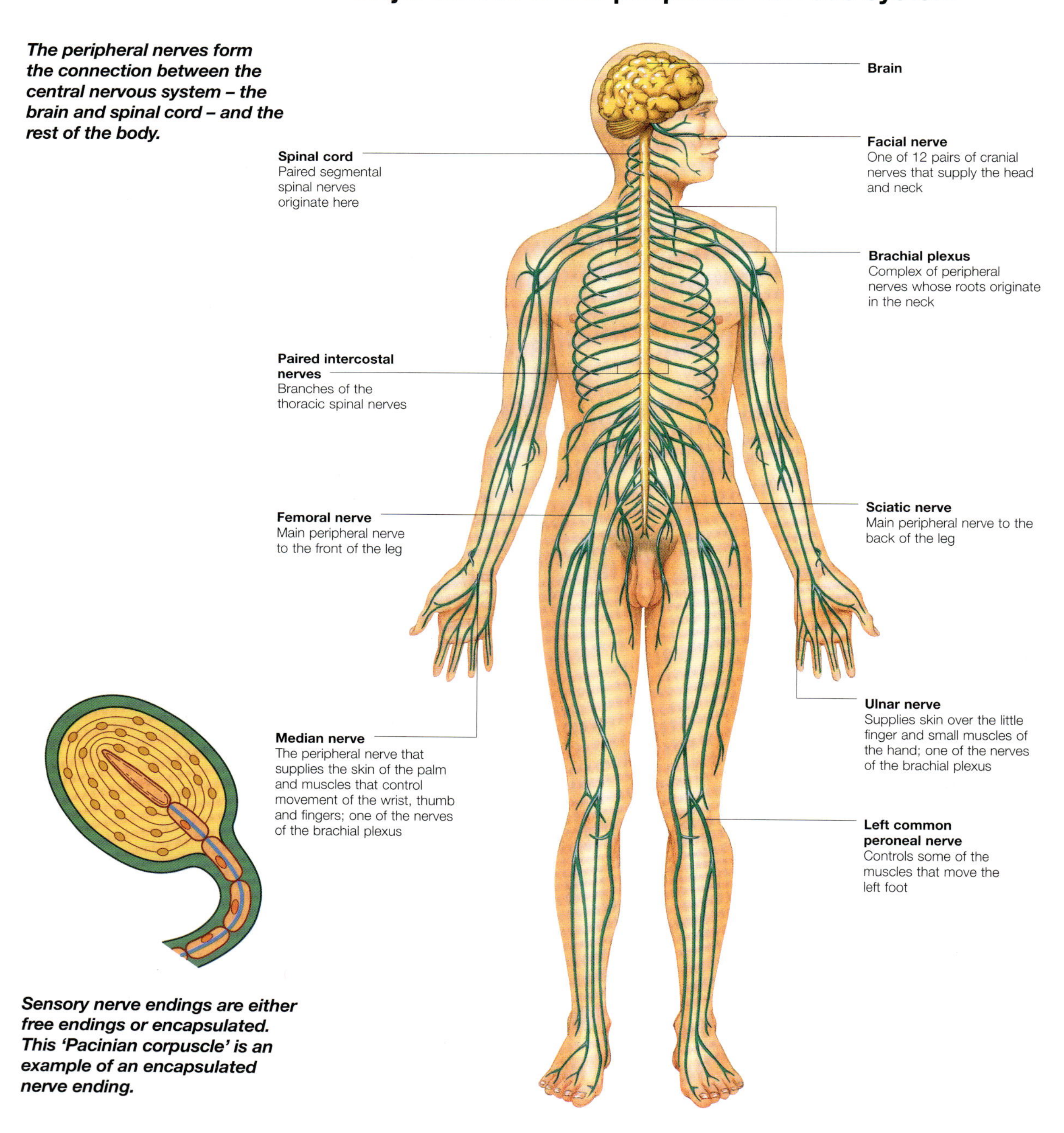

Sensory nerve endings are either free endings or encapsulated. This 'Pacinian corpuscle' is an example of an encapsulated nerve ending.

Structure of the peripheral nerve

Each peripheral nerve consists of a bundle of nerve fibres, a fascicle, some with an insulating layer of myelin, enclosed within connective tissue. The nerve resembles a strong white cord.

The greater part of a peripheral nerve is made up of three connective tissue coverings, without which the fragile nerve fibres would be vulnerable:

- **Endoneurium** – a layer of delicate connective tissue that surrounds the smallest unit of the peripheral nerve, the axon. This layer may also enclose an axon's myelin sheath.
- **Perineurium** – a layer of connective tissue that encloses a group of protected nerve fibres, called fascicles, which are grouped together in bundles.
- **Epineurium** – a tough connective tissue coat that binds the nerve fascicles into a peripheral nerve. The epineurium also encloses blood vessels that help to nourish the nerve fibres and their connective tissue coverings.

Motor nerve endings

Motor nerve endings lie on muscle fibres and secretory cells. They receive signals from the central nervous system via peripheral nerves and enable the muscles to contract or the cells to secrete their products. The point at which the motor nerve ending of a peripheral nerve connects with voluntary muscle is known as the neuromuscular junction.

Nerve function

Most peripheral nerves carry information to and from the central nervous system (sensory and motor functions respectively), and are known as 'mixed' nerves. Purely sensory or purely motor nerves are rare.

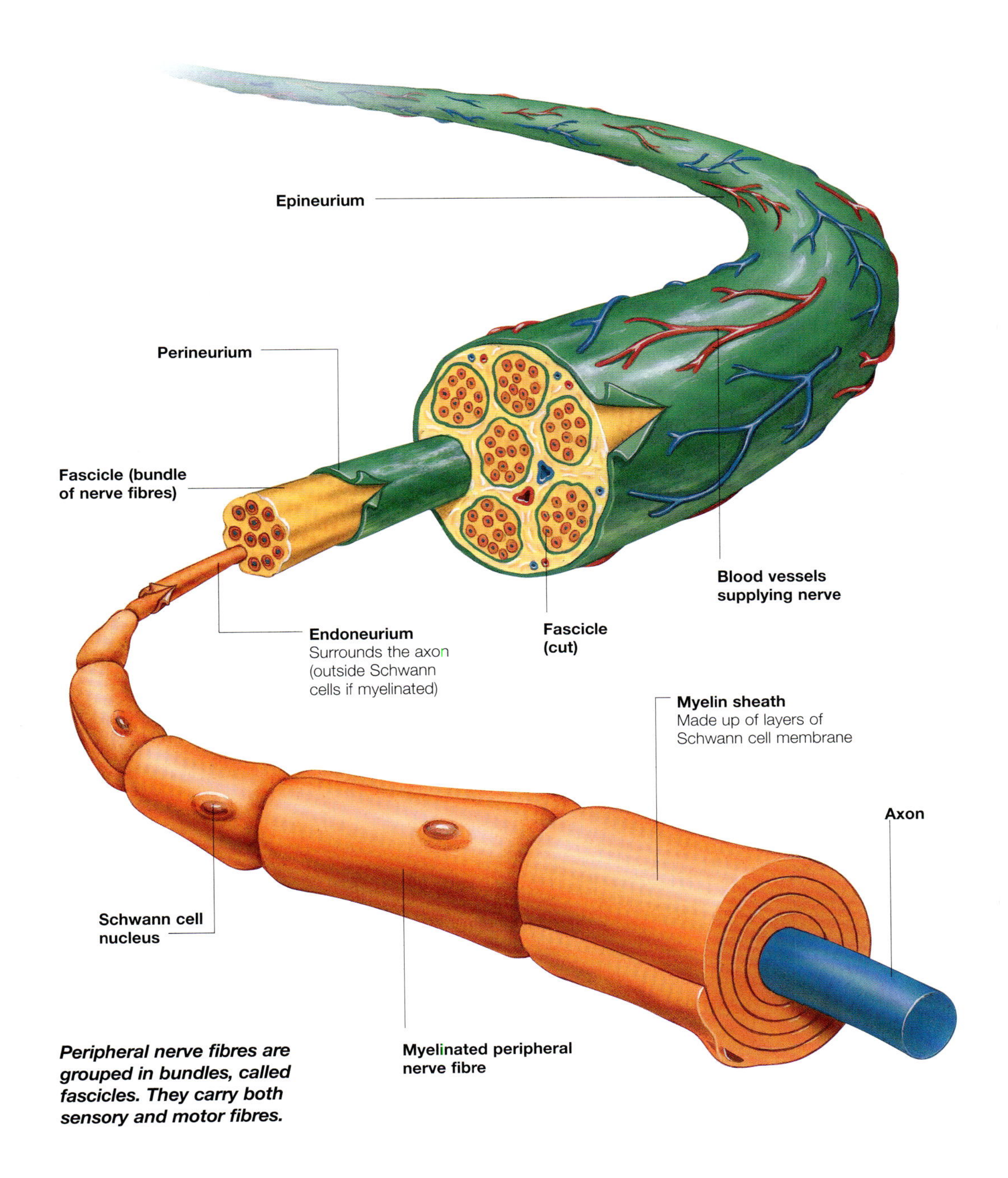

Peripheral nerve fibres are grouped in bundles, called fascicles. They carry both sensory and motor fibres.

Nerves of the Arm

The nerves of the arm supply the skin and muscles of the forearm and hand. There are four main nerves in the arm: the radial, musculocutaneous, median and ulnar nerves.

The nerve supply to the upper limb is provided by four main nerves and their branches. These receive sensory information from the hand and arm, and also innervate the numerous muscles of the upper limb. The radial and musculocutaneous nerves supply muscles and skin of all parts of the arm, while the median and ulnar nerves only supply structures below the elbow.

Radial Nerve

The radial nerve is of great importance as it is the main supplier of innervation to the extensor muscles that straighten the bent elbow, wrist and fingers. It arises as the largest branch of the 'brachial plexus', a network of nerves from the spinal cord in the neck.

Near the lateral epicondyle, the radial nerve divides into its two terminal branches:

- The superficial terminal branch – sensory nerve supply to the skin over the back of the hand, thumb and adjoining two and a half fingers
- The deep terminal branch – motor nerve supply to all of the extensor muscles of the forearm.

Musculocutaneous Nerve

The musculocutaneous nerve supplies both muscles and skin in the front of the upper arm. Below the elbow it becomes the lateral cutaneous nerve of the forearm, a sensory nerve that supplies a large area of forearm skin.

Rear view – nerves of the arm

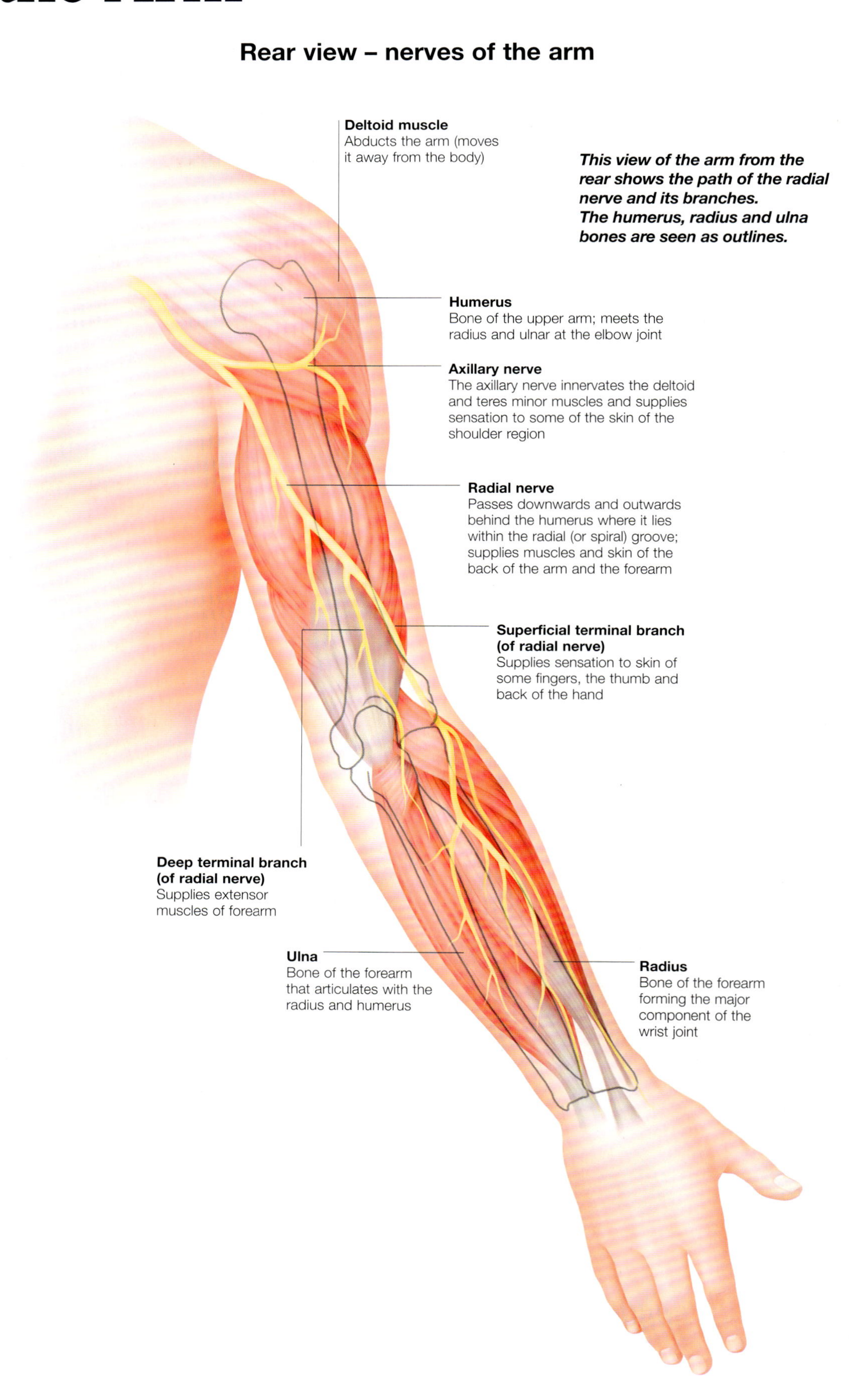

This view of the arm from the rear shows the path of the radial nerve and its branches. The humerus, radius and ulna bones are seen as outlines.

Median and ulnar nerves

The median nerve supplies the forearm muscles enabling the actions of flexion and pronation. The ulnar nerve passes behind the elbow to supply some of the small muscles of the hand.

The median nerve of the upper limb arises from the brachial plexus and runs downwards centrally to the elbow. It is the main nerve of the front of the forearm, and contains the muscles of flexion and pronation.

At the wrist, the median nerve passes through the carpal tunnel. The median nerve ends in branches that supply some of the small muscles of the hand, as well as the skin over the thumb and some neighbouring fingers.

Ulnar nerve

The ulnar nerve passes down along the humerus to the elbow where it loops behind the medial epicondyle, beneath the skin where it can easily be felt. It gives off branches to supply the elbow, two of the muscles of the forearm and several areas of overlying skin before entering the hand. In the hand, the ulnar nerve divides into deep and superficial branches.

Median nerve damage

The median nerve can be damaged by fractures of the lower end of the humerus or compressed by swollen muscle tendons within the carpal tunnel (carpal tunnel syndrome). Median nerve injury can make it difficult to use the 'pincer grip' of the thumb and fingers, as the nerve supplies the small muscles of the thenar eminence (the fleshy prominence below the base of the thumb).

The ulnar nerve is most vulnerable to injury as it passes behind the medial epicondyle of the humerus. The feeling from knocking the 'funny bone' occurs when the nerve is compressed against the underlying bone. Severe damage can lead to sensory loss, paralysis and wasting of the muscles it supplies.

Front view – nerves of the arm

The paths of the ulnar, median and musculocutaneous nerves can be seen in this illustration of the arm from the front.

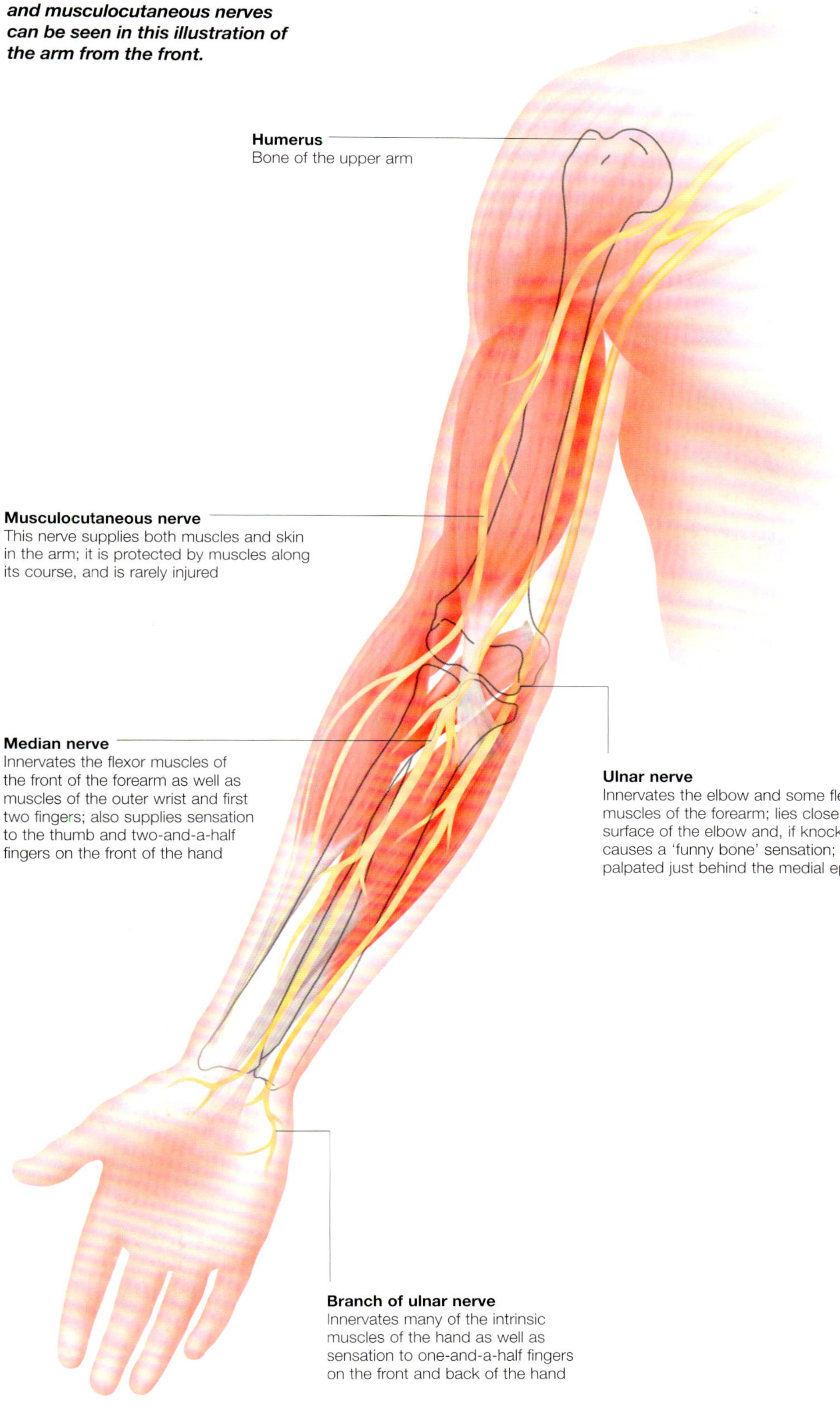

Humerus
Bone of the upper arm

Musculocutaneous nerve
This nerve supplies both muscles and skin in the arm; it is protected by muscles along its course, and is rarely injured

Median nerve
Innervates the flexor muscles of the front of the forearm as well as muscles of the outer wrist and first two fingers; also supplies sensation to the thumb and two-and-a-half fingers on the front of the hand

Ulnar nerve
Innervates the elbow and some flexor muscles of the forearm; lies close to the surface of the elbow and, if knocked, causes a 'funny bone' sensation; it can be palpated just behind the medial epicondyle

Branch of ulnar nerve
Innervates many of the intrinsic muscles of the hand as well as sensation to one-and-a-half fingers on the front and back of the hand

Nerves of the Hand

The structures of the hand receive their nerve supply from terminal branches of the three main nerves of the upper limb: the median, ulnar and radial nerves.

The median nerve enters the hand on the palmar side by passing under the flexor retinaculum within the carpal tunnel. Within the hand the median nerve supplies:

- The three muscles of the thenar eminence – abductor pollicis brevis, flexor pollicis brevis and opponens pollicis
- The first and second lumbrical muscles
- The skin of the palm and palmar surface of the first three-and-a-half digits as well as the dorsal surface (back) of the tips of those fingers.

The branch of the median nerve that supplies the skin of the central palm arises before the median nerve enters the carpal tunnel and passes over the flexor retinaculum so the skin will continue to receive its nerve supply if the median nerve is damaged there.

Ulnar Nerve

The ulnar nerve enters the medial side of the hand by passing over the flexor reticulanum. This nerve supplies:

- The skin on the medial side of the palm, via its palmar cutaneous branch
- The skin of the medial half of the back of the hand, little finger and medial half of the ring finger
- The skin of the palmar side of the little finger and half the ring finger
- The muscles of the hypothenar eminence
- The adductor pollis muscle
- The third and fourth lumbrical muscles and all the interosseus muscles.

Radial Nerve

This nerve runs down the back of the forearm to the dorsal surface of the hand and supplies the skin of the back of the lateral three and a half digits.

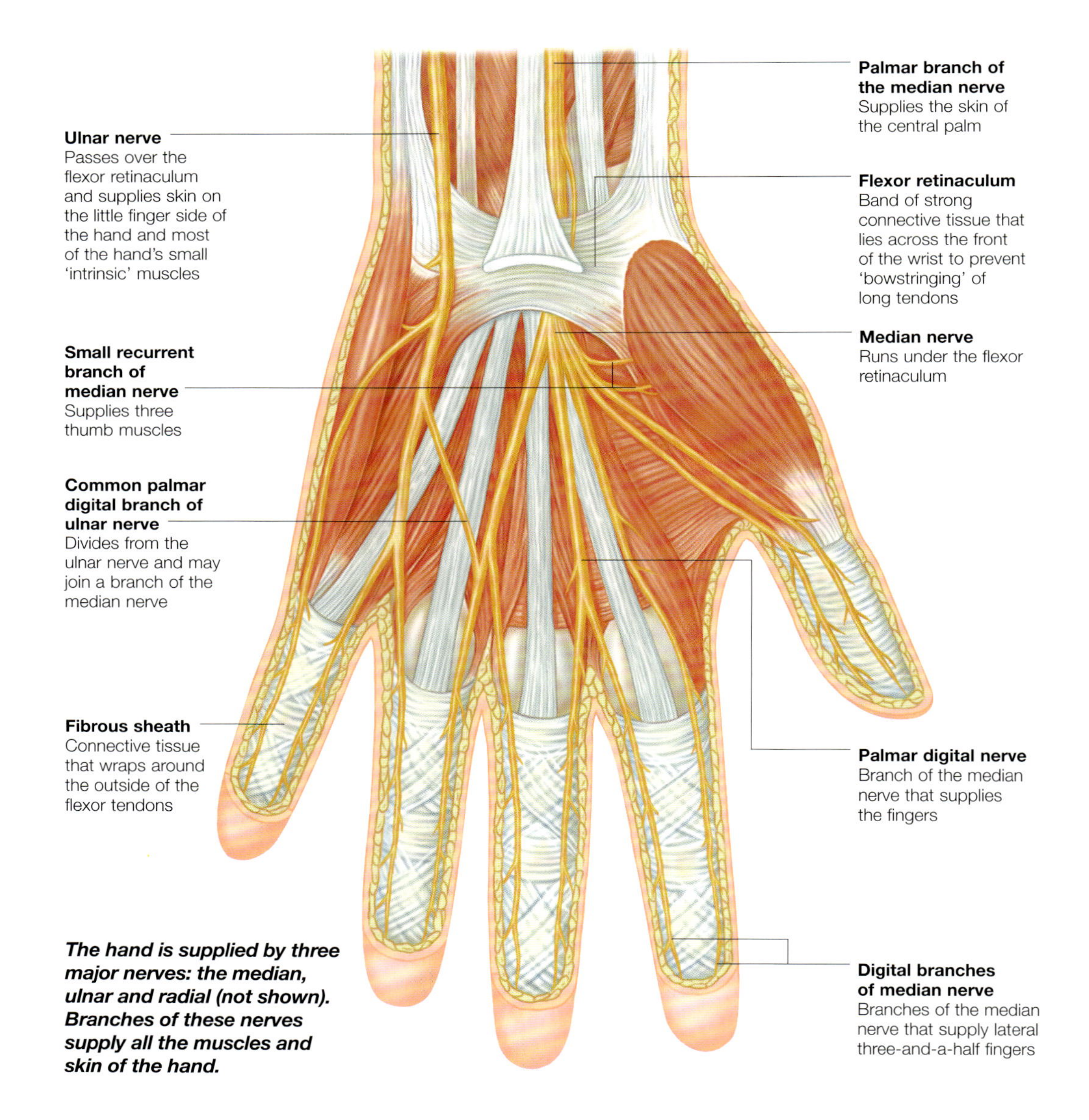

The hand is supplied by three major nerves: the median, ulnar and radial (not shown). Branches of these nerves supply all the muscles and skin of the hand.

Sciatic Nerve

The sciatic nerve is the main nerve of the leg and the largest nerve in the body. Its branches supply the muscles of the hip, many of the thigh and all of the muscles of the lower leg and foot.

The sciatic nerve is made up of two nerves, the tibial nerve and the common peroneal (or fibular) nerve. These are bound together by connective tissue to form a wide band that runs the full length of the back of the thigh.

Origin and Course

The sciatic nerve arises from a network of nerves at the base of the spine, called the sacral plexus. From here, it passes out through the greater sciatic foramen and then curves downwards through the gluteal region under the gluteus maximus muscle (midway between the bony landmarks of the greater trochanter of the femur and the ischial tuberosity of the pelvis).

The sciatic nerve leaves the gluteal region by passing under the long head of the biceps femoris muscle to enter the thigh and runs down the centre of the back of the thigh, branching off into the hamstring muscles (a collective name for the biceps femoris, semitendinosus and semimembranosus muscles). It then divides to form two branches, the tibial nerve and the common peroneal nerve just above the knee.

Higher Division

Rarely, the sciatic nerve divides into two at a much higher level. In this instance the common peroneal nerve may pass above or even through the piriformis muscle in the gluteal region of the lower limbs.

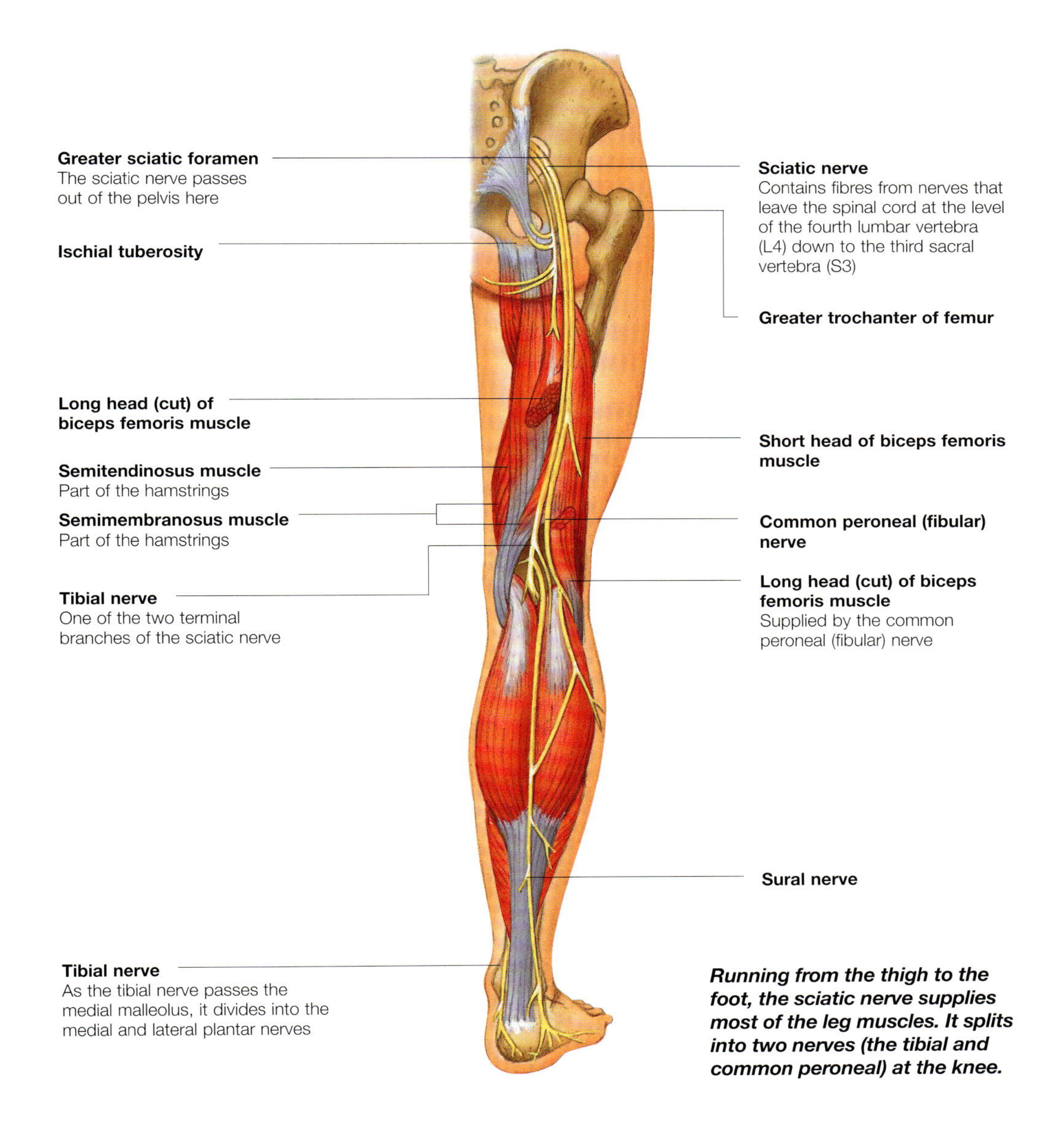

Running from the thigh to the foot, the sciatic nerve supplies most of the leg muscles. It splits into two nerves (the tibial and common peroneal) at the knee.

Terminal branches of the sciatic nerve

The sciatic nerve divides into two branches: the common peroneal (fibular) nerve and the tibial nerve. The common peroneal nerve supplies the front of the leg, while the tibial nerve supplies the back.

The common peroneal nerve leaves the sciatic nerve in the lower third of the thigh and runs down around the outer side of the lower leg before dividing into two just below the knee.

Nerve branches

The two branches of the peroneal nerve include:

- The superficial branch of the peroneal nerve – this supplies the lateral (outer) compartment of the lower leg in which it lies. It then subdivides into smaller branches to supply the muscles around it.
- The deep peroneal nerve – this runs in front of the interosseous membrane between the tibia and the fibula, and then passes over the ankle into the foot. These two terminal branches also supply the knee joint and the skin over the outer side of the calf and the top of the foot.

Injury

As the common peroneal nerve passes around the outer side of the lower leg, it lies just under the skin and very close to the head of the fibula. It is very vulnerable to damage, especially if the fibula suffers a fracture. It is the most commonly damaged nerve in the leg.

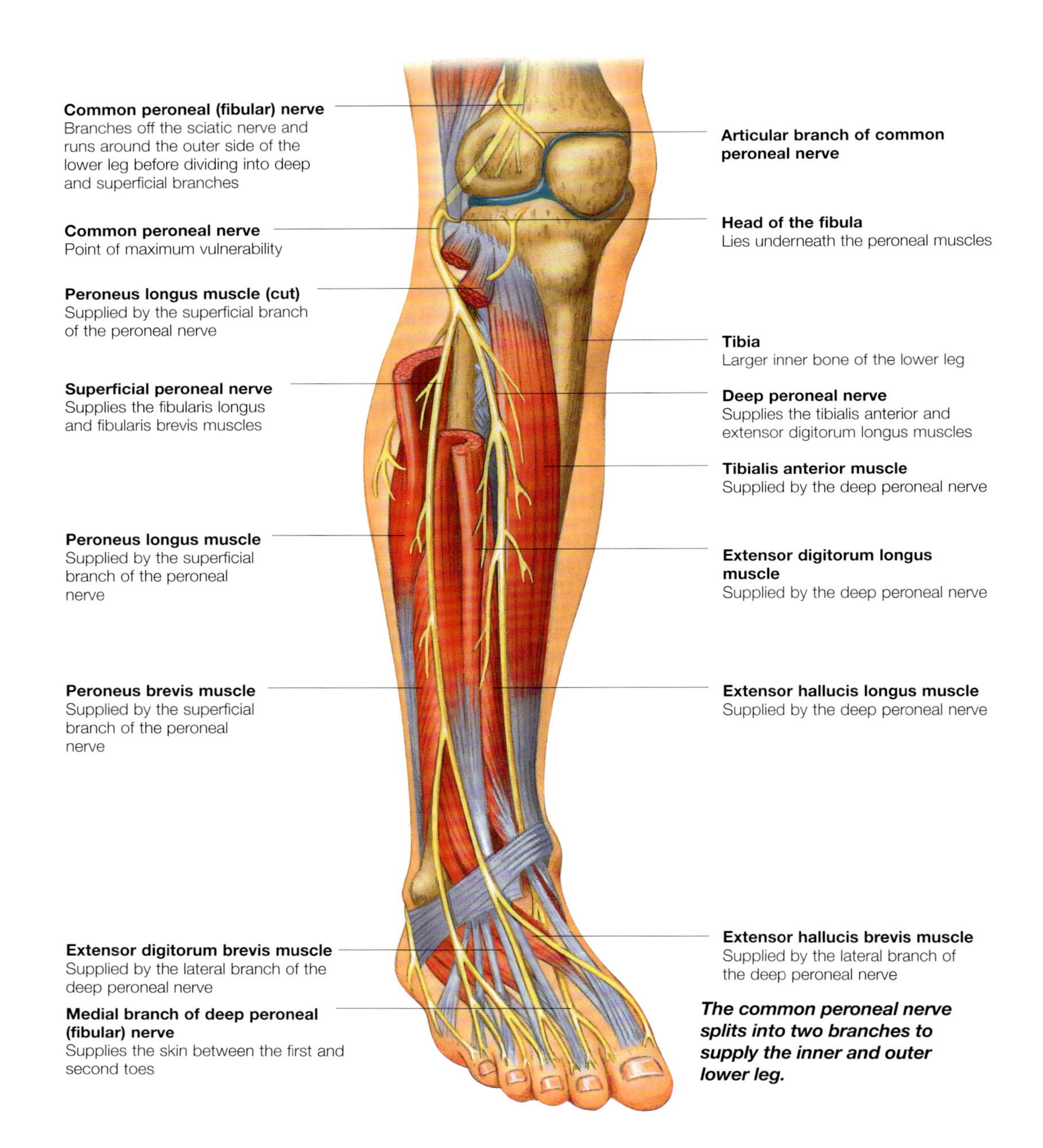

The common peroneal nerve splits into two branches to supply the inner and outer lower leg.

Cutaneous Nerves of the Leg

Cutaneous nerves provide the skin with sensation. There are a number of these within the leg, many of which branch off from the main nerves of the leg: the sciatic, femoral and tibial nerves.

Cutaneous nerves supply the skin. Those of the leg lie within the subcutaneous tissue and often branch off larger nerves that serve muscles and joints.

Nerves of the Thigh

The cutaneous nerves include:

- Ilioinguinal nerve – supplies an area of the front/inner thigh
- Genitofemoral nerve – supplies a small area of skin under the centre of the inguinal ligament
- Lateral cutaneous nerve – supplies the outer thigh
- Obturator nerve – branches of this nerve supply the inner thigh
- Medial/intermediate cutaneous nerves – supply the front of the thigh not supplied by the ilioinguinal nerve
- Posterior cutaneous nerve – supplies the back of the thigh and the popliteal fossa.

Lower Leg and Foot

The following nerves branch off the sciatic and femoral nerves:

- Saphenous nerve – supplies the front and inner lower leg
- Lateral cutaneous nerve – supplies the upper front and outer side of the lower leg
- Superficial peroneal (fibular) nerve – supplies the lower outer calf and the top of the foot
- Sural nerve – a branch of the tibial nerve, supplies the lower outer part of the back of the lower leg and the lateral border of the foot and little toe
- Medial and lateral plantar nerves – supply the sole of the foot
- Tibial nerve – supplies the heel of the foot.

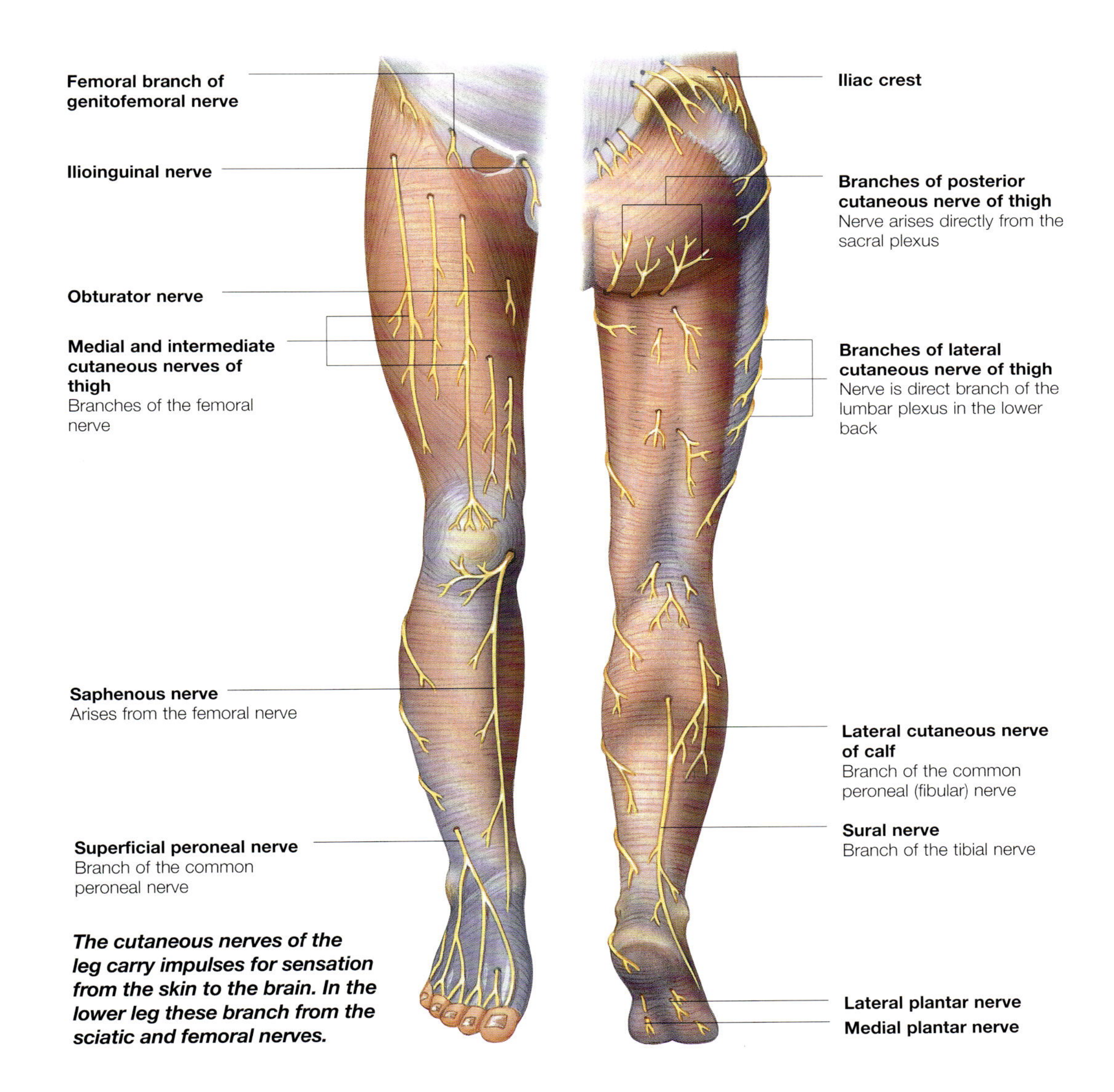

The cutaneous nerves of the leg carry impulses for sensation from the skin to the brain. In the lower leg these branch from the sciatic and femoral nerves.

Femoral Nerve

The femoral nerve branches off from lumbar spinal nerves. It runs through the pelvis and down the front of the thigh, supplying the powerful quadriceps muscles and the skin of the front and inner aspects of the leg.

The femoral nerve is a large nerve arising from lumbar spinal nerves L2, L3 and L4 in the lumbar plexus, a network formed by the spinal nerves that emerge from the lumbar vertebrae. The nerve enters the front of the thigh by passing under the inguinal ligament lateral to the femoral artery and vein. Unlike the blood vessels, it does not lie within the protective femoral sheath.

Branches

The femoral nerve divides into its terminal branches around 3–4cm (1.1–1.5in) below the inguinal ligament. These smaller nerves provide a supply to a number of structures within the lower limb:

- The anterior thigh muscles – the femoral nerve has a very important role in the provision of nerve stimuli to the powerful muscles of the front of the thigh. The four muscles that make up the large quadriceps femoris group are all supplied by the femoral nerve, as are the pectineus and sartorius muscles.
- The hip and knee joints – the femoral nerve sends articular branches to these two large joints, between which it lies
- The skin over the front of the thigh – the cutaneous branches of the femoral nerve provide this area with sensation
- The skin below the knee – this is supplied by another large cutaneous branch, the saphenous nerve. This passes down the leg with the femoral artery from the knee to the toes.

When the spinal roots of the lumbar region are compressed, such as in the case of a herniated (or slipped) disc, the structures supplied by the femoral nerve may be affected.

As the nerve supplies muscles that move both the hip and the knee, this can have a serious effect upon walking. Numbness of the skin over the front of the thigh may also develop.

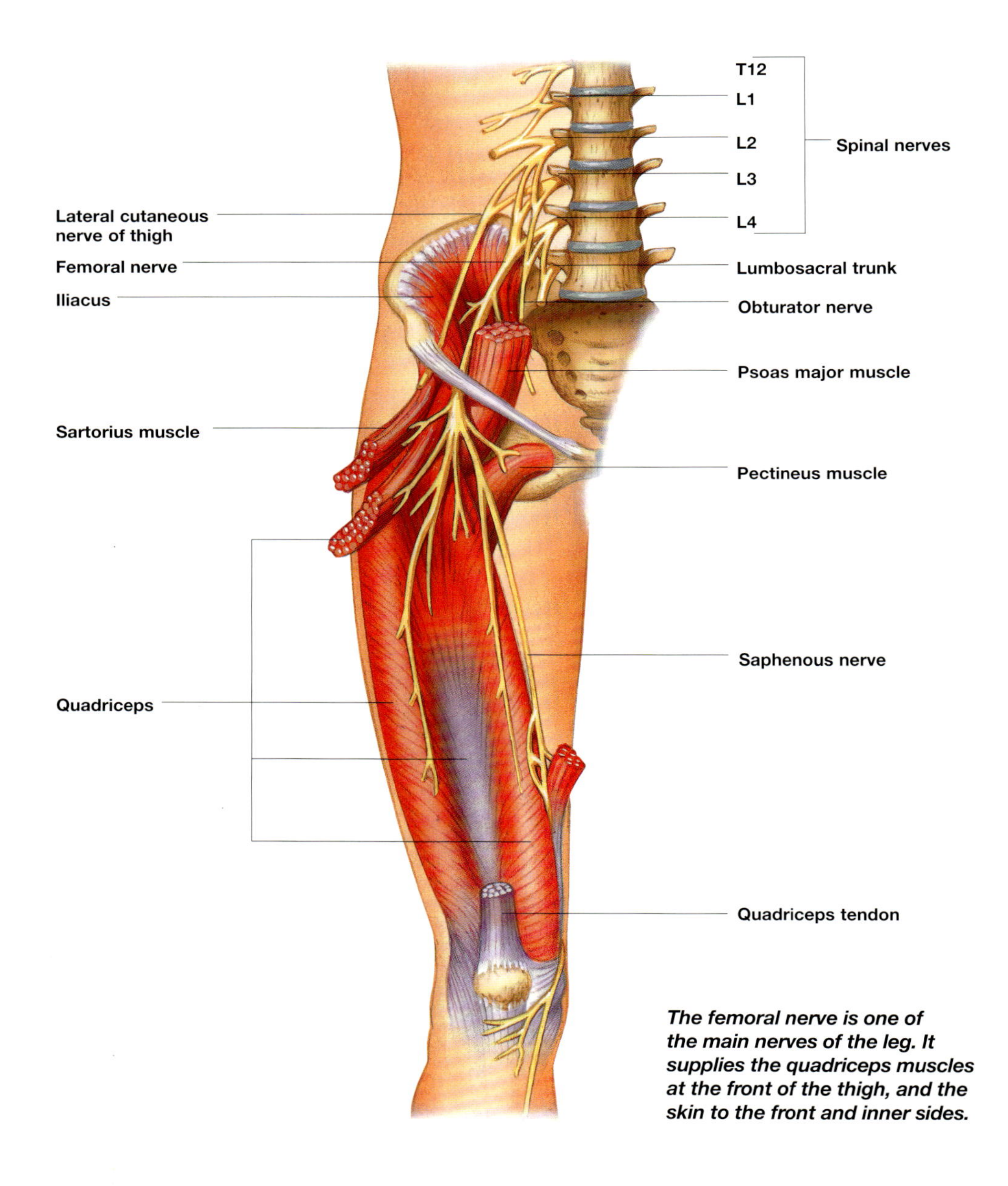

The femoral nerve is one of the main nerves of the leg. It supplies the quadriceps muscles at the front of the thigh, and the skin to the front and inner sides.

Obturator and Tibial Nerves

The obturator nerve originates from the lumbar plexus, innervating the adductor muscles of the thigh. The tibial nerve, a branch of the sciatic nerve, provides motor and sensory function to the lower leg, foot, and plantar muscles.

Like the femoral nerve, the obturator nerve arises from the lumbar plexus. It is formed within the psoas major muscle and passes down through the obturator foramen of the pelvis alongside the obturator artery and vein.

From here, the obturator nerve enters the inner adductor compartment of the thigh. This contains the muscles that adduct the legs (pull them in towards the centre of the body). It lies within the inner aspect of the thigh.

From within this compartment, the obturator nerve supplies:

- All the adductor muscles except the lower part of the adductor magnus
- The skin on the inner aspect of the lower thigh, via a cutaneous branch
- The hip and the knee joints. The small branch that supplies the knee joint descends through the adductor hiatus, a gap in the adductor magnus muscle, to reach the joint.

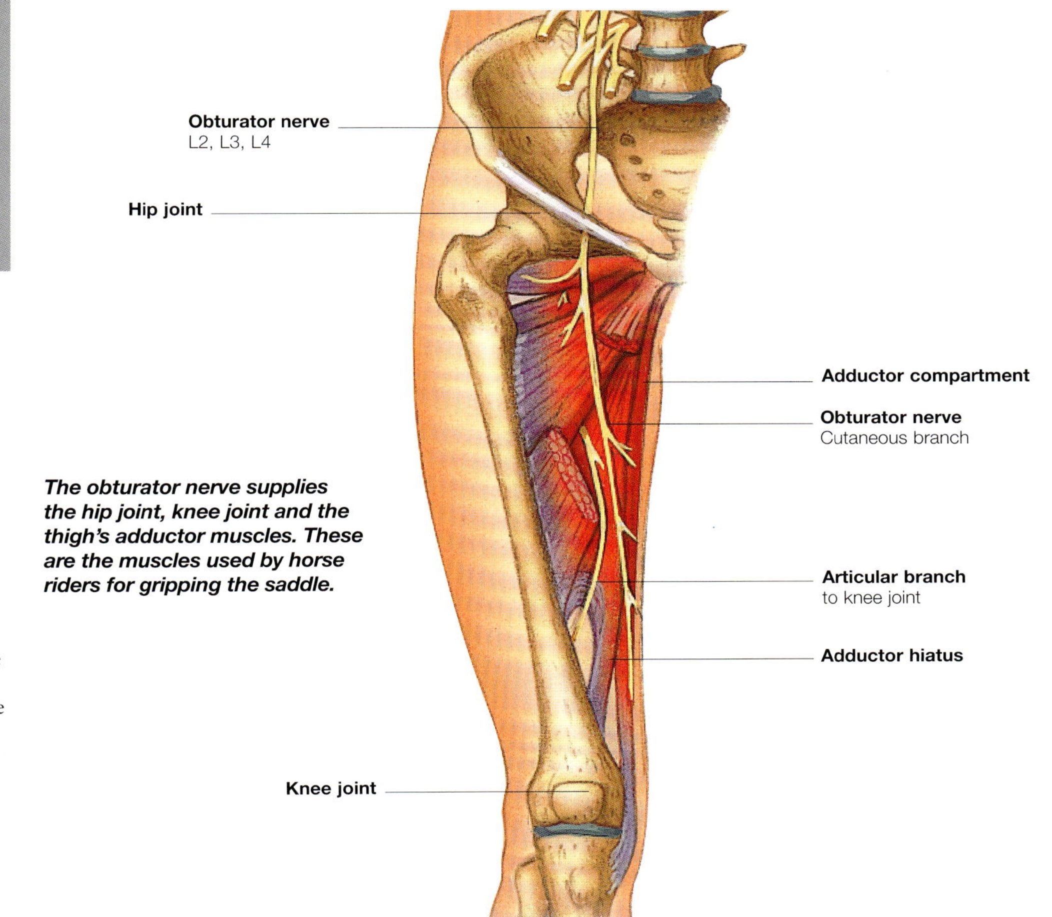

The obturator nerve supplies the hip joint, knee joint and the thigh's adductor muscles. These are the muscles used by horse riders for gripping the saddle.

Tibial nerve

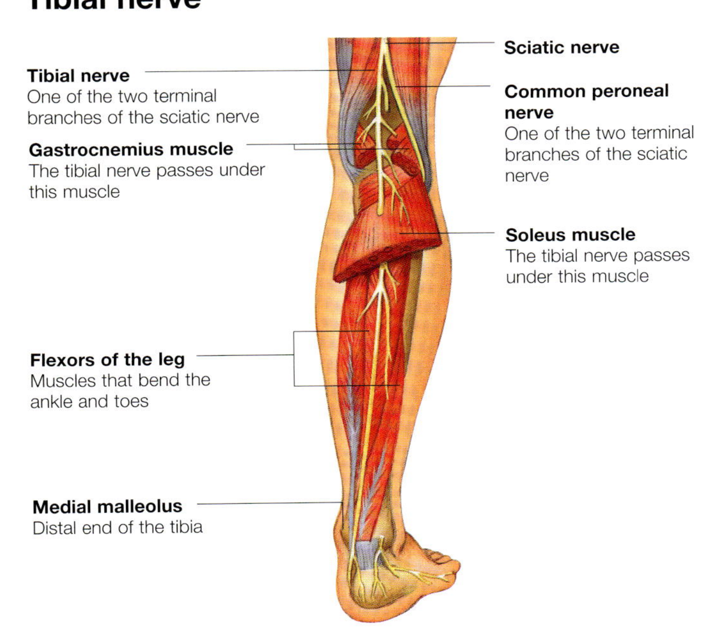

The tibial nerve is the larger of the two terminal branches of the sciatic nerve. It supplies the flexors of the leg: those muscles that bend, rather than straighten, the joints.

Path down the leg

The tibial nerve arises in the lower third of the thigh, where it supplies the hamstring muscles. It then separates from the common peroneal nerve before following a course down the back of the leg:

- It passes through the popliteal fossa (a space behind the knee) alongside the popliteal artery
- It then descends under the large gastrocnemius and soleus muscles
- It reaches the posterior compartment of the lower leg where it gives off branches to the flexor muscles found there
- At the ankle it passes behind the medial malleolus, before dividing into the medial and lateral plantar nerves of the foot.

Branches

The tibial nerve has two cutaneous branches that supply areas of skin: the sural nerve (in the calf) and the medial calcaneal nerve (in the heel).

The tibial nerve splits off from the sciatic nerve to course down the back of the lower leg. Branches supply the muscles and skin with sensation.

Autonomic Nervous System

The autonomic nervous system provides the nerve supply to those parts of the body that are not consciously directed. It can be subdivided into the sympathetic and parasympathetic nervous systems.

The autonomic nervous system has two parts: the sympathetic system and the parasympathetic system. Both generally supply the same organs, but with opposing effects. In each system two neurones make up the pathway from the central nervous system (CNS) to the organ that is being supplied.

Sympathetic Nervous System
The effects of stimulation by the sympathetic system are often referred to as the 'fight or flight' response. In stressful situations, the sympathetic nervous system becomes more active, causing the heart rate to increase and the skin to become pale and sweaty as blood is diverted to muscle.

Structure
The cell bodies of the neurones of the sympathetic system lie within a section of the spinal cord. Fibres from these cell bodies exit the spinal column at the ventral root and pass through the white rami communicantes to reach the paravertebral sympathetic chain. Some fibres that enter the sympathetic chain connect there with the second cell of their pathway. Fibres then exit through the grey rami communicantes to join the ventral spinal nerve.

Adrenal Medulla
The sympathetic nervous system also stimulates the adrenal gland to release the hormones adrenalin and noradrenalin, which amplify the response to stress.

Anatomy of a sympathetic trunk

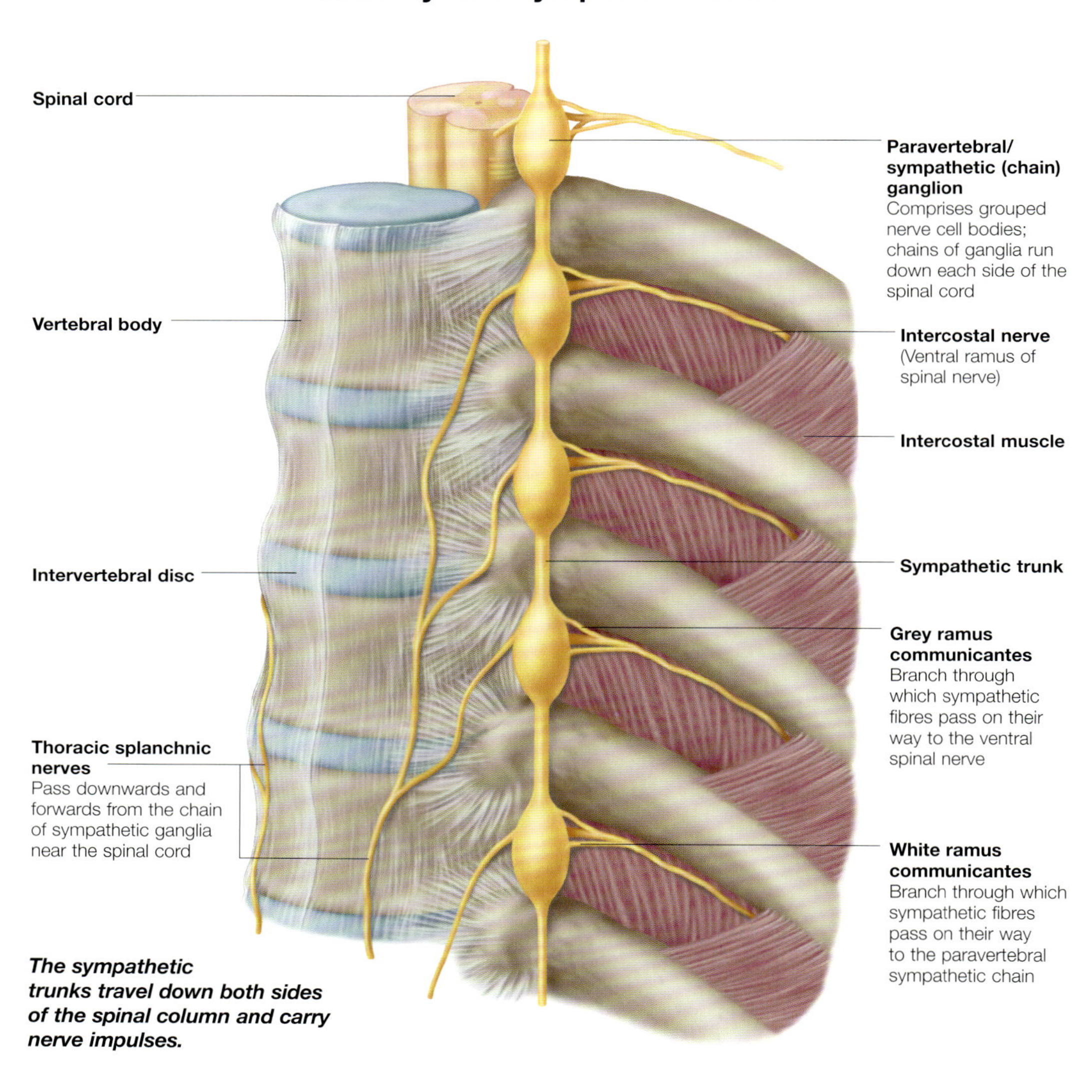

The sympathetic trunks travel down both sides of the spinal column and carry nerve impulses.

Parasympathetic nervous system

The parasympathetic system has opposing effects to the sympathetic system and is most active during periods of rest. It helps the body to conserve energy and to digest food, for example.

The structure of the parasympathetic nervous system is simpler than that of the sympathetic nervous system.

Location of cell bodies

The cell bodies of the first of the two neurones in the pathway are located in only two places:

- The brainstem – fibres from the parasympathetic cell bodies in the grey matter of the brainstem leave the skull as part of a number of cranial nerves. Together, these fibres make up what is known as the cranial parasympathetic outflow
- The sacral region of the spinal cord – the sacral outflow arises from parasympathetic cell bodies that lie within part of the spinal cord. Fibres leave through the ventral root.

Because of the locations of the origins of parasympathetic fibres, the parasympathetic system is sometimes known as the craniosacral division of the autonomic nervous system; the sympathetic system is known as the thoracolumbar division.

Distribution

The cranial outflow provides parasympathetic innervation for the head, and the sacral outflow supplies the pelvis. The area between (the majority of the abdominal and thoracic internal organs) is supplied by part of the cranial outflow that is carried within the vagus (tenth cranial nerve).

Organs controlled by the parasympathetic nervous system

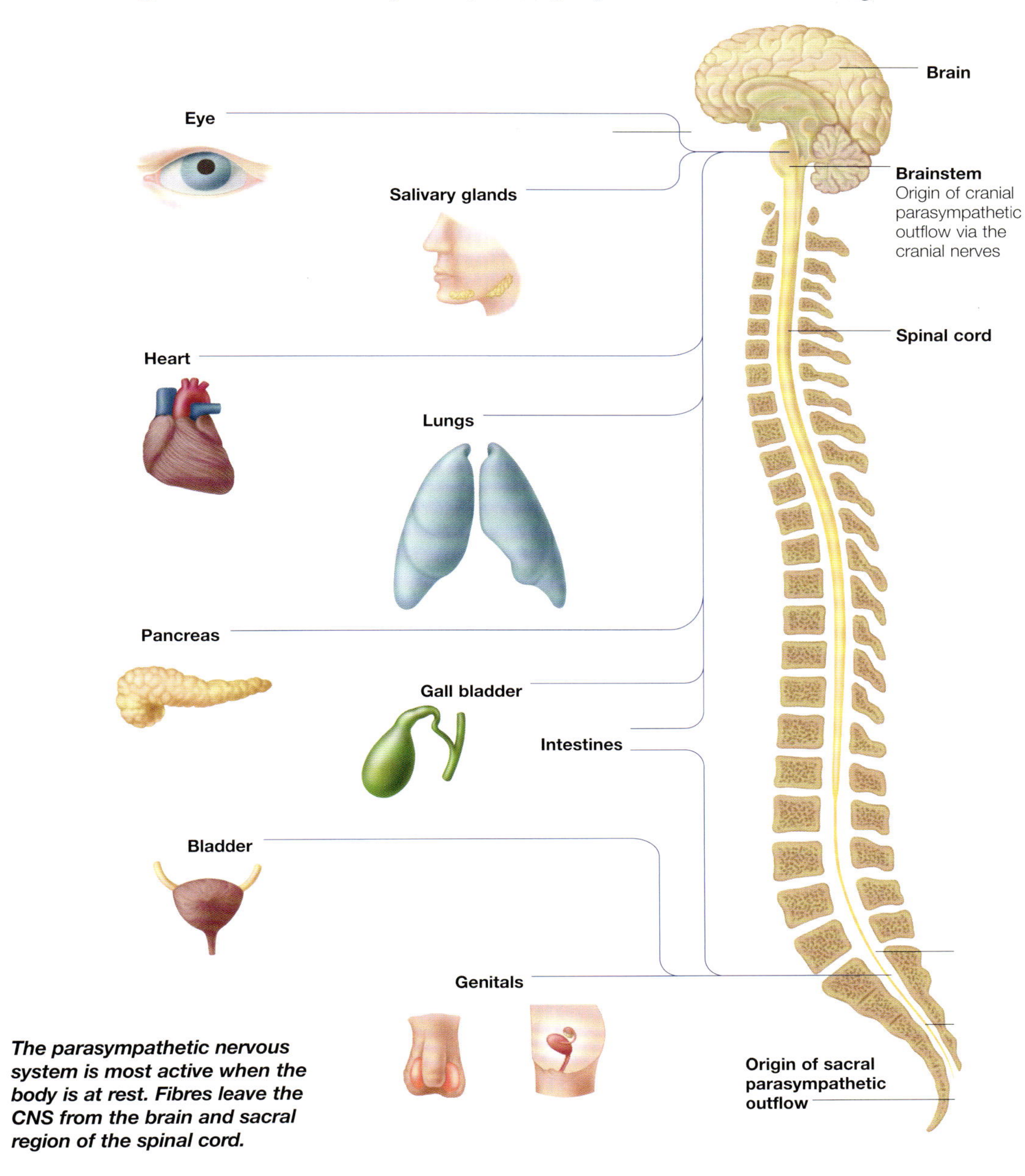

The parasympathetic nervous system is most active when the body is at rest. Fibres leave the CNS from the brain and sacral region of the spinal cord.

Opposing effects

The sympathetic nervous system prepares the body in times of stress or danger, while the parasympathetic system helps the body to rest, digest food and conserve energy. As these tasks are in many ways mutually exclusive, the two systems often have opposite effects upon the body, some of which are:

- Heart – the sympathetic system increases the rate and strength of the heartbeat; the parasympathetic system decreases them
- Digestive tract – the sympathetic system inhibits digestion and reduces blood supply; the parasympathetic system stimulates them
- Liver – the sympathetic system encourages the breakdown of glycogen (a carbohydrate) in the liver to provide energy; the parasympathetic system encourages its formation
- Salivary glands – the sympathetic system reduces the production of saliva, which also becomes thicker; the parasympathetic system promotes a free flow of watery saliva.

How Reflexes Work

Bodily actions that can occur independently of conscious control are called reflexes. They are especially important when a rapid involuntary response is required.

The nervous system is able to perform highly complex tasks, and not all of these require conscious thought. Those actions that are involuntary in nature are called reflexes, preprogrammed and predictable responses to a specific sensory stimulus.

Somatic Reflexes

Somatic reflexes can result in the movement of a muscle or the secretion of a chemical from a gland. The somatic nervous system involves a direct conduction of messages from the central nervous system to the skeletal muscles that enable the body to respond instantly.

For example, touching a hot surface or object triggers an instant withdrawal of the affected body part, therefore bypassing conscious decision-making.

Autonomic Reflexes

The autonomic nervous system is outside conscious control. For example, the baroreceptor reflex corrects a rise in arterial blood pressure without us being aware of the process.

A simple reflex arc

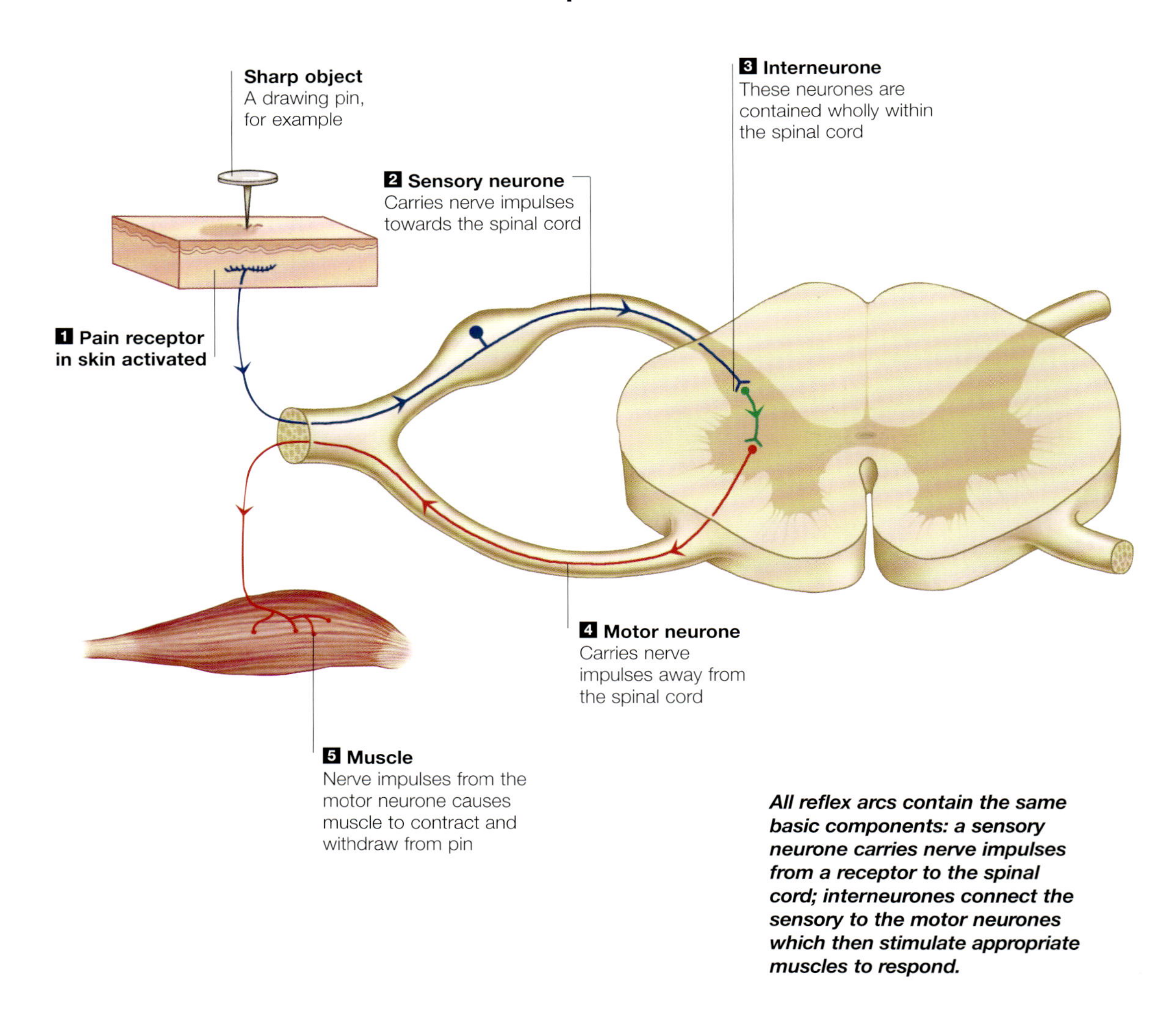

All reflex arcs contain the same basic components: a sensory neurone carries nerve impulses from a receptor to the spinal cord; interneurones connect the sensory to the motor neurones which then stimulate appropriate muscles to respond.

Complex reflexes

Although some spinal reflexes, are relatively simple and involve only a few nerve cells, the spinal cord is capable of carrying out more complicated functions without involving the brain.

An example of a complex reflex is the response to stepping on a sharp object. with the right foot. The crossed extensor reflex is initiated to withdraw the foot and shift the body's weight onto the left leg.

Nerve signals

Initially, the injury stimulates pain receptors in the skin of the foot, causing them to send nerve impulses, via afferent nerve fibres, to the right side of the spinal cord. Neurones in this half of the spinal cord send nerve signals away from the cord via efferent nerve fibres to enable the extensor muscles to relax and the flexors to contract.

Transfer of weight

These events result in the injured leg being moved away from the sharp object. To stabilize the body and avoid a fall, neurones from the right-hand side of the spinal cord cross to the left-hand side and synapse with motor neurones that innervate muscles in the left leg. These motor neurones enable the extensor muscles in the left leg to contract and the flexors to relax, causing the leg to be extended so that it can carry the body's weight.

The crossed extensor reflex

When a bare foot treads on a sharp object, it is rapidly withdrawn and the body's weight is transferred to the other leg.

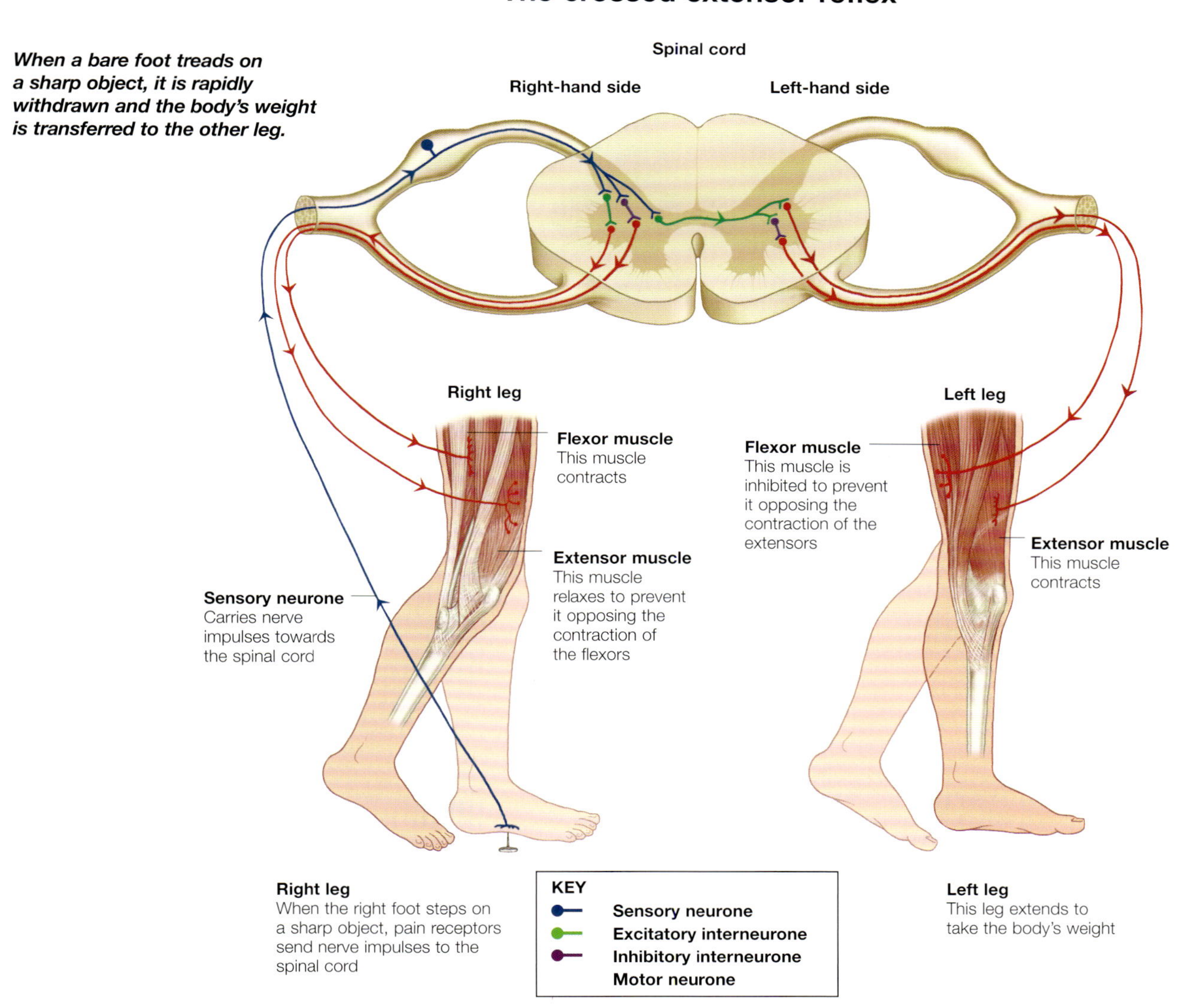

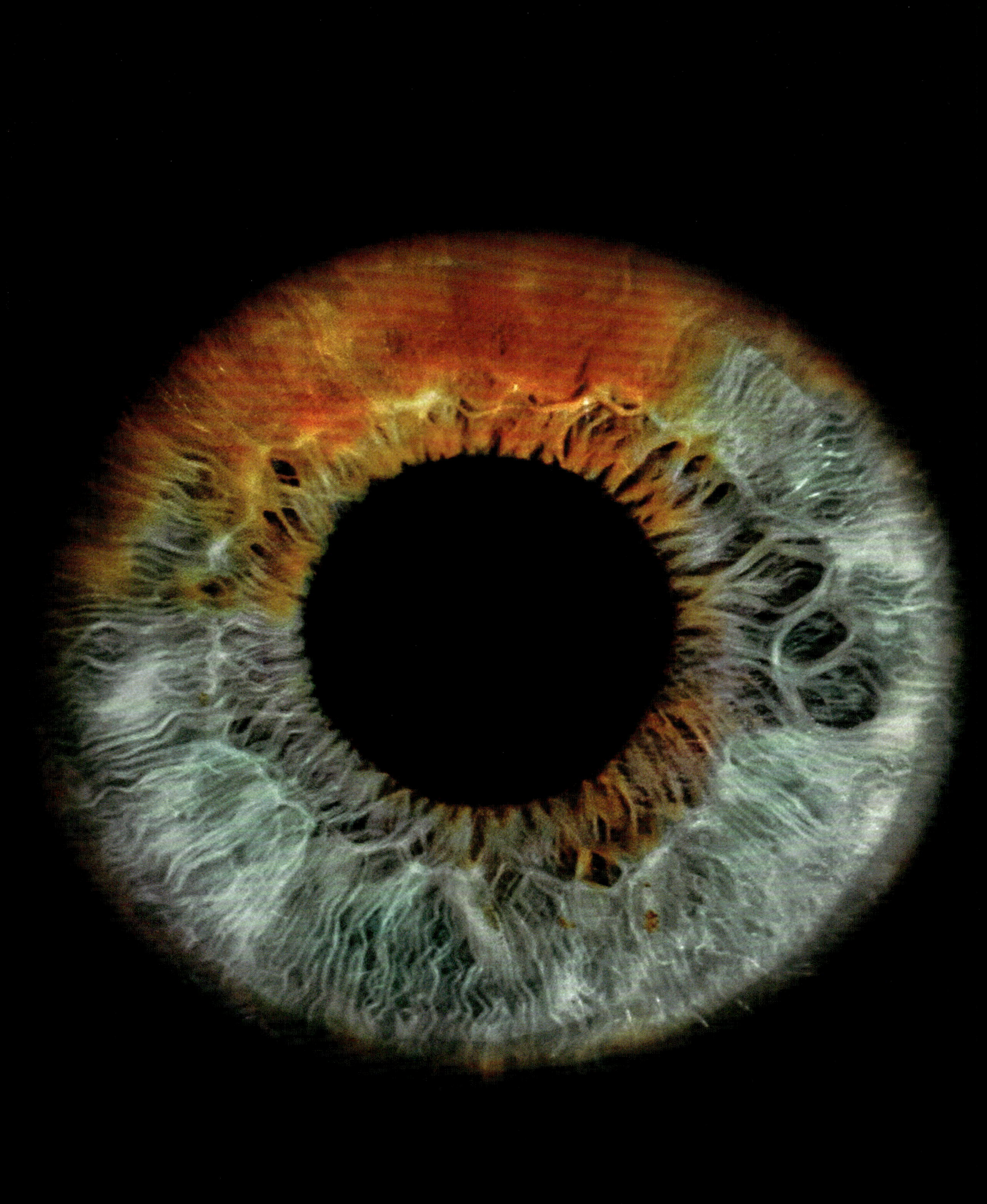

CHAPTER 5

The Senses

The special senses of sight, hearing, smell, taste and touch enable us to assess the world around us. They are the means by which the brain can make sense of the world, making it possible for us to enjoy what we see, hear, touch, smell and taste but also alerting us to danger. Of the five organs of sense, the eye and its associated receptors is the most complex. The eye detects patterns of light in the environment and transfers this information to the brain where it is perceived as images. Whereas the eyes detect light, the ears detect sound waves. They are the vital organs of hearing and balance and transmit sound to the auditory cortex of the brain where it is interpreted and acted on if necessary. Specialized cells in the nasal cavity, the olfactory receptors, convert the chemicals present in an odour to electrical signals that travel to the brain for interpretation. Together with smell, taste is a chemical sense and transmits information from the taste buds on the tongue to the thalamus in the brain.

Finally, touch is the body's ability to perceive sensation in the skin and tissues and to send messages to the brain to identify the source of the sensation and enable us to make decisions. This chapter will explain how the special senses function, and describe their interactions with the central nervous system.

Opposite: The coloured part of the eye, the iris, controls the size of the pupil and the amount of light that passes through the retina.

Sight

The eyes are the specialized organs of sight, designed to respond to light.

The eyes enable us to receive information from our surroundings by detecting patterns of light. This information is communicated to the brain, which processes it so that it can be perceived as images.

Each eyeball is embedded in protective fatty tissue within a bony cavity, known as the orbit. The orbit has a large opening at the front to allow light to enter, and smaller openings at the back, allowing the optic nerve to pass to the brain, and blood vessels and nerves to enter the orbit.

Chambers

The eyeball is divided into three internal chambers. The two aqueous chambers at the front of the eye are the anterior and posterior chambers, and are separated by the iris. These chambers are filled with clear, watery aqueous humour (liquid), which is secreted into the posterior chamber by a layer of cells covering the ciliary body. The aqueous humour passes into the anterior chamber through the pupil, then into the bloodstream via a number of small channels found where the base of the iris meets the margin of the cornea. The largest of the chambers is the vitreous body, which lies behind the aqueous chambers, and is separated from them by the lens and the suspensory ligaments (zonular fibres), which connect the lens to the ciliary body. The vitreous body is filled with clear, jelly-like vitreous humour.

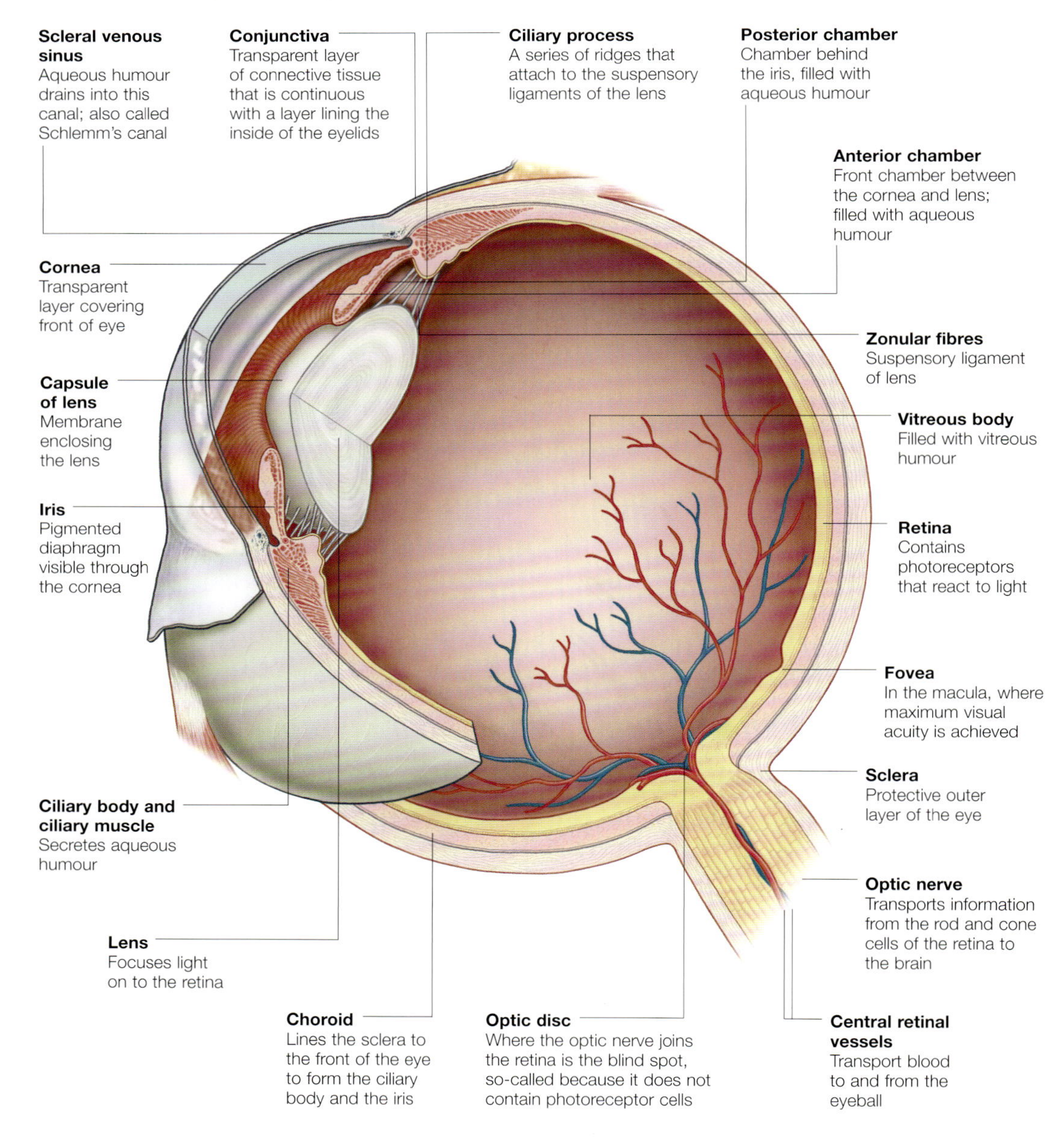

Layers of the eye

The eyeball is covered with three different layers, each of which has a special function.

The outer layer of the eyeball is called the sclera, and is a tough, fibrous, protective layer. At the front of the eye, the sclera is visible as the 'white of the eye'. This is covered by the conjunctiva, a transparent layer of connective tissue. The transparent cornea covers the front of the eyeball, allowing light to enter the eye.

Uvea

The intermediate layer, the uvea, contains many blood vessels, nerves and pigmented cells. The uvea is divided into three main regions: the choroid, the ciliary body and the iris. The choroid extends from where the optic nerve meets the eyeball to the front of the eye, where it forms both the ciliary body and the iris.

Retina

The innermost layer of the eye is the retina, a layer of nerve tissue containing photosensitive (light-sensitive) cells called photoreceptors. It lines all but the most anterior (frontal) part of the vitreous body. There are two types of photoreceptor cells: rod cells detect light intensity and are concentrated towards the periphery of the retina; and cone cells detect colour, and are most concentrated at the fovea at the most posterior part of the eyeball.

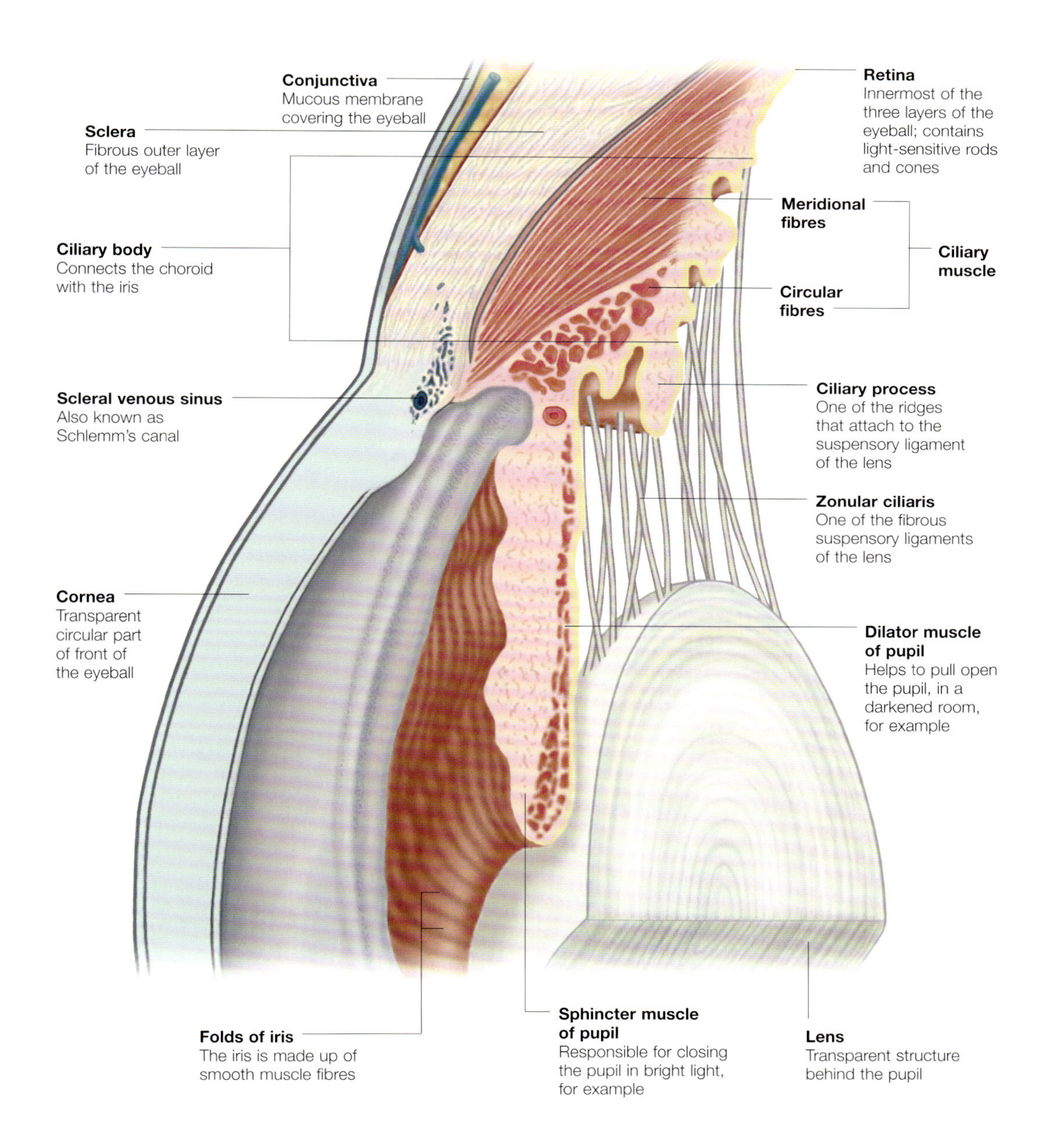

Muscles of the Eye

The rotational movements of the eye are controlled by six rope-like extraocular muscles.

The muscles of the eye can be divided into three groups: the muscles inside the eyeball, the muscles of the eyelids and the extraocular muscles, which rotate the eyeball within its orbit. The six extraocular muscles are rope-like, attaching directly to the sclera. Four of the muscles are rectus (straight) muscles – superior, inferior, lateral (the temple side of the eye) and medial (nasal side). Each rectus muscle arises from connective tissue, the common tendinous ring (annulus) at the back of the orbit that passes forward to insert just behind the junction of the sclera and cornea.

Oblique Muscles

The two extraocular muscles are the oblique muscles. The superior oblique arises from bone near the back of the orbit, and extends to the front of the orbit. There, its tendon loops through the trochlea, a 'pulley' made of fibres and cartilage, and turns back to insert into the sclera.

The inferior oblique arises from the floor of the orbit, passing backwards and laterally under the eyeball to insert towards the back of the eye.

Left eye (side view)

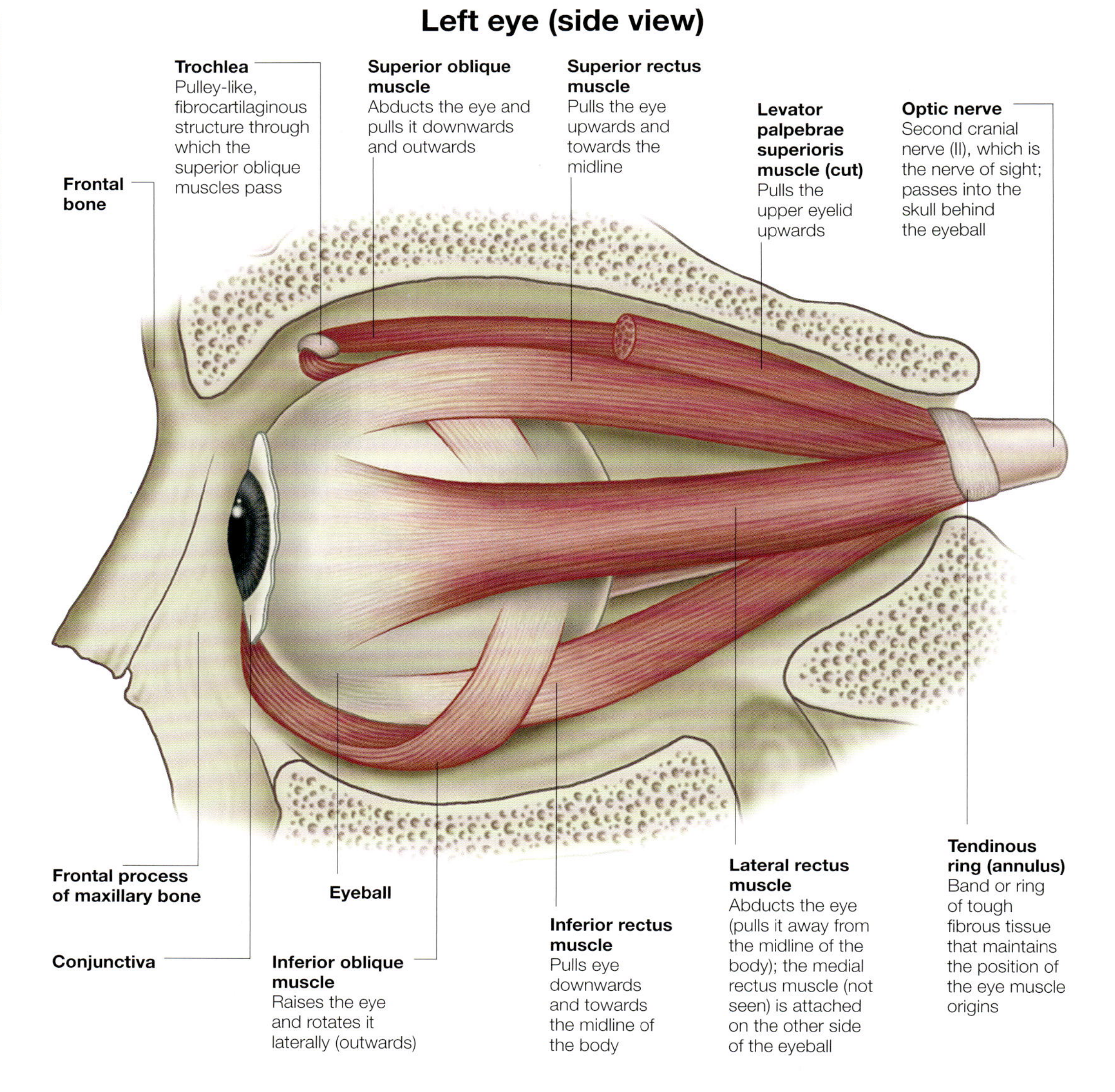

Movement of the eye

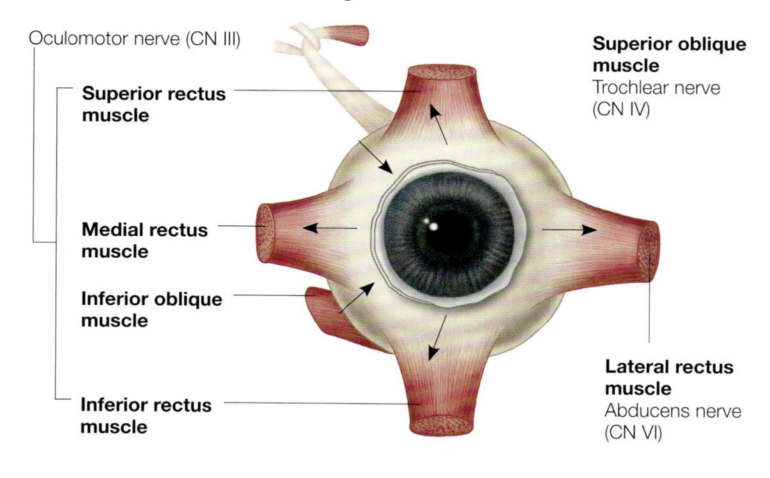

The contraction of the extraocular muscles is controlled by cranial nerves, specifically the trochlear (CN IV), oculomotor (CN III) and abducens (CN VI). The muscles act individually to turn the cornea, though it should be noted that the direction of turn for a particular muscle differs between the right and left eyes. For example, in the right eye, the lateral rectus will turn the cornea to the right, while in the left eye it would turn it to the left. Since eye movements normally occur in parallel, different muscles in each eye act together to turn the eyes.

Therefore to look left, the lateral rectus will turn the left eye and the medial rectus the right eye. The eye movements of a single eye are usually the result of more than one of these muscles acting together.

The extraocular muscles are innervated by cranial nerves. The nerves and muscles serve to move the eye in the directions indicated by the arrows.

Nerves and blood vessels

The eye muscles are served by a series of nerves and blood vessels that help to make sight our dominant sense.

Nerves of the eye enter and leave the orbit through its openings posteriorly (at the back). Cranial nerve (CN) II – the optic nerve, which carries the visual signals from the retina to the brain, passes from the orbit to the cranial cavity through the optic canal. The other nerves – including branches of the ophthalmic nerve, the sensory nerve of the eye – enter the orbit through the orbital fissure. Another nerve important for the eye is the facial nerve (CN VII). This supplies orbicularis oculi (a muscle of facial expression), causes blinking and also controls secretion from the lacrimal glands, which keep the eyes moist. These glands secrete fluid (tears) continuously, which is spread over the surface of the cornea by blinking. Irritation of the cornea can cause an increase in tear production.

Blood vessels of the eye

The main artery of the eye is the ophthalmic artery, which is a branch of the internal carotid artery. The ophthalmic artery enters the orbit within the sheath of the optic nerve, and then branches to the extraocular muscles, the eyeball, the lacrimal gland and surrounding tissues.

The retinal artery remains within the optic nerve stalk until it reaches the optic disc, where it sends out branches supplying the retina. Veins drain the orbit to the cavernous sinus in the cranial cavity and to the facial vein, thus forming a connection between facial blood vessels and the brain.

Left eye (from above)

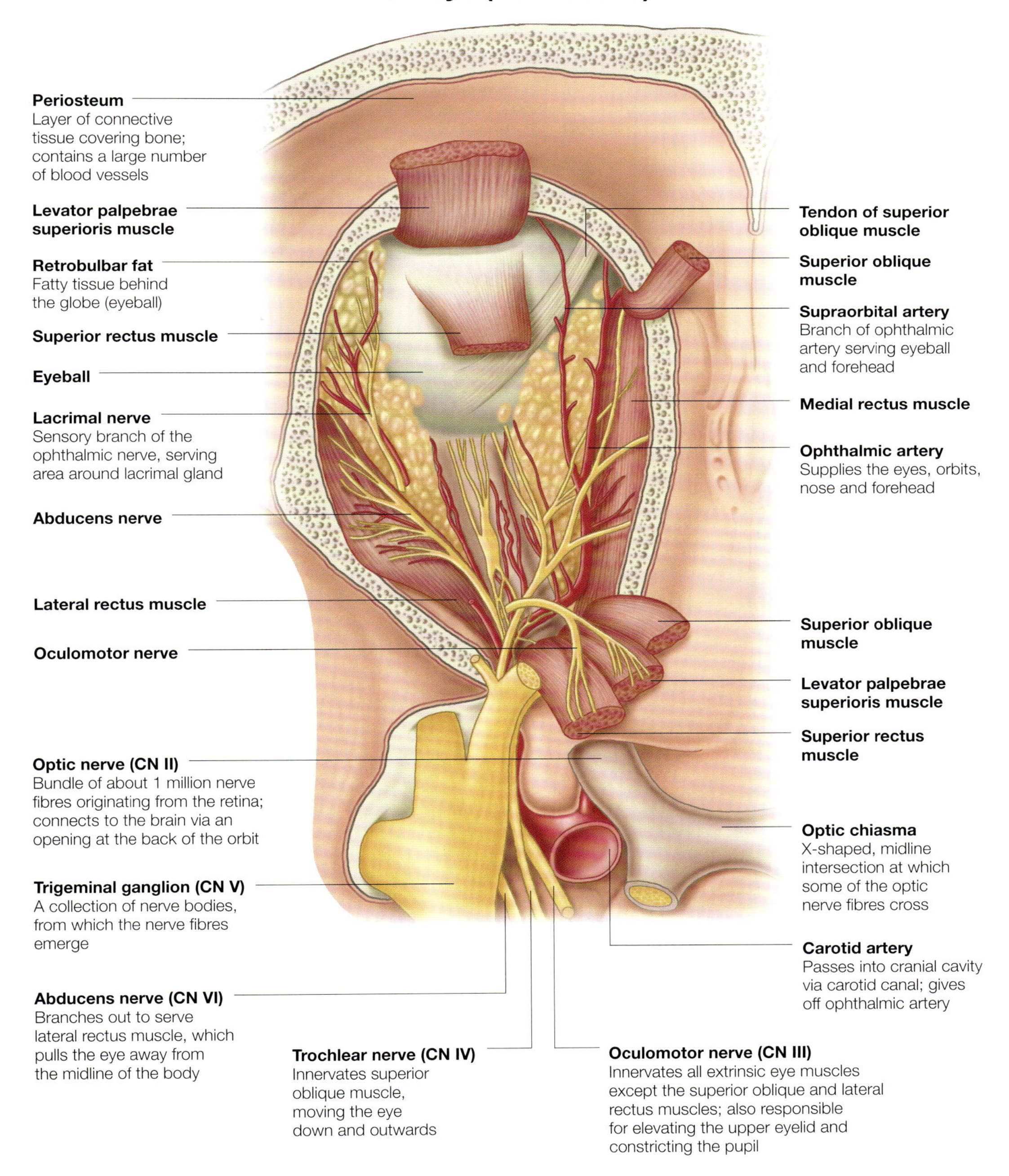

Focusing the Eye

Sight is the principal human sense, and we rely on our eyes for all our visual information. Despite their small size, we can focus on a distant star or a speck of dust, and see in bright sunlight or at dusk.

The human eye works like a camera. Light rays from an object pass through an aperture (the pupil) and are focused by a lens on to the retina, a light-sensitive layer at the back of the eye. The retina – the eye's equivalent of camera film – is a light-sensitive membrane composed of layers of nerve fibres and a pigmented light-sensitive membrane. It contains two kinds of light-sensitive cells: cones and rods.

Cones and Rods

Cones are sensitive to either red, green or blue light, and their signals enable the brain to interpret a colour image. They also give the eye acute vision.

Rods are extremely sensitive to low light but cannot differentiate between colours, which is why objects appear to lose their colour at night. The rods and cones are linked to the brain by nerve cells that all pass out of the back of the eye via the optic nerve.

To see objects clearly, the muscles of the eye must pull on the lens and focus light on the retina. If this process is faulty, or the lens or the eye are the wrong shape, the image will appear blurred.

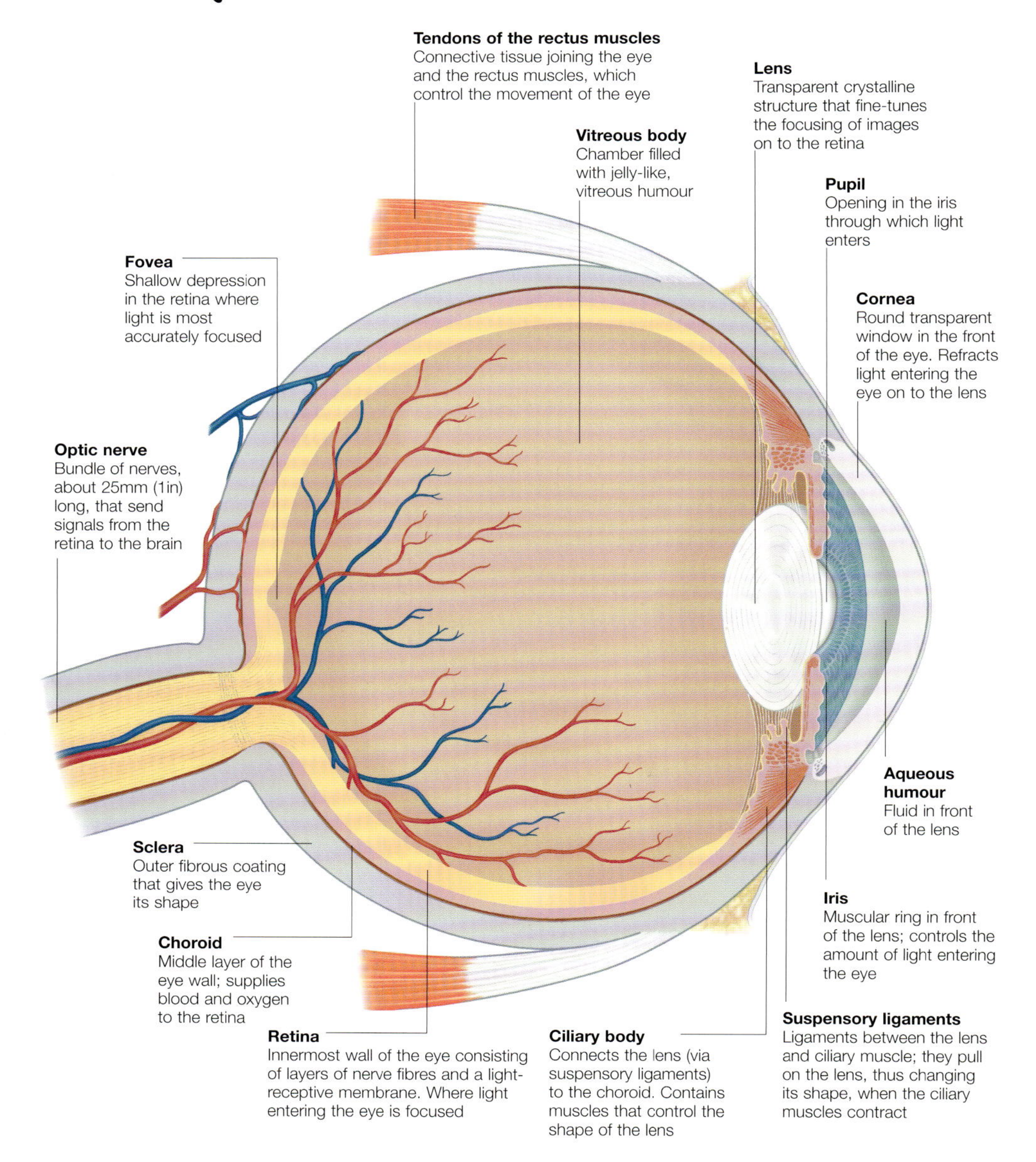

Iris

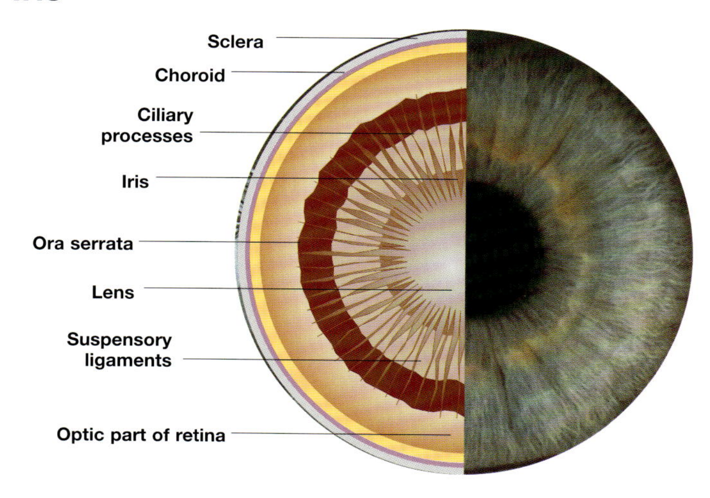

The iris is a muscular, ring-shaped structure with a hole in the middle, called the pupil. The iris contains a distinctive coloured pigment. The muscles of the iris are used to make the pupil larger or smaller, therefore allowing more or less light into the eye according to the surrounding conditions.

The muscles of the iris are found in the ciliary body, which is the part of the eye connecting the choroid (the middle layer of the eye wall) with the iris.

The ciliary body consists of three parts:

- The ciliary ring, adjoining the choroid
- The ciliary processes, 70 radial ridges around the ciliary body
- The ciliary muscle, which controls lens curvature.

This composite picture shows (left) the structure of the eyeball from the inside, with the lens in the centre; and (right) the outer appearance of the eye, where the lens itself is covered by the cornea.

Lens

Light entering the eye passes through the cornea and the aqueous humour, both of which cause refraction (bending) of the light rays inwards.

The cornea of the eye refracts most of the incoming light, and it is the task of the lens to 'fine-tune' the focusing of the rays so that the image falls accurately onto the retina.

The lens is a crystalline structure, made up of several layers. It is attached to the muscular ciliary body by suspensory ligaments. Movements of the ciliary muscle alter the shape of the lens, according to whether the eye needs to focus on a distant or nearby object. The illustrations below (viewing the eye from inside and from the side respectively) demonstrate how the shape of the lens is adjusted as necessary.

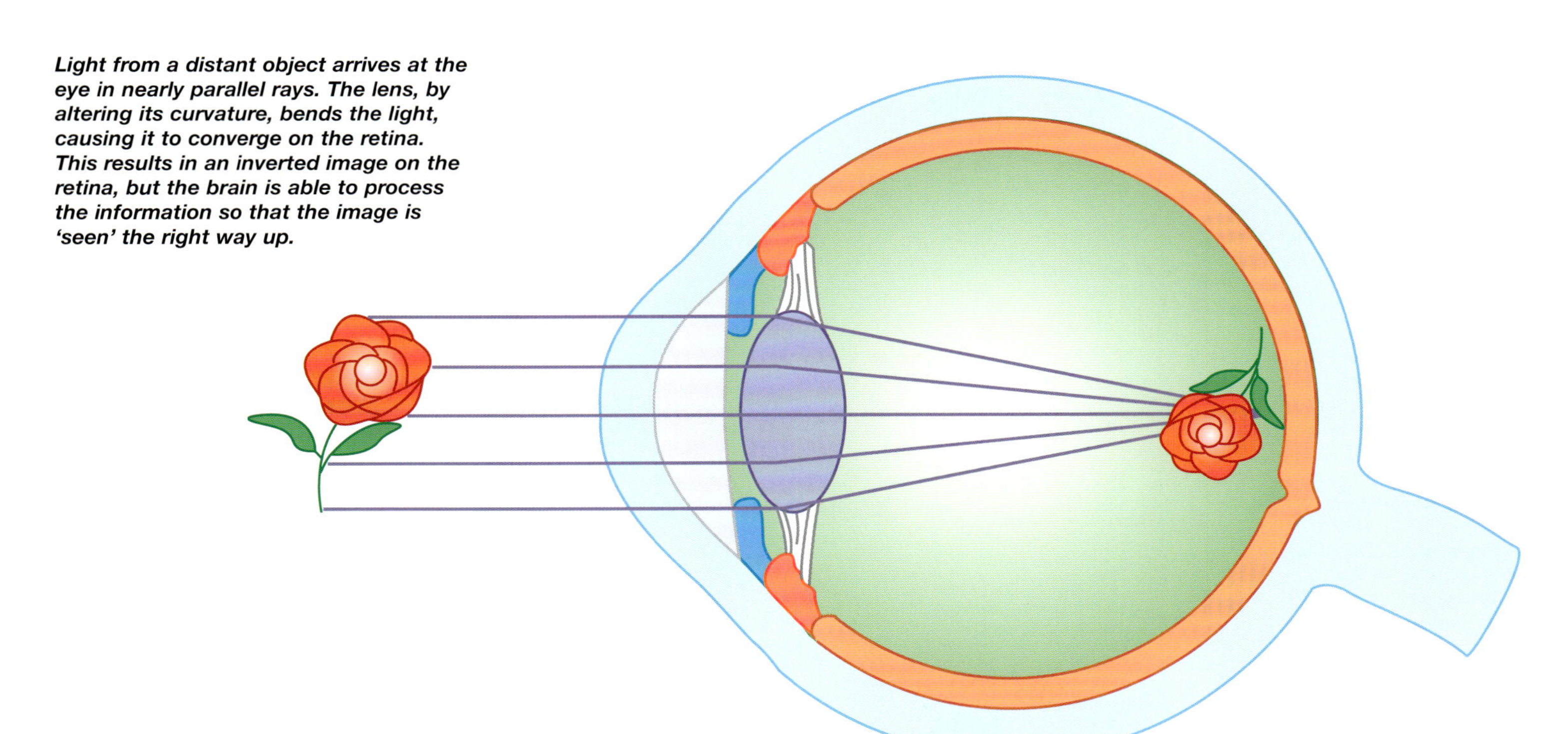

Light from a distant object arrives at the eye in nearly parallel rays. The lens, by altering its curvature, bends the light, causing it to converge on the retina. This results in an inverted image on the retina, but the brain is able to process the information so that the image is 'seen' the right way up.

Looking at close objects

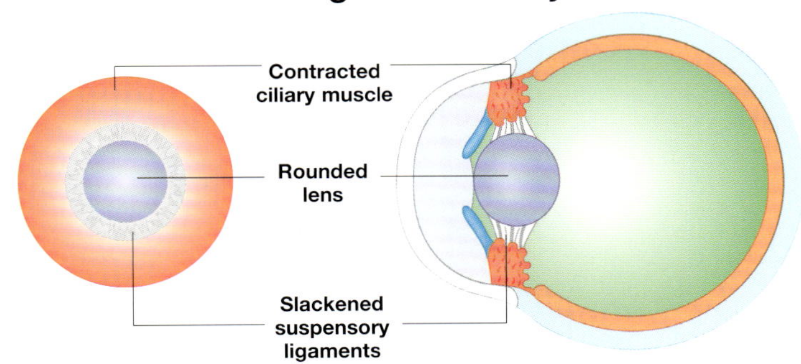

Light rays from a nearby object are more diverged, needing greater refraction. The ciliary muscle contracts, reducing the tension on the suspensory ligaments, and the lens becomes more rounded. As the light rays pass through the rounded lens they are sharply converged on the back of the eye.

Looking at distant objects

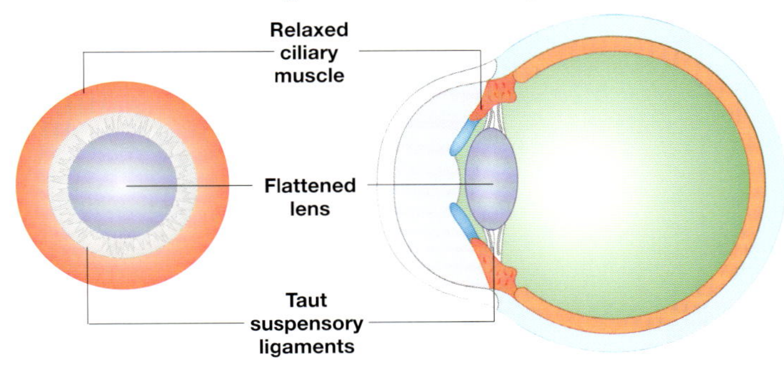

Light rays from a distant object are more parallel when they reach the eye, so require less refraction by the lens. The ciliary muscle relaxes and the tension on the suspensory ligaments pulls the edges of the lens outwards, thus making it thinner and flatter. The rays are focused on the back of the eye.

Retina

The retina, located at the back of the eye, contains specialized cells called photoreceptors that are sensitive to light of different colours. These allow us to see in both the light and the dark.

The eye has adapted through evolution to be extremely sensitive to light. However, the bulk of the eye tissue is not responsive to light. Rather, the muscles that surround the eyeball, as well as the iris, cornea and lens, all act to focus light onto the retina, a relatively small area at the back of the eyeball that contains photoreceptors.

Structure

At its simplest level, the retina consists of four layers of cells:

- At the back of the retina is the outer, pigmented layer – these epithelial cells absorb light (but do not 'detect' it), so preventing it from scattering in the eye
- Next are aligned a layer of photoreceptors, which are able to convert light energy into electrical energy
- The electrical potentials the photoreceptors generate are transmitted to the 'bipolar cells'
- The bipolar cells in turn communicate with 'ganglion cells'; the axons (nerve fibres) of the latter converge and make a right-angled turn before leaving the eye through the optic nerve, which carries information about the visual scene to the brain.

Thus light has first to travel through the ganglion and bipolar cells before it reaches the light-sensitive photoreceptors at the back of the retina. This apparent 'back-to-front' arrangement does not hinder photoreceptors from detecting light.

Structure of the retina

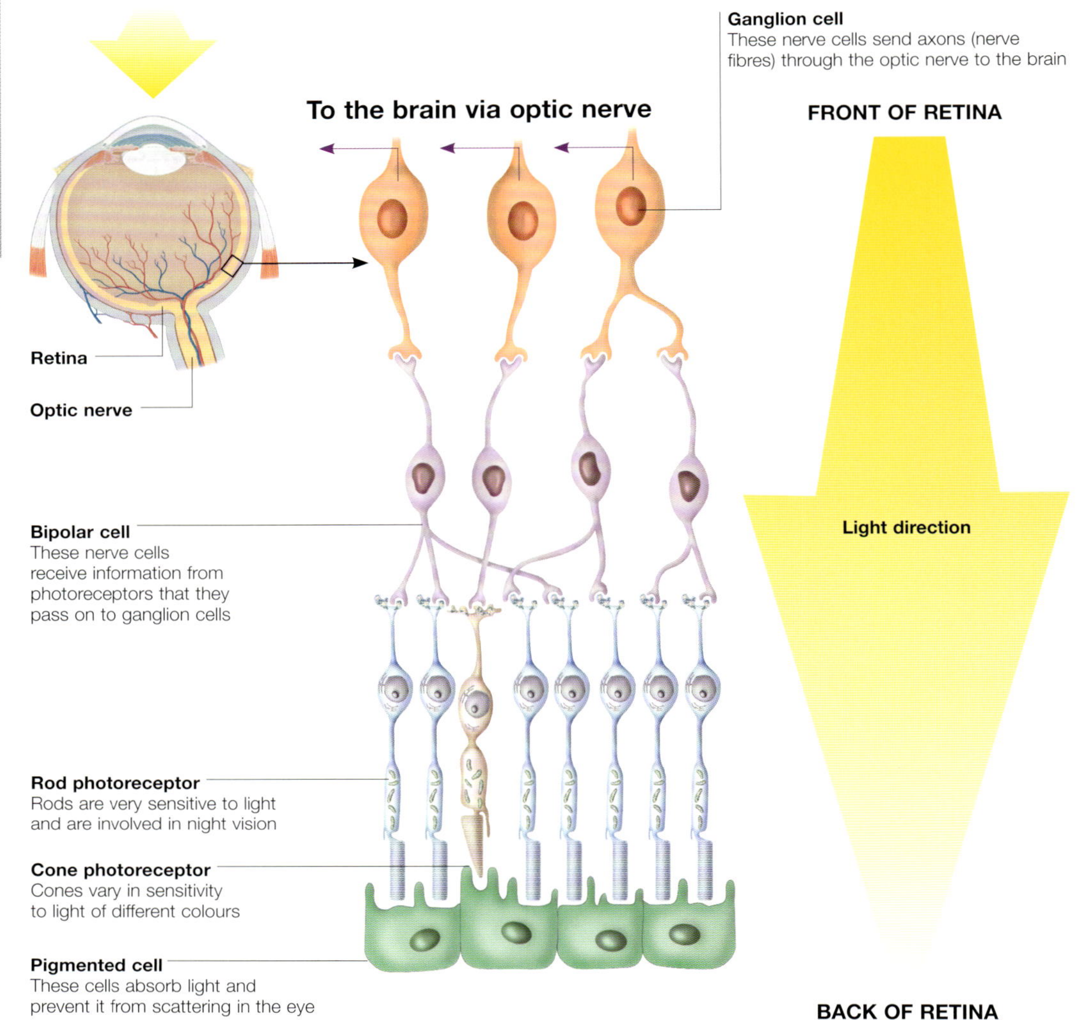

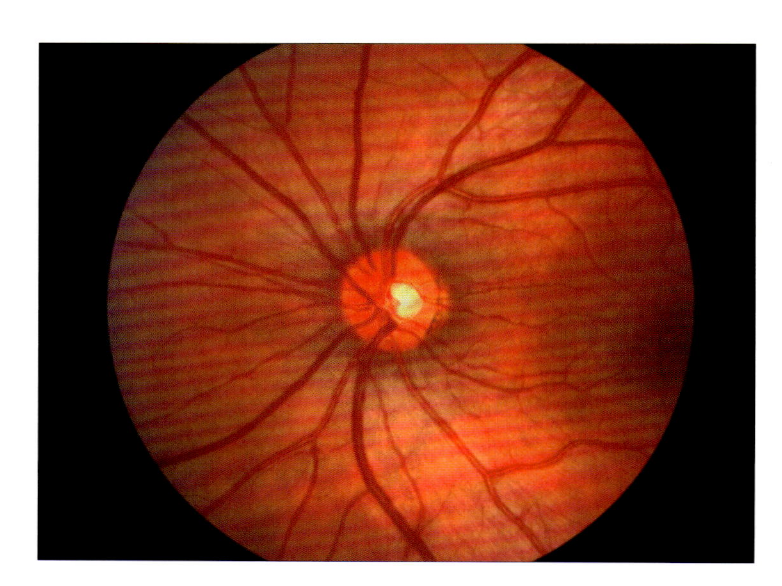

The light-sensitive back of the eye is known as the retina. It acts much like the film in a camera, and sends information to the brain.

Rods and cones

There are two types of photoreceptors: rods, which are sensitive to low levels of light, and cones, which are responsive to light of different colours.

Rods are the most numerous of the two types of photoreceptor; it has been estimated that there are 120 million rod cells compared to only six million cones. Rods are also about 300 times more sensitive to light than cones.

Night vision

As a result of their sensitivity, coupled with their relative abundance, rod cells are ideal for seeing in the dark when light levels are low. However, rod cells provide the brain with only low acuity vision in scales of grey. This is because a rod cell makes connections with more than one bipolar cell, which, in turn, sends electrical impulses to the brain via many ganglion cells. Thus a ganglion cell – which leaves the eye through the optic nerve – provides the brain with information gathered from a large number of rod cells.

Day vision

In contrast to rods, cones operate primarily in strong light and provide the brain with high acuity, colour information on the visual scene. In addition, each individual cone cell has a 'direct line' to the brain; one cone cell is in contact with only one bipolar cell, which in turn communicates with only one ganglion cell. Thus a neuron in the brain can receive information on the activity of a single cone photoreceptor.

Fovea

At the centre of the retina, directly behind the middle of the lens, is a region the size of a pinhead called the fovea. This contains entirely cones and is the only part of the retina that has cones at sufficient density to provide highly detailed colour vision. Hence, only one thousandth of our visual field can be in hard focus at any one time and we need to move our eyes continuously to comprehend a rapidly changing scene.

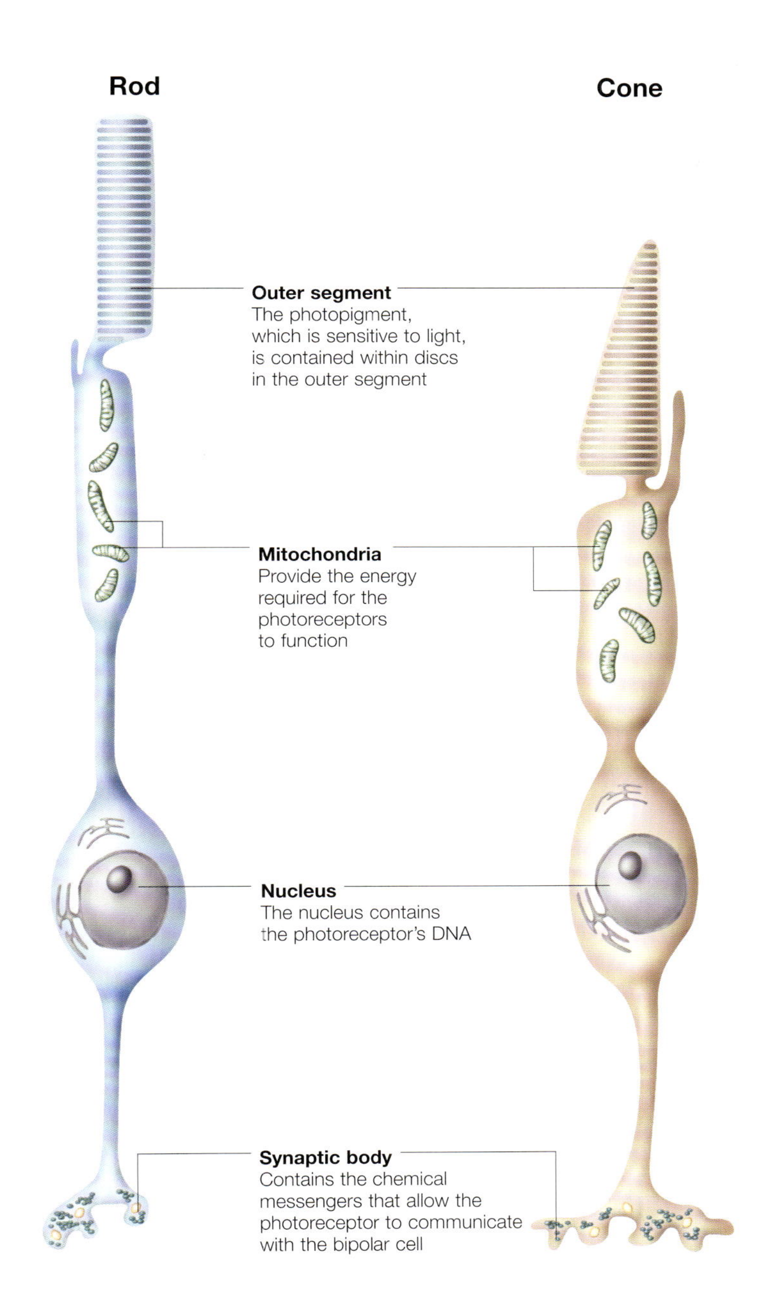

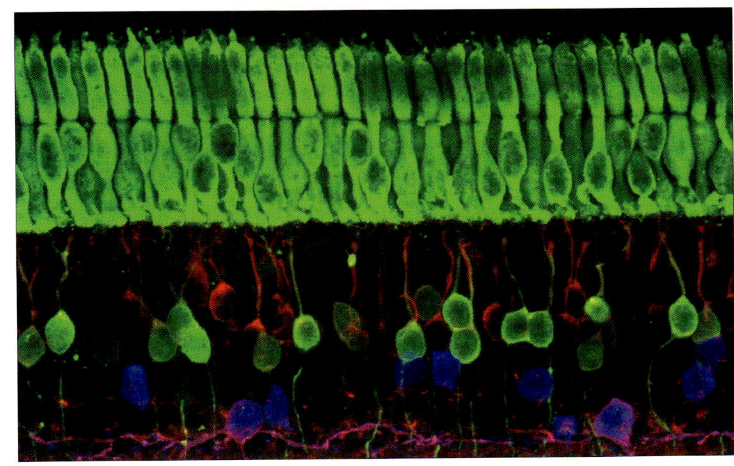

The human eye contains about 120 million rod cells and 6 million cone cells (shown in green). Rods function mainly in strong light, whereas cones enable us to see at night.

Eyelids

The eyelids are thin folds of skin that can close over the eye to protect it from injury and excessive light. The lacrimal apparatus is responsible for producing and draining lacrimal fluid.

Each eyelid is strengthened by a band of dense elastic connective tissue called a tarsal plate. These give the eyelids a curvature that matches that of the eye.

Eyelid Structure

The tarsal plate of the upper eyelid is larger than that of the lower. The inner and outer ends of both tarsal plates are attached to the underlying bone by tiny ligaments. Between the front surface of the tarsal glands and the overlying skin, lie fibres of the orbicularis oculi muscle.

The eyelashes project from the free edge of the eyelids. The follicles of the eyelashes, from which the hairs emerge, have nerve endings that can sense any movement of the lashes.

The tarsal plates contain glands, called meibomian glands, that secrete an oily liquid that prevents the eyelids sticking together. There are also other tiny ciliary glands associated with the eyelash follicles.

Eyelid Movement

The eye closes in response to movement of the upper eyelid. The orbicularis oculi muscle contracts to close the eye, whereas the upper eyelid is opened by the levator palpebrae superioris muscle.

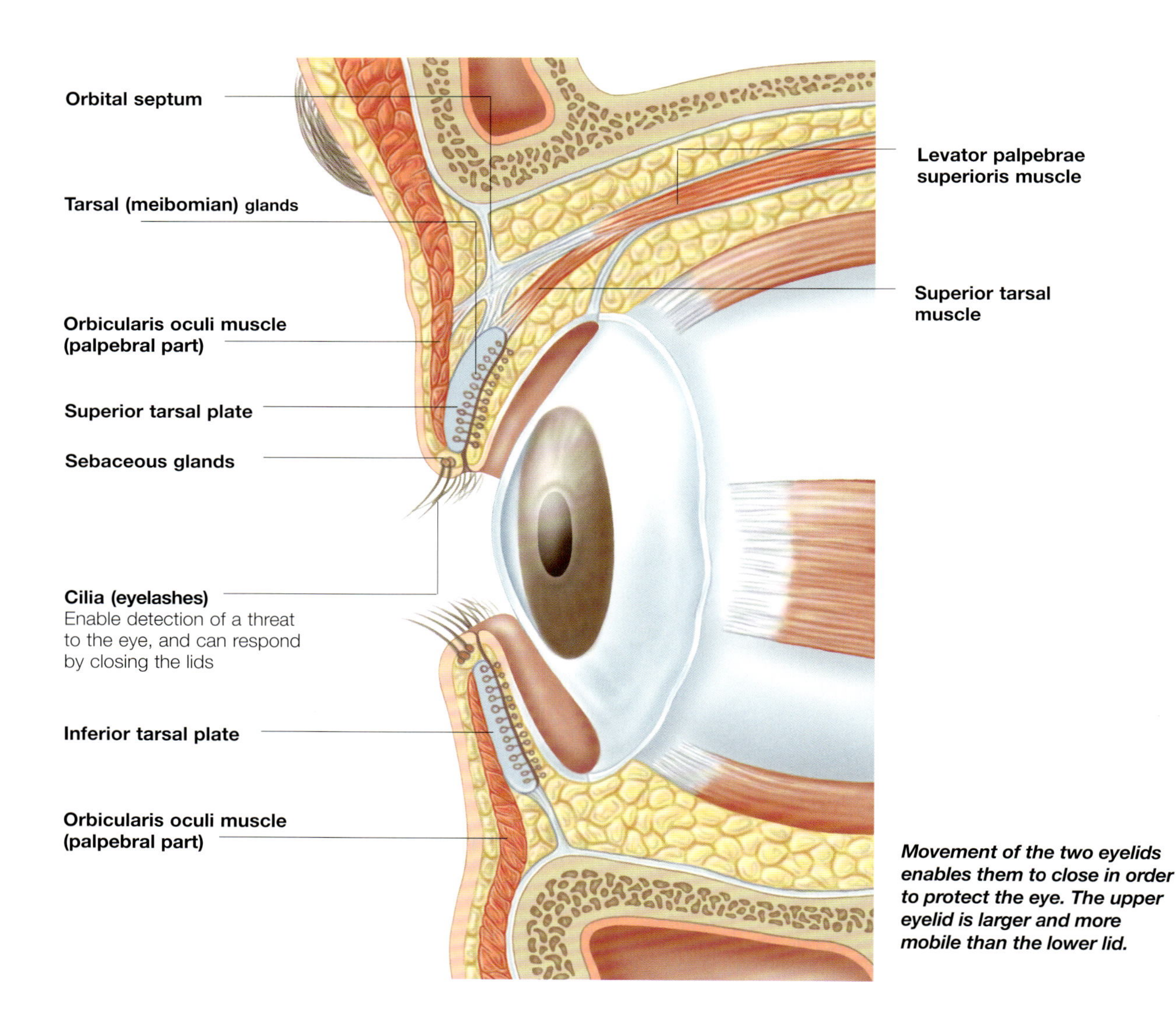

Movement of the two eyelids enables them to close in order to protect the eye. The upper eyelid is larger and more mobile than the lower lid.

Lacrimal apparatus

The eyes are protected and lubricated by lacrimal fluid, our tears. The lacrimal system produces this fluid and drains the excess to the nasal cavity.

The eyes are kept moist by the production of clear, watery fluid by the lacrimal glands. This fluid also contains lysozyme, an antibacterial substance. Most of the fluid, about 2ml (0.07fl oz) per day, is lost through evaporation. The rest is drained by the nasolacrimal ducts to the back of the nose.

Lacrimal gland

The lacrimal gland lies above the outer side of the eye within a recess in the eye socket. The gland has an upper orbital part and a lower palpebral part. There are accessory lacrimal glands that lie in the upper lid. The glands each have 12 tiny ducts that release secretions into the conjunctival sac through openings under the upper lid in the superior fornix. The upper and lower eyelid both have a raised papilla at their inner ends, which has a tiny opening, the lacrimal punctum. Excess fluid enters these openings to be carried away by the lacrimal canaliculi. The fluid then passes into a collecting area known as the lacrimal sac and is carried down to the back of the nasal cavity by the nasolacrimal duct.

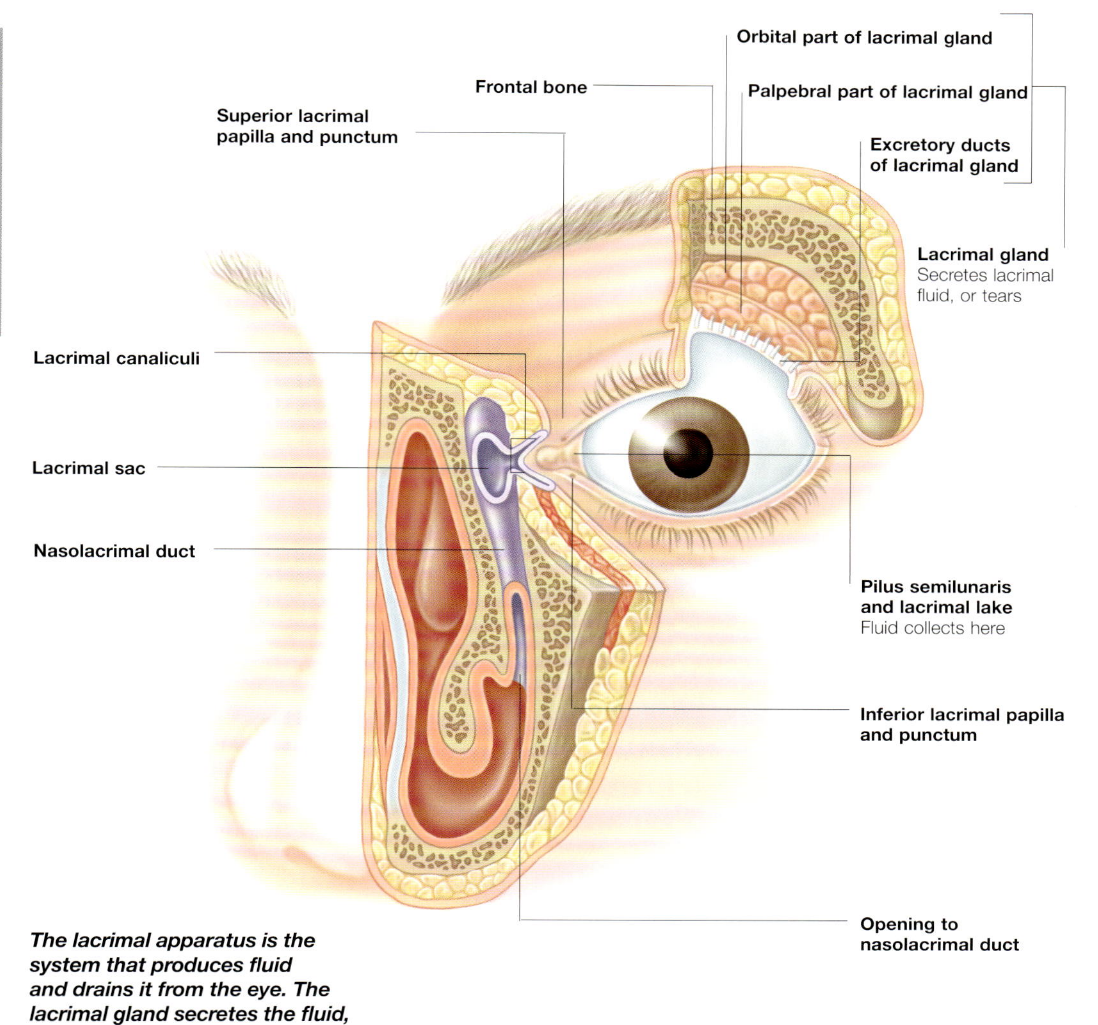

The lacrimal apparatus is the system that produces fluid and drains it from the eye. The lacrimal gland secretes the fluid, which drains via the puncta.

Conjunctiva

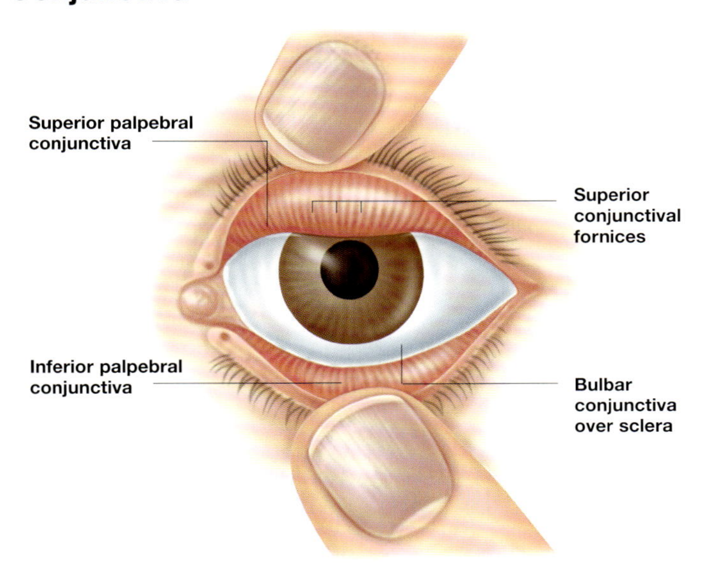

The conjunctiva is a very thin membrane that lines and lubricates the surface of the eyeball and inner surfaces of the eyelid. It can be thought of as having two parts:

- The bulbar conjunctiva – this covers the white of the eye, the sclera. The conjunctiva here is thin and transparent, and is separated from the sclera by loose connective tissue. It does not cover the cornea, which lies over the iris and pupil, but attaches to its periphery.
- The palpebral conjunctiva – this lines the inside of the upper and lower eyelids. Deep recesses, known as conjunctival fornices, are formed where the bulbar and palpebral parts of the conjunctiva meet.

The conjunctiva is a membrane covering the white of the eye and lining the inside of the eyelids. The conjunctiva over the eyeball is transparent.

Tear production

Teardrops are produced when the lacrimal glands are stimulated to secrete increased amounts of fluid that spills out of the eyes. Triggers for tear production include foreign bodies or irritants (such as chemicals). The mechanism of crying or weeping is also a way of communicating our distress and is not fully understood.

Hearing and Balance

The ears are vital sensory organs of hearing and balance. Each ear is divided into three parts – outer, middle and inner ear – each of which is designed to respond to sound or movement in a different way.

The ear can be divided anatomically into three different parts: the external, middle and inner ear. The external and middle ear are important in the gathering and transmitting of sound waves. The inner ear is the organ of hearing and is also vital in enabling us to maintain our balance.

Transmitting Information

The external ear consists of the visible auricle, or pinna, and the canal that passes into the head – the external auditory meatus. At the inner end of the meatus is the tympanic membrane, or eardrum, which marks the border between the external and middle ear.

The middle ear is connected to the back of the throat via the auditory, or Eustachian, tube. Within the middle ear are three tiny bones called the ossicles. These bones are linked together in such a way that movements of the eardrum are transmitted via the footplate of the stapes to the oval window (the opening in between the middle and inner ear).

The inner ear contains the main organ of hearing, the cochlea, and the vestibular system that controls balance. Information from both these parts of the ear passes to specific areas within the brainstem via the vestibulocochlear nerve.

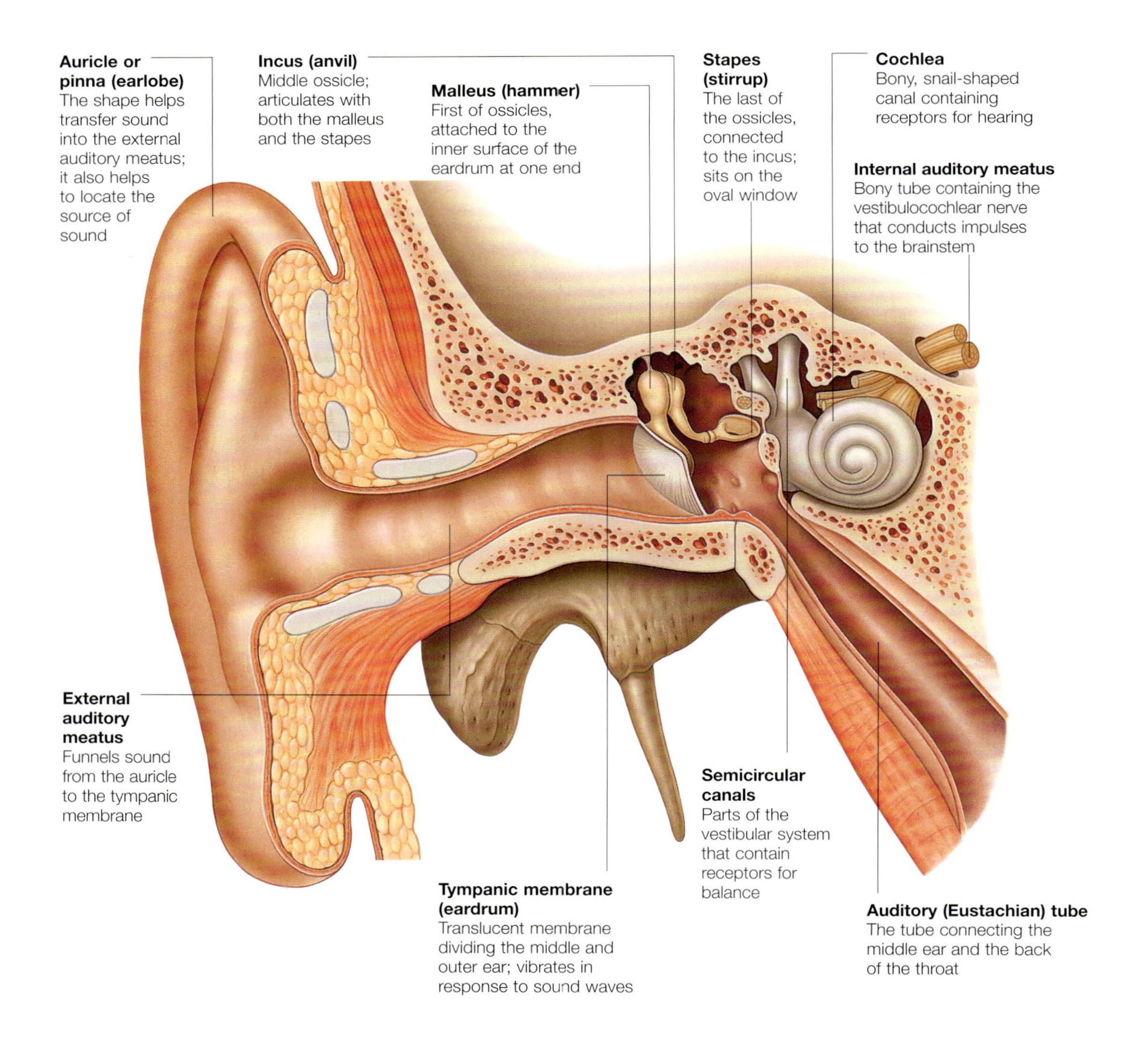

Pinna (external ear)

The pinna is the visible part of the ear on the side of the head consisting of skin and cartilage. It serves to channel sound into the middle ear.

The pinna, or auricle, collects sound from the environment and channels it into the external auditory meatus. It consists of a thin sheet of elastic cartilage and a lower portion called the lobule, consisting mainly of fatty tissue, with a tight covering of skin.

The auricle is attached to the head by a series of ligaments and muscles, and the external ear has a complex sensory nerve supply involving three of the cranial nerves.

External auditory meatus
The external auditory meatus is a tube extending from the tragus to the tympanic membrane and is about 2.5cm (1in) long in adults. The outer third of the tube is made of cartilage (similar to that in the auricle), but the inner two-thirds is bony (part of the temporal bone). In the skin covering the cartilaginous part of the meatus, there are coarse hairs and ceruminous glands that secrete cerumen (earwax).

Usually wax dries up and falls out of the ear, but it can build up and interfere with hearing. The combination of wax and hairs prevents dust and foreign objects from entering.

The boundary between the outer and middle ear is the tympanic membrane, or eardrum. This is a translucent membrane that can be viewed using an auriscope. The tympanic membrane can occasionally perforate due to middle ear infection.

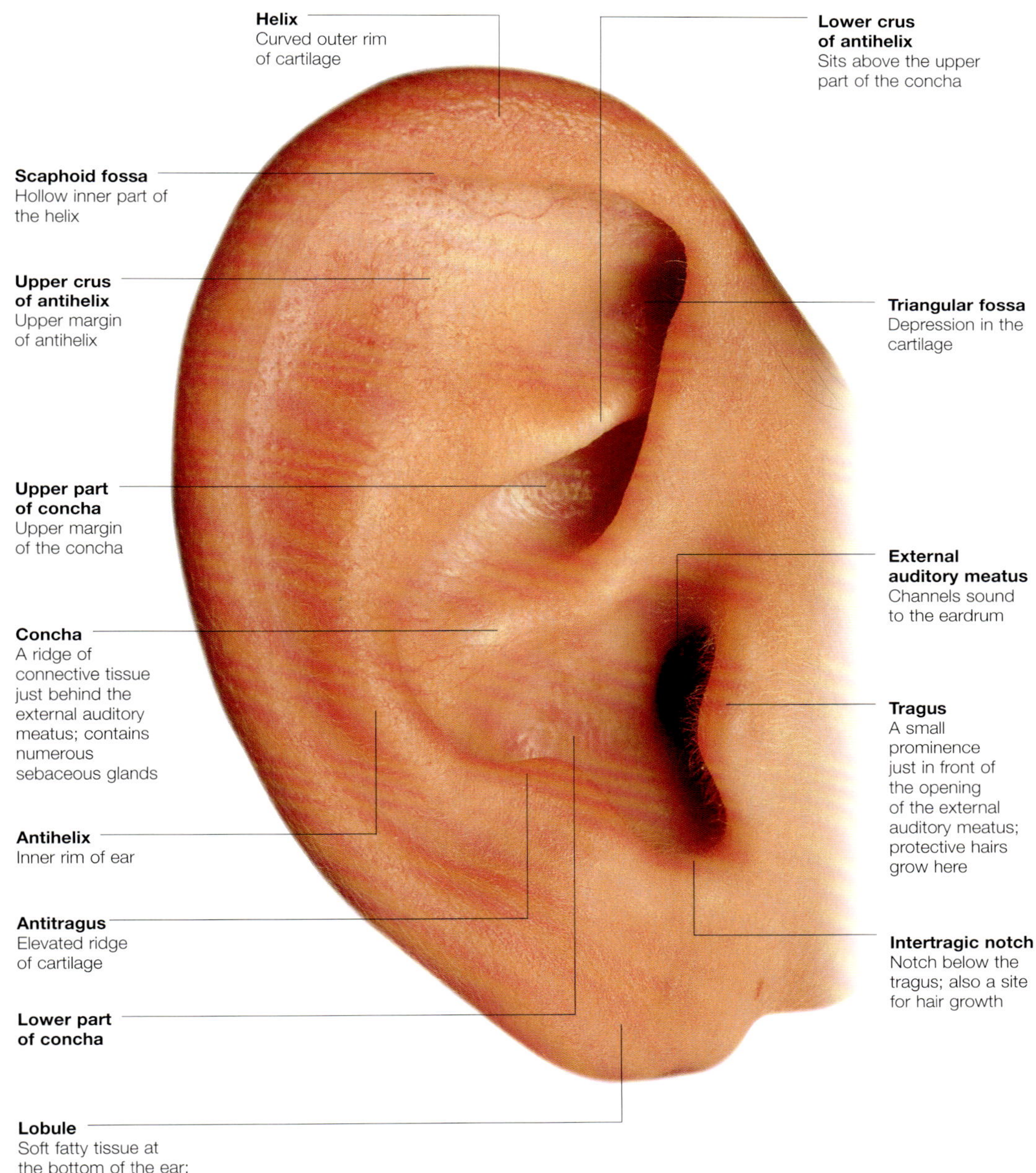

Middle Ear

The middle ear is an air-filled cavity containing the eardrum and three small bones that help transmit sound to the inner ear. It is also connected to the throat via the auditory tube.

The middle ear is an air-filled, box-shaped cavity within the temporal bone of the skull. It contains small bones or ossicles – the malleus, incus and stapes – that span the space between the tympanic membrane (the eardrum) and the medial wall of the cavity.

Tensor Tympani and Stapedius

Two small muscles are also present: the tensor tympani, attached to the handle of the malleus; and the stapedius, attached to the stapes. Both help to modulate the movements of the ossicles. The medial wall divides the middle ear from the inner ear and contains two membrane-covered openings; the oval and round windows.

Auditory Tube

The middle ear is connected to the throat by the auditory (Eustachian) tube. This tube is a possible route of infection. Just below the floor of the middle ear cavity is the bulb of the internal jugular vein, and just in front is the internal carotid artery.

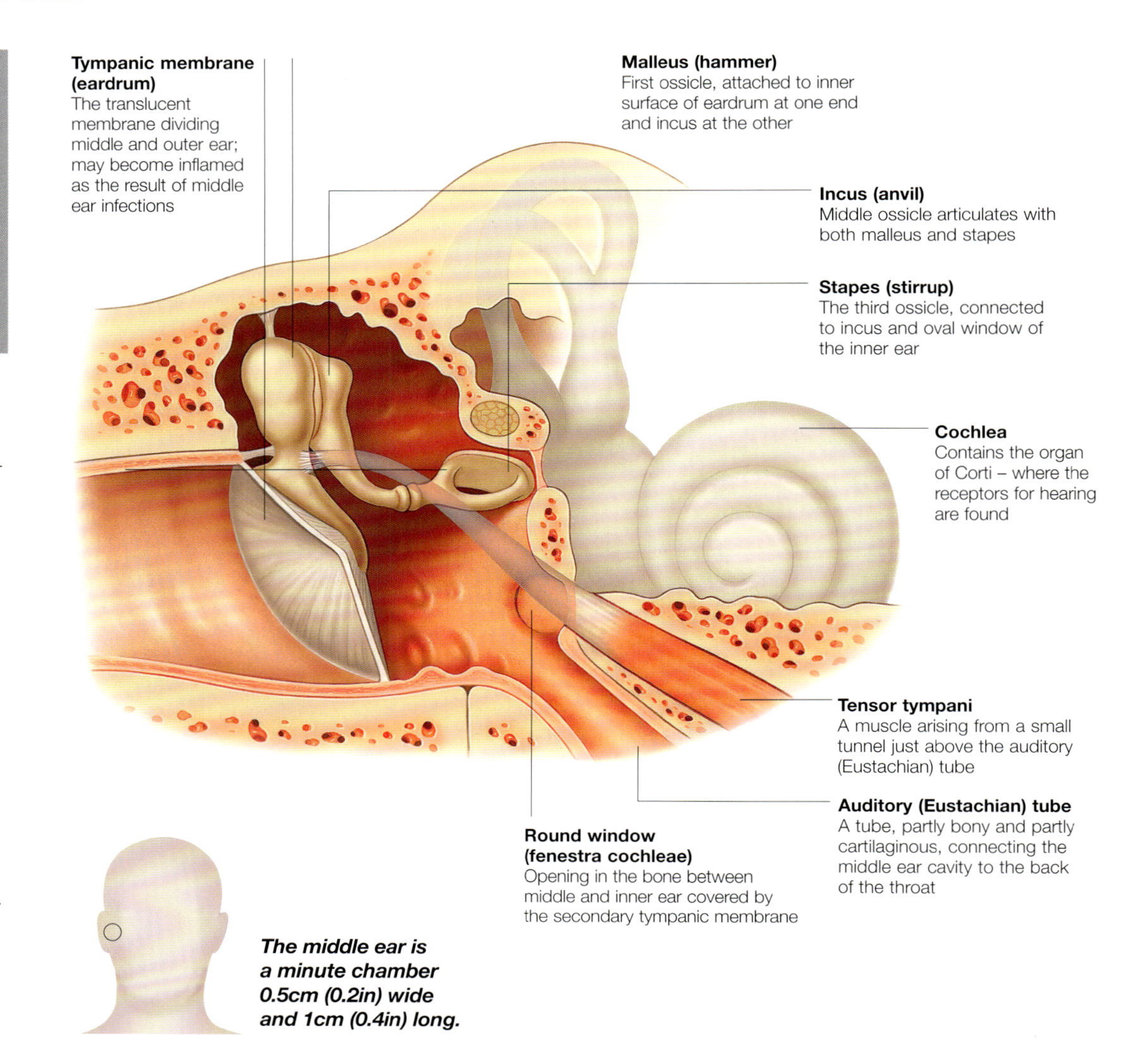

The middle ear is a minute chamber 0.5cm (0.2in) wide and 1cm (0.4in) long.

Ossicles

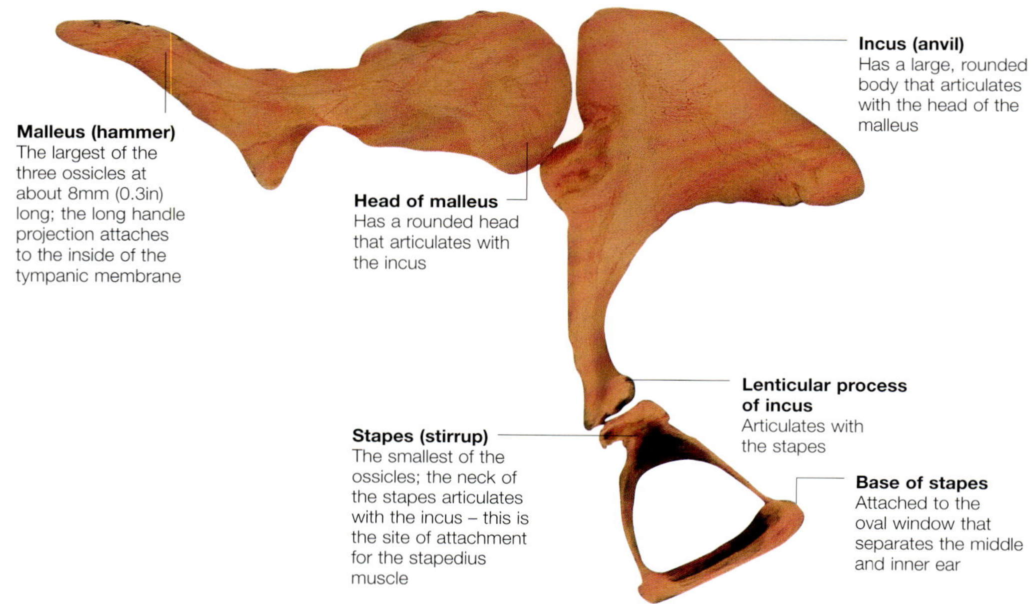

The ossicles are arranged so that vibrations in the tympanic membrane are transmitted across the middle ear to the oval window and to the inner ear. All three bones are held in place by ligaments; in addition, there are two muscles that modulate movement:

* Stapedius, the smallest skeletal muscle in the body, arises from a bony projection called the pyramid and attaches to the neck of the stapes. Contraction of this muscle helps to damp down loud sounds.
* The other muscle, the tensor tympani, has a similar damping effect but it acts by increasing the tension in the tympanic membrane.

The ossicles are three tiny bones in the middle ear. Together, they transmit sound as vibrations from the tympanic membrane to the oval window of the inner ear.

Inner Ear

This part of the ear houses the organs of balance and hearing. It contains the labyrinth that helps us to orientate ourselves, and the cochlea, the organ of hearing.

The inner ear, also known as the labyrinth because of its contorted shape, contains the organ of balance (the vestibule) and of hearing (the cochlea).

It can be divided into an outer bony labyrinth and an inner membranous labyrinth. The bony labyrinth is filled with perilymph and the membranous labyrinth contains a fluid called endolymph, with a different chemical composition.

Orientation

The membranous labyrinth contains the utricle and saccule – two linked, sac-like structures within the bony vestibule. They help detect orientation within the environment.

Related to these are the semicircular ducts lying within the bony semicircular canals. Where they are connected to the utricle, the semicircular canals enlarge to form ampullae, containing sensory receptors. Changes in the movement of the fluid in the ducts provides information about acceleration and deceleration of the head.

The cochlea is a bony spiral canal, wound around a central pillar – the modiolus. Within the cochlea are hair cells, the hearing receptors, that react to vibrations in the endolymph caused by the movement of the stapes on the oval window. They lie within the organ of Corti.

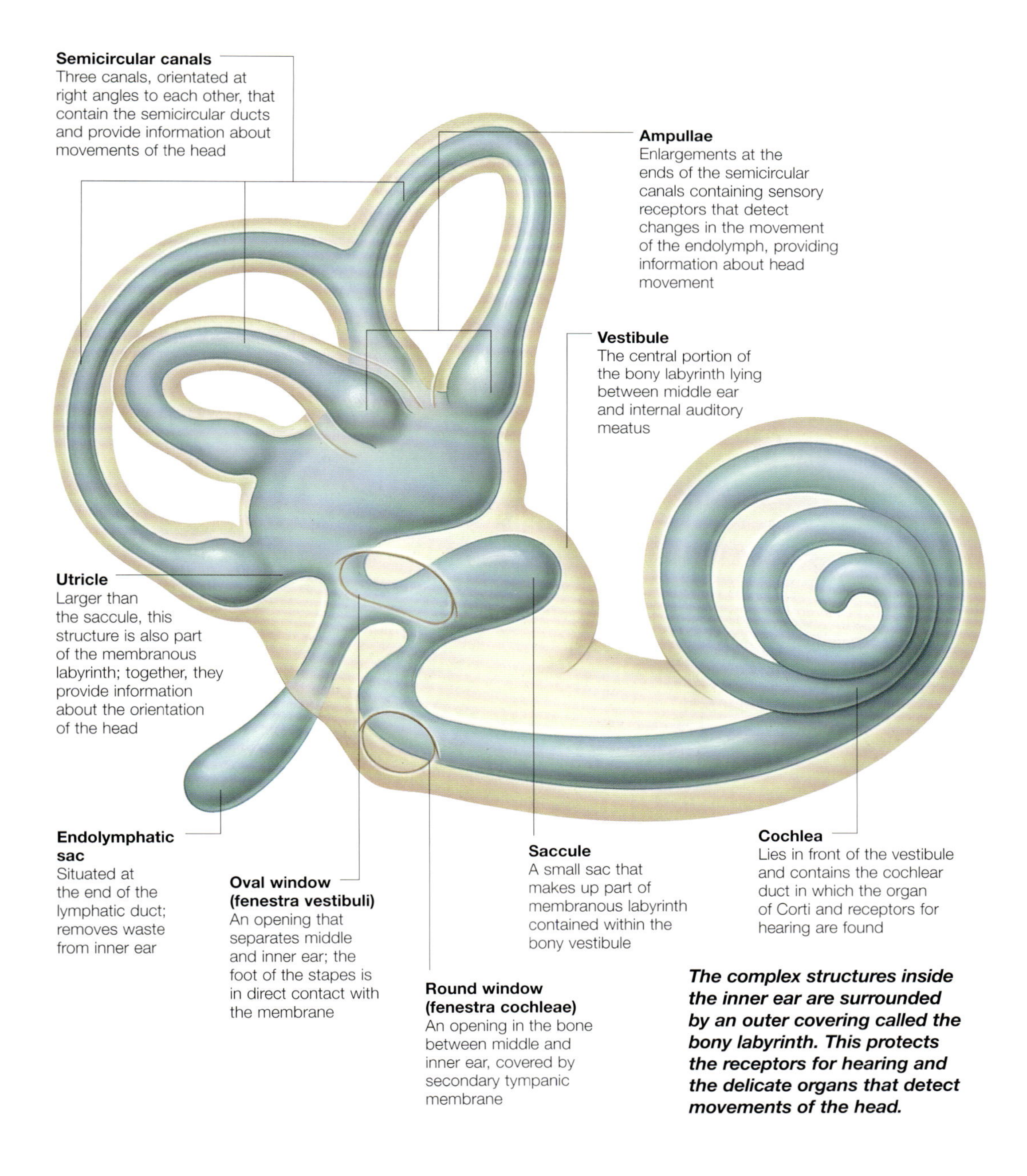

The complex structures inside the inner ear are surrounded by an outer covering called the bony labyrinth. This protects the receptors for hearing and the delicate organs that detect movements of the head.

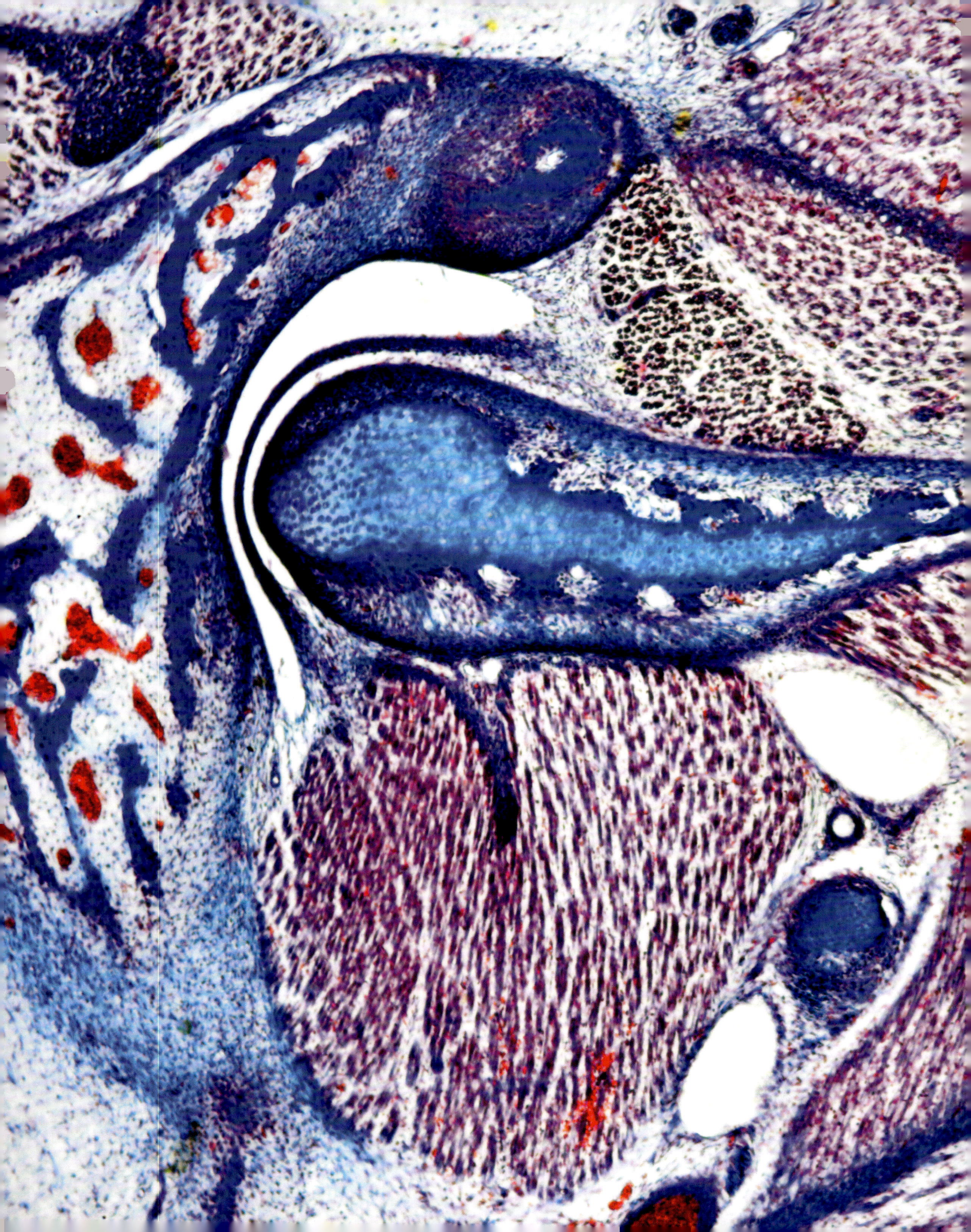

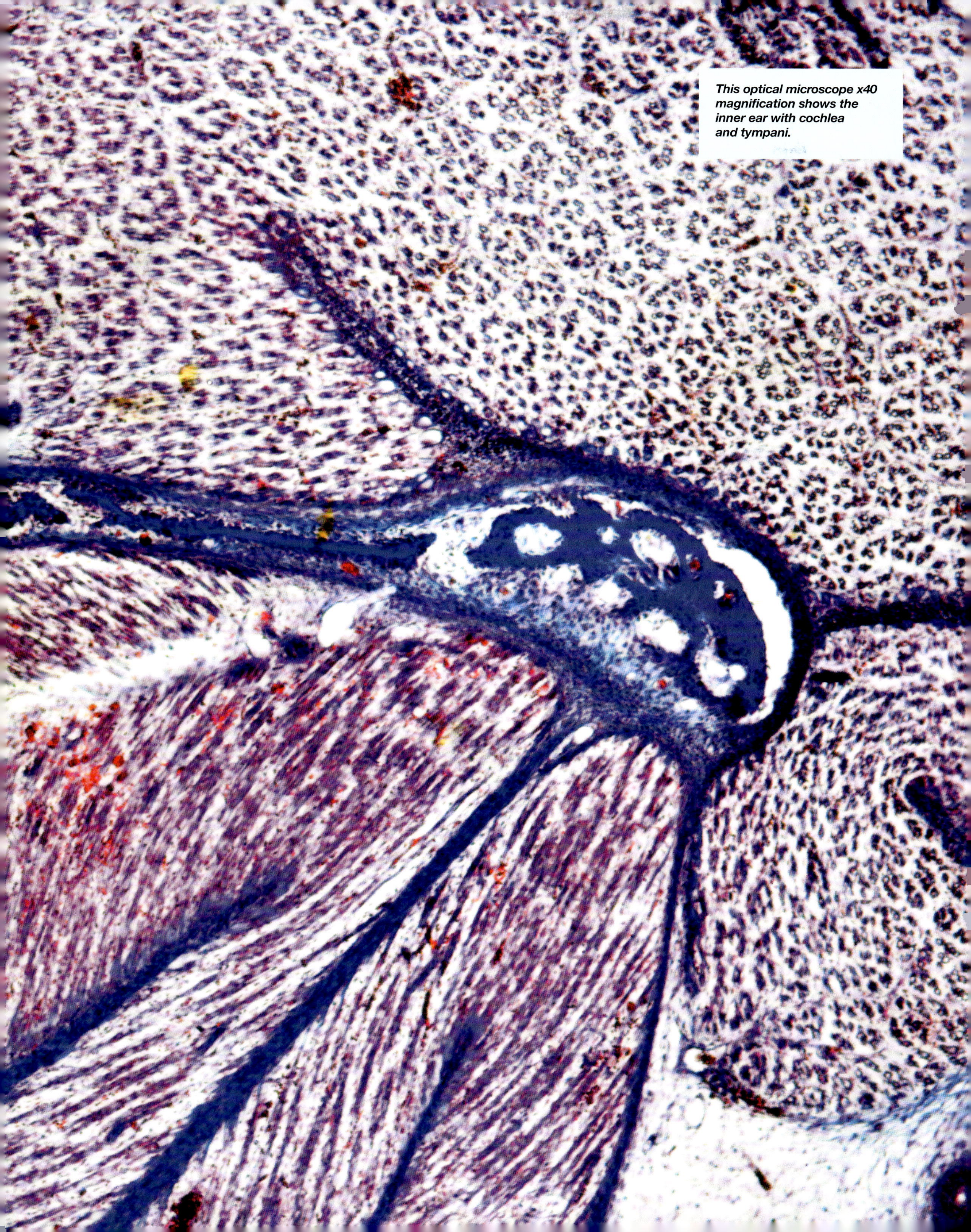

This optical microscope x40 magnification shows the inner ear with cochlea and tympani.

Balance and Orientation

The ear not only facilitates hearing, but is also responsible for maintaining balance when everyday tasks are performed.

The ear is made up of three parts. The outer, visible part of the ear (the pinna and auditory canal) gathers and focuses the sound waves. In the middle ear, the eardrum vibrates, and the three ossicles (small bones) transmit these vibrations to the inner ear.

Inner Ear

The inner ear performs two functions: the cochlea receives the sound waves and transmits them to the brain, where they are interpreted as sound; and the non-auditory or vestibular labyrinth detects changes in the body's position.

Bony Labyrinth

The part of the inner ear concerned with balance is the bony labyrinth. Within this are the vestibule, the semicircular canals and the membranous labyrinth. The membranous labyrinth is surrounded by a fluid called perilymph. Another fluid, endolymph, is contained in the membranous labyrinth. These fluids do not just fill space – they are a vital part of the whole equilibrium system.

The individual parts of the bony labyrinth are sensitive to movement, rotation and orientation of the head.

Structure of the ear

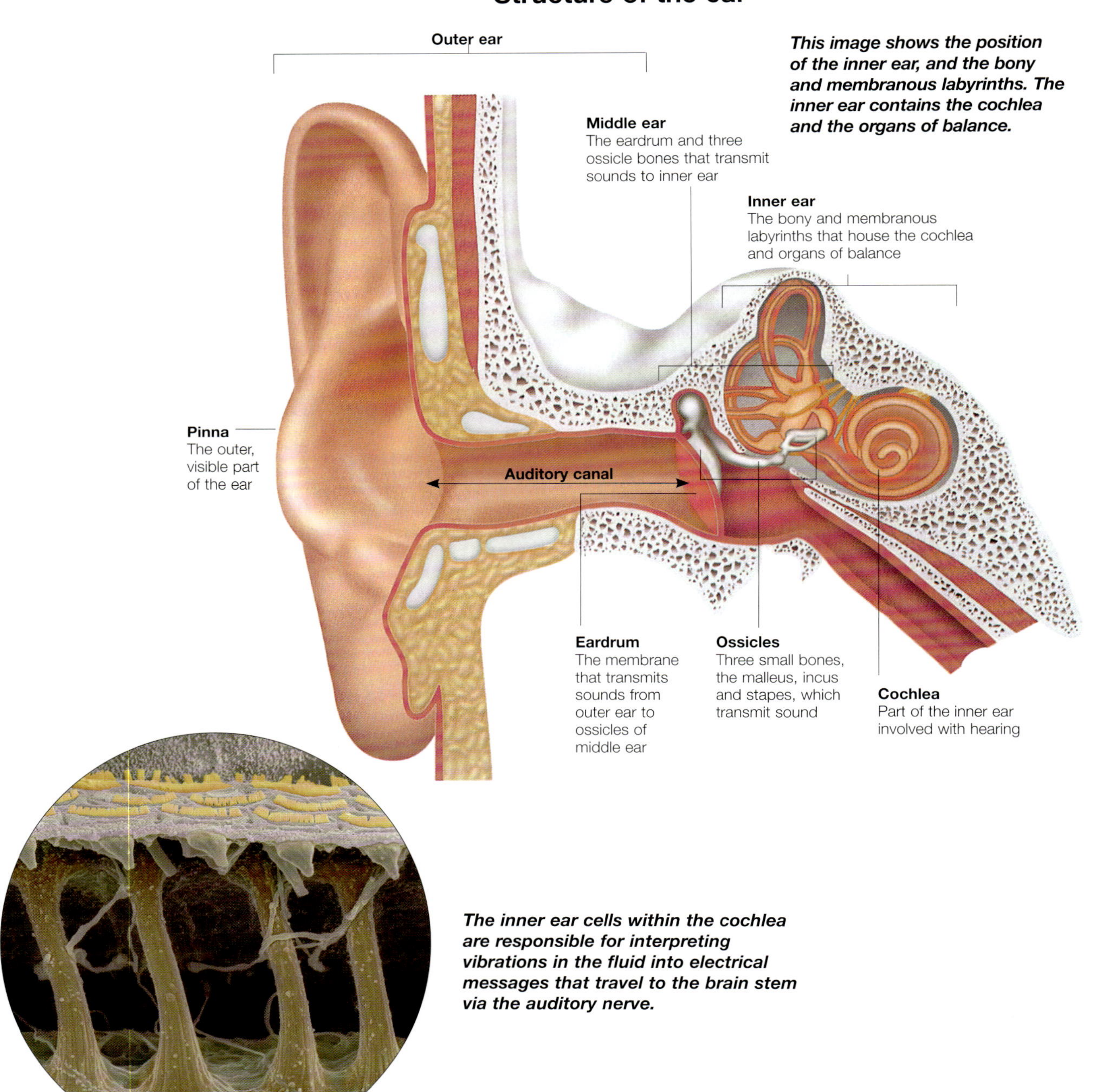

This image shows the position of the inner ear, and the bony and membranous labyrinths. The inner ear contains the cochlea and the organs of balance.

The inner ear cells within the cochlea are responsible for interpreting vibrations in the fluid into electrical messages that travel to the brain stem via the auditory nerve.

Semicircular canals and otolith organs

The tubes and chambers of the bony labyrinth protect the membranous tubes and chambers of the membranous labyrinth, with their fluids and sensors.

The semicircular canals are three bony tubes in each ear that lie at right-angles to each other. Because of their position and structure they are able to detect movement in three-dimensional space and are the parts sensitive to rotation.

Each canal has an expanded end called the ampulla, which is filled with endolymph. Receptor cells located in the ampulla of each canal have fine hairs that project into the endolymph. When we move, these projecting hairs are displaced by the movement of the endolymph. This stimulates the vestibular nerve, which sends signals to the cerebellum of the brain.

Membranous labyrinth

The vestibule contains two membranous sacs (or otolith organs) called the utricle and the saccule, which respond to our orientation. On the inner surface of each sac is a 2mm (0.07in)-wide patch of sensory cells – a macula – that monitors the position of the head.

The utricle maculae lie horizontally and provide information when the head moves from side to side. Less is known about the saccular maculae, but because they are arranged vertically they probably respond to backward and forward tilting of the head. Together, they allow the detection of all the possible positions of the head.

The sensory organs (particularly in the utricle) play a role in controlling the muscles of the legs, trunk and neck to maintain an upright position.

The inner ear

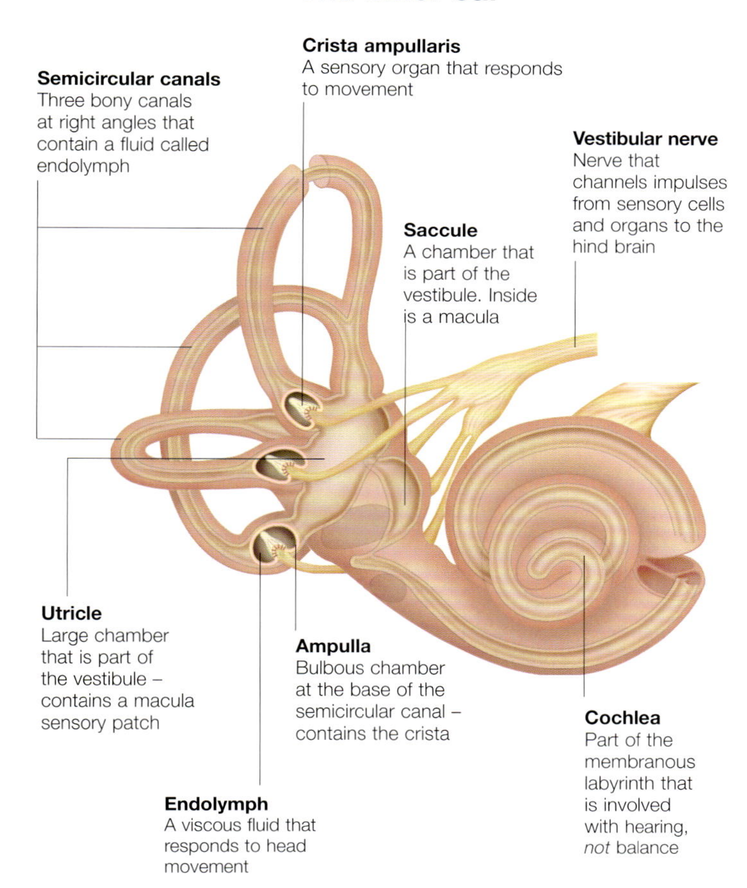

Maculae

Each macula consists of a layer of tissue known as the neuroepithelium. In this layer are sensory cells called hair cells that send continuous nerve impulses to the brain. The hair cells are covered by a gelatinous cap containing small granular particles that weigh against the hairs. When the hair bundles are deflected – because of a tilt of the head, for example – the hair cells are stimulated to alter the rate of nerve impulses being sent. Hair cells near the centre are rounded, and those on the periphery are cylindrical. This may increase sensitivity to a slight tilting of the head.

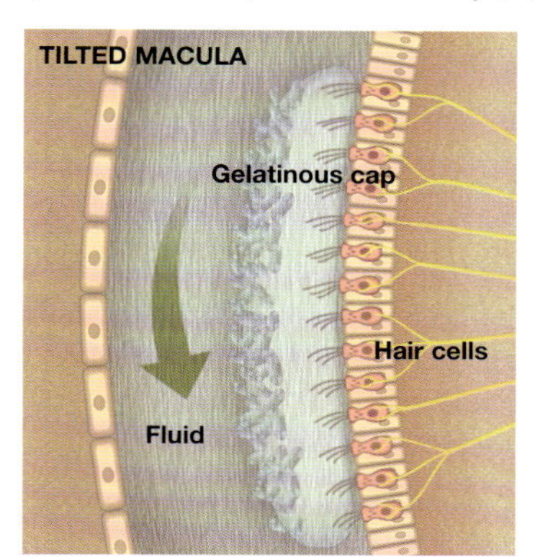

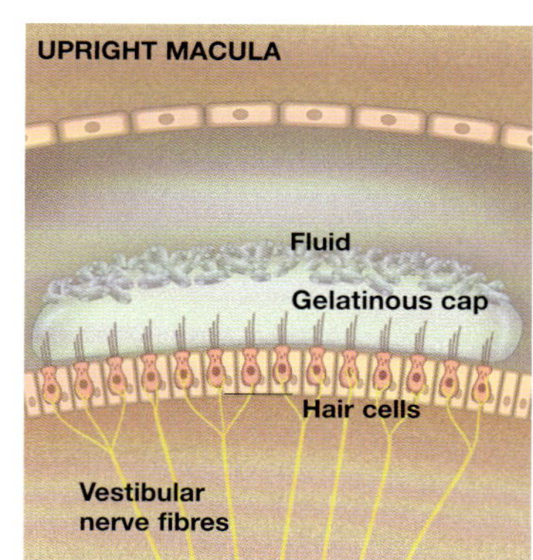

Cristae

The crista is a cone-shaped sensory structure within the ampulla – the swollen base of each semicircular canal. There are six cristae in each ear. Each crista is surrounded by a fluid called endolymph. Each crista responds to changes in the rate of movement of the head, passing information along the vestibular nerve to the brain. Sensitive hair cells are embedded in a gelatinous cone called the cupula. Any kind of head movement causes fluid to swirl past the cupula, bending it and activating the hair cells.

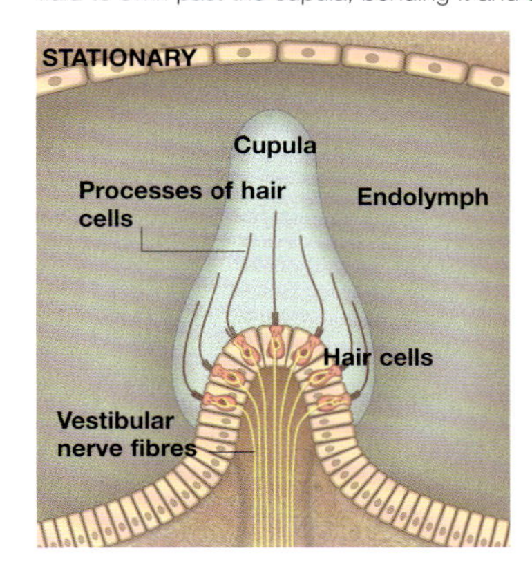

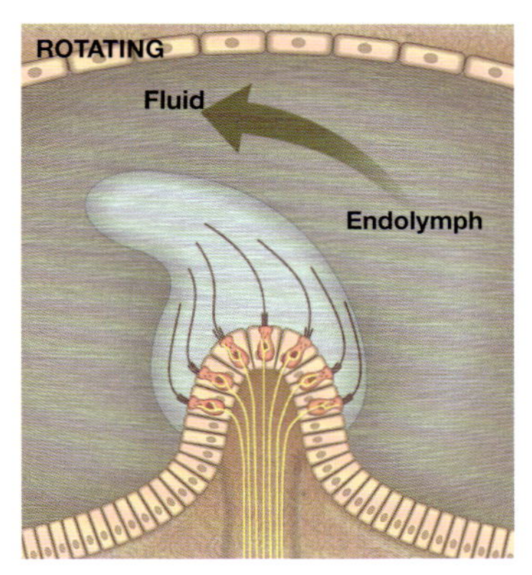

Sound Interpretation

Sound hitting the inner ear is converted into neuronal (nerve) signals. This is a complex and subtle process, which enables the brain to interpret and understand a wide range of sound.

The cochlea – the organ of hearing located in the inner ear – is a coiled bony structure containing a fluid-filled system of cavities.

The central cavity, or cochlear duct, contains the specific structure for hearing, called the organ of Corti. Located on the basilar membrane, this spiral organ contains thousands of sensory hair cells that convert mechanical movement (caused by sound vibrations resonating through the fluid) into electrical nerve impulses that are then transmitted to the cortex of the brain.

Neuronal Pathways
The neuronal pathways of the auditory system are composed of sequences of neurones arranged in series and parallel. The impulses begin in the organ of Corti and ultimately reach the auditory areas of the cerebral cortex known as the transverse temporal gyri of Heschl.

Transit Stations
As neuronal activity is transmitted towards the brain it goes through several 'transit stations'. Some of these transit stations respond in particular ways to various aspects of the auditory signal thus giving the brain more context to the sound. For example, some cochlear neurones have a sharp burst of activity at the start of a sound, called a primary-like response pattern; this informs the auditory cortex of the start of a sound sequence.

The neurones, transit stations and various brain auditory centres are found on both sides of the body. The auditory centres in the brain receive sound from the opposite ear.

Pathway of signals to the brain

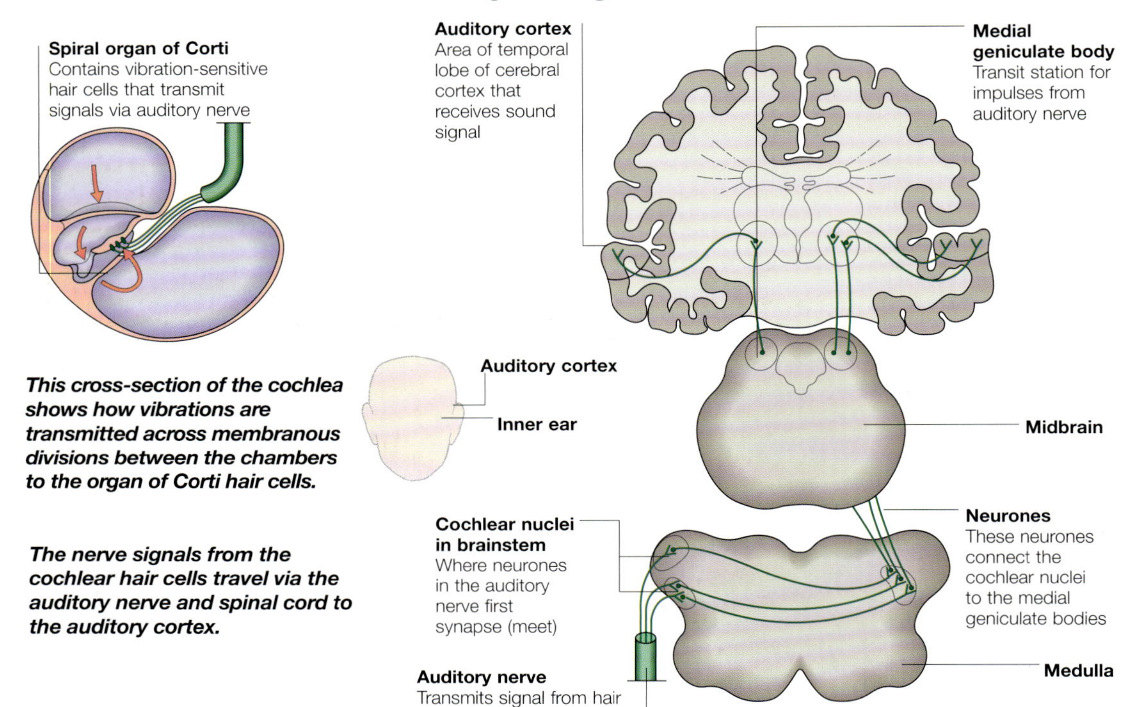

This cross-section of the cochlea shows how vibrations are transmitted across membranous divisions between the chambers to the organ of Corti hair cells.

The nerve signals from the cochlear hair cells travel via the auditory nerve and spinal cord to the auditory cortex.

Interpreting pitch of sound

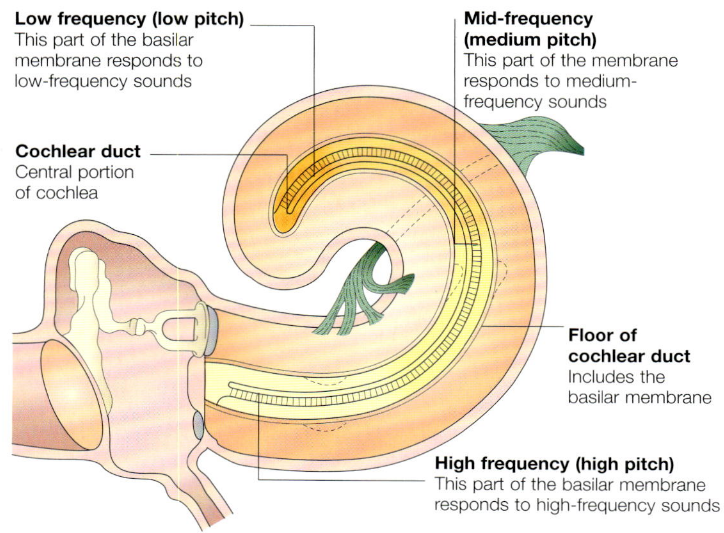

The hair cells in the spiral organ of Corti are able to convey varied tones by responding to different frequencies at locations along the basilar membrane, thereby contributing to the sound-filtering process. Cells at the base of the basilar membrane respond more readily to high-frequency sound waves, while those at the tip are more sensitive to low-frequency sounds. This is equivalent to how a piano emits sounds, with one end of the keyboard producing high notes and the other end low ones.

The stimulation of groups of hair cells at specific locations along the basilar membrane allows the brain to differentiate sounds of different frequencies or pitch.

However, there are additional subtleties used to transduce the different tones.

Imagine a tuning fork that emits the note 'A' is struck. The sound waves reaching the cochlea will all resonate at a frequency of 440 cycles per second (Hertz). This triggers the basilar membrane to vibrate at 440 times a second. Yet there is a particular section of the basilar membrane that is made in such a way that it will vibrate with the largest amplitude at 440 times a second. This will then set the neurones from that region signalling at 440 times a second.

How the brain interprets sound signals

Once nerve impulses are transmitted to the auditory cortex, several areas of the brain are responsible for interpreting the signals.

There are several areas of the temporal lobe on both sides of the brain responsible for interpreting different aspects of sound. These areas receive additional contextual information from the various staging posts as the basic neuronal signals make their way to the auditory cortex.

Identifying sounds

The brain identifies sounds by recognizing essential features of each sound, such as volume, pitch, duration and intervals between sounds. From these elements the brain creates a unique acoustic 'picture' of each sound, in much the same way that a colour television can reproduce the whole spectrum of colours on a screen using dots of just three colours.

The auditory cortex also has to separate many different sounds arriving at the same time, filtering and analyzing them to produce meaningful information. The brain uses the context in which sound is received to make certain assumptions about what it will hear. For example, if the visual cortex tells it that a man is speaking it will expect speech of a certain pitch.

Auditory association cortex

The auditory association cortex is used to process complex sounds, when many sound waves arrive at the same time. This is particularly important in language recognition, and damage to this area results in a person detecting sounds without being able to distinguish between them.

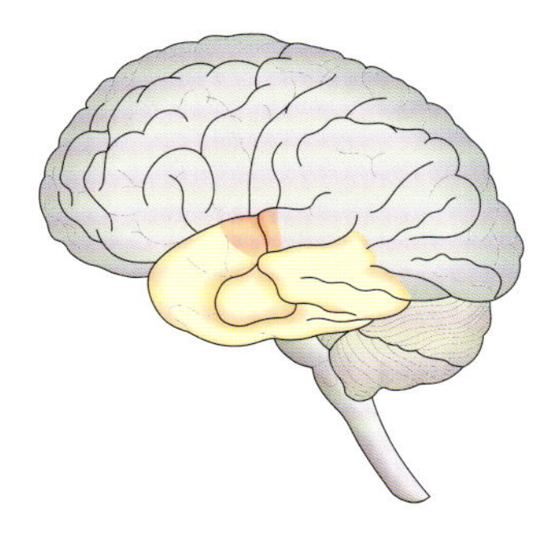

The auditory cortex (pink) recognizes and analyzes sounds. The association cortex (yellow) acts to distinguish more complex features of the sounds.

Locating sound

Listening to sound from behind

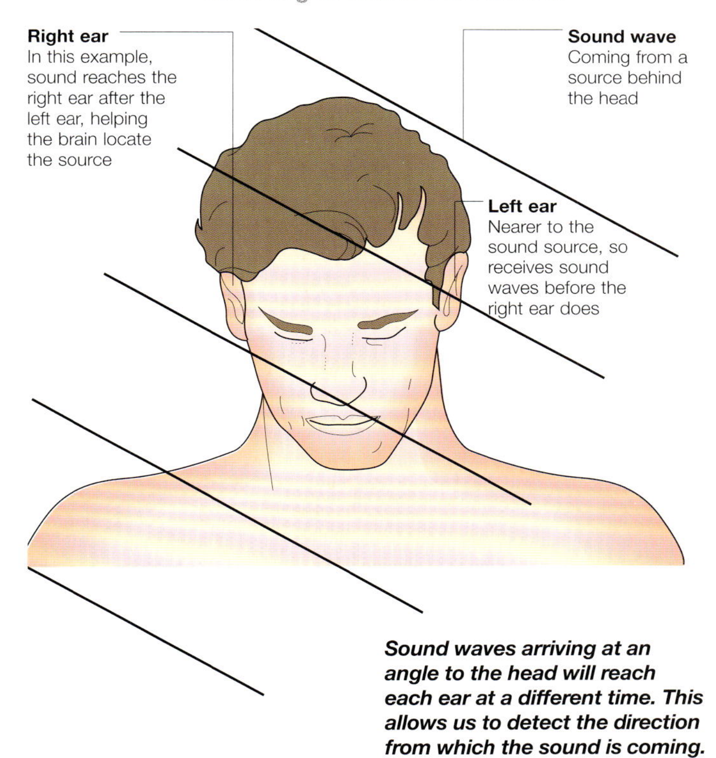

Sound waves arriving at an angle to the head will reach each ear at a different time. This allows us to detect the direction from which the sound is coming.

The brain is very accurate at integrating information in order to locate sound and can detect minute differences in the timing and intensity of the sound reaching the two ears. A sound wave will reach the closer ear a fraction of a second before it reaches the other. The brain can interpret the time difference to distinguish the direction of the sound.

In addition, if sound is coming from the side, the head causes a 'sound shadow' screening one ear, so that it receives less sound than the other ear. Often, our response is to turn our head in the direction of the sound, to enable both ears to hear equally. Locating a sound source also depends on sound waves deflecting off the irregular surface of the pinna, which vary with the angle at which the sound approaches the ear.

As we develop, we learn that particular sound differences are associated with particular directions, and from this can detect the direction of a source.

Nasal Cavity

The nasal cavity runs from the nostrils to the pharynx, and is divided in two by the septum. The roof forms part of the floor of the cranial cavity.

The nasal cavity has several functions:
- Serves as the entrance to the respiratory system, warming the air as it passes to the lungs
- Functions as the organ of smell
- The passages and chambers help to form speech.

The nasal cavity is partitioned into two halves by a vertical plate called the nasal septum, which is part bone and part cartilage. Each half of the nasal cavity is open in front at the nostril, and opens into the pharynx at the back through an opening called the choana.

Nasal Cavity Roof

The roof of the nasal cavity is arched from front to back. The central part of this roof is the cribriform plate of the ethmoid bone, a strip of bone perforated with a number of holes. This forms part of the floor of the cranial cavity, which contains the brain.

Running through the sieve-like cribriform plate from the nasal cavity to the brain is the olfactory nerve, which transmits the sensation of smell.

Olfactory Region

The nasal cavity is lined by mucous membranes. The olfactory receptors, which detect odours, lie in this membrane just above the nasal conchae, at the top of the nasal cavity.

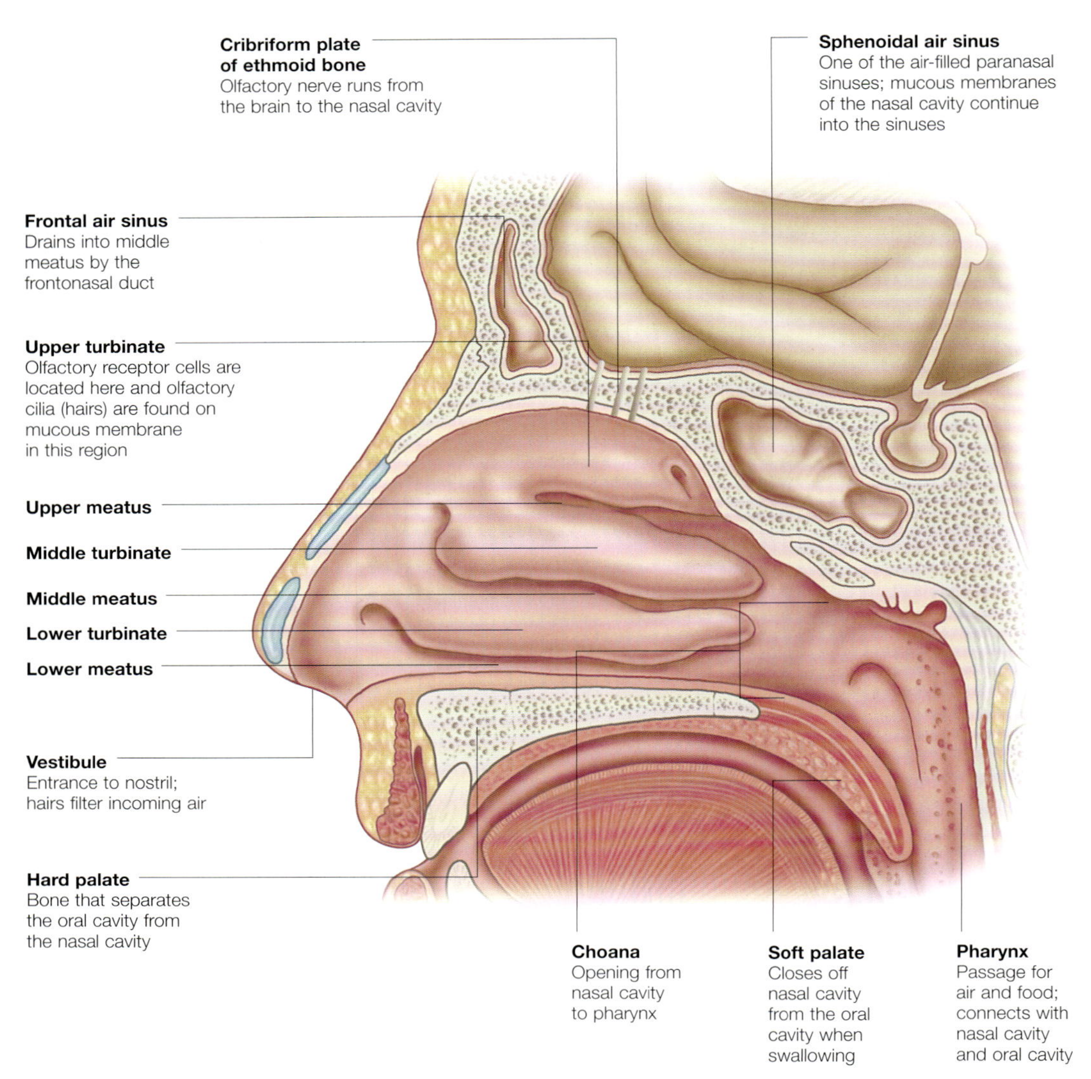

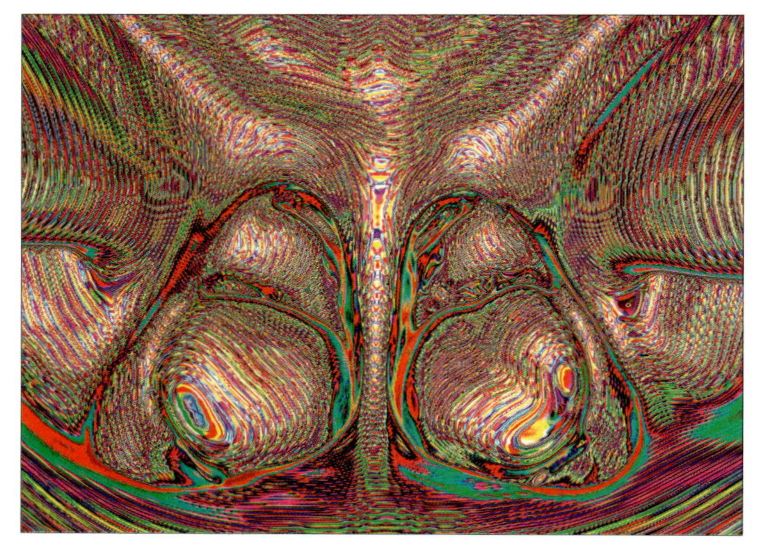

The back of the nose and nasopharynx are shown here in a coloured 3D CT scan. The nasal septum and turbinates are in the centre.

Smell

The nostrils carry air towards specialized cells located just below the front of the skull. These cells are able to detect thousands of different types of odours at very low concentrations.

Our sense of smell is in many ways similar to our sense of taste. This is because both taste and smell rely on the ability of specialized cells to detect and respond to the presence of many different chemicals.

The olfactory (smell) receptors present in the nose 'transduce' (convert) these chemical signals into electrical signals that travel along nerve fibres to the brain.

Olfactory Receptors

Odours are carried into the nose when we inhale, and dissolve in the mucus-coated interior of the nasal cavity. This mucus acts as a solvent, 'capturing' the gaseous odour molecules. It is continuously renewed, ensuring that odour molecules inhaled in each breath have full access to the olfactory receptor cells.

A small patch of mucous membrane located on the roof of the nasal sinuses, just under the base of the brain, contains around 40 million olfactory receptor cells. These are specialized nerve cells that are responsive to odours at concentrations of a few parts per trillion. The tip of each olfactory cell contains up to 20 hairs, known as cilia, that float in the nasal mucus. These greatly increase the surface area of the cell, thereby enhancing its ability to detect chemicals.

When odour molecules bind to receptor proteins on an olfactory cell they initiate a series of nerve impulses. These impulses travel along the cell's axon (a nerve fibre attached to the nerve cell body), which projects through the cribriform plate, the thin layer of skull immediately above the olfactory epithelia. The olfactory cells in turn communicate with other nerve cells, located in the olfactory bulb, which carry information, via the tiny olfactory nerves (also known as cranial nerves I), to the brain.

The olfactory system

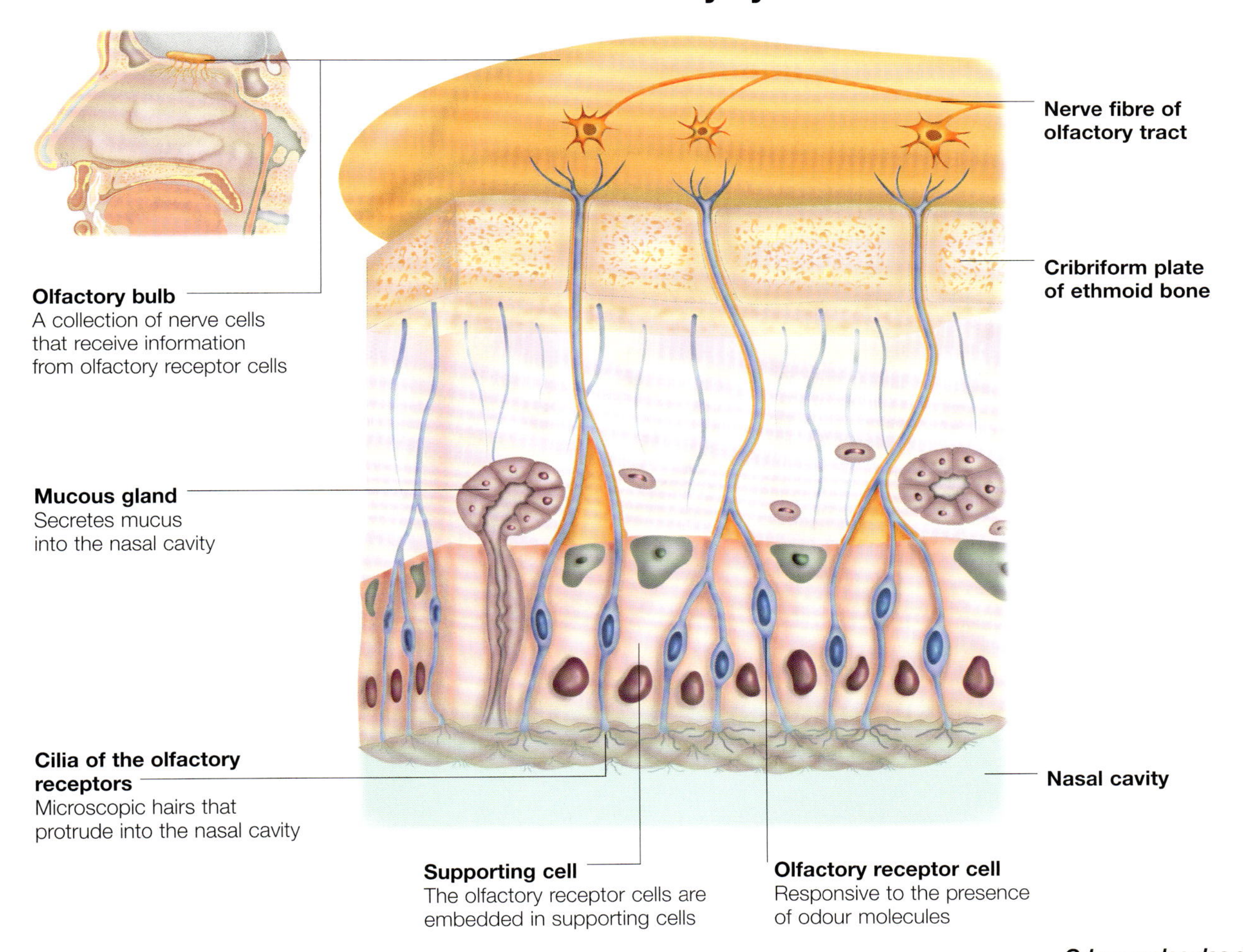

Odour molecules dissolve in mucus secreted into the nasal cavity. Specialized receptors respond to odour molecules by sending nervous impulses to the brain via a structure called the olfactory bulb.

Taste

> We have approximately 10,000 taste buds, located mainly on the surface of the tongue and the soft tissues of the mouth.

Taste, together with smell, is a chemical sense that is reliant on the binding of chemicals from food to receptors located in specific cells, the taste buds. These cells transmit information via nerves to the brain for interpretation as 'tastes'.

The tongue is the main organ of taste, as ingested food must pass through the mouth. The tongue's upper surface is covered in many small projections called papillae, and it is around these structures that most of the taste buds are clustered. However, a few are found elsewhere in the mouth, such as on the pharynx, the soft palate and the epiglottis.

Papillae
There are three major types of papillae. In increasing order of size, these are filiform (cone-like), fungiform (mushroom-shaped) and circumvallate (circular). In humans, most taste buds are the last two types. Fungiform papillae are distributed all over the tongue, with a higher number along the sides and the tip. Circumvallate papillae are the largest and there are between 7 and 12 towards the rear of the tongue, arranged in a shallow inverted 'V' form. Taste buds are found in the sides of the circumvallate papillae and on the upper surfaces of the fungiform papillae.

Cellular Structure
Each taste bud is made up of 40 to 100 epithelial cells, of which there are three types: supporting, receptor and basal cells. Receptor cells are also called the gustatory or taste cells, which give rise to taste sensations. The supporting cells form the major part of the taste bud and separate the receptor cells from each other. Taste bud cells are replaced continually throughout life and the typical lifespan is about 10 days.

Parts of the tongue

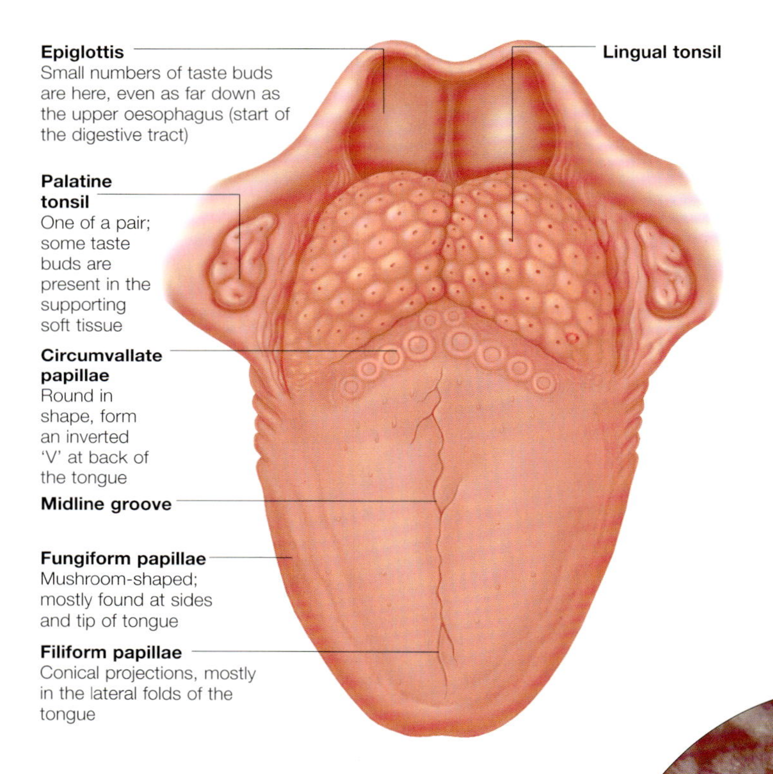

Right: This enlarged image of receptor taste buds show the villae on the tongue.

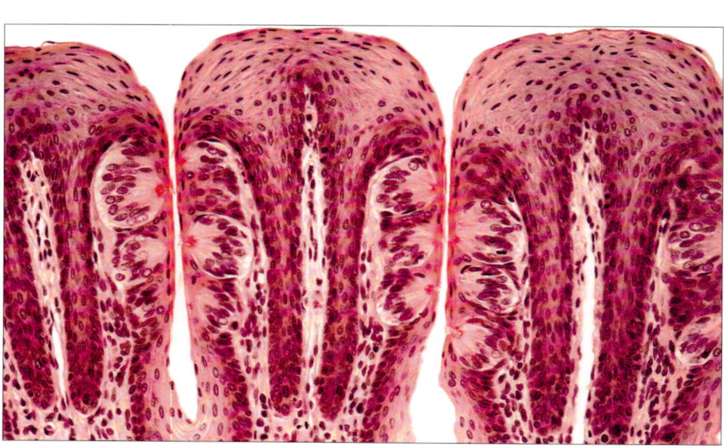

Left: This cross section shows taste buds as round zobes along the grooves between the foliate papillae.

Tasting mechanism

Once food is dissolved by saliva in the mouth, the taste buds on the tongue's surface are stimulated. Gustatory cells then convert the chemical reaction into nerve impulses.

When a food chemical binds to a gustatory cell, nerve impulses are sent to the thalamus, the part of the brain that receives sensory information. The thalamus processes the impulses and transmits them to the part of the brain associated with the sense of taste – the gustatory or taste cortex. The thalamus is unable to discern to any great extent whether the taste experience is positive or negative.

Gustatory cortex

The gustatory cortex identifies the food as good or bad and judges whether to continue eating or not. For a substance to be tasted it must be dissolved in saliva and come into contact with the gustatory hairs. From there, nerve impulses are set up to transmit impulses to the brain.

Information to the brain

A branch of the facial nerve transmits impulses from the taste buds in the front two-thirds of the tongue, and the lingual branch of the glossopharyngeal nerve serves the rear third of the tongue. It is thought that there is a two-way flow of information to the brain regarding taste and the need to eat certain foods to meet the body's requirements.

Taste sensations

The gustatory, or taste, cells in the different regions of the tongue have varying thresholds at which they are activated. In the bitter region of the tongue the cells can detect substances such as poisons in very small concentrations. This explains how the apparent disadvantage of its location is overcome and how its 'protective' nature works. The sour receptors are less sensitive and the sweet and salty receptors are the least sensitive of all. The taste receptors react rapidly to a new sensation, usually within three to five seconds.

Taste depends to a great extent on our sense of smell, which explains why when we have a heavy cold, we are unable to taste food.

Taste bud structure

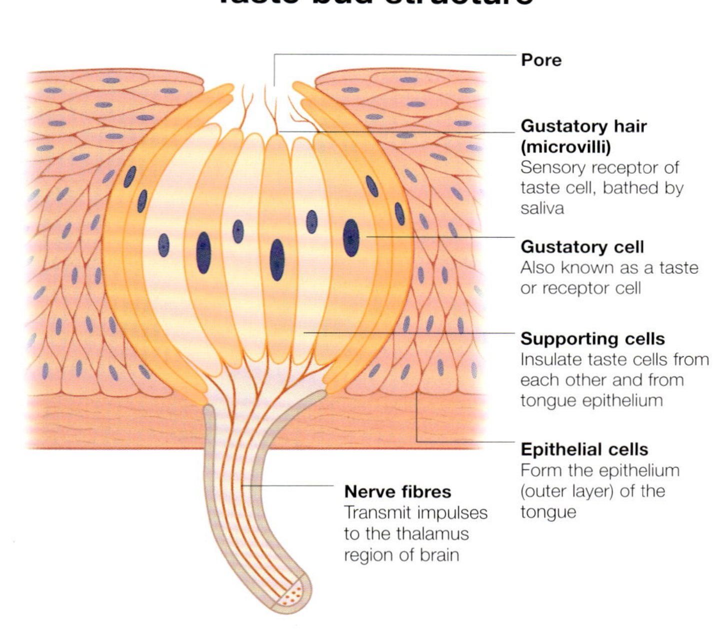

Papilla cross-section

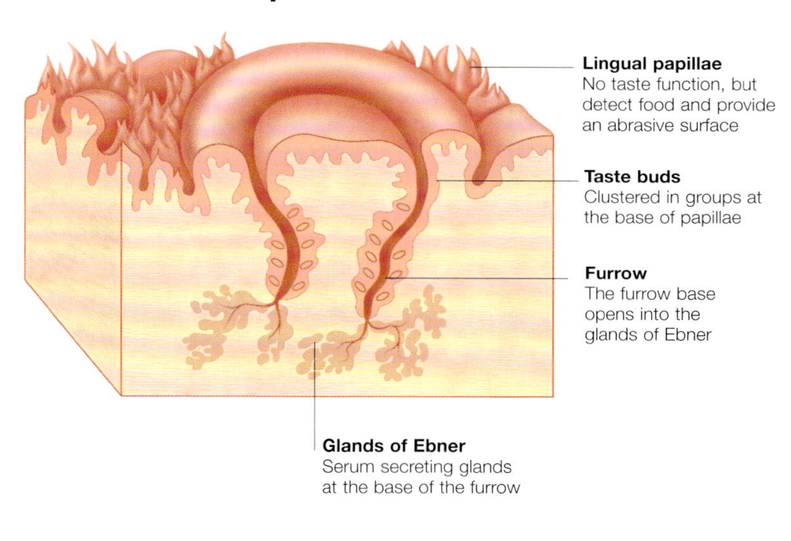

Which part of the tongue tastes what?

Taste sensations can be grouped into four main categories. These are: sweet, sour, salty and bitter. Different parts of the tongue are more sensitive to different taste sensations, although there is no structural difference between the taste buds in the different areas.

The tip of the tongue is most sensitive to sweet and salty tastes, the sides of the tongue are most sensitive to sour tastes and the back of the tongue tastes bitter flavours most strongly. However, these differences are not absolute, as most taste buds can respond to two or three – and sometimes all four – taste sensations. Certain substances seem to change in flavour as they move through the mouth: saccharin, for example, tastes sweet at first, but goes on to develop a bitter aftertaste.

Many natural poisons and spoiled foods have a bitter flavour. It is perhaps likely, therefore, that bitterness receptors are located at the back of the tongue as a protective measure. In other words, the back of the tongue screens for 'bad' food and rejects it.

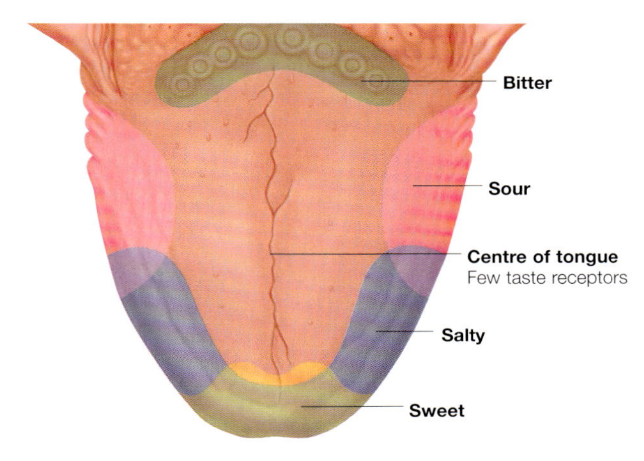

Touch

Touch is the body's ability to perceive sensation on the skin and to send messages to the brain. The brain identifies the source of the sensation and enables us to make decisions.

Touch is made possible by the presence of millions of tiny sensory receptors that are located all over the body, just under the skin. These receptors are specialized nerve endings that receive information about temperature, pain or pressure and convert it into a nerve impulse. The more receptors there are in a particular area, the more sensitive it is. The nerve impulses are conducted along a neural pathway to the somatosensory cortex of the brain, where they are converted into sensations. Sensations enable us to act fast if we are in danger, for example, removing a hand from extreme heat or cold.

Types of Receptors

The sensory receptors linked with touch (temperature, pain, pressure and vibration) are found near the body's surface and are known as exteroceptors. They can be classified according to the external stimuli that they detect:

- Mechanoreceptors – detect stimuli such as gentle touch, pressure or vibration
- Thermoreptors – sensitive to temperature, such as heat or cold
- Nocireceptors – pain detectors that are sensitive to actual tissue damage or extreme pressure.

Mechanoreceptors

There are four main types of mechanoreceptor: Meissner's corpuscles, Merkel's discs, Ruffini's corpuscles and Pacinian corpuscles. All these receptors require only very light stimuli to respond with an impulse.

- Meissner's corpuscles are mainly found in the tips of the fingers, palms of the hands and soles of the feet, genitals, eyelids, tip of the tongue and lips.
- Merkel's discs are located in similar areas to the Meissner's corpuscles.
- Ruffini's corpuscles are located deeper in the dermis and other deeper tissues. They respond to firm, continuous touch.
- Pacinian corpuscles are concerned with pressure and are found in the subcutaneous layer under the skin as well as in other areas of the body, such as around joints, mammary glands and genitals.

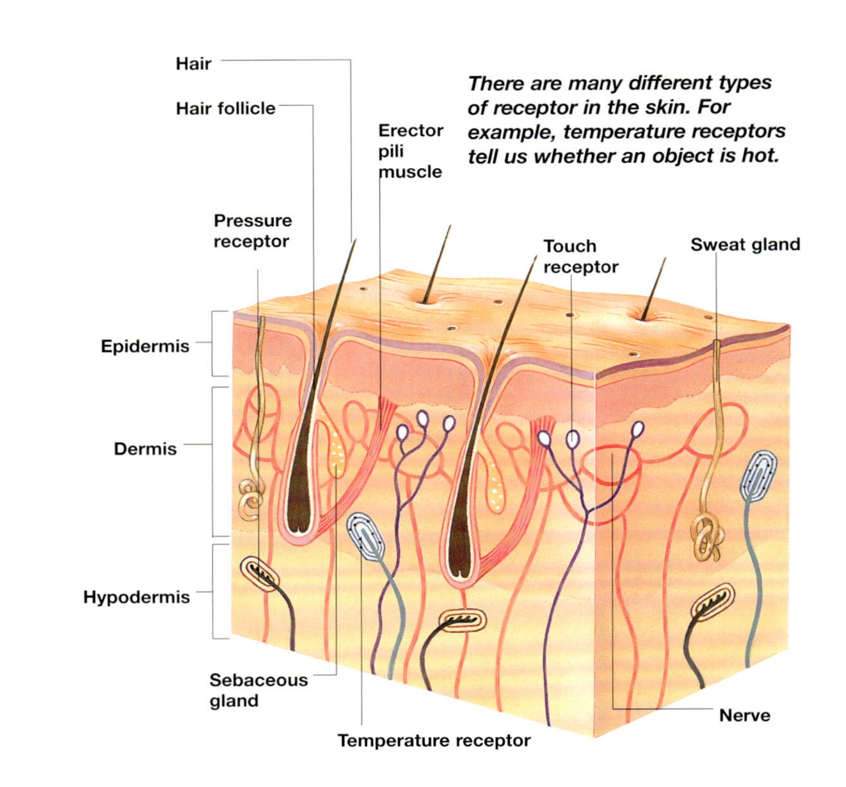

There are many different types of receptor in the skin. For example, temperature receptors tell us whether an object is hot.

Types of receptors

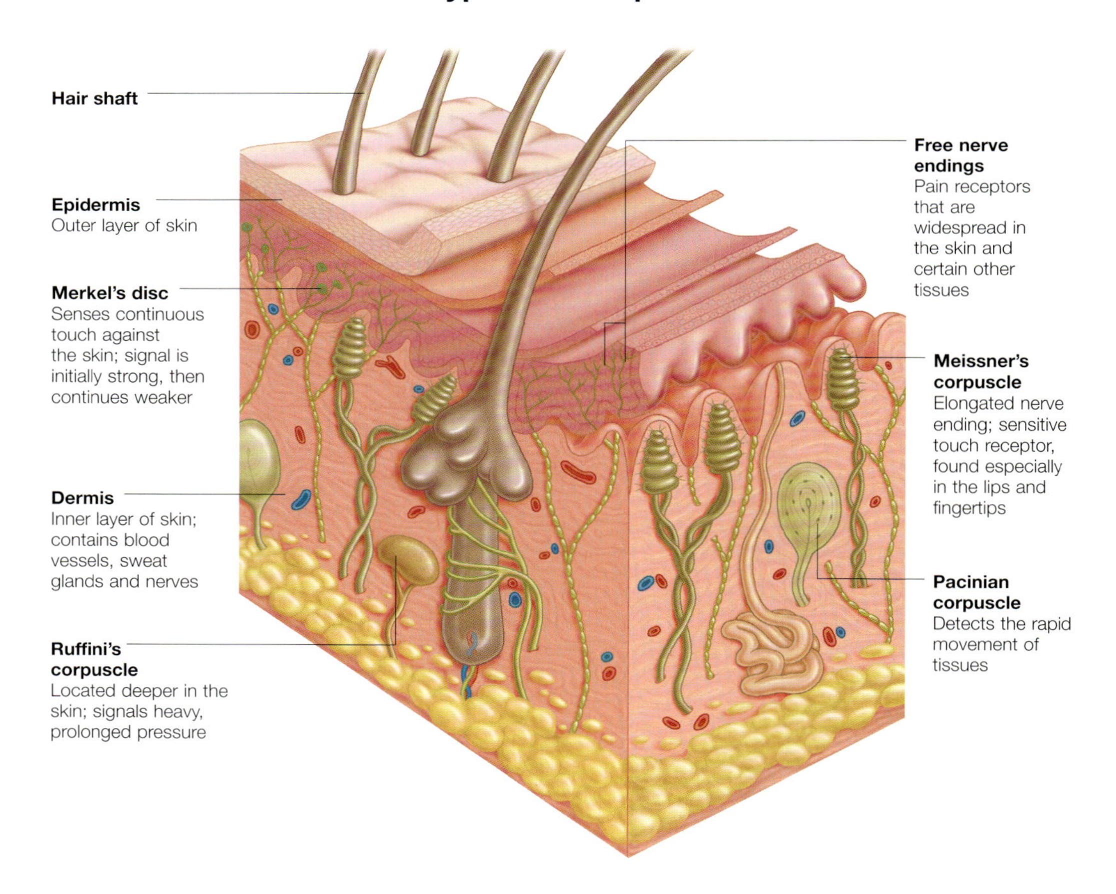

Pathway to the brain

The sensory receptors in the skin receive information, which travels as impulses along the nerve pathway to the brain, either directly or via the spinal cord.

A specialized location in each of the two cerebral hemispheres of the brain, known as the primary somatosensory cortex, receives and processes most of the information from the sensory receptors around the body. Each cerebral hemisphere receives information from the opposite side of the body about touch, pain, temperature and the position of joints and muscles (proprioception).

Pain

An event that causes damage to the tissues brings about the release of chemicals, such as serotonin or histamine. These chemicals are detected by special sensory cells called free nerve endings (also known as nociceptors) located in the superficial layers of skin as well as in internal organs. The sensory nerves send nerve impulses to the brain where the information is analyzed and perceived as pain. In response to pain signals, the brain produces chemicals called endorphins, the body's own painkillers.

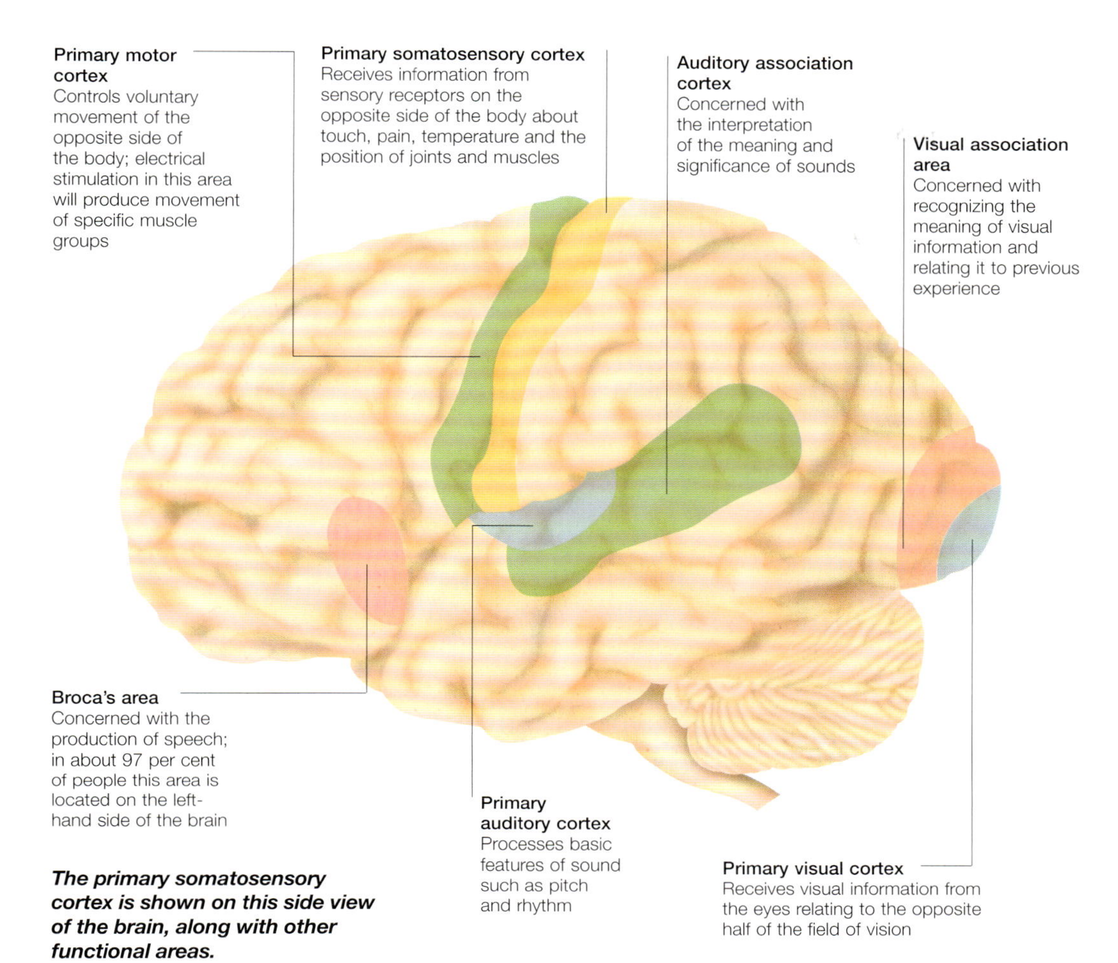

The primary somatosensory cortex is shown on this side view of the brain, along with other functional areas.

Motor and sensory body map

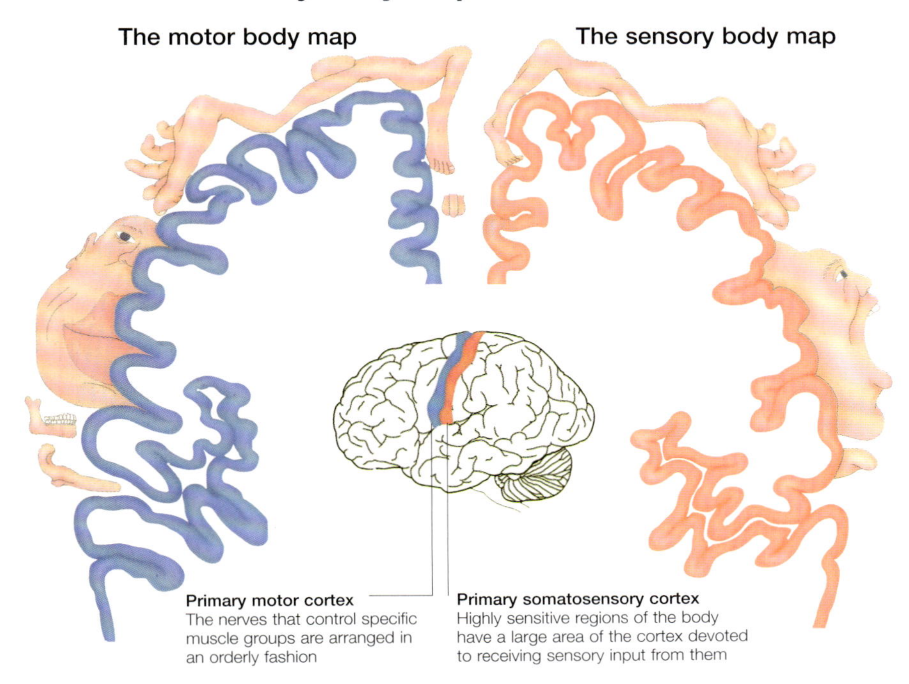

The body surface is represented in the sensory and motor regions of the cerebral cortex in an orderly arrangement.

A Canadian neurosurgeon called Wilder Penfield, working in the 1950s, mapped the regions of the sensory cortex that receive input from the different regions of the body. He stimulated the surface of the brain in locally anaesthetized patients and asked them to describe what they felt.

Penfield discovered that stimulation of regions of the postcentral gyrus produced tactile sensations in specific regions of the opposite side of the body.

Other research has shown that the volume of motor cortex devoted to different areas of the body is related to the degree of fine control and complexity of the movements involved, rather than the muscle bulk.

The cortex does not receive, or send out, the same amount of information from every region of the body.

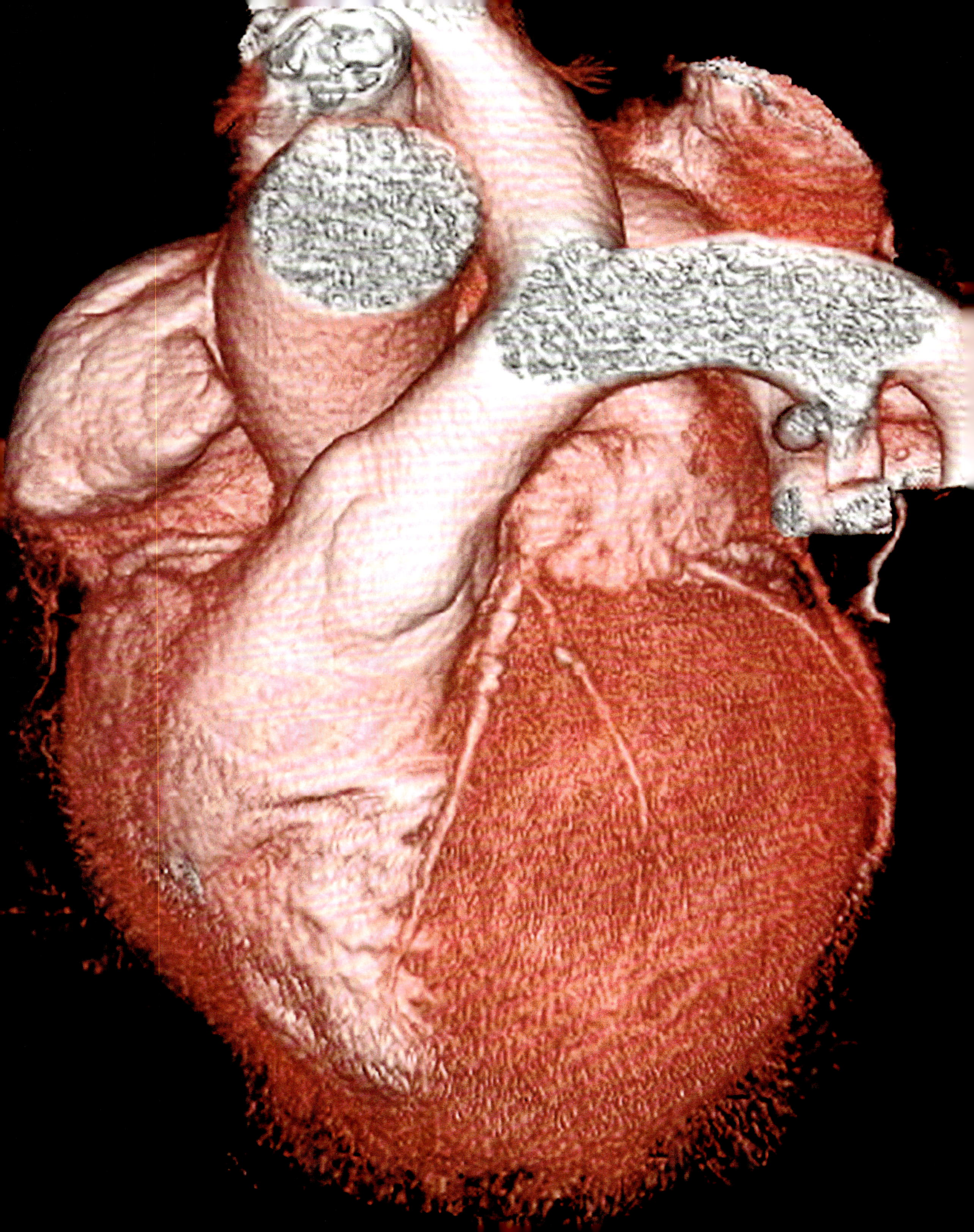

CHAPTER 6

The Cardiovascular System

The blood, heart and blood vessels together constitute the cardiovascular system, the main route of delivery of oxygen, nutrients, chemicals and hormones to all the organs and tissues of the body. This whole transport system (including the heart) can be divided into two halves: the arterial system, which carries oxygenated blood throughout the body keeping organs and tissues functioning; and the venous system, which returns deoxygenated blood to the lungs and carries waste products away from the tissues to be processed. The heart is the working pump for this process, ceaselessly beating to send our total blood volume throughout the entire body about once every minute.

The arterial and venous circulation to the limbs only are described in this section, whereas blood supply to other organs and tissues in the body are outlined in the relevant section of the book.

Opposite: This view of a healthy heart clearly shows the great vessels leaving the chambers.

Overview of Blood Circulation

There are two blood vessel networks in the body. The pulmonary circulation transports blood between the heart and lungs; the systemic circulation supplies blood to all parts except the lungs.

The blood circulatory system can be divided into two parts:

- Systemic circulation – vessels that carry blood to and from the tissues of the body
- Pulmonary circulation – vessels that carry blood through the lungs to take up oxygen and release carbon dioxide.

Systemic Arterial System

The systemic arterial system carries blood away from the heart to nourish the tissues. Oxygenated blood from the lungs is first pumped into the aorta via the heart. Branches from the aorta pass to the upper limbs, head, trunk and the lower limbs in turn. These large branches give off smaller branches, which then divide again and again. The tiniest arteries (arterioles) feed blood into capillaries. It is in the capillaries that fluid, nutrients, hormones and other substances are exchanged with the tissues.

Pulmonary Circulation

With each beat of the heart, blood is pumped from the right ventricle into the lungs through the pulmonary artery (this carries deoxygenated blood). After many arterial divisions, the blood flows through the capillaries of the alveoli (air sacs) of the lung to be reoxygenated. The blood eventually enters one of the four pulmonary veins. These pass to the left atrium, from where the blood is pumped through the heart to the systemic circulation.

Aortic arch

Right pulmonary artery

Left pulmonary artery

The pulmonary circulation involves the flow of blood between the heart and lungs. In the lungs, blood gains oxygen and loses waste carbon dioxide.

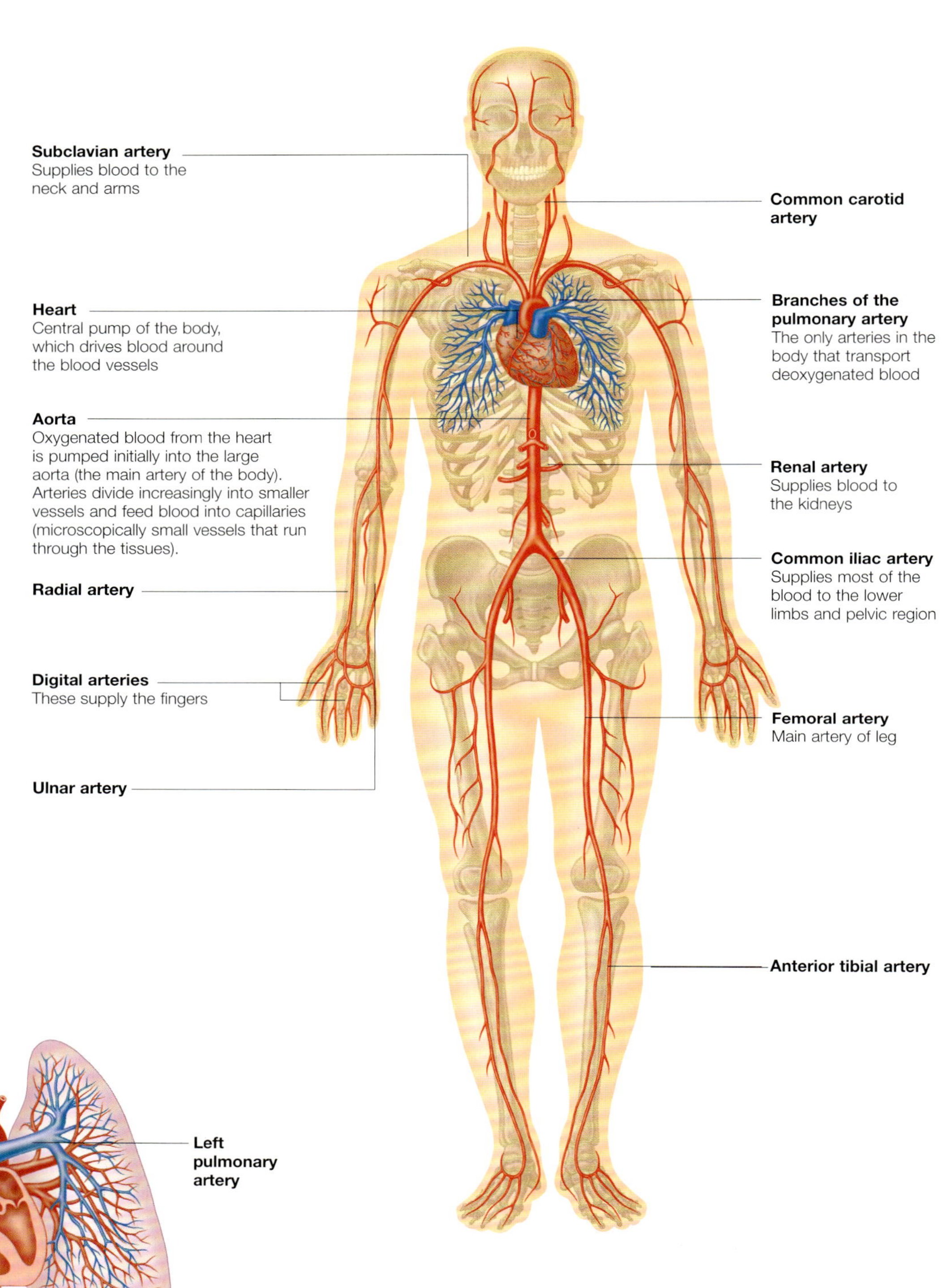

The vessels of the systemic arterial system carry blood from the heart to the tissues. Blood carries oxygen and essential nutrients around the body.

Venous system

The systemic venous system carries blood back to the heart from the tissues. This blood is then pumped through the pulmonary circulation to be reoxygenated before entering the systemic circulation again.

Veins originate in tiny venules that receive blood from the capillaries. The veins converge upon one another, forming increasingly large vessels until the two main collecting veins of the body, the superior and inferior vena cavae, are formed. These then drain into the heart. At any one time, about 65 per cent of the total blood volume is contained in the venous system.

Differences

The systemic venous system is similar in many ways to the arterial system. However, there are some important differences:

- Vessel walls – arteries tend to have thicker walls than veins to cope with the greater pressure exerted by arterial blood.
- Depth – most arteries are deep within the body to protect them, whereas many veins lie just under the skin.
- Portal venous system – the blood that leaves the gut in the veins of the stomach and intestine does not pass directly back to the heart. It first passes into the hepatic portal venous system, which carries the blood through the liver tissues before it can return to the systemic circulation.
- Variations – while the pattern of arteries tends to be the same from person to person, there is far greater variability in the layout of the systemic veins.

Major veins of the body

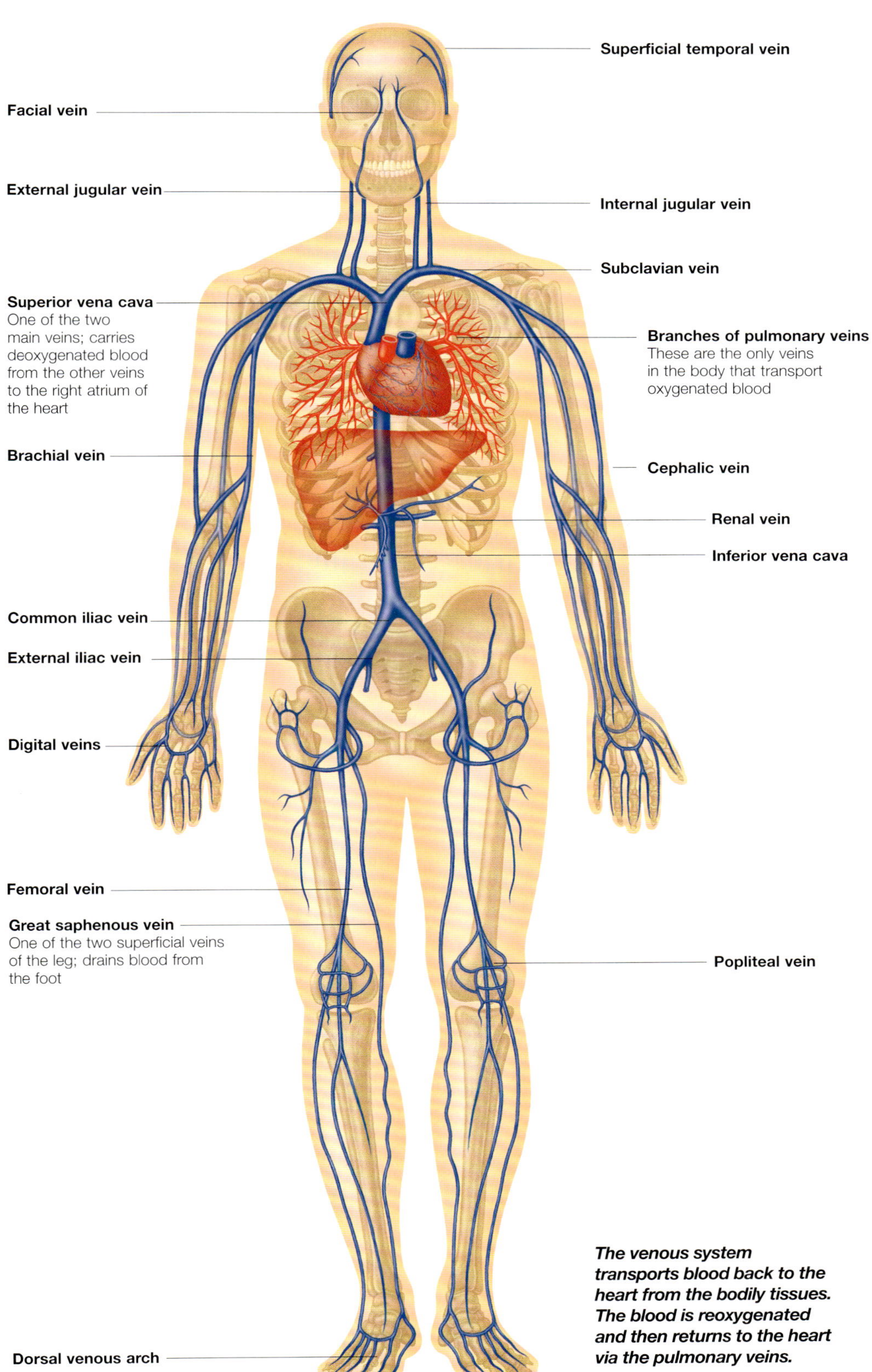

The venous system transports blood back to the heart from the bodily tissues. The blood is reoxygenated and then returns to the heart via the pulmonary veins.

The Function of Blood

Blood transports life-giving oxygen and all the vital nutrients that the cells of our bodies need in order to function. It also carries away the waste products that are produced by our tissues.

Blood makes up about eight per cent of the weight of the human body. The average adult has around 5 litres (8.8 pints) of blood, although the volume varies a great deal, depending mainly on the size of the person. A newborn baby has approximately 80ml (0.03fl oz) of blood per kilogram of body weight.

Blood Circulation

Blood circulates inside a closed system of blood vessels, made up of arteries, capillaries and veins. This complex network transports blood to and from all tissues and organs of the body. At any one time in the average adult, the amount of blood in the various parts of the circulation is approximately as follows:

- Arteries 1200ml (42fl oz)
- Capillaries 350ml (12fl oz)
- Veins 3400ml (120fl oz)

Therefore, most of the blood in circulation is actually in our veins, and very little is in the capillaries.

The blood in the veins (venous blood, returning to the heart) is much darker in colour than arterial blood because it contains relatively little oxygen. Oxygenated blood from the heart, which is found in the arteries, is strikingly scarlet. Capillary blood – which we see when we cut ourselves – has a slightly less bright red colour than arterial blood.

Inside an artery

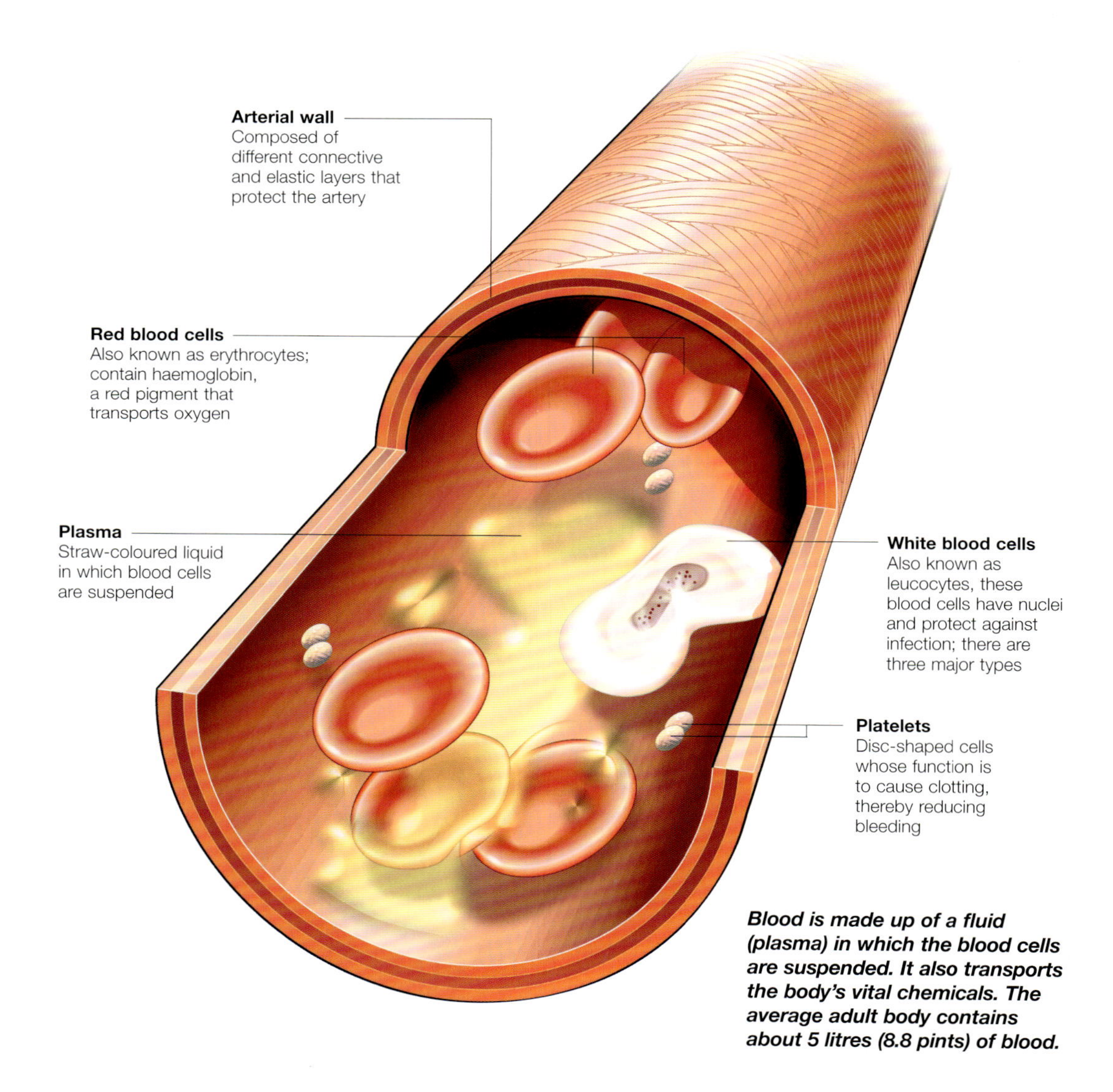

Blood is made up of a fluid (plasma) in which the blood cells are suspended. It also transports the body's vital chemicals. The average adult body contains about 5 litres (8.8 pints) of blood.

Components of blood

The blood that circulates around our bodies consists of several important components. Suspended in plasma are red and white blood cells and platelets; each type of cell has a specific purpose.

Blood consists of various cells suspended in a pale yellow liquid called plasma. Plasma is a sticky fluid containing various chemicals that are in transit from one part of the body to another. Its constituents include:

- Protein 7 per cent
- Salt 0.9 per cent
- Glucose 0.1 per cent

The main proteins in blood plasma are albumin, globulin and fibrinogen. They help to supply nutrition to tissues of the body, and are also important in protecting against infectious diseases. Fibrinogen plays a vital role in blood clotting – it converts into fibrin, a mesh-like material, which helps to stop bleeding after an injury.

Glucose – a form of sugar – is the body's principal fuel. Salt is the body's most important mineral.

Red blood cells

There are three types of cells in blood: red cells, white cells and platelets. Red cells (also known as red corpuscles or erythrocytes) a re the most common cells in the blood. Red cells contain the pigment haemoglobin. This is the iron-containing protein that takes up oxygen in the lungs.

White cells

White cells (also known as white corpuscles or leucocytes) are far fewer in number than red cells. Children have about 10,000 of them in a cubic millimetre of blood, and adults have fewer. White cells are vital in protecting against disease. There are several types:

- Neutrophils: combat bacterial, viral and other infections
- Eosinophils: help defend the body against parasites, and are active in allergic reactions
- Lymphocytes: assist in creating immunity to infection
- Monocytes: capable of engulfing invading particles in the bloodstream
- Basophils: prevent blood clots, release histamine to stimulate inflammation and have an important role in allergic reactions
- Platelets (also known as thrombocytes) are very small cells involved in the process of blood clotting. In a cubic millimetre of blood, there are about a quarter of a million of them. When a blood vessel is cut or damaged, platelets – which are very sticky – immediately adhere to the injured spot, and to each other, and so (along with fibrin) help to plug the gap and stop the bleeding.

Major blood elements

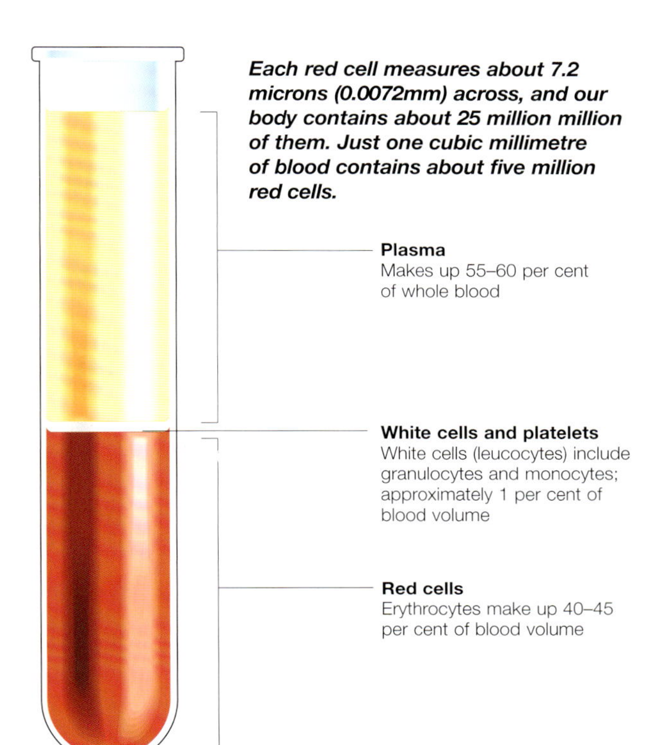

Each red cell measures about 7.2 microns (0.0072mm) across, and our body contains about 25 million million of them. Just one cubic millimetre of blood contains about five million red cells.

White blood cells, known as lymphocytes, help to form immunity against infection.

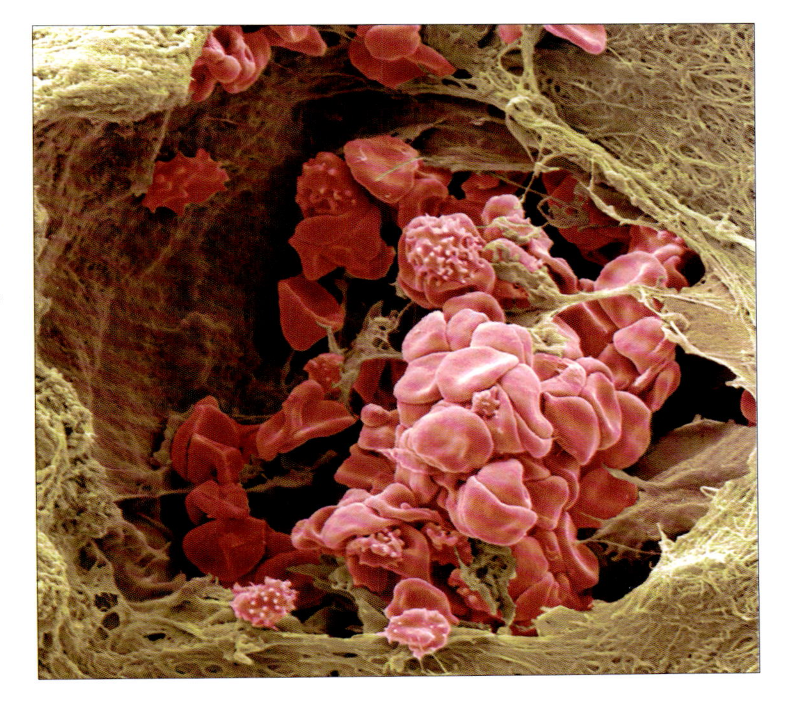

Red blood cells, or erythrocytes, are shown here within a blood vessel. These carry oxygen to all the cells and tissues in the body.

How the Blood Circulates

The circulation transports blood to and from every tissue in the body, maintaining an optimal environment for cell survival and function. It also allows the transport of hormones around the body.

The function of blood circulation is to supply fuel, nutrients and oxygen to the cells of the body. It also carries waste products away from the tissues, transporting them to the kidneys or lungs for disposal.

The Arterial System

Circulation is achieved by the heart pumping blood forcefully through the arterial system. The arteries divide into increasingly smaller branches, and the smallest arteries (arterioles) deliver blood into microscopic capillaries. The capillaries pass through the tissues and anastomose (join) with the smallest veins (venules). The venules join up to form veins, which take blood back to the heart again. The blood on returning to the heart is then pumped to the lungs to be reoxygenated.

Blood Pressure

Blood pressure is force per unit area exerted by the blood in the arterial system. The pressure is higher in the arterial system than in the venous, as it is pumped by the heart. Venous pressure is lower as the venous circulation is dependent on gravity, propulsion by muscles and pressure gradients within the system.

Blood pressure can be monitored using a medical device. It is measured in millimetres of mercury (mm Hg – UK and USA), or kilopascals (other European countries).

Blood pressure is expressed as two figures, for instance 130/70. The first, or upper figure represents the pressure in the arteries when the heart is contracting (systole) – the systolic pressure. The second, or lower figure represents the pressure in the arteries while the heart is relaxing (diastole) – the diastolic pressure.

Circulatory system

The circulatory system is the branching network of blood vessels. The arteries carry oxygenated blood (red) to the tissues, and the veins return the deoxygenated blood (blue) to the heart.

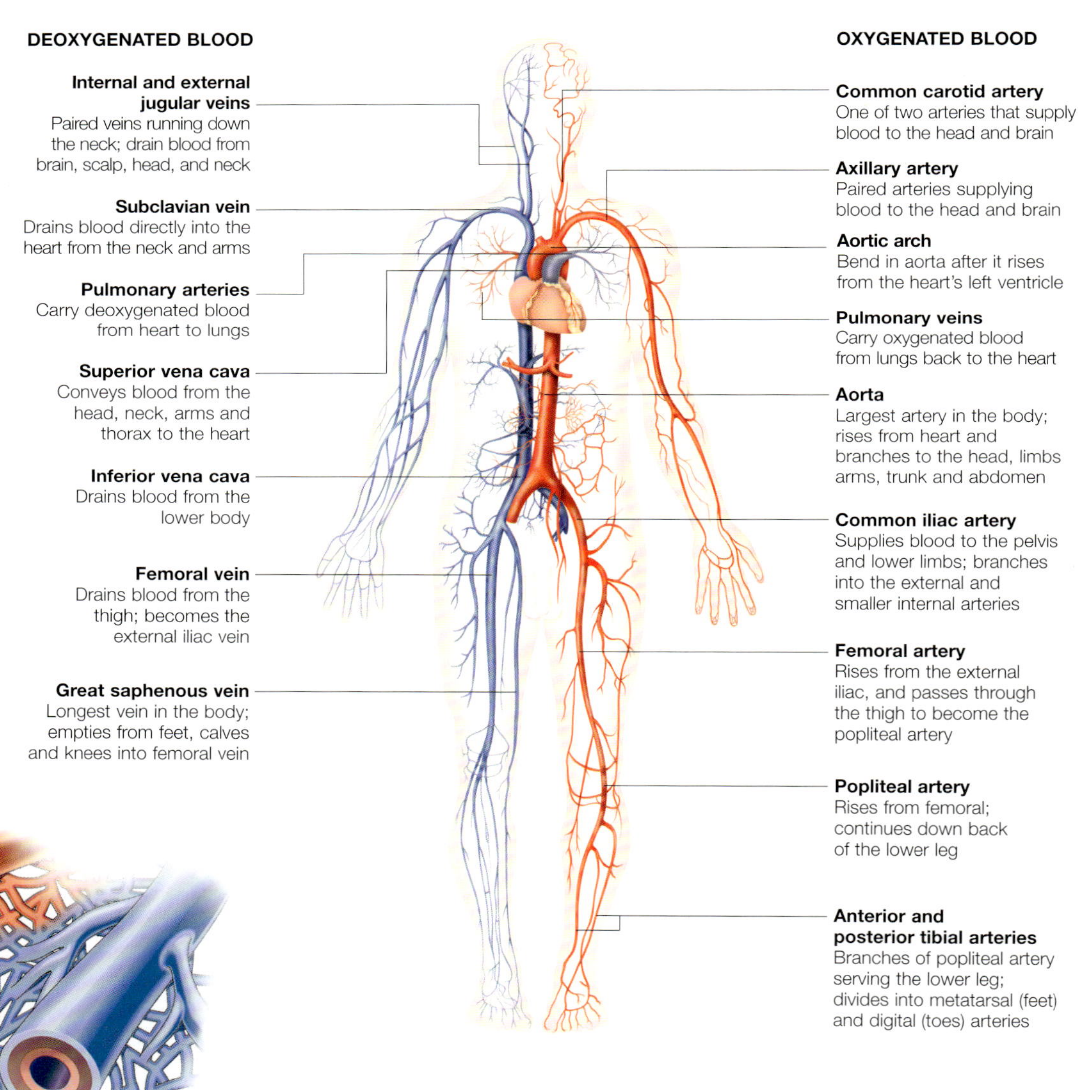

The arteries and veins are linked by a meshwork of capillaries. Over 150,000km (93,000 miles) long, this network allows the exchange of oxygen and nutrients between the arterial and venous systems.

Artery

Vein

Blood flow through the body

Blood flow is the volume of blood flowing through the circulatory system, an organ of the body or an individual blood vessel in a given period of time.

The flow of blood through a blood vessel is determined by a combination of the pressure difference between the two ends of the vessel and the resistance to blood flow through the vessel.

Blood pressure is greatest in the vessels nearest the heart in the aorta and the pulmonary artery. As the blood flows away from the heart, the pressure falls. However, of the two parameters – pressure and resistance – it is resistance that has the greater influence on blood flow. The total blood flow in the the circulation of an adult at rest is about five litres (nine pints) per minute; this is referred to as the cardiac output.

The blood flow to individual tissues is almost precisely controlled in relation to the tissue's needs. When tissues are active, they may require up to 20 or 30 times more blood flow than when they are at rest. However, cardiac output cannot increase more than about four to seven times.

Since the body cannot simply increase total blood flow, local blood flow to specific tissues is controlled by internal monitoring mechanisms. Blood is distributed according to the specific tissues' needs, and redirected away from tissues that do not require nutrients or oxygen at that time.

Venous blood flow

Blood flows back through veins towards the heart by a combination of mechanisms: the contraction of the leg and arm muscles; the presence of efficient valves in the veins that prevent backflow; and the simple process of breathing, which helps to 'suck' the blood through the veins, towards the chest.

Distribution of blood

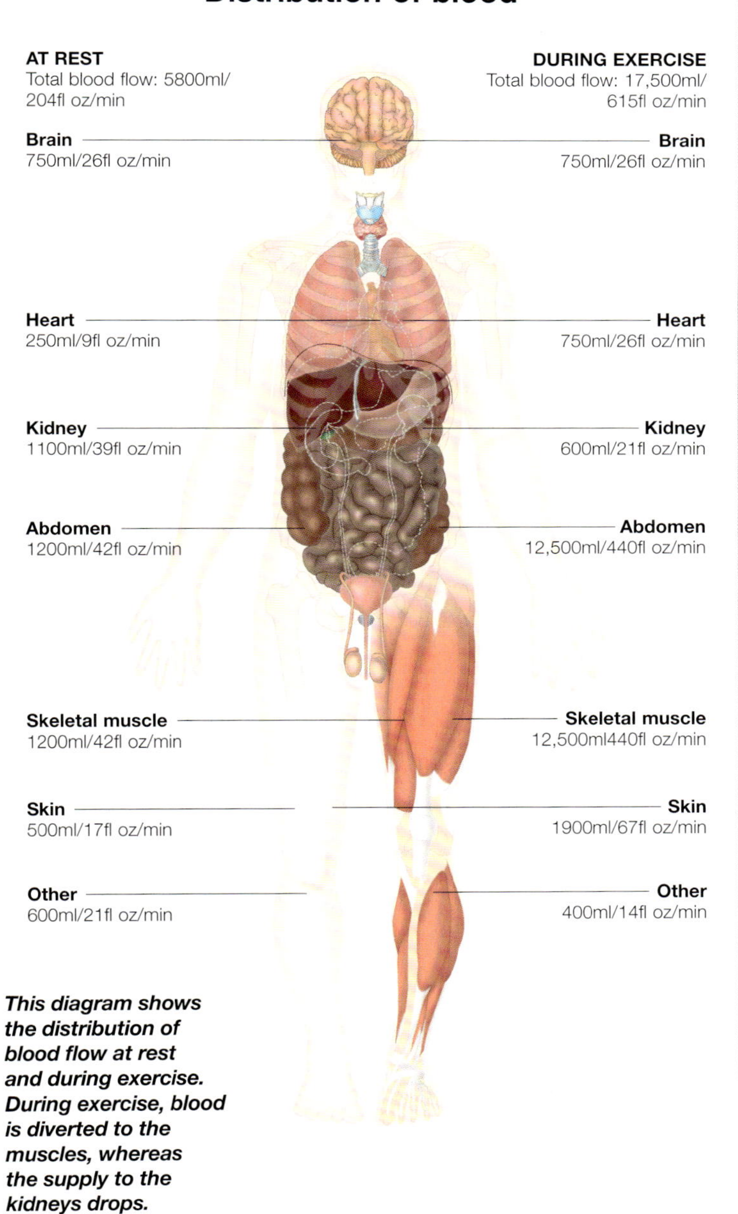

This diagram shows the distribution of blood flow at rest and during exercise. During exercise, blood is diverted to the muscles, whereas the supply to the kidneys drops.

Distribution of blood volume

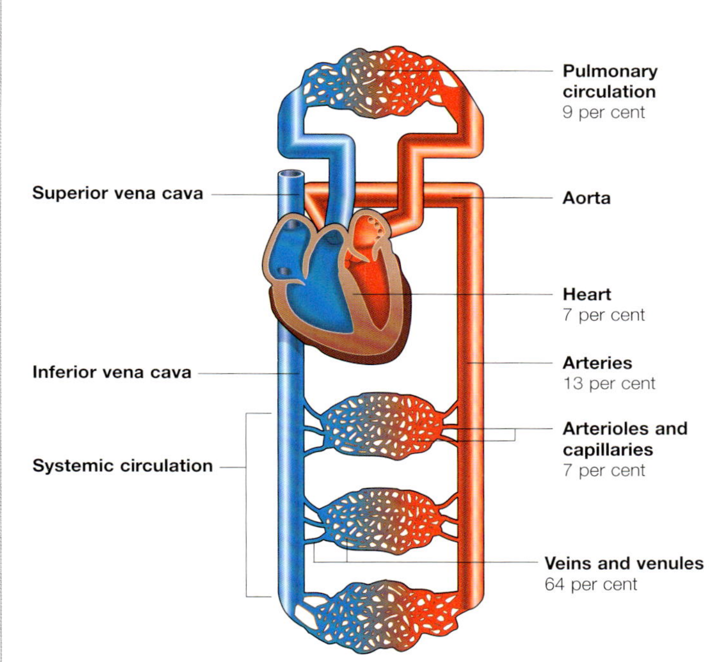

The circulatory system can be divided into two main portions: the pulmonary (lungs) and systemic (entire body). The diagram shows how blood is distributed around these areas.

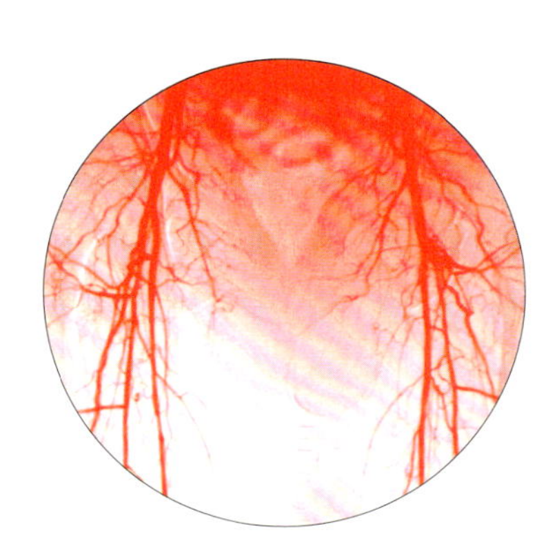

This x-ray of the femoral arteries in the legs show the many branching vessels that take oxygenated blood to the tissues.

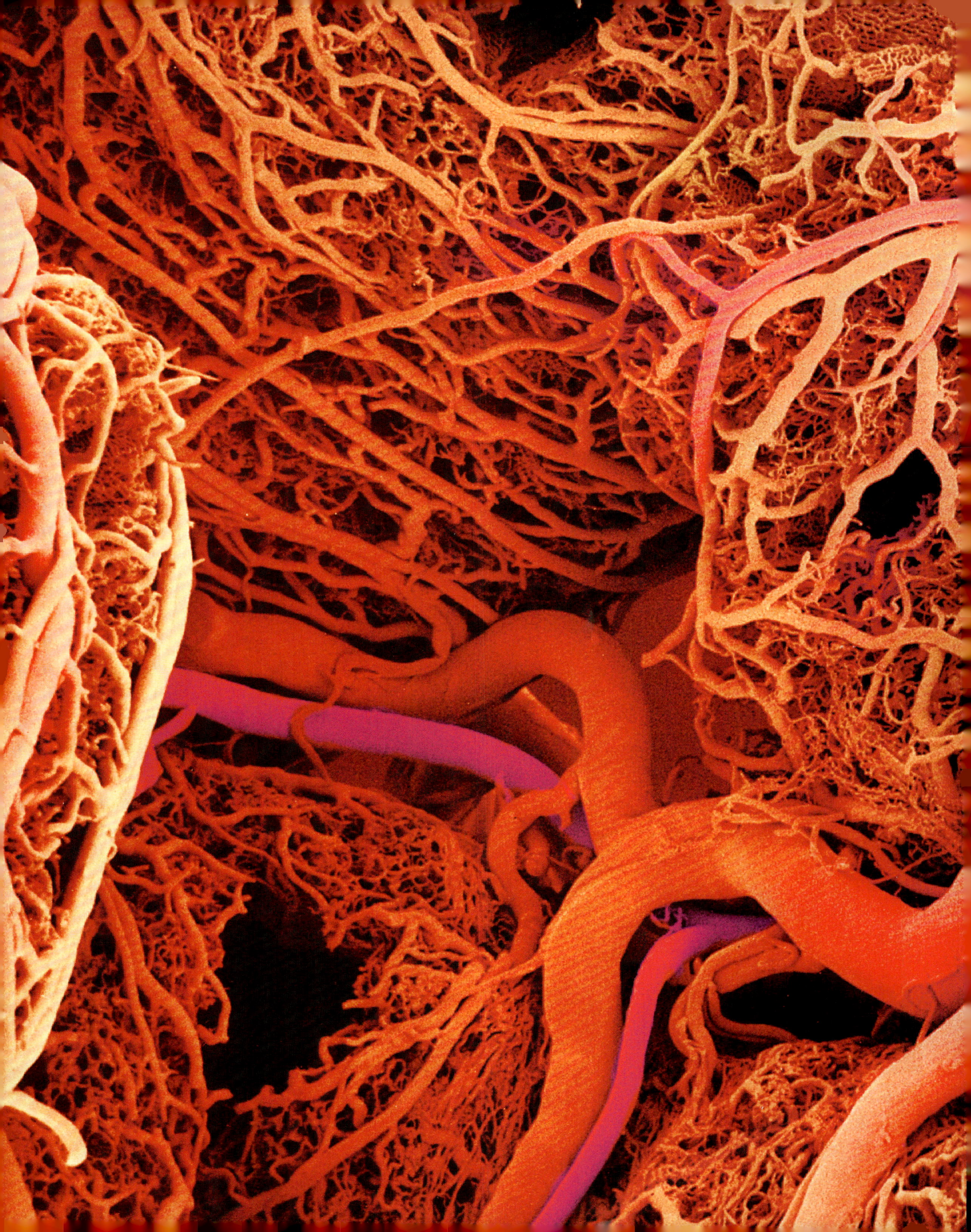

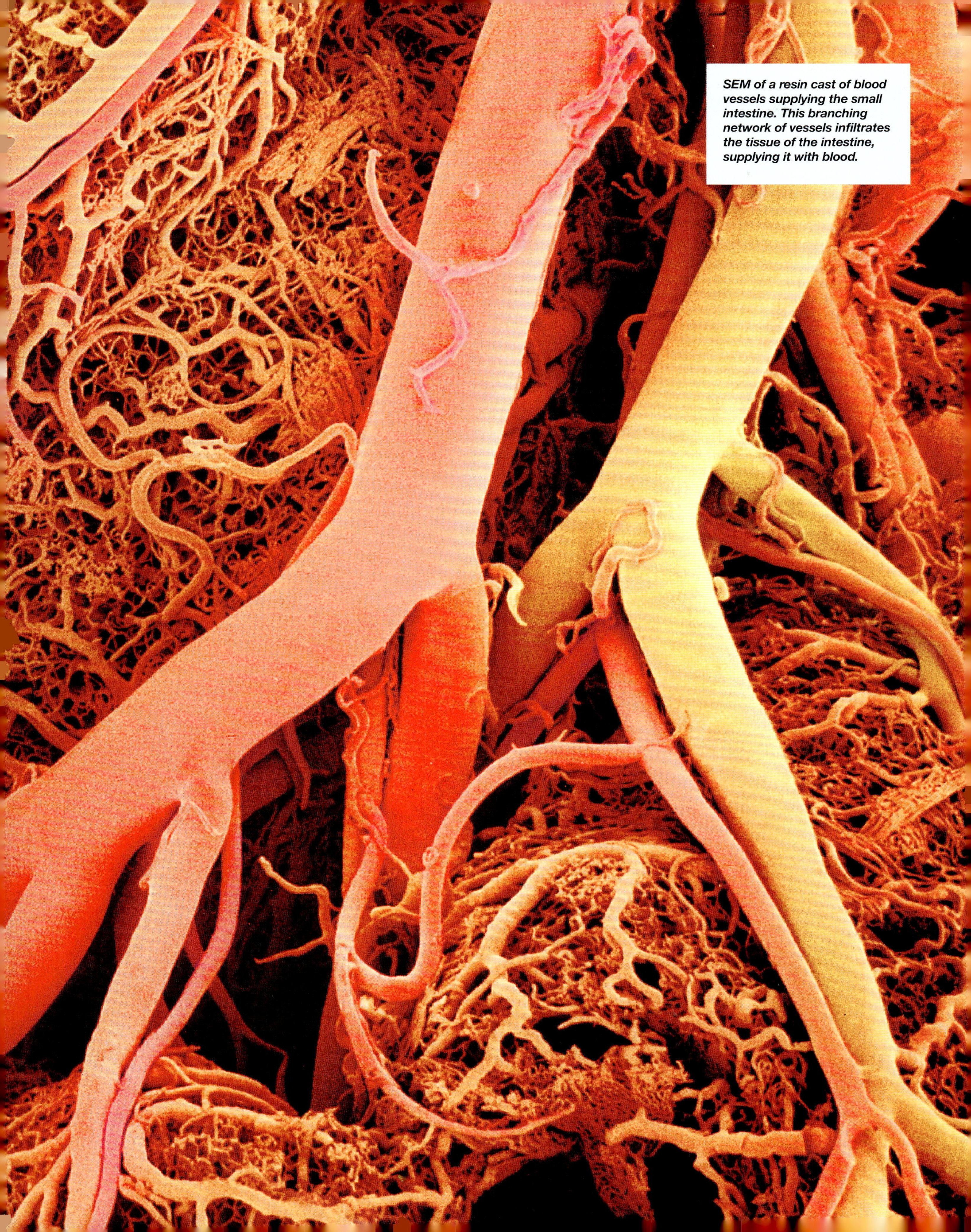

SEM of a resin cast of blood vessels supplying the small intestine. This branching network of vessels infiltrates the tissue of the intestine, supplying it with blood.

How Blood is Transported

Blood vessels are the tubes that carry blood around the body. Arteries carry blood from the heart to the body's tissues. From there veins carry the deoxygenated blood back to the heart.

Blood vessels vary in size according to the amount of blood they carry; the largest vessels are found nearest the heart. Blood destined for body tissues leaves the heart via the aorta, which arches over and behind the heart and carries blood down the trunk. From the aorta, smaller arteries lead to the main organs of the body, where they branch into even smaller vessels called arterioles. These deliver blood to the capillaries, a branching network where the arterioles connect with tiny veins (venules). Here oxygen and nutrients are absorbed into the tissues, and carbon dioxide and waste materials flow back into the circulation.

Blood leaving the tissues collects into the venules, which feed blood into larger and larger vessels, the largest of which, the two vena cavae, return blood to the heart. From the heart, the blood is pumped to the lungs where it is reoxygenated for circulation.

Arteries and Arterioles

Blood leaves the heart under high pressure, so arteries need thick, muscular walls made up of several layers (tunicae). Surrounding the central canal (lumen) is the tunica intima, which consists of a lining of endothelial cells, a layer of connective tissue and a layer of tissue called the internal elastic lamina. The middle layer (tunica media) is formed of smooth muscle cells and sheets of elastic tissue, or elastin. The outer layer (tunica adventitia) is a tough outer coat of fibrous connective tissue.

The largest arteries lead directly from the heart. They are known as elastic or conducting arteries because they contain a relatively high proportion of elastic tissue. This allows them to expand as they fill with blood and then contract again, forcing the blood onwards towards the smaller arteries.

Arteries with a diameter of between 0.3mm (0.01in) and 0.01mm (0.003in) are called arterioles. The largest of these possess all three tunicae, but the tunica media contains only scattered elastic fibres. The smallest have no outer coat and consist only of an endothelial lining surrounded by a single layer of spiralling muscle cells. The flow of blood from arterioles into capillaries is controlled by sympathetic nerves, which cause the muscle cells to contract, thus constricting or dilating the lumen of the arterioles.

Structure of a typical artery

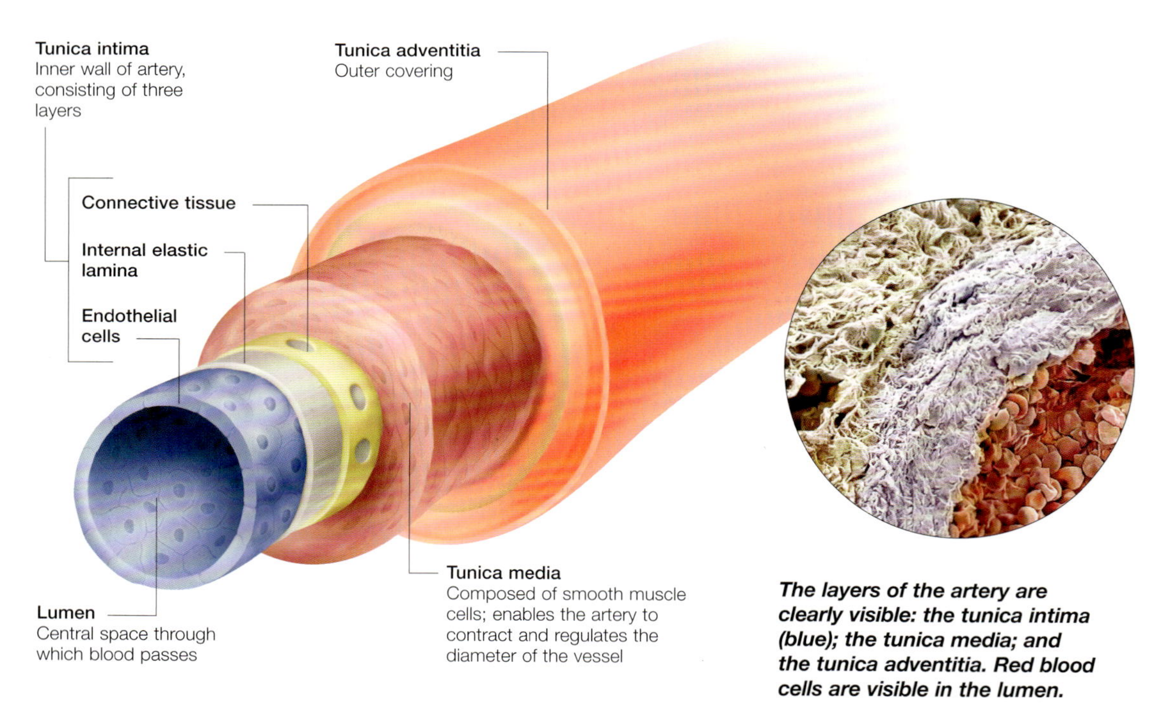

The layers of the artery are clearly visible: the tunica intima (blue); the tunica media; and the tunica adventitia. Red blood cells are visible in the lumen.

This section of a large artery shows the thick walls of the vessel and the many red blood cells in the lumen (central space).

Veins and capillaries

Veins are the vessels that transport deoxygenated blood from around the body to the heart. The capillaries make up the network between veins and arteries in all the tissues.

The structure of veins is very similar to that of arteries, but veins are usually larger with thinner walls, and contain less muscle, elastic and collagenous tissue, so they can be compressed or distended. Venules – the smallest veins – collect blood from the capillaries, which flows into increasingly larger veins. Blood from the lower body flows to the inferior vena cava and drains into the right atrium of the heart. Blood from the upper body is collected by the superior vena cava and also drains into the right atrium.

Many veins, especially those in the limbs, have a system of valves that allow blood to flow in one direction only. These valves are semi-lunar, formed of two half-circles of tissue.

The blood pressure in veins is low. Movement of the blood is helped by the skeletal muscle pump, in which the contraction of surrounding skeletal muscles squeezes the vein and forces the blood along. In veins of less than a millimetre diameter and in regions where muscular activity is more or less continuous, such as the chest and abdominal cavities, there are no valves, as the blood flow is maintained by muscle pressure alone.

Types of capillary

There are at least three different kinds of capillary:

- Continuous capillaries are made up of a single long endothelial cell curved round to form a tube
- Fenestrated capillaries are made up of two or more endothelial cells that have pores (fenestrations), especially near the junctions of the cells
- Discontinuous capillaries, also called sinusoids or vascular sinuses, are made up of a number of cells with large fenestrations.

Continuous capillaries are the least permeable, and liquids are transferred to and from the surrounding tissues by exocytosis and endocytosis, processes by which vesicles (sacs) containing the liquids are moved across the endothelial cells.

In fenestrated capillaries and sinusoids, chemicals pass more easily through the thin membranes that cover the pores. Fenestrated capillaries are common in the endocrine glands and kidneys; sinusoids are found in the liver and spleen.

Skeletal muscle pump

Muscles relaxed

Vein
Each vein is divided into segments by non-return valves to prevent the blood flowing backwards

Muscles contracted

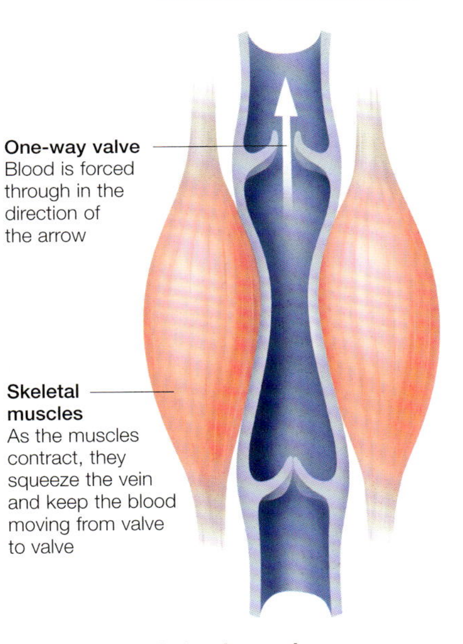

The skeletal muscle pump moves blood through the veins back to the heart. Muscles contract against the flexible vein, forcing the valves to open.

Types of capillary

Structure of a fenestrated capillary

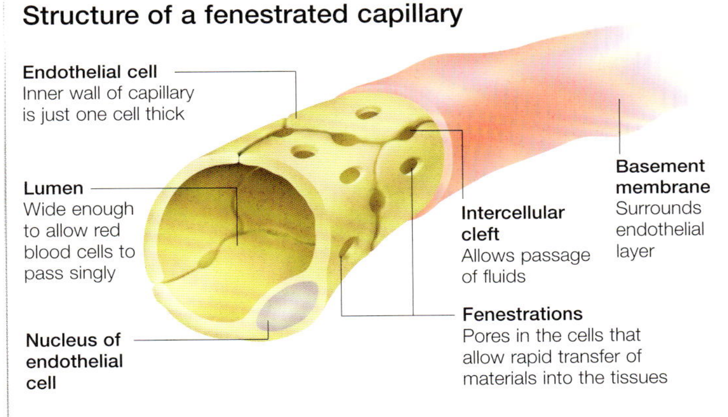

This magnification of a fenestrated capillary shows a fenestration, or pore, in the inner wall of the vessel.

How Blood Clots

Without a mechanism to prevent bleeding, blood loss could prove fatal. Specialized factors in blood can form clots to repair injuries.

Blood flows freely in intact blood vessels due partly to an excess of naturally occurring anticoagulants. However, if the blood vessel wall is damaged, a series of chemical reactions are initiated to prevent blood loss (haemostasis). Without these processes, even the smallest cut could be fatal. Haemostasis involves many blood coagulation factors, which are present in the plasma, as well as chemicals that are released from platelets and injured cells.

Stages of Haemostasis

Haemostasis can be broken down into three main stages, which occur in rapid succession after an injury:

- Vasoconstriction – the first stage involves the constriction of the damaged blood vessel; this can significantly reduce blood loss in the short term
- Platelet plug formation – damage to the blood vessel causes platelets, which are present within the plasma, to become sticky and adhere to each other and to the damaged vessel wall
- Coagulation (blood clotting) – next, the platelet plug is reinforced with a meshwork of fibrin fibres. This fibrin net traps red and white blood cells to form a secondary haemostatic plug, or blood clot.

Coagulation

The formation of a blood clot is a complex process involving over 30 different chemicals. Some of these chemicals, called coagulation factors, enhance clot formation, whereas others, called anticoagulants, prevent clotting (which could in itself be fatal). Maintaining a balance is a vital function of the blood.

Clotting is initiated by a cascade of biochemical reactions involving 13 coagulation factors. The end result is the formation of a complex chemical called prothrombin activator. This compound enables the conversion of a plasma protein called prothrombin into a smaller protein called thrombin. Thrombin, in turn, enables the joining together of fibrinogen molecules present in the plasma to produce a fibrin mesh. It is this mesh that traps blood cells in the hole in the blood vessel wall.

Repair of vessel walls

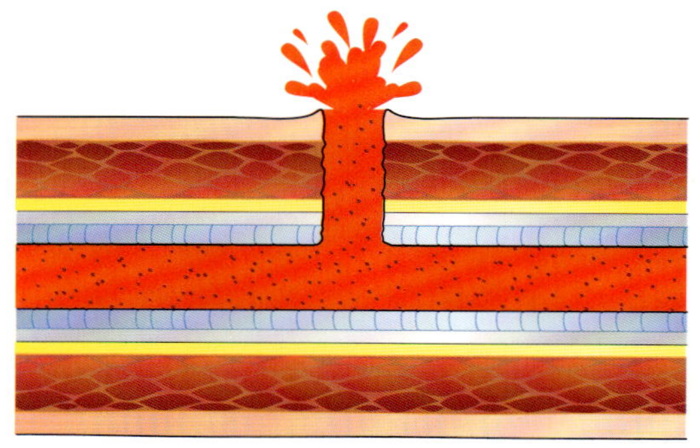

Injury

When a blood vessel is damaged, blood escapes the circulation, reducing blood volume. Excessive blood loss is prevented by haemostasis.

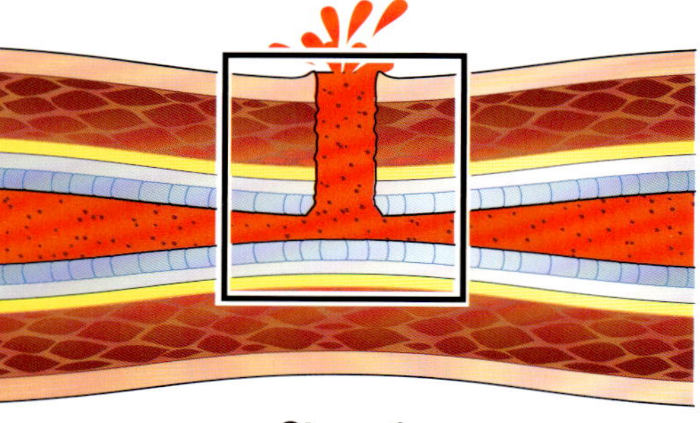

Stage 1

The first stage of haemostasis is vasoconstriction; the damaged blood vessel constricts to reduce the amount of blood flowing through it.

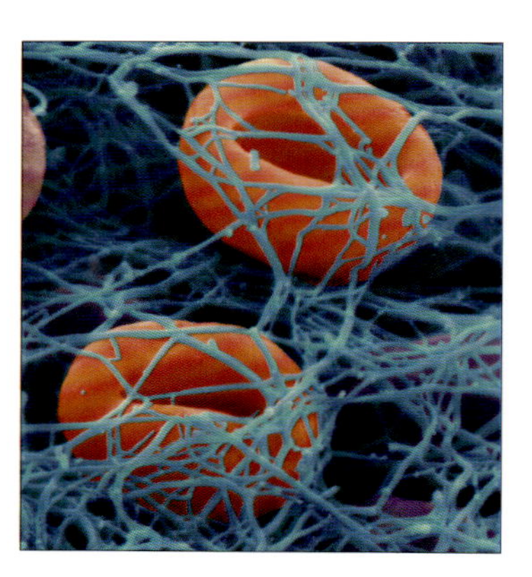

These red blood cells are trapped within a fibrin mesh, which forms a blood clot.

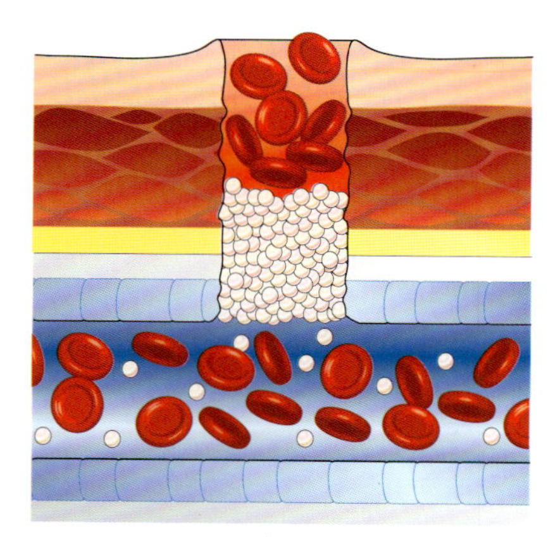

Stage 2

The second stage is the formation of a platelet plug. Platelets (white) stick to one another to temporarily seal the hole in the vessel wall.

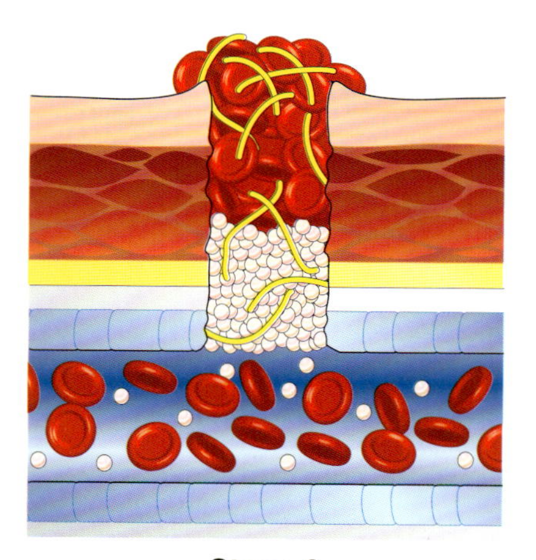

Stage 3

Finally, a blood clot is formed; blood cells are trapped in a fibrin mesh (yellow strands) that seals the hole until it can be permanently repaired.

Clot contraction and repair

About 30–60 minutes after a blood clot has formed, the platelets within the clot contract. As in muscle, platelets contain two contractile proteins called actin and myosin. This contraction pulls on the fibrin strands, bringing the edges of the injured tissue closer together and helping the wound to close.

The blood clot is temporary; at the same time as the clot is contracting, surrounding tissues divide to repair the vessel wall.

Once the tissue has healed (after about two days), the fibrin mesh that holds the clot together is dissolved. This process, called fibrinolysis, is enabled by the enzyme plasmin, which is produced from the plasma protein plasminogen. Plasminogen molecules are incorporated into the blood clot during its formation, where they lie dormant until activated by the healing process. As a result, most of the plasmin is restricted to the clot.

Normally, a fine balance between coagulation and fibrinolysis is maintained in the body.

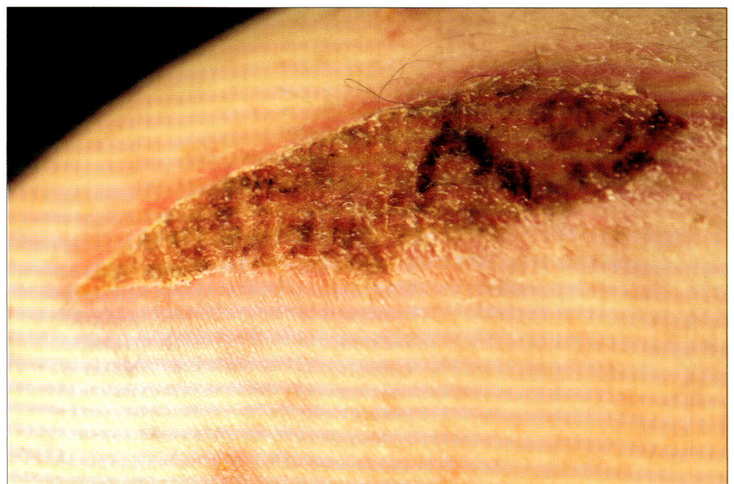

In this photograph the process of healing has begun, and a clot has formed on the skin of the arm.

Platelets

Platelets are cytoplasmic fragments that are able to survive in the blood circulation for up to 10 days. They are formed in bone marrow by extremely large cells called megakaryocytes. They are not classified as cells as they do not have a nucleus and therefore cannot divide.

Electron microscopy reveals three platelet areas:

1 The outer membrane consists of a glycoprotein surface coat that causes it to adhere only to injured tissues. The membrane also contains large numbers of phospholipids that play a number of roles in the blood clotting process.

2 The cytosol (solution inside the cellular membrane) contains contractile proteins (including actin and myosin), microfilaments and microtubules. These are important for clot contraction.

3 Platelet granules contain a variety of haemostatically active compounds that are released when platelets are activated. These compounds are potent aggregating agents that attract more platelets to the wound site. Consequently, the formation of the platelet plug is a self-perpetuating process.

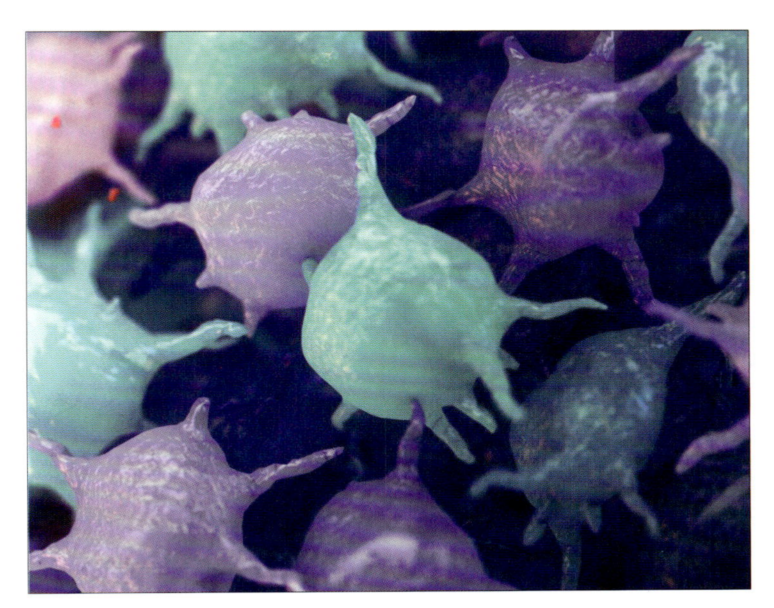

Above: When necessary, platelets stick together and plug holes and tears in damaged blood vessels. They also contain chemicals that aid clotting.

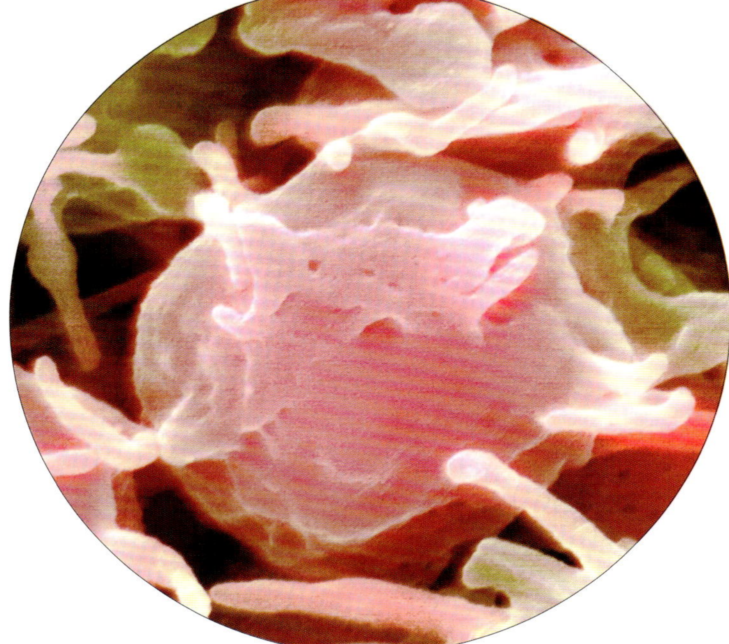

Left: Platelets play a vital role in blood clotting and wound repair. The average lifespan of a platelet is 8–10 days.

Mediastinum

The mediastinum is an area or compartment within the centre of the chest cavity that contains a number of vital organs and structures.

The mediastinum occupies the central space of the thoracic cavity, extending from the base of the neck to the diaphragm, and from the sternum and ribs back to the bones of the spine.

It is lined with the thin, lubricated membranes known as the mediastinal pleura, and contains a number of vital organs and structures that are held loosely together by fatty connective tissue. This arrangement allows for movements due to changes in body posture or by the changes in pressure and volume of the thoracic cavity during breathing.

Structures of the Mediastinum

Contained within the mediastinum are the heart and major blood vessels, the thymus gland, the trachea, the oesophagus and some important nerves, including the vagus and phrenic nerves.

- The heart and its associated large blood vessels. The heart lies in the inferior mediastinum and forms the bulk of the contents of the middle mediastinum. The aorta, the large artery, which carries blood from the heart to the body, lies above the heart within the superior mediastinum, together with the superior vena cava, one of the main veins bringing blood back to the heart from the upper body.
- The thymus gland. This gland, larger in children than in adults, lies within the superior mediastinum just behind the manubrium (upper part of the breastbone).
- The trachea, or windpipe. This descends from the larynx in the neck to enter the superior mediastinum in front of the oesophagus. It lies a little to the right of the mid line and comes to an end by dividing into the right and left main bronchi.
- The oesophagus, or gullet. This muscular tube passes from the pharynx in the neck, down through the length of the whole mediastinum to enter the abdomen via the oesophageal aperture in the diaphragm.
- Nerves. The paired vagus nerves, which arise from the brain, pass down through the mediastinum and then the diaphragm to ultimately reach the stomach. Within the mediastinum the vagus nerves contribute to the innervation of the lungs and the oesophagus. The right and left phrenic nerves, which supply the muscle of the diaphragm, also pass through the mediastinum.
- Lymph nodes and lymphatic vessels. The loose connective tissue of the mediastinum contains many important groups of lymph nodes that receive lymph from the lungs and other mediastinal structures. The main lymphatic vessel, the thoracic duct, passes up into the mediastinum through the aortic opening in the diaphragm and receives lymph from other large lymphatic vessels before emptying into one of the large veins above the heart.

Major vessels of the mediastinum

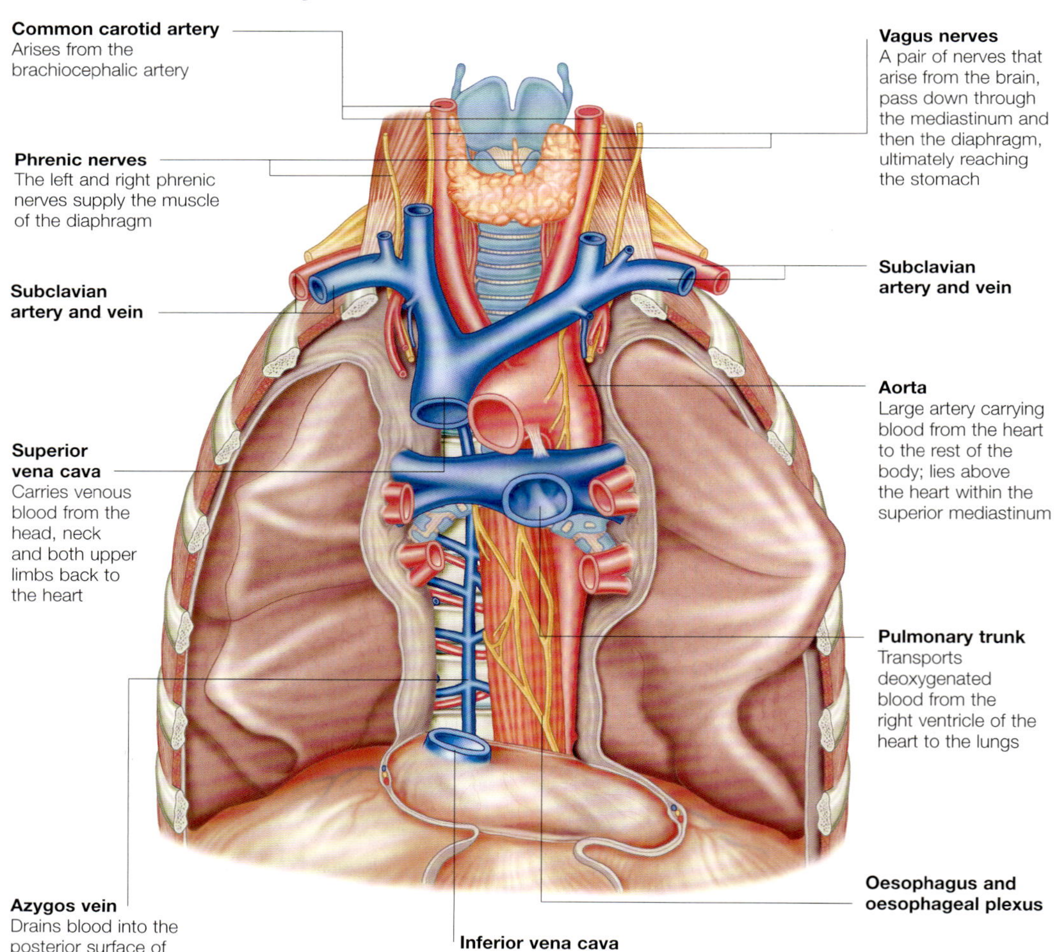

The mediastinum is a moveable compartment extending from the neck to the diaphragm. Its size and shape varies with breathing.

Divisions of the mediastinum

For descriptive purposes, the mediastinum is divided into two main parts; the superior (upper) mediastinum and the inferior (lower) mediastinum. The inferior mediastinum is divided still further into anterior, middle and posterior divisions.

Superior mediastinum
The superior division of the mediastinum extends down as far as an imaginary line that passes from the sternal angle anteriorly (at the front) to the level of the intervertebral disc between the fourth and fifth thoracic vertebrae.

Inferior mediastinum
From this imaginary line down to the diaphragm lies the inferior mediastinum, which is divided into three sections. From front to back, these are:

- Anterior mediastinum – the area in front of the heart and behind the sternum
- Middle mediastinum – containing the heart and great vessels
- Posterior mediastinum – the area behind the heart and in front of the vertebral column.

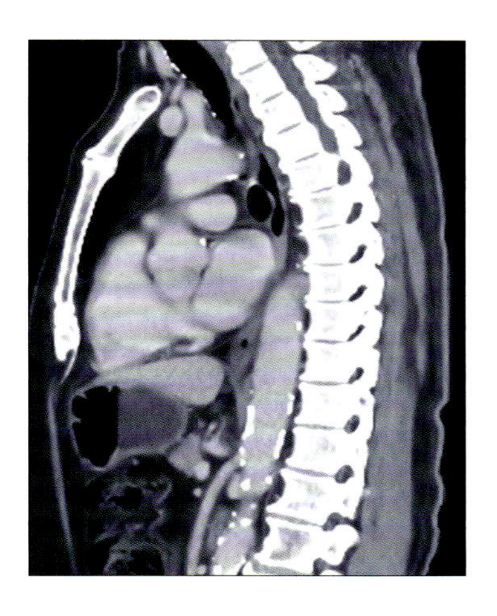

The area of the mediastinum is shown on this CT scan of the chest. It contains the heart and great vessels.

Sagittal section of the mediastinum

This illustration shows the location of the superior mediastinum and, below the dotted line, the three divisions of the inferior mediastinum: anterior, middle and posterior.

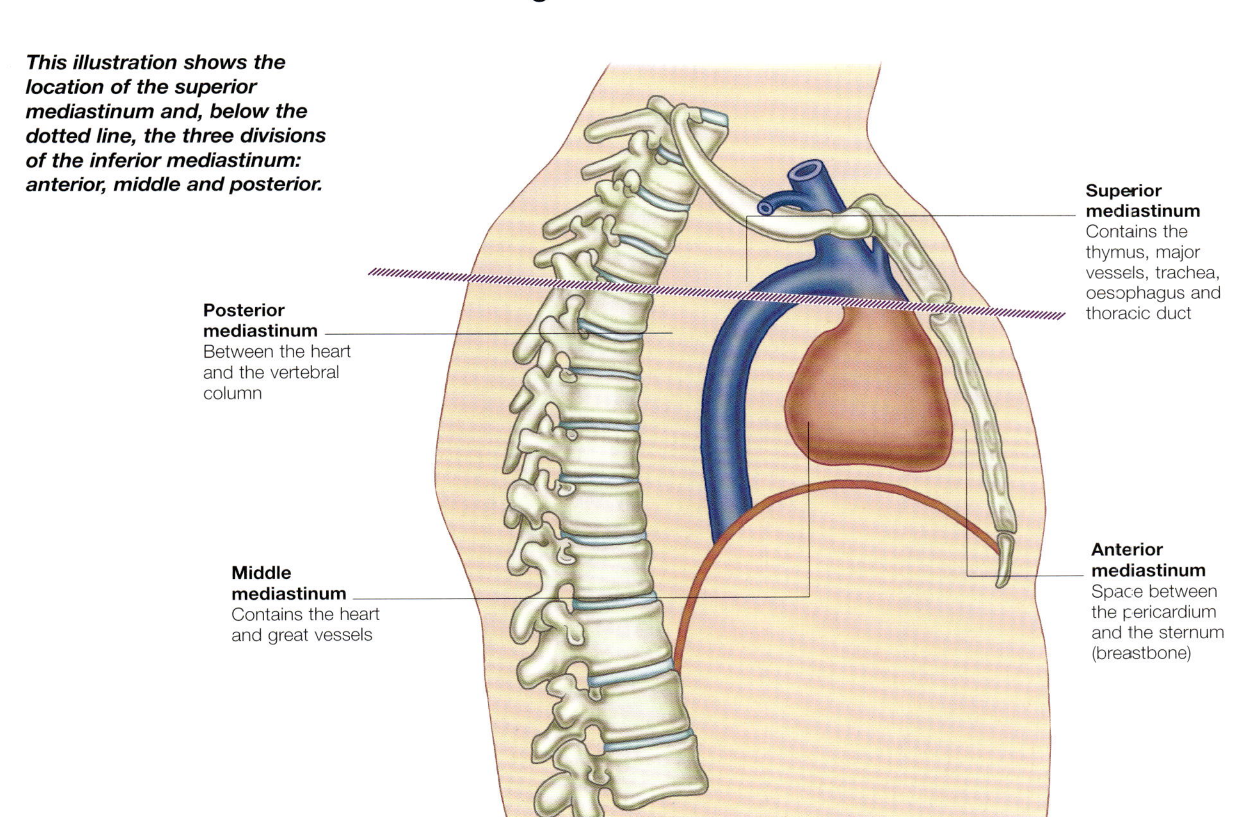

Heart

The adult heart is about the size of a clenched fist and lies within the mediastinum in the thoracic cavity. It rests on the central tendon of the diaphragm and is flanked on either side by the lungs.

The heart is a hollow organ that is composed almost entirely of muscle. It has four chambers – two atria and two ventricles. The typical weight of a normal heart is only about 250–350g (9–12oz) yet it has incredible power and stamina, beating over 70 times every minute to pump blood around the body. Surrounding the heart is a protective sac of connective tissue called the pericardium.

Surfaces of the Heart

Roughly the shape of a pyramid on its side, the heart is said to have a base, three surfaces and an apex:

- The base of the heart lies posteriorly (at the back) and is formed mainly by the left atrium, the chamber of the heart that receives oxygenated blood from the lungs
- The inferior or diaphragmatic surface lies on the underside and is formed by the left and right ventricles separated by the posterior interventricular groove

The right and left ventricles are large chambers that pump blood around the lungs and the body respectively

- The anterior, or sternocostal, surface lies at the front of the heart just behind the sternum and the ribs and is formed mainly by the right ventricle
- The left, or pulmonary, surface is formed mainly by the large left ventricle, which lies in a concavity of the left lung.

Position of the heart

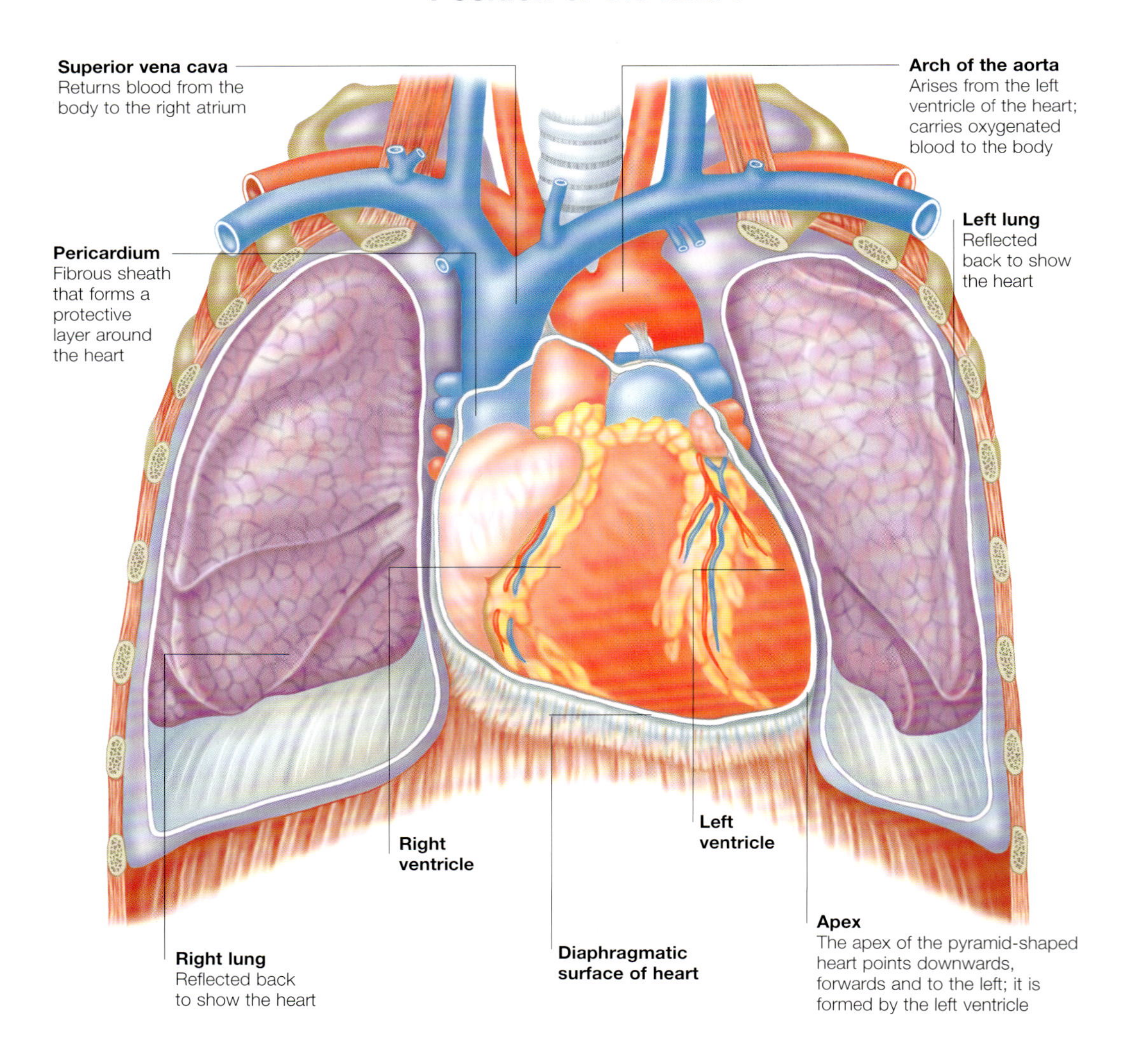

Pericardium

The heart is enclosed within a protective triple-layered sac of connective tissue called the pericardium. The pericardium is formed of two layers – the fibrous pericardium and the serous pericardium.

■ Fibrous pericardium. This is the outer layer of the pericardium and is composed of tough fibrous connective tissue. It has three main functions:
Protection – it is strong enough to provide some protection from trauma.
Attachment – there are fibrous attachments between the pericardium to both the sternum and the diaphragm. The fibrous pericardium also fuses with the walls of the arteries that pass through it from the heart. These attachments help to anchor the heart to its surrounding structures.
Prevention of overfilling of the heart – the pericardium is non-elastic so does not allow the heart to expand with blood beyond a safe limit.

■ Serous pericardium. This covers and surrounds the heart in the same way as the pleura does in the lungs. It is a thin membrane with two layers, the parietal and the visceral. The parietal layer lines the inner surface of the fibrous pericardium and turns back on itself to form the visceral layer. Between them is a small amount of fluid and this allows the chambers of the heart to move freely within the pericardium as the heart beats.

The pericardial sac with the heart removed

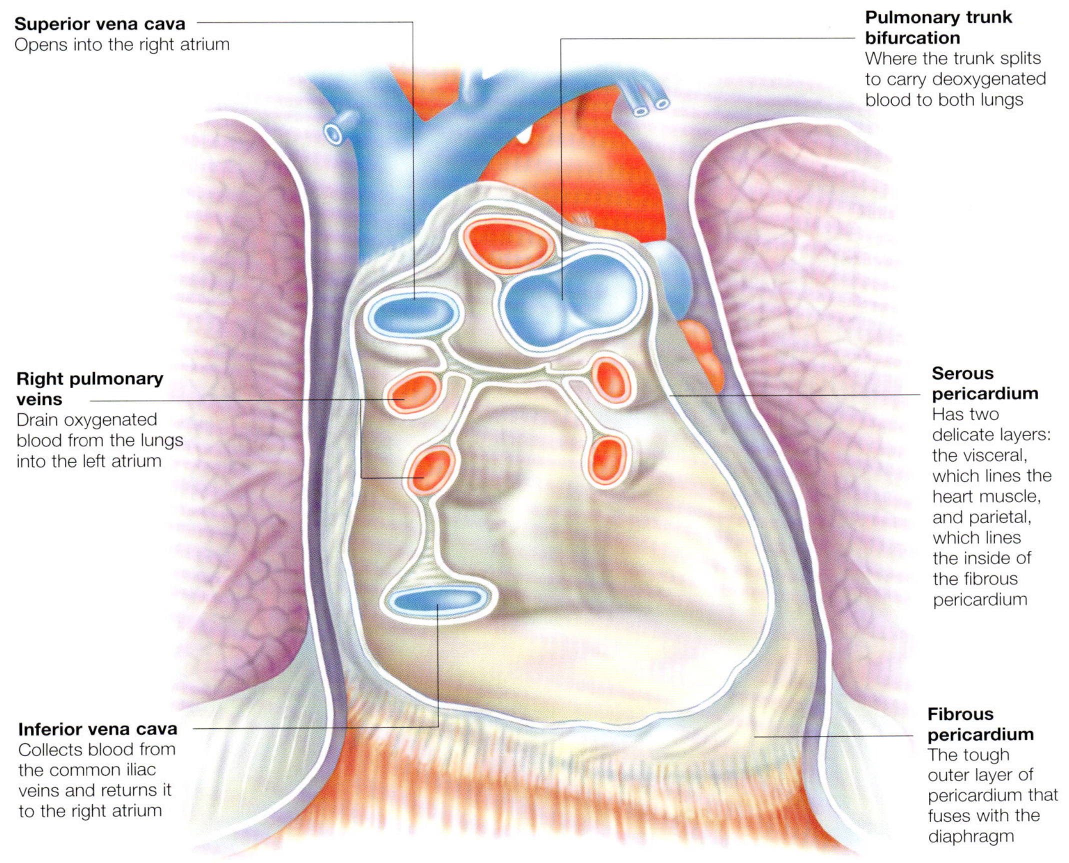

Layers of the heart wall

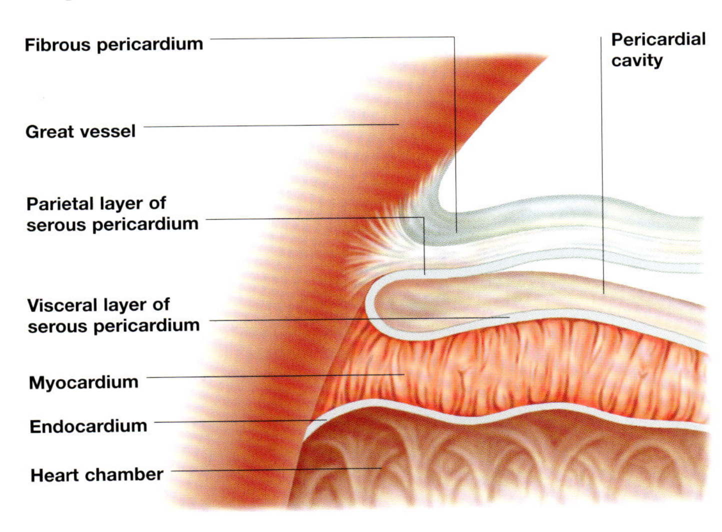

Inside the pericardial cavity, the heart wall is made up of three layers: the epicardium, the myocardium and the endocardium.

■ The epicardium is the visceral layer of the serous pericardium, which covers the outer surface of the heart and is attached firmly to it.

■ The myocardium makes up the bulk of the heart wall and is composed of specialized cardiac muscle fibres. This type of muscle occurs only in the heart and is adapted for the special role it plays there. The muscle fibres of the myocardium are supported and held together by interlocking fibres of connective tissue.

■ The endocardium is a smooth, delicate membrane, formed by a very thin layer of cells, and lines the inner surface of the heart chambers and valves.

The blood vessels entering and leaving the heart are lined by a similar layer – the endothelium – which is a continuation of the endocardium.

A section taken through the heart at the junction of a typical great vessel reveals the different layers of pericardium and heart wall.

Ventricles

The heart is divided into four chambers: two thin-walled atria, which receive venous blood, and two larger, thick-walled ventricles, which pump blood into the arterial system.

The heart is divided into left and right sides, each having an atrium and ventricle. The two ventricles make up the bulk of the muscle of the heart, the left being larger and more powerful than the right. The right ventricle lies in front, forming much of the anterior surface of the heart, while the left lies behind and below, comprising the greater part of the inferior surface. The apex of the heart is formed by the tip of the left ventricle.

The right ventricle receives blood from the right atrium, backflow being prevented by the tricuspid valve. Blood is then pumped by contraction of the ventricular muscle up through the pulmonary valve into the pulmonary trunk and from there into the lungs.

The left ventricle receives blood from the left atrium through the left atrioventricular orifice, which bears the mitral valve. Powerful contractions of the left ventricle then pump the blood up through the aortic valve into the aorta.

The ventricles

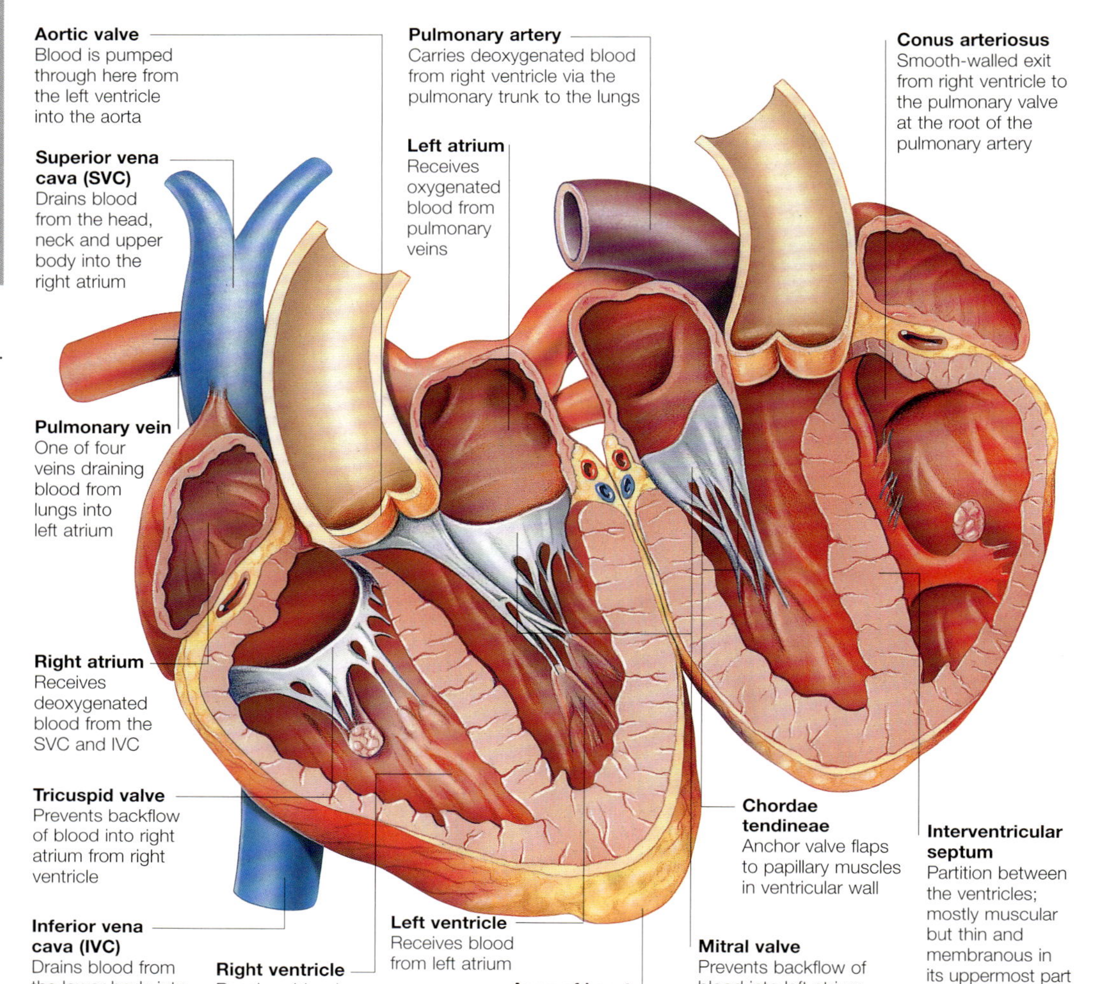

This illustration shows the internal structure of the heart when opened along a plane connecting the root of the aorta and the apex of the heart.

Architecture of the ventricular walls

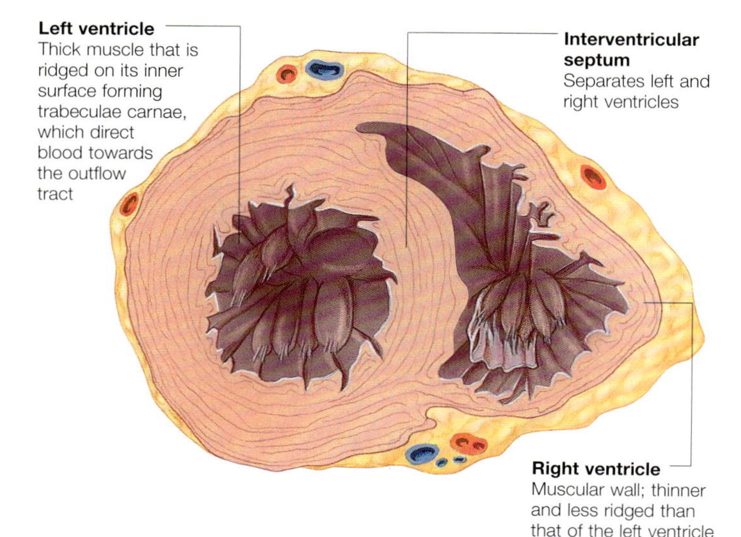

The muscular walls of the left ventricle are twice as thick as those of the right, and form a rough circle in cross-section. The right ventricle forms a crescent in cross-section as it is deformed by the more muscular left ventricle.

The difference in muscle thickness between the chambers reflects the pressure required to empty the relevant chamber when the muscle contracts.

Arising from the walls of both ventricles are the papillary muscles, which taper to a point and bear tendinous chords (chordae tendineae) that attach to the tricuspid and mitral valves to stabilize them during pumping.

The inner surfaces of the ventricular walls, especially where blood enters, are roughened by irregular ridges of muscle, the trabeculae carnae, which give way to smoother walls near the outflow tracts through which blood is pumped out. There is only a small area of smooth wall in the left ventricle, just before the aortic valve. The right ventricle has a larger, funnel-shaped area of smooth wall below the pulmonary valve known as the conus arteriosus, or infundibulum.

A cross-section of the heart through the ventricles shows the difference in thickness of the muscular walls of the left and right ventricles.

Atria

The atria are the two smaller, thin-walled chambers of the heart. They sit above the ventricles and are separated by the atrioventricular valves.

All the venous blood from the body is delivered to the right atrium by the two great veins, the superior and inferior vena cavae (SVC and IVC). The coronary sinus, the vessel that collects venous blood from the heart tissues, also drains into the right atrium.

The interior has a smooth-walled posterior part and a rough-walled anterior section. These two areas are separated by a ridge of tissue known as the crista terminalis.

The roughened anterior wall is thicker than the posterior part, being composed of the pectinate muscles, which give a comb-like appearance to the inner surface. The fossa ovalis is a depression on the wall adjoining the left atrium.

The pectinate muscles extend into a small, ear-like outpouching of the right atrium called the auricle. This conical chamber wraps around the outside of the main artery from the heart – the aorta – and acts to increase the capacity of the right atrium.

The right atrium

The SVC, which receives blood from the upper half of the body, opens into the upper part of the smooth area of the right atrium. The IVC, which receives blood from the lower half of the body, enters the lower part of the right atrium. The SVC has no valve to prevent backflow of blood; the IVC has only a rudimentary non-functional valve. The opening of the coronary sinus lies between the IVC opening and the opening that allows blood through into the right ventricle (the right atrioventricular orifice).

The left atrium

The left atrium is smaller than the right, and forms the main part of the base of the heart. It is roughly cuboid in shape and has smooth walls, except for the lining of the left auricle, which is roughened by muscle ridges. The four pulmonary veins, which bring oxygenated blood back from the lungs, open into the posterior part of the left atrium. There are no valves in these orifices.

In the wall adjoining the right atrium lies the oval fossa, which corresponds to the oval fossa on the right side.

Right atrium of the heart

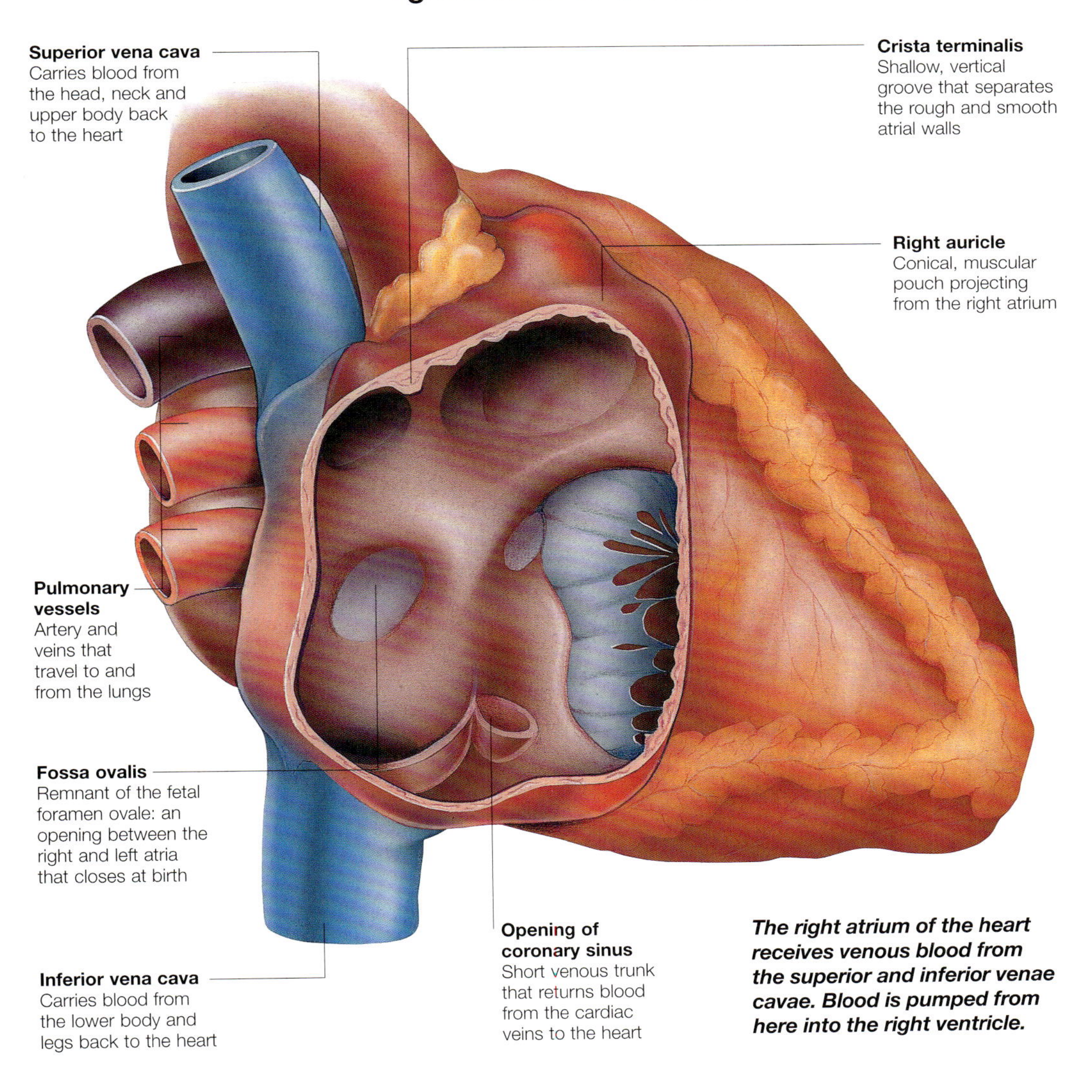

The right atrium of the heart receives venous blood from the superior and inferior venae cavae. Blood is pumped from here into the right ventricle.

Valves of the Heart

The heart is a powerful muscular pump through which blood flows in a forward direction only. Backflow is prevented by the four heart valves that play a vital role in maintaining the circulation.

Each of the two sides of the heart has two valves. On the right side of the heart, the tricuspid valve lies between the atrium and the ventricle, and the pulmonary valve lies at the junction of the ventricle and the pulmonary trunk. On the left side, the mitral valve separates the atrium and ventricle while the aortic valve lies between the ventricle and the aorta.

The tricuspid and mitral valves are also known as the atrioventricular valves as they lie between the atria and the ventricles on each side. They are composed of tough connective tissue covered with endocardium, the thin layer of cells that lines the entire heart. The upper surface of the valves is smooth whereas the lower surface carries the attachments of the chordae tendineae.

The tricuspid valve has three cusps, or flaps. In contrast, the mitral valve has only two and is consequently also known as the bicuspid valve; the name 'mitral' comes from its supposed likeness to a bishop's mitre.

The Heartbeat

During its contraction, the normal heart makes a two-component sound (often described as 'lub-dup'), which can be heard using a stethoscope. The first of these sounds comes from the closure of the atrioventricular valves while the second is due to the closure of the pulmonary and aortic valves.

Cross-section of the heart in diastole (resting)

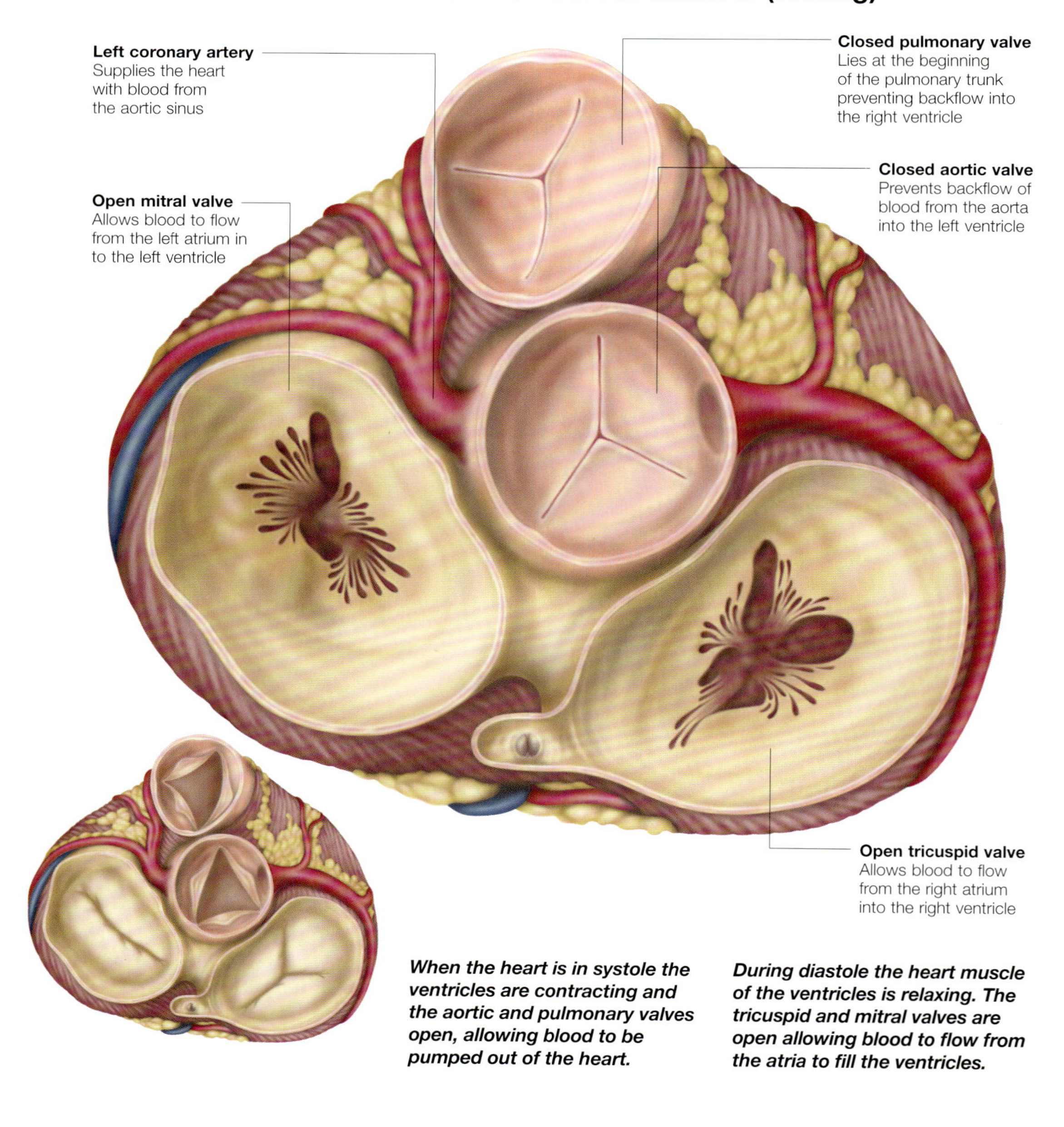

When the heart is in systole the ventricles are contracting and the aortic and pulmonary valves open, allowing blood to be pumped out of the heart.

During diastole the heart muscle of the ventricles is relaxing. The tricuspid and mitral valves are open allowing blood to flow from the atria to fill the ventricles.

Aortic and pulmonary valves

The pulmonary and aortic valves are also known as the semilunar valves. They guard the route of exit of blood from the heart, preventing backflow of blood into the ventricles as they relax after a contraction.

Each of these two valves is composed of three semilunar pocket-like cusps, which have a core of connective tissue covered by a lining of endothelium. This lining ensures a smooth surface for the passage of blood.

Aortic valve

The aortic valve lies between the left ventricle and the aorta, the main artery that carries oxygenated blood to the body. It is stronger and more robust than the pulmonary valve as it has to cope with the higher pressures of the systemic circulation circuit.

Above each cusp of the valve, formed by bulges of the aortic wall, lie the aortic sinuses. From two of these sinuses arise the right and left coronary arteries, which carry blood to the muscle and coverings of the heart itself.

Pulmonary valve

The pulmonary valve separates the ventricle from the pulmonary trunk, the large artery that carries blood from the heart towards the lungs. Just above each cusp of the valve the pulmonary trunk bulges slightly to form the pulmonary sinuses, blood-filled spaces that prevent the cusps from sticking to the arterial wall behind them when they open.

View of the left ventricle opened up

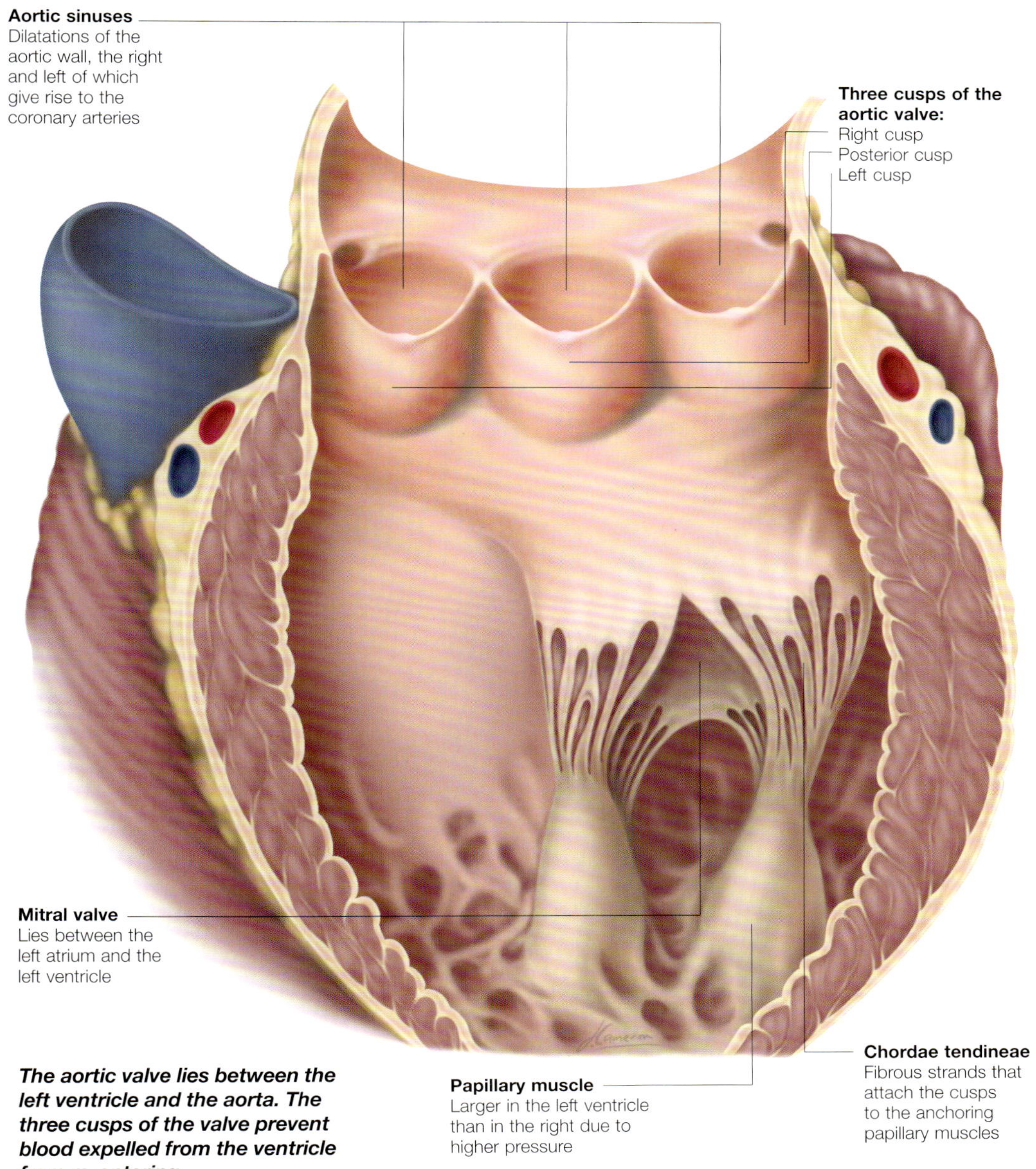

The aortic valve lies between the left ventricle and the aorta. The three cusps of the valve prevent blood expelled from the ventricle from re-entering.

Vessels of the Heart

Blood is delivered to the heart by two large veins – the superior and inferior venae cavae – and pumped out into the aorta. The venae cavae and aorta are collectively known as the great vessels.

The Venae Cavae

The superior vena cava is the large vein that drains blood from the upper body to the right atrium of the heart. It is formed by the union of the right and left brachiocephalic veins which, in turn, have been formed by smaller veins that receive blood from the head, neck and upper limbs.

The inferior vena cava is the widest vein in the body, but only its last section lies within the thorax as it passes up through the diaphragm to deliver blood to the right atrium.

The Aorta

The aorta is the largest artery in the body, having an internal diameter of about 2.5cm (1in) in adults. Its relatively thick walls contain elastic connective tissue that allows the vessel to expand slightly, as blood is pumped into it under pressure, and then recoil, consequently maintaining blood pressure between heartbeats.

The aorta passes upwards initially, then curves around to the left and travels down into the abdomen. It consists of the ascending aorta, the arch of the aorta and the descending (thoracic) aorta. The various sections of the aorta are named for their shape or the positions in which they lie, and each has branches that carry blood to the tissues of the body.

The heart and great vessels

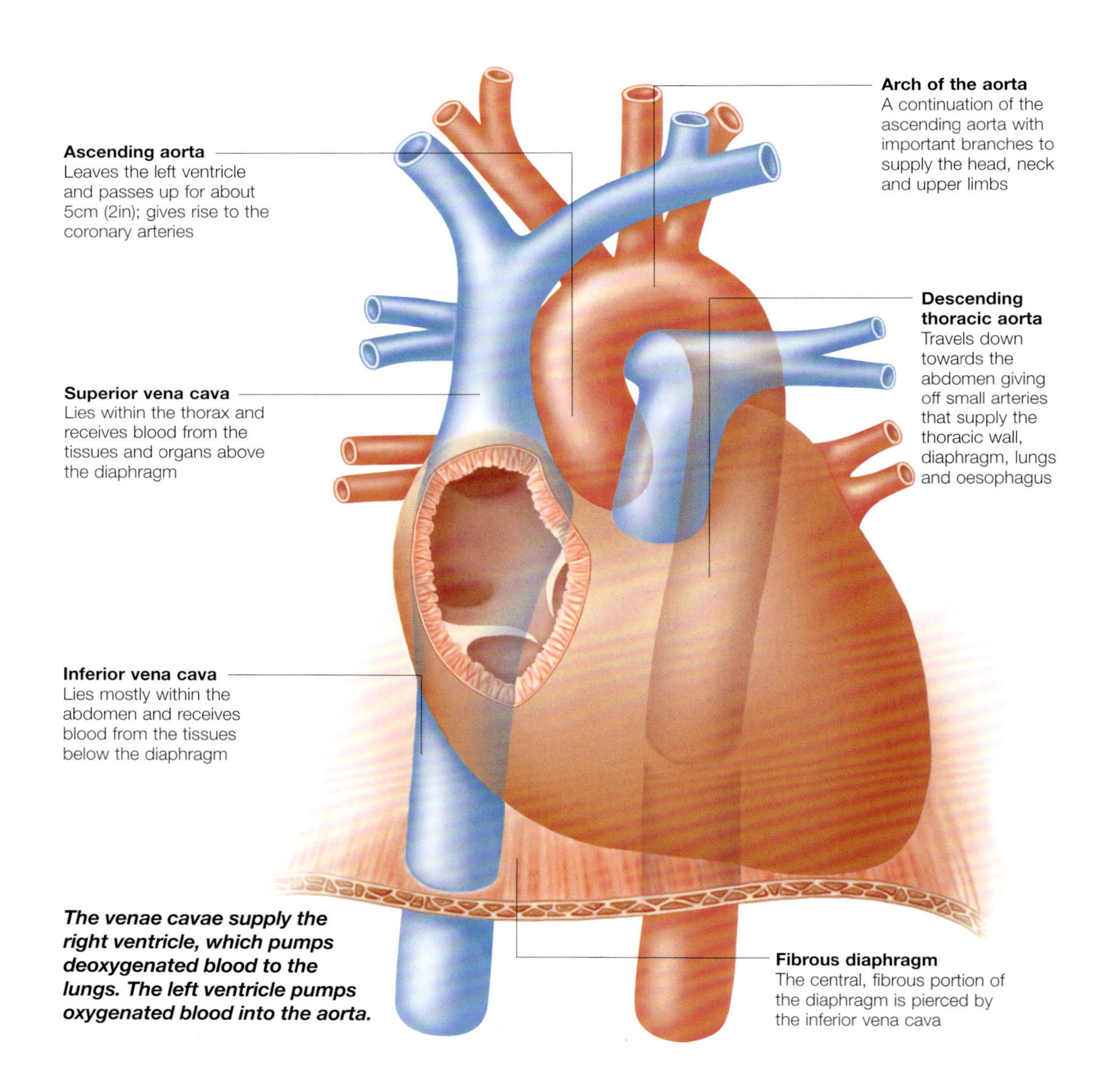

The venae cavae supply the right ventricle, which pumps deoxygenated blood to the lungs. The left ventricle pumps oxygenated blood into the aorta.

Supplying blood to the heart

The heart muscle and the coverings of the heart need their own arterial blood supply, which is provided by the two main coronary arteries and their branches.

There are two coronary arteries: right and left. These arise from the ascending aorta just above the aortic valve and run around the heart just beneath the epicardium, embedded in fat.

The right coronary artery

This arises within the right aortic sinus, a small outpouching of the arterial wall just behind the aortic valve. It runs down and to the right, along the groove between the right atrium and the right ventricle until it lies along the inferior surface of the heart. Here it terminates in an anastomosis (connecting network) with branches of the left coronary artery. The right coronary artery gives off several branches.

The left coronary artery

This arises from the coronary sinus above the aortic valve and runs down towards the apex of the heart. The left coronary artery divides early on into two branches.

Venous drainage

The main vein of the heart is the coronary sinus. It receives blood from the cardiac veins and empties into the right atrium. In general, cardiac veins follow the routes of the coronary arteries.

Coronary arteries

This is a resin cast of the coronary arteries and their branches, as seen from the front. It clearly shows their complex branching network.

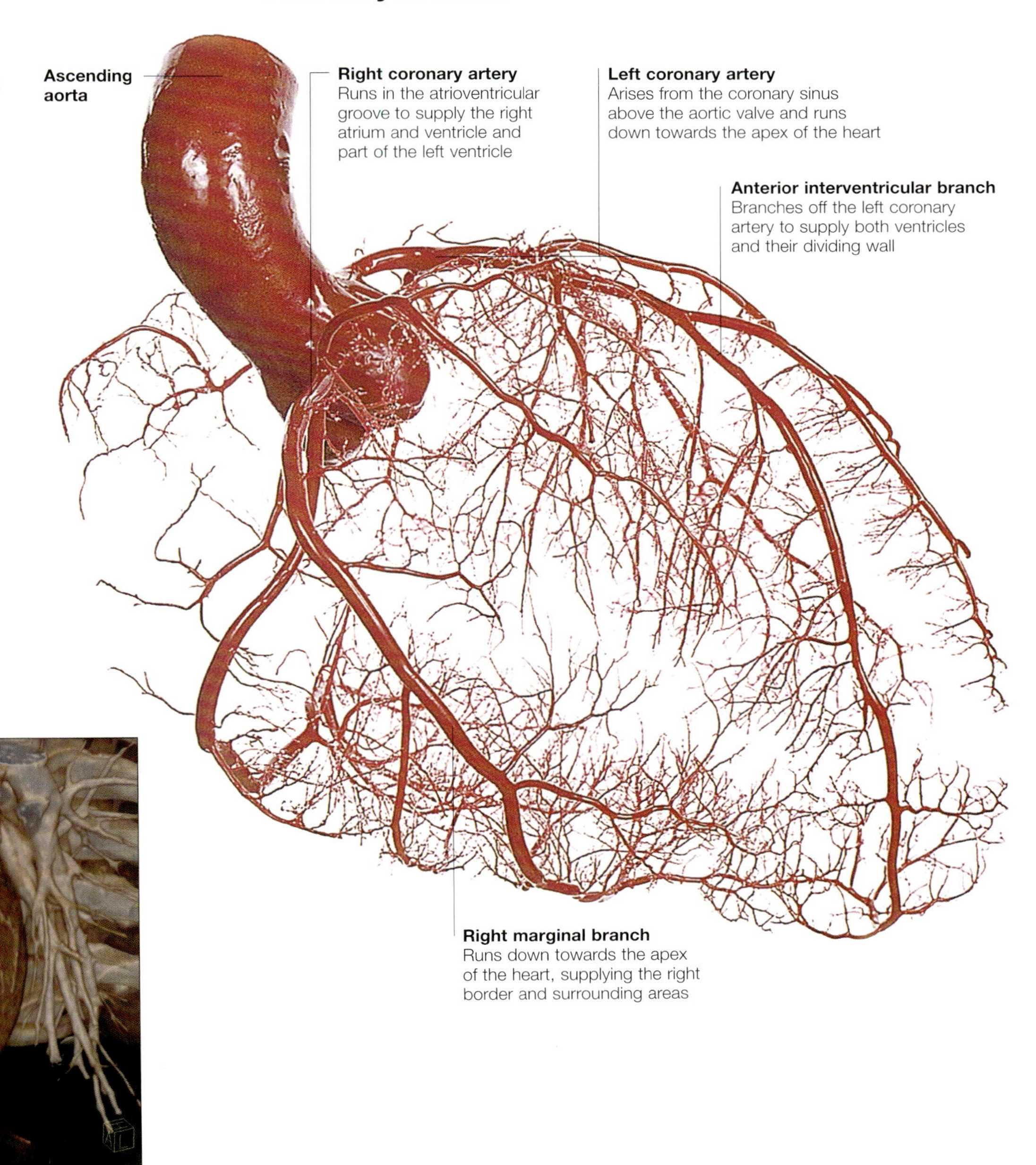

The ascending aorta and some of the coronary arteries are shown in this 3D CT scan of the heart.

Conducting System of the Heart

When the body is at rest, the heart beats at a rate of about 70 to 80 beats per minute. Within its muscular walls, a conducting system sets the pace and ensures that the muscle contracts in a co-ordinated way.

Sinoatrial Node
The sinoatrial (SA) node is a collection of cells within the wall of the right atrium. Each contraction of the cells of the SA node generates an electrical impulse, which is passed to the other muscle cells of the right and left atria and then to the atrioventricular (AV) node.

Atrioventricular Node
The cells of the AV node will initiate contractions of their own, and pass on impulses at a slower rate, if not stimulated by the SA node. Impulses from the AV node are passed to the ventricles through the next stage of conducting tissue.

Atrioventricular Bundle
The AV bundle passes from the atria to the ventricles through an insulating layer of fibrous tissue. It then divides into two parts, the right and left bundle branches, which supply the right and left ventricles respectively.

The intrinsic conduction system of the heart

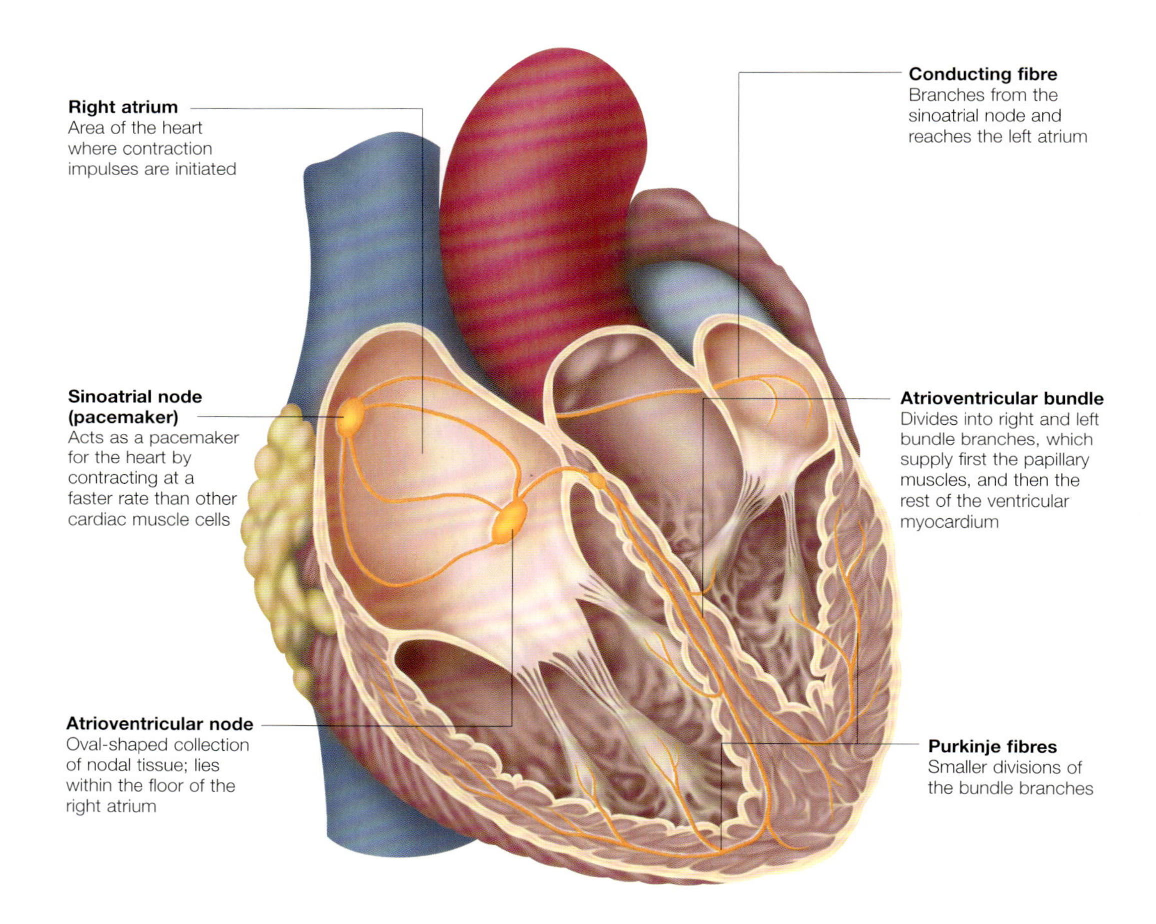

The intrinsic conduction system of the heart carries a wave of nerve impulses, which create synchronized contraction of the heart muscle.

Cardiac cycle

The cardiac cycle is the series of changes within the heart that causes blood to be pumped around the body. It is divided into a period when heart muscle contracts, known as systole and a period when it is relaxed, known as diastole.

Ventricular filling
During diastole the tricuspid and mitral valves are open. Blood from the great veins fills the atria and then passes through these open valves to fill the relaxing ventricles.

Atrial contraction
As diastole ends and systole begins, the sinoatrial node sparks off a contraction of the atrial muscle, which forces more blood into the ventricles.

Ventricular contraction
The wave of contraction reaches the ventricles via the atrioventricular bundles and the Purkinje fibres. The tricuspid and mitral valves snap shut as pressure increases. The blood pushes against the closed pulmonary and aortic valves and causes them to open. As the wave of contraction dies away, the ventricles relax. The cycle begins again with the next sinoatrial node impulse about a second later.

Events of the cardiac cycle

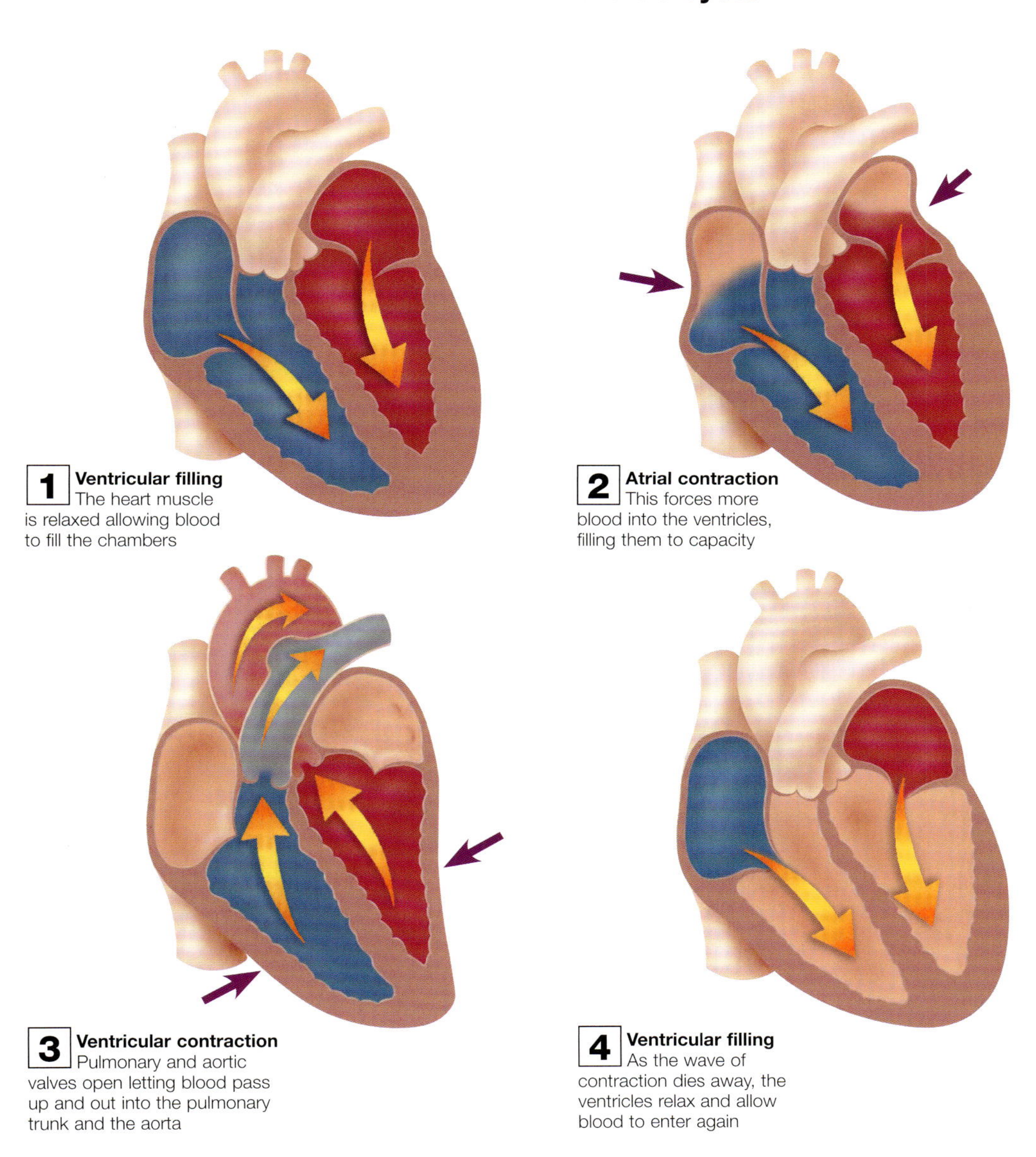

1 Ventricular filling The heart muscle is relaxed allowing blood to fill the chambers

2 Atrial contraction This forces more blood into the ventricles, filling them to capacity

3 Ventricular contraction Pulmonary and aortic valves open letting blood pass up and out into the pulmonary trunk and the aorta

4 Ventricular filling As the wave of contraction dies away, the ventricles relax and allow blood to enter again

The movements of the heart cause the circulation of blood. The sequence of contraction is repeated, in normal adults, about 70 to 90 times a minute.

How the Heart Beats

The heart contains specialized tissue that generates an intrinsic rhythmic beat. The brain controls the heart rate by sending nervous impulses that alter this inherent rhythm.

A remarkable feature of the heart is that as long as it is bathed with a solution containing vital nutrients, it will continue to beat for long periods when removed from the body. This is because the heartbeat originates from within the heart itself, rather than resulting from electrical impulses from the brain.

Specialized pacemaker tissue and an electrical conducting system are responsible for generating the electrical 'spark' underlying the heartbeat and transmitting it in an orderly sequence across the upper and lower heart chambers.

Sinoatrial Node

The primary pacemaker of the heart is called the sinoatrial (SA) node, which is a small area of tissue located in the right atrium (upper chamber) of the heart. The SA node is about 20mm (0.78in) long by 5mm (0.19in) wide. Specialized electrical properties of the cells in the SA node allow it to generate regular 'sparks' of electricity that initiate each heartbeat.

Heart muscle cells are connected to one another in such a way that electrical events pass rapidly from cell to cell. Thus, when the cells of the SA node generate electrical impulses, the emerging wave of electrical excitation spreads very rapidly across both atria. This leads to a synchronized contraction of the atria, which pushes blood into the lower chambers of the heart – the ventricles.

Spread of electrical activity across the heart

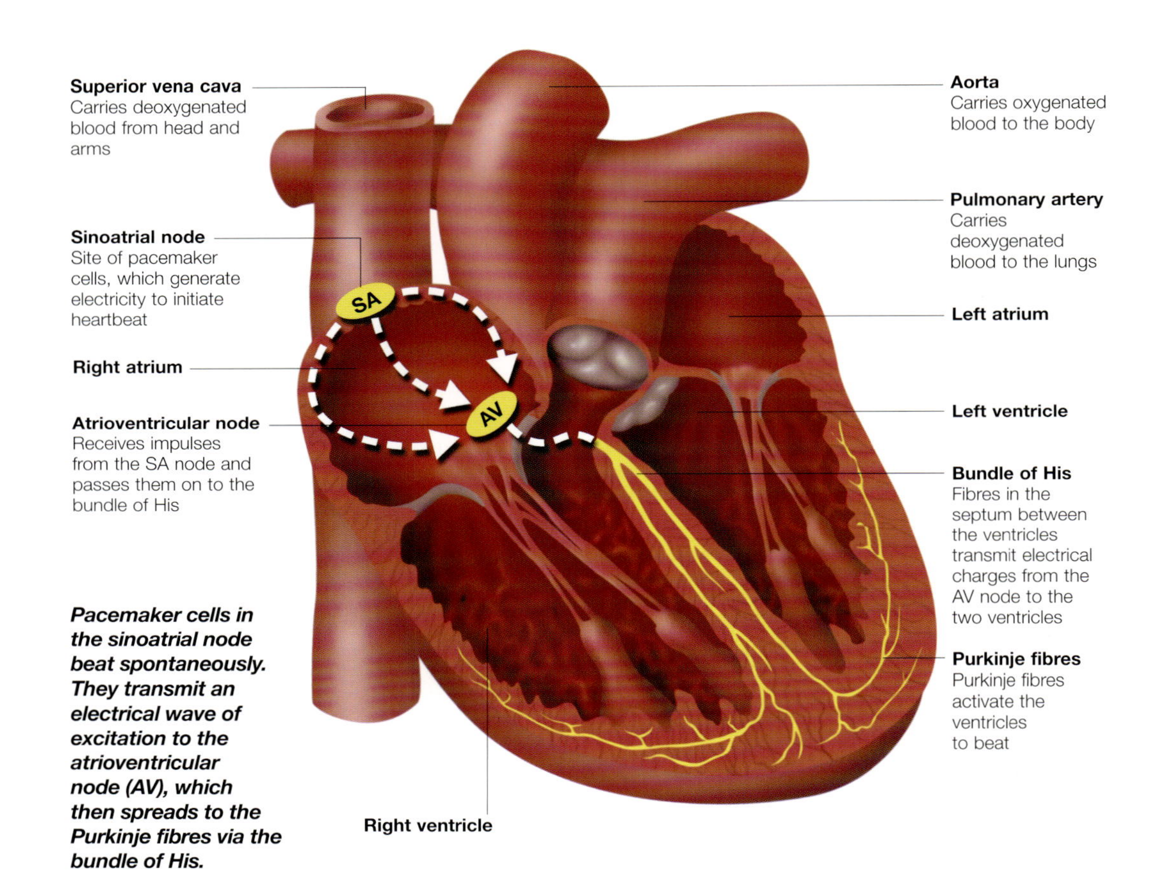

Pacemaker cells in the sinoatrial node beat spontaneously. They transmit an electrical wave of excitation to the atrioventricular node (AV), which then spreads to the Purkinje fibres via the bundle of His.

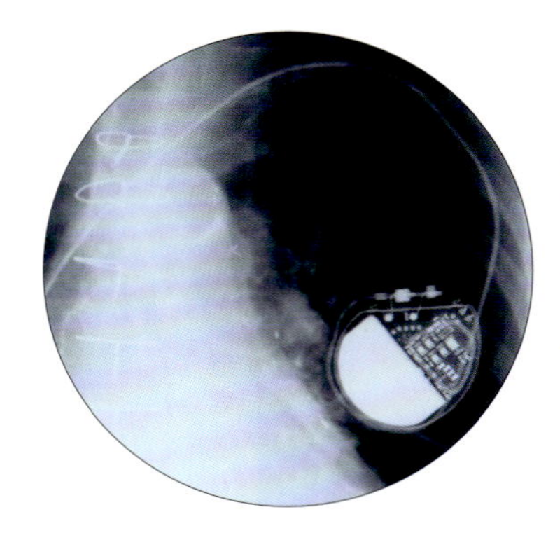

In older age, the electrical conducting system of the heart may become disrupted. A pacemaker device (shown in this x-ray) can be surgically implanted to keep the heart beating regularly.

How the heart rate is controlled

The brain is able to modulate the heart rate via parasympathetic and sympathetic nerve fibres. These adapt the strength and timing of the heartbeat during rest, exercise or emotion.

Although the heartbeat arises in the sinoatrial nodal tissue of the heart, it can be modulated by the brain via a series of nerve fibres. These nerve fibres are subdivided anatomically and functionally into two groups:

- Parasympathetic nerves – these decrease heart rate
- Sympathetic nerves – these increase the rate and strength of the heart's beating.

Parasympathetic control

In the absence of any influence from the nervous system, the inherent rate of SA node impulse generation is approximately 100 beats/minute in humans, higher than the normal resting heart rate, which is around 70 beats/minute. This is because parasympathetic activity (via the vagus nerve) slows the rate of automatic sinoatrial impulse generation.

At rest, therefore, the heart is considered to be under 'vagal tone'; this allows the brain to increase the heart rate by reducing the activity of the vagus nerve.

Sympathetic control

During increased demands on the circulation, such as occurs during exercise, sympathetic fibres release noradrenaline, which speeds up the rate of sinoatrial nodal impulse generation. In addition, sympathetic activity increases the speed of electrical conduction through the AV node allowing the ventricles to be excited and therefore beat more frequently.

As well as exercise, intense emotional states (fear, for example) can increase the heart rate via increased sympathetic activity.

Nervous control of the heart

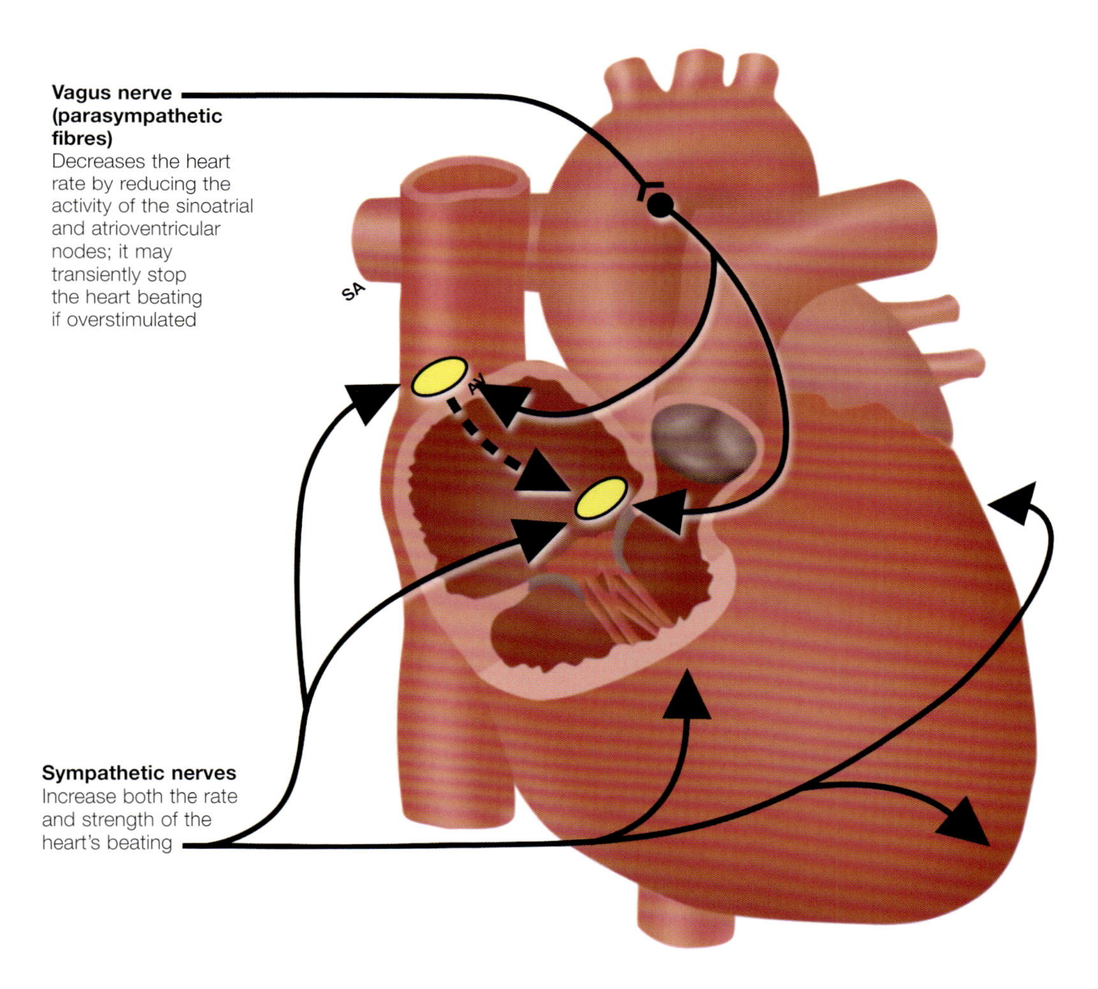

The brain is able to alter the heart's rate and strength of beating by sending nervous impulses along sympathetic and parasympathetic nerve fibres.

Arteries of the Arm

The arteries of the arm supply blood to the soft tissues and bones. The main arteries divide to form many smaller vessels that communicate at networks – anastomoses – at the elbow and wrist.

The main blood supply to the arm is provided by the brachial artery, a continuation of the axillary artery, which runs down the inner side of the upper arm. It gives rise to many smaller branches that supply surrounding muscles and the humerus (upper bone of the arm). The largest of these is the profunda brachii artery, which supplies the muscles that straighten the elbow.

The profunda brachii artery and the other, smaller arteries branching from the lower part of the brachial artery run down around the elbow joint. They then form a network of connecting arteries before rejoining the main arteries of the forearm.

Forearm and Hand

The brachial artery divides below the elbow joint into the radial and the ulnar arteries. The radial artery runs from the cubital fossa along the length of the radius (bone of the forearm). At the lower end of the radius, it lies under the skin and connective tissue; pulsations can be felt here. The ulnar artery runs towards the base of the ulna (the other bone of the forearm).

The hand has a profuse blood supply from the end branches of the radial and ulnar arteries. Branches of the two arteries join together in the palm to form the deep and the superficial palmar arches, from which small arteries arise to supply the fingers.

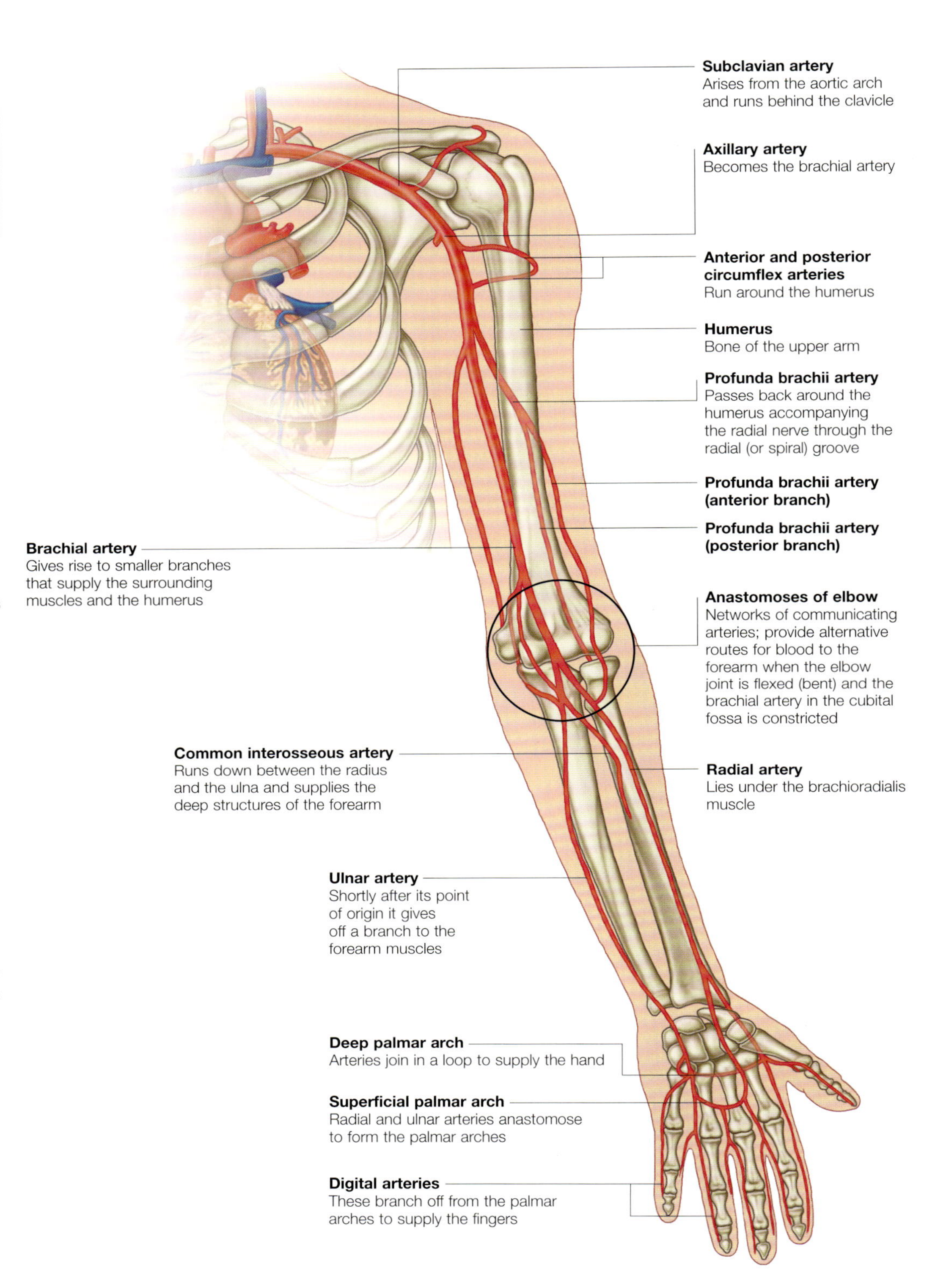

The arterial blood supply to the arm originates in the subclavian artery. Arteries are deeper than veins, protecting them against damage.

Veins of the Arm

The veins of the upper limb are divided into deep and superficial veins. The superficial veins lie close to the skin's surface and are often easily visible.

Venous drainage of the upper limb is achieved by two interconnecting series of veins, the deep and the superficial systems. Deep veins run alongside the arteries, while superficial veins lie in the subcutaneous tissue. The layout of the veins is very variable but usually resembles the pattern detailed right.

Deep Veins

In most cases, the deep veins are paired or double veins (venae comitantes) that lie on either side of the artery they accompany, making frequent anastomoses and forming a network surrounding the artery. The pulsations of blood within the artery alternately compress and release the surrounding veins, helping blood return to the heart.

The radial and ulnar veins arise from the palmar venous arches of the hand and run up the forearm to merge at the elbow, forming the brachial vein. This, in turn, merges with the basilic vein to form the large axillary vein.

Superficial Veins

There are two main superficial veins of the arm, the cephalic and the basilic veins, which originate at the dorsal venous arch of the hand. The cephalic vein runs under the skin along the radial side of the forearm.

The basilic vein runs up the ulnar side of the forearm, crossing the elbow to lie along the border of the biceps muscle. About halfway up the upper arm it turns inwards to become a deep vein.

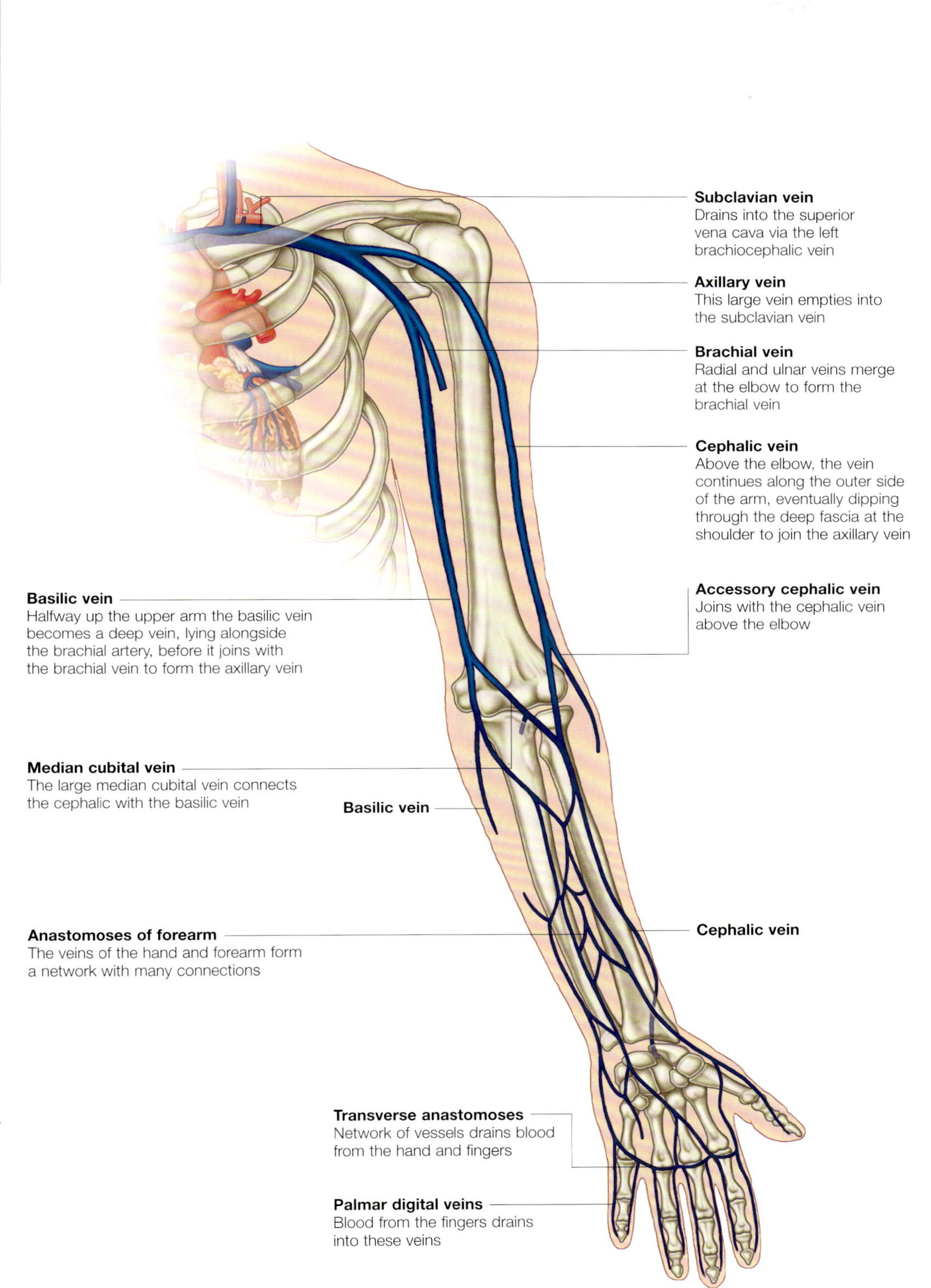

Superficial veins of the arm lie in the subcutaneous tissues (just under the skin) of the arm and are often visible.

Arteries of the Hand

The hand is supplied with numerous arteries and veins. These join to form networks of small, interconnecting blood vessels that ensure a good blood supply to all fingers, even if one artery is damaged.

The hand has a plentiful blood supply from the ulnar and radial arteries. These have many interconnections (anastomoses), maintaining the blood supply even if an artery is damaged.

Superficial palmar arch

The ulnar artery enters the hand on the same side as the little finger and crosses the palm to join with the radial artery to form the 'superficial palmar arch'. This gives off small digital arteries that supply blood to the little, ring and middle fingers.

Deep palmar arch

The deep palmar arch is formed by a continuation of the radial artery. This enters the palm from below the base of the thumb and gives off small arteries that supply the thumb and index finger as well as metacarpal branches that anastomose with the digital arteries.

Back of the hand

An irregular network of small arteries extends over the back of the wrist, which supplies the back of the hand and fingers.

Palmar view of arteries of the left hand

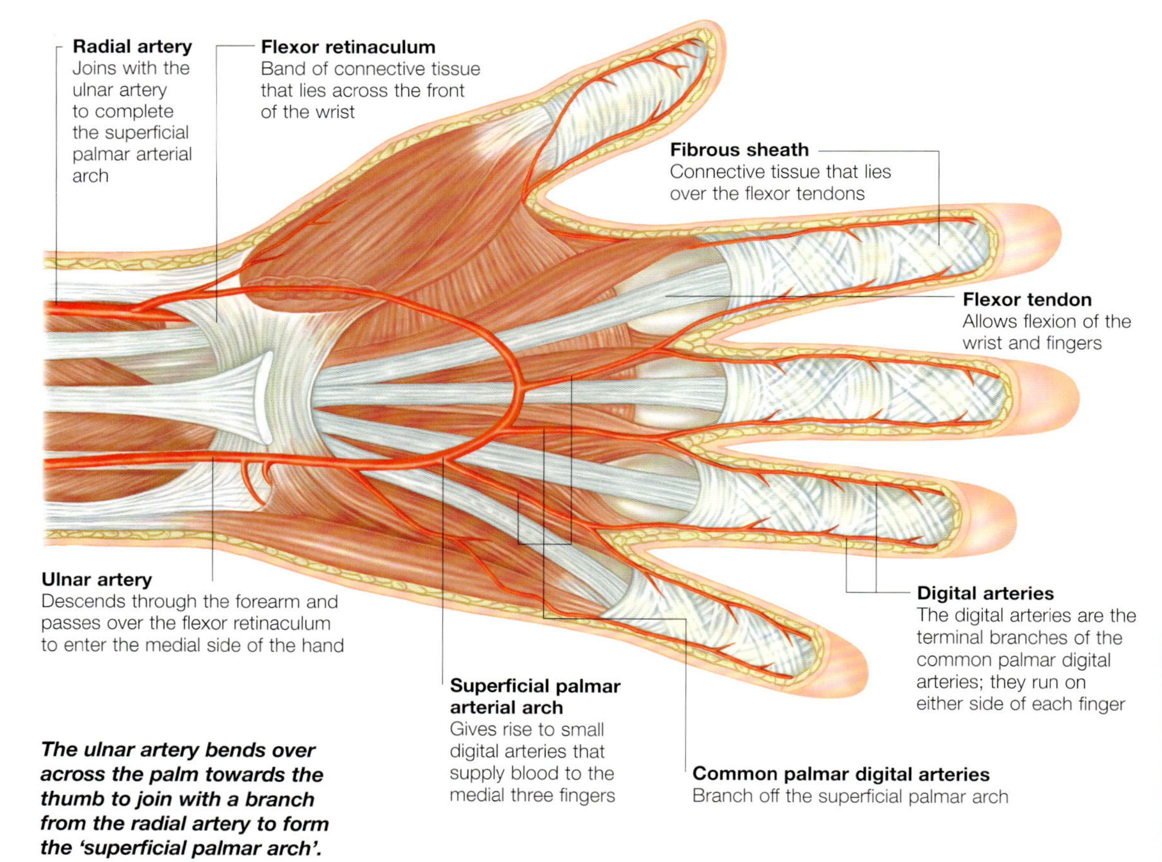

The ulnar artery bends over across the palm towards the thumb to join with a branch from the radial artery to form the 'superficial palmar arch'.

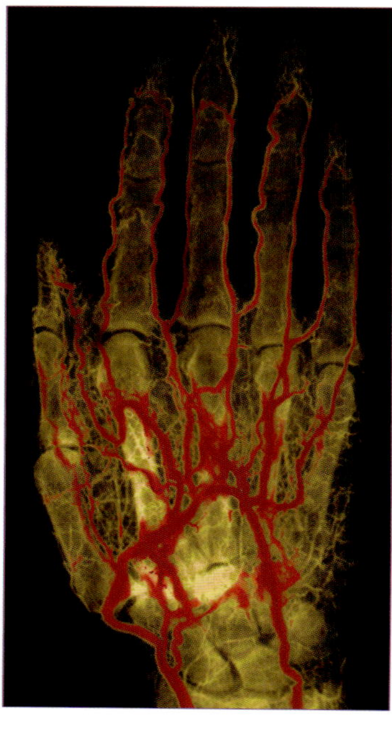

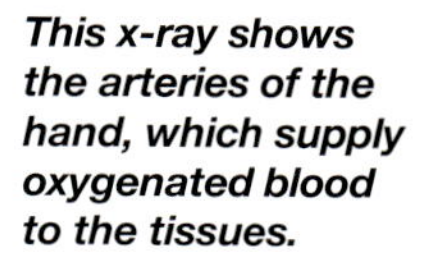

This x-ray shows the arteries of the hand, which supply oxygenated blood to the tissues.

Veins of the hand

The small veins of the hand join together to form the larger veins of the arm, which carry the blood back to the heart.

The veins of the back of the fingers join to form the prominent dorsal venous arch. Blood from this network of veins on the lateral (thumb) side drains into the large cephalic vein. On the other side of the back of the hand, the network drains into the large basilic vein.

Palmar veins

The veins of the palm form interconnecting arches that lie deeply alongside the deep palmar arterial arch and, more superficially, accompany the superficial palmar arch. They receive blood from the fingers via small digital veins running down either side of each finger.

The deep veins of the palm run with the radial and ulnar arteries of the forearm. The superficial palmar veins run with their equivalent arteries.

Dorsal view of the veins of the left hand

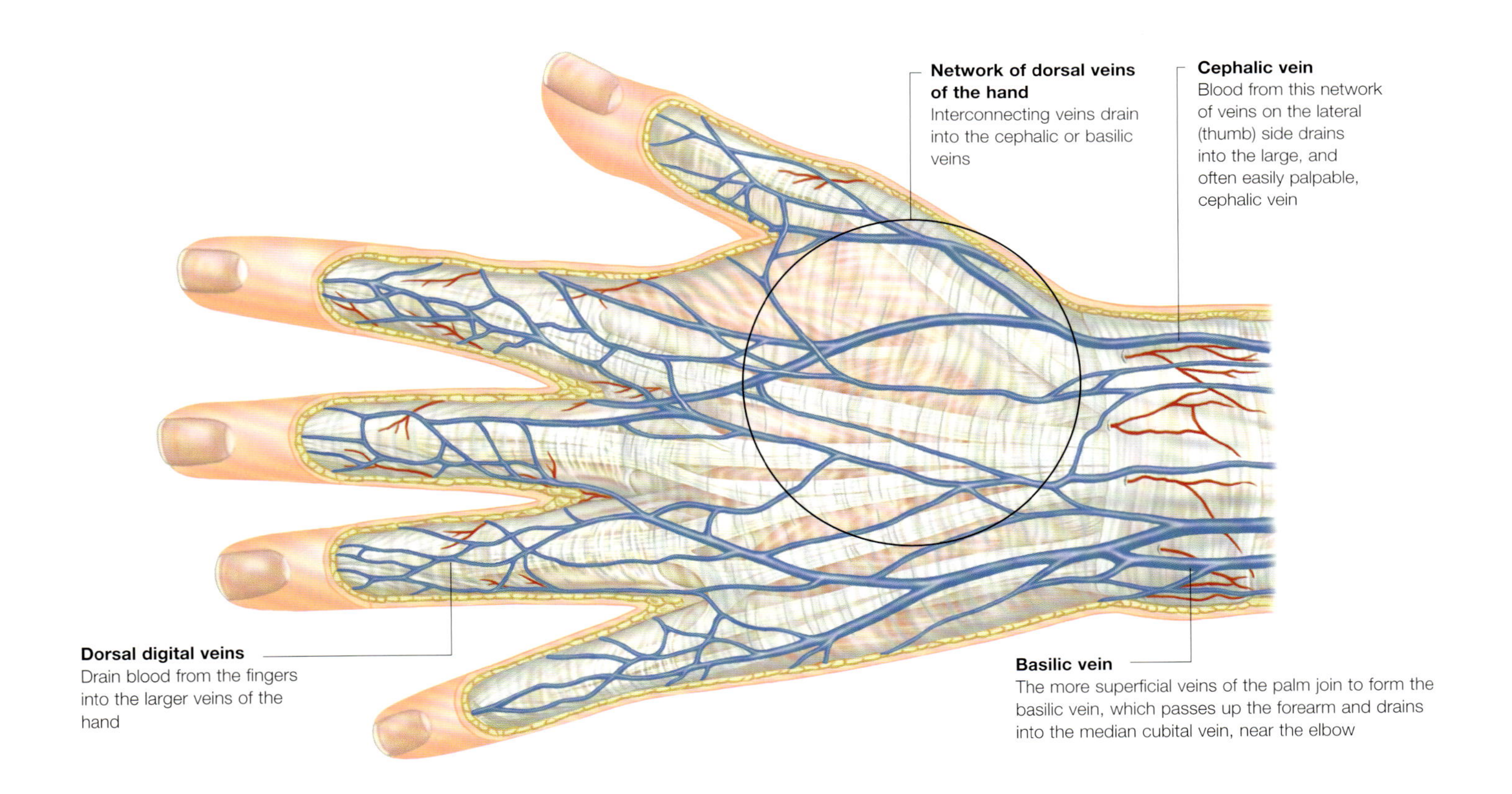

Arteries of the Leg

The lower limb is supplied by a series of arteries that arise from the external iliac artery of the pelvis. These arteries pass down the leg, branching to reach muscles, bones, joints and skin.

A network of arteries supplies the tissues of the lower limb with nutrients. The main arteries give off important and smaller branches to provide nourishment to various joints and muscles.

- **The femoral artery** – the primary artery of the leg. Its main branch is the profunda femoris (deep femoral) artery. Small branches supply nearby muscles before the artery enters a gap in the adductor magnus muscle, the 'adductor hiatus', to enter the popliteal fossa (behind the knee)
- **Profunda femoris** (deep femoral artery) – the main artery of the thigh. It gives off several branches including the medial and lateral circumflex femoral arteries and the four perforating arteries
- **Popliteal artery** – a continuation of the femoral artery. It runs down the back of the knee, giving off small branches to nourish that joint, before dividing into the anterior and posterior tibial arteries
- **Anterior tibial artery** – this supplies the structures within the anterior (front) compartment of the lower leg. It runs downwards to the foot and becomes the dorsalis pedis artery
- **Posterior tibial artery** – this artery remains at the back of the lower leg and, together with the peroneal (fibular) artery, it supplies the structures of the back and outer compartments.

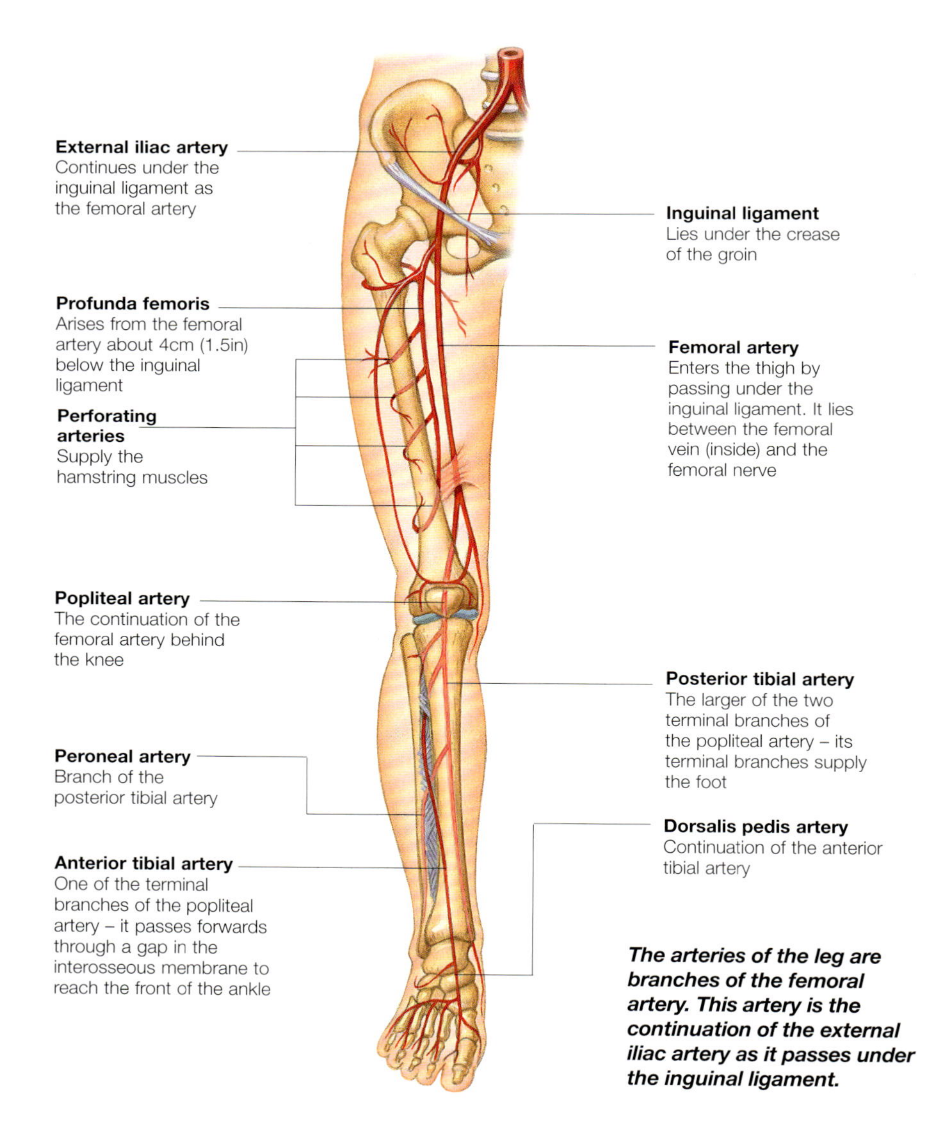

The arteries of the leg are branches of the femoral artery. This artery is the continuation of the external iliac artery as it passes under the inguinal ligament.

Arteries of the Foot

In a pattern similar to that in the hand, the small arteries of the foot form arches that interconnect, giving off branches to each side of the toes. Branches of the arteries give the sole of the foot a particularly rich blood supply.

The arterial supply of the foot is provided by the terminal branches of the anterior and posterior tibial arteries.

Top of the foot

As the anterior tibial artery passes down in front of the ankle it becomes the dorsalis pedis artery. This then runs down across the top of the foot towards the space between the first and second toes, where it gives off a deep branch that joins the arteries on the sole of the foot. Branches of the dorsalis pedis on the top of the foot join to form an arch that gives off branches to the toes.

Sole of the foot

The sole of the foot has a rich blood supply that is provided by branches of the posterior tibial artery. As the artery enters the sole, it divides into two to form the medial and lateral plantar arteries.

- **The medial plantar artery** – this is the smaller of the two branches of the posterior tibial artery. It provides blood for the muscles of the big toe and sends tiny branches to the other toes.
- **The lateral plantar artery** – this artery is much larger than the medial plantar artery and curves around under the metatarsal bones to form the deep plantar arch.

The deep branch of the dorsalis pedis artery joins the inner end of this arch so making a connection between the arterial supply of the top of the foot and the sole.

Plantar aspect of foot (sole)

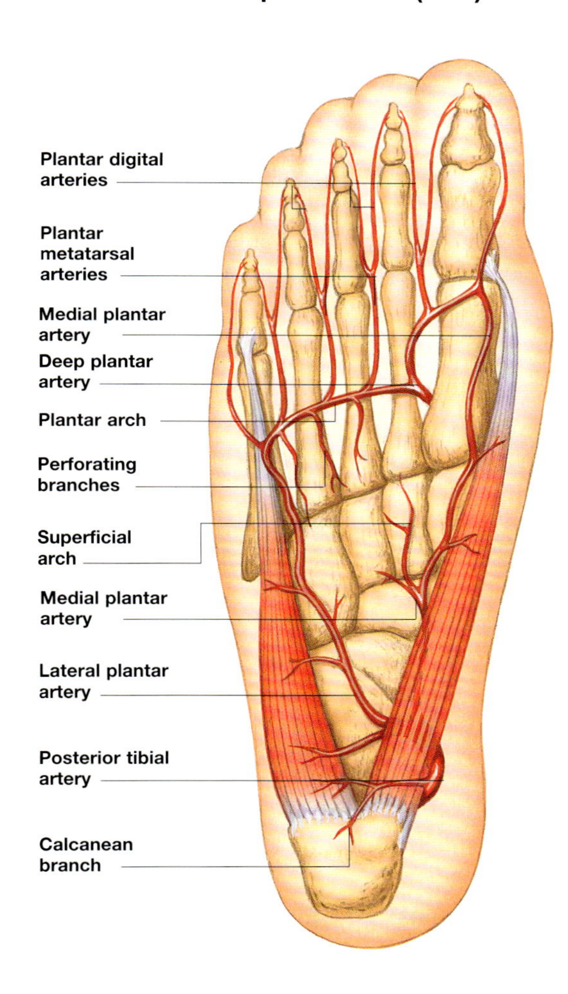

Dorsum of foot (top)

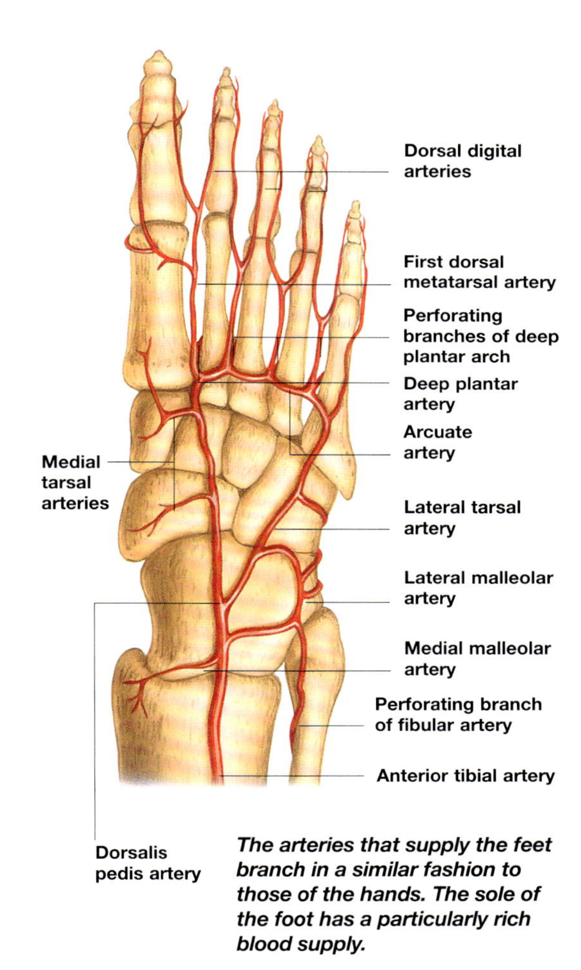

The arteries that supply the feet branch in a similar fashion to those of the hands. The sole of the foot has a particularly rich blood supply.

Veins of the Leg

The lower limb is drained by a series of veins that can be divided into two groups, superficial and deep. The perforating veins connect the two groups of veins.

Lying within the subcutaneous (beneath the skin) tissue, there are two main superficial veins of the leg, the great and small saphenous veins.

Great Saphenous Vein

The great saphenous vein is the longest vein in the body and is sometimes used during surgical procedures to replace damaged or diseased arteries in the heart. It arises from the medial (inner) end of the dorsal venous arch of the foot and runs up the leg towards the groin.

On its journey, the great saphenous vein passes in front of the medial malleolus (inner ankle bone), tucks behind the medial condyle of the femur at the knee and passes through the saphenous opening in the groin to drain into the large femoral vein.

Small Saphenous Vein

This smaller superficial vein arises from the lateral (outer) end of the dorsal venous arch and passes behind the lateral malleolus (outer ankle bone) and up the centre of the back of the calf. As it approaches the knee, the small saphenous vein empties into the deep popliteal vein.

Tributaries

The great and small saphenous veins receive blood along the way from many smaller veins and also intercommunicate freely, or 'anastomose', with each other.

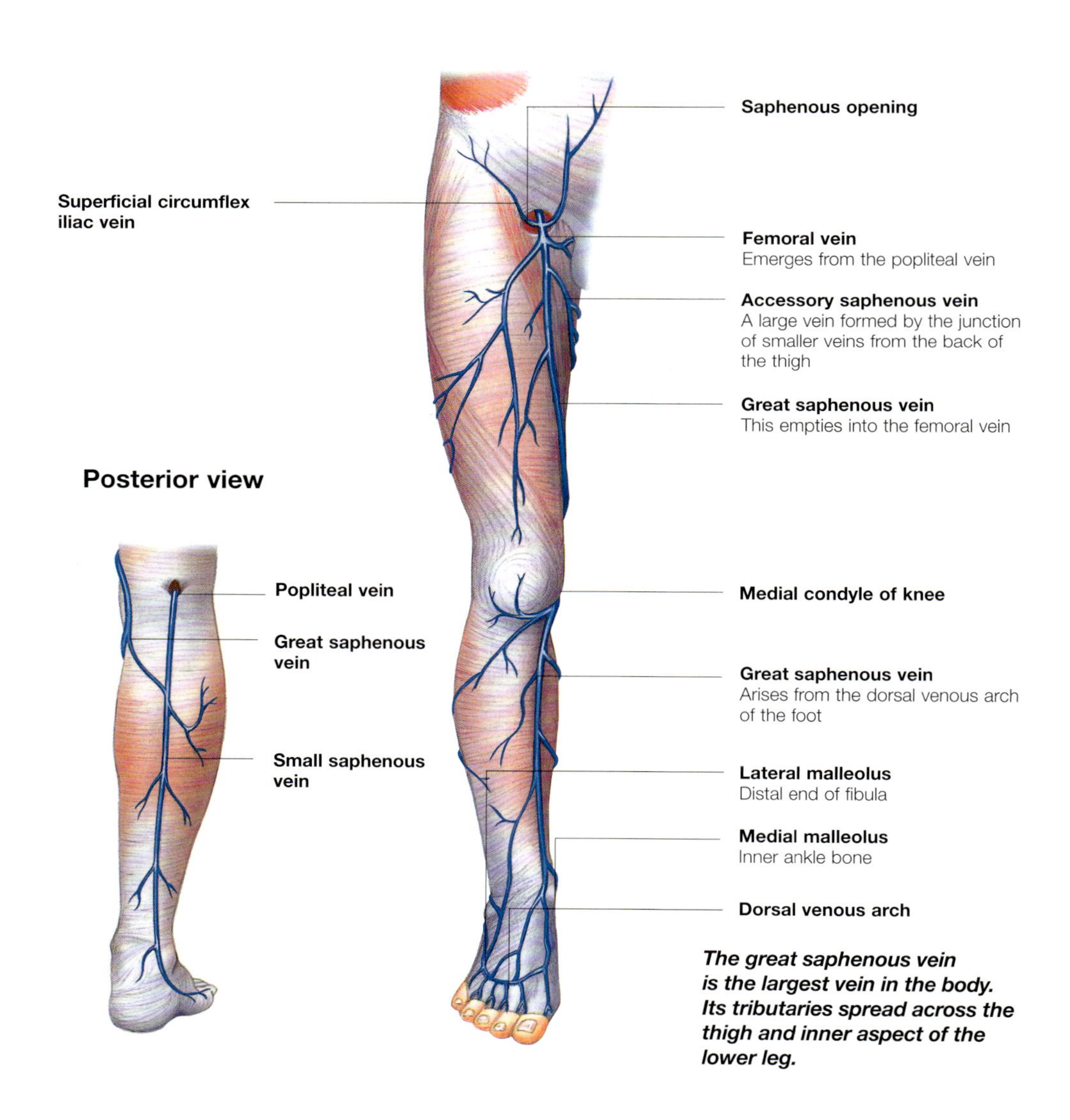

The great saphenous vein is the largest vein in the body. Its tributaries spread across the thigh and inner aspect of the lower leg.

Deep veins of the leg

The deep veins of the leg follow the pattern of the arteries, which they accompany along their length. As well as draining venous blood from the tissues of the leg, the deep veins receive blood from the superficial veins via the perforating veins.

Although the deep leg veins are referred to and illustrated as single veins they are usually, in fact, paired veins that lie either side of the artery. These veins are known as venae comitantes and they are common throughout the body.

- **The posterior tibial vein** – this is formed by the joining together of the small medial and lateral plantar veins of the sole of the foot. As it approaches the knee it is joined from its lateral side by the fibular (peroneal) vein before joining with the anterior tibial vein to form the large popliteal vein
- **The anterior tibial vein** – this is the continuation of the dorsalis pedis vein on the top of the foot. It passes up the front of the lower leg
- **The popliteal vein** – this lies behind the knee and receives blood from the veins that surround the knee joint
- **The femoral vein** – this is the continuation of the popliteal vein as it passes up the thigh. The large femoral vein receives blood from the superficial veins and continues up into the groin to become the external iliac vein of the pelvis.

Deep veins of leg, anterior view

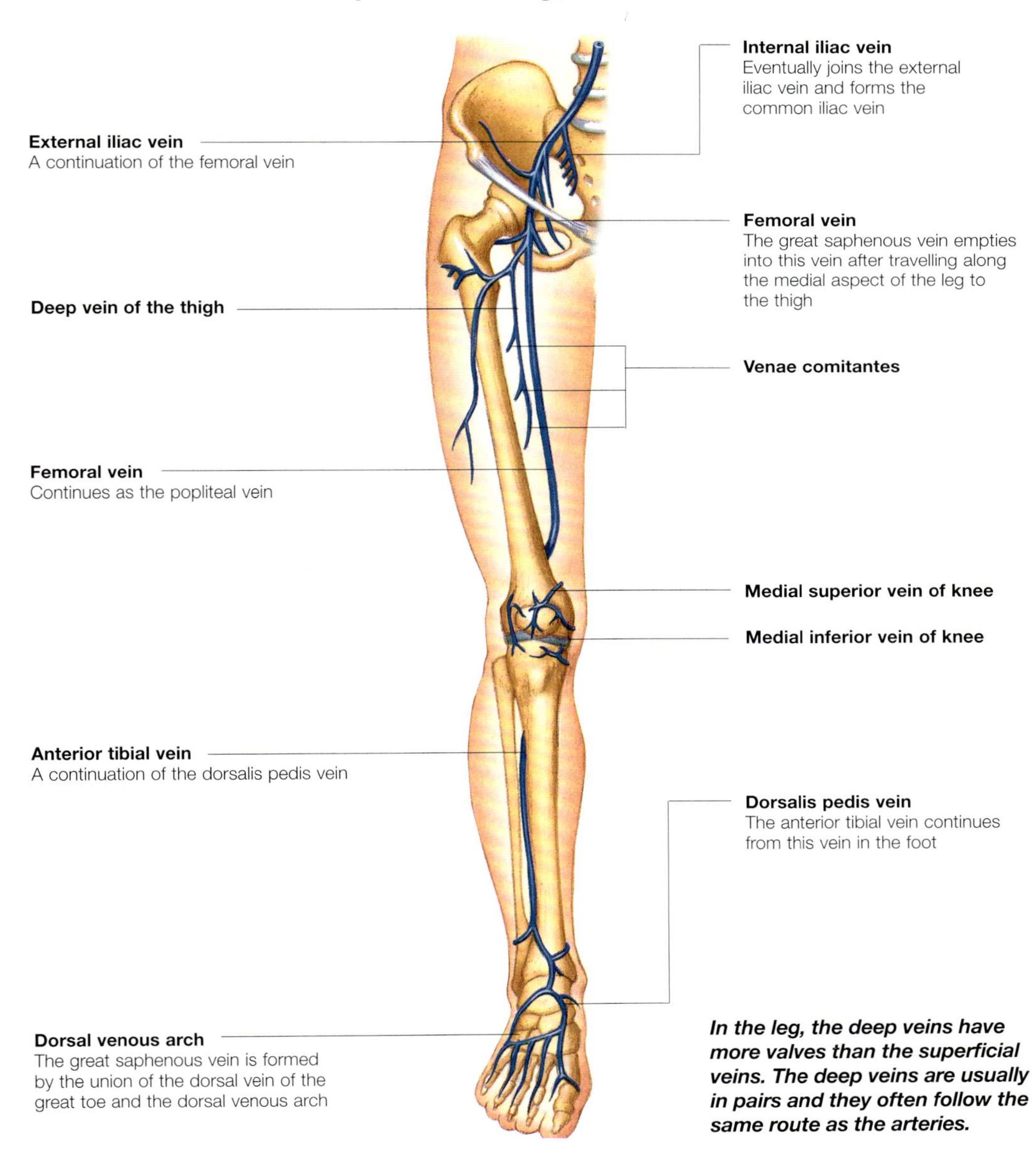

In the leg, the deep veins have more valves than the superficial veins. The deep veins are usually in pairs and they often follow the same route as the arteries.

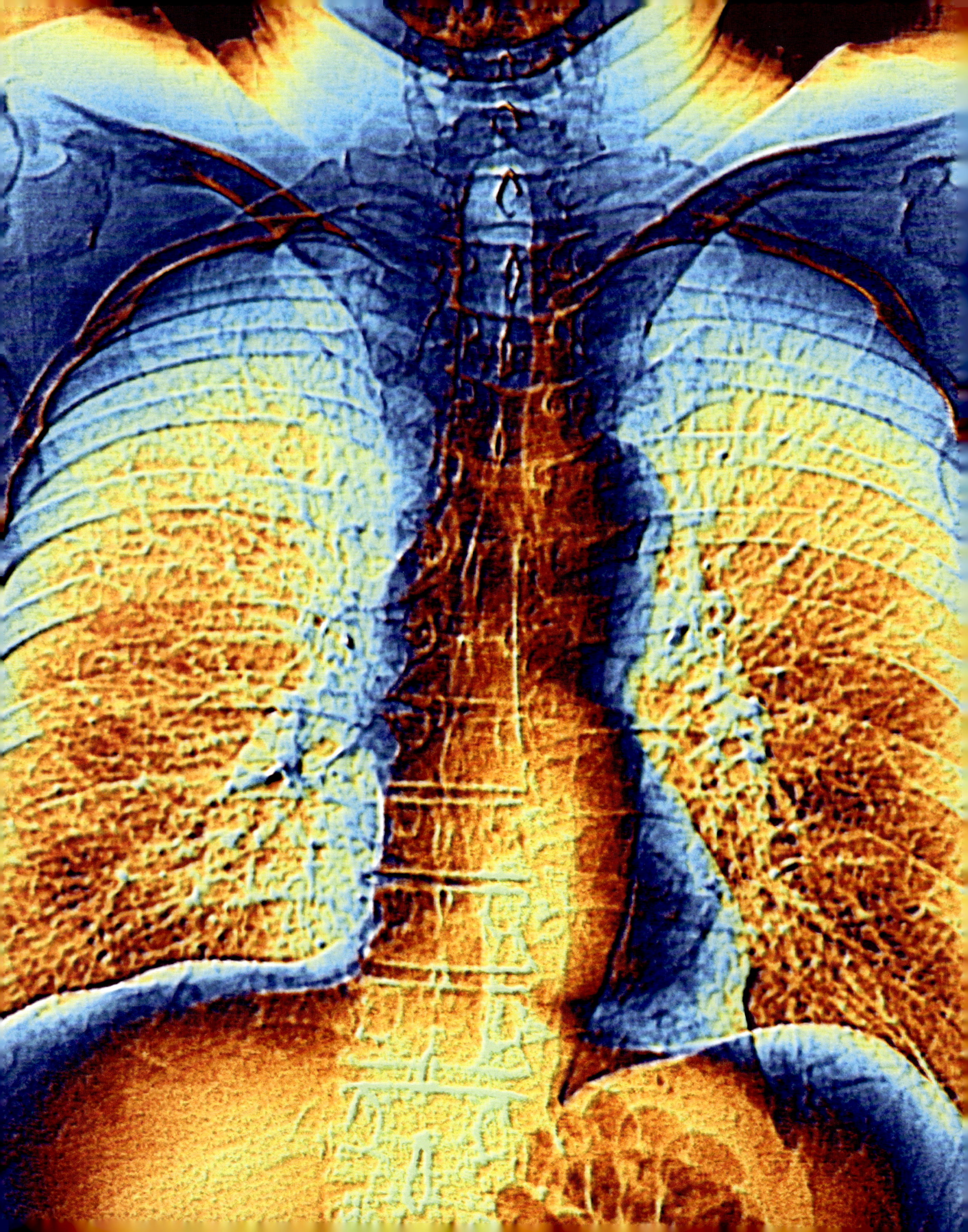

CHAPTER 7

The Respiratory System

Cells require a constant supply of oxygen to carry out their activities, during which they excrete the waste product carbon dioxide. The respiratory system works closely with the cardiovascular system to enable oxygen in the atmosphere to be transferred to the cells, and to ensure that carbon dioxide is successfully excreted. This section covers the airways (the pharynx, larynx, trachea and bronchi) and the lungs themselves, and explains the exchange of gases in the complex alveoli – the functioning parts of the lungs. The lungs are located either side of the heart and are surrounded and protected by the ribcage. Their functioning is dependent on muscles such as the diaphragm and the intercostal muscles to expand the chest to pull air into them, and to relax them on breathing out.

Opposite: The paired human lungs play a vital role in providing all cells and tissues with oxygen and removing carbon dioxide.

Nose

'Nose' commonly implies just the external structure, but anatomically it also includes the nasal cavity. The nose is the organ of smell and, as the opening of the respiratory tract, it serves to warm and filter air.

The external nose is a pyramid-shaped structure in the centre of the face, with the tip of the nose forming the apex of the pyramid. The underlying nasal cavity is a relatively large space and is the very first part of the respiratory tract (air passage).

The nasal cavity lies above the oral cavity (mouth) and is separated by a horizontal plate of bone called the hard palate. Both cavities open into the pharynx, a muscular, tube-like passageway.

External Structure

The external nose is made up of bone in its upper part and cartilage and fibrous tissue in its lower part. The upper part of the skeleton of the nose is mainly made up of a pair of plate-like bones known as the nasal bones. These join, by their upper edges, with the frontal bone (forehead). Joining the outer edge of each nasal bone is the frontal process of the maxilla – a projection from the cheekbone between the nasal bone and the inner wall of the orbit (eye socket).

The bridge of the nose consists almost entirely of the two nasal bones, and is joined to the forehead between the two orbits. Because of their location and their relative fragility, the nasal bones are vulnerable to fracturing.

The lower half of the external nose is made up of plates of cartilage on each side. These join each other, and the cartilages of the other side along the midline of the nose.

Lateral view

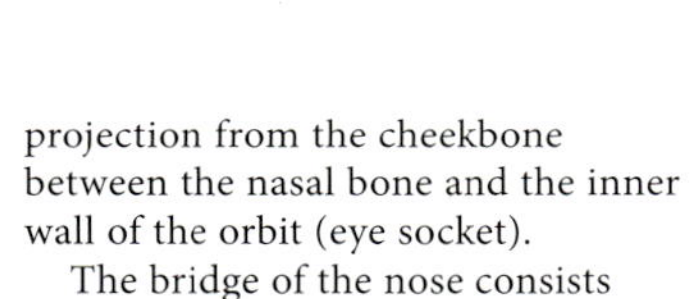

Inside the nostrils

The part of the nasal cavity immediately above the nostril is somewhat flared and is called the vestibule. It is lined with hair-bearing skin. Elsewhere, the nasal cavity has a more delicate inner lining – the mucous membrane.

The lining of the nasal cavity receives a generous blood flow due to the presence of a rich network of blood vessels. This ensures that inhaled air is adequately warmed and moistened in the nasal cavity before it reaches the lungs.

The delicate lining of the nasal cavity is prone to damage. The most common sign of this is bleeding from the membrane, known as a nosebleed (epistaxis). The lining also contains an abundance of cells, which, when inflamed or infected, tend to secrete an excess of viscous fluid (with a cold, for example).

The roof of the nasal cavity has a different kind of lining called the olfactory epithelium. This contains specialized cells that are receptors for the sense of smell.

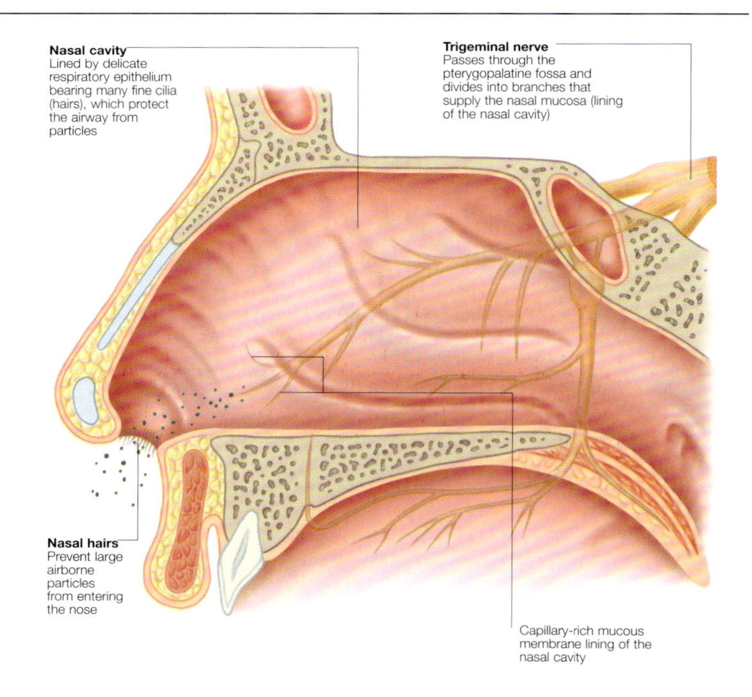

The mucous mebrane lining the nasal cavity warms and moistens air as it is breathed into the nostrils.

Pharynx

The pharynx, situated at the back of the throat, acts as a passage both for food to the digestive system and air to the lungs.

The pharynx, a fibromuscular, 15cm (6in)-long tube at the back of the throat, is a passage for both food and air. The constrictor muscles of the pharynx allow food to be squeezed into the oesophagus. The pharynx is divided into three main parts and the entrance is guarded by the tonsils.

Nasopharynx – the uppermost part of the pharynx, lying above the soft palate. The most prominent feature on each side is the tubal elevation, the end of the auditory (Eustachian) tube that enables air pressure to be equalized between the nasopharynx and the middle ear cavity. Lymphoid (adenoid) tissue is found on the back wall.

Oropharynx – lies at the back of the throat. Its roof is the undersurface of the soft palate; the floor is the back of the tongue. The palatine tonsil lies in the side wall, and is bounded by the palatoglossal fold in front and the palatopharyngeal fold behind.

Laryngopharynx – extends from the upper border of the epiglottic cartilage (which covers the opening of the airway during swallowing) to the lower border of the cricoid cartilage, where it continues into the oesophagus. The inlet of the airway lies in the front section.

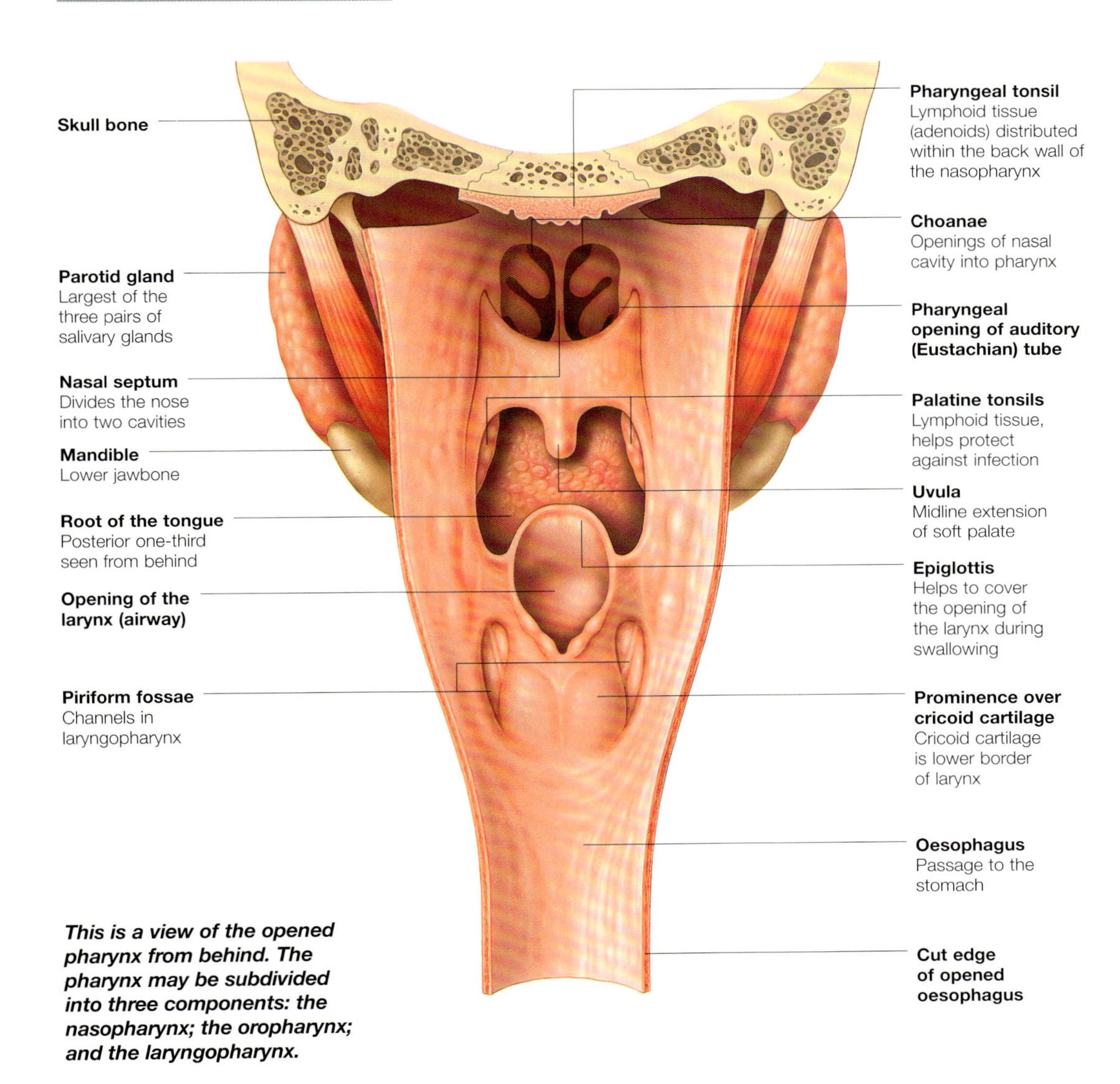

This is a view of the opened pharynx from behind. The pharynx may be subdivided into three components: the nasopharynx; the oropharynx; and the laryngopharynx.

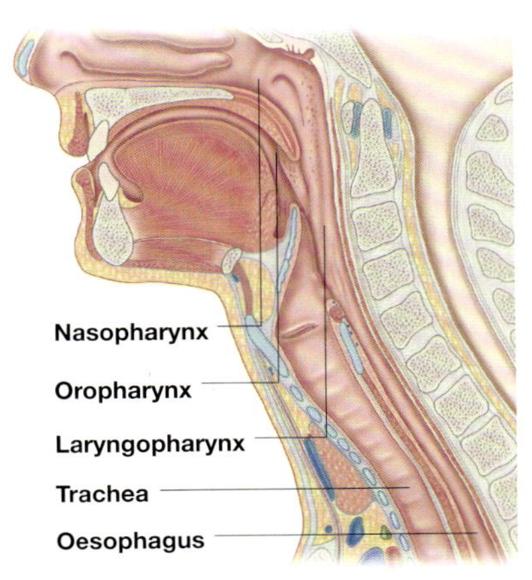

The relative positions of the three major parts of the larynx, the trachea and the oesophagus are evident in this sagittal section.

Larynx

The larynx is situated in the neck below and in front of the pharynx. It is the inlet protecting the lungs, and contains the vocal folds (also known as cords).

The larynx is composed of five cartilages (three single and one paired), connected by membranes, ligaments and muscles. In adult men, the larynx lies opposite the third to sixth cervical vertebrae (slightly higher in women and children), between the base of the tongue and the trachea.

The larynx serves as an inlet to the airways, taking air from the nose and mouth to the trachea. Because air and food share a common pathway, the primary function of the larynx is to prevent food and liquid from entering the airway. This is achieved by three 'sphincters' and by elevation. The larynx has also evolved as an organ of phonation – the act of producing sounds – allowing vocalization.

Laryngeal Cartilages

The laryngeal prominence (thyroid cartilage protrusion, also known as Adam's apple) is readily visible in most men. Its greater protrusion in men compared to women is due to the influence of the hormone testosterone. It typically develops during puberty.

The thyroid cartilage has two rear extensions, a superior and an inferior 'horn'. The cricoid cartilage, the only complete ring of cartilage in the airway, is partly overlapped above by the thyroid cartilage. Above it sit a pair of mobile, pyramid-shaped arytenoid cartilages.

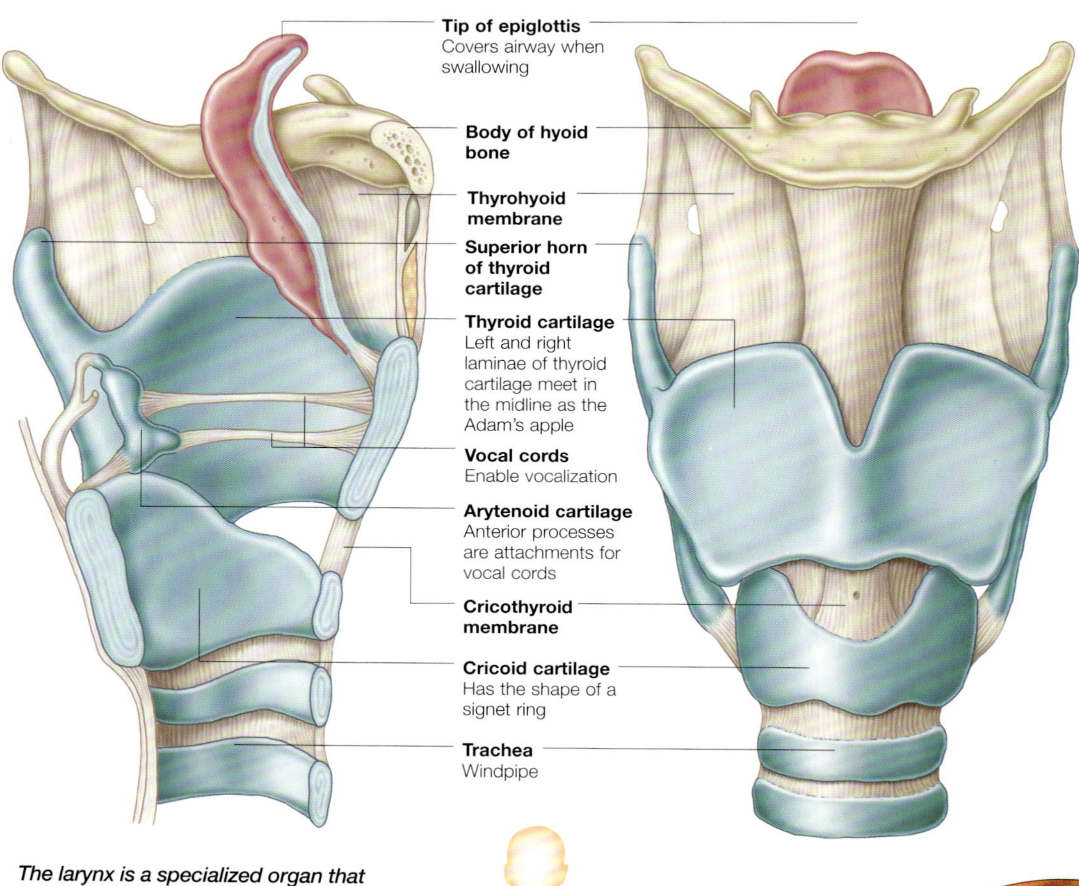

The larynx is a specialized organ that protects the inlet of the air passages and produces the voice. Its intrinsic sphincters close the airway during swallowing.

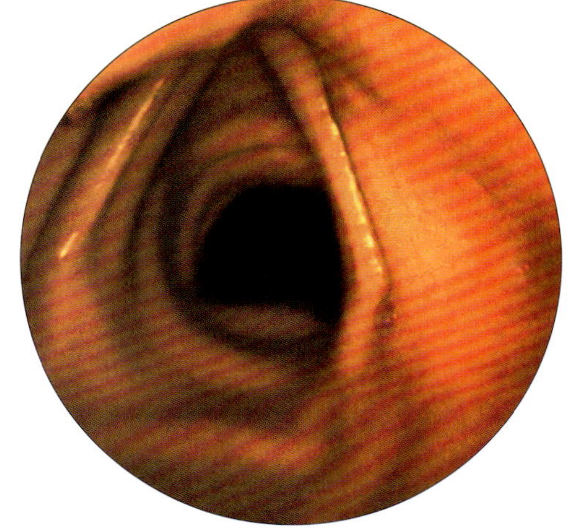

This endoscopic view of the larynx shows the vocal folds in an open position during normal breathing.

Muscles of the larynx

The muscles of the larynx act to close the laryngeal inlet while swallowing and move the vocal cords to enable speech.

During swallowing, the epiglottis, along with the rest of the larynx, is raised. As the front surface hits the rear part of the tongue, it flips backwards over the laryngeal inlet.

Aryepiglottic folds

The aryepiglottic folds of tissue are the free upper margins of the membranes that run between the epiglottis and the arytenoid cartilages. They contain a pair of transverse aryepiglottic and oblique aryepiglottic muscles. These arise from the muscular process of the opposite arytenoid cartilage, and attach to the sides of the epiglottis. They act like a 'purse string', closing the laryngeal inlet. The lower ends of each of the quadrangular membranes form the vestibular folds, or 'false' vocal cords.

Mucous glands

The quadrangular membranes are covered by a mucosa, and a submucosa rich in mucous glands. These are connected to the inner walls of the thyroid and cricoid cartilages. They keep the vocal folds (cords) moist, as the folds have no submucosa themselves, and therefore rely upon these secretions from above.

Piriform fossae

A groove, the piriform fossae, slopes backwards and serves to channel liquids toward the oesophagus and away from the larynx.

Rear view

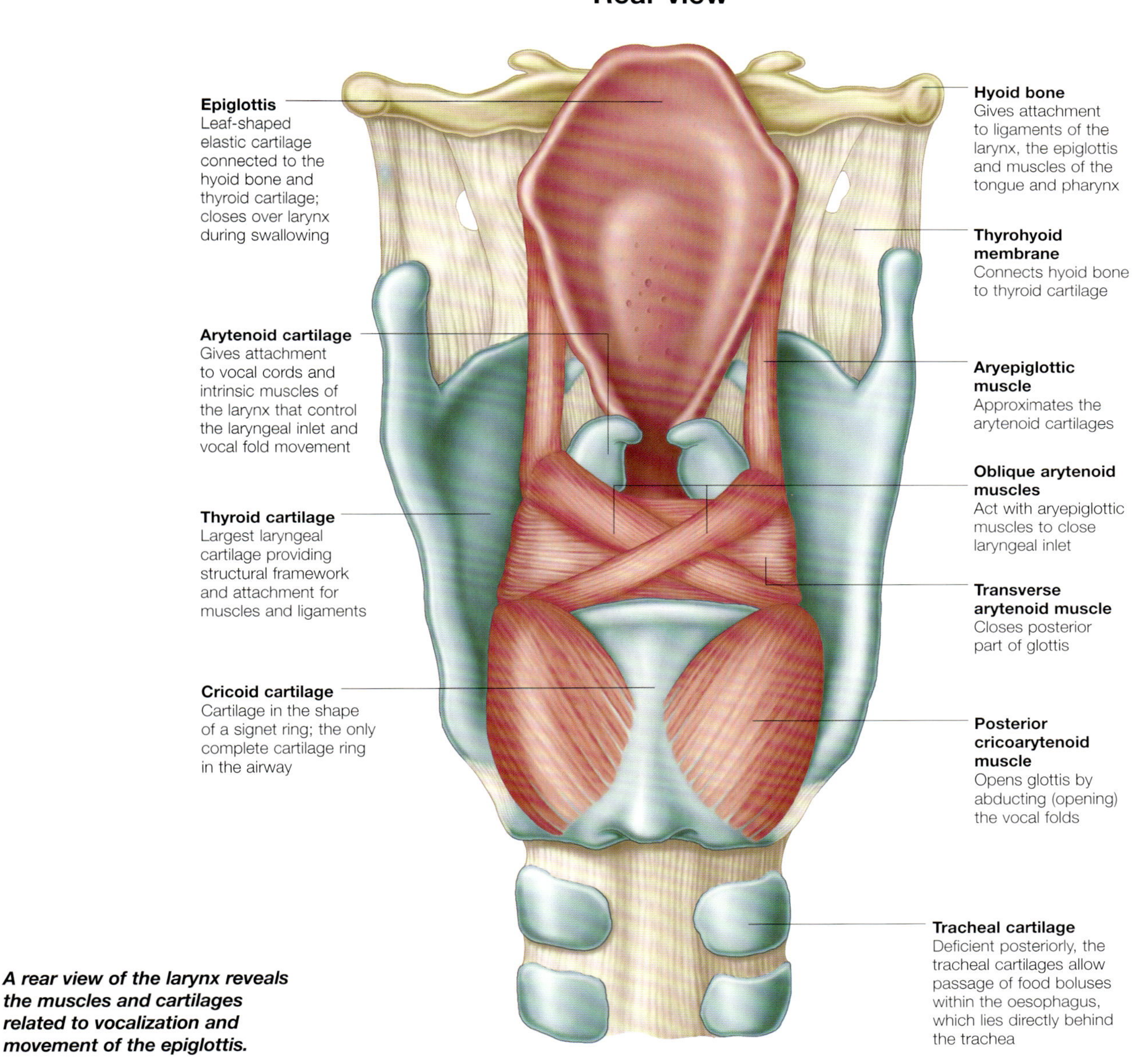

A rear view of the larynx reveals the muscles and cartilages related to vocalization and movement of the epiglottis.

How We Speak

All spoken languages are constructed from a number of separate speech sounds, known as phonemes.

All the speech sounds produced when vocalizing the majority of different languages are the direct result of expelling air from the lungs. In the first instance, air travels from the lungs, via the trachea, into the larynx (the voice box).

Voice Box

The larynx acts as a valve, sealing off the lungs from harmful irritants during coughing, for example. The opening in the larynx is called the glottis, and is covered by two flaps of retractable tissue called the vocal folds (also known as cords).

Vocal Folds

As air rushes through the glottis, the vocal folds resonate, producing a buzzing sound. The pitch of this buzz is determined by the tension and position of the vocal folds. However, not all speech sounds rely on the 'voicing' produced by the vocal folds; for example, the sound 'sssss' lacks voicing, whereas the sound 'zzzzzz' requires the vocal folds to vibrate.

Expulsion of Air

The vibrating air then moves up through the pharynx (throat) before travelling either over the tongue and through the mouth, or behind the soft palate and through the nose.

Speech organs

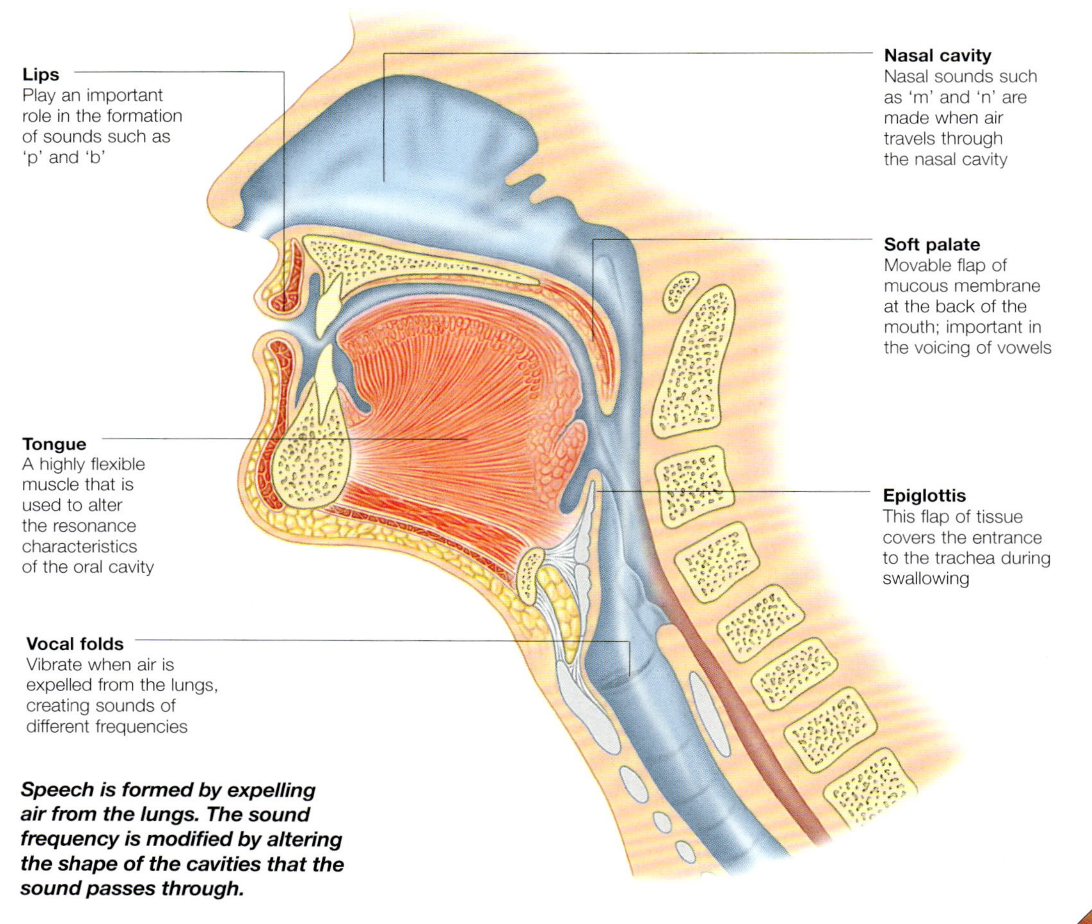

Speech is formed by expelling air from the lungs. The sound frequency is modified by altering the shape of the cavities that the sound passes through.

This view shows the vocal folds shut. The folds tense during speech, and vibrate to form sounds.

Voice sounds

Each of the chambers that air from the lungs passes through is of a different size and shape; the sound wavelength is altered as it travels through these chambers, resulting in a modified sound from the mouth or nose.

Vowels

Vowel sounds are produced when air is able to travel freely from the larynx to the outside. These vowel sounds are generated by altering the dimensions of the chambers the sound has to pass through. For example, when you repeat the vowel sounds in 'bet' and 'but' alternately, you should be able to feel the body of your tongue move backwards and forwards. This motion alters the resonance characteristics of the mouth cavity, thus altering the sound produced.

The lips are also important (note the difference in the lip position when pronouncing the vowel sounds in 'loot' and 'look') in determining the final sound, as is the soft palate (the fleshy flap at the back of the roof of the mouth). If the soft palate is opened, air will be able to flow out through the nose, as well as through the mouth, producing a 'nasal twang'.

Consonants

In contrast to vowels, consonants are produced when a barrier is put in the way of the passing air. When the sound 'sssss' is made, the tip of the tongue is brought up to just behind the teeth; this narrows the passage that the air can flow through, producing a hissing sound. Sounds like this are called 'fricatives' because they are created by the friction of moving air. Other fricative sounds include 'sh', 'th' and 'f', which are all produced by creating turbulence in the airflow.

Other consonant sounds are made by stopping the flow of air entirely, rather than just impeding it. This can be achieved by using the tip of the tongue ('t'), the body of the tongue ('k') or the lips ('p'). Alternatively, the passage of air out through the mouth can be blocked, opening the soft palate to produce sounds like 'm' and 'n'.

Producing consonants

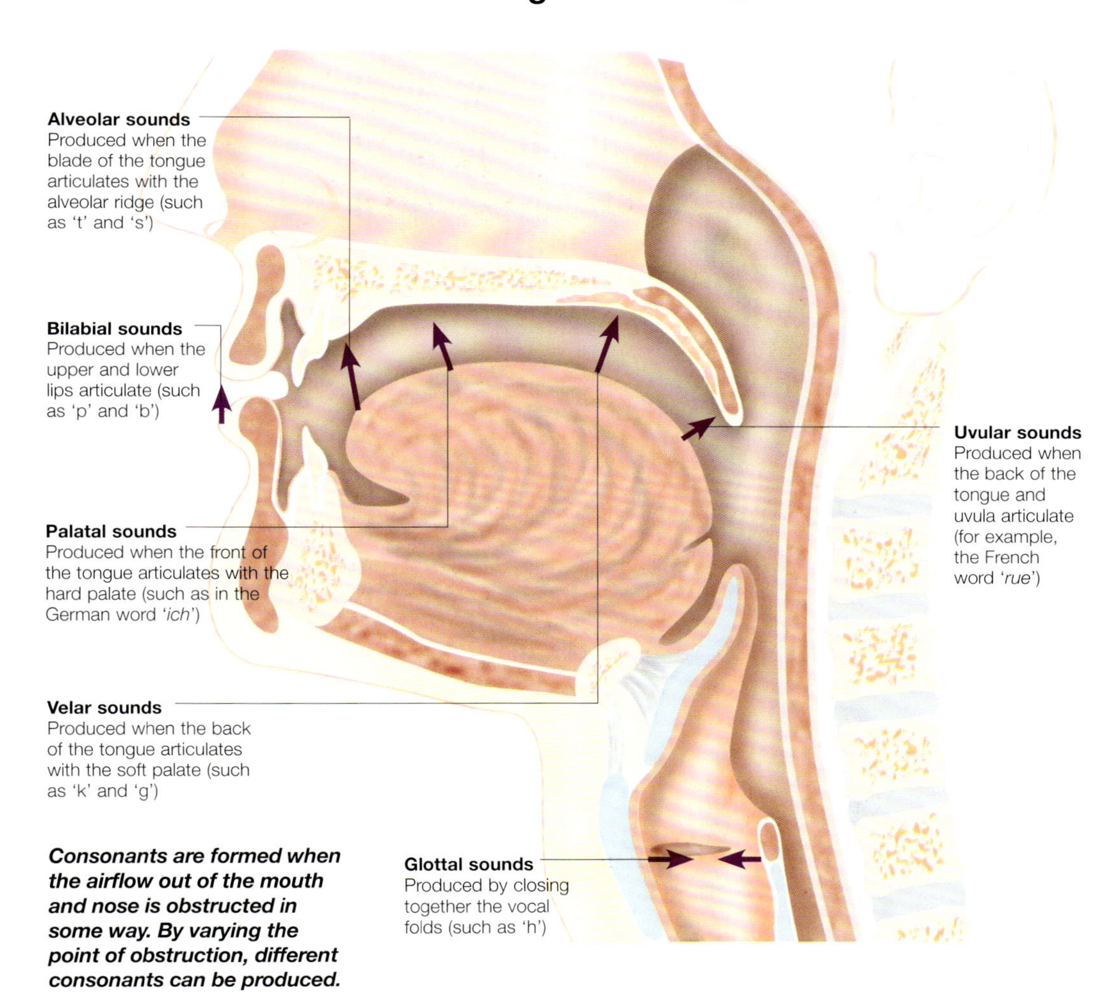

Consonants are formed when the airflow out of the mouth and nose is obstructed in some way. By varying the point of obstruction, different consonants can be produced.

Respiratory Airways

The airways form a network along which air travels to, from and within the lungs. The airways branch repeatedly, each branch narrowing until the end terminals – the alveoli – are reached.

As a breath is taken, air enters through the nose and mouth, then passes down through the larynx to enter the main airway, the trachea (windpipe). The trachea divides into two smaller airways – the bronchi – which divide to form progressively smaller tubes, known as bronchioles, that reach all areas of the lung. These tubes terminate in the alveolar sacs, which form the substance of the lung. It is in these thin-walled sacs that gas exchange with the blood occurs.

Trachea

The trachea extends down from the cricoid cartilage just below the larynx in the neck to the level of the sternal angle where it divides into two branches, the right and left main bronchi. The trachea is formed of strong fibroelastic tissue, within which are embedded a series of incomplete rings of hyaline cartilage, the tracheal cartilages. In adults the trachea is quite wide (approximately 2.5cm/1in), but it is much narrower in infants – about the width of a pencil.

The posterior (back) surface of the trachea has no cartilaginous support and instead consists of fibrous tissue and trachealis muscle fibres. This posterior wall lies in contact with the oesophagus, which is directly behind the trachea.

Tracheal lining

The epithelium (cellular lining) of the trachea contains goblet cells, which secrete mucus onto the surface. Tiny brush-like cilia (hairs) help to catch dust particles and move them back up towards the larynx and away from the lung. Between the epithelium and the rings of cartilage lies a layer of connective tissue containing small blood vessels, nerves, lymphatic vessels and glands that secrete water mucus. There are also many elastic fibres, which help to give the trachea its flexibility.

The major airways

Inhaled air passes down through the larynx into the trachea. From here it is channelled into two streams, each of which supplies a lung via a main bronchus.

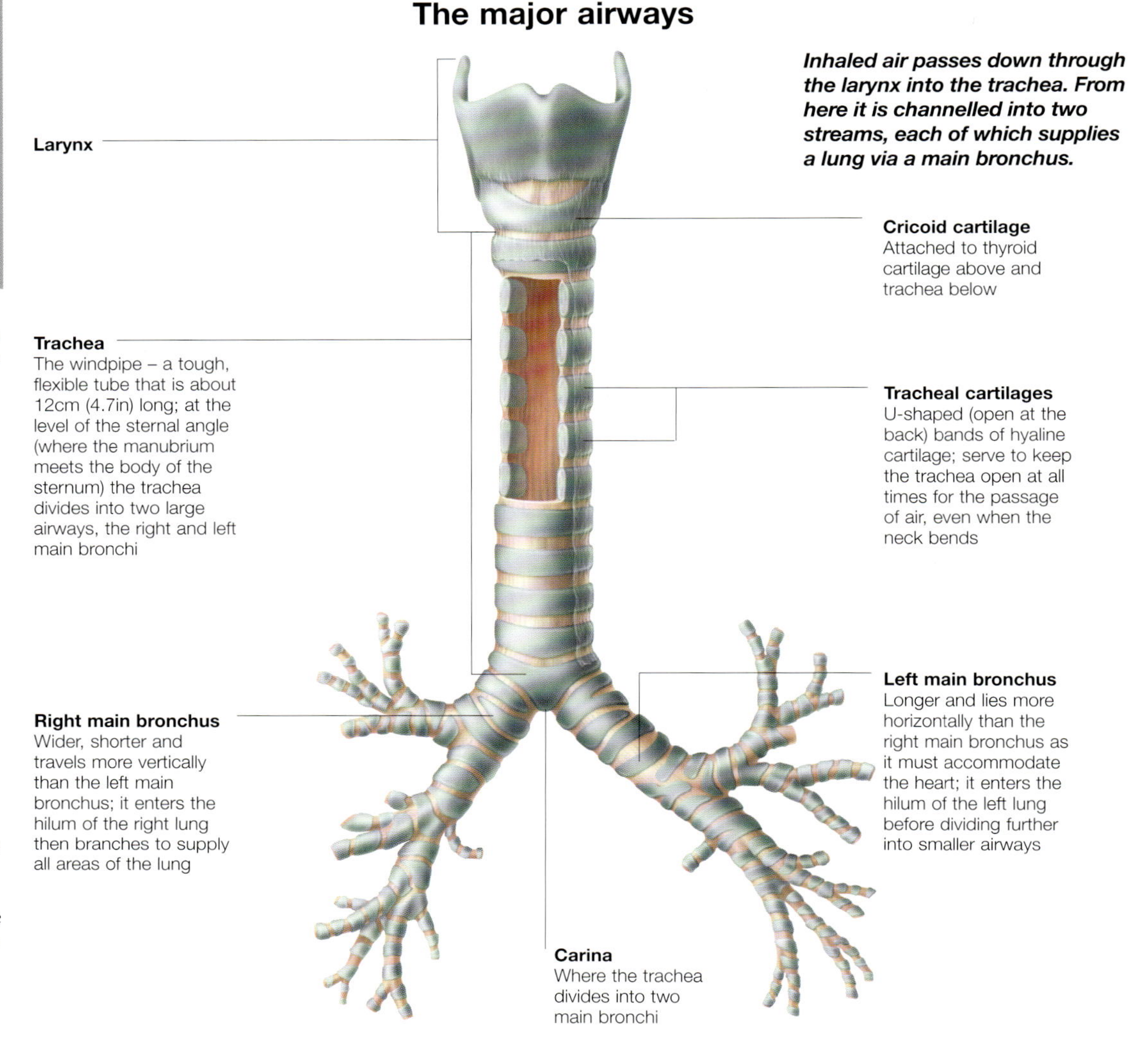

Cross-section through the trachea

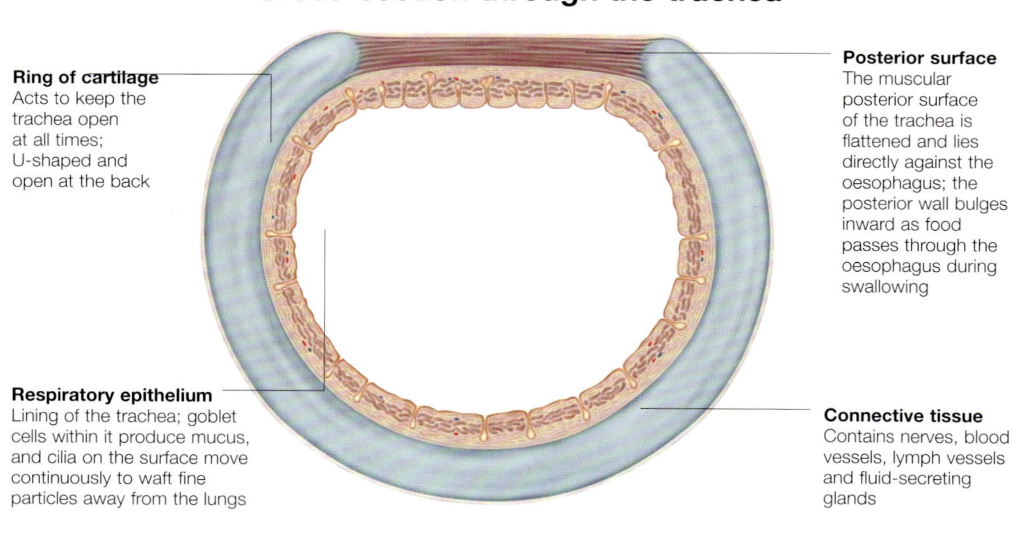

Bronchioles and alveoli

Each of the main bronchi divides again and again, forming the 'bronchial tree', which takes air to all parts of the lung.

The main bronchus first divides into the lobar bronchi, three on the right and two on the left, which each supply one lobe of the lung. These lobar bronchi divide to form smaller bronchi that supply each of the independent bronchopulmonary segments.

Bronchi structure

The bronchi have a similar structure to the trachea, and are very elastic and flexible. They have cartilage in their walls and are lined with respiratory epithelium. There are also numerous muscle fibres, which allow for changes in diameter of these tubes.

Bronchioles

Within the bronchopulmonary segments the bronchi continue to divide, perhaps as many as 25 times, before they terminate in the blind-ended alveolar sacs.

At each division the tubes become smaller, although the total cross-sectional area increases. When the air tubes have an internal diameter of less than 1mm (0.03in) they become known as bronchioles.

Bronchioles differ from the bronchi in that they have no cartilage in their walls nor any mucus-secreting cells in their lining. They do, however, still have muscle fibres in their walls.

Respiratory bronchioles

Further divisions lead to the formation of terminal bronchioles, which in turn divide to form a series of respiratory bronchioles, the smallest and finest air passages. Respiratory bronchioles are so named because they have a few alveoli (air sacs) opening directly into them. Most of the alveoli, however, arise in clusters from alveolar ducts, which are formed from division of the respiratory bronchioles.

Bronchioles

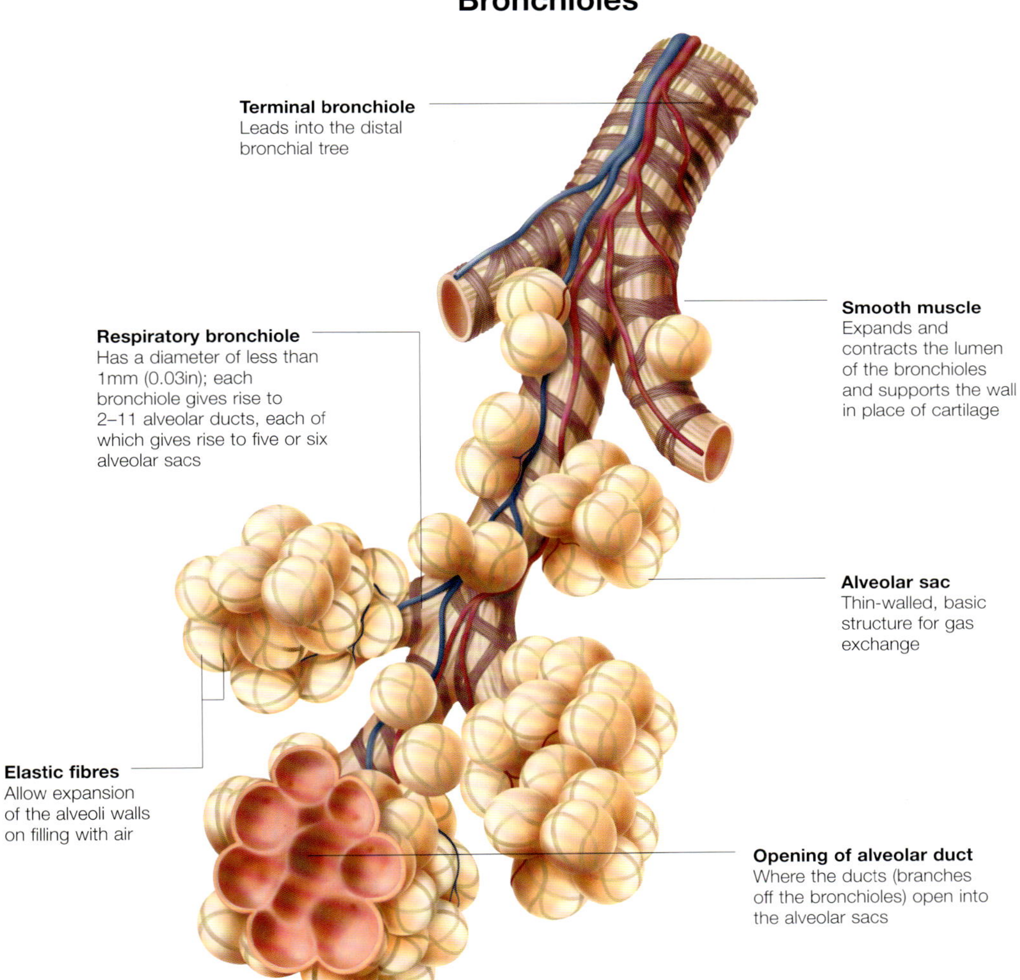

The branching of the bronchial tree ends in numerous terminal bronchioles from which arise the respiratory bronchioles, alveolar ducts and alveoli. The clusters of alveoli provide a large surface area for gaseous exchange.

Alveoli

The alveoli are tiny hollow sacs with extremely thin walls and are the sites of gaseous exchange within the lungs. It is through the alveolar walls that oxygen diffuses from the air into the pulmonary bloodstream, and waste carbon dioxide diffuses out.

There are many millions of alveoli in the human lung that together provide a huge surface area (about 140sq m) for this exchange to take place.

The alveoli lie in clusters like bunches of grapes around the alveolar ducts, each having a narrowed opening into a duct. They also have small holes, or pores, through which they connect with neighbouring alveoli. The alveolar walls are lined by flattened epithelial cells, and are supported by a framework of elastic and collagen fibres.

Two other types of cell are found in the alveoli: macrophages (defence cells), which engulf any foreign particles; and cells that produce surfactant, an important substance that lowers the surface tension in the fluid lining the alveoli, preventing their collapse.

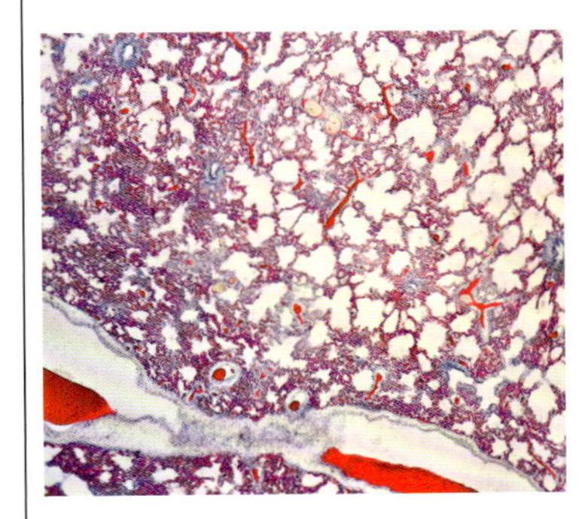

The spaces of the alveoli can be seen here in white. Numerous tiny blood vessels are visible within the tissue.

Lungs

The paired lungs are cone-shaped organs of respiration that occupy the thoracic cavity. They lie either side of the heart, great blood vessels and other structures of the central mediastinum.

The right and left lungs are each enclosed within a membranous bag known as the pleural sac. The lungs are attached to the mediastinum only by a root made up of the main bronchus and large blood vessels.

Lung Tissue

Lung tissue is soft and spongy and has great elasticity. In children the lungs are pink in colour, but they usually become darker and mottled later in life as they are exposed to dust which is taken in by the defence cells of the lining of the airways.

Each lung has:

- An apex that projects up into the base of the neck behind the clavicle (collarbone)
- A base with a concave surface that rests on the superior surface of the diaphragm
- A concave mediastinal surface that lies against various structures of the mediastinum.

Lobes and Fissures

The lungs are divided by deep fissures into sections known as lobes. The right lung has three lobes whereas the left lung, which is slightly smaller due to the position of the heart, has two. Each lobe is independent of the others, receiving air via its own lobar bronchus and blood from lobar arteries. The fissures are deep, extending right through the structure of the lung, and are lined by the pleural membrane.

Anterior view of the lungs

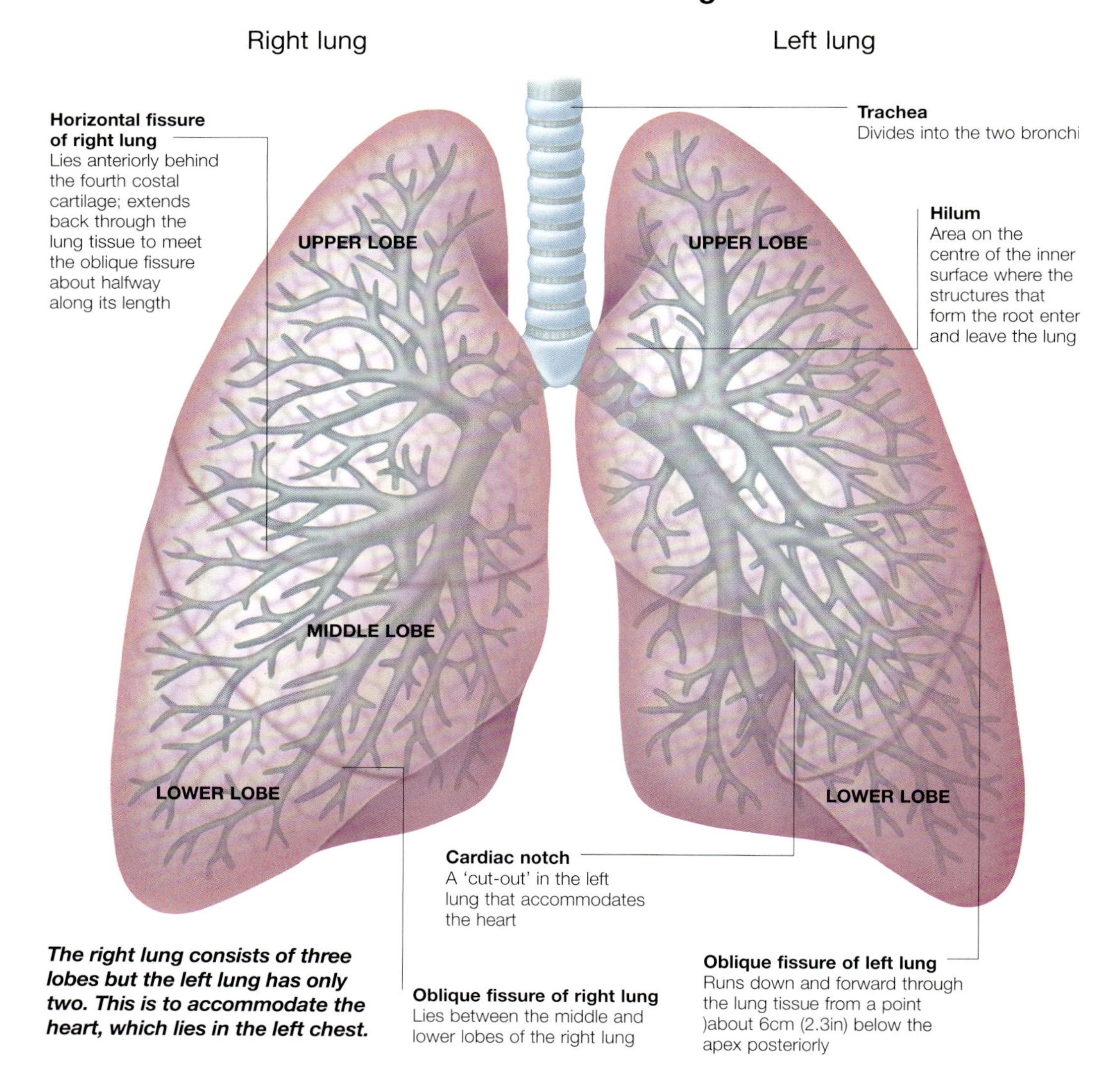

The right lung consists of three lobes but the left lung has only two. This is to accommodate the heart, which lies in the left chest.

Pleurae

Each lung is covered by a thin membrane known as the pleura. The pleurae line both the outer surface of the lungs and the inner surface of the thoracic cage.

The layer of pleura covering each lung is called the visceral pleura, while that lining the thoracic cage is the parietal pleura.

Visceral Pleura

This thin membrane covers the lung surface, dipping down into the fissures between the lobes of the lung.

Parietal Pleura

This is continuous with the visceral pleura at the hilum of the lung. Here, the membrane reflects back and lines all the inner surfaces of the thoracic cavity. The parietal pleura is one continuous membrane divided into areas that are named after the surfaces they cover:

- Costal pleura – lines the inside of the ribcage, the back of the sternum and the sides of the vertebral bodies of the spine
- Mediastinal pleura – covers the mediastinum, the central area of the thoracic cavity
- Diaphragmatic pleura – lines the upper surface of the diaphragm, except where it is covered by the pericardium
- Cervical pleura – covers the tip of the lung as it projects up into the neck.

The pleural cavity, which lies between the visceral and parietal layers of pleura, is a narrow area filled with a small amount of pleural fluid. The fluid lubricates the movement of the lung within the thoracic cavity and also acts to provide a tight seal, holding the lung against the thoracic wall and diaphragm by surface tension. It is this seal that forces the elastic tissue of the lung to expand when the diaphragm contracts and the ribcage lifts during inspiration.

Pleural Recesses

During quiet breathing, the lungs do not completely fill the pleural sacs within which they lie. There is room for expansion in the pleural recesses, areas where the sacs are empty and where parietal pleura comes into contact with itself rather than the visceral pleura overlying the lung tissue. The lungs only expand fully into these recesses during deep inspiration, when lung volume is at a maximum.

Pleural Effusion

At the base of the thoracic cavity, the lowermost parts of the costal pleura come into contact with the diaphragmatic pleura in the costodiaphragmatic recess. This recess is of importance clinically because it provides a potential space that may become filled with fluid – a pleural effusion – in certain medical conditions, such as heart failure.

The costomediastinal recess is smaller and of less clinical importance.

Position of the lungs and pleurae

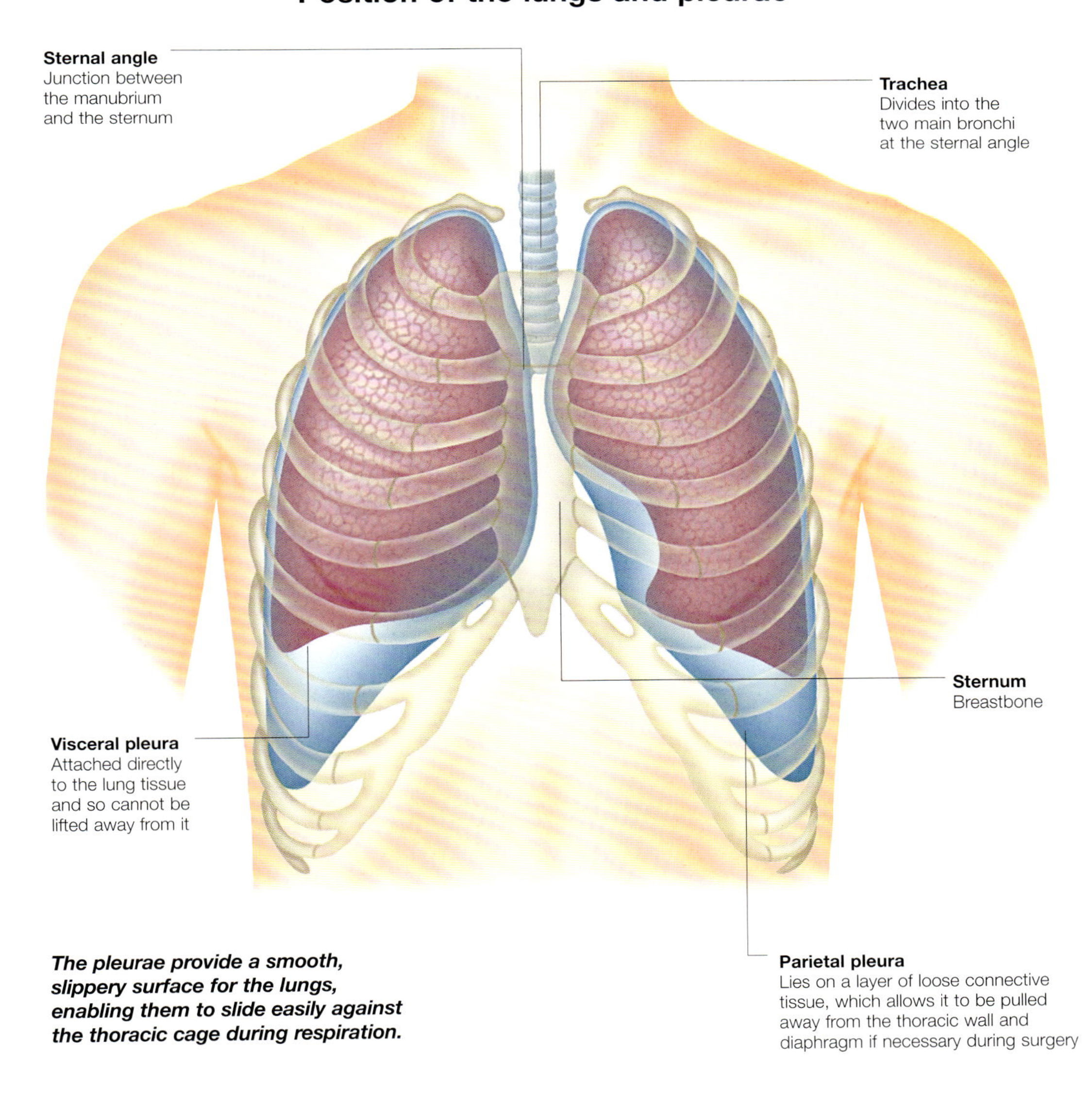

The pleurae provide a smooth, slippery surface for the lungs, enabling them to slide easily against the thoracic cage during respiration.

Blood Vessels of the Lungs

The primary function of the lungs is to reoxygenate the blood used by the tissues of the body and to remove accumulated waste carbon dioxide. This is effected via the pulmonary blood circulation.

Blood from the body returns to the right side of the heart and then passes directly to the lungs via the pulmonary arteries. Once oxygenated in the lungs, blood returns to the left side of the heart via the pulmonary veins and is pumped around the body. Collectively, pulmonary arteries and veins are known as the pulmonary circulation.

Pulmonary vessels

A large artery known as the pulmonary trunk arises from the heart's right ventricle carrying dark-red deoxygenated blood from the body into the lungs. The pulmonary trunk divides into the right and left pulmonary arteries, which run horizontally and enter the lungs at the hilum, alongside the bronchi. Within the lungs the arteries divide to supply each lobe of their respective lung; two on the left and three on the right. The lobar arteries divide further to give the segmental arteries, which supply the bronchopulmonary segments (structural units). Each segmental artery ends in a network of capillaries.

Oxygenated blood then returns to the left side of the heart through the pulmonary veins.

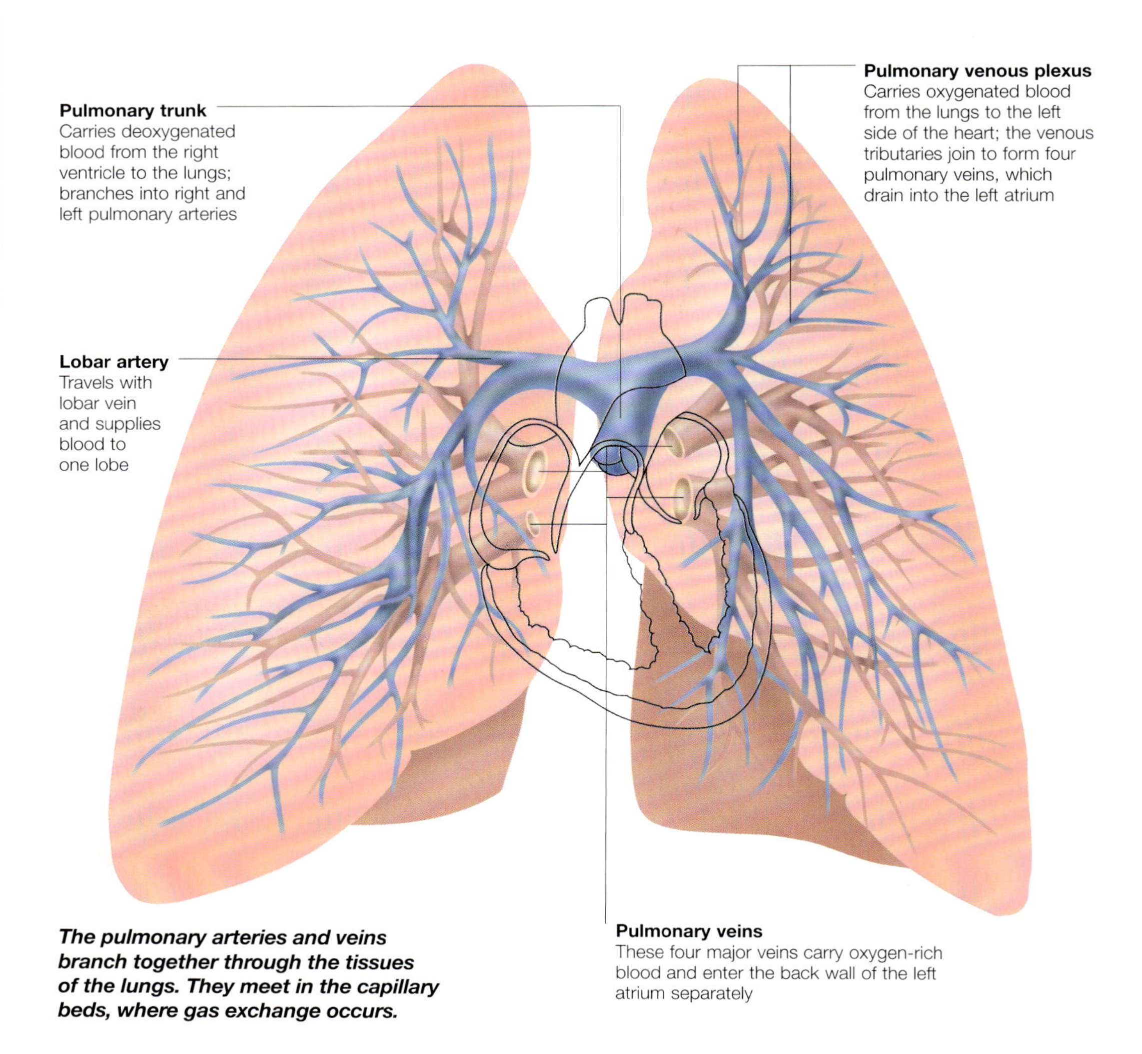

The pulmonary arteries and veins branch together through the tissues of the lungs. They meet in the capillary beds, where gas exchange occurs.

Alveolar capillary plexus

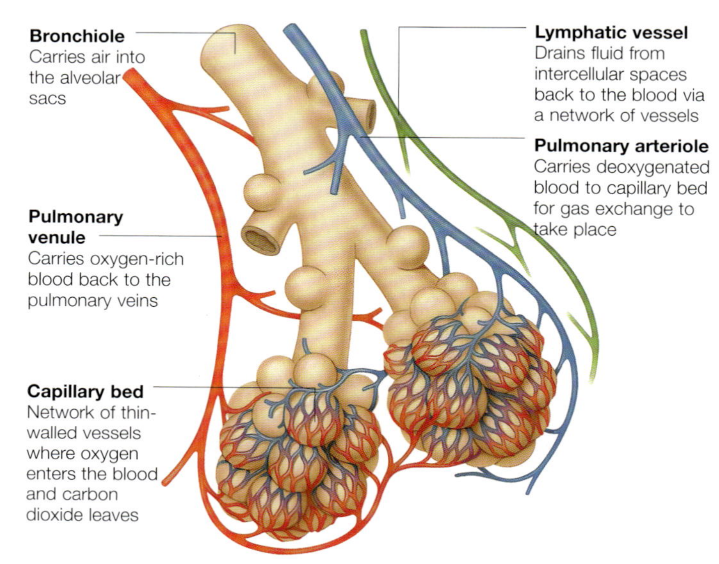

Within the lung, repeated division of the pulmonary arteries results in a network (plexus) of tiny blood vessels (capillaries), around each of the millions of alveolar sacs. The walls of the capillaries are extremely thin, which allows the blood within them to come into close contact with the walls of the alveoli, through which gas exchange takes place.

As oxygen enters and carbon dioxide leaves the pulmonary blood it changes from dark to light red. The newly oxygenated blood is collected into venules that drain each capillary plexus. These venules ultimately join to form the pulmonary veins, which complete the pulmonary circulation by returning the blood to the heart.

Intrinsic blood supply

The tissues of the smallest airways can absorb oxygen from the air they contain, but this is not true for the larger airways, the supporting connective tissue of the lung and the pleura covering the lung. These areas receive their blood supply directly from two small bronchial arteries arising from the thoracic aorta.

Each alveolus is surrounded by a capillary plexus. Deoxygenated blood is oxygenated by gaseous exchange, which occurs through the walls of the alveoli.

Lymphatics of the Lungs

Lymphatic drainage of the lung originates in two main networks, or plexuses: the superficial (subpleural) plexus and the deep lymphatic plexus. These communicate freely with each other.

Lymph is a fluid that is collected from the spaces between cells and carried in lymphatic vessels back to the venous circulation. On its way, the lymph must pass through a series of lymph nodes that act as filters to remove particulate matter and any invading micro-organisms.

Superficial Plexus

This network of fine lymphatic vessels extends over the surface of the lung, just beneath the visceral pleura (covering of the lung). The superficial plexus drains lymph from the lung towards the bronchi and trachea, where the main groups of lymph nodes are found. Lymph from the superficial plexus arrives first at the bronchopulmonary group of lymph nodes, which lie at the hilum of the lung.

Deep Plexus

The lymphatic vessels of the deep plexus originate in the connective tissue surrounding the small airways, bronchioles and bronchi (the alveoli have no lymphatic vessels). There are also small lymphatic vessels within the lining of the larger airways.

Lymph Nodes of the Bronchi

These lymphatic vessels merge and run back along the route of the bronchi and pulmonary blood vessels, passing through intrapulmonary nodes that lie within the lung. From these nodes lymph passes through vessels that drain towards the hilum into the bronchopulmonary lymph nodes.

The bronchopulmonary nodes at the hilum of the lung therefore receive lymph from both superficial and deep lymphatic plexuses.

Lymph Nodes of the Trachea

From the bronchopulmonary nodes, lymph drains to the tracheobronchial (carinal) lymph nodes. From here, lymph passes up through the paratracheal nodes lying alongside the trachea, into the paired bronchomediastinal lymph trunks, which return the fluid to the venous system in the neck.

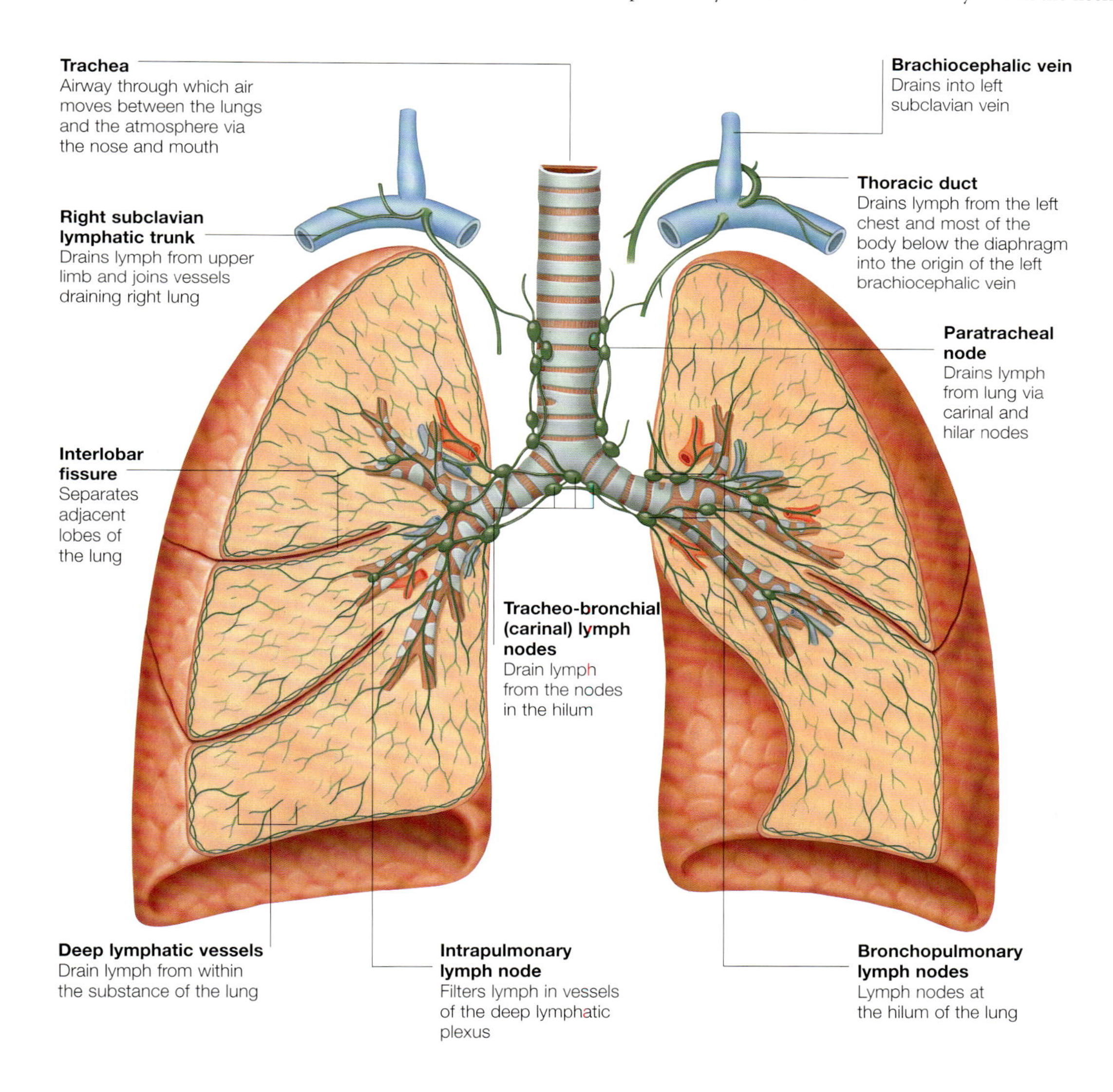

Diaphragm

The diaphragm is a sheet of muscle that separates the thorax from the abdominal cavity. It is essential for breathing as its contraction expands the chest cavity, allowing air to enter.

The diaphragm is the main muscle involved in respiration and has several apertures to allow important structures to pass between the thorax and abdomen.

Muscle of the Diaphragm

The muscle tissue of the diaphragm arises from three areas of the chest wall, merges to form a continuous sheet and converges on the central tendon, which acts as a site of muscular attachment. The three areas of origin of the diaphragm give rise to three separately named parts: the sternal part, the costal part and the lumbar or vertebral part, which arises from the crus and arcuate ligaments.

Central Tendon

The muscle fibres of the diaphragm insert into the central tendon, which has a three-leaved shape. The central part lies just beneath, and is depressed by, the heart. It is attached by ligaments to the pericardium, the membrane surrounding the heart. The two lateral leaves lie towards the back and help form the right and left domes (cupolae) of the diaphragm.

Nerve Supply

The motor nerve supply of the diaphragm, which causes it to contract, comes from the two phrenic nerves that originate from the spinal cord in the neck. These also provide sensory nerve supply, detecting pain and providing information on position.

Abdominal surface of the diaphragm

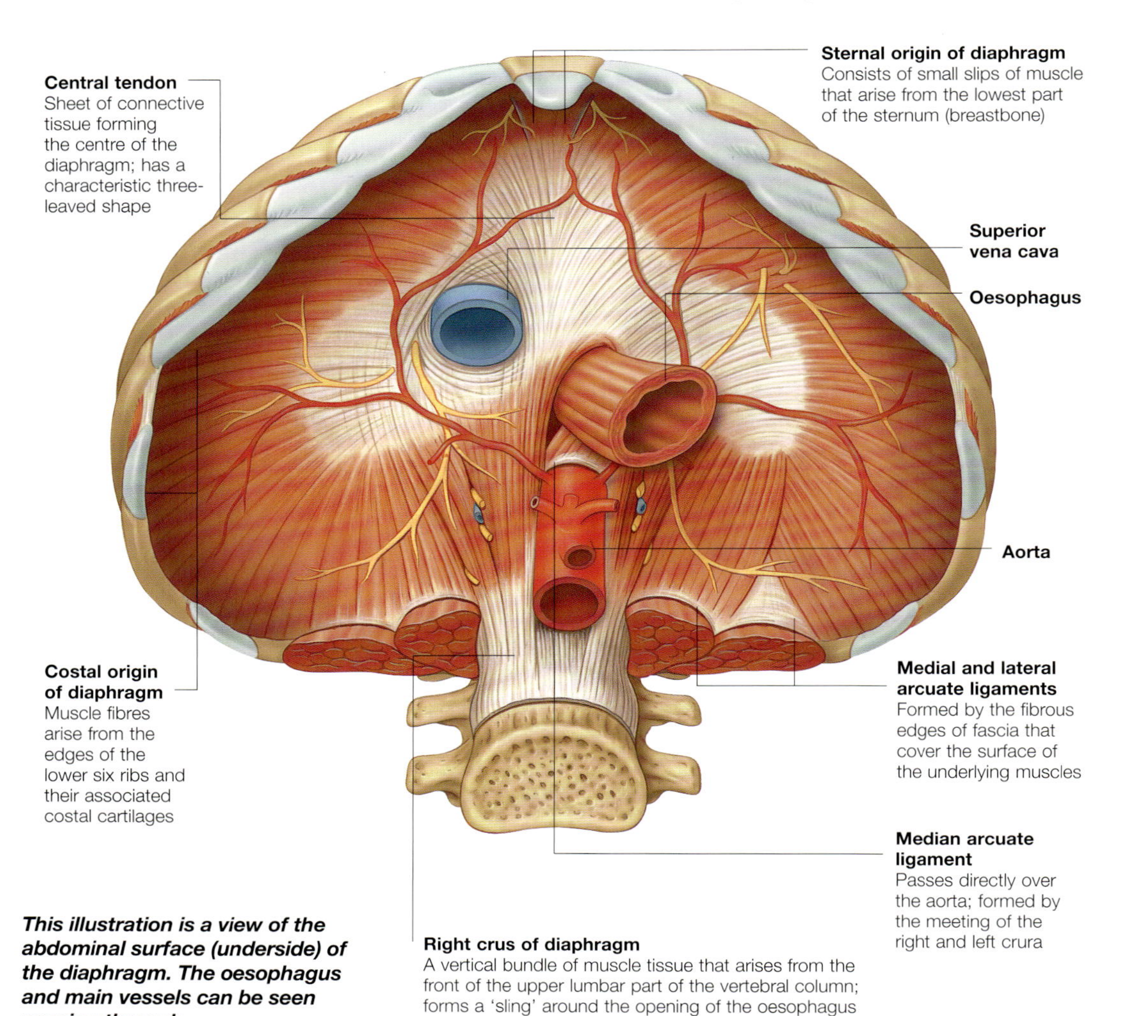

This illustration is a view of the abdominal surface (underside) of the diaphragm. The oesophagus and main vessels can be seen passing through.

Function of the diaphragm

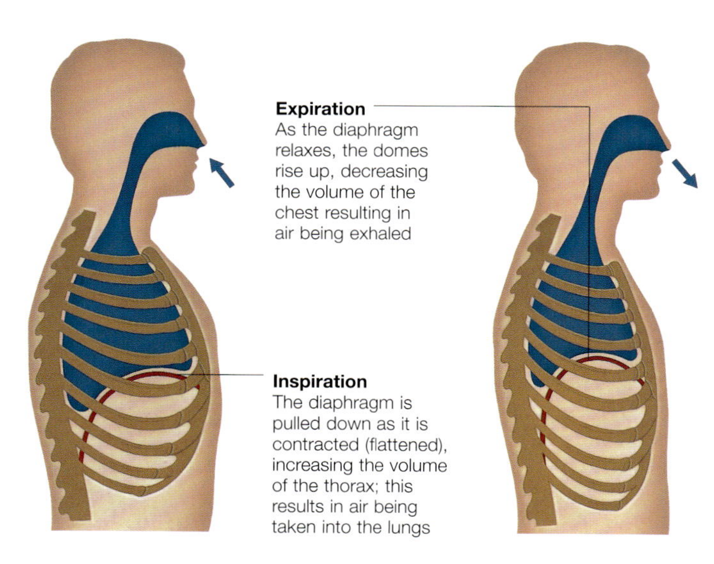

The diaphragm is the main muscle of respiration. Separating the thoracic and abdominal cavities, it curves upwards into two domes, separated by a central depression where the heart rests. By working in conjunction with the changing shape of the ribcage, air is inhaled and exhaled. The periphery of the diaphragm is at a constant level as it is attached to the thoracic wall, but the heights of the domes vary depending on the degree to which the diaphragm muscle is contracted.

Contraction of the muscle fibres cause the domes to be pulled down, which expands the thoracic cavity, and air enters.

Relaxation of the diaphragmatic muscle allows the domes to rise and air is exhaled.

Thoracic surface of the diaphragm

The upper aspect of the diaphragm is convex and forms the floor of the thoracic (chest) cavity. It is perforated by major vessels and structures that must pass through the muscle sheet so as to reach the abdomen.

The central part of the surface of the diaphragm is covered by the pericardium, the membrane that surrounds the heart. To either side, the upper surface of the diaphragm is lined with the diaphragmatic part of the parietal pleura (the thin membrane that lines the chest cavity). This is continuous around the edges of the diaphragm with the costal pleura, which covers the inside of the chest wall.

Diaphragmatic apertures
Although the diaphragm acts to separate the chest and abdominal cavities, certain structures do pass through a series of 'diaphragmatic apertures'. There are three major apertures.

Caval aperture
This is an opening in the central tendon of the diaphragm that allows passage of the inferior vena cava, the main vein of the abdomen and lower limbs. As the opening is in the central tendon rather than the muscle of the diaphragm it will not close when the diaphragm contracts during inspiration; in fact the opening widens and blood flow increases. The opening also contains branches of the right phrenic nerve and lymphatic vessels.

Oesophageal aperture
This allows the passage of the oesophagus through the diaphragm to reach the stomach. Muscle fibres of the right crus act as a sphincter, closing off the oesophageal opening when the diaphragm contracts during inspiration. As well as the oesophagus, the aperture gives passage to nerves, arteries and lymphatic vessels.

Aortic aperture
This opening lies behind the diaphragm rather than within it. As the aorta does not actually pierce the diaphragm, the flow of blood within it is not affected by diaphragmatic contractions while breathing. The aorta emerges under the median arcuate ligament, in front of the vertebral column. The aortic aperture also transmits the thoracic duct (major lymphatic channel) and the azygos vein.

Diaphragm from above

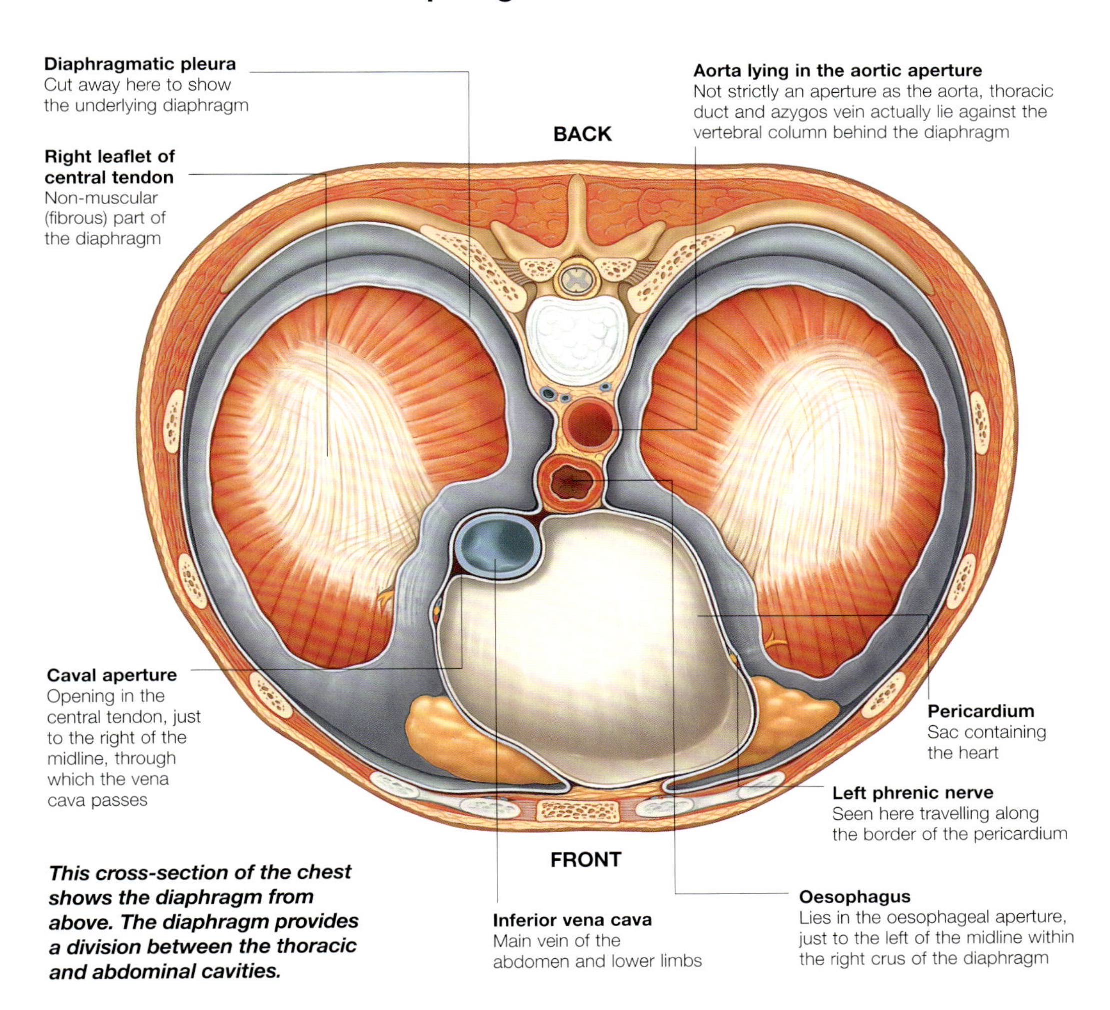

This cross-section of the chest shows the diaphragm from above. The diaphragm provides a division between the thoracic and abdominal cavities.

How the Lungs Work

The lungs, which take up most of the upper chest cavity, have a surface area equivalent to a tennis court. They work tirelessly to sustain life, supplying the body with oxygen and filtering harmful carbon dioxide from the blood.

The lungs remove waste carbon dioxide from the body and exchange it for a fresh supply of oxygen. Air is drawn into the lungs by expanding the chest cavity and then expelled by allowing the cavity to collapse or by forcing the air out.

The upper part of the cavity is bounded by the ribs and the intercostal muscles. The base of the cavity is bounded by the diaphragm – a flat sheet of tissue that forms a wall between the chest and the abdomen. It is these muscles that enable the action of breathing. At rest a person will inhale and exhale about 500ml (17fl oz) of air 13–17 times a minute. The lungs will expand and contract between 15 and 85 times within this time, depending on the body's activity.

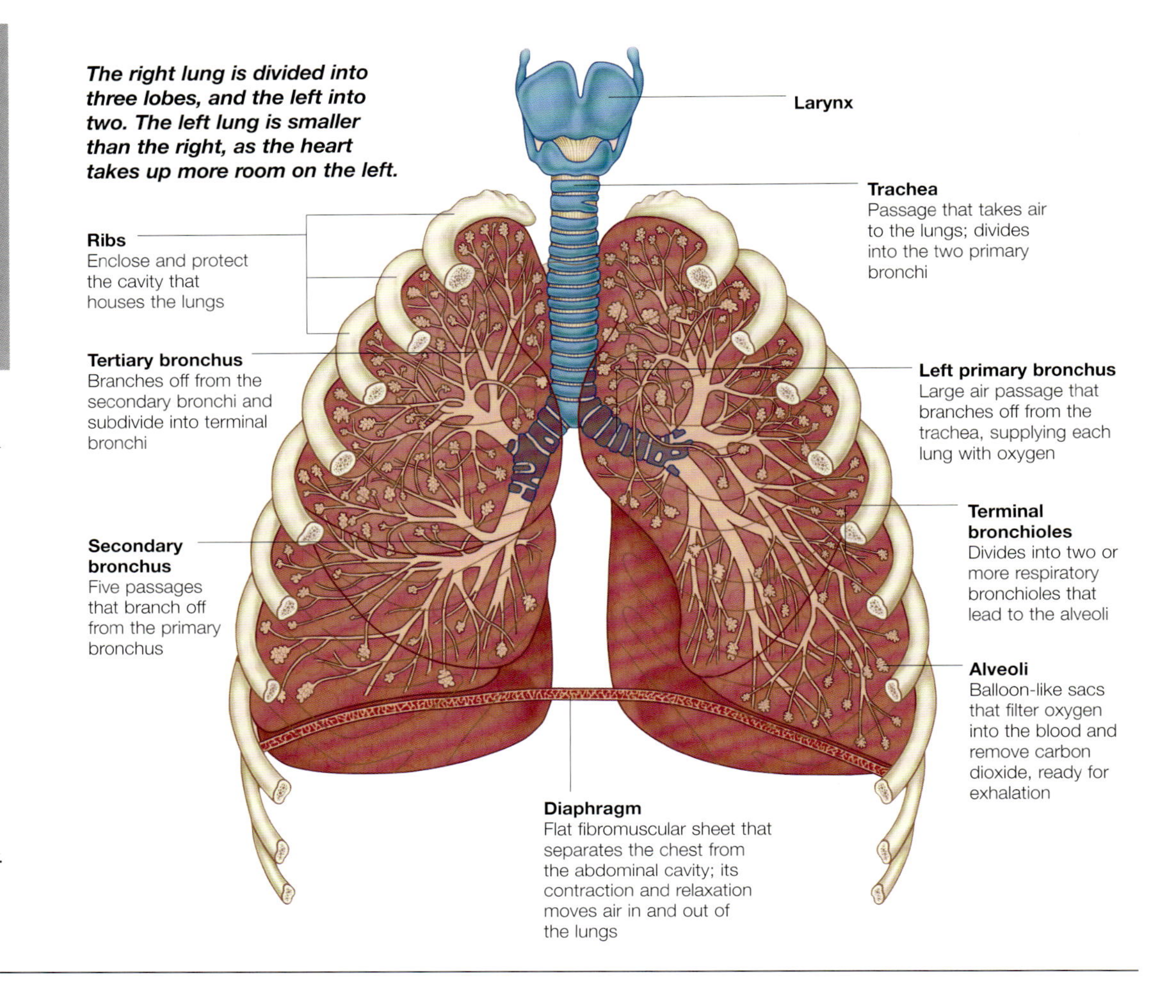

The right lung is divided into three lobes, and the left into two. The left lung is smaller than the right, as the heart takes up more room on the left.

Breathing in and out

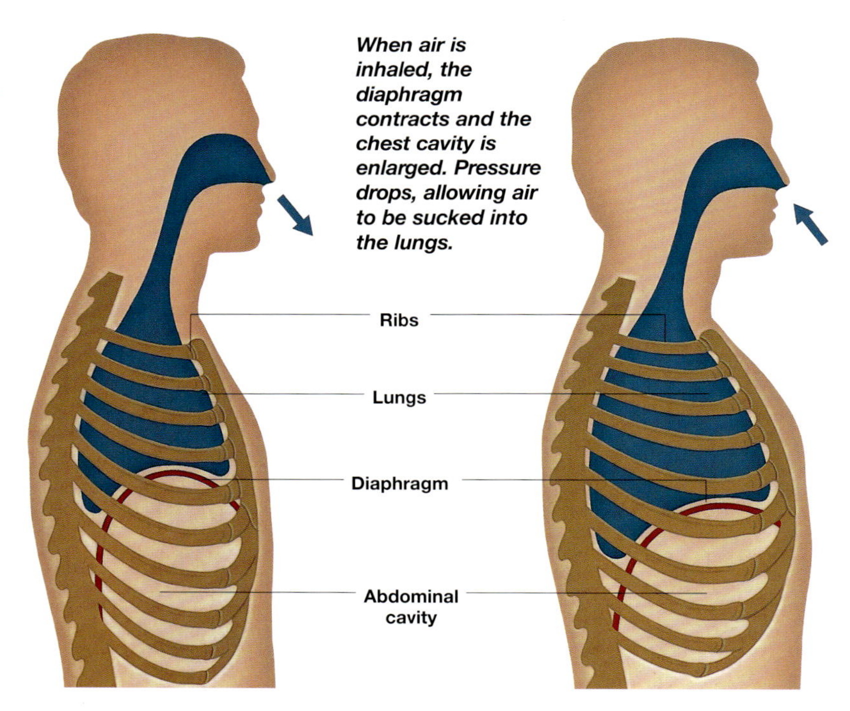

As air is expelled, the diaphragm relaxes and moves up. This causes pressure to rise in the lower chest cavity; exhalation equalizes the pressure.

When air is inhaled, the diaphragm contracts and the chest cavity is enlarged. Pressure drops, allowing air to be sucked into the lungs.

The lungs have a natural tendency to collapse but are held open by surface tension created by fluid produced by the inner pleural membrane.

To draw air into the lungs, the chest cavity is expanded. The muscles of the diaphragm contract, causing it to become flatter. At the same time, the intercostal muscles of the ribs contract, lifting the ribs upwards and outwards. The pressure in the chest cavity is thus reduced and the lungs expand, drawing in air via the mouth or nose.

When the intercostal muscles relax, the ribs fall downwards and inwards and the lungs collapse, forcing the air out. At the same time, the diaphragm relaxes and is pulled up into the chest cavity. In order to force more air out of the chest cavity the abdominal muscles can be used to push the diaphragm further into the chest cavity.

Respiration

An adult's lung capacity is about 5 litres (9 pints), but during normal breathing only 500ml (17fl oz) of air is exchanged. This movement is associated with inside and outside pressure.

Although breathing can be controlled voluntarily, respiratory movements are generally a series of reflex actions. These actions are controlled by the respiratory centre in the hindbrain (the part of the brain that regulates basic bodily systems). This has two regions: an inspiratory centre and an expiratory centre.

Nerve impulses from the inspiratory centre cause the contraction of the intercostal muscles (which move the ribs) and the diaphragm. This is the beginning of an intake of air into the lungs. As the lungs expand, stretch receptors in the walls of the lungs send back signals that begin to inhibit the signals from the inspiratory centre.

At the same time, impulses from the inspiratory centre activate the expiratory centre, which sends back inhibitory signals. The result is the relaxation of the intercostal muscles and the diaphragm, which ceases the intake of air and begins exhalation. The whole process is now ready to start again.

Breathing is also controlled and regulated by the level of carbon dioxide (CO_2) in the blood. Excess CO_2 causes the blood to become more acid. This is detected by the brain, and the inspiratory centre starts to produce deeper breathing until the CO_2 level is reduced.

Controlling breathing

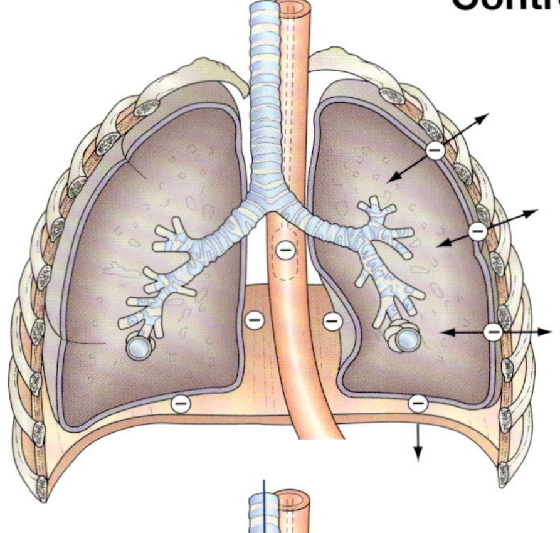

LUNGS AT REST: ***At rest, the muscles associated with respiration are relaxed. The air in the trachea and bronchi is at atmospheric pressure (the standard pressure exerted on the environment) and there is no airflow. Recoil of lung and chest wall are equal but opposite.***

BREATHING IN: ***During inspiration, muscles contract and the chest expands. Pressure in the alveoli becomes less than that outside the lungs, and air flows into the airways.***

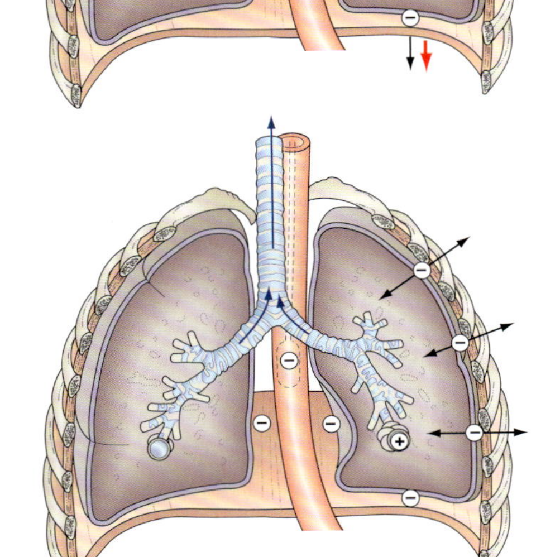

BREATHING OUT: ***During expiration, muscle relaxation causes the lungs to contract. This increases the pressure in the alveoli to a point where it is greater than the pressure at the airway opening. Air now flows out of the lungs.***

KEY

 Pressure within the alveolar greater than external air pressure

 Pressure within the alveolar lesser than external air pressure

Air flow to and from the lungs

 Internal or external forces, resulting in an increase or decrease of pressure

 Muscular contraction or relaxation

Gas exchange

The exchange of the gases oxygen and carbon dioxide takes place within the alveoli, of which there are about 300 million. The alveoli are covered with numerous blood capillaries that can accommodate about 900ml (32fl oz) of blood. These capillaries are very narrow, so that only a single blood cell can travel down them. This provides maximum exposure of the cell to oxygen.

Air that is breathed in contains about 21 per cent oxygen. Inside the alveoli, some of this oxygen dissolves in the surface moisture and diffuses through the thin lining into the blood, where most of it is picked up by the haemoglobin of the red blood cells. At the same time, carbon dioxide, most of which is carried in the blood plasma, diffuses into the lungs, where it is released as a gas ready to be breathed out. Exhaled air contains about 16 per cent oxygen.

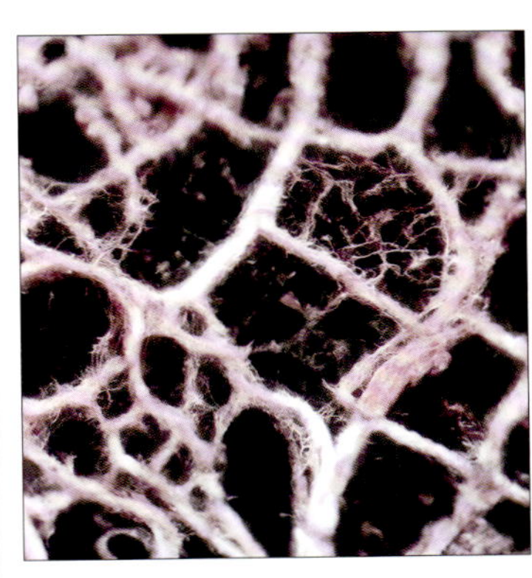

The spaces seen in this meshwork of lung tissue are alveoli, which perform the lung's primary role of gas exchange.

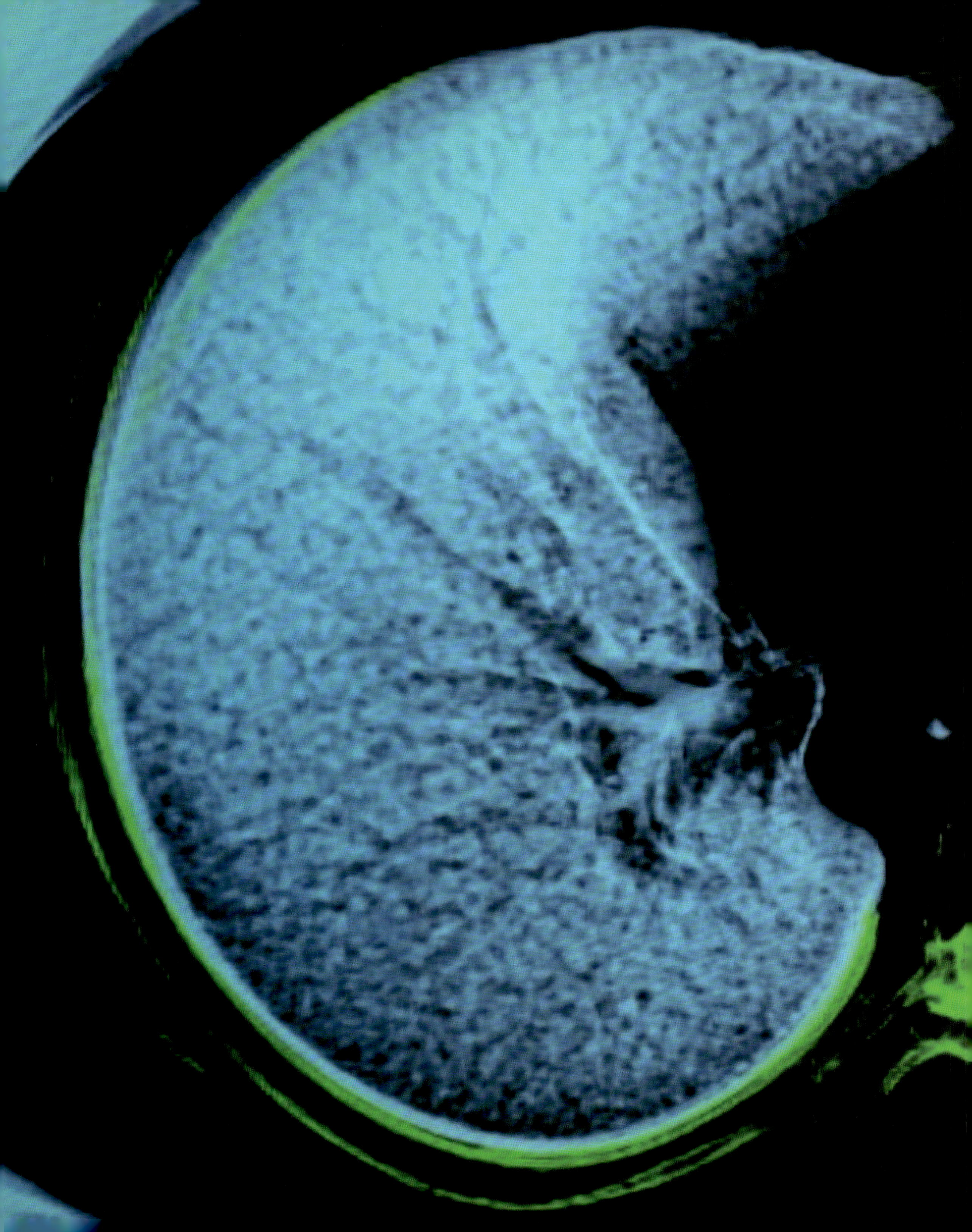

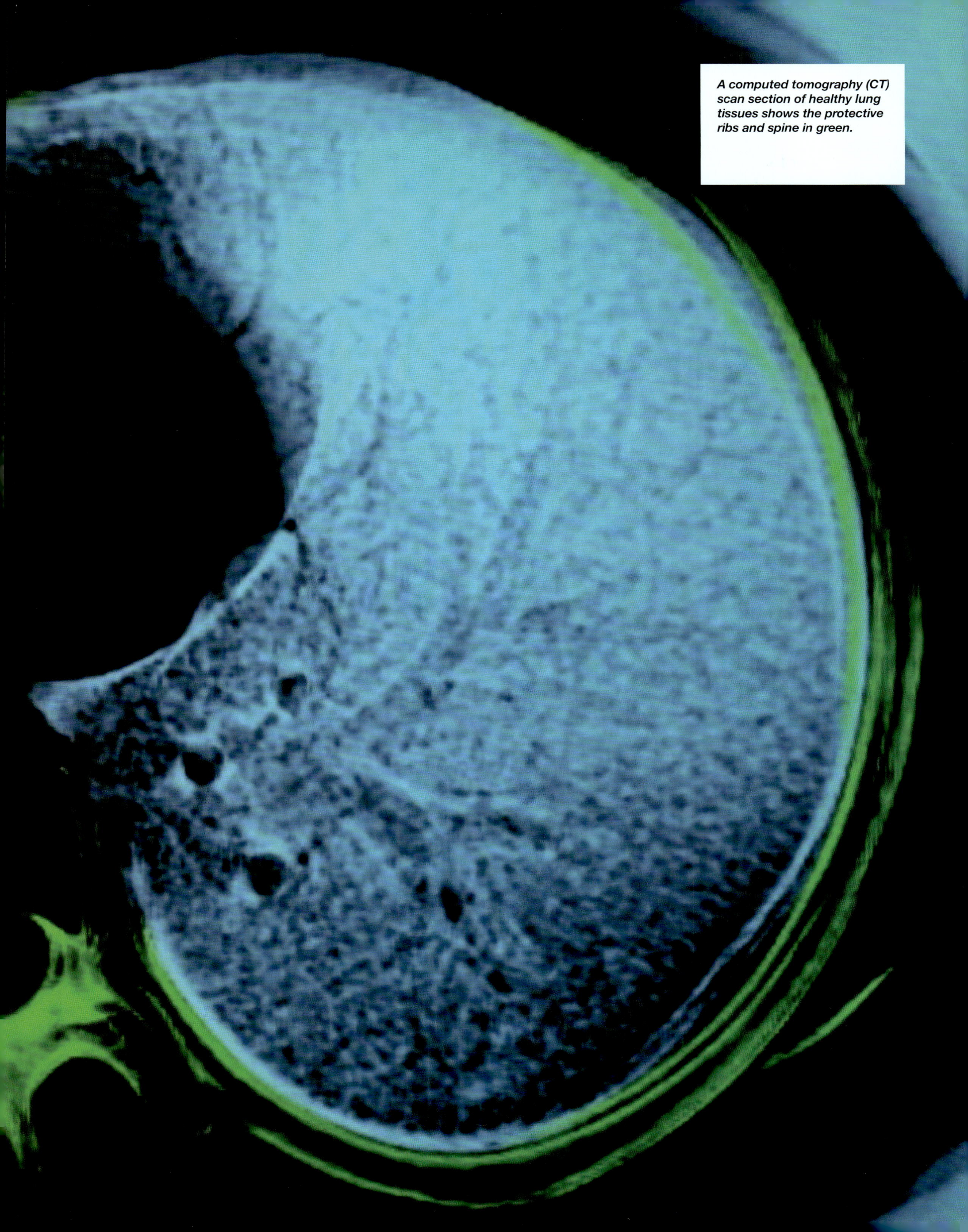

A computed tomography (CT) scan section of healthy lung tissues shows the protective ribs and spine in green.

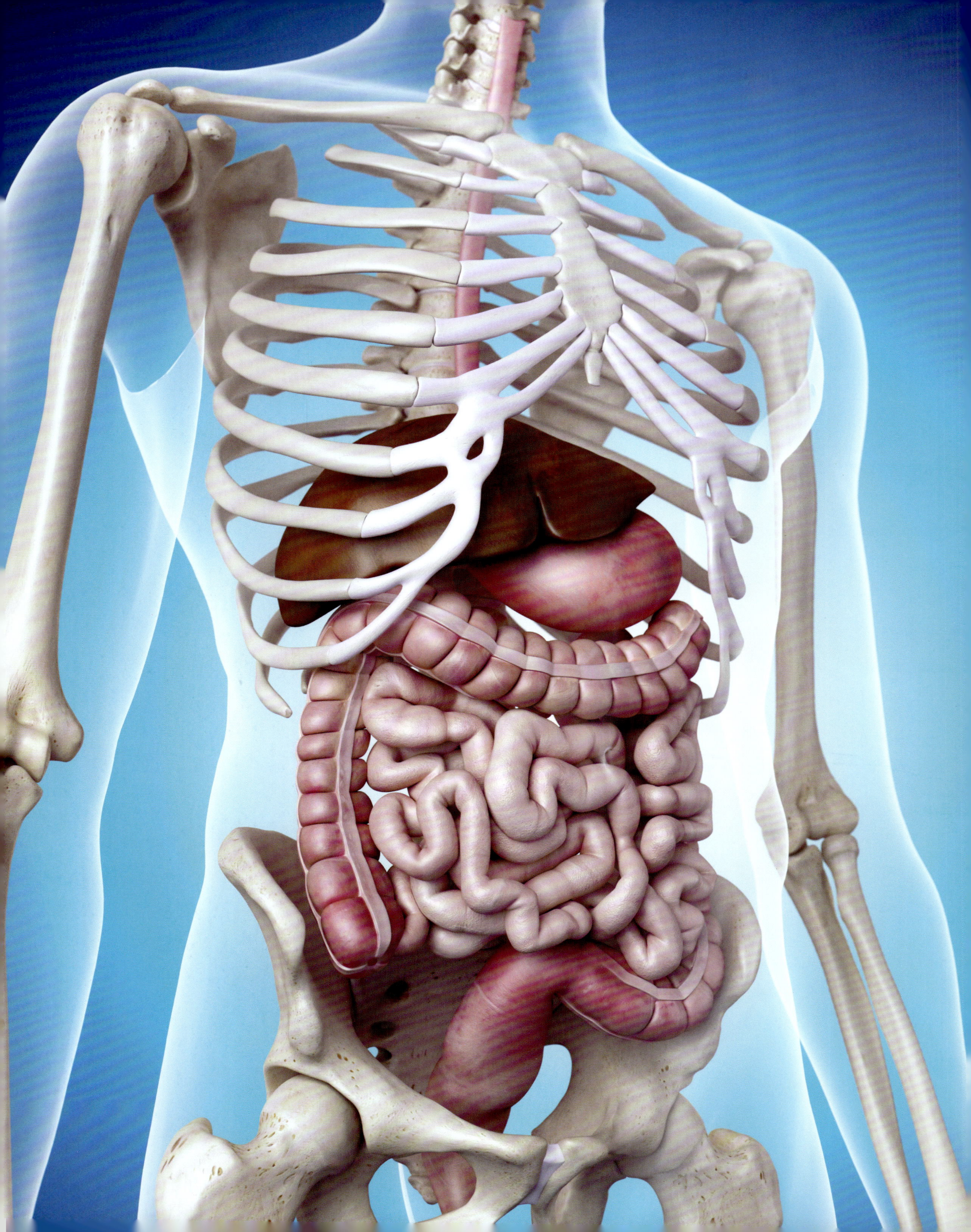

CHAPTER 8

The Digestive System

The cells and tissues of the body require a source of energy to survive and function – this energy is provided by the food we eat. The function of the digestive system is to break down food into smaller components (molecules) that can be easily absorbed into the bloodstream and transported to where they are needed.

The system stretches from the mouth to the anus and is approximately 9m (30ft) long. It includes: the mouth where the teeth grind food and enzymes in the saliva start to chemically break down the food; the stomach, where food is mechanically churned and exposed to gastric enzymes; the small intestine where the process of digestion starts, further breaking down the food and absorbing molecules into the blood; and the large bowel where water is absorbed and from where waste products are expelled as faeces. Along the way, other organs such as the liver, gall bladder and pancreas contribute to the process, producing digestive enzymes and processing nutrients.

Opposite: The digestive system processes the food we eat, absorbing it into the body and excreting waste products.

Teeth

Teeth are the body's toughest and most durable structures. They play a vital role in the digestion of food by helping to break it down into smaller fragments by biting and mastication (chewing).

The teeth are used to chew and grind food into smaller pieces. The action of chewing increases the surface area of food exposed to digestive enzymes, thus speeding up the process of digestion.

Teeth also play an important role in speech – the teeth, lips and tongue allow the formation of words by controlling airflow through the mouth. In addition, teeth provide structural support for muscles in the face as well as helping to form the smile.

Anatomy

Each tooth is composed of the crown and the root. The crown is the visible part of the tooth that emerges from the gingiva or gum (which helps to hold the tooth firmly in place). The crown of each premolar and molar includes projections or cusps that facilitate the chewing and grinding of food. The root is the portion of the tooth embedded in the jawbone.

Structure

Teeth are composed of four distinct types of tissue:

- Enamel – the clear outer layer and the hardest substance in the body. Composed of a densely packed structure, heavily mineralized with calcium salts, this layer helps to protect the inner layers of the teeth from harmful bacteria, and changes in temperature caused by hot or cold food and drink.
- Dentine – encloses and protects the inner core of the tooth, and is similar in composition to bone. It is composed of odontoblast cells, which secrete and maintain dentine throughout adult life.
- Pulp – contains blood vessels that supply the tooth with oxygen and nutrients. It also contains nerves responsible for the transmission of pain and temperature sensations to the brain.
- Cementum – covers the outer surface of the root. It is a calcium-containing connective tissue that attaches the tooth to the periodontal ligament, which anchors the tooth firmly in the tooth socket (alveolus) located in the jawbone.

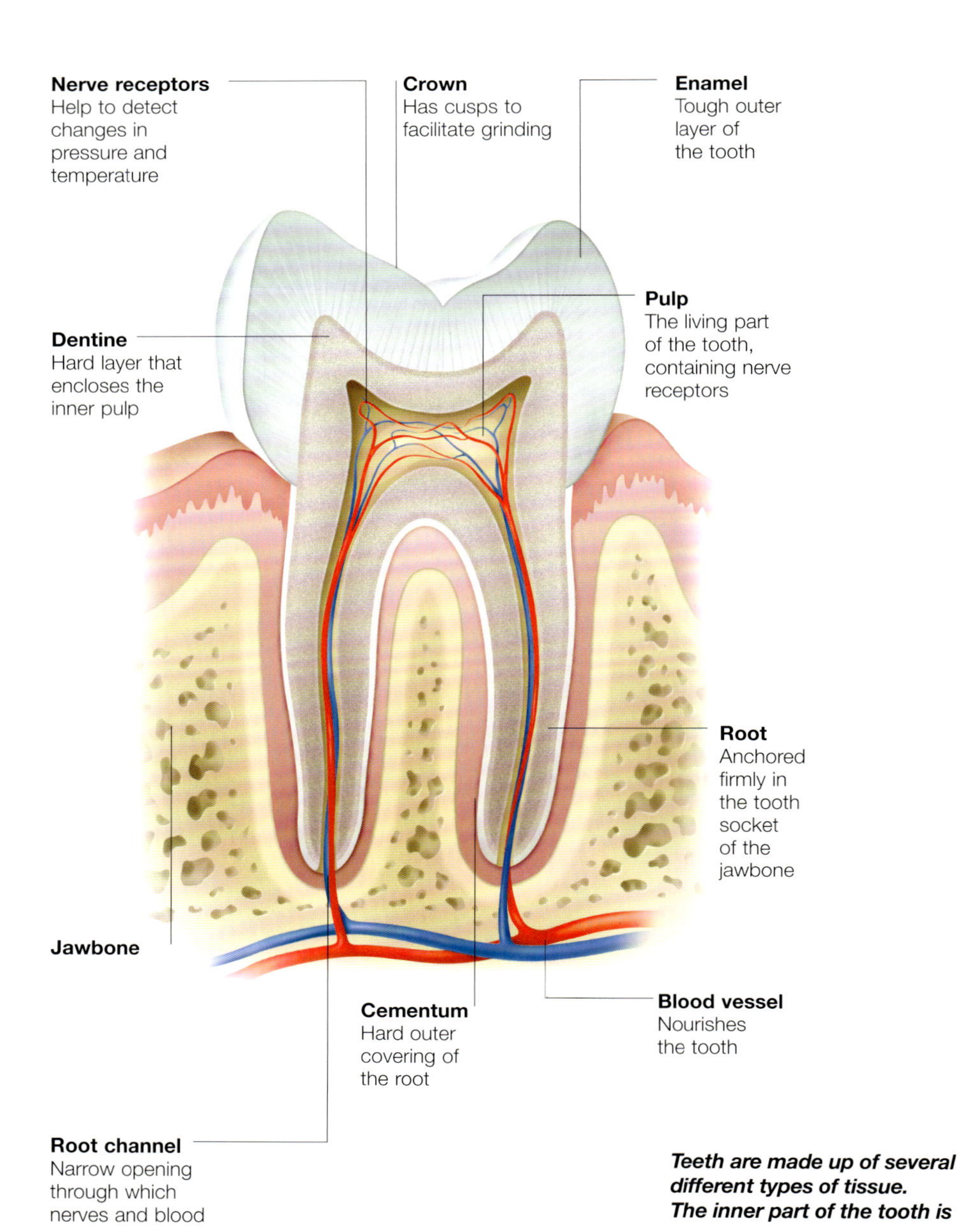

Teeth are made up of several different types of tissue. The inner part of the tooth is nourished and innervated via a small root channel.

Development of teeth

The teeth begin to develop in the human embryo around the sixth week of pregnancy. Six to eight months after birth, root growth pushes the tooth crown through the gum in the process of eruption called teething.

Primary teeth

This first set are the primary or deciduous teeth (milk teeth). These erupt in a specific order, usually the lower central incisors first, followed by the upper central incisors. The deciduous teeth do not include premolars.

Adult tooth growth

The tooth buds for the second wave of tooth production develop at the same time. These permanent teeth remain dormant until the ages of five to seven, when they begin to grow, causing the roots of the deciduous teeth to break down. This breakdown, together with the pressure of the underlying permanent teeth, results in the shedding of the deciduous teeth. The new teeth then start to appear and continue to do so until the ages of 10 to 12.

Eruption of the permanent set follows a similar pattern to the earlier growth (although premolars erupt between the canines and the molars). The permanent set has additional, third molars (wisdom teeth) that tend to appear after between 15 and 25 years.

Types of teeth

Over several years, the milk teeth are replaced by a set of permanent adult teeth. There are 32 adult teeth, including the third molars, known as wisdom teeth.

Around the age of six, the roots of the deciduous teeth are slowly eroded by the pressure of erupting permanent teeth and by the action of specialized bone cells in the jaws. This process, called resorption, allows the permanent teeth to emerge from beneath. If a permanent tooth is missing – a relatively common condition, the corresponding milk tooth is retained.

Complete set

As the milk teeth are replaced, the mouth and jaw lose their childhood shape and take on a more pronounced and adult appearance. Adult teeth are usually darker in colour and differ in size and proportion from the milk teeth. A full complement of permanent teeth is generally present by the end of adolescence, with the exception of the third molars (wisdom teeth), which tend to emerge at around 18–25 years.

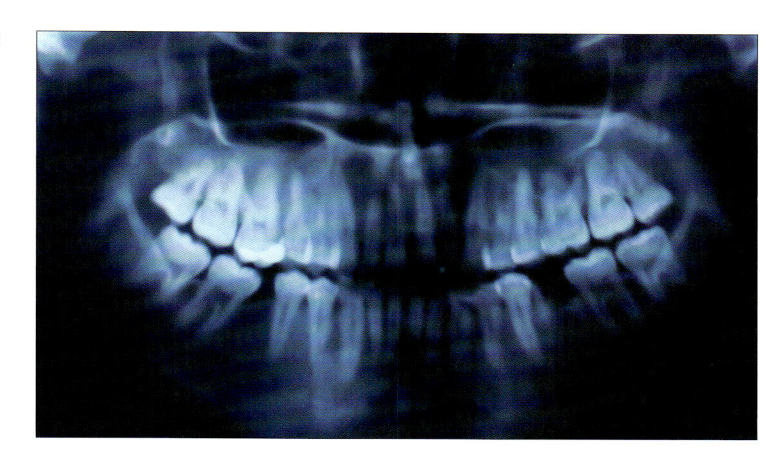

This x-ray shows a healthy set of teeth, with the molars showing prominently as a stronger white colour.

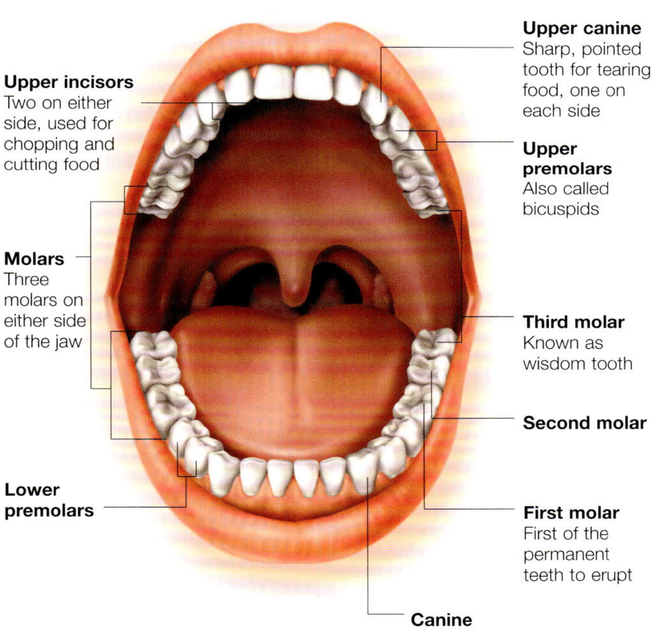

Types of teeth

Adults generally have 32 teeth – 16 on the upper jaw and 16 on the lower – which fit together to bite and chew food. Humans are referred to as heterodonts as they have a variety of different tooth types, with a specific size, shape and function:

- Incisors – adults have eight incisors located at the front of the mouth – four in the upper jaw, and four in the lower jaw. Incisors have a sharp edge used to cut up food.
- Canines – on either side of the incisors are canines, so called because of their resemblance to the sharp fangs of dogs. There are two canines on each jaw, and their primary role is to pierce and tear food.
- Bicuspids – also known as premolars, these are flat teeth with pronounced cusps that grind and mash food. There are four bicuspids in each jaw.
- Molars – behind the bicuspids are the molars, where the most vigorous chewing occurs. There are 12 molars, referred to as the first, second and third molars. Third molars are commonly called wisdom teeth.

Wisdom teeth

Wisdom teeth are remnants from thousands of years ago when the human diet consisted of raw foods that required the extra chewing and grinding power of a third set of molars. Today, wisdom teeth are not required for chewing and, as they can crowd other teeth and cause them to become impacted, they are often removed by dentists.

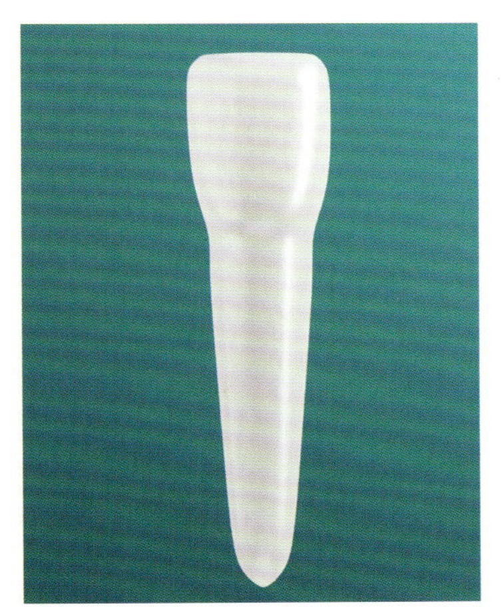

Incisor

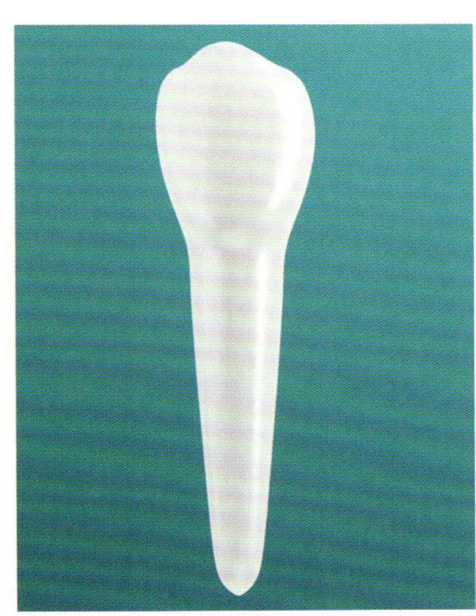

Canine

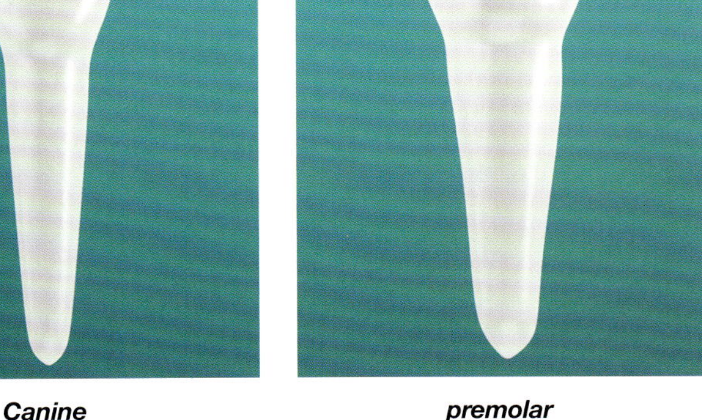

premolar

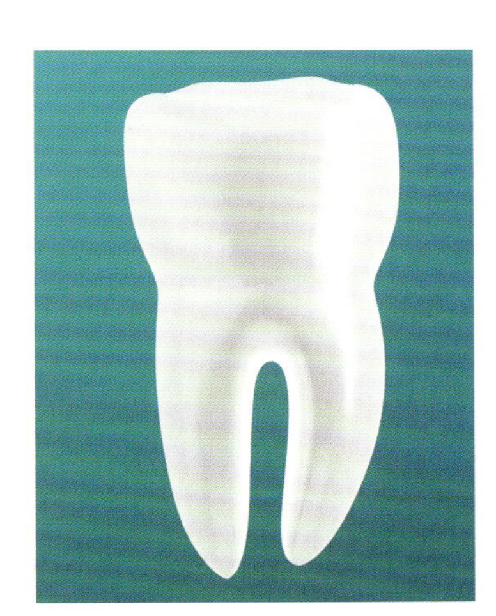

Molar

Oral Cavity

Also known as the mouth, the oral cavity extends from the lips to the fauces, the opening leading to the pharynx.

The roof of the mouth, viewed from below, shows two structures: the dental arch and the palate. The dental arch is the curved part of the maxilla bone at the front and sides of the roof, and the palate is a horizontal plate of tissue separating the mouth from the nose.

Hard and Soft Palate

The front two-thirds of the palate are bony and hard, and are formed by the maxillary bone. The hard palate is covered with a mucous membrane, beneath which run arteries, veins and nerves. These nourish and provide sensation to the palate and the overlying mucous glands, which often form fibrous ridges known as rugae. The mucus secreted by these glands lubricates food to facilitate swallowing.

The rear third of the palate is composed of glandula mucosa, muscle and tendon. Forming much of the soft palate are the tensor and levator palati muscles. These muscles close off the nasal cavity from the mouth during swallowing by tensing and elevating the soft palate. They act with other muscles to open the auditory (Eustachian) tube, which equalizes pressure on either side of the eardrum.

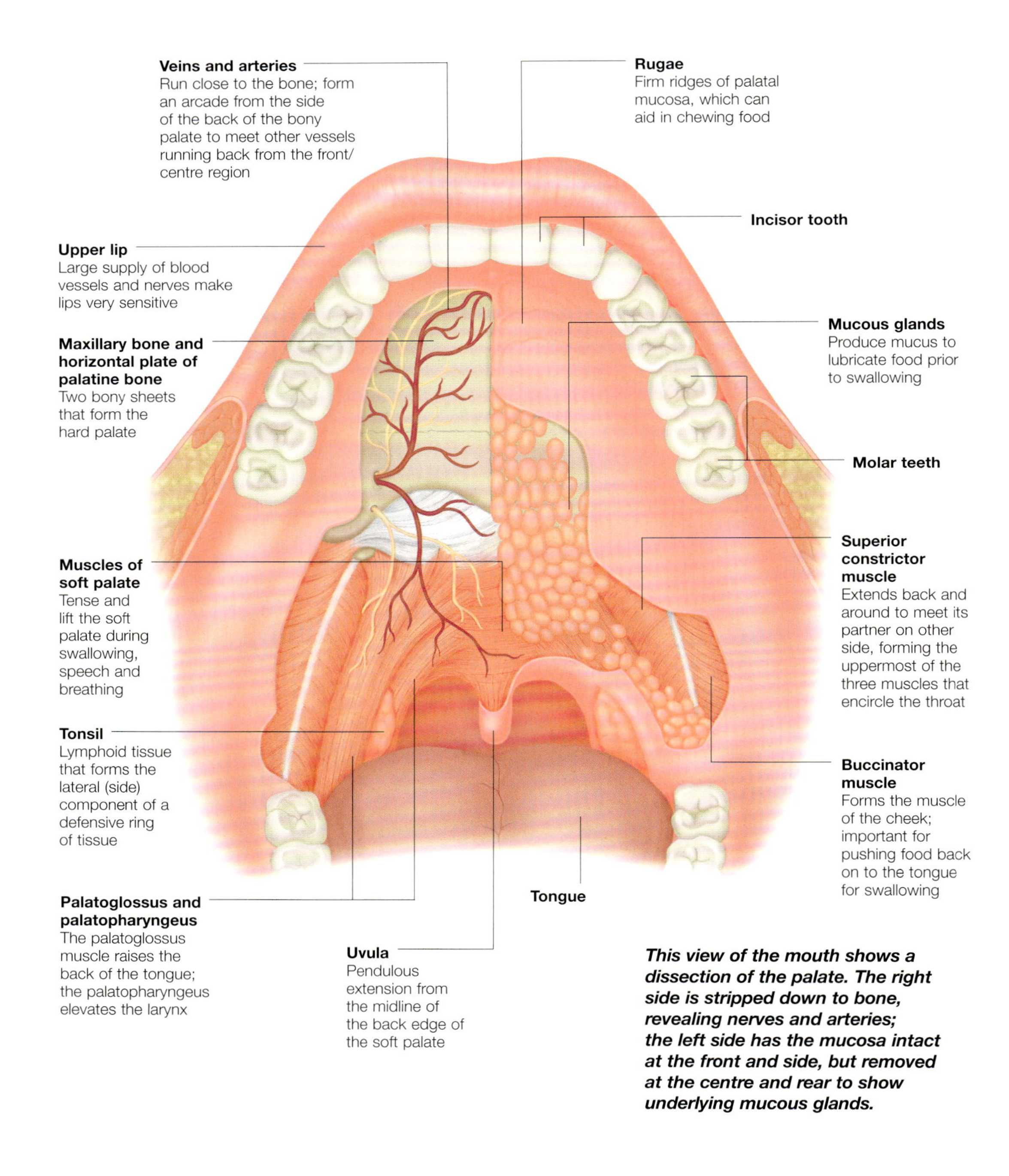

This view of the mouth shows a dissection of the palate. The right side is stripped down to bone, revealing nerves and arteries; the left side has the mucosa intact at the front and side, but removed at the centre and rear to show underlying mucous glands.

Floor of the mouth

The floor of the mouth acts as the foundation for a network of muscles and glands that are essential to its function.

The tongue is situated over the mylohyoid muscle, which forms the muscular floor of the mouth. The hyoglossus muscle anchors the tongue to the hyoid bone and provides extra strength, while the genioglossus muscle stops the tongue from moving back into the throat.

Temporalis muscles are the muscles of mastication. The lingula is a small bony projection of the mandible. The mandibular nerve passes below this, through the mandibular foramen and runs in the body of the mandible to supply sensation to the lower teeth and lip.

Salivary glands

There are a pair each of the sublingual and submandibular salivary glands on either side of the oral floor and, together with the paired parotid glands, they make up the six salivary glands. Saliva, a watery fluid that lubricates the mouth, flows along the submandibular gland duct on the mylohyoid muscle, and emerges in the front of the oral cavity on either side of the tongue, behind the lower front teeth. Saliva from the sublingual glands either runs into the submandibular duct, or flows out through openings in the mucosa to the side of the tongue.

The lingual nerve provides taste and sensation to the front two-thirds of the tongue.

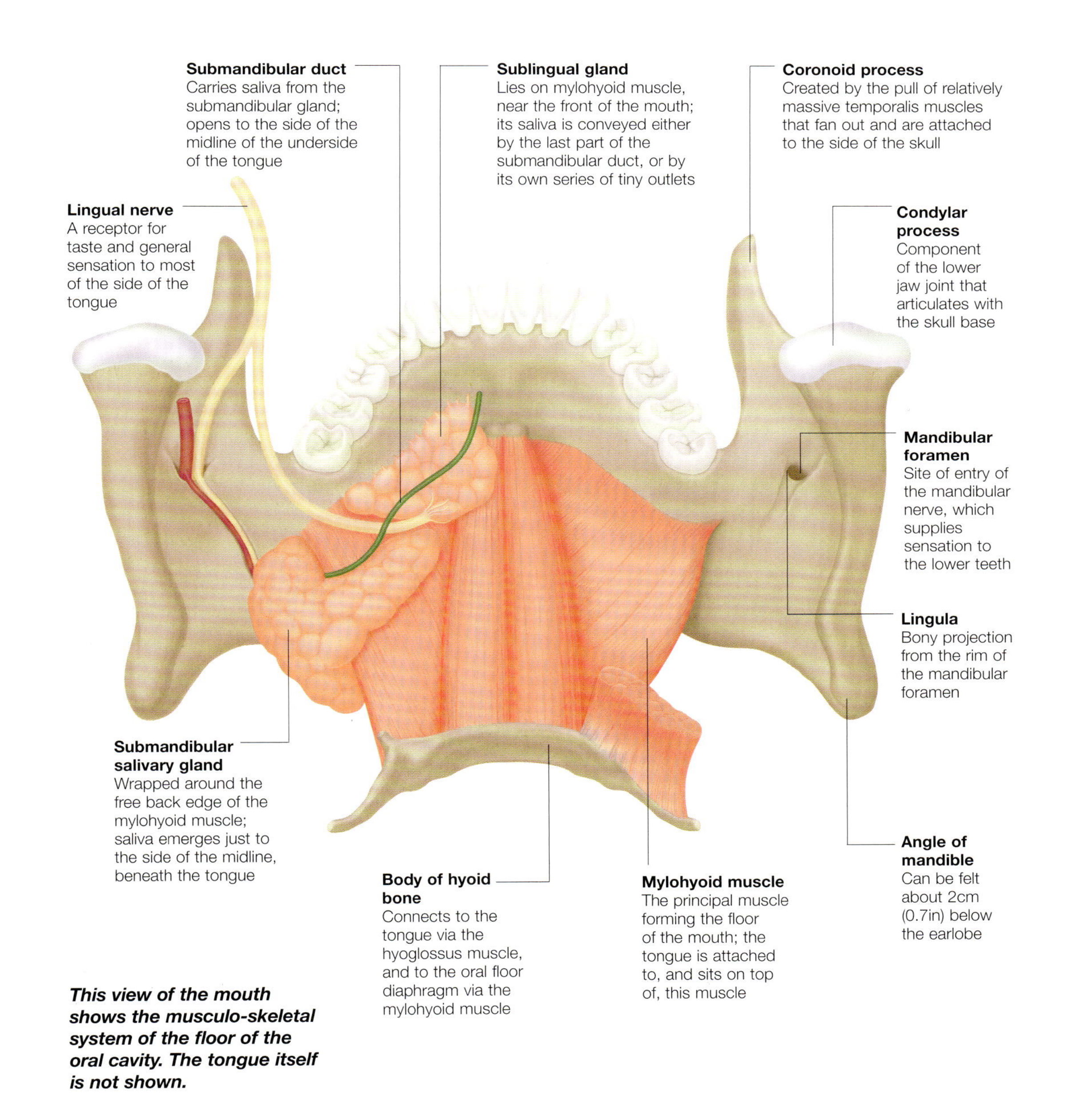

This view of the mouth shows the musculo-skeletal system of the floor of the oral cavity. The tongue itself is not shown.

Salivary Glands

The salivary glands produce about three-quarters of a litre of saliva a day. Saliva plays a major role in lubricating and protecting the mouth and in aiding digestion.

There are three pairs of major salivary glands that produce about 90 per cent of our saliva; the remaining 10 per cent is produced by minor salivary glands located in the cheeks, lips, tongue and palate.

Role of Saliva

The most important roles of saliva are lubrication, which enables mastication, swallowing and speech. It also has a protective function, keeping the mouth and gums moist and limiting bacterial activity.

The cells producing saliva are located in clusters at the end of a branching series of ducts. Two different types of saliva are produced by two distinctive cell types, called mucous and serous cells. The secretory products of mucous cells form a viscous mucin-rich product whereas the serous cells produce a watery fluid containing the enzyme amylase.

Parotid gland

The largest salivary glands are the parotid glands, which secrete serous fluid. Each parotid is superficial, lying just beneath the skin, situated between the mandible (lower jaw) and the ear. Several important structures pass through the parotid gland. The deepest of these is the external carotid artery; the most superficial is the facial nerve, which supplies the muscles of facial expression.

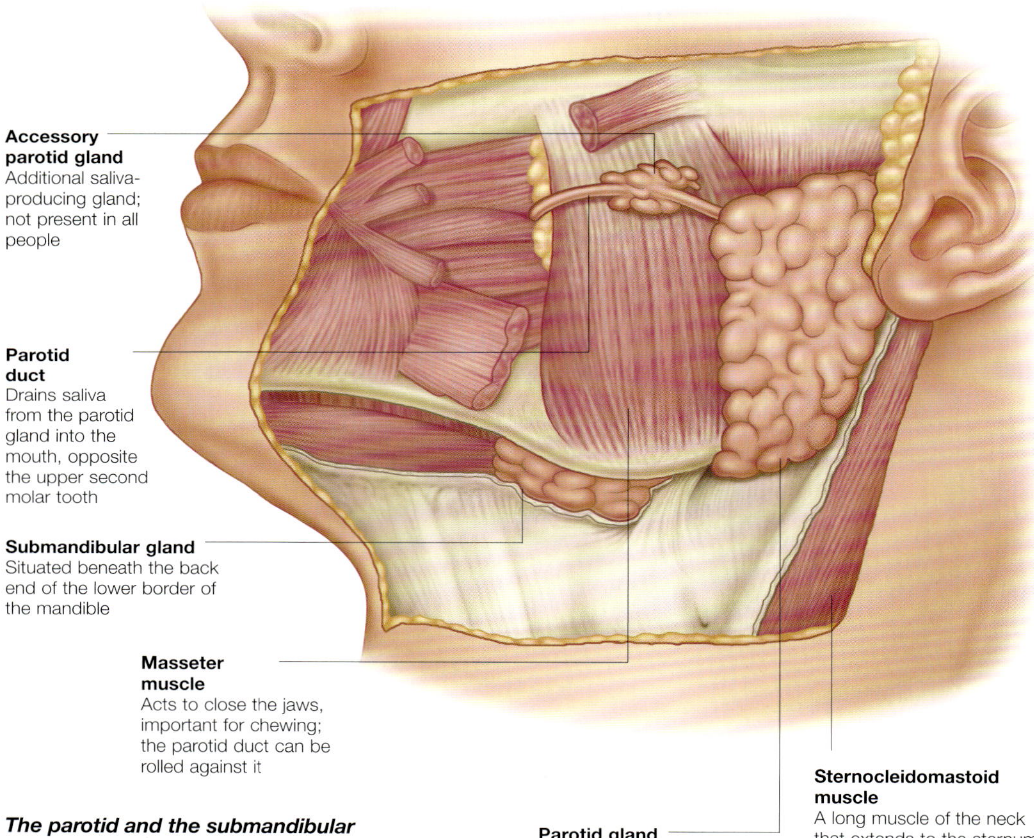

The parotid and the submandibular salivary glands are exposed in this dissection of the side of the face. The deeper surface of the parotid gland lies on the inner surface of the mandible and close to the wall of the pharynx.

Submandibular and sublingual glands

The two smaller pairs of salivary glands are the submandibular and the sublingual glands situated in the floor of the mouth.

The submandibular gland is situated beneath the lower border of the mandible towards the angle of the jaw. It is a mixed salivary gland containing serous cells (about 60 per cent) and mucous cells (about 40 per cent).

Submandibular gland

About the size of a walnut, the gland has two parts: a large, superficial part and a smaller, deep part tucked behind the mylohyoid muscle, which forms the floor of the mouth. The saliva produced by the submandibular gland is carried in the submandibular duct, which opens in the sublingual papilla (protuberance) underneath the tongue.

Sublingual gland

The sublingual gland is the smallest of the three major salivary glands and is almond-shaped. It is composed of about 60 per cent mucous cells and 40 per cent serous cells and lies under the tongue in the sublingual fossa. The two sublingual glands almost meet in the midline, and lie on the mylohyoid muscle.

Behind, the sublingual gland sits close to the deep part of the submandibular gland. Unlike the other glands, the sublingual gland does not have a single major collecting duct, but many smaller ones opening separately into the floor of the mouth or into the submandibular duct.

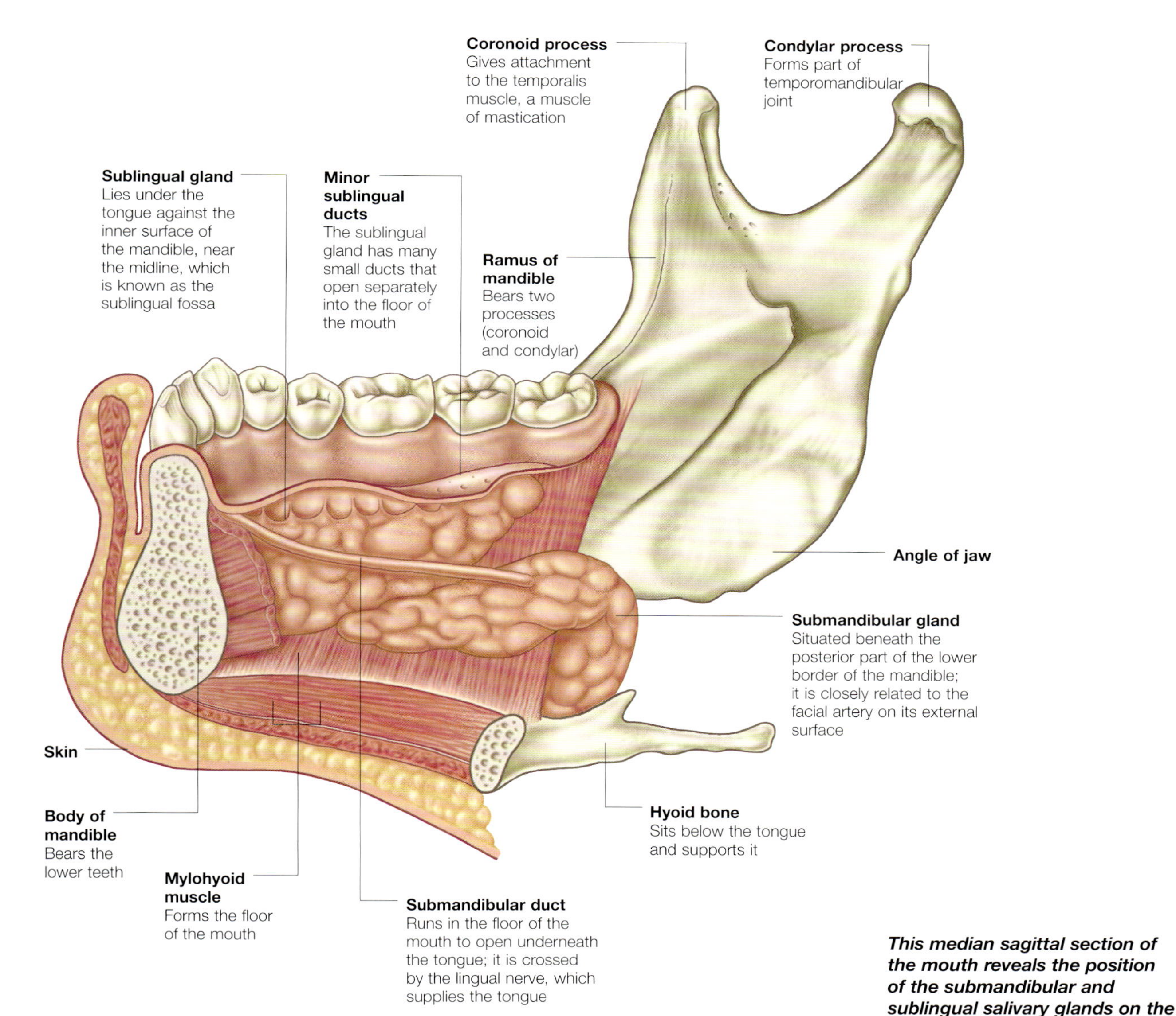

This median sagittal section of the mouth reveals the position of the submandibular and sublingual salivary glands on the inside of the jaw, on the floor of the mouth.

Tongue

The tongue is a mass of muscle, whose complex movement is essential for speech, mastication and swallowing. Its upper surface is lined with specialized tissue that contains taste buds.

The surface of the tongue is covered with an epithelium specialized for the sense of taste. The anterior two-thirds of the tongue lies in the lower dental arcade. The posterior third slopes back and down to form part of the front wall of the oropharynx. Its musculature and movements are described opposite.

Feel of Tongue

The tongue's upper surface is characterized by filiform papillae, tiny protuberances that give the surface a rough feel. The filiform papillae have tufts of keratin which, when elongated, may give the surface a 'hairy' appearance and feel. These 'hairs' can be stained by food, medicine and nicotine.

Larger Papillae

Scattered among them are the larger fungiform papillae. Larger still are the 8–12 circumvallate papillae, which form an inverse V at the junction of the anterior two-thirds and posterior third. These papillae are the major site of taste buds, although they do occur in other papillae and are scattered over the tongue surface, the cheek mucosa and the pharynx.

The posterior third of the dorsal surface has a cobbled appearance due to the presence of 40–100 nodules of lymphoid tissue, which together form the lingual tonsil.

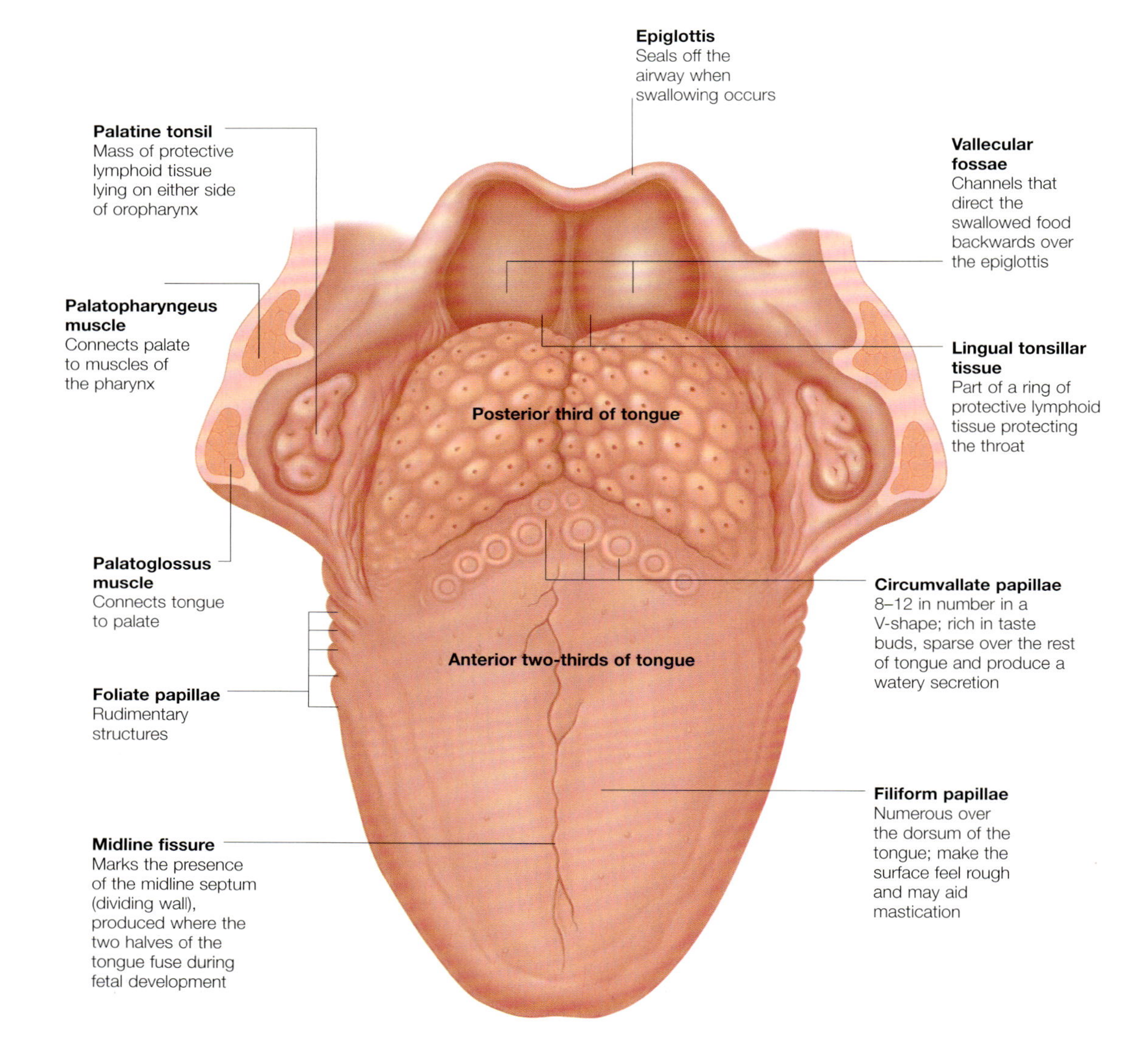

Muscles of the Tongue

The muscles within the tongue (intrinsic muscles) comprise three groups of fibre bundles running the length, breadth and depth of the organ.

The intrinsic muscles of the tongue alter the shape of the tongue to facilitate speech, mastication (chewing) and swallowing. The other muscles attached to the tongue are known as extrinsic muscles and these move the organ as a whole. The names of the extrinsic muscles denote their attachments and the general direction of movement promoted.

Protrusion of the tongue (sticking it out), elevation of its sides and depression of its centre are functions of the intrinsic muscles. They also, together with an intact palate, the lips and the teeth, allow the formation of specific sounds in speech.

Swallowing

When food has been chewed and mixed with lubricating saliva, it is forced up and back between the hard palate and the upper surface of the tongue by contraction of the styloglossus muscles, which pull the tongue up and back.

The palatoglossus muscles then contract, squeezing the food bolus (lump of chewed food) into the oral part of the pharynx. The levator palati muscles lift the soft palate to seal off the nasal passage, whereas the larynx and laryngopharynx are pulled up sealing the airway against the back of the epiglottis while the bolus passes over it.

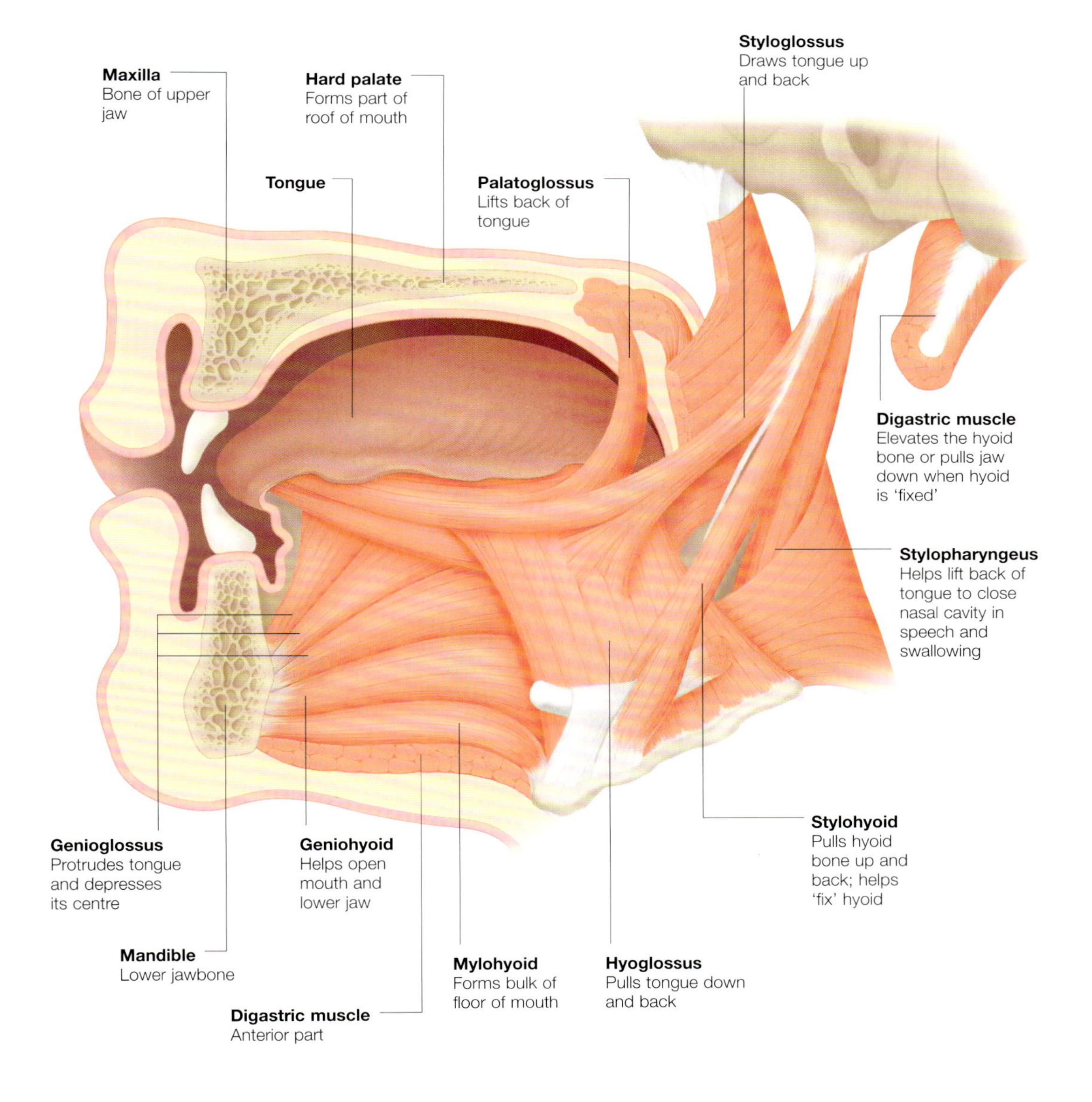

Pharynx

There are six pairs of muscles that together make up the pharynx. These muscles can be divided into two groups.

The first group of pharyngeal muscles are the three pairs of constrictor muscles: the superior, middle and inferior constrictors. These constrict the pharynx, squeezing food downwards into the oesophagus.

The constrictor muscles overlap each other from below upwards (like three stacked cups). The constrictor muscle fibres sweep backwards into a longitudinally running fibrous band in the midline, the pharyngeal raphé, which is attached to the base of the skull.

Pharangeal Muscles

The second group is the three pairs of muscles that run from above and down into the pharynx: the salpingopharyngeus, the stylopharyngeus and the palatopharyngeus. These raise the pharynx during swallowing, elevating the larynx and protecting the airway.

Nerve Supply

Most of the pharynx derives its sensory nerve supply from the glossopharyngeal (ninth cranial) nerve. Stimulation of the oropharynx (middle section of pharynx) at the back of the throat triggers the swallowing and gagging reflexes. The pharynx muscles are supplied predominantly by the 11th cranial (accessory) nerve.

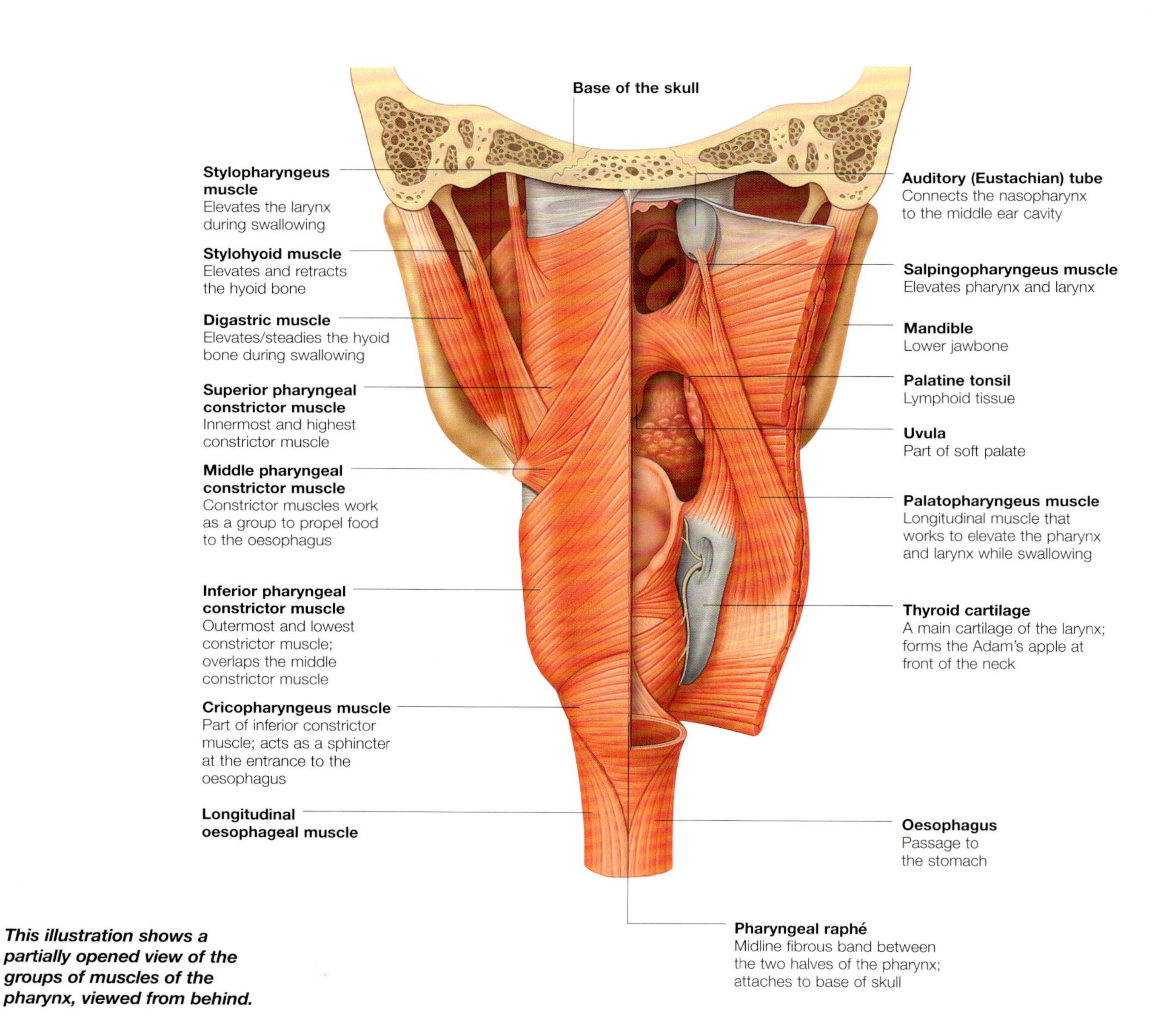

This illustration shows a partially opened view of the groups of muscles of the pharynx, viewed from behind.

Oesophagus

The oesophagus is about 25cm (10in) long in adults and is the muscular tube for the passage of food from the pharynx in the neck to the stomach.

As it is soft and flexible, the contour and path of the oesophagus is not straight; rather it curves around, and is indented by firmer structures, such as the arch of the aorta and the left main bronchus.

When no food is passing through the oesophagus, its inner lining lies in folds which fill the lumen, or central space. As a bolus (lump) of food is swallowed and passes down, it distends this lining and the oesophageal walls. Food is carried down the oesophagus by waves of contractions in a process known as peristalsis.

Layers of Oesophageal wall

In cross-section, the oesophagus has four layers:

- Mucosa – the innermost layer lined by stratified squamous epithelium; it is resistant to abrasion by food
- Submucosa – composed of loose connective tissue; it also contains glands that secrete mucus to aid the passage of food
- Muscle layer – striated muscle (under voluntary control) lines the upper oesophagus; smooth muscle the lower part; and a combination in the mid-region
- Adventitia – a covering layer of fibrous connective tissue.

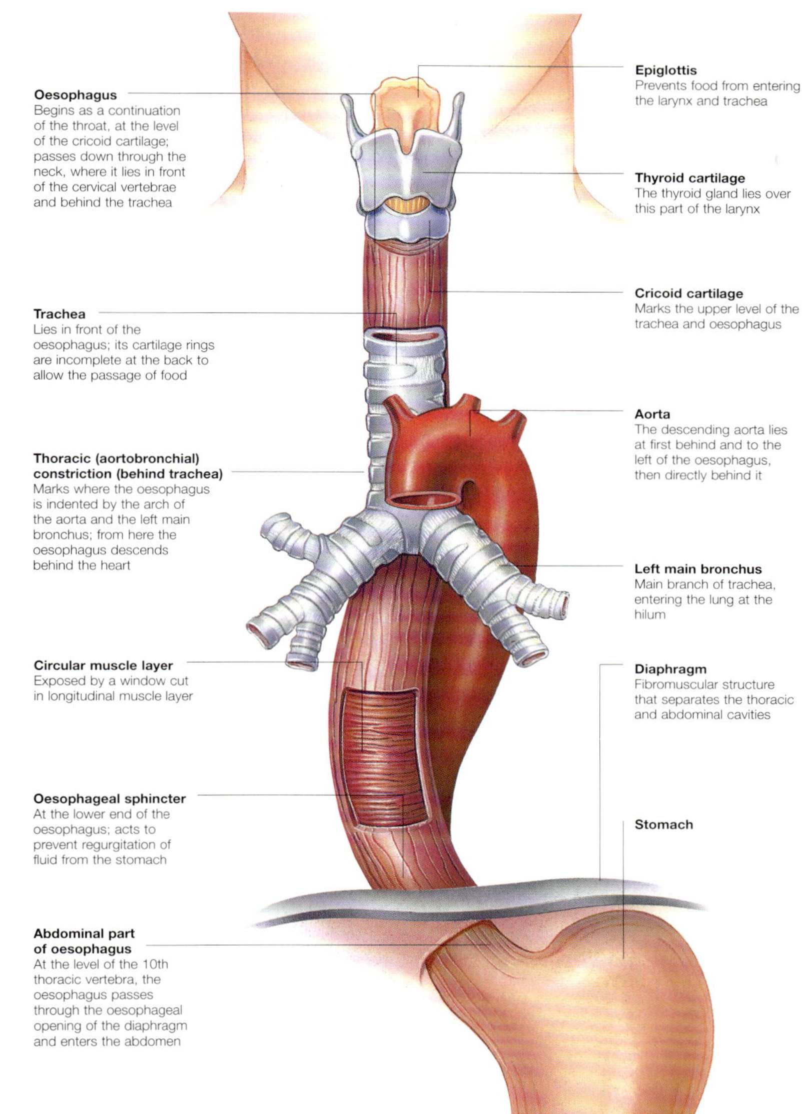

This illustration shows a frontal view of the oesophagus and the structures associated with it. The tube forms a link between the mouth and the stomach.

Nerves of the oesophagus

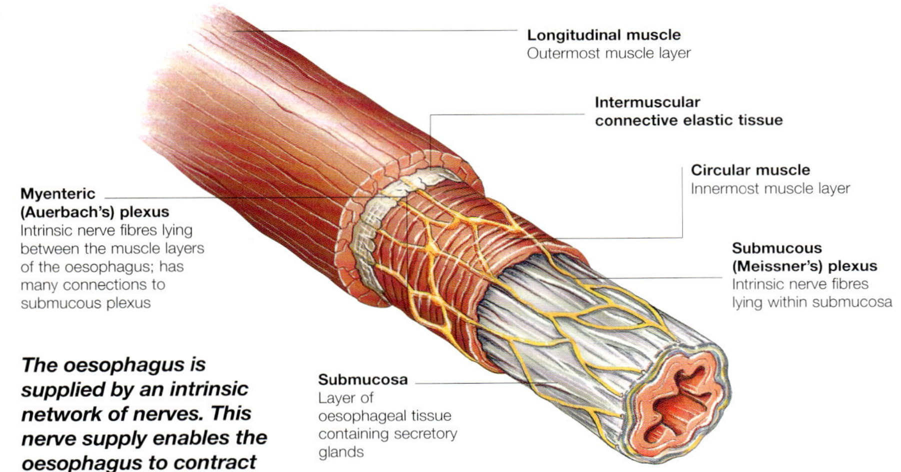

The oesophagus is supplied by an intrinsic network of nerves. This nerve supply enables the oesophagus to contract and relax during peristalsis.

In common with the rest of the gastrointestinal tract, the oesophagus has its own intrinsic nerve supply, which allows it to contract and relax during the process of peristalsis without any external nervous stimulation.

This intrinsic nerve supply derives from two main nerve plexuses within the walls known as the submucous (Meissner's) plexus and the myenteric (Auerbach's) plexus. These connect with each other, and together regulate the glandular secretion and movements of the oesophagus.

External control

The functioning of the intrinsic system can be modified by the autonomic nervous system, which regulates the body's internal environment. External nerve fibres come from the sympathetic trunk and from the vagus (10th cranial) nerve.

Stomach

The stomach is the expanded part of the digestive tract that receives swallowed food from the oesophagus. Food is stored here before being propelled into the small intestine.

The stomach is a distendable muscular bag lined by mucous membrane. It is fixed at two points: the oesophageal opening at the top and at the beginning of the small intestine below. Between these points it is mobile and can vary in position.

Stomach Lining

When empty, the stomach lining lies in numerous folds (or rugae,) which run from one opening to the other. The walls of the stomach are similar to other parts of the gut but with some modifications:

- The gastric epithelium – the layer of cells that lines the stomach; it contains many glands that secrete protective mucus, and others that produce enzymes and acid, which begin the process of digestion
- The muscle layer – this has an inner oblique layer of muscle as well as the usual longitudinal and circular fibres. This arrangement helps the stomach to churn food thoroughly before propelling it on towards the small intestine.

Regions of Stomach

The stomach is said to have four parts, and two curvatures:

- The cardia
- The fundus
- The body
- Pyloric region – outlet area of stomach
- The lesser curvature
- The greater curvature.

Location and structure of the stomach

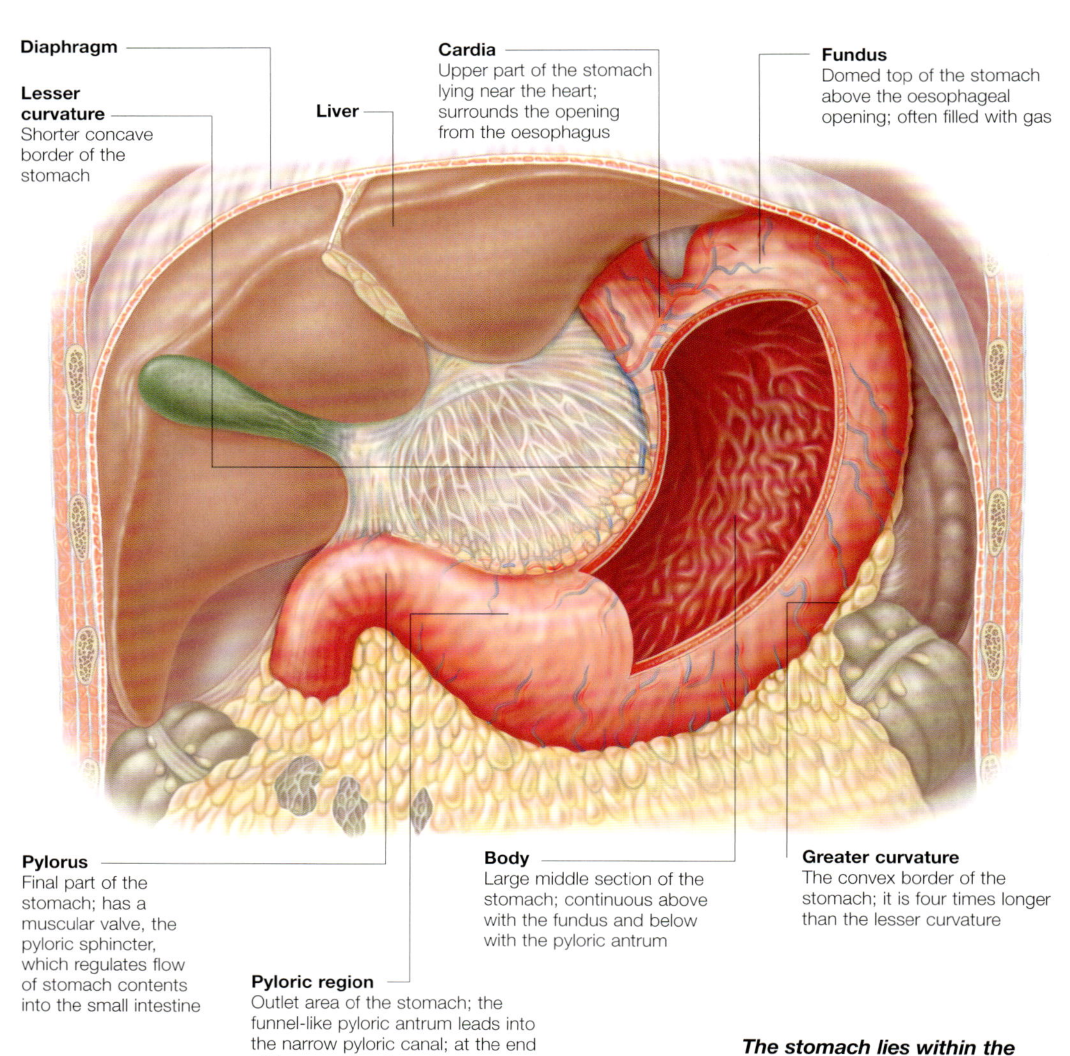

The stomach lies within the epigastric region of the abdomen, below the diaphragm. It lies to the right of the spleen and partly under the liver.

The gastro-oesophageal junction

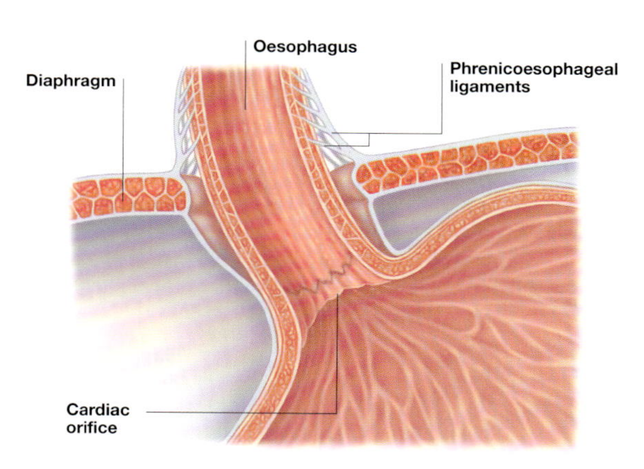

The muscular tube of the oesophagus becomes continuous with the stomach just below the diaphragm. This is where the oesophageal contents enter the stomach.

At the lower end of the oesophagus the epithelium (or lining layer of cells) changes from multilayered, stratified squamous, to the typical gastric mucosa in a zigzag junction.

Connective ligaments

The oesophagus and upper part of the stomach are held to the diaphragm by phrenicoesophageal ligaments. These are extensions of the fascia, a connective tissue covering the diaphragm's surface.

Physiological sphincter

There is no identifiable valve at the top of the stomach to control the passage of food. However, the surrounding muscle fibres of the diaphragm act to keep the tube closed except when a bolus (swallowed mass) of food passes through. This is referred to as the physiological oesophageal sphincter, through which the oesophagus passes.

Blood supply of the stomach

The stomach has a profuse blood supply, which comes from the various branches of the coeliac trunk.

The vessels that supply the stomach are:

- Left gastric artery – a branch of the coeliac trunk
- Right gastric artery – usually arises from the hepatic artery (a branch of the coeliac trunk)
- Right gastroepiploic artery – arises from the gastroduodenal branch of the hepatic artery
- Left gastroepiploic artery – arises from the splenic artery
- Short gastric arteries – arise from the splenic artery.

Veins

The gastric veins run alongside the various gastric arteries. Blood from the stomach is drained ultimately into the portal venous system, which takes blood through the liver before returning it to the heart.

Lymphatics

Lymph collected from the stomach walls drains through lymphatic vessels into the many lymph nodes, which lie in groups along the lesser and greater curvature. It is then transported to the coeliac lymph nodes.

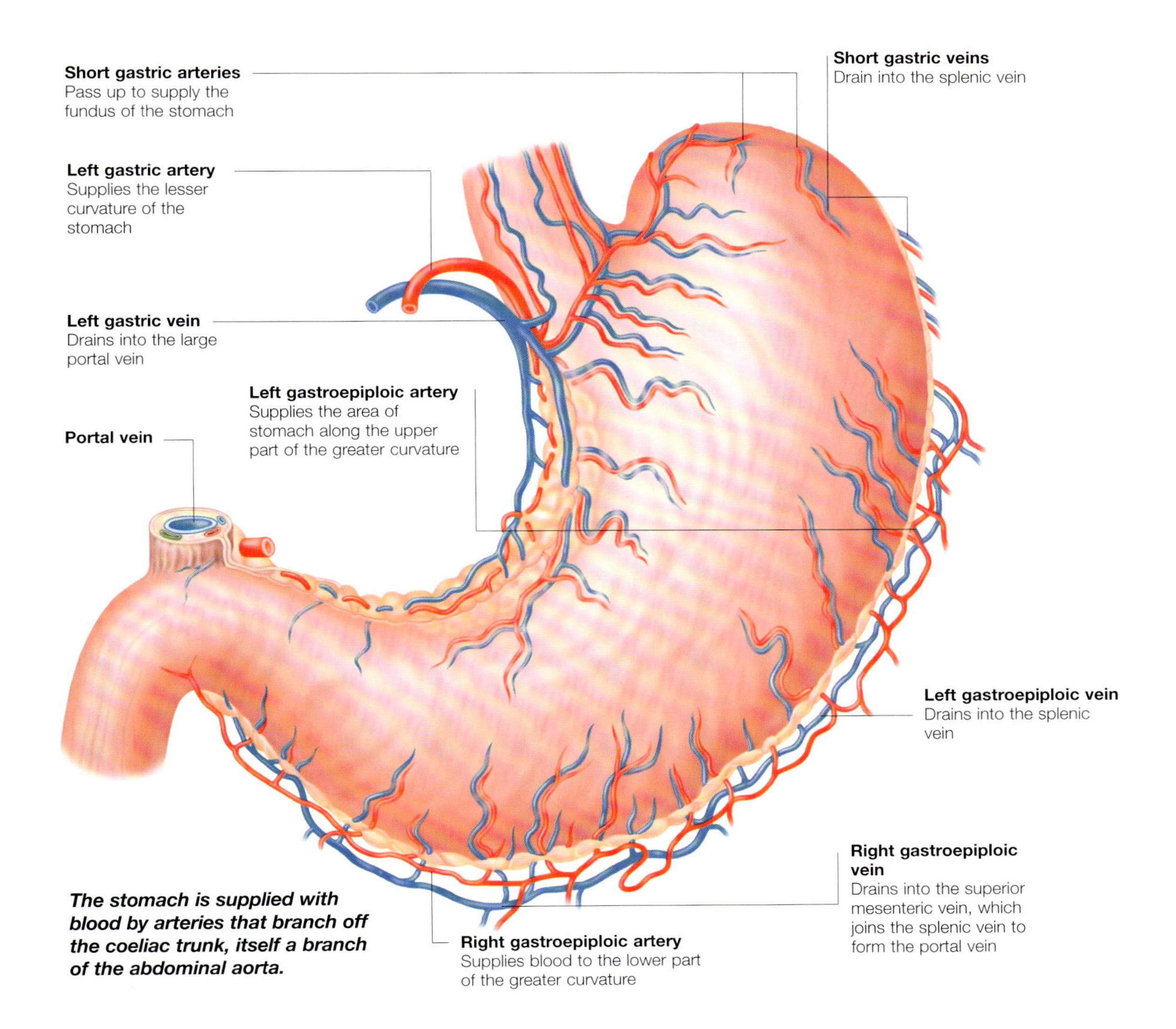

The stomach is supplied with blood by arteries that branch off the coeliac trunk, itself a branch of the abdominal aorta.

Small Intestine

The small intestine extends from the stomach to the junction with the large intestine. It is made up of three parts, and is the main site in the body where food is digested and absorbed.

The small intestine is the main site of digestion and absorption of food. It is about 7m (23ft) in length in adults and extends from the stomach to the junction with the large intestine. It is divided into three parts: the duodenum, the jejunum and the ileum.

Duodenum

The duodenum is the first part of the small intestine and the shortest (about 25cm/10in in length). It receives the contents of the stomach with each wave of contraction of the stomach walls. In the duodenum the contents are mixed with secretions from the duodenal walls, pancreas and gall bladder. The duodenum cannot move, but it is fixed in place behind the peritoneum, the sheet of connective tissue that lines the abdominal cavity.

Blood Supply

The duodenum receives arterial blood from various branches off the aorta, the body's largest artery. These, in turn, give off small branches that provide each part of the duodenum with a rich supply of blood. Venous blood supply mirrors the arterial pattern, returning blood to the hepatic portal venous system.

Structure

The duodenum's walls have two layers of muscle fibres, one circular and one longitudinal.

The mucosa (or lining of the duodenum) is particularly thick. It contains numerous glands (Brunner's glands) that secrete a thick alkaline fluid. This helps counter the acidic nature of the contents that have reached the duodenum from the stomach. The mucosa in the first part of the duodenum is smooth, but thereafter it is thrown into deep, permanent folds of tissue, known as plicae.

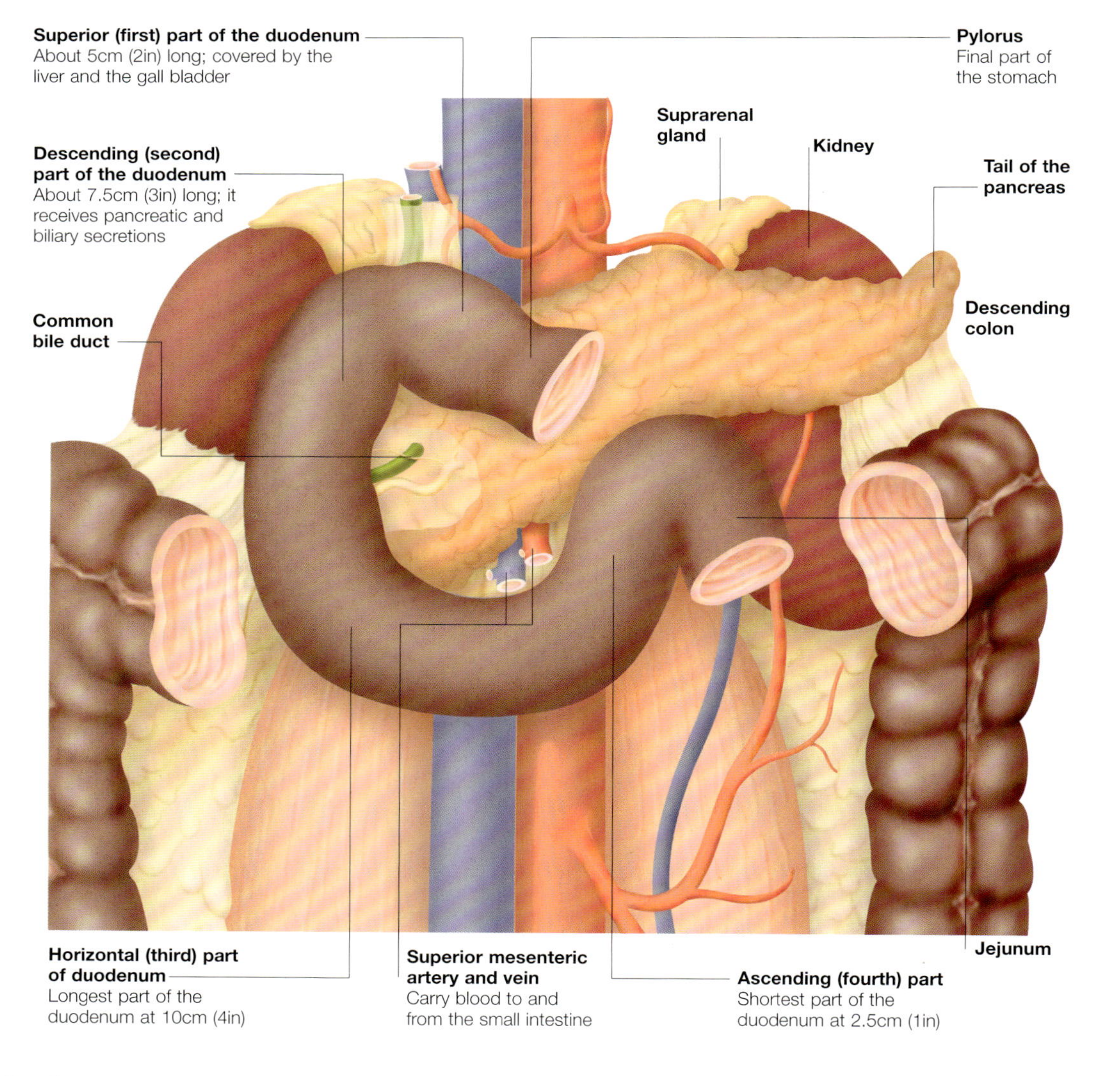

The duodenum is the first part of the small intestine. It is roughly C-shaped and is made up of four parts.

Jejunum and Ileum

The jejunum and the ileum together form the longest part of the small intestine. Unlike the duodenum, they can move within the abdomen.

The jejunum and the ileum comprise the longest part of the small intestine. They are surrounded and supported by a fan-shaped fold of the peritoneum – the mesentery – which allows them to move within the abdominal cavity. The mesentery is 15cm (6in) long.

Blood Supply

The jejunum and the ileum receive their arterial blood supply from 15–18 branches of the superior mesenteric artery. These branches anastomose (join) to form arches, called arterial arcades. Straight arteries pass out from the arterial arcades to supply all parts of the small intestine. Venous blood from the jejunum and ileum enters the superior mesenteric vein. This vein lies alongside the superior mesenteric artery and drains into the hepatic portal venous system.

Role of Lymph

Fat is absorbed from the contents of the small intestine into specialized lymphatic vessels, known as lacteals, which are found within the mucosa. The milky lymphatic fluid produced by this absorption enters lymphatic plexuses (networks of lymphatic vessels) within the walls of the intestine. The fluid is then carried to special nodes called mesenteric lymph nodes.

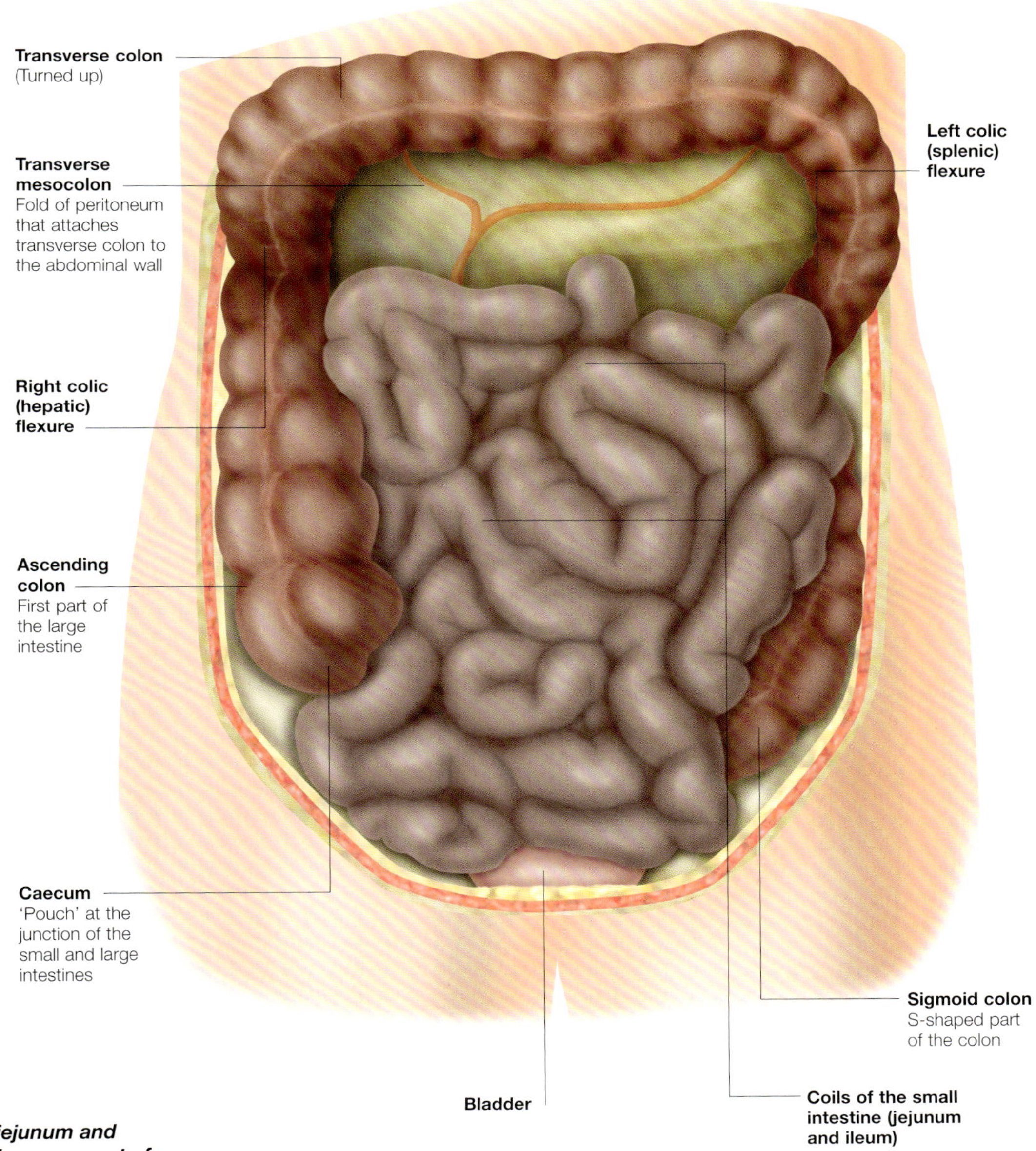

The coils of the jejunum and ileum take up a large amount of central space in the abdominal cavity. They are not fixed and are able to move within the cavity.

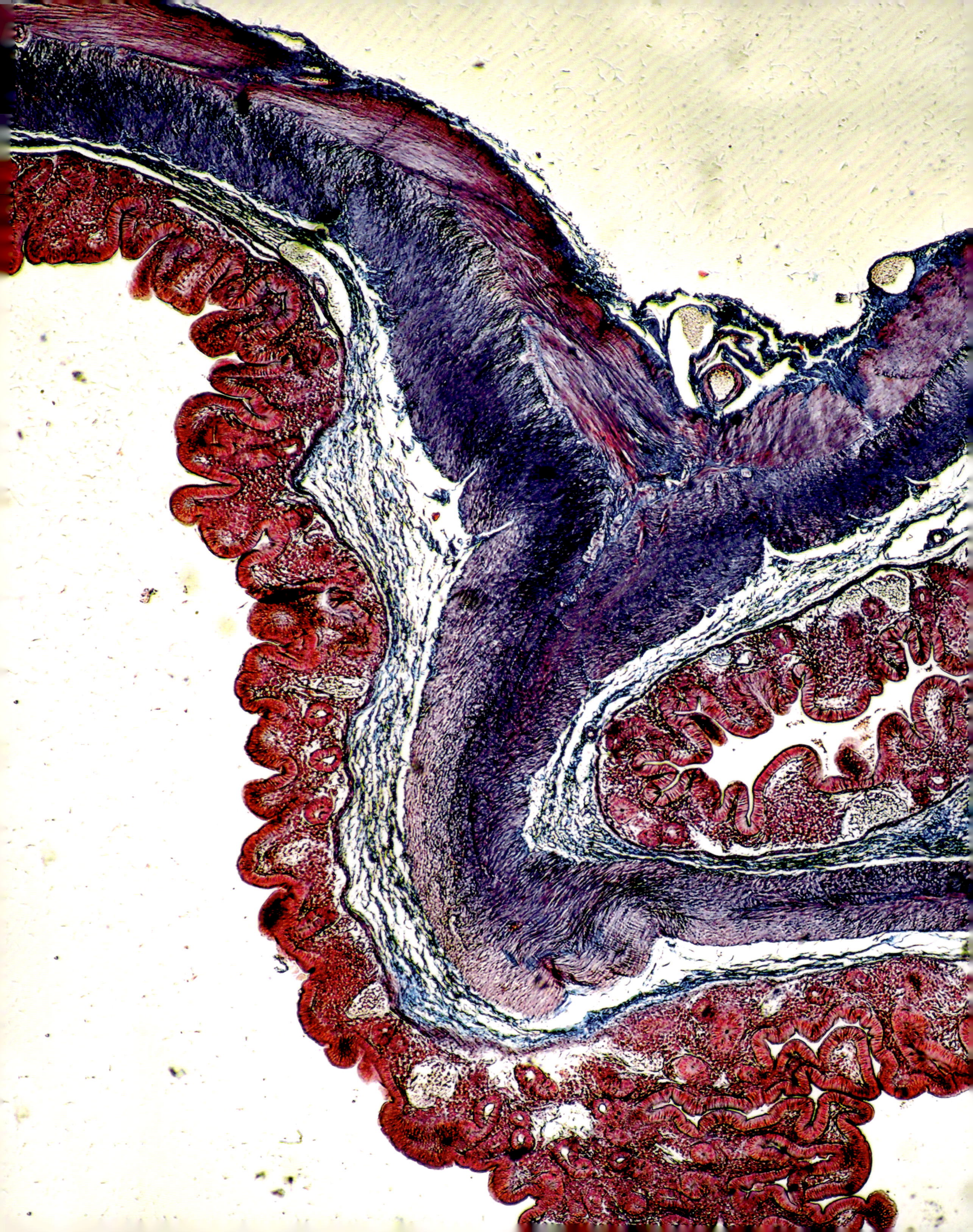

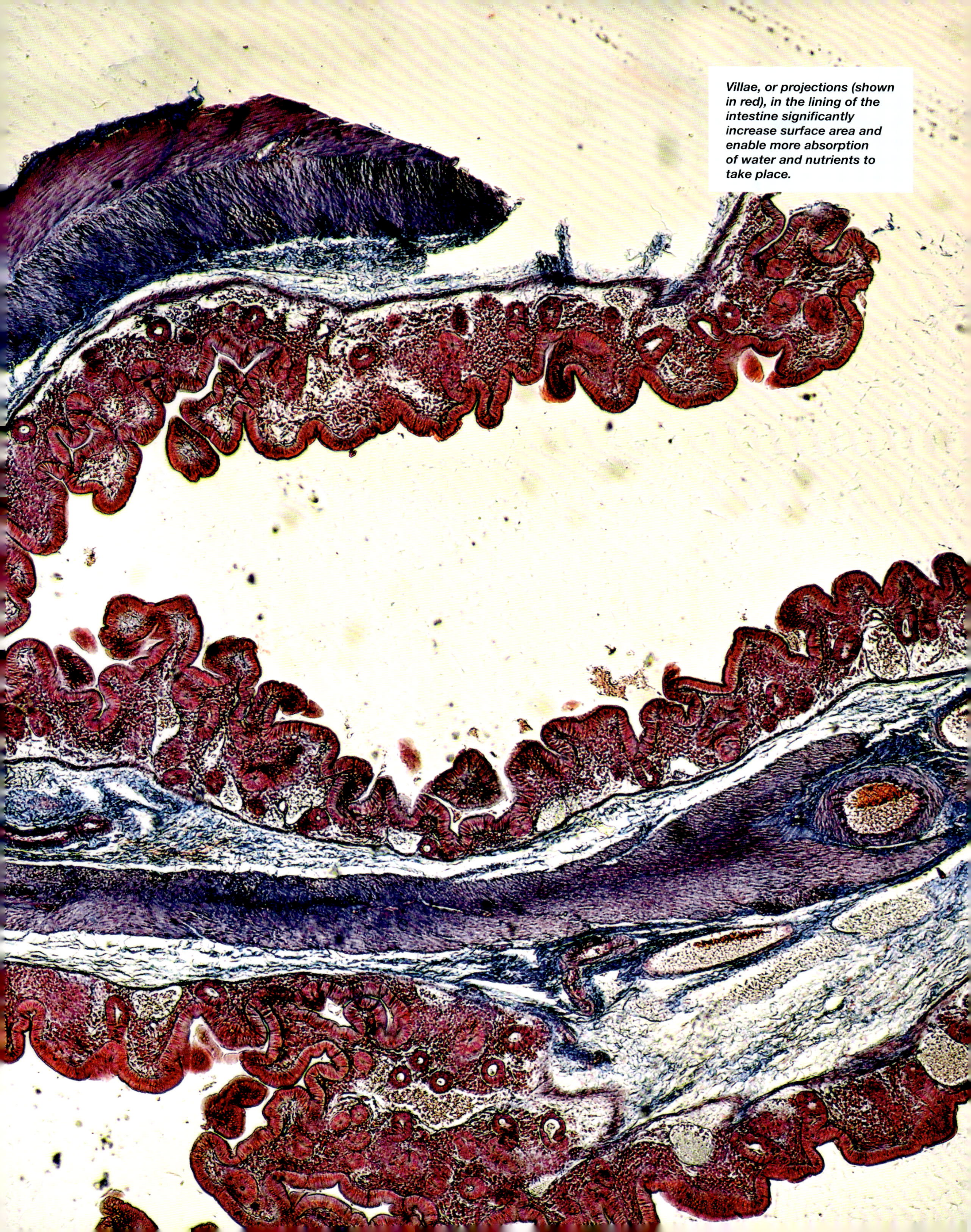

Villae, or projections (shown in red), in the lining of the intestine significantly increase surface area and enable more absorption of water and nutrients to take place.

Caecum

The caecum and appendix lie at the junction of the large and small intestine, an area also known as the ileocaecal region. The caecum, from which the appendix arises, receives food from the small intestine.

The caecum is the first part of the large intestine. Food travels from the terminal ileum, the last part of the small intestine, into the large intestine through the ileocaecal valve; the caecum lies below this valve. The caecum is a blind-ended pouch, about 7.5cm (3in) in length and breadth, which continues as the ascending colon.

Ileocaecal Valve
The ileocaecal valve surrounds the orifice, or opening, through which the liquefied contents of the terminal ileum enter the large bowel. The valve is raised above the caecal wall and is surrounded by a ring of circular muscle fibres, which keep it closed.

Muscle Fibres
The muscular 'coat' of the small intestine is continued in the walls of the large intestine, but here it becomes separated into three bands of muscle, the taeniae coli. As food passes through the intestine, the caecum may become distended with faeces or gas.

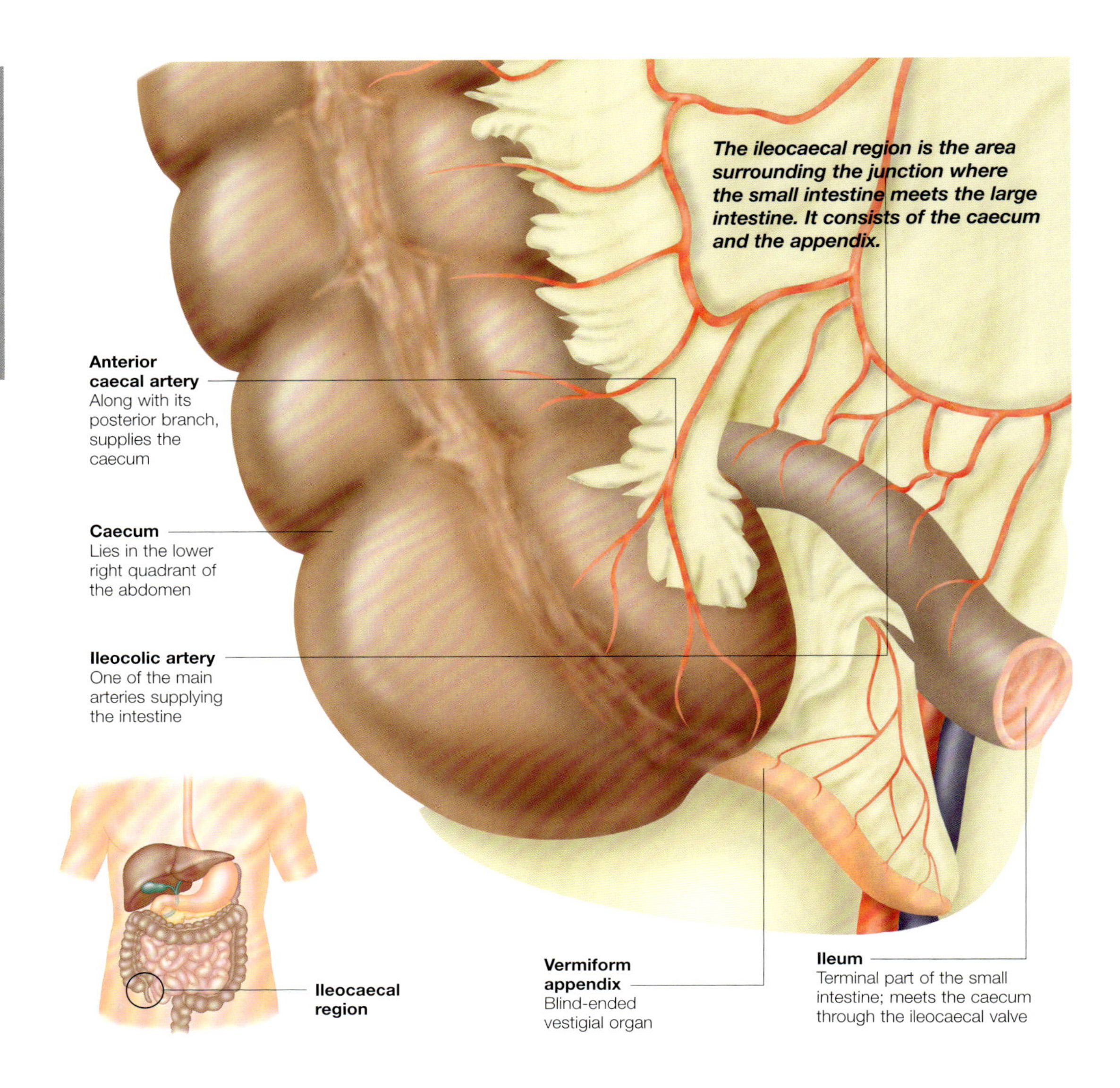

The ileocaecal region is the area surrounding the junction where the small intestine meets the large intestine. It consists of the caecum and the appendix.

The ileocaecal valve

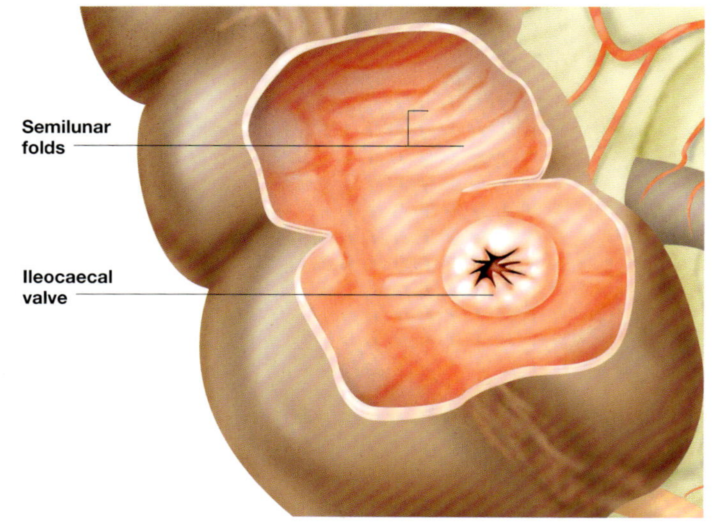

The ileocaecal valve surrounds the orifice, or opening, through which the liquefied contents of the terminal ileum, the last part of the small intestine, enter the caecum.

Anatomical studies
Anatomical studies of cadavers (dead bodies) in the past had shown this orifice to be enclosed between folds or ridges in the caecal wall, which were thought to act like a valve.

Now that it is possible to study this area in living people, using endoscopes, it is clear that the opening looks quite different in life. The ileocaecal orifice is in fact raised above the caecal wall and is surrounded by a ring of circular muscle fibres that help keep it closed.

The ileocaecal valve surrounds the orifice through which intestinal contents pass into the caecum. This valve is not believed to be very effective.

Barium studies
Although the contents of the caecum do not easily pass back into the ileum when the caecal walls contract, the ileocaecal valve is not very effective. Barium X-ray studies of the large intestine commonly show leakage of contents backwards from the caecum into the terminal ileum.

Appendix

The appendix is a narrow, muscular outpouching of the caecum. It is usually between 6–10cm (2–4in) in length, and arises from the back of the caecum, its lower end being free and mobile.

The vermiform (or 'worm-like') appendix is attached to the caecum at the beginning of the large intestine. The walls of the appendix contain lymphoid tissue. The lymphoid tissue of the appendix, and that within the walls of the small intestine, protects the body from microorganisms within the gut.

Muscle layer

Whereas the longitudinal muscle in the walls of the rest of the large intestine is present only in three bands – the taeniae coli – the appendix has a complete muscle layer. This is because the three taeniae coli converge on the base of the appendix and their fibres join to cover its entire surface.

Peritoneum

The appendix is enclosed within a covering of peritoneum that forms a fold between the ileum, the caecum and the first part of the appendix. This fold is known as the mesoappendix.

Base of appendix

The base of the appendix, where it arises from the caecum, is usually in a fixed position. The corresponding area on the surface of the abdomen is known as McBurney's point.

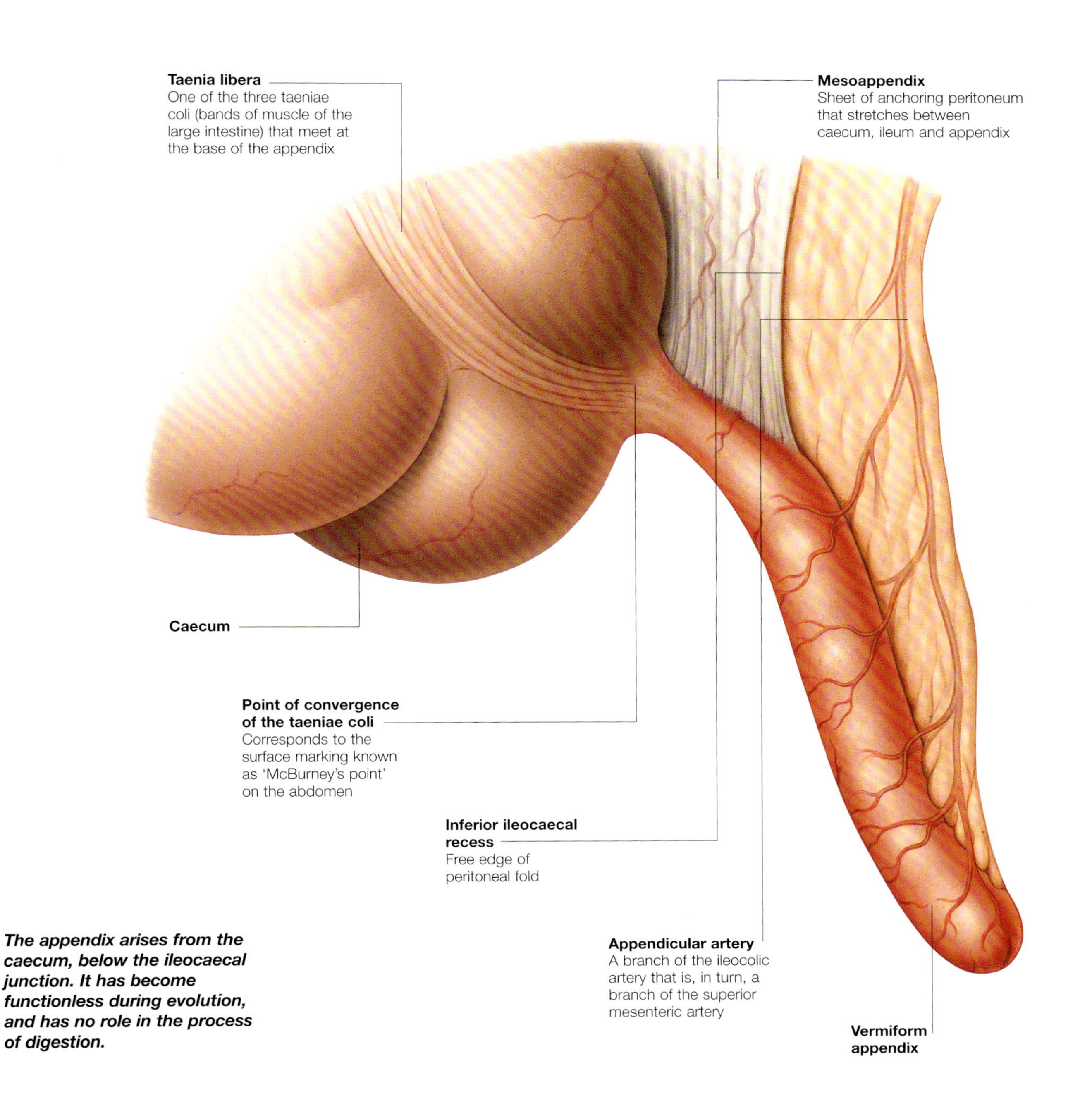

The appendix arises from the caecum, below the ileocaecal junction. It has become functionless during evolution, and has no role in the process of digestion.

Colon

The colon forms the main part of the large intestine. Although a continuous tube, the colon has four parts: the ascending colon, the transverse colon, the descending colon and the sigmoid colon.

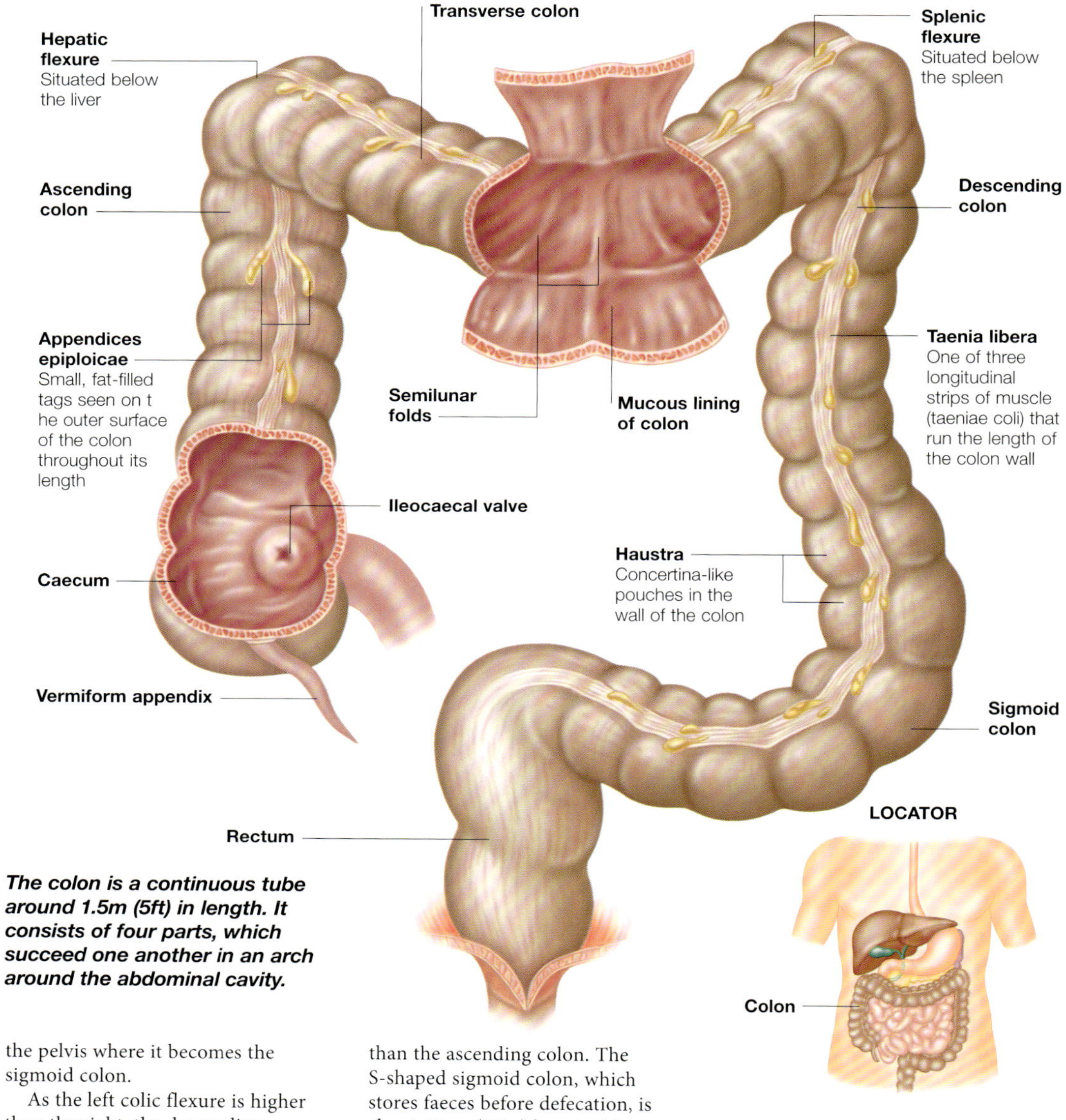

The colon is a continuous tube around 1.5m (5ft) in length. It consists of four parts, which succeed one another in an arch around the abdominal cavity.

The colon receives the liquefied contents of the small intestine and reabsorbs the water to form semisolid waste. This is then expelled through the rectum and anal canal as faeces. There are two sharp bends, or flexures, in the colon known as the right colic (or hepatic) flexure and the left colic (or splenic) flexure.

Ascending Colon

The ascending colon runs from the ileocaecal valve up to the right colic flexure, where it becomes the transverse colon. It is about 12cm (4.7in) long and lies against the posterior (back) abdominal wall and is covered on the front and sides by the peritoneum, a thin sheet of connective tissue that lines the abdominal organs.

Transverse Colon

The transverse colon begins at the right colic flexure, under the right lobe of the liver, and runs across the body towards the left colic flexure next to the spleen. At about 45cm (18in), it is the longest, most mobile part of the large intestine, as it hangs down suspended within a fold of peritoneum.

Descending Colon

The descending colon runs from the left colic flexure down to the brim of the pelvis where it becomes the sigmoid colon.

As the left colic flexure is higher than the right, the descending colon is consequently longer than the ascending colon. The S-shaped sigmoid colon, which stores faeces before defecation, is about 40cm (16in) long and leads into the rectum.

Sigmoid colon and the colon lining

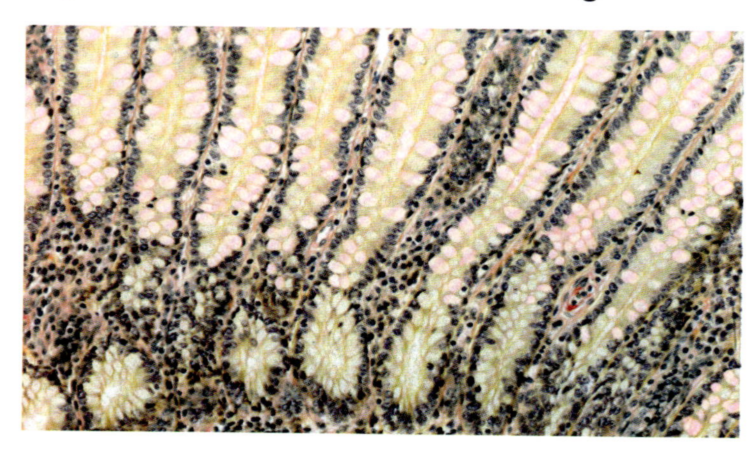

The sigmoid – 'S-shaped' – colon is the continuation of the descending colon, starting at the pelvic brim.

Characteristics

It is about 40cm (16in) long and, unlike the descending colon, quite mobile as it lies within its mesentery, or fold of peritoneum. At its far end the sigmoid colon leads into the rectum. The function of the sigmoid colon is to store faeces before defecation, and so its size and position vary depending on dietary intake and whether it is full or empty.

The colon lining – the mucosa (green) – contains glands (yellow). Cells in these glands are involved in water absorption and mucus secretion.

Lining of the colon

The lining of the colon has a simple layer of cells with many deep depressions, or crypts, which contain mucus-secreting cells. The mucus is important for lubricating the passage of faeces and protecting the walls from acids and gases produced by the intestinal bacteria.

Blood supply of the colon

Like the rest of the intestine, each of the parts of the colon is readily supplied with, and drained of, blood from a network of arteries and veins. Venous drainage mirrors the pattern of the arteries.

The arterial supply to the colon comes from the superior and inferior mesenteric branches of the aorta, the large central artery of the abdomen. The ascending colon and first two-thirds of the transverse colon are supplied by the superior mesenteric artery, whereas the last third of the transverse colon, the descending colon and the sigmoid colon are supplied by the inferior mesenteric.

Pattern of the arteries

As in other parts of the gastrointestinal tract, there are connections between the branches of these two major arteries. The superior mesenteric artery gives off the ileocolic, right colic and middle colic arteries, which anastomose (join) with each other and with the left colic and sigmoid branches of the inferior mesenteric artery.

In this way an 'arcade' of arteries is formed around the wall of the colon, supplying all parts with arterial blood.

Arterial system of the colon

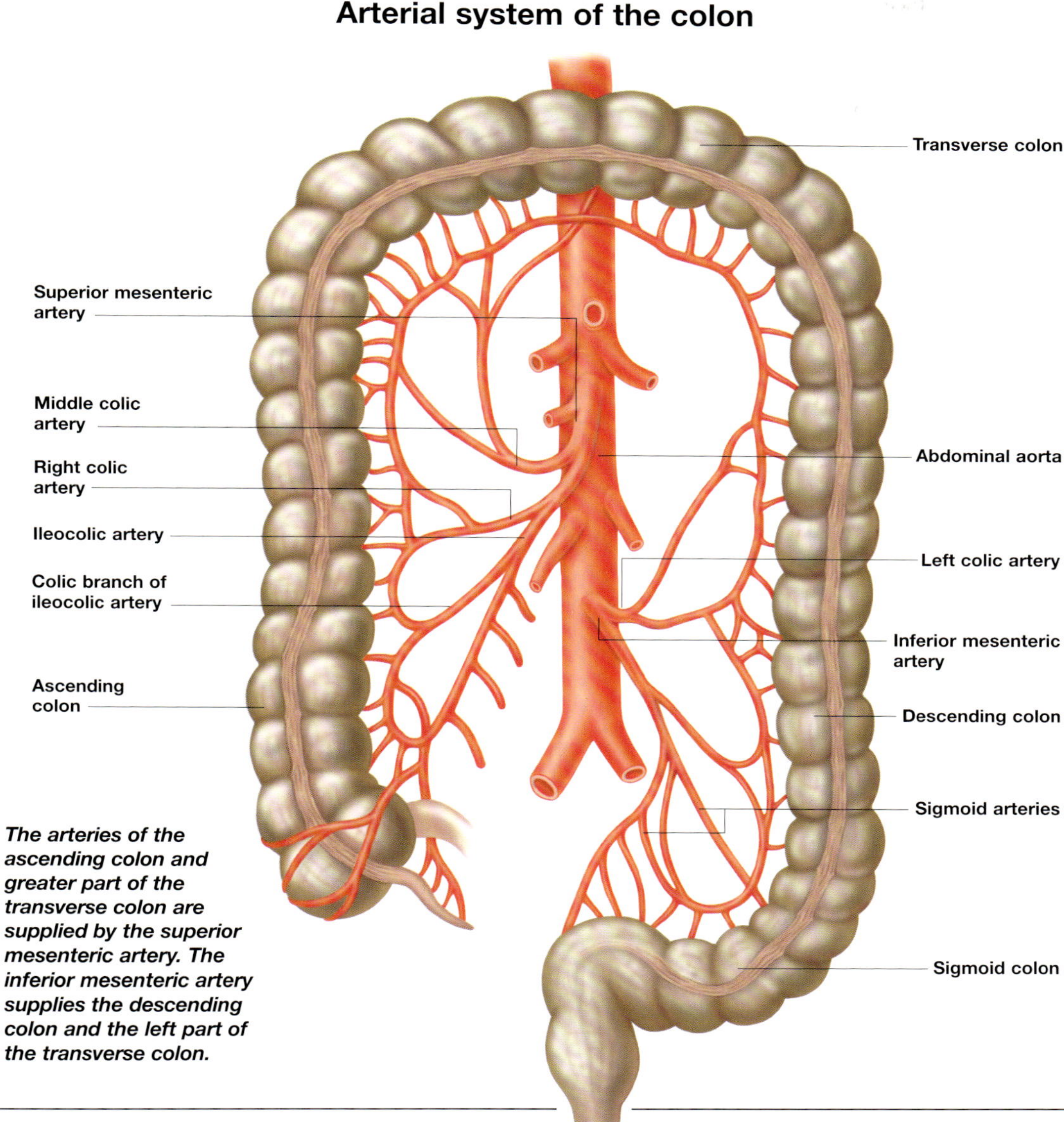

The arteries of the ascending colon and greater part of the transverse colon are supplied by the superior mesenteric artery. The inferior mesenteric artery supplies the descending colon and the left part of the transverse colon.

Venous drainage of the colon

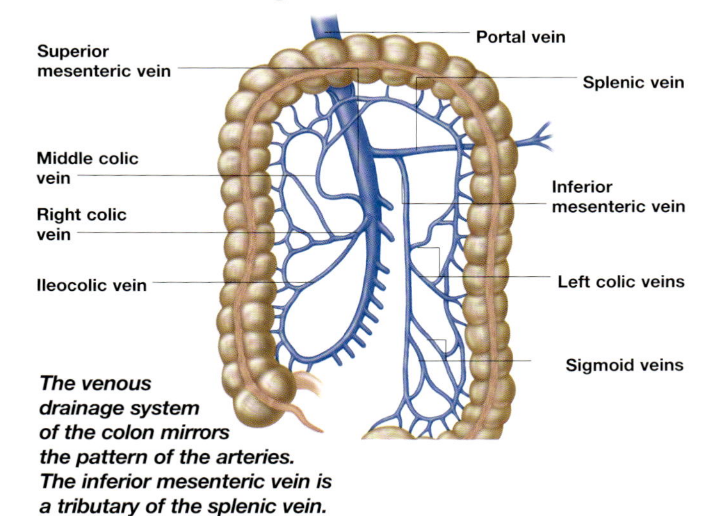

The venous drainage system of the colon mirrors the pattern of the arteries. The inferior mesenteric vein is a tributary of the splenic vein.

Venous blood from the colon is collected ultimately in the portal vein. In general, blood from the ascending colon and first two-thirds of the transverse colon runs into the superior mesenteric vein, with blood from the remainder of the colon being drained by the inferior mesenteric vein.

The inferior mesenteric vein drains into the splenic vein, which then joins with the superior mesenteric vein to form the portal vein. The portal vein then carries all the venous blood through the liver on its way back to the heart.

Lymphatic drainage

Lymph collected from the walls of the colon travels in lymphatic vessels back alongside the arteries towards the main abdominal lymph-collecting vessel, the cysterna chyli. There are many lymph nodes that filter the fluid before it is returned to the venous system.

Lymph passes through the lymph nodes on the wall of the colon, through the nodes adjacent to the small arteries supplying the colon and then through the superior and inferior mesenteric nodes.

Haustra

Unlike the small intestine, the walls of the colon are puckered into concertina-like pockets, or haustra, which show up quite clearly on direct examination, although this pattern can be absent if there is chronic inflammation such as in colitis.

Rectum and Anal Canal

The rectum and anal canal together form the last part of the digestive tract. They receive waste matter in the form of faeces and enable it to be passed out of the body.

The rectum continues on from the sigmoid colon. It follows the curve of the sacrum and coccyx, which form the back of the bony pelvis.

The lower end of the rectum joins to the anal canal with an 80–90 degree change in direction, which prevents faeces passing into the anal canal until necessary. The longitudinal muscle of the rectum is in two broad bands that run down the front and the back surfaces. There are three horizontal folds in the rectal wall known as the superior, middle and inferior transverse folds. Below the inferior fold the rectum widens into the ampulla.

Anal Canal

The anal canal runs from the anorectal flexure to the anus. Except during defecation, the canal is empty. The lining of the anal canal changes along its length. The upper part has longitudinal ridges called anal columns that begin at the ano-rectal junction and end at the pectinate line. At the lower end of these are the anal sinuses and valves. The sinuses produce lubricating mucus when faeces are being passed. The valves prevent the passage of mucus out of the anal canal at other times.

Coronal section through the rectum and anal canal

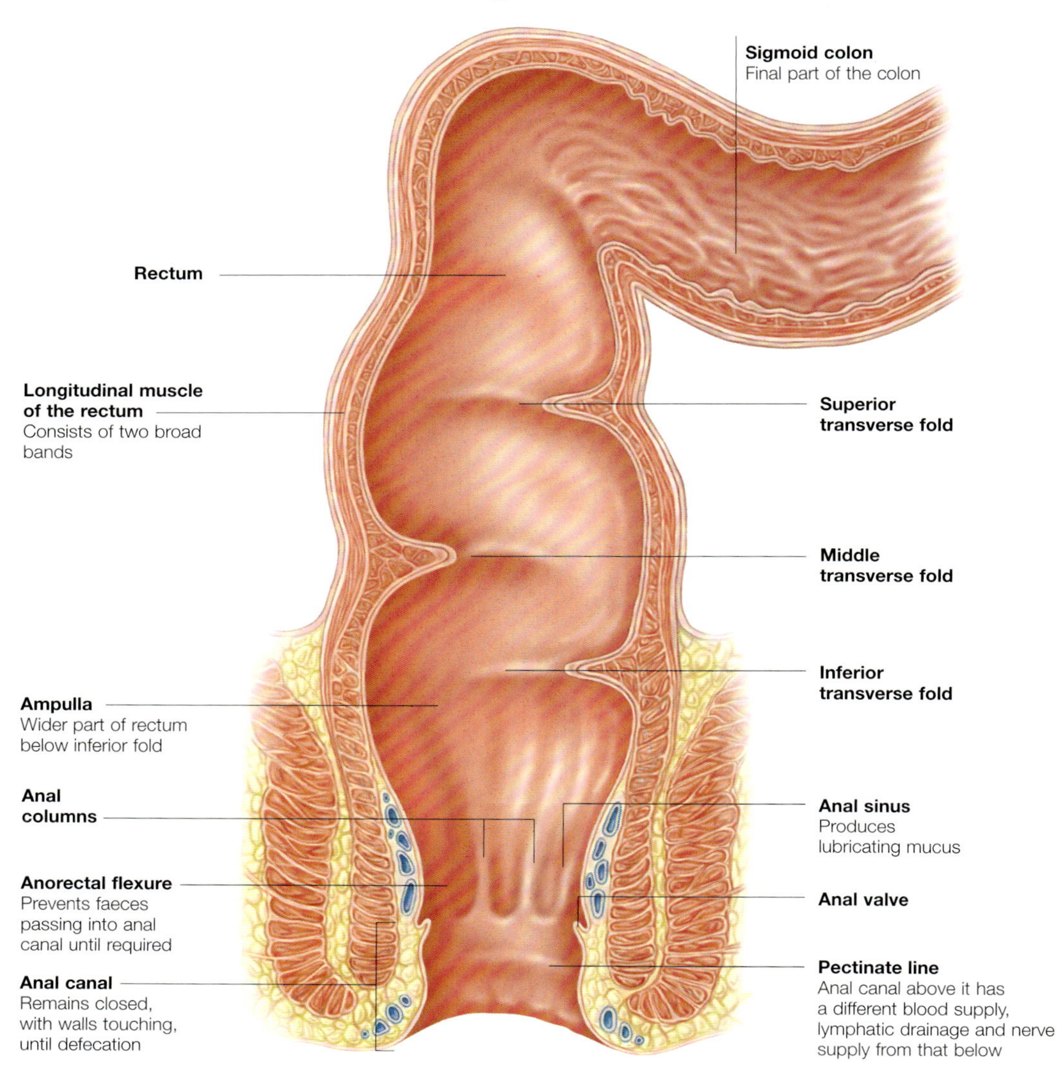

The anal sphincter

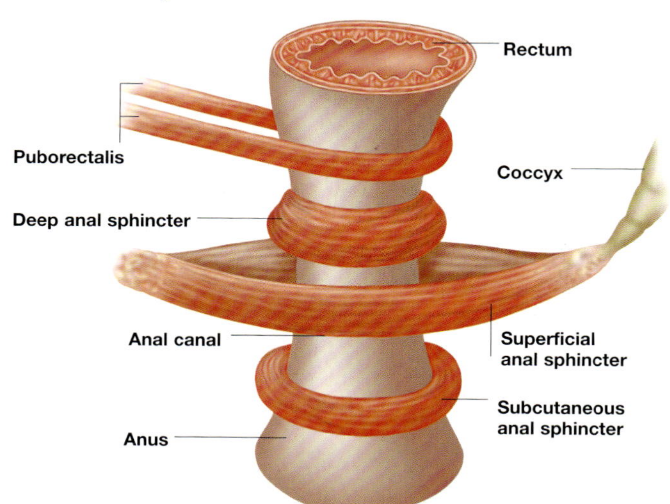

The contents of the intestines are constantly being moved on to the next stage without our conscious awareness.

However, it is important that there is control of the movement of faeces at the final stage. This control is achieved through the function of the anal sphincter, which is made up of several parts:

- The internal anal sphincter
A thickening of the normal circular muscle layer of the bowel in the upper two thirds of the anal canal. It is not under voluntary control.
- The puborectalis muscle
A sling of muscle that loops around the anorectal junction forming an angle and preventing passage of the contents of the rectum into the anal canal.
- External anal sphincter
In three parts, deep, superficial and subcutaneous, this sphincter is under voluntary control, and so can be relaxed by an act of will when convenient.

The anal sphincter, which consists of several parts, controls the release of faeces from the body. Only the external anal sphincter is under voluntary control.

Vessels of the rectum and anus

The rectum and anal canal have a rich blood supply. A network of veins drains blood from this area.

Beneath the lining of the rectum and anal canal lies a network of small veins, the rectal venous plexus:

- The internal rectal venous plexus lies just under the lining
- The external rectal venous plexus lies outside the muscle layer.

These receive blood from the tissues and carry it to the larger veins that drain the area: the superior, middle and inferior rectal veins that drain the corresponding parts of the rectum. The internal venous plexus of the anal canal drains blood in two directions on either side of the pectinate line region. Above this level blood drains into the superior rectal vein while from below it drains into the inferior rectal vein.

The rectum receives its arterial blood supply from three sources. The upper part is supplied by the superior rectal artery, the lower portion is supplied by the middle rectal arteries whereas the anorectal junction receives blood from the inferior rectal arteries.

Within the anal canal the superior rectal artery travels down to provide blood above the pectinate line. The two inferior rectal arteries, branches of the pudendal, supply the anal canal below the pectinate line.

Venous drainage system of the rectum and anus

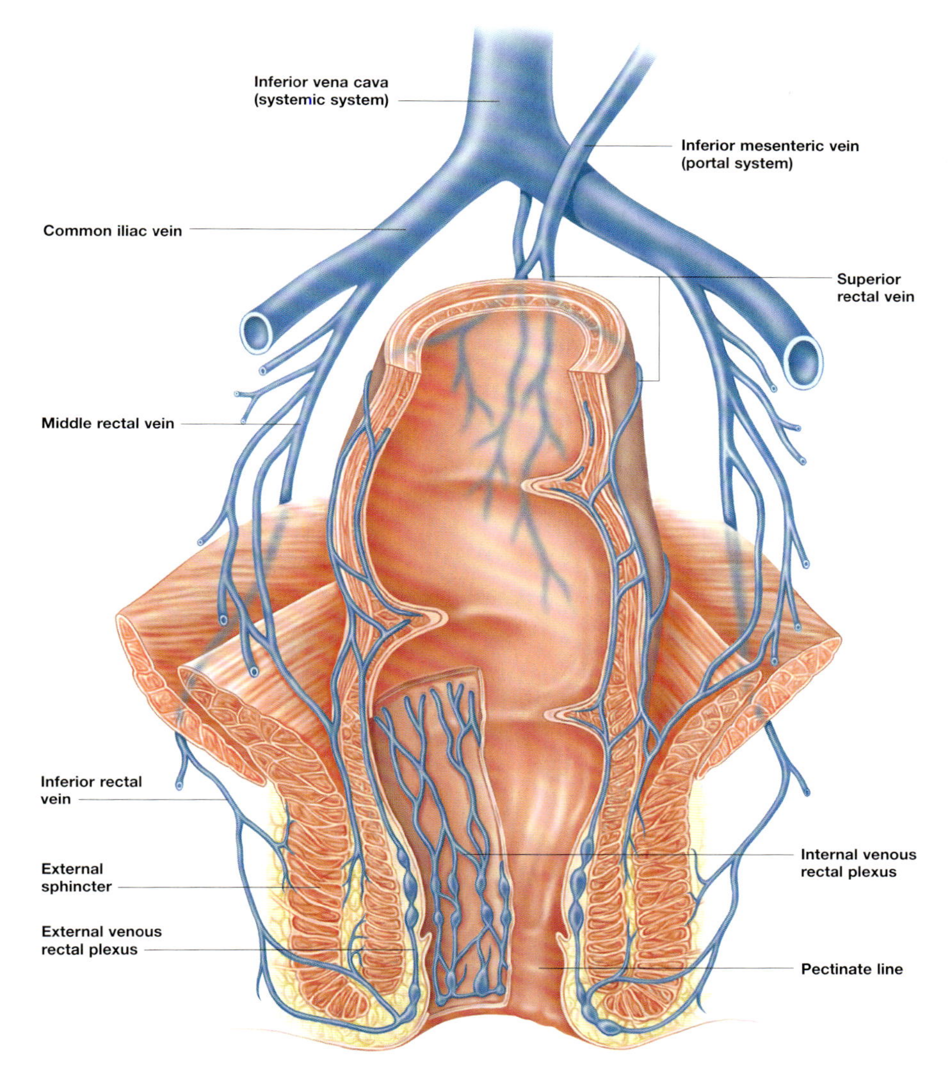

Nerves of the rectum and anal canal

Nerve supply

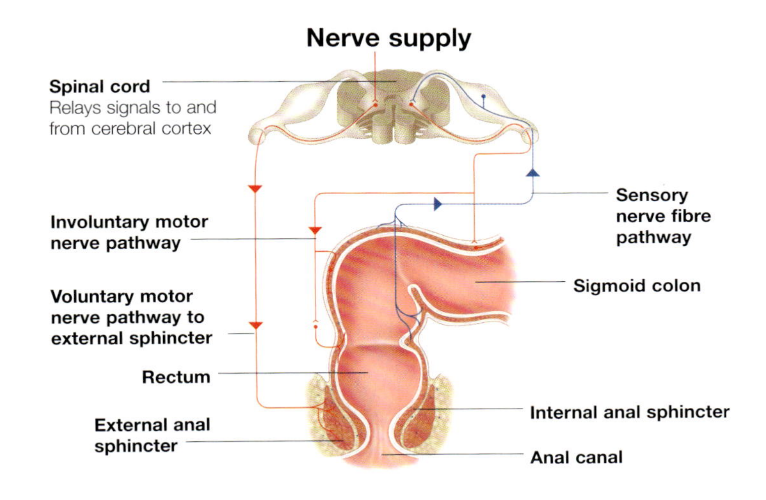

Like the rest of the gastrointestinal tract, the walls of the rectum and anal canal have a nerve supply from the body's autonomic nervous system. This system works 'in the background', usually without us being aware of it, to regulate and control the body's internal functions.

Reflex contraction

These nerves can sense the filling of the rectum and can then cause the reflex contraction of the rectal walls to push the faeces into the anal canal and the relaxation of the internal anal sphincter. However, the anal canal, or more specifically the external anal sphincter, also has a nerve supply from the 'voluntary' nervous system.

These nerves, which originate from the second, third and fourth sacral spinal nerves, allow us to contract the sphincter muscle by an act of will and so prevent filling of the anal canal until an appropriate time for defecation.

When the rectum is full, a defecation reflex is triggered in the spinal cord. Signals are sent to the rectal muscles to start contracting.

Liver

The liver is the largest abdominal organ, weighing about 1.5kg (3lb) in adult men. It plays an important role in digestion and also produces bile, which is secreted into the duodenum.

The liver lies under the diaphragm in the abdominal cavity, on the right side, protected largely by the ribcage. The tissue of the liver is soft and pliable, and reddish brown in colour. It has a rich blood supply from both the portal vein and the hepatic artery.

Lobes of the Liver

Although it has four lobes, functionally, the liver is divided into two parts, right and left, each receiving its own separate blood supply. The two smaller lobes, the caudate and the quadrate, can only be seen on the underside of the liver.

The lobes of the liver are made of small hexagonal lobules, which are the functioning units. Each lobule is made of hepatic cells that are arranged like the spokes of a wheel around a central vein – a tributary of the hepatic vein.

Sinusoids

Blood flows past the hepatocytes and into this central vein through tiny vessels known as sinusoids. The sinusoids receive blood from the vessels of the portal triad, groupings of three vessels that lie at the six points of the lobule. The portal triad is made up of a small branch of the hepatic artery, a small branch of the portal vein and a small biliary duct that collects the bile made by the liver cells.

The sinusoids within each lobule contain tiny specialized cells known as Kupffer cells. These remove debris and worn out blood cells from the blood before it is taken to the heart.

Peritoneal Coverings

The greater part of the liver is covered with the peritoneum, a sheet of connective tissue that lines the walls and structures of the abdomen. Folds of the peritoneum form the various ligaments of the liver.

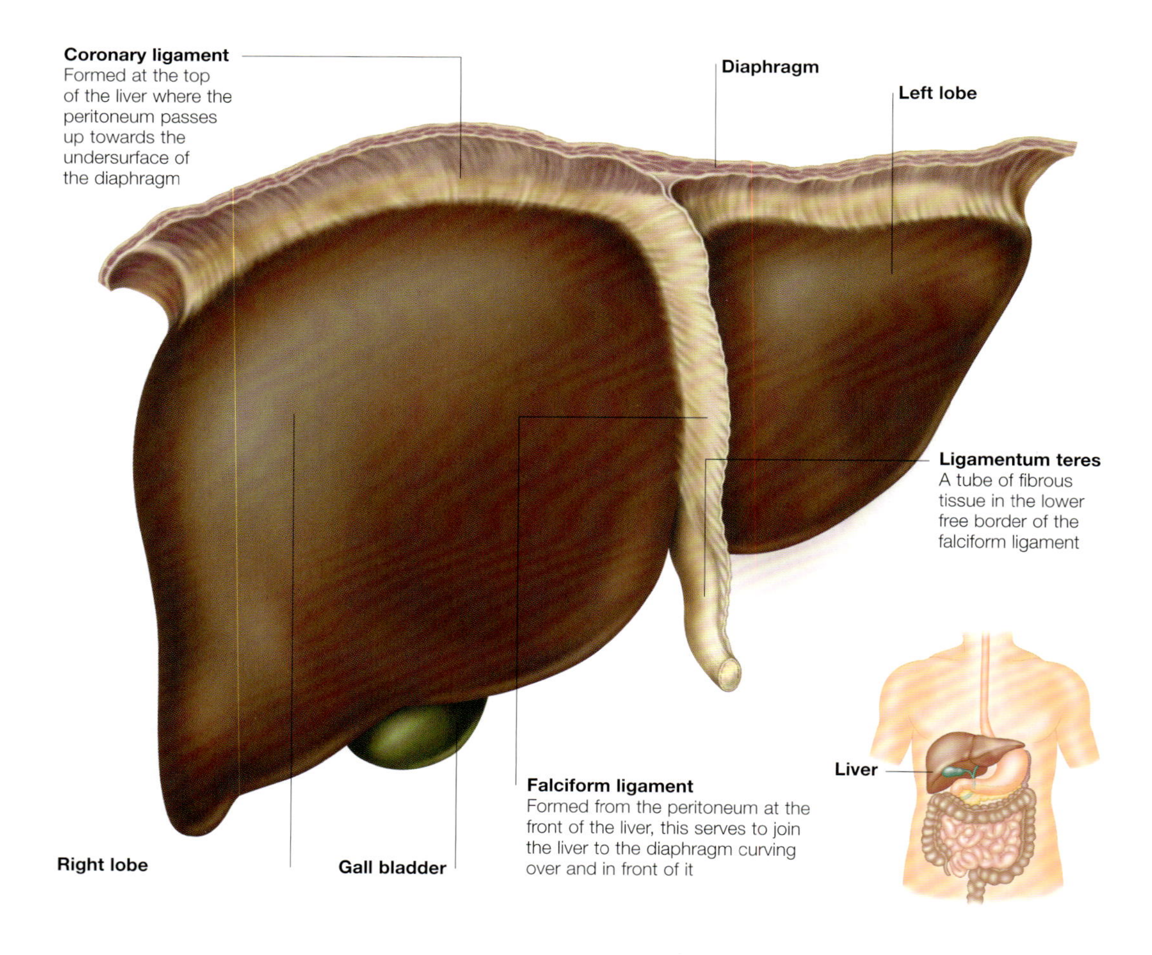

The position of the liver may vary during respiration. It is pushed down on inhalation, and rises again as the breath is exhaled.

Visceral surface of the liver

The underside of the liver is called the visceral surface as it lies against the abdominal organs, or viscera. The impressions of adjacent organs, the related vessels and the positions of the inferior vena cava and gall bladder can be seen.

The liver lies closely against many other organs in the abdomen. Because the tissue of the liver is soft and pliable, these surrounding structures may leave impressions on its surface. The largest and most obvious impressions are seen on the surfaces of the right and left lobes.

Porta hepatis

The porta hepatis is an area that is similar to the hilum of the lungs, in that major vessels enter and leave the liver together clothed in a sleeve of connective tissue, in this case peritoneum.

Structures that pass through the porta hepatis include the portal vein, the hepatic artery, the bile ducts, lymphatic vessels and nerves.

Blood supply

The liver is unusual in that it receives blood from two sources:

- The hepatic artery

This conveys 30 per cent of the liver's blood supply. It arises from the common hepatic artery and carries fresh oxygenated blood. On entering the liver it divides into right and left branches. The right branch supplies the right lobe whereas the left branch supplies the caudate, quadrate and left lobes.

- The hepatic portal vein

This conveys 70 per cent of the liver's blood supply. This large vein drains blood from the gastrointestinal tract, from the stomach to the rectum. Portal blood is rich in nutrients that have been absorbed after digestion in the gut. Like the hepatic artery, it divides into right and left branches with similar distributions. Venous blood from the liver is returned to the heart via the hepatic vein.

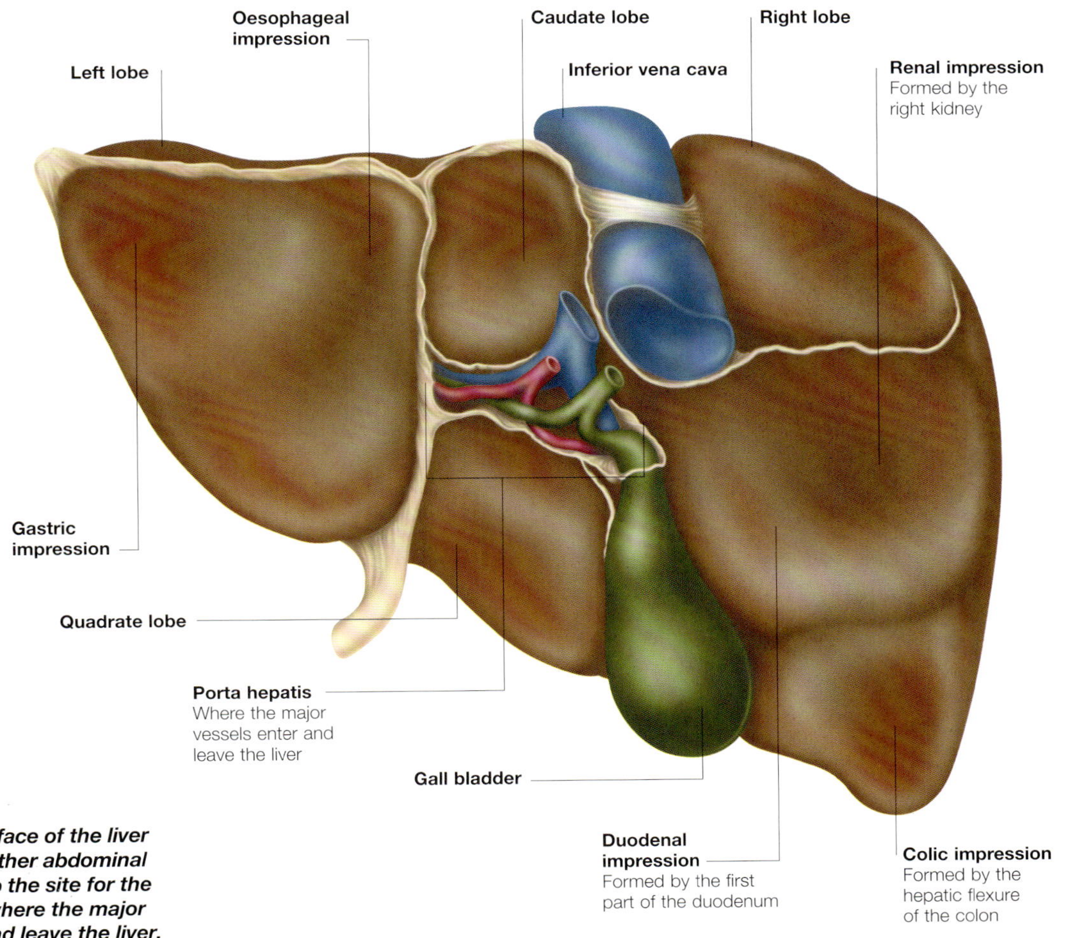

The visceral surface of the liver is indented by other abdominal organs. It is also the site for the porta hepatis, where the major vessels enter and leave the liver.

Biliary System

The biliary system ensures the delivery and storage of bile – a substance that breaks down dietary fats within the small intestine.

Bile, which is essential for the absorption of vitamins D and E and the breakdown of fats, is produced by liver cells and stored in the gall bladder. Bile consists of bile salts and bile pigments, which come from the breakdown of red blood cells, cholesterol and lecithin.

Gall Bladder

This is a pear-shaped sac, about 8cm (3in) long, that lies underneath and extends just below the ninth rib. The lining of the gall bladder is very similar to that of the stomach, as it is formed of mucous membrane that lies in microscopic folds, or rugae. The function of the gall bladder is to store bile and to concentrate it by absorbing water. The presence of fats in the stomach causes the gall bladder to contract, which squeezes bile through the common bile duct and into the duodenum. Here, it emulsifies fats, making them easier to digest.

Passage of Bile

Bile passes from the liver into small bile ducts that merge to form the right and left hepatic ducts. These ducts pass out of the liver through the porta hepatis then unite to form the common hepatic duct.

Common Bile Duct

The common hepatic duct is joined by the cystic duct, forming the common bile duct. This continues down towards the duodenum where, together with the duct carrying secretions of the pancreas, it empties through the major duodenal papilla (or ampulla of Vater) into the small intestine.

When there is no food in the small intestine, a valve around the ampulla of Vater closes and the biliary flow is diverted to the gall bladder for storage.

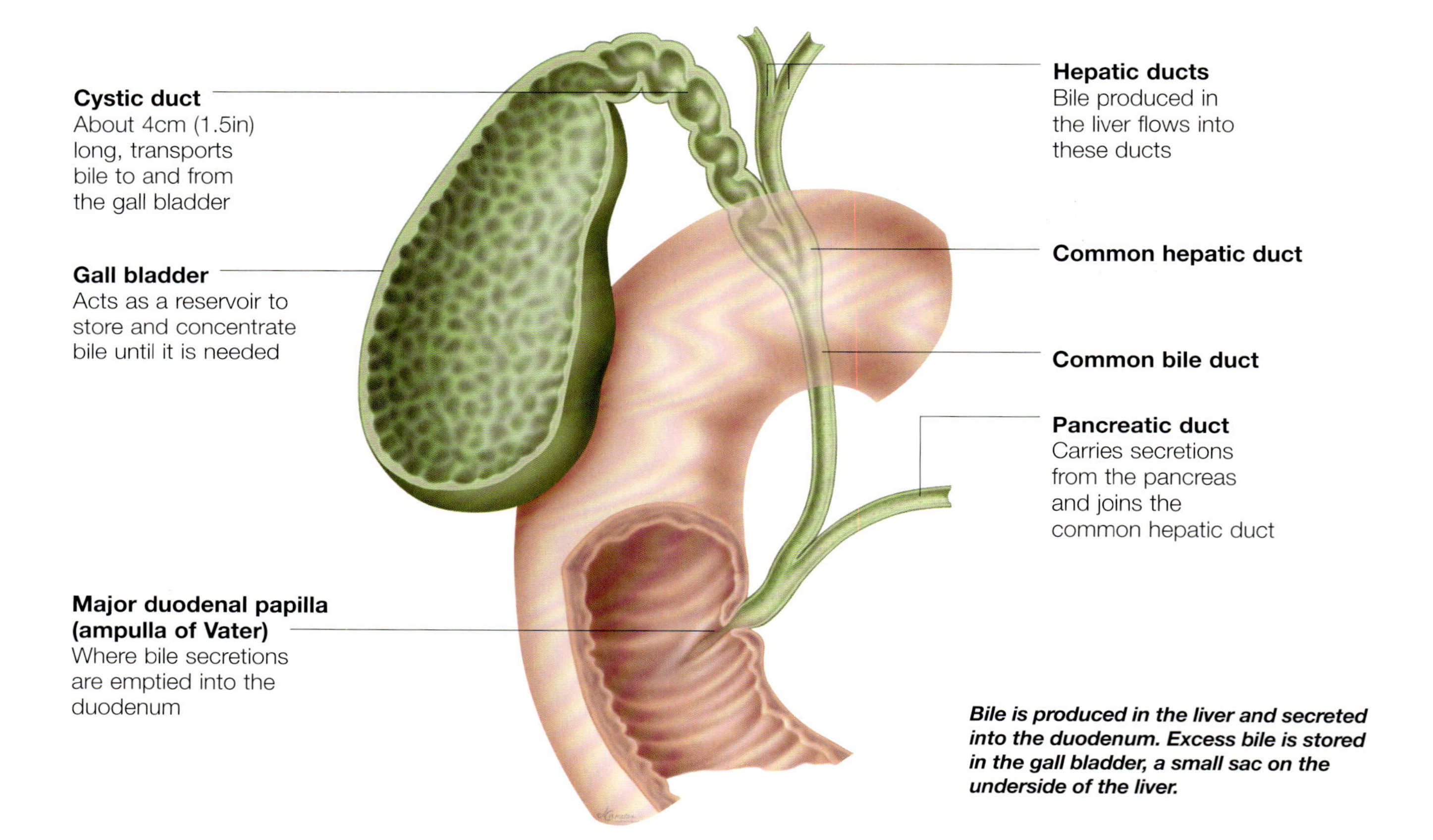

Bile is produced in the liver and secreted into the duodenum. Excess bile is stored in the gall bladder, a small sac on the underside of the liver.

Functions of the Liver

The liver is one of the most complex organs in the body. It controls more than 500 chemical reactions, and manufactures and stores substances that are vital to sustaining life.

The liver consists of four lobes, each one made up of hexagonally shaped areas called lobules, which consist of a central vein surrounded by liver cells.

Hepatic System

The whole structure is permeated by a network of veins, arteries and ducts. The ducts are channels that collect bile, which is produced by liver cells. The veins and arteries form the liver's own circulation system, known as the hepatic portal system. The portal vein collects blood from the digestive tract and delivers it to the liver cells for processing, whereas the hepatic artery branches off the aorta to supply nutrients to the liver cells.

Functions

One of the important functions of the liver is to remove harmful toxins from the general circulation and digestive organs and process them. In addition the liver processes and stores glycogen and fat, and can convert these into sugar when necessary.

The liver cells produce bile, manufacture anti-clotting factors, for example heparin, and manufacture plasma proteins, such as albumin. Old red and white blood cells and some bacteria, are destroyed in the liver. Together with the kidneys, the liver has a role in activating vitamin D. Other functions are listed below.

Microscopic anatomy

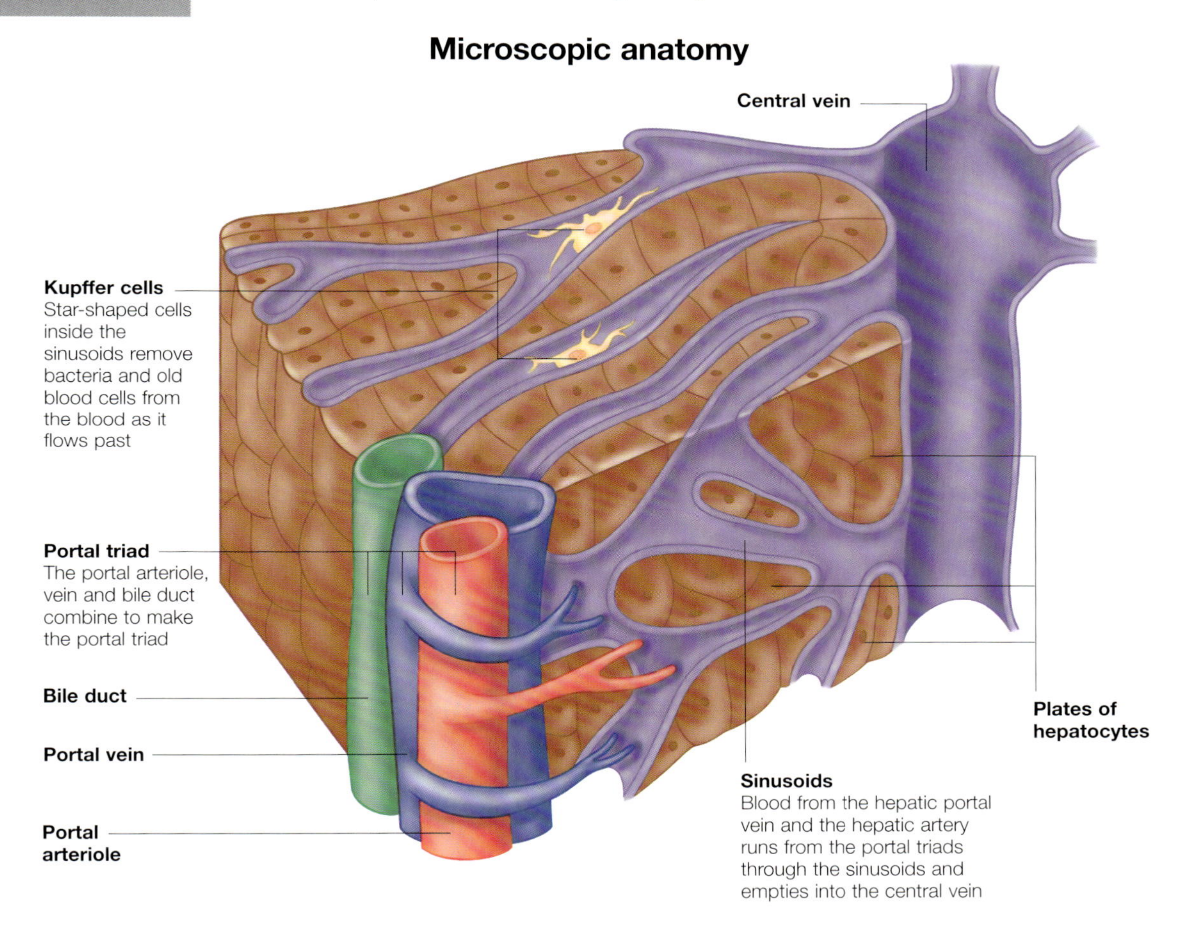

Processes in the liver

The functions of the liver involve the control of more than 500 chemical reactions, making the liver the most important organ of metabolism – that is, the process whereby chemicals are changed in the body.
These include:

- Storing carbohydrates. The liver breaks down glucose, the form in which carbohydrates are carried in the blood, and stores it as glycogen. The process is reversed when blood glucose levels fall or there is a sudden demand for extra energy.
- Disposal of amino acids. The liver breaks down surplus amino acids, which make up proteins, and turn the ammonia that is produced into urea, a constituent of urine.
- Using fat to provide energy. When there is insufficient carbohydrate in the diet to fulfil energy needs, the liver breaks down stored fat into chemicals (ketones), which are used to produce energy and heat.
- Manufacturing cholesterol. Naturally produced cholesterol is essential for the production of bile and hormones such as cortisol and progesterone.
- Storing minerals and vitamins. The liver stores sufficient minerals, such as iron and copper (needed for red blood cells), and vitamins A (which it synthesizes as well as stores), B12 and D to meet the body's requirements.
- Processing blood. The liver breaks down old red blood cells, using some of their constituents to make bile pigments. It also manufactures prothrombin and heparin – proteins that affect blood clotting.

Pancreas

The pancreas is a large gland that produces both enzymes and hormones. It lies in the upper abdomen behind the stomach, one end in the curve of the duodenum and the other end touching the spleen.

The pancreas secretes enzymes into the duodenum, the first part of the small intestine, to aid the digestion of food. Lying across the posterior wall of the abdomen, the pancreas is said to have four parts:

- The head – lies within the C-shaped curve of the duodenum. It is attached to the inner side of the duodenum; a small, hook-like projection, the uncinate process, projects towards the midline.
- The neck – narrower than the head, due to the large hepatic portal vein behind; it lies over the superior mesenteric blood vessels.
- The body – triangular in cross-section and lies in front of the aorta; it passes up and to the left to merge with the tail.
- The tail – comes to a tapering end within the concavity of the spleen.

Blood Supply

The pancreas has a very rich blood supply. The pancreatic head is supplied from two arterial arcades that are formed from the superior and inferior pancreaticoduodenal arteries. The body and tail of the pancreas are supplied with blood by branches of the splenic artery.

Venous blood from the pancreas travels to the liver via the portal venous system; the vein positions mirror the arterial supply.

Pancreatic Enzymes

The pancreas produces about 1.5 litres (53fl oz) of pancreatic juice every 24 hours. The production of juice and its release into the duodenum is regulated by nerve impulses from the stomach. The small intestine also releases hormones that affect its secretion. Pancreatic juice contains water, salts, sodium bicarbonate and enzymes. The alkaline nature of the pancreatic juice helps to neutralize the stomach acids. The enzymes include pancreatic amylase, which digests carbohydrate, trypsin that digests protein and pancreatic lipase, which is the main fat-digesting enzyme in the body.

Location of pancreas

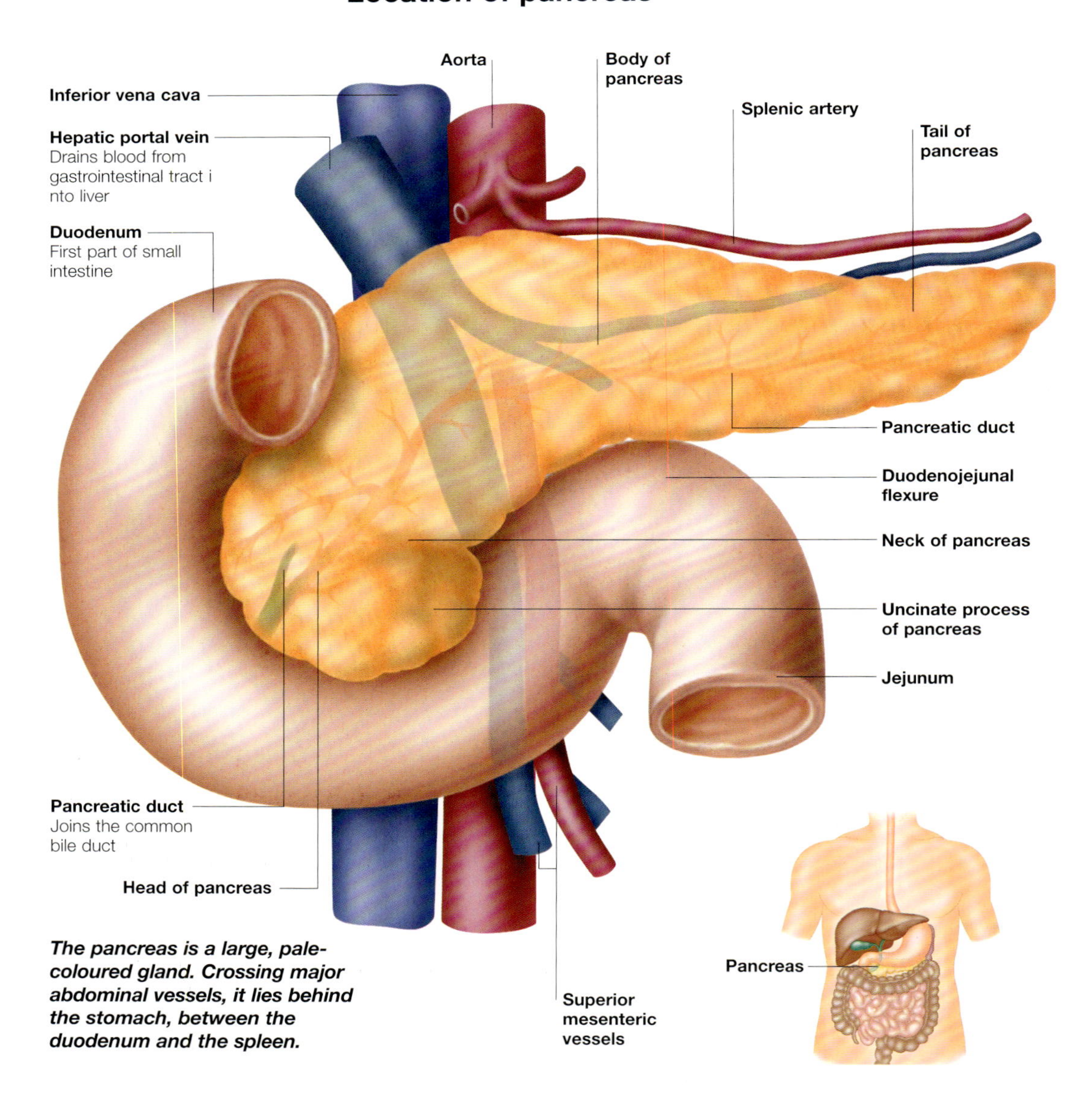

The pancreas is a large, pale-coloured gland. Crossing major abdominal vessels, it lies behind the stomach, between the duodenum and the spleen.

Pancreatic duct and duodenal papilla

Digestive juices produced by cells in the pancreas travel down the pancreatic duct into the small intestine to aid digestion.

The main pancreatic duct, known as the duct of Wirsung, runs the length of the pancreas from the tail to the head, receiving smaller tributaries as it goes. The pancreatic duct transports digestive juices from the secreting cells of the pancreas to the small intestine where they aid digestion.

At the head of the pancreas the pancreatic duct joins with the common bile duct to form a short, dilated tube known as the hepatopancreatic ampulla, or ampulla of Vater. This duct opens to discharge its contents into the duodenum at the tip of the major duodenal papilla, raised areas of the lining of the duodenum, about 10cm (4in) below the pylorus of the stomach.

Muscle fibres

Involuntary muscle fibres run around the walls of the two ducts, and around the wall of the combined duct they form, to provide sphincters (specialized rings of muscle that surround an orifice), which regulate the flow of the ducts' contents into the duodenum.

Accessory duct

Most people have a single pancreatic duct but in some cases there is an accessory pancreatic duct, also known as the duct of Santorini. This accessory duct may have its own, smaller opening into the duodenum, known as the minor duodenal papilla.

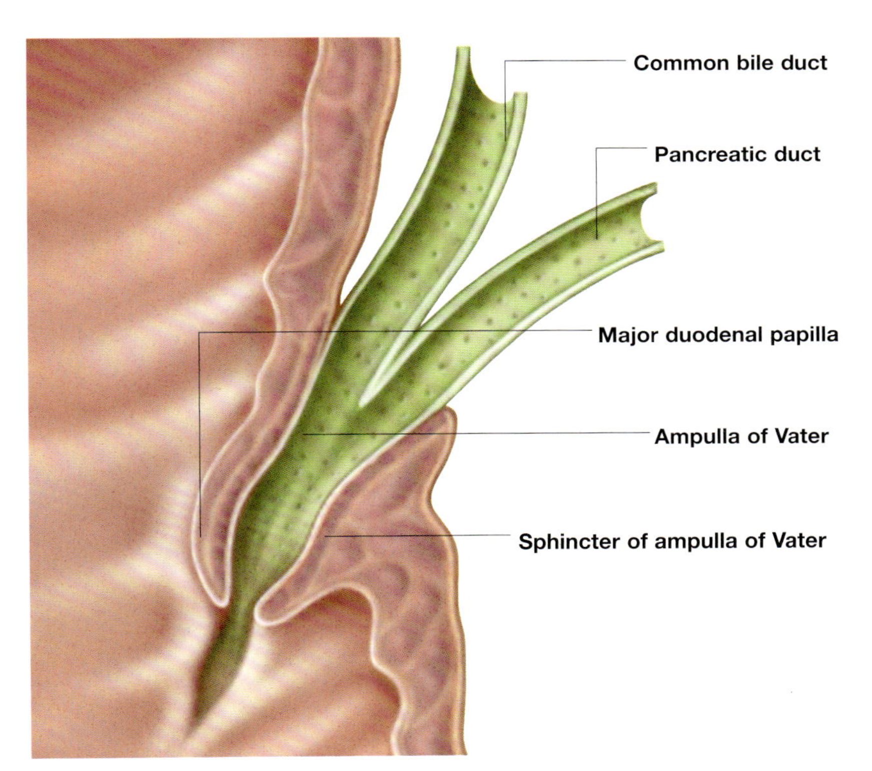

The pancreatic duct and common bile duct join within the head of the pancreas to form the ampulla of Vater. This empties into the duodenum.

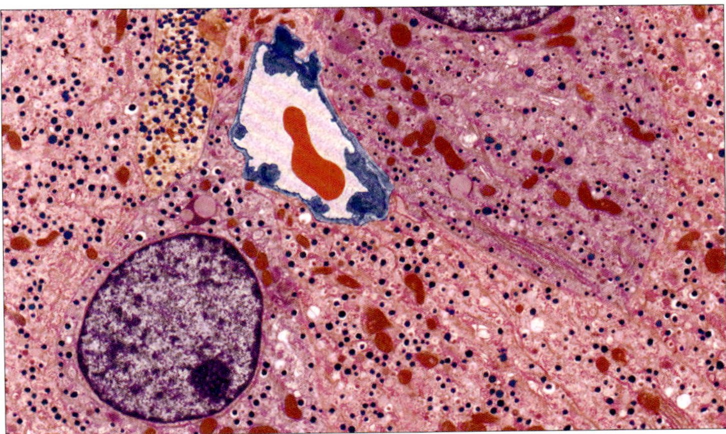

Pancreatic tissue contains many tiny ducts that carry digestive juices towards the duodenum.

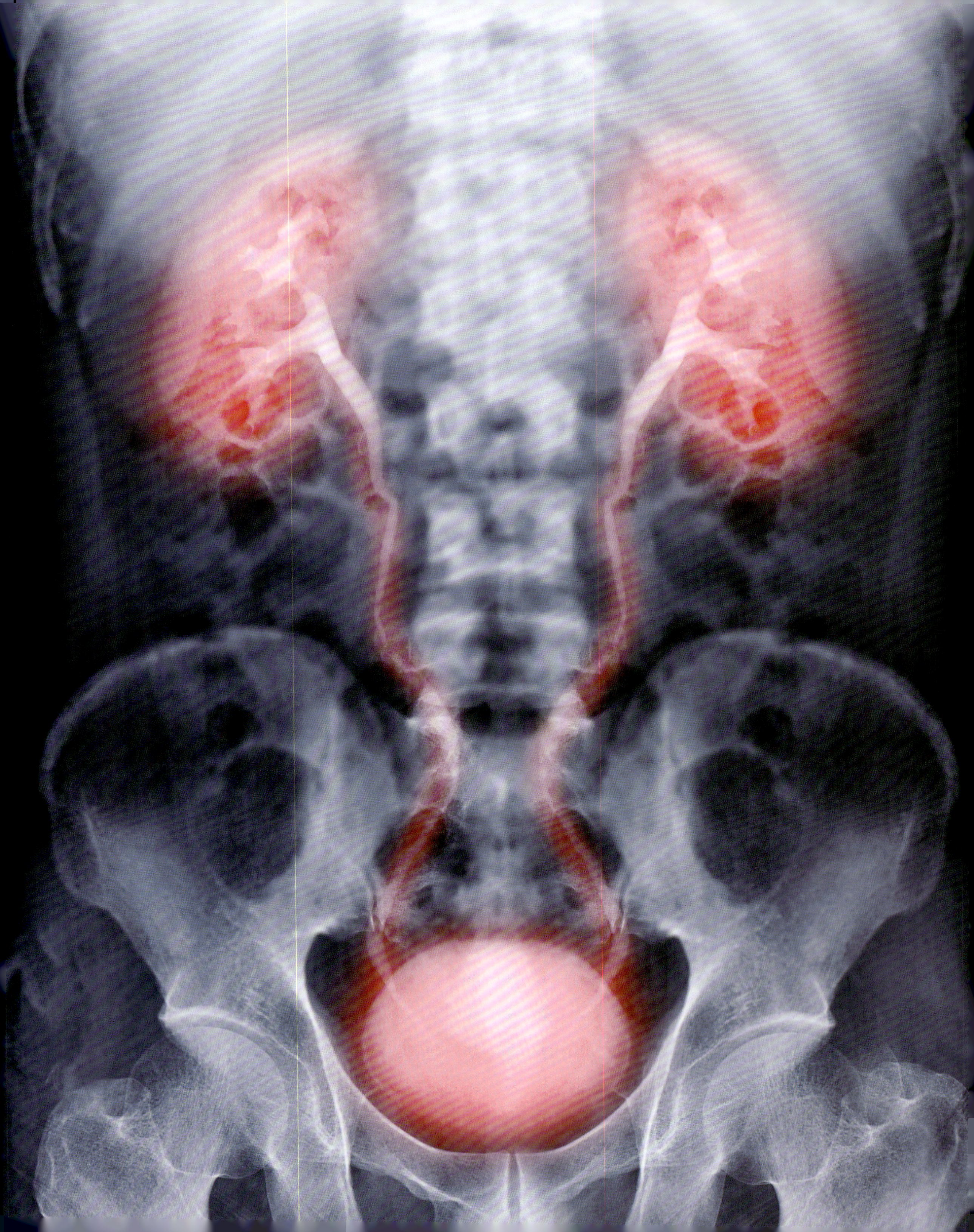

CHAPTER 9

Kidneys and Urinary System

The paired, bean-shaped kidneys are responsible for fluid and electrolyte balance in the body. They also influence other systems and have a role to play in the control of blood pressure and the production of red blood cells. Each kidney contains approximately one million nephrons, the functioning units of the urinary system. These filter about 180 litres (40 gallons) of blood per day, removing metabolic waste products, excess electrolytes and water, which leaves the body as urine. The two ureters transport urine to the bladder where it is stored and then expelled, under conscious control, via the urethra. The delicate process of maintaining an ideal fluid and electrolyte balance is known as homeostasis; in the nephrons, absorption of water and essential nutrients back into the body occurs according to need.

This chapter provides a detailed description of the anatomy of the kidneys, ureters, bladder and urethra and describes the functions of the urinary system.

Opposite: This specialized x-ray of the urinary tract clearly shows the kidneys, ureters and bladder.

Urinary Tract

The urinary tract consists of the kidneys, ureters, urinary bladder and urethra. Together, these organs are responsible for the production of urine and its expulsion from the body.

The kidneys filter the blood to remove waste chemicals and excess fluid, which they excrete as urine. Urine passes down through the narrow ureters to the bladder, which stores it temporarily before it is expelled through the urethra.

Ureters
From the hilus of each of the two kidneys, emerge the right and left ureters. These long narrow tubes receive the urine that is produced continually by the kidney and transport it to the bladder.

Bladder
Urine is received and stored temporarily in the bladder, a collapsible, balloon-like structure that lies within the pelvis.

Urethra
When appropriate, the bladder contracts to expel its contents through the urethra, a thin-walled muscular tube.

The Urinary Tract

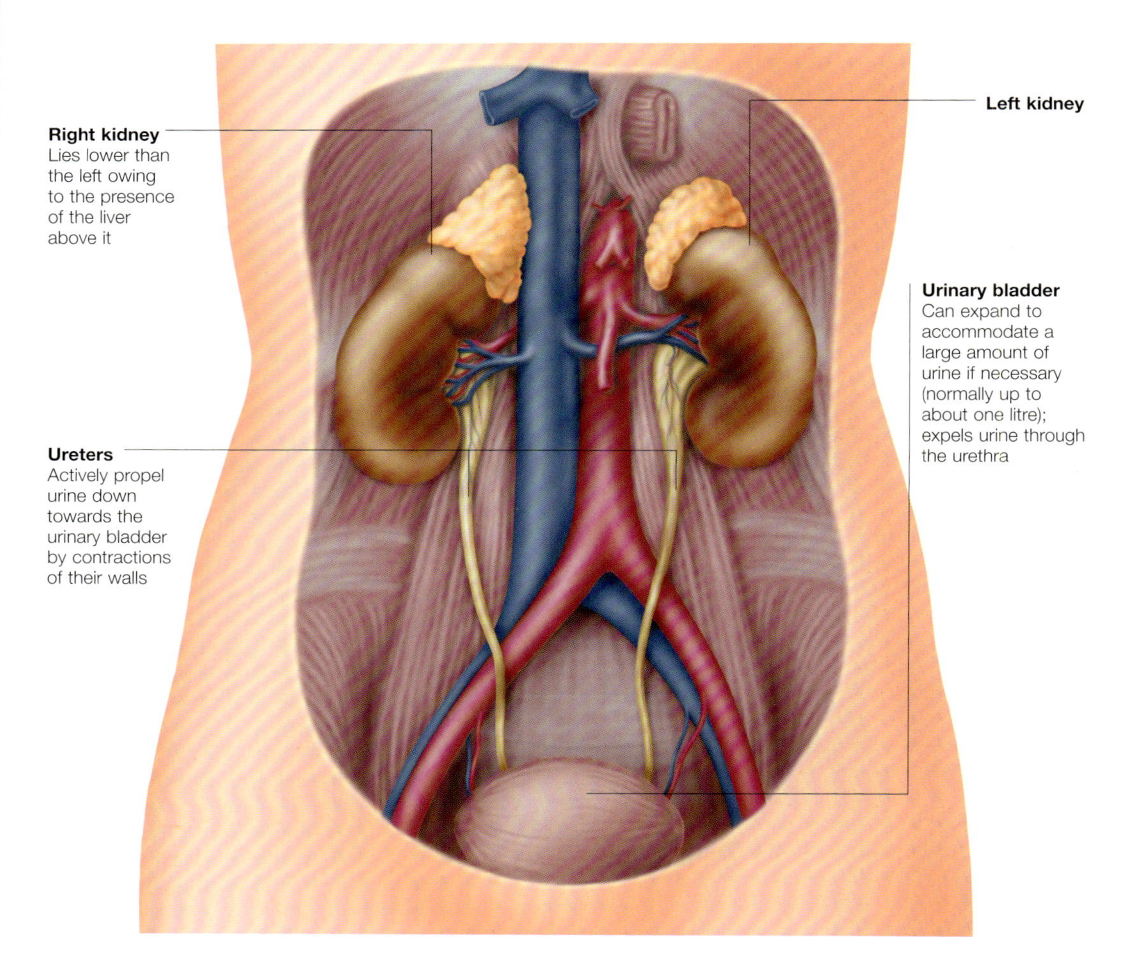

The urinary tract consists of structures involved in the, production, storage and expulsion of urine. It extends from the abdomen into the pelvis.

Rear view of the urinary system

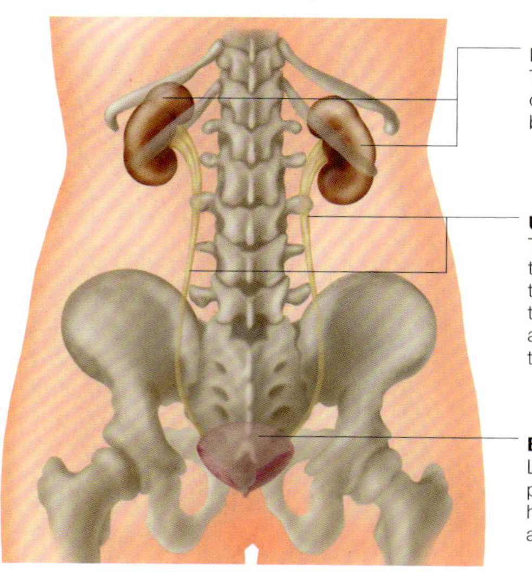

The kidneys lie against the posterior (back) wall of the abdomen, just above the waist, their upper poles lying under the 11th and 12th ribs. They lie outside the peritoneum, the lining of the abdominal cavity. The right kidney lies approximately 2.5cm (1in) lower than the left to allow for the large liver. Both kidneys move up and down during respiration and with changes in posture.

Ureters
These long tubes (one for each kidney) descend down the rear of the abdomen towards the bladder. They are well cushioned as they are buried deep within a dense mass of tissue for protection.

Bladder
The bladder lies protected within the pelvic cavity behind the symphysis pubis. In males, it is situated in front of the rectum whereas in females it lies in front of the vagina.

The urinary tract extends down the length of the abdomen and pelvis. The kidneys lie behind the lower ribs and the bladder is situated on the pelvic floor.

Kidneys

The kidneys are a pair of solid organs situated at the back of the abdomen. They act as filtering units for blood and maintain the balance and composition of fluids within the body.

The paired kidneys lie within the abdominal cavity against the posterior abdominal wall. Each kidney is about 10cm (4in) in length, reddish brown in colour and has a characteristic 'bean' shape. On the medial, or inward facing, surface lies the hilus of the kidney from which the blood vessels enter and leave. The hilus is also the site of exit for the ureters, via which urine is transported to the bladder.

Regions of the Kidney
The kidney has three regions, each of which plays a role in the production or collection of urine:

- Renal cortex – the most superficial layer; it is pale and has a granular appearance
- Renal medulla – composed of dark reddish tissue, it lies within the cortex in the form of 'pyramids'
- Renal pelvis – the central, funnel-like area of the kidney that collects urine and is continuous with the ureters at the hilus.

Outer Layers
Each kidney is covered by a tough, fibrous capsule. Outside the kidney lies a layer of fat that is contained within the renal fascia – a dense connective tissue that anchors the kidneys and adrenal glands to surrounding structures.

Cross-section through the kidney

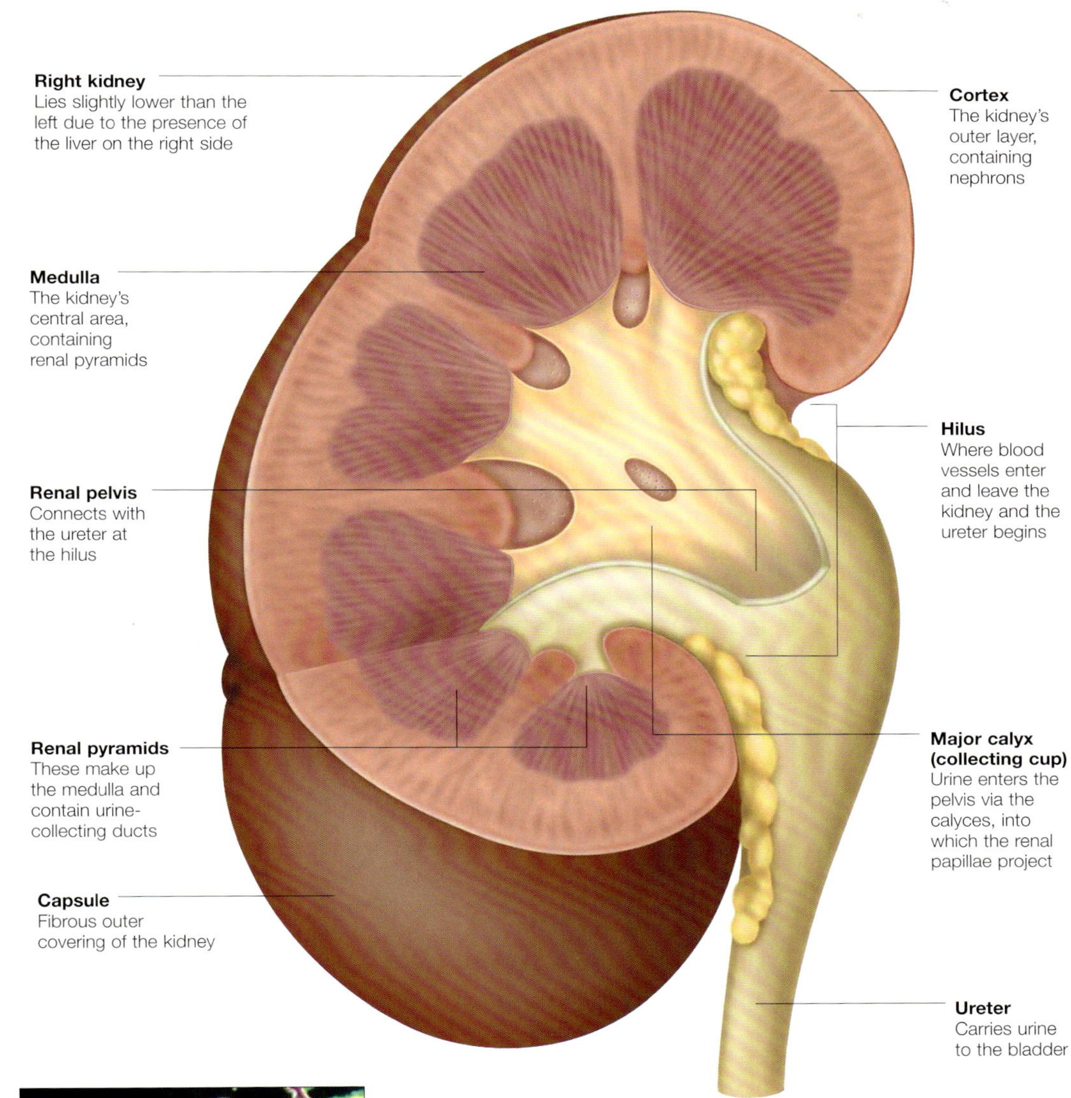

The kidneys are responsible for the excretion of waste from the blood. Each has three regions: cortex, medulla and renal pelvis.

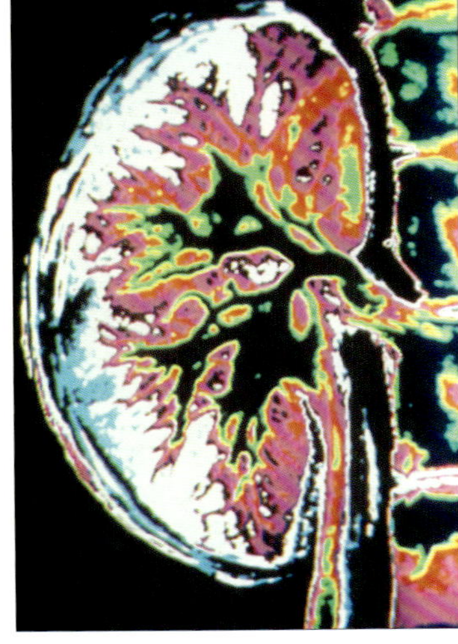

This scan shows the anatomy of the kidney, including the renal pelvis and calyx.

Nephron

There are over one million tiny nephrons, or urine-producing units within the kidney, and from all of these, one millilitre of urine is produced every minute.

Each nephron contains a renal corpuscle within the cortex of the kidney and a long loop of renal tubule, mainly in the medulla:

■ Renal corpuscle
The renal corpuscle is composed of a clump of tiny arterioles, the glomerulus, surrounded by an expanded cup of renal tubule known as Bowman's capsule. Fluid filters out of the blood to enter the renal tubule for processing.

■ Renal tubule
The renal tubule meanders from the Bowman's capsule into the cortex and back again as the loop of Henle. It is here that reabsorption of water and solutes takes place according to need.

Excretion of Waste Products

The waste products of metabolism, such as urea, uric acid, bilirubin and creatinine, are eliminated by the kidneys. The nephron works by a process of secretion followed by reabsorption. Nutrients and waste products flow freely out of the blood in the glomerulus into the Bowman's capsule to be excreted by the body. However, these waste chemicals are accompanied by water and essential nutrients, which must be reclaimed by the body. Most of the reabsorption back into the blood takes place in the distal convoluted tubule and the loop of Henle.

The nephron and its blood supply

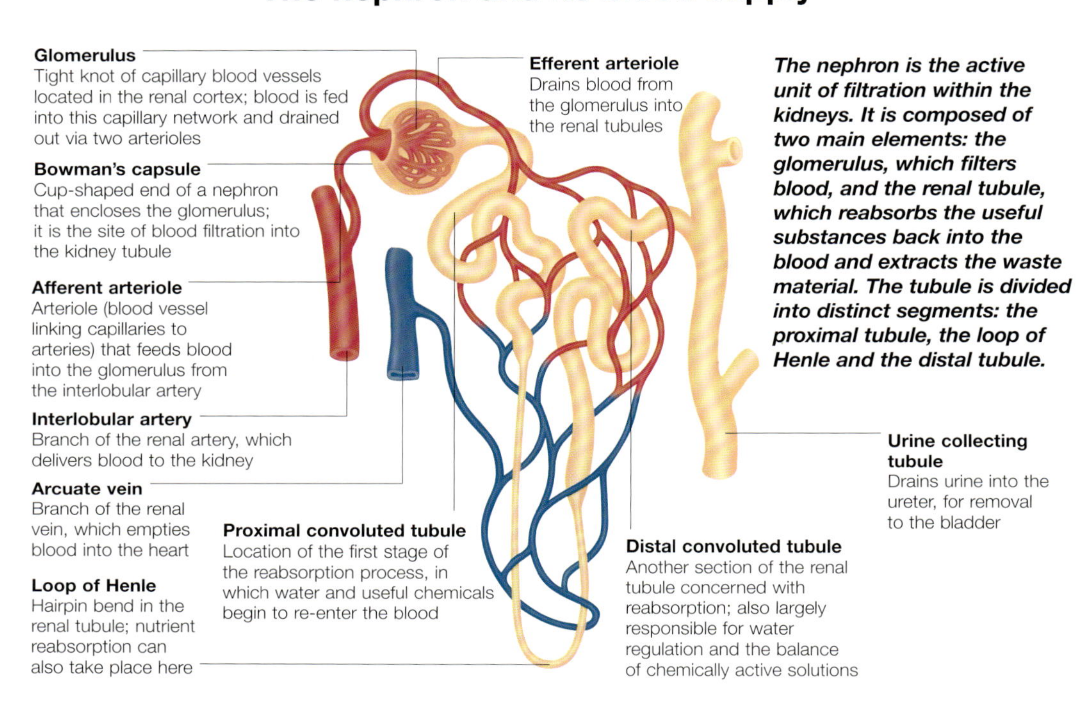

The nephron is the active unit of filtration within the kidneys. It is composed of two main elements: the glomerulus, which filters blood, and the renal tubule, which reabsorbs the useful substances back into the blood and extracts the waste material. The tubule is divided into distinct segments: the proximal tubule, the loop of Henle and the distal tubule.

Capillary networks

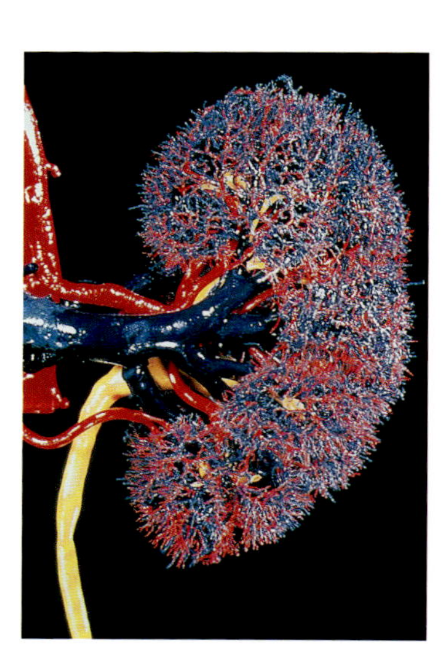

On entering the kidney, the renal artery divides into several branches, each radiating towards the cortex. In the cortex, the branches subdivide repeatedly into smaller and smaller vessels. The final sub-branch is called an arteriole. Each arteriole supplies blood to one nephron.

The anatomy of the arterial blood supply to the kidney nephrons is unique, in that each nephron is supplied by two, rather than one, capillary beds. The arteriole supplying the nephron is known as the efferent arteriole. It is the tight knotting of the resulting capillaries that forms the glomerulus.

On leaving the capillary tuft, the microvessels join together to form the outgoing arteriole, known as the afferent arteriole. This arteriole then redivides into the peritubular capillaries – a second network of microvessels surrounding the urine-collecting tubule further down its length. These capillaries empty into the vessels of the venous system, eventually draining into the renal vein.

The pressure within the glomerulus is high, forcing fluid, nutrients and waste products out of the blood into the nephron capsule. The pressure in the peritubular capillaries is low, allowing fluid reabsorption. Adjustments to the pressure differences between the two capillary beds control the excretion and reabsorption of water and chemicals within the blood.

A cast of a normal kidney shows the complex capillary networks within the organ. There are approximately one million arterioles in each kidney.

Blood supply to the kidneys

The function of the kidneys is to filter blood, for which they receive an exceedingly rich blood supply. As with other parts of the body, the pattern of drainage of venous blood mirrors the arterial supply.

Arterial blood is carried to the kidneys by the right and left renal arteries, which arise directly from the main artery of the body, the aorta. The right renal artery is longer than the left as the aorta lies slightly to the left of the midline. One in three people have an additional, accessory, renal artery.

Renal arteries

The renal artery enters the kidney at the hilus and divides into between three and five segmental arteries, each of which further divides into lobar arteries that enter the kidney as interlobar arteries. There are no connections between branches of neighbouring segmental arteries.

Interlobar arteries

The interlobar arteries pass between the renal pyramids and branch to form the arcuate arteries, which run along the junction of the cortex and the medulla. Numerous interlobular arteries pass into the tissue of the renal cortex to carry blood to the glomeruli of the nephrons, where it is filtered to remove excess fluid and waste products.

Venous drainage

Blood enters the interlobular, arcuate and then interlobar veins before being collected by the renal vein and returned to the inferior vena cava, the main collecting vein of the abdomen.

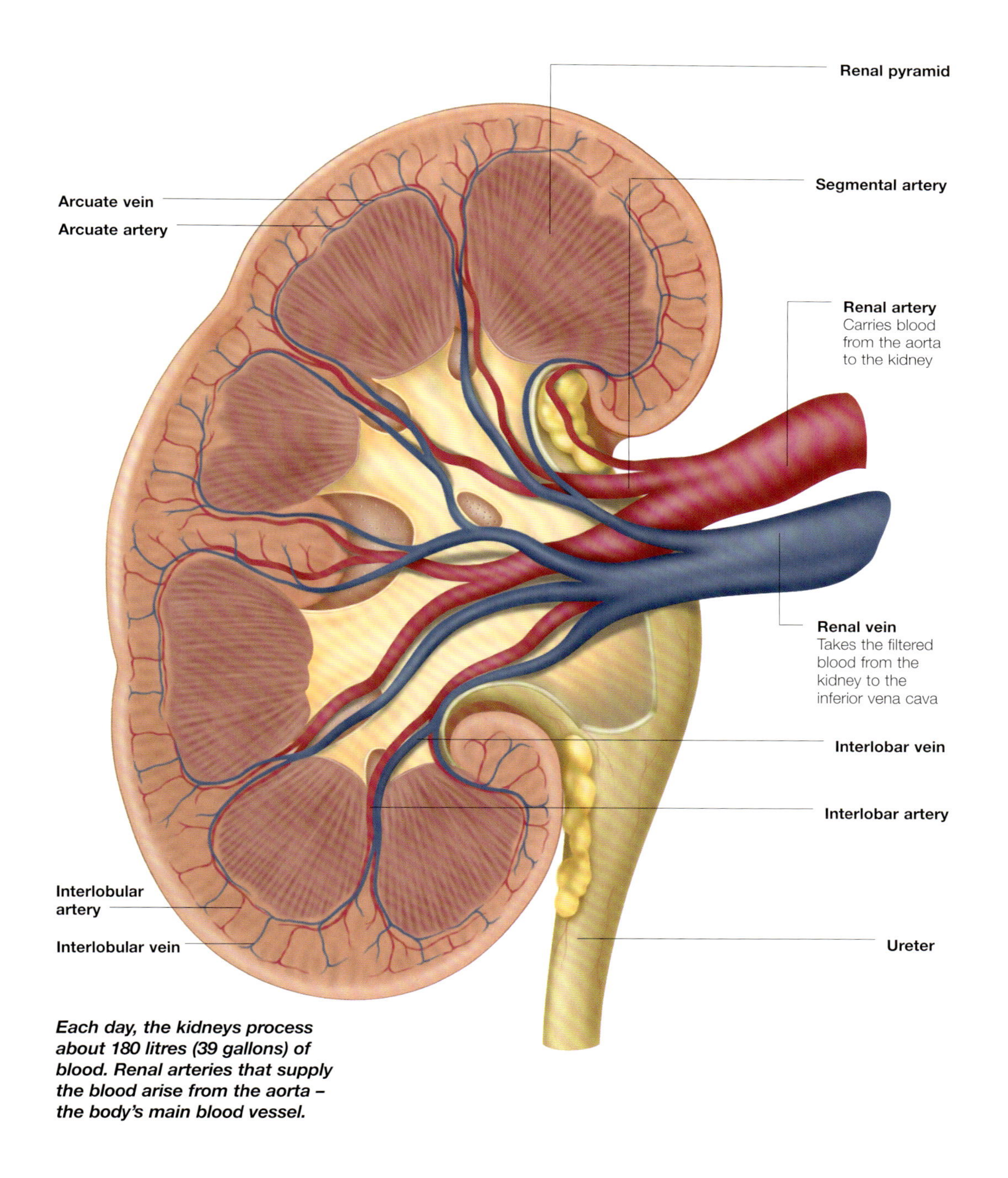

Each day, the kidneys process about 180 litres (39 gallons) of blood. Renal arteries that supply the blood arise from the aorta – the body's main blood vessel.

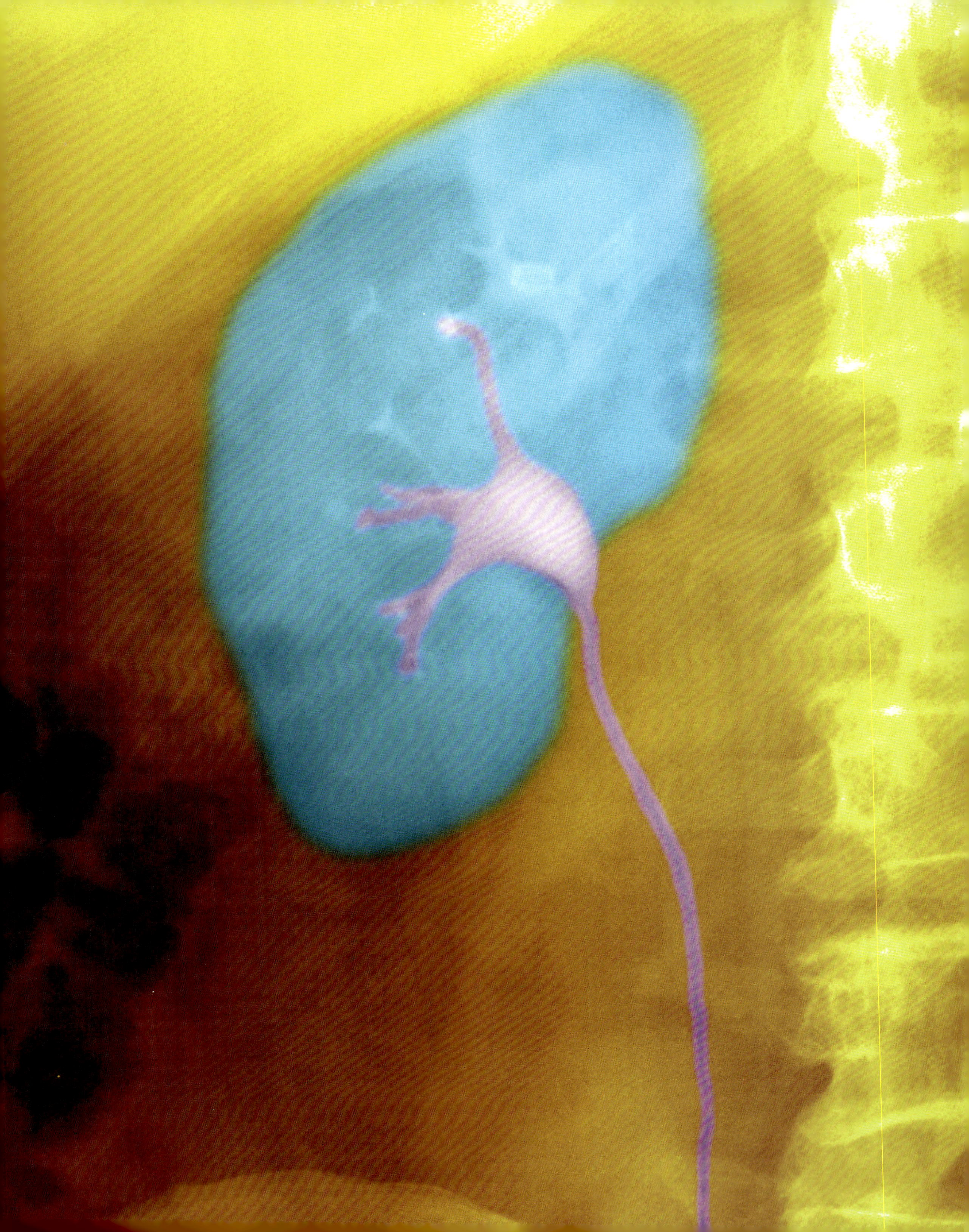

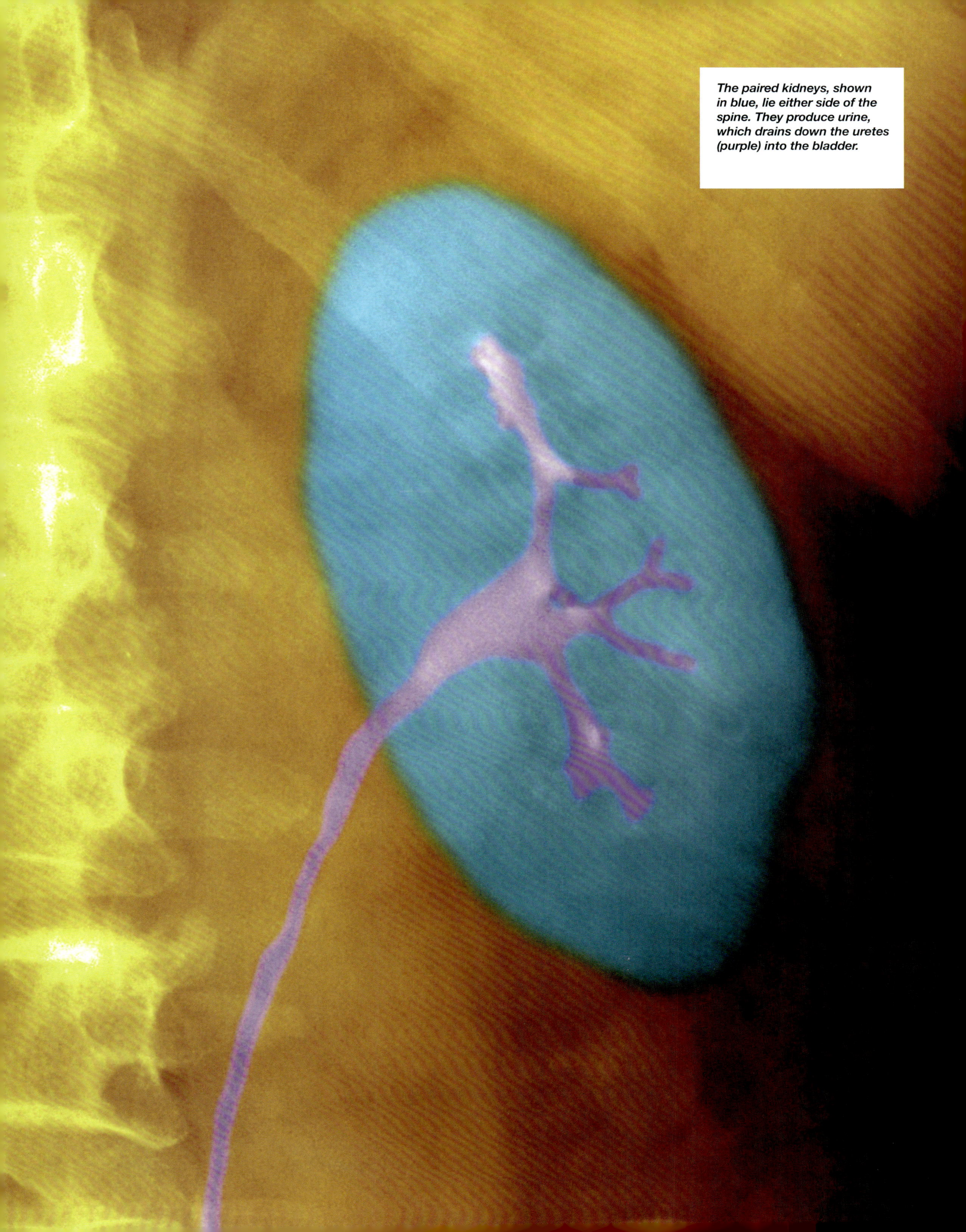

The paired kidneys, shown in blue, lie either side of the spine. They produce urine, which drains down the uretes (purple) into the bladder.

Ureters

The ureters are tubular and propel the urine towards the bladder. Each ureter squeezes and contracts its muscles to encourage the free flow of urine.

The ureters are narrow, thin-walled muscular tubes that actively propel urine from the kidneys and on to the urinary bladder.

Each of the two ureters is 25–30cm (9.8–11.8in) in length and about 3mm (0.11in) wide. They originate at the kidney and pass down the posterior abdominal wall to cross the bony brim of the pelvis and enter the bladder by piercing its posterior wall.

Parts of Ureter

Each ureter consists of three anatomically distinct parts:

- Renal pelvis

This is the first part of the ureter, which lies within the hilum of the kidney. It is funnel-shaped as it receives urine from the major calyces and then tapers to form the narrow ureteric tube. The junction of this part of the ureter with the next is one of the narrowest parts of the whole structure.

- Abdominal ureter

The ureter passes downwards through the abdomen and then slightly towards the midline until it reaches the pelvic brim and enters the pelvis. During its course through the abdomen the ureter runs behind the peritoneum, the membranous lining of the abdominal cavity.

- Pelvic ureter

The ureter enters the pelvis just in front of the division of the large common iliac artery. It runs down the back wall of the pelvis before turning to enter the posterior wall of the bladder.

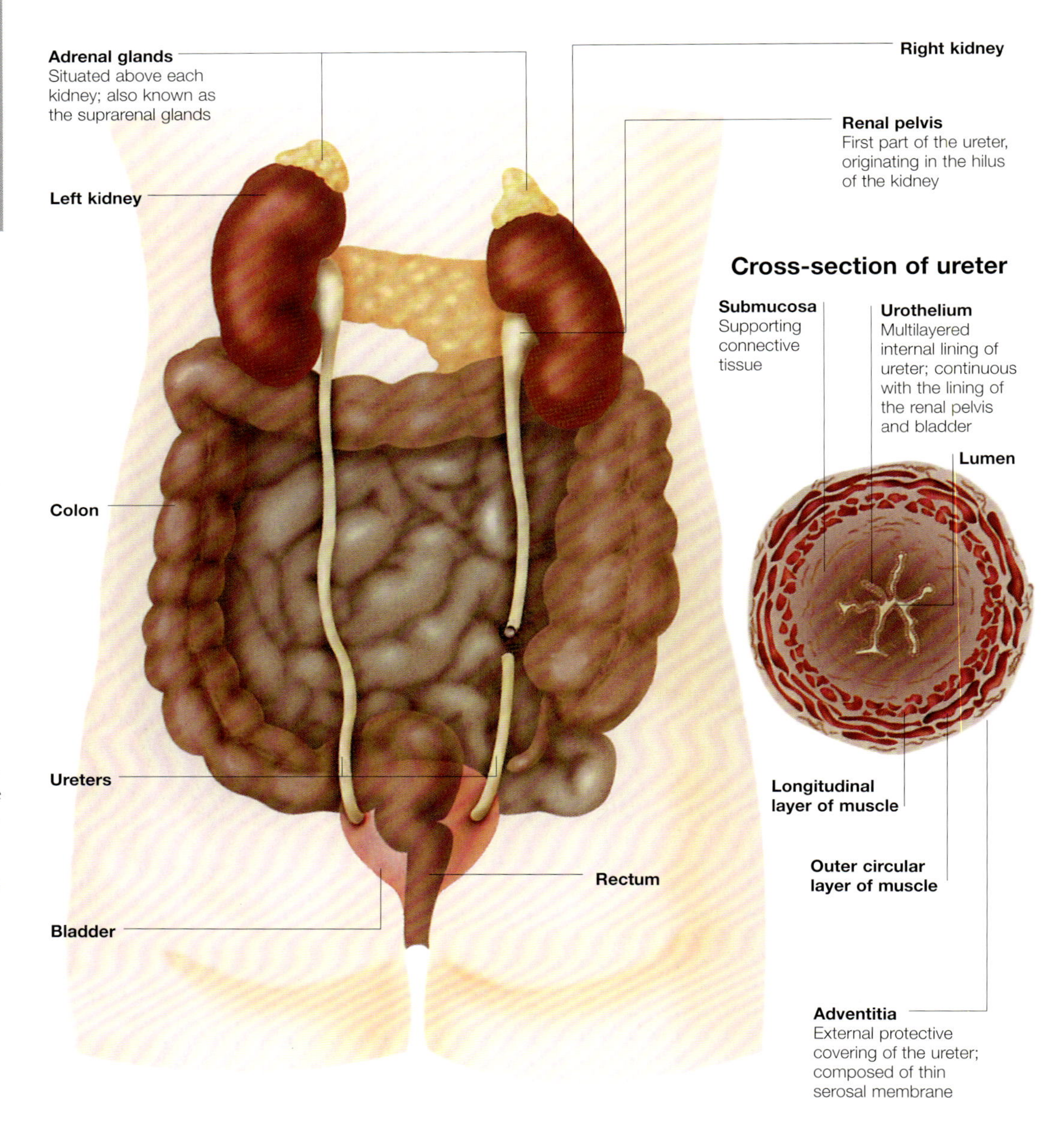

Urine is actively propelled along the ureters to the bladder by contraction of the muscular walls. This is the action known as 'peristalsis'.

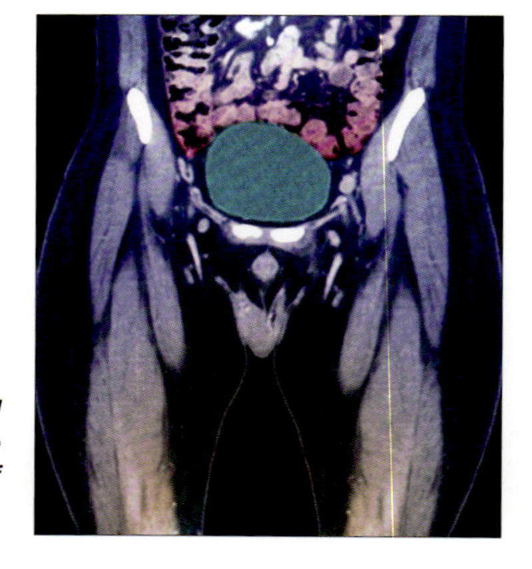

Coloured computed tomography (CT) scan of the healthy full bladder (green) of a 45-year-old patient.

Bladder and Urethra

The ureters channel urine produced by the kidneys down their length and into the urinary bladder. Urine is stored in the bladder until it is expelled from the body via the urethra.

Urine is continuously produced by the kidneys and is carried down to the urinary bladder by two muscular tubes, the ureters.

Bladder

The bladder stores urine until it is passed out via the urethra. When the bladder is empty, it is pyramidal in shape, its walls thrown into folds, or rugae, which flatten out on filling. The position of the bladder varies:

- In adults the empty bladder lies low within the pelvis, rising up into the abdomen as it fills
- In infants the bladder is higher, being within the abdomen even when empty
- The bladder walls contain many muscle fibres, collectively known as the detrusor muscle, which allow the bladder to contract and expel its contents.

Trigone

The trigone is a triangular area of the bladder wall at the base of the structure. The wall here contains muscle fibres that act to prevent urine from ascending the ureters when the bladder contracts. A muscular sphincter around the urethral opening keeps it closed until urine is passed out of the body.

Coronal section of female bladder and urethra

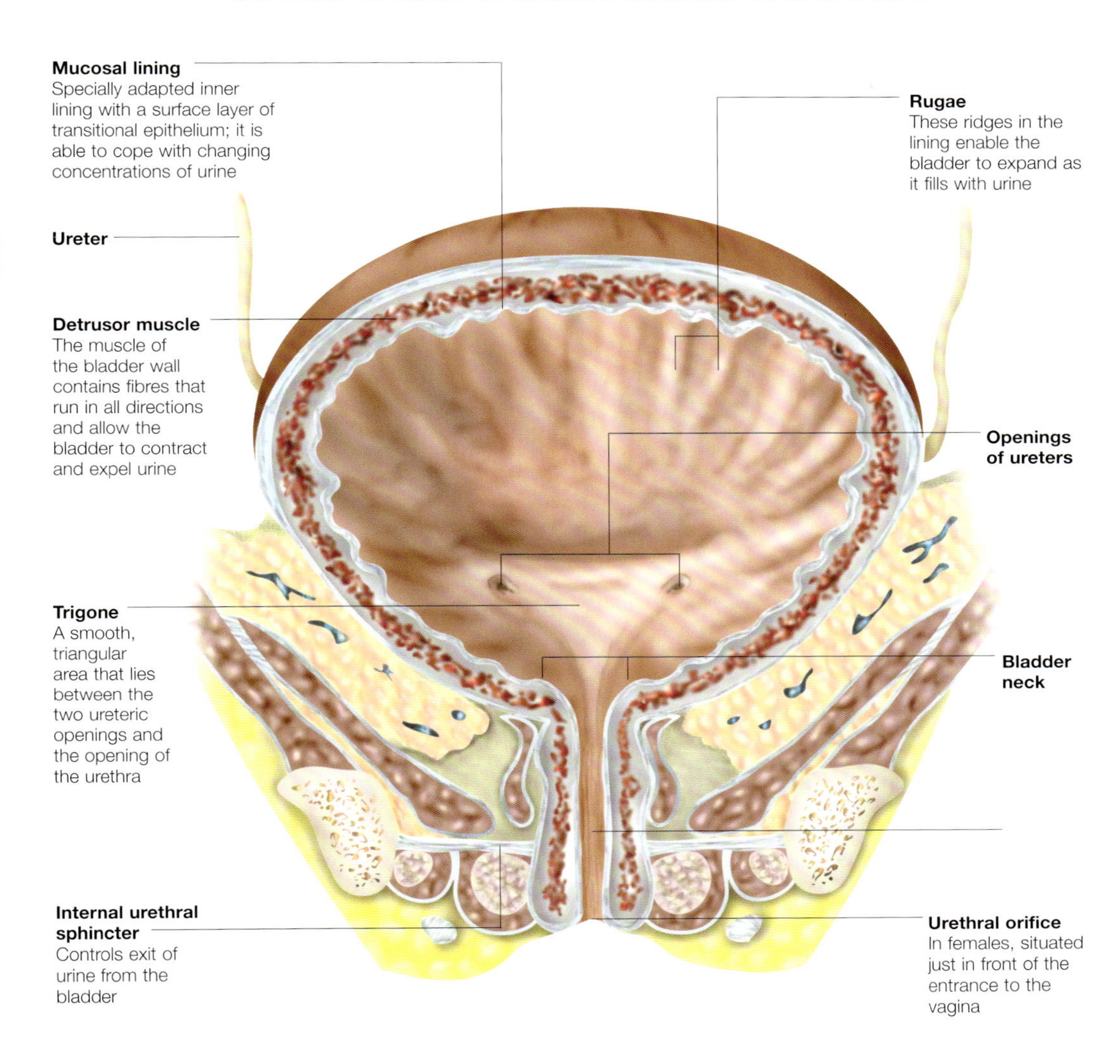

The urinary bladder is flexible enough to expand as it fills. It is made from strong muscle fibres that facilitate the expulsion of urine when necessary.

Differences in male and female anatomy

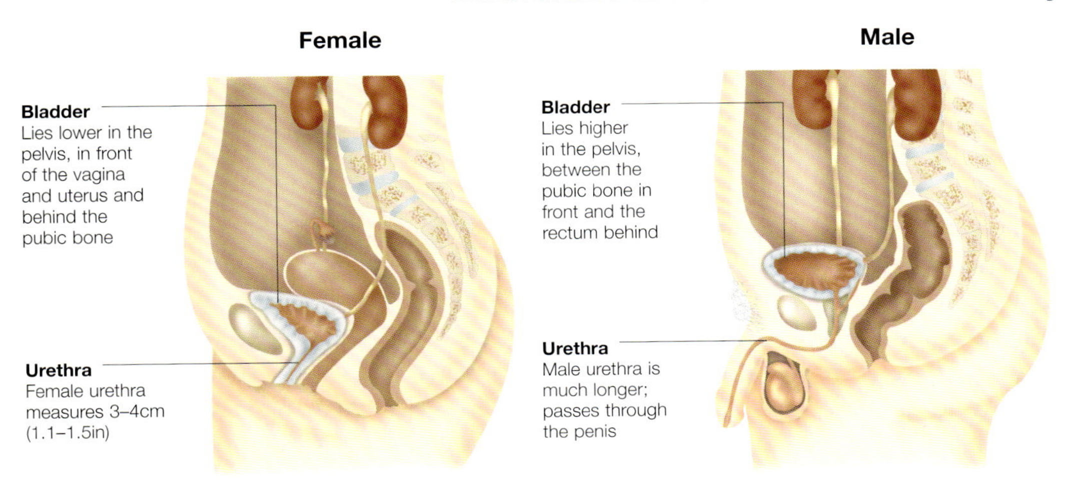

Due to the presence of reproductive organs the bladder position, and size, shape and position of the urethra vary between males and females:

- In men the urethra is about 20cm (7.8in) long, passing through the prostate gland and then running along the penis before opening at the external urethral orifice
- In women the urethra is 3–4cm (1.1–1.5in) in length and opens at the urethral orifice, which lies just in front of the vaginal opening.

The major difference in male and female urinary tract anatomy is the length of the urethra. An adult male urethra is five times the length of a female one.

Maintaining Homeostasis

The kidneys maintain the volume and chemical composition of bodily fluids. They do this by filtering impurities from the blood and excreting excess water and metabolic by-products as urine.

The kidneys are responsible for maintaining the constancy of body fluids by filtering toxins, metabolic waste products and excess ions from the blood. At the same time, the kidneys also maintain blood volume (the correct balance of water and salts) and the correct acidity of body fluids. This complex process is known as homeostasis.

The production of urine is a three-stage process: filtration, reabsorption and secretion. Once the required water and essential nutrients have been reabsorbed into the body, the remaining fluid is urine, which empties into the ureters to be excreted.

Functioning Units

There are over one million blood processing units within the kidney that are called nephrons. It is within the nephrons that the complex processes of filtering the blood, reabsorbing essential nutrients and water and disposing of waste products takes place. These tiny units 'straddle' the renal cortex (outermost layer) and the middle zone, the medulla. The glomeruli of the nephrons are found in the cortex, while their collecting tubules drain processed urine into the renal pelvis via the pyramids in the medulla.

Internal structure of the kidney

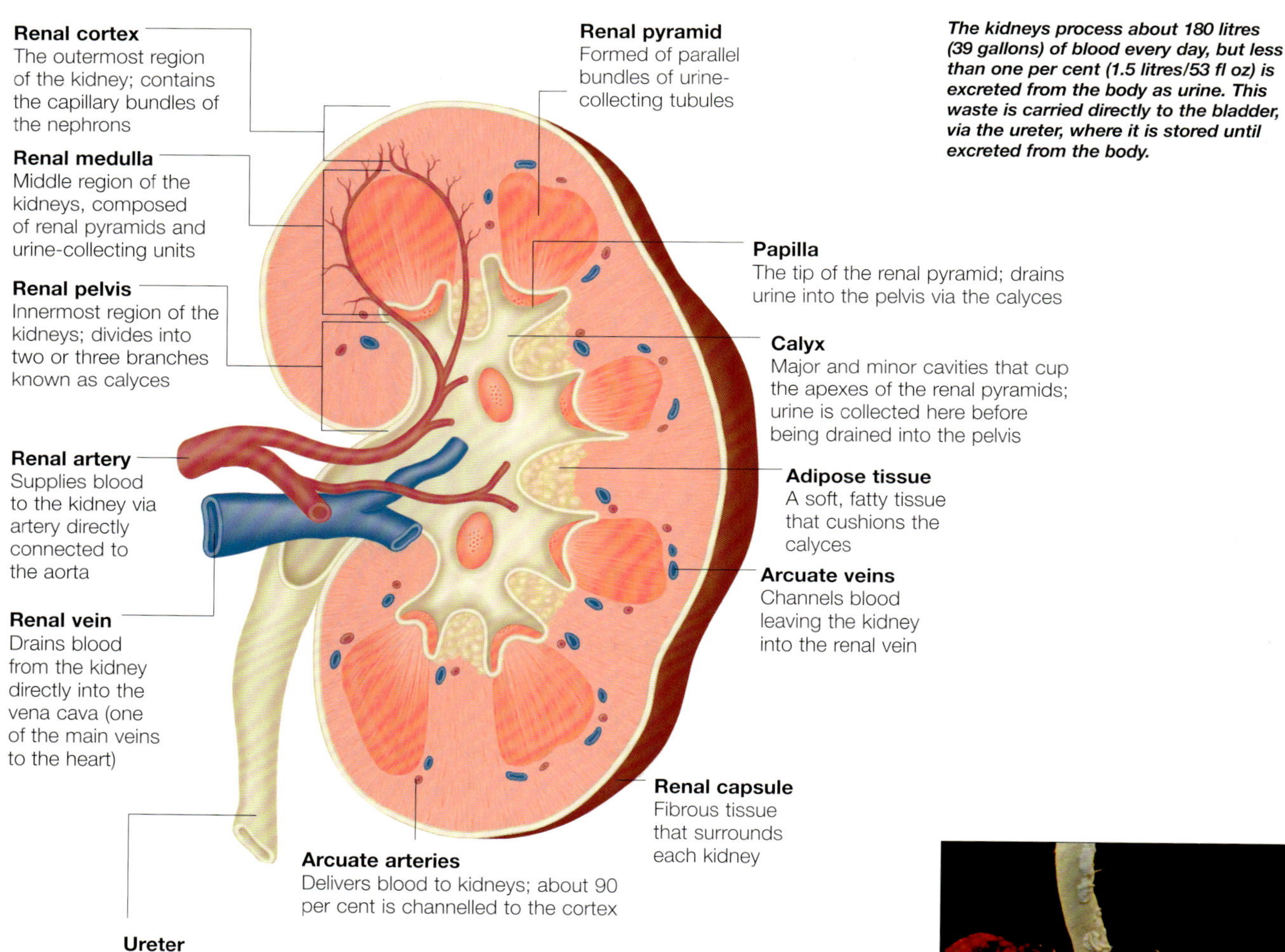

The kidneys process about 180 litres (39 gallons) of blood every day, but less than one per cent (1.5 litres/53 fl oz) is excreted from the body as urine. This waste is carried directly to the bladder, via the ureter, where it is stored until excreted from the body.

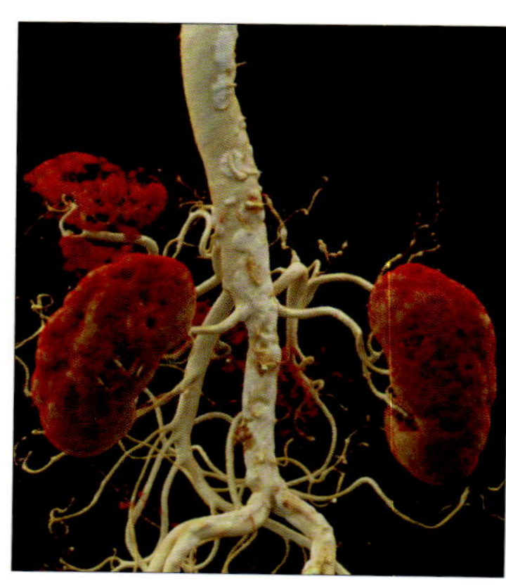

This CT angiography shows the abundant arterial blood supply to the kidneys.

Control of blood pressure

Blood volume is a direct indicator of blood pressure. The kidneys continually monitor blood absorption and sodium levels to maintain an even pressure.

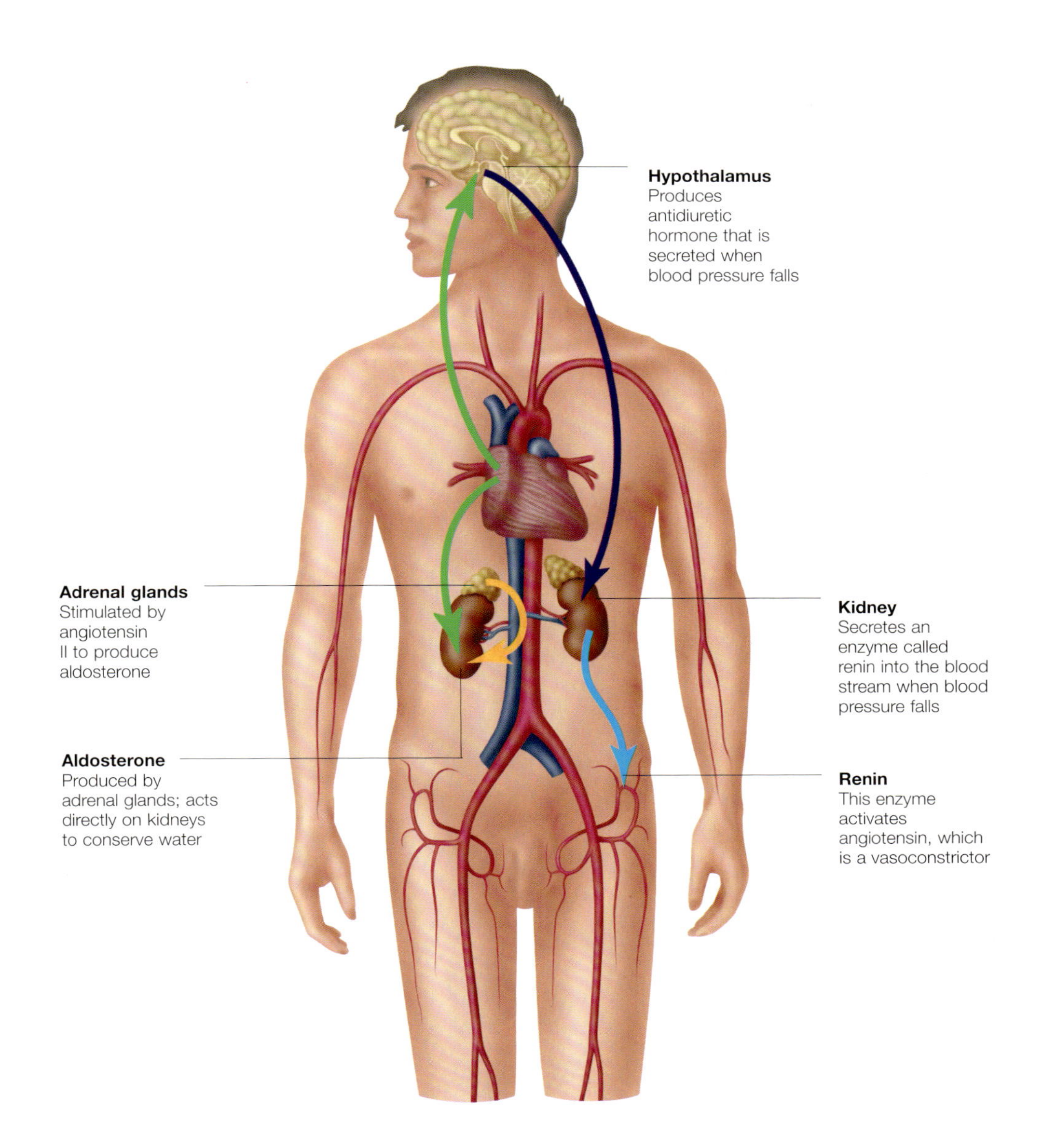

The kidneys regulate blood pressure by altering the amount of urine passed, thereby regulating blood volume. When blood pressure is low, the kidneys conserve water in the circulation and when it is raised, they ensure that greater volumes of water are passed as urine.

Filtration rate

Within each nephron (functional unit of the kidney) is a bundle of arterioles (blood vessels) called the glomerulus. Water and solutes are 'pushed' out of the blood into the collecting tubules by the higher blood pressure in the glomerulus. An average person filters about 125ml (4.3fl oz) of filtrate per minute.

If the blood pressure is too low, water will be retained in the circulation to help boost the blood pressure. If the blood pressure is high, more water is forced into the tubules and passed as urine.

Feedback mechanism

The walls of the blood vessels supplying the nephrons contain specialized cells that are able to detect blood pressure. It is these cells that set into motion additional processes needed to rectify abnormal pressure:

- The blood pressure falls below normal limits and the specialized cells detect this change
- A hormone called renin is secreted into the blood stream
- Renin converts a substance called angiotensin into angiotensin I, which then becomes angiotensin II as it passes through the lungs in the blood
- Angiotensin II stimulates the adrenal glands (located on the top of the kidneys) to produce aldosterone
- Aldosterone acts directly on the nephrons in the kidneys so that more salt and water are reabsorbed back into the blood circulation. This results in an increase in blood pressure
- In addition to this mechanism, angiotensin II constricts blood vessels, thus increasing the pressure within them.

Antidiuretic hormone

The hypothalamus in the brain also has a role to play. When the water concentration in the blood is low, potentially leading to a drop in blood pressure, the hypothalamus secretes antidiuretic hormone (ADH). This acts on the tubules in the nephrons, making them more permeable so that more water is reabsorbed into the blood.

Influencing factors

The normal blood pressure of a resting adult is usually around 120/80 mmHg, but this can be influenced by a wide range of factors:

- Age

Blood pressure naturally increases throughout life. This is because the arteries lose the elasticity that, in younger people, absorbs the force of heart contractions

- Gender

Men generally experience higher blood pressure than women or children

- Lifestyle choices

Being overweight, consuming high levels of alcohol or enduring a long period of stress can all contribute to high blood pressure.

Hypertension

Abnormally high blood pressure (hypertension) may be caused by a number of factors, but is commonly caused by atherosclerosis, a disease that causes narrowing of the blood vessels. When the disease affects the arteries of the kidney (renal arteries), it may cause long-term problems with blood pressure regulation.

Hypotension

Abnormally low blood pressure (hypotension) is usually due to reduced blood volume or increased blood vessel capacity. This can happen in the case of severe burns or dehydration, which both lower blood volume, or through an infection such as septicaemia, which causes a widening of the blood vessels.

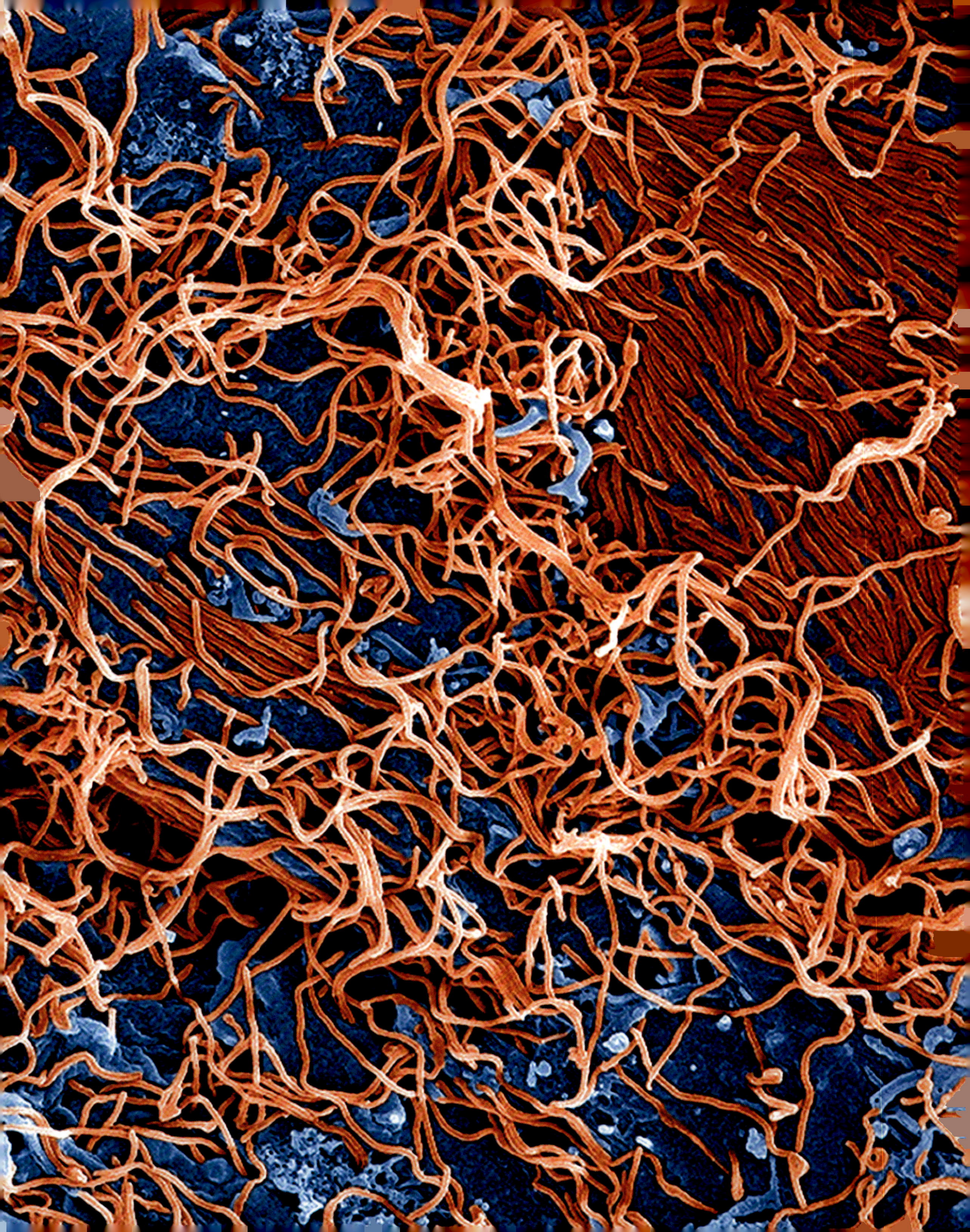

CHAPTER 10

The Immune and Lymphatic Systems

Our immune system protects us from infection, providing us with the means to ward off invading micro-organisms that could potentially cause infection or cancers. Many organs contribute to this defence system, including the skin, which provides overall protection, and the lymphatic organs and tissues.

The body's immune system acts in several ways: some blood cells engulf invaders, whereas others secrete antibodies that eliminate them or destroy them with toxins. The lymphatic system has an important immune function in that it contains lymphocytes – white cells that recognize foreign cells and mount an attack. The lymphatic system is also responsible for draining fluid from the space between the cells and transporting fats around the body.

Opposite: Our immune system protects us against potentially life threatening infections, such as Ebola virus (shown here in red).

Protection from Disease

As well as carrying nutrients to and waste products away from all tissues of the body, blood contains components that are a vital part of the human immune response to infection.

Blood is the great defensive fluid of our bodies. It is constantly present in the circulatory system, ready to respond to any microbial threat that may present.

Bone Marrow
All blood cells begin life in the bone marrow – the jelly-like substance contained within the cavities of bones. All types of blood cells are derived from a single type of cell called a stem cell, which goes on to form either red blood cells, platelets or the white blood cells of the immune system. Cells can then migrate to other parts of the immune system, such as the spleen or thymus gland where they mature into other cell types.

Lymphatic System
The functioning of the immune system is facilitated by the lymphatic system. This system circulates a solution called lymph around the whole of the body, but it is different to the circulatory system, which carries blood. Importantly, the lymphatic system carries white blood cells around the body.

In capillaries – the smallest of the blood vessels – pressure causes fluid and small molecules to be forced into the spaces between cells. This is called interstitial fluid, which bathes and feeds surrounding tissues. This is subsequently drained into the lymphatic system, where it circulates and eventually drains back into the bloodstream. It is not actively pumped, but relies on the vessels being squeezed by surrounding muscles.

Defending against infection

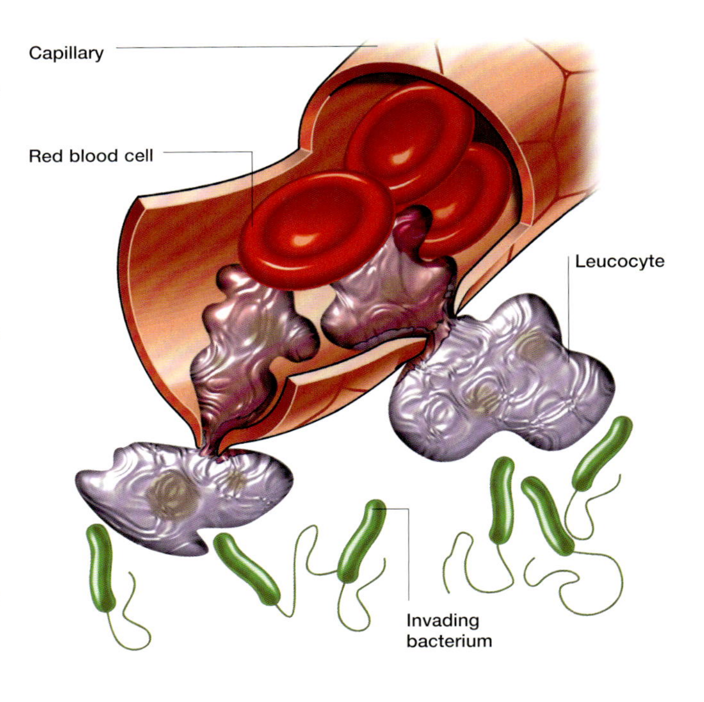

When the body is infected by bacteria, chemical signals are released. These cause white blood cells, called leucocytes, to leave the capillaries and attack the invading bacteria.

Viruses

Because viruses are so small (only 0.00001mm/in in diameter), they are very efficient at entering the respiratory and gastrointestinal tracts. They can only replicate by invading a 'host' cell. Lymphocytes mount an immune response to attack the invaders.

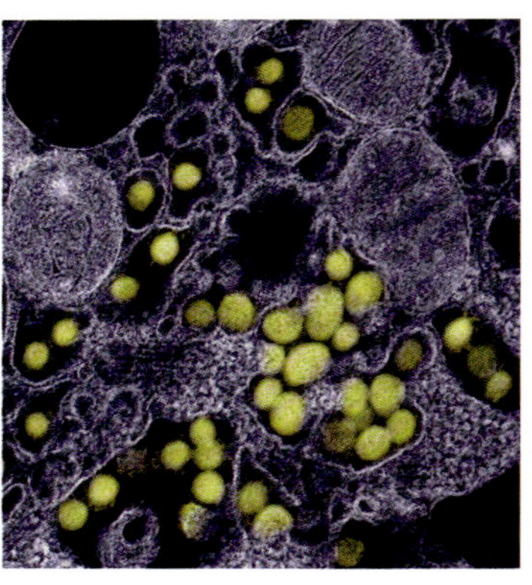

Right: Covid viruses are visible here (yellow). They invade host cells and replicate within them.

Single-cell invaders

Bacteria and protozoans are sought out, engulfed and killed by white blood cells called phagocytes. Invading microbes trigger the production of factors that attract phagocytes to the infected area; they are then coated with antibodies and ingested.

Right: E. coli bacteria are associated with food poisoning. Phagocytes in the blood are capable of ingesting such microbes.

Multi-cellular invaders

Helminths are parasitic worms, commoner in warmer countries. The blood attacks them with specialized white cells called eosinophils – so named because they stain red when exposed to eosin laboratory dye.

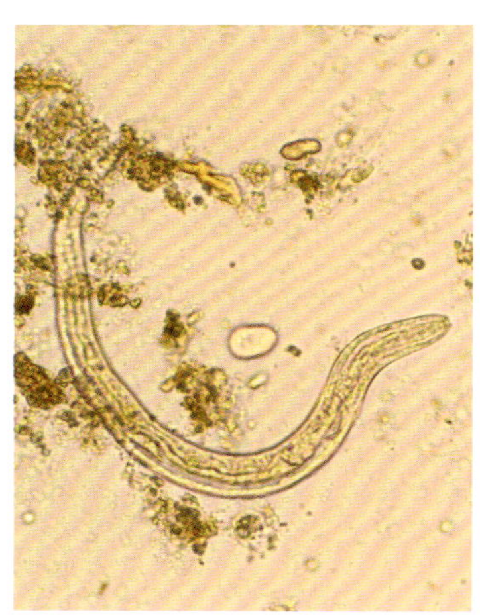

Right: Parasites, such as this hookworm, are found in the intestines. Eosinophils in the bloodstream are capable of attacking some of these invaders.

Fungi

Fungal organisms invade moist, warm areas of the human body, such as between the toes. The body mounts an immune response against the fungi by transporting antibodies to the site, via the blood.

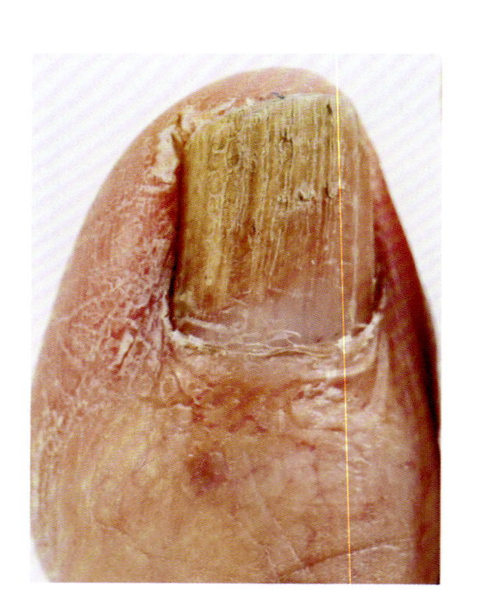

Right: The body responds to many fungal infections by producing antibodies. These are carried in the blood to the relevant area.

Defensive components of blood

Although some infections can overcome our defences, the various components of the blood successfully fight back against most invaders.

The components of the blood that combat infections are:

■ Phagocytes.

If a microbe enters the body it will almost certainly encounter specialized white blood cells known as neutrophils and monocytes. Their function is to engulf invading particles (phagocytose) and break them up through a process of intracellular digestion. Phagocytes do not live exclusively in the blood. Instead, they spread out from the blood vessels and into the tissues where they are best placed to attack invading microbes.

Of the two types of phagocyte, neutrophils are relatively short-lived, while monocytes are longer-lasting and turn into another group of cells, termed macrophages. Macrophages create a zone of inflammation around microbes, helping to limit their spread. Where possible, they engulf them.

■ Lymphoid cells.

These white cells come in three forms:

T-lymphocytes. These are very effective at attacking viruses. Virologists classify them into various groups (helper T-cells; suppressor T-cells; cytotoxic T-cells; and hypersensitivity-mediating T-cells), which all combine to attempt to destroy viruses.

B-lymphocytes. These are involved in the production of antibodies against microbes.

Killer cells and natural killer (NK) cells. These are often able to recognize human cells that have been taken over by viruses as intracellular 'factories' and destroy them.

■ Interferons. These are chemical agents that are produced by cells that have been infected by viruses and T-lymphocytes. Interferons flow through the bloodstream, activating NK cells and providing defence against viruses.

■ Complement. This blood component consists of about 20 proteins. When infection occurs, they work together to attack bacteria and organize inflammation around the infected area.

■ Acute phase proteins. These are blood proteins with the ability to attach to certain bacteria and disable them in the early stages of an infection.

■ Eosinophils. These are specialized white blood cells that play a role in fighting off infection by helminths. They are capable of inactivating some of these parasites by binding to them and releasing a toxic protein.

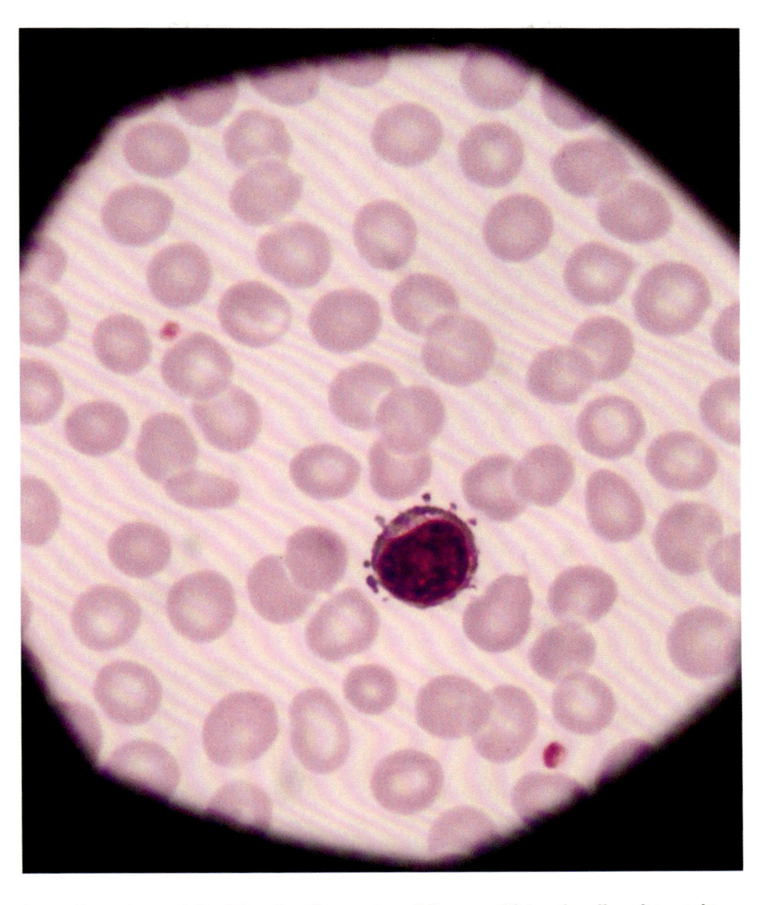

Lymphocyte – white blood cells neutrophil on red blood cells, shown in the background.

Blood antibodies

Antibodies are vital components of the blood. They are complex molecules called immunoglobulins, which are formed in response to infection. There are various types of immunoglobulin:

■ IgG makes up about three-quarters of the immunoglobulin in normal blood. It is effective in neutralizing the toxins (poisons) produced by certain microbes.

■ IgM makes up about one-fourteenth of the serum immunoglobulins. It activates complement so that it can attack foreign cells.

■ IgA makes up about a fifth of the blood's immunoglobulin load. It is mainly delivered to areas such as the mouth, air passages and intestine, where germs are likely to attack. It acts as an antiseptic secretion, helping to keep microbes from penetrating the mucous surfaces of the body.

■ IgE is thought to play a part in defending the body against helminths, by creating defensive inflammations. It is often produced in excessive amounts in people who have allergies, causing inappropriate inflammation. This inflammantion is associated with the symptoms of asthma, hay fever and allergic skin reactions.

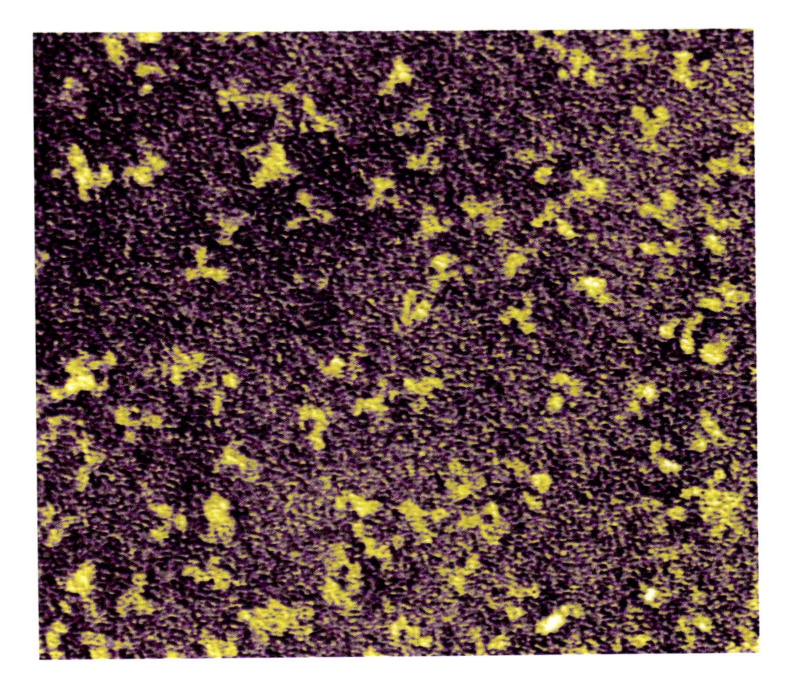

The structure of an antibody is shown on this computer-generated image. Antibodies are able to bind to foreign cells or toxins and neutralize them.

Lymphatic System

The lymphatic system consists of a network of lymph vessels and organs and specialized cells throughout the body. It is an essential part of the body's defence against invading micro-organisms.

The lymphatic system is the lesser known part of the circulatory system, which works together with the cardiovascular system to transport a fluid called lymph around the body. The lymphatic system plays a vital role in the defence of the body against disease.

Lymph Fluid

Lymph is a clear, watery fluid containing electrolytes and proteins that is derived from blood and bathes the body's tissues. Lymphocytes – specialized white blood cells involved in the body's immune system – are found in lymph as well as blood. They attack and destroy foreign micro-organisms, thereby maintaining the body's health. This is known as an immune response.

Although the vessels of the lymphatic system carry lymph, the fluid is not pumped around the body as blood is; instead, contractions of muscles surrounding the lymph vessels move the fluid along.

Parts of Lymphatic System

The lymphatic system is made up of a number of interrelated parts:

- Lymph nodes – lie along the routes of the lymphatic vessels and filter lymph
- Lymphatic vessels – small capillaries leading to larger vessels that eventually drain lymph into the veins
- Lymphatic cells (lymphocytes) – cells through which the body's immune response is mounted
- Lymphatic tissues and organs – scattered throughout the body, these act as reservoirs for lymphoid cells and play an important role in immunity.

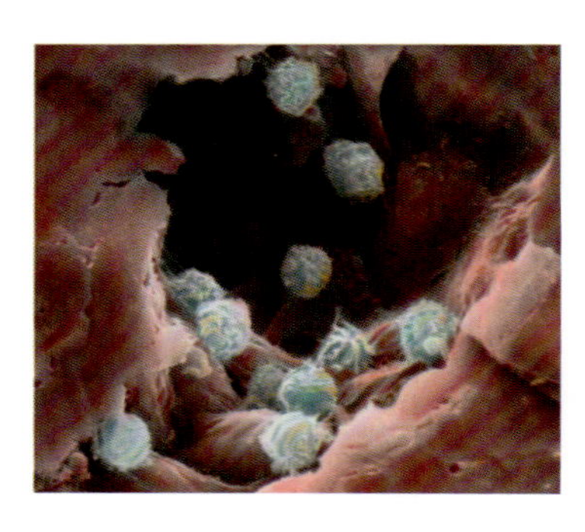

Lymphocytes, white blood cells involved in the body's immune response, appear blue on this false-colour electron micrograph.

Lymphatic system

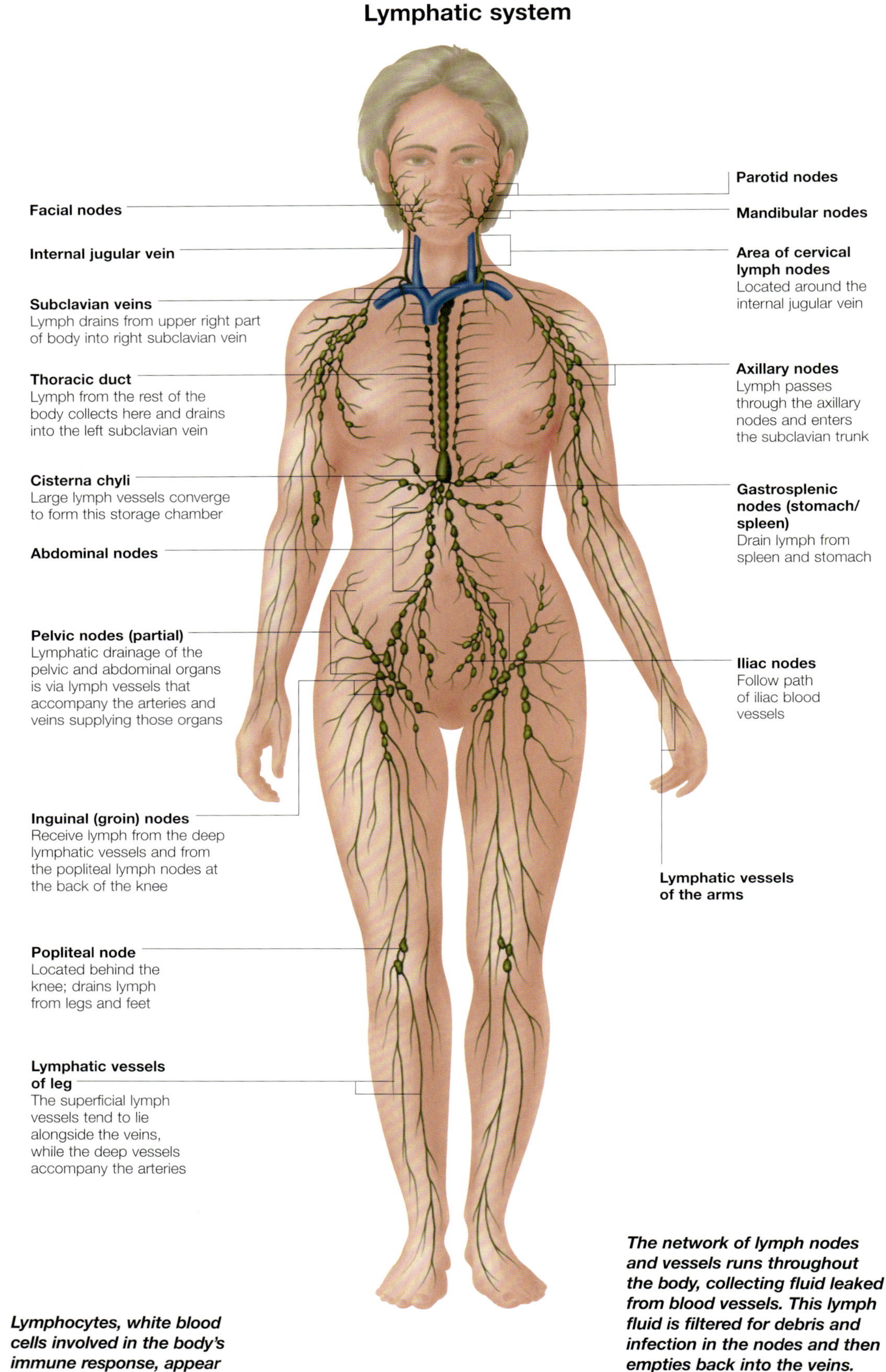

The network of lymph nodes and vessels runs throughout the body, collecting fluid leaked from blood vessels. This lymph fluid is filtered for debris and infection in the nodes and then empties back into the veins.

Lymph nodes

Lymph nodes lie along the route of the lymphatic vessels. They filter the lymph for invading micro-organisms, infected cells and other foreign particles.

Lymph nodes are small, rounded organs that lie along the course of the lymphatic vessels and act as filters for the lymph. The nodes vary in size, but they are mostly bean-shaped, 1–25mm (0.03–1in) in length, surrounded by a fibrous capsule and embedded in connective tissue.

Lymph node function

As well as fluid, the tiny lymphatic vessels in the tissues pick up other items, such as parts of broken cells, bacteria and viruses. Within the lymph node, fluid slows and comes into contact with lymphatic cells that ingest any solid particles and recognize foreign micro-organisms. To prevent these particles from entering the bloodstream – and to allow the body to mount a defence against invading organisms – lymph is filtered through a number of lymph nodes before draining into the veins.

Some lymph nodes are grouped together in regions and given names according to their position and the region in which they are found (for example, the axillary nodes in the axilla), the blood vessels they surround (such as the aortic nodes around the large central artery, the aorta), or the organ they receive lymph from (pulmonary nodes in the lungs).

Structure of a lymph node

Afferent lymph vessel
Transports lymph fluid to node

Lymphocyte

Capsule
Collagen and elastin make up the fibrous capsule that surrounds the lymph node

B-cell
Produces antibodies to invading micro-organisms

Germinal centre
Releases lymphocytes to fight invading micro-organisms; these lymphocytes mature into B- and T-lymphocyte cells

Arteriole

Venule

Macrophage
A large scavenger cell

Sinus
Acts to slow the passage of lymph through the node, so that macrophages can attack invading micro-organisms

Vein

Artery

Trabeculae
Columns of fibrous tissue that divide the lymph node into segments

T-cell
Kills micro-organisms and infected cells

Efferent vessel
One vessel carries the filtered lymph fluid from the lymph node

The internal structure of a lymph node slows the passage of lymph fluid, so that specialized lymphocyte cells can filter out any micro-organisms.

Lymph vessels

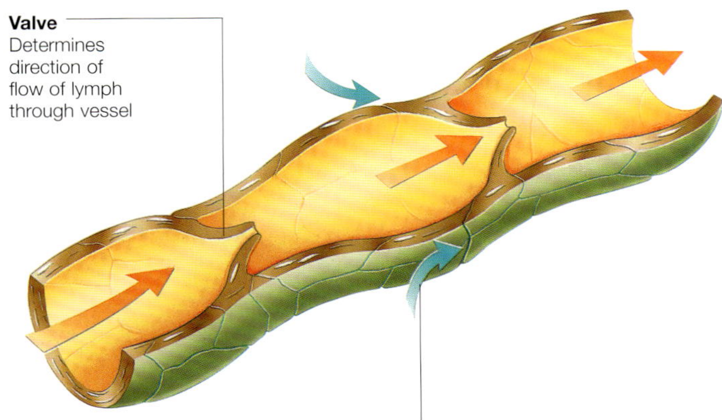

The fluid circulating around the cells in tissues drains into lymph capillaries. From here, it flows through valves in these vessels to the lymph nodes.

Arteries supply blood to the body's tissues under pressure. This has the effect of causing fluid and proteins to leak out of the tiny capillaries and into the spaces around the cells of those tissues.

Much of this leaked fluid will pass back into the capillaries, which gradually converge to form veins that carry blood back to the heart for further circulation. However, some of the fluid – and the proteins – remain behind and would accumulate in the tissues were it not for the network of tiny lymphatic vessels in the tissue spaces.

The lymph fluid travels up the converging lymphatic vessels, which eventually join to form the main lymphatic trunks. These unite to form the two large lymphatic ducts – the thoracic duct and the right lymphatic duct. These drain into the large veins above the heart, returning the retrieved fluid and proteins to the bloodstream.

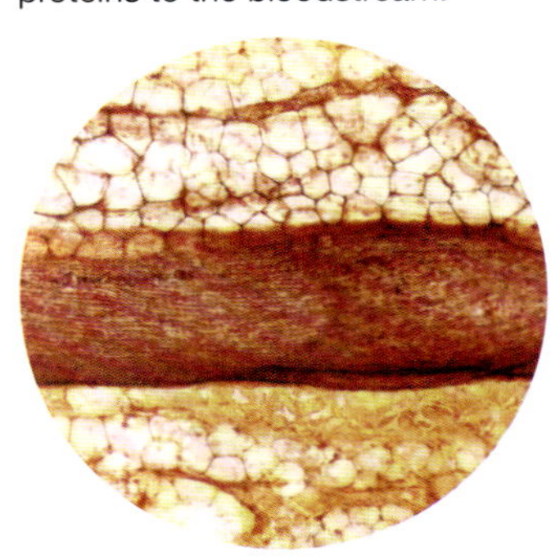

The 'non-return' value can be seen halfway down this lymph vessel.

Lymphatic tissue

In addition to lymph nodes and vessels there are clusters of cells and organs that form part of the immune system. These include the spleen, the thymus and lymphoid tissue in the gut.

Scattered throughout the body are discrete groups of lymphatic tissue, which play an important role in the immune system:

- The spleen – an organ that produces B-lymphocytes and destroys bacteria and damaged red blood cells
- The thymus – a small gland that lies in the chest just behind the upper part of the sternum (breastbone). It receives newly formed lymphocytes from the bone marrow, which mature into T-lymphocytes, an important group of lymphatic cells
- Lymphatic tissue of the gastrointestinal tract – lymphoid tissue lying beneath the lining of the gut, the ring of lymphatic tissue (tonsils and adenoids) at the back of the mouth and some discrete clumps of lymphoid nodules known as Peyer's patches, found in the walls of the last part of the small intestine. These are thought to be the site of maturation of B-lymphocytes, another important set of lymphocytes.

The large amount of lymphatic tissue in the gut wall helps to protect against infection caused by organisms entering through the mouth.

Lymphatic organs and tissues

Lymphatic organs are found throughout the body. Within these organs, specialized cells screen the lymph fluid for foreign particles.

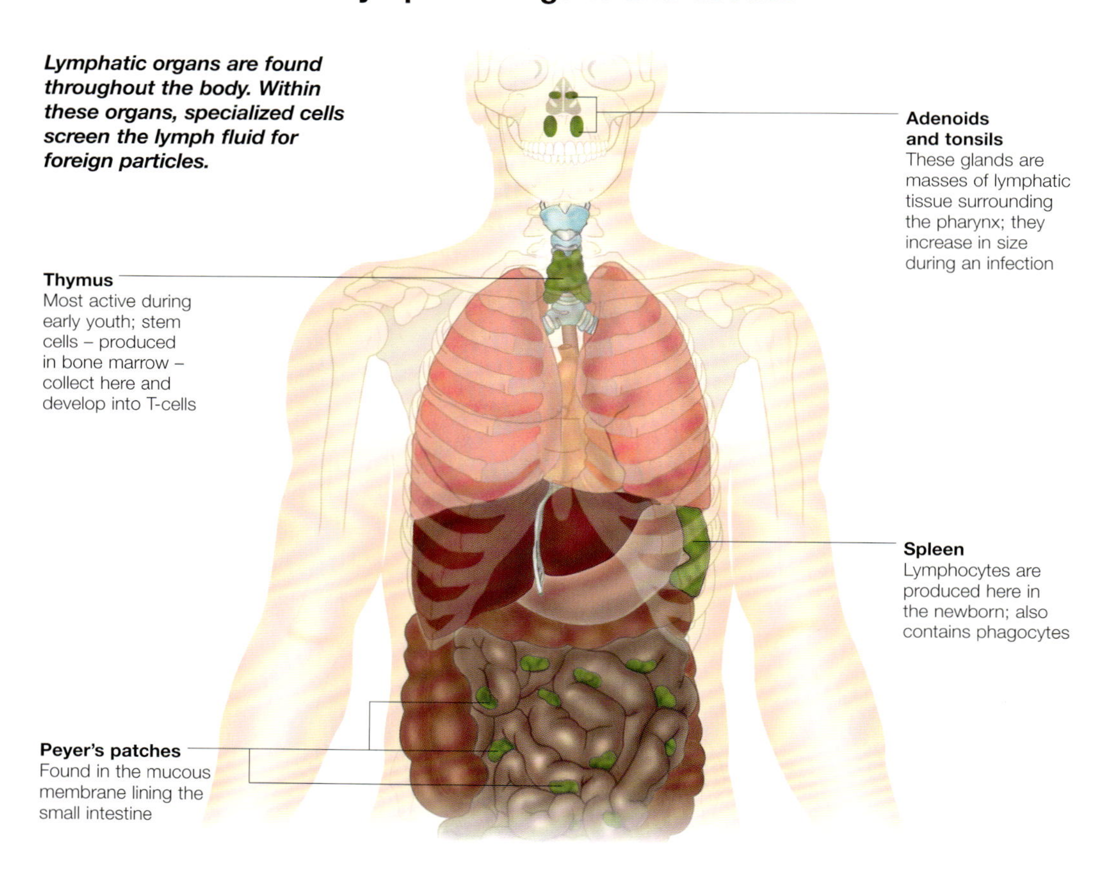

Lymphocytes

The cells of the immune system, lymphocytes, can recognize foreign proteins, such as those found on the surface of invading micro-organisms or on the cells of transplanted organs.

In response, the lymphocyte cells multiply and mount an immune response, some (T-cells) by directly attacking foreign cells and some (B-cells) by manufacturing antibodies that attach to foreign proteins, allowing them to be found and destroyed.

Lymphocytes originate in the bone marrow and circulate freely in the bloodstream. As they circulate, they can quickly mount a response to infections.

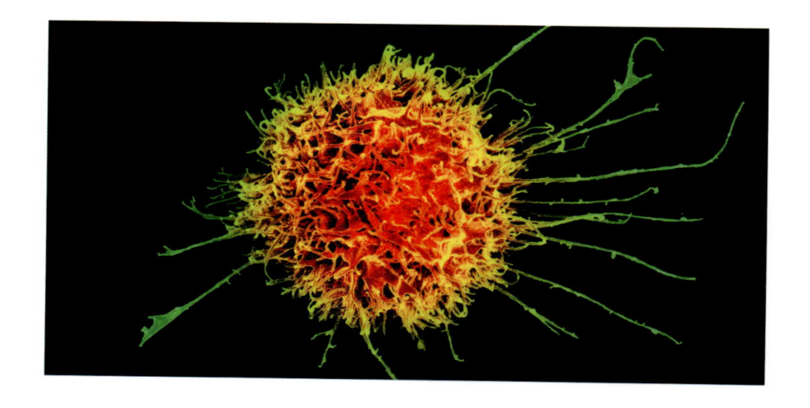

'Natural killer cells' are a type of lymphocyte. They are able to destroy cancer cells and cells infected with viruses.

Lymph drainage vessels

The lymphatic vessels form a network that runs through the tissues. These vessels converge and empty into the veins.

Drainage of chest

Of the lymph nodes that lie in the chest, the most important are the internal mammary nodes on either side of the sternum. They receive 25 per cent of the lymph from the breast and may be a site for spread of breast cancer. Within the chest, the largest group of lymph nodes lie around the base of the trachea (windpipe) and the bronchi. Other lymph node groups within the chest lie alongside the major blood vessels.

Upper and lower limbs

In the limbs, there are superficial and deep lymph vessels; the superficial vessels tend to lie alongside the veins whereas the deep vessels accompany the arteries.

The axillary (armpit) group of nodes receives lymph from the whole of the upper limb, the trunk above the umbilicus and the breast. The inguinal (groin) lymph nodes receive lymph from the superficial vessels and the deep lymph vessels that run alongside the arteries.

Lymph travels up from the inguinal nodes to the nodes alongside the aorta and eventually joins the lumbar lymph trunks.

Lymphatic ducts and nodes

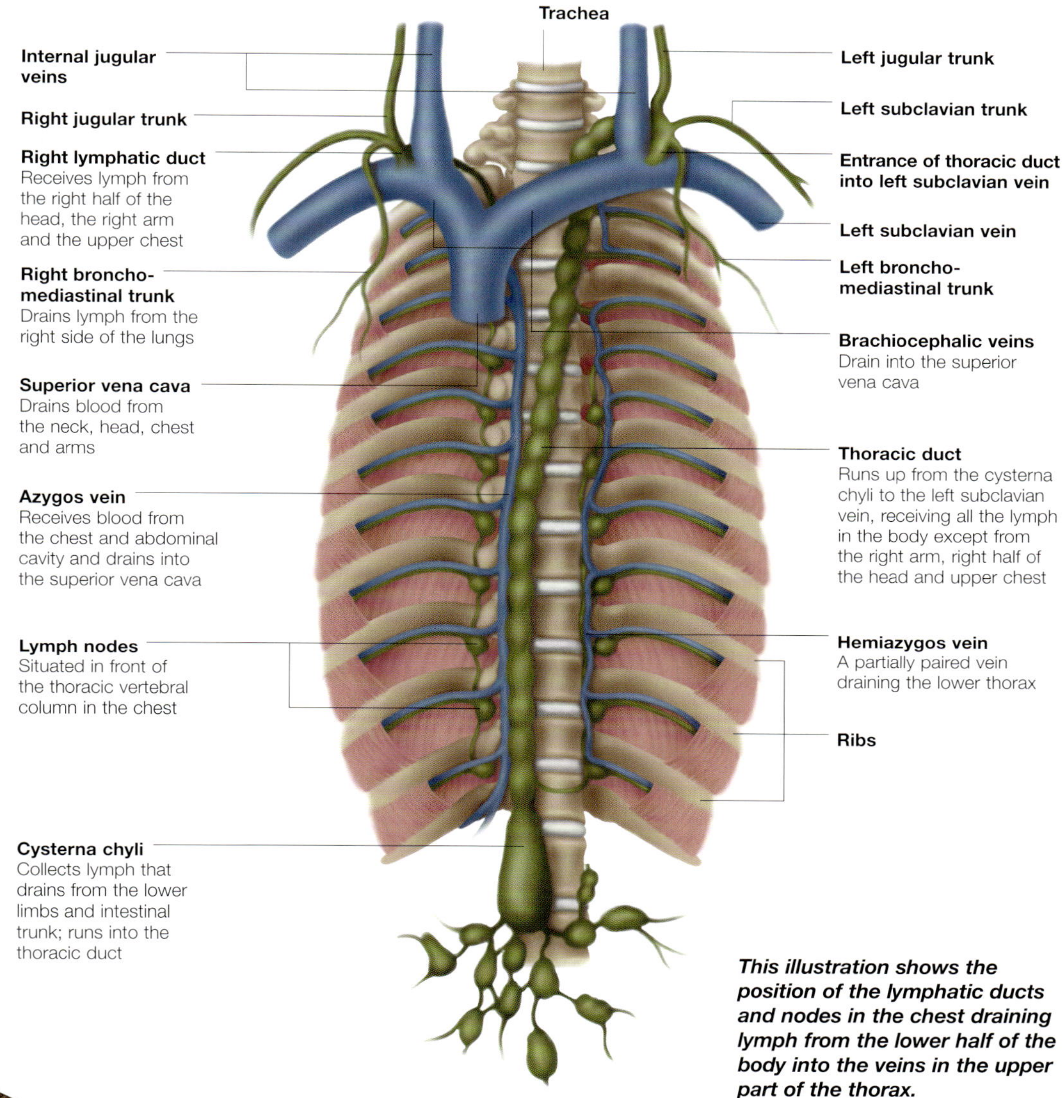

This illustration shows the position of the lymphatic ducts and nodes in the chest draining lymph from the lower half of the body into the veins in the upper part of the thorax.

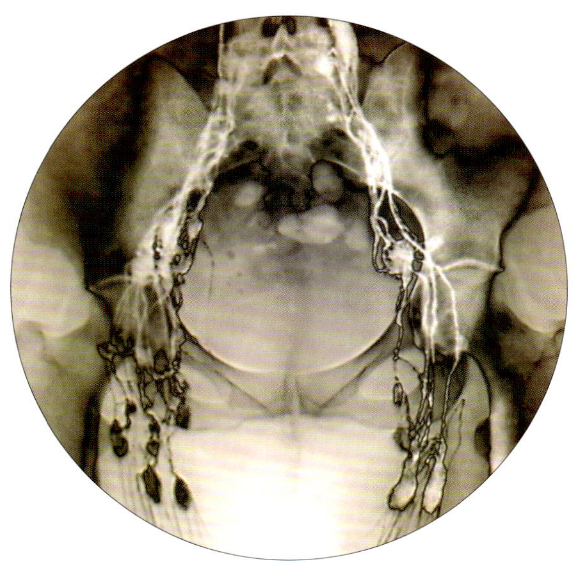

Normal lymph vessels and nodes can be seen in this x-ray of the pelvic area.

Regional Lymphatic Drainage

Lymph from the body returns to the bloodstream via a series of lymph nodes. An understanding of the lymph drainage pattern is vital in monitoring the spread of cancers or infection.

Lymph is the fluid present within the vessels of the lymphatic system. The main function of the lymphatic vessels is to collect excess tissue fluid and return it to the blood circulation.

Lymph from each part of the body follows a specific path on its way back to rejoin the blood circulation, passing through lymph node groups – which have a filtering role – along the way.

Head and Neck Nodes

The lymph node groups of the structures of the head and neck are named according to their positions. The important lymph node groups include the:

- Occipital
- Mastoid, or retroauricular (behind the ear)
- Parotid
- Buccal
- Submandibular (under the jaw)
- Submental (under the chin)
- Anterior cervical
- Superficial cervical
- Deep within the neck lie other groups of nodes, which surround and drain the pharynx, larynx and trachea.

Deep Cervical Nodes

These lymph nodes all drain ultimately into the deep cervical group of nodes that lie in a chain alongside the major blood vessels of the neck.

Lymph nodes of the head and neck

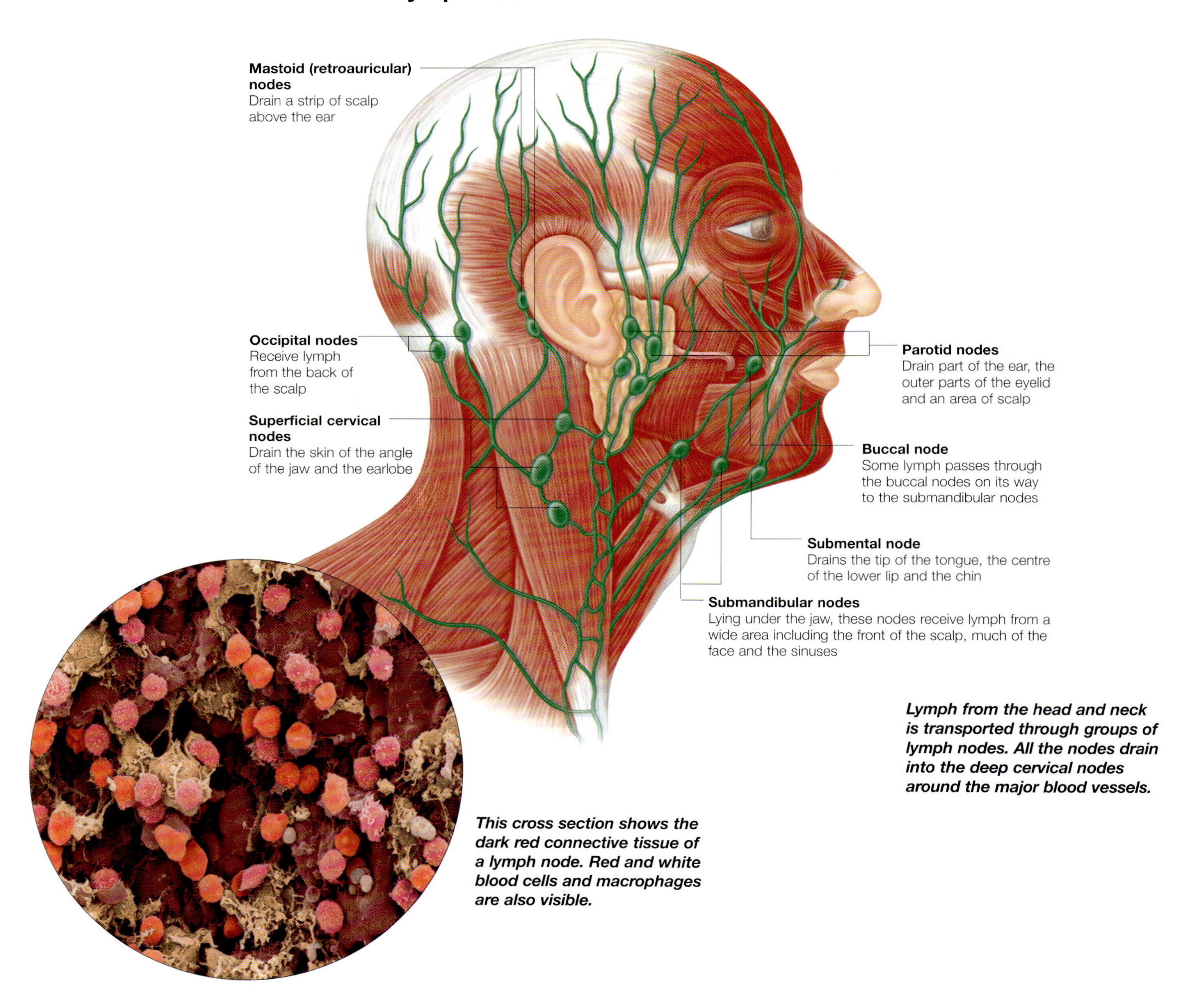

Lymph from the head and neck is transported through groups of lymph nodes. All the nodes drain into the deep cervical nodes around the major blood vessels.

This cross section shows the dark red connective tissue of a lymph node. Red and white blood cells and macrophages are also visible.

Lymphatic drainage of the intestines

The lymph vessels and nodes that make up the lymphatic drainage of the gastrointestinal system follow the pattern of arteries that supply the gut with blood. Lymph from the small intestine transports fats from food into the bloodstream.

Much of the gut is enclosed and suspended within a fold of connective tissue, known as the mesentery. The blood vessels that supply the gut lie within this mesentery, forming arcades that connect with each other to reach all parts of this lengthy structure.

Site of nodes

The lymph nodes that initially receive lymph from the intestine are found within the mesentery in a number of places:

- By the wall of the intestine
- Among the arterial arcades
- Alongside the large superior and inferior mesenteric arteries.

These mesenteric groups of nodes are, in some cases, named according to their positions in relation to the intestine or to the artery they accompany. From the intestinal wall, lymph drains through these nodes in turn to eventually reach the pre-aortic nodes, which lie next to the large central artery, the aorta.

Absorption of fat

In addition to its normal function, the lymph that leaves the small intestine has a further role – that of transporting the fats absorbed from food.

The lining of the small intestine bears numerous microvilli. These tiny projections of the mucous membrane greatly increase the surface area of the intestine to help absorption.

Central vessels

Within each microvillus lies a central lymph vessel, called a lacteal. The function of the lacteals is to carry away fat particles absorbed from food that are too big to enter the blood capillaries.

These fats travel through the lymphatic system to be delivered into the bloodstream with the rest of the lymph.

Gastrointestinal lymph nodes

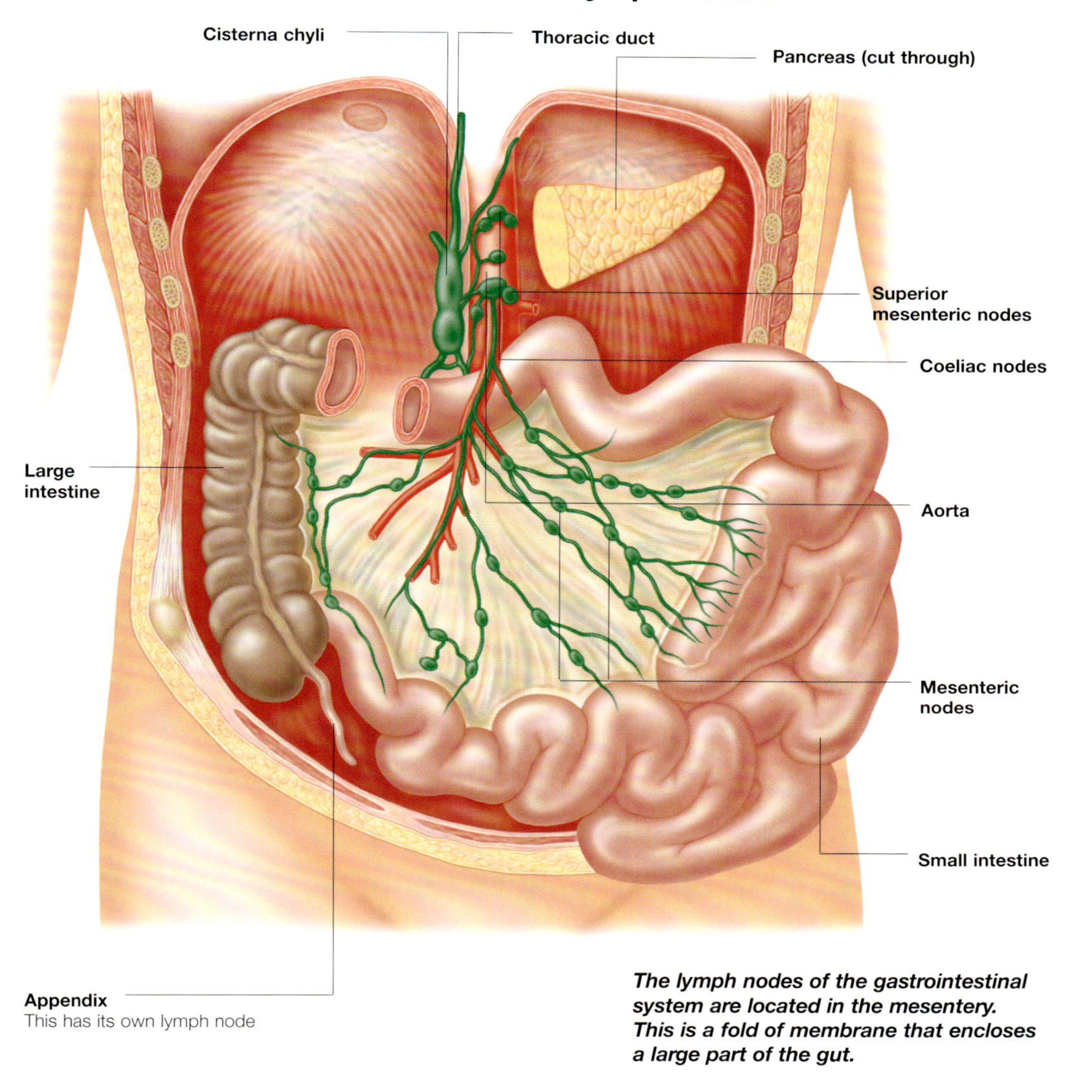

The lymph nodes of the gastrointestinal system are located in the mesentery. This is a fold of membrane that encloses a large part of the gut.

Lymphatic drainage of the stomach

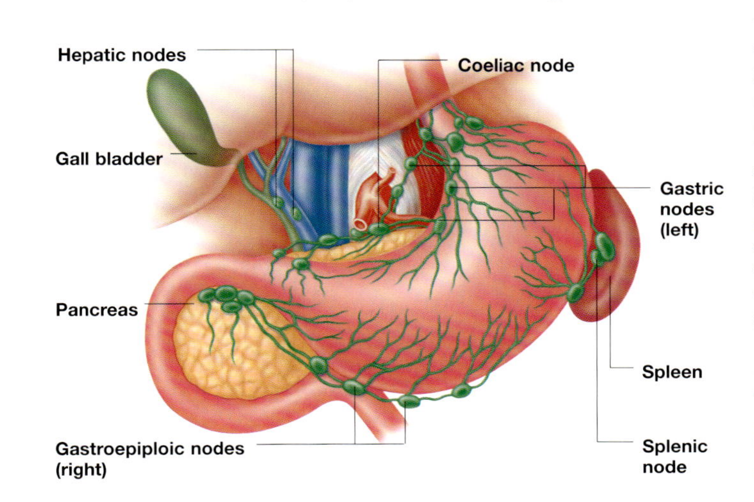

There are four main groups of nodes that receive lymph from the stomach:

- The left and right gastric nodes receive lymph from the area supplied by the left and right gastric arteries. They lie along the lesser curve of the stomach.
- The splenic nodes lie at the hilum of the spleen on the left side of the stomach. They receive lymph from the area of the stomach supplied by the short gastric arteries.
- The left and right gastroepiploic nodes lie along the greater curve of the stomach and receive lymph from areas supplied by the gastroepiploic arteries.
- All the lymph received from the stomach by these groups travels on to drain into the coeliac nodes.

Spleen

The spleen is the largest of the lymphatic organs. It is dark purple in colour and lies under the lower ribs on the left side of the upper abdomen.

The spleen is usually about the size of a clenched fist; in older age, the spleen naturally atrophies and reduces in size.

The hilum of the spleen contains its blood vessels (the splenic artery and vein) and some lymphatic vessels. The hilum also contains lymph nodes and the tail of the pancreas, all enclosed within the lienorenal ligament – a fold of peritoneum.

Surface of the Spleen

The spleen shows indentations of the organs that surround it. The surface that lies against the diaphragm is curved smoothly, whereas the visceral surface carries the impressions of the stomach, the left kidney and the splenic flexure of the colon.

Spleen Coverings

The spleen is surrounded and protected by a thin capsule, which is composed of irregular fibroelastic connective tissue. Contained within the tissue of the capsule are muscle fibres that allow the spleen to contract periodically. These contractions expel the blood the spleen has filtered back into the circulation. Outside the capsule the spleen is completely enclosed by the peritoneum, the thin sheet of connective tissue that lines the abdominal cavity and covers the organs within it.

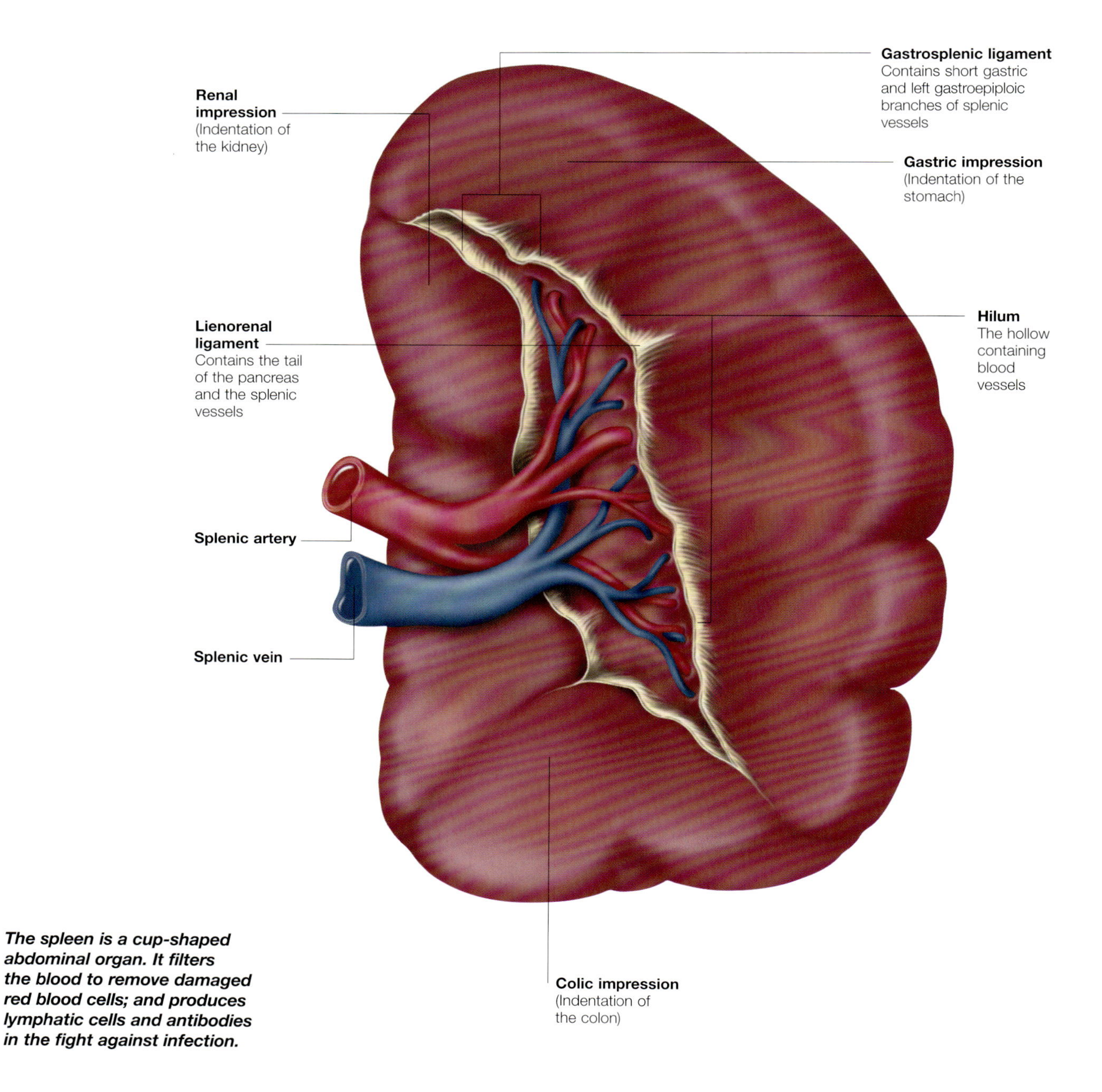

The spleen is a cup-shaped abdominal organ. It filters the blood to remove damaged red blood cells; and produces lymphatic cells and antibodies in the fight against infection.

Functions of the spleen

The spleen is an important organ that is part of the immune system. It is responsible for both creating new lymphocytes and for destroying worn-out and damaged red cells and platelets.

The spleen is about 12cm (4.7in) in length and is the largest lymphatic organ in the body. There are two different types of tissue in the spleen, known as white pulp and red pulp.

Fetal development

In the fetus the spleen contributes to the manufacture of haemoglobin in the blood. After birth, however, it does not have this function.

White pulp

The white pulp is composed mainly of lymphatic cells lying clustered around the small branches of the splenic artery, which brings blood into the spleen. The main function of this area of the spleen is to initiate an immune response. It does this by manufacturing and activating B-lymphocytes, which themselves become antibody-producing plasma cells. The spleen also stores lymphocytes and releases them when necessary. It is thought that at any one time, around 25 per cent of the body's white cells are in the spleen.

Red pulp

The red pulp (within which the islands of white pulp lie) consists of connective tissue. This connective tissue contains red blood cells and macrophages, cells that can engulf and destroy other cells.

Cells in the red pulp filter the blood and remove damaged red cells and platelets from the bloodstream and destroy them. They are able to recycle the iron from the destroyed red cells and either store it, or export it into the bloodstream as ferritin.

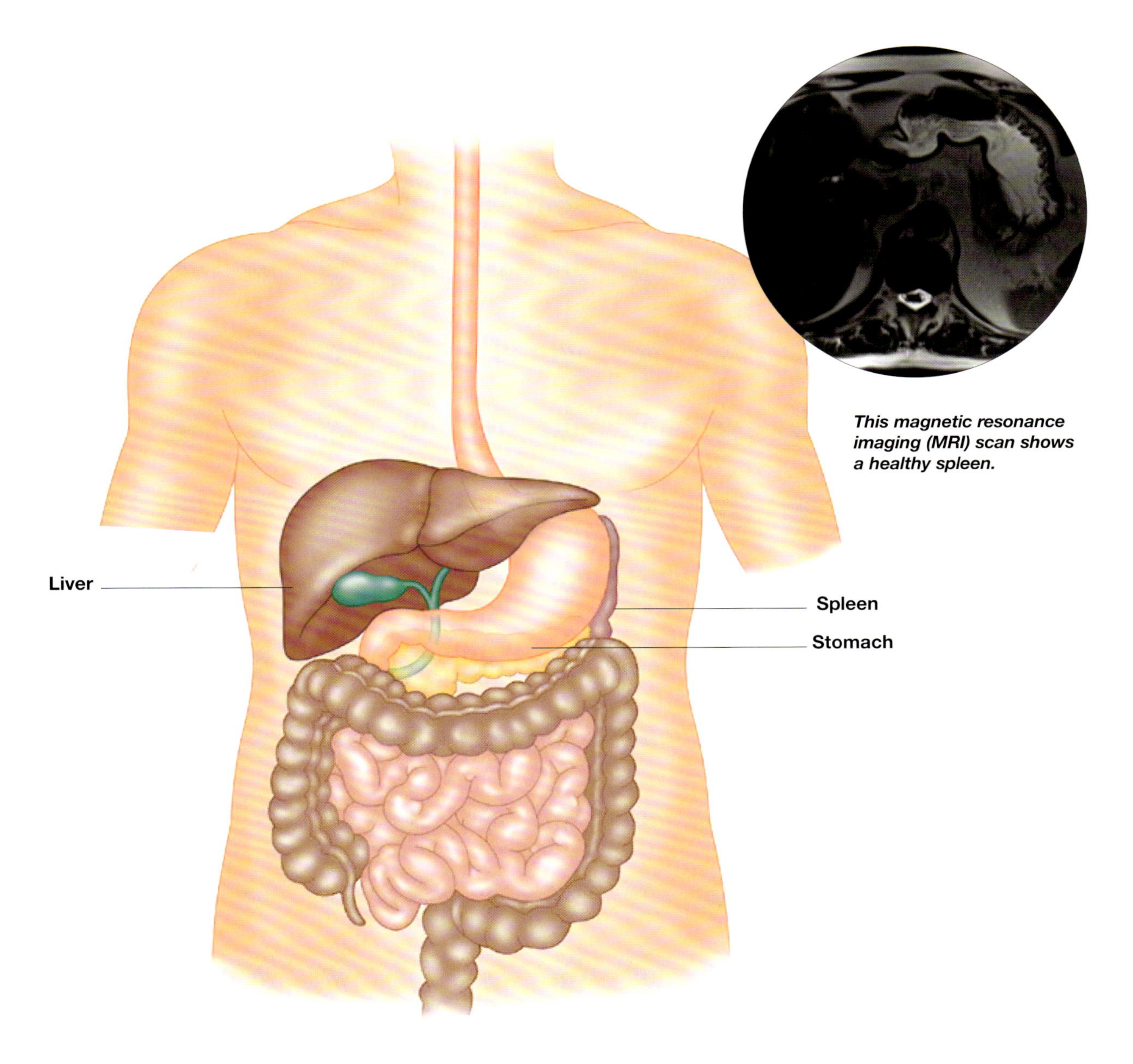

This magnetic resonance imaging (MRI) scan shows a healthy spleen.

Thymus

The thymus gland forms a vital part of the body's immune system and is the site of development of important specialized immune cells called T-lymphocytes. It lies in the anterior mediastinum.

The thymus is a pink, flattened, often bilobed gland that lies within the anterior mediastinum and extends up to the superior mediastinum. It is situated in front of the large blood vessels and the trachea, just behind the manubrium (the upper part of the breastbone).

During childhood, when the gland is at its largest, it may also extend upwards into the root of the neck and downwards into the anterior mediastinum in front of the heart.

The thymus is surrounded by a fibrous capsule from which extensions enter the thymus tissue itself to divide the gland into lobules.

Blood Supply

There is a generous blood supply from the internal thoracic arteries via their anterior intercostal and anterior mediastinal branches. Venous drainage is via the thymic veins, which drain into the brachiocephalic vein close to the heart.

Growth of the Thymus

The thymus gland is at its largest during infancy and childhood. Occasionally, in some newborn babies, it may be so large that it projects up into the neck, compressing the trachea and causing breathing difficulties.

Similar to other lymphatic tissue throughout the body, the thymus becomes relatively less significant in size as a child grows into adulthood. At birth the thymus may weigh 10–15g (0.3–0.5oz), but this increases to only 30–40g (1–1.4oz) at puberty, despite the fact that the body weight as a whole has increased tenfold.

Changes through Life

After puberty, the functional tissue of the thymus gradually becomes increasingly smaller and is replaced by fat until, in old age, the gland may be very difficult to find at all.

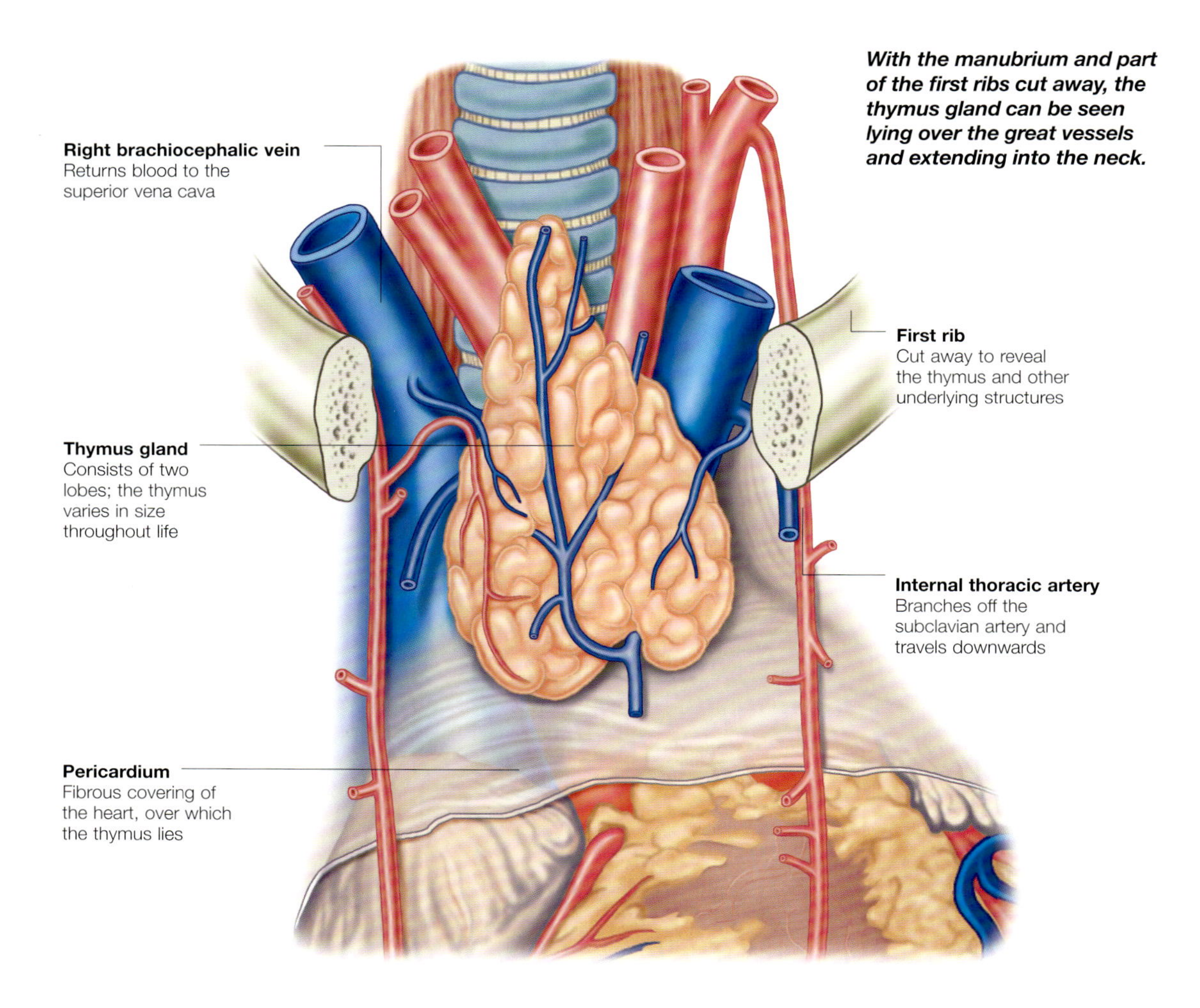

With the manubrium and part of the first ribs cut away, the thymus gland can be seen lying over the great vessels and extending into the neck.

Functions of the thymus

The thymus has an important role to play in helping to produce T-cells, so named because they are produced in the thymus.

The primary function of the thymus gland is to process lymphocytes to become T-cells. Lymphocytes can be broadly classified into two types:

- B-cells that turn into plasma cells, which produce antibodies to attack microbial invaders. This is known as humoral immunity and is mainly effective against bacteria and viruses.
- T-cells that are responsible for cellular immunity – this involves the formation of lymphocytes that are able to recognize and attach to an invading organism and kill it. This type of immunity is effective against fungi, parasites, cancer cells and viruses.

T-cells

All lymphocytes start their lives in the embryonic bone marrow. Just before birth, about half of the c ells travel to the thymus gland where they are 'processed' and given the ability to perform very specific immune reactions.

Once processed, the cells leave the thymus gland and become embedded in lymphatic tissue throughout the body.

Activation of T-cells

There are many different types of T-cells, each with a particular role and each with the ability to respond to specific antigens.

For the majority of the time, most T-cells remain inactive but if an antigen (a marker of a foreign body) enters the body, the T-cell that is programmed to react with that particular antigen will activate. The T-cell then grows and divides and mounts an attack on the invader.

Hormones

The thymus also produces the hormones thymosin, thymic humoral factor, thymic factor and thymopoietin, which enable T-cells to mature.

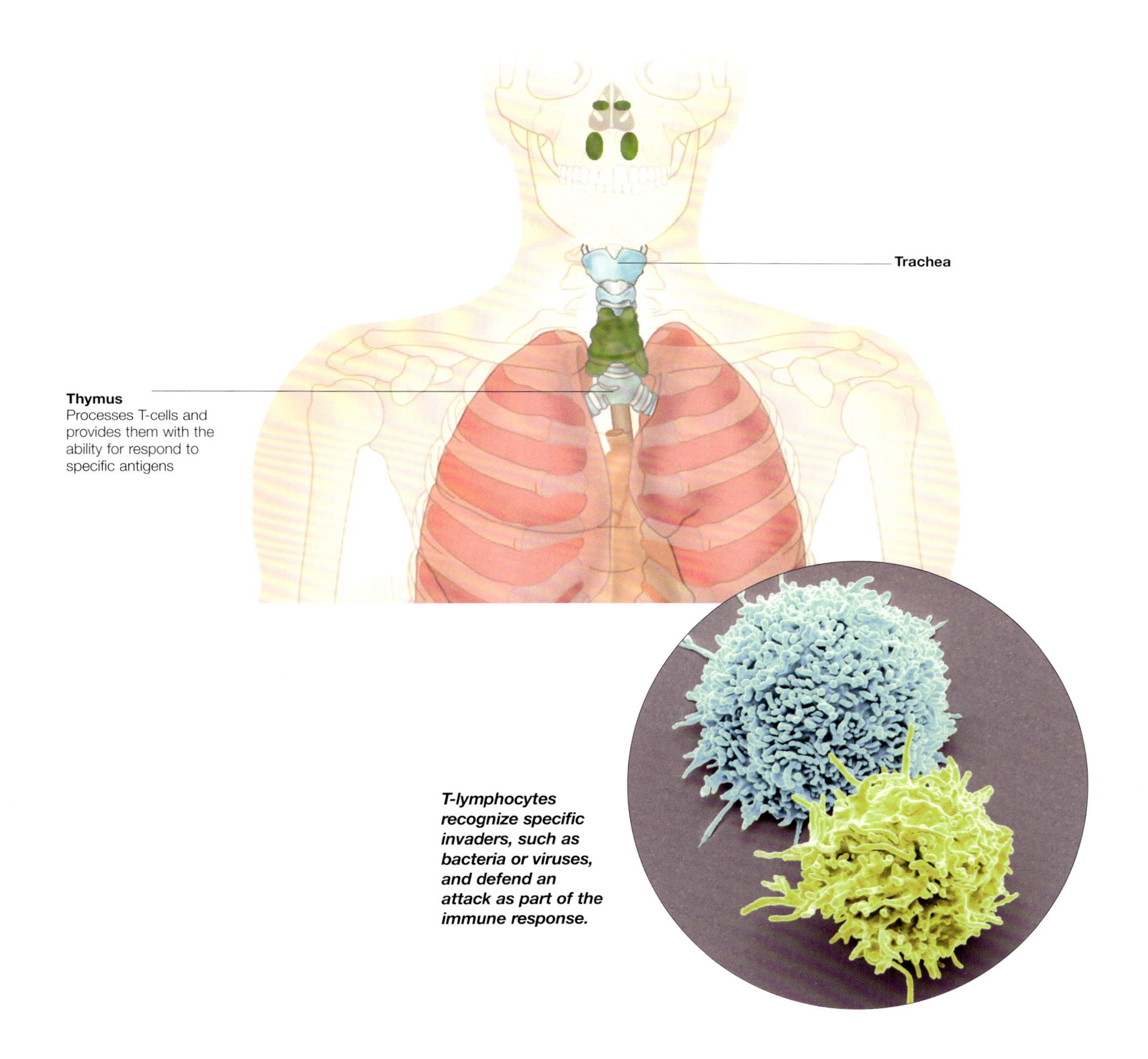

T-lymphocytes recognize specific invaders, such as bacteria or viruses, and defend an attack as part of the immune response.

Immune Responses

Infections develop when a bacterium or virus manages to break through the immune defences and replicate inside the body. The body usually responds efficiently to these invaders but they sometimes cause illness.

The body is exposed to a myriad of micro-organisms every day. In fact, it plays host to millions of bacteria, all living in a state of coexistence. Most bacteria are harmless as long as they stay in protected places such as the surface of the skin, intestines, nose, mouth or vagina. However, if these surfaces become damaged through injury or disease, and micro-organisms are allowed to enter the normally sterile internal tissues of the body, infection can occur.

Alternatively, bacteria or viruses can spread from another source such as an infected person or an animal or insect.

Protective Barriers

Fortunately the body has a number of protective barriers that act as a first line of defence against infection, including:

- The skin – this provides a physical barrier to pathogens (disease-causing organisms), helping to maintain a sterile internal environment
- The nose – this contains sticky mucus and hairs to trap potentially harmful micro-organisms
- Saliva – this contains antibodies that combat pathogens
- Tears – these contain antibodies to prevent infection of the eyes
- Throat – this is protected by the reflex reaction of coughing
- Stomach – this produces a strong acid that destroys any ingested pathogens.

Local Infection

If pathogens manage to breach the body's first line of defence, they can multiply in the tissues, causing infection. The body's response to this is to trigger inflammation and produce antibodies to fight the infective organisms and prevent the spread of infection.

Redness

If a sufficient number of pathogens invade the body, they will release harmful toxins or cause enough damage to cells for local blood vessels to dilate, resulting in an increased blood flow to the affected area. This gives rise to the redness and warmth typical of an inflamed area. In addition, a watery fluid leaks out of the blood vessels, causing the surrounding area to swell visibly.

The increase in blood flow enables cells of the immune system, including phagocytes (a type of white blood cell that engulfs and destroys pathogens) to reach the area and attack the organisms present. This is usually sufficient to prevent infection spread, and the swelling eventually subsides as the pathogens are destroyed.

If the infection is particularly severe, then the body will also form a wall of fibrous tissue around the infected area. This wall serves to keep the infection localized while it is brought under control by the immune system.

Within the fibrous walls, a build-up of pus may occur; this contains dead white cells, body cells and bacteria, and cell debris.

Incubation Period

After a pathogen has invaded the body, there is an interval of time before there is any evidence of disease. This is because all pathogens undergo an incubation period during which they multiply. Once there are sufficient pathogens, they will cause noticeable effects or symptoms in the patient.

Variable Length

Incubation periods vary greatly, from only a few hours to a number of years. Cholera, for example, can develop within a couple of hours of drinking contaminated water, but AIDS may not develop until many years after the HIV virus has been acquired.

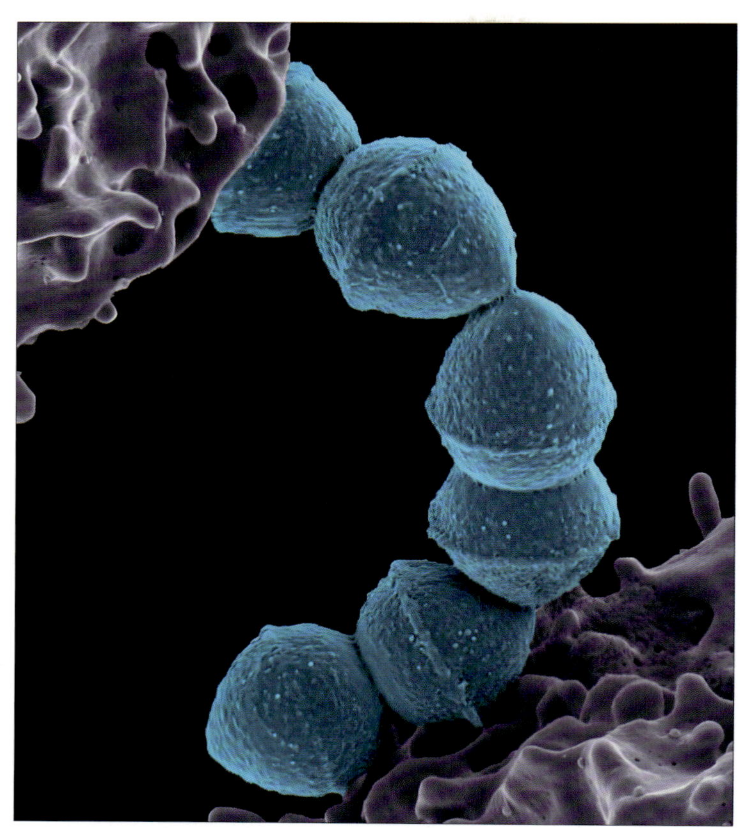

Pathogens go through an incubation period in which they multiply in the body. Here, Streptococcus bacteria are dividing.

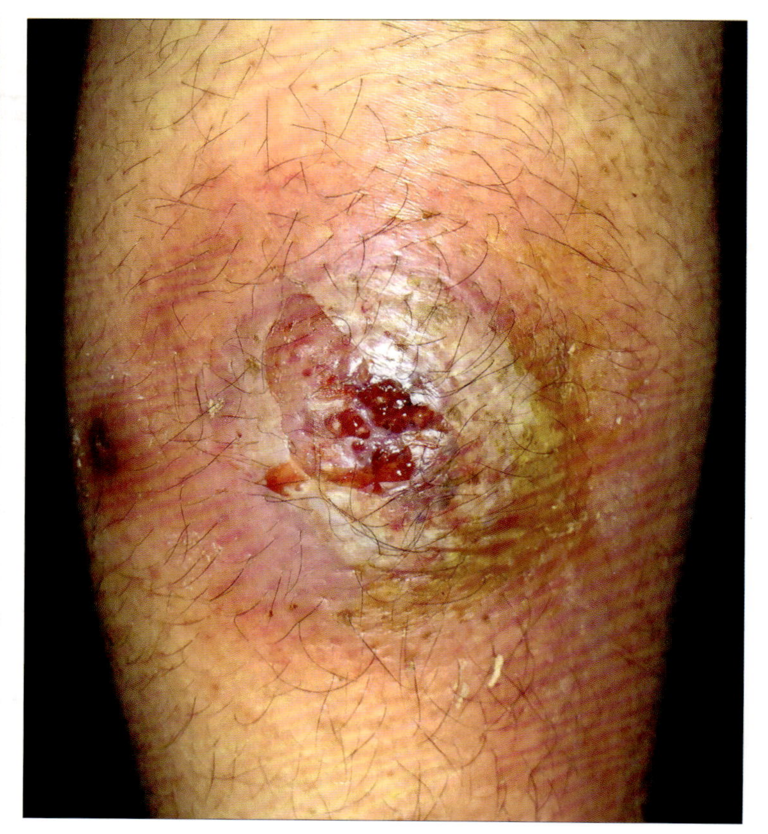

In some infections, a wall of fibrous tissue may form around the inflamed area. Pus can then build up within this wall to form an abscess.

Spread of infection

Some micro-organisms enter the bloodstream and spread quickly to engulf the whole body. Common signs of systemic infection are a fever or rash.

In some cases, infecting organisms, or the toxins they produce, enter the bloodstream and rapidly spread throughout the entire body. This is known as systemic infection and can lead to some characteristic symptoms, such as fever or rash.

Fever

Fever occurs when the immune system cells are damaged by invading pathogens, causing them to release substances called cytokines. These affect the body's 'thermostat' (controlled by the brain), effectively adjusting it to a higher setting. As a result, the normal body temperature is perceived by the brain to be too low, causing shivering to occur, which automatically produces extra heat. This causes the body temperature to rise to a level that is fatal to most invading micro-organisms.

Rash

Skin rashes in systemic infections are caused by multiple areas of skin inflammation as a result of the micro-organisms or their toxins. Rashes indicate that similar damage may be occurring within the body.

Spread of infection

The majority of infections are acquired, directly or indirectly, from other people, and may be spread in the following ways:

- Skin-to-skin contact – if the dose of micro-organisms is large or virulent enough, skin infections may be spread by contact. Some organisms, such as staphylococci, penetrate the sweat glands and hair follicles, causing pustules and boils. For example, impetigo, a bacterial skin infection, can very easily be spread through contact with infected skin.
- Transfer to eye – pathogens may be spread from the fingers to the eye, causing infections such as conjunctivitis. This may be spread from one eye to the other, and can even be transmitted by infected towels or make-up products.
- Transfer to nose – pathogens are often picked up by the fingers and spread to the nose through rubbing. In fact, the rhinoviruses that cause the common cold are more readily transmitted by hand-shaking than by sneezing.
- Inhalation – a number of infections are spread through the inhalation of airborne droplets released during coughing or sneezing, for example respiratory infections like Covid-19, or measles.
- Ingestion – although stomach acid destroys the majority of ingested pathogens, some manage to survive and pass through to the intestines. The consumption of contaminated food or water can spread infection in this way, causing gastroenteritis. Food poisoning can also be caused by food contaminated with material from the infected hands of food handlers.
- Faecal contamination – this is a common cause of infection, since faeces can contain pathogens that are transmitted to food prepared by a person who has not washed their hands properly (such as in salmonella poisoning). Certain viruses (enteroviruses) can be spread by the ingestion of faecal traces, for example the viruses that cause polio and hepatitis A.
- Pregnancy – infection can be spread directly from mother to baby during pregnancy via the placenta, for example, toxoplasmosis. During birth, babies may also contract infections such as herpes or streptococcus through contact with an infected vagina.
- Blood – pathogens in blood can be transmitted by the use of an infected syringe, or through tattooing and ear-piercing with unsterile instruments. HIV and hepatitis B can be spread in this way.
- Sexually-transmitted infection – many diseases (such as herpes, chlamydia and shigella) can be spread during sexual activity due to intimate contact and exchange of bodily fluids.

Animal contact

A few infections develop through contact with animals and insects. Some, such as rabies, may be acquired from infected animals; others, like malaria, are transmitted by insects that act as vectors for a disease.

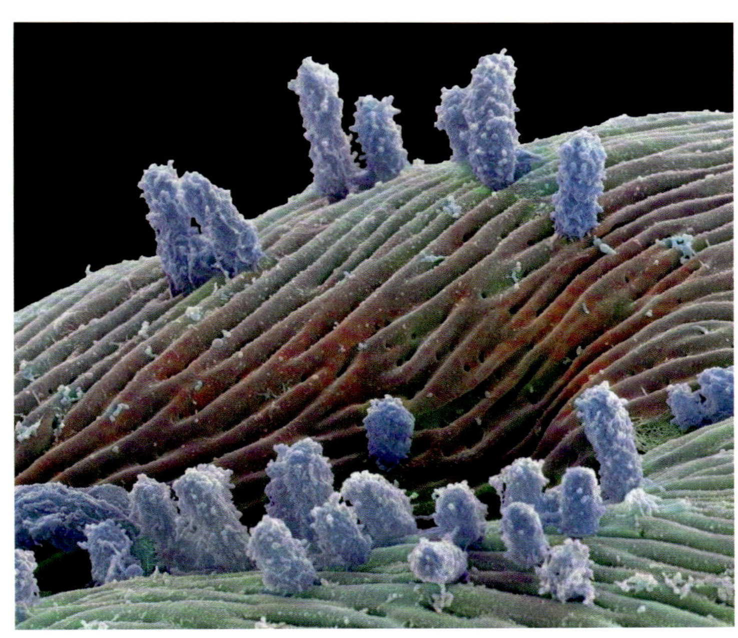

Infection can develop by drinking water that is contaminated by bacteria such as pseudomonas or E.coli.

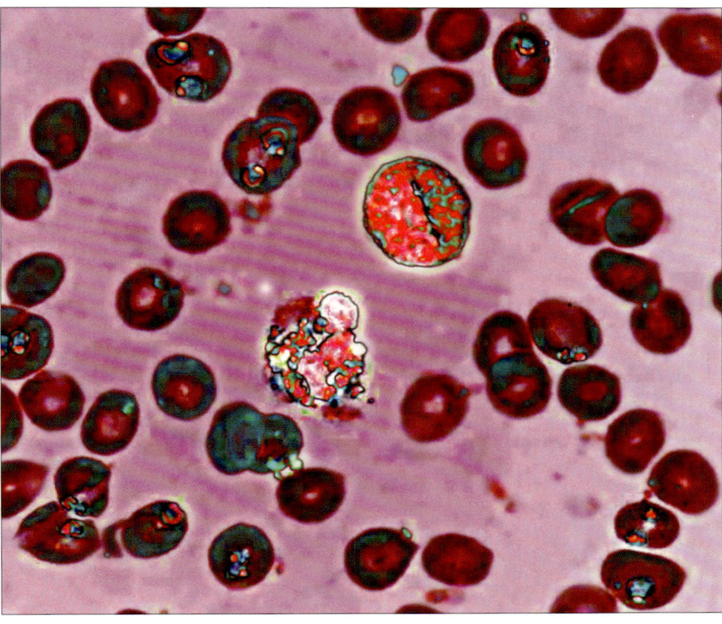

Merozoites – malaria-causing organisms – are shown here invading red blood cells, where they multiply.

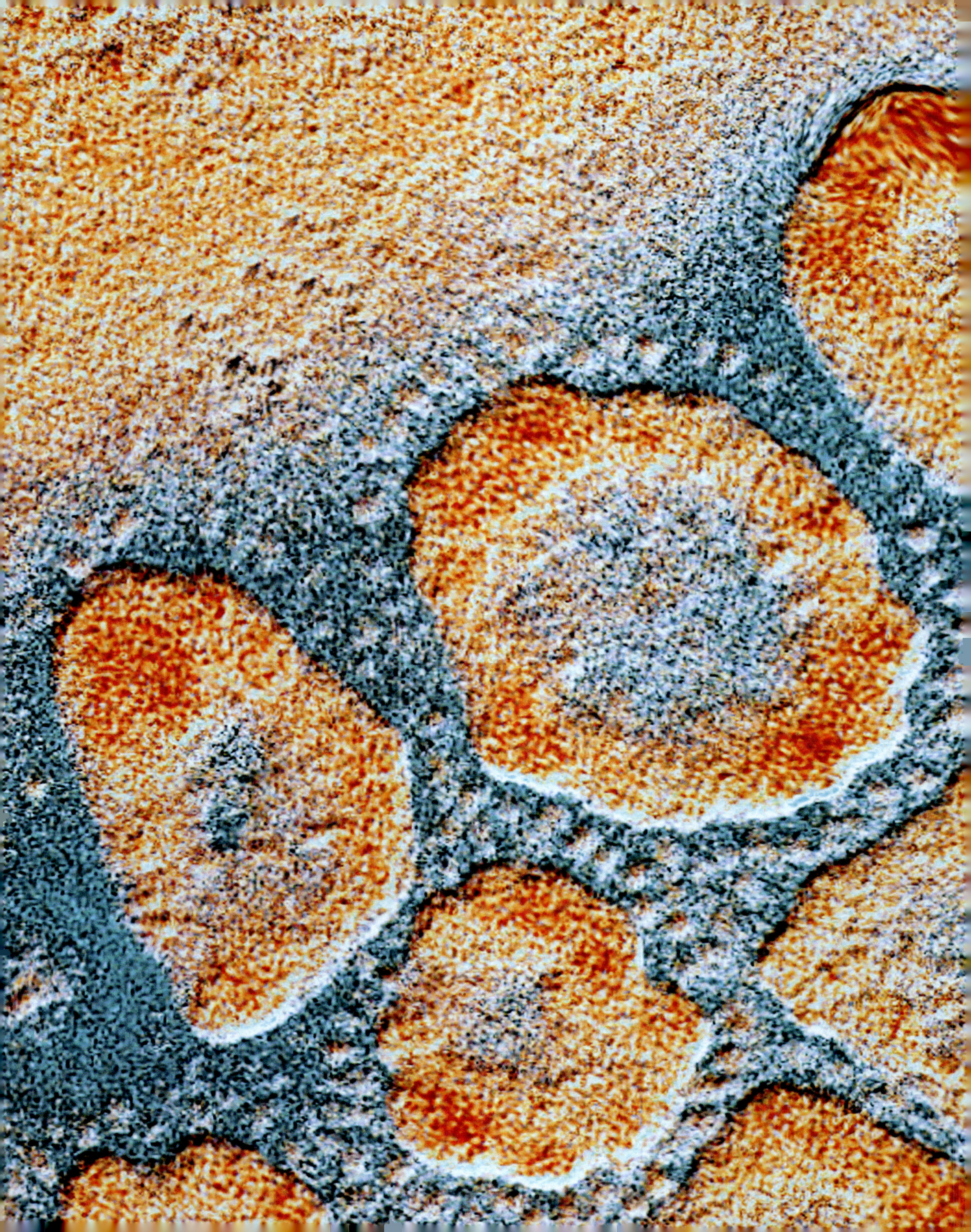

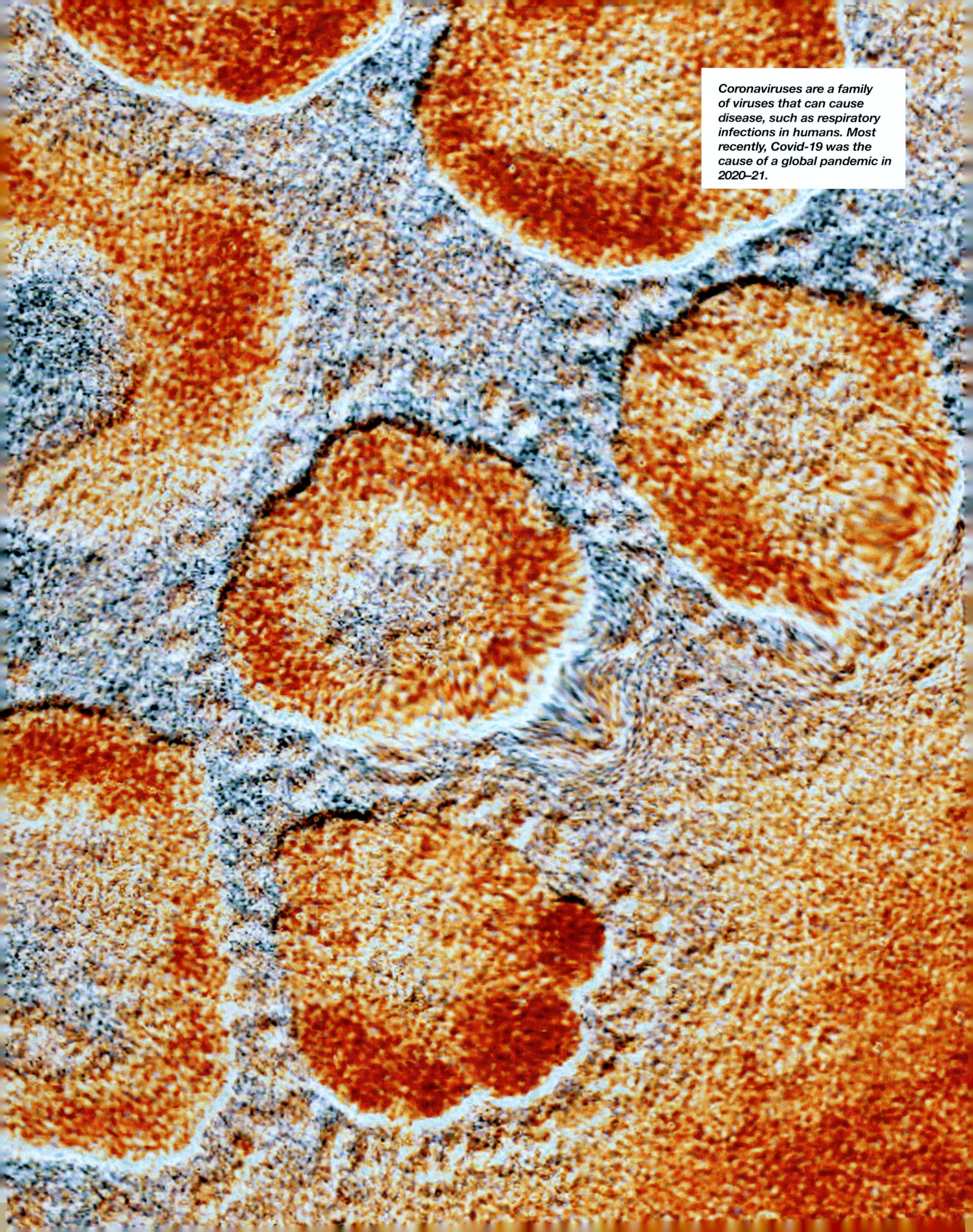

Coronaviruses are a family of viruses that can cause disease, such as respiratory infections in humans. Most recently, Covid-19 was the cause of a global pandemic in 2020–21.

Allergy

An allergy is an inappropriate response by the body's immune system to a normally harmless substance. Allergies vary from hay fever and asthma to life-threatening anaphylactic shock.

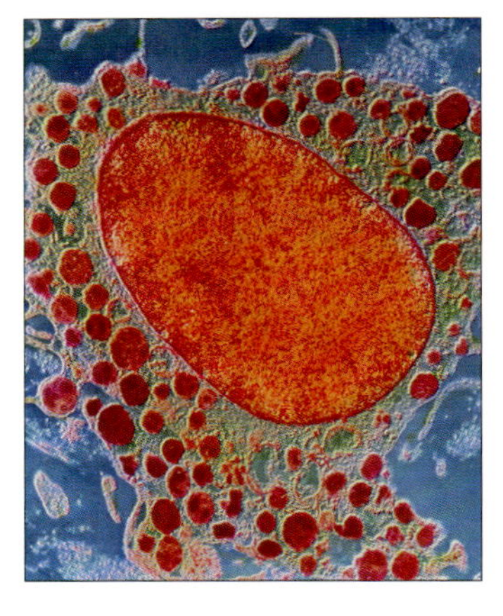

Above: Mast cells are large cells found in connective tissue. Histamine (which helps the body to fight infection) is produced in the cells' granules (shown in black).

An allergy is a hypersensitivity of the body to a particular substance. If the body comes into contact with this substance, unpleasant and even life-threatening symptoms may occur.

Immune Reaction

Allergies occur when the immune system – the body's defence against infection – misidentifies an innocuous substance as being harmful and overreacts to it. This can result in mild symptoms such as a rash or runny nose or, in some cases, life-threatening shock. Allergies can be caused by any substance, but typical allergens are pollen, penicillin, latex, peanuts and shellfish.

The main components of the body's immune system are lymphocytes (white blood cells). B-cells are able to identify foreign particles (antigens) and form appropriate antibodies (immunoglobulins) specifically engineered to fight them. There are five basic types of antibodies: IgA, IgD, IgE, IgG and IgM. The immunoglobulin responsible for allergic reaction is IgE.

Allergies can be inherited, so that the gene responsible for producing the protein that enables lymphocytes to distinguish between threatening and non-threatening proteins is faulty. This means that in a person allergic to shellfish, for example, a B-cell is unable to recognize that a protein ingested as part of a meal containing shellfish is not invading the body. As a result, the B-cell produces large quantities of IgE antibodies.

Sensitization

These antibodies subsequently attach themselves to basophils (a type of white blood cell) and mast cells (found in connective tissue) in the body, causing the body to become sensitized to the allergenic protein.

Basophils and mast cells both produce histamine, an important weapon in the body's defence against infection. When released in extreme quantities, histamine can have a devastating effect upon the body.

Over a period of around 10 days from initial exposure to the allergen, all the body's basophils and mast cells are primed with IgE antibodies and the body becomes sensitized to that allergen. If the body then comes into contact with the allergen for a second time, it will be prepared to attack immediately and a cascade reaction occurs, in which a domino effect is triggered.

Allergic Cascade

The allergic cascade occurs as follows:

1) The body and the allergen come into contact.

2) The cells of the immune system are stimulated.

3) The IgE antibodies within the body are alerted.

4) The IgE antibodies, bound to the surface of the mast cells and basophils, recognize the allergen by the specific protein markers on its surface.

5) The IgE antibodies, still attached to the mast cells and basophils, attach to the surface proteins of the allergen. The healthy mast cells and basophils are destroyed (degranulation). Histamine is released, which causes the surface blood vessels to dilate, leading to a drop in blood pressure; the spaces between surrounding cells fill with fluid.

6) Depending on the allergen, and where the reaction occurs, this may result in immediate symptoms. For example, if the reaction occurs in the mucous membrane of the nose, it may cause symptoms of hay fever, such as sneezing.

Non-allergic Reaction

In a non-allergic person, the allergic cascade fails to progress because the allergen is destroyed. A group of around 20 proteins that are present in the blood bind, one by one, to the allergen/antibody site. When the string of proteins is complete, the allergen is destroyed.

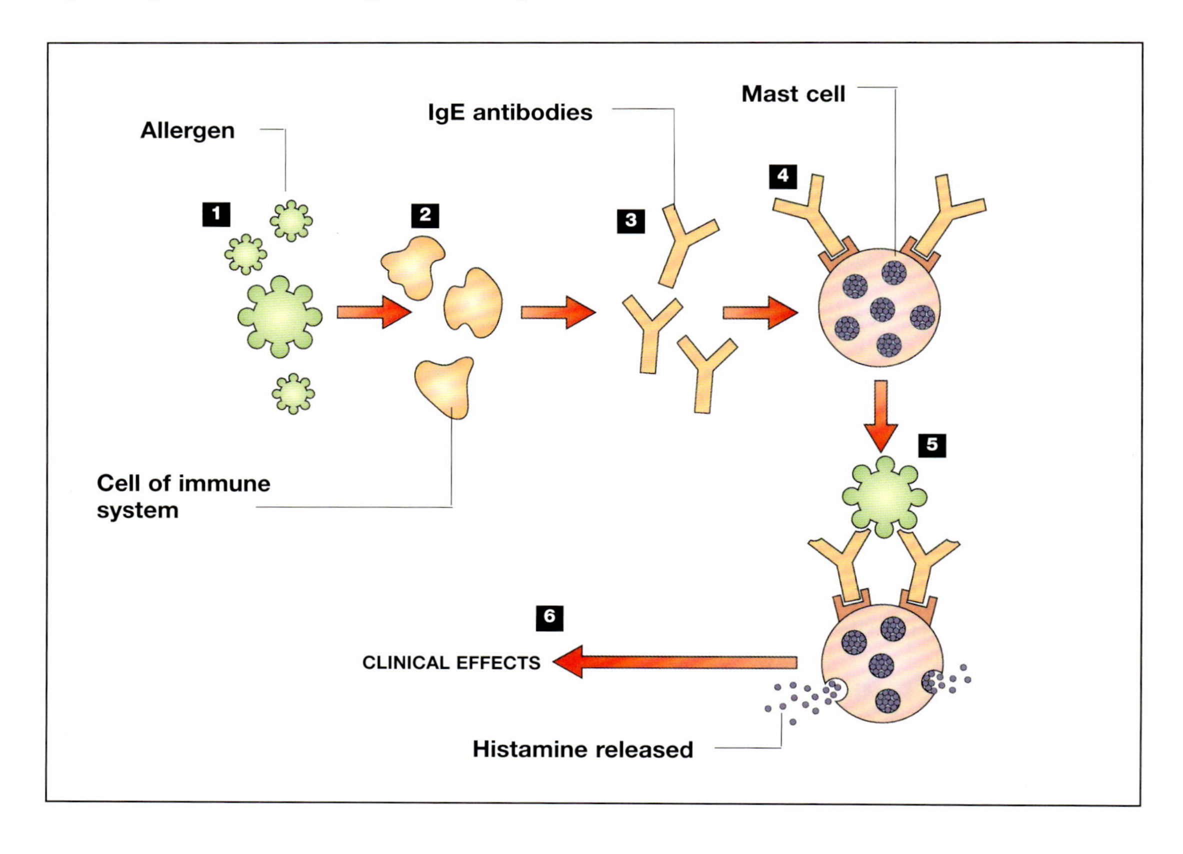

Right: When an individual develops an allergic response to a substance, a domino reaction occurs. A chain of events is set in motion, known as the allergic cascade.

Anaphylaxis

Anaphylactic shock is an extreme allergic reaction that affects the whole body. Without treatment with adrenaline, the condition may be fatal.

In some cases, an allergic reaction can involve the entire body; this is known as a systemic reaction. During this reaction, histamine is released throughout the body, causing capillaries in many tissues to dilate. Anaphylaxis occurs when the reaction is so severe that the blood pressure becomes dangerously low. In extreme cases, the blood pressure drops so low that the body goes into shock. This is known as anaphylactic shock, and is often a fatal condition.

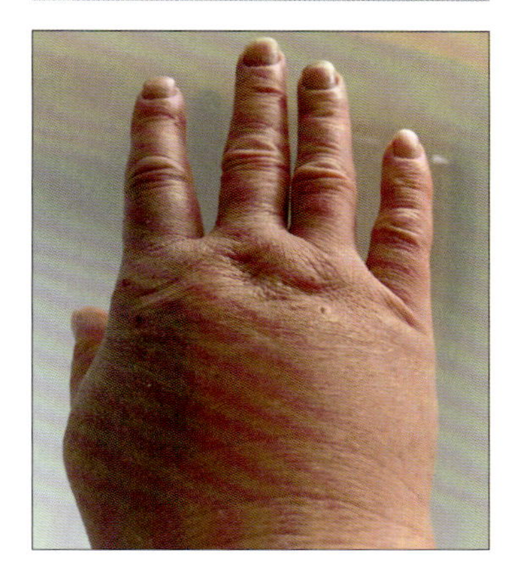

A severe allergic reaction can cause localized swelling, known as oedema, in the tissues. This man has been stung by a bee on his lip resulting in inflammation.

Severe reaction

Anaphylaxis develops very suddenly and presents in a number of ways. A person may rapidly develop a rash and the throat may swell as cells release fluid into surrounding tissue, causing breathing difficulties. A dangerous and rapid drop in blood pressure accompanies this as the blood vessels throughout the body dilate. The brain and other vital organs become starved of oxygen and, within a matter of minutes, the person may die. Even if the victim survives this form of allergic reaction, the brain and kidneys may be permanently damaged.

Adrenaline

The only effective treatment for anaphylaxis is an intramuscular injection of adrenaline, a hormone naturally produced by the adrenal glands. Adrenaline counteracts the symptoms caused by excess histamine by constricting the body's blood vessels and opening the airways. It is vital that the injection is administered correctly at the onset of symptoms for it to be effective.

People who are aware of a serious allergy usually carry an injection for self-administration.

Diagnosis for allergies

If a person suspects that they have an allergy, they can request tests to determine its exact nature. The skin prick test is a common means of determining the cause of an allergy. This involves applying a diluted extract of a possible allergen to the skin (usually of the forearm) and then pricking the skin under the allergen with a needle. If swelling or redness develops in the pricked area, it indicates that IgE antibodies to that allergen are present.

Blood tests are also used to diagnose an allergy, especially in young children, since exposing a child to even minute amounts of allergen during a scratch test could trigger an anaphylactic reaction.

Neither technique is completely accurate, but a combination of both tests along with a patient's medical history can aid diagnosis and the formulation of a treatment plan.

Managing the allergy

Once identified, many allergens, such as dog hair or shellfish, can simply be avoided. However, some allergens, such as pollen, mould or dust, that are present in the environment cannot be avoided. The resulting allergies are kept in check using antihistamines, decongestants, corticosteroids and, in the case of anaphylaxis, adrenaline.

Immunotherapy

For people with severe allergies that cannot be avoided or managed with medication, immunotherapy may be their only hope of leading a normal life. This involves a number of injections of the specific allergen, starting with a very weak dilution and building up to a higher dose that can be maintained over time.

These injections allow the immune system to adjust and desensitize to the allergen over time, so that it produces fewer IgE antibodies. Immunotherapy also stimulates the production of IgG antibodies, which block the effects of IgE. The treatment is expensive, time-consuming and entails risk (such as severe allergic reaction).

Skin prick tests are often used to identify allergens responsible for allergic conditions. Allergens include pollen, fungal spores and dust.

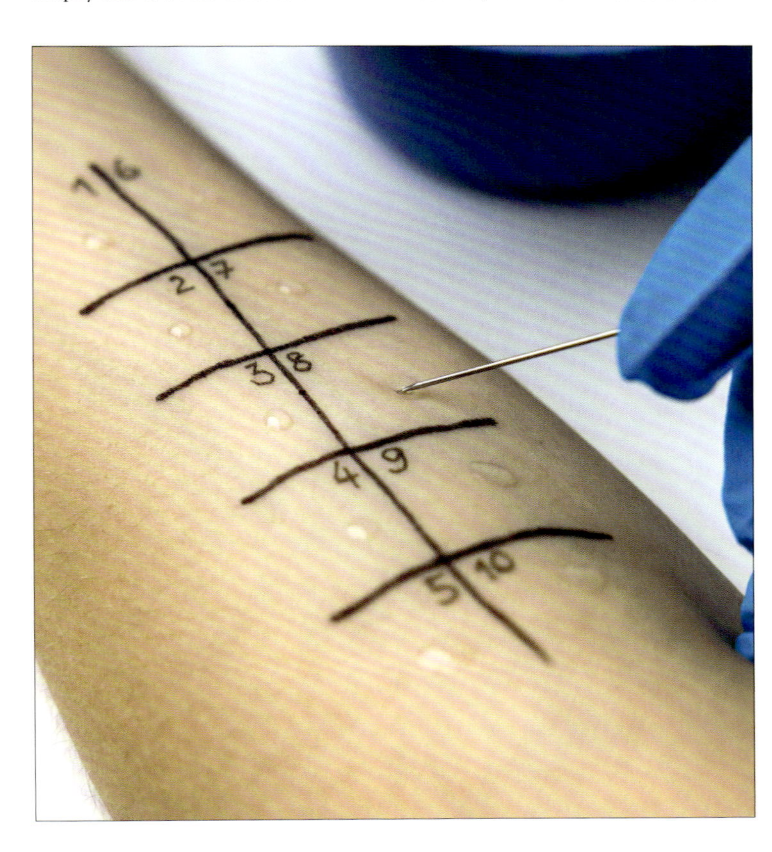

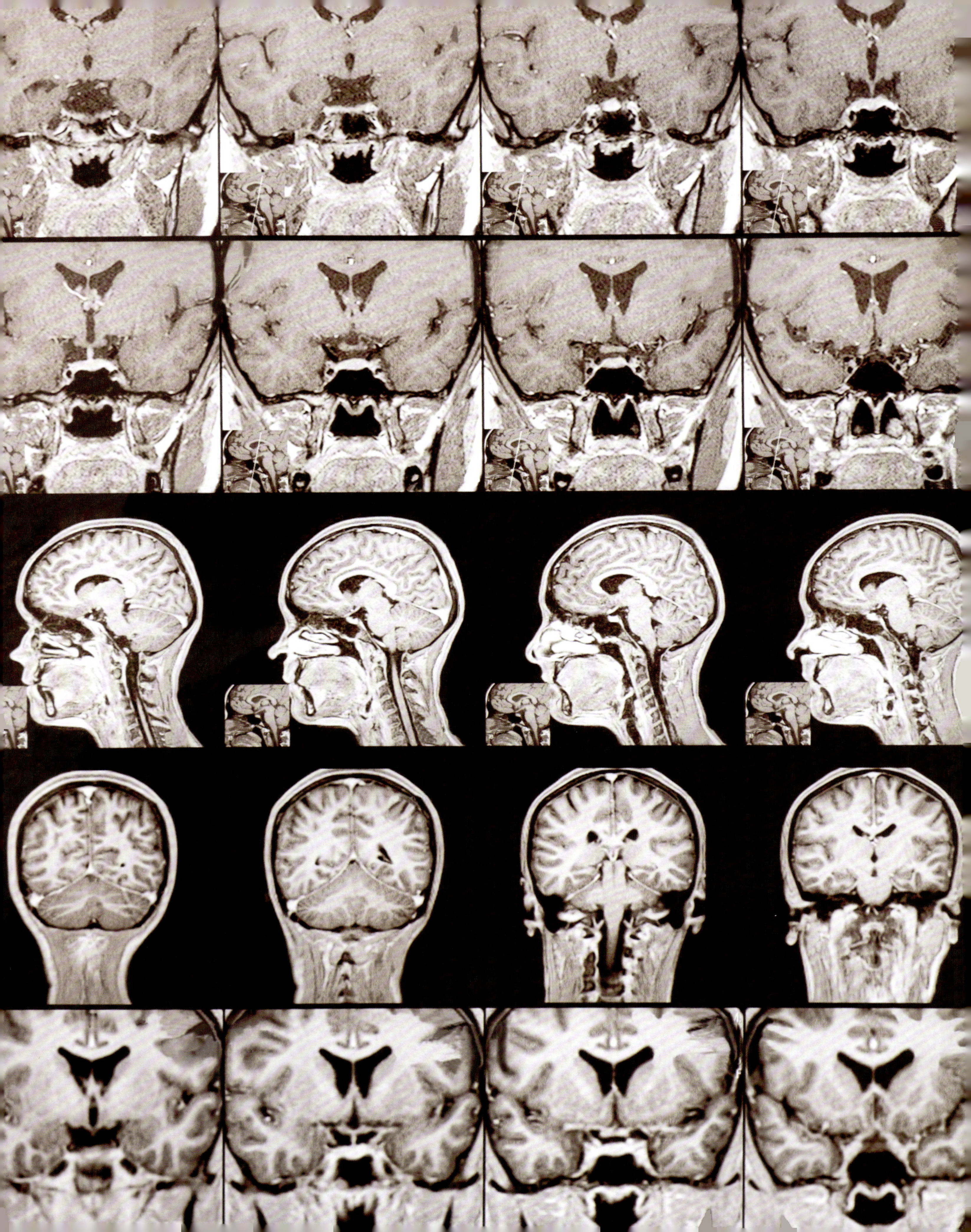

CHAPTER 11

The Endocrine System

Like the nervous system, the endocrine system's main function is to transmit messages. The endocrine system influences the 'behaviour' of distant organs and tissues using chemical messengers known as hormones. There are many different hormones, each with specific responsibilities, but all vital to maintaining homeostasis – a stable, functioning environment within the body.

Hormones are produced by a group of glands in the body, the most important of which is the pituitary gland, which manufactures eight hormones as well as receiving two from the nearby hypothalamus. The pituitary, thyroid and parathyroid glands and adrenal glands purely produce hormones, while the pancreas, ovaries and testes have other functions outside the endocrine system, such as manufacturing digestive enzymes and producing eggs and sperm for reproduction.

This chapter explains the vital functions of hormones and describes the anatomy of the endocrine glands. Where glands have other functions, these will be discussed in the appropriate chapters.

Opposite: A magnetic resonance image (MRI) of the head of a healthy person, showing the pituitary gland.

Endocrine System

The endocrine system is vital in maintaining homeostasis in the body by controlling the release of chemicals called hormones that 'instruct' cells how to behave.

The glands and tissues of the endocrine system control many of the body's functions by secreting substances called hormones that are released into the bloodstream and have specific effects on cells, enabling homeostasis (a constant and stable internal environment). The rate at which hormones are secreted is carefully regulated by feedback mechanisms and by need and demand. As a result, there should never be over- or under-production of hormones in a healthy person.

Endocrine and Exocrine Glands
Glands are clusters of tissue that secrete chemicals and other substances. There are two types of gland: exocrine and endocrine.

- Exocrine glands secrete through ducts directly onto a surface or into ducts. For example, sweat glands produce sweat that is carried through a duct to the surface of the skin. Likewise, lacrimal glands manufacture tears that are released onto the surface of the eye through a tear duct.

- Endocrine glands produce hormones, substances that pass directly into the bloodstream to be carried to target organs or cells. The endocrine glands include the hypothalamus and pituitary glands, thyroid and parathyroid glands, adrenal glands and pineal gland.

Other Hormone Glands
There are other important organs that have different functions but which also contain endocrine tissue that secretes hormones. These include the pancreas and the ovaries and testes.

Nervous System
The endocrine system works closely with the nervous system; both systems coordinate to stimulate and release hormones. The hypothalamus in the brain is the main link between the nervous system and the endocrine system as it controls the pituitary gland.

The response to stress is triggered by the sympathetic nervous system. The hormones adrenaline and noradrenaline are secreted by the adrenal medulla and are carried in the bloodstream to many different organs, reinforcing the effects of the sympathetic nervous system.

Feedback Mechanisms
The production of hormones is regulated by negative feedback control. An example of this mechanism is blood calcium control. When the body's level of calcium drops, the parathyroid glands are stimulated to increase production of a parathyroid hormone, which acts on other parts of the body to release calcium into the bloodstream.

Response to stress

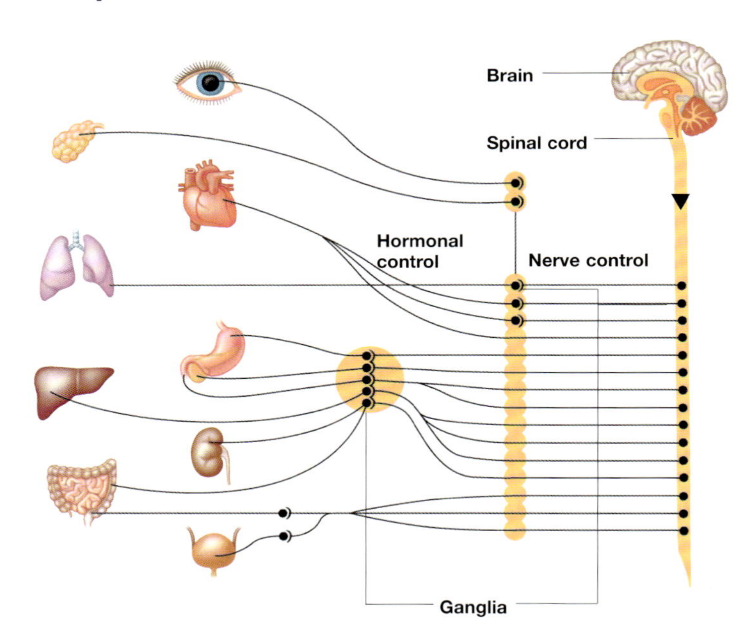

Under conditions of stress, the sympathetic nervous system triggers production of hormones in the adrenal and pituitary glands.

Sweat gland

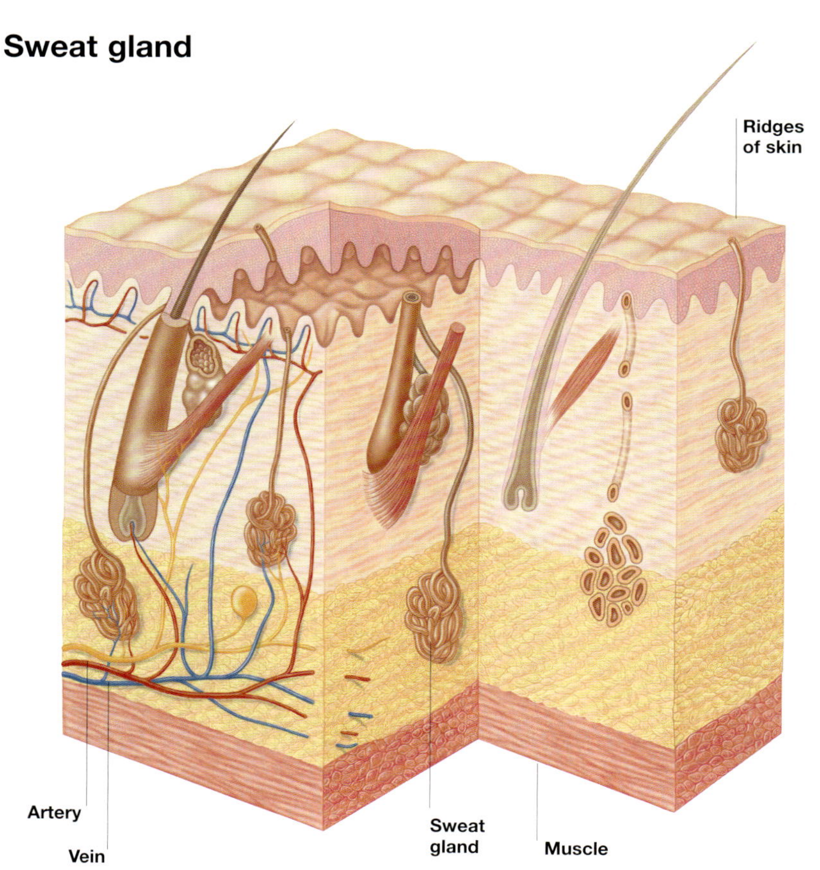

Sweat glands are an example of an exocrine gland; glands that secrete into ducts and onto surfaces. Endocrine glands release hormones directly into the blood.

Hormones

Hormones are chemicals that are released by an endocrine gland into the bloodstream, which then carries them to distant sites throughout the body.

Hormones may have a specific site of action or can affect a wide variety of different cells, simultaneously regulating a large number of different bodily processes.

For a hormone to affect the internal biochemistry of a cell (known as a cell's 'behaviour'), the cell must have an appropriate protein receptor embedded within its cell membrane.

Glandular tissue
The cells within a gland release certain hormones into the bloodstream

Blood vessel
The hormone travels through the bloodstream to reach target cells

Target cells
Only target cells are affected by the hormone

Hormones are chemical messengers that are released by a gland into the bloodstream to affect distant tissues.

Categories of hormone

There are three categories of hormones:

- Amines, for example, those secreted by the thyroid gland and adrenals
- Proteins and peptides, made from chains of amino acids, which include insulin (to control blood sugar) and oxytocin (to stimulate labour)
- Steroids, such as cortisol, the sex hormones and alsdosterone.

All have one common function and that is to alter the activities of cells.

Receptor cells

There are about 50 different hormones produced in the body, all of which are linked to specific cells, known as target cells. When triggered by feedback systems, hormones are able to locate their target cells by recognizing their receptors. The receptors are protein molecules that are embedded in the plasma membrane of the cell and are only able to recognize certain hormones.

The hormone is able to bind to its specific receptor, deliver its message and activate the cell to behave in a certain way. This might be to manage a stressful or dangerous situation or to release chemicals to affect metabolism.

Hypothalamus-pituitary complex

Of all the endocrine glands in the body, the pituitary gland is the most influential. This tiny gland in the brain manufactures 10 hormones that have far-reaching actions. The pituitary gland works together with an area of the brain called the hypothalamus, which regulates levels of chemicals in the body via feedback systems and issues instructions to increase or reduce the production of specific hormones accordingly.

Neurohormones

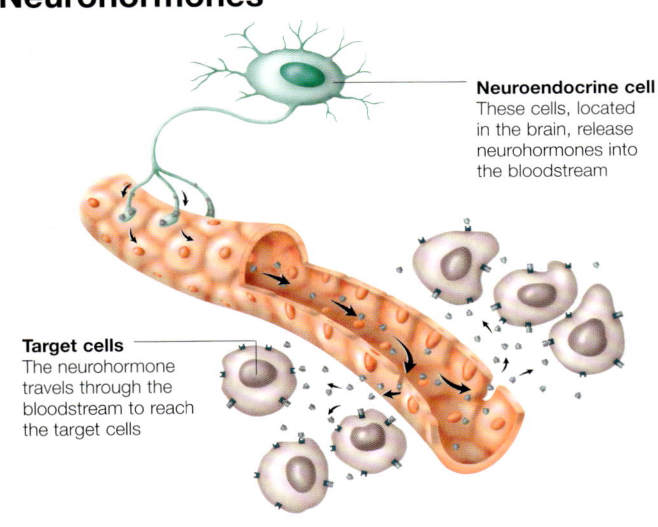

Most neurones communicate with each other by releasing a chemical messenger that diffuses between the gap (called a synapse) that separates them. However, some neurones do not synapse with another nerve cell.

Rather, their synaptic terminals are located near blood vessels; when these neurones are stimulated they release a neurohormone into the bloodstream, which is then carried to distant target organs in much the same way that a hormone is released from a gland.

Neurohormones are released by specialized nerve cells called neuroendocrine cells. These chemicals are carried in the bloodstream to the target cells.

Oxytocin

Oxytocin is a neurohormone that is released into the bloodstream by neuroendocrine cells located in the hypothalamus. This occurs in response to the stimulation of sensory nerves in the mother's nipple by a suckling infant. The blood carries the neurohormone to the mammary gland where it causes milk to be ejected from the nipple.

Hypothalamus

The hypothalamus is located in the deep core of the brain. It regulates fundamental aspects of body function, and is critical for homeostasis – the maintenance of equilibrium in the body's internal environment.

The hypothalamus lies below the thalamus in the floor of the diencephalon, the structures that surround the third ventricle of the brain. It is part of the endocrine system in that it controls the pituitary gland, which descends from its base. A hollow structure, known as the infundibulum, or pituitary stalk, connects the hypothalamus to the posterior part of the gland.

Regulating Factors

The hypothalamus regulates a wide range of basic processes and is the main link between the central nervous system and the endocrine system, controlling pituitary gland function. Chemicals known as regulating factors are secreted by the hypothalamus and delivered directly to the pituitary gland via the blood circulation, where they control the levels of hormone production in the body depending on need. An example of these regulating factors is growth hormone-releasing factor (stimulates release) and growth hormone-inhibiting factor, or somatostatin (inhibits release).

Hypothalamic nuclei

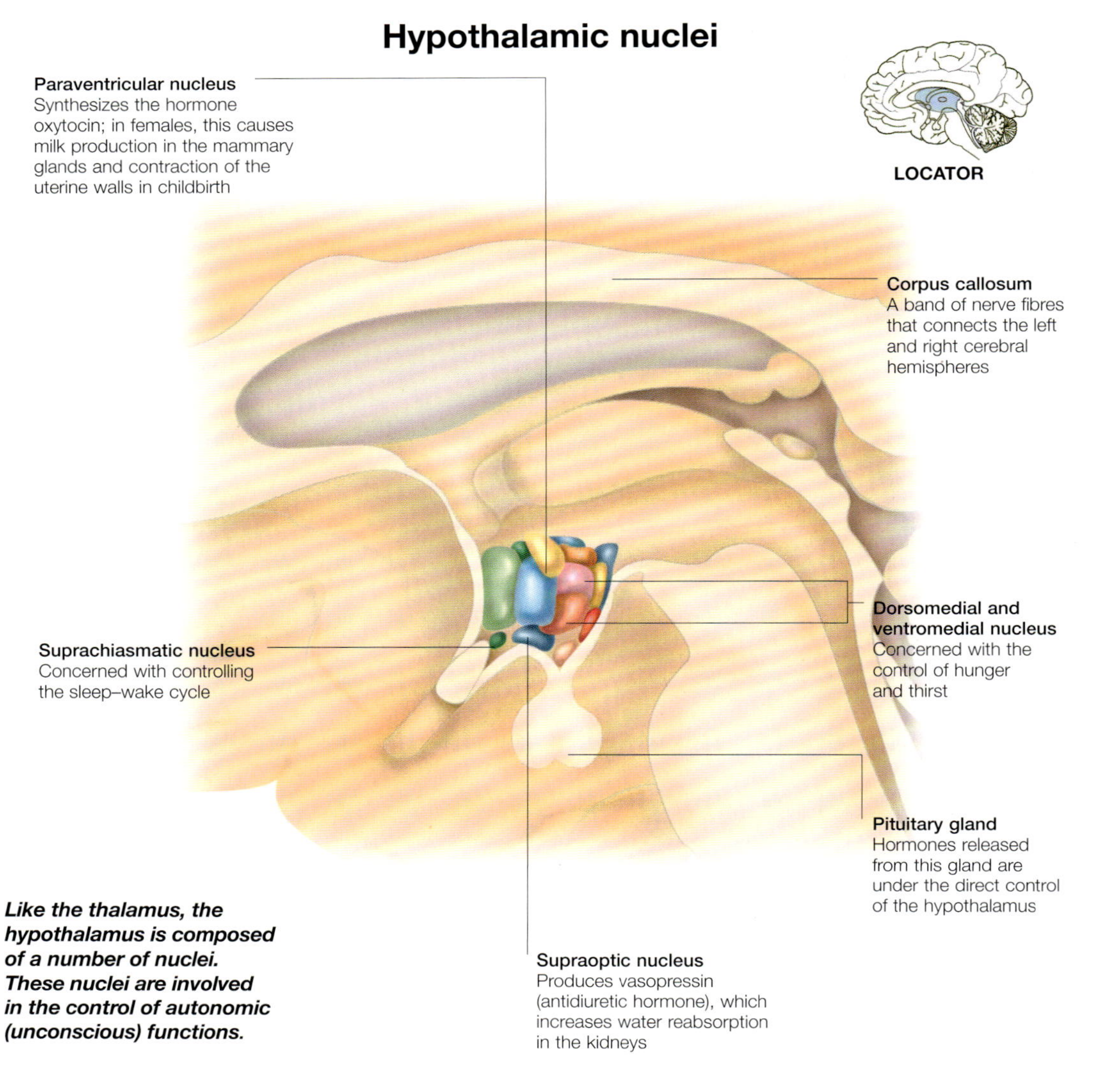

Like the thalamus, the hypothalamus is composed of a number of nuclei. These nuclei are involved in the control of autonomic (unconscious) functions.

Hypothalamic control of the pituitary

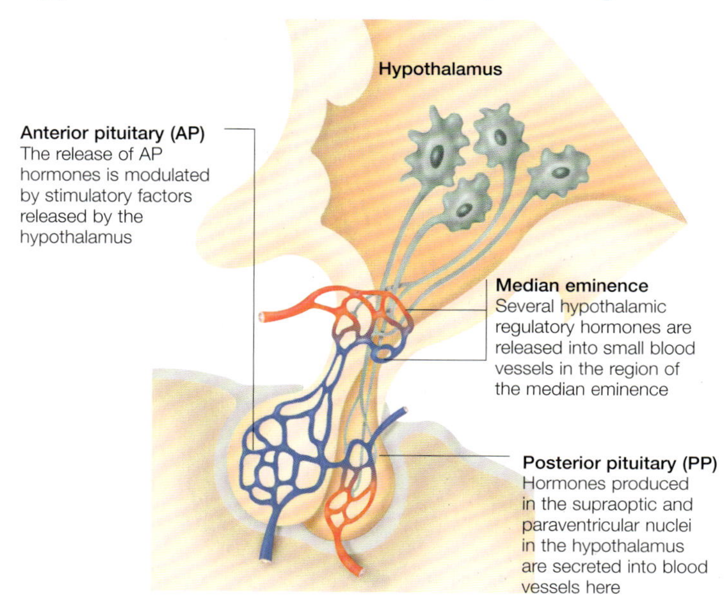

Among its many functions, the hypothalamus is the main link between the central nervous system and the endocrine system, controlling pituitary gland function. It lies just above the pituitary gland and is responsible for detecting changes in the body's environment.

Regulating factors

In response to these changes, the hypothalamus releases chemicals known as regulating factors that have a direct influence on the pituitary gland. These regulating factors either inhibit or stimulate the production of hormones in direct reponse to the feedback messages that have been received. An example of these regulating factors are those hormones that inhibit or stimulate growth. Overproduction of growth hormone could result in abnormally long bones, whereas underproduction could stunt growth, so a balance of regulating factors is vital.

The hypothalamus controls the pituitary gland via nerve fibres innervating the posterior pituitary, and blood capillaries supplying the anterior pituitary.

ADH and oxytocin

As well as influencing the pituitary gland to produce hormones, the hypothalamus produces two hormones of its own. These are antidiuretic hormone, which acts on the kidneys to preserve water, and oxytocin, which stimulates the uterus to contract during labour. Both hormones are stored in the pituitary and only released when necessary.

Pituitary gland

The pituitary gland, or 'master gland' is situated in the brain directly below the hypothalamus. It is controlled by the hypothalamus and is involved in many bodily activities.

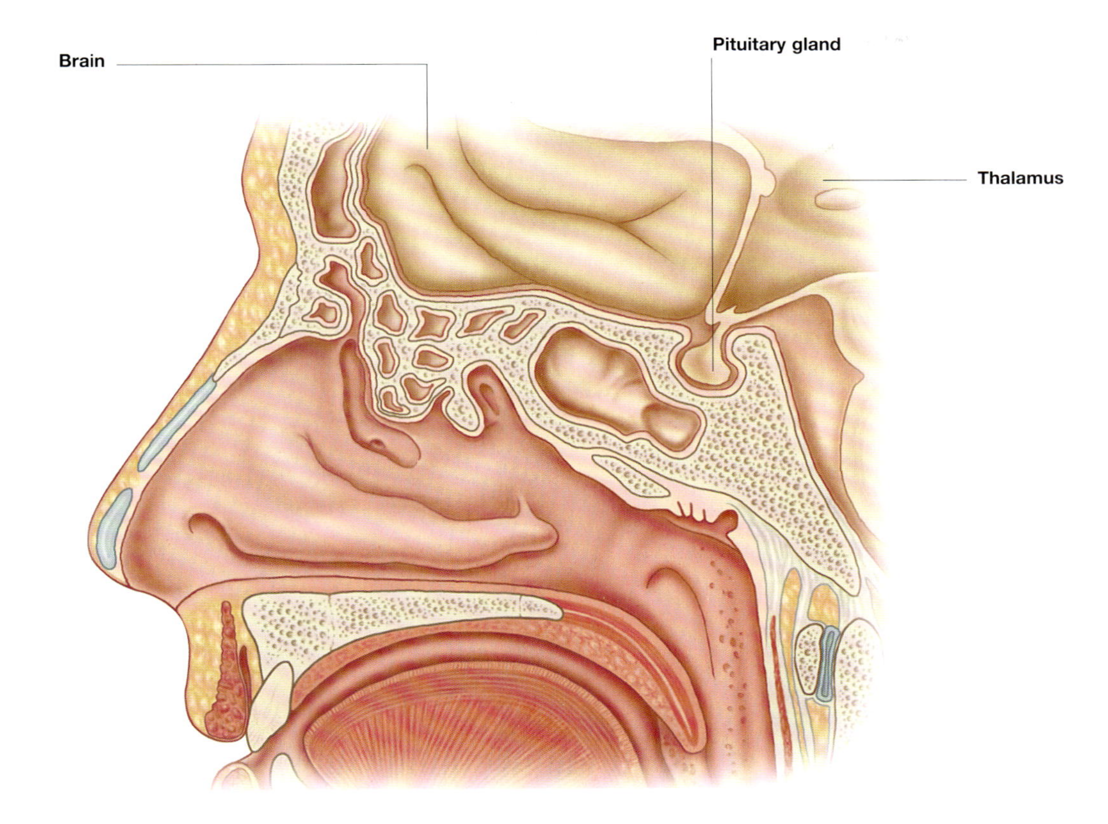

The pituitary gland has two lobes:

- The anterior lobe, also known as the adenohypophysis, is the largest lobe and contains glandular epithelial cells that secrete hormones.
- The posterior lobe, also known as the neurohypophysis, is a smaller lobe. Its neurones are directly connected to the hypothalamus.

Pituitary hormones

There are seven hormones secreted by the anterior lobe of the pituitary gland, many of which stimulate other endocrine glands to act. They include:

- Growth hormone – 'switches on' body cells so that they divide, mainly in the skeleton and muscles
- Prolactin – stimulates the production of breast milk
- Adrenocorticotrophic hormone – stimulates the cortex of the adrenal gland to produce its own hormones
- Melanocyte-stimulating hormone – influences the colour of the skin
- Thyroid-stimulating hormone – controls the thyroid gland
- Follicle-stimulating hormone – plays a role in sexual development, stimulating the production of eggs in the ovaries and sperm in the testes
- Luteinizing hormone – stimulates the ovary to release an egg and acts on the lining of the uterus to prepare it for implantation. It also plays a role in breast milk production. In males, the luteinizing hormone stimulates the testes to produce testosterone.

The posterior lobe of the pituitary produces antidiuretic hormone (ADH) that increases blood pressure and decreases urine production and oxytocin, which cause uterine contractions.

The pineal gland

This tiny gland produces melatonin, which plays an important role in the regulation of the sleep cycle. The production of this hormone is triggered by signals that it receives from the retina in the eye (mediated by the hypothalamus). In darkness the pineal gland secretes melatonin, while the presence of light suppresses production.

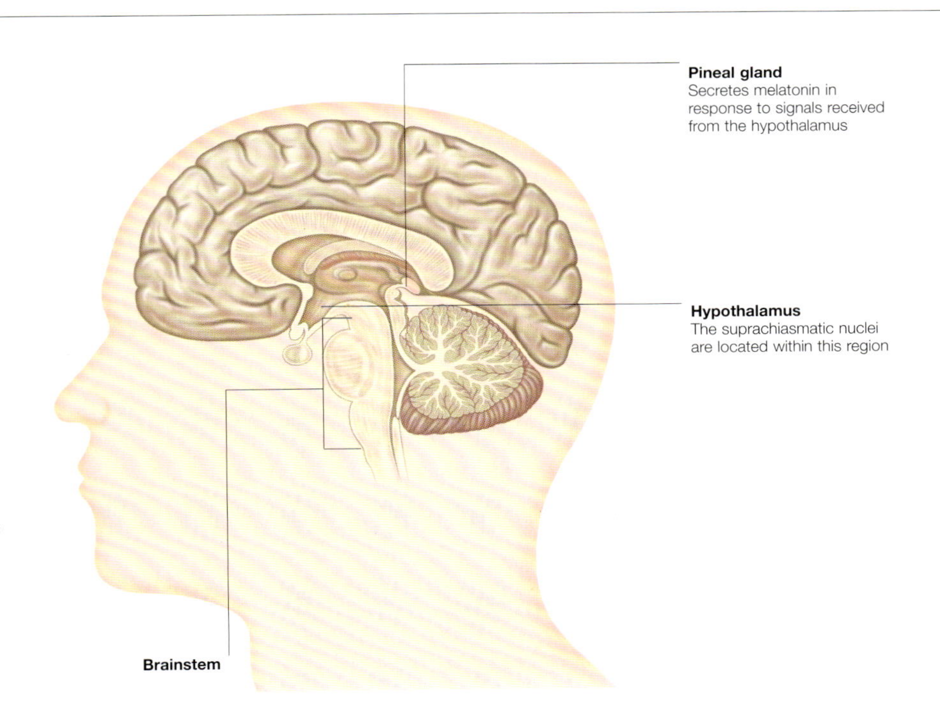

The pea-sized pineal gland is located deep in the centre of the brain. It is linked by neurones to the retina and, in darkness, produces the hormone melatonin to aid sleep.

Thyroid Gland

The thyroid and parathyroid glands are situated in the neck. Together, they produce important hormones responsible for regulating growth, metabolism and calcium levels in the blood.

The thyroid gland is an endocrine gland situated in the neck, lying to the front and side of the larynx and trachea. It is similar to a bow-tie in shape, and it produces two iodine-dependent hormones: tri-iodothyronine and thyroxine.

Thyroid Hormones

These hormones are responsible for controlling metabolism through the promotion of metabolic enzyme production. They also regulate growth, and influence the activity of the nervous system. In addition, the gland secretes calcitonin, which is involved in the regulation of calcium levels in the blood. In children, growth is dependent upon the thyroid gland, through its stimulation of the metabolism of carbohydrates, proteins and fats.

The gland has two conical-shaped lobes connected by an 'isthmus' (a band of tissue connecting the lobes), which usually lies in front of the second and third tracheal cartilage rings. The entire gland is surrounded by a thin connective tissue capsule and a layer of deep cervical fascia.

Pyramidal Lobe

There is often a small third lobe, the pyramidal lobe, which extends upwards from near the isthmus and lies over the cricothyroid membrane and median cricothyroid ligament.

ANTERIOR VIEW OF THE THYROID GLAND

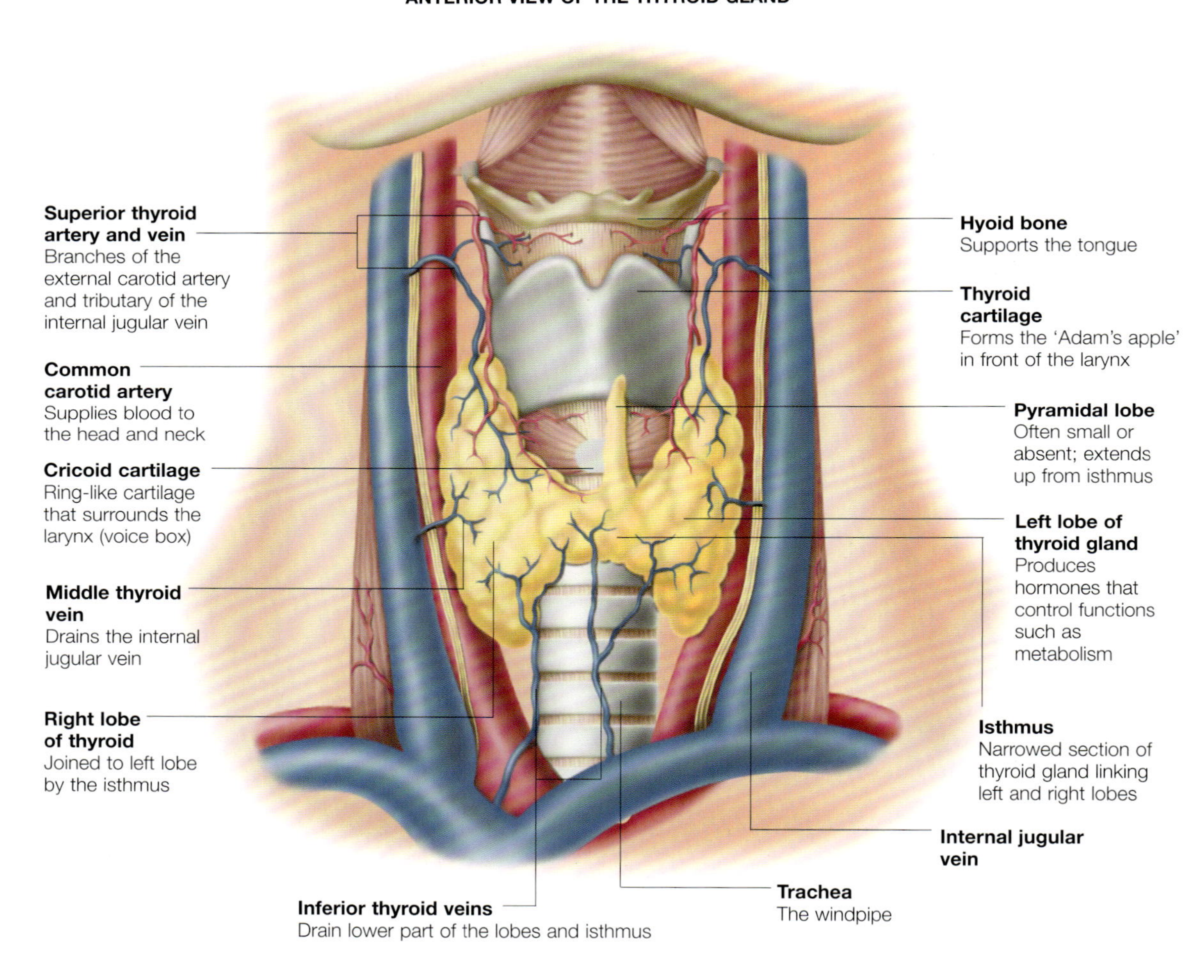

Blood and nerve supply

The posterior view of the thyroid reveals the small parathyroid glands, embedded within the lobes. A rich network of vessels supply the glands.

The thyroid gland is well supplied by blood vessels. The upper pole receives arterial blood from the superior thyroid artery, a branch of the external carotid artery. The lower pole is supplied by the inferior thyroid artery, a branch of the thyrocervical trunk. The lobes of the gland are directly related to the common carotid arteries.

The hormones are distributed to the bloodstream via a network (plexus) of veins in and around the gland, which ultimately drain into the internal jugular and brachiocephalic veins.

Relation to nerves

In addition to the blood vessels, the gland is closely related to nerves. Posteriorly, the most important relation is the pair of recurrent laryngeal nerves from the vagus nerve. These ascend in the groove between the oesophagus and trachea, heading towards the larynx where they supply motor nerves to all laryngeal muscles (except the cricothyroid), and sensory nerves to the sub-glottic larynx.

POSTERIOR VIEW OF THE THYROID GLAND

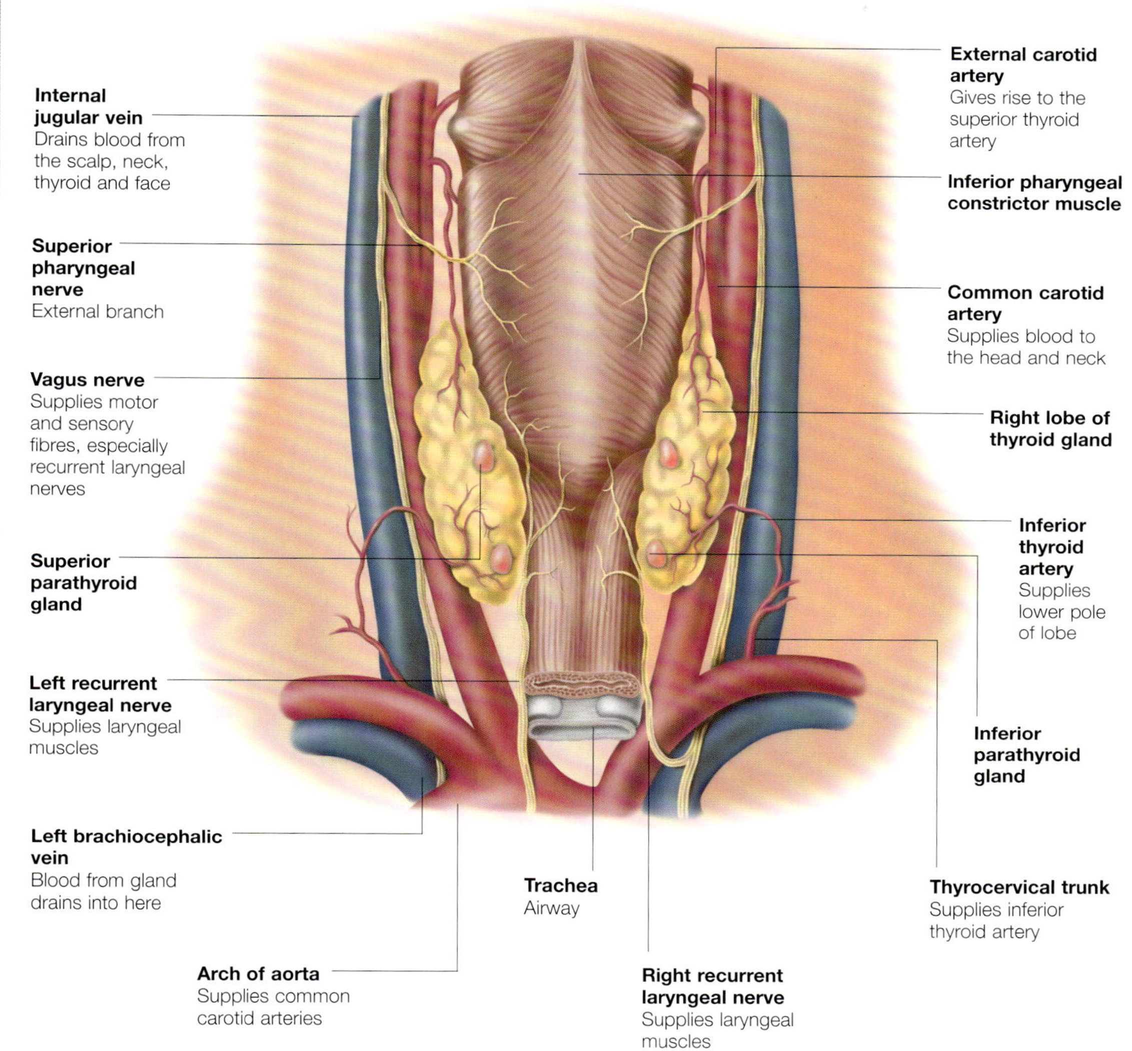

Parathyroid glands

The pea-sized parathyroid glands (superior and inferior) are embedded within the rear tissue of the thyroid gland. They secrete parathyroid hormone (PTH) which, together with calcitonin and vitamin D, controls calcium metabolism.

Parathyroid hormone

The hormone PTH is produced by the principal (chief) cells in the glands. It is responsible for controlling levels of calcium and phosphate in the blood.

When calcium levels are low, PTH stimulates the body to absorb more calcium, magnesium and phosphate from the digestive tract into the bloodstream.

In addition, PTH activates osteoclasts (bone-destroying cells) to break down bone tissue and release calcium and phosphate into the blood.

Kidney changes

The parathyroid hormone also has an effect on the kidneys, increasing the amount of calcium and magnesium that is reabsorbed from the urine and returned to the blood.

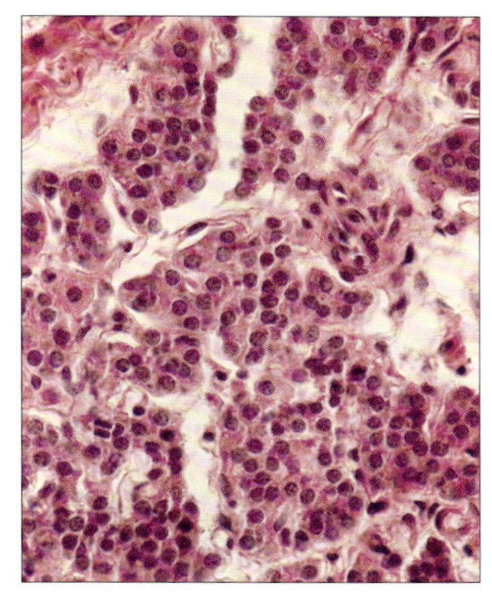

This electron micrograph shows part of a parathyroid gland cell. These glands secrete chemicals vital for calcium metabolism.

Adrenal Glands

The adrenal glands are situated above the kidneys, but are not part of the urinary tract. Each gland consists of two separate parts: a medulla surrounded by a cortex.

Lying on top of the kidneys are the paired adrenal glands, also known as the suprarenal glands. Although these glands are physically close to the kidneys, they play no part in the urinary system: they are endocrine glands that produce hormones vital to the healthy functioning of the body.

Surrounding Tissues

The yellowish adrenal glands lie above the kidneys and under the diaphragm. They are surrounded by a thick layer of fatty tissue and are enclosed by renal fascia although they are separated from the kidneys themselves by fibrous tissue.

This separation allows a kidney to be surgically removed without damaging these delicate and important glands.

Gland Differences

Due to the position of the surrounding structures the soft adrenal glands differ in appearance:

- The right adrenal gland

The right adrenal gland is pyramid-shaped and sits on the upper pole of the right kidney. It lies in contact with the diaphragm, the liver and the inferior vena cava, the main vein of the abdomen.

- The left adrenal gland

The left adrenal gland has the shape of a half moon and lies along the upper surface of the left kidney from the pole down to the hilus. It lies in contact with the spleen, the stomach, the pancreas and the diaphragm.

Blood Supply

As with other endocrine glands, which secrete their hormones directly into the bloodstream, the adrenal glands have a very rich blood supply. They receive arterial blood from three sources – the superior, middle and inferior adrenal arteries – which arise from the inferior phrenic artery, the aorta and the renal artery respectively.

Near the adrenal glands these arteries branch repeatedly, numerous tiny arteries thereby entering the glands over their entire surface.

A single vein leaves the adrenal glands on each side to drain blood into the inferior vena cava on the right and the renal vein on the left.

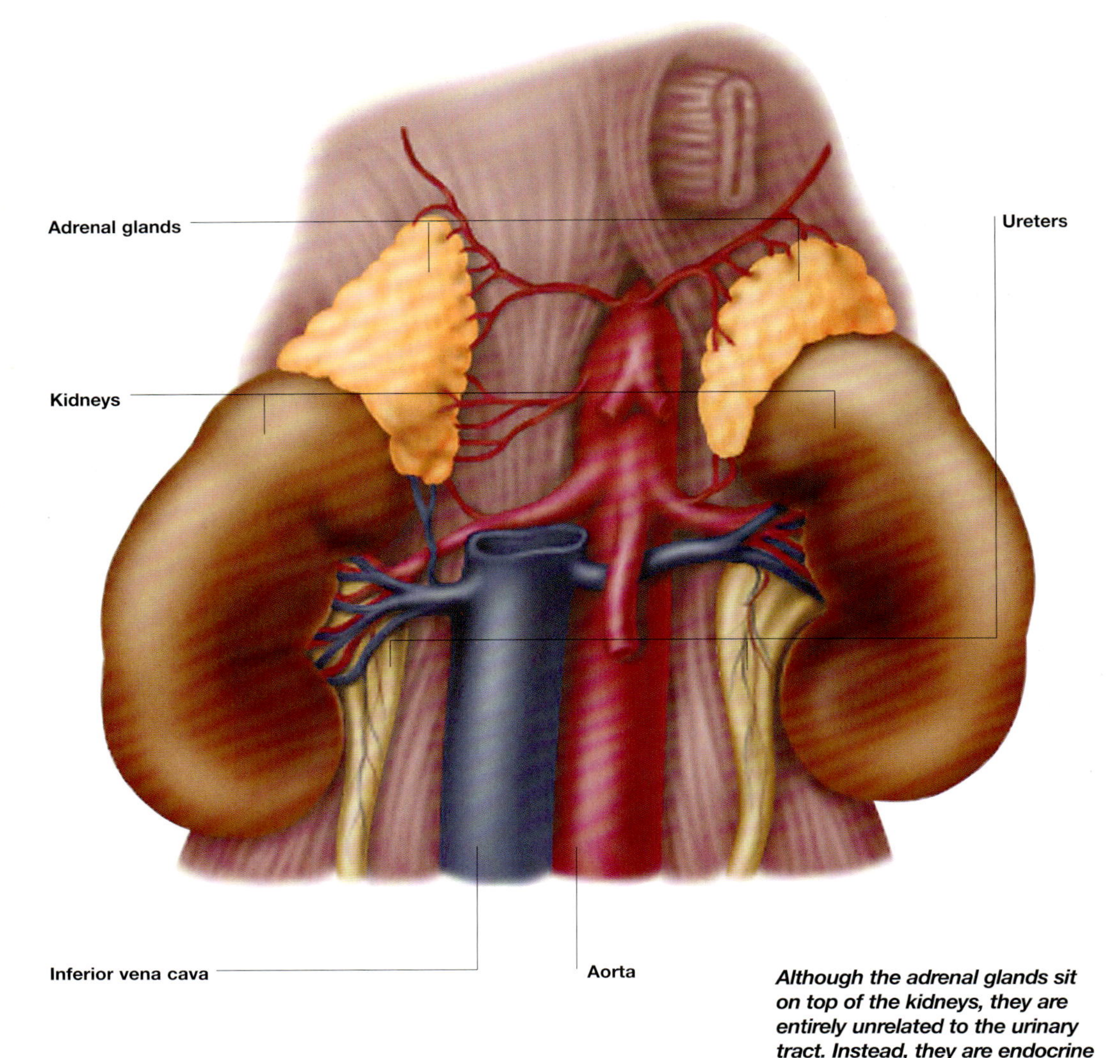

Although the adrenal glands sit on top of the kidneys, they are entirely unrelated to the urinary tract. Instead, they are endocrine glands that secrete hormones into the bloodstream.

Adrenal medulla

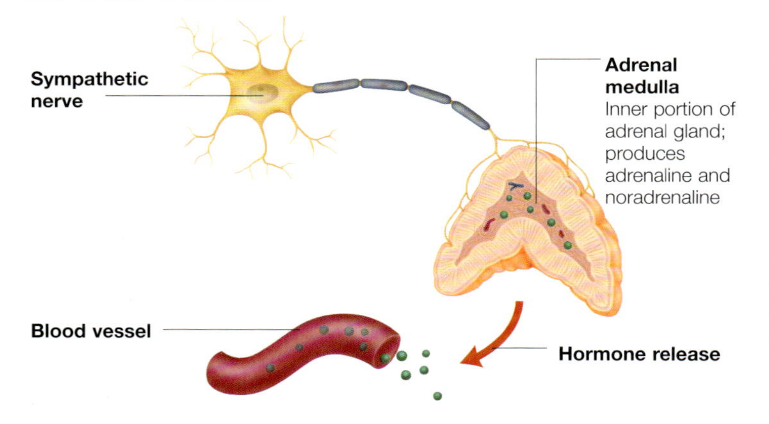

As a reaction to stress, the adrenal medulla is stimulated to release hormones into the bloodstream. These hormones prepare the body for action.

Each adrenal gland has an outer cortex and an inner medulla. The medulla is made up of a 'knot' of nervous tissue and is the site of production of the hormones adrenaline and noradrenaline, which are responsible for the 'fight-or-flight' response.

When triggered by the hypothalamus, adrenaline and noradrenaline act to amplify the effects of the sympathetic nervous system. They increase the heart rate and blood pressure, constrict the blood vessels, increase the breathing rate, slow down digestion and act to increase the effectiveness of muscle contractions. They also increase the blood sugar.

Pancreas

Sugar is an important source of energy that exists naturally in the blood as glucose. The correct balance of blood-sugar levels is vital to life, and this is regulated by hormones secreted from the pancreas.

The pancreas lies in the upper abdomen behind the stomach, one end lying in the curve of the duodenum and the other end touching the spleen. It has two main functions: to produce digestive enzymes that break down food in the stomach (this function is discussed in the digestive system); and to secrete hormones that regulate the levels of sugar in the body directly into the bloodstream. This latter function is part of the endocrine system.

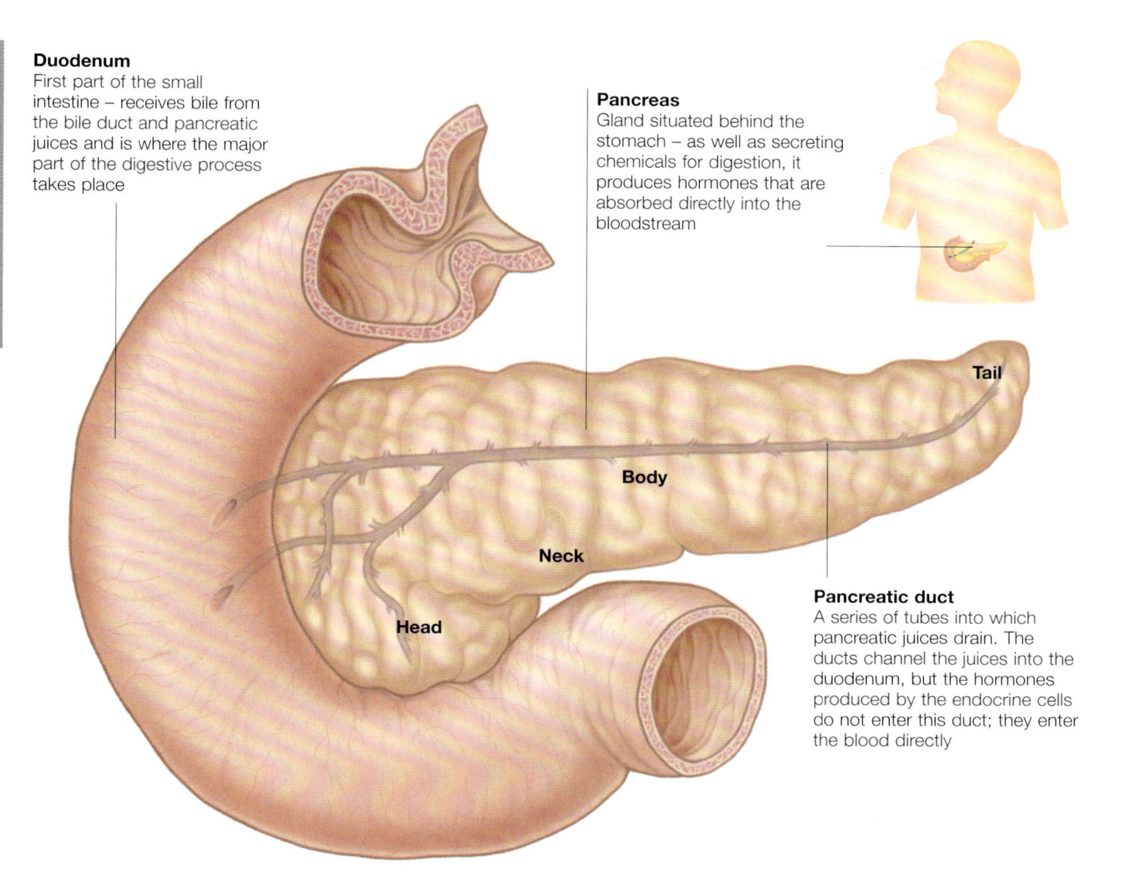

Pancreatic Hormones

The endocrine part of the pancreas is formed of clusters of cells called the islets of Langerhans, which are scattered through the pancreas. There are three types of cells in the clusters:

- Alpha cells – these are responsible for producing the hormone glucagon, which regulates the blood sugar level
- Beta cells – these cells secrete insulin, an important hormone that also regulates blood sugar level
- Delta cells – these produce growth hormone-inhibiting factor (GHIF), otherwise known as somatostatin.

Glucose

Glucose is a simple sugar that is vital for brain function and an important source of energy for the rest of the body. It is stored in the body in the form of glycogen – long chains of sugar molecules found in the liver and muscles and transported round the body in the blood.

Glucagon

Glucagon is a hormone that is responsible for increasing the levels of glucose in the blood. It achieves this by causing glycogen and other nutrients present in the liver, such as amino acid and lactic acid, to convert into glucose. Once converted, the glucose is released by the liver into the bloodstream. Hormones respond to feedback mechanisms and only act if stimulated; if the blood glucose drops, then specialized cells in the islets of Langerhans are stimulated to secrete glucagon. When blood sugar levels rise again, glucagon production decreases.

Insulin

This hormone is produced by the beta cells in the islets of Langerhans and acts in an opposite way to glucagon, by lowering the blood glucose level. It achieves this by encouraging glucose to return to the cells and convert back to glycogen. The production of insulin is also regulated by a feedback mechanism so that when blood sugar levels are high, insulin secretion is stimulated and when low, inhibited. A deficiency of insulin results in the condition diabetes mellitus.

Growth Hormone-inhibiting Factor

GHIF is also known as somatostatin and is secreted by the alpha cells in the islets of Langerhans, as well as in other parts of the body. Somatostatin prevents the release of pancreatic hormones, including insulin, and glucagon, as part of the hormonal feedback mechanism.

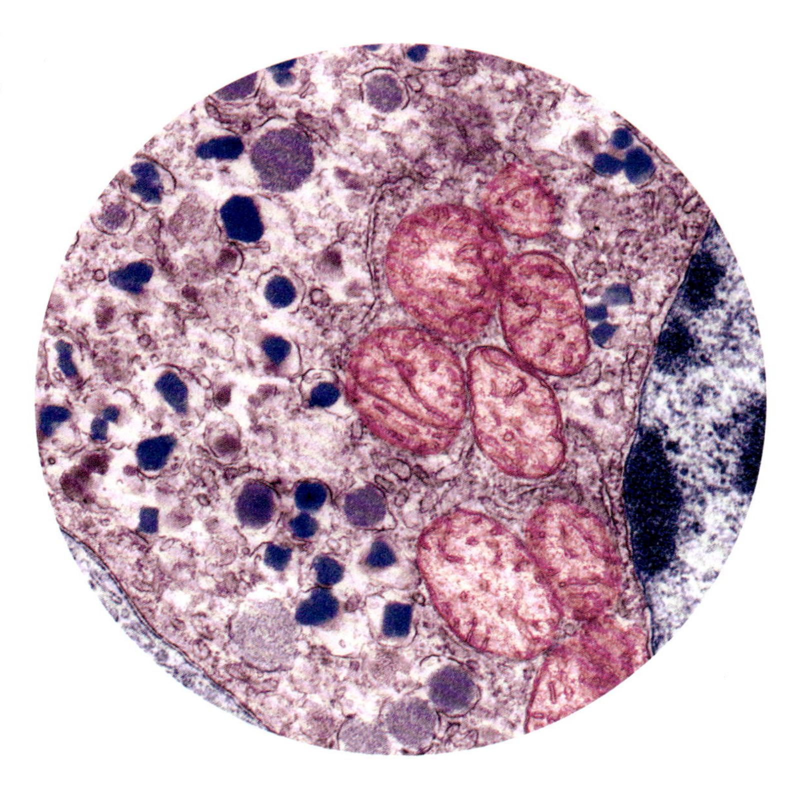

Pancreatic islet cells. Coloured transmission electron micrograph (TEM) of a section through an Islets of langerhans from a human pancreas.

Ovaries

Ovaries are integral to the endocrine system as they produce vital sex hormones that influence sexual development and reproduction.

The ovaries are paired glands that are situated in the pelvic cavity close to the uterus and attached to it by the uterine (Fallopian) tubes. The two female sex hormones, oestrogen and progesterone, are produced by the ovaries. These maintain the female characteristics, and act to influence the growth of an embryo and childbirth. In partnership with the hormones produced by the pituitary gland, oestrogen and progesterone control and regulate the menstrual cycle and prepare the female body for pregnancy.

Relaxin is another hormone produced by the ovaries and placenta that relaxes the symphysis pubis and dilates the cervix ready for childbirth. It also has a part in increasing sperm motility.

Corpus luteum
Growing primary follicle
Secondary follicle
Follicular vein and artery
Mature Graafian follicle
Egg released from Graafian follicle

The ovaries are paired glands that produce three important hormones: oestrogen, progesterone and relaxin.

The menstrual cycle

The oestrus, or menstrual cycle, refers to the cyclical changes that take place in the female reproductive system during the production of eggs.

These changes are controlled by hormones released by the pituitary gland and ovaries: oestrogen, progesterone, luteinizing hormone and follicle-stimulating hormone.

Uterine changes

Following menstruation the uterine lining (endometrium) thickens and becomes more vascular under the influence of oestrogen and follicle-stimulating hormone.

During the first 14 days of the menstrual cycle a Graafian follicle matures. Ovulation occurs around day 14 when the secondary oocyte is expelled and swept into the uterine tube.

The ruptured follicle becomes a hormone-secreting body called the corpus luteum. This secretes progesterone, stimulating further thickening of the endometrium in which the fertilized ovum will implant.

If fertilization does not occur, the levels of progesterone and oestrogen decrease. This causes the endometrium to break down and be excreted into the menstrual flow.

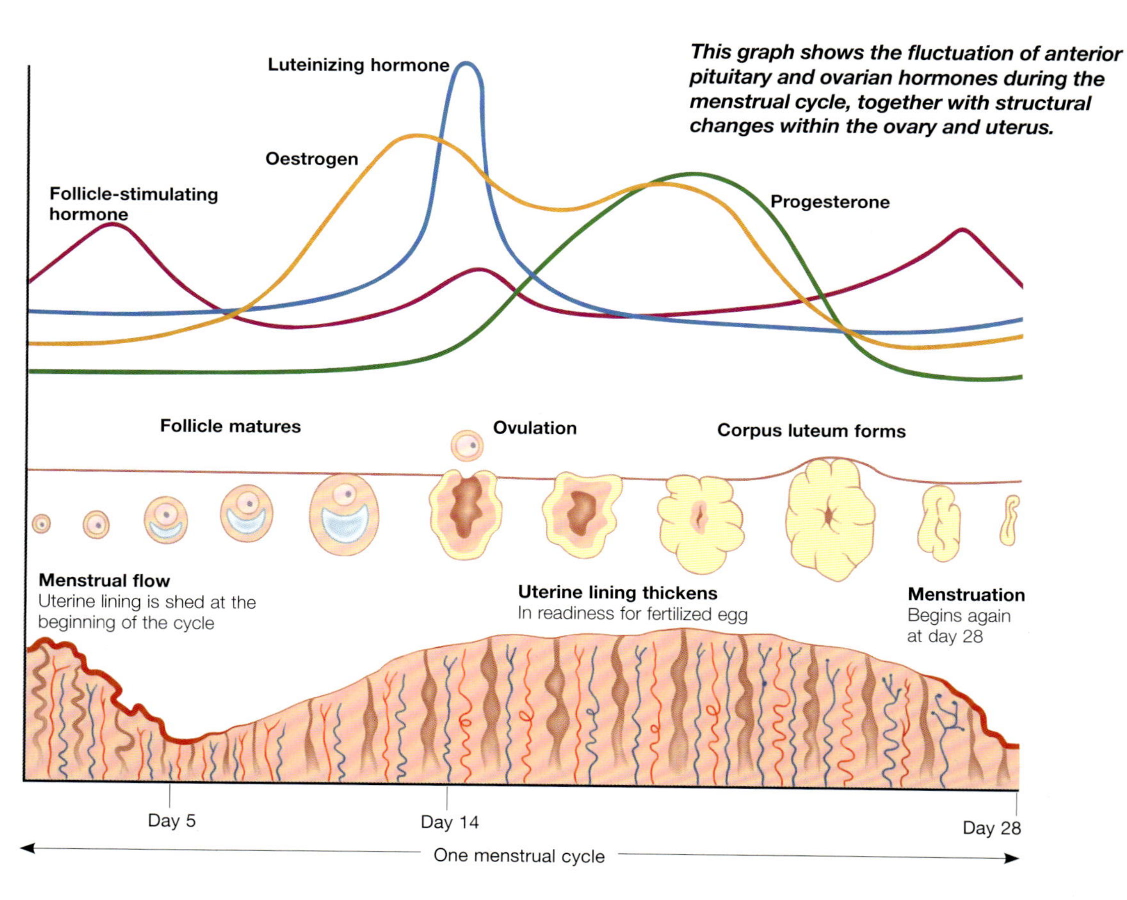

This graph shows the fluctuation of anterior pituitary and ovarian hormones during the menstrual cycle, together with structural changes within the ovary and uterus.

Testes

The testes produce testosterone, the primary male hormone that enables the development of male sex characteristics.

The paired testes are firm, oval-shaped structures that lie within the scrotum, a bag formed as an outpouching of the anterior abdominal wall.

Hormones

Male hormones are known as androgens and they are responsible for the male sex characteristics. There are two hormones produced by the testes: testosterone, which is secreted by the interstitial endocrinocytes – clusters of cells between the seminiferous tubules – and inhibin.

Testosterone

This important androgen has several effects on the body. It is responsible for the development and growth of the male sex organs and secondary sex characteristics (such as body hair and deepening of the voice). Testosterone also has a role to play in the development of sperm. The production of testosterone, like many other hormones, is stimulated by the pituitary gland.

Inhibin

This protein hormone acts directly on the pituitary gland to inhibit the production of follicle-stimulating hormone (FSH) by the pituitary gland. FSH acts on the seminiferous tubules to produce sperm. As soon as there are adequate numbers of mature sperm to reproduce, inhibin feeds back this information to the pituitary gland, which slows down production of FSH.

Sagittal section of the scrotum

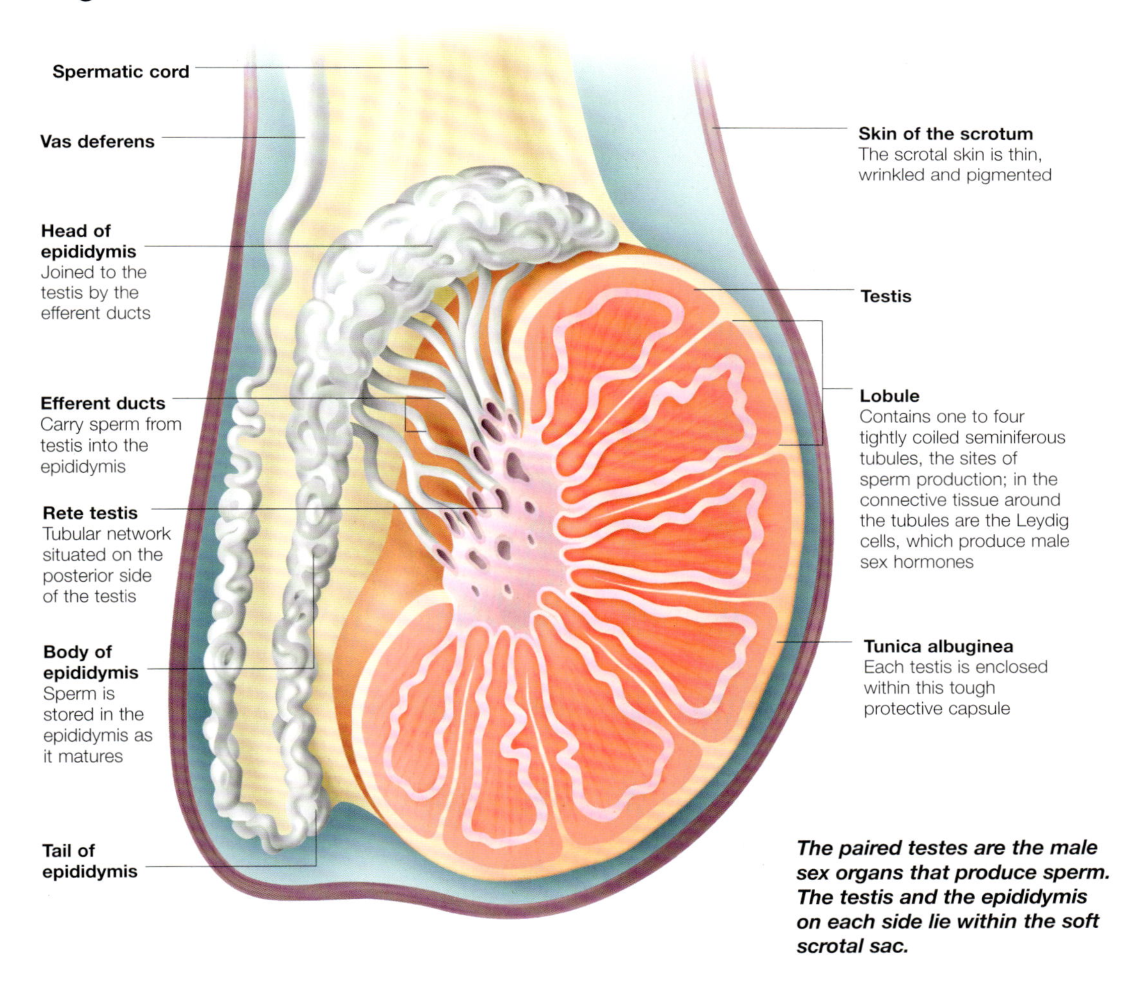

The paired testes are the male sex organs that produce sperm. The testis and the epididymis on each side lie within the soft scrotal sac.

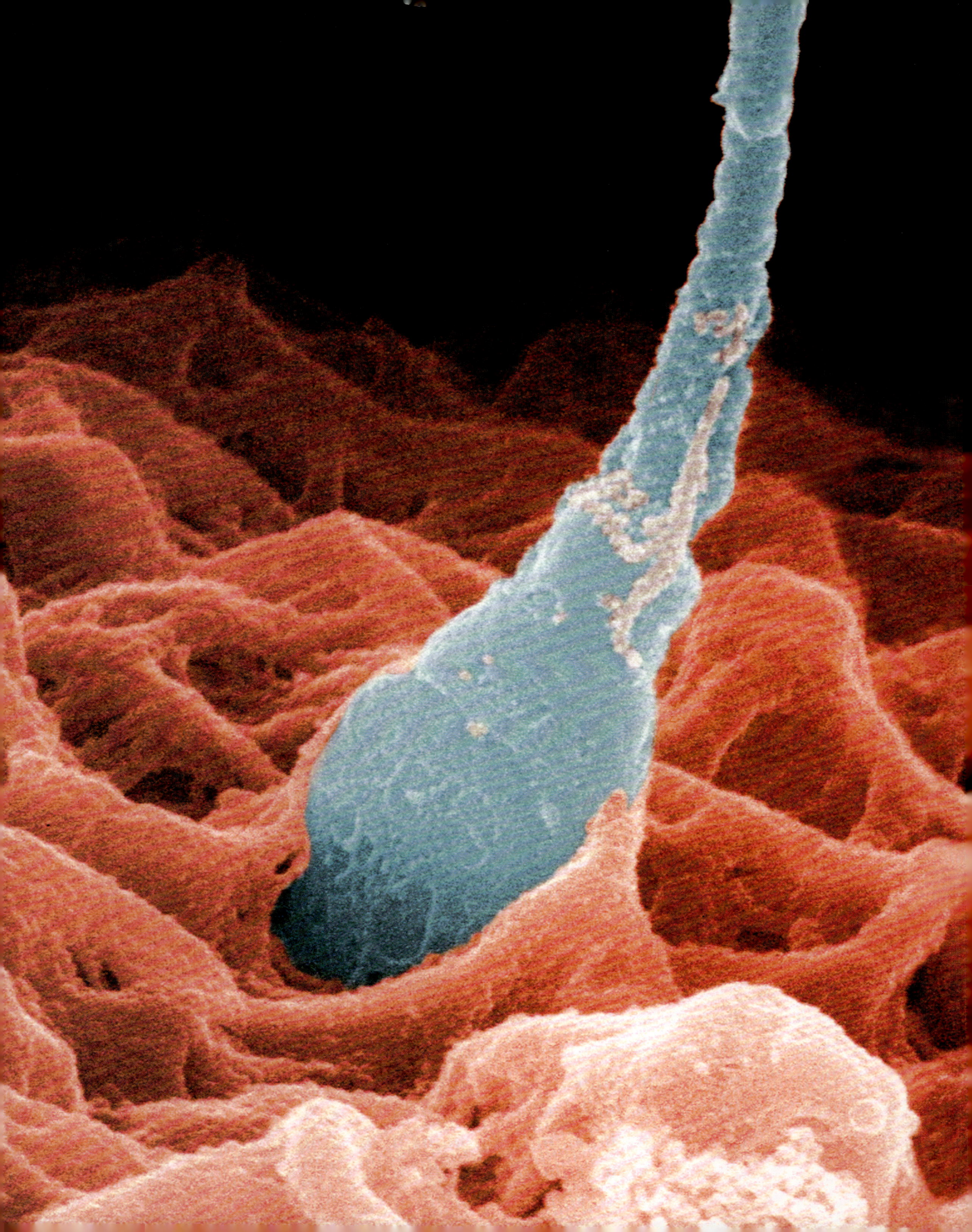

CHAPTER 12

The Reproductive System

From the moment we are born, our bodies are equipped with the organs and tissues that make up the male and female reproductive systems. These remain inactive until we reach puberty, when hormone levels change and our bodies become sexually mature in preparation for reproduction.

This chapter describes the anatomical layout of our reproductive systems and explores each stage of the sexual and reproductive journey – from the onset of puberty to the development of the foetus and the process of childbirth.

Opposite: Coloured scanning electron micrograph (SEM) of a sperm (blue) penetrating a human egg (ovum). Each sperm (spermatozoan) has a rounded head (acrosome) and a long tail with which it swims. Women usually release one egg per month, whereas men release millions of sperm in each ejaculation.

Puberty

During puberty, both boys and girls undergo enormous physical and emotional change. This is due to the production of sex hormones, triggering development essential to fertility.

Puberty is the period of physical change that occurs during adolescence and results in sexual maturity. In girls it tends to occur between the ages of 10 and 14, while in boys it is likely to start between 10 and 14 and continue until about 17.

Secondary Sexual Characteristics

The physical changes that take place during puberty are manifested in the appearance of secondary sexual characteristics, such as a deepening of the voice in boys and the growth of the breasts in girls.

Accelerated Growth

During puberty there is a striking growth spurt of around 8–10cm (3–4in) per year in height. Since boys reach full maturity later than girls, their growth period is extended, with the result that they tend to be significantly taller.

This accelerated growth spurt affects different parts of the body at any one time, so the body may appear to be out of proportion during this period.

Growth acceleration usually affects the feet first, followed by the legs and the torso. Finally, the face, particularly the lower jaw, undergoes development. Body weight can almost double during this time. In girls this is largely due to increased fat deposition in response to changing hormone levels, while in boys it is due to an increase in muscle bulk.

Trends

Studies have shown that menarche (the onset of menstruation) appears to be occurring at an increasingly early age in girls, at a rate of four to six months earlier every decade. This is thought to be due to improved nutrition. It is likely that boys are also maturing at an earlier age.

Hormonal Triggers

Puberty is triggered by the production of a gonadotrophin-releasing hormone from a region of the brain known as the hypothalamus. It is not clear what triggers the release of this hormone. There is speculation that it may be controlled by the interaction between the pineal gland and the hypothalamus, acting as a biological clock.

Sex Gland Stimulation

Gonadotrophin-releasing hormone stimulates a small gland in the brain known as the pituitary. This triggers the release of a group of hormones known as gonadotrophins (sex gland stimulators), at around the age of 10–14 years. Gonadotrophins stimulate the ovaries to secrete oestrogens, and the testes to produce testosterone. It is these hormones that are responsible for the development of the secondary sexual characteristics during puberty.

Emotional Changes

The many physical changes that take place during puberty are accompanied by a number of emotional changes. The individual may have difficulty coming to terms with the many physical changes taking place in the body, such as the onset of menstrual periods in girls and the deepening of the voice in boys. In addition, fluctuating levels of hormones during puberty can affect mood.

Below: Girls start puberty at different ages. However, most girls will have reached the same level of sexual maturity by the time they are 16.

Below right: In addition to undergoing major physical changes, teenagers also suffer the emotional consequences of hormonal fluctuations.

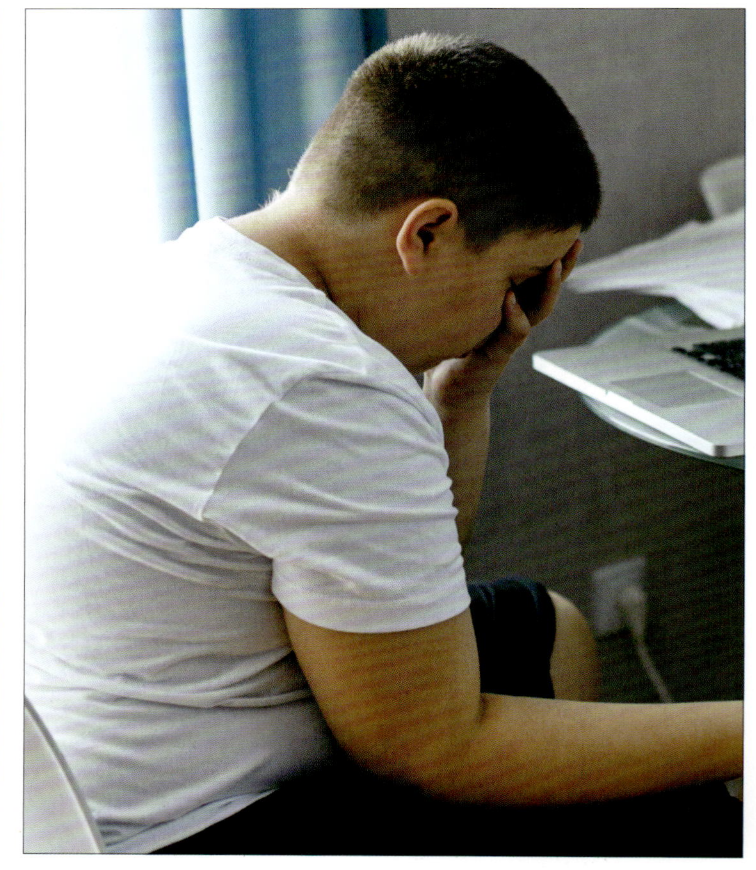

Physical changes during puberty

Testosterone is a key hormone during the period of puberty, causing complex and profound change in both boys and girls.

Puberty in boys

Boys start to go through puberty between the ages of 10 and 14. The physical changes that occur during this time are brought about by the male sex hormone, testosterone. This is a growth-promoting hormone that is produced by cells within the testes.

Sperm production

Before puberty, the testes contain numerous solid cords of cells. With the onset of puberty, the cells at the centre of the cords die, so that the cords become the hollow tubes called seminiferous tubules, in which sperm cells develop. The production of testosterone within the testes in turn triggers:

- The onset of sperm production. Large numbers of sperm cells are produced – around 300–600 per gram of testicle every second
- Growth of the testes, scrotum and penis
- Spontaneous erections; present since birth, these can now be psychologically induced
- Maturation of the sperm-carrying ducts and enlargement of the seminal vesicles (sperm-storing sacs)
- Enlargement of the prostate gland, which starts to secrete fluid that makes up part of the seminal fluid
- Ejaculation – first occurs around a year after the penis undergoes accelerated growth.

Changes continue until the age of around 17. The voice box enlarges, the vocal cords lengthen, and the voice deepens and becomes more resonant. Body hair begins to grow in the pubic region, in the armpits and on the face, chest and abdomen. Testosterone also accelerates muscular development.

Puberty in girls

Puberty in girls starts between the ages of 10 and 14, but varies from person to person, and so some girls reach sexual maturity before others. By the age of 16, most girls will have reached the same level of maturity and will have undergone body growth, alterations in proportions, and changes in the sexual and reproductive organs:

- The first sign of puberty in girls is usually breast budding. Hormones, such as oestrogen and progesterone, trigger the nipples to enlarge and the breast tissue to grow as milk glands and ducts develop. After this time breast growth is rapid.
- The adrenal glands start to produce male sex hormones, such as testosterone that cause a sudden surge of physical growth and the development of pubic and underarm hair.
- Hip development – changes take place in the bones of the pelvis, making it wider in relation to the rest of the skeleton.
- Menstruation usually begins around a year after these hormones are released.

Abnormal puberty

Abnormal changes in the hypothalamus or adrenal glands, such as a tumour, can cause the process of puberty to occur at a much earlier age. This rare phenomenon is known as precocious puberty, and can result in full sexual development in young children.

Puberty in both sexes can be delayed by malnutrition or constant physical exertion. Many athletes and gymnasts do not develop sexual characteristics until they have more relaxed training regimes. A number of genetic disorders (such as cystic fibrosis) can also affect puberty.

Male physical changes

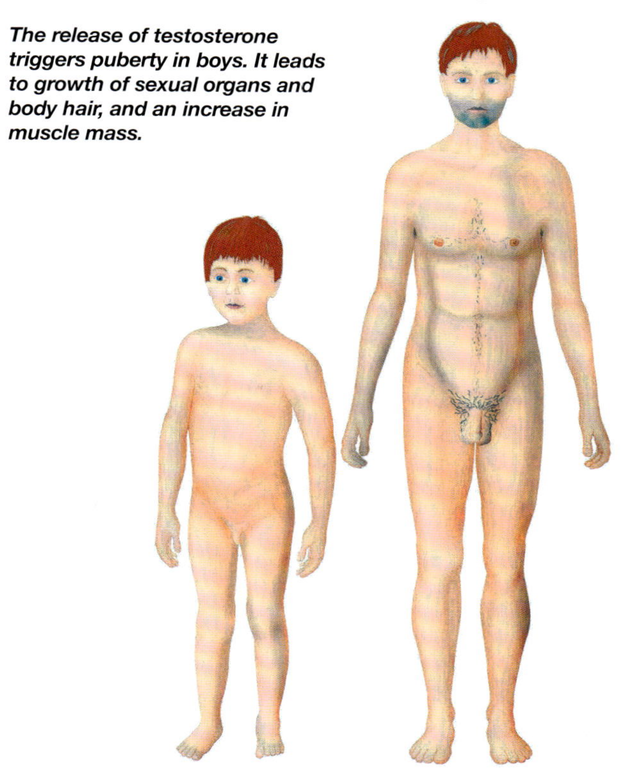

The release of testosterone triggers puberty in boys. It leads to growth of sexual organs and body hair, and an increase in muscle mass.

Female physical changes

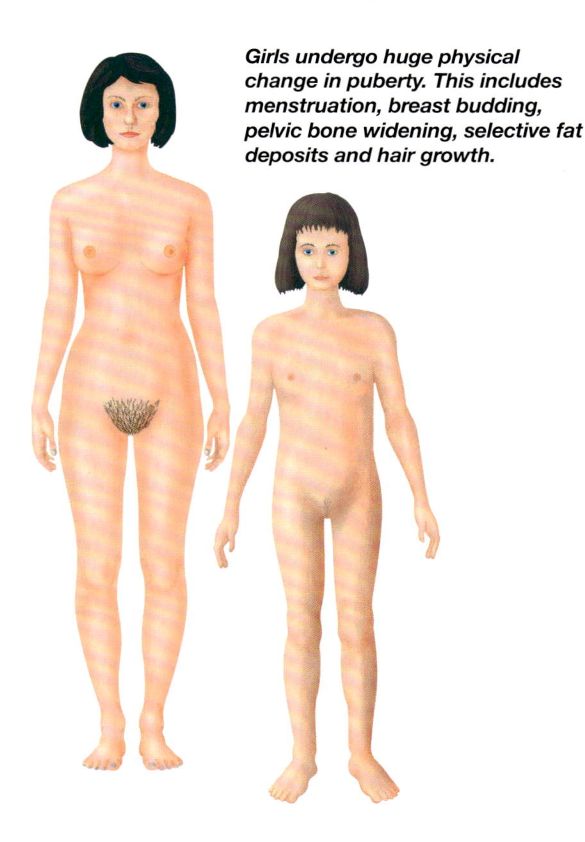

Girls undergo huge physical change in puberty. This includes menstruation, breast budding, pelvic bone widening, selective fat deposits and hair growth.

Male Reproductive System

The male reproductive system includes the penis, scrotum and the two testes (contained within the scrotum). The internal structures of the reproductive system are contained within the pelvis.

The structures constituting the male reproductive tract are responsible for the production of sperm and seminal fluid and their carriage out of the body. Unlike other organs it is not until puberty that they develop and become fully functional.

Constituent Parts
The male reproductive system consists of a number of interrelated parts:

- Testis – the paired testes lie suspended in the scrotum. Sperm are carried away from the testes through tubes or ducts, the first of which is the epididymis.
- Epididymis – on ejaculation sperm leave the epididymis and enter the vas deferens.
- Vas deferens – sperm are carried along this muscular tube to the prostate gland.
- Seminal vesicle – on leaving the vas deferens sperm mix with fluid from the seminal vesicle gland in a combined 'ejaculatory' duct.
- Prostate – the ejaculatory duct empties into the urethra within the prostate gland.
- Penis – on leaving the prostate gland, the urethra then becomes the central core of the penis.

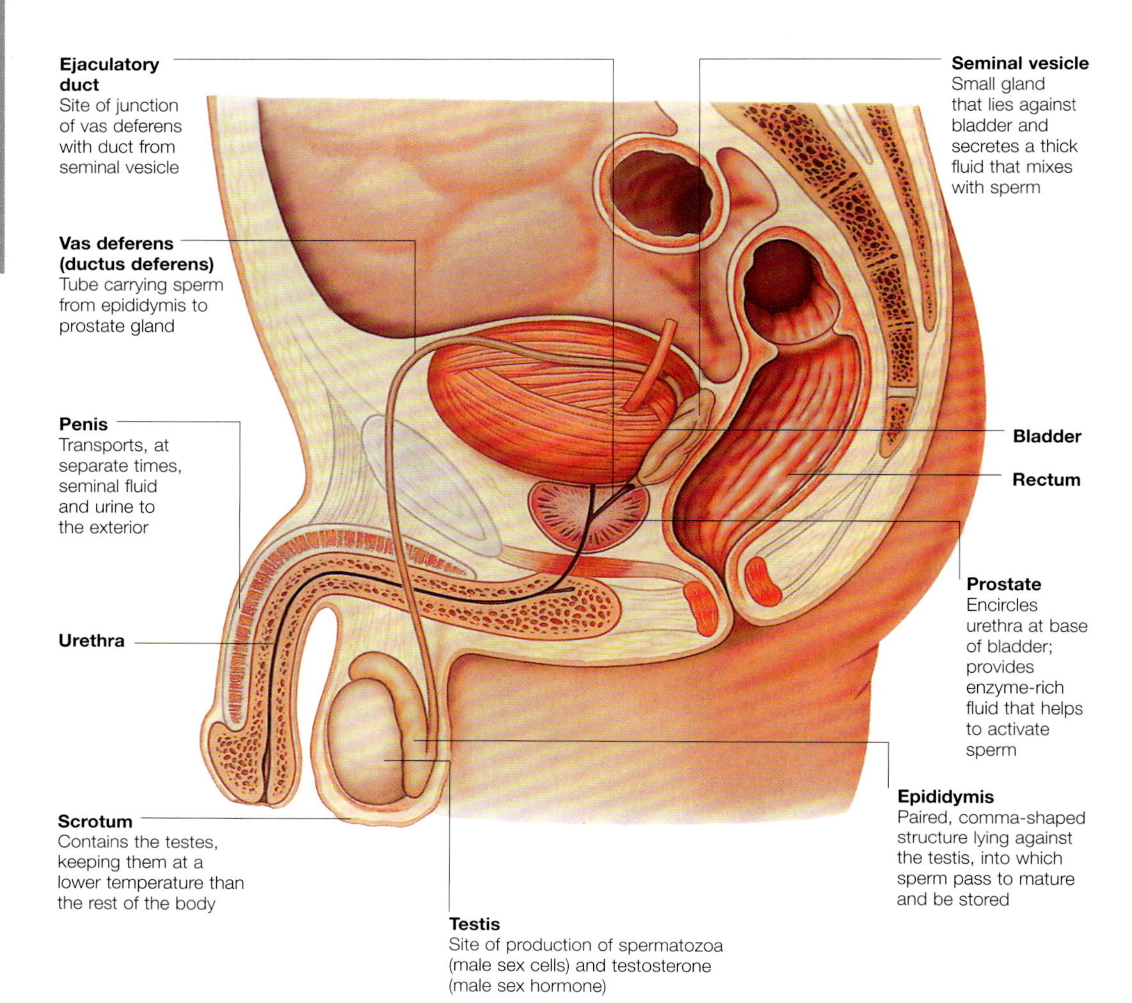

External male gentitalia

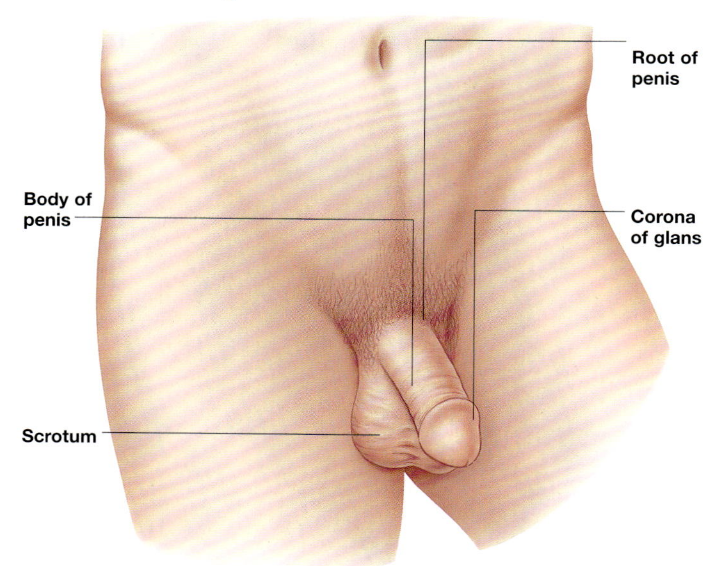

The external genitalia are those parts of the reproductive tract that are visible in the pubic region, while other parts remain hidden within the pelvic cavity.
Male external genitalia consists of:

- The scrotum
- The penis.

In adults, these are surrounded by coarse pubic hair.

Scrotum
The scrotum is a loose bag of skin and connective tissue that holds the testes suspended within it.

The external male genitalia consist of the scrotum and the penis, which are situated in the pubic area. In adults, pubic hair surrounds the root of the penis.

There is a midline septum, or partition, that separates the testes from each other. The testes are located in a potentially vulnerable position outside the body cavity so that the sperm can be kept as cool as possible.

Penis
Most of the penis consists of erectile tissue, which becomes engorged with blood during sexual arousal, causing the penis to become erect. The urethra, through which urine and semen pass, runs through the penis.

Prostate gland and seminal vesicles

The prostate gland is a vital part of the male reproductive system, providing enzyme-rich fluid. It produces up to a third of the total volume of the seminal fluid.

About 3cm (1.18in) in length, the prostate gland lies just under the bladder and encircles the first part of the urethra. Its base lies closely attached to the base of the bladder, its rounded anterior surface lying just behind the pubic bone.

Capsule

The prostate is covered by a tough capsule made up of dense fibrous connective tissue. Outside this true capsule is a further layer of fibrous connective tissue known as the prostatic sheath.

Internal structure

The urethra, the outflow tract from the bladder, runs vertically through the centre of the prostate gland, where it is known as the prostatic urethra. The ejaculatory ducts open into the prostatic urethra on a raised ridge, the seminal colliculus.

The prostate gland is said to be divided into lobes although they are not distinct:

- Anterior lobe – lies in front of the urethra and contains mainly fibromuscular tissue
- Posterior lobe – lies behind the urethra and beneath the ejaculatory ducts
- Lateral lobes – these two lobes lie on either side of the urethra and form the main part of the gland
- Median lobe – lies between the urethra and the ejaculatory ducts.

Location of the prostate gland

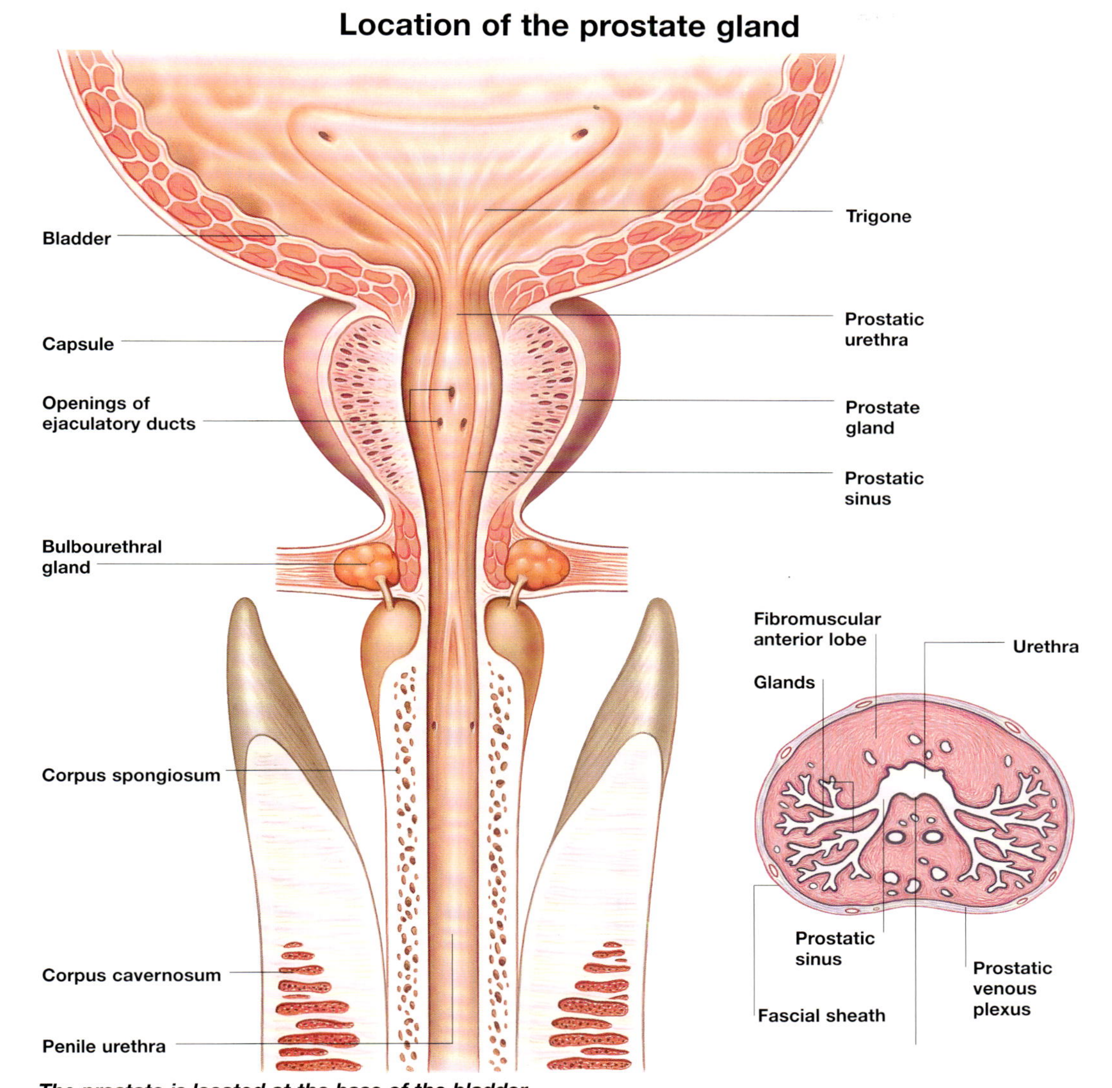

The prostate is located at the base of the bladder and surrounds the urethra. It is a firm, smooth organ, approximately the size of a walnut.

Seminal vesicles

The paired seminal vesicles are accessory glands of the male reproductive tract and produce a thick, sugary, alkaline fluid that forms the main part of the seminal fluid.

Structure and shape

Each seminal vesicle is an elongated structure about the size and shape of a little finger and lies behind the bladder and in front of the rectum, the two forming a V-shape.

Prostate secretions

The prostate gland is sac-like, with a volume of approximately 10–15ml (0.3–0.5fl oz). It consists internally of coiled secretory tubules with muscular walls.

The secretions leave the gland in the duct of the seminal vesicle, which joins with the vas deferens just inside the prostate to form the ejaculatory duct.

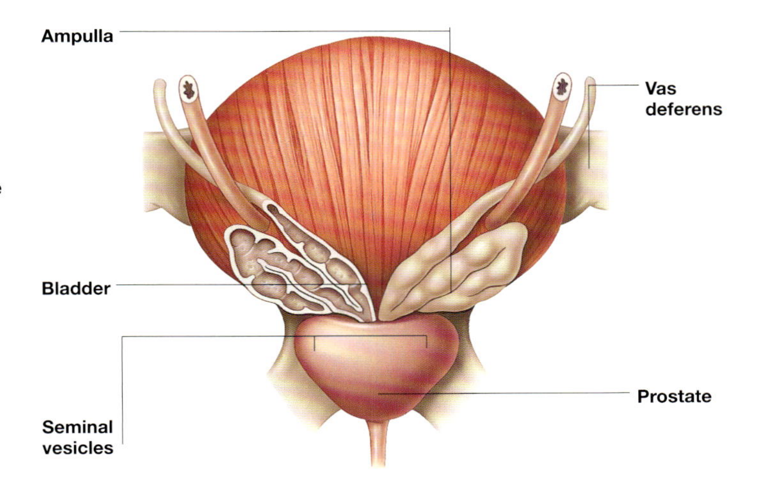

The seminal vesicles are situated at the back of the bladder. Secretions pass into the vas deferentia, which empty into the prostatic urethra.

Testes, Scrotum and Epididymis

The two testes, which lie suspended within the scrotum, are the sites of sperm production. The scrotum also contains the two epididymides – long, coiled tubes, which connect to the vas deferens.

The paired testes are firm, mobile, oval-shaped structures about 4cm (1.5in) in length and 2.5cm (1in) in width. They lie within the scrotum, an outpouching of the anterior abdominal wall, and are attached above to the spermatic cord, from which they hang.

Temperature Control
Normal sperm can only be produced if the temperature of the testes is about three degrees lower than the internal body temperature. Muscle fibres within the spermatic cord and walls of the scrotum help to regulate the scrotal temperature by lifting the testes up towards the body when it is cold, and relaxing when the ambient temperature is higher.

Epididymis
Each epididymis is a firm, comma-shaped structure that lies closely attached to the upper pole of the testis, running down its posterior surface. The epididymis receives the sperm made in the testis and is composed of a highly coiled tube which, if extended, would be 6m (20ft) in length.

From the tail of the epididymis emerges the vas deferens. This tube will carry the sperm back up the spermatic cord and into the pelvic cavity on the next stage of the journey.

Sagittal section of the contents of the scrotum

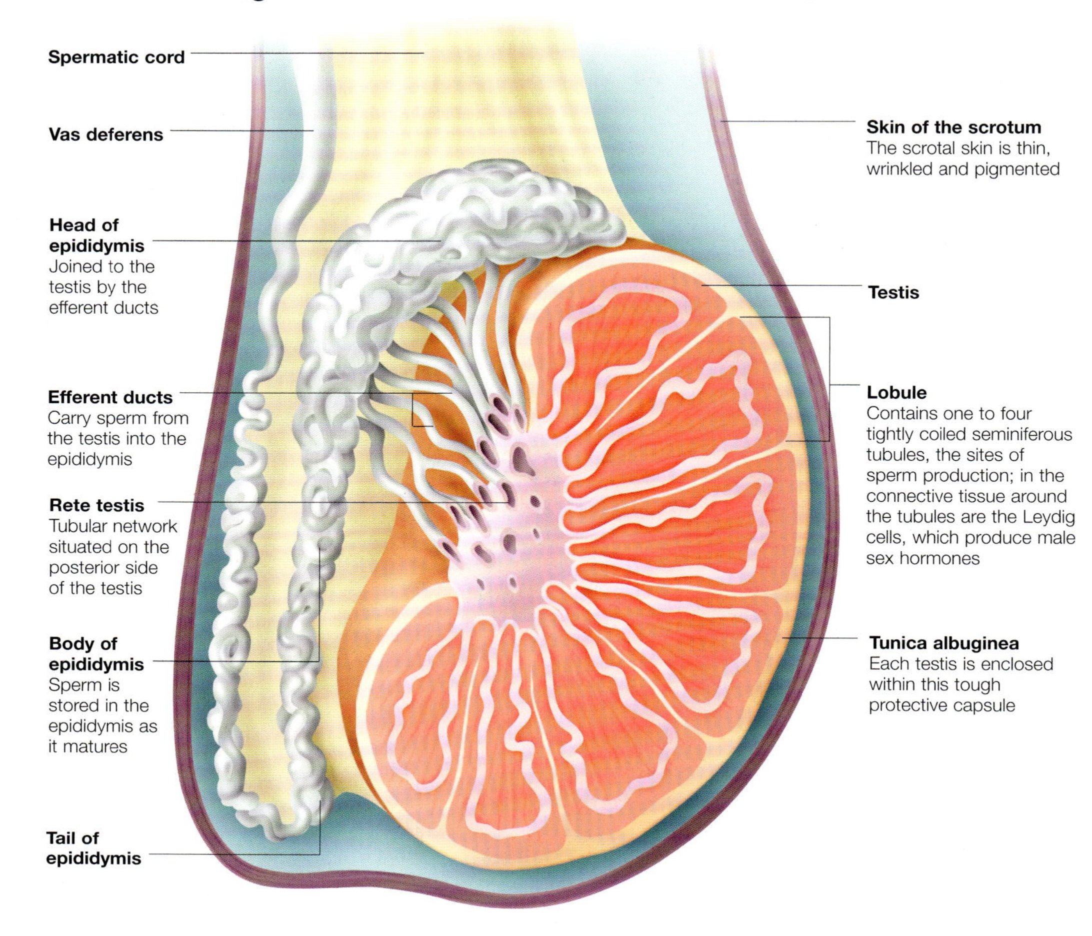

The paired testes are the male sex organs that produce sperm. The testis and the epididymis on each side lie within the soft scrotal sac.

Walls of the scrotum

The scrotum originated as an outpouching of the anterior abdominal wall, and has a number of layers. These include:

- Skin, which is thin, wrinkled and pigmented
- Dartos fascia, a layer of connective tissue with smooth muscle fibres
- Three layers of fascia derived from the three muscular layers of the abdominal wall, with further cremasteric muscle fibres
- Tunica vaginalis, a closed sac of thin, slippery, serous membrane, like the peritoneum in the abdomen. The tunica vaginalis contains a small amount of fluid to lubricate movement of the testes against surrounding structures.

Unlike the abdominal wall, there is no fat in the coverings around the testes, which is believed to help keep them cool.

Cross-section of the scrotum

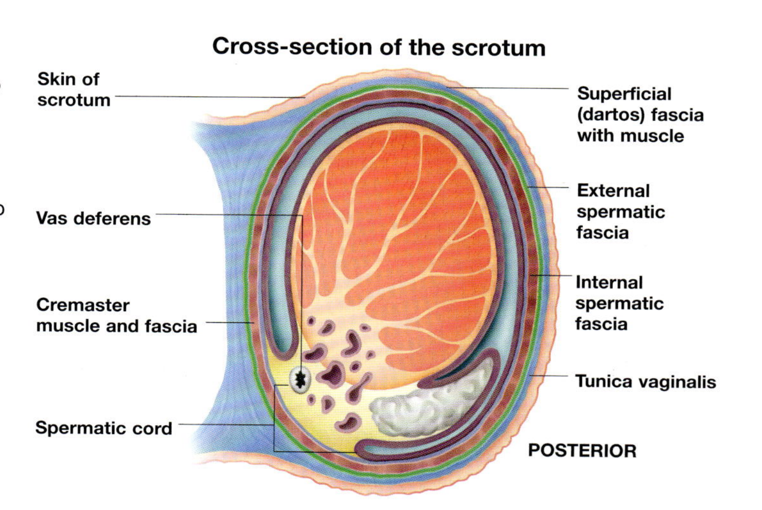

The scrotum contains the testes and hangs outside the body. It consists of an outer covering of skin, which surrounds several protective layers.

Blood supply of the testes

The arterial blood supply of the testes arises from the abdominal aorta, and descends to the scrotum. Venous drainage follows the same route in reverse.

During embryonic life, the testes develop within the abdomen; it is only at birth that they descend into their final position within the scrotum. Because of this the blood supply of the testes arises from the abdominal aorta, and travels down with the descending testis to the scrotum.

Testicular arteries

The paired testicular arteries are long and narrow and arise from the abdominal aorta. They then pass down the posterior abdominal wall, crossing the ureters as they go, until they reach the deep inguinal rings and enter the inguinal canal.

As part of the spermatic cord they leave the inguinal canal and enter the scrotum where they supply the testis, also forming interconnections with the artery to the vas deferens.

Testicular veins

Testicular veins arise from the testis and epididymis on each side. Their course differs from that of the testicular arteries within the spermatic cord where, instead of a single vein, there is a network of veins, known as the pampiniform plexus.

Further up in the abdomen, the right testicular vein drains into the large inferior vena cava, while the left normally drains into the left renal vein.

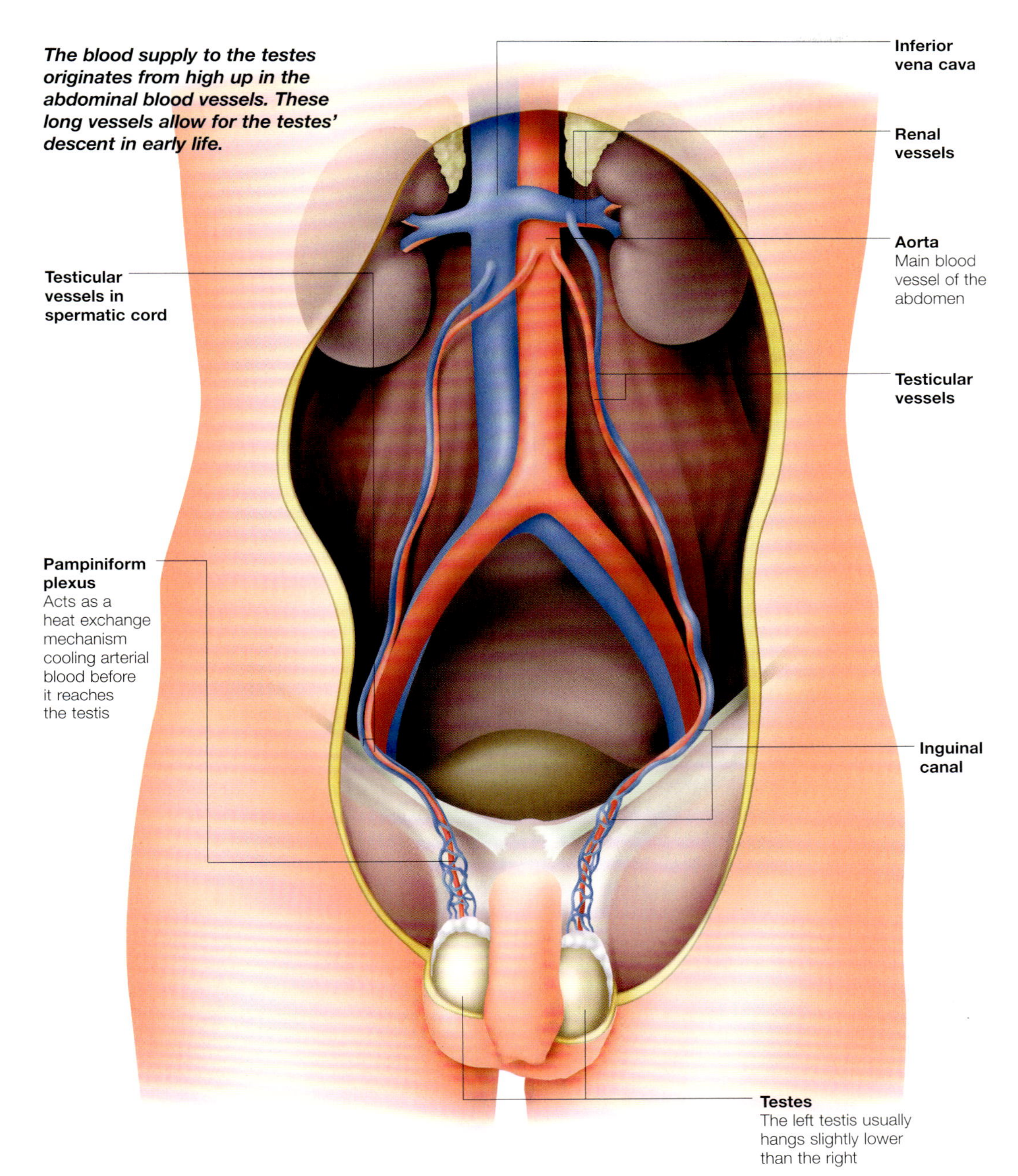

The blood supply to the testes originates from high up in the abdominal blood vessels. These long vessels allow for the testes' descent in early life.

Internal structure of the testis

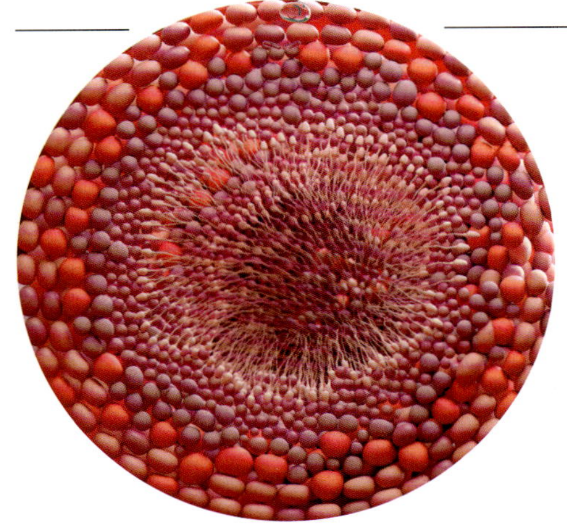

This micrograph shows a sectioned seminiferous tubule. Developing sperm (red) are inside the tubule, which is surrounded by Leydig cells (green).

Each testis is enclosed within a tough, protective capsule, the tunica albuginea, from which numerous septa, or partitions, pass down to divide the testis into about 250 tiny lobules. Each wedge-shaped lobule contains one to four tightly coiled seminiferous tubules, which are the actual sites of sperm production. It has been estimated that there is a total of 350m (1140ft) of sperm-producing tubules in each testis.

Tubules

Sperm are collected from the coiled seminiferous tubules into the straight tubules of the rete testis and from there into the epididymis.

Between the seminiferous tubules lie groups of specialized cells, known as the interstitial or Leydig cells, which are the site of production of hormones such as testosterone.

Penis

The penis is the male sex organ, which, when erect, conveys sperm into the vagina during sexual intercourse. To enable this, the penis is largely composed of erectile tissue.

The penis is mostly composed of three columns of sponge-like erectile tissue, the two corpora cavernosa and the corpus spongiosum. These are able to fill and become engorged with blood, causing an erection. There is only a small amount of muscular tissue associated with the penis, and what there is lies in its root. The shaft and glans have no muscle fibres. The main components of the penis are:

- Root – this first part of the penis is fixed in position and is made up of the expanded bases of the three columns of erectile tissue covered by muscle fibres
- Shaft – this hangs down in the flaccid condition and is made up of erectile tissue, connective tissue, and blood and lymphatic vessels
- Glans – the tip of the penis, this is formed from the expanded end of the corpus spongiosum and carries the outlet of the urethra, the external urethral orifice
- Skin – this is continuous with that of the scrotum and is thin, dark and hairless. It is attached only loosely to the underlying fascia and lies in wrinkles when the penis is flaccid.

At the tip of the penis the skin extends as a double layer, which covers the glans; this is known as the prepuce, or foreskin.

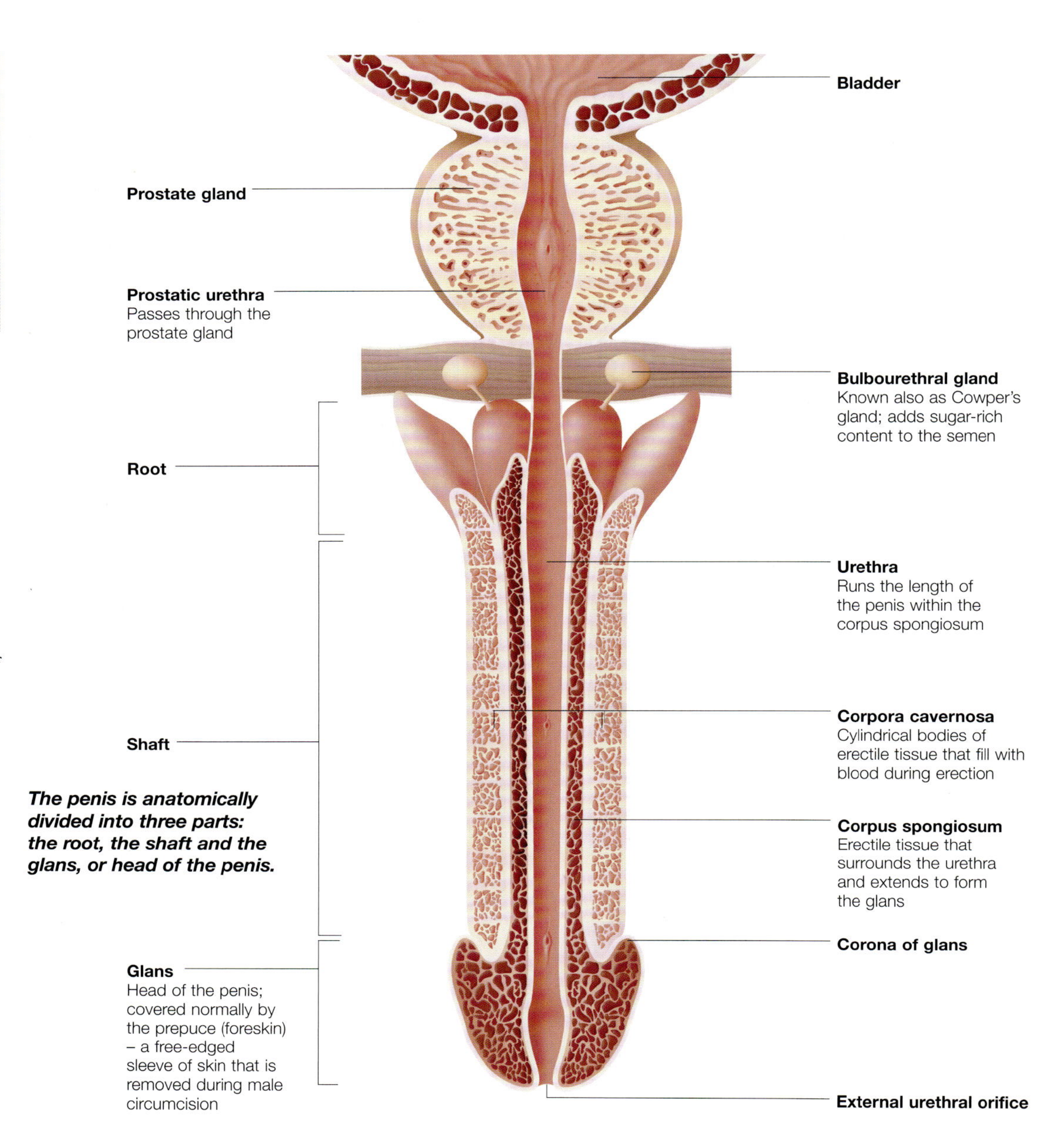

The penis is anatomically divided into three parts: the root, the shaft and the glans, or head of the penis.

Cross-section of shaft

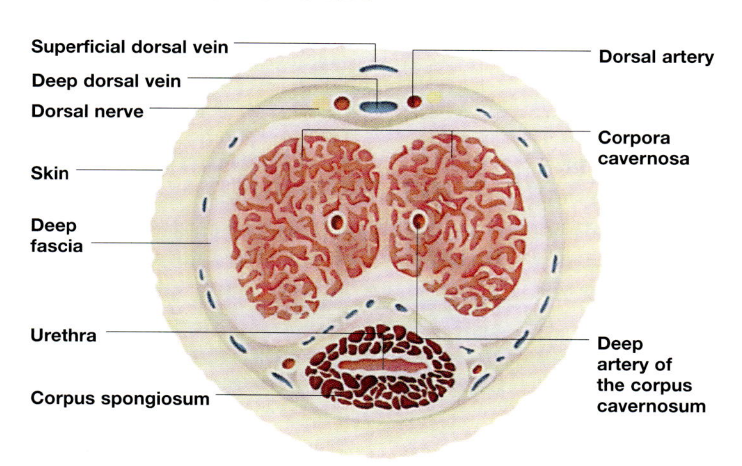

In a cross-section of the shaft of the penis, the relationship of erectile tissue, blood vessels and fascia can be seen more easily. The main bulk is made up of the three masses of erectile tissue, the smaller corpus spongiosum containing within its length the urethra. Each corpus cavernosum carries a central deep artery, which supplies the blood needed for erection.

The main body of the penis, the shaft, consists of three bodies of erectile tissue. These fill with blood during sexual stimulation, resulting in an erection.

Connective tissue

A sleeve of connective tissue, the deep fascia, encloses the erectile tissue and the deep dorsal vein and dorsal arteries and nerves. Outside the deep fascia is a layer of loose connective tissue that contains the superficial veins. The skin that overlies this loose connective tissue layer is firmly attached to the underlying structures only at the glans.

Associated muscles

Several muscles are associated with the penis. Their fibres are confined to the root and structures around the penis, rather than to the shaft or glans.

These muscles are known collectively as the superficial perineal muscles. They lie in the perineum, the area around the anus and external genitalia. There are three main muscles:

- Superficial transverse perineal muscle – This narrow, paired muscle lies just under the skin in front of the anus. It runs from the ischial tuberosity of the pelvic bone on each side right across to the midline of the body.
- Bulbospongiosus – This muscle acts to compress the base of the corpus spongiosum, and thus the urethra, to help expel its contents. It originates in a central tendon or raphe, which unites the two sides and passes round to encircle the root of the penis.
- Ischiocavernosus – This muscle originates from the ischial tuberosity of the pelvic bone to surround the crura or bases of the corpora cavernosa on each side. Contraction of this muscle helps to maintain erection of the penis.

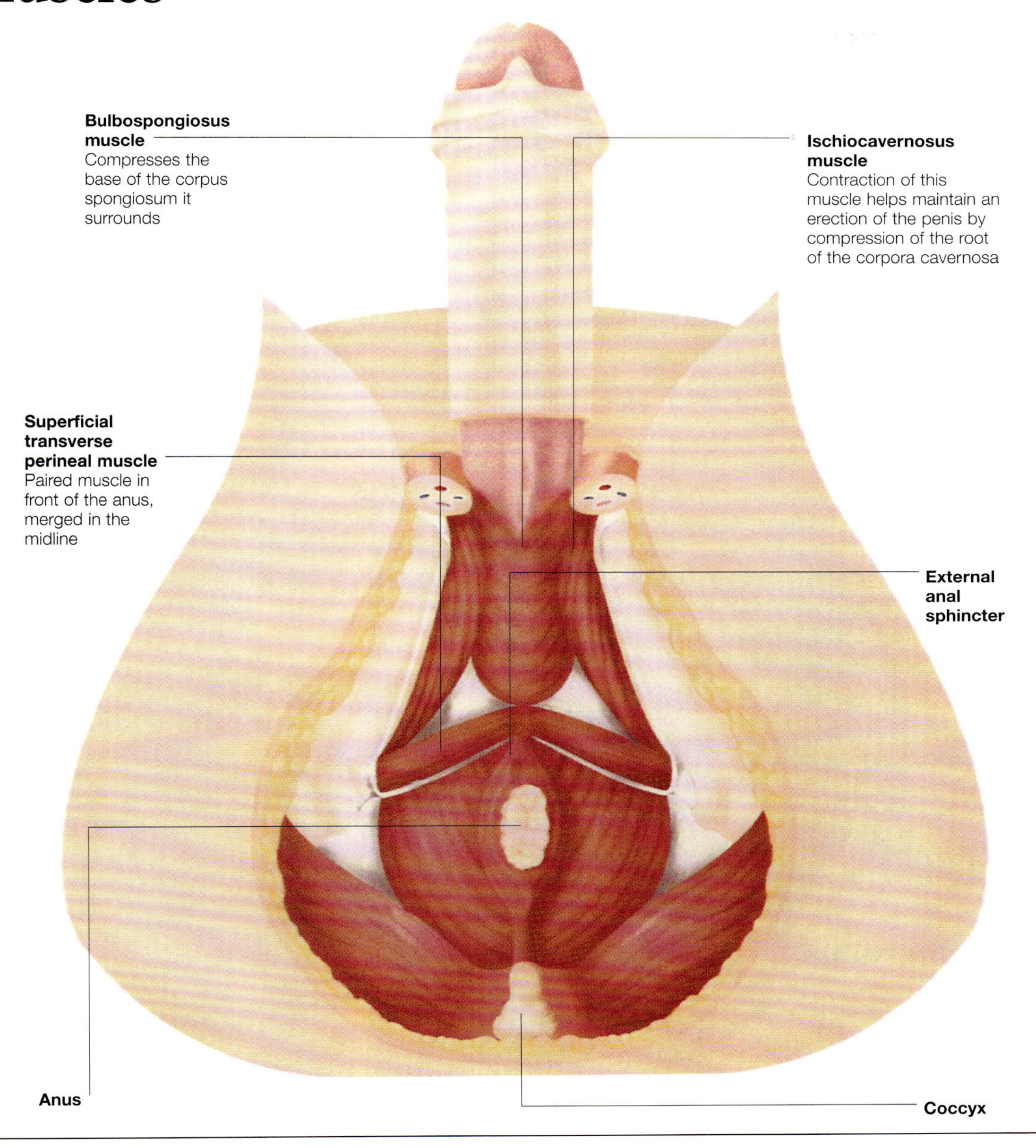

Blood supply

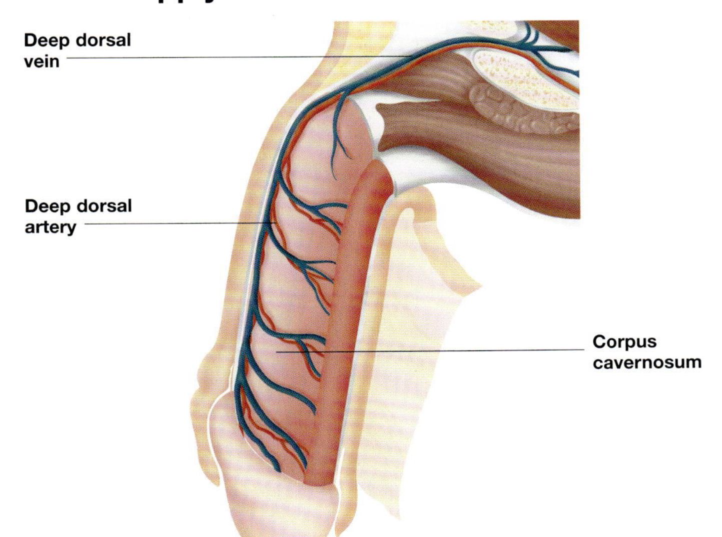

The arterial supply of the penis has two functions. As with any organ, it has to provide the necessary oxygenated blood for the tissues of the penis. It must also, however, provide an additional supply to allow engorgement of the spongy erectile tissues for erection.

Arteries

All the arteries supplying the penis originate from the internal pudendal arteries of the pelvis.

The blood supply of the penis originates from the internal pudendal arteries. The deep arteries supply the corpora cavernosa during an erection.

The dorsal arteries lie on each side of the midline deep dorsal vein, and supply connective tissue and skin. The deep arteries run within the spongy tissue of the corpora cavernosa to supply tissue there and to allow flooding of that tissue during erection.

Venous drainage

The deep dorsal vein of the penis receives blood from the cavernous spaces while blood from the overlying connective tissue and skin is drained by the superficial dorsal vein. Venous blood drains ultimately into the pudendal veins within the pelvis.

Sperm

Sperm are the male sex cells, produced and stored in the testes. Owing to the process of meiosis, a specialized division of the cell nucleus, each cell contains a unique set of genes.

Sperm are mature male sex cells, vital to fertilization. They are produced in the testes, two walnut-sized organs located in the scrotum. The scrotum is the pouch that hangs below the penis, and is around two degrees cooler than the core temperature of the body, so providing the optimum temperature for the production of sperm.

In order to maintain this temperature, the scrotum can pull up closer to the body when the surrounding temperature is low, and can drop farther away as the temperature rises.

Sexual Organs

The testes are the primary producers of testosterone (the male sex hormone). These specialized organs each contain around 1,000 seminiferous tubules, which are responsible for the manufacture and storage of sperm. The tubules are lined by small cells known as spermatogonia. From puberty onwards spermatogonia cells start to divide to produce cells that eventually develop into sperm.

Alternating with the spermatogonia are much larger cells, the Sertoli cells, which secrete nutrient fluid into the tubules.

Spermatogenesis

Spermatogenesis (the formation of sperm) is a complex process that involves the constant proliferation of spermatogonia cells to form primary spermatocytes. These cells have a full set of genes, identical to those in other body cells.

Meiosis

The primary spermatocytes then undergo a specialized division known as meiosis, in which they split twice to produce cells with a random half (haploid) set of genes. These cells, known as spermatids, develop and grow to produce mature, motile (moving) sperm.

Formation of sperm

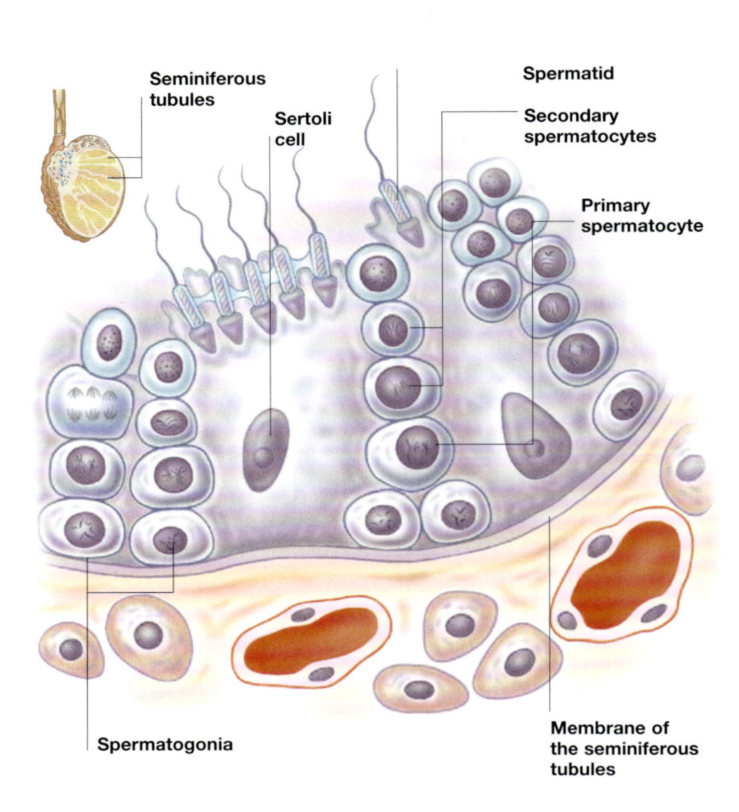

The seminiferous tubules of the testes are lined with small cells, called spermatogonia. These cells divide to produce primary spermatocytes.

Division of genetic information

Each primary spermatocyte contains a set of chromosomes, arranged in 23 pairs (diploid set). It then undergoes the specialized process of meiosis by which it splits in two, to form a pair of spermatids, each containing only half a set of chromosomes (haploid set). This process occurs in two stages to produce four spermatids.

First stage

During the first stage, the chromosomes within the spermatocyte's nucleus replicate (double up) and then pair off. The chromosomes exchange random blocks of genes within each pair.

This exchange is nature's way of 'shuffling' the gene pool and introducing variation within the offspring. The paired chromosomes separate as the cell divides, each cell receiving two copies of one member of each chromosome pair. The spermatocytes then divide again.

Second stage

During the second stage of meiosis, the 23 replicated chromosomes within each nucleus split up and the spermatocytes divide again.

The end result of the meiosis process is the production of spermatids that contain half the number of chromosomes of a spermatocyte. The genetic make-up of each resulting spermatid is unique, due to the mixing process and the chances of any two being identical are virtually zero.

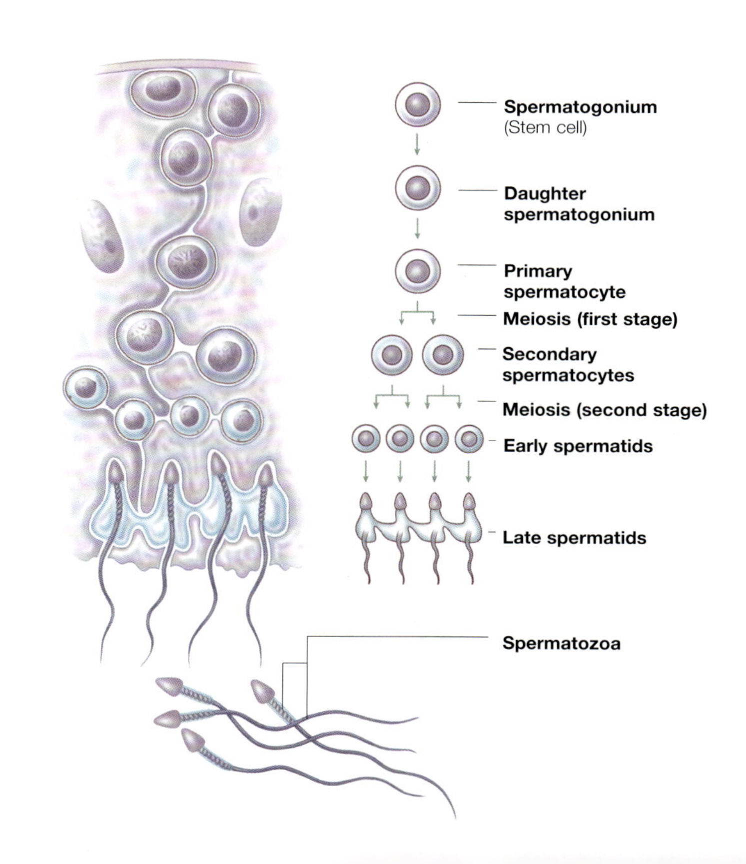

As a result of meiosis, each spermatocyte divides into four spermatids. Each spermatid contains half the genetic material found in the spermatocyte.

Structure of sperm

Mature sperm cells are specially designed to facilitate the swimming movement used to propel them towards the female's egg.

The immature spermatids move towards the nearest Sertoli cell, where they receive nourishment in the form of glycogen, proteins, sugars and other nutrients. This provides them with energy and helps them to mature into spermatozoa.

Spermatozoa are among the most specialized cells in the body. Each sperm (spermatozoon) measures 0.05mm (0.001in) in length and consists of a head, neck and tail.

Spermatozoa

The head of the sperm is shaped like a flattened teardrop, and contains a sac of enzymes known as the acrosome. These enzymes are vitally important to the sperm's ability to break down and penetrate the protective outer layer of the female's egg during fertilization.

Behind the acrosome is the cell nucleus, which contains a random half set of male genetic material (DNA) tightly coiled within 23 chromosomes. Thanks to the process of meiosis, each sperm possesses a unique set of genetic information.

The neck is a fibrous area where the middle part of the sperm joins the head. It is a flexible structure and allows the head to swing from side to side, facilitating the swimming movement.

Tail structure

The sperm tail consists of a pair of long filaments surrounded by two rings each containing nine fibrils. At the front end of the tail are a further ring of outer dense fibres and also a protective tail sheath. The tail is divided into three sections:

- The middle piece – the fattest part of the tail, due to an additional spiral layer full of energy-producing units known as mitochondria. These produce energy that fuels the sperm, allowing it to swim.
- The principal piece – consists of the 20 filaments, along with the outer dense fibres and tail sheath.
- The end piece – here the dense fibres and tail sheath thin out, with t he result that this part of the tail is enclosed only by a thin cell membrane. This gradual tapering is what produces the sperm's characteristic whiplash-like swimming motion, driving the sperm towards the egg.

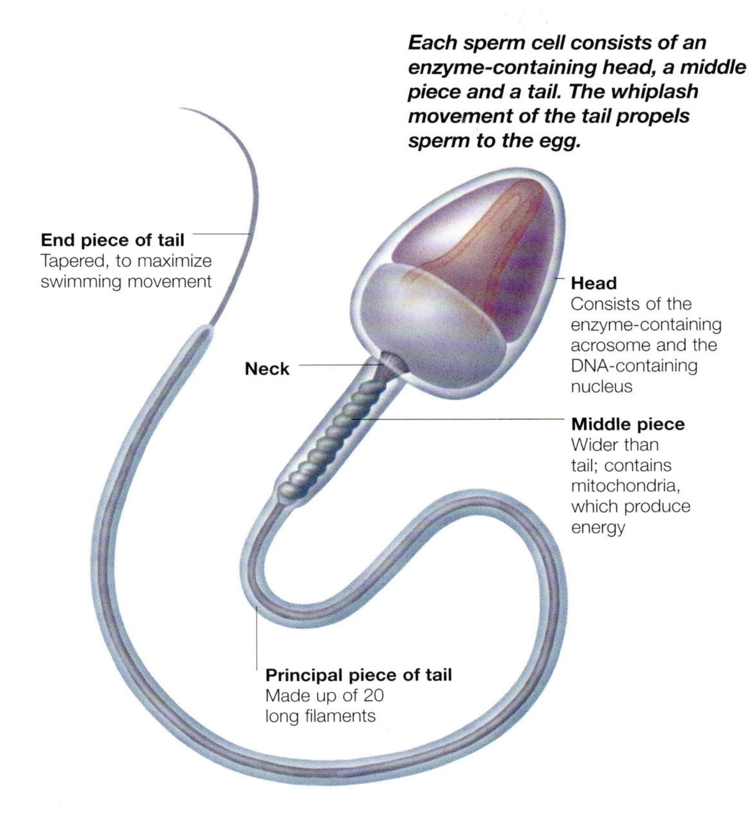

Each sperm cell consists of an enzyme-containing head, a middle piece and a tail. The whiplash movement of the tail propels sperm to the egg.

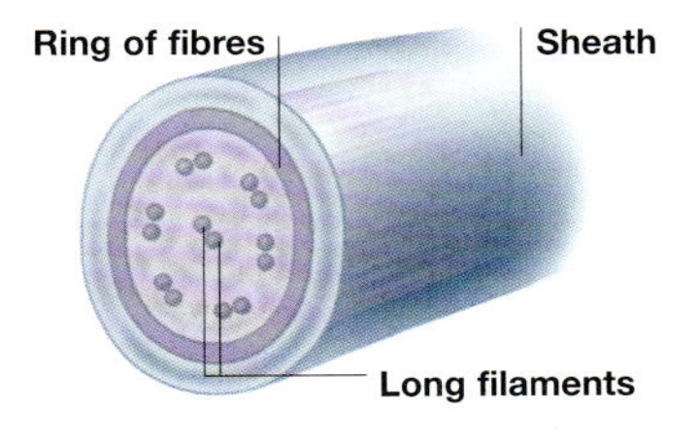

The tail consists of a central pair of filaments surrounded by an outer ring of nine paired filaments. At the front end of the tail is a further ring of outer dense fibres and a protective sheath.

Production of semen and ejaculation

Once their tails are fully developed the sperm are released by the Sertoli cells into the seminiferous tubule. As fluid is secreted into the tubule by the Sertoli cells, a current is produced that washes the sperm cells towards the epididymis. This is a long tube coiled against the testes in which the mature sperm are stored.

Ejaculation

The sperm are propelled from the epididymis during sexual stimulation and up the vas deferens via a wave of muscular contractions within the ducting system.

They travel to the ejaculatory duct, through the prostate and into the urethra. Here they are bathed in secretions from the prostate gland and seminal vesicles (small sacs that hold constituents of semen). The result is a thick, yellowish-white fluid, known as semen. The average discharge of semen (ejaculate) contains approximately 300 million sperm.

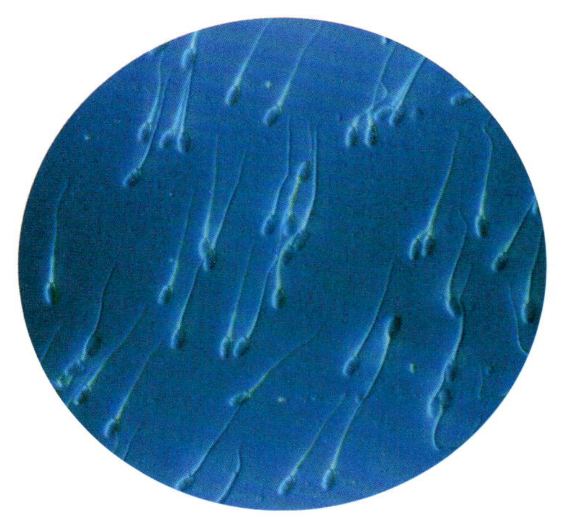

Light micrograph of a field of swimming human sperm (spermatozoa). These tiny male sex cells produced in the testes are responsible for fertilisation of the female's ovum (egg). Sperm take 74 days to form and a further 26 days to mature and pass through the epididymis and vas deferens. They can survive in the female for up to five days.

Female Reproductive System

The role of the female reproductive tract is two-fold. The ovaries produce eggs for fertilization, and the uterus nurtures and protects any resulting fetus for its nine-month gestation.

The female reproductive tract is composed of the internal genitalia – the ovaries, uterine (Fallopian) tubes, uterus and vagina – and the external genitalia (the vulva).

Internal Genitalia

The almond-shaped ovaries lie on either side of the uterus, suspended by ligaments. Above the ovaries are the paired uterine tubes, each of which provides a site for fertilization of the oocyte (egg), which then travels down the tube to the uterus.

The uterus lies within the pelvic cavity and rises into the lower abdominal cavity as a pregnancy progresses. The vagina, which connects the cervix to the vulva, can be distended greatly, as occurs during childbirth when it forms much of the birth canal.

External Genitalia

The female external genitalia, or vulva, is where the reproductive tract opens to the exterior. The vaginal opening lies behind the opening of the urethra in an area known as the vestibule. This is covered by two folds of skin on each side, the labia minora and labia majora, in front of which lies the raised clitoris.

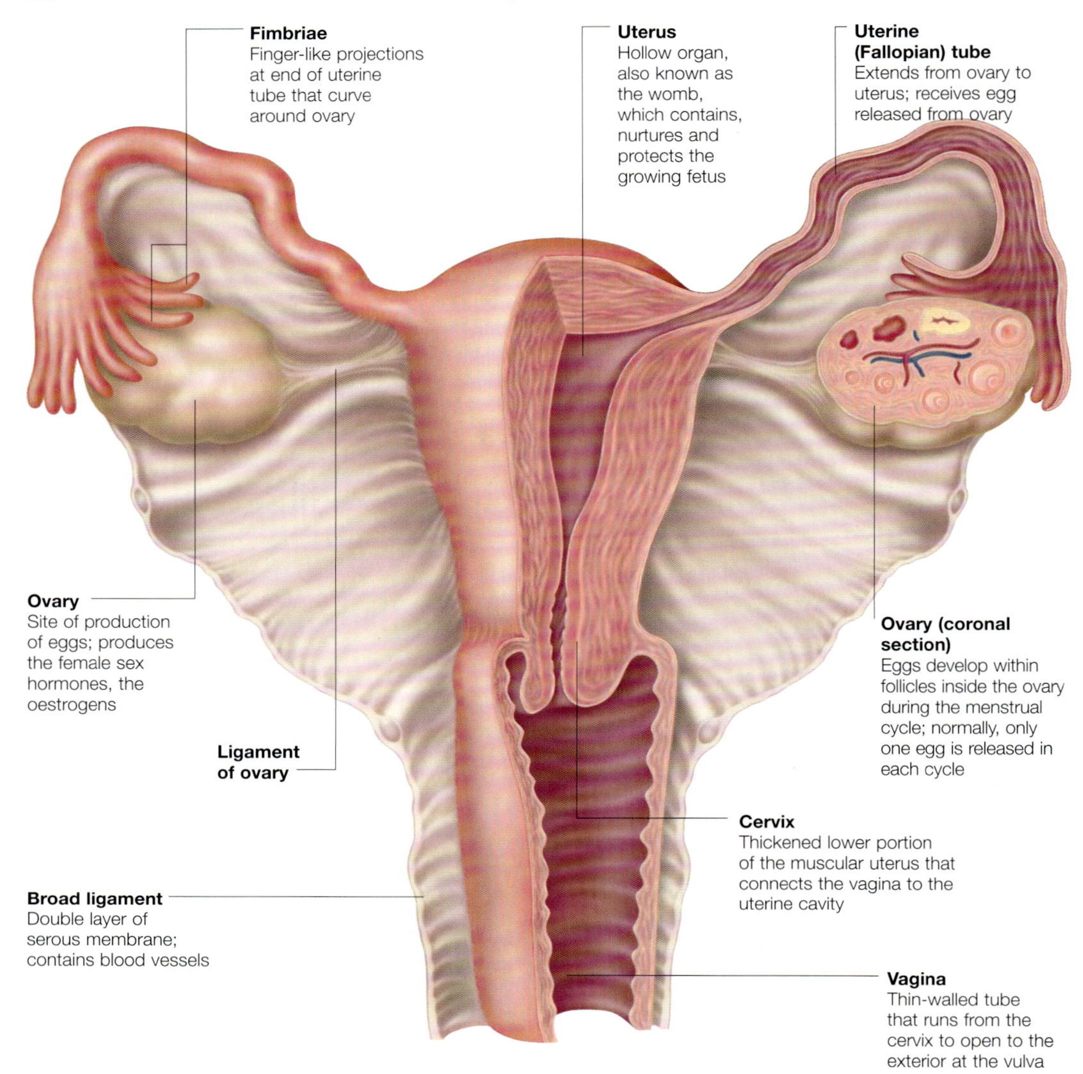

The female reproductive system is composed of internal and external organs. The internal genitalia are T-shaped and lie within the pelvic cavity.

Position of the reproductive duct

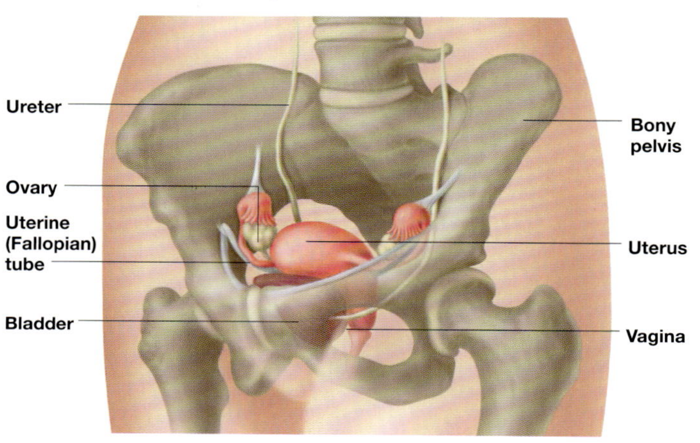

In adult women the internal genitalia (which, apart from the ovaries, are tubular in structure) are located deep within the pelvic cavity. They are thus protected by the presence of the circle of bone that makes up the pelvis. This is in contrast to the pelvic cavity of young children, which is relatively shallow. A female child's uterus, therefore, like the bladder behind which it sits, is located within the lower abdomen.

The internal reproductive organs in adult women are positioned deep within the pelvic cavity. They are therefore protected by the bony pelvis.

Broad ligaments

The upper surface of the uterus and ovaries is draped in a 'tent' of peritoneum, the thin lining of the abdominal and pelvic cavities, forming the broad ligament that helps to keep the uterus in its position.

Blood supply of the reproductive system

The female reproductive tract receives a rich blood supply via an interconnecting network of arteries. Venous blood is drained by a network of veins.

The four principal arteries of the female genitalia are:

- Ovarian artery – this runs from the abdominal aorta to the ovary. Branches from the ovarian artery on each side pass through the mesovarium, the fold of peritoneum in which the ovary lies, to supply the ovary and uterine (Fallopian) tubes. The ovarian artery in the tissue of the mesovarium connects with the uterine artery.
- Uterine artery – this is a branch of the large internal iliac artery of the pelvis. The uterine artery approaches the uterus at the level of the cervix, which is anchored in place by cervical ligaments. The uterine artery connects with the ovarian artery above, while a branch connects with the arteries below to supply the cervix and vagina.
- Vaginal artery – this is also a branch of the internal iliac artery. Together with blood from the uterine artery, its branches supply blood to the vaginal walls.
- Internal pudendal artery – this contributes to the blood supply of the lower third of the vagina and anus.

Veins

A plexus, or network, of small veins lies within the walls of the uterus and vagina. Blood received into these vessels drains into the internal iliac veins via the uterine vein.

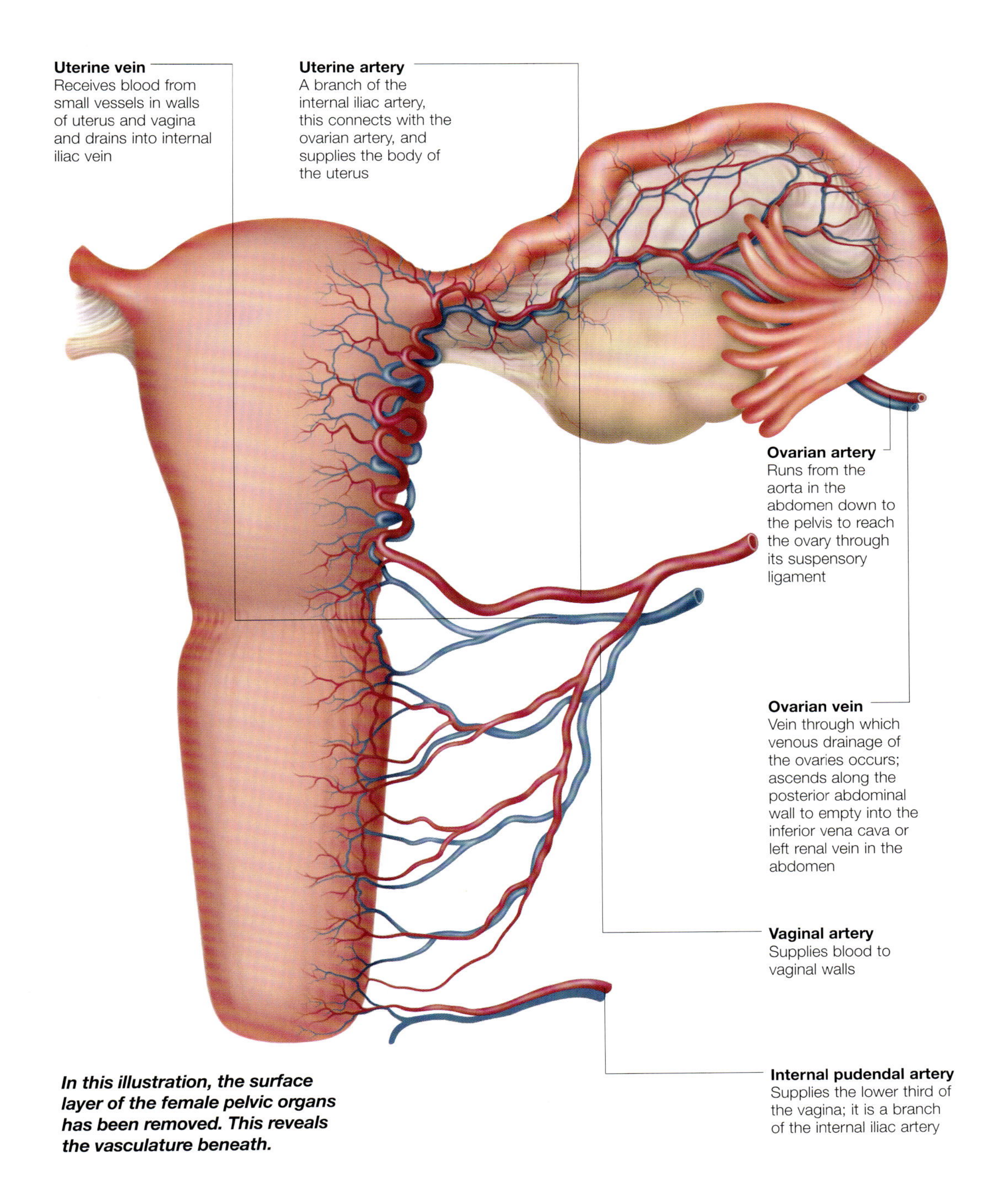

In this illustration, the surface layer of the female pelvic organs has been removed. This reveals the vasculature beneath.

Female Breast

The breast undergoes structural changes throughout the life of a woman. The most obvious changes occur during pregnancy as the breast prepares for its function as the source of milk for the baby.

Men and women both have breast tissue, but the breast is normally a well-developed structure only in women. The two female breasts are roughly hemispherical and are composed of fat and glandular tissue that overlie the muscle layer of the front of the chest wall on either side of the sternum.

Breast Structure

The base of the breast is roughly circular in shape and extends from the level of the second rib above to the sixth rib below. In addition, there may be an extension of breast tissue towards the axilla (armpit), known as the 'axillary tail'.

The mammary glands consist of 15 to 20 lobules – clusters of secretory tissue from which milk is produced. Milk is carried to the surface of the breast from each lobule by a tube known as a 'lactiferous duct', which has its opening at the nipple.

The nipple is a protruding structure surrounded by a circular, pigmented area, called the areola. The skin of the nipple is very thin and delicate and has no hair follicles or sweat glands.

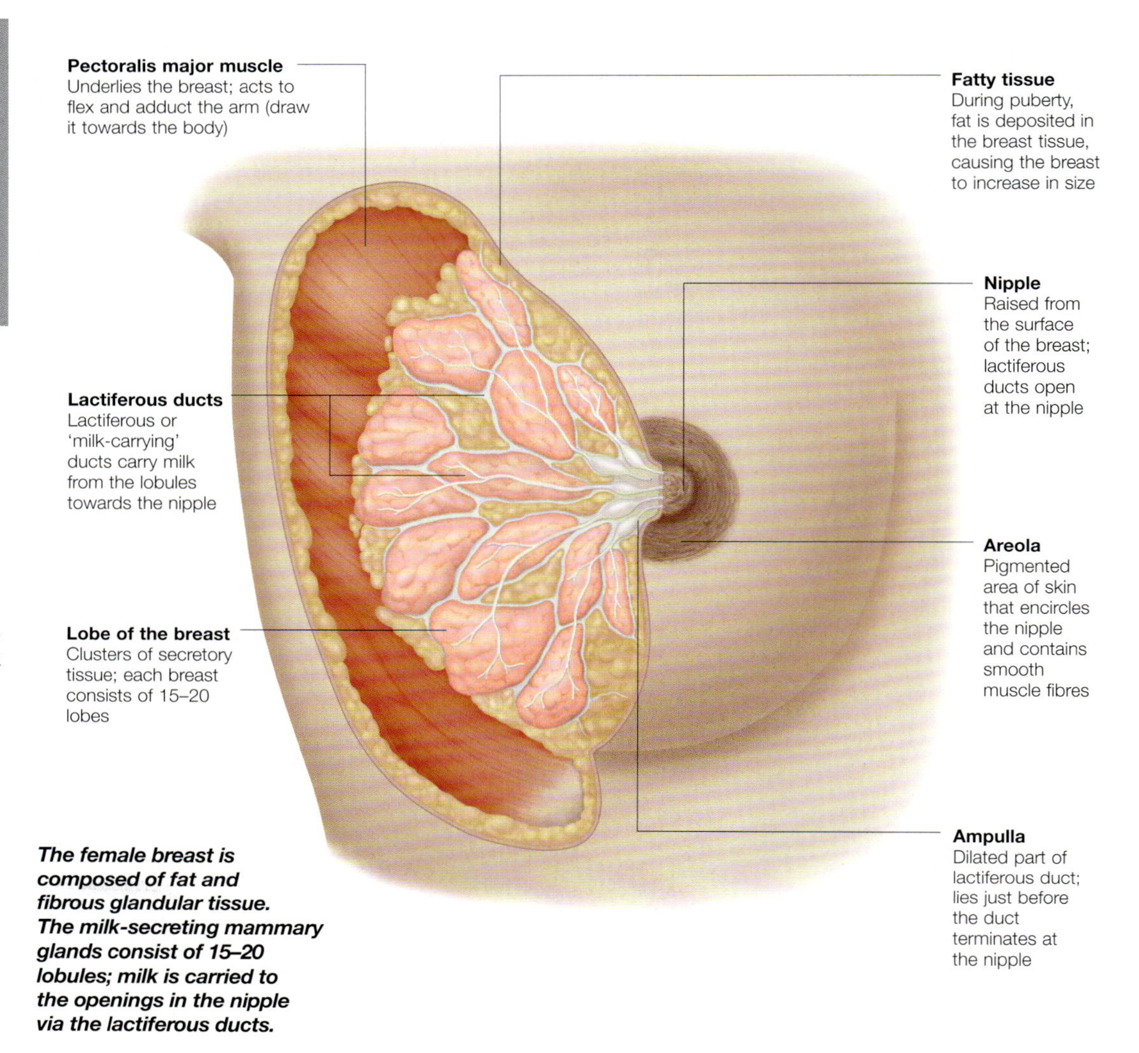

The female breast is composed of fat and fibrous glandular tissue. The milk-secreting mammary glands consist of 15–20 lobules; milk is carried to the openings in the nipple via the lactiferous ducts.

Blood vessels of the breast

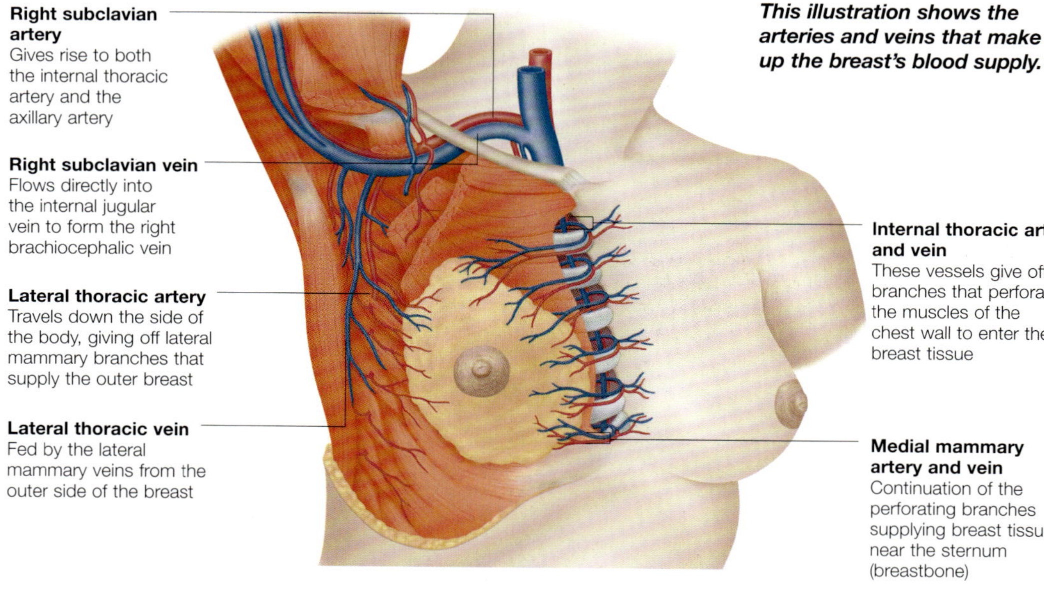

This illustration shows the arteries and veins that make up the breast's blood supply.

The blood supply to the breast comes from a number of sources; these include the internal thoracic artery, which runs down the length of the front of the chest, and the lateral thoracic artery, which supplies the outer part of the breast and some of the posterior intercostal arteries.

A network of superficial veins underlies the skin of the breast, especially in the region of the areola, and these veins may become very prominent during pregnancy.

The blood collected in these veins drains in various directions, following a similar pattern to the arterial supply, travelling via the internal thoracic veins, the lateral thoracic veins and the posterior intercostal veins to the large veins that return blood to the heart.

Lymphatic drainage of the breast

Lymphatic fluid, is returned to the blood circulation by the lymphatic system. Lymph passes through a series of lymph nodes, which act as filters to remove bacteria, cells and other particles.

Tiny lymphatic vessels arise from the tissue spaces and converge to form larger vessels that carry the (usually) clear lymph away from the tissues and into the venous system.

Lymph drains from the nipple, areola and mammary gland lobules into a network of small lymphatic vessels, the 'subareolar lymphatic plexus'. From this plexus the lymph may be carried in several different directions.

Pattern of drainage

About 75 per cent of the lymph from the subareolar plexus drains to the lymph nodes of the armpit, mostly from the outer quadrants of the breast. The lymph passes through a series of nodes in the region of the armpit draining into the subclavian lymph trunk, and ultimately into the right lymphatic trunk, which returns the lymph to the veins above the heart.

Most of the remaining lymph, mainly from the inner quadrants of the breast, is carried to the 'parasternal' lymph nodes, which lie towards the midline of the front of the chest. A small percentage of lymphatic vessels from the breast take another route and travel to the posterior intercostal nodes.

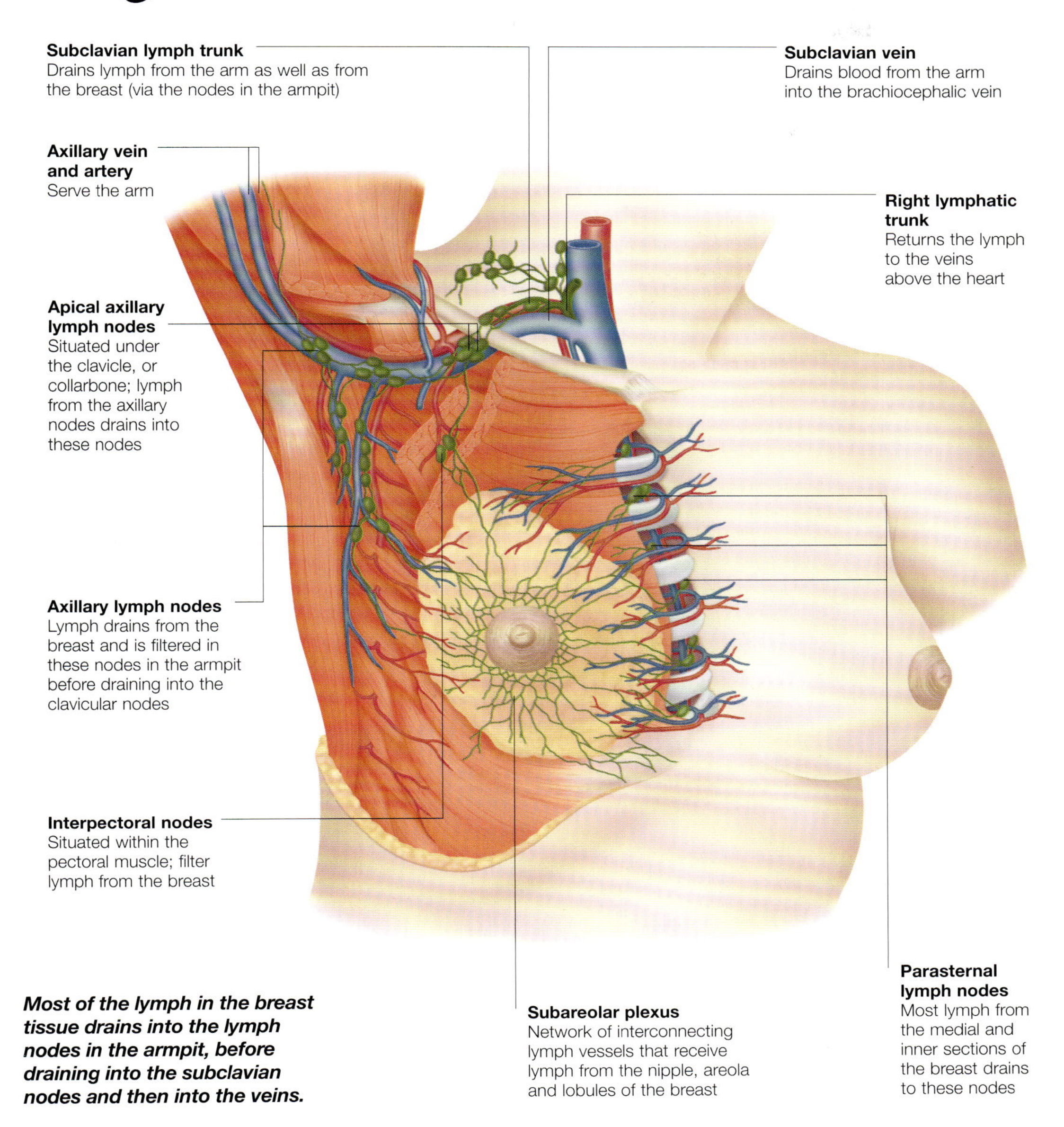

Most of the lymph in the breast tissue drains into the lymph nodes in the armpit, before draining into the subclavian nodes and then into the veins.

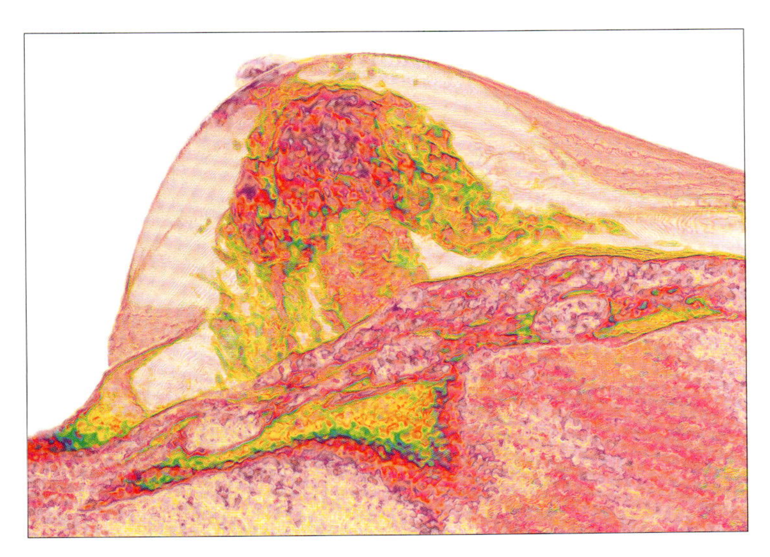

Coloured 3D magnetic resonance imaging (MRI) scan of a normal female breast. This axial view shows the breast parenchyma tissue (pink). The parenchyma tissue is the bulk tissue of the breast, excluding the adipose (fatty) tissue.

Uterus

The uterus, or womb, is the part of the female reproductive tract that nurtures and protects the fetus during pregnancy. It lies within the pelvic cavity and is a hollow, muscular organ.

In the non-pregnant state, the uterus is about 7.5cm (3in) long and 5cm (2in) across at its widest point. During pregnancy, the uterus expands to accommodate the fetus.

Structure

The uterus is made up of two parts:

- The body, forming the upper part of the uterus – this is fairly mobile as it must expand during pregnancy. The central triangular space, or cavity, of the body of the uterus receives the openings of the paired uterine tubes.
- The cervix, the lower part of the uterus – this is a thick, muscular canal that is anchored to the surrounding pelvic structures for stability.

Uterine Walls

The body of the uterus has a thick wall that is composed of three layers:

- Perimetrium – the thin outer coat that is continuous with the pelvic peritoneum
- Myometrium – forming the great bulk of the uterine wall
- Endometrium – the delicate lining that is specialized to allow implantation of an embryo should fertilization occur.

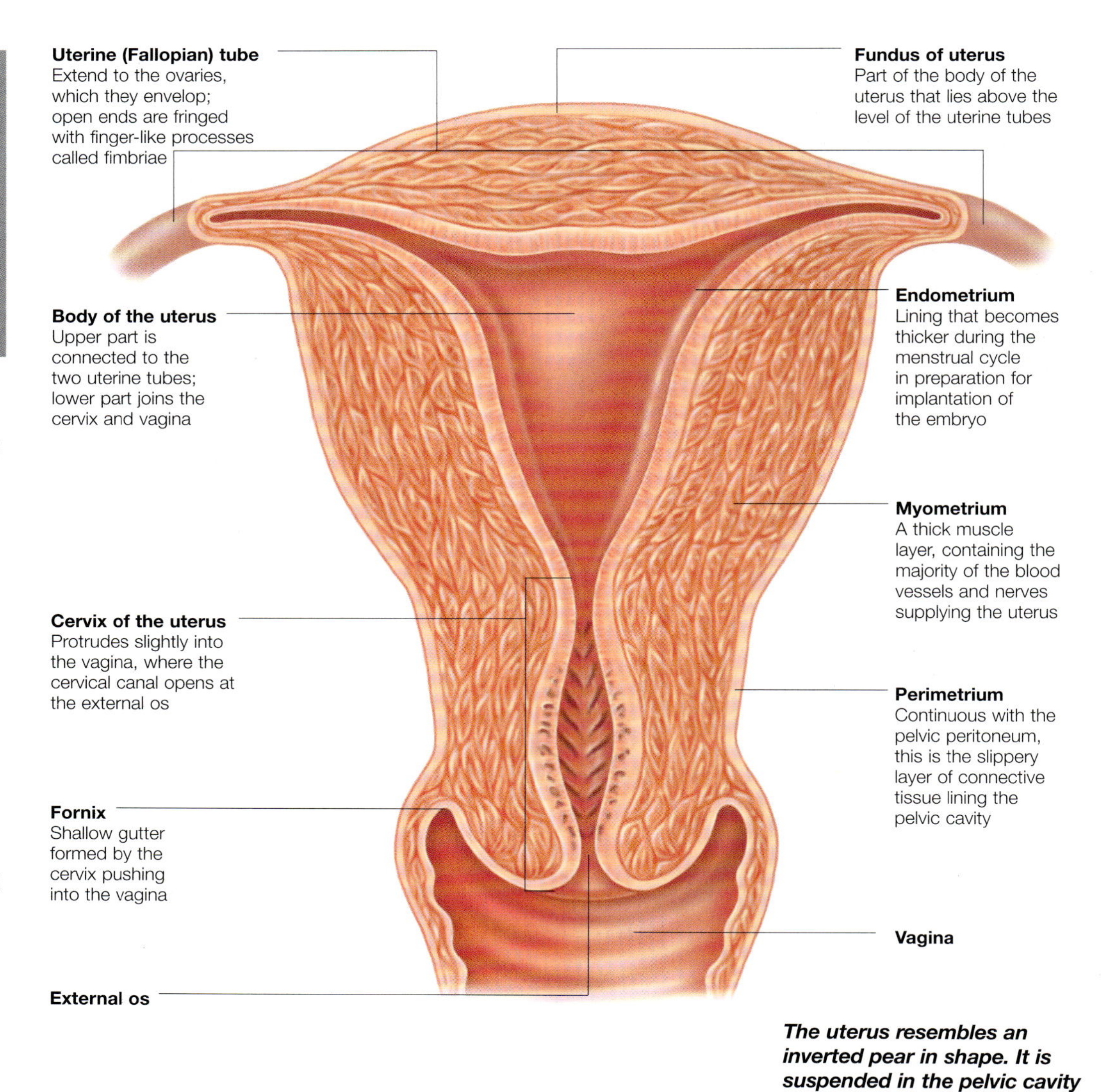

The uterus resembles an inverted pear in shape. It is suspended in the pelvic cavity by peritoneal folds or ligaments.

Position of the uterus

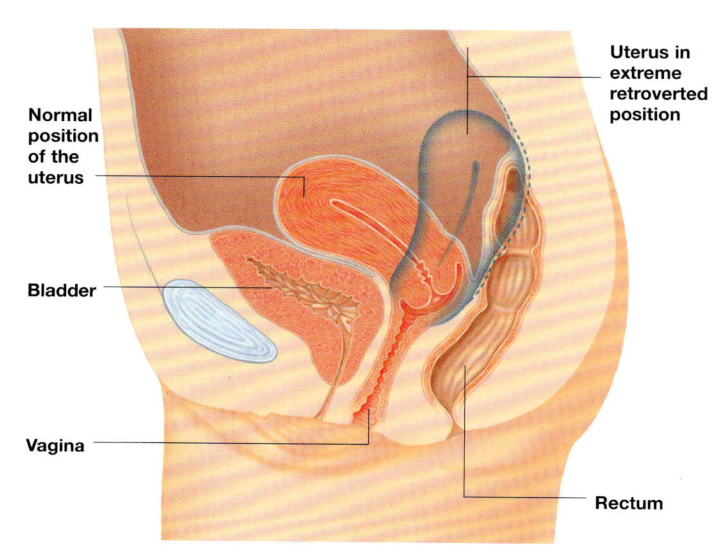

The uterus lies in the pelvis between the bladder and the rectum. However, its position changes with the stage of filling of these two structures and with different postures.

Normal position

Normally the long axis of the uterus forms an angle of 90 degrees with the long axis of the vagina, with the uterus lying forwards on top of the bladder. This usual position is known as anteversion.

In most women the uterus lies on the bladder, moving backwards as the bladder fills. However, it can lie in any position between the two extremes shown.

Anteflexion

In some women, the uterus lies in the normal position, but may curve forwards slightly between the cervix and fundus. This is termed anteflexion.

Retroflexion

In some cases, however, the uterus bends not forwards but backwards, the fundus coming to lie next to the rectum. This is known as a retroverted uterus.

Regardless of the uterine position, it will normally bend forwards as it expands in pregnancy. A pregnant retroverted uterus, however, may take longer to reach the pelvic brim, at which point it becomes palpable abdominally.

Uterus in pregnancy

In pregnancy the uterus must enlarge to hold the growing fetus. From being a small pelvic organ, it increases in size to take up much of the space of the abdominal cavity.

Pressure of the enlarged uterus on the abdominal organs pushes them up against the diaphragm, encroaching on the thoracic cavity and causing the ribs to flare out to compensate. Organs such as the stomach and bladder are compressed to such an extent in late pregnancy that their capacity is greatly diminished and they become full sooner.

After pregnancy, the uterus rapidly decreases in size again, although it always remains slightly larger.

Height of fundus

During pregnancy the enlarging uterus is accommodated within the pelvis for the first 12 weeks, at which time the uppermost part, the fundus, can just be palpated in the lower abdomen. By 20 weeks, the fundus reaches the region of the umbilicus, and by late pregnancy it has reached the xiphisternum, the lowest part of the breastbone.

Weight of uterus

In the final stages of pregnancy the uterus increases in weight from a pre-pregnant 45g (1.5oz) to around 900g (32oz). The myometrium (muscle layer) grows as the individual fibres increase in size (hypertrophy). In addition, the fibres increase in number (hyperplasia).

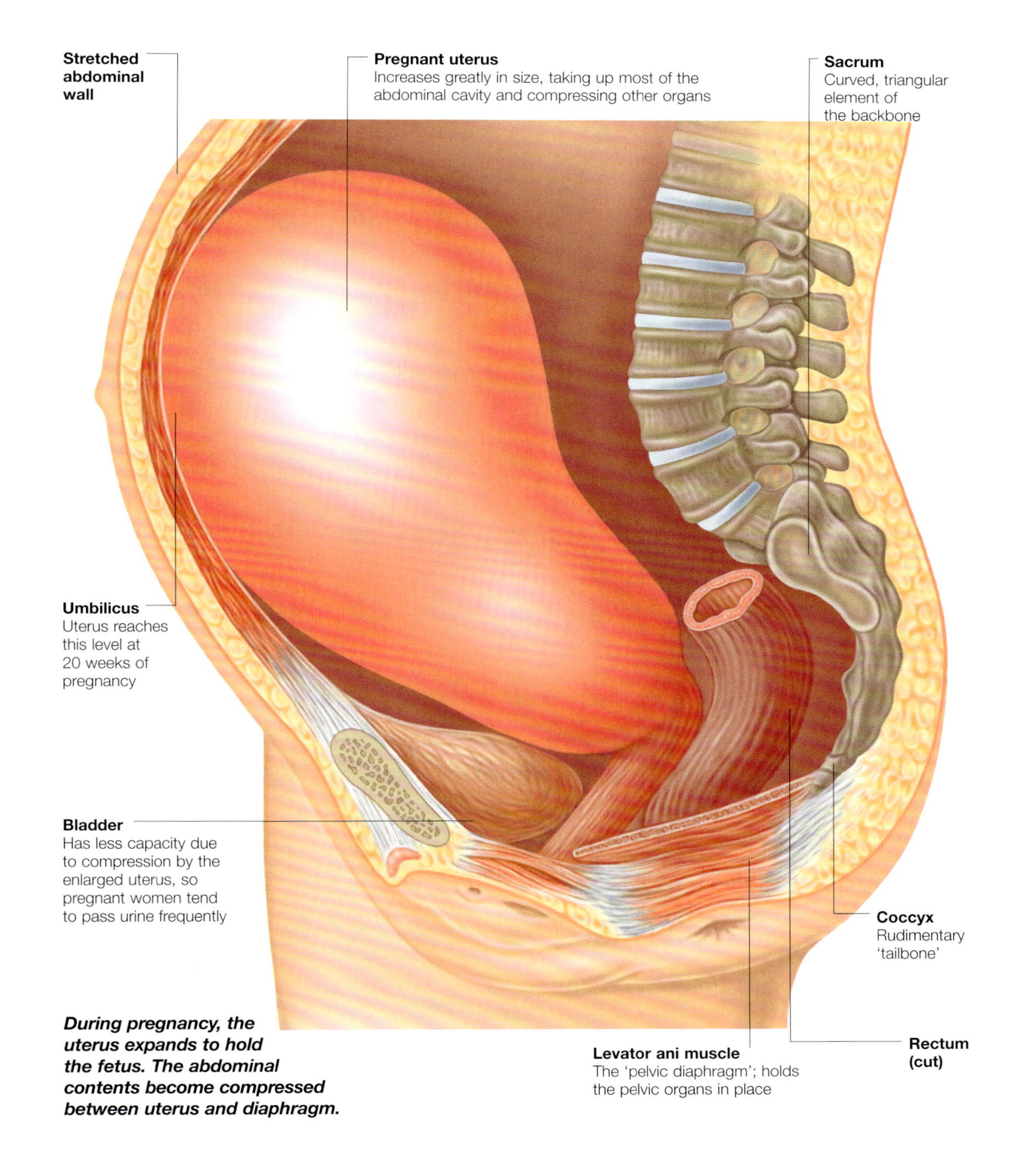

During pregnancy, the uterus expands to hold the fetus. The abdominal contents become compressed between uterus and diaphragm.

Vagina

The vagina is the thin-walled muscular tube that extends from the cervix of the uterus to the external genitalia. The vagina is normally closed but is designed to stretch during intercourse or childbirth.

The vagina is approximately 8cm (3in) in length and lies between the bladder and the rectum. It forms the main part of the birth canal and receives the penis during sexual intercourse.

Structure

The front and back walls of the vagina normally lie in contact with one another, closing the lumen (central space), although the vagina can expand greatly, as happens in childbirth.

The cervix, the lower end of the uterus, projects down into the lumen of the vagina at its upper end. Where the vagina arches up to meet the cervix, it forms recesses known as the vaginal fornices. These are divided into anterior, posterior, right and left fornices, although they form a complete ring.

The thin wall of the vagina has three layers:

- Adventitia – outer layer composed of fibroelastic connective tissue that allows distension when necessary
- Muscularis – the central muscular layer of the vaginal wall
- Mucosa – the inner layer of the vagina; this is thrown into many rugae (deep folds), and has a layered, stratified squamous (skin-like) epithelium (cell lining), which helps to resist abrasion during intercourse.

Coronal section through the vagina

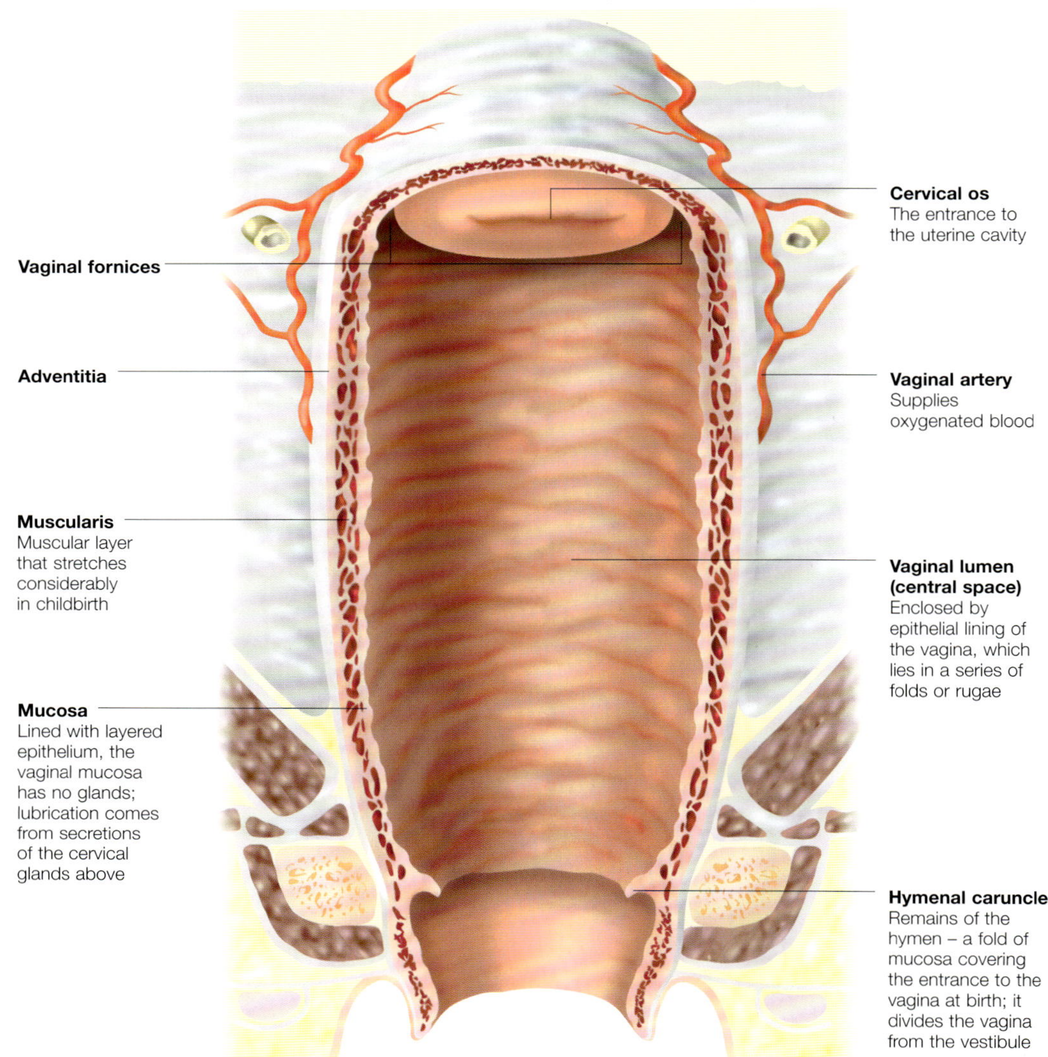

The vagina is a muscular, tubular organ designed to expand during sexual intercourse and childbirth. It is approximately 8cm (3in) in length.

Female exterior genitalia

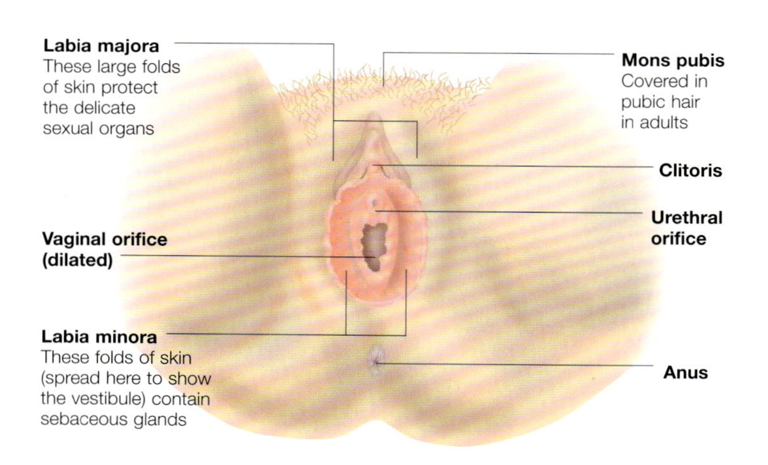

The female external genitalia, or vulva, are those parts that lie at the surface of the body, external to the vagina. They include:

- Mons pubis – the rounded, hairy, fatty area that lies above the pubic bone
- Labia majora – the two outer fatty folds of skin that lie across the vulval opening
- Labia minora – the two smaller folds of skin that lie inside the cleft of the vulva
- Vestibule – area into which the urethra and vagina open
- Clitoris – a structure composed of erectile tissue and containing a rich sensory nerve supply; it is analogous to the penis in males.

The vulval opening is partially closed off by a fold of mucosa, the hymen; this may rupture at first intercourse, with tampon use or during a pelvic examination.

The external genitalia include four folds of skin, known as the labia. These cover and protect the clitoris and the vaginal and urethral orifices.

Cervix

The cervix, or neck of the uterus, is the narrowed, lower part of the uterus that projects down into the upper vagina.

The cervix is fixed in position by the cervical ligaments, anchoring the mobile uterine body above. It has a narrow canal about 2.5cm (1in) long in adult women. The walls of the cervix are tough, containing fibrous tissue as well as muscle, unlike the body of the uterus, which is mainly muscular.

The central canal of the cervix is the downwards continuation of the uterine cavity that opens at its lower end, the external os, into the vagina. The canal is widest at its central point, narrowing slightly at the internal os at the upper end and the external os below.

Lining of the Cervix

The lining of the cervix is of two types:

- Endocervix – this is the lining of the cervical canal, inside the cervix. The epithelium is a single layer of columnar cells that overlies a surface thrown into many folds containing glands.
- Ectocervix – this covers the portion of the cervix that projects down into the vagina; it is composed of squamous epithelium and has many layers.

Cervical Os

The opening of the cervical canal into the upper vagina is known as the cervical os. In a woman who has not given birth the opening appears round in shape. After childbirth, the os becomes slit-like in appearance.

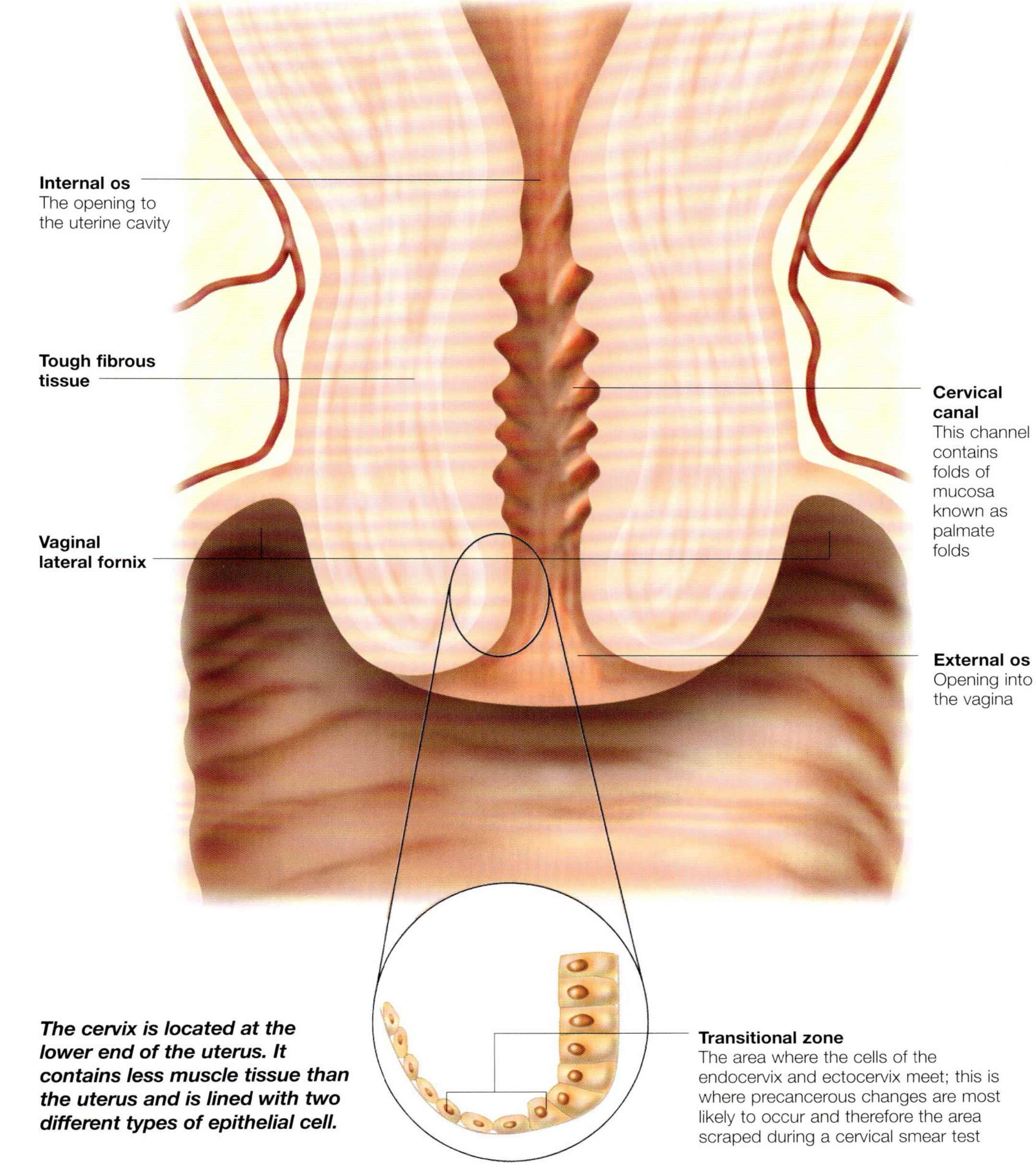

The cervix is located at the lower end of the uterus. It contains less muscle tissue than the uterus and is lined with two different types of epithelial cell.

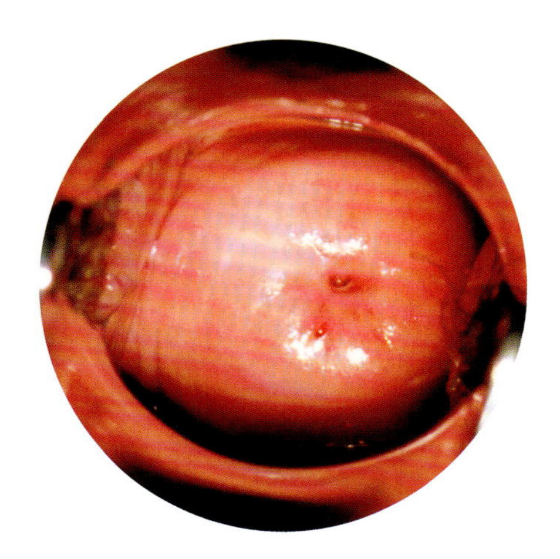

This healthy cervix is viewed through a metal speculum.

Ovaries

The ovaries are the site of production of oocytes, or eggs, which are fertilized by sperm to produce embryos.

The paired ovaries are situated in the lower abdomen and lie on either side of the uterus. Their position is variable, especially after childbirth, when the supporting ligaments have been stretched. Each ovary consists of:

- Tunica albuginea – a protective layer of fibrous tissue
- Medulla – a central region with blood vessels and nerves
- Cortex – within which the oocytes develop
- Surface layer – smooth before puberty but becoming more pitted in the reproductive years.

Blood Supply

The arterial supply to the ovaries comes via the ovarian arteries, which arise from the abdominal aorta. After supplying the uterine tubes, the ovarian and uterine arteries overlap.

Blood from the ovaries enters a network of tiny veins, the pampiniform plexus, within the broad ligament. From here it enters the right and left ovarian veins, which ascend into the abdomen to drain ultimately into the large inferior vena cavae and the renal vein respectively.

Cross-section of an ovary

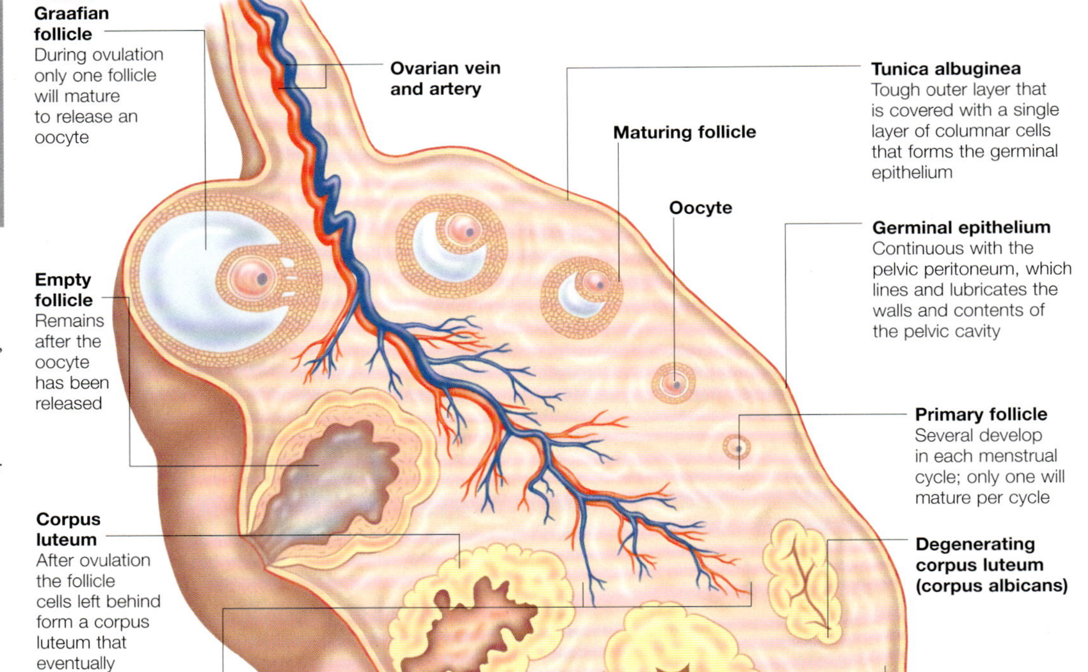

This cross-section shows the follicles situated in the cortex of the ovary. Each follicle contains an oocyte at a different stage of development.

Supporting ligaments

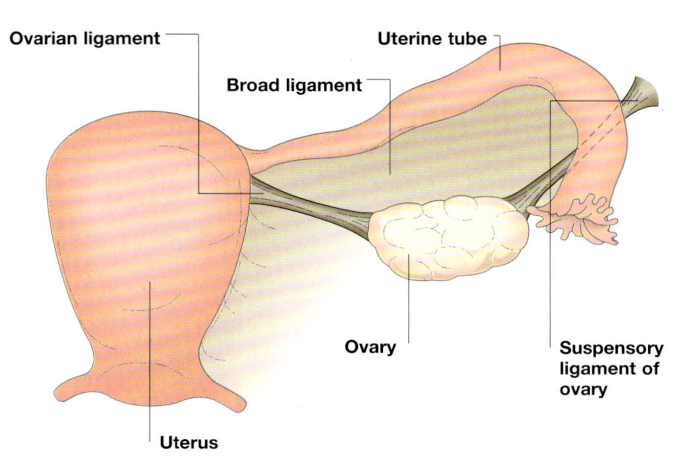

Each ovary is held in its position relative to the uterus and uterine tubes by several ligaments.

Main ligaments

These ligaments include the following:

- Broad ligament – the tent-like fold of pelvic peritoneum that hangs down on either side of the uterus, enclosing the uterine tubes and ovaries
- Suspensory ligament of the ovary – the part of the broad ligament that anchors the ovary to the side wall of the pelvis and carries the ovarian vessels and lymphatics
- Mesovarium – the fold of the broad ligament within which the ovary lies
- Ovarian ligament – attaches the ovary to the uterus and runs within the broad ligament.

These ligaments may become stretched in women following childbirth, which in many cases means that the position of the ovary may be more variable than before pregnancy.

Each ovary is suspended by several ligaments to hold it in position. However, the position varies, especially if the ligaments have stretched.

Uterine (Fallopian) Tubes

The uterine, or Fallopian, tubes collect the oocytes (eggs) released from the ovaries and transport them to the uterus. They also provide a site for fertilization of the oocyte by a sperm to take place.

Each uterine tube is about 10cm (4in) long and extends outwards from the upper part of the uterus towards the lateral wall of the pelvic cavity. The tubes run within the upper edge of the broad ligament and open into the peritoneal cavity near the ovary.

Structure

The tubes are divided into four parts:

- Infundibulum – the funnel-shaped outer end of the uterine tubes that opens into the peritoneal cavity
- Ampulla – the longest and widest part and the most usual site for fertilization of the oocyte
- Isthmus – a constricted region with thick walls
- Uterine part – the shortest part of the tube.

A layer of smooth muscle fibres within the walls allows the tubes to contract rhythmically to propel the oocyte to the uterus. In addition, the walls are lined with cells that bear cilia, tiny brush-like projections that beat to 'sweep' the oocyte towards the uterus. Non-ciliated cells in deep crypts in the lining of the tubes produce secretions that keep the oocyte, and any sperm that may be present, nourished.

Blood Supply

The uterine tubes have a very rich blood supply from both the ovarian and uterine arteries; these overlap to form an arterial arcade. Venous blood drains from the tubes in a pattern that mirrors the arterial supply.

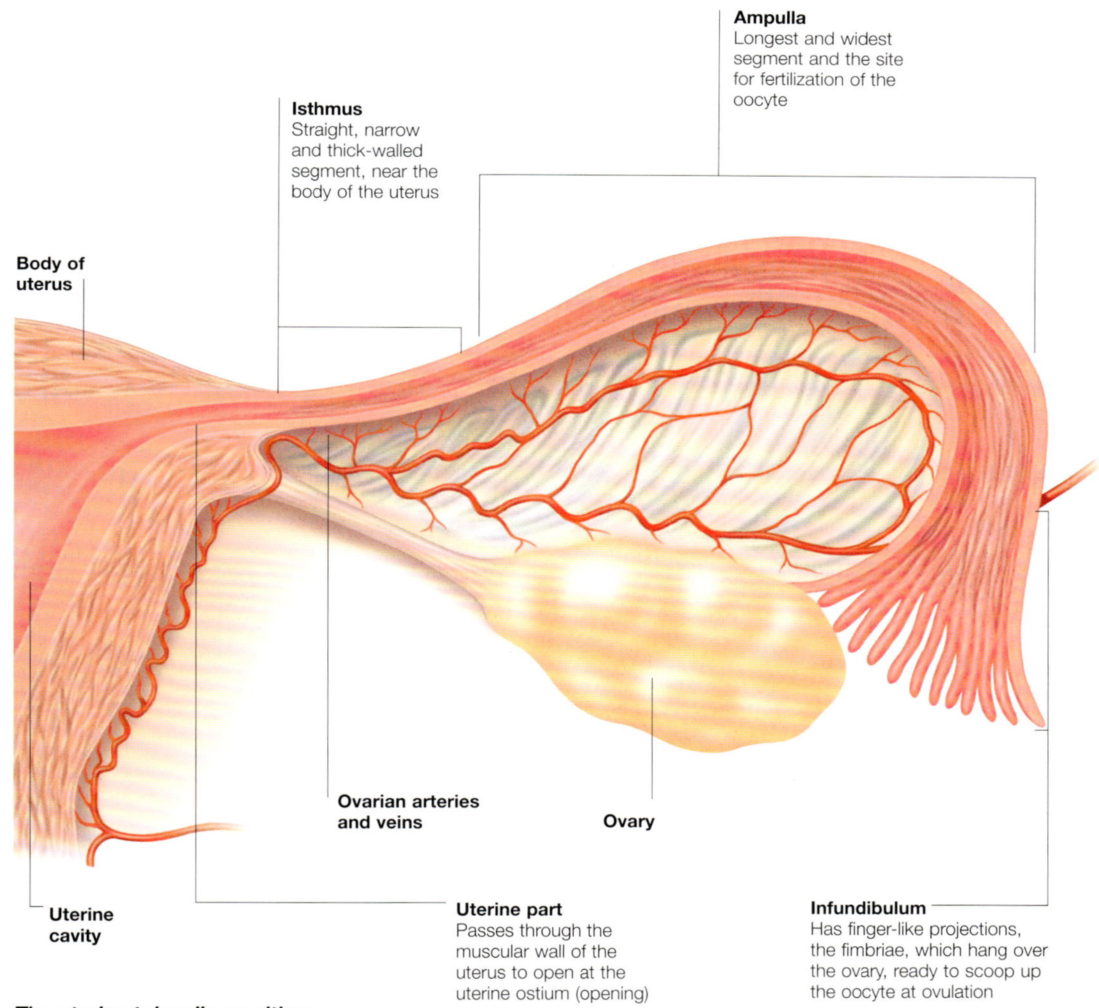

The uterine tubes lie on either side of the body. The outer part of each tube lies near the ovary, its end opening there into the abdominal cavity.

Menstrual Cycle

The menstrual cycle is the process by which an egg is released from an ovary in preparation for pregnancy. This occurs about every four weeks from the time of a woman's first period.

The menstrual cycle is characterized by the periodic maturation of oocytes (cells that develop into eggs) in the ovaries and associated physical changes in the uterus. Reproductive maturity occurs after a sudden increase in the secretion of hormones during puberty, usually between the ages of 11 and 15.

Cycle Onset

The time of the first period, which occurs at about the age of 12, is called the menarche. After this, a reproductive cycle begins, averaging 28 days. This length of time may be longer, shorter or variable, depending on the individual. The cycle is usually continuous, apart from during pregnancy or hormonal upset.

Menstruation

Each month, if conception does not occur, oestrogen and progesterone levels fall and the blood-rich lining of the uterus is shed at menstruation (menses). This takes place every 28 days or so, but the time can range from 19 to 36 days.

Menstruation lasts for about five days. Around 50ml/1.75fl oz (about an eggcup) of blood, uterine tissues and fluid is lost during this time, but again this volume varies from woman to woman. Some women lose only 10ml (0.3fl oz) of blood, while others lose 110 ml (3.8fl oz).

Excessive menstrual bleeding is known as menorrhagia; temporary cessation of menstruation – such as during pregnancy – is called amenorrhoea. The menopause is the complete cessation of the menstrual cycle, and usually occurs between 45 and 55.

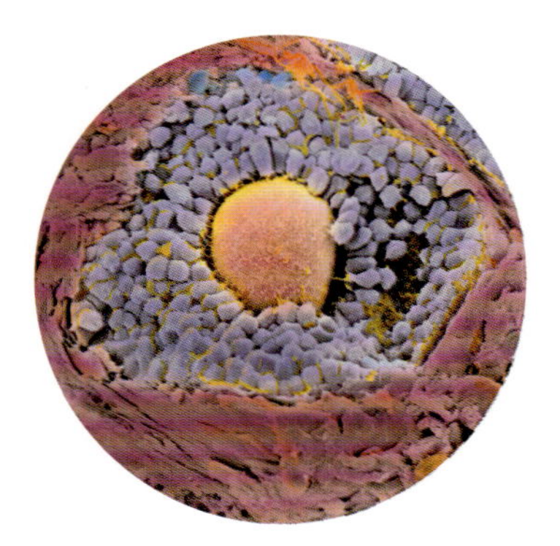

A developing egg in the centre of its follicle. The number of eggs is finite, and they are usually used up by the age of 50.

Monthly physiological changes

This diagram illustrates the changes during the cycle. Between days one and five, the lining is discharged, while another follicle is developing. The uterine lining thickens, and at day 14 the egg is released (ovulation).

FSH: Follicle-stimulating hormone
LH: Luteinizing hormone

Point of ovulation
This occurs at around day 14 of the cycle. The production of LH peaks, causing the mature follicle to rupture, releasing the egg. The uterine lining is at its thickest, ready to receive the egg should fertilization take place. If implantation does not take place, the corpus luteum degenerates and the lining is shed

Gonadotrophic hormones
Released by the pituitary gland to promote the production of the egg and of sex hormones in the gonads (ovaries)

Ovarian activity
Each month, one follicle develops to maturity, then releases an egg at ovulation; the surviving tissue in the ovary forms the corpus luteum, a temporary hormone-producing gland

Ovarian hormones
Secreted by the ovary to encourage the lining to grow; extra progesterone is produced by the corpus luteum after ovulation to prepare the uterus for pregnancy

Lining of uterus
Progressively thickens to receive the fertilized egg; if the egg does not implant, the lining is shed (menses) during the first five days of the cycle

LH
FSH
Follicle matures
Ovulation
Ruptured follicle forms corpus luteum
Oestrogen
Progesterone
Menses
Proliferation
Days
5
14
28
One menstrual cycle

Egg development

The process of developing a healthy egg for release at ovulation takes around six months. It occurs throughout life until the stock of oocytes is exhausted.

One to two million eggs are present at birth, distributed between the two ovaries, and about 400,000 are left by the time of the first period. During each menstrual cycle, only one egg – from a pool of around 20 potential eggs – develops and is released. By the time menopause is reached, the process of atresia (cell degeneration) in the ovaries is complete and no eggs remain.

Eggs develop within cavity-forming secretory structures called follicles. The first stage of follicle development occurs when an egg becomes surrounded by a single layer of granulosa cells and is called a primordial (primary) follicle. The genetic material within the egg at this stage remains undisturbed – but susceptible to alteration – until ovulation of that egg occurs, up to 45 years after it first developed. This helps to explain the increase in abnormal chromosomes in eggs and offspring of women who conceive later in life.

Primordial follicles develop into secondary follicles by meiotic (reductive) division and then into tertiary (or antral, meaning 'with a cavity') follicles. As many as 20 primary follicles will begin to mature, although 19 will eventually regress. If more than one follicle develops to maturity, twins or triplets may be conceived.

Ovulation

The final 14 day period of follicular development takes place during the first half of the menstrual cycle and depends on the precise hormonal interplay between the ovary, pituitary gland and the hypothalamus.

The trigger for selecting a healthy egg for development at the start of each cycle is a rise in the secretion of follicle-stimulating hormone (FSH) by the pituitary gland. This occurs in response to a fall in the hormones oestrogen and progesterone during the luteal phase (second 14 days) of the previous cycle if conception has not occurred.

Egg selection

At the time of the FSH signal, there are about 20 secondary follicles, 2–5mm (0.1–0.2in) in diameter, distributed between the two ovaries. A single follicle is selected from this pool, while the others undergo atresia. Once a follicle is selected, the development of further follicles is prevented. A typical 5mm (0.2in) secondary follicle will then require 10–12 days of sustained stimulation by FSH to grow to a diameter of 20mm (0.8in) before rupturing, releasing the egg into the uterine (Fallopian) tube.

As the follicle enlarges, there is a steady rise in oestrogen production, triggering a mid-cycle rise in luteinizing hormone (LH) by the pituitary, which in turn causes release and maturation of the egg. The interval between the LH peak and ovulation is relatively constant (about 36 hours). The ruptured follicle (corpus luteum) that remains after ovulation becomes a very important endocrine gland, secreting oestrogen and progesterone.

Hormone regulation

Progesterone levels rise to a peak about seven days after ovulation. If fertilization takes place, the corpus luteum maintains the pregnancy until the placenta takes over at about three months' gestation. If no conception takes place, the gland has a lifespan of 14 days, and oestrogen and progesterone levels decline in anticipation of the next cycle.

In the first half of the cycle, oestrogen secreted by the developing follicle (stage before corpus luteum) enables the lining of the uterus (endometrium) to proliferate and increase in thickness ready to nourish the egg should it become fertilized. Once the corpus luteum is formed, progesterone converts the endometrium to a more compact layer in anticipation of an embryo implanting.

Under a light microscope, a secondary oocyte (mature egg) can be seen surrounded by the cells of the corona radiata that support it during development.

The fully developed egg is surrounded by a protein coating called the zona pellucida. This serves to trap and bind a single sperm during the process of fertilization.

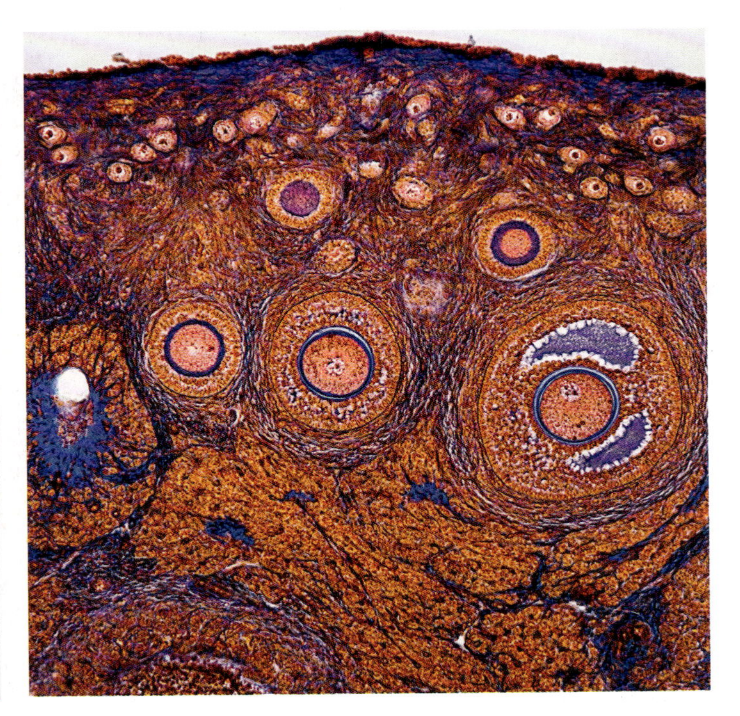

The follicles are located in the cortex of the ovary. This micrograph shows the follicle separated by connective tissue.

Ovulation

The total supply of eggs for a woman's reproductive years is determined before she is born. These immature eggs are stored in the ovary until puberty, after which one is released every month.

An ovum (egg) is the female gamete, or sex cell, which unites with a sperm to form a new individual. Eggs are produced and stored in the ovaries, two walnut-sized organs connected to the uterus via the uterine (Fallopian) tubes.

Ovaries

Each ovary is covered by a protective layer of peritoneum (abdominal lining). Immediately below this layer is a dense fibrous capsule, the tunica albuginea. The ovary itself consists of a dense outer region, called the cortex, and a less dense inner region, the medulla.

Gamete Production

In females, the total supply of eggs is determined at birth. Egg-forming cells degenerate from birth to puberty and the timespan during which a woman can release mature eggs is limited from puberty until menopause.

The process by which ova are produced is known as oogenesis, which literally means 'the beginning of an egg'.

Germ cells in the fetus produce many oogonia cells. These divide to form primary oocytes, which are enclosed in groups of follicle cells (support cells).

Genetic Division

The primary oocytes begin to divide by meiosis (a specialized nuclear division) but this process is interrupted in its first phase and is not completed until after puberty.

At birth, a lifetime's supply of primary oocytes, numbering between 700,000 and two million, will have been formed. These specialized cells will lie dormant in the cortical region of the immature ovary and slowly degenerate, so that by puberty only 40,000 remain.

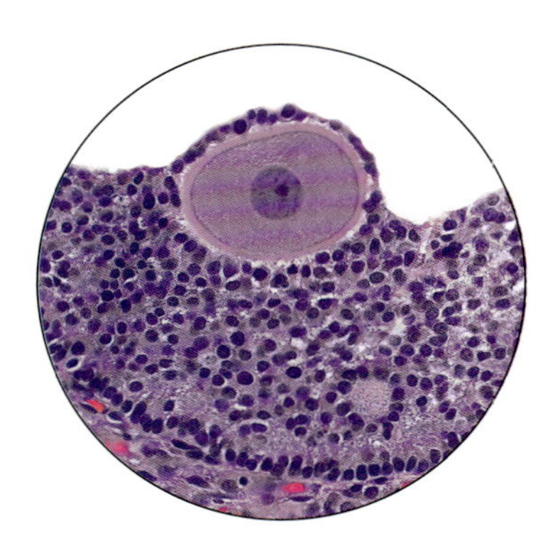

This micrograph shows an ovary with several large follicles (white). During ovulation, up to 20 follicles begin to develop, but only one matures to release an egg.

Egg development

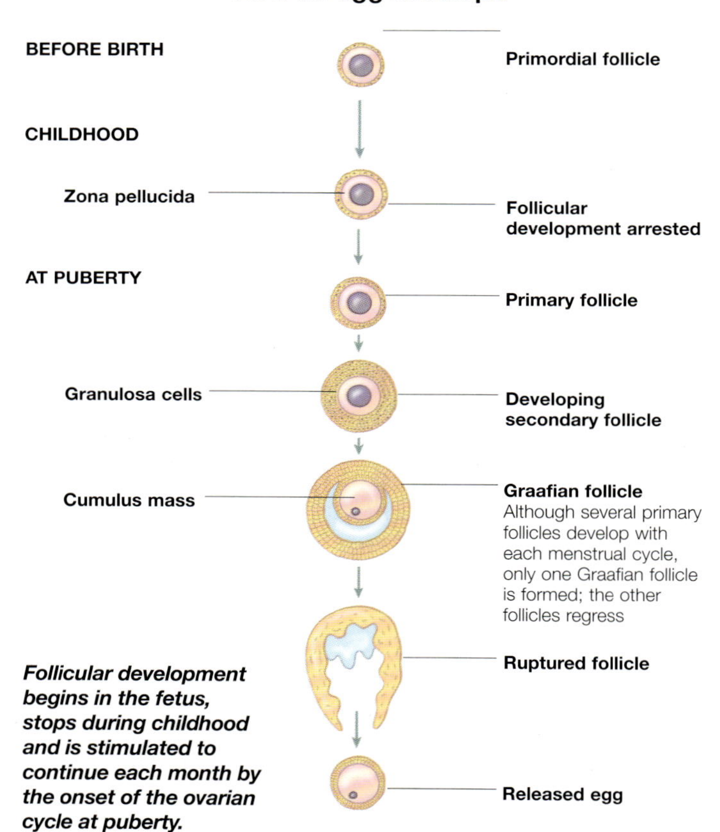

Follicular development begins in the fetus, stops during childhood and is stimulated to continue each month by the onset of the ovarian cycle at puberty.

Before puberty the primary oocyte is surrounded by a layer of cells (the granulosa cells), forming a primary follicle.

Puberty

With the onset of puberty, some of the primary follicles are stimulated each month by hormones to continue development and become secondary follicles:

- A layer of clear viscous fluid, the zona pellucida, is deposited on the surface of the oocyte.
- The granulosa cells multiply and form an increasing number of layers around the oocyte.
- The centre of the follicle becomes a chamber (the antrum) that fills with fluid secreted by the granulosa cells.
- The oocyte is pushed to one side of the follicle, and lies in a mass of follicular cells called the cumulus mass.

A mature secondary follicle is called a Graafian follicle.

Meiosis

The first meiotic division produces two cells of unequal size – the secondary oocyte and the first polar body. The secondary oocyte contains nearly all the cytoplasm of the primary oocyte. Both cells begin a second division; however, this process is halted, and is not completed until the oocyte is fertilized by a sperm.

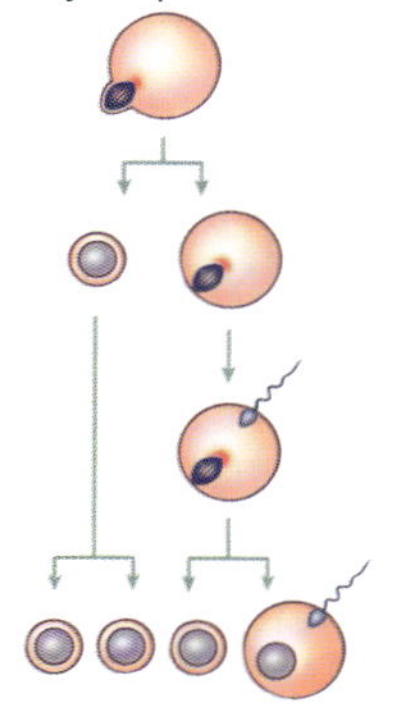

Meiosis, a specialized nuclear division, occurs in the ovaries, giving rise to a female sex cell and three polar bodies.

Egg release

Ovulation occurs when a follicle ruptures, releasing a mature oocyte into the uterine tube. It is at this stage in the menstrual cycle that fertilization may occur.

As the Graafian follicle continues to swell, it can be seen on the surface of the ovary as a blister-like structure.

Hormonal changes

In response to hormonal changes, the follicular cells surrounding the oocyte begin to secrete a thinner fluid at an increased rate, so that the follicle rapidly swells. As a result, the follicular wall becomes very thin over the area exposed to the ovarian surface, and the follicle eventually ruptures.

Ovulation

A small amount of blood and follicular fluid is forced out of the vesicle, and the secondary oocyte, surrounded by the cumulus mass and zona pellucida, is expelled from the follicle into the peritoneal cavity – the process of ovulation.

Women are generally unaware of this phenomenon, although some experience a twinge of pain in the lower abdomen. This is caused by the intense stretching of the ovarian wall.

Fertile period

Ovulation occurs around the 14th day of a woman's menstrual cycle, and it is at this time that a woman is at her most fertile. As sperm can survive in the uterus for up to five days, there is a period of about a week when fertilization can occur.

In the event that the secondary oocyte is penetrated by a sperm cell and pregnancy ensues, the final stages of meiotic division will be triggered. If, however, the egg is not fertilized, the second stage of meiosis will not be completed and the secondary oocyte will simply degenerate.

The ruptured follicle forms a gland called the corpus luteum that secretes progesterone. This hormone prepares the uterine lining to receive an embryo.

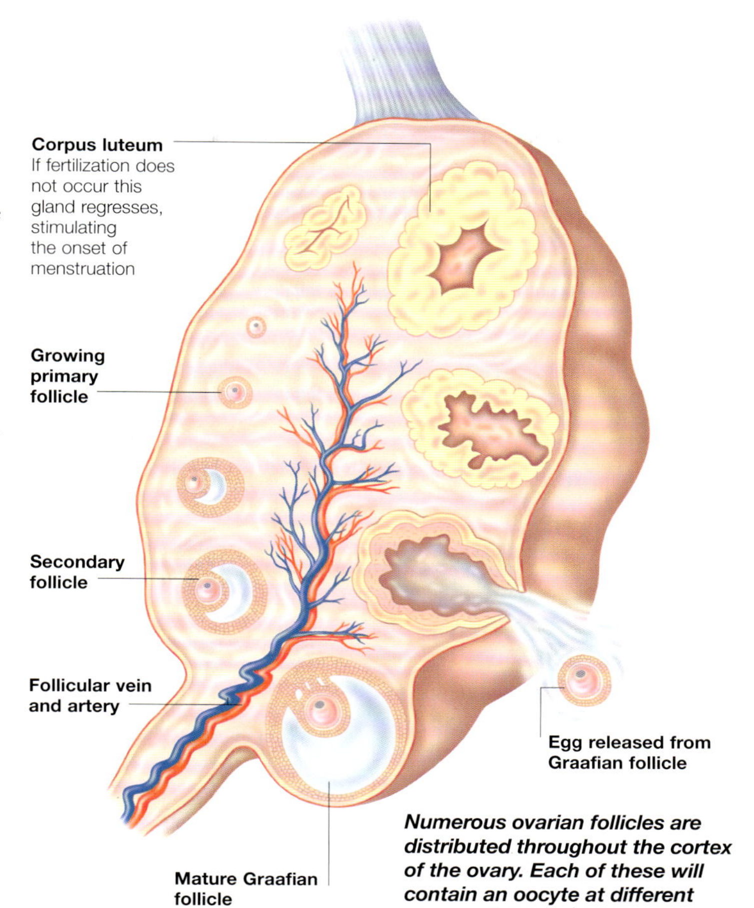

Numerous ovarian follicles are distributed throughout the cortex of the ovary. Each of these will contain an oocyte at different stages of development.

The menstrual cycle

The oestrus, or menstrual cycle, refers to the cyclical changes that take place in the female reproductive system during the production of eggs.

These changes are controlled by hormones released by the pituitary gland and ovaries: oestrogen, progesterone, luteinizing hormone and follicle-stimulating hormone.

Uterine changes

Following menstruation the endometrium thickens and becomes more vascular under the influence of oestrogen and follicle-stimulating hormone. During the first 14 days of the menstrual cycle a Graafian follicle matures. Ovulation occurs around day 14 when the secondary oocyte is expelled and swept into the uterine tube.

The ruptured follicle becomes a hormone-secreting body called the corpus luteum. This secretes progesterone, stimulating further thickening of the uterine lining in which the fertilized ovum will implant.

If fertilization does not occur, progesterone and oestrogen levels decrease. This causes the endometrium to break down and be excreted into the menstrual flow.

This graph shows the fluctuation of anterior pituitary and ovarian hormones during the menstrual cycle, together with structural changes within the ovary and uterus.

Conception

Millions of sperm cells travel up the female reproductive tract in search of the oocyte (egg). It takes hundreds of sperm to break down the outer coating of the oocyte, but only one will fertilize it.

Fertilization occurs when a single male gamete (sperm cell) and a female gamete (egg or oocyte) are united following sexual intercourse. Fusion of the two cells occurs and a new life is conceived.

Sperm

Following sexual intercourse, the sperm contained in the semen travel up to the uterus. Along the way they are nourished by the alkaline mucus of the cervical canal. From the uterus the sperm continue their journey into the uterine (Fallopian) tube. Although the distance involved is only around 20cm (7.8in), the journey can take up to two hours.

Survival

Although an average ejaculation contains around 300 million sperm cells, only a fraction of these (around 10,000) will manage to reach the uterine tube where the oocyte is located. Even fewer will actually reach the oocyte. Many sperm will be destroyed by the acidic vaginal environment, or become lost in other areas of the reproductive tract.

Sperm do not become capable of fertilizing an oocyte until they have spent some time in the woman's body. Fluids in the reproductive tract activate the sperm, so that the whiplash motion of their tails becomes more powerful.

The sperm are also helped on their way by contractions of the uterus, which force them upwards into the body. The contractions are stimulated by prostaglandins contained in the semen, and which are also produced during female orgasm.

Oocyte

Once it has been ejected from the follicle (during ovulation), the oocyte is pushed towards the uterus by the wave-like motion of the cells lining the uterine tube. The oocyte is usually united with the sperm about two hours after sexual intercourse in the outer part of the uterine tube.

Following sexual intercourse, millions of sperm cells make their way up the reproductive tract in search of the oocyte.

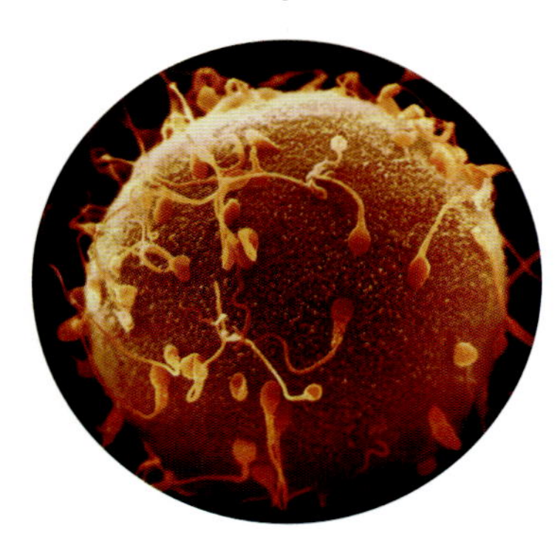

The path to fertilization

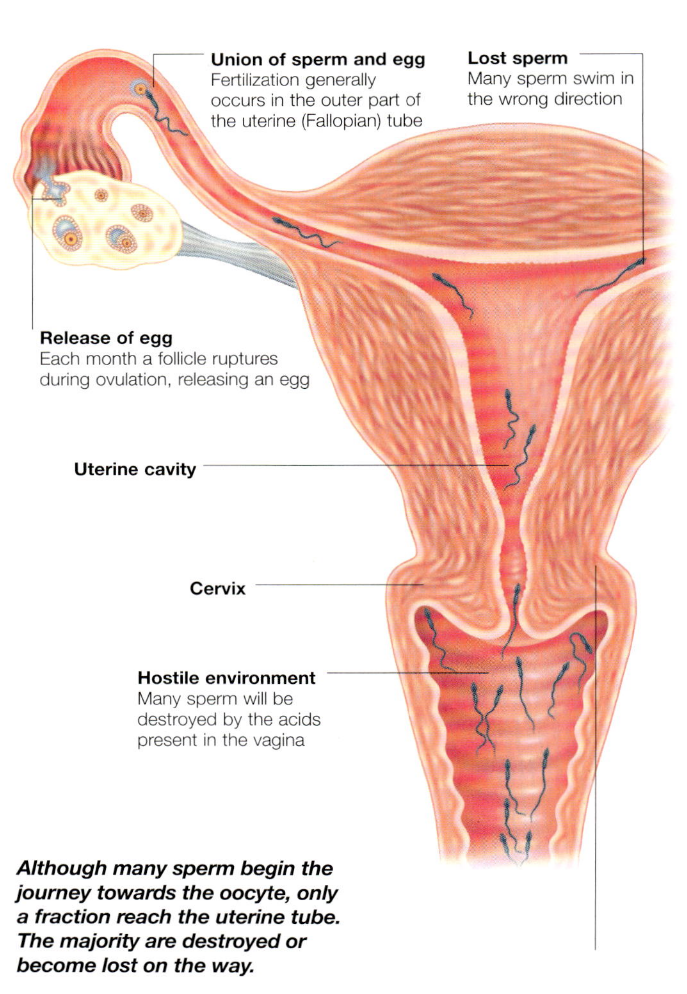

Although many sperm begin the journey towards the oocyte, only a fraction reach the uterine tube. The majority are destroyed or become lost on the way.

Reaching the oocyte

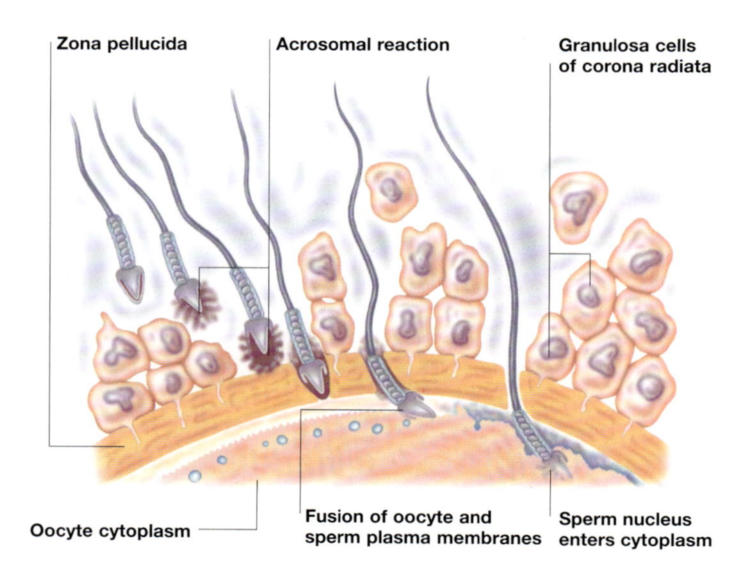

On the journey towards the oocyte, secretions present in the female reproductive tract deplete the sperm cells' cholesterol, thus weakening their acrosomal membranes. This process is known as capacitation, and without it fertilization could not occur.

Once in the vicinity of the oocyte, the sperm are chemically attracted to it. When the sperm cells finally come into contact with the oocyte, their acrosomal membranes are completely stripped away, so that the contents of each acrosome (the enzyme-containing compartment of the sperm) are released.

Penetration

The enzymes released by the sperm cells cause the breakdown of the cumulus mass cells and the zona pellucida, the protective outer layers of the oocyte. It takes at least 100 acrosomes to rupture in order for a path to be digested through these layers and for a single sperm to enter. In this way the sperm cells that reach the oocyte first sacrifice themselves, to allow penetration of the cytoplasm of the oocyte by another sperm.

When sperm cells reach the oocyte they release enzymes. These enzymes break down the protective outer layers of the ovum, allowing a sperm to enter.

Fertilization

When a single sperm has entered the oocyte, the genetic material from each cell fuses. A zygote is formed, which divides to form an embryo.

Once a sperm has penetrated the oocyte, a chemical reaction takes place within the oocyte, making it impossible for another sperm to enter.

Meiosis II

Entry of the sperm nucleus into the oocyte triggers the completion of nuclear division (meiosis II) begun during ovulation. A haploid oocyte and the second polar body (which degenerates) are formed. Almost immediately, the nuclei of the sperm and oocyte fuse to produce a diploid zygote, containing genetic material from both the mother and father.

Determination of sex

It is at the point of fertilization that sex is determined. It is the sperm, and therefore the father, that dictates what sex the offspring will be. Sex is determined by a combination of the two sex chromosomes, the X and the Y. The female will contribute an X chromosome, whereas a male may contribute either an X or a Y. Fertilization of the oocyte (X), will either be by a sperm containing an X or a Y to give a female (XX) or a male (XY).

Cell division

Several hours after fertilization the zygote undergoes a series of mitotic divisions to produce a cluster of cells known as a morula. The morula cells divide every 12 to 15 hours, producing a blastocyst comprised of around 100 cells.

The blastocyst secretes the hormone human chorionic gonadotrophin. This prevents the corpus luteum from being broken down, thus maintaining progesterone secretion.

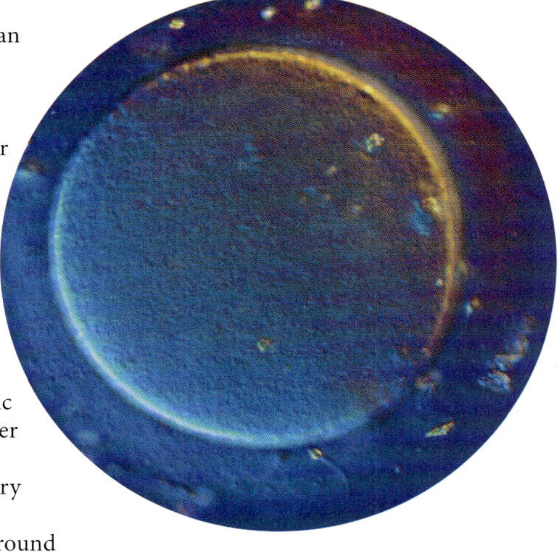

Once a sperm has penetrated the oocyte, the nuclei of both cells fuse. A diploid zygote forms, containing both the mother's and father's genes.

Implantation and development

Around three days after fertilization, the blastocyst begins its journey from the uterine (Fallopian) tube to the uterus. Normally the blastocyst would be unable to pass through the sphincter muscle in the uterine tube. However, the increasing levels of progesterone triggered by fertilization cause the muscle to relax, allowing the blastocyst to continue its journey to the uterus.

A damaged or blocked uterine tube preventing the blastocyst from passing at this stage could result in an ectopic pregnancy in which the embryo starts to develop in the tube.

Multiple births

In most cases a woman releases one oocyte every month from alternate ovaries. Occasionally, however, a woman produces an oocyte from each ovary, both of which are fertilized by separate sperm. This results in the development of non-identical twins, when each fetus is nourished by its own placenta.

Very occasionally a fertilized oocyte splits spontaneously to produce two embryos. This results in identical twins that share exactly the same genes, and even the same placenta.

Implantation

Once it has reached the uterus, the blastocyst implants itself in the thickened lining of the uterine wall. Hormones released from the blastocyst mean that it is not identified as a foreign body and expelled. Once the blastocyst is safely implanted, gestation begins.

Imperfections

About one third of fertilized oocytes fail to implant in the uterus and are lost. Of those embryos that do implant, many contain imperfections in their genetic material, such as an extra chromosome. These imperfections can cause the embryo to be lost soon after implantation. This can occur even before the first missed period, so that a woman will not even have known that she was pregnant.

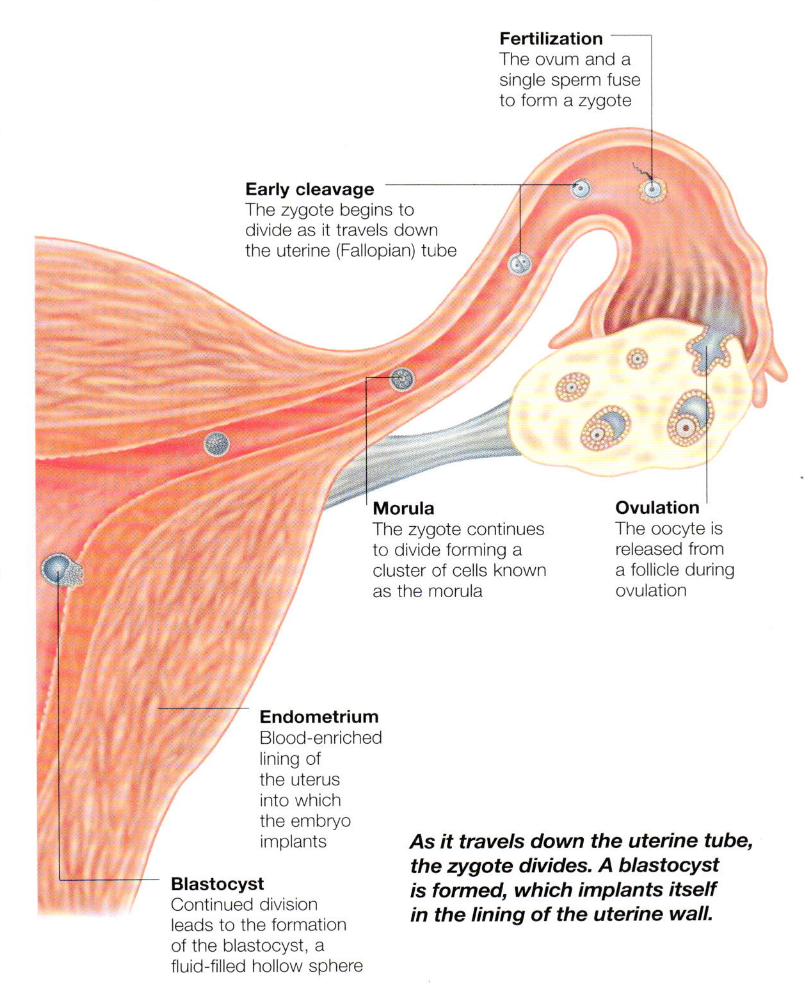

As it travels down the uterine tube, the zygote divides. A blastocyst is formed, which implants itself in the lining of the uterine wall.

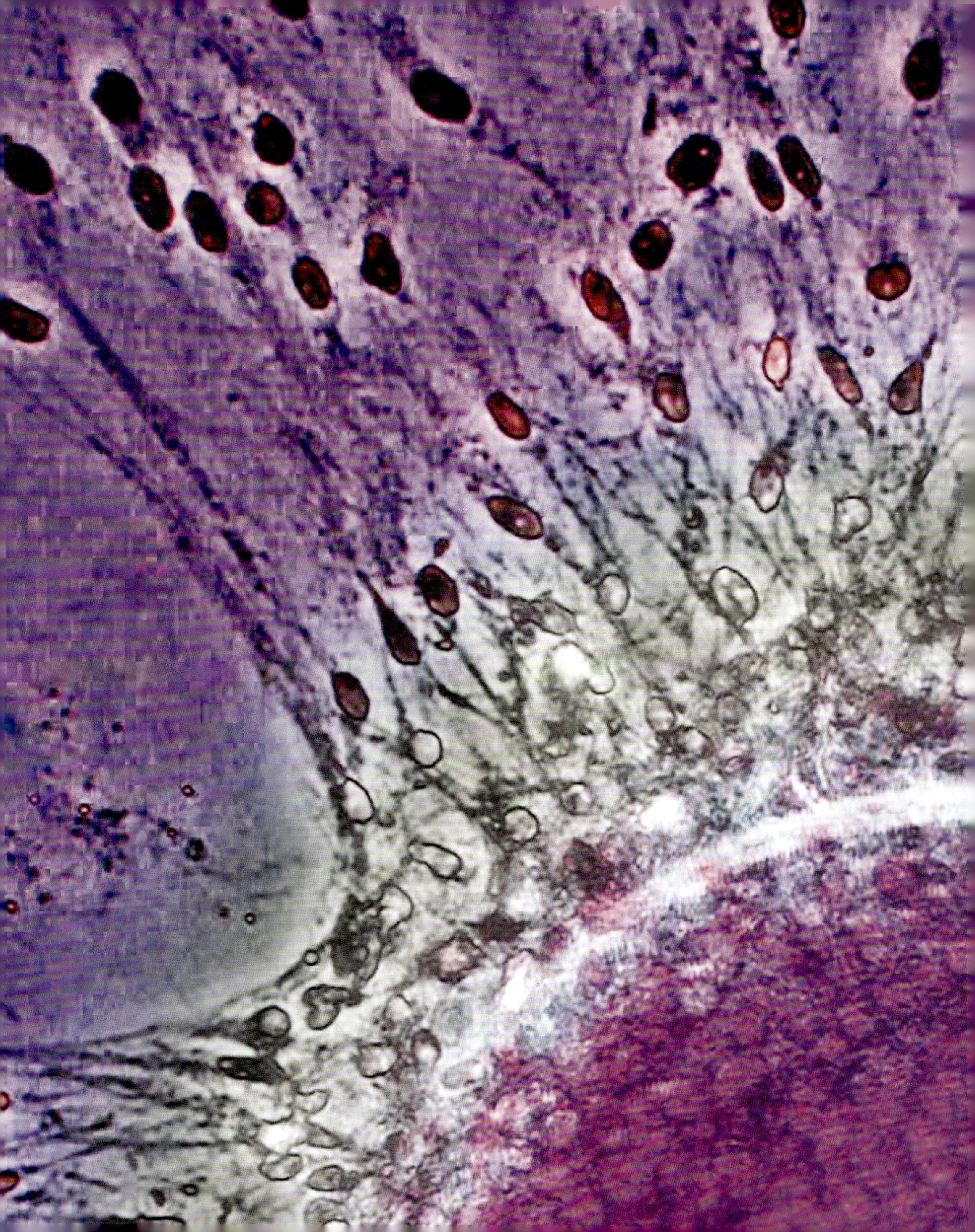

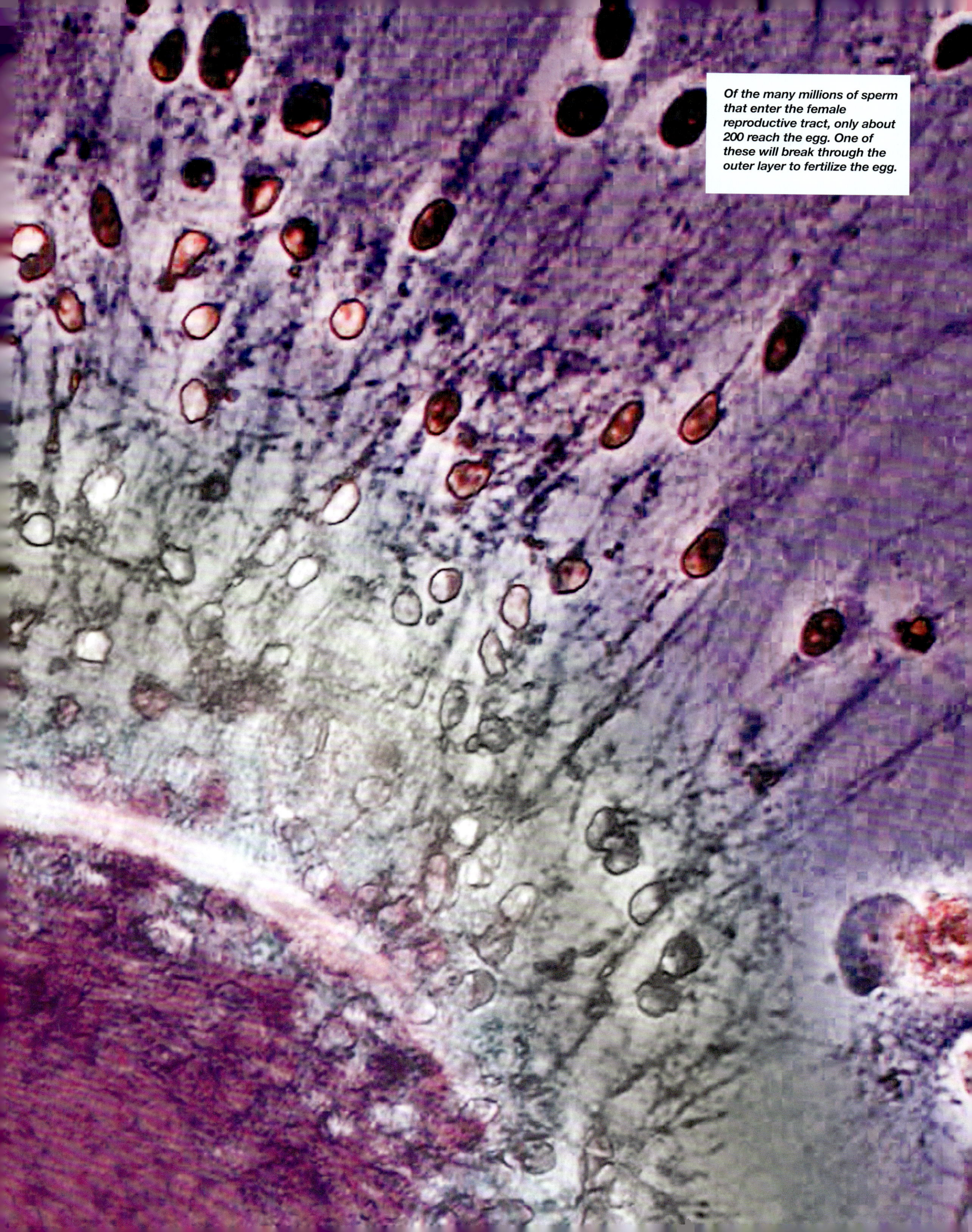

Of the many millions of sperm that enter the female reproductive tract, only about 200 reach the egg. One of these will break through the outer layer to fertilize the egg.

Placenta

The placenta is the organ that provides the developing fetus with all the nutrients it needs. It is a temporary structure formed in the uterus during pregnancy from fetal and maternal tissues.

The placenta takes on the role of the lungs and the intestine for the developing fetus. It achieves this by exposing the blood of the fetus to the maternal blood within its internal structure, allowing the fetus to take up oxygen and nutrients, while waste products are carried away.

The placenta becomes detached at birth and is delivered after the baby in what is known as the third stage of labour. It is then examined to check that it is complete and shows no evidence of abnormality or disease that may have affected the fetus.

Placental Appearance

At full term the placenta is a deep red, round or oval flattened organ. It normally weighs about 500g (1.1lb), or one sixth of the weight of the fetus it nourishes. There are two sides to the delivered placenta:

- The maternal aspect (which is attached to the lining of the womb) – this shows subdivisions where the placental tissue is divided by fibrous bands (septa). It is deep red and feels spongy.
- The fetal aspect (from which the umbilical cord arises) – this is covered in fetal membranes. Its surface is shiny and smooth with large umbilical vessels.

Fetal aspect of placenta

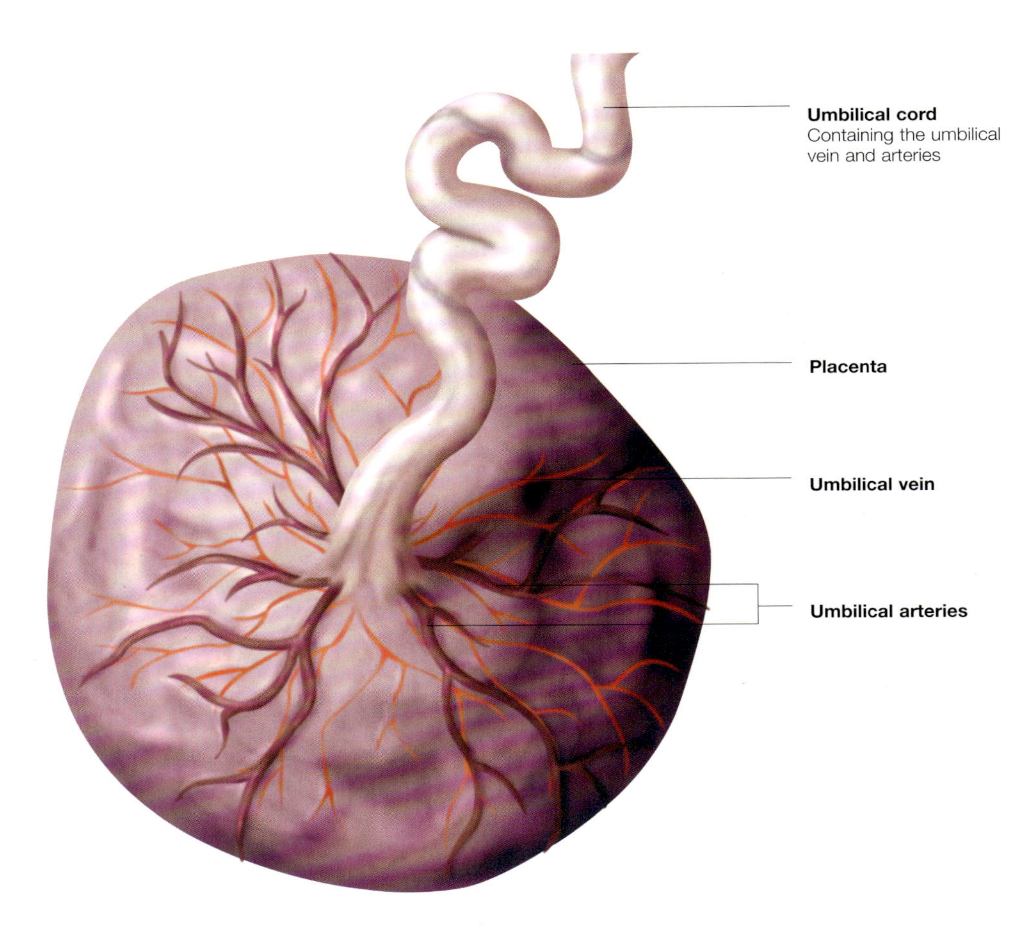

The placenta has two aspects: the maternal and the fetal. The fetal aspect of the placenta (shown here) is identifiable by the large umbilical vessels.

Variations in the placenta

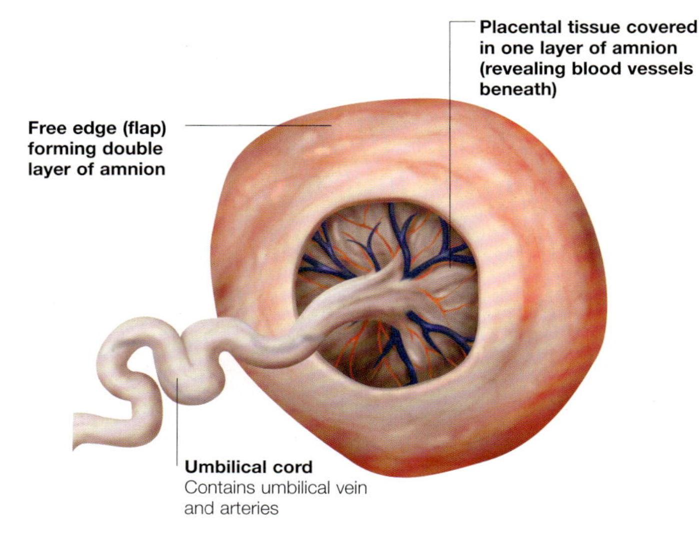

In a circumvallate placenta the amnion (the membranous sac containing the baby) folds in on itself, creating a double layer over most of the placenta.

A number of variations can occur in the form of the placenta. These are usually of little clinical significance, posing no threat to mother or fetus, although occasionally they may present problems.

Placental variations

Possible variations include:

- Succenturiate placenta – an extra lobe of the placenta that lies within the fetal membranes a short distance away from the main placenta.
- Battledore placenta – this is the name given to a placenta in which the cord inserts at one edge rather than centrally, as is normal.
- Velamentous insertion of the cord – the unusual arrangement whereby the umbilical cord itself does not reach the placenta but inserts into the fetal membranes a little distance away. The umbilical vessels then divide on their way to the placenta.
- Circumvallate placenta – this occurs when there is extensive folding back of the membranes, which may be linked to bleeding during birth.

Inside the placenta

As the placenta develops, the fetal blood vessels form chorionic villi (finger-like projections) within it, to absorb nutrients and oxygen from the incoming maternal blood vessels. Waste is also passed back to the maternal blood.

The placenta provides the means by which the growing fetus can receive oxygen and nutrients from the maternal blood circulation and, at the same time, dispose of its waste products. To allow these transfers to take place, the placenta has a very rich blood supply from both the mother and the fetus.

A cross-section of the placenta reveals that this organ is made up partly from maternal tissue and partly from fetal tissue. The spiral arteries that arise from the maternal uterine arteries bring blood into the base of the placenta. This blood then leaves the arteries and fills wide 'pools' (intervillous spaces) in which the fetal villi are suspended. The maternal blood then returns to the mother's circulation through numerous veins.

The fetal villi are finger-like projections that contain blood vessels connected to the fetus through the umbilical cord. They branch again and again to create the maximum amount of surface area for the transfer of oxygen, nutrients and waste substances to and from the maternal blood.

Although the two circulations come close to each other, maternal and fetal blood do not mix, being divided by the thin walls of the villi.

Functions of the placenta

The placenta has a number of functions that are vital to fetal growth and development:

- Respiration – fetal blood is supplied with oxygen from the maternal circulation via the placenta, which also carries away waste carbon dioxide
- Nutrition – nutrients which circulate in the maternal bloodstream are passed to the fetus through the placenta
- Excretion – waste products from the fetus are passed from the two umbilical arteries to the villi, and ultimately the maternal circulation, for disposal
- Hormone production – the placenta is an important source of hormones, especially oestrogen and progesterone. These hormones not only help to maintain the pregnancy, but also prepare the mother for the birth of her child.

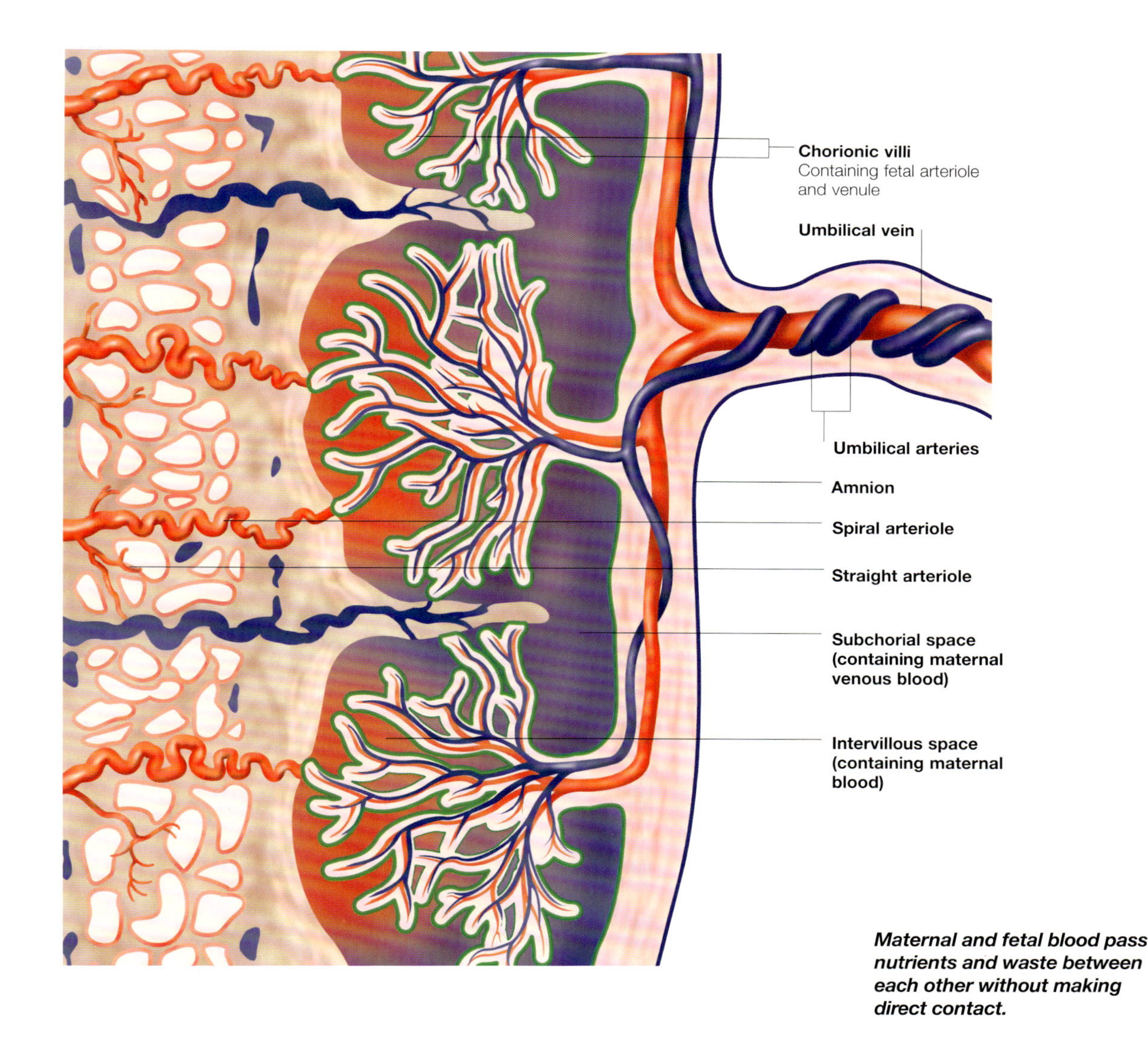

Maternal and fetal blood pass nutrients and waste between each other without making direct contact.

Childbirth

Towards the end of pregnancy physiological changes occur in both mother and fetus. Hormonal triggers cause the muscles in the uterine wall to contract, delivering the baby and placenta.

Parturition is the final stage of pregnancy and usually occurs 280 days (40 weeks) from the last menstrual period. The series of physiological events that lead to the baby being delivered from the mother's body are referred to collectively as labour.

Initiation of Labour

The precise signal that triggers labour is not known, but many factors that play a role in its initiation have been identified. Progesterone is a hormone that is responsible for maintaining the uterine lining during pregnancy and has an inhibitory effect on the smooth muscle of the uterus, preventing it from contracting. During pregnancy, levels of progesterone secreted by the placenta into the mother's circulation are high.

Hormonal Triggers

Towards the end of the pregnancy, there is increasingly limited space in the uterus and the fetus' oxygen supply becomes more restricted as it outgrows the placenta. The anterior lobe of the fetus' pituitary gland increases the secretion of ACTH (adrenocorticotropic hormone). In turn, the fetus' adrenal cortex is triggered to produce chemical messengers (glucocorticoids) that inhibit progesterone secretion from the placenta. Meanwhile the levels of oestrogen released by the placenta into the mother's circulation reach a peak. This causes the myometrial cells of the uterus to produce a higher number of oxytocin receptors, making the uterus more sensitive to oxytocin, a hormone that helps the uterus contract and the cervix dilate.

Contractions

Eventually the inhibitory influence of progesterone on the smooth muscle cells of the uterus is overcome by the stimulatory effect of oestrogen.

The inner lining of the uterus (myometrium) weakens, and the uterus begins to contract irregularly. These contractions, known as Braxton Hicks, help to soften the cervix in preparation for the birth and are often mistaken by pregnant mothers for the onset of labour.

As the pregnancy reaches full term, stretch receptors in the uterine cervix activate the mother's hypothalamus (a region of the brain) to stimulate her posterior pituitary gland to release the hormone oxytocin. Certain cells in the fetus also begin to release this hormone. Elevated levels of oxytocin trigger the placenta to release prostaglandins and together they stimulate the uterus to contract.

Labour

As the uterus is weakened due to suppressed levels of progesterone and is more sensitive to oxytocin, the contractions become stronger and more frequent, and the rhythmic contractions of labour begin.

A 'positive feedback' mechanism is activated whereby the greater the intensity of the contractions the more oxytocin is released, which in turn causes the contractions to become more intense. This chain is broken when the cervix is no longer stretched after delivery and oxytocin levels fall.

Hormonal changes before delivery

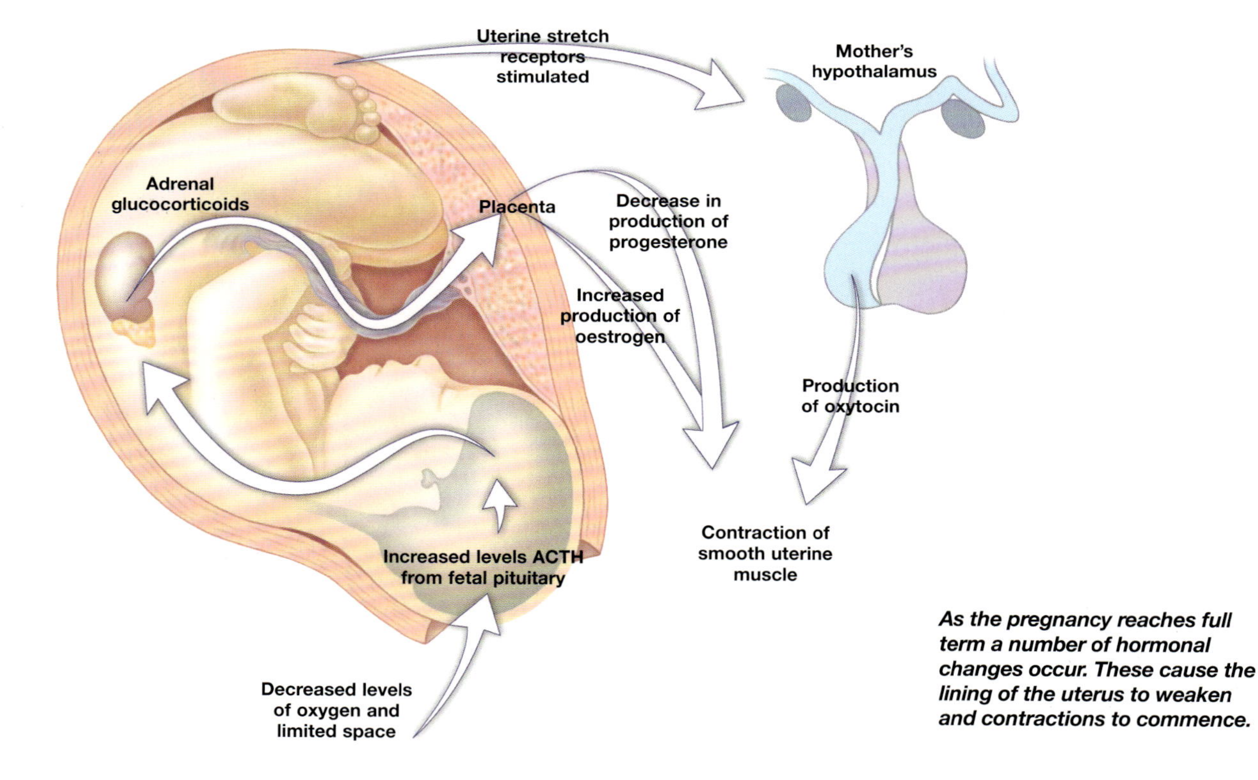

As the pregnancy reaches full term a number of hormonal changes occur. These cause the lining of the uterus to weaken and contractions to commence.

Stages of labour

Childbirth can be divided in to three distinct stages: dilatation of the cervix, birth of the baby and delivery of the placenta.

In order for the baby's head to pass through the birth canal, the cervix and vagina must dilate to around 10cm (4in) in diameter. As labour commences, weak but regular contractions begin in the upper part of the uterus.

These initial contractions are 15–30 minutes apart and last around 10–30 seconds. As the labour progresses, the contractions become faster and more intense, and the lower part of the uterus begins to contract as well.

The baby's head is forced against the cervix with each contraction, causing the cervix to soften, and gradually dilate.

Eventually the amniotic sac, which has protected the baby for the duration of the pregnancy, ruptures and the amniotic fluid is released.

Engagement

The dilatation stage is the longest part of labour and can last from 8 to 24 hours. During this phase the baby begins to descend through the birth canal, rotating as it does so, until the head engages, entering the pelvis.

Dilatation is the longest stage of labour. It can take up to 24 hours for the cervix to dilate sufficiently to allow delivery.

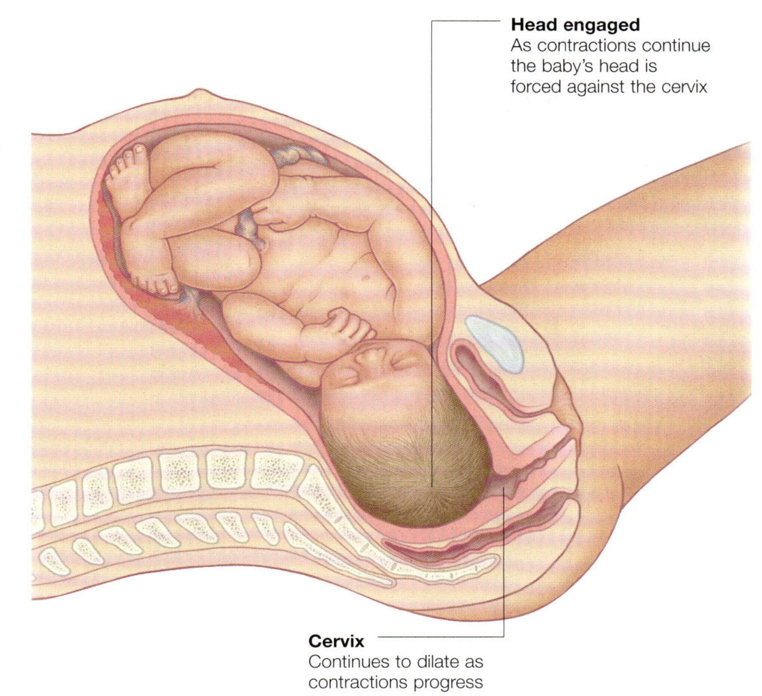

Second stage

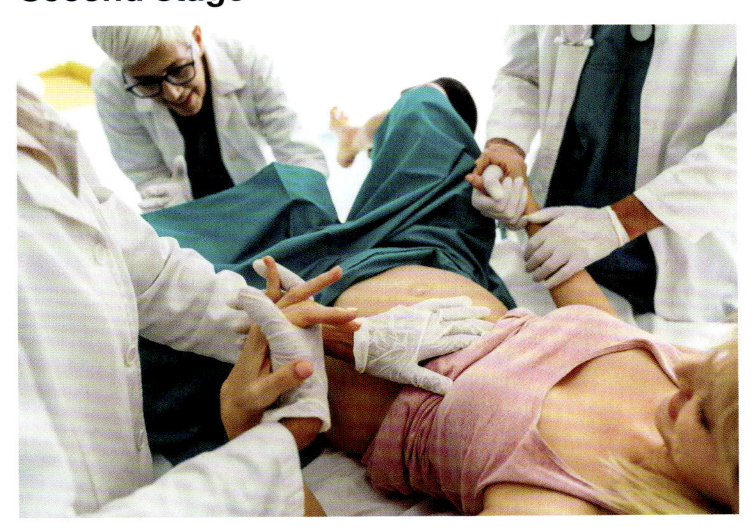

The second stage of labour lasts from full dilatation to the birth of the child. Usually by the time the cervix is fully dilated, strong contractions occur every 2–3 minutes and last around a minute.

Urge to push

At this point the mother will have an overwhelming urge to push or bear down with the abdominal muscles. This phase can take as long as two hours, but is generally much quicker in subsequent births.

Once the cervix is fully dilated the baby is ready to be delivered. The mother will feel a strong urge to push, expelling the baby through the cervix.

Delivery

Crowning takes place when the largest part of the baby's head reaches the vagina. In many cases the vagina will distend to such an extent that it tears. Once the baby's head has exited, the rest of the body is delivered much more easily.

When the baby emerges head first, the skull (at its widest diameter) acts as a wedge to dilate the cervix. This head-first presentation allows the baby to breathe before it is completely delivered.

Delivery of the placenta

The final stage of labour, when the placenta is delivered, can take place up to 30 minutes after the birth. Once the baby has been delivered, the uterine contractions continue, and act to compress the uterine blood vessels to limit bleeding. The contractions also cause the placenta to break away from the wall of the uterus.

Afterbirth

The placenta and attached fetal membranes (the afterbirth) are then easily removed by maintaining gentle manual traction on the umbilical cord. All placental fragments must be removed to prevent continued uterine bleeding and infection after birth.

Hormone levels

Oestrogen and progesterone levels fall once their source, the placenta, has been delivered. During the four weeks after birth the uterus shrinks but remains larger than it was before pregnancy.

Contractions continue after the birth, which helps the placenta to detach from the uterine wall. Gentle traction on the cord may be used to help deliver the placenta.

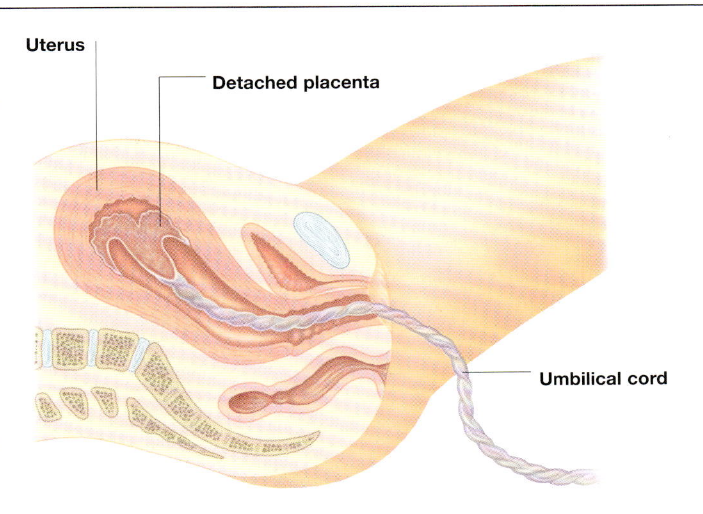

CHAPTER 13

Ageing and Death

In about the third decade of life, the cells and tissues of the body begin to 'age'. Biological ageing is the result of long-term, irreversible cell damage over many years, which leads to a gradual decrease in mental and physical ability and, ultimately, death. Research into the factors that cause cell and tissue deterioration has not yet revealed a solution, but it is evident that ageing can be influenced by lifestyle factors. The biological or physiological age of a person is not always the same as their chronological age (years since birth) and a lifestyle that includes regular exercise and a healthy diet does seem to have an impact on life expectancy. This chapter describes the ageing process at a cellular level and describes the external and internal changes that take place as the years progress. The actual process of death is also explained, as are the changes that the body undergoes afterwards.

Opposite: The ageing process in the cells and tissues of the body is inevitable. However, a healthy lifestyle can delay the onset of poor health in older age.

Science of Ageing

Ageing is the gradual degeneration of the molecules and cells of the body over time, resulting in decreased physical and mental ability, and a greater susceptibility to disease.

Ageing is the term used to describe the physiological changes that take place in the body as it slowly degenerates with time. This process occurs gradually over a number of years, beginning in the third decade of life (age 20–30).

Life Expectancy

The longest lifespan recorded is currently 122 years, although on average in the UK, life expectancy is about 78 years for men and 82 years for women. Worldwide, pre-Covid pandemic life expectancy was increasing, but has now slowed.

Extensive research has been carried out into the biological mechanisms behind ageing, particularly into the reasons for difference in chronological age (actual age in years) and biological age (the physiological age of the cells, tissues and organs), which can differ greatly between individuals. A greater understanding of the factors that slow down the ageing process have enabled us to live healthier and longer lives and although these factors are not yet fully understood, there do appear to be some consistent factors: lifestyle; diet; environmental factors; and genetics.

Research confirms that a supportive and social lifestyle throughout life can impact health in older age.

Cellular ageing

These cells are undergoing apoptosis, or cellular death.

Cells are the units that form all the tissues in the body. These cells are replenished through the process of replication (cell division). Once cells are no longer able to divide, they are known as 'senescent' cells, and are usually effectively removed by the immune system. As the body ages, the immune system becomes less effective at disposing of these cells, and this leads to inflammation and damage to other tissues, contributing to ageing.

Cell death

Research has shown that cells divide a finite number of times before undergoing programmed cell death, or apoptosis. In addition, the remaining cells may not function as well in old age. Cellular repair enzymes may be less active, so more time is required for essential chemical reactions to occur. As the cells fail to reproduce, an organ becomes less efficient until it can no longer fulfil its biological role.

DNA damage

Damage to, or mutations (change in the base sequence) of, DNA is thought to be a primary cause of ageing. The damage can result from complex internal repair or cell replication mechanisms failing later in life. Damage can also be the direct result of external factors such as long-term exposure to tobacco smoke or sunlight. If repair mechanisms are not functioning, then damaged cells accumulate and perform badly. Damaged cells also affect neighbouring tissues and may give rise to cancer.

Telemere attrition

Telemeres are an example of a DNA repair mechanism. At each end of a chromosome is a telemere, a long section of DNA that protects the ends of chromosomes from becoming damaged. Every time a cell divides, the telomeres shorten slightly, a process known as attrition. Eventually, they become so short that the cell is unable to divide further, and the cell becomes senescent.

Environmental factors

Environmental damage to DNA is constant, with numerous assaults a day from pollution, sunlight, chemicals, drugs, among others. Over time, these factors can permanently damage DNA. A visible example is UV light damage on the skin of an older person, which leads to discolouration and occasionally cancers.

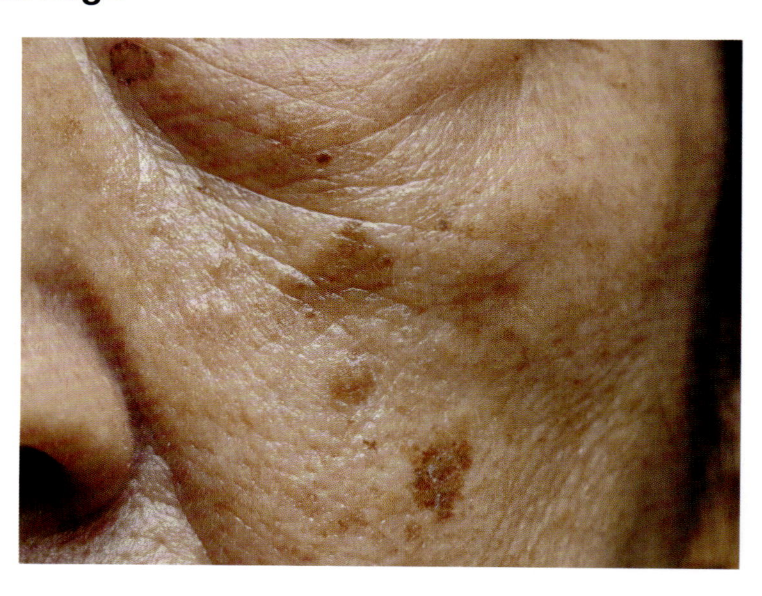

Long-term exposure to UV light damages skin cells in the skin. These 'solar lentigo' develop in areas most exposed to the sun.

Research into ageing

For decades, scientists have been researching the causes of growing old with the purpose of slowing or reversing the ageing process, and enabling a healthy old age.

People are living longer and, in many countries, there is a shift in the age profile with more people in the older age groups. The data suggest that although people are living longer, they are often not healthy in later life and many have two or more medical conditions that require physical care and drugs. Mobility may be poor, resulting in people becoming less active and increasingly isolated. Research is therefore focused on improving quality of life and delaying the deterioration of health and the onset of chronic illnesses, such as arthritis, heart disease and dementia.

Biological versus chronological age

Researchers are looking at biological age (the physiological age of the body) and chronological age (a person's actual age in years) and why they can be different. For example, a person may have a biological age of 65 and a chronological age of 80. Knowing the reasons for this discrepancy would help experts to unlock the secrets of a healthy old age.

Hallmarks of ageing

To establish clear causes of ageing, researchers have developed certain hallmarks to measure biological age. These hallmarks, such as cellular attrition and telemere senescence (see opposite page) have provided a framework for scientists to continue to identify the factors that cause ageing. However, progress towards solutions is slow, with most focus on the diseases of old age rather than ensuring a healthy older age.

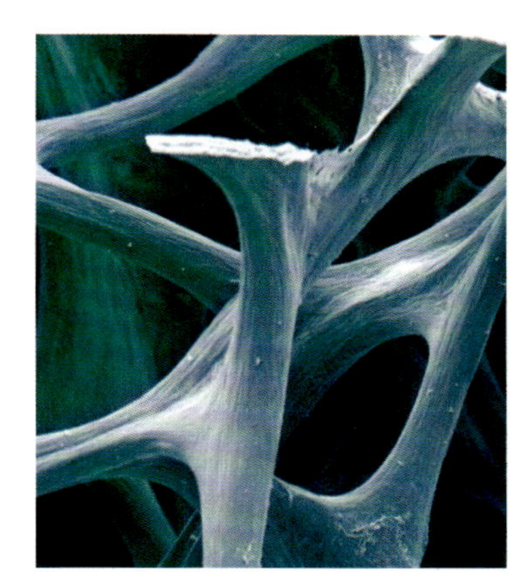

Healthy spongy bone tissue is shown here, greatly magnified.

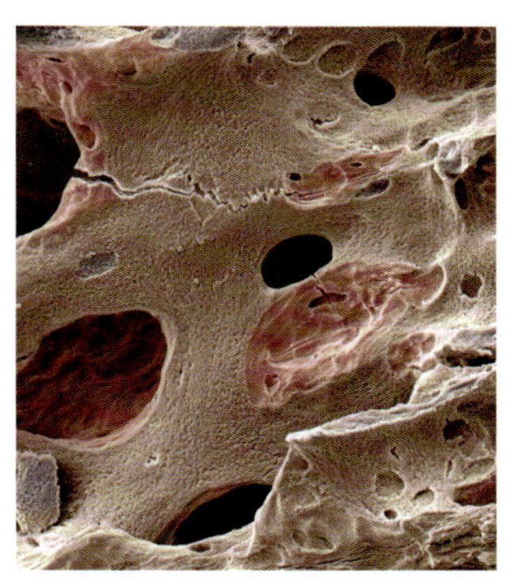

This image depicts developing osteoperosis, a decrease of bone density. Areas of the bone tissue appear degraded (shown in darker pink).

Healthy older age

Research into why some older people are healthier than others in later life does appear to show that several factors contribute.

Determinants of healthy ageing

The determinants identified for healthy ageing include physical activity, diet, self-awareness, outlook/attitude, lifelong learning, faith, social support, financial security, community engagement and independence.

Positive actions

Some of these determinants have a direct effect on physical health. For example, osteoporosis of the bones can be delayed by regular weight-bearing exercise. Exercise will also strengthen muscles, ligaments and joints, which in turn improves bone health. Other determinants improve psychological health and reduce stress and its harmful effects.

Social determinants

There is increasing evidence that determinants such as social support and community engagement can have a significant impact on health in older age. For example, loneliness, isolation and poverty can lead to a deterioration in both mental and physical health.

Brain ageing

Research continues into factors that affect our cognitive ability, and in particular dementia. In general, although the number of brain cells decreases throughout life, this represents only a small percentage of the total number of cells in the brain. There is no conclusive evidence that intelligence deteriorates with age; rather it seems that there is an association with education and lifestyle.

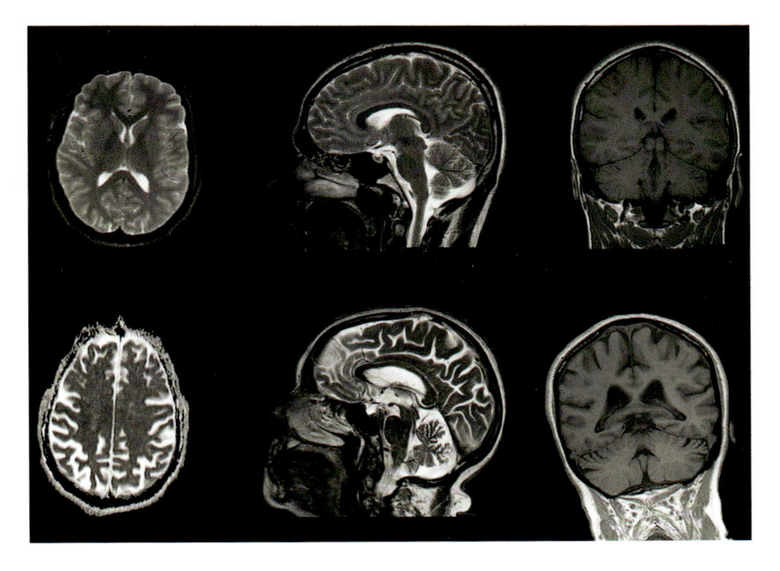

These MRI scans show the brains of healthy 22 (top) and 96 year olds (bottom). The older woman's brain shows brain shrinkage (white areas).

Assessing health in old age

The outward signs of ageing are obvious: changes in skin; greying and loss of hair; arthritic joints and loss of mobility. There are also a number of medical biological markers and tests that are used to assess internal health in old age:

* Blood tests look at the function of organs, such as the liver and kidneys, and will show any deterioration.
* Blood tests can also indicate a rise in cholesterol or sugar levels, and are used to assess risk of heart disease, stroke, or diabetes.
* Brain function can be assessed using specialized cognitive tests, for example those for dementia or memory loss.
* Imaging techniques show the extent of bone tissue deterioration.
* Tools are also available to test frailty in older people – that is the ability to cope with falls and infections. These tools assess a person's balance, their ability to walk unaided or to stand up from a seated position.

External Changes

Outward signs of ageing take place gradually over several decades. Although these changes are inevitable, factors such as lifestyle, environment and genetics can influence the rate at which they occur.

The first visible signs of ageing can be detected in the body while people are still young. A few grey hairs start to appear, wrinkles form at the corners of the eyes, and the flesh becomes less firm.

The changes progress so slowly that, at first, people are not really aware of them. However, by the time people have reached their late 50s and 60s, the cumulative effects of external body changes have become impossible to ignore.

Rate of ageing

Changes occur in hair colour, hair thickness and consistency, skin quality, body fat distribution, height and, sometimes, the shape of the spine.

A person's precise rate of ageing depends on their genetic inheritance, and is modified by circumstances such as their early upbringing, lifestyle and environment. For example, a person who drinks, smokes and sunbathes is likely to look older than one who does not.

Effects on hair

Hair begins to grey when the production of pigment in the hair follicles starts to slow down. It is a variable process that can start as early as the 20s for some people and as late as the 60s for others. Genes play an important role in determining the age at which grey hair replaces the natural colour.

Both men and women undergo characteristic localized changes in the quality and quantity of their hair. In men, hair in the nostrils, eyebrows and ears that was once fine, short and colourless becomes coarser, longer and darker. In women over the age of 65, hair on the chin and above the lips may undergo a similar transformation.

Baldness

In male pattern baldness, hair on the scalp changes in the opposite way. Here, dark, coarse hair is replaced by fine, short, colourless hair. The familiar male-balding pattern with receding hairline may also affect women, but this tends to happen only if women live into their 80s and 90s.

Skin changes

In later life, skin has a slower cell turnover and the cells are smaller in size and number. As a result, skin tends to be thinner, drier and more fragile and this increased laxity leads to the formation of wrinkles. These dramatic changes in the appearance of the skin are largely due to a progressive destruction of the delicate architecture of the connective tissue

Right: After the age of 40, muscle tissue is no longer replaced when it dies. As a result, there is a distinctly reduced muscle mass in older age.

Below: Healthy collagen fibres are shown in this micrograph of tendon tissue. As we age, the collagen degenerates and stiffens, so we become more susceptible to injury.

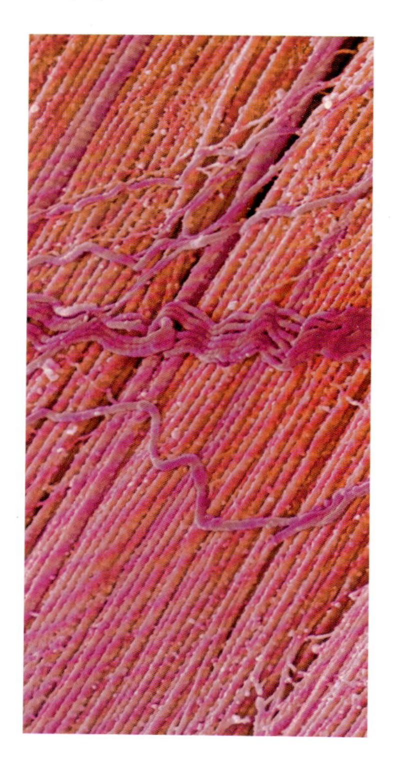

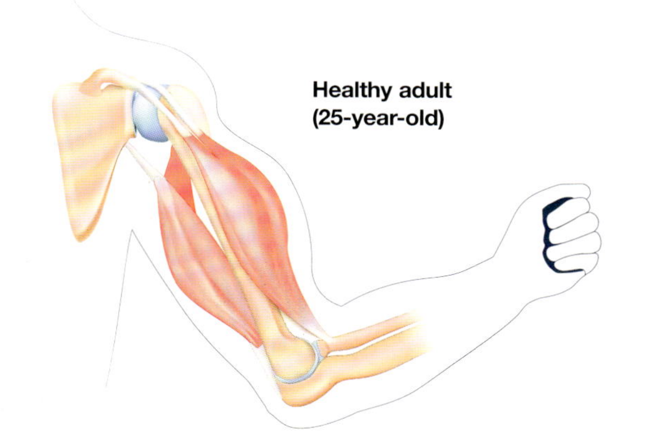

Young and active

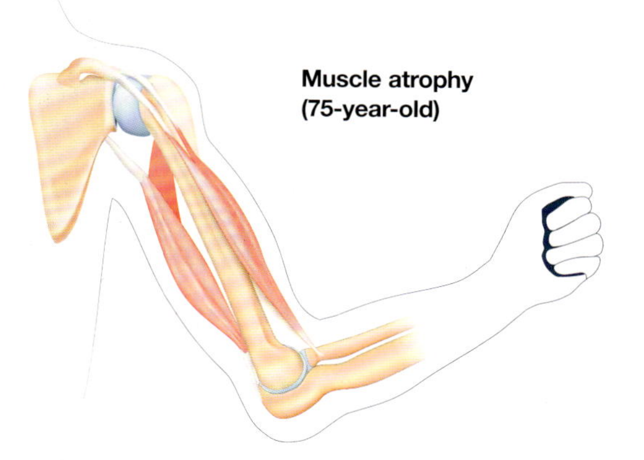

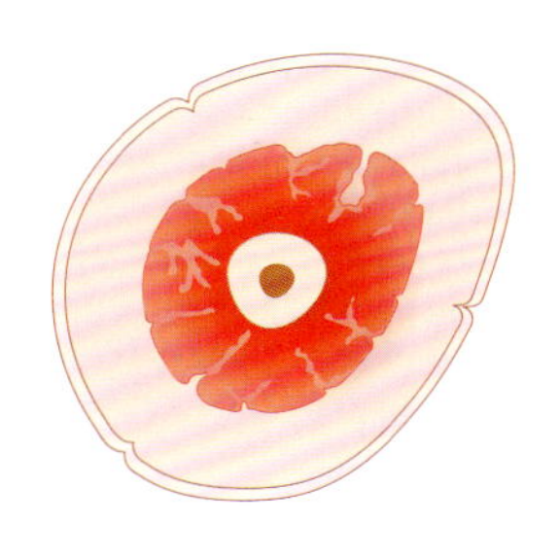

Old and passive

components – collagen and elastin (a tissue rich in elastic fibres) – that are found in the deeper layer of the skin (dermis). Elastic fibres maintain normal skin tension, but as people age, these fibres decrease in number, which means that the skin loses its elasticity.

Inhaled free radicals – highly reactive molecules in the atmosphere – have the unfortunate effect of attacking the skin's collagen. Their effects on the collagen contribute to an older appearance.

Ageing skin can be exacerbated by sunbathing and smoking because both activities produce greater quantities of free radicals.

Facial appearance

The facial appearance of older people changes because the amount of subcutaneous fat – the fat just beneath the skin – steadily decreases. When combined with the reduced thickness of the dermis and the diminished elasticity of the skin, this causes the skin to sag and make wrinkles more pronounced.

In addition, the contraction of facial bone allows skin to fold away from underlying muscles. The effect of gravity pulling this skin down can then lead to drooping features.

Wrinkles

Wrinkles develop gradually over many years. First to appear are horizontal furrows, or frown lines, across the forehead during the 20s, followed by crow's feet around the eyes at 40, wrinkles at the corner of the mouth from 50, and creases around the lips from 60.

Pigmentation

Although pigment-containing cells in the skin decrease with age, those that remain increase in size and sometimes cluster to form age spots, also known as liver spots.

The ageing of skin in women begins about 10 years earlier than in men. This is because the oil-producing glands in women's skin begin to atrophy (waste away) after the menopause when the levels of the hormone oestrogen decline. In men, the oil-producing glands continue to function, and as a result male skin is thicker and oilier, postponing the appearance of wrinkles.

Altered muscle mass

Unlike skin cells, muscle cells cannot replace themselves. Muscle density increases until about the age of 39; muscle mass then begins to shrink as the muscle fibres decrease in number and diameter.

Fat replaces some of the muscles as it starts to infiltrate between them. This means that even if a person's weight is the same at age 60 as it was at 25, their body will generally contain more fat than it once did. The loss of muscle mass is accompanied by the loss of muscle strength, tone, flexibility and speed of movement.

After the menopause, when levels of oestrogen and progesterone are low, women start to notice that fat becomes more concentrated around the waist, abdomen and breasts. As a result, with increasing age, the female shape may become less curvy.

Bone density and loss of height

Osteoporotic changes in bone density can lead to the development of a curved appearance to the upper spine. As the bones of the spine lose their density, the collapse of the vertebrae causes the ribcage to tilt downwards towards the hips.

A curvature in the upper spine creates a second curve in the lower spinal column, pushing the internal organs outwards. Due to the compressed spinal column, several inches in height can be lost. Internal functions are impaired as the organs shift position and obstruct other organs and systems.

Even for older people who do not have osteoporosis, there is a loss in height of just over an inch in men and two inches in women. Loss of height is due to altered posture and a thinning of the cartilage discs that provide cushioning between the spinal vertebrae.

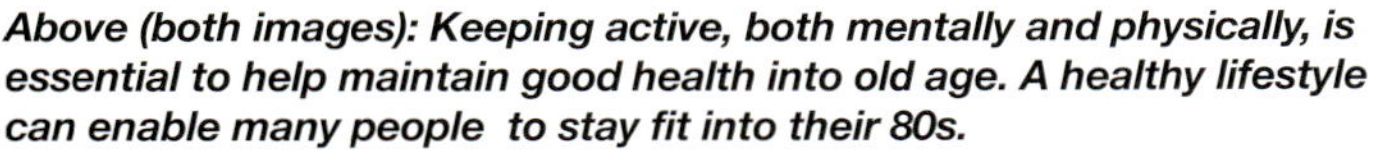

Above (both images): Keeping active, both mentally and physically, is essential to help maintain good health into old age. A healthy lifestyle can enable many people to stay fit into their 80s.

Internal Body Changes

As people age, a number of physical changes take place. Some are inevitable, but others depend on factors such as lifestyle and the state of health during youth and middle age.

As the body ages, its cells degenerate and vital cellular processes slow down. The internal organs become less efficient, affecting the body functions. The exact cause of these internal changes is unknown, but there are a number of factors that contribute, including the physiological ageing processes discussed earlier in this section as well as social determinants of health.

Greater understanding of these issues means that we can take action to prevent or mitigate many of them, and enable older people to enjoy good health in later life.

Brain
As the brain ages, the number of neurones and their supporting cells decreases and by the age of 90, the total brain mass has decreased by approximately 11 per cent. However, research has indicated that this may not be an inevitable consequence of ageing. In healthy elderly people, researchers have found no neuronal loss with age, only a shrinkage of neurone size.

The autonomic nervous system, which regulates body systems, can become slow or weakened with age. The result is that the body takes longer to react to changing conditions – for example, the increase in heart rate during exercise – and may not respond as effectively to stress.

Organs and metabolism
Many organs, such as the liver, kidneys, spleen, pancreas, lungs and liver, shrink in size and therefore function less efficiently as their cells gradually degenerate. There is a general decline in regulating mechanisms, resulting in the body being less adaptable to external changes. Older people are more sensitive to extremes of temperature. They may also take longer to recover from illness and infections than younger people.

Appetite and digestion
Ageing has an effect on appetite and can lead to older people losing weight. The hormones that boost appetite are decreased and the overall elasticity of the digestive system is reduced so that food moves more slowly through the gut. The organs vital for digesting food – the liver and pancreas – may not be functioning as well as they should. A dry mouth, gum disease and dental problems can also affect the appetite.

Right: All men have an enlarged prostate in old age. An enlarged prostate is shown in this CT scan in pink.

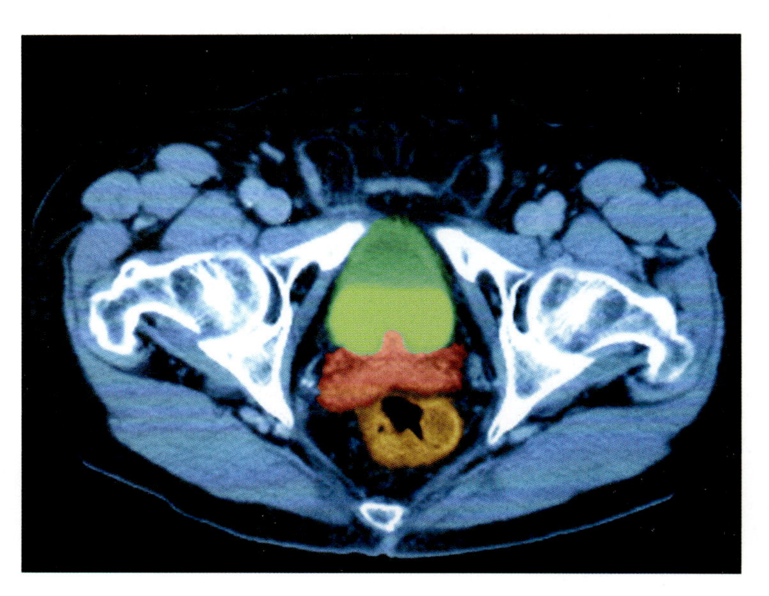

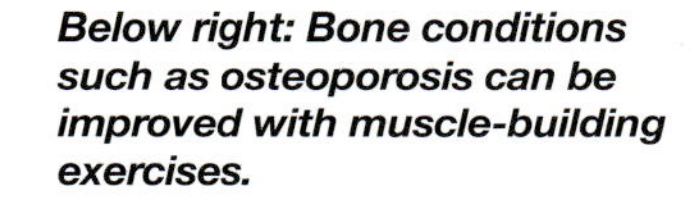

Below right: Bone conditions such as osteoporosis can be improved with muscle-building exercises.

Below: Osteoarthritis of the hip seen on a frontal x-ray.

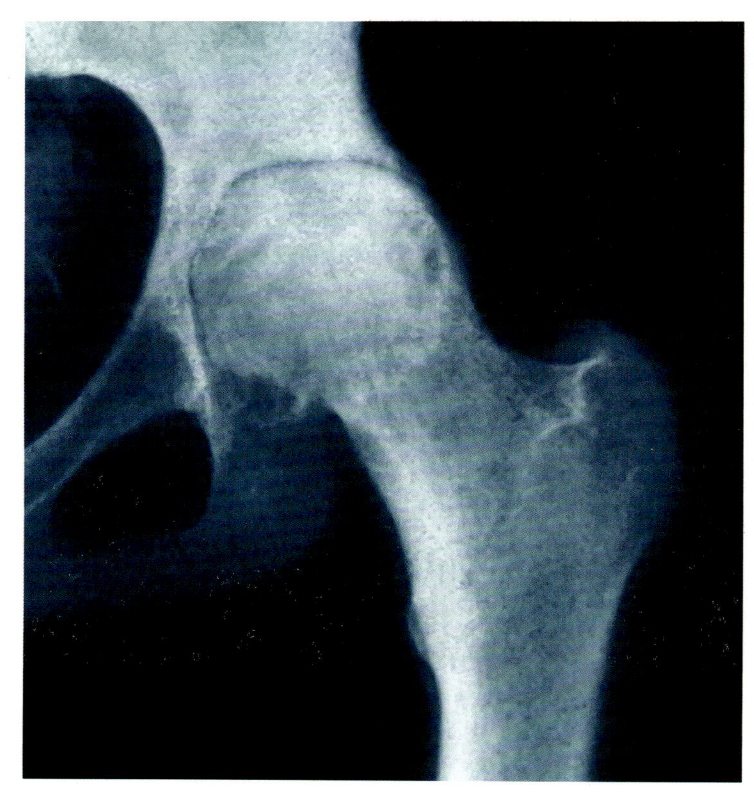

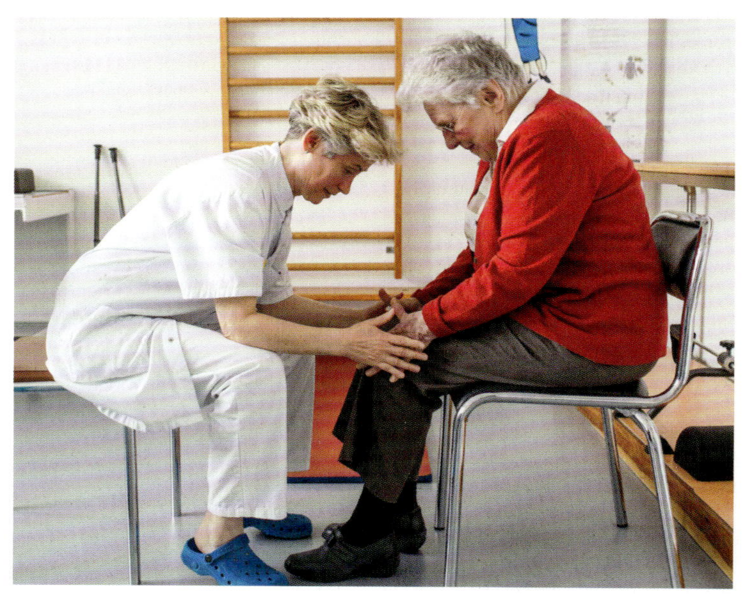

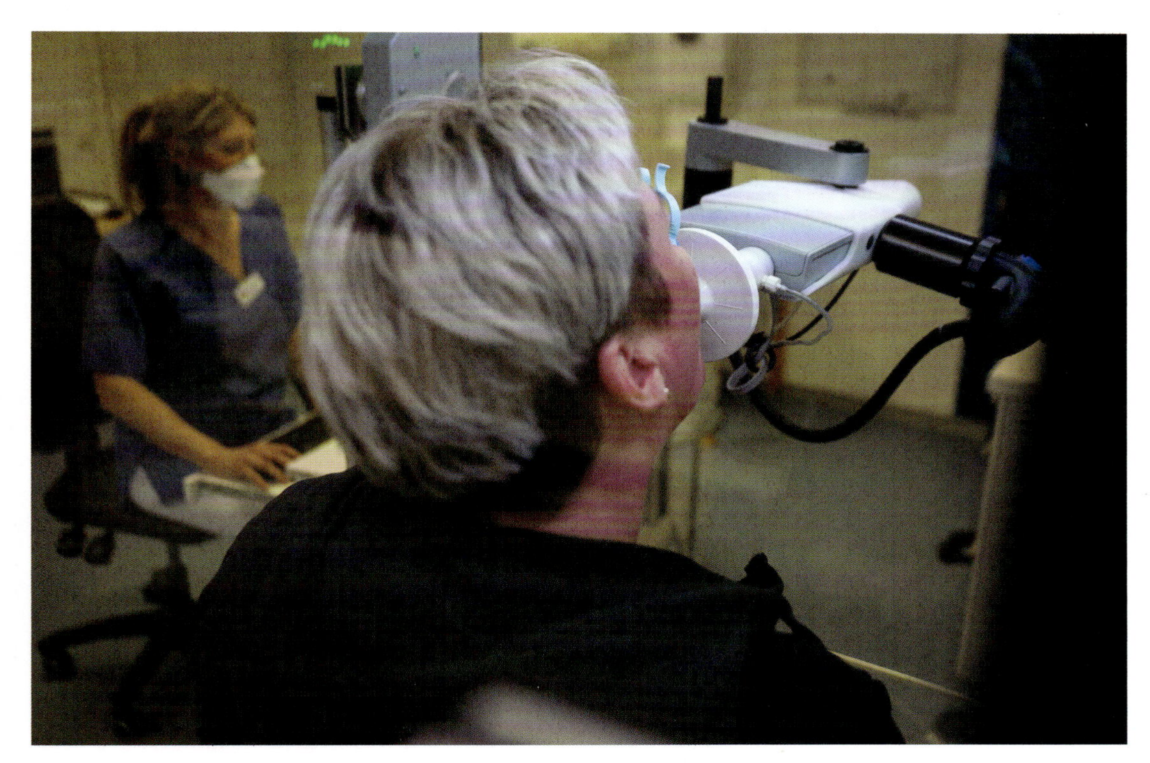

Left: Lung function tests assess how well the lungs work and can help to diagnose long-term conditions.

Cardiovascular system
Age has some important effects on the heart. The collagen that surrounds the heart's fibres loses its elasticity. Under conditions of stress, even healthy hearts are less effective than they once were. For example, under stress or during physical activity, the heart rate is unable to beat as rapidly. The blood vessels lose their elasticity and their walls thicken, particularly the lining of the arteries. This leads to a greater resistance to blood flow so that the heart has to work harder to pump blood around the body. Blood pressure may increase because of the thickened arteries.

Lungs
With increasing age, the lungs and chest wall become less elastic, and maximum breathing capacity is reduced. In addition, the mechanisms protecting the lungs from infection – such as ciliary action (tiny hair-like cells preventing foreign particles from entering the lungs) and normal mucus production – become less effective. This can result in an increased susceptibility to developing chest infections.

Older people may find that, during physical activity, they have to breathe harder to maintain satisfactory blood oxygen levels.

Bladder and kidney function
The capacity of the bladder diminishes with age. This seems to be the result of muscle weakness and changes in the connective tissue. The result is that the average bladder of an older person holds less than half as much as that of a young person.

Signals from the ageing bladder also change. When young, the sensation of bladder fullness, which produces the urge to urinate, is felt when the bladder is about half full. However, among older adults, the urge is not felt until the bladder is nearly full.

Moreover, each kidney contains millions of tiny filtering units called nephrons, which cleanse the blood of impurities. The function of the kidneys declines as nephrons die off and the blood supply to the kidneys is reduced.

Immune system
The thymus – an endocrine gland that is a major component of the immune system – is important in the ageing process. As the thymus shrinks, there is an increase in the bloodstream of immature T-cells (white blood cells responsible for fighting infection), and the functioning of mature T-cells becomes less efficient, whereas other white blood cells, the B-cells, produce fewer antibodies. Consequently, the body is less capable of mounting an immune response against invading organisms, making the body more vulnerable to infection. Whether such changes are an inevitable consequence of ageing remains uncertain.

Musculoskeletal system
After the age of 40, there is a gradual loss of muscle mass and this process accelerates after the age 60. As a result, older people lose muscle strength, have difficulty with mobility and are prone to falls. In addition, bones lose calcium and density leading to a greater risk of fractures and conditions such as osteoporosis and osteoarthritis. Bone disease in the vertebrae and thinning in the intervertebral discs can result in a stooping posture and loss of height.

Endocrine system
The endocrine system is a group of glands throughout the body that secretes hormones. These hormones are responsible for a range of functions that maintain homeostasis. In older age, some hormone levels are reduced. An example is the reduction in melotonin, a hormone that is synthesized by the pineal gland and is linked to regulating sleep patterns. In some people this can cause disturbed sleep or even insomnia.

Fertility in men and women
Women lose their fertility with the onset of menopause, which usually occurs in the late 40s or early 50s. Among men in their 50s, 68.5 per cent have active sperm in their semen, but among men in their 70s the proportion has decreased to 48 per cent.

In most men, normal ageing of the prostate gland is difficult to monitor as abnormal but harmless cell growth enlarges the gland. Among men in their 70s, it is not uncommon for the prostate gland to have doubled in size. Sometimes, the gland becomes so enlarged that it interferes with the function of the bladder and has to be removed.

Below: Atheroma, shown in yellow, can build up in the lining of the arteries and cause cardiovascular conditions, such as stroke.

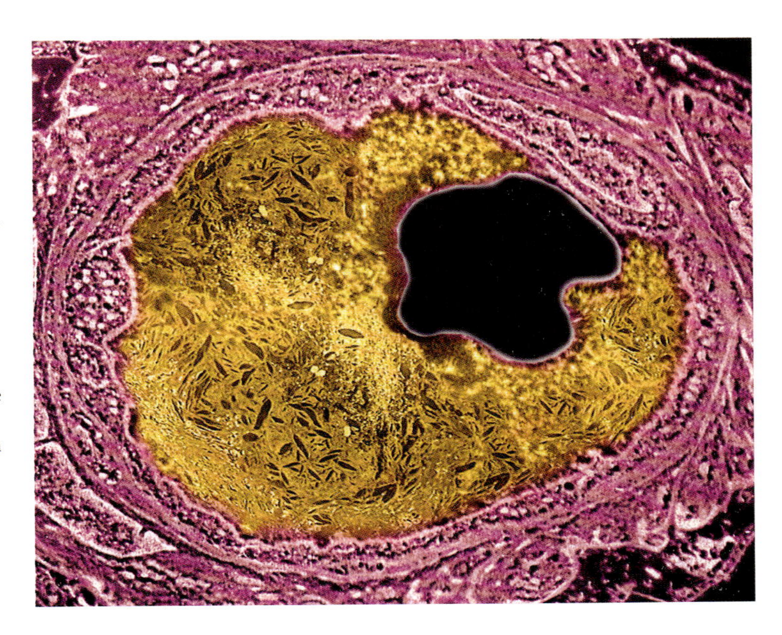

Sensory Changes in Old Age

Many of the sensory changes experienced in older age are due to the deterioration of the bodily structures governing sight, hearing, smell, taste and touch.

During the ageing process, the senses, which give us information about the world around us, tend to become less acute. Some of these changes can have a tremendous impact on the lives of older people.

Hearing problems

Many people lose the ability to hear clearly as they get older. It is estimated that 30 per cent of people over the age of 65 have significant hearing impairment. Total or partial deafness can have a major effect on a person's life, and can be an isolating experience.

As people get older, the structures that make up the ear deteriorate. The eardrum often becomes thicker and there are changes in the bones of the middle ear, which are responsible for transmitting sound to the inner ear. The semicircular canals that control balance may also be affected.

Auditory nerves

Sensorineural hearing loss involves changes in the inner ear, auditory nerve or brain. High-frequency sounds can become more difficult to hear as the part of the cochlea that responds to these frequencies ceases to work properly. This condition is called presbyacusis.

Hearing speech while there is background noise becomes more difficult. This is because the ear is not able to detect such a wide range of frequencies, making it more difficult to distinguish sounds from different sources. The overall effect can be one of jumbled sounds.

Auditory acuity

Changes in the auditory nerve can also cause loss of auditory acuity (sharpness), whereas changes in the brain will lessen its ability to process information from the ears and 'translate' it into meaningful information.

Tinnitus (ringing in the ears) occurs when the inner ear transmits signals to the brain without being first stimulated by sound from outside. Often, however, there is no obvious cause for this and it can affect people of all ages.

There is currently no effective treatment for these conditions, although a modern hearing aid can ease some of the problems associated with presbyacusis.

Conductive hearing loss

Conductive hearing loss occurs when the structures of the outer and middle ear no longer transmit sound efficiently to the inner ear. A hearing aid or, sometimes, surgery can help with this type of hearing loss, depending on the specific cause.

Vision

From middle age onwards, most people will notice that they have more difficulty reading. This is because of a condition called presbyopia in which the lens of the eye becomes thicker, loses elasticity and changes shape, resulting in deteriorating close vision. By age 60, these changes stabilize and glasses usually resolve the problem.

Cataracts

Cataracts, in which there is a change in the protein in the lens, causing it

Right: Regular eye examinations can detect visual problems early, so opticians can correct abnormalities with glasses or lenses.

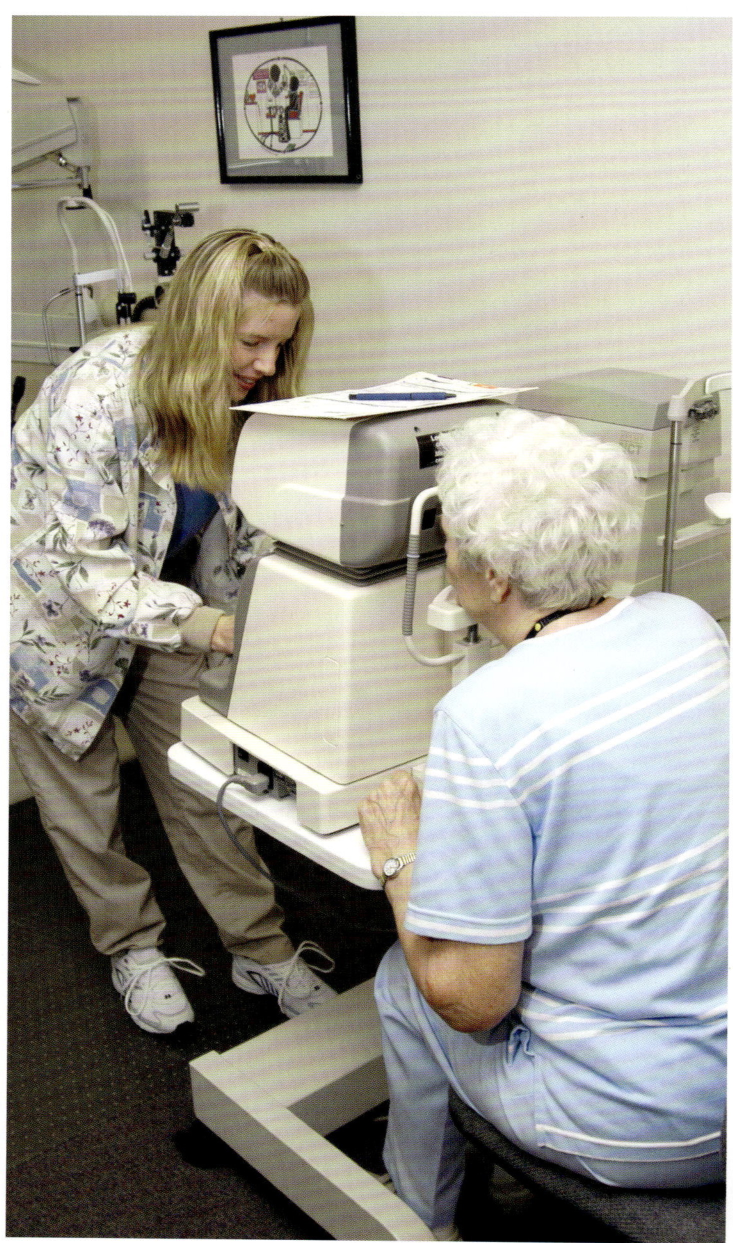

Below: Cloudy lenses, or cataracts, can be easily replaced under local anaesthetic. A new artificial lense can restore clarity of vision.

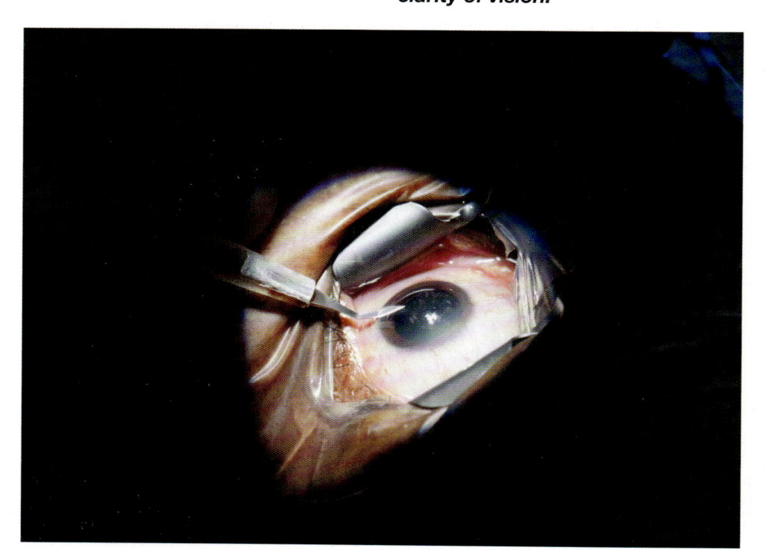

to become cloudy and eventually opaque, are common in older people and easily treated with lens replacement surgery.

Some more serious eye conditions, such as glaucoma (in which the pressure within the eyeball increases) and macular degeneration (damage to the macula, an area in the eye that enables detailed vision) are less common, but are ususally only seen in older people. They are treatable with surgery or medication if detected promptly.

Loss of smell

According to some research, people start to lose their sense of smell at the age of 30. However, the deterioration takes place so slowly that people are rarely aware that it is happening.

The causes are not fully understood, but it is known that the patch of nerve endings that act as smell receptors in the nose begin to thin. At the same time, the olfactory bulb (the part of the brain responsible for interpreting the signals from the smell receptors) becomes smaller. Consequently, older people can find it harder to distinguish smells.

Loss of taste

At the same time, food may start to appear more bland. This is probably partly because the sense of taste is to some extent dependent on the ability to smell and partly because the taste buds also change.

A young person has around 9000 taste buds in the tongue. From the age of about 40–50 years in women and 50–60 years in men, the number of taste buds starts to decrease. At the same time, the remaining taste buds start to lose mass (atrophy).

Even so, the sensitivity to the four different taste sensations (sweet, salty, sour and bitter) does not seem to decrease until after the age of 60, if at all. There is also some doubt as to whether loss of taste and smell is actually related to ageing itself. The loss, if it occurs, may be more concerned with other factors, such as disease, smoking and a lifetime's exposure to chemicals in the environment.

Touch and pain

The skin of the body is covered in receptors that are sensitive to touch, pressure, vibration, cold, heat and pain. There are also receptors inside the body that inform the brain about the condition and position of internal organs. As people age, their sensitivity to these sensations tends to be reduced. The cause is uncertain. It may be that the reduced circulation of old age leads to a corresponding reduced blood flow to the receptors, making them less efficient. Changes in the brain or spinal cord may also play a part. Equally, the loss of sensitivity may be a result of the disorders that occur more often in the elderly. Minor dietary deficiencies, such as a decreased thiamine level, may play a part.

The sense of fine touch decreases up until the age of about 70, when it seems actually to increase again, due to the skin becoming thinner.

A reduced ability to detect touch, pain and heat means that there is a greater risk of injury.

Above: Grandparents share the joys of gardening with their grandchildren.

Life Expectancy

Human beings are not likely to achieve immortality. However, scientists are looking for ways of slowing the ageing process, and future generations may have normal life expectancies of over 100 years.

Life expectancy depends largely on a person's general health and the care that they receive when they are sick. In most parts of the world, living conditions and medical care have improved greatly over the last 50 years and most people now live longer than their ancestors did.

Medical Advances

In 1900, the average life expectancy worldwide was just 48 years. By 2019, life expectancy had increased to 73 but then the COVID pandemic reversed the trend back to the 2012 level of 62. The overall increase in life expectancy is due to improved access to nutritious food, improvements in primary healthcare, easier access to safe water, improved sanitation, immunization, and antibiotics and other medicines.

However, relatively few people live to see their 100th birthday. Even in Japan where, statistically, people live longer than anywhere else, the average lifespan for men is only 81.7 years and 87.2 years for women. In the UK, these figures are 81.9 for women and 78.3 for men, while in the USA they are 79.1 and 73.7 years (WHO data).

Other Challenges

In low-income countries, such as sub-Saharan Africa, life expectancy is lower due to multiple challenges, for example: infectious diseases such as tuberculosis; an increase in non-infectious diseases , such as cancer and deaths due to conflict and violence.

Female Longevity

In parts of the world where life expectancy is highest, women tend to outlive men by five to eight years. The reason for this is not clear, but it may be linked to the female menopause. This involves reduced hormone levels due to the cessation of egg production, compared to hormone levels in men, which remain high and are associated with increased chances of heart disease and prostate cancer.

Below: Few people live to reach a 100, but this number is increasing. Jeanne Calment died at the age of 122 in France in 1997.

Reaching Older Age

Life expectancy actually increases with age. Children who reach the age of five have a better chance of surviving into adulthood because their bodies are better able to fight disease. And, the longer a person lives, the greater their chances of reaching older ages. For example, those who live beyond the age of 70 have a greater chance of surviving into their 80s or 90s. Furthermore, anyone who reaches the age of 100 is more likely to reach the age of 101 than someone in their 90s, even though they have a greater and increasing chance of dying in subsequent years.

DNA

Human beings are not actually programmed to die. Ageing and death are, it seems, the accidental consequence of errors in cellular DNA. These errors are the result of damage caused by chemical processes and subsequent errors in replication and protein production. Although the body has repair mechanisms that deal with much of the damage, some errors persist, with the result that permanent errors gradually accumulate. During the reproductive years, errors normally have no marked effect and thus are not selected against by the evolutionary process. However, as a person ages, the accumulation of DNA errors in genes that would otherwise prolong life expectancy causes these genes to fail, and humans therefore become more prone to disease and disability.

Lifespan

Humans already live to ages well beyond the point at which they are capable of producing and rearing children. In fact, the maximum lifespan for a human being appears to be about 120 years. The oldest recorded person is Jeanne Calment of France, who was born in 1875 and died at the age of 122. However, this is exceptional; according to at least one researcher, the average human lifespan is unlikely to reach 100 for several centuries to come.

Free Radicals

Scientists have established that free oxygen radicals (potentially damaging molecules that are introduced into the body through exposure to UV radiation and inhaling pollutants and cigarette smoke) are one of the main causes of ageing. Free oxygen radicals are impossible to avoid. We need oxygen to survive, and free radicals are created during many of the body's metabolic processes. It has been shown that using antioxidant drugs to remove these free radicals could potentially slow down the ageing process. Certain foods, such as fruit, fresh vegetables, tea, red wine and even chocolate, also contain antioxidants that can block the action of free oxygen radicals.

Melatonin

Other research has looked at the effects of melatonin, a hormone that is secreted by the pineal gland in the brain and which has a rejuvenating effect on the body. Some researchers believe that, as the pineal gland calcifies with age, melatonin levels decrease, and it is claimed that injections of melatonin could provide protection against almost every disease of old age.

Telomeres

Another line of research is the study of telomeres, the rounded tips of chromosomes. It has been shown that, as cells grow older, telomeres become shorter, until a point is reached at which the chromosomes cannot replicate. Cancer and reproductive cells, on the other hand, have intact telomeres, due to the presence of the enzyme telomerase. Researchers are looking at ways of activating the telomerase gene in body cells. If this can be achieved, it could prevent cellular ageing, and consequently increase the lifespan of the body.

Increasing Lifespan

In time, therefore, the human lifespan may increase. However, even if the ageing process is eventually overcome, true immortality is virtually impossible. Accidents would still happen, after all, and it is estimated that few people would survive longer than about 600 years.

And, even extending average life expectancy by a shorter period – say 50 years – would create new challenges, putting pressure on population levels, housing, and the worlds of work and leisure. Furthermore, in a world in which an increasing proportion of the population would be over 70, social attitudes to old age would have to change dramatically.

What is death?

Eventually, either because of old age or illness, death occurs in which the body's organs cease to function and the complex cells and tissues decompose into simpler organic matter.

The final stage of life, or death, is inevitable as the damage that occurs to cells and tissues throughout life becomes unsustainable and the body is unable to function. Death is the process that occurs when the heart stops beating, closely followed by breathing cessation. The cells are therefore no longer receiving oxygen and nutrients and start to undergo the process of decomposition.

The stages of death

There are five stages of death: pallor mortis; livor mortis; algor mortis; rigor mortis and decomposition. These terms refers to the physiological processes that occur in the body once the heart has stopped beating and cells and tissues are dying.

■ **Pallor mortis** simply refers to the pale skin that is evident immediately after the heart has stopped beating.

■ **Livor mortis** (or hypostasis) describes the process of blood pooling in the body according to gravity as damaged blood cells leak out of faulty blood vessels and pool in the capillaries. This process usually occurs about 2–4 hours after death.

■ **Algor mortis** refers to the body temperature. After death the overall body temperature reduces significantly, but it takes about 24 hours for it to be equivalent to the environmental temperature.

■ **Rigor mortis** is the stiffening of muscles that develops a few hours after death. The muscles of the body work by contracting and relaxing. Muscle fibres are formed from many long thin strands, known as myofibrils, which in turn consist of two types of overlapping protein filaments. These proteins called actin and myosin, enable the contraction of the fibres. Energy is provided to actin and myosin in the form of adenosine triphosphate (ATP). After death, there is no ATP and therefore no energy sources for cells and so actin and myosin cease to function leading to muscle rigidity and locked joints. This condition is known as rigor mortis and starts to develop a few hours after death. After about 12 hours, enzyme activity in the cell enables the muscles to relax again.

■ **Decomposition** describes the process of the breaking down of tissues, which becomes evident around four days after death. There are two aids to decomposition, enzymes and bacteria. As the cells are deprived of oxygen, the acid content within them increases and chemical reactions occur. Enzymes are produced that begin to digest the membranes of the cells. This process starts earlier in some organs, for example, the brain and the liver.

In life, the body is teeming with bacteria (particularly in the gut) that usually cause no harm and often are beneficial. After death, however, there is no immune system to keep the bacteria at bay and they multiply and spread throughout the body, digesting soft tissue and breaking it down into gases, fluids and salts. Fly larvae (maggots) also assist with this process by eating dead tissue.

■ **Skeletonization**

The final stage of decomposition is skeletonization, in which the soft tissues and organs have completely broken down and only the bones (and teeth) remain. The amount of time it takes to reach this stage is variable and depends on many factors such as the environment, insect activity, soil composition, depth of burial, ambient temperature and cause of death. Depending on the conditions, the process to skeletonization takes from weeks to years.

The skeleton itself takes much longer to decay – sometimes many decades in a moderate climate. However, in conditions that preserve bone tissue, for example, peat and ice, it can take thousands of years for the skeleton to break down, as evidenced by archaeological specimens.

In the first stage after death – decomposition – tissues are broken down by enzymes and bacteria.

This body is in the final stage of decomposition – skeletonization. Only the bones and teeth remain.

Index

Picture credits

Alamy: 25 (Connect Images), 26 top right (Science History Images), 32 (Alf Jacob Nilsen), 34 (BSIP SA), 40 (Mediscan), 81 (Prakaymas Vitchitchalao), 176 (Science History Images), 177 (Adrian Weston), 236 (Olga Kramer), 244 (Tim Mainiero), 245 (Science History Images), 252/253 (Jose Maria Barres Manuel), 260 right (Science History Images), 271 (Scott Camazine), 274 left & 275 (BSIP SA), 277 top (Przemyslaw Nieprzecki), 277 middle (nobeastsofierce), 279 (Samunella), 290 (Spencer Grant), 294 (Science History Images), 304 (Mediscan), 309 (Jose Maria Barres Manuel), 317 (Stu Gray), 336/337 (blickwinkel), 350 (Samunella), 353 (Black Star), 356/357 (Science History Images), 360 (Samunella), 362 (Stocktrek Images), 364 bottom left and top right (BSIP SA), 366 (Custom Medical Stock Photo), 368 (BSIP SA), 376 left (Imago), 382, 417 bottom right (Juan Carlos Juarez), 422/423 (Deco Images II), 427 (Andor Bujdoso), 434 top right (Science History Images), 434 bottom right (Phanie - Sipa Press), 435 top (dpa picture alliance), 436 left (Julia Catt Photography), 436 right (Dennis MacDonald), 439 left (Medicimage Education)

Alamy/Science Photo Library: 38/39 (Juan Gaertner), 80 (Rajaaisya), 170 (Sebastian Kaulitzki), 264 (Miriam Maslo), 276 & 277 bottom (Dennis Kunkel Microscopy), 300 (Alfred Pasieka), 318/319 (Rajaaisya), 358 (Zephyr), 378/379 (Alfred Pasieka), 380 (CNRI), 401 (Design Cells)

Alamy/Science Photo Library/Steve Gschmeissner: 14, 20 top, 30/31, 41, 50 top, 254, 269 top, 269 bottom, 274 right, 340, 349, 375, 377 left, 432 left

Dreamstime: 12 (Sebastian Kaulitzki), 51 (Puwadol Jaturawutthichai), 90/91 (Ivansmuk), 133 (Skyhawk911), 163 (Jakup Cejpek), 260 left (Jlcalvo), 320 (Sebastian Kaulitzki), 323 top (Guasor), 323 bottom (Dreamerb), 381 bottom (Ronstik), 396 left (Elena Elisseeva), 396 right (Albert Shakirov), 430 top (Pressmaster), 430 middle (Justlight), 430 bottom (Zay Nyi Nyi), 433 left (Drx), 433 right (Roman Chazov), 437 (Wavebreakmedia)

Getty Images: 22 (Ed Reschke), 43 (De Agostini), 48 (Steve Gschmeissner/Science Photo Library), 50 bottom (QAI Publishing), 55 (BSIP), 287 & 377 right (BSIP), 389 (De Agostini), 431 top both (Steve Gschmeissner/ Science Photo Library), 434 left & 435 bottom (BSIP), 438 (Eric Fougere), 439 right (Global Images Ukraine)

Public Domain: 10, 364 top left

Science Photo Library: 16 top (Dennis Kunkel Microscopy), 16 bottom (Professors P. Motta & T. Naguro), 18 top (Professors P.M. Motta & J. Van Blerkom), 18 bottom (Steve Gschmeissner), 19 top left (Dr. Juan F. Gimenez-Abian), 19 top right (Thomas Deerinck, NCMIR), 19 middle left (Dr Torsten Wittmann), 19 middle right (Dr Alexey Khodjakov), 19 bottom (Lennart Nilsson, TT), 20 bottom (Jose Calvo), 26 top left (Power & Syred), 26 bottom (A. Barrington Brown, © Gonville & Caius College /Coloured by SPL) , 29 (Eye of Science), 174/175 (Dr Torsten Wittmann), 217 top (Jose Calvo), 217 bottom (Steve Gschmeissner), 258 (K H Fung), 272/273 (Susumu Nishinaga), 306 (CNRI), 365 bottom (Dr Klaus Boller), 367 (M.I. Walker), 369 (Zephyr), 370 (Steve Gschmeissner), 373 (Steven Needell), 391 (Steve Gschmeissner), 394 (Clouds Hill Imaging), 405 (Science Pictures), 409 (K H Fung), 413, 416 (Professor P.M. Motta, G. Macchiarelli, S.A. Nottola), 417 top (Professor P.M. Motta et al), 417 bottom left (Dr Yorgos Nikas), 418 (Ziad M. El-Zaatari), 420 (D. Phillips), 421 (Lennart Nilsson, TT), 431 bottom (Dr P. Marazzi)

Shutterstock: 11 (VesnaArt), 36 top (sruilk), 36 bottom (tanitost), 44 (natchaprommee), 142/143 (Somsak Thapthimthong), 364 bottom right (Warren Price Photography), 365 top (S. Toey), 376 right (Zay Nyi Nyi), 381 top (Joanne Vincent), 428 (NDAB Creativity), 432 right (Tetiana Zhabska)

Wellcome Collection: 9

All artworks courtesy Bright Star Publishing plc.